D0164428

Introduction to Modern Inorganic Chemistry

6th edition

Introduction to Modern Inorganic Chemistry

6th edition

K.M. Mackay

R.A. Mackay

W. Henderson

Department of Chemistry
University of Waikato
Hamilton
New Zealand

First published in 1968
Second edition 1972
Third edition 1981
Fourth edition 1989
Fifth edition 1996

Reprinted in 2002 by:
Nelson Thornes Ltd
Delta Place
27 Bath Road
CHELTENHAM
GL53 7TH
United Kingdom

02 03 04 05 06 / 10 9 8 7 6 5 4 3 2 1

A catalogue record for this book is available from the British Library

ISBN 0 7487 6420 8

Page make-up by Alden Bookset Ltd, Exeter

Printed in Croatia by Zrinski

Contents

PREFACE XIII

SI UNITS AND NAMES 1

CHAPTER 1

INTRODUCTION 8

1.1 INORGANIC CHEMISTRY AND THE DISCOVERY OF THE ELEMENTS 8

1.2 DEVELOPMENT 8

1.3 RECENT ADVANCES 10

1.4 INORGANIC NOMENCLATURE 13

1.5 APPROACH TO INORGANIC CHEMISTRY AND FURTHER READING 14

PROBLEMS 17

CHAPTER 2

THE ELECTRONIC STRUCTURE AND THE PROPERTIES OF ATOMS 18

INTRODUCTION 18

2.1 BACKGROUND 18

2.2 ISOTOPES 19

2.3 RADIOACTIVE ISOTOPES AND TRACER STUDIES 20

THEORY OF THE ELECTRONIC STRUCTURE OF HYDROGEN 21

2.4 INTRODUCTION 21

2.5 THE DUAL NATURE OF THE ELECTRON 22

2.6 THE HYDROGEN ATOM 23

MANY-ELECTRON ATOMS 28

2.7 THE APPROACH TO THE WAVE EQUATION 28

2.8 THE ELECTRONIC STRUCTURES OF ATOMS 29

SHAPES OF ATOMIC ORBITALS 32

2.9 THE s ORBITAL 32

2.10 THE p ORBITALS 37

2.11 THE d ORBITALS 38

2.12 THE PERIODIC TABLE 40

FURTHER PROPERTIES OF THE ELEMENTS 44

2.13 IONIZATION POTENTIAL 44

2.14 ELECTRON AFFINITY 46

2.15 ATOMIC AND OTHER RADII 48

2.15.1 Covalent species 48

2.15.2 Ionic species 50

2.15.3 Metals 52

2.16 ELECTRONEGATIVITY 53

2.17 COORDINATION NUMBER, VALENCY AND OXIDATION STATE 55

PROBLEMS 58

CHAPTER 3

COVALENT MOLECULES: DIATOMICS 60

GENERAL BACKGROUND 60

3.1 INTRODUCTION 60

3.2 BOND FORMATION AND ORBITALS 62

DIATOMIC MOLECULES 63

3.3 THE COMBINATION OF s ORBITALS 63

3.4 THE COMBINATION OF p ORBITALS 66

3.5 BOND ORDERS OF DIATOMIC MOLECULES 70

3.6 ENERGY LEVELS IN DIATOMIC MOLECULES 74

3.7 SUMMARY 78

PROBLEMS 79

CHAPTER 4

POLYATOMIC COVALENT MOLECULES 81

4.1 INTRODUCTION 81

4.2 THE SHAPES OF MOLECULES AND IONS CONTAINING σ BONDS ONLY 81

4.2.1 The arrangement of σ bonds 81

4.2.2 The effect of lone pairs 82

4.2.3 Three pairs 84

4.2.4 Four pairs 84

4.2.5 Five pairs 84

4.2.6 Six pairs 85

4.3 THE SHAPES OF SPECIES CONTAINING π BONDS 86

4.3.1 Electron counting procedure for π electrons 88

4.3.2 Summary 91

4.4 GENERAL APPROACHES TO BONDING IN POLYATOMIC SPECIES 91
4.5 BONDING IN POLYATOMICS: THE TWO-CENTRE BOND APPROACH 92
4.6 TWO-CENTRED ORBITALS: HYBRIDIZATION 93
4.6.1 Equivalent hybrids 93
4.6.2 Nonequivalent hybrids 95
4.7 DELOCALIZED, OR MULTI-CENTRED, σ ORBITALS 97
4.7.1 Summary of Sections 4.5 to 4.7 101
4.8 π BONDING IN POLYATOMIC MOLECULES 101
4.8.1 The π orbitals of benzene 103
4.8.2 The π orbitals of the nitrite ion, NO_2^- 104
4.9 AN EXAMPLE OF THE APPROACH USING DELOCALIZED BONDING THROUGHOUT 105
4.9.1 Carbon dioxide, CO_2 105
4.10 EXTENSION TO OTHER MOLECULES 107
4.10.1 Ozone, O_3 107
4.10.2 The nitrate ion, NO_3^- 107
4.10.3 Sulphur trioxide, SO_3 107
4.10.4 Summary 107
PROBLEMS 108

CHAPTER 5
THE SOLID STATE 110

SIMPLE IONIC CRYSTALS 110
5.1 THE FORMATION OF IONIC COMPOUNDS 110
5.2 THE BORN–HABER CYCLE 115
5.3 THE LATTICE ENERGY 116
5.4 THE ENDOTHERMIC TERMS IN THE FORMATION OF AN IONIC SOLID 118
5.5 BONDING WHICH IS NOT PURELY IONIC 119
5.6 METALLIC BONDING 121
5.7 COMPLEX IONS 127
5.8 THE CRYSTAL STRUCTURES OF COVALENT COMPOUNDS 128
5.9 DEFECT STRUCTURES AND NONSTOICHIOMETRIC SOLIDS 130
PROBLEMS 131

CHAPTER 6
SOLUTION CHEMISTRY 133

AQUEOUS SOLUTIONS 133
6.1 SOLUBILITY 133
6.1.1 Ionic substances 133
6.1.2 Covalent substances 135
6.2 ACIDS AND BASES 136
6.2.1 Strengths of oxyacids 138
6.3 OXIDATION AND REDUCTION 138
NONAQUEOUS SOLVENTS 145
6.4 SOLUBILITY AND SOLVENT INTERACTION IN NONAQUEOUS SOLVENTS 146
6.5 ACID–BASE BEHAVIOUR IN NONAQUEOUS SOLVENTS 148
6.6 GENERAL USES OF NONAQUEOUS SOLVENTS 150
6.7 LIQUID AMMONIA 151
6.8 ANHYDROUS ACETIC ACID 153
6.9 'SUPERACID' MEDIA 155
6.10 BROMINE TRIFLUORIDE 156
6.11 SUPERCRITICAL FLUIDS 158
PROBLEMS 159

CHAPTER 7
EXPERIMENTAL METHODS 160

SEPARATION METHODS 160
7.1 ION EXCHANGE 160
7.2 CHROMATOGRAPHY 161
7.3 SOLVENT EXTRACTION 162
STRUCTURE DETERMINATION 164
7.4 DIFFRACTION METHODS 164
7.5 SPECTROSCOPIC METHODS AND THE ELECTROMAGNETIC SPECTRUM 166
7.6 ELECTRONIC SPECTRA 168
7.7 VIBRATIONAL SPECTRA 169
7.8 NUCLEAR MAGNETIC RESONANCE (NMR) 171

7.9 FURTHER METHODS OF MOLECULAR SPECTROSCOPY 175
7.9.1 Electron spin resonance (esr) spectroscopy 175
7.9.2 Mössbauer spectroscopy 175
7.9.3 Mass spectrometry 176

7.10 FOURIER TRANSFORM METHODS 177

7.11 OTHER METHODS 178
7.11.1 Magnetic measurements 178
7.11.2 Dipole moments 179
7.11.3 Scanning electron microscopy (SEM) 180
7.11.4 Scanning tunnelling microscopy (STM) 180

DETERMINATION OF ENERGY LEVELS 182

7.12 PHOTOELECTRON SPECTROSCOPY 182
7.12.1 The energy source 182
7.12.2 Vibrational structure 183

PROBLEMS 184

CHAPTER 8

GENERAL PROPERTIES OF THE ELEMENTS IN RELATION TO THE PERIODIC TABLE 185

8.1 VARIATION IN ENERGIES OF ATOMIC ORBITALS WITH ATOMIC NUMBER 185

8.2 EXCHANGE ENERGY 188

8.3 STABLE CONFIGURATIONS 190

8.4 ATOMIC AND IONIC SIZES 193

8.5 CHEMICAL BEHAVIOUR AND PERIODIC POSITION 194

8.6 METHODS OF SHOWING THE STABILITIES OF OXIDATION STATES 196

8.7 THE ABUNDANCE AND OCCURRENCE OF THE ELEMENTS 200

8.8 THE EXTRACTION OF THE ELEMENTS 203

PROBLEMS 207

CHAPTER 9

HYDROGEN 209

9.1 GENERAL AND PHYSICAL PROPERTIES OF HYDROGEN 209

9.2 CHEMICAL PROPERTIES OF HYDROGEN 211

9.3 IONIC HYDRIDES 213
9.3.1 Complex hydride anions 214

9.4 METALLIC HYDRIDES 215

9.5 COVALENT HYDRIDES 218
9.5.1 Preparation 218
9.5.2 Properties 220

9.6 ELECTRON DEFICIENT HYDRIDES 222
9.6.1 Boron hydrides 222
9.6.2 Wade's rules 224
9.6.3 Other electron deficient hydrides 226

9.7 THE HYDROGEN BOND 228

PROBLEMS 233

CHAPTER 10

THE 'S' ELEMENTS 234

10.1 GENERAL AND PHYSICAL PROPERTIES, OCCURRENCE AND USES 234

10.2 COMPOUNDS WITH OXYGEN AND OZONE 237

10.3 CARBON COMPOUNDS 238

10.4 COMPLEXES OF THE HEAVIER ELEMENTS 239

10.5 CROWN ETHERS, CRYPTATES AND ALKALI METAL ANIONS 240

10.6 SPECIAL FEATURES IN THE CHEMISTRY OF LITHIUM AND MAGNESIUM 241

10.7 BERYLLIUM CHEMISTRY 243

PROBLEMS 245

CHAPTER 11

THE SCANDIUM GROUP AND THE LANTHANIDES 246

11.1 GENERAL AND PHYSICAL PROPERTIES 246

11.2 CHEMISTRY OF THE TRIVALENT STATE 248

11.3 THE SEPARATION OF THE ELEMENTS 250

11.4 OXIDATION STATES OTHER THAN III 252
11.4.1 Cerium(IV) 252
11.4.2 Europium(II) 252
11.4.3 Other IV states 253
11.4.4 Other II states 253

11.5 PROPERTIES ASSOCIATED WITH THE PRESENCE OF f ELECTRONS 255
PROBLEMS 256

CHAPTER 12
THE ACTINIDE ELEMENTS 257

12.1 SOURCES AND PHYSICAL PROPERTIES 257
12.2 GENERAL CHEMICAL BEHAVIOUR OF THE ACTINIDES 258
12.3 THORIUM 261
12.4 PROTACTINIUM 263
12.5 URANIUM 264
12.6 NEPTUNIUM, PLUTONIUM AND AMERICIUM 268
12.7 THE HEAVIER ACTINIDE ELEMENTS 270
PROBLEMS 272

CHAPTER 13
THE TRANSITION METALS: GENERAL PROPERTIES AND COMPLEXES 273

13.1 INTRODUCTION TO THE TRANSITION ELEMENTS 273
13.2 THE TRANSITION ION AND ITS ENVIRONMENT: LIGAND FIELD THEORY 277
13.3 LIGAND FIELD THEORY AND OCTAHEDRAL COMPLEXES 279
13.4 COORDINATION NUMBER FOUR 285
13.5 STABLE CONFIGURATIONS 288
13.6 COORDINATION NUMBERS OTHER THAN FOUR OR SIX 289
13.7 EFFECT OF LIGAND ON STABILITY OF COMPLEXES 292
13.8 ISOMERISM 296
13.9 MECHANISMS OF TRANSITION METAL REACTIONS 297
13.10 STRUCTURAL ASPECTS OF LIGAND FIELD EFFECTS 302
13.11 SPECTRA OF TRANSITION ELEMENT COMPLEXES 304
13.12 π BONDING BETWEEN METAL AND LIGANDS 307
PROBLEMS 310

CHAPTER 14
THE TRANSITION ELEMENTS OF THE FIRST SERIES 312

14.1 GENERAL PROPERTIES 312
14.2 TITANIUM, $3d^2 4s^2$ 314
14.2.1 Titanium(IV) 315
14.2.2 Titanium(III) and lower oxidation states 317
14.3 VANADIUM, $3d^3 4s^2$ 318
14.3.1 Vanadium(V) 319
14.3.2 Vanadium(IV) 322
14.3.3 Vanadium(III) 323
14.4 CHROMIUM, $3d^5 4s^1$ 324
14.4.1 Chromium(VI) 325
14.4.2 Chromium(V) and chromium(IV) 325
14.4.3 Chromium(III) 327
14.4.4 Chromium(II) and lower oxidation states 328
14.5 MANGANESE, $3d^5 4s^2$ 329
14.5.1 The high oxidation states, manganese(VII), (VI) and (V) 329
14.5.2 Manganese(IV) and manganese(III) 330
14.5.3 Manganese(II) and lower oxidation states 332
14.6 IRON, $3d^6 4s^2$ 333
14.6.1 The iron–oxygen system 334
14.6.2 The higher oxidation states of iron 334
14.6.3 The stable states, iron(III) and iron(II) 335
14.7 COBALT, $3d^7 4s^2$ 338
14.7.1 Cobalt oxidation states greater than (III) 338
14.7.2 Cobalt(III) 339
14.7.3 Cobalt(II) 340
14.7.4 Lower oxidation states of cobalt 341
14.8 NICKEL, $3d^8 4s^2$ 342
14.8.1 Higher oxidation states of nickel 342
14.8.2 Nickel(II) 343
14.9 COPPER, $3d^{10} 4s^1$ 344
14.9.1 Copper(IV) and copper(III) 345
14.9.2 Copper(II) 345
14.9.3 Copper(I) 347

14.10 THE RELATIVE STABILITIES OF THE DIHALIDES AND TRIHALIDES OF THE ELEMENTS OF THE FIRST TRANSITION SERIES 348
PROBLEMS 350

CHAPTER 15

THE ELEMENTS OF THE SECOND AND THIRD TRANSITION SERIES 351

15.1 GENERAL PROPERTIES 351
15.2 ZIRCONIUM, $4d^25s^2$, AND HAFNIUM, $5d^26s^2$ 352
15.3 NIOBIUM, $4d^45s^1$, AND TANTALUM, $5d^36s^2$ 354
15.3.1 The V state 355
15.3.2 The IV state 357
15.3.3 The lower oxidation states of niobium and tantalum 357
15.4 MOLYBDENUM, $4d^55s^1$, AND TUNGSTEN, $5d^46s^2$ 359
15.4.1 Molybdenum(VI) and tungsten(VI) 360
15.4.2 The V state 363
15.4.3 The IV state 364
15.4.4 The lower oxidation states 365
15.5 TECHNETIUM, $4d^65s^1$, AND RHENIUM, $5d^56s^2$ 366
15.5.1 The VII oxidation state 367
15.5.2 The VI state 369
15.5.3 The V state 370
15.5.4 The IV state 370
15.5.5 The lower oxidation states 371
15.6 RUTHENIUM, $4d^75s^1$, AND OSMIUM $5d^66s^2$ 372
15.6.1 The higher oxidation states 372
15.6.2 The lower oxidation states 374
15.7 RHODIUM, $4d^85s^1$, AND IRIDIUM, $5d^96s^0$ 376
15.8 PALLADIUM, $4d^{10}5s^0$, AND PLATINUM, $5d^96s^1$ 379
15.8.1 The VI, V and IV oxidation states 379
15.8.2 The III, II and lower oxidation states 380
15.9 SILVER, $4d^{10}5s^1$, AND GOLD, $5d^{10}6s^1$ 382
15.9.1 Silver 382
15.9.2 Gold 384
15.10 THE ZINC GROUP 385
15.10.1 THE I STATE AND SUBVALENT COMPOUNDS 386
15.10.2 THE II STATE 388
PROBLEMS 390

CHAPTER 16

TRANSITION METALS: SELECTED TOPICS 391

16.1 COPPER OXIDE CERAMIC SUPERCONDUCTORS 391
16.2 CARBONYL COMPOUNDS OF THE TRANSITION ELEMENTS 395
16.2.1 Formulae 395
16.2.2 Bonding 396
16.2.3 Structures 397
16.2.4 Related species 398
16.3 METAL–ORGANIC COMPOUNDS 399
16.3.1 Metal–carbon σ bonding 399
16.3.2 Metal–carbon multiple bonding 401
16.4 π-BONDED CYCLOPENTADIENYLS AND RELATED SPECIES 402
16.5 THE ORGANOMETALLIC CHEMISTRY OF THE LANTHANIDES 403
16.6 ACTINIDE ORGANOMETALLIC CHEMISTRY 405
16.7 MULTIPLE METAL–METAL BONDS 406
16.7.1 Bonding 408
16.8 TRANSITION METAL CLUSTERS 409
16.8.1 Halide clusters 409
16.8.2 Carbonyl and related clusters 409
16.8.3 Gold clusters 411
16.8.4 Very large clusters 412
16.9 METAL–DIOXYGEN SPECIES 413
16.10 COMPOUNDS CONTAINING M-N_2 UNITS AND THEIR RELATIONSHIP TO NITROGEN FIXATION 415
16.10.1 Metal compounds with coordinated dinitrogen species 416
16.10.2 Bonding between N_2 and M 418
16.11 METAL–DIHYDROGEN COMPLEXES 419
16.11.1 Discovery 419
16.11.2 Properties and bonding 420
16.12 POST-ACTINIDE 'SUPERHEAVY' ELEMENTS 421
16.12.1 Background 421
16.12.2 Synthesis and properties 421

16.13 RELATIVISTIC EFFECTS 424
PROBLEMS 426

CHAPTER 17

THE ELEMENTS OF THE 'p' BLOCK 427

17.1 INTRODUCTION AND GENERAL PROPERTIES 427
17.2 THE FIRST ELEMENT IN A p GROUP 431
17.3 THE REMAINING ELEMENTS OF THE p GROUP 433
17.4 THE BORON GROUP, ns^2np^1 435
17.4.1 The elements, general properties and uses 435
17.4.2 The III state 439
17.4.3 The I oxidation state and mixed oxidation state compounds 441
17.5 THE CARBON GROUP, ns^2np^2 445
17.5.1 General properties of the elements, uses 445
17.5.2 The IV state 449
17.5.3 Hydride and organic derivatives 450
17.5.4 The II state 452
17.5.5 Reaction mechanisms of silicon 454
17.6 THE NITROGEN GROUP, ns^2np^3 456
17.6.1 General properties 456
17.6.2 The V state 461
17.6.3 The III state 464
17.6.4 Other oxidation states 468
17.7 THE OXYGEN GROUP, ns^2np^4 470
17.7.1 General properties 470
17.7.2 Oxygen 473
17.7.3 The other elements: the VI state 476
17.7.4 The IV state 478
17.7.5 The II state: the —II state 479
17.7.6 Compounds with an S-S or Se-Se bond 481
17.7.7 Sulfur–nitrogen ring compounds 482
17.8 THE FLUORINE GROUP, ns^2np^5 (THE HALOGENS) 483
17.8.1 General properties 483
17.8.2 The positive oxidation states 485
17.8.3 The —I oxidation state 493
17.9 THE HELIUM GROUP 494
17.9.1 Xenon compounds 495
17.9.2 Preparation and properties of simple compounds 496
17.9.3 Structures 499
17.9.4 Reactions with fluorides 500
17.9.5 Other rare gas species 502
17.10 BONDING IN MAIN GROUP COMPOUNDS: THE USE OF d ORBITALS 502
PROBLEMS 504

CHAPTER 18

SELECTED TOPICS IN MAIN GROUP CHEMISTRY AND BONDING 506

18.1 THE FORMATION OF BONDS BETWEEN LIKE MAIN GROUP ATOMS 506
18.2 POLYSULFUR AND POLYSELENIUM RINGS AND CHAINS 506
18.2.1 Polysulfur rings 507
18.2.2 Polysulfur ions 508
18.2.3 X-S_n-X species 509
18.2.4 Polyselenium and polytellurium rings and chains 510
18.3 NETS AND LINKED RINGS 512
18.3.1 Polyphosphorus compounds and related species 512
18.3.2 Polyarsenic compounds and other analogues 514
18.3.3 Mixed systems 515
18.4 CLUSTER COMPOUNDS OF THE p BLOCK ELEMENTS 516
18.4.1 General 516
18.4.2 Skeletal electron pairs 516
18.4.3 Boron subhalides 518
18.4.4 Naked metal cluster ions 518
18.4.5 Group 14 prismanes and related structures 520
18.5 POLYNUCLEAR IONS AND THE ACID STRENGTH OF PREPARATION MEDIA 521
18.6 SILICATES, ALUMINOSILICATES AND RELATED MATERIALS 523
18.6.1 Simple silicate anions: SiO_4^{4-} and $Si_2O_7^{6-}$ 523
18.6.2 Rings and chains: $[SiO_3]_n^{2-}$ 525
18.6.3 Sheet structures: $Si_4O_{10}^{4-}$ 526
18.6.4 Three-dimensional structures 527
18.6.5 Zeolites 527
18.6.6 Metal phosphates and metal phosphonates 529
18.7 MULTIPLE BONDS INVOLVING HEAVIER MAIN GROUP ELEMENTS 530
18.8 COMMENTARY ON VSEPR 531
18.8.1 Experimental electron densities 531
18.8.2 Limiting cases 532
18.8.3 AB_2 dihalides 532
18.8.4 The problem of s^2 configurations 533
18.9 BONDING IN COMPOUNDS OF THE HEAVIER MAIN GROUP ELEMENTS 534

CHAPTER 19

GENERAL TOPICS 536

19.1 ELECTRON DENSITY DETERMINATIONS 536

19.2 METAL-POLYCHALCOGENIDE COMPOUNDS 538
- 19.2.1 Metal–polysulfur compounds 538
- 19.2.2 Metal–polyselenide and –telluride complexes 539

19.3 FULLERENES, NANOTUBES AND CARBON 'ONIONS'–NEW FORMS OF ELEMENTAL CARBON 541
- 19.3.1 Fullerenes and their metal derivatives 541
- 19.3.2 Carbon nanotubes and giant fullerenes 543
- 19.3.3 Polyhedral structures formed by other materials: transition metal chalcogenides and metallacarbohedranes 543

19.4 DENDRIMERIC MOLECULES 544
- 19.4.1 Introduction and dendrimer synthesis 544
- 19.4.2 Applications 545

CHAPTER 20

BIOLOGICAL, MEDICINAL AND ENVIRONMENTAL INORGANIC CHEMISTRY 546

20.1 BIOLOGICAL INORGANIC CHEMISTRY 546
- 20.1.1 Introduction 546
- 20.1.2 The s elements in biochemistry 546
- 20.1.3 The p elements in biochemistry 547
- 20.1.4 The transition metals in biochemistry 549

20.2 MEDICINAL INORGANIC CHEMISTRY 556
- 20.2.1 Overview 556
- 20.2.2 Inorganic diagnostic (imaging) agents in medicine 556
- 20.2.3 Technetium imaging agents 557
- 20.2.4 Magnetic resonance imaging 558
- 20.2.5 Boron neutron capture therapy (BNCT) 559
- 20.2.6 Platinum-based drugs in cancer chemotherapy 559
- 20.2.7 Therapeutic gold drugs 560
- 20.2.8 Chelation therapy for metal poisoning 560

20.3 ENVIRONMENTAL INORGANIC CHEMISTRY 561
- 20.3.1 Phosphates in detergents: eutrophication 562
- 20.3.2 Chlorofluorocarbons: the ozone hole 563
- 20.3.3 The greenhouse effect 564
- 20.3.4 Acid rain 566
- 20.3.5 Other inorganic environmental problems 566

APPENDIX A
FURTHER READING 568

1 BOOKS 568

2 REVIEWS AND JOURNALS 572

3 BIBLIOGRAPHIES FOR PARTICULAR SECTIONS OF THE TEXT 572

4 ELECTRONIC ACCESS TO CHEMICAL INFORMATION 585

APPENDIX B
SOME COMMON POLYDENTATE LIGANDS 586

APPENDIX C
MOLECULAR SYMMETRY AND POINT GROUPS 589

INDEX 599

RELATIVE ATOMIC MASSES 609

PERIODIC TABLE OF THE ELEMENTS 610

Preface

Inorganic Chemistry continues to fascinate! The diversity of chemistry covered by the "inorganic" elements brings new and sometimes surprising compounds, insights into the chemistry of biological systems and applications in fields as diverse as polymers and medicine. The overall format of the book closely follows that of previous editions, with the earlier Chapters (1–8 and 13) concentrating on principles, Chapters 9–12, 14 and 15 covering the systematic chemistry of the elements, and the later Chapters (16–20) covering more specialised topics. The Periodic Table continues to be a vital tool in Inorganic Chemistry for the rationalisation and prediction of chemical trends.

In this edition, we have concentrated on a thorough update of material from previous editions, reordering some material, and presenting the material in a modern and readable style by means of boxes, margin notes and many redrawn diagrams. Additional Problems have been added at the end of many chapters as an aid-in-learning for the reader. Topics such as the use of supercritical fluids as solvents, the application of mass spectrometry in inorganic chemistry, carbonyl complexes of metals in high oxidation states, very large metal clusters and the isolation and chemistry of the superheavy elements are an illustration of some of the new and exciting areas which have been added or updated in this edition.

The amount of Inorganic Chemistry knowledge continues to grow at a substantial pace and this is reflected in the large number of journals which report Inorganic Chemistry research. At first, the student may be daunted by the task of finding out information on any particular Inorganic Chemistry topic. To aid this process, we have comprehensively updated the Bibliography, and added a new section on Electronic Sources of Information. The accessibility and power of electronic search tools means that time-consuming literature searches are now a thing of the past; journals themselves are being increasingly accessed in electronic form. Additionally, many excellent books covering general and specialised aspects of Inorganic Chemistry have been published in recent years, and many of these have been included in the reading list.

We are grateful to colleagues, students and readers who have provided us with feedback on previous editions, and are especially grateful to those inorganic chemists whose research is included in this book. We are also very grateful to Professor Martin Ystenes of the Norwegian Institute of Technology, Trondheim.

Ken Mackay
Bill Henderson
Ann Mackay

SI UNITS AND NAMES

UNITS USED IN THIS EDITION 3
THE NAMING OF GEOMETRICAL SHAPES 5

SI Units and Names

Many international scientific and engineering bodies have recommended a unified system of units and the SI (Système International d'Unités) has been generally adopted for use in scientific journals. The full scheme is not universally accepted, and some variant units are found. There are seven basic units:

Physical quantity	*Symbol for quantity*	*Name of unit*	*Symbol for unit*
length	l	metre	m
mass	m	kilogram	kg
time	t	second	s
electric current	I	ampere	A
temperature	T	kelvin	K
amount of substance	n	mole	mol
luminous intensity	I_v	candela	cd

Two supplementary units are radian (rad) for plane angles and steradian (sr) for solid angles. The SI system differs from the old metric system by replacing the centimetre and the gram by the metre and the kilogram.

Multiples of these units are normally restricted to steps of a thousand, and fractions by steps of a thousandth, i.e. multiples of $10^{\pm 3}$, but with the multiples 10,100,1/10 and 1/100 retained.

Fraction	*Prefix*	*Symbol*	*Multiple*	*Prefix*	*Symbol*
10^{-1}	deci	d	10	deca	da
10^{-2}	centi	c	10^2	hecto	h
10^{-3}	milli	m	10^3	kilo	k
10^{-6}	micro	μ	10^6	mega	M
10^{-9}	nano	n	10^9	giga	G
10^{-12}	pico	p	10^{12}	tera	T
10^{-15}	femto	f	10^{15}	peta	P
10^{-18}	atto	a	10^{18}	exa	E
10^{-21}	zepto	z	10^{21}	zetta	Z
10^{-24}	yocto	y	10^{24}	yotta	Y

It is important to note that 1 km^2 implies 1 $(km)^2$, that is 10^6 m^2 and *not* 10^3 m^2.

A range of units derive from these basic units, or are supplementary to them. Those commonly used by the chemist are shown in Table 1. The unit of force, the newton, is independent of the Earth's gravitation and avoids the introduction of g (the gravitational acceleration) into equations. The unit of energy is the joule (newton × metre) and of power, the joule per second (watt).

The gram will be used until a new name is adopted for the kilogram as the basic unit of mass, both as an elementary unit to avoid 'millikilogram', and with prefixes, e.g. mg.

All units not compatible with SI are being abandoned progressively, and the majority of such units are steadily, if slowly, disappearing from the literature. A few units with particular

TABLE 1 Units derived from the basic SI units, or supplementary to them

Physical quantity	*Name of unit*	*Symbol and definition*
force	newton	$N = m\ kg\ s^{-2}$ or $J\ m^{-1}$
pressure	pascal	$Pa = m^{-1}\ kg\ s^{-2}$ or $N\ m^{-2}$
energy, work, heat	joule	$J = m^2\ kg\ s^{-2}$ or $N\ m$
power	watt	$W = m^2\ kg\ s^{-3}$ or $J\ s^{-1}$
electric charge	coulomb	$C = s\ A$
electric potential	volt	$V = m^2\ kg\ s^{-3}\ A^{-1}$ or $J\ s^{-1}\ A^{-1}$
resistance	ohm	$\Omega = m^2\ kg\ s^{-3}\ A^{-2}$ or $V\ A^{-1}$
capacitance	farad	$F = m^{-2}\ kg^{-1}\ s^4\ A^2$ or $s\ A\ V^{-1}$
frequency	hertz	$Hz = s^{-1}$
temperature, t	degree Celsius (centigrade)	°C where $t/°C = T/K - 273.15$
area	square metre	m^2
volume	cubic metre	m^3
density	kilogram per cubic metre	$kg\ m^{-3}$
velocity	metre per second	$m\ s^{-1}$
angular velocity	radian per second	$rad\ s^{-1}$
acceleration	metre per (second)^2	$m\ s^{-2}$
magnetic flux density	tesla	$T = kg\ s^{-2}\ A^{-1}$ or $V\ s\ m^{-2}$
time	hour, year, etc. will continue to be used	

advantages will probably persist, as any system of units should combine logical consistency with a reasonable degree of convenience. Among such traditional chemical units are:

Litre. If multiples are restricted to $10^{\pm 3}$, then lengths would be confined to the metre and millimetre, giving volume units of m^3 and mm^3. It seems unlikely that a system restricted to values differing by 10^9 will find favour and the cm^3 and dm^3 are likely to be long with us. It is probable that the litre will survive as a convenient name for the cubic decimetre.

Ångström. The ångström unit ($10^{-8}\ cm = 10^{-10}\ m = 10^{-1}\ nm = 100\ pm$) was originally introduced as a unit of length for use on the interatomic scale. It continues to be widely used by crystallographers, though some replacement by the picometre, pm, is occurring.

Atomic mass. This is expressed relative to $^{12}C = 12.0000$ and is termed relative atomic mass, symbol A_r. *Relative molecular mass*, M_r, is the sum of the A_r values for each atom in the molecule.

Energy. As the joule is of the same order of magnitude, there are no reasons for retaining the calorie. Energies of chemical processes commonly fall into the convenient range of 10 to 10^3 kJ mol^{-1}, and kcal mol^{-1} will disappear. Chemists also use two other units in energy measurement: the *wave number*, *reciprocal centimetre* or *Kayser*, written as cm^{-1}, and the *electron volt*, eV. The latter is strictly the energy acquired by one electron falling through a potential of one volt, but eV is commonly used for the molar unit found by multiplying the strict value by Avogadro's constant. The main advantages of the electron volt are its close relation to the methods used to measure certain parameters, such as ionization potentials, and its larger size. At about 10^2 kJ mol^{-1}, it is convenient for the larger chemical energies and its use is likely to continue for many years. The cm^{-1} is not a unit of energy but is used in spectroscopy. The relations underlying this application are described in Chapter 7.

The SI system is for reporting precise measurements in a fundamentally self-consistent way, and does not require that other units should never be used. It is not intended to preclude the use of 'working units' and units used in a nonrigorous context. Laboratory workers continue to fractionally distil at pressures measured in millimetres of mercury, autoclave at hundreds of atmospheres and read temperatures in degrees Celsius.

In Table 1 are listed the units which are derived from the basic SI units or their supplements. In Table 2 we list those units which are contrary to SI and are being phased

TABLE 2 Commonly occurring units which are contrary to SI

Unit	*Quantity*	*Equivalent*
(A) Units differing from SI units by powers of 10		
ångström (Å or A)	length	10^{-10} m = 10^{-1} nm = 10^2 pm
litre (l or L)	volume	10^{-3} m^3 = dm^3
dyne (dyn)	force	10^{-5} N
erg	energy	10^{-7} J
mho (siemens or reciprocal ohm)	conductance	Ω^{-1}
bar	pressure	10^5 Pa
poise	viscosity	0.1 Pa s
tonne (t)	mass	10^3 kg
(B) Other units		
calorie (cal)	energy	I.T. cal = 4.186 8 J; 15° cal = 4.185 5 J; thermochemical cal = 4.184 J
electron volt (eV)	energy	1.602 1 × 10^{-19} J
(electron volt per mole, also symbolized eV = 96.484 kJ mol^{-1})		
atmosphere (atm)	pressure	101.325 kN m^{-2}
millimetre of mercury (mmHg) or torr (Torr)	pressure	133.322 N m^{-2}
atomic mass number (amu or u = 1/12 mass of ^{12}C)	mass	1.660 41 × 10^{-27} kg

out. These have conversion factors which are exactly defined, apart from the electron volt and the atomic mass unit which are given in terms of the best experimentally determined factors. The values of physical constants in SI units are listed in Table 3.

Units used in this edition

A textbook written entirely in SI units faces some disadvantages, particularly as many literature sources still in use are not in SI units and scientists need to be able to convert existing data between systems. With this in mind, the following convention has been adopted in this edition.

Length. The SI system of m, mm, and smaller fractions is used. Notice that many interatomic distances (given originally in ångströms = 10^{-10} m) were known to an accuracy expressed by two decimal places in ångströms, e.g. 1.07 Å. It follows that the most convenient SI-allowed multiple is the picometre, pm, so that values become whole numbers of picometres, e.g. 107 pm, and more accurate values are distinguished by having figures after the decimal point as in 107.6 pm = 1.076 Å. This usage is more convenient than using fractions of the nanometre, as in 0.107 6 nm.

Energy. Most values are given in kJ mol^{-1} and kcal is not used. Where appropriate, cm^{-1} and eV are also used and conversion factors are included in tables in these units. Occasionally, particularly when dealing with magnetic resonance, frequencies in Hz are found in place of wave numbers in cm^{-1}. As 1 Hz is one wave per second and 1 cm^{-1} is one wave per cm, the two are connected by the speed of light. Table 4 shows the interconversion factors for all these units.

Temperatures. Temperatures are quoted in degrees Celsius (°C), or in Kelvin (K)—*note* no degree symbol is used.

Electrical units. These do not occur widely in this text. Note that the interaction between charges is modified by the permittivity of a vacuum, ε_0. Thus the non-SI factors e^2/r, which occur in the wave equation, for example, now become $e^2/4\pi\varepsilon_0 r$.

Other parameters. SI units are used where exact values are stated. Note that the litre is used as the name of the strict SI dm^3. There is some discussion that the gram-molecule

SI UNITS AND NAMES

TABLE 3 Values of physical constants

Physical constant	*Symbol*	*Recommended value*
Speed of light in a vacuum	C_0	$299\,729\,458$ m s^{-1}
Atomic mass unit	u	10^{-3} kg mol^{-1}
		$1.660\,565\,5 \times 10^{-27}$ kg
Mass of proton	m_p	$1.672\,623\,1 \times 10^{-27}$ kg
Mass of neutron	m_n	$1.674\,954\,3 \times 10^{-27}$ kg
Mass of electron	m_e	$9.109\,389\,7 \times 10^{-31}$ kg
Charge on proton or electron (−)	e	$1.602\,177\,33 \times 10^{-19}$ C
Boltzmann constant	k or k_B	$1.380\,658 \times 10^{-23}$ J K^{-1}
Planck constant	h	$6.626\,075\,5 \times 10^{-34}$ J s
Permeability of a vacuum	μ_0	$4\pi \times 10^{-7}$ J s^2 C^{-2} m^{-1}
Rydberg constant	$R_\infty = \dfrac{\mu_0^2 m_e e^4 c^3}{8h^3}$	$1.097\,373\,177 \times 10^7$ m^{-1}
Bohr magneton	$\mu_B = \dfrac{eh}{4\pi m_e}$	$9.274\,078 \times 10^{-24}$ J T^{-1}
Avogadro constant	N_A or L	$6.022\,136\,7 \times 10^{23}$ mol^{-1}
Gas constant	R	$8.314\,510$ J K^{-1} mol^{-1}
'Ice-point' temperature	T_{ice} or T_0	273.150 K
		($RT_0 = 2.271\,081 \times 10^3$ J mol^{-1})
Permittivity of a vacuum	$\varepsilon_0 = (\mu_0 c^2)^{-1}$	$8.854\,187\,82 \times 10^{-12}$ F m^{-1}
Faraday constant	$F = Le$	$9.648\,456 \times 10^4$ C mol^{-1}
RT ln $10/F$ at 298 K		5.916×10^{-2} V
Bohr radius	a_0	$5.291\,770\,6 \times 10^{-11}$ m
Molar volume of ideal gas (273.15 K, 1 atm)	$V_0 = RT/P_0$	$2.241\,038\,3 \times 10^{-2}$ m^3 mol^{-1}

TABLE 4 Conversion factors

	kJ mol^{-1}	cm^{-1}	eV	MHz	kcal mol^{-1}
kJ mol^{-1}	1	83.626	$1.036\,4 \times 10^{-2}$	$2.506\,2 \times 10^6$	0.2394
cm^{-1}	$1.195\,7 \times 10^{-2}$	1	$1.239\,4 \times 10^{-4}$	$2.997\,9 \times 10^4$	2.858×10^{-3}
eV	96.484	8068.3	1	$2.418\,8 \times 10^8$	23.063
MHz	$3.990\,3 \times 10^{-7}$	$3.335\,6 \times 10^{-5}$	$4.134\,4 \times 10^{-9}$	1	$9.534\,5 \times 10^{-8}$
kcal mol^{-1}	4.184	349.83	$4.335\,9 \times 10^{-2}$	$1.048\,7 \times 10^7$	1

might be replaced by the kilogram-molecule but most chemists would find this unacceptable.

In the text, numerical values for physical properties, such as atomic radii or oxidation–reduction potentials, are chosen from consistent sets and are quoted mainly for purposes of comparison. Much more accurate values are often available for particular data, and any calculations of physical significance should use such values, which can be found in a number of critical compilations of data or in the original literature.

Formulae are commonly used in place of chemical names where the former are clearer and less clumsy. Equations which are written unbalanced are used either to show the major product of a reaction, or to indicate the variety of products without being definite about their relative proportions.

In addition to the symbols for fundamental constants given in Table 3, a number of other symbols and abbreviations are to be found in the text. These are:

S	entropy
H	enthalpy (heat content)
K	equilibrium constant
ccp	cubic close-packed
hcp	hexagonal close-packed
bcc	body-centred cube
m or m.p.	melts or melting point
b or b.p.	boils or boiling point
d	decomposes
subl	sublimes

The naming of geometrical shapes

Since the geometry of solids was first worked out by the ancient Greeks, much of the terminology comes from Greek and occasionally causes misunderstanding. A regular solid is named for the number of faces, and the Greek root is *hedron*. Thus the tetrahedron is named for its four faces. The plural is formed by turning *-ron* into *-ra*, and similarly for the adjective. Thus 'tetrahedra' is the plural and 'tetrahedral' the adjective. Since we normally deal with atoms arranged in a geometrical form around a central atom, the chemist is usually more interested in the number of points or *vertices* (note the singular is *vertex*) in a figure, rather than the number of faces. While a tetrahedron has four points and four faces, the number of points and faces differs for all other solids. Thus an octahedron, familiar as the shape found for 6-coordination, has *eight* faces (hence octa-) and six vertices.

A further relation of interest is that of *capping* which means placing a further atom above a face so that it is at equal distances from each of the atoms which define that face. If we cap planar figures we get *pyramids*: capping a square gives a square pyramid and capping it again on the other side gives a square *bipyramid*. The capping atom is often termed the *apex* (plural *apices*). It may be placed at any distance from the face, but if the distance is equal to the length of the edge of the capped figure, then a special *regular* shape results. Thus a regularly capped triangle (or regular trigonal pyramid) is a tetrahedron and the octahedron is a regular square bipyramid. If regular figures are capped, further relationships emerge. Thus the eight atoms which regularly cap the faces of an octahedron, themselves form a cube, and if the six faces of a cube are regularly capped, the caps form an octahedron.

We also note some interrelationships involving the cube. If every second vertex of a cube is selected, these four points define a tetrahedron, so a cube may be seen as two identical interpenetrating tetrahedra (see Fig. 13.9). If these two tetrahedra are now altered regularly so that one is elongated and one is compressed (equivalent to pulling two pairs of atoms defining opposite face diagonals out of the face, forming triangles bent about the diagonal) the result is a *dodecahedron*. If a tetrahedron is regularly capped on all four faces, the eight atoms lie at the corners of a cube, the figure resulting corresponds to the formula A_4B_4 and is called a *cubane*.

Finally, one class of less regular figures also occurs among chemical structures—the *prisms*. A prism is formed by connecting two regular plane figures held in an eclipsed position. For example, a general pentagonal prism has two regular pentagonal faces and five rectangular ones. Again, special figures result if all the edges are equal—a cube is a special case of the square prism, for example. The *antiprism* is formed if the regular plane figures are placed in a staggered configuration, forming for example a square antiprism. The regular trigonal antiprism is the octahedron.

Examples of all these figures will be found in later chapters. It is extremely valuable to study solid models, or a good computer simulation program, as an aid to understanding the relationships between them.

Table 5 summarizes the properties of the commoner shapes met in chemistry.

SI units and names

Table 5 Some common shapes

Shape	*Name*	*Significant geometrical properties*	*Examples (Figures)*
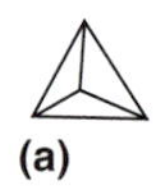 (a)	Tetrahedron (a) (for relationship with a cube, see Fig. 13.9)	4 faces; 4 points; 6 edges	4.3a; 5.1c
	Trigonal pyramid	as above, 3 edges longer than others	17.66c
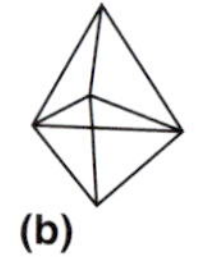 (b)	Trigonal bipyramid (b)	6 faces; 5 points; 9 edges	4.4a; 9.8a; 14.29
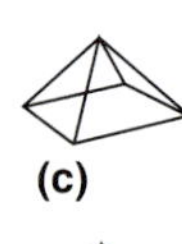 (c)	Square pyramid (c)	4 triangular + 1 square face; 5 points; 8 edges	13.11b; 14.13; 17.66d
 (d)	Octahedron (d)	8 faces; 6 points; 12 edges	4.5a; 5.1a; 15.11
	face-bridged	8 bridging atoms	15.19
	edge-bridged	12 bridging atoms	15.9
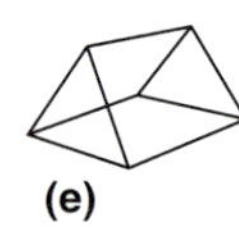 (e)	Trigonal prism (e)	2 triangular + 3 rectangular faces; 6 points; 9 edges	5.10c 13.11a
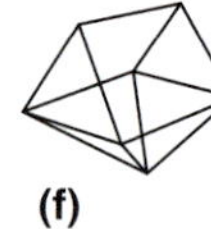 (f)	monocapped (f)	cap on rectangular face	15.1
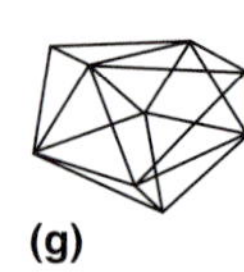 (g)	tricapped (g)	caps on all 3 rectangular faces	15.23; 17.40c
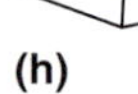 (h)	Pentagonal pyramid (h)	1 pentagonal + 5 triangular faces; 6 points; 10 edges	9.8b; 14.17a
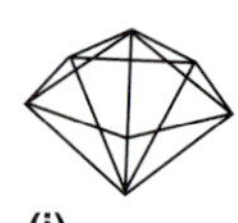 (i)	Pentagonal bipyramid (i)	10 faces; 7 points; 15 edges	14.12; 14.17c
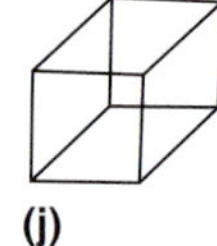 (j)	Cube (j)	6 faces; 8 points; 12 edges	5.1b; 5.12g; 15.16
	cubane	A_4B_4 at alternate vertices	15.15
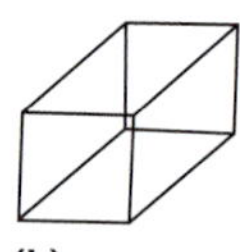  (k)	Square prism (k)	as cube, four edges different length	14.33
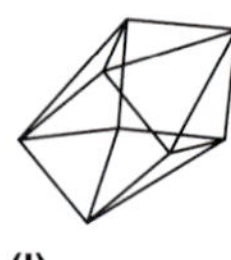 (l)	Square antiprism	2 square + 8 triangular faces; 8 points; 16 edges	12.5; 15.5b
	monocapped (l)	cap on square face	18.21

TABLE 5 (Contd.)

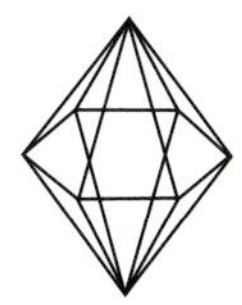

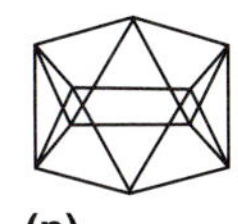
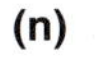

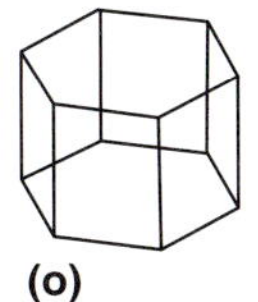
(o)

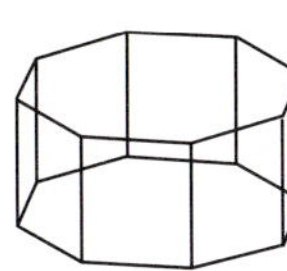
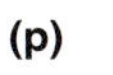

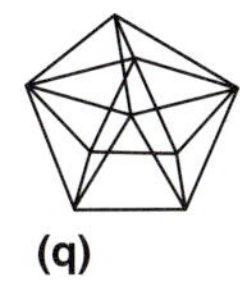
(q)

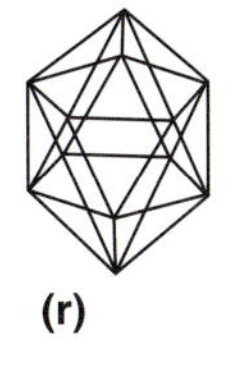
(r)

Name	*Significant geometrical properties*	*Examples (Figures)*
Hexagonal bipyramid (m)	12 faces; 8 points; 18 edges	12.6
Pentagonal antiprism (n)	2 pentagonal + 10 triangular faces; 10 points; 20 edges	16.7a; 16.9
Hexagonal prism (o)	2 hexagonal + 6 rectangular faces; 12 points; 18 edges	5.12e (outline); 16.7b
Octagonal prism (p)	2 octagonal + 8 rectangular edges; 16 points; 24 edges	16.11a
Dodecahedron (q)	12 faces; 8 points; 16 edges	15.2; 15.18
Icosahedron (r)	20 faces; 12 points; 30 edges	17.7d

Note: examples are chosen to give a variety of representations of the shapes.

Introduction

CHAPTER 1

1.1	INORGANIC CHEMISTRY AND THE DISCOVERY OF THE ELEMENTS	8
1.2	DEVELOPMENT	8
1.3	RECENT ADVANCES	10
1.4	INORGANIC NOMENCLATURE	13
1.5	APPROACH TO INORGANIC CHEMISTRY AND FURTHER READING	14
PROBLEMS		17

1.1 Inorganic chemistry and the discovery of the elements

Chemistry is one of the oldest and most wide-ranging of the sciences, and hence of human knowledge and endeavour. It had already grown sufficiently by the end of the 19th century to be conveniently divided into the three classical branches of *inorganic, organic* and *physical* chemistry. Inorganic chemistry covers the properties and reactions of all the chemical elements apart from carbon—now exceeding 110.

A major theme of inorganic chemistry over the last two millennia has been the discovery and characterization of the elements themselves. This continues to the present day in the synthesis of ultra-high atomic number elements by high energy bombardment (Section 16.12).

The discovery of the elements is summarized in Table 1.1. If the pattern of discovery is plotted against time, a curve is obtained which mirrors the pattern of development in many sciences. A long slow period of completely empirical advance in the ancient world was followed by a phase mainly of preservation and rediscovery through the Arab alchemists and in India and China. For the century up to AD 1750, some of the basic ideas of what we now call chemistry were developed from more deliberate investigations. From AD 1750 up to the first half of the 20th century, there was a sharply accelerating pattern of discovery as theory and technique advanced in parallel. Within this period we see individual spurts reflecting specific advances, like the 18th-century studies of gases, the early 19th-century use of electrolysis to isolate the very active metals, or the recognition of the Rare Gas Group which gave five new elements in five years. Eventually the pace slowed, in the decade to 1940, because there were 'no new worlds to conquer' and all the elements up to uranium had been identified. This was not the end of the story, as it turned out that further post-uranium elements could be synthesized. This phase is now slowing down, reflecting the decreasing intrinsic stability of the nuclei. Whether this is finally the end of the story of the elements is not yet clear (compare Section 16.12). The overall pattern, found in many other developing fields, is of empirical discovery, acceleration fuelled by the interaction of greater understanding and improved methods, then maturity when the pace of change slows. Often, new accelerations start up from the mature phase, as unexpected observations or new ideas trigger off further developments.

1.2 Development

While we have followed the tale of the elements, the growth of inorganic chemistry as a whole followed a similar pattern. Inorganic chemistry was the first of the chemical sciences to flower in the course of the Scientific Revolution, and most of the work leading to the formulation of the atomic theory was carried out on inorganic systems, especially the gases and simple compounds like the nitrates, carbonates or sulfates. A critical advance in technique was the development of ever more accurate measures of the quantity of material—both by weighing and by measurement of gases. Once it could be established that a particular substance had the same composition when prepared by different routes (for example, an oxide prepared from the metal and air, from heating the carbonate, from precipitating the hydroxide from solution and igniting) the way was open to following changes quantitatively, to formulating generalizations like 'the Law of Constant Composition', and ultimately to Dalton's atomic theory. It is worth remarking that even the most sophisticated modern experiment depends ultimately on accurate measurement of weight changes.

TABLE 1.1 The discovery of the elements

Date range	*Number of elements discovered*	*Comments*
Prehistoric	3	C, S, Au which occur native, i.e. uncombined
ca. 3000 BC	5	Ag, Cu, Pb, Sn, Hg with readily processed ores
ca. 1000 BC	1	Fe requiring higher temperature reduction
ca. 500 BC	1	Zn ca. 90% pure
Up to 1650	4	As, Sb, Bi: Zn rediscovered
1650–1700	1	First dated discovery: P in 1669
1700–50	3	Co, Ni and native Pt
1750–75	7	First gases H, N, O, Cl and Ni, Mn, Bi*
1775–1800	5	Cr, Mo, W, Te, Ti (finally pure in 1910)
1800–25	18	Active metals, Li, Na, K, Mg to Ba: heavier metals Ce, Ir, Os, Pd, Rh, Zr: also B, Cd, I, Se
1825–50	9	Br, Si, Be, Al, V, La, Ru, Th, U
1850–75	5	Rb, Cs, Ga, Tl, Nb {He seen in solar spectrum}
1875–1900	approx. 11	5 inert gases: F, Ge: radioactive Po, Ra, Ac: some lanthanides
1900–25	approx. 10	Rn, Ta, In, Hf, Re, Pa, lanthanides
1925–50	11	2 lanthanides: radioactive Tc, Pm, Fr, At: man-made post-uranium Np, Pu, Am, Cm, Bk
1950–75	10	Last 2 purified lanthanides: 8 man-made
1975–today	approx. 12	Man-made (a few atoms only)

Notes: 1. Compare Tables 2.5 and 2.6 for the names and periodic positions of the elements. 2. Many dates of discovery are approximate, as the existence of many elements was recognized anything from a few months up to a century before final purification. The very similar lanthanides present particular difficulties of definition. 3. *Bismuth known earlier but confused with lead. 4. See the reference by Ringnes (Appendix A) for a very readable account of the origins of the names of the chemical elements.

In the first half of the 19th century, not only had more than half of the elements been isolated but a great many of their simpler compounds had been studied. It is remarkable that explosive nitrogen trichloride or highly corrosive hydrogen fluoride were under study around 1800. By contrast, only a few simple organic compounds were known by 1820 and little progress was being made in organic chemistry as much of the effort was directed to extremely complex materials like milk or blood. By the middle of the 19th century came the period of

BRANCHES OF CHEMISTRY

Since the chemistry of one element, carbon, is so enormously ramified—probably similar in extent to the total chemistry of the remaining 109 put together—it has traditionally made up the separate field of organic chemistry. The bridging discipline covering the organic chemistry of the inorganic elements, *organo-element* or *organometallic* chemistry, is an extensive field to which we shall make substantial reference. In addition, part of the chemistry of carbon, covering the element and simpler compounds like its oxides and oxyions or the carbides, is traditionally covered in inorganic chemistry. No attempt is made to draw rigid boundaries.

The detailed study of energy changes, reaction mechanisms, much of bonding theory, the chemistry of polymers, chemistry which occurs at surfaces and interfaces, the behaviour of metallic systems—all these fall into physical chemistry. Again, there are no rigid demarcations, and much of the most exciting work is done at the points of overlap. Other long-standing subdivisions include *analytical* chemistry and *theoretical* chemistry.

As might be expected, the huge expansion of chemistry in the last few decades has led to further subdivision within inorganic chemistry—such as phosphorus chemistry or transition metal chemistry—as well as the defining of new fields with substantial overlap with inorganic chemistry. The latter include materials science, catalysis, inorganic biochemistry, computational chemistry and many more.

THE ROOTS OF INORGANIC CHEMISTRY

The origins of inorganic chemistry are ancient. Observation followed by what we now describe as empirical experiments led to the slow development of new materials from the early stages in human history. Thus beads of glass and of ceramics are found in ancient Egyptian burials and pottery was made by the earliest civilizations. Considerable control was achieved: black or red pottery was made by reducing or increasing the proportion of air, and colours and glazes had developed to a high degree of sophistication by 500 BC.

The discovery of metal extraction and processing was a most important theme and led to substantial mastery of the technologies by ancient craftsmen. Gold is usually found naturally as the free metal and, as shown by grave goods, has always been highly valued. Modern analysis of ancient artefacts shows that it was understood that the addition of small proportions of silver or copper gave a harder, more wear-resisting metal, and gave desirable variations in colour. The metal contents of gold coins held to highly consistent standards which correlated through many countries in a chain of related weights and gold contents—for example, from Macedonia to India in the 5th century BC. Copper has been known for about 7000 years and has been extracted from sulfide ores for around 5500 years. Small metal items like beads and bracelets are found in graves for several centuries before larger products like tools or weapons, suggesting a period where manufacture was difficult and not well understood. Analysis on ancient kilns shows that temperatures up to 1200°C were achieved in Bronze age copper smelting. Once production methods had evolved to the larger scale, it was rapidly found that alloying copper with tin to give bronze, or with zinc to give brass, produced metals of superior properties.

Tin, which is fairly inert to air, was probably prepared in a pure form many centuries before the more reactive zinc. Iron requires much more sophisticated treatment, and can be extracted only at the high temperatures achieved with air blown through a bed of charcoal. The carbon is also necessary to remove the oxygen from the ore. Even so, early temperatures did not reach the melting point, so casting was not possible and the metal was shaped by hammering. As iron weapons are much superior to bronze, it is thought that the initial discovery was kept secret for a millennium and knowledge of iron only became widespread after the destruction of the Hittite empire about 1200 BC (recent archaeology suggests this picture is oversimplified). Compared with iron, isolation of the other elements found up to the 18th century is relatively easy, so iron-working represents a peak in technological achievement which lasted for something like 4000 years.

spectacular advance in organic chemistry, followed around 1900 by a great upsurge of interest in physical chemistry. These advances meant nearly a century of comparative neglect of inorganic chemistry.

Of course, very important advances were made, including the formulation of the Periodic Table, the discovery and exploitation of radiochemistry, and the classical work on non-aqueous solvents and on the complex chemistry of the transition elements, but it was not until the 1930s that the modern upsurge of interest in inorganic chemistry got under way. Among the seeds of this renaissance were the work of Stock and his school on volatile hydrides of boron and silicon, of Werner and others on the chemistry of transition metal complexes, of Kraus and Walden on non-aqueous solvents, and the work of a number of groups on radioactive decay processes. At the same time, the theories which play an important part in modern inorganic chemistry were being formulated and applied to chemical problems. The discovery of the fundamental particles and the structure of the atom culminated in the development of wave mechanics, which is the basis of all modern approaches to valency and bonding. This theory is outlined in Chapter 2, and its application to molecular structure is given in Chapters 3 and 4. A little later, the effect of an atom's environment on the energy of its d electrons was brought into the treatment of transition metal compounds in the crystal field theory which is discussed in Chapter 13.

1.3 Recent advances

All these developments prepared the ground for the expansion of inorganic chemistry, starting in the 1950s, which was stimulated both by developments on the academic side in experiment and theory, and by the demand for new materials and for knowledge of many elements hitherto scarcely studied.

The advent of atomic energy focused attention on heavy transition elements and lanthanides (for example, the chemically very similar Zr and Hf have quite different neutron absorption properties). Similarly, the growth of electronics, followed by computers, led to growth in the chemistry of lesser-known Main Group elements involved in semiconductors, such as Ge, Ga, In and Se. A further significant change in the latter half of the 20th century was the very rapid growth in the number of working scientists and technologists, allowing simultaneous growth throughout chemistry. In earlier times, fields expanded only at the price of relative neglect in other areas.

Starting in the 1950s, transition element chemistry grew from 10% of Honours courses to become the dominant area of interest for several decades, largely as a result of the strong mutual stimuli of experimental and theoretical advances. More recent growth has encompassed fields such as low oxidation state chemistry, organometallic compounds, metal cluster chemistry, dendrites and other macromolecules, and multiple metal–metal bonding (compare Chapter 16). A decade or so later came a similar expansion in Main Group chemistry, building from the more traditional compounds into a substantial range of new species, including rare gas compounds, compounds containing chains, rings or clusters of like atoms, and unusual oxidation states stabilized by specially designed ligands (compare Section 17.9 and Chapter 18). All these developments were led by advances in preparative chemistry and in efficient methods for separating and rapidly characterizing new compounds (see Chapter 7).

Far from slackening, the pace has further increased. Through the turn of the millenium, we live with a continuing headlong expansion in inorganic chemistry, fuelled by new methods, new theories, new fields of interest like metals in biological systems, the search for new materials, new catalysts, more output for less pollution, and many other driving forces. Even unstable elements like technetium are finding uses in medicine, and highly radioactive isotopes, such as americium-241, find a variety of applications, including household smoke detectors. Such interests have generated research in every corner of the Periodic Table, and there are now no elements, apart from the very unstable heavy ones, where there is not a very substantial body of knowledge available.

This rapid growth of inorganic chemistry continues to make it a very lively and exciting subject in which to work and teach but it does lead to problems from the student's point of view. Textbooks tend to be out of date by the time they are published and the treatment of each subject changes as new discoveries are made.

The *critical temperature* of a superconductor is that temperature below which the phenomenon of resistance-free conduction occurs.

Particularly striking examples arise when the advance arouses widespread interest outside the specialist field. An illustration was provided by the announcement, late in 1986, of a superconductor whose critical temperature was around 40 K. This followed a long period when the highest critical temperature found had risen only very slowly from about 5 K to around 23 K. The new superconductors were oxide phases involving copper and elements like the lanthanides and the alkaline earth metals. Excitement was enormous, as higher temperature superconductors have tremendous potential for all electrical devices. There was a very rapid exploration of the chemistry. A superconducting phase of major interest is $YBa_2Cu_3O_{7-x}$, where x is around 0.1, and the pattern of exploration is exactly what any inorganic chemist in the last hundred years would have followed—basically study of complex oxides of related elements, guided by the Periodic Table (see Section 16.1 for a full review including other recently reported classes of superconducting materials).

Another example was the discovery, leading to the 1997 Nobel Prize, of new allotropic forms of carbon—the polyhedral carbon species commonly known as fullerenes, discussed in greater depth in Section 19.3. Coming after at least a century when the only established allotropes of carbon were diamond and graphite, this discovery initiated a wave of studies of these materials, producing new inorganic and organic derivatives, novel materials with interesting physical and electronic properties and many physical and theoretical studies.

Such examples, where the driving forces range from the excitement of new types of molecule through to expectations of major practical applications, are representative of many advances. Some novel and unexpected discovery triggers a period of intense interest, where rapid and widespread exploration occurs, *heavily based on the pattern of previous knowledge.* To a substantial extent inorganic developments are rooted in the relationships of the Periodic Table and established systematic chemistry. The incidence and progress of the more novel new discoveries and growing points are unpredictable, and this should give pause to those who would attempt too closely to guide the development of science into areas deemed to be more 'relevant' to the problems of the day.

A textbook must attempt to reflect both the steadily growing core of basic material, and the areas of current interest and excitement. An introductory text can only sample, while an advanced text will make a valiant attempt to cover all areas of current interest.

Theory. In the area of theory, the inorganic chemistry student is presented with a number of approaches at different levels of sophistication. Because chemical entities and their interactions are relatively complex, chemistry is much less 'theory-led' than is physics. The power and sophistication of chemical computation is steadily increasing, but it is still the case that only quite simple inorganic chemical observations can be described *exactly* by theory. Much insight comes from approximate methods, and currently we are faced with a range of theoretical approaches varying in power and degree of approximation. While there have been arguments about which of several alternatives is the best, in most cases we are content to use overlapping and even apparently conflicting theories, depending on the specific application. For example, many species may be described in electrostatic terms—as charged ions, dipoles, etc.—and the energy changes calculated by electrostatics. The same species may equally well be described in terms of covalent bonding, with a theoretical approach which has its base in quantum mechanics. Often, neither approach gives an exact answer because approximations are needed to bring the calculation within the compass of even the most powerful computer. In general, that approach is chosen which gives the most convenient answer to a specific problem, and different methods may be used to tackle different parts of the same problem. It follows that there is no one answer to a question like 'is this compound ionic or covalent?', but rather an understanding that either description is more or less useful, and more or less of an approximation.

Difficulties can arise where a relatively simple approach allows the rationalization and systematization of a particular body of data. Because the approach is fairly simple, it is usually not complete—that is, there will be exceptions and anomalies. If the model is reasonably wide-ranging, it is worth retaining and using it even after cases appear which are not covered. On occasion, two different partial models will be used even though they overlap and are not fully compatible. Such situations are quite common, and usually do not greatly disturb the scientist working in the field. They can be confusing to the student on first acquaintance, as there is the feeling that only one can be 'right', and we tend to use fairly high-powered words like 'law', 'theory' or 'principle' to describe them. If they are seen as *partial* descriptions or models, to be used as convenient, many of these problems disappear for the chemist (however much they disturb the philosopher of science!).

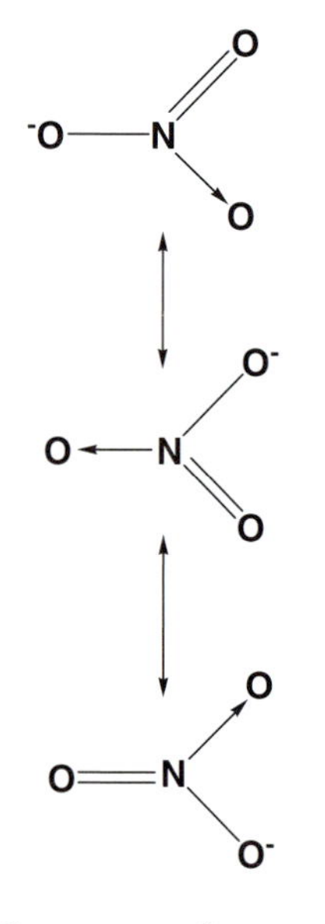

Resonance forms of the nitrate ion

In applying wave mechanics to chemistry, the two common approaches have been the *valence bond* and *molecular orbital* methods. Each is a different approximation to the wave equation for a system, and they converge to the same answer for very simple systems. For polyatomic molecules, approximation is essential and the theories are used side by side. Older preferences were for the valence bond approach which is closer to the classic picture of a molecule as linked together by discrete electron pair bonds. For example, the partial double-bonding in a species such as the nitrate ion was described in terms of 'resonance' between contributing forms, each of which was described in terms of single and double bonds.

This way of thinking is now favoured rather less than the molecular orbital approach, which discusses such species in terms of *delocalized* bonds extending over all the molecule (see Section 4.4). In other areas, such as properties of excited states and of species with extended multiple bonding, the molecular orbital theory is more satisfactory. In this text, the structures of molecules and ions are described largely in terms of the molecular orbital theory.

Similarly, in transition metal chemistry, *ligand field theory,* which deals with compounds which have valency electrons in the d orbitals, subsumes the electrostatic *crystal field theory* and also wave-mechanical aspects giving a molecular orbital treatment of transition metal compounds with multi-centred bonds. Again the treatments are at a number of levels of generality and approximation, and are often used in tandem.

All these changes are reflected in the succeeding pages, but the reader will realize that the rapid rate of growth means that any text is somewhat out of date by the time it is published and the review literature should be consulted for recent advances.

1.4 Inorganic nomenclature

The nomenclature of inorganic chemistry was put on a definitive basis by the publication of the IUPAC (International Union of Pure and Applied Chemistry) Rules in 1957, 1970 and 1990. These rules define a systematic method for naming all inorganic compounds, but they also allow the retention of a number of trivial names which are well established. (See Section 2.12 and Table 2.7 for Periodic Table nomenclature.)

The principles of systematic naming are straightforward and are tabulated in the box.

Notice that, according to rule (iii), all polyatomic ions have names ending with -ate. This must not be confused with the trivial naming of oxygen anions where the endings -ate, -ite, etc., are used to indicate the oxidation state (see below). In systematic naming, all such anions end with -ate and the oxidation state is shown by the stoichiometry or by Roman numerals. Thus, $SnCl_6^{2-}$ is hexachlorostannate(IV), $SnCl_3^-$ is trichlorostannate(II), SO_4^{2-} is tetraoxosulfate (or tetraoxosulfate(VI)) and SO_3^{2-} is trioxosulfate. Some examples are given in Table 1.2.

The system outlined above gives a means of providing an unambiguous systematic name for any inorganic compound, but see the box titled 'Chemistry and communication' on p. 15. The system is intended to be flexible and a considerable number of familiar names are retained for common use. However, it is important to have unique and unambiguous names for many purposes, such as safety, international trade and legal definition.

The following points may be particularly noted.

(i) The terminations -ous and -ic may be retained for cations of elements with only two oxidation states, as in ferrous and ferric or stannous and stannic.

(ii) In anions, the termination -ite to distinguish a lower oxidation state is retained in such cases as nitrite, sulfite, phosphite and chlorite. Similarly the hypo- ... -ite method

TABLE 1.2 Examples of systematic inorganic nomenclature. Of the detailed rules for the order of citation of ligands in complexes we need only note that anionic ligands come before neutral and cationic ones and 'oxo' is often dropped from the names of familiar oxyions

Formula	Name	Comment
$NaCl$	Sodium chloride	Rules i, ii and v: note that mono- is usually omitted as a prefix
SiC	Silicon carbide	
As_4S_4	Tetra-arsenic tetrasulfide	
Cl_2O	Dichlorine oxide	
OF_2	Oxygen difluoride	
$KICl_4$	Potassium tetrachloroiodate	
$FeCl_2$	Iron dichloride (or iron(II) chloride)	Rules i, iii, iv and v. Notice that the use of roman numerals in rule iv, and the prefixes of rule v, often provide alternative names; superfluous information is avoided
$Pb_2^{II}Pb^{IV}O_4$	Trilead tetroxide (or dilead(II) lead(IV) oxide)	
$K_4[Fe(CN)_6]$	Potassium hexacyanoferrate(II)	
$K_3[Fe(CN)_6]$	Potassium hexacyanoferrate(III)	
$Na(SO_3F)$	Sodium trioxofluorosulfate	
NaH_2PO_4	Sodium dihydrogen tetraoxophosphate	
$Na(NH_4)HPO_4.4H_2O$	Sodium ammonium hydrogen phosphate tetrahydrate	Note that multiple groups are written separately (often all run together) in an order which is defined by the rules. Notice also that oxo- is often dropped from the names of familiar oxyanions
$BiOCl$	Bismuth oxide chloride	
$VOSO_4$	Vanadium(IV) oxide sulfate	
$ZrOCl_2.8H_2O$	Zirconium oxide dichloride octahydrate	
$Li(AlH_4)$	Lithium tetrahydroaluminate	
$NH_4[Cr(SCN)_4(NH_3)_2]$	Ammonium tetrathiocyanatodiamminechromate(III)	
$[Co(CO_3)(NH_3)_4]Cl$	Carbonatotetra-amminecobalt(III) chloride	
$[Be_4O(CH_3COO)_6]$	(See Fig. 10.9) μ_4-oxo-hexa-μ-acetatotetraberyllium	

BASIC RULES OF INORGANIC CHEMISTRY NOMENCLATURE

(i) The cation, or electropositive component, of a compound has its name unmodified.

(ii) If the anion, or electronegative constituent, is monatomic, its name is modified to end in -ide.

(iii) If the anion is polyatomic, its name is modified to end in -ate.

(iv) Where oxidation states are to be indicated, they are shown by means of Roman numerals following the name of the element.

(v) The stoichiometric proportions of constituents are denoted by Greek prefixes (mono, di, tri, tetra, penta, hexa, hepta, octa, nona, deca, undeca and dodeca). Alternatively, numerals may be used, as in $B_{10}H_{14}$—decaborane-14. In addition, the multiplicative prefixes (bis, tris, tetrakis, etc.) may be used to indicate a multiplicity of complex groups, especially when these already contain a numeral, as in $Ni(PPh_3)_4$—tetrakis(triphenylphosphine) nickel.

(vi) In extended structures, a bridging group is indicated by the prefix μ, for example $[(NH_3)_5Cr\text{-}OH\text{-}Cr(NH_3)_5]Cl_5$ which is μ-hydroxo-bis[penta-ammine-chromium(III)] chloride. A group bridging three atoms, e.g. the face-bridging Cl atoms in Fig. 15.19, is labelled μ^3. A further term which is now commonly used is hapto, symbol η. Its use is best illustrated by considering cyclopentadiene. If this bonds equally through all five C atoms, as in ferrocene, Fig. 16.7a, it is labelled penta-hapto, symbol η^5. A single C-element bond would be (mono) hapto, η^1, while if only one diene group bonded, involving two of the five carbons, the pentadiene group would be described as dihapto, η^2.

(vii) A further terminology will be found for Main Group compounds where organic and organometallic nomenclature overlaps. Hydrides are now given names ending in '-ane' and compounds are often named as hydride derivatives. Thus PH_3 is now *phosphane* instead of phosphine, AsH_3 is *arsane*, and we now often find the ligand Ph_3P called triphenylphosphane. While NH_3 and H_2O are systematically named nitrane or oxane respectively, it is unlikely that terms based on ammonia and water will disappear.

of showing an even lower oxidation state is retained for hyponitrite, hypophosphite and hypochlorite. Corresponding acid names end in -ous and hypo- . . . -ous.

(iii) The term thio- is used to denote the replacement of an oxygen atom by a sulfur one, as in $PSCl_3$—thiophosphoryl chloride. Similar use is made of seleno- and telluro-.

(iv) The terms ortho- and meta- are retained to indicate different 'water contents' of acids, as in orthophosphoric acid, H_3PO_4, and metaphosphoric acid, $(HPO_3)_n$, or H_5IO_6 orthoperiodic acid and HIO_4 periodic acid.

(v) As the last example shows, the prefix per- is used to indicate an oxidation state above the one indicated by the normal -ate or -ic termination of an anion or acid. (Per should not be used for metals and cations.) This usage should be confined to the cases of perchlorate, perbromate, periodate, permanganate and per-rhenate. The prefix peroxo- should be used to denote the presence of the -O-O-group derived from hydrogen peroxide, as in peroxodisulfuric acid (HO_3SOOSO_3H), although the old 'per' is still often found, especially in commercial use. The term 'perfluoro' is widely used in the organic literature to denote a complete substitution of H atoms by F atoms. Thus, the compound $C_{10}F_{18}$ is widely known as perfluorodecalin (decalin$=C_{10}H_{18}$) which is of interest as a blood substitute because of its high capacity for dissolving oxygen gas.

On the whole, the tendency is to use the simplest name available for well-known compounds, as long as this is accurate. New classes of compounds tend to be given their systematic name unless, as in the case of ferrocene for example, the discoverer happens to have hit on a suitable trivial name that has become widely accepted. Inorganic nomenclature need not cause any difficulty to the beginner who only has to translate the systematic name into a more familiar form, as long as the details above, especially about the various uses of the termination -ate, are familiar.

1.5 Approach to inorganic chemistry and further reading

The starting point for an understanding of inorganic chemistry lies in the electronic structure of the atom. The basis of this is provided by quantum theory and wave mechanics and is outlined in Chapter 2, using a relatively pictorial and qualitative approach.

Combination of atoms into diatomic species is covered in Chapter 3, setting the basis for polyatomics discussed in Chapter 4. This chapter also covers the shapes of covalent polyatomic molecules and ions. Chapter 5 deals with compounds in the solid state, both ionic and covalent materials, while Chapter 6 deals with solutions in water and in non-aqueous solvents. Chapter 7 gives a brief review of the experimental methods used in inorganic chemistry. The latter three-quarters of the book covers the chemistry of the elements. After placing this chemistry in the context of the Periodic Table in Chapter 8, and covering hydrogen as a forerunner in Chapter 9, Chapters 10, 11, 12, 14, 15 and 17 cover the different blocks of the Periodic Table in a Group-by-Group pattern. Chapter 13 provides a general context for the transition element chemistry of Chapters 14 to 16, while Sections 17.1 to 17.3 serve a similar function for the Elements of the Main Groups covered in the rest of Chapters 17 and 18. Finally, some themes of strong current interest are selected for more detailed discussion—for transition elements in Chapter 16, for Main Groups in Chapter 18, and for more general topics in Chapters 19 and 20.

Clearly, the arrangement sketched above is only one possible ordering, and does not bind the reader. You will find that any topic may be taken as a starting point, and will eventually lead through most of the text. To illustrate, one possible network of relationships is sketched in Figure 1.1 for two of the topics mentioned in Section 1.4. Many others are possible, and you will start to feel some mastery of the subject once you have followed through a number of themes for yourself.

An introductory text can only provide a framework on which a more detailed knowledge may be built. We have tried to give a basic survey by concentrating on relatively well-known compounds, particularly the oxides, halides and hydrides. Added to these, we have tried to signal many other areas of great interest by relatively brief mention, and have extended the treatment of these in a limited number of cases.

Any textbook of inorganic chemistry nowadays covers only a few percent of established knowledge. Likewise, even the most dedicated student can only become familiar with a tiny part of the whole. It is therefore important to become efficient in searching out material when you turn to a new area of the subject. What do you do when, with only a broad idea of copper chemistry and a shadowy recollection of the perovskite structure, your manager or pupils or others demand to know about these new superconductors which are in the news? What if chromium contamination suddenly becomes a problem, or your company decides to study indium telluride as a semiconductor when all you know about is silicon? Clearly,

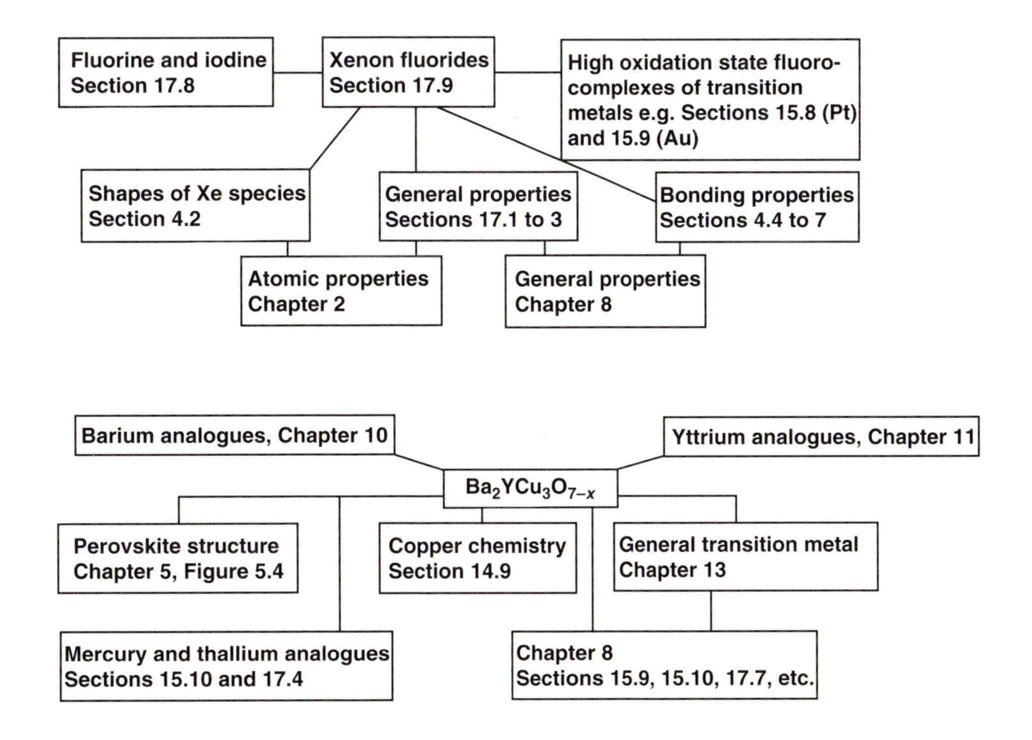

FIG. 1.1 Some correlated study areas

CHEMISTRY AND COMMUNICATION

Even the brief sketch of Section 1.4 highlights the power, but also the problems, of chemical nomenclature. The current system was developed from the time of Lavoisier and his contemporaries at the end of the 18th century—at least 2000 years after observations in what we now call chemistry were first made! With names arising from empirical observations in many crafts such as metallurgy and glassmaking, and especially from alchemy, there were many cases where the same substance had several different names, and the same name was applied to several different substances. While many names such as *oil of vitriol* for sulfuric acid gave some information about source (*vitriol* was applied to sulfates such as those of copper or iron), others such as the alchemical terms *azoch* or *sol* or *red dragon* were often deliberately obscure. Thus modern chemical nomenclature developed to be precise, specific and informative, and the systems for inorganic, organic and other nomenclatures are very powerful. Modern systemic nomenclature works extremely well with compounds of moderate complexity where a chemist can attribute a formula and structure to a named compound.

However, with the number of identified chemical compounds rising rapidly towards 10 000 000 molecules of ever-increasing complexity, producing an unambiguous and generally acceptable name demands ever more complicated rules of nomenclature whose formulation, agreement and promulgation inevitably lags behind the pace of discovery. Commonly research results are reported using structure diagrams, leaving it to experts, such as journal editors or nomenclature committees, to determine exact IUPAC names for new compounds.

The nomenclature problem is one example of the communication difficulties which arise from the huge development of modern chemistry (and indeed of science and technology as a whole). There has also arisen a large specialist vocabulary to describe phenomena—for example, most of the index entries in this book. If society as a whole is to gain a general understanding of current science, there is a vital need for translation between the active research frontier and the public. Major communicators are science journalists and the authors of popular science books, and there are often several layers of interpretation and generalization between the researcher and the layperson (and we are all laypersons outside our specialities!).

The problem is compounded by the dual function of many words. Technical terms are often common words used in a specialized sense—*compound, complex, coordination*. Conversely, we find public use in a wide sense of words which have a restricted meaning to chemists. For example, *rust* which refers to iron is used for general corrosion products so we hear of rusting copper or even plastics, and of course the ubiquitous *chemical* to describe any manufactured product which the speaker does not like. Other communication problems include the widespread use of pseudo-scientific jargon to give an air of authority to dubious claims.

From all this the student needs to be aware of the communication problem and the need to alleviate it as much as possible. It also offers interesting career opportunities within the communication network which links the specialist researcher to the general public.

no-one expects to have an in-depth knowledge of the whole of inorganic chemistry, but you do need to know how to start finding out.

Since the selection and approach of any author is an individual one, your first step is to consult similar sources—find three or four other broad texts and build up a basic picture. Then, to get more depth, go to the more advanced texts, those written for British Honours students or for American graduate courses—a number of these are listed in Appendix A, and these will often give further references to reviews or papers. You should beware of becoming too entangled in specific detail early on in a survey, so you will normally focus on reviews. Inorganic chemistry is well provided with review series (Appendix A) and most major topics will be found. Again, be aware that the depth of presentation may vary considerably, even within a single volume, and it is probably best to follow the references back to earlier treatments and start there. Among the more general presentations of inorganic topics are the Royal Society of Chemistry *Monographs for Teachers,* a number of articles in the journals, *Education in Chemistry* and *Journal of Chemical Education*, and articles in the 'news' publications *Chemistry in Britain* (Royal Society of Chemistry) and in *Chemical and Engineering News* (American Chemical Society). Websites of these societies and publications are listed in Appendix A, together with more advanced reviews.

Some relatively recent multi-volume compendia give expert overviews but the date of publication may be a year or two later than the terminus of the individual articles.

Titles include *Comprehensive Inorganic Chemistry, Comprehensive Coordination Chemistry,* and the two editions of *Comprehensive Organometallic Chemistry.* The *Dictionary of Inorganic Compounds* and the *Dictionary of Organometallic Compounds* are useful sources of information on specific individual compounds.

Problems

1.1 Choose a simple broad topic, such as 'the chemistry of manganese'. Read the introduction to Appendix A.

(a) List all the pages in this book which give relevant information

(b) Similarly list the pages in any three other textbooks

(c) Find three reviews which cover some aspect of the topic (try to find an example of a broad survey and a detailed account of a specific area)

(d) Find three papers published in the last two years covering different parts of the topic.

1.2 Draw up the time plot suggested in Section 1.2 for the discovery of the elements.

1.3 Plot the elements discovered at different times on the Periodic Table. Are there any patterns? Are there any other correlations, e.g. with reactivity as shown by ionization potentials, bond energies to oxygen, etc.?

CHAPTER 2

INTRODUCTION		18
2.1	BACKGROUND	18
2.2	ISOTOPES	19
2.3	RADIOACTIVE ISOTOPES AND TRACER STUDIES	20
THEORY OF THE ELECTRONIC STRUCTURE OF HYDROGEN		21
2.4	INTRODUCTION	21
2.5	THE DUAL NATURE OF THE ELECTRON	22
2.6	THE HYDROGEN ATOM	23
MANY-ELECTRON ATOMS		28
2.7	THE APPROACH TO THE WAVE EQUATION	28
2.8	THE ELECTRONIC STRUCTURES OF ATOMS	29
SHAPES OF ATOMIC ORBITALS		32
2.9	THE s ORBITAL	32
2.10	THE p ORBITALS	37
2.11	THE d ORBITALS	38
2.12	THE PERIODIC TABLE	40
FURTHER PROPERTIES OF THE ELEMENTS		44
2.13	IONIZATION POTENTIAL	44
2.14	ELECTRON AFFINITY	46
2.15	ATOMIC AND OTHER RADII	48
2.15.1	Covalent Species	48
2.15.2	Ionic Species	50
2.15.3	Metals	52
2.16	ELECTRONEGATIVITY	53
2.17	COORDINATION NUMBER, VALENCY AND OXIDATION STATE	55
PROBLEMS		58

The Electronic Structure and the Properties of Atoms

Introduction

2.1 Background

Although some 200 different sub-atomic particles have been discovered by the physicists, only three, the proton, the neutron and the electron, are of direct interest to the chemist. The masses and charges of these different particles are so minute that it is convenient to define much smaller units than the gram and the coulomb which are used on the macroscopic scale. The proton and neutron are very nearly equal in mass at about 10^{-24} g, and this is used as the unit of mass on the atomic scale—the *atomic mass unit* or *a.m.u.* The mass of the electron is very much less than that of the other two particles and is of the order of one two-thousandth a.m.u. The charges on the proton and electron are equal in size though opposite in sign, that on the electron being negative. (There is also a short-lived particle of the same mass as the electron, but with unit positive charge, called the positron or positive electron.) This electronic charge, which equals about -1.6×10^{-19} C, is taken as the unit of charge on the atomic scale and is given the symbol $-e$. The neutron has no charge. Thus, the proton has unit mass and unit positive charge, the neutron has unit mass and zero charge while the electron has negligible mass and unit negative charge. The exact values are given in Table 2.1, but this approximation suffices for most purposes.

The foundation of the modern theory of atomic structure was laid by the work of Rutherford on the scattering of α-particles by very thin metal targets. He found that, when a beam of α-particles (mass = 4, charge = $+2e$) was directed at a target of thin metal foil, nearly all the particles passed through the target with scarcely any deflection, a few were deflected through large angles, and an even smaller proportion were reversed along their paths. These observations suggested a model of the atom as a small, dense, positively charged core surrounded by a much larger and more tenuously occupied region of electrons. The α-particles are so much more massive than the electrons that they would be relatively unaffected when they passed through the outer regions of the atoms and deflected only if they came close to the core. Since most of the α-particles passed straight through the target, which was about a thousand atoms thick, it followed that the cores must be very small. When the α-particle passed close to a core, it was strongly deflected by the charge repulsion, while the occasional particle which happened to be heading straight at the core (with a mass and charge about thirty times its own) was repelled back along its path. The effect is illustrated in a very diagrammatic form in Fig. 2.1 on p. 18.

This work has been fully substantiated by later investigations. In the modern view, the atom consists of a tiny, dense, positively charged nucleus containing protons and neutrons, surrounded by a much larger, more tenuous, cloud of electrons. Since the electron mass is so much smaller than the masses of the other particles, the relative atomic mass is approximately equal to the nuclear mass number, A; which equals the sum of the number of protons, Z, and the number of neutrons $(A–Z)$ or N. Since there are Z protons, the nucleus bears a charge of $+Ze$ and this is balanced by having Z electrons surrounding

TABLE 2.1 Properties of the fundamental particles

*Particle**	*Charge* (C)	*Relative mass*
Proton ^1_1p	$+e = 1.6022 \times 10^{-19}$	1.007 29
Neutron ^1_0n	Zero	1.008 660
Electron $^{\ 0}_{-1}e$ or β^-	$-e = -1.6022 \times 10^{-19}$	0.000 5486

*The symbols are explained in Section 2.2. The positron has the symbol $^{\ 0}_{+1}e$ or β^+.

Note that *e* is also used as a symbol for the electron.

the nucleus so that the atom is electrically neutral. *Z* is termed the *atomic number* and is the most important single property of an atom.

The volume of an atom is essentially the space occupied by its electron cloud, the nucleus filling only a minute proportion of the whole (about 10^{-15} of the atomic volume). The radius of an atom is of the order of 100 pm, while the nuclear radius is about 10^{-3} pm. The radii of molecules extend to about 1000 pm, or even more for very complex or polymeric species. For comparison, the present-day electron microscope can resolve down to about 1000 pm and can thus be used to 'see' molecules of moderate size.

2.2 Isotopes

Chemical behaviour is determined by the interaction between the electron clouds of atoms whose character, in turn, depends on *Z*, the number of electrons present. That is, the atomic number determines the chemical properties of an element, and all the atoms of a particular element have *Z* protons in their nuclei and *Z* electrons. It is found, however, that these *Z* protons may be accompanied in the nucleus by varying numbers of neutrons so that atoms of the same atomic number may have different nuclear mass, *A*. For example, chlorine with $Z = 17$ has two forms, one with eighteen neutrons and nuclear mass $A = 35$, and the other with twenty neutrons and nuclear mass 37. Such atoms with the same atomic number *Z*, but with different nuclear mass *A*, are termed *isotopes.*† Since the chemical properties of an element depend only on *Z*, all the isotopes of an element undergo identical chemical reactions although the rates of reaction (and other effects which depend on mass) may show small differences. These mass effects are negligible except for the lightest elements such as hydrogen. However, the enrichment of uranium (for use in nuclear weapons and nuclear reactors) relies on the very small difference in the rates of diffusion between $^{235}UF_6$ and $^{238}UF_6$ molecules in the gas phase. The number of naturally occurring isotopes of an element varies widely. Some elements such as ^9_4Be, $^{31}_{15}P$ or $^{197}_{79}Au$ are found in only one isotopic form while others may form up to ten stable isotopes. For example:

Element	*Z*	*A*
Cadmium	48	106, 108, 110, 112, 113, 114, 116
Tin	50	112, 114, 115, 116, 117, 118, 119, 120, 122, <u>124</u>
Tellurium	52	120, 122, 123, <u>124</u>, 125, 126, 128, 130
Xenon	54	<u>124</u>, 126, 128, 130, 131, 132, 134, 136

This illustrates that, as well as isotopes with the same *Z* value and different *A* values, there also exist atoms with the same mass numbers but differing atomic numbers. Such atoms are called *isobars*. One example is provided by the isotopes of tin, tellurium and xenon of mass 124 which are underlined in the table. Isobars are of less importance to the chemist than are isotopes.

The nuclear mass, as determined chemically, is the mean weight of the naturally occurring mixture of isotopes. Thus chlorine, which consists of 75.4 percent ^{35}Cl and 24.6 percent ^{37}Cl, has a relative atomic mass of 35.453. Very extensive studies have shown that, where elements are not involved in natural radioactive decay processes, the isotopic composition and relative atomic mass of samples from widely varied sources is nearly always

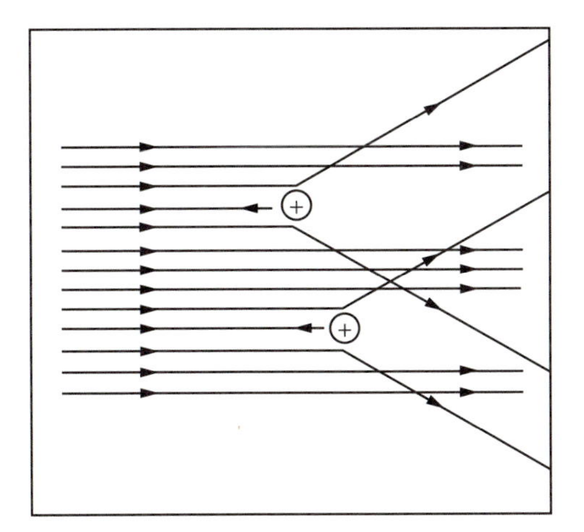

FIG. 2.1 Rutherford's experiment. Diagrammatic representation of the deflections observed when a beam of α-particles hits a metal target

†These nuclear properties are commonly shown by left sub- and superscripts on the element symbol: ^A_ZX. Thus the α-particle, which is the helium nucleus, is written $^4_2He^{2+}$. (The alternative form, $_ZX^A$ is also found, especially in American publications.)

constant. The exceptions are those elements, e.g. copper and lead with variable isotopic composition in naturally obtained samples, and also lithium, boron and uranium in which an isotopic separation may result from methods used in commercial preparation of a sample.

2.3 Radioactive isotopes and tracer studies

Only certain combinations of protons and neutrons form stable isotopes. If the neutron/proton ratio becomes too large or too small, the nucleus is unstable and radioactive, and some nuclear process takes place to restore the balance. For example, the isotope of hydrogen of mass 3 called tritium, ${}^{3}_{1}H$, has too many neutrons to be stable. It spontaneously converts a neutron into a proton by emitting a negative electron to form a helium isotope, ${}^{3}_{2}He$. This type of process is written out as a nuclear equation; in this case

$$ {}^{3}_{1}H \rightarrow {}^{0}_{-1}e + {}^{3}_{2}He $$

(In nuclear equations, both the charge on the nucleus and the mass are shown and each of these quantities must balance.)

The converse case is illustrated by magnesium-23 which has excess protons and converts a proton to a neutron by the emission of the positive electron (positron) to form a sodium isotope:

$$ {}^{23}_{12}Mg \rightarrow {}^{0}_{+1}e + {}^{23}_{11}Na $$

Nuclei are not stabilized only by electron emission: other ways include the emission of an α-particle or the capture of an orbital electron (K capture). Sometimes a nucleus is so far from a stable ratio of protons to neutrons that the daughter nucleus produced by the first radioactive decay step is itself unstable and rearranges, and this process continues until a stable nucleus results.

All the atoms in a sample of an unstable material do not undergo a nuclear reaction simultaneously and instantaneously. It is found that a particular radioactive isotope is characterized by a half-life, given the symbol $t_{\frac{1}{2}}$, which is the time in which half the atoms present in a given sample have undergone transformation. Half-lives range from minute fractions of a second to millions of years. The half-life is a constant, characteristic of the isotope, and is not varied by any change in the environment of the atom. For tritium, $t_{\frac{1}{2}} = 12.4$ years and for magnesium-23 it is 11.6 seconds. A very long half-life is found for uranium-238 ($t_{\frac{1}{2}} = 4.5 \times 10^{9}$ years).

Since all the isotopes of an element have identical chemical properties, it is possible to study the course of many reactions by using an element in an isotopic form other than the natural one. To choose a very simple example, it could be shown that an acid ionizes in water by dissolving it in 'heavy water' formed from the isotope of hydrogen of mass two (heavy hydrogen or deuterium, ${}^{2}_{1}H$, often given the special symbol D). When the acid is recovered from solution all the ionizable hydrogen atoms in the molecule will have been replaced by deuterium ions which are present in great excess in the solution. For example, acetic acid—CH_3COOH—is recovered from heavy water as CH_3COOD, showing that only the carboxylic hydrogen atom is acidic.

Radioactive atoms may be detected, by means of their characteristic radiations, in extremely small amounts and are therefore valuable for tracer studies such as these. One illustration is provided by a study of hypophosphorous acid, H_3PO_2. When this is dissolved in heavy water, only one of the three hydrogen atoms is replaced rapidly by a deuterium atom and this confirms that the other two hydrogen atoms are directly bonded to the phosphorus atom and do not ionize:

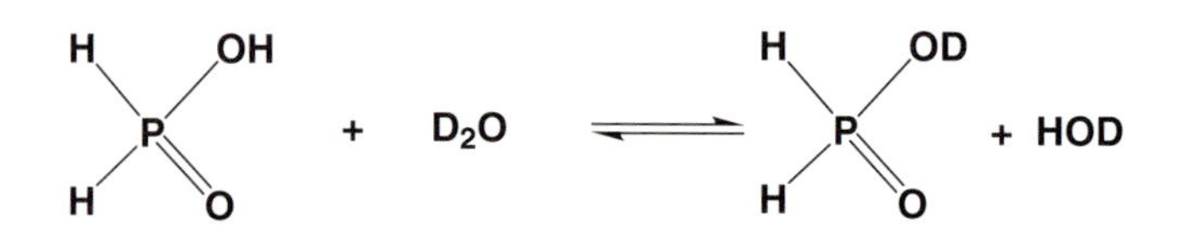

SOME USES OF ISOTOPES

In many natural kinetic processes, partial fractionation of *stable isotopes* of light elements occurs. Examples include the effect of temperature on processes involving water or carbon dioxide, or because different pathways to biological products are followed in different organisms which affect nitrogen ratios as well as carbon and oxygen. Thus the ratios of D to H; ^{13}C to ^{12}C; ^{15}N to ^{14}N; and ^{18}O to ^{16}O give valuable biological or geochemical information, such as the sea temperature when a limestone deposit formed. Oxygen isotope ratios in ice, determined from annual layers in the long cores bored in Greenland and the Antarctic, are one of the main lines of evidence about climate change in the last hundred thousand years. A longer timescale is accessible through the oxygen and carbon ratios in the skeletons of marine microorganisms recovered from deep sea cores. Similarly, carbon and nitrogen ratios can give information about diet from ancient skeletal material where land plants, meat and sea products give different signatures. For example, such studies have traced the spread of the domestication of wheat or barley in Europe, rice in East Asia and maize in the Americas.

Since *radioactive isotopes* can be detected with high sensitivity, they have many applications. Thus, tritium, ^{14}C or ^{32}P have been widely used to trace biochemical pathways in organisms. A further important application is in dating. The very long half-life of uranium and thorium minerals allows the dating of rocks up to a billion years old, while more recent deposits can be dated using shorter-lived isotopes such as potassium/argon for ages around 100 000 years and ^{14}C ($t_{\frac{1}{2}} = 5730$ years) for dating historical and prehistorical materials. Recent spectacular examples include the Turin Shroud and the mummified body of the 'Iceman' which emerged from a glacier on the Italian–Austrian border.

A further interesting application arises where naturally occurring ores have variable isotopic composition. This is particularly the case for lead, where different lead isotopes are formed in decay series which start with different isotopes of uranium or thorium. Thus many lead sources have characteristic isotopic signatures. Apart from lead itself, many ancient metals have significant lead contents—for example, much ancient silver was recovered from lead ores. In a favourable case, an ancient metallic object can be analysed for its lead isotope pattern and traced to a particular ore source, or at least to a region. This

However, when the radioactive form of hydrogen, tritium, was used to study the exchange, the much higher sensitivity allowed the detection of a further, very slow, reaction in which the hydrogen bonded to the phosphorus atom exchanged with the solvent by means of an isomerization reaction:

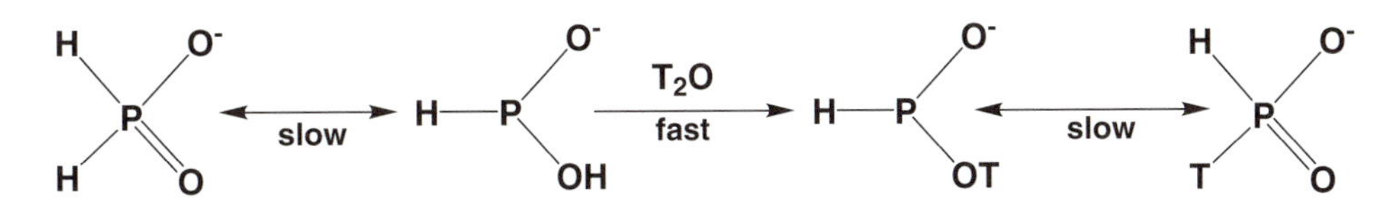

Theory of the electronic structure of hydrogen

2.4 Introduction

As the chemical properties of elements depend on the interaction of the electron clouds of their atoms, any fundamental theory of chemistry must start by examining the electronic structure of atoms. The remainder of this chapter is devoted to this theme while the succeeding ones discuss the ways in which the electron clouds interact in the formation of molecules and ions. The modern theory of atomic structure—based on the work of Heisenberg and Schrödinger—attempts to describe the shape and arrangement of the electron clouds and to calculate the energy of any given configuration of electrons. If this aim could be carried out completely and accurately, a complete description of chemical phenomena could be derived purely from theory. The course of a reaction, or the shape of a molecule, would be determined by calculating the energy of each alternative and then choosing the most favourable. At present, theoretical chemistry does not reach this idealized position. It is not possible to carry out calculations using theory alone for any systems other than the simplest. The difficulties are formidable as the energy of a reaction is a relatively small difference between two large quantities. For example, hydrogen atoms combine to form hydrogen molecules in an extremely vigorous and exothermic reaction:

$$2H \rightarrow H_2$$

The theory yields a value for the *total electronic energy* of the atoms or molecules. This is the gain in energy when, in this case, two protons and two electrons are brought from infinite separation to form the two hydrogen atoms on the left of the equation or else the hydrogen molecule on the right. The energy of formation of hydrogen molecules from hydrogen atoms is the difference between these total electronic energies: 2652 kJ mol^{-1} for 2H and 3109 kJ mol^{-1} for H_2, equal to 434.3 kJ mol^{-1} after allowing for minor effects. It can be seen that, even in this simple case, an error of one percent in each of the total electronic energies can produce an error of about thirteen percent in the resultant energy of formation.

In general, the total electronic energies of the reactants and products are much greater, and the energies of reaction are smaller, than in the case of hydrogen. Consider, for example, the question whether carbon and hydrogen combine as:

$$\text{(a) } C_{(solid)} + H_{2(gas)} \rightarrow CH_{2(gas)}$$
$$\text{or (b) } C_{(solid)} + 2H_{2(gas)} \rightarrow CH_{4(gas)}$$

The difference in the heats of the reactions (a) and (b) is under 42 kJ mol^{-1} while the total electronic energy of small, light molecules like CH_2 and CH_4 is of the order of 100 000 kJ mol^{-1}. It follows that if we try to predict the outcome of such reactions from absolute values of the total electronic energy, we must be able to calculate values which are accurate to 1 in 10^5. For larger molecules and those containing heavier elements, the demand is even higher. While calculations have improved rapidly in the last decades, such limits are unobtainable. For most calculations of interest to chemists, however, it is not necessary to achieve absolute values. The largest energy contributions are those of the tightly bound inner electron shells (see Section 2.8). Since these will be common to the alternative products they cancel out, leaving the focus on differences in the energies of the outermost electrons which arise from the different bonds and shapes of alternative structures, as in (a) and (b) above. Modern calculations, at the level of packages available on personal computers, do provide useful explanations of observed stabilities and structures. Perhaps fortunately, while calculations can be a valuable guide, the day when we can replace the laboratory with the computer is still distant.

To generalize broadly, the great value of the theory of the electronic structure of atoms and molecules in its present form lies, not in its use for absolute calculations, but in the correlation and rationalization it provides for a great mass of experimental results. In current theory, much use is made of experimental data (such as bond lengths and angles) to help to simplify and solve the equations, and these solutions in turn help to suggest further experimental work. This provision of a wide, stimulating and flexible theoretical framework and the strong interaction between theory and experiment has been one of the most exciting and vital aspects of chemistry in the last few decades. In the next few sections an attempt has been made to outline the basic steps in this theory.

2.5 The dual nature of the electron

Before examining the electronic structure of the atom, two important properties of the electron itself must be discussed.

These are:

(i) the dual nature of the electron which partakes of the properties both of a particle and of a wave, and

(ii) the effect of Heisenberg's Uncertainty Principle when applied to the electron.

The classical picture of the electron is of a tiny particle whose position in space can be accurately defined by its co-ordinates x, y and z. Its motion in an atom is then described by the variation of (x, y, z) with time. This was the way in which Rutherford and Bohr described the atom on a 'planetary' model with the massive central nucleus and the light electrons moving in 'orbits' around it. However, it was shown in the 1920s that moving particles should behave in some ways as waves and that this effect should be particularly marked for a particle as light as the electron. Experimental support for this prediction was soon found. It was shown, for example, that a beam of electrons could be diffracted by a suitable grating in exactly the same way as a beam of light. These wave properties were introduced into the theory of atomic structure by Schrödinger in his *wave mechanics* in which the electron in an atom is described by a wave equation.

The Uncertainty Principle places an absolute limit on the accuracy with which the position and motion of a particle may simultaneously be known. It must be noted that this limit is a fundamental property of the particles, not a problem with measurements. A formal statement is that the product of the uncertainty in the position Δx and of the uncertainty in the momentum Δp of a particle cannot be less than the modified Planck's constant, $h/2\pi$. (Planck's constant $h = 6.6261 \times 10^{-34}$ J s.) That is $\Delta x \Delta p \nless h/2\pi$. This limit is so small that the uncertainty is negligible for normal bodies but it is large for a particle as light as the electron. As a result, the idea that the electron has a definite position (for example, of its following a definite orbit) must be replaced by the concept of a probability distribution for the electron. In other words, the answer to the question, 'where is the electron in an atom?', becomes a statistical one.

When both these concepts—the wave nature of the electron and the Uncertainty Principle—were taken into account, it was found that a satisfactory description of an atom resulted from Schrödinger's Wave Equation whose solutions, named *wave functions*, were given the symbol ψ. The value of the square of the wave function at any given point $P(x, y, z)$—written $\psi^2_{(x,y,z)}$—gave the probability of finding the electron at P. ψ^2 is sometimes termed the probability density of the electron. Put in rather crude terms, the old particle-in-an-orbit picture of classical atomic theory is replaced in wave mechanics by an electron 'smeared out' into a charge cloud, and the function ψ^2 describes how the density of the charge cloud is distributed in space.

2.6 The hydrogen atom

The theory of wave mechanics may now be examined in more detail for the simplest case, the hydrogen atom, where there is only one electron moving in the field of a singly charged nucleus. Once the hydrogen atom is understood, the results for more complex systems may be derived similarly. The box titled 'Approach to the wave equation for hydrogen-like atoms' shows details of the approach. The key result is the derivation of the quantum numbers which define the complete set of allowed solutions to the wave equation.

When this set of solutions is examined, it becomes apparent that the orbitals fall into families. The first type, of which ψ_1 is an example, is of the form $\psi = f(r)$. In other words, the value of the wave function, and hence of its square, the probability of finding the electron, depends only on the distance from the nucleus and is the same in all directions in space. These orbitals are spherical and they correspond to the cases where the quantum number l (and therefore m also) is zero. Orbitals are usually designated by a number equal to the value of n, and a letter corresponding to the value of l as follows:

l = 0, 1, 2, 3, 4, 5, ...
s p d f g h..........

where the rather odd selection of letters at the beginning arises for historical reasons. Thus these orbitals which are functions of r only and have $l = 0$, are s orbitals and ψ_1 is the 1s orbital. There is an s orbital for each value of n and they increase in energy as n increases (see Table 2.3 and Fig. 2.3).

A second type of solution to the wave equation has the form $\psi = f(r)f(x)$. Clearly these orbitals now have directional properties and will have different magnitudes in the $\pm x$

APPROACH TO THE WAVE EQUATION FOR HYDROGEN-LIKE ATOMS

The time independent form of the Schrödinger wave equation for the hydrogen atom looks like this:*

$$\frac{-h^2}{8\pi^2 m}\nabla^2\psi - \frac{e^2}{4\pi\varepsilon_0 r}\psi = E\psi \qquad (2.1)$$

In this expression, e and m are the charge and mass, respectively, of the electron, h is Planck's constant, r is the distance from the nucleus which is taken as the origin of coordinates, ψ is the wave function, ε_0 is the permittivity of a vacuum and E is the total energy of the system. This equation looks rather formidable but it is just a statement in wave-mechanical terms of the principle of conservation of energy. The del-squared term corresponds to the kinetic energy, the term $-e^2\psi/r$ is the potential energy (the attraction between the charges on the electron ($-e$) and on the nucleus ($+e$) at a separation r), while the right hand side of the equation is the total energy.

A number of wave functions which satisfy Equation (2.1) may be found. These are written ψ_1, ψ_2, ψ_3, etc., and are called *orbitals* by analogy with the orbits of the old planetary theory. To each solution or orbital there corresponds a certain value—E_1, E_2, E_3, etc.—of the total energy of the system. That orbital, say ψ_1, which corresponds to the lowest value of the total energy describes the electron distribution in the normal, most stable state of the hydrogen atom (called the *ground state*), and this is the orbital where the electron density is concentrated most closely to the nucleus. The expression for the ground state orbital is:

$$\psi_1 = \frac{1}{\sqrt{(\pi a_0^3)}}\exp(-r/a_0) \qquad (2.2)$$

where $a_0 = 0.529$ Å $= 52.9$ pm is the atomic length unit or Bohr radius. The corresponding energy E_1 is:

$$E_1 = -e^2/8\pi\varepsilon_0 a_0 = -1310\,\text{kJ mol}^{-1} = -13.60\,\text{eV} \qquad (2.3)$$

The other solutions to the wave equation describe states of hydrogen of higher energies than the ground state (called *excited states*) where the electron is in one of the orbitals concentrated further out from the nucleus. The hydrogen atom will be excited from the ground state to one of these higher-energy states if it absorbs energy, equal in amount to the difference between E_l and the value of E of the higher orbital, which will promote the electron to this higher orbital. This energy absorption may be observed in the electronic spectrum of the hydrogen atom and the frequencies of electromagnetic radiation absorbed give experimental values for the various energy states E_1, E_2, E_3, etc., of the atom, which agree very well with the calculated E values. Thus Fig. 2.2 shows that

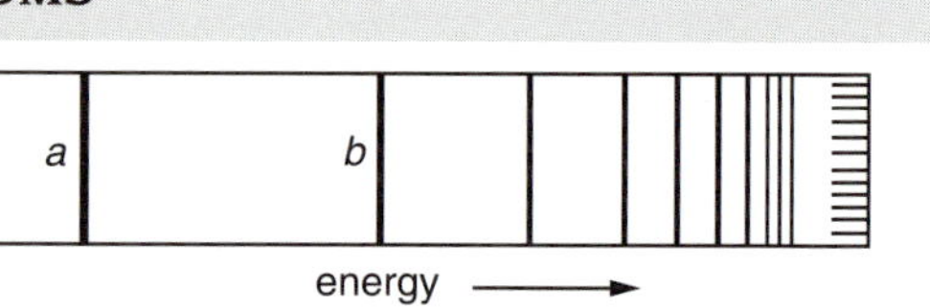

FIG. 2.2 The spectrum of hydrogen. This diagram shows the Balmer series in the visible spectrum. Transitions to upper states become increasingly close together until the continuum on the right of the diagram is reached. The line labelled a, for example, results from the transition between E_2 and E_3 while that labelled b is the transition between E_2 and E_4

part of the hydrogen atom spectrum which involves transitions between E_2 and higher levels E_3, E_4, E_5, etc.

The wave equation (2.1) for hydrogen holds for all other 'hydrogen-like' species, that is for those with only one electron such as He^+ or Li^{2+}. The only modification to Equation (2.1) required is to allow for the different nuclear charge Z in the potential energy term. Thus the general form of the wave equation for hydrogen and hydrogen-like atoms is:

$$\frac{-h^2}{8\pi^2 m}\nabla^2\psi - \frac{Ze^2}{4\pi\varepsilon_0 r}\psi = E\psi \qquad (2.4)$$

and the solutions are exactly the same as for hydrogen except that the value of Z will carry through the working.

Since the value of ψ^2 times unit volume at a given point P(x, y, z) is the probability of finding the electron in that volume at P, it follows that acceptable solutions of the wave equation must have certain properties. Suitable functions must, for example, be single-valued at all points P as there cannot be two or more answers to the question 'what is the probability of finding the electron at P?' Similarly, ψ must be continuous and finite. Further, the total value of ψ^2 summed over all the points in space $\left(\text{i.e. } \int_{-\infty}^{+\infty}\psi^2 \text{d}x\,\text{d}y\,\text{d}z\right)$ must equal one, since the probability of finding the electron somewhere in space must be certainty (which is unity by definition). The result of these restrictions is to limit the acceptable solutions of the wave equation to those which can be determined by three quantum numbers n, l and m which may take only those values shown in Table 2.2. A set of three quantum numbers is required to describe each orbital. Thus the ground state of the hydrogen atom (Equation 2.2) has the electron in

TABLE 2.2 The quantum numbers, n, l and m

Quantum number	*Allowed values*
n	1, 2, 3, 4, 5,. . ..
l	$(n-1)$, $(n-2)$,. . . 2, 1, 0
m	$+l, +(l-1), +(l-2), \ldots 2, 1, 0, -1, -2, \ldots -(l-2), -(l-1), -l$

* ∇^2 is an abbreviation for $\frac{\partial^2}{\partial x^2} + \frac{\partial^2}{\partial y^2} + \frac{\partial^2}{\partial z^2}$, where (x, y, z) are the space coordinates of the electron. It is read 'del-squared'.

the orbital where $n = 1$, $l = 0$, $m = 0$. These quantum numbers arise naturally in the course of the mathematics because of the requirement that acceptable solutions are well-behaved functions. This is in contrast to the older theory where the quantum numbers had to be added, apparently arbitrarily, to the classical description. The detail of these calculations is too complicated to be shown here but the interested reader is referred to the sources cited at the end of the book.

TABLE 2.3 Atomic orbitals with n values up to four

n	*l*	*m*	*Symbol*	*Number of levels with this n value*
1	0	0	1s	1
2	0	0	2s	4
	1	$\pm 1, 0$	2p	
3	0	0	3s	9
	1	$\pm 1, 0$	3p	
	2	$\pm 2, \pm 1, 0$	3d	
4	0	0	4s	16
	1	$\pm 1, 0$	4p	
	2	$\pm 2, \pm 1, 0$	4d	
	3	$\pm 3, \pm 2, \pm 1, 0$	4f	

direction than in the rest of space. There are two other exactly similar types of orbitals, $\psi = f(r)f(y)$ and $\psi = f(r)f(z)$ which are concentrated in the y and z directions respectively. The most stable representatives of these orbital types for hydrogen are:

$$\begin{aligned} \psi_x &= kx \exp(-r/2a_0) \\ \psi_y &= ky \exp(-r/2a_0) \\ \psi_z &= kz \exp(-r/2a_0) \end{aligned} \tag{2.5}$$

where k is a constant compounded of a number of fundamental constants. These three orbitals are equal in energy and for each of them:

$$E = -e^2/32\pi\varepsilon_0 a_0 \tag{2.6}$$

This equality of energy for orbitals of this second type, in sets of three, is generally true. The three orbitals are entirely equivalent except for their direction in space.* They have $l = 1$ and are therefore p orbitals. The set of lowest energy in hydrogen, ψ_x, ψ_y and ψ_z, are 2p orbitals.† It follows from Table 2.2 that the p orbitals must occur in sets of three for each value of n as, when $l = 1$, m may take the three values $+1$, 0, -1.

The theory requires only that there should be three independent p orbitals and any set of three may be chosen. It is usually most convenient to choose the above set, ψ_x, ψ_y and ψ_z, which coincide with the coordinate axes but different sets of three p orbitals may be useful on occasion. For example, a set making different angles with the axes may be chosen and these will be formed from appropriate combinations of ψ_x, ψ_y and ψ_z. In particular, it should be noted that ψ_x, ψ_y and ψ_z do not correspond directly to the m values, ± 1 and 0. On the normal convention of axial directions, the z orbital is that for which $m = 0$ but the two orbitals for which $m = \pm 1$ have to be rearranged to give the two orbitals ψ_x and ψ_y.‡ There are further sets of higher energy p orbitals for the higher n values. These 3p, 4p, 5p, etc., sets of orbitals again contain three independent orbitals of equal energy.

*Such a set of equal energy levels is termed *degenerate*.
†Since $l = n - 1$, there are no p orbitals for $n = 1$.
‡When the direction of a p orbitals is to be distinguished a subscript is added to the symbol. For example, the orbitals of Equation (2.5) are respectively, p_x, p_y and p_z.

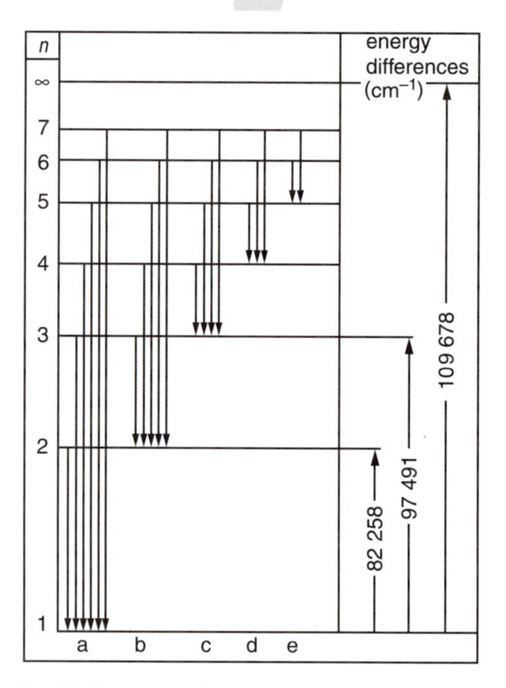

FIG. 2.3 Energy level diagram for hydrogen, showing the transitions corresponding to the various series of lines in the electronic spectrum of hydrogen. The transitions are lettered corresponding to the different series of lines in the hydrogen spectrum: (a) Lyman, (b) Balmer, (c) Paschen, (d) Brackett, (e) Pfund. Values shown are experimentally determined: compare the values calculated from the formula given in Table 2.4

The next type of solution of the wave equation to be considered is the set of orbitals which are functions of two directions as well as of r, of the type $\psi = f(x)f(y)f(r)$. Such orbitals have l values equal to two and there are five independent orbitals of equal energy in a set, corresponding to the five m values, ± 2, ± 1, 0. These are d orbitals and, since five values of m can only arise when $n = 3$ or more (Table 2.2), the lowest energy d orbitals are the 3d set. The 4d, 5d, 6d, etc., orbitals are of progressively higher energy and each set contains five orbitals of equal energy.

When $l = 3$, the 7 possible m values give rise to the 7 f orbitals and these arise when n is 4 or more. Likewise g, h, orbitals etc. are possible, but the known elements are accounted for without such orbitals being occupied. The primary interest in introductory chemistry lies in the s, p and d orbitals. Table 2.3 lists all the orbitals with n values up to four.

It is convenient to change from Cartesian coordinates (x, y, z) to polar coordinates (r, θ, ϕ). This does not alter the principles but simplifies the mathematics and makes it possible to separate the wave functions expressed in polar coordinates into a radial part involving only r, the distance from the nucleus, and an angular part $f(\theta, \phi)$ which expresses the directional properties of the orbitals. Conventionally, θ is the angle from the z axis and ϕ is the angle from the x axis. The relations are

$$x = r \sin\theta \cos\phi$$
$$y = r \sin\theta \sin\phi$$
$$z = r \cos\theta$$
$$r^2 = x^2 + y^2 + z^2$$

Equations (2.5) then have the form

$$\begin{aligned} \psi_z &= k' r(\exp - r/2a_0)\cos\theta \\ \psi_x &= k' r(\exp - r/2a_0)\sin\theta\cos\phi \\ \psi_y &= k' r(\exp - r/2a_0)\sin\theta\sin\phi \end{aligned} \tag{2.7}$$

The k' values are products of fundamental constants. If the nuclear charge is Z, but only one electron is present, that is, if the atom is hydrogen-like, then the exponential term becomes $\exp(-Zr/2a_0)$.

In hydrogen and hydrogen-like atoms, the energy of the electron depends only on the n value of the orbital in which it is found. That is, the order of energies is:

$$1s < 2s = 2p < 3s = 3p = 3d < 4s = 4p = 4d = 4f < 5s = 5p = 5d = 5f < \ldots$$

The electron in the ground state is in the 1s orbital and is found in one of the higher energy orbitals only if the atom has been excited by the absorption of energy. The lowest excited state is where the electron is in an orbital of n value $= 2$, and this corresponds to the absorption of energy equal to $(e^2/8\pi\varepsilon_0 a_0 - e^2/32\pi\varepsilon_0 a_0)$, (see Equations 2.3 and 2.6) which equals 10.20 eV (984.3 kJ mol^{-1}) or the absorption of radiation of wave number 82 273 cm^{-1}. The energies required for excitation to higher states can be calculated similarly.

These excitations can also be observed experimentally in the spectrum of hydrogen, and the first major test of the theory to be applied was to see if the experimental and theoretical energies matched. The electronic spectrum of hydrogen was observed many years before the development of Schrödinger's theory, or before Bohr's earlier theory, and both theories do give energy levels which agree with the observed ones. Schrödinger's theory also gives close agreement for many-electron atoms where Bohr's theory is less successful.

If electromagnetic radiation consisting of a continuous range of frequencies—'white' radiation—is shone on to an absorbing system, then those frequencies which correspond to the difference, ΔE, in energy between two states of the system will be absorbed. (The relation between energy and frequency, ν, is $E = h\nu$.) Thus the absorption spectrum shows lines at frequencies which correspond to energy level differences in the irradiated species. Lines at the same frequencies may be *emitted* when the species return from the higher to the lower energy level and it is easier, for atoms, to observe this *emission spectrum*. When the hydrogen spectrum is examined several series of lines, like that in Fig. 2.2, are

TABLE 2.4 The electronic spectrum of hydrogen

Series	m	m'	$\tilde{\nu}$	Corresponding orbital description
Lyman	1	2	82 303	$n=1 \rightarrow n=2$
		3	97 544	$n=1 \rightarrow n=3$
		4	102 879	$n=1 \rightarrow n=4$
		∞	109 737	Total electronic energy when electron is in the 1s orbital: dissociation energy for the ground state
Balmer	2	3	15 241	$n=2 \rightarrow n=3$
		4	20 576	$n=2 \rightarrow n=4$
		∞	27 434	Total electronic energy when electron is in 2s or 2p orbitals
Paschen	3	4	5 334	$n=3 \rightarrow n=4$
		5	7 803	$n=3 \rightarrow n=5$
		∞	12 193	Total electronic energy when electron is in 3s, 3p or 3d orbitals
Brackett	4	5	2 469	$n=4 \rightarrow n=5$
		∞	6 859	Total electronic energy when electron is in any orbital with $n=4$
Pfund	5	6	1 341	$n=5 \rightarrow n=6$
		∞	4 390	Total electronic energy when electron is in any orbital with $n=5$

observed. Each series is named after its discoverer. These series occur in different regions of the spectrum—the Lyman series in the ultraviolet, the Balmer series in the visible, the Paschen series in the near-infrared, and the Brackett and Pfund series in the far-infrared. Each is similar to Fig. 2.2 in showing lines which become progressively less intense and closer together towards higher frequencies. They all show an upper frequency limit for line absorption, above which continuum is observed. In each series the wave numbers $\tilde{\nu}$ of the lines fit the general formula:

$$\tilde{\nu} = R(1/m^2 - 1/m'^2)$$

In this formula, m has an integral value which is different for each series of lines while $m' = (m+1), (m+2), (m+3),\ldots,\infty$. R is a constant, the Rydberg constant, which equals 109 740 cm^{-1} approximately. The values of m and examples of the transitions for the various series are shown in Table 2.4. These various series may be portrayed on an energy level diagram, as in Fig. 2.3, where the zero value for the energy is taken as the start of the continuum corresponding to $m' = \infty$. The correlation between this diagram and the electronic levels of the atom, as derived from the wave equation, is obvious. The states whose line spectra correspond to $m = 1, 2, 3, \ldots$ are those whose n quantum number takes the values, $1, 2, 3, \ldots$ respectively. The transition in the Lyman series corresponding to

$$\tilde{\nu} = R(1/1^2 - 1/2^2)$$

(i.e. with $m = 1$, $m' = 2$) is the transition from the ls level to the 2s (or 2p) level. The Lyman series corresponds to transitions from the ground state, 1s, to levels with higher n values; while the Balmer, Paschen, etc., series, for which $m = 2$, 3, etc., respectively, correspond to transitions when electrons which were already excited are further excited to higher levels. Thus the frequencies of the Lyman series correspond to the energy gaps between the 1s level and the higher levels and, in particular, the start of the continuum corresponding to $m' = \infty$ corresponds to the complete removal of the 1s electron. Similar transitions from the 2s level are observed in the Balmer series, from the 3s level in the Paschen series, and so forth. As an illustration of the correlation between the calculated and observed energy differences, it is worth calculating the transitions in the Lyman series $m' = 2$ (equal to the 1s—2s energy gap) and $m' = \infty$ (equal to the energy of the 1s orbital with respect to complete dissociation). For 1s to 2s the value is $\tilde{\nu} = R(1/1^2 - 1/2^2) = 3/4 \times 109\,740\ \text{cm}^{-1} = 82\,303\ \text{cm}^{-1}$ (10.20 eV). The dissociation transition is $\tilde{\nu} = R(1/1^2 - 1/\infty^2) = R = 109\,737\ \text{cm}^{-1}$ (13.61 eV). The values calculated from the wave equation agree exactly.

The Bohr theory gives an equally good correlation with experiment for the hydrogen atom but wave mechanics is to be preferred for dealing with many-electron atoms.

Many-electron atoms

2.7 The approach to the wave equation

When the wave equation for helium with two electrons and $Z = 2$ is considered, it is found to be more complicated than the wave equation for the hydrogen atom both in the kinetic energy term and in the potential energy term. The box titled 'Approach to the wave equation for many-electron atoms' indicates the line of attack.

The self-consistent field approximation discussed in the box, or other methods of approximation, use hydrogen-like fields. Thus most of the results for the exactly solved hydrogen atom apply in a qualitatively similar form. The electron distributions in many-electron atoms are described by wave functions ψ similar to those for hydrogen, and the square

APPROACH TO THE WAVE EQUATION FOR MANY-ELECTRON ATOMS

As both electrons contribute to the kinetic energy there are now two del-squared terms, one applying to each electron. The potential energy term consists of three parts in place of the single term in Equation (2.1). In helium there are two attractions—between each electron and the nucleus—and there is also a repulsion term between the two electrons.
The wave equation for helium is:

$$k(\nabla_1^2 + \nabla_2^2)\psi - \left(\frac{Ze^2}{4\pi\varepsilon_0 r_1} + \frac{Ze^2}{4\pi\varepsilon_0 r_2} - \frac{e^2}{4\pi\varepsilon_0 r_{12}}\right)\psi = E\psi \quad (2.8)$$

where k is a product of universal constants, r_1 and r_2 are the distances of each electron from the nucleus and r_{12} is the interelectronic distance. This increase in complexity adds considerably to the difficulty of the calculation. Particular difficulty arises when dealing with the repulsion between the two electrons. These problems become even more complex as the analysis is extended to atoms with larger numbers of electrons and it is at present impossible to solve the wave equation of a many-electron atom directly. Methods of approximation have to be sought and many of these methods attempt to get round the problem of the interelectron terms by considering only one electron at a time. In other words, the many-electron atom is treated as a series of problems based on hydrogen-like situations and this, in turn, means that many of the results of the last section carry over for many-electron atoms with only minor modifications.

One method of approximation, for example, starts with the wave equation for only one of the electrons moving in a potential field determined by the nuclear charge and the averaged-out field of all the other electrons. This case is then like that of the hydrogen atom with a modified value for the nuclear charge and may be solved to give a 'first approximation' distribution function for this electron. Each electron is considered in turn and a set of first approximation distribution functions is obtained. These functions are then used to refine the value of the potential field of the nucleus plus electrons and then the analysis is repeated giving 'second approximation' functions. The whole process is continued until self-consistent results are obtained and a solution to the many-electron problem is found using the method for the hydrogen-like atom at each step in the calculation. The above method is called the *self-consistent field method*. Other, more accurate, approaches are available, some of which start off from the self-consistent field answers, but these cannot be followed out here. An introduction is given in Coulson's book listed in the references. The process of reiteration outlined for the *self-consistent field method* is greatly enhanced by the power of modern computers.

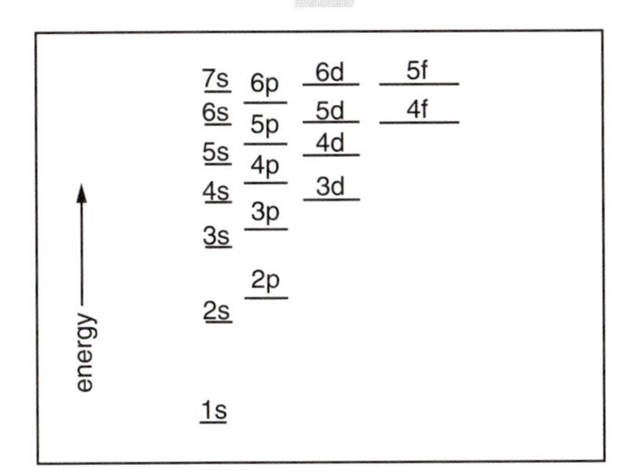

FIG. 2.4 Energy levels in many-electron atoms. This diagram shows the relative energy levels of the orbitals at the Z values where they are about to be filled

of the wave function again describes the probability distribution of the electron in one of these orbitals. These atomic orbitals are defined by the three quantum numbers n, l and m which are restricted to certain integral values by the same rules as for hydrogen, as in Table 2.2. One difference between many-electron atoms and the hydrogen atom is that the energy of an electron in an orbital depends on l as well as on n: that is, the order of energies now includes s<p<d<f as well as $n = 1<2<3\ldots$. As a result, when there are d and f levels corresponding to an n value, these overlap in energy with the s and p levels of higher n values. The order of energies for a many-electron atom is approximately 1s<2s<2p<3s<3p<4s=3d<4p<5s=4d<5p<6s=5d=4f<6p<7s=6d=5f. This is illustrated in Fig. 2.4. Two points are of importance in this figure: first, that for higher n values the ns level is approximately equal in energy to the $(n-1)$d level and to the $(n-2)$f level, and second, that the energy gaps between successive levels with the same l value become smaller as n values increase. This has important effects on the chemistry of the heavier elements.

Although the energy of an orbital in a many-electron atom depends on both the n and l values, there is no dependence on the value of m in the free atom, and the three p orbitals or the five d orbitals of a given n value are equal in energy as they are in the hydrogen atom. The total electronic energy of an atom is the sum of the energies of each electron in its orbital, with a correction for the interaction between the electrons. The spectra of many-electron atoms are much more complicated than that of hydrogen, as there are more energy levels and some overlap. However, most atomic spectra have now been successfully analysed and these give experimental values of the energy levels which agree with the calculated ones.

STELLAR SPECTRA

Atoms can exist in the cooler outer regions of the Sun and other stars, and thus hydrogen can be identified through its Lyman or other series by spectroscopic examination of starlight. While the spectra of many-electron atoms are more complicated than that of hydrogen, they show series of lines of similar origin which can be matched with spectra measured experimentally on Earth. This technique allows elements to be identified in the stars and in interstellar gases. Further, because stars are moving relative to the Earth, the Doppler effect shifts the spectra in proportion to their relative velocity—for receding sources the shift is to lower frequencies. Such redshift studies show that stars, and indeed galaxies, are receding with a velocity that increases in proportion to their distance and are the basis of current models of the expanding Universe. Such studies are the basis of cosmochemistry, see Section 8.7

2.8 The electronic structures of atoms

The electronic structure of an atom may be built up by placing the electrons in the atomic orbitals. Clearly, the orbitals of lowest energy will be the first to be filled and the questions which have to be answered in order to carry out the building-up, or *aufbau*, process are:

(i) how many electrons in each orbital?
(ii) what happens when there are a number of orbitals of equal energy, such as three 2p orbitals.

The answer to (i) depends on one other property of the electron, its *spin*. This property was discovered during a study of the spectra of the alkali metals, when it was found that the absorption lines had a fine structure which could be explained only if the electron was regarded as spinning on its axis and able to take up one of two orientations with respect to a given direction. This spin is included in the description of the state of an electron by introducing a fourth quantum number, m_s, which may take one of the two values $\pm\frac{1}{2}$. The spin is usually indicated in diagrams by using an arrow, ↑ or ↓. The distribution of electrons in an atom is then determined by Pauli's Principle that no two electrons may have all four

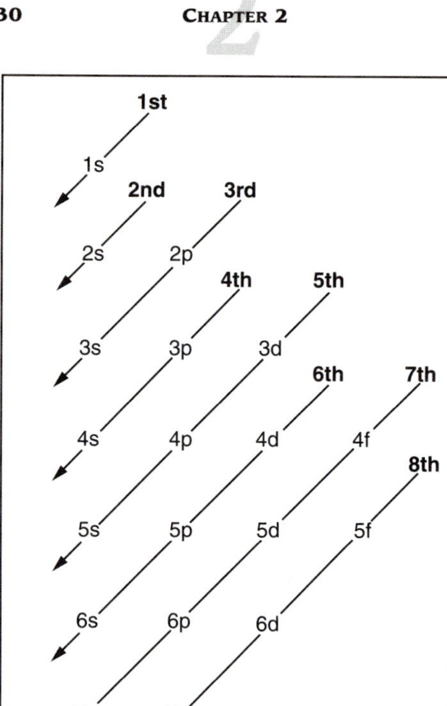

FIG. 2.5 The order of filling energy levels in an atom. The levels are writen out in a square array and filled in the order in which they are cut by a series of diagonals as shown.

quantum numbers the same. Thus the an answer to (i) is that each orbital, which is defined by the three quantum numbers n, l and m, can hold only two electrons and these will have $m_s = +\frac{1}{2}$ and $-\frac{1}{2}$. Such a pair of electrons of opposite spin in one orbital is termed 'spin-paired'.

The answer to (ii) is given by Hund's Rules which may be stated as:

(a) electrons tend to avoid as far as possible being in the same orbital,
(b) electrons in different orbitals of the same energy have parallel spins.

The electronic structure of any atom may be worked out taking account of these factors as follows:

(1) Orbitals are filled in order of increasing energy. This may be remembered from the diagram shown in Fig. 2.5 where the orbitals corresponding to a given value of n are written out in horizontal rows and then filled in the order in which they are cut by the series of diagonal lines as shown. This device gives an order which is substantially correct, though there are a few slight anomalies for heavier atoms. In particular, one electron enters the 5d level before the 4f level is filled and some uncertainty exists about the distribution of electrons between the 6d and 5f levels in the heaviest atoms.

(2) Each orbital (which is defined by specific values of n, l and m) may hold only two electrons with $m_s = \pm\frac{1}{2}$. In other words, each electron is described by the four quantum numbers, n, l, m and m_s, which may not all have the same values.

(3) Where a number of orbitals of equal energy is available, the electrons fill each singly, keeping their spins parallel, before spin-pairing starts.

The use of these rules to build up the electronic structures of the elements may be illustrated by a few examples, but first two useful notations for describing these structures must be defined. A convenient way of showing the electronic structures by a diagram is to place cells on each level of the appropriate part of Fig. 2.4, corresponding to the number of orbitals in each level. The electrons are indicated by arrows and a cell with a pair of arrows shows a filled orbital. The diagrams for a number of the lighter elements are shown in Fig. 2.6. A second notation is to write out the orbitals in order of increasing energy and to indicate the number of electrons by superscripts. To save this latter notation becoming too clumsy, it is convenient to show the configuration of inner, completed levels by the symbol of the appropriate rare gas element. Thus, helium is written, $He = 1s^2$, lithium is $Li = (1s)^2(2s)^1$ or $Li = [He](2s)^1$ and sodium, which, in full, is $(1s)^2(2s)^2(2p)^6(3s)^1$, is shortened to $Na = [Ne](3s)^1$.

The electron in the hydrogen atom is described by the four values of the quantum numbers $(1, 0, 0, \frac{1}{2})$. Then, in helium, the second electron enters the lowest energy orbital (rule 1) which is the 1s, orbital singly occupied in the hydrogen atom. This second electron must have its spin antiparallel with the first (rule 2) and corresponds to the quantum

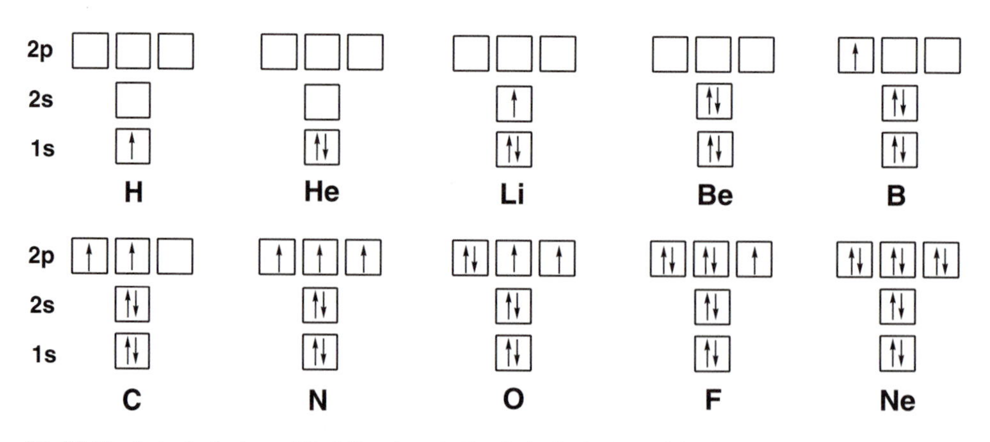

FIG. 2.6 The electronic structures of the lighter elements. The electronic structures of the elements where the 1s, 2s and 2p levels are being filled, illustrating the *aufbau* process

numbers $(1, 0, 0, -\frac{1}{2})$. In the next element, lithium, the third electron must enter the next lower orbital, 2s, as in Fig. 2.6.

The operation of the third rule is illustrated by the case of carbon with $Z = 6$. Starting from the three electrons in lithium, the fourth completes the 2s orbital and the fifth enters the next lowest level, 2p. The sixth electron then enters one of the two remaining empty 2p orbitals with its spin parallel to that of the previous electron. The third 2p level is occupied in the next element, nitrogen, and not until there is a fourth electron to be accommodated in the 2p level—at oxygen—does spin-pairing occur in that level. The orbitals with $n = 2$ are completely filled at neon, $Z = 10$. Sets of orbitals with the same value of the n quantum number are known as quantum *shells* and the electrons in that shell which is partly filled in a particular atom are termed the *valency electrons*.

The rules for deriving atomic structures are further illustrated below for some of the heavier elements.

Consider first iron, Fe with $Z = 26$, whose electronic structure is shown in Fig. 2.7. The first ten electrons fill up to the neon structure while the next eight fill the 3s and 3p levels, repeating the pattern of the second shell, to form the argon core. This accounts for eighteen electrons. Fig. 2.4 shows that the level which comes next in energy to the 3p level is 4s which is a little more stable than 3d. The nineteenth and twentieth electrons fill the 4s level, leaving six electrons to be accommodated in the five 3d orbitals. The first five electrons enter these orbitals singly with all their spins parallel while the last electron pairs up with one of these. Thus the configuration is $Fe = [Ar](3d)^6(4s)^2$ with four unpaired spins. Note that 3d is now more stable than 4s.

As a second example, take gadolinium, Gd with $Z = 64$, see Fig. 2.7. Continuing from the configuration of iron, the next ten electrons fill the 3d and 4p levels to give

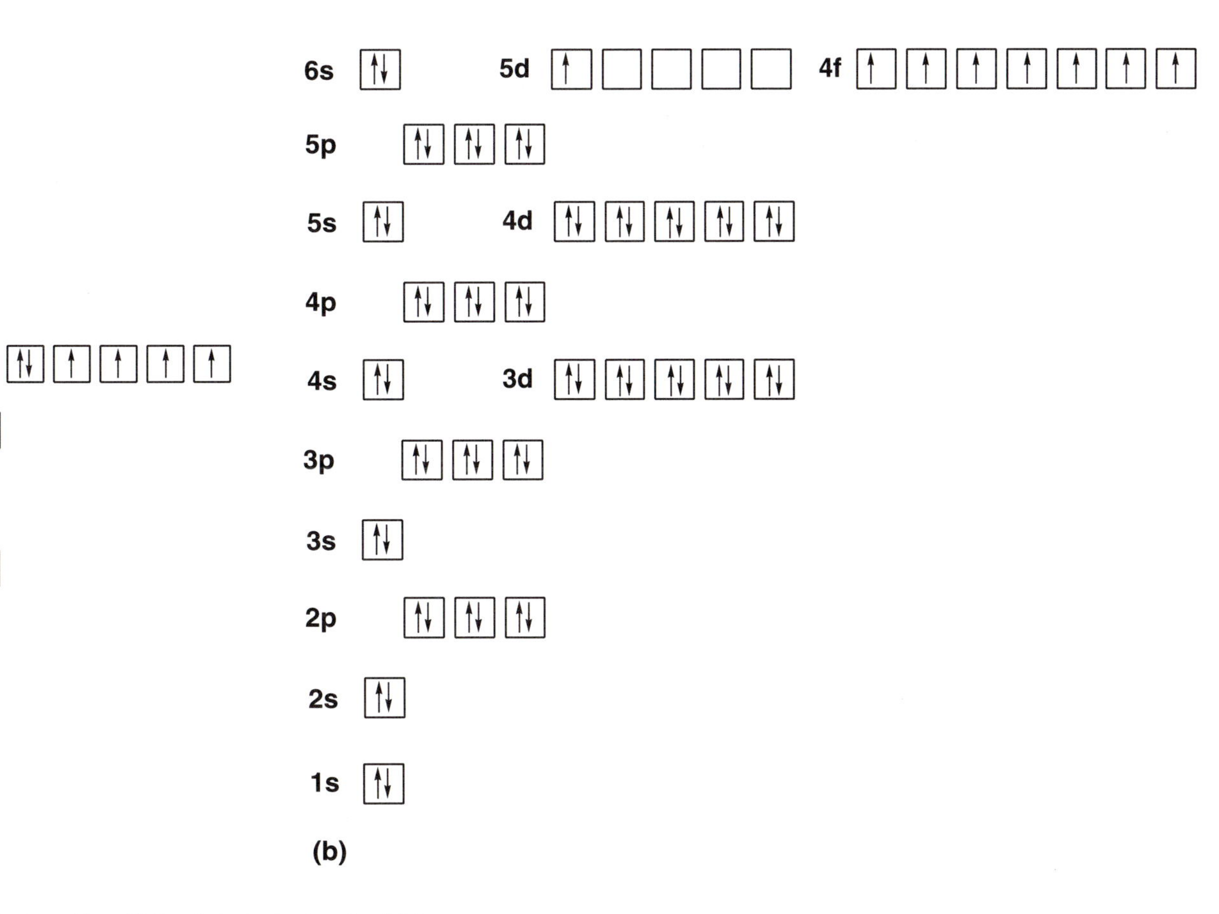

FIG. 2.7 (a) The electronic structure of iron. (b) The electronic structure of gadolinium

the configuration of krypton, $Z = 36$. The next eighteen repeat this pattern in the 5s, 4d and 5p levels giving the xenon ($Z = 54$) configuration. The next level in energy is 6s which is filled by the next two electrons and then come the 5d and 4f levels which are very close in energy, being almost identical in the range of Z values around $Z = 60$. In the event, the first electron enters the 5d level and the last seven electrons in gadolinium enter the 4f orbitals. As there are seven orbitals in an f level, each one is singly occupied and gadolinium has the configuration Gd = $[Xe](4f)^7(5d)^1(6s)^2$ with eight unpaired electrons (the single d electron and the seven parallel f electrons). As the 5d and 4f orbitals are so similar in energy, the elements in this region of the Periodic Table vary in the distribution of their electrons between the two. There is never more than one electron in the 5d level, but often there is none at all. Thus, europium with $Z = 63$, which precedes gadolinium, has the configuration Eu = $[Xe](4f)^7(5d)^0(6s)^2$.

The rest of the elements up to $Z = 86$ complete the 4f, 5d and 6p levels to give the configuration of the heaviest rare gas, radon. The next two electrons fill the 7s level and then there is a close correspondence between the energies of the 5f and 6d levels, similar to that between the 4f and 5d levels. Here the energy gap is even smaller and there has been considerable difficulty and confusion about the levels being occupied in the heaviest elements. These elements are best regarded as paralleling the 4f ones in filling mainly the 5f level, but the first members of the set make more use of the d level than their lighter congeners. The full electronic structure of all the elements is given in Table 2.5.

Shapes of atomic orbitals

In the last section, the electronic structures of the elements were built up from a knowledge of the energy levels of the orbitals derived from the wave equation. Since the chemist is primarily concerned with the outermost electrons and these are found in s, p or d orbitals, the shape and extension in space of these orbitals is of primary importance. Indeed, the general aim of understanding and predicting the shape and reactions of ions and molecules may be carried quite a long way by qualitative reasoning which is based largely on diagrams of the relevant atomic orbitals, as the next chapters show.

2.9 The s orbital

The solution of the wave equation for the 1s orbital of hydrogen is the wave function given in Equation (2.2):

$$\psi_1 = \sqrt{(1/\pi a_0^3)\exp(-r/a_0)}$$

where r is the distance from the nucleus. The probability density of the electron distribution is the square of this wave function: $\psi_1^2 = (1/\pi a_0^3)\exp(-2r/a_0)$.

As these expressions are functions only of r, the distance from the nucleus, they are spherically symmetric around the nucleus. As they are exponential functions, the values of the wave function and of the electron density fall off smoothly and rapidly with increasing distance from the nucleus (see Fig. 2.8a). The contour diagram which results from joining points in space with the same value of ψ or ψ^2 has the general appearance of Fig. 2.8b (where the values decrease outwards from the nucleus). As exponential functions never quite fall to zero, there is always a finite electron density outside any given one of these contour lines but it is possible to include the major part of the electron cloud within a boundary surface close to the nucleus, and it is common to represent the orbital by a single boundary contour (as in Fig. 2.8c) enclosing an arbitrary fraction—say 90 percent—of the electron density. Such a boundary diagram may be used to represent the electron density, ψ^2, or it may be the corresponding diagram of the orbital, ψ. Since both functions are

See Figs 2.10 and 2.11 for examples, the 1s orbital has the same sign throughout.

TABLE 2.5 The electronic configurations of the elements

Element	*Symbol*	*Z*	A_r	*Electronic configuration*			
				Inner shells	*Valency shell*		
						1s	
Hydrogen	H	1	1.007 94			1	
Helium	He	2	4.002 602			2	
						2s	2p
Lithium	Li	3	6.941	He		1	
Beryllium	Be	4	9.012 182	He		2	
Boron	B	5	10.811	He		2	1
Carbon	C	6	12.0107	He		2	2
Nitrogen	N	7	14.006 7	He		2	3
Oxygen	O	8	15.999 4	He		2	4
Fluorine	F	9	18.998 403 2	He		2	5
Neon	Ne	10	20.179 7	He		2	6
						3s	3p
Sodium	Na	11	22.989 770	Ne		1	
Magnesium	Mg	12	24.305 0	Ne		2	
Aluminium	Al	13	26.981 538	Ne		2	1
Silicon	Si	14	28.085 5	Ne		2	2
Phosphorus	P	15	30.973 761	Ne		2	3
Sulfur	S	16	32.065	Ne		2	4
Chlorine	Cl	17	35.453	Ne		2	5
Argon	Ar	18	39.948	Ne		2	6
					3d	4s	4p
Potassium	K	19	39.098 3	Ar		1	
Calcium	Ca	20	40.078	Ar		2	
Scandium	Sc	21	44.955 910	Ar	1	2	
Titanium	Ti	22	47.867	Ar	2	2	
Vanadium	V	23	50.941 5	Ar	3	2	
Chromium	Cr	24	51.996 1	Ar	5	1	
Manganese	Mn	25	54.938 049	Ar	5	2	
Iron	Fe	26	55.845	Ar	6	2	
Cobalt	Co	27	58.933 200	Ar	7	2	
Nickel	Ni	28	58.693 4	Ar	8	2	
Copper	Cu	29	63.546	Ar	10	1	
Zinc	Zn	30	65.39	Ar	10	2	
Gallium	Ga	31	69.723	Ar	10	2	1
Germanium	Ge	32	72.64	Ar	10	2	2
Arsenic	As	33	74.921 60	Ar	10	2	3
Selenium	Se	34	78.96	Ar	10	2	4
Bromine	Br	35	79.904	Ar	10	2	5
Krypton	Kr	36	83.80	Ar	10	2	6
					4d	5s	5p
Rubidium	Rb	37	85.467 8	Kr		1	
Strontium	Sr	38	87.62	Kr		2	
Yttrium	Y	39	88.905 85	Kr	1	2	
Zirconium	Zr	40	91.224	Kr	2	2	
Niobium	Nb	41	92.906 38	Kr	4	1	
Molybdenum	Mo	42	95.94	Kr	5	1	
Technetium	Tc	43	98.906 3*	Kr	6	1	
Ruthenium	Ru	44	101.07	Kr	7	1	
Rhodium	Rh	45	102.905 50	Kr	8	1	
Palladium	Pd	46	106.42	Kr	10	0	
Silver	Ag	47	107.868 2	Kr	10	1	

TABLE 2.5 (Contd.)

Element	Symbol	Z	A_r	Electronic configuration				
				Inner shells	Valency shell			
						4d	5s	5p
Cadmium	Cd	48	112.411	Kr		10	2	
Indium	In	49	114.818	Kr		10	2	1
Tin	Sn	50	118.710	Kr		10	2	2
Antimony	Sb	51	121.760	Kr		10	2	3
Tellurium	Te	52	127.60	Kr		10	2	4
Iodine	I	53	126.904 47	Kr		10	2	5
Xenon	Xe	54	131.293	Kr		10	2	6
					4f	5d	6s	6p
Caesium	Cs	55	132.905 45	Xe			1	
Barium	Ba	56	137.327	Xe			2	
Lanthanum	La	57	138.905 5	Xe		1	2	
Cerium	Ce	58	140.116	Xe	2		2	
Praseodymium	Pr	59	140.907 65	Xe	3		2	
Neodymium	Nd	60	144.24	Xe	4		2	
Promethium	Pm	61	146.915 1*	Xe	5		2	
Samarium	Sm	62	150.36	Xe	6		2	
Europium	Eu	63	151.964	Xe	7		2	
Gadolinium	Gd	64	157.25	Xe	7	1	2	
Terbium	Tb	65	158.925 34	Xe	9		2	
Dysprosium	Dy	66	162.50	Xe	10		2	
Holmium	Ho	67	164.930 32	Xe	11		2	
Erbium	Er	68	167.259	Xe	12		2	
Thulium	Tm	69	168.934 21	Xe	13		2	
Ytterbium	Yb	70	173.04	Xe	14		2	
Lutetium	Lu	71	174.967	Xe	14	1	2	
Hafnium	Hf	72	178.49	Xe	14	2	2	
Tantalum	Ta	73	180.947 9	Xe	14	3	2	
Tungsten	W	74	183.84	Xe	14	4	2	
Rhenium	Re	75	186.207	Xe	14	5	2	
Osmium	Os	76	190.23	Xe	14	6	2	
Iridium	Ir	77	192.217	Xe	14	7	2	
Platinum	Pt	78	195.078	Xe	14	9	1	
Gold	Au	79	196.966 55	Xe	14	10	1	
Mercury	Hg	80	200.59	Xe	14	10	2	
Thallium	Tl	81	204.383 3	Xe	14	10	2	1
Lead	Pb	82	207.2	Xe	14	10	2	2
Bismuth	Bi	83	208.980 38	Xe	14	10	2	3
Polonium	Po	84	209.982 4*	Xe	14	10	2	4
Astatine	At	85	209.987 1*	Xe	14	10	2	5
Radon	Rn	86	222.017 6*	Xe	14	10	2	6
					5f	6d	7s	
Francium	Fr	87	223.019 7*	Rn			1	
Radium	Ra	88	226.025 4*	Rn			2	
Actinium	Ac	89	227.027 7*	Rn		1	2	
Thorium	Th	90	232.038 1*	Rn		2	2	
Protactinium	Pa	91	231.035 9*	Rn	2	1	2	
Uranium	U	92	238.028 91	Rn	3	1	2	
Neptunium	Np	93	237.048 2*	Rn	5		2	
Plutonium	Pu	94	244.064 2*	Rn	6		2	
Americium	Am	95	241.061 4*	Rn	7		2	
Curium	Cm	96	247.070 4*	Rn	7	1	2	

TABLE 2.5 (Contd.)

Element	Symbol	Z	A_r	Electronic configuration				
				Inner shells	Valency shell			
					5f	6d	7s	7p
Berkelium	Bk	97	247.070 3*	Rn	8	1	2	
Californium	Cf	98	251.079 6	Rn	10		2	
Einsteinium	Es	99	(252)	Rn	11		2	
Fermium	Fm	100	(257)	Rn	12		2	
Mendelevium	Md	101	(258,260)	Rn	13		2	
Nobelium	No	102	(259)	Rn	14		2	
Lawrencium	Lr	103	(262)	Rn	14	1?	2	
Rutherfordium	Rf	104	(261)	Rn	14	1?	2	1?
Dubnium	Db	105	(262)	Rn	14	3	2	
Seaborgium	Sg	106	(265,266)	Rn	14	4	2	
Bohrium	Bh	107	(264)					
Hassium	Hs	108	(277)					
Meitnerium	Mt	109	(268)					
(Ununnilium	Uun	110)	(281)					
(Unununium	Uuu	111)	(272)					
(Ununbium	Uub	112)	(285)					
(Ununquadium	Uuq	114)	(289)					

(a) Values for relative atomic masses are based on 1999 revision. Recent studies of isotopic composition have shown variations from different sources (H, Li, B, C, O, Si, S, Ar, Cu, Pb) or variations introduced by commercial isolation (Li, B, U) which limits the accuracy of the atomic mass or requires that the isotopic composition of a particular sample should be determined. Elements involved in radioactive decay processes (Ar, Sr, Xe, Er, Pb, Ra) may have different isotopic compositions in different geological specimens. The other changes are mainly in precision. (b) Weights marked with an asterisk are those of the commonest long-lived isotope of a radioactive element: those in brackets indicate the most accessible isotope of the heavier elements. (c) Relative atomic masses based on $^{12}C = 12.000\,0$: in SI, $^{12}C = 12.000\,0 \times 1.660\,57 \times 10^{-27}$ kg. (d) Elements 110, 111, 112 and 114 have been reported but are not yet fully established (see Section 16.12).

exponential ones, the boundary diagrams for ψ and ψ^2 are similar in appearance, that for ψ being distinguished by showing the variation in the sign of the wave function in different regions of space.

As discussed in the next chapter, a bond is formed by the combination of atomic orbitals, and the way in which they may combine depends on the signs of the various wave functions. For this reason, the chemist usually works initially with boundary diagrams of the wave functions, ψ. However, we must always remember that we are interested in the electron distribution around the nuclei—that is, having decided how the wave functions, ψ, combine, we square the resultant to get ψ^2, the probability density of the electron distribution.

A further important function used in representing an orbital is the *radial density function*, $4\pi r^2\psi^2(r)$. The volume of a spherical shell of thickness dr at a distance r from the nucleus is $4\pi r^2 dr$. Hence, $4\pi r^2\psi^2(r)dr$ is the probability of the electrons being found at a distance between r and $(r+dr)$ from the nucleus. The plot of the variation of this function with distance from the nucleus for the hydrogen ls orbital is shown in Fig. 2.9a. The radial density in hydrogen is at a maximum at that distance, a_0, from the nucleus that Bohr calculated as the radius of the most stable orbit in the planetary theory. Note that the radial density function is the product of an r^2 function (increasing from zero at the nucleus) and the exponentially decreasing wave function; hence, it starts at zero at the nucleus and passes through a maximum.

The s orbitals of higher n value resemble the ls orbital, being spherically symmetrical and having the same sign for the wave function in all directions. They extend further into space and there are changes close in towards the nucleus where spherical nodes appear across

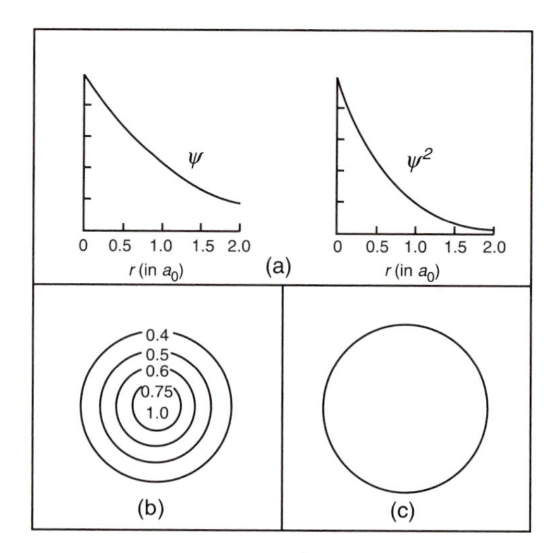

FIG. 2.8 The 1s orbital of hydrogen. (a) Plots of ψ_{1s} and ψ_{1s}^2 against r. (b) Contour representation of the 1s orbital. (c) Boundary contour representation

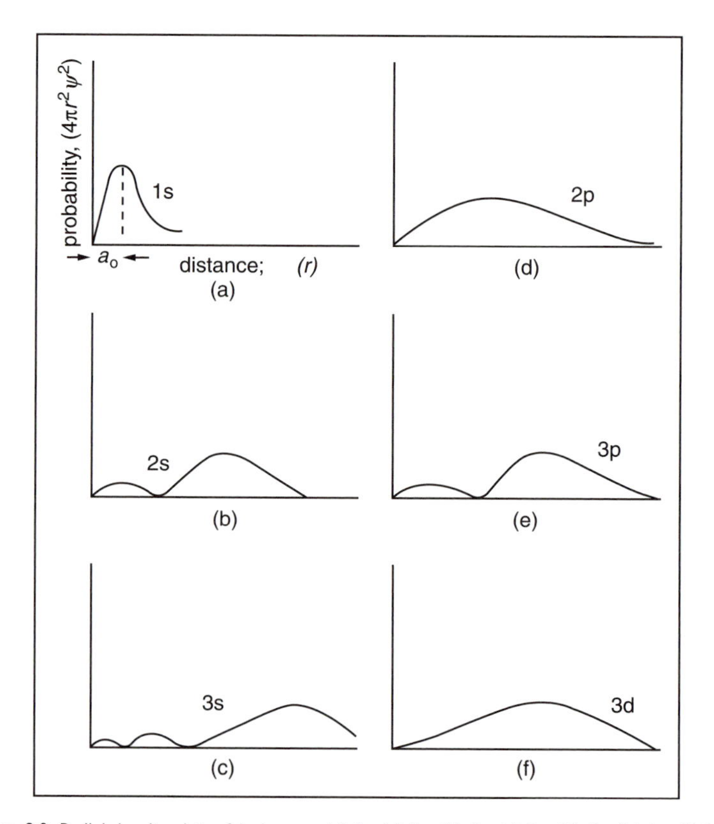

FIG. 2.9 Radial density plots of hydrogen orbitals. (a) 1s, (b) 2s, (c) 3s, (d) 2p, (e) 3p, (f) 3d

which the wave function changes sign. As only the outer regions are of chemical interest, these nodes are rarely important. They appear as minima in the radial density plot; compare for example, the hydrogen 2s orbital in Fig. 2.9b. The distance from the nucleus of the maximum probability increases rapidly to about $5.3a_0$ for 2s and nearly $14a_0$ for 3s so that these orbitals are much more diffuse (recall that the total area under the curve equals a probability of 1 for each orbital).

2.10 The p orbitals

From Table 2.3, the lowest-energy p orbitals are those where $n = 2$, given by Equations (2.5) or (2.7) for hydrogen. These orbitals do have directional properties, in contrast to the s orbitals which are alike in every direction from the nucleus. Equations (2.7) allow us to see this directional property most readily, as the total wave function may be separated into two parts, one a function only of r (the radial part), and one a function of θ and ϕ (the angular part). The radial part falls off exponentially from the nucleus, just as for the s orbitals, and this mainly determines the energy of the electron in the p orbital. For our purposes, the more important component is the angular part. For p_z this is simply $\cos\theta$, the angle with the z axis, while p_x varies as $\sin\theta\cos\phi$ and p_y as $\sin\theta\sin\phi$, both involving the angle to the x axis as well. In Fig. 2.10a are shown the boundary contour representations of these angular parts of the 2p orbitals. Only two regions of space are occupied, aligned along the plus and minus directions of one axis. The function changes sign across the nucleus, and there is a nodal plane through the nucleus where the value is zero. The three equations of (2.7) are chosen so that the three orbitals are identical apart from their orientation—along the z, x and y axes respectively.

Several different conventions are in use to display atomic orbitals, and the use of shading to indicate the + and − parts of the wave function is becoming common. An alternative version of the p orbitals is shown in Fig. 2.10b, using this shading convention.

Multiplying by the radial part of the wave function does not change the fundamental property that the p wave functions describe two lobes of opposite sign separated by a nodal plane, although the lobes are elongated into a fat tear-shape.

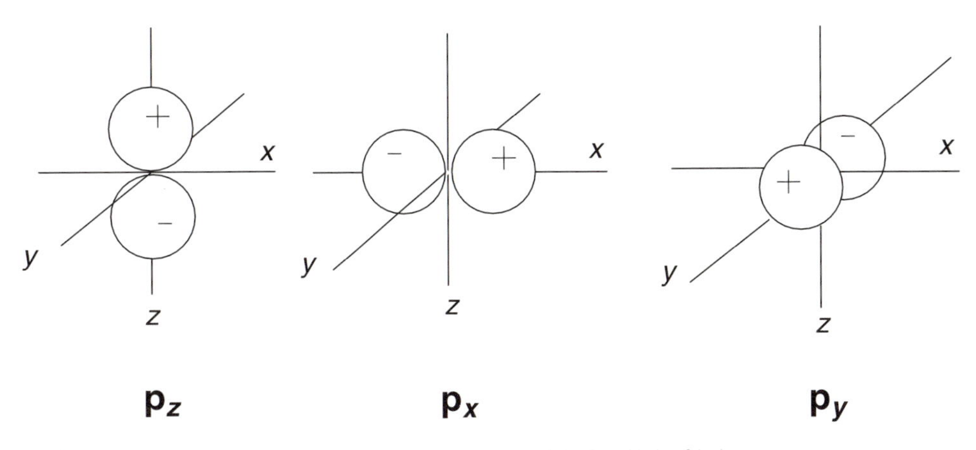

FIG. 2.10a Boundary contour representations of the angular parts of the 2p orbitals of hydrogen

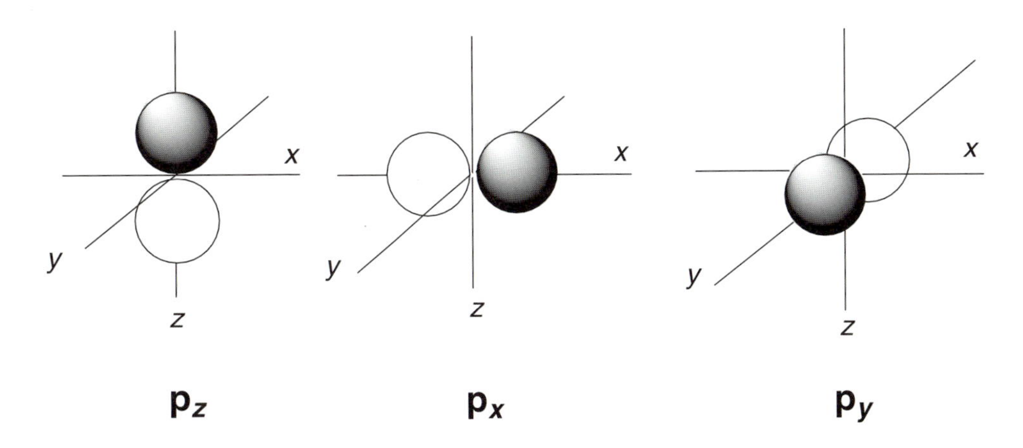

FIG. 2.10b Boundary contour representation of the 2p orbitals of hydrogen, using the convention in which the shaded and unshaded lobes represent opposite signs of the wave function

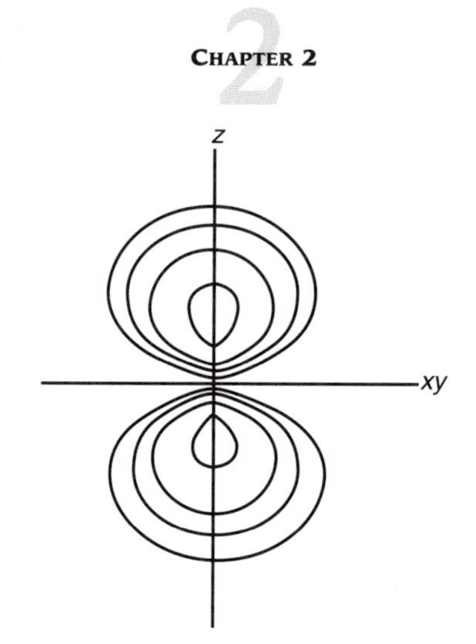

FIG. 2.11 Contour representation of a 2p orbital

Thus when the wave function is squared to get the electron density, this is concentrated in two lobes (as sketched in Fig. 2.11) and there is zero electron density in the nodal plane passing through the nucleus perpendicular to the axis of the lobes. Contours join points where the electron density decreases outwards by a factor of 4 from one to the next. For effective nuclear charges appropriate to B to F, the maximum electron density is found at distances ranging from about 0.3 to $1.3a_0$ from the nucleus.

Thus the p orbitals contrast with the s orbitals in being directional and having zero electron density at the nucleus. When the radial density function is plotted for the 2p orbital of hydrogen, we find the maximum at $4a_0$ (i.e. n^2a_0 as predicted also by the Bohr theory). For higher values of n, the p wave function is similarly oriented in two lobes with a nodal plane through the nucleus and there are also radial nodes of zero electron density, as shown by Fig. 2.9e. As for the higher s orbitals, these radial nodes are too close in to the nucleus to be of significance in a qualitative treatment and we may disregard them.

Modern calculations are able to give detailed descriptions of the electron density in p orbitals for the majority of the elements, though there are still difficulties in dealing with the largest numbers of electrons. Some consequences are discussed in Section 18.9: here we need only the qualitative picture and concentrate on the sign changes of Fig. 2.10.

2.11 The d orbitals

In a similar way, the angular parts of the d orbitals may be separated and are shown in Fig. 2.12. Multiplying in the radial part does not change the basic general character of a four-lobed figure with alternating signs, and squaring the total wave function to get

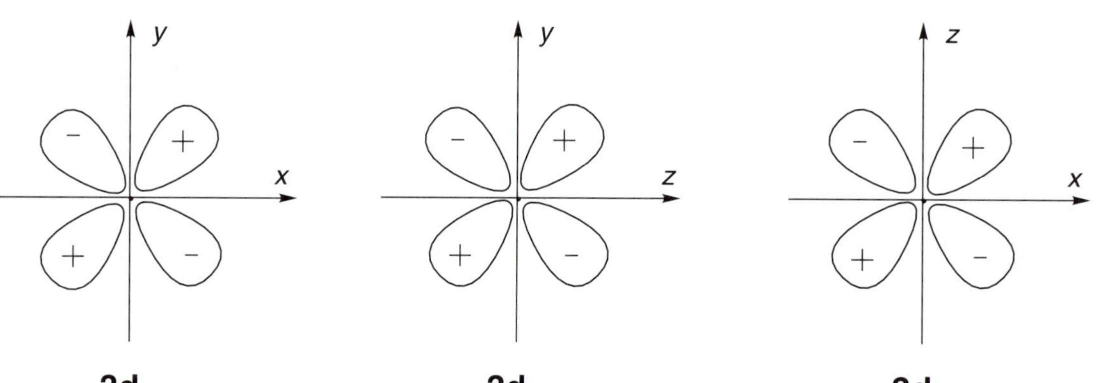

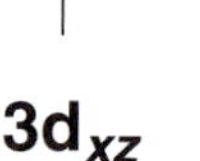

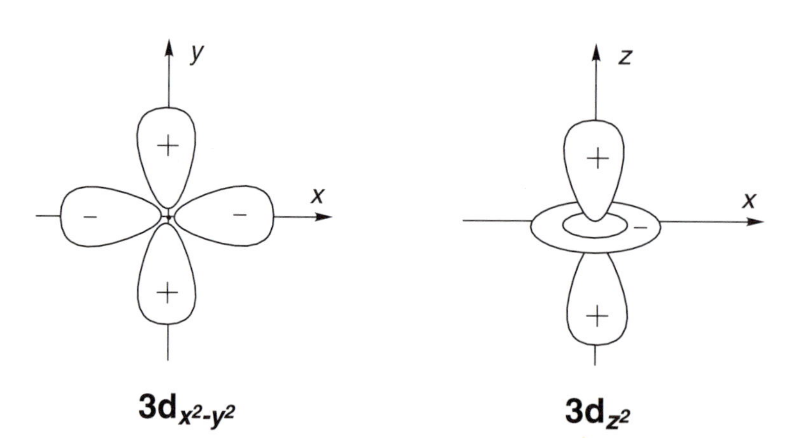

FIG. 2.12a Boundary contour representations of the angular parts of the 3d orbitals (note the changes of axes)

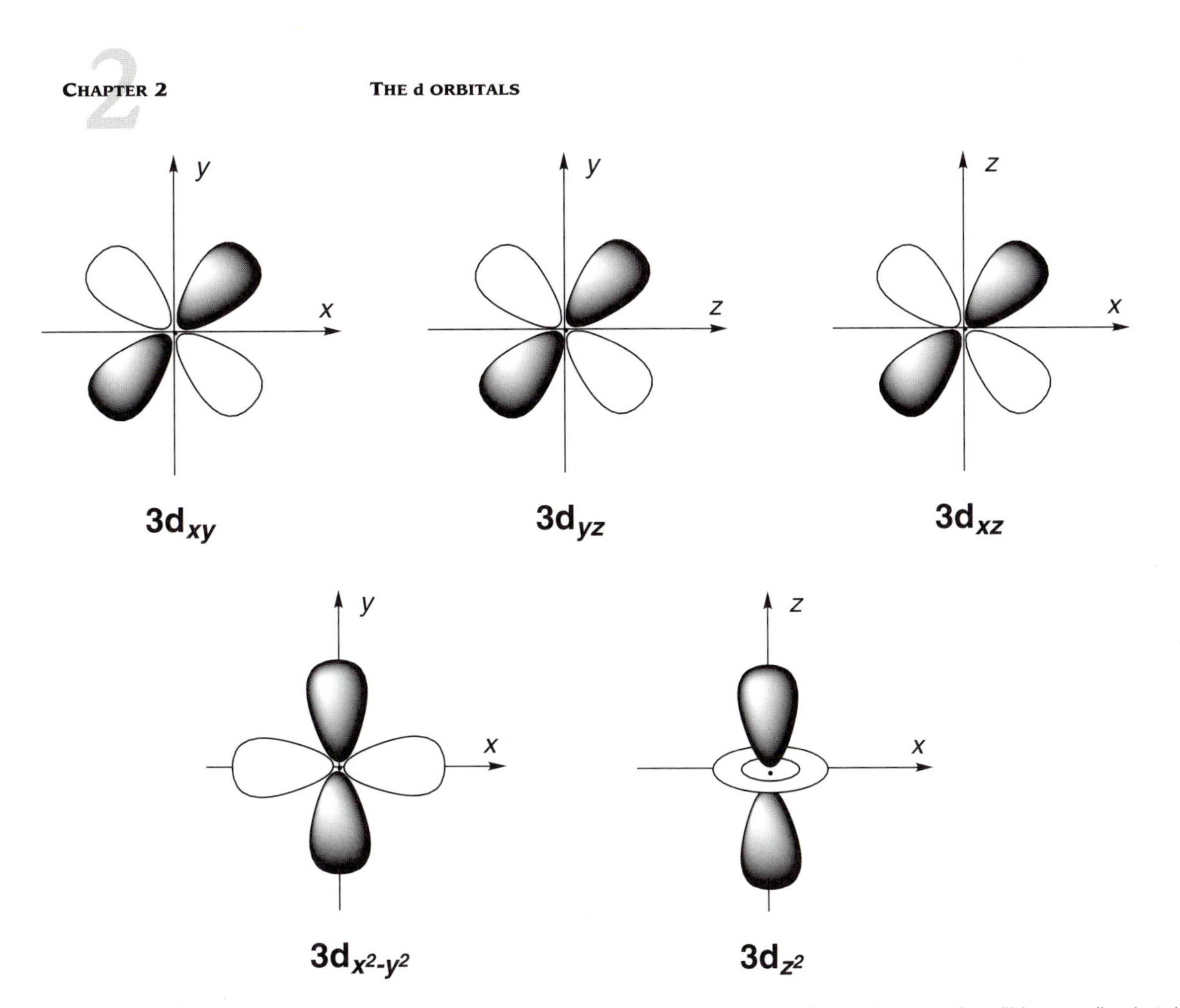

FIG. 2.12b Representation in which the shading indicates opposite signs. Note: in subsequent diagrams the shading convention will be generally adopted

the electron density changes the shapes of the lobes but does not change the basic four-lobed character. The radial density function for hydrogen, shown in Fig. 2.9f, has a maximum at $9a_0$. As for the p orbitals, this maximum contracts rapidly with nuclear charge and is found about one Bohr radius from the nucleus in the first row transition metals.

In a d orbital, there are two nodal planes at right angles through the nucleus and therefore zero electron density at the nucleus. The d_{z^2} orbital does not fit this description and it arises in the following way. The three orbitals d_{xy}, d_{yz} and d_{zx} are identical except for their orientation in space, and the $d_{x^2-y^2}$ orbital is the same as the d_{xy} orbital but rotated through 45°. There are two other possible orbitals corresponding to the $d_{x^2-y^2}$ orbital and directed along the other two pairs of axes (in an obvious notation, $d_{y^2-z^2}$ and $d_{z^2-x^2}$) but these would give six d orbitals in all, although there are only five independent solutions of the d type of the wave equation. It is easy to show that the three solutions of the d_{xy} type are independent while the three of the $d_{x^2-y^2}$ type are not. (If the orbital diagrams of this latter set of three are superimposed, taking account of the signs, the lobes all cancel out.) Any two of the three would be acceptable along with the three orbitals of the d_{xy} type, but it is convenient to use one of them, say $d_{x^2-y^2}$, together with a combination of the other two, d_{z^2}, where:

$$\psi_{d_{x^2}} = 1/\sqrt{3(\psi_{d_{x^2-y^2}} - \psi_{d_{y^2-x^2}})}$$

This gives a satisfactory set of five independent d orbitals and this is the set in common use. Naturally, the z direction may be chosen to suit the situation. For example, if the atom is in an external magnetic field, it is useful to have the z axis parallel to the field. The higher d orbitals are of the same basic shape as the 3d orbitals.

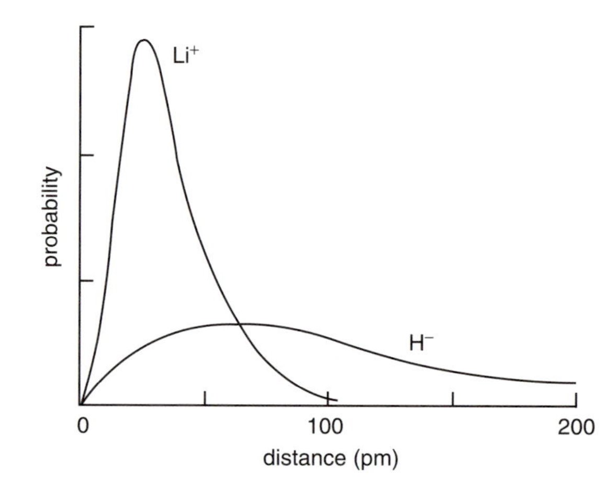

FIG. 2.13 Radial density curves for ions containing two electrons and with different nuclear charges. This plot for isoelectronic ions illustrates the effect of increasing the nuclear charge from $Z = 1$ to $Z = 3$ on the spatial distribution of the electron probability density

It might be noted that the five d orbitals described above do not all correspond directly to particular m values. While $m = 0$ gives d_{z^2}, d_{xz} and d_{yz} are formed from combinations of $m = \pm 1$, and d_{xy} and $d_{x^2-y^2}$ from combinations of $m = \pm 2$.

The f orbitals have shapes reminiscent of the d orbitals but with a further nodal plane through the nucleus. For example f_{xyz} is an 8-lobed figure with three planes at right angles. As these orbitals are little used in bonding they need not be discussed further.

The basic shapes of orbitals are unchanged by changes in the nuclear charge, although their extensions into space depend inversely on the value of Z. Fig. 2.13 shows the radial density curve for the ls orbital of helium-like species with different values of the nuclear charge. The radius of the electron cloud around an atom is the resultant of this contraction of the charge cloud with increasing Z, and the fact that orbitals further and further out from the nucleus are being occupied as the atomic number rises. There is a general increase in atomic size as Z increases, but this change occurs in a very irregular manner (compare the plot of atomic radius against atomic number in Fig. 8.8). The greatest jumps in radius come when the outermost electron starts to fill the s level of a new quantum shell and there are less sharp increases when any new level starts to be occupied and when spin-pairing occurs at p^4 and d^6 configurations.

2.12 The Periodic Table

As chemical behaviour depends on the interaction of the electron clouds of atoms, and especially on the interaction of the outermost parts of these clouds, atoms which have their outer electrons in the same type of orbital should have similar chemical behaviour. For example, a configuration such as s^2p^3 implies the same shape of electron cloud, whatever the n values of the orbitals, although the extension of the electron cloud, and hence the atomic radius, clearly depends on the atomic number. If the elements are arranged so that those with the same outer electron configuration fall into Groups, the result is the Periodic Table of the elements. This is one of the most important of all scientific generalizations and grew from an increasing realization through the nineteenth century that elements fell into families whose properties followed a trend as their weight increased—for example, the 'triad' Cl, Br, I where the properties of Br are intermediate between Cl and I. As atomic weights became more accurate, more and more elements were fitted in and chemists like Newlands and Olding produced tabular arrays of elements which were close to the modern table for the lighter elements. By 1869, with some 60–70 elements more or less firmly identified, this development was brought to a climax simultaneously by Mendeléef and by

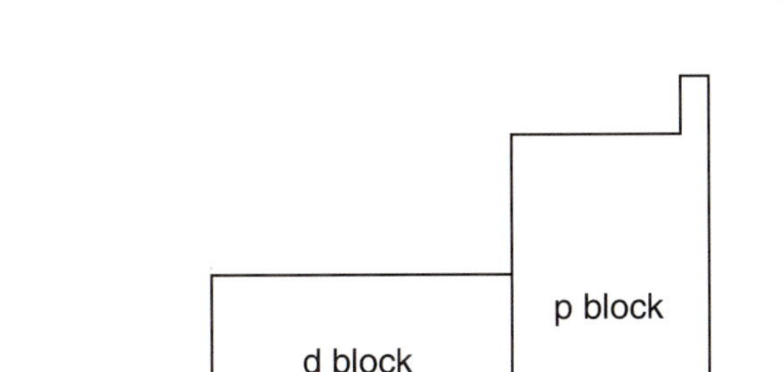

FIG. 2.14 Block diagram of the Periodic Table

THE PERIODIC TABLE

It is difficult to realize, nowadays when all this has become accepted, just how important a contribution a knowledge of electronic structure made to the chemist's understanding and use of the Periodic Table. In particular, the table produced by Mendeléef, based on valency considerations, was in the 'short' form where the transition elements were introduced as sub-groups into the main groups of the s and p blocks. This practice was justifiable at the time as there are some resemblances, particularly when the transition elements are showing their maximum valencies, and there were many blanks in the knowledge of the chemistry of these elements. As such knowledge increased, it became apparent that more anomalies than analogies were introduced by the short form of presentation. The long form removes many of these difficulties, and the case for its adoption became overwhelming when the electronic structures of the elements were worked out. There are still a number of minor anomalies. In particular, no form of the table completely reflects the differences in the ground state electron distribution between the s and d levels among the transition elements, nor those between the d and f levels among the inner transition elements (see Table 2.5), but these differences have no effect on the chemistry of these elements as the anomalies disappear in the valency states. In fact, all the problems and objections to the long form of the Periodic Table disappear when it is regarded as a very successful broad generalization about properties, based on the electronic structures of the elements but not reflecting them in every detail.

Meyer, in the Periodic Law which defined the basic Periodic Table. The new table served as the focus for the discovery and tabulation of new elements, and for much other chemical development, in the sixty years up to the development of quantum approaches. The electron configurations derived from the electronic theory of atoms then provided the theoretical rationalization and refinement of the Periodic Table.

The Periodic Table reflects the order of energy levels in the atoms as they are derived from the wave equation, and the form of the Periodic Table follows from the allowed values of the quantum numbers as given in Table 2.2. This is illustrated by the block form of the Periodic Table shown in Fig. 2.14. The different 'blocks' hold sets of two, six, ten and fourteen elements; these being, respectively, the elements where the s, p, d, and f levels are filling and these blocks thus follow from the number of orbitals of each type which are allowed by the quantum rules. The order of the blocks across the table from left to right follows from the general order of energy levels: $n\mathrm{s} < (n-1)\mathrm{d} \leqslant (n-2)\mathrm{f} < n\mathrm{p}$. The modern 'long' form of the Periodic Table is given as Table 2.6.

A number of special names are given to particular sections of the Periodic Table. There is, unfortunately, some confusion in nomenclature and Table 2.7 lists both the special names approved for general use and some of the cases where conflicting usages appear in text-books. In this book, Groups which do not have a trivial name listed in the Table will be named by the lightest element of the group—carbon Group, titanium Group, etc.

TABLE 2.6 Periodic Table of the elements

	Groups																				*Rare gas electrons*
	1	*2*		*3*	*4*	*5*	*6*	*7*	*8*	*9*	*10*	*11*	*12*		*13*	*14*	*15*	*16*	*17*	*18*	
Period 1 1s	H																		H	He	2
2 2s	Li	Be												2p	B	C	N	O	F	Ne	2, 8
3 3s	Na	Mg												3p	Al	Si	P	S	Cl	Ar	2, 8, 8
4 4s	K	Ca	3d	Sc	Ti	V	Cr	Mn	Fe	Co	Ni	Cu	Zn	4p	Ga	Ge	As	Se	Br	Kr	2, 8, 18, 8
5 5s	Rb	Sr	4d	Y	Zr	Nb	Mo	Tc	Ru	Rh	Pd	Ag	Cd	5p	In	Sn	Sb	Te	I	Xe	2, 8, 18, 18, 8
6 6s	Cs	Ba	5d	La*	Hf	Ta	W	Re	Os	Ir	Pt	Au	Hg	6p	Tl	Pb	Bi	Po	At	Rn	2, 8, 18, 32, 18, 8
7 7s	Fr	Ra	6d	Ac†	Rf	Db	Sg	Bh	Hs	Mt	110	111	112	7p		114					
*Lanthanides				4f	Ce	Pr	Nd	Pm	Sm	Eu	Gd	Tb	Dy	Ho	Er	Tm	Yb	Lu			
†Actinides				5f	Th	Pa	U	Np	Pu	Am	Cm	Bk	Cf	Es	Fm	Md	No	Lr			

TABLE 2.7 Nomenclature of the elements

	General sections of the Periodic Table	
Main Group elements (or typical elements or representative elements)	Those elements with the outermost electrons in the s or p levels–the Groups headed by Li, Be, B, C, N, O, F, and He, plus H. Sometimes the two lighter elements in each group are excluded.	
Transition elements	Those elements where the d or f levels are filling. On the fuller definition of Main Group elements, this class then includes the rest of the elements.	
Inner transition elements	Those elements where an f shell is filling. With this usage, 'transition elements' is then confined to elements of the d block.	
A and B subgroups	This terminology is derived from the older short form. The Main and Transition elements were divided into subgroups, A and B. In Groups I and II the Main Group elements (the alkali and alkaline earth metals respectively) were termed A and the transition elements (copper and zinc Groups) were termed B. But in the other Groups conflicting usages exist; in one system, the Main Group was termed A throughout while in the other,* more common one, it was termed B. Thus Group IVB might be the carbon Group of Main elements or the titanium Group of Transition elements, depending on the author, and the reader has to check which convention was used.	
M (Main) and T (Transition) subgroups	This was an attempt to replace the A/B usages by an unambiguous nomenclature. It has not found official favour since the initials apply only to the English terms.	
The 18 Group system*	Now recommended by IUPAC to finally eliminate the problems of ambiguity. At present subject to lively criticism as the simpler relations between Group number and maximum valency are lost. Use is indicated in Table 2.6.	
In this text we stick by our original view that memorizing the group of elements that belong to a particular number is no easier than memorizing a group of elements! Thus we refer to the Periodic Groups either by the accepted trivial names listed below or by the name of the lightest element in the Group.		
	Trivial names for Groups of elements	
Alkali metals*	Li, Na, K, Rb, Cs, Fr	
Alkaline earth metals*	Ca, Sr, Ba, Ra	
Pnictogens	N, P, As, Sb, Bi	
Chalcogens* or calcogens	O, S, Se, Te, Po	
Halogens*	F, Cl, Br, I, At	
Inert gases or rare or noble* gases	He, Ne, Ar, Kr, Xe, Rn	
Rare earth elements*	Sc, Y, La to Lu inclusive	These divisions are not usually strictly observed and these three names tend to be used interchangeably.
Lanthanum series* or lanthanons	La to Lu inclusive	
Lanthanides or lanthanoids*	Ce to Lu inclusive	
Actinium series*	Ac onwards	
Actinides or actinoids*	Those elements where the 5f shell is being filled (see Chapter 11).	
Transuranium elements*	The elements following U	
Coinage metals	Cu, Ag, Au	
Platinum metals	Ru, Os, Rh, Ir, Pd, Pt	
Noble metals	An ill-defined term applied to the platinum metals, Au, and sometimes includes Ag, Re, and even Hg.	
Metal and non-metal	These two terms are widely used but it is not clear precisely where the boundary between them comes. The term metalloid or semi-metal is often applied to elements of intermediate properties such as B, Si, Ge or As.	

*These usages are approved by the International Union of Pure and Applied Chemistry Rules for Nomenclature of Inorganic Chemistry.

HEAVY METALS

The term 'Heavy Metal' is often found, but in such a wide variety of contexts that it is almost useless. At least twenty different usages are found mostly

in terms of density or specific gravity, but with no agreed cut-off

in terms of atomic number or atomic weight, but again with no agreed starting point

in terms of toxicity and environmental effects.

In current common parlance, *heavy metal* implies toxicity. This use probably stems from concerns about elements such as lead or mercury. However, there is no relationship between toxicity and metallic character—consider arsenic or tellurium—or between toxicity and 'weight'. Thus one of the lightest elements, beryllium, is also one of the most poisonous. In general, toxicity is highly dependent on the compound formed and there is no easy generalization.

The term 'heavy metal' is thus best avoided and replaced by something more precise.

In most chemistry texts, 'heavy' is used in the obvious sense of lower members of the Group.

Further properties of the elements

In this section a number of important atomic properties are defined and discussed.

2.13 Ionization potential

If sufficient energy is available, it is possible to detach one or more electrons from an atom, molecule or ion. The minimum amount of energy required to remove one electron from a gaseous atom, leaving both the electron and the resulting ion without any kinetic energy, is termed the *ionization potential*. Since energy has to be provided to remove the electron against the attraction of the nucleus, ionization potentials are always positive. The energies required to remove the first, second, third, etc., electrons from an atom are its first, second, third, etc., ionization potentials. It is clear that the successive ionization potentials will increase in size as it becomes increasingly difficult to remove further electrons from the positively charged ions. Ionization potentials of molecules and ions are defined in a similar way to those for atoms.

The ionization potentials of atoms reflect the binding energies of their outermost electrons and will be lowest for those elements where the valency electrons have just started to enter a new quantum level. If Table 2.8 of ionization potentials and Figs 8.4 and 8.5 are examined, it will be seen that the lowest first ionization potentials are shown by the alkali metals where the last electron has entered a new quantum shell and has its main probability density markedly further out from the nucleus than the preceding electrons. As a quantum shell fills, on going across a Period from the alkali metal, the outermost electrons become more and more tightly bound and the ionization potentials rise to a maximum at the rare gas element where that quantum shell is completed. The stability of the completed shell is also shown when the successive ionization potentials of any particular element are examined. As electrons are removed from a partly filled shell the successive ionization potentials rise steadily, but there is a great leap in the energy required to remove an electron when all the valency electrons have been removed and the underlying complete quantum shell has to be broken. For example, it requires 518 kJ mol^{-1} to remove the single 2s electron from lithium but, when the closed $1s^2$ group has to be broken in order to remove a second electron, the second ionization potential shoots up to 7280 kJ mol^{-1}. Similarly, the successive ionization potentials of aluminium are:

1st $Al[Ne](3s)^2(3p)^1 \rightarrow Al^+[Ne](3s)^2$ needs 577 kJ mol^{-1}
2nd $Al^+[Ne](3s)^2 \rightarrow Al^{2+}[Ne](3s)^1$ needs 1815 kJ mol^{-1}
3rd $Al^{2+}[Ne](3s)^1 \rightarrow Al^{3+}[Ne]$ needs 2740 kJ mol^{-1}
but 4th $Al^{3+}[Ne] \rightarrow Al^{4+}[He](2s)^2(2p)^5$ needs 11 590 kJ mol^{-1}

TABLE 2.8 Ionization energies of the elements

Element	*Ionization energies (electron volts, 1 eV = 96.48 kJ mol^{-1})*							
	1st	*2nd*	*3rd*	*4th*	*5th*	*6th*	*7th*	*8th*
H	13.60							
He	24.59	54.4						
Li	5.39	75.6						
Be	9.32	18.2	154					
B	8.30	25.2	37.9	259				
C	11.26	24.4	47.9	64.5	392			
N	14.53	29.6	47.5	77.5	97.9	552		
O	13.62	35.1	54.9	77.4	114	138	739	
F	17.42	35.0	62.7	87.1	114	157	185	954
Ne	21.56	41.0	63.5	97.1	126	158	207	239
Na	5.14	47.3						
Mg	7.65	15.0	80.1					
Al	5.99	18.8	28.5	120				
Si	8.15	16.3	33.5	45.1	167			
P	10.49	19.7	30.2	51.4	65.0	220		
S	10.36	23.3	34.8	47.3	72.7	88.0	281	
Cl	12.97	23.8	39.6	53.5	67.8	97.0	114	348
Ar	15.76	27.6	40.7	59.8	75.0	91.0	124	143
K	4.34	31.6						
Ca	6.11	11.9	50.9					
Sc	6.54	12.8	24.8	73.5				
Ti	6.82	13.6	27.5	43.3	99.2			
V	6.74	14.7	29.3	46.7	65.2	128		
Cr	6.77	16.5	31.0	49.1	69.3	90.6	161	
Mn	7.44	15.6	33.7	51.2	72.4	95	119	196
Fe	7.87	16.2	30.7	54.8	75.0	99	125	151
Co	7.86	17.1	33.5	51.3	79.5	102	129	157
Ni	7.64	18.2	35.2	54.9	75.5	108	133	162
Cu	7.73	20.3	36.8	55.2	79.9	103	139	166
Zn	9.39	18.0	39.7	59.4	82.6	108	134	174
Ga	6.00	20.5	30.7	63.6				
Ge	7.90	15.9	34.2	45.7	93.5			
As	9.81	18.6	28.4	50.1	62.6	128		
Se	9.75	21.2	30.8	42.9	68.3	81.7	155	
Br	11.81	21.8	36	47.3	59.7	88.6	103	193
Kr	14.00	24.5	37.0	52.5	64.7	78.5	111	126
Rb	4.18	27.3						
Sr	5.70	11.0	43.6					
Y	6.38	12.2	20.5	61.8				
Zr	6.84	13.1	23.0	34.3	80.4			
Nb	6.88	14.3	25.0	38.3	50.6	103		
Mo	7.10	16.2	27.2	46.4	61.2	68	127	
Tc	7.28	15.3	29.5	43	59	76	94	162
Ru	7.37	16.8	28.5	46.5	63	81	100	119
Rh	7.46	18.1	31.1	45.6	67	85	105	126
Pd	8.34	19.4	32.9	49.0	66	90	110	132
Ag	7.58	21.5	34.8	52	70	89	116	139
Cd	8.99	16.9	37.5					
In	5.79	18.9	28.0	54				
Sn	7.34	14.6	30.5	40.7	72.3			
Sb	8.64	16.5	25.3	44.2	55.5	108		
Te	9.01	18.6	28.0	37.4	58.8	70.7	137	

TABLE 2.8 (Contd.)

Element	*Ionization energies (electron volts, 1 eV = 96.48 kJ mol^{-1})*							
	1st	*2nd*	*3rd*	*4th*	*5th*	*6th*	*7th*	*8th*
I	10.45	19.1	33.0					
Xe	12.13	21.2	32.1					
Cs	3.89	23.1						
Ba	5.21	10.0						
La	5.58	11.1	19.2					
Hf	6.65	14.9	23.3	33.3				
Ta	7.89	16.2	22.3	33.1	45			
W	7.98	17.7	24.1	35.4	48	61		
Re	7.88	16.6	26.0	37.7	51	64	79	
Os	8.7	16.9	25	40	54	68	83	99
Ir	9.1	16	27	39	57	72	88	104
Pt	9.0	18.6	28.5	41.1	55	75	92	109
Au	9.23	20.5	30.5	43.5	58	73	96	114
Hg	10.44	18.8	34.2					
Tl	6.11	20.4	29.8	50.5				
Pb	7.42	15.0	32.0	42.3	68.8			
Bi	7.29	16.7	25.6	45.3	56.0	88.3		
Po	8.42							
At	9.2							
Rn	10.75							
Fr	3.98							
Ra	5.28	10.2						
Ac	5.17	12.1						
Th	6.08	11.5	20.0	28.8				

Other examples are shown in the table of ionization potentials. For all atoms the energies required to remove electrons from filled shells below the valence shells are so great that they cannot be provided in the course of a chemical reaction, and ions such as Li^{2+} or Al^{4+} are found only in high energy discharges. Thus the energy gap between the valence shell and the underlying filled shell, which is reflected in the successive ionization potentials, puts an effective upper limit on the valency of an element and this, of course, follows from the electronic structures of the elements as derived from the wave equation in the early part of this chapter.

2.14 Electron affinity

The electron affinity is the energy of the reverse process to ionization: the uniting of an electron with a gaseous atom or ion or molecule. The energy change in this process is the electron affinity of the species. Electron affinities are difficult to measure experimentally and only a few have been directly determined with accuracy. Zollweg has compiled a tabulation of first electron affinities corresponding to

$$M_{(g)} + e^- \rightarrow M^-_{(g)}$$

The convention used here is the accepted thermodynamic one of endothermic changes being positive and exothermic changes negative. The opposite convention may be found in older texts.

by examining measured and interpolated values normalized to the accurately known first electron affinities of elements like the halogens and oxygen. These first electron affinities are listed in Table 2.9.

The stability of the rare gas configuration is reflected in the high electron affinities of the halogens which are forming the anion with the rare gas electronic structure. Conversely,

TABLE 2.9 Electron affinities of the elements (after Zollweg) (kJ mol^{-1}) (exothermic changes are taken as negative)

H																	He
−74.5																	+21.2
Li	Be											B	C	N	O	F	Ne
−59.8	−36.7											−17.3	−122.3	+20.1	−141.3	−337.4	+28.9
Na	Mg											Al	Si	P	S	Cl	Ar
−52.2	+21.2											−19.3	−131	−68.5	−196.8	−349.2	+35.7
K	Ca	Sc	Ti	V	Cr	Mn	Fe	Co	Ni	Cu	Zn	Ga	Ge	As	Se	Br	Kr
−45.4	+186	+70.5	+1.93	−60.8	−93.5	+93.5	−44.5	−102	−156	−173	−8.7	−35.3	−139	−103	−203	−324.1	+40.5
Rb	Sr	Y	Zr	Nb	Mo	Tc	Ru	Rh	Pd	Ag	Cd	In	Sn	Sb	Te	I	Xe
(−37.6)	+145	+38.6	−43.5	−109	−114	−95.5	−145	−162	−98.5	−193	+26.1	−19.3	−99.5	−90.5	−189	−295.2	+43.5
Cs	Ba	La	Hf	Ta	W	Re	Os	Ir	Pt	Au	Hg	Tl	Pb	Bi	Po	At	Rn
(−36.7)	+46.4	−53.1	+60.8	−14.4	−119	−36.7	−139	−190	−247	−270	+18.6	−30.4	−99.5	−91.5	−127	−270	

the electron affinities for the rare gases, where the extra electron in the anion starts a new quantum shell, are all endothermic.

While most of the first electron affinities, shown in Table 2.9, are exothermic, the second electron affinities for elements forming doubly charged anions are always large and positive (i.e. energy has to be provided to add the second electron) with the result that the formation of doubly (or higher) charged anions requires the net addition of energy. For example:

$$O_{(gas)} + 2e^- \rightarrow O^{2-}_{(gas)}$$

requires 703 kJ mol^{-1}. Such values are usually derived indirectly from the Born–Haber cycle (see Section 5.4).

It should be noted that the largest exothermic electron affinity, that of chlorine, is smaller than the ionization potential of caesium which is the least endothermic of any atom. The result is that any electron transfer between a pair of atoms to form a pair of ions is an endothermic process:

$$Cl_{(gas)} + Cs_{(gas)} \rightarrow Cs^+Cl^-_{(gas)} \quad \Delta H = +12.1\ \text{kJ mol}^{-1}$$

and usually a considerable amount of energy is required: e.g.

$$I_{(gas)} + Na_{(gas)} \rightarrow Na^+I^-_{(gas)} \quad \Delta H = +180\ \text{kJ mol}^{-1}$$

It follows that the formation of an ionic compound from its component elements occurs only because of the additional energy provided by the electrostatic attractions between the ions in the solid. This is further discussed in Chapter 5.

2.15 Atomic and other radii

As the probability density distribution of an electron decreases exponentially from the nucleus, it never exactly equals zero. There is therefore no unambiguous definition of the radius of an isolated atom, though it may be taken as the radius of, say, the 95% contour. In a molecule, and even more so in a liquid or solid, the electrons are subject to the fields of all the neighbouring atoms and their distribution depends on the detailed chemical environment. The radius will differ from that of the single atom, and will also differ from compound to compound, though the latter variation is usually relatively small. Furthermore, only the distances between atoms (bond lengths) can be measured experimentally and the radii of the atoms forming the bond have to be deduced from these bond lengths.

Although the atomic radius is not an exact concept, it is possible to compile sets of atomic radii which reproduce most of the observed interatomic distances to within ten percent or so. These sets of atomic radii are valuable, as any marked discrepancy between the observed bond length and that calculated from the atomic radii suggests that there is some change in the type of bond or some other effect which should be investigated further. Moreover, when working with closely related compounds the bond lengths should agree much more closely than to ten percent so that quite small discrepancies are meaningful and worth further study. A number of sets of values for radii are required depending on the nature of the bonding—covalent, ionic or metallic.

2.15.1 Covalent species

A set of covalent radii may be derived by starting from the experimentally measured bond lengths in the elements. If these bond lengths are divided by two they give reasonable values for the radii of the atoms, and then the atomic radii of elements which do not form single bonds in the elemental state may be deduced from the bond lengths in suitable compounds with elements of known radii. A few simple examples of the process of building up a set of atomic radii are shown below.

Element	*Bond length (pm)*	*Atomic radius (pm)*
F_2	142	F = 71
Cl_2	199	Cl = 99
Br_2	228	Br = 114
I_2	267	I = 134
C (diamond)	154	C = 77

		Bond length (pm)		
Molecule		*Found*	*Calculated from above*	*Difference*
CF_4	C-F	132	148	16
CCl_4	C-Cl	177	176	1
CBr_4	C-Br	191	191	0
CI_4	C-I	214	211	3

The agreement between experimental and calculated values is excellent for the heavier halogens but less good for fluorine. It is found from a wide number of fluorine compounds that better general agreement with experiment is found if the atomic radius of fluorine is taken as F = 64 pm. This is a purely empirical correction chosen to give the best fit with experimental data. In a similar way, the value used for the hydrogen radius is H = 29 pm although the bond length in the H_2 molecule is 74 pm. With these and similar empirical adjustments to the experimental values the table of atomic radii shown in Table 2.10a was built up.

Most values calculated from Table 2.10a will agree to within 20–30 pm with the experimental bond lengths. The discrepancies are often wider when hydrogen or fluorine are involved as is illustrated by the measured values shown in Table 2.10b for the halides of the carbon Group elements.

A self-consistent and semi-empirical set of values of this type is the best that can be done with a single set of figures for atomic radii. A number of suggestions have been made for modifying the calculated bond length to allow for environmental effects. One example is the Schomaker–Stevenson correction which allows for the polarity of the bond. This is:

$$r_{A-B} = r_A + r_B - 0.09|x_A - x_B|$$

TABLE 2.10a Atomic radii in covalent molecules (pm)

Be	B	C	N	O	F	H
89	80	77	70	66	64	29
	Al	Si	P	S	Cl	
	126	117	110	104	99	
Zn	Ga	Ge	As	Se	Br	
131	126	122	121	117	114	
Cd	In	Sn	Sb	Te	I	
148	144	140	141	137	133	
Hg	Tl	Pb	Bi			
148	147	146	151			

	B	C	N	O
Double bond radii	71	67	62	62
Triple bond radii	64	60	55	

TABLE 2.10b Bond lengths of halides of the heavier elements of the carbon Group (pm)

	Silicon	*Germanium*	*Tin*
MF_4	154	167	?
MCl_4	201	208	231
MBr_4	215	231	244
MI_4	243	250	264

TABLE 2.11 Van der Waals' radii (pm)

H		N	O	F
100		155	152	147
	Si	P	S	Cl
	210	180	180	175
	Ge	As	Se	Br
	195	185	190	185
	Sn	Sb	Te	I
	210	205	206	198

where r_{A-B} is the bond length, r_A and r_B are the covalent radii, and x_A and x_B are the electronegativities (see Section 2.16). This formula does improve the agreement between calculated and experimental values in many cases, especially for fluorides, but the discrepancies are still significant and it is probably better to accept the purely empirical nature of the atomic radius and seek other evidence to establish the existence of special effects within the bond.

The discussion above applies only to single bonds. When double or triple bonds are present, the bond length is shortened and appropriate values of the atomic radius must be used. For example, in ethylene the $C=C$ distance is 135 pm and in acetylene $C \equiv C$ is 120 pm, corresponding to a double bond radius for carbon of 67 pm and a triple bond radius of 60 pm. Approximate values for other double and triple bond distances may be calculated by using the radii which are given in the last two lines of Table 2.10a. It will be noticed that the variations in bond lengths due to multiple bonding are larger than the uncertainties associated with the empirical nature of the set of atomic radii, but not by a very great margin. This means that attempts to deduce the bond order from variations in the bond length are legitimate but should be treated with some reserve unless closely similar compounds are being discussed.

In addition to the covalent radius just discussed, a further, much larger radius called the Van der Waals' radius is characteristic of atoms in covalent compounds. This radius represents the shortest distance to which atoms which are not chemically bound to each other will approach before repulsions between the electron clouds come into play. The Van der Waals' radius therefore governs steric effects between different parts of a molecule. Some values of Van der Waals' radii are shown in Table 2.11.

2.15.2 Ionic species

For ionic radii, we look for a set of self-consistent values to reproduce the observed interionic distances in ionic solids, in the same way as covalent radii do in molecular compounds. It is more difficult, however, to devise a set of ionic radii as there is no obvious way of dividing the observed internuclear distances, r_{MX}, in ionic compounds into cationic (r_+) and anionic (r_-) radii. This is in contrast to the case of covalent or metallic radii where there are distances between like atoms which can be divided into two. In addition, ionic radii may be expected to vary with the environment. This is particularly the case with anions as the electrons are generally less tightly held. (Consider for example K^+ and Cl^-. Both have the same number of electrons but the positive change on the potassium nucleus is two higher than that on the chlorine nucleus.) Careful studies and calculations suggest cations are relatively invariable in size, with less than 3% contraction between the free ion and that ion in a symmetric crystal environment. By contrast anions vary quite substantially with, for example, the radius of Cl^- decreasing from 187 pm in Cs^+Cl^- to 125 pm in Cu^+Cl^-. There are grounds for questioning how good the ionic approximation is for compounds like CuCl (compare Chapter 5), but for compounds of the s element cations M^+ and M^{2+} (apart from Be^{2+}), a range of about 17 pm or around 10% is appropriate for halide ion radii. The H^- ion, with two electrons and a single nuclear charge, exhibits extreme variations and is discussed in Chapter 9. Overall, we can expect to produce a usable set of cation radii for species with rare gas configurations but not for those with d electron populations. To go with these, a set of anion radii may be chosen which can vary within 10%. As with covalent radii, these values are both empirical and experimental. They are useful to (a) systematize a large number of experimental observations of inter-ionic distances, and (b) indicate anomalies which might suggest unusual bonding or other interesting phenomena worth further study.

There have been two quite different types of attack on the problem of dividing up the measured interionic distances into anion and cation radii. The older, classical, approach was to find some acceptable assumption to act as a starting point, and two sets of values, produced respectively by Goldschmidt and by Pauling in the period 1926–28, are widely used.

Goldschmidt's method was to assume that in a compound with a large anion and a small cation, such as LiI, the anions would be in contact. Then half the I-I distance equals the radius of I^-. This value is then used, in compounds with larger cations such as NaI, KI, etc., to calculate cation radii for Na^+, K^+ etc., and these in turn allow the calculation of the radii of other anions such as Cl^- or O^{2-}. Finally, the radius of Li^+ is derived from some compound with a small anion such as LiCl or Li_2O. This method, with further refinements, was used to compile the set of empirical ionic radii shown in Table 2.12. The value of 145 pm for oxygen is quoted in this set although the lower values of 140 pm or 135 pm are usually more compatible with transition metal values.

An alternative approach, used by Pauling, was to assume that the radii of isoelectronic ions, such as Cl^- and K^+, varied inversely as their effective nuclear charge. This then gave

TABLE 2.12 Ionic radii (pm)

	Symmetric ion radii		*Crystal radii* (Shannon)			
Ion	*Goldschmidt*	*Johnson*	*C.N. 4*	6	8	12
Li^+	68	92	73	90	106	
Na^+	98	118	113	116	132	153
K^+	133	145	151	152	165	178
Rb^+	148	156		166	175	186
Cs^+	167	168		181	188	202
Be^{2+}	30	69	41	59		
Mg^{2+}	65	102	71	86	103	
Ca^{2+}	94	126	114	126	148	
Sr^{2+}	110	138		132	140	158
Ba^{2+}	129	140		149	156	175
Al^{3+}		92	53	68		
Sc^{3+}		106		89	101	
Y^{3+}		114		104	116	
La^{3+}		120		117	130	150
F^-	133	112	117	119		
Cl^-	181	164		167		
Br^-	195	179		182		
I^-	216	202		206		
O^{2-}	145(135)	116	124	126	128	
S^{2-}	190	158		170		
Se^{2-}	202	174		184		
Te^{2-}	222	192		207		
Cu^+	96		74	91		
Ag^+	126		114	129	142	
Au^+	137			151		
Zn^{2+}	83		74	88	104	
Cd^{2+}	103		92	109	124	145
Hg^{2+}	112		110	116	128	
Tl^+	149			164	173	184
Pb^{2+}				133	143	163

a way of dividing the experimentally observed M-X distances and allowed a different set of internally consistent ionic radii to be built up. Either the Pauling or the Goldschmidt radii allow a reasonable prediction of interatomic distances in crystals but, of course, the two sets of values must not be mixed.

Recent evaluations of ionic radii use experimental evidence that was not available in the 1920s, particularly modern X-ray diffraction. As X-rays are scattered by electrons, the major contribution (see Section 7.4) is from the highly concentrated inner electrons. Positions of the centres of these (and thus of the nuclei) can be observed accurately. The small number of outer electrons in the valence levels make only small contributions to the electron density map, and superimposed on these are a number of experimental uncertainties and errors, such as those arising from thermal motion.

In a number of cases, careful X-ray analysis has allowed these errors to be minimized. When the electron density from the inner electrons is subtracted out, positions of minimum electron density between the cation and anion may be determined and these are used to define cation and anion radii. For example, the minima suggest the following 'experimental' radii:

$$
\begin{aligned}
Li^+ &= 92\ \text{pm}, \quad & F^- &= 109\ \text{pm in LiF} \\
Na^+ &= 118\ \text{pm}, \quad & Cl^- &= 164\ \text{pm in NaCl} \\
Mg^{2+} &= 102\ \text{pm}, \quad & O^{2-} &= 109\ \text{pm in MgO} \\
Ca^{2+} &= 126\ \text{pm}, \quad & F^- &= 110\ \text{pm in } CaF_2.
\end{aligned}
$$

It will be seen that these values are internally self-consistent, and they add up to give good agreement with other experimental interatomic distances (e.g. $Ca^{2+} + O^{2-} = 240$ pm: experimental value in CaO = 240 pm).

On such a basis, Ladd formulated a set of 'experimental' radii which Johnson has extended beyond directly measured cation radii by using the interatomic distance in metals, a, in the simple formula for cation radii r_+:

$$r_+ = 0.64 \times a/2$$

The 0.64 is an empirical constant (0.61 is better for first-row elements like lithium). Using the directly measured values for Li^+, Na^+, K^+, Mg^{2+} and Ca^{2+} together with the formula gives the Johnson values listed in Table 2.12 for 'spherical potential ions', broadly covering ions in symmetric environments. By difference, the corresponding anion radii were derived but it should be noted that these are average values from a relatively wide range and apply only in a symmetric environment.

In what is essentially an update of the classical approach, Shannon has thoroughly considered all the factors which influence ionic radii including effects of coordination number, charge of oxidation state, covalent and metallic contributions and distortions, and crystal vacancies. He has defined a set of *crystal radii,* based on $O^{2-} = 126$ pm in 6-coordination. These relate directly to more traditional radii based on $O^{2-} = 140$ pm by subtraction of 14 pm from the crystal radius. We tabulate crystal radii because, in Shannon's words, they 'correspond more closely to the physical size of ions in a crystal'.

Listed in Table 2.12 are Goldschmidt radii representing the early lists, Johnson values based on direct measurements of electron density minima and usable only for symmetric environments, and Shannon values for representative coordination numbers. In each case the values are optimized to give the best fit to observed data. Values from any one set may be used to predict new interionic distances, but values from different sets must not be mixed. See also Chapter 5 for a discussion of the ionic model of solids.

2.15.3 Metals

In metals, the environment of each atom is the same, so a set of metallic radii may be derived by halving the interatomic distances. The structures of metals (see Section 5.6) are usually close-packed with a coordination number of twelve, and the metallic radii listed in Table 2.13 are for 12-coordination. Some metal structures involve 8-coordination, and Goldschmidt proposed to take 0.97 of the 12-coordinate radius as an estimate of

TABLE 2.13 Metallic radii (pm)

Li		Be							
157		112							
Na		Mg		Al					
191		160		143					
K		Ca		Ga		Ge			
235		197		153		139			
Rb		Sr		In		Sn	Sb		
250		215		167		158	161		
Cs		Ba		Tl		Pb	Bi		
272		224		171		175	182		
Sc	Ti	V	Cr	Mn	Fe	Co	Ni	Cu	Zn
164	147	135	129	137	126	125	125	128	137
Y	Zr	Nb	Mo	Tc	Ru	Rh	Pd	Ag	Cd
182	160	147	140	135	134	134	137	144	152
La	Hf	Ta	W	Re	Os	Ir	Pt	Au	Hg
188	159	147	141	137	135	136	139	144	155

the 8-coordinate one. Use of metallic radii should be confined to metals and alloys, and similar provisos apply to their use as for covalent or ionic radii.

When dealing with all these sets of radii it is essential to keep in mind that the experimentally determined quantities are the inter-atomic or inter-ionic distances which can be measured to high accuracy (usually to within a few tenths of a picometre). The values listed for the atomic or ionic radii are empirical and chosen to give the best fit over the widest range of experimental data. As a result, small deviations between calculated and measured values (of up to 5 pm or so) are significant only if very critically examined. Larger differences (of the order of 10 pm) suggest the presence of abnormal bonding, either multiple bonds or strong polarization effects in covalent compounds, or polarization and covalent contributions in ionic compounds.

2.16 Electronegativity

One parameter which is widely used in general discussion of the chemical character of an element is its electronegativity. This is defined as *the ability of an atom in a molecule to attract an electron to itself*. There is no direct way of measuring this ability though a number of indirect methods have been suggested, such as the proposal of Mulliken who defined the electronegativity of an atom as the average of its electron affinity and ionization potential (as the electron affinity is a measure of the tendency of the atom to gain an electron, and the ionization potential indicates its tendency to lose an electron). This is the most fundamental of a number of proposed definitions of electronegativity and it may be applied where electron affinity values are known. However, the values available (such as Table 2.9) are not all directly measured, and other measures of electronegativity are used. The classic one is the Pauling electronegativity, based on bond energies. Despite all attempts at improvement, the Pauling values are still the most generally used (Table 2.14a). A major difficulty is that the attraction for an electron is clearly not expected to be the same for different valencies of an element. Zhang has proposed a set of values, based on covalent radii and ionization potentials and geared to Pauling values, which are defined for each of the main oxidation states (see Section 2.17) of the element. These values are one of the more general sets available, though they have some deficiencies for the Main Group elements. They are listed in Table 2.14b.

There has been a great deal of discussion, argument, and often confusion, about the significance of electronegativity values, largely because various authors have used the concept with different degrees of sophistication. The electronegativity is extremely valuable as a brief summary, within one parameter, of the general chemical behaviour of an atom but it must be used in a general way and little significance attaches to small differences in

TABLE 2.14a Pauling's values of the electronegativity of elements

Li	Be											B	C	N	O	F
1.0	1.5											2.0	2.5	3.0	3.5	4.0
Na	Mg					H = 2.1						Al	Si	P	S	Cl
0.9	1.2											1.5	1.8	2.1	2.5	3.0
K	Ca	Sc	Ti	V	Cr	Mn	Fe	Co	Ni	Cu	Zn	Ga	Ge	As	Se	Br
0.8	1.0	1.3	1.5	1.6	1.6	1.5	1.8	1.9	1.9	1.9	1.6	1.6	1.8	2.0	2.4	2.8
Rb	Sr	Y	Zr	Nb	Mo	Tc	Ru	Rh	Pd	Ag	Cd	In	Sn	Sb	Te	I
0.8	1.0	1.2	1.4	1.6	1.8	1.9	2.2	2.2	2.2	1.9	1.7	1.7	1.8	1.9	2.1	2.5
Cs	Ba	La	Hf	Ta	W	Re	Os	Ir	Pt	Au	Hg	Tl	Pb	Bi	Po	At
0.7	0.9	1.0	1.3	1.5	1.7	1.9	2.2	2.2	2.2	2.4	1.9	1.8	1.9	1.9	2.0	2.2
Fr	Ra	Ac														
0.7	0.9	1.1														
						Lanthanides range from 1.0 to 1.2										
						Actinides range from 1.3 to 1.4										

TABLE 2.14b Electronegativity values after Zhang

H																
1 2.25																
Li	Be											B	C	N	O	F
1 0.95	**2** 1.45											**3** 1.95	**4** 2.55	3.05	3.65	4.2
Na	Mg											Al	Si	P	S	Cl
1 0.95	**2** 1.2											**3** 1.5	**4** 1.75	**5** 2.1	**6** 2.45	**7** 2.85
														3 1.7	**4** 2.0	**5** 2.35
K	Ca	Sc	Ti	V	Cr	Mn	Fe	Co	Ni	Cu	Zn	Ga	Ge	As	Se	Br
1 0.9	**2** 1.05	**3** 1.3	**4** 1.6	**5** 2.0	**6** 2.3	**7** 2.5	**6** 2.4	**3** 1.75	**2** 1.5	**2** 1.5	**2** 1.45	**3** 1.55	**4** 1.8	**5** 2.05	**6** 2.3	**7** 2.55
			3 1.4	**4** 1.85	**4** 1.9	**6** 2.4	**3** 1.7	**2** 1.45		**1** 1.25		**1** 1.1	**2** 1.4	**3** 1.6	**4** 1.85	**5** 2.1
			2 1.2	**3** 1.6	**3** 1.65	**4** 1.95	**2** 1.45									
				2 1.35	**2** 1.4	**2** 1.45										
Rb	Sr	Y	Zr	Nb	Mo	Te	Ru	Rh	Pd	Ag	Cd	In	Sn	Sb	Te	I
1 0.9	**2** 1.0	**3** 1.2	**4** 1.5	**5** 1.75	**6** 2.0	**7** 2.3	**4** 1.9	**4** 1.85	**4** 1.85	**1** 1.15	**2** 1.3	**3** 1.45	**4** 1.6	**5** 1.75	**6** 1.95	**7** 2.15
			3 1.35	**4** 1.65	**4** 1.8	**4** 1.8	**3** 1.65	**3** 1.65	**2** 1.45			**1** 1.1	**2** 1.25	**3** 1.45	**4** 1.6	**5** 1.8
			2 1.2	**3** 1.5	**2** 1.4	**2** 1.4	**2** 1.45	**2** 1.45								
Cs	Ba	La	Hf	Ta	W	Re	Os	Ir	Pt	Au	Hg	Tl	Pb	Bi	Po	At
1 0.9	**2** 1.0	**3** 1.2	**4** 1.55	**5** 1.9	**6** 2.15	**7** 2.35	**8** 2.6	**4** 1.9	**4** 1.9	**3** 1.7	**2** 1.35	**3** 1.5	**4** 1.55	**5** 1.7	**6** 1.9	**7** 2.05
			3 1.45	**4** 1.75	**5** 2.0	**6** 2.2	**6** 2.3	**3** 1.7	**2** 1.5	**1** 1.25	**1** 1.2	**1** 1.1	**2** 1.25	**3** 1.4	**4** 1.6	**5** 1.75
			2 1.3	**3** 1.55	**4** 1.85	**4** 1.9	**4** 1.95	**2** 1.5								
Fr	Ra	Ac														
1 0.9	**2** 0.95	**3** 1.25														
					Lanthanides		**4** 1.4 to 1.5									
							3 1.2 to 1.35									
							2 1.05 to 1.2									

Note: Values rounded to 0.05. Oxidation states in **boldface**.

values between two atoms. The most electronegative elements occur in the top right-hand corner of the Periodic Table and electronegativity falls on going down a Group towards the heavier elements or on going to the left along a Period towards the alkali metals.

Electronegativities are most useful in the guidance they give to the electron distribution in a bond. In a bond A-B between two atoms, the electron density in the bond may lie evenly between the two atoms or be concentrated more towards one atom, say B, than towards the other, when the bond is said to be *polarized*. In the limiting case, when the electron density of the bonding electrons is entirely on B, an electron has been fully transferred from A to B and an ionic compound, A^+B^-; forms. The electron density distribution in the bond may be predicted from the electronegativities of A and B. If A and B have the same electronegativities, it follows from the definition that A and B attract

ELECTRONEGATIVITY AND HARDNESS

We have suggested the use of *electronegativity* as a semi-quantitative concept which is useful as a general rationalizing principle in thinking about reacting molecules. Using the Mulliken definition, electronegativity is related to *hardness* as, respectively, the sum and the difference of the basic atomic properties, *ionization energy* (Table 2.8) and *electron affinity* (Table 2.9). A number of workers, including Pearson and Sanderson (see reading lists), have carried these concepts further and have developed approaches from quantum mechanics.

For readers wishing to follow up this more in-depth approach, a useful starting point is the review by D. Bergmann and J. Hinze in *Angewandte Chemie, International Edition*, 1996, vol 35 pages 150–163.

the electrons in the bond equally and no polarization results. If B is more electronegative than A, its attraction for the bond electrons is the stronger and polarization results, the degree of polarization being proportional to the difference in electronegativity. A large electronegativity difference favours the formation of ions and, as a rough guide, an ionic compound forms between A and B if they differ in electronegativity by more than two units. Thus elements with very high or very low electronegativities are more likely to form ionic compounds than those with intermediate values.

The electronegativity of an element depends on the other atoms attached to the one in question. Thus, carbon in H_3C-X is less electronegative than carbon in F_3C-X, as the highly electronegative fluorine atoms in the trifluoromethyl compound remove more electron density from the carbon in the C-F bonds than do the hydrogen atoms in the C-H bonds of the methyl compounds. As a result, the carbon atom in F_3C-X has more tendency to attract the electrons in the C-X bond than has the carbon atom in H_3C-X. It follows that the electronegativity values given in Table 2.14 represent the behaviour of the elements in an 'average' chemical environment and the effective electronegativity of an element in any particular compound depends in detail on its environment.

Another parameter related and complementary to electronegativity is *hardness*. Whereas the Mulliken electronegativity is defined as the average of the ionization potential (I) and electron affinity (A), hardness η can be approximated as half the difference between these two values:

$$\eta = (I - A)/2$$

The hardness of an atom is a parameter which attempts to quantify the ability of electrons to redistribute themselves within the atom and thus is a measure of the *polarizability* of the atom, as described earlier. Atoms with small ionization energies and small electron affinities, such as the heavy halogens and oxygen Group elements (i.e. those elements on the bottom right-hand side of the p-block), are termed 'soft'. Small atoms, such as sodium, oxygen and fluorine, are termed 'hard'. The hardness of the donor atoms of a ligand bonding to a metal atom is of great consequence in determining the strength of the bonding interaction and this topic is discussed in greater detail for transition metal complexes in Section 13.7. The general rule is that 'like bonds to like', i.e. soft metal centres such as Hg(II) and Ag(I) have a strong preference for binding to soft donor atoms such as P, S, Se and I.

2.17 Coordination number, valency and oxidation state

The three terms, coordination number, valency and oxidation state, are used to describe the environment and chemical state of an atom in a compound. The three overlap somewhat in meaning and application, but the use of each has advantages in certain circumstances.

The simplest term to describe an atom in a compound is its *coordination number,* which is the number of nearest neighbours to the given atom, whatever the bonding between them. The coordination number is a purely empirical property of the element determined from the structure of the compound. This simplicity is the main advantage in the use of the term, as a compound may be described by the coordination numbers of its constituent atoms, however difficult it may be to determine the bonding between these atoms. The only difficulty in determining the coordination number comes when all the distances between like substituents and the central atom are not the same. In some cases, it may be difficult to decide whether some distance which is distinctly longer than the rest is part of the coordination number or not. However, few such cases cause any real problem, and causes of asymmetry in coordination are well understood.

When more information about the atom is required, the valency or the oxidation state must be determined. *Valency* is a familiar term which is used to describe the bonding of the atom. Confusion sometimes arises because the value of the valency depends on the theory chosen to describe the bonding. While this is based on experimentally determined properties there are many cases where alternative theories each have validity, giving rise to different values for the valency. This problem is often disguised by familiarity but it may arise in acute form with the discovery of a new class of compound with unexpected structures or properties. Until an adequate theoretical description of the bonding is agreed, the valency remains undefined or ambiguous. One historical example is nickel carbonyl, $Ni(CO)_4$, which was known for many years before there was an adequate theory of its bonding. Its structure has been written at various times with the nickel-carbon monoxide bond as Ni=C=O, Ni-C≡O, Ni←C≡O and Ni ⇌ CO implying that the nickel is, respectively, eight-, four-, zero- and zero-valent.

Apart from this type of problem, the valency nomenclature is sometimes clumsy (just because it gives a more complete picture of the molecule). For example, cobalt in the ion $[Co(NH_3)_6]^{3+}$ has to be described as having a covalency of six and an electrovalency of three. There are also occasions when the term 'valency' conceals differences in properties. An example is given by ammonia, NH_3, and nitrite ion, NO_2^-. In both compounds the nitrogen atom is properly described as trivalent and yet it has to be oxidized to pass from one compound to the other, and it is more useful in some contexts to discuss ammonia and related compounds such as the amines, R_3N, separately from the nitrites and other trivalent oxy-compounds.

Considerations such as the above, led to the introduction of a narrower, more empirical term, *oxidation number* (or *oxidation state*). The oxidation number of an element in a compound may be simply determined from a number of empirical rules and it is quite independent of the nature of the bonding. Obviously, it gives less information about the chemical state of the element than does an accurate description in terms of valency but it is useful and convenient when that extra information is not required or available.

The oxidation number of an atom in a compound is defined by the following rules:

In nearly all compounds, rules (i) to (iii) are equivalent to taking O=−II (except in peroxides, where O=−I, and in OF_2, where O=+II), H=+I (except in ionic hydrides) and halogens=−I (except in their oxygen compounds, not including OF_2, as mentioned above).

(i) The oxidation number of an atom in the element is zero.
(ii) The oxidation number of an atom in an ionic compound is equal to the charge on that atom (with the sign).
(iii) The oxidation number of an atom in a covalent compound is equal to the charge which it would have in the most probable ionic formulation of the compound.

The first two rules are perfectly clear but a little experience is required to find the artificial ionic form required by rule (iii). The electronegativities of the elements in the compound usually serve to make the most probable ionic formulation clear, as illustrated by the examples given below:

Compound	*More electronegative element*	*Ionic formulation*	*Oxidation numbers*
BCl_3	Cl	$B^{3+}(Cl^-)_3$	B=III, Cl=−I
SO_2	O	$S^{4+}(O^{2-})_2$	S=IV, O=−II
NH_3	N	$N^{3-}(H^+)_3$	H=I, N=−III
NH_4^+	N	$[N^{3-}(H^+)_4]^+$	H=I, N=−III
NO_2^-	O	$[N^{3+}(O^{2-})_2]^-$	N=III, O=−II
CrO_4^{2-}	O	$[Cr^{6+}(O^{2-})_4]^{2-}$	Cr=VI, O=−II
$Cr_2O_7^{2-}$	O	$[(Cr^{6+})_2(O^{2-})_2]^{2-}$	Cr=VI, O=−II

Notice, in the last column, that the sum of the oxidation numbers of the atoms equals the overall charge on the species. Although atoms may be shown with large charges, e.g. Cr^{6+} or S^{4+}, this by no means implies the existence of such unlikely ions. To make this clear, it is usual to indicate the oxidation state by Roman numbers —Cr(VI) or S(IV).

Oxidation and reduction are very simple to define in terms of oxidation numbers. Oxidation is any process which increases the oxidation number of an element while reduction corresponds to a decrease in the oxidation number. For example, the conversion of ammonia to nitrogen involves an increase from −III to zero in the oxidation number of the nitrogen and is an oxidation by three steps, the conversion of ammonia to nitrite involves an oxidation by six steps, while the conversion to nitrate involves a change of eight steps to nitrogen(V). On the other hand, the change from ammonia, NH_3, to ammonium ion, NH_4^+, involves no change in the oxidation numbers and is not an oxidation. The same applies to the change from chromate to dichromate which sometimes causes trouble in analytical calculations. Further examples are provided by the range of nitrogen compounds below:

Oxidation number of the nitrogen	*Examples*
−III	NH_3, or NH_4^+
−II	N_2H_4
−I	NH_2OH
0	N_2
I	N_2O or $N_2O_2^{2-}$
II	NO
III	N_2O_3 or NO_2^-
IV	N_2O_4
V	N_2O_5 or NO_3^-

In complex ions, if the ligand is a neutral molecule like ammonia in $[Co(NH_3)_6]^{3+}$ or water in $[Cu(H_2O)_4]^{2+}$, the metal has an oxidation number equal to the charge, Co(III) and Cu(II) respectively. Similarly, nickel in nickel carbonyl, $Ni(CO)_4$, has an oxidation number of zero. If the ligand is charged, then the oxidation number of the metal must balance with the total charge on the ion: Fe(II) in ferrocyanide, $[Fe(CN)_6]^{4-}$, Fe(III) in ferricyanide, $[Fe(CN)_6]^{3-}$, or Co(III) in $[Co(NH_3)_3Cl_3]$ and in $[CoF_6]^{3-}$.

The use of oxidation numbers simplifies the calculations involved in oxidation–reduction titrations. In the overall reaction, the change in oxidation state of the reductant must balance that of the oxidant. The reaction stoichiometry is thus readily worked out from the oxidation state changes of the reactants. A full account of the method is given in the standard analytical textbooks but the following examples illustrate the approach. Compare also Section 6.3.

The oxidation of arsenite by permanganate in acid solution

$$MnO_4^- + AsO_3^{3-} \text{ to } Mn^{2+} + AsO_4^{3-}$$

The manganese change is from MnO_4^-, where the Mn=VII to Mn^{2+} with Mn=II; change in manganese oxidation state = −5.

The arsenic change is from AsO_3^{3-}, where the As=III to AsO_4^{3-} with As=V; change in arsenic oxidation state=+2. The reaction stoichiometry is therefore:

$$2MnO_4^- + 5AsO_3^{3-}$$

The equation may then be balanced by introducing hydrogen ions and water molecules in the usual way to give:

$$2MnO_4^- + 5AsO_3^{3-} + 6H^+ \rightarrow 2Mn^{2+} + 5AsO_4^{3-} + 3H_2O$$

The reaction between iodate and iodide

$$IO_3^- + I^- \text{ to } I_2$$

In this case, the oxidant and the reductant end up in the same form. In iodate, the iodine is in the V oxidation state so that the change in going from iodate to iodine is by −5. The change in oxidation state from the −I in iodide to the element is by +1 and the reaction stoichiometry is therefore:

$$IO_3^- + 5I^-$$

The balanced equation is:

$$IO_3^- + 5I^- + 6H^+ \rightarrow 3I_2 + 3H_2O$$

If this reaction is carried out in concentrated hydrochloric acid, instead of in dilute acid as above, the final product is not iodine but iodine monochloride, ICl, in which the iodine has an oxidation state of +I. In this case, the change from iodate to ICl is −4 and the change from iodide is +2 so that the balanced equation becomes:

$$IO_3^- + 2I^- + 6H^+ \rightarrow 3I^+ + 3H_2O$$

As far as most calculations are concerned, only the reaction stoichiometry has to be known and the use of oxidation numbers in the calculation gives this very rapidly and easily.

The oxidation state concept breaks down in those cases where an ionic formulation is ambiguous. One example is in the case of the metal nitrosyls which contain groups, M-NO, which could quite validly be formulated in three ways—as $(NO)^+$, $(NO)^-$ or with neutral NO groups. Similar difficulties are encountered in, for example, the hydrides of boron or phosphorus where the electronegativities (B = 2.0, P = 2.05, H = 2.1) are so close that doubts arise whether to write H^+ or H^-: in fact, the hydrogen is negatively polarized in most boron–hydrogen compounds and positively polarized in most phosphorus–hydrogen ones. In organic chemistry, also, the oxidation state concept is not very useful; it is more convenient to discuss reactions such as $CH_4 \rightarrow CH_3Cl$ in terms of substitution rather than in terms of a change in the carbon oxidation state.

The three concepts, coordination number, oxidation state and valency, become less empirical and convey increasing amounts of information in that order.

Problems

Readers may best test and reinforce their understanding of this chapter by applying the various formulae, and manipulating numerical data. A number of cases should be worked out, and the questions given below are mostly illustrations of the type of example which you can make up.

2.1 How many protons, neutrons and electrons are present in ^{24}Mg, $^{24}Na^+$, ^{99}Mo, ^{99}Tc, ^{129}Xe, $^{127}I^-$, $^{195}Pt^{2+}$, $^{197}Au^{3+}$

For example, $^{37}Cl^-$: from the Periodic Table, Cl is element 17, hence has 17 protons, leaving 20 neutrons to make up the atomic weight. There will be 18 electrons, 17 to balance the protons plus one extra for the negative charge.

2.2 What is the minimum uncertainty in the position of (a) a mass of 1 mg, (b) a molecule of UF_6, (c) a molecule of H_2, (d) a neutron, (e) an electron where each is moving at half the velocity of light?

In this and the other calculations, you need values from the tables in SI Units and names, pages 1 to 6. Be careful to use compatible units. You need work only to 3 or 4 significant figures. See Table 2.5 or the end paper table for relative atomic masses.

2.3 Show by substitution that ψ_1 (Equation 2.2) satisfies the Schrödinger equation.

2.4 Calculate the value of the minimum energy, E_1, of He^+ (see Equations 2.1 to 2.4).

2.5 Calculate and plot out the values of ψ_1 and ψ_1^2 for the hydrogen 1*s* orbital for distances from the nucleus of 0, 25, 50, 100 and 200 pm. Compare with Fig. 2.8.

2.6 Choose various values of the atomic number, *Z*, and decide the electron configuration. Compare your answers with Table 2.5. Decide the number of unpaired electrons in each case.

For the larger values, it is convenient to start off with the rare gas configuration next below the target *Z*. For example, for *Z* = 58 the rare gas is Xe with *Z* = 54 (filling 5s, 4d and 5p levels) leaving 4 electrons. From Section 2.7 and Fig. 2.5, two of these fill 6s, and then the last two have a choice of 4f and 5d which are of very similar energy. In actuality, the first enters 5d and the last one enters 4f, giving the configuration of cerium, the first of the lanthanide elements.

2.7 Calculate the excitation and dissociation energies from various excited states of the H atom, e.g. the dissociation energy for $n = 4$ or the excitation energy from $n = 3$ to $n = 6$. Work out the values in kJ mol^{-1}, eV and cm^{-1}.

2.8 Which of the following sets of quantum numbers represent permissible solutions of the Schrödinger wave equation?

n	l	m	m_s
3	1	0	$-\frac{1}{2}$
3	1	1	1
3	2	4	$\frac{1}{2}$
5	4	−3	$\frac{1}{2}$
2	−1	−1	$-\frac{1}{2}$

2.9 Write out all the permissible sets of quantum numbers for

(a) 4 electrons in 3p orbitals,
(b) 4 electrons in 5d orbitals.

2.10 For a p orbital, plot the general form of the radial part of the wave function, $r(\exp{-r/2a_0})$. Multiply this by the angular part to generate the shape of the orbital. Confirm that the change of sign and the nodal plane of the angular part remain in the full wave function.

2.11 Find from the references the electron density contour plot of

(a) an electron in a 2p orbital: in (i) H and (ii) Be
(b) an electron in a 3p orbital in (i) H and (ii) Al
(c) an electron in two different 3d orbitals.

2.12 Plot electron affinity (Table 2.9) against Period position (parallel to Figs 8.4 to 8.7). Discuss any relationship between these curves and the ionization energy ones.

2.13 Many alternative presentations of the Periodic Table have been suggested, particularly ones using three dimensions. Authors have also proposed plotting a further parameter, such as electronegativity or some function reflecting valence shell electron energies, in the third dimension.

Create two alternative Periodic Tables, and compare your ideas with published ones (see references in the reading list).

2.14 Calculate Mulliken electronegativities for a Group of elements and compare with the listed values. Which set correlate best with the chemistry?

> In Section 2.16, the Mulliken electronegativity is defined as $1/2(I-A)$ (refer to Tables 2.8, 2.9, 2.14 and the appropriate section on systematic chemistry).

2.15 Look up the references listed on electronegativity. Use these to find earlier references: can you find six different sets of electronegativity values? Assess the value of (a) the general idea and (b) each specific set of values. How many significant figures do you think should be used in expressing electronegativities?

2.16 Draw to scale the covalent and ionic radii of the halogens. Draw scale diagrams of X_2, HX (X = halogen) including the Van der Waals' radii. For ClF_3 (see Fig. 4.4; bond lengths are 170 pm, axial, and 160 pm, equatorial) do the Van der Waals' radii of axial and equatorial F atoms overlap?

2.17 (a) Determine the stoichiometry of the oxidation of NH_3 to *each* of the other nitrogen oxidation states with MnO_4^-

(i) in acid, forming Mn^{2+}
(ii) in alkali, forming MnO_2.

(b) Write balanced equations for NH_3 going

(i) to N_2, and
(ii) to NO_3^-

by reacting with MnO_4^- in acid.

2.18 De Broglie proposed that the electron wave may be described by the same equations that apply to a photon, that is, $h\nu' = E = mv^2$, where v is the velocity of the electron.

(a) From the values of the constants, and using the relationship of Equation (7.2), calculate the wavelength of an electron moving at 1%, 10% and 90% of the speed of light.
(b) What would be the velocity of an electron whose wavelength was

(i) equal to
(ii) one fifth of

the circumference corresponding to the Bohr radius of hydrogen?

CHAPTER 3

GENERAL BACKGROUND 60

3.1 INTRODUCTION 60

3.2 BOND FORMATION AND ORBITALS 62

DIATOMIC MOLECULES 63

3.3 THE COMBINATION OF s ORBITALS 63

3.4 THE COMBINATION OF p ORBITALS 66

3.5 BOND ORDERS OF DIATOMIC MOLECULES 70

3.6 ENERGY LEVELS IN DIATOMIC MOLECULES 74

3.7 SUMMARY 78

PROBLEMS 79

Covalent Molecules: Diatomics

General background

3.1 Introduction

In the last chapter, a picture of the electron structure of atoms was built up from the theory of wave mechanics taken together with experimental data. This knowledge of atomic structure will now be used to examine the process of combining atoms to form molecules.

One factor in the ordering of elements into families with related properties, which culminated in the formulation of the Periodic Table, was the grouping together of elements with the same pattern of valencies. It is well known that Mendeléef left gaps in his table, to be filled by hitherto-undiscovered elements, and the valency of these new elements (e.g. in the form of formulae of oxides or halides) was one of their properties which he predicted.

As we saw in Chapter 2, once the electron configuration of the elements was understood, the Periodic Table was found to result directly. The question then arises whether bonding and valency may also be explained in terms of electron configurations. At this point, we adopt the convenient subdivision of valency into two phenomena—*covalency*, where electrons are shared between bonded atoms, and *electrovalency*, where electrons are transferred from one atom to another to form *ions* which are then bonded by electrostatic forces. The formation of covalent bonds is discussed in this chapter and the next, while ionic compounds are the subject of Chapter 5.

The first widely accepted electronic description of a covalent bond is that due to Lewis, of two electrons shared between two atoms and binding them together. Furthermore, for light elements at least, stable configurations were those where a total of eight electrons, either wholly owned or shared, surrounded an atom. This was the 'octet rule', and gave a good guide to the probable formulae of stable species. Thus the idea of electron sharing and the octet rule went a long way to rationalize the observed valencies. In this way, carbon, with four outer electrons, and fluorine with seven, can achieve octets in CF_4

$$\cdot\dot{\underset{\cdot}{C}}\cdot + 4\cdot\ddot{\underset{\cdot\cdot}{F}}: \rightarrow \begin{array}{c} :\ddot{F}: \\ :\ddot{F}:C:\ddot{F}: \\ :\ddot{F}: \end{array}$$

Hydrogen is stable with two electrons, hence

$$2H\cdot + \dot{\underset{\cdot\cdot}{O}}: \rightarrow H:\ddot{\underset{\cdot\cdot}{O}}:H$$

Double bonds are readily explained by the sharing of four electrons

$$\cdot\dot{\underset{\cdot}{C}}\cdot + 2\ddot{\underset{\cdot\cdot}{O}}: \rightarrow \dot{O}::C::\dot{O}$$

and so on. Species may be charged, as in

$$4H\cdot + .\ddot{N}: - 1e \rightarrow \left[\begin{array}{c} H \\ H:\ddot{\underset{\cdot\cdot}{N}}:H \\ H \end{array}\right]^+$$

CHEMICAL COMPOSITION AND CHEMICAL FORMULAE

A major step in the understanding of substances was made by the 18th century chemists who started to measure the changes in weight, or the corresponding changes in gas volumes, of substances which were undergoing transformations such as burning or reacting with acids. Such experiments established regularities in the composition of materials which were expressed by the Laws of Constant Composition, Multiple Proportions and the like. These generalizations in turn could be understood in terms of Dalton's atomic hypothesis, and were evidence for it.

Thus the idea grew in the 19th century that atoms formed a fixed number of links and hence combined to give molecules of fixed composition. As wider ranges of compounds were examined, it was concluded that the number of links characteristically formed by a particular element was limited to a small number of values, often only a single one. This number is the *valency*, see Section 2.17. For example, the first great rationalization of organic chemistry came from the realization that carbon was nearly always tetravalent. The concept of the link between atoms was refined into the notion of a *bond* and then molecules were formulated so that the number of bonds matched the valencies of the atoms. The whole process is illustrated in the figure.

All this is basic revision for the reader but it summarizes an evolution of thinking about substances which took well over a century to refine from the first quantitative observations. These steps amounted to a major revolution in thinking about the material world, and all the chemistry from the middle of the nineteenth century to today is built thereon.

Given the idea of constant valency, the picture of bonds became elaborated. Thus, if carbon is tetravalent and oxygen is divalent, then a molecular formula CO_2 can be understood by writing two bonds between C and each O. Using the convention whereby a line between atoms signifies a bond, we write CO_2 as O=C=O, and similarly HCN satisfies known valencies when written H-C≡N with a triple bond between C and N.

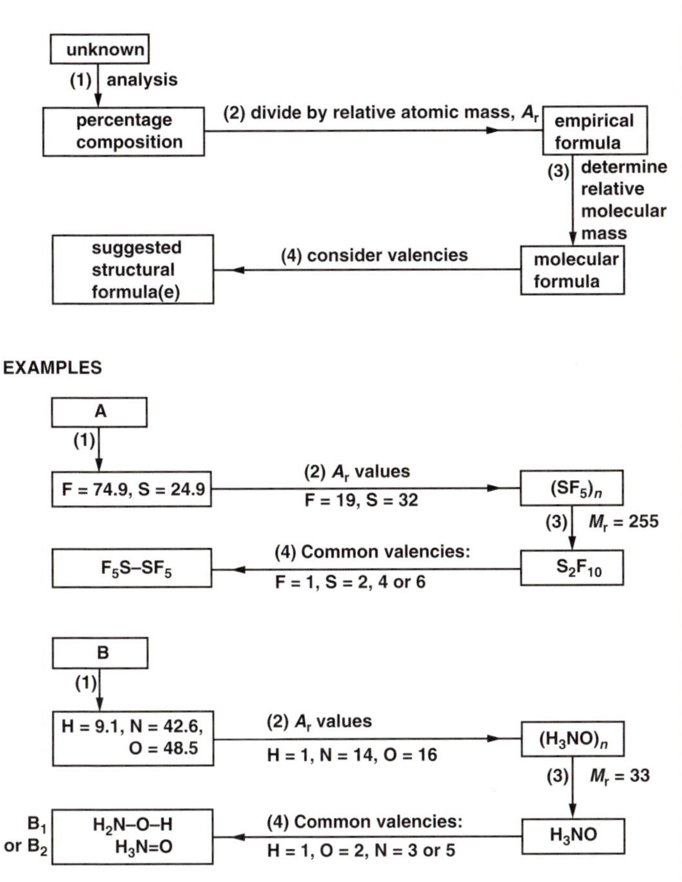

Formulation of an unknown. Stages shown by vertical steps involve experimental determinations, those shown by horizontal steps use known values. Note that in (2), as the atom ratios must be whole numbers, the calculations can be done with rounded-off values. While steps (1) to (3) are unambiguous, it may be possible to write more than one acceptable formula at step (4)

Some species do not form an octet, but do share all the electrons available to them, such as boron or beryllium in their halides

$$\cdot\dot{B}\cdot + 3\cdot\ddot{\underset{\cdot\cdot}{Cl}}: \rightarrow :\ddot{\underset{\cdot\cdot}{Cl}}:\underset{}{\overset{:\ddot{Cl}:}{B}}:\ddot{\underset{\cdot\cdot}{Cl}}:$$

$$\cdot Be \cdot + 2\cdot\ddot{\underset{\cdot\cdot}{F}}: \rightarrow :\ddot{\underset{\cdot\cdot}{F}}:Be:\ddot{\underset{\cdot\cdot}{F}}:$$

and these species, though stable, can complete their octets by co-ordinating with other molecules where unshared pairs are present as in

$$:\ddot{\underset{\cdot\cdot}{F}}:\overset{:\ddot{F}:}{B}:\ddot{\underset{\cdot\cdot}{F}}: + :\underset{H}{\overset{H}{\ddot{N}}}:H \rightarrow :\ddot{\underset{\cdot\cdot}{F}}:\underset{:\underset{\cdot\cdot}{F}:}{\overset{:\ddot{F}:}{B}}:\underset{H}{\overset{H}{N}}:H$$

Once again, these concepts are revision for the reader, but are important in representing the introduction of electron configuration into thinking about molecule formation. The octet rule is, of course, equivalent to emphasizing the stability of the rare gas configurations—those where the valence level s and all three p orbitals are filled with two electrons each. The octet rule is formally broken (a) if there are insufficient electrons to give an octet around each atom, as in BF_3, or (b) where there is a further energy level fairly close to the p level which can accept extra electrons. Thus we find that nitrogen forms only NF_3

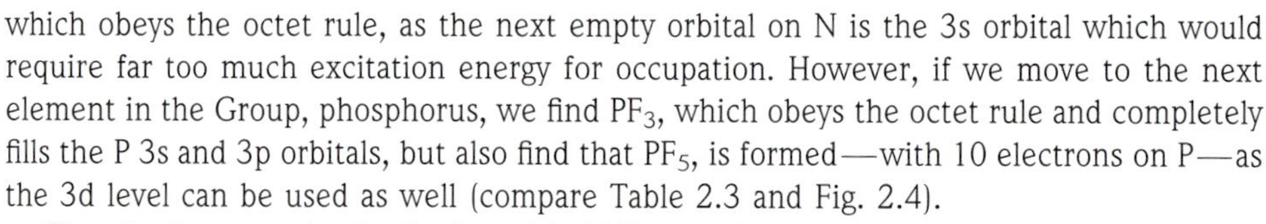

which obeys the octet rule, as the next empty orbital on N is the 3s orbital which would require far too much excitation energy for occupation. However, if we move to the next element in the Group, phosphorus, we find PF_3, which obeys the octet rule and completely fills the P 3s and 3p orbitals, but also find that PF_5, is formed—with 10 electrons on P—as the 3d level can be used as well (compare Table 2.3 and Fig. 2.4).

For the heavier Main Group elements with coordination numbers greater than 4, an explanation in terms of d orbital participation is generally accepted at the basic level. However, a more complex picture emerges from modern calculations, see Section 18.9.

Thus in the examples in the box titled 'Chemical composition and chemical formulae' structure A contains S with 6 bonds and therefore 12 electrons around it. Similarly, while B_1 obeys the octet rule, B_2 needs 10 electrons on N which is not permissible.

3.2 Bond formation and orbitals

We must now examine the idea of a bond at a greater level of sophistication. One way of translating the Lewis idea into terms of wave mechanics will be discussed here, though a number of other approaches are possible and these are presented in the references.

Let us start with the simplest of all molecules, the positive ion of the hydrogen molecule, H_2^+. If we generalize Equation (2.1) to cover one electron moving in the field of two nuclei, there is only one del-squared term in the kinetic energy part of the wave equation, while the potential energy part contains three terms—the attractions between each nucleus and the electron and the repulsion between the nuclei. The equation is similar to that of a hydrogen-like atom (see Equation 2.4), but now the nuclear field is not spherically symmetrical. Although the problem is more difficult than that of the hydrogen atom, the wave equation for the hydrogen molecule ion can be solved exactly and the resulting energy levels and energy of formation agree with the experimental values.

Just as with atoms, major difficulties arise in the case of molecules as soon as more than one electron is involved. For example, the presence of the second electron in the hydrogen molecule, H_2, makes it a molecular analogue of the helium atom. In the wave equation for the hydrogen molecule (compare Equation 2.8) there will be two del-squared terms in the kinetic part describing the two electrons and there are six components of the potential energy term. These are the four attractions between each electron and each nucleus, the repulsion between the nuclei, and finally the inter-electron repulsion term which provides the main source of difficulty in these calculations.

It is clear that such complexities increase rapidly as the number of electrons rises, and methods of approximation have to be found. Before turning to these, it is of interest to record the results of the more rigorous calculations of electronic energies for some simple molecules. These are shown in Table 3.1.

The agreement is very close in these cases and gives grounds for confidence in the general approach. The calculations are very long and complex, however, and may require substantial computer resources. In addition, the quantity most readily derived from the calculation is the total electronic energy of the molecule, while the changes involved in a reaction are dependent on the small differences between such total energies. At present, these total energies cannot be calculated to the very high order of accuracy required for direct predictions of reaction paths or molecular structures. Simplifications must be introduced in order to solve the equations and the overall theory becomes a semi-quantitative guide. Even in this relatively modest form, the wave mechanical approach to molecular structure has wrought an impressive change in the way in which chemists think about molecules.

TABLE 3.1 Calculations of total electronic energies

Molecule	Calculated value (kJ mol^{-1})	Experimental value (kJ mol^{-1})
H_2^-	2887.2	2893.8
H_2	3081.2	3081.1
CH_4	104 850	106 210
Other diatomic molecules like CO, N_2 or HF	Differences between calculated and experimental values lie in the range 0.5–1.5%	

The total electronic energy of a species is the energy evolved when all its constituent particles (nuclei and electrons) are brought from infinite separation and combined to form the molecule or ion in its equilibrium configuration.

In this text, the theory will be used as a general guide, and diagrams rather than calculations will be used to describe the processes of molecule formation. The reader is asked to keep in mind that these diagrams do mirror the calculations and that definite values may be found for the parameters—such as bond lengths and bond energies—which are qualitatively described here. Fuller accounts are given by Coulson for example (see references).

Diatomic molecules

One well-known approximation method for the wave equation for molecules is the method of molecular orbitals. In this approach, the aim is to construct orbitals analogous to the atomic orbitals of Chapter 2 but centred on both nuclei. Then the electrons are fed in—two into each orbital in order of increasing energy—to build up the electronic structure of the molecule, just as the electronic structures of the atoms were built up.

The first problem is to find a way of constructing these molecular orbitals. One starting point is to consider that, when the electron in a molecule is close to one nucleus, it is in almost the same environment as in the free atom. This suggests that the molecular orbitals may be derived from some combination of atomic orbitals. The simplest way of combining the orbitals is additively and a simple *linear combination of atomic orbitals* (LCAO) has been widely used.

3.3 The combination of s orbitals

In the case of H_2^+, the molecular orbitals are formed from the linear combinations (Equation 3.1) of the atomic orbitals ψ_X and ψ_Y on the two hydrogen atoms, X and Y:

$$\begin{aligned} \phi_B &= \psi_X + \psi_Y \\ \phi_A &= \psi_X - \psi_Y \end{aligned} \tag{3.1}$$

Consider the case where the wave functions ψ_X and ψ_Y represent 1s orbitals on the two hydrogen atoms. The result of combining these two atomic orbitals into the molecular orbital ϕ_B may be shown diagrammatically, using the curves for the 1s orbital which were given in Chapter 2 in Fig. 2.8.

In Fig. 3.1, the curves giving the complete cross-section through each nucleus, X and Y, of their respective 1s orbitals are drawn out and the process of addition to give ϕ_B is indicated to the right. The two nuclei are brought together to a distance equal to the interatomic distance in the molecule, and the wave function curves are summed to give the wave function of the molecule. If we square, to get the electron probability density, the corresponding curve is a cross-section through the two nuclei of the electron distribution in the molecule.

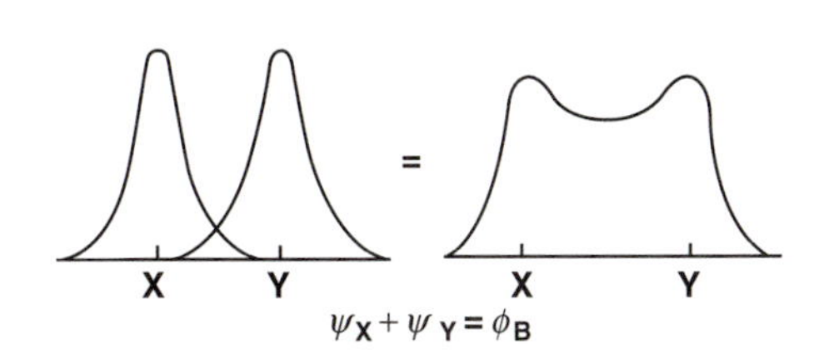

FIG. 3.1 The formation of the molecular orbital $\phi_B(\sigma_s)$. The nuclei X and Y are placed at the measured internuclear distance in the molecule and the wave function curves are superimposed to give the summation curve shown

The parts of ϕ_B outside the internuclear region follow an exponential curve and the electron density falls off rapidly on moving away from the nuclei. The orbital ϕ_B may be shown in projection as a contour map as in Fig. 3.2a, or in the more convenient boundary diagram of Fig. 3.2b where the contour line which encloses 90 or 95% of the electron density is drawn (compare with atomic orbitals as in Fig. 2.8c).

In equations such as (3.1) and (3.2), a fully rigorous discussion requires that the RHS is multiplied by a factor N, called the *normalizing constant*. This is a numerical factor which adjusts the equation so that $\int \phi^2 d\tau = 1$; i.e. so that the probability of finding the electron described by the molecular orbital ϕ somewhere in space is unity (compare the requirement for atomic orbitals in Chapter 2). In this particular case, as the atomic orbitals ψ_X and ψ_Y are normalized (i.e. $\int \psi_X^2 d\tau = 1 = \int \psi_Y^2 d\tau$) $N = 1/\sqrt{(2)}$ and the full version of Equation (3.1) is

$$\phi_B = \frac{1}{\sqrt{(2)}}(\psi_X + \psi_Y)$$

However, as N is a numerical constant, its actual value rarely affects the qualitative discussion of molecular orbitals which we are giving here and we shall normally omit it.

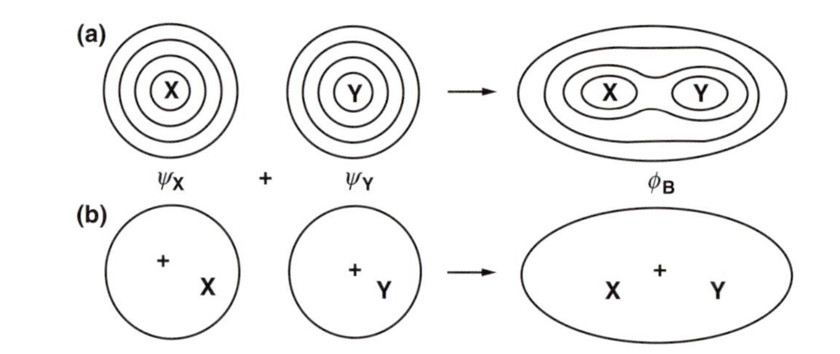

FIG. 3.2 (a) Contour representation of $\phi_B(\sigma_s)$. (b) Boundary contour representation $\phi_B(\sigma_s)$. The countour line enclosing 90% of the electrons density is drawn. Such diagrams represent either the wave function or its square, the probability density function. The wave function diagram is the more useful and shows the sign of the wave function, as here

It is clear from these diagrams that, in the molecular orbital ϕ_B, there is an accumulation of electron density in the region between the two nuclei. This markedly decreases the repulsion between the two nuclear charges. At the same time, the electron has a high probability of being in the region where it experiences the attraction of both the nuclear charges, and is more strongly held than if it was in one of the contributing atomic orbitals and attracted by only one nuclear charge. The result is that the presence of electrons in this molecular orbital holds the two atoms together and a bond is formed. The orbital ϕ_B is termed a *bonding molecular orbital*.

The molecular orbital ϕ_A may be treated in the same way. The combination of atomic 1s orbitals $(\psi_X - \psi_Y)$ is shown in Fig. 3.3, and the contour representation and the boundary line diagram are given in Fig. 3.4. In this molecular orbital, ϕ_A changes sign at the mid-point of XY and the electron probability density, ϕ_A^2, falls to zero here across a *nodal plane* perpendicular to the internuclear axis. Thus, if an electron is placed in the orbital ϕ_A, electron density is removed from the region between the nuclei and accumulated on the remote side of the atoms. The internuclear repulsion has full effect and no bond results. ϕ_A is termed an *antibonding molecular orbital*.

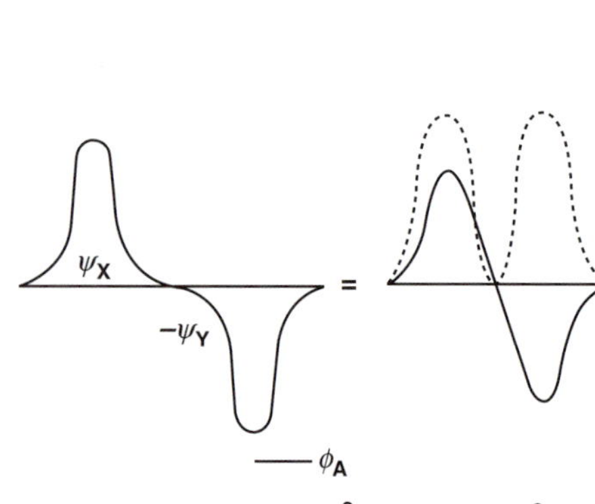

FIG. 3.3 The formation of the molecular orbital $\phi_A(\sigma_s^*)$

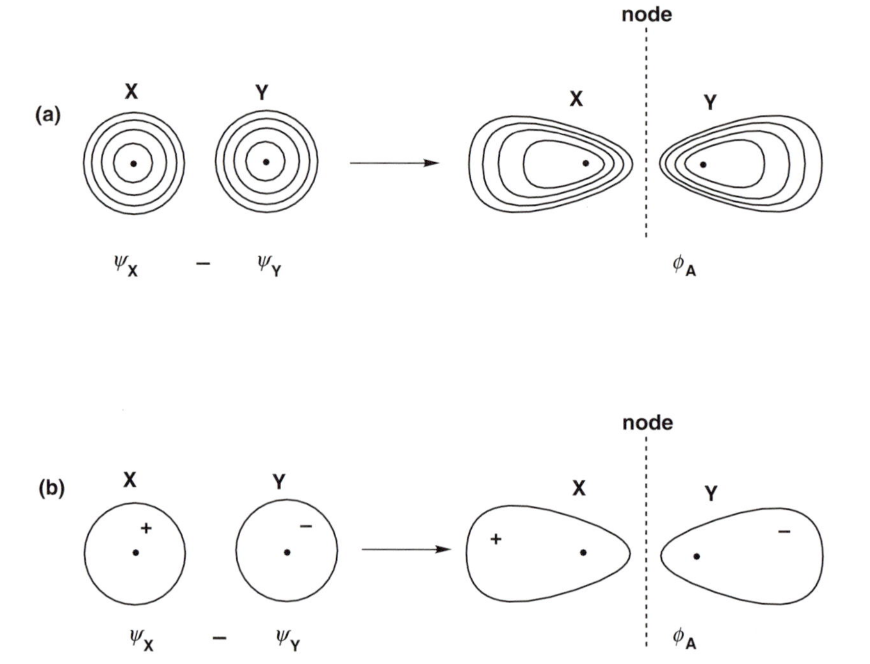

FIG. 3.4 (a) Contour representation of $\phi_A(\sigma_s^*)$. (b) Boundary contour representation of $\phi_A(\sigma_s^*)$. (b) shows the molecular orbital wave function which changes sign across the nodal plane perpendicular to the internuclear axis

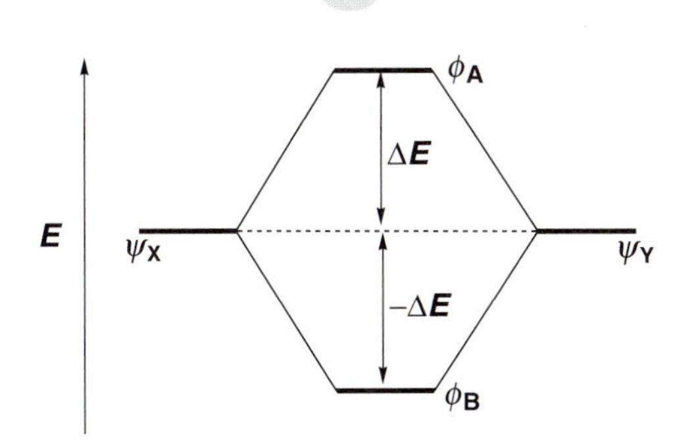

FIG. 3.5 Energy level diagram for ϕ_B and ϕ_A (σ_s and σ_s^*). This shows the energies of the molecular orbitals, ϕ_B and ϕ_A, relative to the energies of the constituent atomic orbitals. The atomic orbital energy levels are indicated by the horizontal lines to left and right while the molecular orbitals are shown in the centre of the diagram. Those atomic orbitals which contribute to a particular molecular orbital are connected to it by the inner sloping lines. This convention is followed in all diagrams of this type. The bonding orbital, ϕ_B, is more stable than the atomic orbitals by the same amount of energy as the antibonding orbital, ϕ_A, is less stable than the atomic orbitals

> This is a first approximation. In a more quantitative treatment, the destabilization is somewhat greater than the stabilization so that complete occupation of the antibonding orbital (compare He_2 below) is *less* stable than reversion to filled atomic orbitals. A convenient treatment is given in the *Journal of Chemical Education* papers by Smith quoted in the reading list. See also 'The simplest molecule' reference by McNab for a discussion of more detailed calculation methods and their very good match with experimental evidence.

When the energies, E_B and E_A, of an electron in the molecular orbitals ϕ_B and ϕ_A are calculated, it is found that E_B is less than the energy of the electron in the constituent atomic 1s orbitals while E_A is greater than the atomic orbital energy by the same amount, ΔE (see box). That is, the molecular orbital ϕ_B is stabilized, and the molecular orbital ϕ_A is destabilized, relative to χ_{1s}, by the energy ΔE. This may be shown on an energy level diagram as in Fig. 3.5. The two atomic orbitals thus combine to form two molecular orbitals, one of which is more stable than the atomic orbitals while the other is less stable by the same amount of energy. The process of deriving the electronic structure of a molecule can then be formalized in steps similar to those used for atoms. The nuclei are first placed together at the appropriate distance, then molecular orbitals are constructed from the atomic orbitals, and finally the electrons are fed into the molecular orbitals in order of increasing energy. Just as with atomic orbitals, a molecular orbital holds no more than two electrons and, when there are a number of molecular orbitals of equal energy, the electrons enter them singly with parallel spins. The number of molecular orbitals formed must exactly equal the number of atomic orbitals used in their construction.

Although, in the above outline, it has been assumed that the inter-nuclear distance (the bond length) is a known factor—and it will usually be known experimentally—it should be noted that in a full theoretical treatment it is possible to derive the optimum bond length by finding the value which gives the minimum total energy of the system. This has been done for a number of simpler cases with results that agree with the experimentally determined distances.

The formation of molecules can now be followed by using the energy level diagram of Fig. 3.5. For the hydrogen molecule ion, H_2^+, the electronic structure is shown in Fig. 3.6a. The orbital of lowest energy in the molecule is ϕ_B and the electron goes into this, being stabilized by energy, ΔE, relative to its energy in the atom. Thus, when two hydrogen nuclei and one electron combine to form a molecule, this is more stable by ΔE than the separate atom plus ion.

In the hydrogen molecule, H_2, (Fig. 3.6b) there are two electrons to be considered. These both enter ϕ_B and will have their spins paired. The energy stabilization of the molecule over the two isolated atoms is $2\Delta E$ (less a relatively small term due to the inter-electron repulsion) and this bond energy, in the hydrogen molecule, is regarded as that of a normal single bond. It follows that the bond in the ion, H_2^+, which has about half the energy of formation, may be regarded as a 'half-bond'. This accords with experiment. H_2^+ exists as a transient species in electric discharges and its bond energy, which may be determined from its spectrum, is about half the energy of H_2.

Next, consider a three-electron molecule such as the positive ion of diatomic helium, He_2^+, whose energy level diagram is shown in Fig. 3.7. Since ϕ_B is filled by the first two electrons, the third must be placed in the antibonding molecular orbital ϕ_A. The energy of formation of He_2^+ is therefore $-2\Delta E + \Delta E$. The net stabilization from re-arranging two helium nuclei and three electrons as a molecule is thus ΔE, corresponding to a 'half-bond' again.

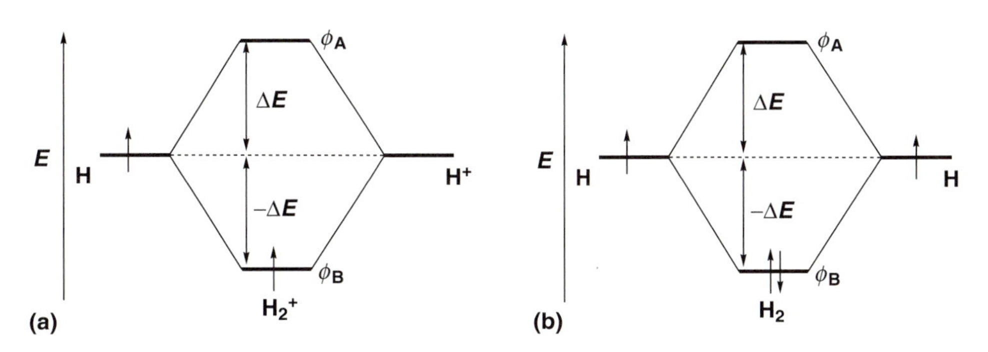

FIG. 3.6 (a) The electronic structure of H_2^+. (b) The electronic structure of H_2

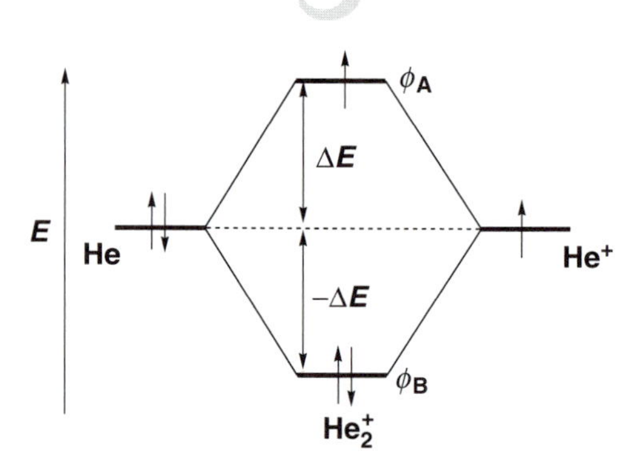

FIG. 3.7 The electronic structure of He_2^+

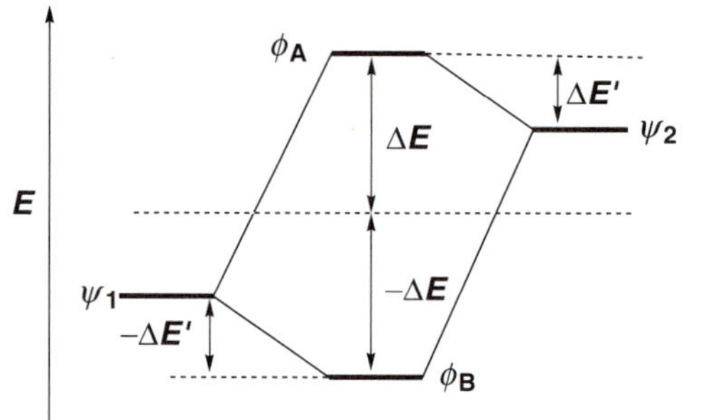

FIG. 3.8 Energy level diagram for the combination of s orbitals of unlike atoms. Stabilization and destabilization is relative to the average energy level of the two atomic orbitals. Note that the differences ΔE in energy from the atomic orbitals are also equal

Finally, in a four-electron molecule such as He_2, two electrons would be placed in the bonding orbital ϕ_B and two in the antibonding orbital ϕ_A. The energy of formation is $-2\Delta E + 2\Delta E$ equal to zero, and no bond results. In fact, helium exists as a monatomic gas. Exact calculation shows that the four electrons in the configuration $\phi_B^2\phi_A^2$ are actually less stable than as $(1s)^2$ on each He atom.

An analysis such as this may be extended in an exactly similar way to all other s orbitals with higher n values, and also to the more general case of diatomic molecules where the two atoms are not the same. In this case, the two atomic orbitals which will combine to form the molecular orbital will be of different energies, and the most favourable combination will not be in the 1:1 ratio of Equation (3.1), but some more general expression of the form:

$$\phi_B = \psi_1 + c\psi_2 \tag{3.2}$$

will be needed. The value of the mixing coefficient c which will give the optimum energy has to be found in the course of the calculation. The energy level diagram which corresponds to this more general case is shown in Fig. 3.8. Here the bonding orbital is stabilized, and the antibonding orbital destabilized, by equal amounts of energy, ΔE, calculated from the mean energy of the contributing atomic orbitals ψ_1 and ψ_2. The gain in energy when an electron is taken from the more stable of the two atomic orbitals is the smaller amount labelled $-\Delta E'$. Clearly, this decreases as the energy difference between the atomic orbitals ψ_1 and ψ_2 increases. If $\Delta E'$ is too small no molecule results. It follows that useful molecular orbitals are formed only when the combining atomic orbitals are of similar energy. As a general rule, this limitation implies that only orbitals in the valency shells of atoms will combine to form molecular orbitals. Thus, in hydrogen chloride, the hydrogen 1s orbital is of too high an energy to combine with 1s or 2s orbitals on the chlorine atom (whose energy levels are greatly stabilized relative to those of hydrogen by the attraction of the nuclear charge of 17), and it is too stable to interact with the chlorine 4s, or higher, orbitals. The hydrogen 1s orbital is comparable in energy with the chlorine 3s or 3p orbitals and could form molecular orbitals with these.

It should be remarked that the actual electron energies both in the atomic orbitals and in the molecular orbitals of helium are different from the energies of electrons in the corresponding orbitals in hydrogen, because of the difference in the nuclear charge. The difference between the atomic and molecular orbital energies, ΔE, will however, be of the same order of size.

3.4 The combination of p orbitals

If the molecular axis in a diatomic molecule is taken as the z axis, then the p_z orbitals on the two atoms may combine to form molecular orbitals similar in type to those formed by s orbitals. Just as with the s orbitals, the p_z orbitals on the two atoms X and Y combine to form two molecular orbitals, ϕ_1 and ϕ_2, which are similar to those given by Equations (3.1).

$$\begin{aligned}\phi_1 &= \psi_X + \psi_Y\\ \phi_2 &= \psi_X - \psi_Y\end{aligned} \tag{3.3}$$

where the atomic orbitals ψ_X and ψ_Y are here the p_z orbitals. The process of combination is illustrated using the boundary contour method in Fig. 3.9. The electron density in the first

We are adopting here the usual convention that the *unique* direction in a molecule is taken as the z axis. For example, the molecular axis in a linear molecule or the axis perpendicular to the molecule in a planar species would normally be labelled z.

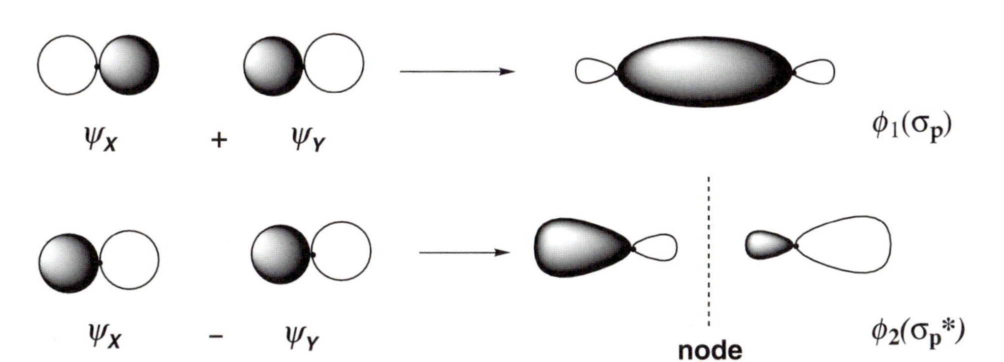

FIG. 3.9 The combination of atomic p_z orbitals. Two atomic p_x orbitals give rise to bonding and antibonding molecular orbitals which, respectively, concentrate or remove electron density between the nuclei. Note that a *plus* combination is defined here as the in-phase one, and *minus* as the out-of-phase one, even though this involves reversing the signs of some of the p orbitals

NOMENCLATURE FOR MOLECULAR ORBITALS

For an atom, the nucleus is at a unique point, and similarly the bond between the two atoms in a diatomic molecule is centred on a unique line. Orbitals in a molecule, such as those of Figs 3.4 and 3.9, which are symmetrical about the bond, are thus analogues of the s orbital in the atom which is symmetrical about the nucleus. The nomenclature adopted is to use sigma, σ, which is the Greek letter s, to label such orbitals. Then subscripts are used to show which atomic orbitals compose the molecular orbital, and antibonding orbitals are starred. Thus, for the molecular orbitals of Figs 3.2 and 3.4, ϕ_B becomes σ_S, and ϕ_A becomes σ_S^*, while for those of Fig. 3.9, $\phi_1 = \sigma_p$ and $\phi_2 = \sigma_p^*$.

In a similar way, molecular orbitals which have one nodal plane containing the bond axis across which the wave function changes sign, such as those in Fig. 3.12 or in Fig. 3.14c, are called pi (π—the Greek p) molecular orbitals by analogy with atomic p orbitals.

Likewise, two d orbitals may overlap sideways-on to form a molecular orbital with two nodal planes containing the bond axis, as in Fig. 3.14d. This is named a delta (δ—the Greek d) molecular orbital, analogous to the atomic d orbital. While π orbitals are common, δ bonds are found mainly in dimeric metal–metal bonded complexes of the heavier transition metals, see Section 16.7.

orbital, ϕ_1^2, which results from the addition of the two atomic p_z orbitals, is concentrated in the region between the two nuclei and this molecular orbital is bonding. In the second orbital, there is a nodal plane between the nuclei and the electron density, ϕ_2^2, is concentrated in the regions remote from the internuclear region, falling to zero between the nuclei. This orbital is therefore antibonding. Thus the combination of the two atomic p_z orbitals gives two molecular orbitals, one of which is bonding and one antibonding, just as in the case of the s orbitals. These molecular orbitals, ϕ_B, ϕ_A, ϕ_1 and ϕ_2, are termed sigma (σ) molecular orbitals.

The formation of sigma molecular orbitals is not restricted to combinations of like atomic orbitals. The only necessary condition is that the two components are symmetrical in sign about the bond axis. Then, if they are of similar energy, they may combine to form a molecular orbital. For example, an s and a p_z orbital may combine to form a bonding sigma molecular orbital as shown in Fig. 3.10. The corresponding sigma antibonding orbital also exists.

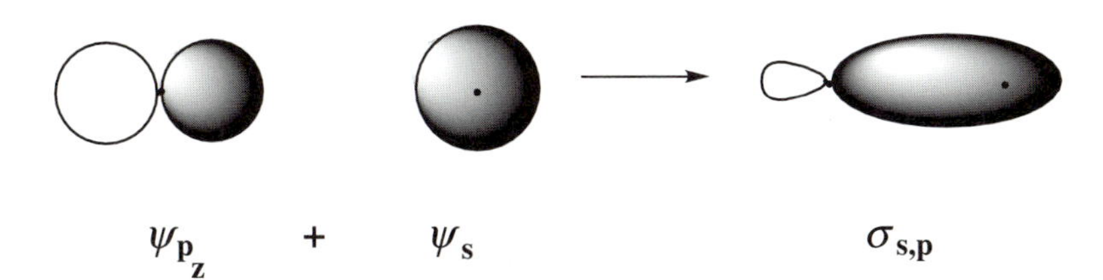

FIG. 3.10 The combination of an s and a p_z orbital. Such an orbital is symmetrical around the inter-nuclear axis and concentrates electron density between the nuclei. The corresponding antibonding orbital may be formed in a similar manner

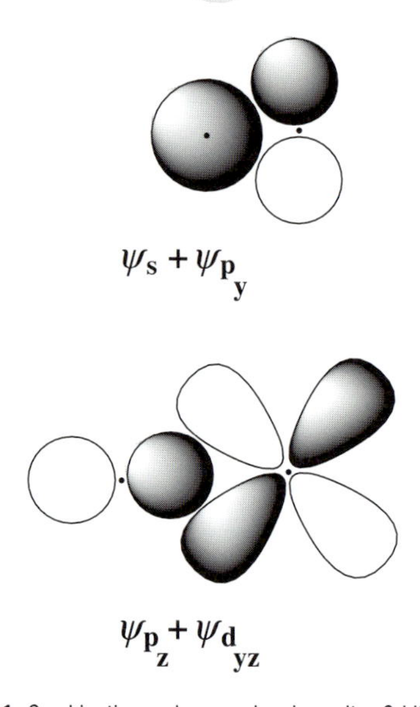

FIG. 3.11 Combinations where no bond results. Orbitals can only combine to give molecular orbitals if both are of the same symmetry with respect to the molecular axis; that is, if the signs of the wave functions change across the axis in the same way

Orbitals with different symmetries about the bond axis cannot combine to form molecular orbitals. This is illustrated in Fig. 3.11 for the cases of s with p_y, and p_z with d_{yz}. It can be seen that the 'plus-plus' and 'plus-minus' areas of overlap cancel each other out.

Although the p_x and p_y orbitals cannot enter into sigma bonding because they are antisymmetric (i.e. change in sign) across the bond axis, they can form a different type of bond, called a pi (π) bond (see the box titled 'Nomenclature for molecular orbitals'), by overlapping 'sideways-on'—see Fig. 3.12. The first combination shown in Fig. 3.12 accumulates charge in the region between the nuclei and is therefore bonding. The second mode of combining the atomic p orbitals has a nodal plane lying between the nuclei and perpendicular to that containing the bond axis. This second molecular orbital is antibonding. These are named, systematically, as π_p and π_p^* respectively.

The bonding π_p molecular orbital is of lower energy than the contributing atomic p orbitals, while the antibonding π_p^* molecular orbital is of higher energy by an equal amount. Because of the existence of the nodal plane containing the nuclei in the π orbitals, the π_p interaction has rather less effect on the electron density between the nuclei than has the σ_p interaction, and the energy gap between the bonding and the antibonding π molecular orbitals is less than that between the bonding and antibonding σ orbitals. The result of this is that the strength of a double bond, composed of a σ plus π contribution, is somewhat less than that of two σ bonds. This is of particular importance when accounting for the reduced stability of homonuclear double bonds between heavy p block elements (e.g. As=As, Sb=Sb, etc.) where the long bond distance and diffuse p orbitals result in poor overlap and a small π-bond energy, and thus σ-bonded compounds are more highly favoured.

Since the p_x and p_y atomic orbitals are both perpendicular to the molecular axis and identical with each other apart from their orientation, they both form π orbitals (with π_{p_x} perpendicular to π_{p_y}) in exactly the same way. The π_x and π_y levels are equal in energy as are the π_x^* and π_y^* levels. The combination of all the p atomic orbitals on two atoms to give the molecular orbitals of a diatomic molecule may therefore be shown on a composite energy level diagram, as in Fig. 3.13.

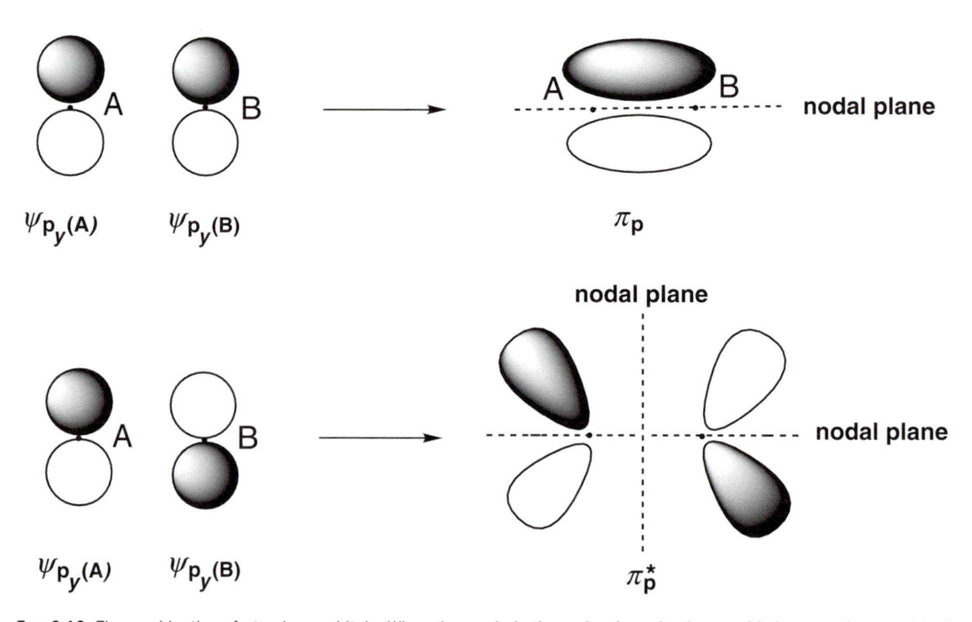

FIG. 3.12 The combination of atomic p_y orbitals. When the z axis is the molecular axis, the p_y orbitals are anti-symmetrical with respect to this axis—that is, the wave function changes sign on crossing the axis. Such orbitals combine to give bonding, π, and antibonding, π^*, orbitals which are also anti-symmetrical with respect to the molecular axis. The bonding orbital accumulates electron density between the nuclei, although less effectively than a σ orbital. The antibonding orbital has a nodal plane between the nuclei and perpendicular to the axis

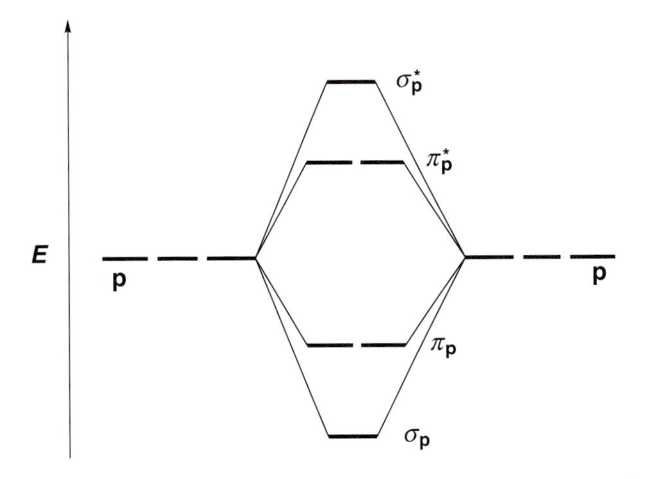

FIG. 3.13 Energy level diagram for the combination of p orbitals. As the π interaction is weaker than the σ one, the π and π^* levels lie between the σ and σ^* ones

This discussion may readily be extended to include d atomic orbitals which may combine with each other or with s or p orbitals to give σ or π molecular orbitals. It is also possible for two d orbitals to overlap 'face-to-face' with each other to form a delta (δ) molecular orbital (see the box titled 'Nomenclature for molecular orbitals'). Figure 3.14 illustrates molecular orbitals involving d orbitals.

The discussion of covalent bonding between a pair of atoms given above may be summarized:

(a) Two atomic orbitals may combine to form two molecular orbitals centred on the nuclei, one of which concentrates electrons in the region between the nuclei and is bonding, while the other removes electron density from this region and is anti-bonding.

(b) Only those atomic orbitals which are of similar energy and of the same symmetry with respect to the inter-atomic axis may combine to form molecular orbitals.

(c) Any atomic orbitals which are symmetrical with respect to the bond axis may combine to form σ molecular orbitals.

(d) Any atomic orbitals which are anti-symmetrical with respect to the bond axis may combine to form π molecular orbitals.

Our next step is to try to describe the electronic structures of diatomic molecules in terms of these ideas. We must recognize that so far our discussion has been entirely qualitative. We have said nothing about the sizes of the ΔE terms, nor about the energy differences between, say, σ_p, and π_p levels in Fig. 3.13. Nor do we have any guidance about combining the energy levels of the orbitals of Fig. 3.5 and 3.13 into a composite diagram. If full and accurate calculations could be carried out, all these questions would be answered from the wave equation. While such calculations are becoming increasingly more feasible as computer power grows and methods improve, we still turn to experimental evidence to help us to complete the picture.

There are three properties we are attempting to rationalize:

(1) Bond orders, reflected in measured bond lengths and heats of formation.

(2) Unpaired electrons, reflected in the magnetic properties (see Section 7.10 for definitions).

(3) Energy levels, reflected in ionization energies.

It is convenient to break the discussion into two parts. First we discuss bond orders and magnetic properties in Section 3.5 which we can do very successfully from a purely qualitative energy level diagram. Then we can get a more quantitative, accurate and sophisticated description by considering energy levels as given, especially, by photoelectron spectroscopy. This is taken up in Section 3.6.

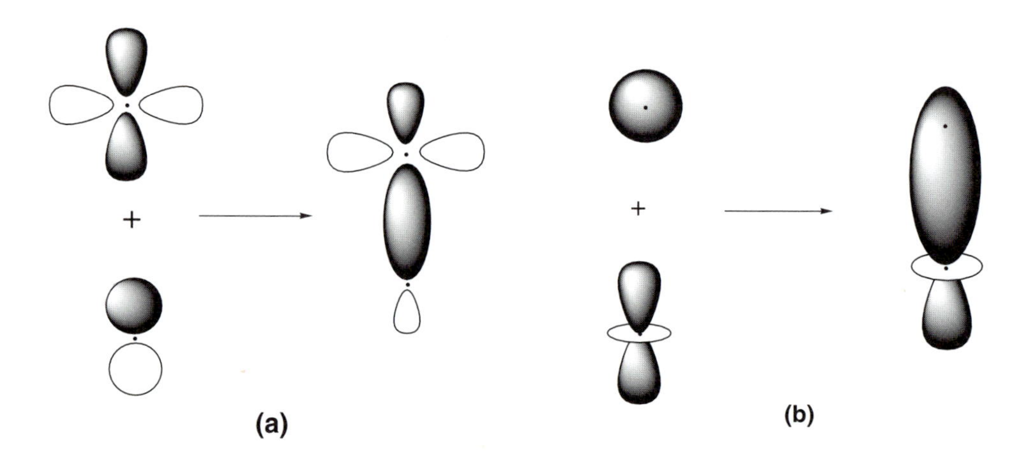

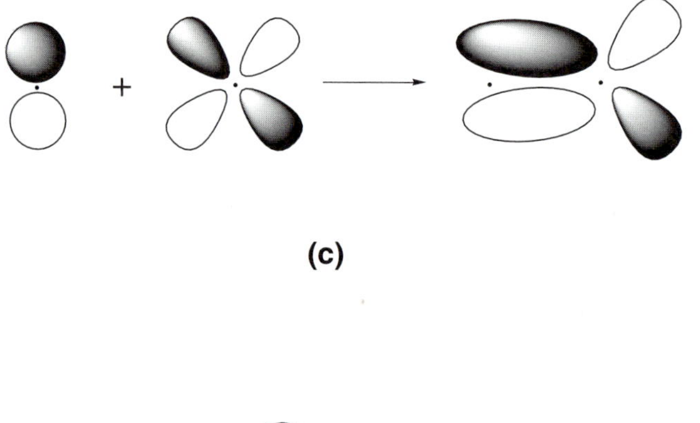

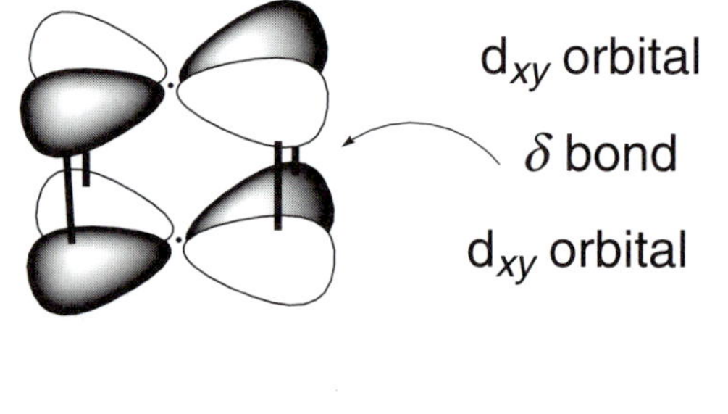

FIG. 3.14 Molecular orbitals involving d orbitals. These diagrams indicate the formation of σ, π and δ bonds, involving d orbitals. (a) σ bond between p_y and $d_{x^2-y^2}$, (b) σ bond between s and d_{z^2}, (c) π bond between p_y and d_{xy}, (d) δ bond between two d_{xy} orbitals

3.5 Bond orders of diatomic molecules

We can assign the valency electrons in a diatomic molecule, and hence predict the bond order and magnetic properties, if we simply combine Figs 3.5 and 3.13 by assuming the s diagram lies at lower energy than the p diagram, and there is no overlap or interaction. This gives us Fig. 3.15.

We shall build up the electron configuration of the molecule by taking all the valency electrons from the atoms and feeding them into the molecular energy level diagram. We obviously fill the most stable level first: each orbital holds a maximum of two electrons with opposite spins, and orbitals of equal energy are first populated singly.

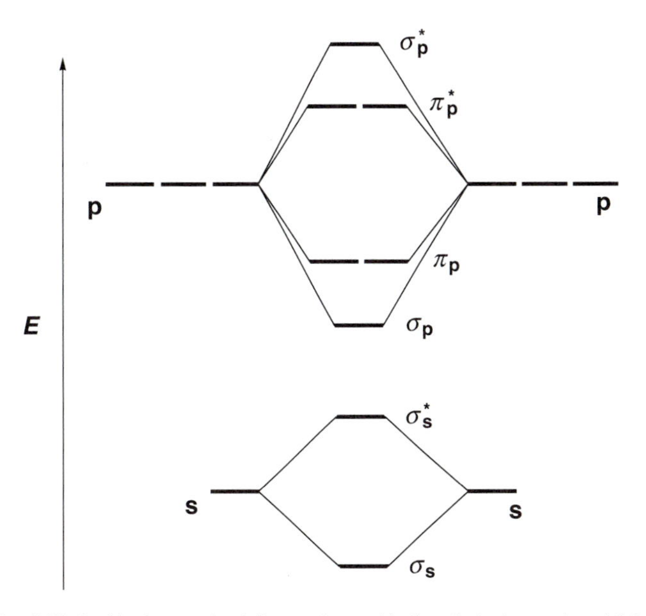

FIG. 3.15 Combined energy level diagram for combination of atomic s and p orbitals

Consider first the case of the fluorine molecule, F_2. The fluorine atom has the electronic configuration $(1s)^2(2s)^2(2p)^5$, and the inner $(1s)^2$ shell is too tightly held to the nucleus to play any part in the bonding. We write the configuration $[He](2s)^2(2p)^5$ to emphasize this (compare Section 2.8). There are thus seven valency electrons from each atom to be fitted into the molecular orbitals of the fluorine molecule and, when these are filled in order of increasing energy, the arrangement shown in Fig. 3.16 results. This may be written as an equation:

$$2F[He](2s)^2(2p)^5 \rightarrow F_2[He][He](\sigma_s)^2(\sigma_s^*)^2(\sigma_p)^2(\pi_p)^4(\pi_p^*)^4$$

To determine the bond order, we determine how many pairs of electrons in bonding orbitals there are in excess of the pairs in antibonding orbitals. In the fluorine molecule

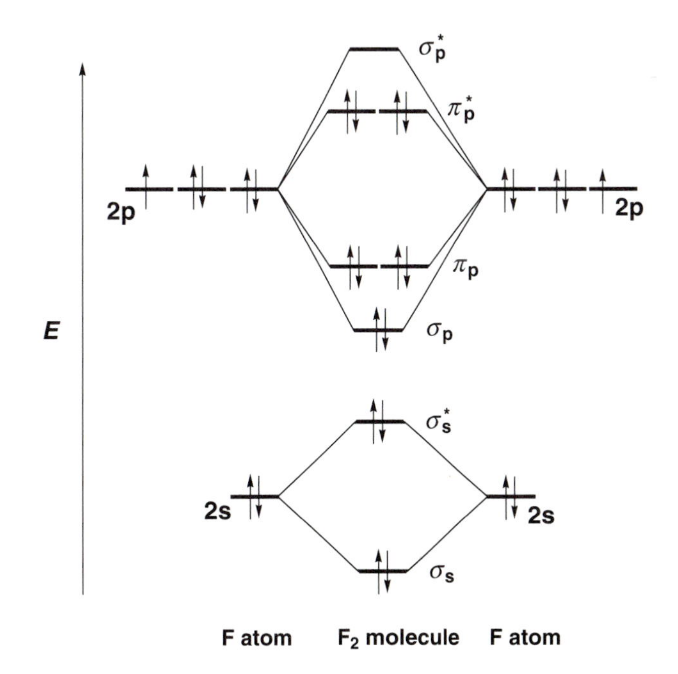

FIG. 3.16 Electronic structure of F_2. An energy level diagram of the molecular orbitals is formed and all the valency electrons are placed in the molecular orbitals, according to Hund's rules

there are eight electrons in bonding orbitals (σ_s, σ_p, π_p) and six in antibonding sigma and pi orbitals. This gives an excess of two bonding over antibonding electrons and corresponds to the single bond in the fluorine molecule. All the electrons are paired, so fluorine is diamagnetic.

The molecular orbital energy level diagram for oxygen is shown in Fig. 3.17. The outer electronic structure of the oxygen atom is $(2s)^2(2p)^4$, so that the oxygen molecule has two fewer electrons than the fluorine molecule and these will be missing from the highest level, the π^* one, leaving only two antibonding π electrons. As there are two electrons to enter the two π^* orbitals, which are of equal energy, they enter these orbitals singly, keeping their spins parallel in accordance with Hund's rules, giving the oxygen molecule two unpaired electrons. The structure is:

$$2O[He](2s)^2(2p)^4 \rightarrow O_2[He][He](\sigma_s)^2(\sigma_s^*)^2(\sigma_p)^2(\pi_p)^4(\pi_p^*)^2$$

There are eight electrons in bonding orbitals and four in the antibonding orbitals σ_s^* and π_p^*, giving a net count of four bonding electrons corresponding to the double bond. Since there are two unpaired electrons, O_2 is *paramagnetic*.

If this description of O_2 is compared with that derived from the Lewis electron pair theory, it will be seen that the molecular orbital treatment has the advantage. In order to complete the octet on oxygen, O_2 would be described as $\dot{\underset{\cdot}{:}}\mathrm{O} :: \mathrm{O}\dot{\underset{\cdot}{:}}$ thus predicting the double bond. However, the Lewis theory gives no reason for expecting unpaired electrons. This ready explanation of the paramagnetism of oxygen was one of the major early successes of the molecular orbital theory.

If diatomic neon, Ne_2 were to form, two further electrons would have to be added to the F_2 structure and they could go only into the highest energy antibonding orbital, σ_p^*. The number of antibonding electrons and the number of bonding electrons both equal eight, no bond results, and neon exists as a monatomic gas.

In a similar way, simple electron counting gives, for example

$$2N[He](2s)^2(2p)^3 \rightarrow N_2[He][He](\sigma_s)^2(\sigma_s^*)^2(\sigma_p)^2(\pi_p)^4$$

(diamagnetic, triple bond)

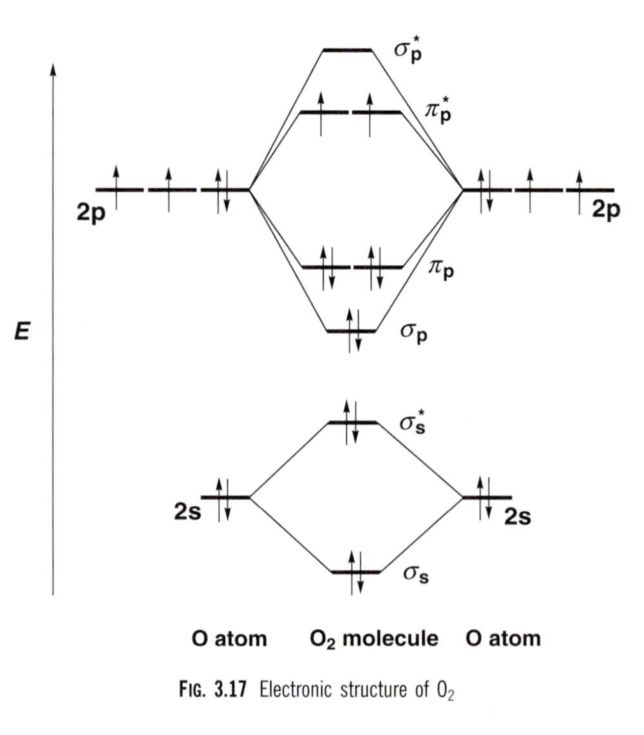

FIG. 3.17 Electronic structure of O_2

$$2C[He](2s)^2(2p)^4 \rightarrow C_2[He][He](\sigma_s)^2(\sigma_s^*)^2(\sigma_p)^2(\pi_p)^2$$
(paramagnetic, double bond)

As the example of C_2 shows, the mere fact that a diatomic species would have a high bond order does not prove that the element will exist in a stable form as a diatomic gas.

The electron counting procedure is simply extended to charged species by adding or subtracting electrons. For example, the superoxide ion O_2^- would be described by

$$2O[He](2s)^2(2p)^4 + e \rightarrow O_2^-[He][He](\sigma_s)^2(\sigma_s^*)^2(\sigma_p)^2(\pi_p)^4(\pi_p^*)^3$$
(paramagnetic, bond order $=1\frac{1}{2}$)

Electron counting may also readily be applied to molecules where the atoms have higher n values than two for their valency levels. For example, the electronic structures of the other halogen molecules are exactly the same as for fluorine:

$$X_2(\sigma_s)^2(\sigma_s^*)^2(\sigma_p)^2(\pi_p)^4(\pi_p^*)^4$$

but here the atomic symbol X is to indicate that all the inner non-valency electron shells remain held by the nuclear attraction and play no part in the bonding. For example, for bromine

$$2Br[Ar](4s)^2(4p)^5 \rightarrow Br_2[Ar][Ar](\sigma_s)^2(\sigma_s^*)^2(\sigma_p)^2(\pi_p)^4(\pi_p^*)^4$$
(diamagnetic, single bond)

where the molecular orbitals are constructed from the 4s and 4p atomic orbitals and all the inner shells of the first, second and third levels remain held by the atomic nuclei.

The analysis may be extended to heteronuclear molecules where the two atoms are in different periods. For example, in ClF, the larger nuclear charge of the chlorine atom lowers the energy of all its orbitals compared with the corresponding fluorine ones. As a result, the 3s and 3p levels of chlorine are approximately equal to the 2s and 2p levels of fluorine (and the Cl 2s and 2p levels are so tightly bound that they play no part in the bonding). The electronic diagram of ClF is then very similar to that of F_2 shown in Fig. 3.16, except that the molecular orbitals are formed by overlap of F orbitals of the second shell with the corresponding Cl orbitals of the third shell:

$$Cl[Ne](3s)^2(3p)^5 + F[He](2s)^2(2p)^5 \rightarrow ClF[Ne][He](\sigma_s)^2(\sigma_s^*)^2(\sigma_p)^2(\pi_p)^4(\pi_p^*)^4$$
(diamagnetic, single bond)

The structure of hydrogen halides follows similarly. For example, in HCl the H 1s orbital is approximately equal to the Cl 3s and 3p orbitals in energy and is of correct symmetry to form a sigma bond either with the 3s or the $3p_z$ orbital (taking z as the H-Cl direction). The overlap is best with the $3p_z$ orbital, and a sigma bonding orbital is formed as in Fig. 3.10, together with the corresponding antibonding orbital. The remaining valency orbitals of the chlorine are non-bonding and hold electron pairs. Thus HCl is described by

$$H(1s)^1 + Cl[Ne](3s)^2(3p)^2 \rightarrow HCl[Ne](\sigma_{s,p})^2(\sigma_{s,p}^*)^0(3s)^2(3p_x)^2(3p_y)^2$$

If we now turn to consider heteronuclear species where the two atoms are in the same period, the only difference is in the relative levels of the atomic orbitals, with those of the atom with highest nuclear charge lying lowest in energy. Consider nitric oxide and its cation, NO^+ (the nitrosonium ion). The energy level diagram for nitric oxide is given in Fig. 3.18. The atomic orbital energy levels are no longer equal because of the difference in nuclear charges. The nitric oxide molecule has an odd number of electrons and the one of highest energy occupies the antibonding π orbital.

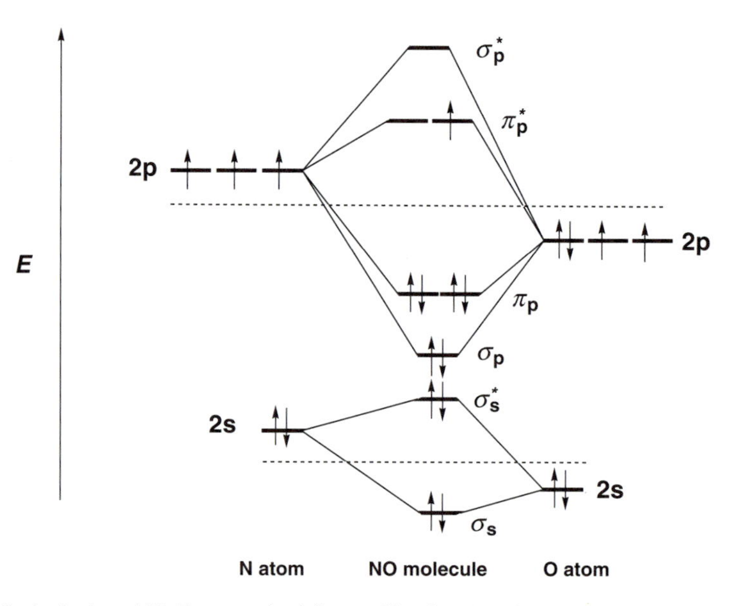

Fig. 3.18 Electronic structure of NO. The energy level diagram differs from that of a homonuclear diatomic molecule, such as O_2, only in having the atomic levels at different energies. The splitting of the bonding and antibonding molecular orbitals of any particular type is symmetrical about the average energy of the constituent atomic orbitals

$$\mathrm{N[He](2s)^2(2p)^3 + O[He](2s)^2(2p)^4}$$
$$\rightarrow \mathrm{NO[He][He]}(\sigma_s)^2(\sigma_s^*)^2(\sigma_p)^2(\pi_p)^4(\pi_p^*)^1$$

There are eight bonding electrons, in the sigma s and p levels and in the pi p level and three antibonding electrons, $(\sigma_s^*)^2$ plus $(\pi_p^*)^1$, leaving a net excess of five bonding electrons corresponding to a bond order of two and a half. If this molecule is ionized, the highest energy electron is the one removed and this is in the antibonding pi orbital. The excess of bonding over antibonding electrons goes up to six and the bond order rises from two and a half to three. (NO^+ is isoelectronic with N_2.) The cation is therefore more strongly bonded than the parent molecule. This can be seen experimentally in the infrared spectrum where the N-O stretching frequency occurs at higher energy for NO^+ than for NO.

In all these cases, and many more, the observed magnetism matches with that predicted. Bond order is more difficult to quantify but all the observed changes are in the directions indicated by the predictions. For example, for oxygen species

	O_2^+	O_2	O_2^-	O_2^{2-}
Predicted bond order	$2\frac{1}{2}$	2	$1\frac{1}{2}$	1
Observed bond length (pm)	112.3	121.1	134	154.1

Thus the simple approach of feeding all the valency electrons into the levels of Fig. 3.15, or related energy level diagrams like Fig. 3.18, successfully rationalizes the observed bond orders and magnetism.

However, Fig. 3.15 is not always an adequate description of the *energies* of the electrons in the molecular orbitals.

3.6 Energy levels in diatomic molecules

Closer consideration of Fig. 3.15 raises two questions.

(1) Remembering that He_2, with a possible configuration $(\sigma)^2(\sigma^*)^2$, does in fact exist only as 2He atoms with $(1s)^2$, the general query arises whether distinct completely filled (bonding+antibonding) levels will persist, or will the better description be as non-bonding pairs on the atoms. That is, does the configuration $(\pi)^4(\pi^*)^4$ (implying *two* different orbital

energies) describe the system better or worse than two filled p orbitals on each atom (implying only *one* orbital energy when the atoms are identical)? Similarly, is the description $(\sigma_s)^2(\sigma_s^*)^2$ better or worse than $2 \times (s)^2$?

(2) We assumed that σ_s and σ_s^* were more stable than σ_p, and further that there was no interaction between them. Theoretical treatments show that orbitals of the same symmetry and similar energies, such as s and p_z, may interact (or 'mix'). How significant is this effect in the molecules discussed?

In fact, the separation of sigma molecular orbitals formed from atomic s orbitals and those formed from the p_z orbitals is a simplification: in principle, all orbitals of sigma symmetry on the atoms are expected to make some contribution to all the molecular sigma orbitals. That is, the two s orbitals and the two p_z orbitals may be combined in various proportions to give four molecular orbitals of sigma symmetry which we shall relabel σ_1, σ_2, σ_3 and σ_4, in order of increasing energy. The two extreme levels are similar to those in the simpler scheme with σ_1 close to σ_s and σ_4 close to σ_p^* in character. The major difference comes with σ_2 and σ_3, which both have substantial s and p contributions. The major effect of the mixing is to stabilize σ_2 below σ_s^* and to destabilize σ_3 relative to σ_p^*. The relation between the two schemes for the sigma molecular orbitals is indicated in Fig. 3.19 parts (a) and (b).

The answers to queries (1) and (2) above come most readily from experimental data. While a number of techniques give information about energy levels, the most readily interpreted is photoelectron spectroscopy (see Section 7.12).

To a good approximation, a particular photoelectron (PE) band gives the ionization energy of an electron in a particular orbital in the molecule. Thus we can use the ionization energies from the PE spectrum to put an energy scale on the molecular orbital energy level diagrams. Further, the fine structure of the PE band gives us the vibrational energy of the product ion A_2^+. If this is less than the vibrational energy of the parent molecule, A_2, it implies the ion has a lower bond order, and hence that an electron has been lost from a bonding molecular orbital in the ionization. (cf. discussion of H_2 in Section 7.12)

$$A_2 + h\nu \rightarrow A_2^+ + e$$

Conversely, equal or greater vibrational energies in A_2^+ implies the loss of a non-bonding or an antibonding electron.

As a simple example, let us consider the PE spectrum of HCl, shown schematically in Fig. 3.20. Two bands are found using He(I) radiation. The one at 12.75 eV has the short vibrational series characteristic of the loss of a nonbonded electron while the long

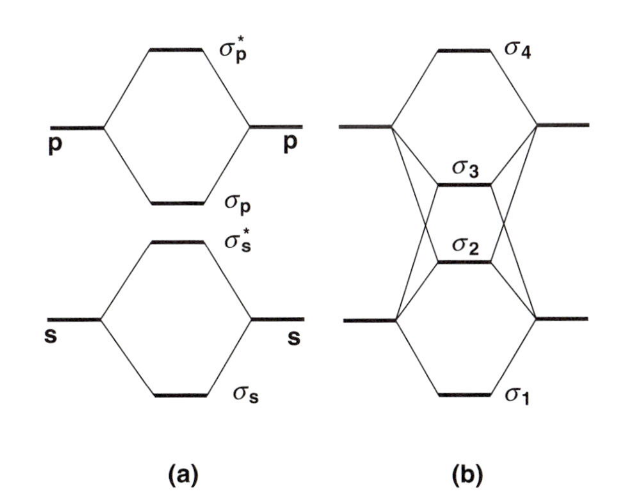

FIG. 3.19 (a) σ combinations without s–p mixing. (b) σ combinations with s mixed with p. This shows the relation between the $\sigma_s\sigma_s^*\sigma_p$ and σ_p^* levels (a) and the levels σ_1 to σ_4 (b) which result when s and p contributions are allowed to mix. The main effect is to stabilize σ_2 relative to σ_s^* and to destabilize σ_3 relative to σ_p

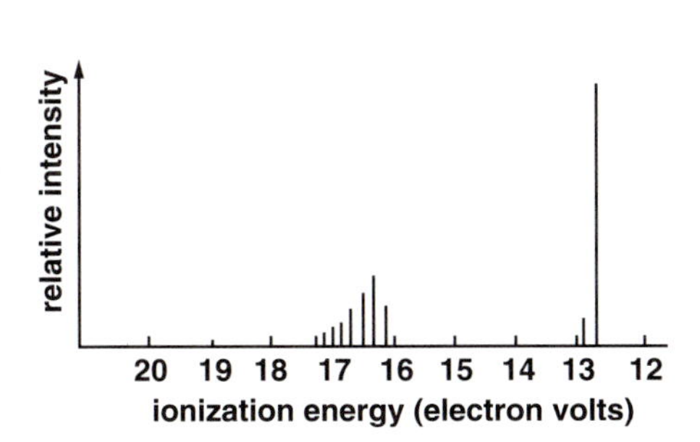

FIG. 3.20 The photoelectron spectrum of HCl

TABLE 3.2 Photoelectron bonds of the hydrogen halides

	Molecular stretching frequency (cm^{-1})	*photoelectron energy* (*eV*)	*Vibrational features*		*Assignment*
			Sequence	*Separation* (cm^{-1})	
HCl	2886	12.75	short	2660	non-bonding
		16.25	long	1610	bonding
HBr	2560	11.68	short	2420	non-bonding
		15.28	long	1290	bonding
HI	2230	10.2	short	2100	non-bonding
		13.8 approx	broad, bands overlap		bonding

Notes: (a) p non-bonding band is actually doubled due to Jahn–Teller effect.
(b) Excitation from 3s level is too energetic to be seen using He(I) radiation.

sequence of the 16.25 eV bond is that expected where a bonding electron is lost. These assignments are supported by the vibrational separations (Table 3.2). Thus the 12.75 eV ionization must be from the least strongly bound non-bonding level, the Cl(3p) level, while the 16.25 eV bond corresponds to ionization from the $\sigma_{s,p}$ orbital. The third occupied orbital, the Cl(3s) level, is too stable for excitation using He(I) radiation. We can present the results in a composite energy level/PE spectrum diagram as in Fig. 3.21. Note that the levels of the empty orbitals, e.g. $\sigma^*_{s,p}$, cannot be located by photoelectron spectroscopy.

For a second example, consider O_2. Here an additional complication arises because of the unpaired electrons in the π^* orbital. If the ionization occurs from, say, the σ_p orbital, then the electron remaining in σ_p in the O_2^+ ion may have its spin parallel or antiparallel to the π^* electrons. Thus each energy state is doubled apart from that arising from π^* ionization. The analysis is indicated in Table 3.3. It will be seen that these observations could be fitted on to Fig. 3.17.

However, if we turn to nitrogen, N_2, we find that we are forced to consider mixing among the sigma states, as in Fig. 3.19b.

The He(I) photoelectron spectrum of N_2 is shown in Fig. 3.22, and the data are listed in Table 3.4. It is immediately obvious that the 15.6 eV ionization is from an orbital that is only weakly bonding, and cannot therefore be from the π level. Further, we expect the N_2 π level to be similar in energy to the O_2 π level which is 17.0 eV from Table 3.3. Thus the 16.69 eV level in N_2 has the expected attributes of the bonding π level. This leads to the conclusion that the weakly bonding level at 15.57 eV and the weakly antibonding level at 18.75 eV arise from the two higher sigma orbitals, and the least stable of these lies *above*

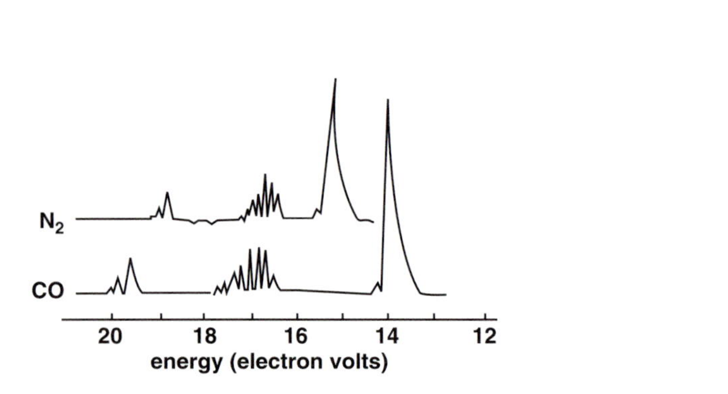

FIG. 3.22 Photoelectron spectra of N_2 and CO

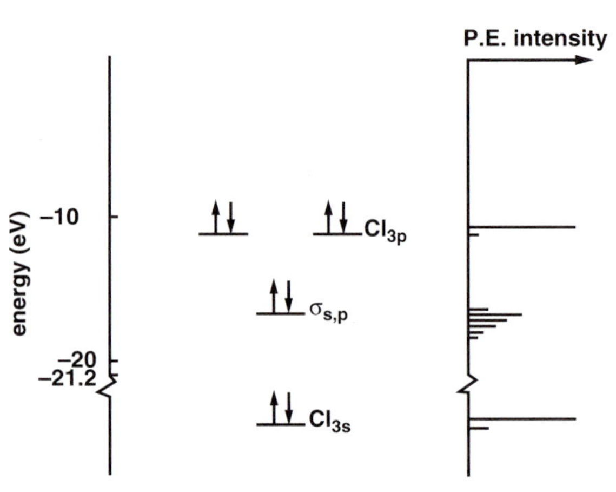

FIG. 3.21 Energy levels and photoelectron spectrum of HCl

TABLE 3.3 Photoionization of oxygen, O_2. (Stretching frequency of neutral $O_2 = 1555\ cm^{-1}$)

	Photoelectron energy (eV)	*Vibrational energy (cm^{-1})*	*Assigned to loss of electron*	
			which is	*from*
For O_2^+	12.07	1780	antibonding	π^*
	17.0	1010	bonding	π
	19.3	1110	bonding	σ_p
	26.9		antibonding	σ_s^*
	40.6		bonding	σ_s

Notes: (a) Last two excitations used He(II) radiation.
(b) For comparison, loss of 1s electrons needs > 540 eV.
(c) Levels doubled by spin effects have been averaged.

the π level. This can only happen if σ_s and σ_s^*, and σ_p and σ_p^*, interact with each other, as discussed earlier in this section. The net effect of this s–p mixing is to (a) raise σ_3 and lower σ_2, and (b) average out their bonding and antibonding character to give relatively non-bonding combinations. It must be emphasized, however, that the π levels are unaffected by this s–p mixing since they are orthogonal to the molecular axis. Thus the energy level diagram for N_2 has the π bonding level lying between σ_2 and σ_3 of Fig. 3.19b. The electronic structure of N_2 is therefore given as in Fig. 3.23.

This interpretation is supported by the data for CO, Table 3.4. Carbon monoxide has the same number of valency electrons as N_2 (such species are termed *isoelectronic*) but its energy level diagram is a skew one, reflecting the unequal atomic levels, Fig. 3.23b. We see that the π level of CO is of almost exactly the same stability as for N_2, but the electrons in σ_3 are less tightly bound (reflecting the greater contribution from $C2p_z$ in σ_3) while those in σ_2 are more tightly bound (O contribution greater) than in the respective orbitals of N_2.

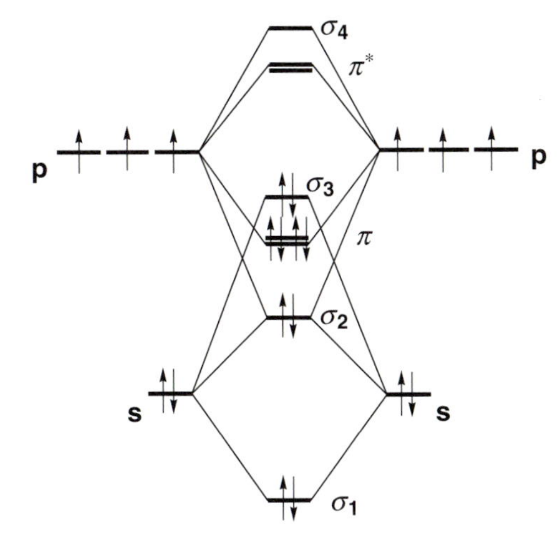

FIG. 3.23(a) The electronic structure of N_2. In the structure of N_2, the bonding π levels lie below σ_3. The bonding and antibonding contributions of σ_2 and σ_3 cancel, leaving σ, and the two π levels to make up a total bond order of three

In general, the importance of the mixing between s and p orbitals depends inversely on the difference in energy between them. With atoms to the right of the Periodic Table, like O or F, the separation into σ_s and σ_p orbitals is a good approximation, and the schemes for F_2 or O_2 shown in Figs 3.16 and 3.17 are satisfactory. Further to the left, the atomic s and p orbitals are closer in energy and the interaction between them is significant. Thus nitrogen is described as in Fig. 3.23a:

$$2N[He]2s^2(2p)^3 = N_2[He][He](\sigma_1)^2(\sigma_2)^2(\pi)^4(\sigma_3)^2$$

The bonding character added to σ_2 relative to σ_s^* exactly equals the loss of bonding character in σ_3 relative to σ_p and there is no net bonding effect when σ_2 and σ_3 are both filled. Thus N_2 has six more bonding than antibonding electrons ($\sigma_1^2\pi^4$) making a triple bond.

TABLE 3.4 Photoionization of N_2 and CO. (Stretching frequency of $N_2 = 2345\ cm^{-1}$, $CO = 2157\ cm^{-1}$)

	Photoelectron energy (eV)	*Vibrational separation* (cm^{-1})	*Assignment of lost electron*
For N_2^+	(1) 15.57	2150	weakly bonding
	(2) 16.69	1810	strongly bonding
	(3) 18.75	2390	weakly antibonding
	(4) 37.3		strongly bonding
	(5) 410		ls electron
For CO^+	(1) 14.01	2160	non-bonding
	(2) 16.53	1610	bonding
	(3) 19.68	1690	bonding
	(4) 38.3		strongly bonding
	(5) 293		C ls electron
	(6) 542		O ls electron

EXCITED STATES OF OXYGEN WITH THE π^* SPINS PAIRED

It is worth taking a closer look at the oxygen molecule and considering how its electronic state influences its properties. As we discussed earlier in Sections 3.5 and 3.6, the ground state of this molecule is that which places one unpaired electron in each of the π_p^* orbitals to give a total of two unpaired electrons, thereby accounting for the observed paramagnetism of the O_2 molecule—and obeying Hund's rule for the most stable configuration.

However, there are states of higher energy (called *excited states*) where the electron spins are anti-parallel. Two configurations are possible for placing such electrons in the π^* orbitals—the electrons can both be placed into one π^* orbital with opposed spins or singly into the two π^* orbitals with opposed spins—respectively (a) and (b) below:

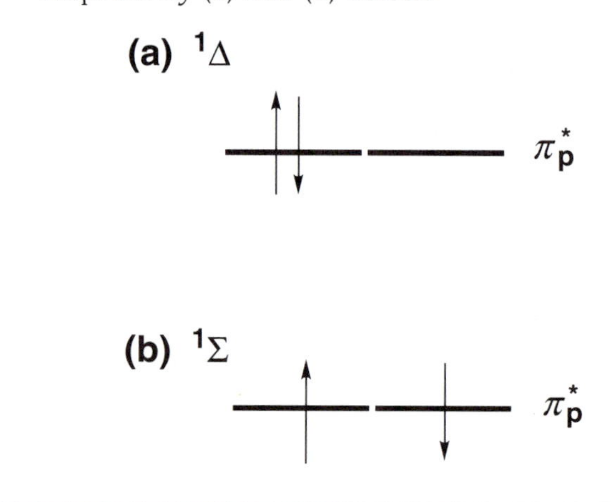

These two diamagnetic forms of oxygen have been detected in the gas phase. The paramagnetic ground state form of O_2, with two parallel spins, is termed 'triplet' oxygen, and given the spectroscopic symbol $^3\Sigma$, whereas the other two are termed 'singlet' oxygen $^1\Delta$ and $^1\Sigma$ (the terms 'singlet' and 'triplet' refer to the spin multiplicity, $2m_s + 1$). The $^1\Delta$ form of singlet oxygen is a chemically important species since it undergoes two-electron reactions, such as Diels–Alder type cyclizations, similar to those undergone by the isoelectronic molecule ethylene. In contrast, the triplet form of oxygen undergoes radical reactions. Singlet oxygen is also responsible for the degradation of rubbers and other synthetic polymers by air, and perhaps the best known 'occurrence' of singlet oxygen is in the chemiluminescence demonstrations involving the compound luminol. The bond lengths of all three forms of the O_2 molecule have been determined and, as would be expected, all point to there being a net double bond, but with the singlet excited states having slightly longer O-O bonds.

In carbon monoxide and in the cyanide ion, the σ_3 level also lies above the bonding π orbitals. The species are triply bonded and the formation equations may be written:

$$\mathrm{C[He](2s)^2(2p)^2 + O[He](2s)^2(2p)^4 \rightarrow CO[He][He](\sigma_1)^2(\sigma_2)^2(\pi)^4(\sigma_3)^2}$$

and

$$\mathrm{C[He](2s)^2(2p)^2 + N[He](2s)^2(2p)^3 + e^- \rightarrow CN^-[He][He](\sigma_1)^2(\sigma_2)^2(\pi)^4(\sigma_3)^2}$$

3.7 Summary

By combining atomic orbitals into molecular orbitals, we can describe the bonding in diatomic molecules in a way that gives greater depth to the early ideas of electron-pair bonds.

Using simple combinations of the s and p orbitals, as in Fig. 3.15, we can feed in the valency electrons and arrive at correct indications of the bond orders and numbers of unpaired electrons for any diatomic species A_2 or AB, or any of their cations or anions. For O_2 or F_2, such an approach also gives the correct relative ordering of the energies of the molecular orbitals.

For the general case, particularly with molecules of atoms to the left of oxygen in the Periodic Table, a more accurate description of the relative energies of the molecular orbitals allows for mixing of s and p contributions to the σ orbitals. This is often sufficiently large to alter the relative position of the π bonding orbital, as in Fig. 3.23.

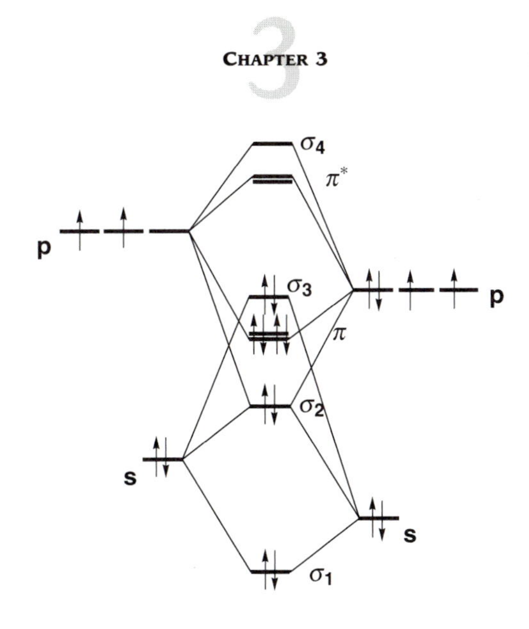

FIG. 3.23(b) The electronic structure of CO

There are thus two stages in the description of a diatomic species:

(1) determine the bond order and magnetic properties by filling the levels of Fig. 3.15;
(2) find the relative energy levels, and the best description of the orbitals, by reference to the photoelectron spectrum or other experimental data. In absence of such data, be prepared to use Fig. 3.23 for molecules containing atoms to the left of oxygen.

The molecular energy level diagrams may also be used in the discussion of the excited states of the molecules. If the molecule absorbs energy equal to the difference between the energy of an occupied orbital and that of one of the higher, empty orbitals, an electron will undergo a transition into the upper orbital, e.g. $\pi \rightarrow \pi^*$ in N_2. The energy differences between molecular orbitals can thus be observed in the electronic spectrum of a diatomic molecule and such experimental values compared with the calculated energy levels, just as in the case of atomic orbitals. The electronic transitions for the light diatomic molecules, such as nitrogen, are of relatively high energy and occur in the far-ultraviolet region of the spectrum.

Thus the photoelectron spectrum gives the absolute energies of the occupied orbitals, while the visible–UV spectrum can give the energy differences between occupied and empty orbitals. By combining both sets of data, the diatomic molecular energy level diagrams can be made quantitative.

Problems

In this chapter, we start to combine atoms into chemically much more significant entities—molecules. All the atomic properties discussed in Chapter 2 are significant in the resulting molecules as later chapters show. However, we emphasize here the atomic orbitals and show how we can think of them combining—considering them purely as waves with one crest reinforced by another or nullified by a trough. We turn the resulting wave into an electron density distribution and decide whether the attractions between the positive nuclear charges and this distribution will be greater or less than in the separate atoms. This brings us qualitatively to the idea of the energy of stabilization, and then to a quantitative picture using data from photoelectron spectra.
Question I revises basic ideas (see page 60).

Many variants of the other questions will suggest themselves. Thus the bond order of any diatomic A_2 or AB (real or hypothetical) or of any ions derived from these can be predicted. You can look for evidence of bond order changes in bond lengths, bond strengths (reflected by heats of formation or vibrational stretching frequencies) or from photoelectron spectra.

3.1 (a) The relative molecular mass of XCl_3 is 181. What is the relative atomic mass of X?

(b) A substance contains 16.1% H, 38.7% C and 45.1% N. If the relative molecular mass is 31, decide its structural formula.

3.2 Using your values from question 2.5, draw out the plots corresponding to Figs 3.1 and 3.3 for two H atoms whose nuclei are separated by (a) 50 pm and (b) 200 pm.

3.3 Using z as the internuclear axis in AB, write down all the combinations of d orbitals on A with s, p or d orbitals on B which will give

(a) σ bonds,
(b) π bonds.

3.4 Work out the electron configurations, bond orders and numbers of unpaired electrons for the species O_2^+, O_2, O_2^- and O_2^{2-}. (Compare your answers with Section 3.5.)

3.5 Using Fig. 3.15, determine the configurations of the diatomic molecules Li_2, Be_2, B_2, C_2, N_2, O_2 and F_2. Arrange in decreasing order of bond strength.

Question 3.4 to 3.8 apply the methods of Section 3.5. The commonest error is to include the electrons in the underlying 1s shells.

The fact that the predicted electron configuration is bonding does not guarantee that the diatomic will be the stable form for the system. One obvious case is where a more complex species is formed. Thus B, C or BN, for example, are all very stable solids and, while diatomics like B_2 are found in the gas phase, condensation to solids is greatly favoured.

Another case is illustrated by BNe. Counting all 8 outer shell electrons from Ne, there are 11 electrons which will give the same configuration as NO, Fig. 3.18. However, the 'skewing' of the diagram becomes extreme as the high nuclear charge on Ne lowers the levels of its 2s and 2p levels compared with O, while the much lower charge on B means that its levels are much higher than those of N. As a result, the overlaps are poor and there is no significant stabilization of the molecular orbitals relative to the Ne atomic levels. Thus the Ne remains as single atoms and cannot be persuaded into combination—with the exception of very short-lived species.

The larger rare gas atoms have less tightly held electrons so we do find diatomics with longer lifetimes, such as XeF.

3.6 Repeat the exercise in Question 3.5 with all the possible mixed diatomics AB.

3.7 If Fig. 3.17b applies, which of the unstable species in Questions 2.5 and 3.6 might be stabilized?

3.8 Which of the species CN, CO, NO, NF and OF would be stabilized by
(a) loss of an electron,
(b) gain of an electron?

3.9 The (simplified) photoelectron spectrum of NO shows the following bands

Ionization energy (eV)	*Vibrational separation (cm^{-1})*
9.26	2260
16.4	1550
18.1	1180

Assign and compare with CO and O_2. (The stretching frequency of NO is 1890 cm^{-1}.)

CHAPTER 4

4.1 INTRODUCTION 81
4.2 THE SHAPES OF MOLECULES AND IONS CONTAINING σ BONDS ONLY 81
4.2.1 The arrangement of σ bonds 81
4.2.2 The effect of lone pairs 82
4.2.3 Three pairs 84
4.2.4 Four pairs 84
4.2.5 Five pairs 84
4.2.6 Six pairs 85
4.3 THE SHAPES OF SPECIES CONTAINING π BONDS 86
4.3.1 Electron counting procedure for π electrons 88
4.3.2 Summary 91
4.4 GENERAL APPROACHES TO BONDING IN POLYATOMIC SPECIES 91
4.5 BONDING IN POLYATOMICS: THE TWO-CENTRE BOND APPROACH 92
4.6 TWO-CENTRED ORBITALS: HYBRIDIZATION 93
4.6.1 Equivalent hybrids 93
4.6.2 Nonequivalent hybrids 96
4.7 DELOCALIZED, OR MULTI-CENTRED, σ ORBITALS 97
4.7.1 Summary of Sections 4.5 to 4.7 101
4.8 π BONDING IN POLYATOMIC MOLECULES 101
4.8.1 The π orbitals of benzene 103
4.8.2 The π orbital of the nitrite ion, NO_2^- 104
4.9 AN EXAMPLE OF THE APPROACH USING DELOCALIZED BONDING THROUGHOUT 105
4.9.1 Carbon dioxide, CO_2^- 105
4.10 EXTENSION TO OTHER MOLECULES 107
4.10.1 Ozone, O_3 107
4.10.2 The nitrate ion, NO_3^- 107
4.10.3 Sulfur trioxide, SO_3 107
4.10.4 Summary 107
PROBLEMS 108

Polyatomic Covalent Molecules

4.1 Introduction

Once more than two atoms are present in a molecule, the problem of molecular shape arises. For example, an AB_4 molecule might be a tetrahedron, a square plane, or some less symmetrical shape. Since the molecule in its ground state adopts the shape which minimizes the total energy, a complete bonding theory would produce the shape of the molecule as one of its results. Unfortunately, such high accuracy calculations are only possible for simple species with few electrons. Thus the shape, or at least the symmetry, has to be determined by experiment, and we then require of our bonding theory a *description* of the shape rather than a prediction.

While the shape may always be determined experimentally (see Chapter 7), it is convenient to be able to decide on the most likely shape of a polyatomic molecule or ion. It turns out that this can be done, quite simply, by considering the repulsive forces between electron pairs in the valency shell of the central atom (or atoms) of the species. This VSEPR theory (*v*alence *s*hell *e*lectron *p*air *r*epulsion) gives a good qualitative prediction of the shape of a molecule which turns out to be accurate for about 95% of all main group compounds and is also basically correct for transition metal complexes.

We outline the VSEPR theory and show the shapes predicted by it in the next two sections before we turn to a discussion of bonding in polyatomics.

4.2 The shapes of molecules and ions containing σ bonds only

The case of σ-bonded species is the simplest and is discussed first.

4.2.1 The arrangement of σ bonds

When two atoms are bound together, electron density is concentrated in the region of space between them (Sections 3.3 and 3.4). If a central atom is bonded to a number of others, it is reasonable to expect the bonds from the central atom to be as far apart as possible in order to reduce the electrostatic repulsions between the electron-dense regions in the bonds. For a triatomic molecule, such as $BeCl_2$, a linear configuration Cl-Be-Cl will minimize the repulsions between the electrons in the two Be-Cl bonds. Similarly, when there are three attached atoms, as in BCl_3, we expect an equilateral triangle with the Cl-B-Cl angles all 120°. Compare Table 5 on page 6 for a list of examples of common shapes. The formal description of the shape of a molecule is in terms of its *symmetry* and *point group*—see Appendix C for a full treatment. Table 4.1 shows the expected configurations for molecules of the types AB_n. In each case the configuration is the one of highest symmetry.

Coordination numbers greater than six are uncommon. Iodine is 7-coordinate in the pentagonal bipyramidal IF_7 and tellurium, for example, is 7- and 8-coordinate in complexes derived from the hexafluoride. There are many examples of the cases shown in Table 4.1, among ions as well as among molecules, and the tetrahedron and octahedron are especially common shapes.

If more than one type of atom is bonded to the central one, the configuration becomes less symmetrical although retaining the basic shape shown in the table. (The tetrahedral

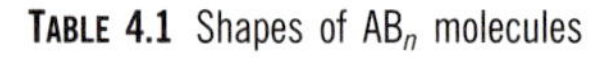

TABLE 4.1 Shapes of AB_n molecules

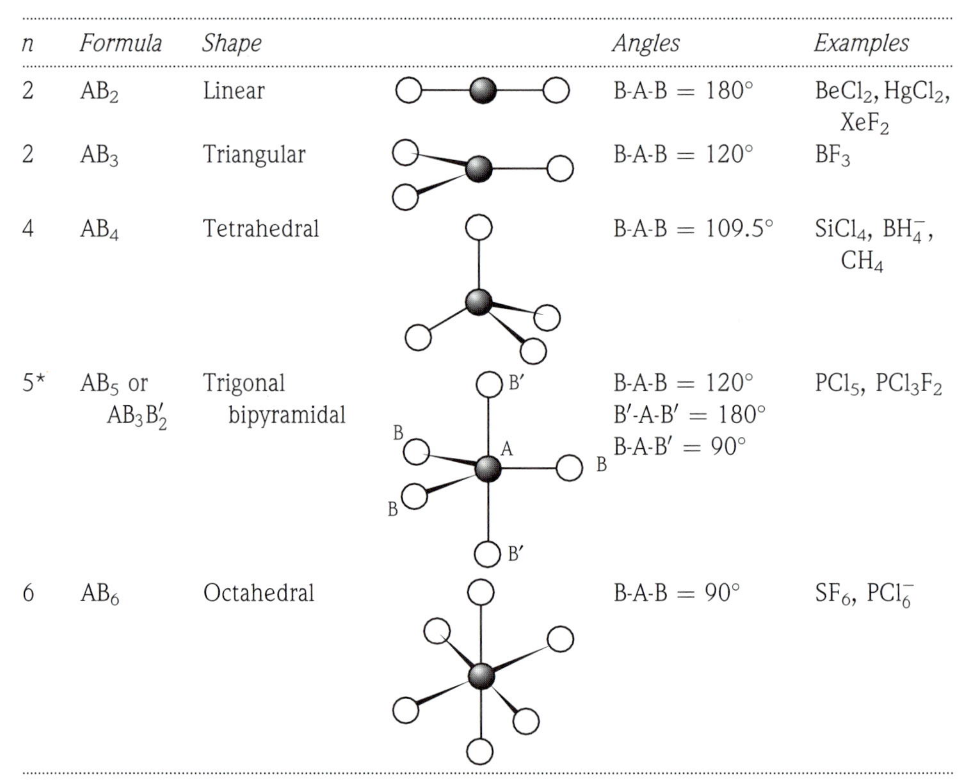

n	*Formula*	*Shape*	*Angles*	*Examples*
2	AB_2	Linear	B-A-B = 180°	$BeCl_2$, $HgCl_2$, XeF_2
2	AB_3	Triangular	B-A-B = 120°	BF_3
4	AB_4	Tetrahedral	B-A-B = 109.5°	$SiCl_4$, BH_4^-, CH_4
5*	AB_5 or $AB_3B'_2$	Trigonal bipyramidal	B-A-B = 120° B′-A-B′ = 180° B-A-B′ = 90°	PCl_5, PCl_3F_2
6	AB_6	Octahedral	B-A-B = 90°	SF_6, PCl_6^-

* All the B positions in all these configurations are equivalent, except in the trigonal bipyramid where the two apical positions (B′) are not equivalent to the three equatorial (B) ones.

configuration of all kinds of organic molecules is an obvious example.) In a molecule AB_rX_s, if the A-X bond is shorter and stronger than the A-B bond, the electron density near A in the X directions will be greater than in the B directions. As a result, the BAB angles close up and the XAX angles open out relative to the values given in Table 4.1 for symmetrical AB_n cases. In addition, the AX bonds will tend to be as far apart in the molecule as possible; for example an octahedral AB_4X_2 species would be expected to be most stable in the *trans* configuration, compare Fig. 13.14.

While any compound with different substituents will be distorted from the symmetrical configuration, the amount and sense of the distortion will not always be obvious. Bond lengths, bond polarities and steric effects may all come into play and tend in opposite directions. However, most changes in bond angles due to unsymmetrical substitution are relatively small—cf. the parameters of the silyl halides below—and the basic shape remains that of Table 4.1.

$$\text{H-Si-H angle in silyl halides, } SiH_3X:$$
$$X = F,\ \text{angle} = 109\tfrac{1}{2}^\circ : X = Cl,\ \text{angle} = 110\tfrac{1}{2}^\circ :$$
$$X = Br,\ \text{angle} = 111\tfrac{1}{2}^\circ$$

4.2.2 The effect of lone pairs

The most important case of distortion by unsymmetrical substitution comes when an atom in the molecule is replaced by an unshared pair of electrons on the central atom. For example, ammonia, NH_3, is not a plane triangular molecule as a casual glance at Table 4.1 would suggest, but is pyramidal with an H-N-H angle of 107°. This angle is near the tetrahedral value and a count of the electrons on the nitrogen shows that there is a lone or unshared pair of electrons in addition to the three bond pairs. If the lone pair occupies a specific direction in space, the four electron pairs would still be expected to have a basically tetrahedral arrangement. Thus ammonia is an extreme case of an unsymmetrically substituted tetrahedron, AB_3X. Since a lone pair is subject to the attraction only of the central atom nucleus, rather than being shared between the central atom and a bonded

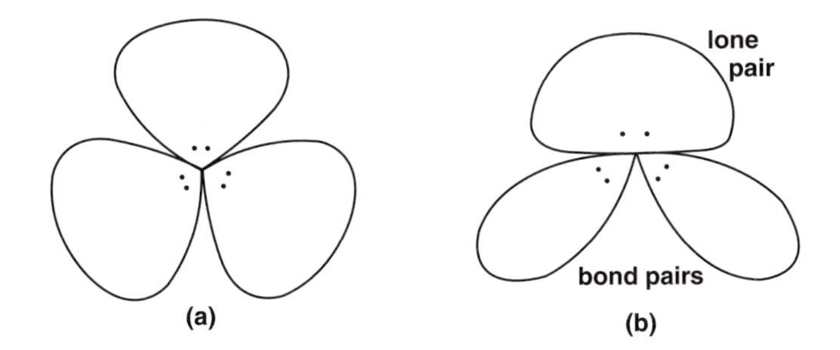

FIG. 4.1 Three electron pairs on a central atom: (a) three equal pairs, (b) two bond pairs and a lone pair. As the lone pair is influenced only by the central nucleus, its density is concentrated much more closely to the central atom than the bond pairs and it thus dominates the stereochemistry

atom, the electron density of a lone pair is concentrated close to the central atom, as shown schematically in Fig. 4.1. Thus the lone pair electrons exert greater repulsion than the bond pairs, and the bond angles close up when a lone pair is present. For example, the X-M-X angles of all the nitrogen group trihalides such as NF_3 or SbI_3 lie within the range 97–104° compared with the tetrahedral angle of $109\frac{1}{2}$°. Again, where geometrical isomerism is possible, lone pairs will tend to be as far apart as possible. Thus a central atom with two lone pairs and four bond pairs around it, is expected to from a square planar molecule, with the two lone pairs in the *trans* positions of Fig. 4.5c.

When lone pairs are taken into account, the structure of any σ-bonded species may be formulated in terms of two simple rules:

(1) The basic shape of a molecule or ion depends on the number of electron pairs surrounding the central atom (lone pairs plus bond pairs) and is that which follows from Table 4.1, assuming that lone pairs occupy positions in space. It must be emphasized that the shape of the *molecule* is determined by the arrangement of the *atoms* themselves. Thus, the NH_3 molecule, while it has four electron pairs in a distorted tetrahedral arrangement, is properly described as trigonal pyramidal and *not* as tetrahedral.

(2) Repulsions decrease in the order: lone pair–lone pair > lone pair–bond pair > bond pair–bond pair, with the results that (a) lone pairs tend to be as far apart from each other as possible and (b) bond angles close up compared with those given in Table 4.1 for the regular structure of the same total number of electron pairs.

It is convenient to use the symbol E for a lone pair of electrons. The main types are listed in Table 4.2 and discussed below. In all these cases the ligands (a general term for any atom or group attached to the central atom) may be radicals like alkyl groups, or ions like cyanide,

TABLE 4.2 Shapes of species with lone pairs of electrons

Total number of electron pairs in the valence shell	*Basic shape*	*Number of lone pairs*	*Formula*	*Shape*	*Examples*
3	Triangle	1	AB_2E	V-shape	$SnCl_2$ (gaseous)
4	Tetrahedron	1	AB_3E	Trigonal pyramid	NH_3, PF_3
		2	AB_2E_2	V-shape	H_2O, SCl_2
5	Trigonal bipyramid	1	AB_4E	See Fig. 4.4	$TeCl_4$
		2	AB_3E_2	T-shape	ClF_3
		3	AB_2E_3	Linear	$(ICl_2)^-$
6	Octahedron	1	AB_5E	Square pyramid	IF_5
		2	AB_4E_2	Square plane	$(ICl_4)^-$

In this table the basic shapes are listed as nouns, while in Table 4.1 adjectives are used—compare 'Naming of geometrical shapes', page 5.

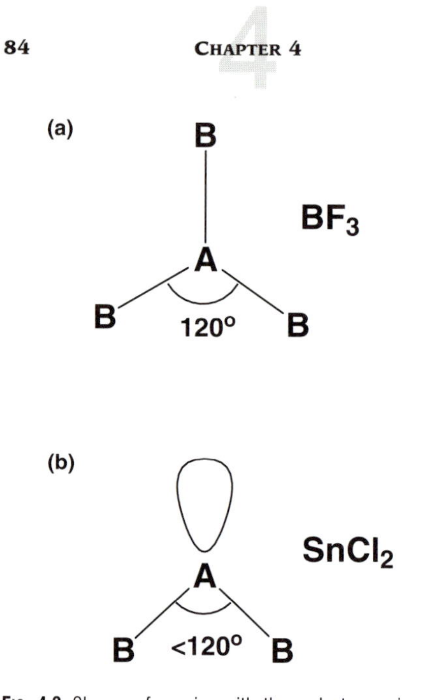

FIG. 4.2 Shapes of species with three electron pairs around the central atom: (a) three bonds, (b) two bonds and one lone pair

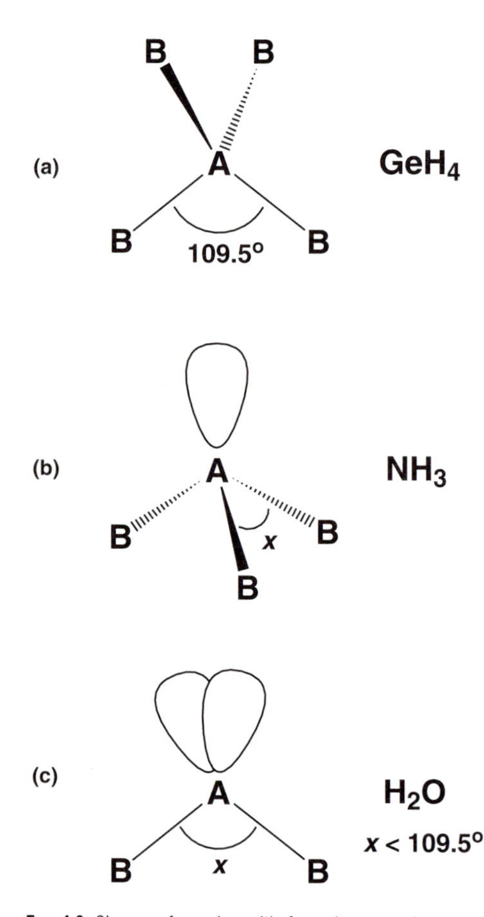

FIG. 4.3 Shapes of species with four electron pairs around the central atom: (a) four bonds, (b) three bonds and one lone pair, (c) two bonds and two lone pairs

or neutral groups donating lone pairs like water or ammonia, as well as atoms as in the table. Thus, $[Zn(H_2O)_4]^{2+}$ is a zinc ion (with no valency electrons of its own) surrounded by four electron pairs donated by the four coordinated water molecules and the shape is therefore tetrahedral.

4.2.3 Three pairs

Figure 4.2 shows the case where there are three electron pairs in the valence shell. When one of these pairs is unshared, the V-shaped molecule of Fig. 4.2b results. (It must be noted that lone pair, or other valency, electron density is rarely found experimentally. The atom centres are normally determined and the electron positions deduced from the resulting structure. In this and the following figures, the lone pairs are indicated schematically to show the relation to the basic structure.) Because of the greater repulsion of the lone pair, the bond angle is reduced from the value of 120°.

4.2.4 Four pairs

Shapes derived from the tetrahedron are shown in Fig. 4.3 and illustrated by methane, ammonia, water and their heavier analogues. CH_4, SiH_4 and GeH_4 are regular tetrahedra with bond angles of $109°28'$ (Fig. 4.3a). In ammonia, one position is occupied by the lone pair of electrons, and the molecular shape is the trigonal pyramid of Fig. 4.3b, with the H-N-H angle reduced to $107°18'$ by the repulsion of the lone pair. Water has two unshared electron pairs on the oxygen atom, so the molecule is V-shaped as shown in Fig. 4.3c and the increased repulsion reduces the H-O-H angle to $104°30'$.

The bond angles in the heavy element analogues of ammonia and water all show considerable reduction from the tetrahedral angle, Table 4.3. The angles of the compounds of the oxygen Group are all smaller than those of the corresponding compounds of the nitrogen Group elements, reflecting the enhanced repulsion effect of the two lone pairs.

The variations in angle shown in Table 4.3 may be rationalized in terms of the electron densities in the bonds of these compounds: the higher the bond electron density at the central atom, the more resistance there is to the bond-closing effect of the lone pairs. Thus, comparing NH_3 and PH_3, the P-H bond is longer than the N-H bond and therefore the electron density is less. Furthermore, as phosphorus is less electronegative than nitrogen, the electron density is concentrated nearer the hydrogen in P-H than in N-H. Both these factors mean that there is less electron density at the phosphorus atom in the P-H bonds than is the case at nitrogen in NH_3, and therefore the $\widehat{HPH}$ angle in PH_3 is much more drastically reduced by the lone pair repulsion. A similar effect occurs if the phosphorus, in turn, is replaced by the rather larger and less electronegative atoms, As or Sb, and the $\widehat{HAsH}$ and $\widehat{HSbH}$ angles show further contraction. The biggest relative change in size and electronegativity comes between nitrogen and phosphorus (compare Tables 2.10 and 2.14) and this is the point where the biggest change in bond angle occurs. Parallel changes are found for H_2O, H_2S, H_2Se and H_2Te.

If NH_3 and NF_3 are now compared, changes in bond length play only a minor role as fluorine is only a little bigger than hydrogen. However, the electronegativities fall in the order $F > N > H$ so that the electron density in the N-F bond is concentrated towards the fluorine—the reverse of the polarization of the N-H bond. Thus the lone pair repulsion closes up the $\widehat{FNF}$ angles more than the $\widehat{HNH}$ ones. Again, a parallel effect is observed in H_2O and OF_2.

The angles in the remaining halogen compounds of these elements are all around 100° and therefore larger than in the hydrides while electronegativity effects would lead us to predict angles smaller than in the hydrides. This may be a result of steric effects with the large halogen atoms coming into contact, or it may be an indication that π bonding is occurring between the halogen and the central atom (see Section 4.3) which returns electron density to the central atom.

4.2.5 Five pairs

Five electron pairs around the central atom give the trigonal bipyramid as the basic shape, Fig. 4.4. For this shape, the two apical (sometimes termed axial) positions (marked B′) are

TABLE 4.3 Bond angles in some AX_3E and AX_2E_2 compounds

		(XAX angles)			
NH_3	107.3°	NF_3	102.5°		
PH_3	93.8°	PF_3	97.8°	PCl_3	100.1°
AsH_3	91.8°	AsF_3	96.2°	$AsCl_3$	98.7°
SbH_3	91.3°	SbF_3	87.3°	$SbCl_3$	99.5°
H_2O	104.5°	OF_2	103.2°	OCl_2	111°
H_2S	92.2°	SF_2	98°	SCl_2	100°
H_2Se	91.0°				
H_2Te	89.5°				

All values measured in molecules on the gas phase. Most values are accurate to ±0.5°.

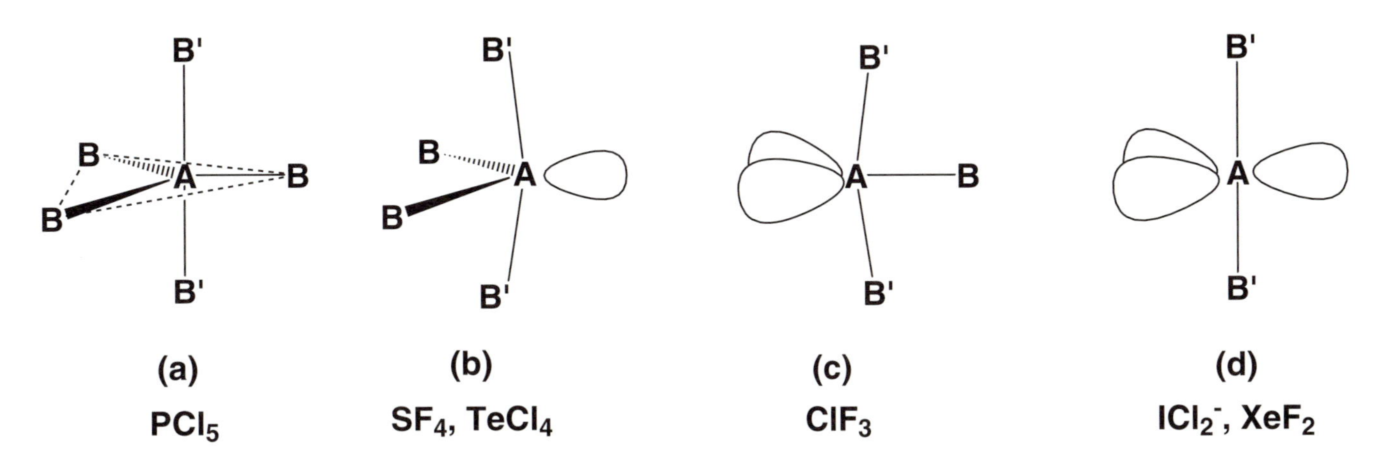

FIG. 4.4 Shapes of species with five electron pairs around the central atom: (a) five bonds, (b) four bonds and one lone pair, (c) three bonds and two lone pairs, (d) two bonds and three lone pairs. The axial and equatorial positions are not equivalent in the trigonal bipyramid. All examples known so far with lone pairs have these lone pairs in equatorial positions

not equivalent to the three equatorial (B) positions. When the ligands are different, for example in PF_3Cl_2, the more electronegative one occupies the axial position.

For trigonal bipyramids with one lone pair, this occupies the equatorial position giving the $AB_2EB'_2$ configuration because this minimizes the repulsions. In this configuration, the lone pair makes two angles of 90° to the A-B′ bond pairs and two angles of 120° to the A-B bonds. If the lone pair was in the axial position, there would be three 90° angles, to the A-B bonds, and one 180° angle to A-B′. As the electron density falls off exponentially, the repulsion effect is much greater at small angles than at large angles. The dominating factor is the additional 90° angle and the equatorial position of the lone pair is the only one found.

For two, or for three, lone pairs in the trigonal bipyramidal configuration, there are again a number of alternative structures possible. However, the only ones found are those where the lone pairs are equatorial. Thus, for two lone pairs, the structure is the T-shaped $ABE_2B'_2$ one of Fig. 4.4c, while three lone pairs give linear $AE_3B'_2$.

In $AB_2EB'_2$ and $ABE_2B'_2$ the repulsions from the lone pairs close up the bond angles, as shown for $TeCl_4$, and ClF_3 in Figs 4.4b and c. The repulsions of the three, symmetrically placed, lone pairs in $AE_3B'_2$ balance out, so these species are strictly linear.

4.2.6 Six pairs

When six electron pairs are present around the central atom, the basic shape is the regular octahedron shown in Fig. 4.5a. In this case, unlike the trigonal bipyramid, all the positions are equivalent and all the BAB angles are 90°.

If one unshared pair is present, the structure becomes the square pyramid of Fig. 4.5b. When there are two lone pairs, the configuration which minimizes the lone pair–lone pair interaction is the square planar configuration in which the lone pairs are *trans* to each other at an angle of 180°, as in Fig. 4.5c. There are no examples of more than two lone pairs in the octahedral case.

In the conventional diagram of the octahedron the equatorial positions are commonly linked up to clarify the figure, and this tends to suggest that the apical and equatorial positions are non-equivalent, so care must be taken until such diagrams become familiar.

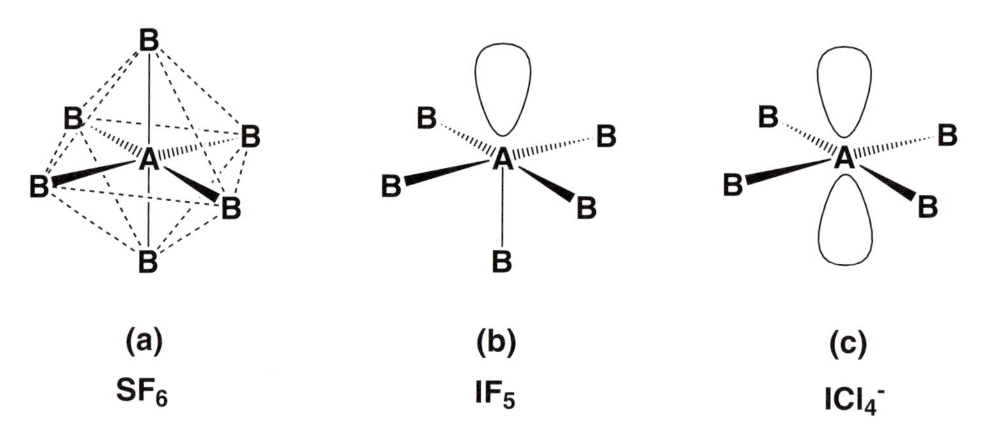

FIG. 4.5 Shapes of species with six electron pairs around the central atom: (a) six bonds, (b) five bonds and one lone pair, (c) four bonds and two lone pairs

4.3 The shapes of species containing π bonds

The extension of the discussion of the preceding section to molecules and ions containing π bonds is basically simple. As the π electrons in a double bond follow the same direction in space as the σ electrons of that bond, the electrons used in π bonding have no major effect on the shape of the molecule.

(a) VSEPR—SUMMARY AND EXAMPLES

With these ideas in mind, it is possible to put the process of determining the shape of any singly-bonded species on a formal basis in terms of simple rules.

(1) Determine which atom is the central one. This is usually the least electronegative one, e.g. Cl in ClO_3^-, O in OF_2. If there are equally plausible choices for the central atom, then the shape should be worked out for each.

(2) Determine the number of electron pairs on the central atom by adding up the number of valency electrons on the central atom plus one electron for each bond (this is the electron contributed by the second element to the bond) plus one electron for each negative charge, or minus one for each positive charge.

(3) The shape then follows from the cases listed in Tables 4.2 and 4.3. Two other rules may be added to allow for special classes of compound:

(4) When a coordinate bond is present, two electrons are added to the total around the central atom since the donor atom or group is providing both the electrons of the bond.

(5) Transition elements follow the rules above but their valency shell d electrons are not always included when determining the number of electrons around the central atom. The shapes of species with partly filled d shells is discussed in Chapter 13; here we need only note that the electron pair analysis applies rigorously to those cases, d^0, d^5 and d^{10}, where the d orbitals are symmetrically occupied and also, in practice, to all configurations, except d^4, d^7 and d^9 (which are distorted) and d^8, which is commonly square planar.

This approach is illustrated in the examples below.

(a) *The 2-coordinated species, SCl_2, ICl_2 and $Ag(CN)_2^-$.*

(i) SCl_2: sulfur has the electron configuration [Ne] $(3s)^2$ $(3p)^4$. The electron count is

valency electrons from S = 6 electrons

add 1 bond electron per Cl = 2 electrons

total = 8 electrons = 4 pairs

SCl_2 has therefore four electron pairs and a V-shaped AB_2E_2 structure:

S
Cl Cl

(ii) ICl_2^-: similarly I = 7 electrons

2Cl = 2 electrons

add for negative charge = 1 electron

ICl_2^- has thus five electron pairs and the structure is the linear $AB_2'E_3$ one: $(Cl\text{-}I\text{-}Cl)^-$.

(iii) $Ag(CN)_2^-$: silver has the electronic configuration $[Kr](4d)^{10}(5s)^1$ of which only the s electron is used. The number of electrons around the silver is thus:

Ag s electron = 1

add bond electron from each CN = 2

add negative charge electron = 1

total = 4 electrons

Thus there are two electron pairs on the Ag and the structure is linear AB_2: $(NC\text{-}Ag\text{-}CN)^-$.

(b) *The rare gas fluorides, XeF_2, XeF_4, and XeF_6.* The xenon atom is regarded as using all eight s^2p^6 electrons. The electron count then proceeds as follows:

XeF_2: Xe = 8 electrons

bond electrons from 2F = 2 electrons

XeF_2 has thus five electron pairs around the xenon atom and the structure is the linear AB'_2E_3 type.

XeF_4: Xe = 8 electrons

4F = 4 electrons

XeF_4 is therefore the square planar AB_4E_2 structure derived from an octahedral arrangement of six electron pairs.

XeF_6: Xe = 8 electrons

6F = 6 electrons

In XeF_6 there are thus seven electron pairs around the xenon. Infrared spectroscopy and electron diffraction on XeF_6 show the molecule to be a distorted octahedron, with the distortion opening up one of the triangular XeF_3 faces in order to accommodate the lone pair of electrons on Xe. A number of other seven-electron pair structures are also known, such as IF_7, TeF_7^- and IOF_6^-, and all of these are pentagonal bipyramids.

(c) *PCl_5 and its dissociation ions, PCl_4^+ and PCl_6^-.* Phosphorus pentachloride exists as monomeric PCl_5 units in the gas phase but forms PCl_4^+ PCl_6^- in the solid by transfer of a chloride ion.

PCl_5: Valency electrons P = 5 electrons

bond electrons from 5Cl = 5 electrons

PCl_5 thus forms a trigonal bipyramid in the gas phase.

PCl_4^+: P = 5 electrons

4Cl = 4 electrons

Charge = −1 electron for the plus charge

Thus there are four electron pairs around the phosphorus in PCl_4^+ and the shape is tetrahedral.

PCl_6^-: P = 5 electrons

6Cl = 6 electrons

charge = 1 electron

There are thus six electron pairs in PCl_6^- and the shape is a regular octahedron.

The ionization in the solid state probably derives from the impossibility of packing the 5-coordinate structure regularly.

(d) *As an example of coordinate bonds, consider the adduct $BF_3.NH_3$.* Boron has three valency electrons, therefore BF_3 has three electron pairs around the boron and is a planar molecule with FBF angles of 120°. Ammonia is a trigonal pyramid with a lone pair on the nitrogen as already seen. As the boron has four valency shell orbitals, one remains vacant in BF_3 and this can accept the electron pair of the ammonia to form the adduct $F_3B \leftarrow NH_3$.

The electron count at the boron atom in the adduct is

valency electrons B = 3 electrons

one bond electron from 3F = 3 electrons

two bond electrons from N = 2 electrons (as the electron pair is donated)

The boron has therefore four electron pairs around it and the configuration (of the three F atoms and the N) is tetrahedral: the FBF angle is decreased towards the tetrahedral value of $109\frac{1}{2}°$.

The electron count at the nitrogen atom remains at four electron pairs as the two donated electrons remain in the nitrogen valence shell. The nitrogen is thus surrounded tetrahedrally by the three H atoms and the B atom.

(e) *Transition metal compounds. $TiCl_4$.* Titanium has four valency electrons, d^2s^2. The electron count is

valency electrons Ti = 4 electrons

bond electrons from 4Cl = 4 electrons

Thus $TiCl_4$ is tetrahedral.

For the majority of transition metal compounds, this approach must be modified as the d electrons contribute less directly to the shape. To construct a general approach, ML_4 or ML_6 complexes of transition elements, M, and any ligand, L (such as halide, cyanide, water, ammonia) are regarded as formed from an ion M^{n+} and four, or six, ligands donating a pair of electrons each. Any d electrons on M^{n+} are not used and the shapes are ML_4 = tetrahedral and ML_6 = octahedral. (Notice that $TiCl_4$ can be treated in this way as Ti^{4+} and $4Cl^-$: this gives the correct shape and electron count, although it is misleading in implying ionic bonding.)

Compounds with other coordination numbers are treated similarly, leading to shape predictions following Table 4.1. This approach will be correct in about three-quarters of all transition metal compounds. However, the underlying d^n configuration does affect the shape, especially when n = 4, 7, 8 or 9, and it is better to tackle transition metal compounds by the different approach of Chapter 13.

(b) VSEPR—ALTERNATIVE FORMULATIONS

You might prefer to formulate a molecule or complex ion in a different way from those illustrated. For example, $Ag(CN)_2^-$ could equally well be regarded as Ag^+ and $2CN^-$ with the cyanide ions acting as lone pair donors to the silver cation. Any valid alternative will not affect the VSEPR count. In this formulation the count is

Ag^+[the d shell is not used] = 0 electrons

$2CN^-$ [each a lone pair donor] = 4 electrons

total = 4 electrons, as before

Alternative formulations are possible for many molecules, and each can be perfectly acceptable. For example, many ligands, such as the halogens, can be regarded as covalently bound and donating one electron to a neutral central atom, or as anions donating two electrons but the central atom then becomes a cation to keep the overall charge the same. Compare also $TiCl_4$ in example (e).

While we tend to choose the formulation that seems most realistic (for example, avoiding high charges such as P^{5+} for PCl_5) we should be clear that all that the VSEPR calculation does is to produce the number of electron pairs, and hence predict the shape. It tells us nothing about the bonding—and conversely, no conclusion about bonding, such as ionic or covalent in these examples, can be drawn from a successful prediction of shape. Notice, however, that if the observed shape differs from the VSEPR prediction, this is a clear indication of some additional bonding or steric effect which should be investigated—for a classical example see trisilylamine, Figures 17.21 and 17.23.

There is another way in which alternative structures may occur. This is when there is more than one acceptable way of linking up the atoms (compare step 1 of box on p. 84).

For example, POF_3 might be formulated with P=O, like $POCl_3$ discussed in 4.3.1. However, a perfectly acceptable alternative is that containing a P-O-F unit, and such oxyfluoride substituents are known in other molecules. The OF group is a univalent ligand, just like OH. Once we have thought of such alternatives, we can predict the probable shape by carrying out the VSEPR count for each. (Such predictions would guide experimental studies to decide the actual formulation.)
For F_2POF, there are two central atoms, P and O, and the electron count is carried out for each.

At P

P = 5 electrons

2F = 2 electrons

OF = 1 electron

total = 8 electrons = 4 pairs

so the shape at P is an AB_3E structure with a lone pair. Thus the 2 PF bonds and the P-OF bond form a trigonal pyramid.

At O

O = 6 electrons

F = 1 electron

F_2P = 1 electron

total = 8 electrons = 4 pairs

so the shape at O is an AB_2E_2 structure with two lone pairs, and the P-O-F link is V-shaped.

Since phosphorus is commonly either 5- or 3-valent, F_3P=O and $F_2P(OF)$ are both acceptable formulations. The VSEPR count predicts structures for each formulation. Once again, VSEPR allows no conclusion to be drawn about which structure is adopted—and indeed, in several cases analogous isomers are both known.
A special case of alternative chemical formulation is that of dimers or polymers indicated in Section 4.3.2.

4.3.1 Electron counting procedure for π electrons

The molecular shape is determined mainly by the σ bonds and lone pairs and the presence of pi bonding causes only minor changes due to the additional electron density. Thus, to determine the shape of a molecule or ion which contains one or more π bonds, the electrons which the central atom uses to form the π bonds must be subtracted from the total of electrons. Then the shape of the species is determined by the number of lone pairs and sigma bond pairs around the central atom. This is best understood in terms of an example. In phosphorus oxychloride, $POCl_3$, the central atom is phosphorus which forms a double bond to the oxygen, $C1_3P$=O. The two electrons of this π bond come one from the phosphorus and one from the oxygen. Thus, when counting the number of electrons around the phosphorus atom, this electron used in the π bond must be subtracted as it has no steric effect. The calculation proceeds

valency electrons at P = 5 electrons

sigma bond electrons from O + 3Cl = 4 electrons

less electron used by P in π bond = −1 electron

Therefore there are four electron pairs around the phosphorus atom which are shape determining. $POCl_3$ is therefore tetrahedral. (The effect on the bond angles of the π bond is discussed later.)

Next consider the carbonate ion, CO_3^{2-}. The most reasonable formulation for this ion in the normal valency form is $O{=}C(O^-)O^-$ Then the electron count at the carbon atom is

> Note that in charged species involving terminal oxygen atoms it is usually more realistic to place the negative charge on the more electronegative ligand atom and not on the central atom.

valency electrons at C = 4 electrons
σ bond electrons from 3O = 3 electrons
less electron used in 1 π bond = −1 electron

Thus there are three electron pairs to determine the shape and the carbonate ion is planar.

As a final example, take the sulfite ion, SO_3^{2-}. This would be formulated as $O{=}S(O^-)O^-$ and therefore:

valency electrons at S = 6 electrons
σ bond electrons from 3O = 3 electrons
less electron used in 1 π bond = −1 electron

Hence four electron pairs and the sulfite ion has an unshared pair on the S atom and is a trigonal pyramid.

> A list of cases involving π bonds is given in Table 4.4.

The main difficulty in dealing with π-bonded species lies in deciding how many electrons the central atom is using for π bonds. It will usually be found satisfactory to write down a formula with single and double bonds which satisfies normal ideas of valency, as in the examples above, and then count up the number of double bonds from the central atom.

In a few cases, notably ozone, O_3, and the nitrate ion, NO_3^-, it will be found that it isnecessary to write some of the bonds as coordinate bonds to keep normal valencies or to avoid violating the octet rule. These two are best written as $O{=}O{\rightarrow}O$ and $O{=}N(O^-){\rightarrow}O$.

All the bonds are, despite this formalism, equivalent. The electron count then proceeds:

O_3	NO_3^-
O = 6 electrons (the central O)	N = 5 electrons
1O = 1 electron (the covalently bound O)	2O = 2 electrons (the two covalently bound O atoms)
1 π bond = −1 electron	1 π bond = −1 electron
total = 3 pairs	total = 3 pairs

Thus O_3 is a V-shaped molecule and NO_3^- is planar. Notice that, as the coordinate bonds originate from the central atom, the oxygen bound by the coordinate bond contributes no electrons to the count of those around the central atom.

Although a formula with localized double bonds is used to assist these calculations, it must not be thought that this type of formula gives a true description of the electron distribution in the species. If the ligand atoms are all equivalent, double bonds and charges are delocalized over the whole molecule or ion. For example, the three configurations:

TABLE 4.4 Structures of species with π bonds

Number of valency electrons on the central atom	*Number of σ bonds*	*Number of π bonds*	*Shape*	*Examples*
Case 1: no lone pairs (Shape follows from column 2)				
4	2	2	Linear	CO_2, HCN
	3	1	Triangular	CO_3^{2-}
5	3	2	Triangular	NO_3^-
	4	1	Tetrahedral	POX_3, PO_4^{3-}, VO_4^{3-}
6	3	3	Triangular	SO_3
	4	2	Tetrahedral	CrO_4^{2-}, SO_4^{2-}
7	4	3	Tetrahedral	IO_4^-, MnO_4^-
	6	1	Octahedral	$IO(OH)_5$
Case 2: one lone pair (Shape is that due to the lone pair plus the number of bonds in column 2)				
5	2	1	V-shaped	NOCl, NO_2^-
6	3	1	Trigonal pyramidal	$SOCl_2$, SO_3^{2-}
	2	2	V-shaped	SO_2
7	3	2	Trigonal pyramidal	IO_3^-, XeO_3
	4	1	Distorted (see Fig. 4.6)	$IO_2F_2^-$
Case 3: two lone pairs (Shape is that due to the two lone pairs plus the bonds)				
7	2	1	V-shaped	ClO_2^-

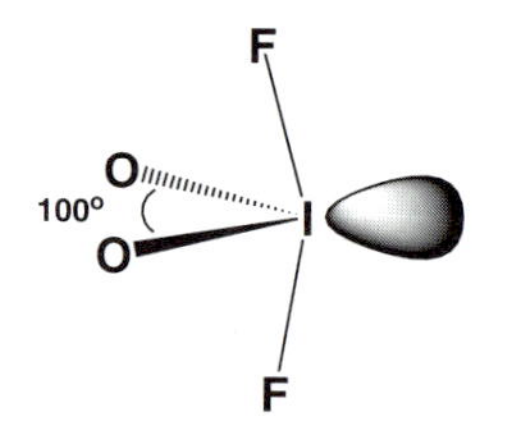

FIG. 4.6 The structure of $IO_2F_2^-$

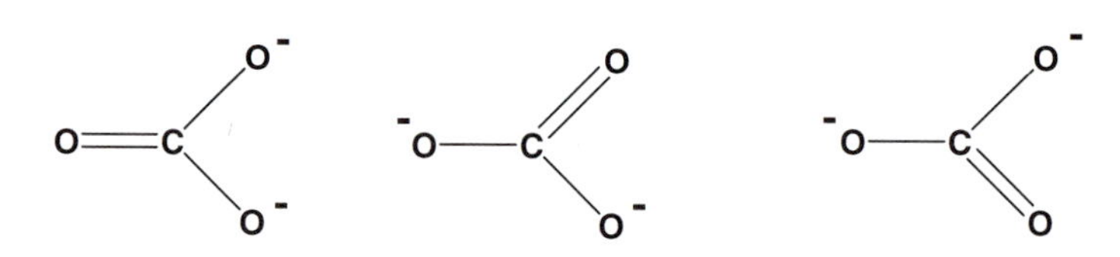

are all equally likely for the carbonate ion, and it is found experimentally that all the C-O distances are identical and that there is a charge of $-\frac{2}{3}e$ on each oxygen.

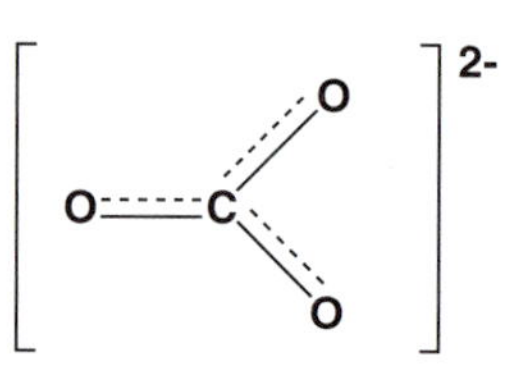

This delocalization is discussed in the next section.

Where π bonding is delocalized, as in this case, its only effect on the stereochemistry is to shorten all the bonds compared with the single bond lengths. If, however, the π bond is localized, as in $Cl_3P{=}O$ or $Cl_2S{=}O$ for example, then it exerts a steric effect. As the double bond region contains two electron pairs and the bond is also shorter than a single bond, the electron density is high and so the repulsion between two double bonds, or between a double bond and a single bond is greater than that between two single bonds. The π bond thus has a steric effect similar to that of a lone pair and the bond angles between single-bonded ligands will be closed up because of the greater repulsion. For example, in $POCl_3$ the Cl-P-Cl angle is reduced from the tetrahedral value to 103.5°. The similarity of the steric effects of a lone pair and of a double bond can be seen in the molecule $XeOF_4$ where all the bond angles are very close to 90°.

4.3.2 Summary

The application of the VSEPR theory, as discussed in the last two sections, may be summarized:

(i) Empirical ideas of the repulsion of the electron-dense regions represented by chemical bonds leads to the principle that bonds around a central atom will be arranged in the most symmetrical manner possible for the coordination number. These are the configurations given in Table 4.1.

(ii) Lone or unshared pairs of electrons on the central atom must be taken into account when deciding the shapes. These lone pairs fill coordination positions in the basic configurations to give the shapes detailed in Table 4.2 and in the commentary thereon.

(iii) The direction of distortion from the basic shapes in unsymmetrically substituted species may be predicted by considering the electron density and polarization within the bonds. When lone pairs are present, they have a dominating effect and the order of repulsions is lone pair–lone pair > lone pair–bond pair > bond pair–bond pair.

(iv) The same analysis applies to species with π bonds. The electrons contributed by the central atom to the π bonds are subtracted and the configuration depends on the number of bonds plus lone pairs left on the central atom as in the cases in Table 4.4. Delocalized π bonds have little steric effect but localized ones are regions of high electron density and cause distortion.

Although it is at first necessary to work out each structure systematically using the rules above, it will be found that the process is readily short-circuited with practice. For example, a major question is always whether there are lone pairs present, and these are nearly always found in compounds where the central element is showing an oxidation state less than its Group maximum one. Similarly, structures of isoelectronic species are usually the same, e.g., CO_3^{2-} and NO_3^-, or SO_2 and O_3, and this often helps to link up with known compounds.

Finally, one warning: the methods above apply on the assumption that the molecular formula is known and does not allow for polymerization. Thus, beryllium dichloride exists as $BeCl_2$ only in the gas phase, in the solid it is the chloride-bridged polymer

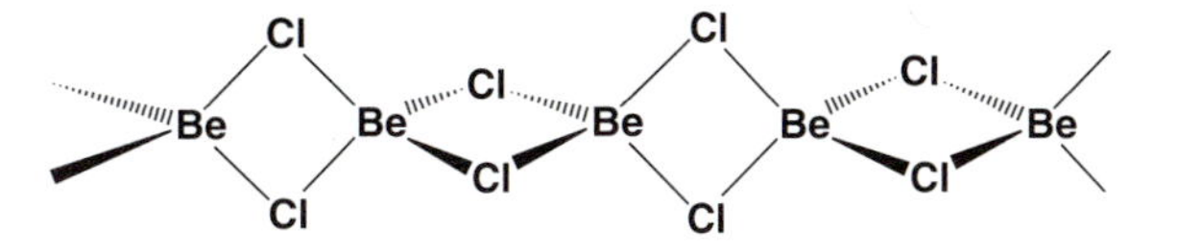

Of course, given this structure, the geometry about the Be atoms can be predicted by VSEPR. There are now four electron pairs at each Be, made up from two donor pairs and the two pairs of each basic $BeCl_2$ unit, giving tetrahedral coordination.

Similarly, aluminium tribromide exists as the dimer Al_2Br_6, and SiO_2 is a three-dimensional polymer instead of a discrete molecule like CO_2. VSEPR gives no information about such alternative molecular forms. However, once the molecular formula is known, the methods given above lead to the structures of the large majority of simple molecules and ions, and also give the probable geometries in polymeric materials.

4.4 General approaches to bonding in polyatomic species

Once we know the shape of a polyatomic species, whether from experiment or by VSEPR theory, or other prediction, we wish to discuss its bonding. The molecular orbital approach may be extended from diatomic molecules to cover molecules with more than two atoms in two different ways:

(1) Just as we took two nuclei and formed orbitals centred on both in the diatomic molecule, we can take the n nuclei of an n-atomic molecule and form n-centred orbitals in a polyatomic molecule.

Specifically we would take the n nuclei with their inner non-bonding electron shells and place them in their positions in the molecule. We then combine all the valency shell atomic orbitals on all the atoms to form polycentred molecule orbitals embracing all the nuclei. As for diatomic molecules, the total number of such *delocalized* molecular orbitals equals the total number of atomic orbitals used in their construction. Each may hold up to two electrons. The valency electrons would then be fed into these n-centred molecular orbitals, in order of increasing energy, to form the molecule in a process which is exactly analogous to the building-up of atomic and diatomic molecular structures which has just been discussed. Thus we would place the atoms of water in their V-shape, and seek to form three-centre molecular orbitals for H_2O from the hydrogen 1s and the oxygen 2s, $2p_x$, $2p_y$ and $2p_z$ atomic orbitals.

(2) Alternatively, we can treat the polyatomic molecule as built up of a set of two-centred units. For each pair of bonded atoms we form two-centre bonds just as for diatomic molecules. The whole molecule is thus built up of two-centred localized molecular orbitals. That is, we describe H_2O in terms of two O-H bonds.

This latter approach corresponds to the long-familiar ideas of electron pair bonds and it is the one which is mainly used to describe sigma bonding in polyatomic molecules. The picture of delocalized many-centred bonds is less familiar. It is particularly useful when discussing π-bonded species where the delocalized picture is the most satisfying, and also best for some σ-bonded cases, such as the 'electron-deficient' molecules typified by the boron hydrides (Section 9.6). In fuller treatments, the two approaches converge and they are often used interchangeably.

We shall start by discussing σ-bonded species in terms of localized two-centred orbitals, as this conforms to the traditional chemist's view of bonding. We then illustrate the alternative multi-centred approach to σ-bonding in a few simple molecules. Next we discuss experimental evidence from photoelectron spectroscopy which allows us to choose which picture is most satisfactory. In the final section we describe π-bonded species.

4.5 Bonding in polyatomics: the two-centre bond approach

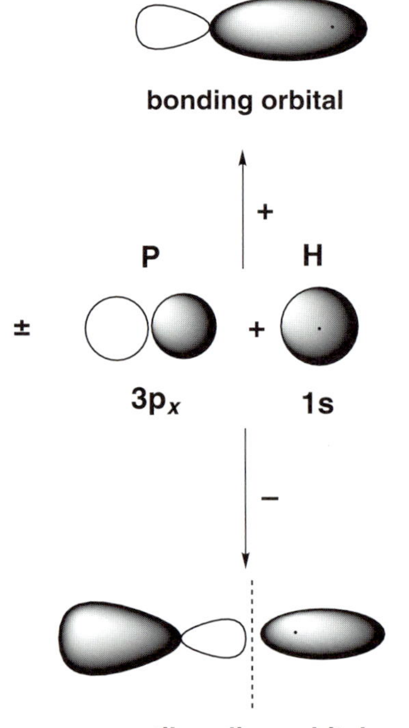

FIG. 4.7 The combination of the phosphorus $3p_x$ orbital with the 1s orbital of the hydrogen on the x axis to give the molecular orbitals of a P-H bond in PH_3. The H atom in the x direction combines with a suitable phosphorus orbital to give bonding and antibonding two-centred molecular orbitals

In tackling the shapes of σ-bonded species, as described in Tables 4.1 to 4.3, we first note that there are already some steric properties in the atomic orbitals. For example, the three p orbitals are at 90° to each other. It is thus possible to describe bond angles of 90° in terms of overlap with p orbitals of the central atom and suitable orbitals on the ligands. Thus phosphane, PH_3, where the H-P-H angle is just above 90°, could be described as having each P-H bond formed by overlap of a phosphorus 3p orbital with the ls orbital of the hydrogen to give a situation very similar to that in a diatomic molecule. The phosphorus atom has the valency shell configuration $(3s)^2(3p_x)^1(3p_y)^1(3p_z)^1$. In the x direction, for example, the phosphorus $3p_x$ orbital (which will be written ψ_{p_x}) is of the correct symmetry to form a σ overlap with the 1s orbital on that hydrogen atom lying in the x direction (written ψ_s). The resulting two-centred molecular orbitals are of the form given in Equation (3.2):

$$\sigma_{s,p_x} = \psi_s + c\psi_{p_x}$$
$$\sigma^*_{s,p_x} = c\psi_s - \psi_{p_x}$$

where the value of c which gives the most favourable energy of combination would have to be determined in the course of the calculation (compare Section 3.3). There are two electrons available; the hydrogen one and the one from the phosphorus p_x orbital, and these fill σ_{s,p_x}, leaving the antibonding orbital empty, and giving a single P-H bond. This step is represented in Fig. 4.7 and may be written as a formation equation:

$$P(3p_x)^1 + H(1s)^1 \rightarrow P\text{-}H(\sigma_{s,p_x})^2(\sigma^*_{s,p_x})^0$$

Thus the whole process of forming this P-H bond is just the same as forming the bond in a heteronuclear diatomic molecule.

The P-H bonds in the y and z directions are formed in the same way giving the phosphane molecule (Fig. 4.8):

$$P(3s)^2(3p_x)^1(3p_y)^1(3p_z)^1 + 3H(1s)^1$$
$$\rightarrow PH_3(3s)^2(\sigma_{s,p_x})^2(\sigma_{s,p_y})^2(\sigma_{s,p_z})^2(\sigma^*_{s,p_x})^0(\sigma^*_{s,p_y})^0(\sigma^*_{s,p_z})^0$$

The molecule may be built up in this way by considering each pair of bonded atoms in turn as if they were in a diatomic molecule and using the methods of Section 3.5. The lone pair of electrons is accommodated in the s orbital (and they will tend to be concentrated in that spatial segment of the s orbital remote from the bond directions) and the H-P-H bond angle follows as a consequence of the 90° angle between the p orbitals.

A similar discussion applies to SbH_3, AsH_3 and to the hydrides of S, Se, and Te. In H_2S (Fig. 4.9) the bonds are formed using two of the sulfur 3p orbitals, giving an H-S-H angle of 90°, and the third p orbital and the 3s orbital hold the two lone pairs. This analysis does not explain the small departures from the 90° angle in these molecules and cannot account for the much larger bond angles in water and ammonia.

4.6 Two-centred orbitals: hybridization

The use of pure atomic s and p orbitals fails to account for the shape of most of the molecules mentioned in the previous sections. Linear, triangular or tetrahedral shapes, for example, cannot be explained using simple atomic orbitals and the concept of mixed or *hybrid* orbitals has to be introduced.

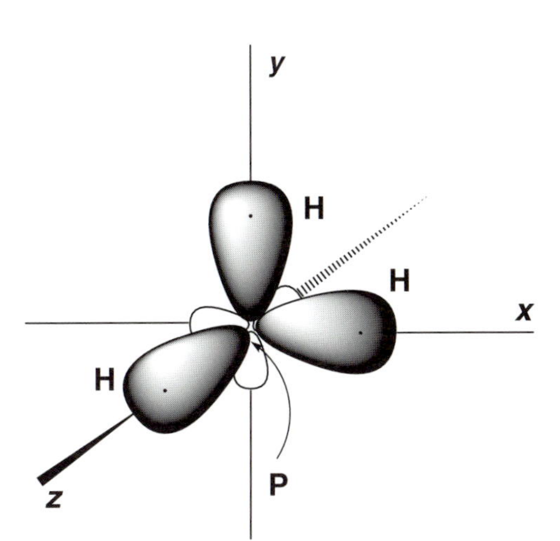

FIG. 4.8 The bonding orbitals in a phosphane molecule. The overlap of the three phosphorus p orbitals with hydrogen s orbitals given three P-H bonds at 90° to each other, reflecting the angle between the p orbitals

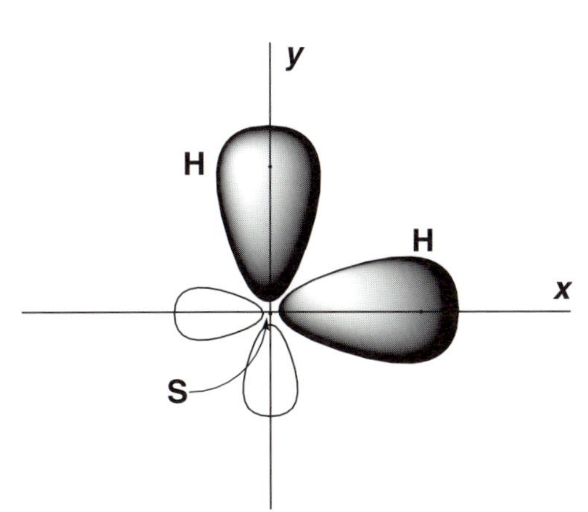

FIG. 4.9 The bonding orbitals in hydrogen sulfide (sulfane). The 2 H and S are linked by two bonds formed by s + p overlap (compare Fig. 4.8)

4.6.1 Equivalent hybrids

Consider first the monomer of beryllium dichloride (which is found in the gas phase). The electronic configuration of the beryllium atom is Be = [He] $2s^2$ $(2p)^0$. The two electrons in the valency shell are paired in the s orbital, and the p orbitals are empty. In order to obtain divalency, the s orbital and a p orbital must be used and, to have two bonds at 180°, this s + p configuration must be rearranged to form two equivalent orbitals. This is done by combining the two into two sp hybrids, as shown in Fig. 4.10 If the s orbital is superimposed on the p one as shown, the positive lobe of the p wave function is reinforced and the negative lobe is diminished. This description reflects diagrammatically the mathematical process:

$$sp = \frac{1}{\sqrt{(2)}}(s + p)$$

which gives the orbital shown in Fig. 4.10a. The second hybrid, which is directed in the reverse sense, is $(1/\sqrt{2})(s - p)$ and is given in Fig. 4.10b. The factors $1/\sqrt{2}$ arise as the total electron density in the two sp hybrids (found by squaring the two wave functions above and adding) must equal s^2+p^2, the electron density in the two constituent atomic orbitals. The total process is: s+p = 2 sp hybrids, two atomic orbitals giving two hybrid orbitals on the central atom. (Note that this process of mixing orbitals on the *same* atom, to give hybrid atomic orbitals, must be distinguished from the process of combining an s and a p orbital on different atoms to form σ molecular orbitals, which was shown in Fig. 3.11). The two valency electrons are then placed one into each sp hybrid and each of these may overlap with a ligand orbital to form molecular orbitals.

The two sp hybrid orbitals are equivalent to each other and directed at 180°. They are used to form the bonds in a linear molecule. For example, the description of $BeCl_2$, given below, resembles the description of PH_3 apart from the use of hybrid orbitals.

$BeCl_2$. The chlorine atom has seven valency electrons and four valence shell orbitals. Six of the electrons are paired up in three of these orbitals, leaving the fourth singly occupied. In this molecule, the unique direction is the Cl-Be-Cl axis and, by convention, this is taken as the *z* direction. The singly occupied chlorine orbital must be of σ symmetry with respect to the Be-Cl bond, and could be the chlorine s or p_z atomic orbitals or a hybrid such as sp pointing in the *z* direction. Molecular orbitals are formed by this chlorine σ orbital and the sp hybrid on the beryllium to give bonding and antibonding molecular orbitals, and the same

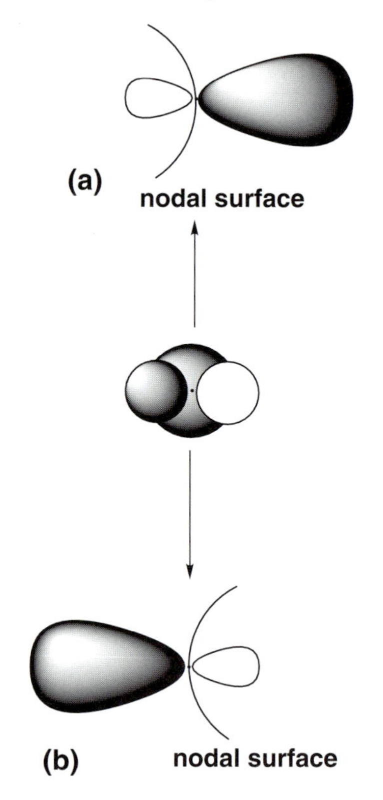

FIG. 4.10 The formation of sp hybrid orbitals: (a) the orbital $(1/\sqrt{2})$ (s+p), (b) the orbital $(1/\sqrt{2})$ (s−p)

happens for the second chlorine atom and the other sp beryllium hybrid. The complete process is first.

$$\mathrm{Be(s)^2(p)^0 \rightarrow Be(sp_A)^1(sp_B)^1} \text{(designating the two chlorine atoms A and B)}$$

that is, the beryllium valency electrons are placed one into each hybrid orbital.

Then $\mathrm{Be(sp_A)^1 + Cl_A(\sigma)^1 \rightarrow Be\text{-}Cl_A(\sigma_{Be\text{-}Cl})^2(\sigma^*_{Be\text{-}Cl})^0}$

where $\sigma_{\mathrm{Be\text{-}Cl}}$ is of the form $\psi_{\mathrm{sp}}+\psi_{\mathrm{p}}$ and, for the antibonding orbital, $\sigma^*_{\mathrm{Be\text{-}Cl}}$ is $\psi_{\mathrm{sp}}-\psi_{\mathrm{p}}$. Similarly for the bond to the second chlorine

$$\mathrm{Be(sp_B)^1 + Cl_B(\sigma)^1 \rightarrow Be\text{-}Cl_B(\sigma)^2(\sigma^*)^0}$$

These are shown in Fig. 4.11

Each beryllium–chlorine bond is a normal single bond, just as in the diatomic molecules or in phosphane, except that it is formed using a hybrid atomic orbital on the beryllium instead of a simple one. The Cl-Be-Cl angle of 180° follows from the angle between two sp hybrids.

As well as providing orbitals oriented at the correct angle, sp hybrids overlap more effectively than simple s or p orbitals, and therefore give stronger bonds. Pauling has proposed the *S* parameter, listed in Table 4.5, as a measure of this relative overlap for various orbitals.

A similar mixing process can be carried out with two p orbitals and the s orbital to give three coplanar equivalent sp^2 hybrid orbitals directed at 120°. These are shown in Fig. 4.12. In the case of a planar molecule the unique direction is perpendicular to the plane and this is taken as *z* by convention: hence the plane is *xy*. If one of the three in-plane orbitals is directed along the x-axis, this has contributions from the s and p_x orbitals. The other two hybrids are composed of the s, p_x and p_y orbitals. The expressions for the three hybrids are, respectively:

$$\sqrt{(\tfrac{1}{3})}\mathrm{s} + \sqrt{(\tfrac{2}{3})}\mathrm{p}_x$$
$$\sqrt{(\tfrac{1}{3})}\mathrm{s} - \sqrt{(\tfrac{1}{6})}\mathrm{p}_x + \sqrt{(\tfrac{1}{2})}\mathrm{p}_y$$
$$\sqrt{(\tfrac{1}{3})}\mathrm{s} - \sqrt{(\tfrac{1}{6})}\mathrm{p}_x - \sqrt{(\tfrac{1}{2})}\mathrm{p}_y$$

The coefficients are again chosen so that the total electron density in the three hybrid orbitals adds up to $s^2+p_x^2+p_y^2$, the electron density in the three constituent orbitals.

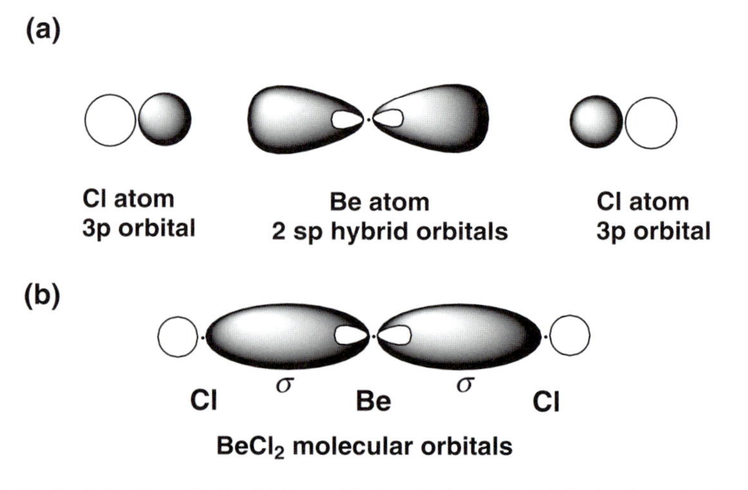

FIG. 4.11 Bonding in beryllium chloride: (a) the contributing atomic orbitals, (b) the bonding molecular orbitals

TABLE 4.5 Values of Pauling's *S* parameter for various atomic orbitals

Orbital	s	p	sp	sp^2	sp^3
Relative strength	1.0	1.73	1.93	1.99	2.00

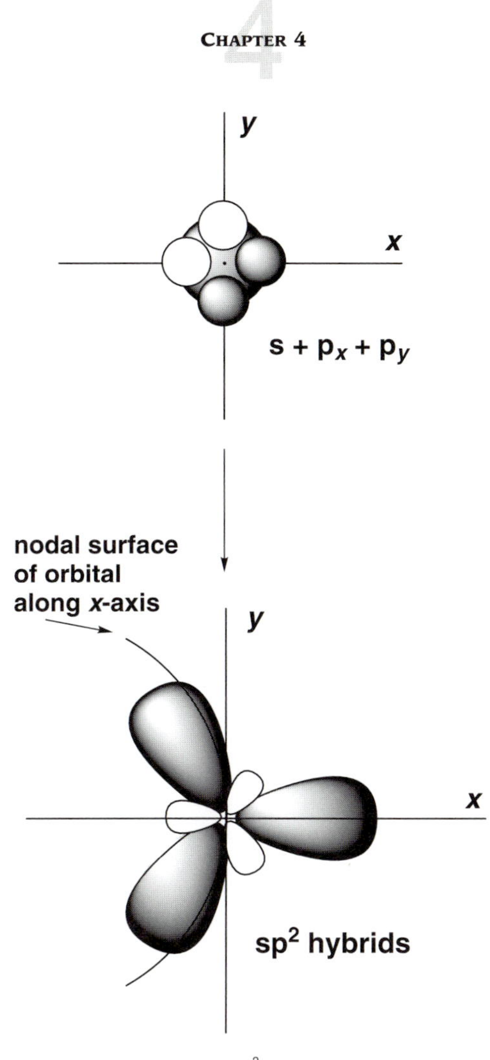

FIG. 4.12 The formation of sp^2 hybrid orbitals. One of the three nodals surfaces is shown. The small lobes are the negative part of these three orbitals

The sp^2 hybrids are then used to form the bonds in a plane triangular molecule, such as BCl_3. The process may be described as:

(i) Formation of the sp^2 hybrids and placing one valency electron in each

$$B(2s^2)(2p)^1 \rightarrow B(sp^2)^1(sp^2)^1(sp^2)^1$$

(ii) Each sp^2 hybrid is combined with a chlorine orbital of σ symmetry to give one bonding and one antibonding two-centre molecular orbital. These are of the form $sp^2_B + c\sigma_{Cl}$ (bonding) and $sp^2_B - c\sigma_{Cl}$ (antibonding). The electrons go into the bonding orbitals:

$$B3(sp^2)^3 + 3Cl(\sigma)^1 \rightarrow BCl_3 3(\sigma_{B\text{-}Cl})^6 3(\sigma^*_{B\text{-}Cl})^0$$

The whole process is that three atomic orbitals on the boron, combined into three atomic hybrid orbitals, combine with three atomic orbitals on the three chlorines to give six molecular orbitals, three bonding and three antibonding.

The six valency electrons fill the bonding orbitals, giving the boron trichloride molecule as shown in Fig. 4.13.

In tetrahedral configurations, all three p orbitals and the s orbital combine to form four sp^3 hybrids, which are shown in Fig. 4.14. When 5- or 6-coordinated structures have to be formed, it becomes necessary to use d orbitals as well as s and p orbitals. 5-coordination arises from sp^3d hybridization and the appropriate d orbital is the d_{z^2} orbital. (This can be seen qualitatively; the three equatorial positions in the trigonal bipyramid correspond to three sp^2 hybrids and the two orbitals in the $\pm z$ direction arise by mixing the atomic orbitals lying on the z axis, that is p_z and d_{z^2}. In a similar manner, octahedral hybridization involves six sp^3d^2 orbitals and, here, the appropriate d orbitals are d_{z^2} and $d_{x^2-y^2}$. d orbitals may also contribute to other shapes. For example, sd^3 (using d_{xy}, d_{yz} and d_{zx}) is also tetrahedral. The above treatment of higher coordination numbers is useful as an initial approach. In more advanced treatments, the extent to which d orbitals are involved in 5- and 6-coordinated compounds is subject to discussion (compare Section 18.9).

For linear, triangular, tetrahedral and octahedral shapes, all the hybrid orbitals in a set are equivalent, have equal angles between them and extend equally in space. For example, each sp^3 hybrid is equivalent to the other three and has one-quarter s character and three-quarters p character.

When hybrids are formed, energy is required to promote the electrons into the configuration required for the bonds, as in the s^2p^0 to sp^1sp^1 configuration discussed above for $BeCl_2$. If the energy of such a step can be calculated, and the energies of the remaining steps are known, it is possible to calculate the energy of formation of a particular molecule and perhaps predict the preferred product(s) of a reaction. Such an analysis is often not possible for lack of data, but the case of the reaction of carbon and hydrogen to form the CH_4 molecule has been worked out and it may be compared with the possible alternative reaction to form CH_2 (Fig. 4.15). It will be noted that the difference in heats of formation favours CH_4 by a factor of 4/3 compared with CH_2 but the difference is only about one per cent of the total energies involved, which again highlights the difficulties of predicting chemical behaviour by direct calculation.

The carbon atom reacts in such a way that all its valency electrons and all its valency orbitals are involved in molecule formation, and this is generally the favoured mode of reaction, at least for lighter atoms. Where there are empty valence orbitals as in $BeCl_2$ or BCl_3, further reaction to use these orbitals is likely and such molecules act as electron pair acceptors: similarly, where there are one or more lone pairs, as in ammonia or water, these tend to form donor bonds to suitable acceptors. Many examples of such behaviour will be found in the later chapters.

4.6.2 Nonequivalent hybrids

In the above cases, a given number of simple atomic orbitals were combined to give the same number of hybrid orbitals which were all equivalent to each other. These modes of

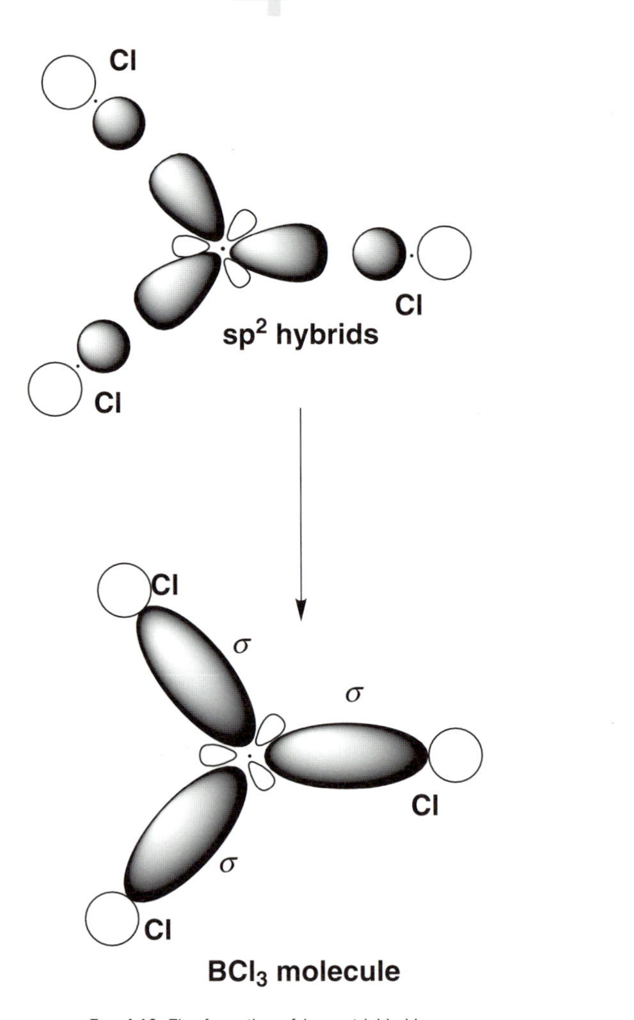

FIG. 4.13 The formation of boron trichloride

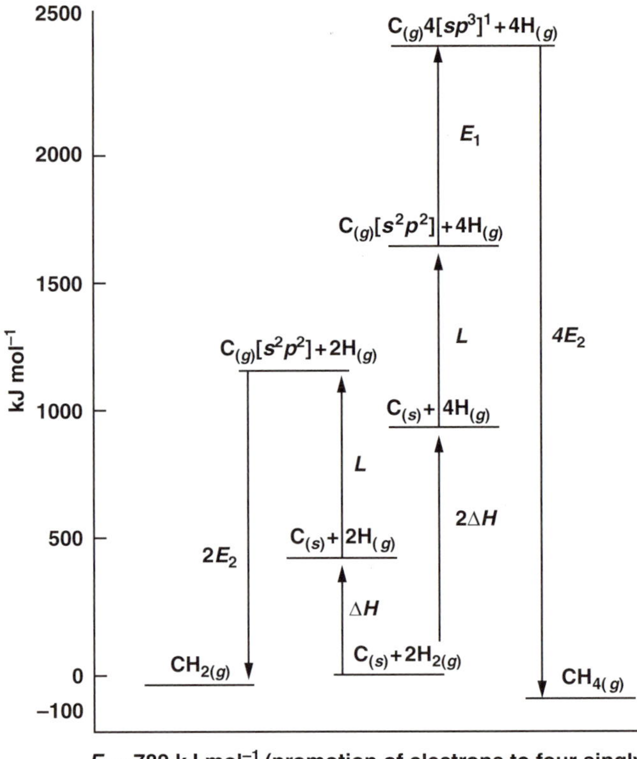

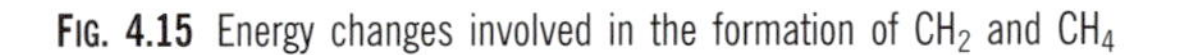

FIG. 4.15 Energy changes involved in the formation of CH_2 and CH_4

FIG. 4.14 The four sp^3 hybrid orbitals. The four small negative lobes (compare Fig. 4.12) have been omitted for clarity. The sp^3 hybrids are tetrahedrally directed

hybridization correspond to the basic shapes of Table 4.1. It is quite possible, however, to form nonequivalent hybrids. For example, an unsymmetrical beryllium compound, BeXY, would almost certainly have a more favourable bonding structure if nonequivalent 's+p' hybrids were used with, say, the Be-X bond formed by a hybrid with rather more s character and the Be-Y bond from one with more p character.

In the same way, trigonal or tetrahedral hybrids may be nonequivalent. For example, in chloromethane, CH_3Cl, the C-Cl bond will be formed by an '$s+p^3$' hybrid which has a different amount of s character from the other three '$s+p^3$' hybrids which form the C-H bonds. The optimum amount of s character in such hybrids will vary from molecule to molecule, although the shape and basic hybrid structure is still tetrahedral.

Such nonequivalent hybrids become even more necessary when molecules containing lone pairs are under discussion. To get the bond angles of 107° in ammonia, the p character of the N-H bonds has to be significantly increased over the 3/4 p of the equivalent sp^3 hybrid, and the orbital holding the lone pair has proportionately more s character. Similar remarks apply to the case of water with a bond angle of 105°.

There is indeed a complete range of possible hybrids from the four equivalent sp^3 hybrids at $109\frac{1}{2}^\circ$ in methane, through the nonequivalent '$s+p^3$' hybrids in CH_3Cl, NH_3 and H_2O, down to the hybrids in PH_3 and H_2S which correspond to bond angles of just over 90° and are mainly p orbitals with a little s character for the bonds and s plus a little p for the lone pair. Moving the other way, towards larger angles, there is the same continuous variation from equivalent sp^3 hybrids at $109\frac{1}{2}^\circ$, through nonequivalent arrangements, to the case of equivalent sp^2 hybrids at 120° plus an unused atomic p orbital.

There is thus a complete continuity of bond angle versus percentage p character in sp^n hybrids, as indicated in Fig. 4.16.

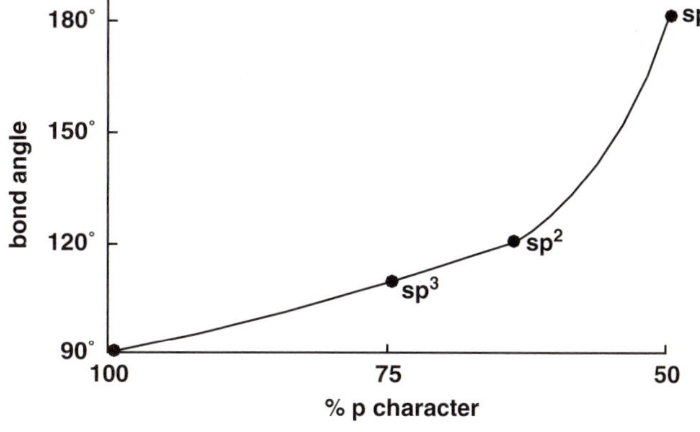

FIG. 4.16 Bond angle and p character

In general terms, any of the structures, symmetrical or not, predicted by the methods of Sections 4.3 and 4.4, may have their electronic structures explained in terms of molecular orbitals formed by appropriate equivalent or nonequivalent hybrids on the central atom together with appropriate ligand orbitals. Hybrids can be constructed to fit any set of bond angles. It is also possible, in a more detailed treatment, to derive theoretical heats of formation by calculations involving hybridization schemes, although the results are too inaccurate in general to give a direct guide to the course of a reaction.

It should be clear from the above discussion that the idea of hybridization is very valuable in providing a *description* of the electron density in a molecule but that it does not provide any *explanation* of the electron arrangements or of the shapes of molecules. It is common to see statements such as 'methane is tetrahedral because it is sp^3 hybridized' but this description is incorrect.

As with other wave-mechanical calculations, hybridization presents a potential method of predicting molecular shapes. It is possible, in principle, to find the optimum hybridization scheme for a molecule such as water which will give the most favourable energy of formation and hence the bond lengths and angles. However, just as in the cases discussed earlier, the necessity of introducing approximate methods to solve the wave equations adds so much uncertainty to the values which result that the predictions are rather crude in the present state of the art.

An interesting discussion by Smith (see references) has used Pauling's S parameter (a measure of hybrid orbital overlap and therefore bond strength) to predict structures and explains some observations which are exceptions to the VSEPR theory.

4.7 Delocalized, or multi-centred, σ orbitals

Instead of constructing a set of two-centre bonds we could describe a polyatomic species by constructing orbitals centred over all the atoms of the molecule. A full discussion is beyond our scope, but the general approach may be illustrated.

Consider first a very simple example, the BeH_2 monomer. The orbital combinations are shown in Fig. 4.17. The beryllium 2s orbital combines, in phase, with the s orbitals on hydrogens A and B to form the 3-centre orbital, ψ_1. The electron probability density in this orbital, ψ_1^2, will be concentrated in the regions of highest field between the nuclei, and will experience more attraction than in the isolated atoms. Thus ψ_1 is bonding. Similarly ψ_2, formed by the combination s (on A) $+p_z-$s (on B), is bonding. The corresponding out-of-phase combinations ψ_1^* and ψ_2^* are clearly anti-bonding. Since both the p_x and p_y orbitals on beryllium have nodes in the plane of the two hydrogens, no combinations between these and the hydrogen s orbitals are possible. Thus we can draw up a qualitative energy level diagram, Fig. 4.18. Notice, firstly, that six atomic orbitals (4 on Be, 1 on each H) give six BeH_2 orbitals ($\psi_1, \psi_1^*, \psi_2, \psi_2^*$ together with p_x and p_y). Secondly, the four electrons are placed in the two bonding levels giving two bonds holding the three atoms together. This is

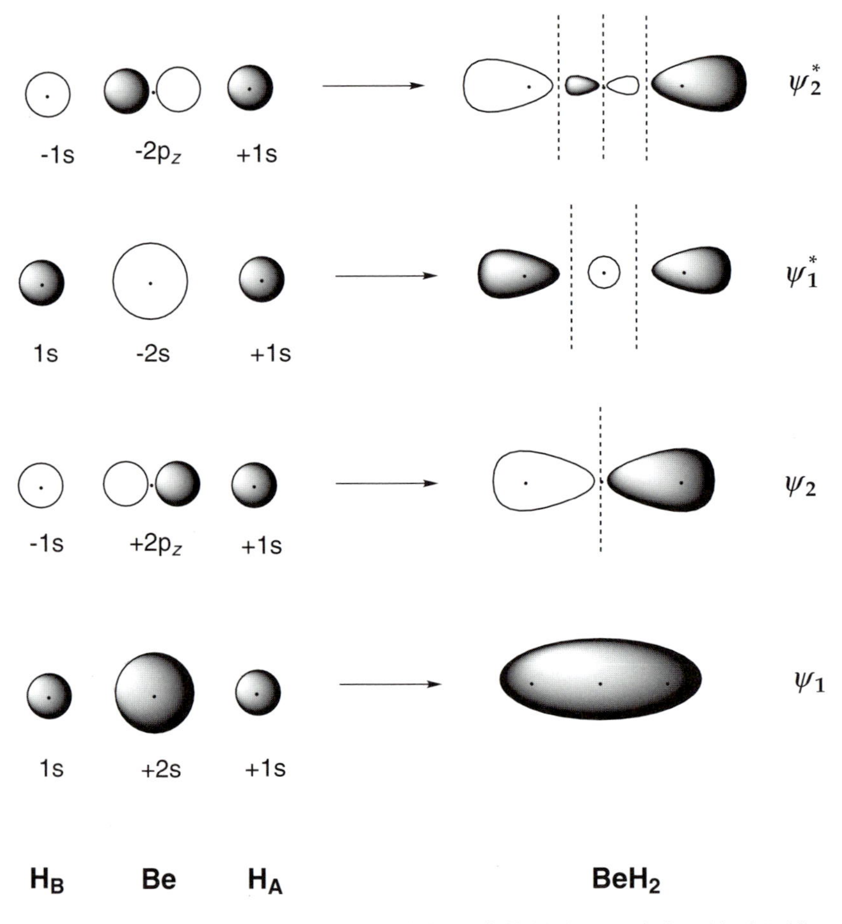

FIG. 4.17 Formation of molecular orbitals in BeH_2 (schematic). Nodal planes are indicated by dotted lines

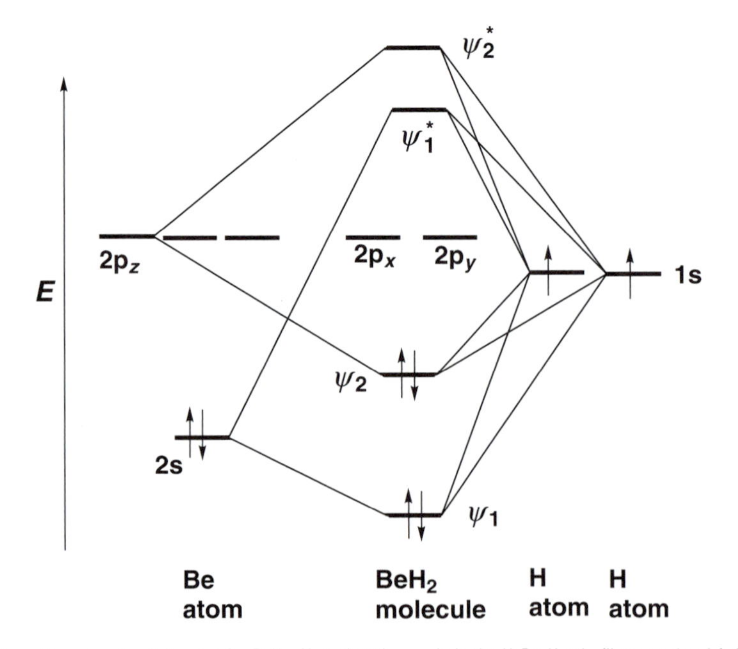

FIG. 4.18 Schematic energy level diagram for BeH_2. Note that the z axis is the H-Be-H axis (linear molecule). The p_x and p_y orbitals on Be are not involved in molecular orbital formation and are simply repeated in the BeH_2 column of the diagram

equivalent to two Be-H single bonds, but the bonding electrons are spread over all three atoms. Finally, the energy separation of ψ_1 and ψ_2 reflects approximately the separation of the beryllium 2s and 2p atomic levels.

For a second example, we can compare the water molecule, H_2O, with BeH_2. Let us take the change in two steps. If we replace Be by an atom of higher nuclear charge, such as O, the electrons will be more tightly bound and the energy levels of the 2s and 2p orbitals will move down in the diagram relative to the hydrogen 1s orbitals. In terms of Fig. 4.17, the 2s orbital would be smaller and the overlaps in ψ_1 would be less. Thus the energy level diagram for a hypothetical linear OH_2 molecule would have ψ_1 and 2s at about the same level and the oxygen 2p would give the closest match to the hydrogen orbitals.

If we now bend the H_2O molecule we have to change the alignment of z since the unique direction is now the bisector of the HOH angle. Let us take the plane of the molecule as xz and thus the x axis is in-plane at right angles to z (i.e. in the position of the z axis of BeH_2) and the y axis is perpendicular to the plane. While this change to sustain the 'z as unique direction' convention may seem pedantic, it avoids confusion when symmetry is introduced more formally in detailed treatments.

With this change, the three-centre orbital in the bent molecule corresponding to ψ_2 in Fig. 4.17

(i) uses the p_x orbital on O in place of p_z and
(ii) is a less effective overlap since the hydrogens have moved down. In compensation, the hydrogens can also overlap on the other in-plane oxygen orbital, p_z, as indicated in Fig. 4.19.

Combining all these ideas, we can describe the interactions in the V-shaped H_2O molecule thus:

(1) because of the increased nuclear charge, the overlap of the oxygen 2s orbital with the hydrogen 1s orbitals is minor so that the ψ_1 orbital in water is essentially the oxygen 2s orbital and makes only a minor contribution to the bonding.
(2) the orbital corresponding to ψ_2 in Fig. 4.17 now involves, by convention, the oxygen $2p_x$ orbital and is less strongly bonding as the overlap is less. Correspondingly, the antibonding combination ψ_2^* is less destabilized.
(3) a new interaction involving the second in-plane orbital on oxygen provides a bonding combination ψ_3 shown in Fig. 4.19, and the corresponding out-of-phase antibonding combination ψ_3^*.

Thus the energy level diagram for water is that of Fig. 4.20. The eight valency electrons fill the two bonding orbitals, ψ_2 and ψ_3, and the two nonbonding orbitals, ψ_1 and p_y. Thus there are two bonds and two nonbonding pairs.

Let us compare this description with the photoelectron spectrum of H_2O, shown in Fig. 4.21. First, we note that the vibrational structure is more complicated than in Fig. 3.22, for example, since there are three modes of vibration for a triatomic molecule (cf. Chapter 7). Band (1) at 12.6 eV has the sharp profile characteristic of the ionization of a nonbonded electron and the vibrational structure is analysed as the overlap of progressions of about 3200 cm^{-1} and 1400 cm^{-1} separations (compare frequencies in H_2O of 3650 cm^{-1} and 1595 cm^{-1}). Band (2), at 13.7 eV, shows a major drop in vibrational frequency, to 975 cm^{-1},

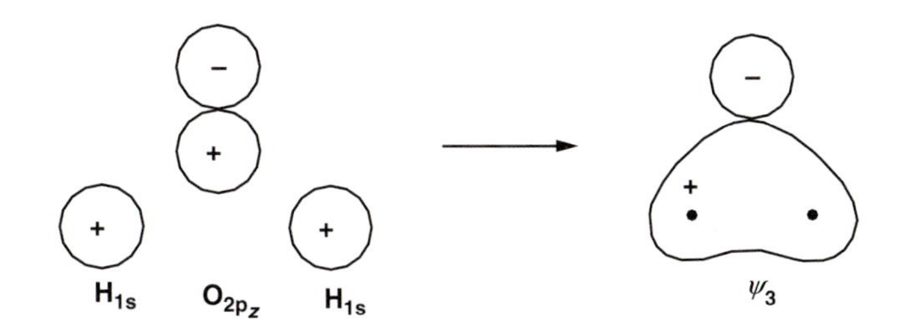

FIG. 4.19 Bonding orbital from in-phase overlap of $2p_z$ orbital on O with H 1s orbitals

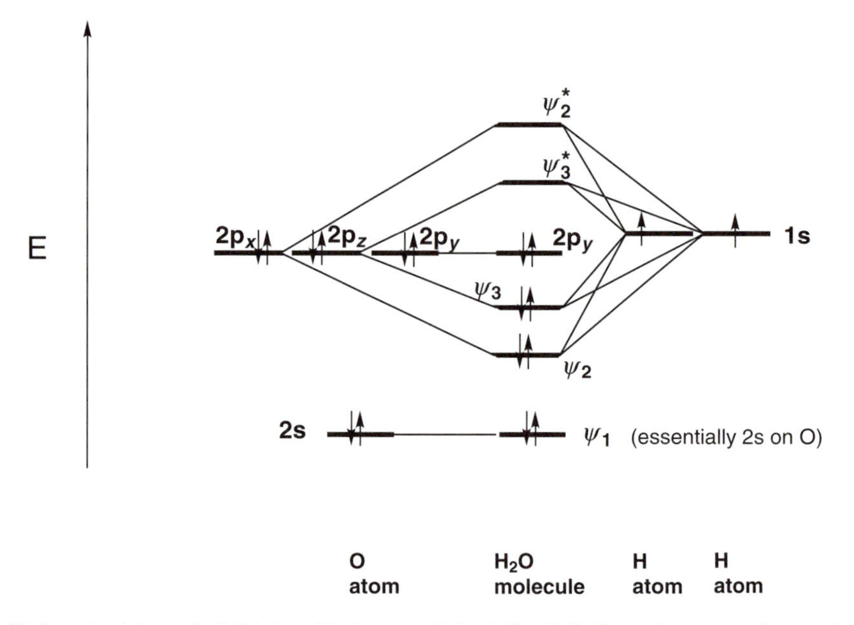

FIG. 4.20 Energy level diagram for H_2O (schematic). Compare with Fig. 4.18 for BeH_2. The nonlinear shape allows overlap with both the p orbitals in the plane of the molecule. As the oxygen s orbital is very tightly bound, ψ_1 is essentially nonbonding

as expected for the loss of a bonding electron. Band (3) at 17.2 eV also corresponds to ionization of a bonding electron. There is then a large energy gap to band (4) at 36 eV.

All this is compatible with the H_2O energy level diagram of Fig. 4.20. Band (4), corresponding to loss of the most tightly bound electron, is near the energy level of the 2s electron in oxygen and is assigned to ionization from ψ_1. (Note: ionization of the oxygen ls electron needs 580 eV.) The gap from 36 to 17 eV corresponds to the s–p energy gap. The two bonding levels come next, ionization from ψ_2 requiring 17.2 eV and from ψ_3 needing 13.7 eV. Finally the 12.6 eV ionization corresponds to the loss of a nonbonding electron from the oxygen $2p_y$ orbital (compare 12.1 eV for the loss of a π^* electron from O_2).

A rather similar picture emerges for ammonia, NH_3. The lowest ionization energy, at 10.2 eV, appears as a sharp band which shows a vibrational sequence under high resolution similar to the bending frequency of the parent molecule. The second band is of about twice the intensity, at 14.8 eV with a poorly resolved vibrational spacing of about 1800 cm^{-1} compared with the molecular stretching frequencies of 3340 cm^{-1} and 3440 cm^{-1}. The final band is weaker and very broad, centred about 27.5 eV. These can be assigned respectively to nonbonding (10.2 eV) and to doubly degenerate bonding (14.8 eV)

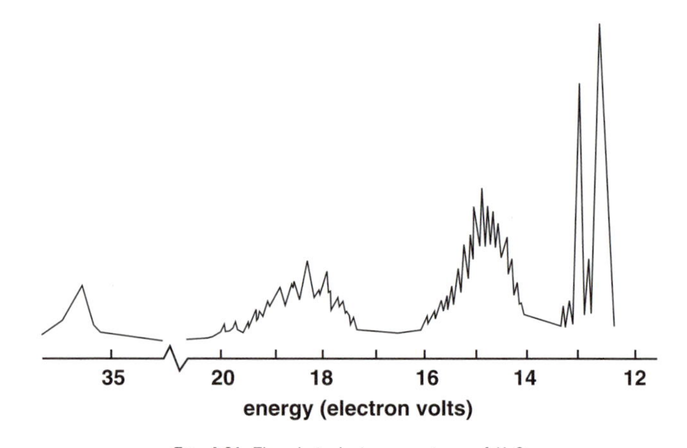

FIG. 4.21 The photoelectron spectrum of H_2O

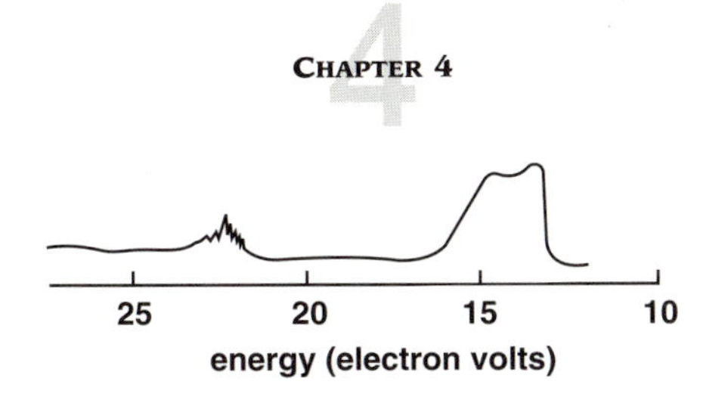

FIG. 4.22 The photoelectron spectrum of CH_4 (low resolution)

molecular orbitals involving the nitrogen p orbitals and to a bonding orbital (27.5 eV) involving the nitrogen s orbital.

Finally, consider the photoelectron spectrum of CH_4, shown in Fig. 4.22. One band at 12.7 eV is very broad and would match ionization from a triply degenerate bonding orbital using the three carbon p orbitals. The weaker band, about 23 eV, would correspond to ionization of an electron from a bonding orbital from the carbon 2s orbital.

In these last two cases, we note that the observed spectrum is *not* that predicted from the two-centre bond model using hybridization, where we would expect the following:

for CH_4: four identical sp^3 hybrids on C forming four identical CH bonding orbitals. Thus we would expect only one ionization in the photoelectron spectrum.

for NH_3: nonequivalent sp^3 hybrids, one (with extra s character) holding the lone pair and the three others forming three identical NH bonds. Thus we would expect two ionizations in the photoelectron spectrum.

We can easily understand the observed spectra if we think of four-centred (for NH_3) or five-centred (for CH_4) molecular orbitals formed by the central atom s and p orbitals and reflecting the s–p energy differences in N or O. The identity of the two bonding combinations involving the nitrogen p orbitals, or of the three bonding combinations involving the carbon p orbitals, follows from symmetry in a more detailed calculation.

4.7.1 Summary of Sections 4.5 to 4.7

σ bonding in polyatomics may be described in terms of hybrid orbitals on the central atom and two-centre bonds, *or* in terms of multi-centre bonds. The two-centre description is easier to visualize, accords with classical ideas, and has some advantages in calculations. The multi-centre description appears to give the best explanation of the observed photoelectron spectra.

It should be emphasized that these are alternative models, each capable of considerable extension and refinement, and either may be used according to the problem being examined. Other models of bonding also exist and some are discussed in the references.

4.8 π bonding in polyatomic molecules

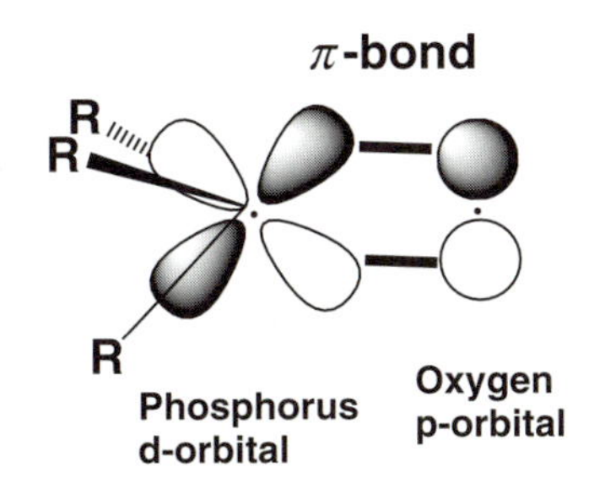

FIG. 4.23 Localized π bonds in a phosphane oxide

Polyatomic molecules which contain isolated localized π bonds, such as the P=O bond in Cl_3PO or R_3PO or the S=O bond in sulfoxides R_2SO, may be treated similarly to the π-bonded diatomic molecules of Section 3.5. These π bonds, which are localized between two atoms, are formed by sideways overlap of atomic orbitals of suitable symmetry. In the case of phosphorus oxychloride or the phosphane oxides, we can form the σ bonds from sp^3 hybrids on the phosphorus atom (using all the 3p orbitals on the phosphorus), so that the π bond requires the use of a suitable phosphorus 3d orbital to overlap with a p orbital on the oxygen. This bond is shown in Fig. 4.23. The sulfoxide is similar with a lone pair on the sulfur and a d_π–p_π bond (Fig. 4.24).

Similarly, the triple bond in hydrogen cyanide, H-C≡N, is localized between the carbon and nitrogen atoms and formed by the overlap of the p_x and p_y orbitals of the carbon and nitrogen atoms.

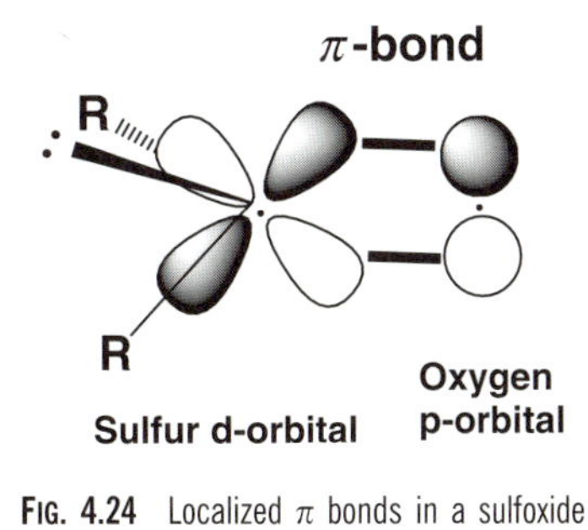

FIG. 4.24 Localized π bonds in a sulfoxide

Such localized π bonds are obviously similar to those in diatomic molecules, but they are relatively uncommon in inorganic chemistry. Most π bonds are *delocalized* over a number of atoms and are best discussed in terms of many-centred atomic orbitals.

For example, in the carbonate ion, CO_3^{2-}, all the experimental evidence shows that all three oxygen atoms are equivalent, and so a structure with a localized π bond, such as

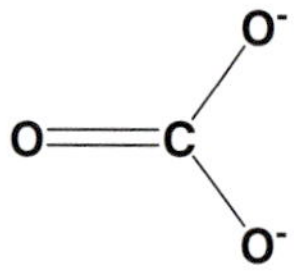

O^-
$O{=}C$, is incorrect and a description of the ion must be found which keeps the
O^-

equivalence of the oxygens. This is done by first deriving the shape of the molecule by the methods of Section 3.7. We then describe the σ bonding. This may be done in terms of hybrid orbitals and two-centre bonds or in terms of delocalized σ orbitals. Since we are going to use delocalized orbitals for the π bonding it seems more logical to do so for the σ bonds also. We shall indeed do this for later examples, (Figs 4.28 and 4.30), but for this first example we shall used the localized, two-centre model to illustrate what is still the commoner approach. In this, once we have accounted for the electrons and orbitals used (a) in σ bonding and (b) to account for lone pairs, we then have available the remaining orbitals and electrons to form the π system.

The carbonate ion is planar with no unshared pair on the carbon. Let the molecular plane be the *xy* plane. Then the carbon uses its 2s, $2p_x$ and $2p_y$ orbitals to form the σ bonds (using sp^2 hybrids), and the oxygen atoms must also use their 2s, $2p_x$ and $2p_y$ orbitals either in the σ bonds or in accommodating nonbonding electrons (there is no chance of sideways overlap between oxygen orbitals in the plane of the molecule as the distances are too large). This accounts for all the p, p_x and p_y orbitals and for eighteen electrons, six in the three σ bonds and four unshared electrons on each oxygen atom. The twelve s and p orbitals form the three bonding σ orbitals and two orbitals on each oxygen to hold the nonbonding electrons, and all these are filled. The remaining three orbitals are the antibonding σ orbitals and these remain empty and are of very high energy (see Fig. 4.25). The ion has a total of twenty-four electrons (including two for the charge) so that six have still to be accommodated and the p_z orbitals on the four atoms have yet to be used. These four p_z orbitals are combined to form four, four-centred, π orbitals, each holding two electrons, Fig. 4.26b. The π orbital of lowest energy is of the form:

$$\psi_1 = (p_C + p_1 + p_2 + p_3)$$

where the constants have been omitted and the orbitals referred to are the carbon p_z (p_C) and the three oxygen p_z orbitals respectively. This orbital is shown in Fig. 4.26a. It has no nodes between C and O, concentrates electron density in the regions between the atomic nuclei, and is bonding. The orbital of highest energy, by contrast, has a node between the carbon atom and each oxygen and is antibonding, Fig. 4.26b. This orbital is of the form:

$$\psi_4 = (p_C - p_1 - p_2 - p_3)$$

As there are four atomic p_z orbitals, there must be four molecular orbitals formed by them. The remaining two orbitals, ψ_2 and ψ_3, are, in this case, of equal energy and lie between ψ_1 and ψ_4. They are nonbonding (see Fig. 4.25).

The six remaining electrons enter ψ_1, ψ_2 and ψ_3, leaving the antibonding ψ_4 empty. The four electrons in ψ_2 and ψ_3 have no bonding effect, and thus the carbonate ion is left with one effective π bond over the whole molecule with a resulting C-O bond order of $1\frac{1}{3}$ (one for the σ bond and $\frac{1}{3}$ for the π bond), which corresponds with that implied by the simple formula.

A more detailed discussion of π orbitals, especially of the orbitals of intermediate energy such as ψ_2 and ψ_3, is beyond the scope of this text, but the general properties of such π systems are readily recognized. Four generalizations about many-centred π orbitals are possible.

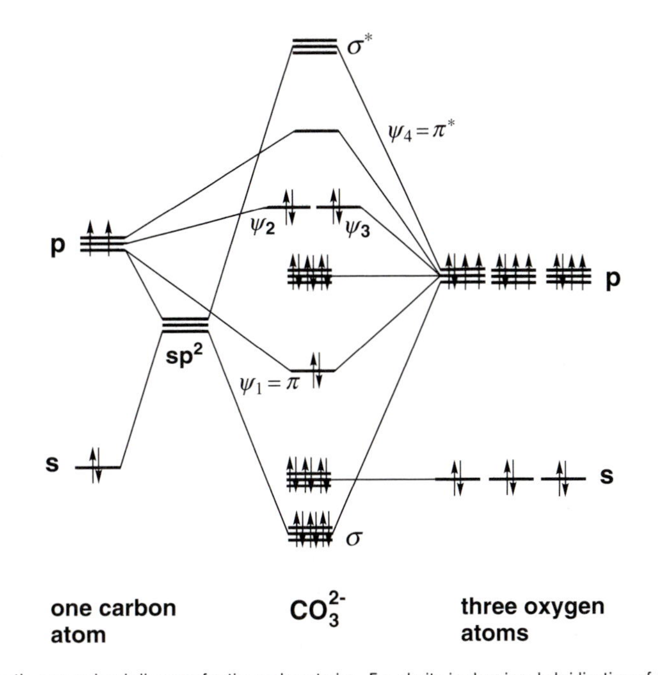

FIG. 4.25 Schematic energy level diagram for the carbonate ion. For clarity in drawing, hybridization of the oxygen orbitals has been neglected. Each oxygen atom has two pairs of nonbonded electrons—shown as the pair in the s orbitals and the pair of nonbonding p electrons. The delocalized π orbitals are ψ_1, ψ_2, ψ_3 and ψ_4. The bonding and antibonding σ C-O orbitals are shown as σ and σ^*

(1) The number of many-centred molecular orbitals equals the number of component atomic orbitals (in the case of the carbonate ion this is four). Each molecular orbital may hold up to two electrons.

(2) The molecular π orbital of lowest energy is that obtained by combining all the atomic orbitals with the same sign so that there are no nodes between pairs of atoms in the resulting π orbital. In the carbonate ion, this orbital is ψ_1.

(3) The molecular π orbital of highest energy is generally one where there is a node between each pair of atoms, that is, where the sign of the wave function is reversed between each pair of atoms as in ψ_4 of the carbonate ion. Such an orbital is strongly antibonding and no π bonding can occur in a case where electrons have to be placed in this type of orbital.

(4) The remaining molecular π orbitals are intermediate in energy and their energies fall symmetrically about the mean energy of the strongly bonding and the strongly antibonding orbitals. Thus, in the carbonate case, ψ_2 and ψ_3 are degenerate and nonbonding. In other cases there will be equal numbers of weakly bonding and weakly antibonding orbitals. Note that where there is an odd number of contributing orbitals there must be at least one nonbonding level.

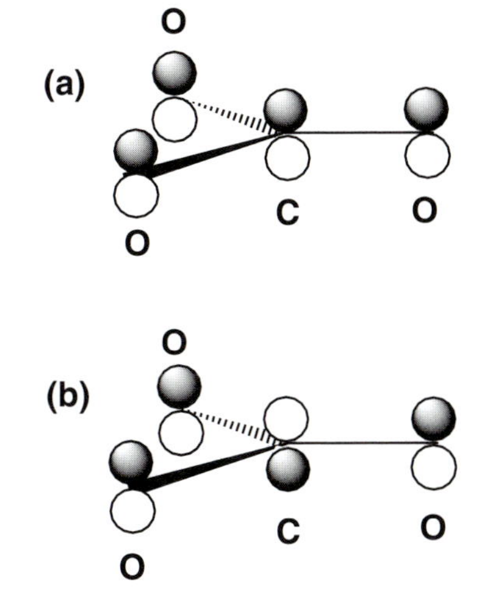

FIG. 4.26 Delocalized π orbitals in the carbonate ion: (a) the most strongly bonding orbital, (b) the most strongly antibonding orbital

The form of the intermediate molecular orbitals is not always clear from simple considerations, but these generalizations make it possible to work out whether there will be any π bonding in a molecule without knowing any more about the formation of the intermediate orbitals. The one necessary condition is that it should be possible to leave the highest antibonding orbital empty and there will then be a net π bonding effect. If other, more weakly antibonding π orbitals also remain empty, the bonding effect is enhanced.

4.8.1 The π orbitals of benzene

The classical example of a delocalized π system is, of course, the case of benzene. The π orbitals here are enumerated below and illustrated in Fig. 4.27 and the reader can see how they fit the generalizations above.

(1) There are six component atomic orbitals and six π orbitals result.

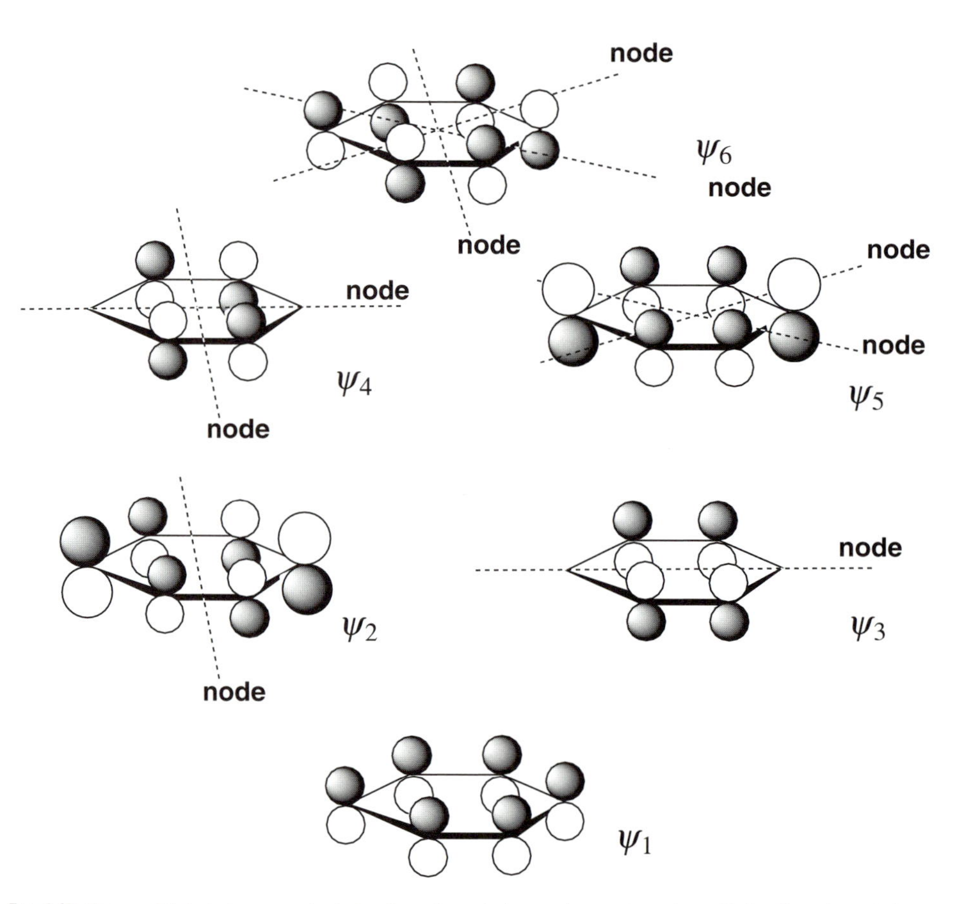

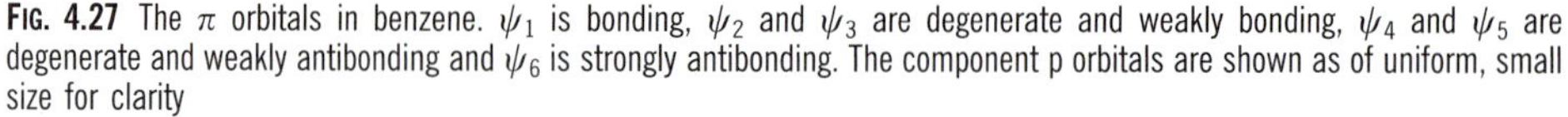

FIG. 4.27 The π orbitals in benzene. ψ_1 is bonding, ψ_2 and ψ_3 are degenerate and weakly bonding, ψ_4 and ψ_5 are degenerate and weakly antibonding and ψ_6 is strongly antibonding. The component p orbitals are shown as of uniform, small size for clarity

(2) The orbital of lowest energy has the form:

$$\psi_1 = k(\mathrm{p}_1 + \mathrm{p}_2 + \mathrm{p}_3 + \mathrm{p}_4 + \mathrm{p}_5 + \mathrm{p}_6)$$

This has no nodes between atoms and is strongly bonding.

(3) The orbital of highest energy has the form:

$$\psi_6 = k(\mathrm{p}_1 - \mathrm{p}_2 + \mathrm{p}_3 - \mathrm{p}_4 + \mathrm{p}_5 - \mathrm{p}_6)$$

Each pair of atoms is separated by a node and the orbital is strongly antibonding.

(4) The other four molecular π orbitals fall into two sets of degenerate pairs. One pair has one node cutting the ring and is weakly bonding while the other pair of orbitals has two nodes cutting the ring and is weakly antibonding. The six π electrons fill the three bonding levels, leaving the three antibonding orbitals empty.

4.8.2 The π orbitals of the nitrite ion, NO_2^-

In this ion, the two oxygen atoms are equivalent and the methods of Section 4.3 show that the ion is bent with a lone pair on the nitrogen atom. Again, let the molecule lie in the xy plane, then only the p_z orbitals on each atom will be involved in π bonding. Of the eighteen electrons in the valency shells of the component atoms, four will form the σ bonds, two give the nitrogen lone pair and there are four nonbonding electrons on each oxygen atom leaving four electrons to go into π bonds. There are three atomic p_z orbitals and thus there are three molecular, three-centred, π orbitals. The lowest energy orbital extends evenly over the molecule and is bonding, while the highest energy π orbital has a node between the nitrogen atom and each oxygen and is antibonding. The third π orbital is nonbonding and has its electron density largely on the oxygens. The four electrons enter the bonding and nonbonding π orbitals, giving a net effect of one π bond over the molecule or an N-O bond order of $1\frac{1}{2}$.

4.9 An example of the approach using delocalized bonding throughout

4.9.1 Carbon dioxide, CO_2

This is a linear molecule with no lone pairs on the carbon, by the method of Section 4.3. If the molecular axis is taken as the z axis, the s and the p_z orbitals on the three atoms will form the σ bonds or hold nonbonding electrons on the oxygen atoms. The carbon orbitals are higher in energy than the corresponding oxygen ones. To a reasonable approximation, however, we can take the s orbital on carbon combining with the s orbitals on the two oxygens to give a three-centred σ bonding orbital ψ_s with no nodes, and a three-centred σ antibonding orbital with two nodes ψ_s^*, similar to ψ_1 and ψ_1^* depicted in Fig. 4.17. However, in contrast to BeH_2 we shall not use the central orbital for anything else, so we have to form a third orbital from these three atomic s orbitals. This will be a nonbonding orbital with a single node at carbon, which we can label ψ_s^0.

In a similar way, we can take the p_z orbital on the carbon with the p_z orbital on each oxygen and form bonding, nonbonding and antibonding three-centred orbitals ψ_p, ψ_p^* and ψ_p^0, as shown in Fig. 4.28.

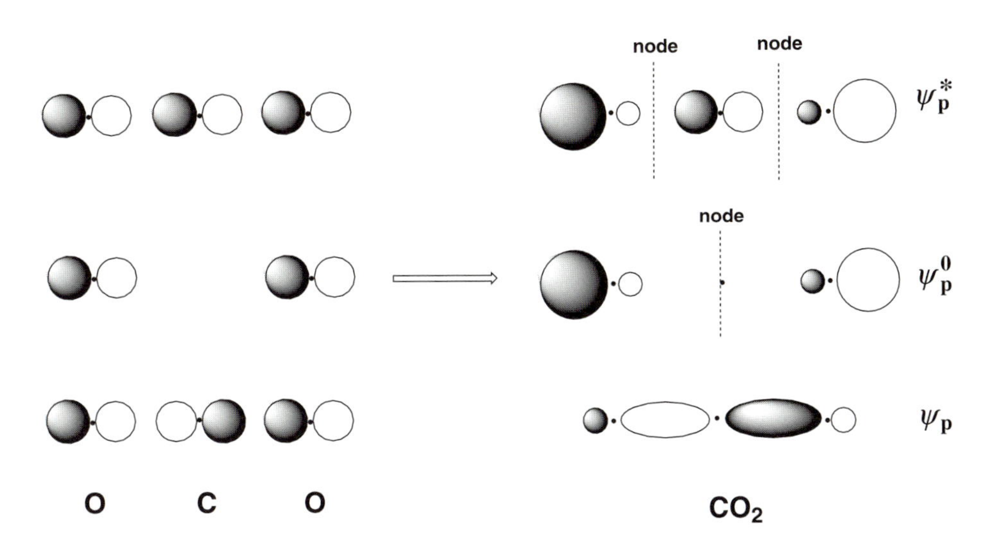

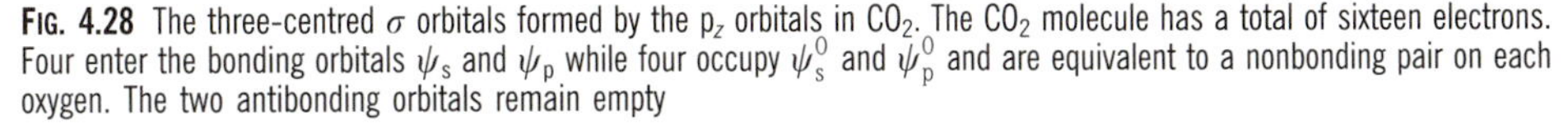

FIG. 4.28 The three-centred σ orbitals formed by the p_z orbitals in CO_2. The CO_2 molecule has a total of sixteen electrons. Four enter the bonding orbitals ψ_s and ψ_p while four occupy ψ_s^0 and ψ_p^0 and are equivalent to a nonbonding pair on each oxygen. The two antibonding orbitals remain empty

This leaves eight electrons for π bonding. The p_y and p_x orbitals on each atom are available to form π bonds. Consider first the p_y orbitals. There are three of these and therefore three delocalized molecular π orbitals will be formed. The lowest energy orbital will have no interatomic nodes and be of the form (omitting constants):

$$\psi_1 = (p_O + p_C + p_O)$$

while the highest energy orbital will have nodes between the carbon and each oxygen and have the form:

$$\psi_3 = (p_O - p_C + p_O)$$

The third molecular orbital, ψ_2, will be nonbonding, Fig. 4.29. In the case of the p_x orbitals, exactly the same combinations will occur, as these orbitals are identical with

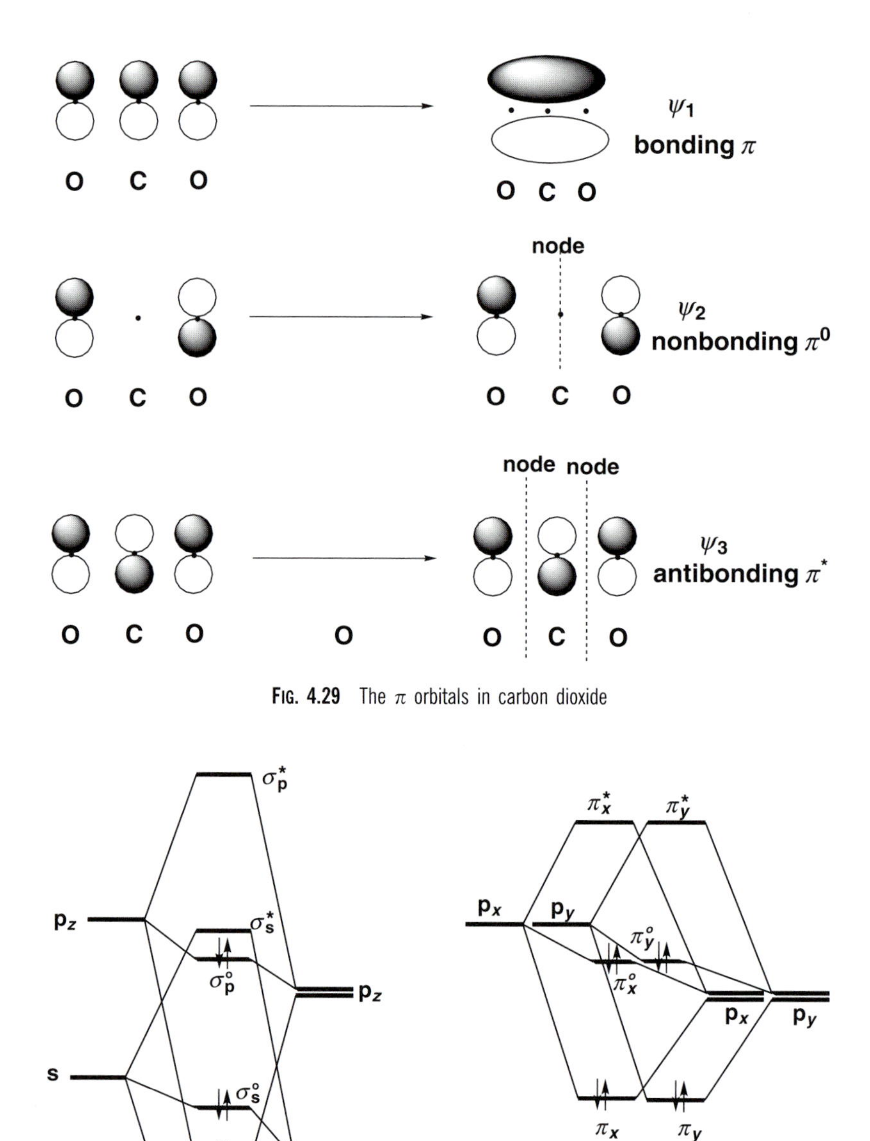

FIG. 4.29 The π orbitals in carbon dioxide

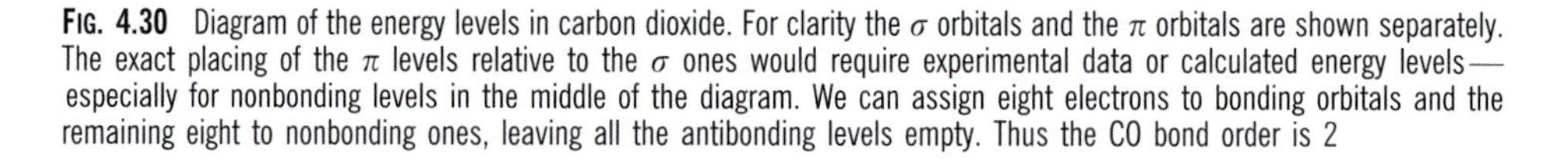

FIG. 4.30 Diagram of the energy levels in carbon dioxide. For clarity the σ orbitals and the π orbitals are shown separately. The exact placing of the π levels relative to the σ ones would require experimental data or calculated energy levels—especially for nonbonding levels in the middle of the diagram. We can assign eight electrons to bonding orbitals and the remaining eight to nonbonding ones, leaving all the antibonding levels empty. Thus the CO bond order is 2

the p_y ones apart from their direction in space. The resulting molecular orbitals are also identical; that is, the six atomic p_y and p_x orbitals combine to form six molecular orbitals, two of which are bonding and of the form of ψ_1, two nonbonding like ψ_2, and two antibonding like ψ_3. The eight valency electrons fill the two bonding π orbitals and the two nonbonding π orbitals, giving a net effect of two π bonds over the molecule. There are thus four bonding σ electrons and four bonding π electrons, adding up to a C-O bond order of two. The energy level diagram for the molecule is shown schematically in Fig. 4.30.

4.10 Extension to other molecules

By comparison with the detailed treatments above, we can draw provisional conclusions about other molecules.

4.10.1 Ozone, O_3

It is interesting to compare the cases of ozone and carbon dioxide. There are two more valency electrons in ozone and, if these are added to the energy level diagram of carbon dioxide in Fig. 4.30, they would have to be placed in the antibonding π orbitals, reducing the net π bonding over the molecule from two to one and, if both electrons were placed in one of the π^* orbitals, say the one in the y direction, the whole π system in the y direction would have zero bonding effect. All this implies that the equivalence of the y and x directions in CO_2 will disappear in O_3, i.e., the molecule is no longer linear. This conclusion may also, of course, be derived by the methods of Section 4.3. Ozone is isostructural and isoelectronic with the nitrite ion and has a bent structure, with a nonbonding pair on the central oxygen atom and two nonbonding pairs on each of the two terminal oxygens. There are three, three-centred, π orbitals of which the bonding and nonbonding ones are occupied giving an O-O bond order of $1\frac{1}{2}$.

Compare Figs 4.25, 4.26

4.10.2 The nitrate ion, NO_3^-

This ion is planar and there are no unshared electrons on the nitrogen atom. The total number of electrons is twenty-four. The ion is isoelectronic and isostructural with carbonate and has the same π electron configuration, with two electrons in a strongly bonding π orbital delocalized over the whole molecule and four electrons in the degenerate pair of nonbonding orbitals. There is a net effect of one π bond in the ion and a N-O bond order of $1\frac{1}{3}$.

4.10.3 Sulfur trioxide, SO_3

This is a plane triangular molecule with twenty-four valency electrons like the nitrate ion. It might therefore be described in the same way if the sulfur atom used only its 3s and 3p orbitals. There would then be six electrons in π orbitals but with a net bonding effect of only one π bond. If, however, the sulfur atom makes use of its 3d orbitals in addition, it is possible to construct π orbitals from six atomic orbitals (S $3p_z$, $3d_{xz}$, $3d_{yz}$, plus three O $2p_z$) instead of four. There would then be three bonding π orbitals and three antibonding ones, and the six π electrons can all be placed in bonding orbitals, giving three delocalized π bonding orbitals and a total S-O bond order of about two (the approximation comes in as all three π orbitals are not equally stable).

In a similar way, SO_2 is isoelectronic and isostructural with O_3 but the possibility exists of the sulfur using d orbitals to increase the number of bonding π orbitals.

Whenever d orbitals of relatively low energy are available, there is the possibility that they will contribute to the bonding. This applies especially to molecules containing very electronegative elements bonded to the elements in the third and higher periods and thus to all such oxygen compounds. The extent of d orbital participation will depend on the relative energies and the number of electrons to be accommodated and may be decided only by calculation (see Section 18.9).

The use of d orbitals by elements of the 3p series has been a hotly debated area in recent years. A recent article has concluded that the bonding in SO_2 can indeed be best described by means of two S=O double bonds and that resonance structures of the type

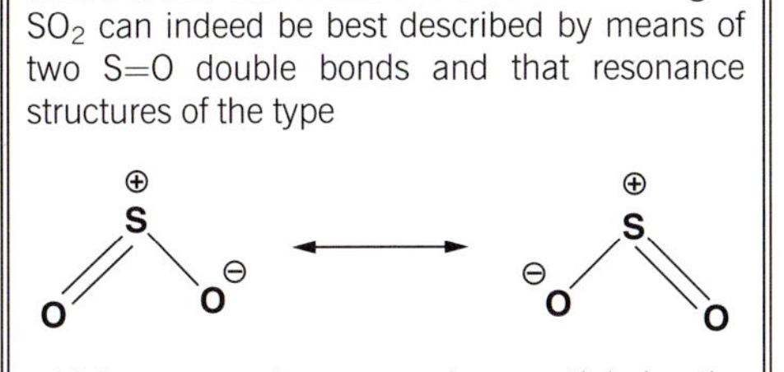

which are analogous and essential in the description of the bonding in O_3, are unimportant for SO_2.

4.10.4 Summary

Shapes of polyatomic molecules may be treated as follows:

(1) Where only σ bonds and unshared pairs of electrons are involved, the shape of a simple molecule is determined by the number of electron pairs around the central atom. The bonding may be treated in terms of localized two-centred molecular orbitals formed between each pair of bonded atoms, or may, with more accuracy but less convenience, be treated in terms of many-centred, delocalized orbitals. These treatments may be extended to more complex species (for example, those with more than one 'central atom') but the predictions become less secure as the complexity rises.

(2) The basic shapes of simple species with π bonds again depend only on the number of σ and lone pairs on the central atom and this can be calculated by the methods of

Section 4.3. Localized π bonds between pairs of atoms may be treated in terms of two-centred molecular orbitals just as in the case of diatomic molecules.

(3) Where a number of equivalent atoms occur in a species with π bonding, the π electrons must be treated as delocalized over the whole molecule. Many-centred molecular π orbitals are constructed from atomic orbitals of suitable symmetry, and these have the property that the molecular orbital of lowest energy and that of highest energy are readily distinguishable, and, as long as the latter remains unoccupied, a net π bonding effect results. The treatment of delocalized π bonding by relatively simple ideas is less complete than that of σ bonding, and problems arise particularly in the cases where d orbital participation is possible. These more complicated cases can only be fully treated by detailed calculation which is beyond the scope of this text. However, the relatively simple ideas outlined in the earlier parts of this chapter allow a reasonably accurate description of the shapes of the large majority of simple polyatomic molecules and ions.

Problems

Many examples of molecular shapes can be found in Chapter 9 onwards. You should list a few examples and then work them out without immediate reference to the text. It is also important to comprehend molecular shapes in three dimensions. You should take every opportunity of looking at, or making, models of structures.

4.1 Determine the most probable shape of the fluorides of the Main Group Elements as listed in Table 17.2b, p. 429

4.2 Determine similarly the shapes of all the interhalogens and all the polyhalide ions given in Table 17.19, p. 489

4.3 Determine the most likely structures of all the xenon oxides, oxyfluorides and anions given in Table 17.22, p. 496

4.4 Discuss the following bond angles: 117° in O_3, ~120° in SO_2, 101° in SiF_2, 180° in CO_2, 180° in N_3^-.

For ozone, it is tricky to formulate a simple structure. Have a go, and then see Section 4.3.1.

4.5 Arrange the following species in order of increasing OSO angle: SO_2, SO_3, SO_3^{2-}, SO_4^{2-}, H_2SO_3, H_2SO_4, SO_2Cl_2, SO_2F_2.

It is easiest to put the charges on the O atoms.

4.6 Discuss the ONO bond angles, 180° in NO_2^+, 134° in NO_2, 115° in NO_2^-.

Bonding in polyatomics should be worked out *both* on the two-centred *and* on the many-centred orbital basis. Again, the starting point is to consider the in-phase overlap of orbitals, thought of as waves.

The bonding material in this chapter should be correlated with the molecules described in the second half of the book. For example, you should go on to consider the bonding description of the compounds in Questions 4.1 to 4.6 above. The following questions will suggest some approaches.

4.7 In a series of molecules PXH_2, the HPH angle was 90°, 95°, 100°, $109\frac{1}{2}$°, 115°, 120° and 135°. What hybridization of phosphorus s and p orbitals is implied in each case? Use Table 4.5 to decide the relative strengths of these PH bonds.

4.8 Imagine a square-planar molecule MH_4. Assume M has valence shell s, p and d orbitals and take the molecular plane as *xy*. Write down all the combinations of M orbitals with hydrogen s orbitals which give delocalized sigma bonds (Compare Section 4.7.)

4.9 By analogy with Figure 4.26, draw out the molecular π orbitals formed by the out-of-plane p orbitals in a square planar AB_4 species.

4.10 In Question 4.9, if A can also use d orbitals, which of these π levels would be further stabilized? (Take the plane as *xy*, and put the B atoms on the *x* and *y* axes.)

The three intermediate orbitals should be taken as weakly bonding, nonbonding and weakly antibonding. (Square symmetry does not allow all three to be equal and nonbonding.)

4.11 Assuming the species remain planar, decide the electron configuration of CO_3^{3+}, CO_3^{2+}, CO_3^+, CO_3, CO_3^-, and CO_3^{3-} by comparison with CO_3^{2-} (Figure 4.25). Which of these would be:

(a) paramagnetic, (b) more strongly bonded?

Which species would be stabilized by becoming nonplanar, making ψ_2 more stable than ψ_3?

4.12 The photoelectron spectrum of formaldehyde H_2CO shows four bands below 21 eV:

	Energy (eV)	*Vibrational sequence*	Vibrational separations (cm^{-1})	
1	10.8	Short	2560	1590
2	14.1	Long	2400	1210
3	15.9	Long	–	1270
4	16.3		Broad envelope	

The stretching frequencies for H_2CO are 2780 and 1740 cm^{-1}.

(a) Decide the probable shape of H_2CO.

(b) Discuss the σ bonds, nonbonding electrons and π bonds.

(c) Discuss the photoelectron spectrum and draw up an energy level diagram.

4.13 (a) The He(I) photoelectron spectrum of CO_2 shows the following bands: 13.8 eV (nonbonding): ca. 17.6 eV (bonding), 18.1 eV (approx. nonbonding): 19.4 eV (approx. nonbonding).

(b) Other evidence indicates that the 13.8 and 17.6 eV ionizations are from π levels.

(c) It is calculated that there are two further ionizations at about 39 and 41 eV, both bonding.

Is this evidence compatible with Fig. 4.30?
Do these results support the three-centre σ bond approach for CO_2 (compare Section 4.7)?

CHAPTER 5

SIMPLE IONIC CRYSTALS 110
5.1 THE FORMATION OF IONIC COMPOUNDS 110
5.2 THE BORN–HABER CYCLE 115
5.3 THE LATTICE ENERGY 116
5.4 THE ENDOTHERMIC TERMS IN THE FORMATION OF AN IONIC SOLID 118
5.5 BONDING WHICH IS NOT PURELY IONIC 119
5.6 METALLIC BONDING 121
5.7 COMPLEX IONS 127
5.8 THE CRYSTAL STRUCTURES OF COVALENT COMPOUNDS 128
5.9 DEFECT STRUCTURES AND NONSTOICHIOMETRIC SOLIDS 130
PROBLEMS 131

The Solid State

When we turn from single molecules to solids, the structures and energies depend on a wider range of forces.

The most direct continuation from our discussions of the small single molecule are those species where normal covalent forces are present throughout the crystal—the crystal is a giant molecule. The classical example is diamond where each C atom is bound by a normal C-C single bond to four neighbours arranged tetrahedrally and this structure extends indefinitely. Such crystals are typically hard and high-melting. An example of a compound of this type is SiO_2 in which each Si is bonded to 4 O and each O to 2 Si in an infinite array.

At the other extreme we find molecular solids where the individual compounds are strongly bound covalent molecules, such as CH_4, but only very weak forces hold the molecules together in the solid. Such crystals are typically soft and low-melting.

We find a third class of solids where the components are ions and the major binding forces in the crystal are the electrostatic ones between the ions, as in NaCl. Such crystals are relatively hard and high-melting.

Finally we may pick out the class of metals which are related to the giant molecules but where the outer electrons are free to move throughout the crystal. Such crystals show a range of hardness and melting points, and are highly reflecting and conducting.

These four classes are idealized cases and a range of interactions apply in most solids (compare also the discussion based on Fig. 5.11 in Section 5.6). We start this chapter with ionic crystals, where the basic interactions are well understood, and we go on to relate other types of solid to these.

Simple ionic crystals

5.1 The formation of ionic compounds

In the last two chapters, the Lewis electron pair theory was extended on the basis of wave mechanics to give a full description of the covalent bond. In this section, ionic bonding will be examined and the various factors which determine the formation and stability of ionic compounds will be discussed.

The basic process in the formation of an ionic compound is the transfer of one or more electrons from one type of atom to another: the resulting ions are then held together by electrostatic attraction.

Isolated ionic species, such as the two-atom entity M^+X^-, do not exist under ordinary conditions. In ionic compounds, we are dealing with an array of ions, which extends in three dimensions to the edges of the crystallite. If the ionic solid dissolves, the ions are indeed separated, but they are stabilized by interaction with the solvent molecules (see Chapter 6) and the free ion has only a transient existence.

The arrangement of the ions in the solid is the one which gives the highest electrostatic energy. To see what factors determine this arrangement, consider the process of bringing up successive anions around a given cation. If there are already n anions surrounding the cation, the addition of a further anion produces an extra attraction between its charge and the cation charge, and also produces a number of repulsions between its charge and the charges on the n anions already present. There are thus two opposing tendencies. One is to increase the attractive forces by making the coordination number of the cation as large as possible and this is balanced by the increase in the repulsive forces as more and more anions are added. When the two tendencies balance, the final structure results. An exactly similar argument holds, of course, for the number of cations to be found around an anion.

The repulsions are at a minimum if the distribution of ions is as symmetrical as possible. Thus ions which are 3-coordinated have their neighbours at 120° in a triangular arrangement, 4-coordinated ions have a tetrahedral, 6-coordinated ions an octahedral, and 8-coordinated ions a cubic arrangement. In addition, some coordination numbers, such as five, which do not pack regularly in a solid are not observed in ionic crystals.

The coordination numbers in a solid of given formula, such as AB, depend on the number of the larger ions which may be packed around the smaller one. The stoichiometry—in this example, 1:1—then determines the coordination number of the larger ion. As the formation of a cation involves the removal of electrons, cations are always smaller than the parent atoms (for example, the atomic radius of K is 203 pm, while the radius of K^+ is 133 pm). Conversely, the addition of electrons to atoms to form anions involves an increase in radius (for example, F = 71 pm but F^- = 133 pm in radius). As a result, anions are generally larger than cations and it is the number of anions which can pack around a cation which usually determines the coordination numbers and the structures. For example, sodium chloride crystallizes in a structure where the sodium ion is surrounded by six chloride ions (Fig. 5.1a), and the chloride ion has, of course, six sodium ions around it. The larger caesium cation allows a coordination number of eight in the structure of caesium chloride (Fig. 5.1b).

The number of anions which can pack around a given cation may be determined from the ratio of the radii of the cation and anion. This *radius ratio*, r_+/r_-, may thus be used to give an indication of the likely coordination number for a salt of given formula type. (Notice that the argument is exactly the same if the anion is smaller than the cation, except that it is the ratio of anion radius to cation radius, r_-/r_+, that is the important one.)

We can calculate the radius ratio value at which one coordination number is expected to change into another from purely geometrical grounds if we make two assumptions:

(1) that ions are charged incompressible spheres of definite radius (see Section 5.5 for further discussion of this assumption)

(2) that the central ion adopts the highest coordination number which allows it to remain in contact with each neighbour. (This second assumption is a good approximation for the balance of attractive and repulsive forces referred to above.) An example is shown in the box titled 'Determining the radius ratio limit for 6-coordination'.

NOTE ON CRYSTAL STRUCTURE DIAGRAMS

It follows from the discussion in Section 5.1 that ions, or atoms, in crystals are expected to be as close together as possible and that the available space will be filled as completely as possible. However, if such a 'spacefilling' situation is represented directly in a diagram it is very difficult to see the arrangement of the atoms. As a result, most diagrams give an exploded view of the crystal. This convention has been adopted here. In accurate diagrams, the centres of the atoms are positioned exactly but their diameters would be reduced. In most of the diagrams in this chapter, further slight distortions or alterations in perspective have been introduced to make the structural arrangement easier to interpret.

In a number of cases, further conventions are used to assist understanding. Atoms or ions in the top layer may be distinguished from more distant layers by the thickness of the circles used to represent them or, as for example in Fig. 5.1, by using darts to represent the bonds joining them with the broad end of the dart on the atom nearest the front of the unit cell. Thus in the Cs Cl diagram, Fig. 5.1b, the Cs atom at the body centre has four Cl atoms in front of it and four behind it. It is often useful to show the connected ions from the neighbouring cell to emphasize coordination numbers as in Figs. 5.1a or 5.3b.

In more elaborate structures, there is a conflict between showing clearly the coordination of each atom and avoiding an extremely elaborate diagram showing many repeat units. Any diagram is necessarily a compromise and a formalized representation of a three-dimensional, space-filling structure and the reader should search for as many representations as possible of difficult structures in the references given. The study of models in three dimensions greatly clarifies the more complex structures and every opportunity of examining such models should be taken (see references).

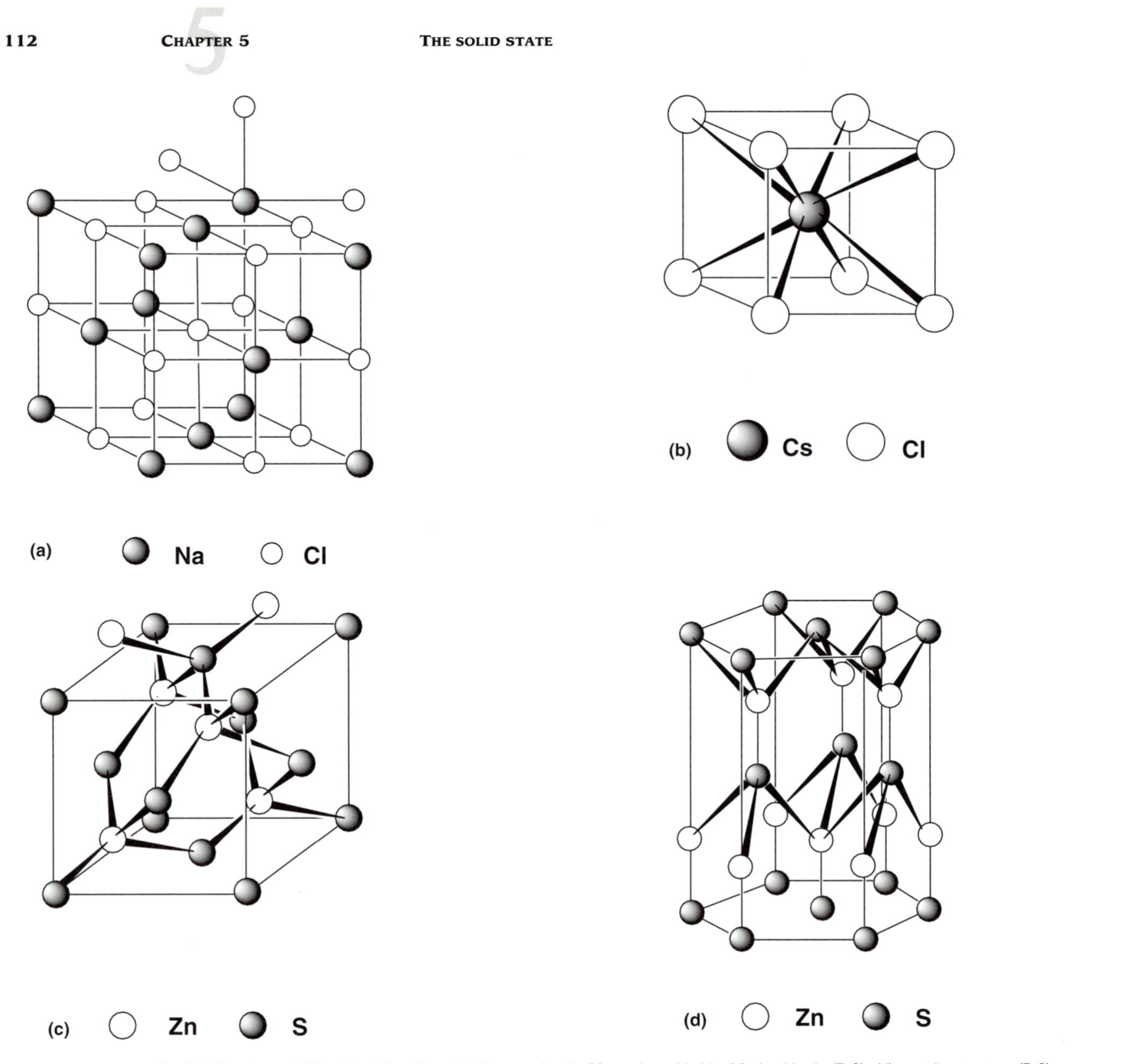

FIG. 5.1 Structures of AB solids: (a) sodium chloride or rock salt, (b) caesium chloride, (c) zinc blende (ZnS), (d) wurtzite structure (ZnS)

DETERMINING THE RADIUS RATIO LIMIT FOR 6-COORDINATION

In 6-coordination, a cross-section through a site in the lattice appears as in Fig. 5.2.
When the anions just touch, BAB′ = 90°. Then

$$AB = AB' = (r_+ + r_-) \text{ and } BB' = 2r_-$$

$$\frac{AB}{BB'} = \frac{r_+ + r_-}{2r_-} = \cos 45° = 1/\sqrt{2}$$

$$\therefore r_+ = \sqrt{2}r_- - r_-$$

$$\text{or } r_+/r_- = \sqrt{2} - 1 = 0.41$$

This represents the lower limit to the radius ratio for 6-coordination. If the cation is any smaller, it is not in contact with the anions, the attractive forces are less, and the structure would revert to a lower coordination.

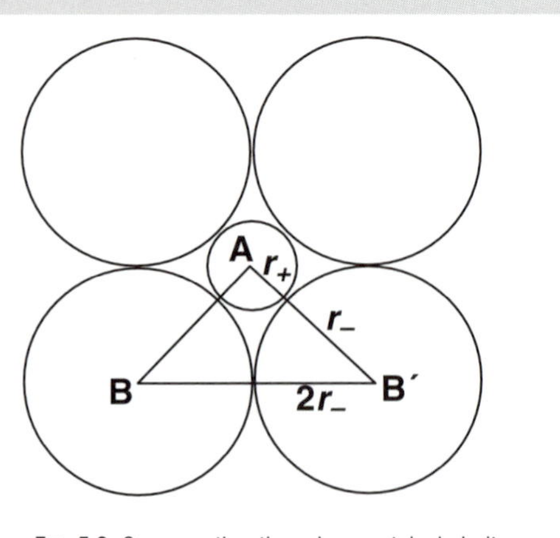

FIG. 5.2 Cross-section through an octahedral site

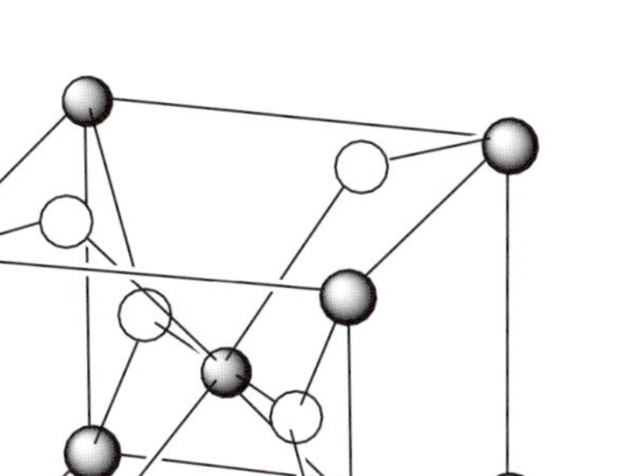

(a) Ti O

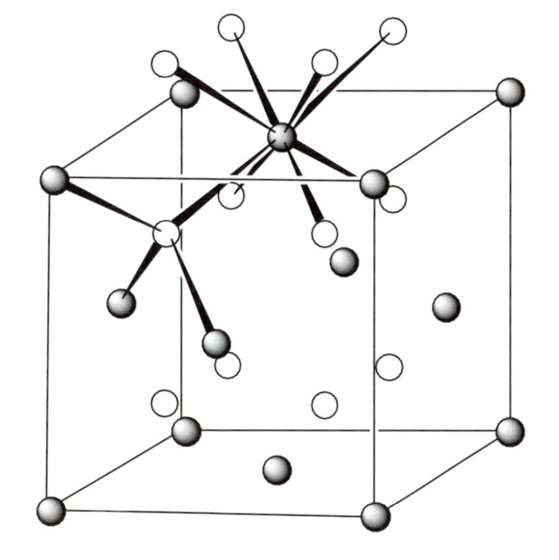

(b) Ca F

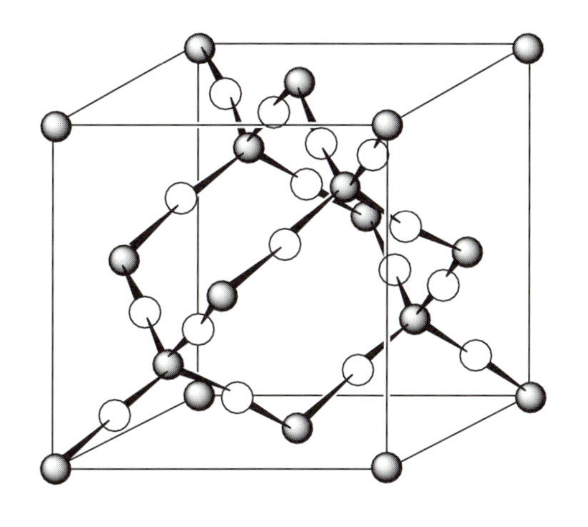

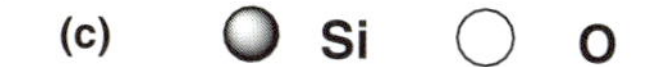

FIG. 5.3 The structures of AB_2 solids: (a) rutile (TiO_2), (b) fluorite (CaF_2) showing the cubic 8-coordination around calcium and the tetrahedral 4-coordination around fluorine, (c) β-cristobalite (SiO_2)

Similar calculations may be carried out for all the coordination numbers. The results indicate the range of values for the radius ratio within which different coordination numbers should be stable. These are shown below.

r_+/r_- :	0.155 to	0.23 to	0.41 to	0.73 to higher values
C.N. :	3	4	6	8

The validity of this simple method of predicting the coordination number may be assessed by examining the structures of some AB and AB_2 compounds.

There are three structures of AB formula where the ions are all in sites of high symmetry and these are shown in Fig. 5.1. By far the commonest is the sodium chloride (rock salt) structure where both ions are octahedrally coordinated. For larger ions, cubic coordination is found in the caesium chloride structure. Smaller ions show tetrahedral coordination in the ZnS structures. These each have both Zn and S atoms in tetrahedral coordination but the overall symmetry of the lattice is cubic in the zinc blende structure, Fig. 5.1(c), and hexagonal in the wurtzite structure, Fig. 5.1(d).

For AB_2 species, the standard structures (Fig. 5.3) are:

(a) titanium dioxide (rutile) with coordinations of 6 (octahedral) and 3 (trigonal).

(b) calcium fluoride (fluorite) with coordinations of 8 (cubic) and 4 (tetrahedral).

(c) β-cristobalite (one form of SiO_2) with coordinations 4 (tetrahedral) and 2 (linear). This is the type compound for the most regular 4:2 structure, though SiO_2 is a giant covalent molecule, not an ionic species.

The structures and radius ratios of some ionic halides and oxides are listed in Table 5.1 calculated using the ionic radii of Table 2.12. Among the AB_2 structures, where the 8-coordinated fluoride structure is expected to be replaced by the 6-coordinated rutile structure at a radius ratio of 0.73, the agreement with prediction is remarkably good. The lower limit for the stability of 6-coordination comes at 0.41, and germanium dioxide, with a ratio of 0.38, occurs in two forms, one isomorphous with SiO_2 and the second with the rutile structure.

In the AB structures, the agreement is less good and the sodium chloride structure persists through a wider range of radius ratios than predicted, both at the upper end as shown in the table and at the lower end of the range as shown by the lithium halides of radius ratios down to 0.28 for LiI, all of which have the rock salt structure.

The very simple model of ions as hard spheres may thus be used as a reasonable first approximation to suggest the structure expected for an ionic crystal. A clear indication of its limitations is given in the discussion of ionic radii in Section 2.15, which should be re-examined.

A number of other structures where the coordination sites are of regular symmetry are shown in Fig. 5.4. Of particular interest is the perovskite ($CaTiO_3$) structure, Fig. 5.4c, since many materials with very interesting magnetic, electrical, optical and catalytic properties adopt this structure. The reader is referred to Section 16.1 for a discussion of the copper oxide superconductors, many of which have structures related to perovskite.

Less symmetrical configurations adopted by compounds in which the bonding is not purely ionic are discussed in Section 5.5.

TABLE 5.1 Radius ratios and structures of some AB and AB_2 solids

Compound	r_+/r_-	*AB structure*	*Compound*	r_+/r_-	*AB_2 structure*
KF	1.00	Sodium chloride	CaF_2	0.71	Fluorite
KCl	0.73	Sodium chloride	SrF_2	0.83	Fluorite
KBr	0.68	Sodium chloride	BaF_2	0.97	Fluorite
KI	0.62	Sodium chloride	MgF_2	0.49	Rutile
RbF	0.92*	Sodium chloride	TiO_2	0.48	Rutile
RbCl	0.82	Sodium chloride (and caesium chloride)	SnO_2	0.51	Rutile
RbBr	0.76	Sodium chloride	GeO_2	0.38	Rutile and also a 4:2 form as in SiO_2
RbI	0.69	Sodium chloride	CeO_2	0.72	Fluorite
CsF	0.81*	Sodium chloride			
CsCl	0.93	Caesium chloride (and sodium chloride)			
CsBr	0.87	Caesium chloride			
CsI	0.76	Caesium chloride			

*These values are r_-/r_+ as the cations are larger than the anion.

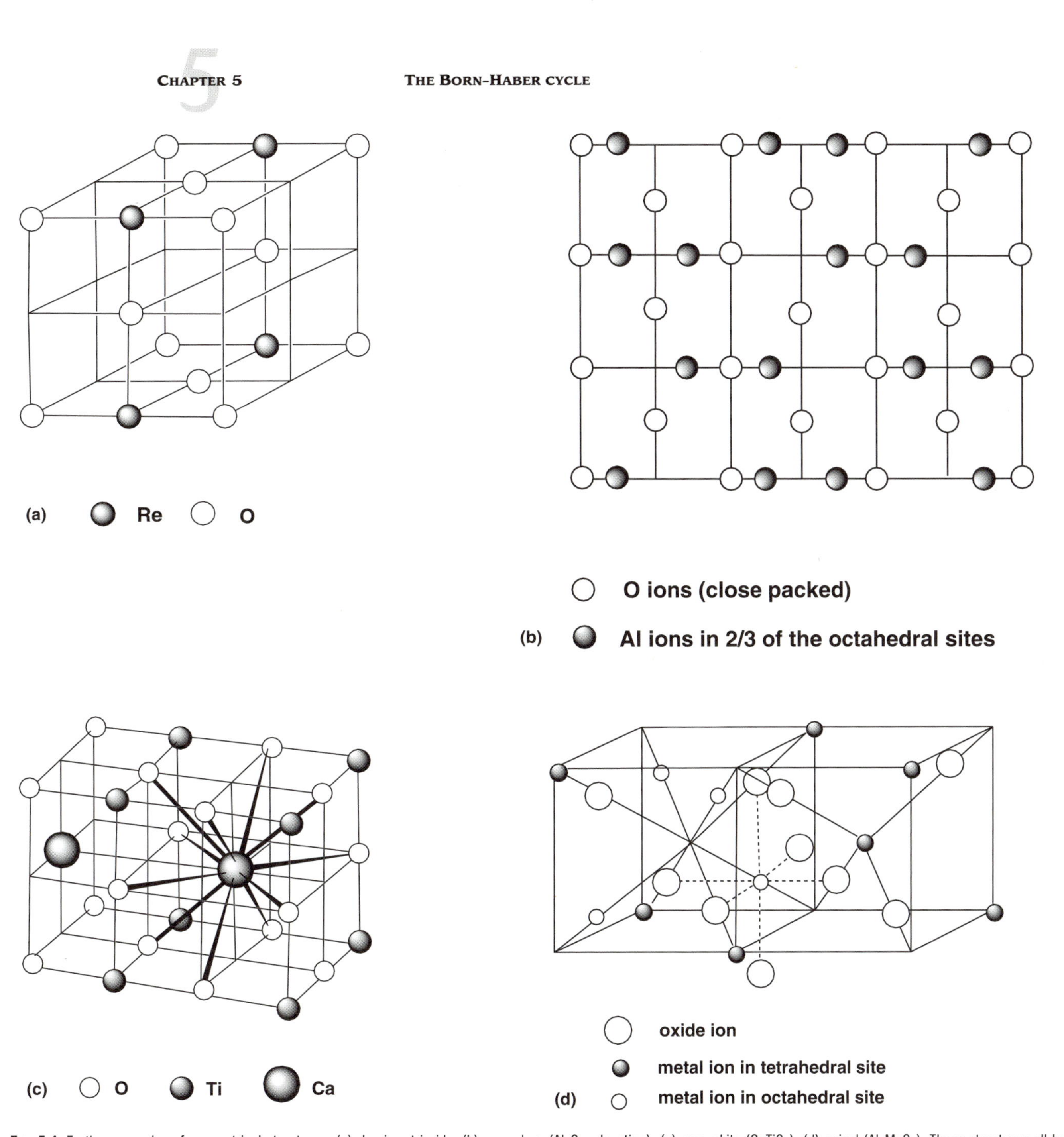

FIG. 5.4 Further examples of symmetrical structures: (a) rhenium trioxide, (b) corundum (Al_2O_3, elevation), (c) perovskite ($CaTiO_3$), (d) spinel (Al_2MgO_4). These structures all have the ions in positions of maximum symmetry with neighbours disposed regularly around them, with the exception of corundum and spinel where the oxygen positions are in a close-packed arrangement but the nearest neighbours are not of highest symmetry

5.2 The Born–Haber cycle

What factors determine whether given elements combine to form an ionic solid? These may be found by considering the energy changes involved in the formation of an ionic solid from the elements and we shall use sodium chloride as an example. This has a measured heat of formation (H_f) of $-410.9\,\text{kJ mol}^{-1}$; i.e. for the reaction

$$Na_{(metal)} + \tfrac{1}{2}Cl_{2(gas)} \rightarrow Na^+Cl^-_{(solid)},\ H_f = -410.9\,\text{kJ mol}^{-1}$$

This reaction may be broken down into simpler steps whose energies are known, as in the diagram of Fig. 5.5 by a method due to Born and Haber.

Known as the Born–Haber cycle, this gives the net energy change calculated from the five simpler steps shown in the table below the diagram, and hence a calculated heat of formation.

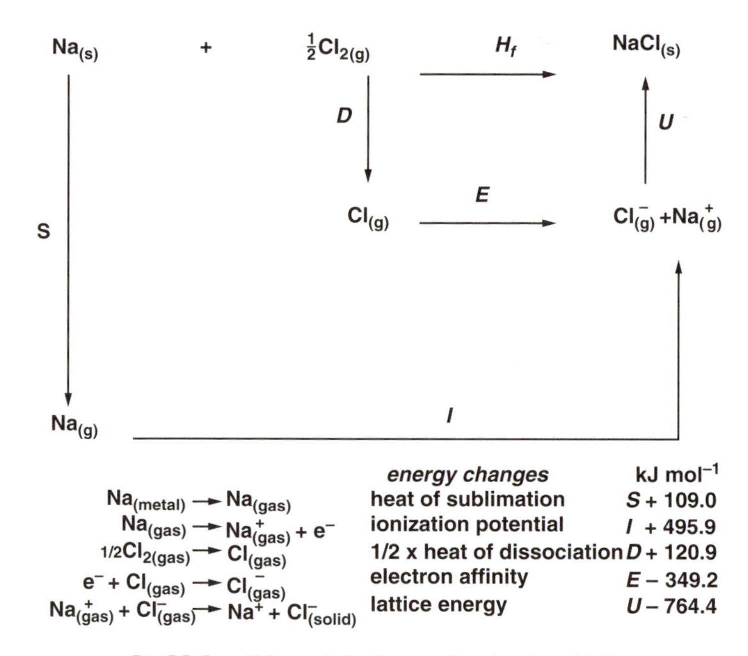

FIG. 5.5 Born–Haber cycle for the formation of sodium chloride

In the case of sodium chloride, the values of all the quantities in the cycle are known so that a calculated heat of formation (equal to $-387.8\ \text{kJ mol}^{-1}$) may be found and compared with the experimental value. The close agreement confirms that the model of ions as incompressible spheres is a reasonable one in this case. Similar agreement between cycle values and experimental ones is found in many cases and this gives grounds for confidence in using the cycle in other ways. The most useful of these is the evaluation of one of the cycle quantities when the others are known. In particular, the electron affinity E which is difficult to measure is usually determined from the Born–Haber cycles of a series of appropriate salts. It can be seen from Fig. 5.5 that:

$$H_f = S + I + D + E + U$$
$$\text{hence } E = H_f - U - S - I - D$$

The Born–Haber cycle may also be used to determine whether or not the bonding in a compound is purely ionic. As we assume that the forces are ionic in order to calculate the lattice energy, if the calculated value does not agree with the experimental one, the assumption is invalid and other effects must be occurring. This aspect is discussed further in Section 5.5.

Our main reason for discussing the formation of an ionic solid in terms of the Born–Haber cycle is to allow a more detailed assessment of the contribution of each component of the energy cycle to the heat of formation of the ionic solid. The factors which determine whether a compound is ionic or covalent may thus be isolated and discussed. These factors will be examined in turn starting with the lattice energy U which is the most important exothermic term in the cycle. Then the endothermic terms—the ionization potential, the electron affinity and the heats of atomization—will be examined.

5.3 The lattice energy

The energy change when the gaseous ions are brought together from infinite separation to their equilibrium distances in the solid is the lattice energy, U. That is, U is the heat of the reaction (for an AB solid)

$$A^{z+}_{(gas)} + B^{z-}_{(gas)} \rightarrow A^{z+}B^{z-}_{(solid)} \quad U$$

This energy arises from electrostatic interactions between the ions.

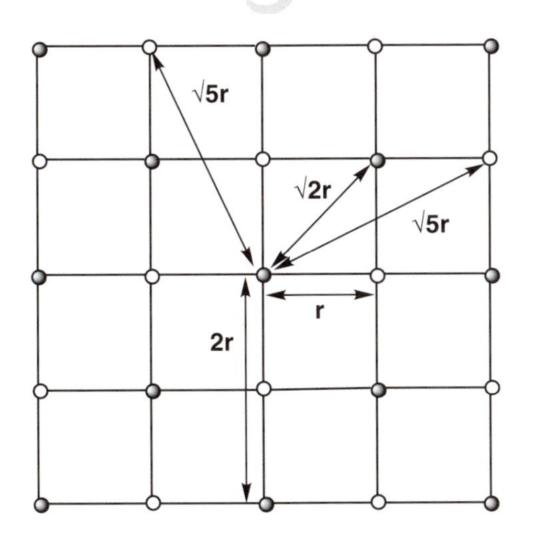

FIG. 5.6 Two-dimensional analogue of a crystal lattice

When the geometry of the solid array of ions is known, the lattice energy may be calculated. The method may be illustrated by considering the square two-dimensional array of ions shown in Fig. 5.6. Any one cation in a square array of interatomic distance r has four anions at a distance r as nearest neighbours. The electrostatic potential energy between these ions is $4Z^+Z^-e^2/4\pi\epsilon_0 r$, where Z^+ and Z^- are the positive and negative charges on the ions (assuming an AB stoichiometry), e is the charge on the proton and ϵ_0 is the permittivity of a vacuum. The next-nearest neighbours to the given cation are four cations at a distance $\sqrt{2}r$ and a repulsion exists between the given cation and these ions of $4(Z^+)^2e^2/4\pi\epsilon_0\sqrt{2}r$. Then come four more cations at $2r$, eight anions at $\sqrt{5}r$ and so forth. The total electrostatic energy of a cation in this square array is thus:

$$E = 4e^2(Z^+)(Z^-)/4\pi\epsilon_0 r + 4e^2(Z^+)^2/4\pi\epsilon_0\sqrt{2}r$$
$$+ 4e^2(Z^+)^2/8\pi\epsilon_0 r + 8e^2(Z^+)(Z^-)/4\pi\epsilon_0\sqrt{5}r + \ldots$$
$$= -Z^2e^2/4\pi\epsilon_0 r(4 - 2\sqrt{2} - 2 + 8/\sqrt{5}\ldots)$$

where $Z^+ = -Z^-$ has been put equal to Z for an AB crystal.

The convergent infinite series in the bracket may be evaluated from the geometrical properties of the array and its sum may be found. This sum is called the *Madelung constant*, A, after its first evaluator. A different geometrical array, for example a rectangular one, would have a different value of the Madelung constant and the whole analysis may be extended to three dimensions. For example, for a rock salt lattice the Madelung constant

$$A_{\mathrm{NaCl}} = (6 - 12/\sqrt{2} + 8/\sqrt{3} - 6/2 + 24/\sqrt{5}\ldots)$$
$$= 1.748\ldots$$

The Madelung constant is the factor relating the electrostatic forces and the spatial arrangement of the ions in a crystal. It depends only on the geometry of the crystal and is independent of the nature or charge of the ions.

The electrostatic energy of a cation in an AB lattice may therefore be written:

$$E = \frac{-Z^2e^2A}{4\pi\epsilon_0 r}$$

In detail, the energy is $N \times (E_{\mathrm{cation}} + E_{\mathrm{anion}}) \div 2$: the division by two is to avoid counting each attraction twice over.

The electrostatic energy of the anion in an AB crystal is the same as that for the cation. If a mole is considered, the electrostatic energy is E times Avogadro's constant, N. Thus, the electrostatic energy of a mole of an AB ionic compound is

$$NE = \frac{-Z^2e^2AN}{4\pi\epsilon_0 r}$$

So far, only the attractive electrostatic forces between the ions have been considered but Born introduced a second, repulsive, energy term into the equation to take account of the repulsion which arises at very short interionic distances when the ions start to interpenetrate (otherwise the expression for NE tends to infinity as r tends to zero). The repulsive force rises steeply as the interionic distance decreases and is represented by a term $E_{\mathrm{rep}} = B/r^n$ where B is a constant similar to the Madelung constant and n is also a constant of the order of nine for sodium chloride. The total energy of the crystal, which is the lattice energy U, is then:

$$U = NE + NE_{\mathrm{rep}} = \frac{-Z^2e^2AN}{4\pi\epsilon_0 r} + \frac{NB}{r^n}$$

The constant B can be eliminated, since the lattice energy is a minimum at the equilibrium value of $r = r_0$, so that by setting $\mathrm{d}U/\mathrm{d}r = 0$ for $r = r_0$, B can be expressed in terms of the other constants. Then:

$$U = \frac{-Z^2e^2NA}{4\pi\epsilon_0 r_0}(1 - 1/n)$$

All the quantities on the right-hand side of this expression are known or may be found; r_0 comes from direct experimental evidence and n from calculation or experiment. The lattice

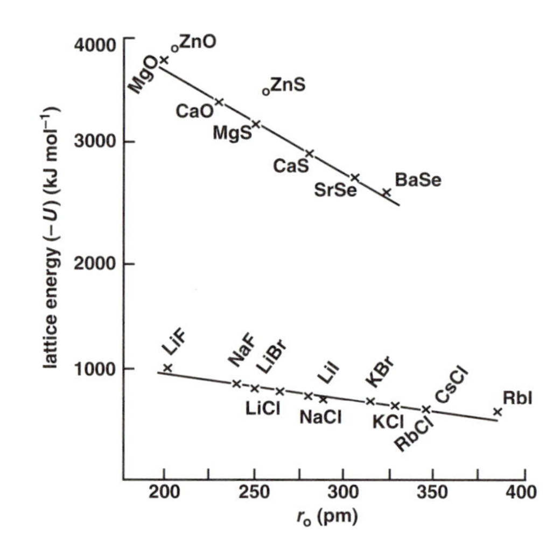

FIG. 5.7 Lattice energies (plotted as $-U$). Lattice energies are plotted as a function of r_0 for a number of cases of the sodium chloride structure and for CsCl, ZnO and ZnS. The marked effect of the ionic charge is clear: it will also be noted that the variation in structure to 8:8 or 4:4 coordination has only a minor effect

energy U can thus be found by a combination of calculation and experimental determination of parameters.

It can be seen that the properties of an ionic solid which determine the lattice energy are the geometry, as reflected in the Madelung constant, the interionic distance and the charges on the ions. The most important effect is that of the charge which appears as a squared term. The lattice energy of an $A^{2+}B^{2-}$ solid is four times that of an $A^{+}B^{-}$ solid of the same geometry.

Of the geometrical factors, the Madelung constant has only a minor effect on the lattice energy as the values of the constant for different structures of the same stoichiometry are similar. For example, in AB structures the Madelung constants are as follows:

8-coordinated, CsCl structure, $A = 1.763$
6-coordinated, NaCl structure, $A = 1.748$
4-coordinated, wurtzite structure, $A = 1.641$

When the stoichiometry changes, the Madelung constant changes to a greater extent, for example, $A = 5.04$ for the fluorite structure; but changes in the stoichiometry imply changes in the charges and it is difficult to make comparisons between different formula types.

Variations in the second geometrical factor, the interionic distance, have a more important effect on the size of the lattice energy. The lattice energy depends inversely on r_0 so that, within a given structure type, the lattice energy will fall as the ionic sizes increase. It will be clear from Section 5.1 that the ratio of anion to cation sizes should not vary too widely as r_0 varies or some other structure will become the more stable one. The effect of variations in charge and interionic distance is best seen by comparing the lattice energies of a series of compounds of the same structure formed by related elements, Fig. 5.7. The marked effect of changes in charge is very obvious, as well as the linear fall in lattice energy as the ions (and therefore the interionic separation) increase in size. The values for AB compounds of different structures (ZnO and ZnS having the wurtzite structure) are included in the figure to show the effect of the Madelung constant.

SUMMARY FOR THE EXOTHERMIC CONTRIBUTIONS TO THE LATTICE ENERGY

The largest values of the lattice energy, the main factor favouring the formation of an ionic solid, will result from

(1) increasing the charges on the ions,
(2) decreasing the interionic separation, i.e. by having small ions,
(3) by changes in the Madelung constant, though this effect is relatively small.

5.4 The endothermic terms in the formation of an ionic solid

In general, the largest of the energy contributions which have to be added to the system before an ionic solid is formed is the ionization potential of the cation, I. If the formation of an ionic solid is to be favoured, the ionization energy should be as small as possible. From Table 2.8, it will be seen that the ionization potentials of the elements have the following characteristics:

(i) The ionization potentials increase from left to right across a Period.
(ii) In any Group, the ionization potential decreases with increasing size down the Group.
(iii) For a given element, the first ionization potential is less than the second, which in turn is less than the third, etc.

It thus requires least energy to form cations of large atoms on the left of the Periodic Table, that is of the larger elements of the alkali and alkaline earth elements. The higher the charge on the ion, the greater is the energy required in its formation.

The energy of formation of the anion is the electron affinity. The electron affinities (compare Section 2.14) may represent an endothermic or an exothermic contribution to the heat of formation of an ionic solid, but even the most exothermic electron affinity is less than the smallest ionization potential so that the formation of gaseous cation plus anion from the gaseous atoms is endothermic for any pair of elements. Only a few electron affinities are exothermic and these are all for the formation of singly charged anions. The formation of doubly charged anions is always a strongly endothermic process and this is true *a fortiori* for anions of higher charge.

The heats of atomization of the elements in their standard states depend very much on the form in which the element exists. Little in the way of generalization can be said except that where elements in a Group occur in the same form—as for example, the halogens—there is a tendency for the heat of atomization to decrease with increasing atomic weight, but often with the lightest element anomalous. As the values for sodium chloride show, the heats of atomization commonly represent only a minor contribution to the energy balance.

If these factors are compared with those which lead to a high lattice energy, it is seen that the two main requirements for low endothermic energies—large ions of low charge—are exactly opposed to those which favour high lattice energies. The small ions of high charge which give the highest exothermic contribution are precisely those which require the highest endothermic energies of formation from the elements in their standard states. The formation of an ionic solid therefore depends on the detailed balance of all the energy contributions in each individual case. Once again, the direction of the chemical change depends on small differences between large values and is difficult to predict *a priori*.

SUMMARY FOR THE ENDOTHERMIC CONTRIBUTIONS TO THE LATTICE ENERGY

The least energy will be required, and thus the formation of an ionic solid will be most favoured, from this aspect when the ions are
(1) of low charge,
(2) large, giving large interionic separations,
(3) formed from elements at the extremes of the Periodic Table.

In general, ionic solids are formed by the s metals and the transition elements in the II or III oxidation states (though often as complexes) with anions from the halogen or chalcogen Groups. Simple ions with charges greater than two are less common, but the lanthanide elements form M^{3+} ions, and the existence of Th^{4+} is well established.

If the values of the ionization potentials are examined in detail, it will be seen that successive ionization potentials for an element rise in a fairly regular manner as long as only electrons in the valence shell are removed. Since lattice energies vary as the square of the charge on the ions, it might seem that, on balance, ion formation would be most favoured in cases where the charges are high. Unfortunately, a further complication appears which upsets this conclusion, in that highly charged ions are those most likely to cause polarization and a departure from purely ionic bonding.

5.5 Bonding which is not purely ionic

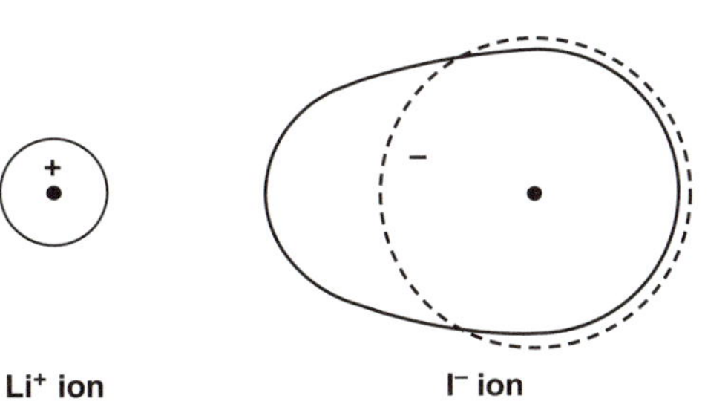

FIG. 5.8 Diagrammatic representation of polarization in LiI

In the discussion in Section 5.1, the assumption was made that ions were hard spheres and that the only forces (except at very short distances) were electrostatic ones between the charges. That this is a reasonable approximation in many cases is shown by the agreement between the calculated and measured heats of formation of the solids. However, in a solid such as lithium iodide where the cation is very small and the anion is large (Fig. 5.8) the high charge density on the cation distorts the rather diffuse electron cloud of the anion as indicated in the figure. The centre of negative charge in the anion no longer coincides with the centre of positive charge and an induced dipole is present in the anion. In such a case, the anion is termed *polarizable* and the cation, *polarizing*. This polarization represents a departure from purely ionic bonding in the compound. Thus the elementary idea of ions

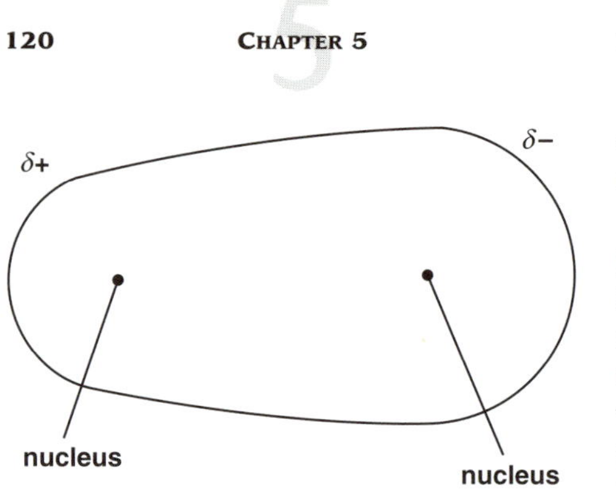

FIG. 5.9 Polarization in a covalent bond

as hard spheres has to be modified by allowing for distortions of the spherical electron clouds where ions have high charge densities.

In a similar way, the purely covalent bond discussed in the last chapter is uncommon. The pair of electrons which form a covalent bond are equally shared between the constituent atoms only if these are identical or, by coincidence, of the same electronegativity. If the two atoms differ in electronegativity, the bonding electrons are more strongly attracted by the more electronegative and a dipole is created in the bond (Fig. 5.9). The pure covalent bond with the electrons equally shared between the two atoms and the pure ionic bond with the electron completely transferred from one atom to the other are the two extreme cases. There is a complete range of bond types lying between the two, the ionic bond distorting in the manner of Fig. 5.8, and the covalent bond distorting in the manner of Fig. 5.9, till the polarized covalent bond and the polarized ionic bond become indistinguishable.

There is one reservation about this picture: the pure covalent bond does exist (between two atoms of the same element) but the pure ionic bond is an abstraction as there is bound to be some polarization between any pair of ions. However, since the environment of an anion in a crystal is symmetrical, the slight polarizations in a strongly ionic compound have only a minor effect on the bond energy and can be regarded as a normal attribute of ionic solids.

The polarizing power of the smallest ion (which is generally the cation) depends on the density of charge on it. The polarizing power is thus greatest for small, highly charged ions like Mg^{2+} or Al^{3+}: so much so that the smaller congeners of these ions, Be^{2+} and B^{3+}, do not exist and beryllium and boron are covalently bound in their compounds. Polarizability is greatest for large ions like I^-, and especially for those with diffuse electron clouds like H^- or N^{3-}.

A marked degree of polarization shows up in the lattice energy values. When a dipole is induced in one ion, there is an ion–dipole attraction to be added to the ion–ion attraction which is used in the calculation of lattice energies on the purely ionic model. The actual lattice energy of a polarized solid should thus be higher than that calculated on a purely ionic model. The values in Table 5.2 illustrate this.

The differences for the alkali halides are less than one percent and the direction of deviation is random. The alkali hydrides have experimental values which differ from the values calculated on an ionic model by up to eight percent and the experimental values are all low.

TABLE 5.2 Calculated and experimental lattice energies, $-U$

	Lattice energy (kJ mol^{-1})		
Compound	*Experimental*	*Calculated*	*Difference*
LiF	1009.2	1019.2	−10.0
NaF	903.9	900.8	+3.1
LiH	905.4	979.1	−73.7
NaH	810.9	845.2	−34.3
KH	714.2	741.4	−27.2
AgF	954	920	+34
AgCl	904	833	+71
AgBr	895	816	+79
AgI	883	778	+105
KF	801.2	805.4	−4.2
KCl	697.9	702.5	−4.6
KBr	672.4	674.9	−2.5
KI	631.8	637.6	−5.8
TlCl	732	686	+46
TlBr	720	665	+55
TlI	695	636	+59
RbCl	677.8	677.8	0
RbBr	649.4	653.1	−3.7
RbI	613.0	619.2	−6.2

The difference is greatest for the lithium compound, showing the large polarizing effect of the small lithium ion. The sign of the deviation probably results from the unusual compressibility of the hydride ion (compare Section 9.1).

The values of the silver and thallium(I) halides are given for comparison with those of the alkali metals of most similar radius, e.g. potassium and rubidium. The figures illustrate the effect of the filled d shell in such ions. The values for the silver and thallous halides (which have the sodium or caesium chloride structures) show marked discrepancies between the calculated and experimental lattice energies and these differences increase from the fluoride to the iodide. In these cases the experimental values are all higher than the calculated ones, showing the presence of the additional energy term due to polarization, which is largest for the iodides. It is interesting to note that gold(I) iodide, AuI, which has a markedly higher lattice energy ($-1050\,\mathrm{kJ\,mol^{-1}}$) than silver iodide, has had its structure determined. This consists of chains ··· Au-I-Au-I- ··· and is clearly not ionic at all. The silver salt thus provides an intermediate case between the ionic alkali metal iodides and the covalent aurous iodide. This example illustrates that the structure of a solid may provide an additional criterion for a departure from ionic bonding.

It is characteristic of an ionic solid that the forces are equal in all directions so that a symmetrical arrangement of ions results. At the other extreme, in a molecular solid formed by a covalent compound, there are two quite different kinds of force: very strong interactions in the bonds between the atoms of the molecule, and very weak interactions between molecules. When the solid is intermediate between these extreme types, there is often stronger bonding in some directions than in others and this shows up in a lowered symmetry of the structure. Thus chains may be formed as in gold(I) iodide above, and another common deviation from ionic bonding is the formation of layer lattices as in CdI_2 (Fig. 5.10a). In general, departure from purely ionic bonding is shown by the adoption of lattices where ions do not have their neighbours in the most symmetrical possible environments. One example is nickel arsenide which is an AB structure with 6:6 coordination but, although the six neighbours of the nickel atom are disposed octahedrally, the six neighbours of the arsenic atom lie at the corners of a trigonal prism. (see entry (e) in Table 5 on page 6). Some of the commoner less-regular structures are shown in Fig. 5.10.

It must be noted that, although structures where the atoms are in environments of low symmetry indicate the presence of nonionic contributions to the bonding, the converse is not true. Many compounds with symmetrical structures are not ionic. For example, many compounds of the transition metals with small atoms, the so-called 'interstitial' compounds, such as TiC or CrN, have the sodium chloride structure but there is no question of these compounds containing C^{4-} or N^{3-} ions and their bonding has appreciable metallic character. Similarly, the silver halides have the sodium chloride structure although there is appreciable covalent character in the bonding as just discussed.

5.6 Metallic bonding

While elements to the right of the Periodic Table favour electron pair sharing and are covalently bound in their elemental form, those to the left, which readily lose their valency electrons, are metallic. A metal has been picturesquely described as 'an array of cations in a sea of electrons'. The cations usually assume one of three simple arrangements described below and the valency electrons become completely delocalized over the whole structure. The electrons are mobile, accounting for the typical metallic properties of high electrical and thermal conductivity, and there are no underlying directed bonds.

From one point of view, metallic bonding is the limit of the process of delocalizing σ electrons. Consider, for example, a Li_{10} unit. From the 10 s orbitals, we can form ten 10-centre orbitals by an extension of the steps outlined in Section 4.7. Only five of the ten orbitals will be needed to hold the ten valency electrons. Furthermore, the spread of energies between the most and least stable of the ten orbitals is limited, so that the energy gap between the highest filled orbital and the lowest empty one will be relatively small. If we now go to a Li_{1000} unit, and form orbitals delocalized over the whole 1000 atoms, the separation between successive orbitals becomes tiny. In the terminology used in metal

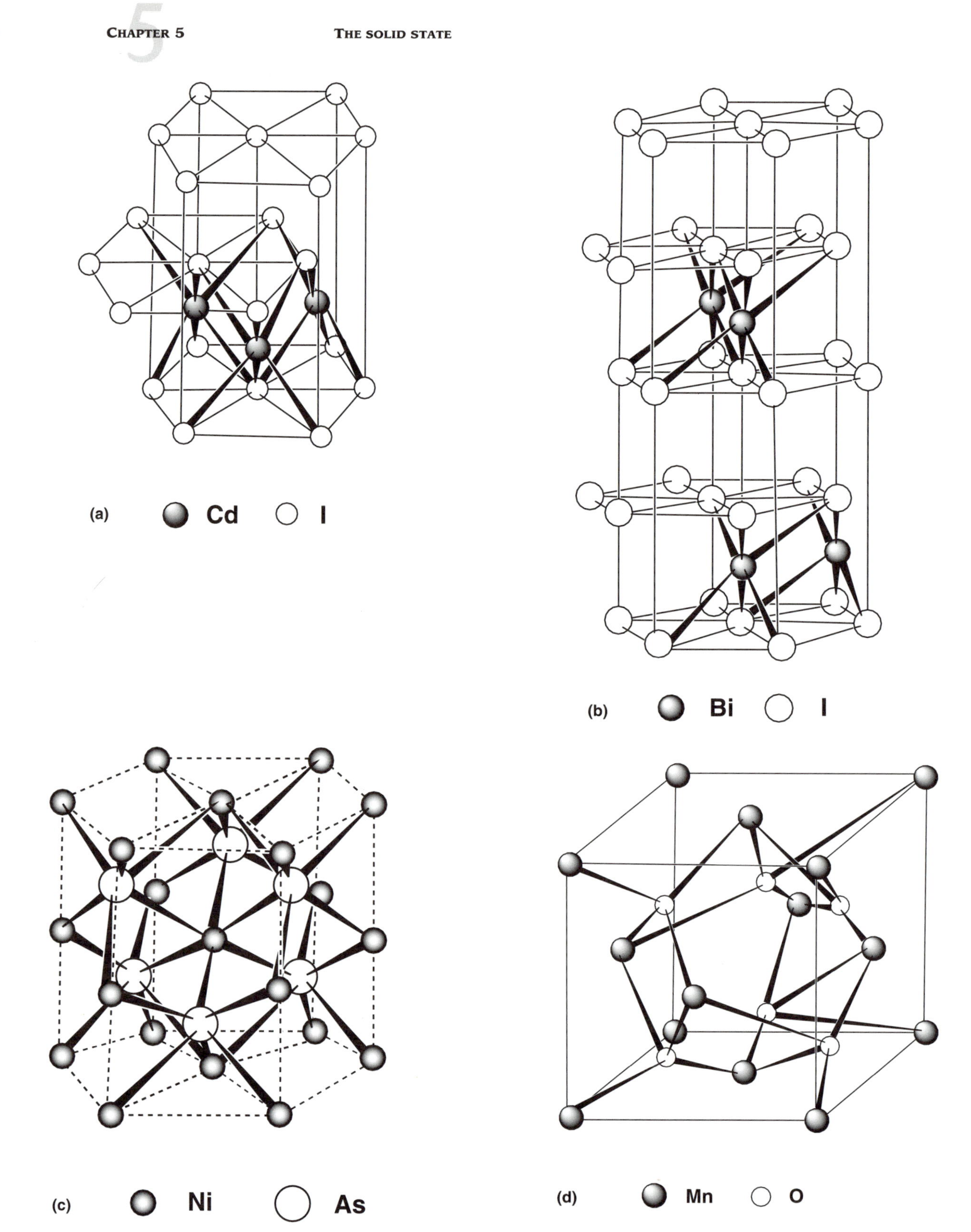

FIG. 5.10 Less-regular structures: (a) cadmium iodide, (b) bismuth triiodide, (c) nickel arsenide, (d) Mn_2O_3. In these structures some or all the ions have their neighbours in coordination positions which are not of the highest possible symmetry and this reflects the directional, nonionic character of part or all of the bonding forces

theory, we are creating a *band* of orbitals, here an s *band*. There are sufficient electrons to half-fill this band. However, the energy gap between this highest filled orbital (the 500th) and the lowest unfilled one is tiny. The population of electrons will have statistical spread of energies, so a few will be found in the 501st and higher orbitals, with corresponding gaps in the orbitals up to 500. Entropy effects will act similarly. These electrons are unpaired and in orbitals which interact with all 500 Li centres. In the case of a crystal, with the number of centres of the order of Avogadro's Number, delocalized orbitals will form over all these centres. Electrons are placed in these orbitals and the higher energy ones can interact with all the metal cations. The result is a wave-mechanical description of a metal where the electrons are standing waves over the whole crystal.

A band need not involve only s electrons. Thus, in calcium, for example, the p orbitals overlap to form a p *band*, whose range of energies overlaps with that of the s band. Thus, though there are 2 electrons per Ca atom, they do not simply fill the s band (which would give an insulator), but partly fill the overlapping p band.

Because the electrons at the top of any partly filled band are free to move through the whole crystal, this accounts for the high electrical and thermal conductivity of metals. Similarly, since excitations of electrons into the upper part of the band can occur with a wide range of energies, electrons interact with all wavelengths of light, giving rise to metallic lustre. When transition elements are involved, d *bands* also occur.

Without going into further detail it can be summarized that there are three basic types of bonding: ionic, involving complete transfer of an electron from one atom to another; covalent, with the sharing of a pair of electrons between two atoms; and metallic, where the electrons are completely delocalized over the crystal. If these three extreme types are thought of as being placed at the corners of a triangle, then some compounds will be represented by points near the vertices of this 'bond triangle', with bonding predominantly of one type. Some compounds will be represented by points along an edge of the triangle, with bonding intermediate between two types. Finally, the majority of compounds would be represented by points within the area of the triangle, showing that the bonding had some of the characteristics of all three types. This idea is illustrated schematically in Fig. 5.11.

There are three common metallic structures which are illustrated in Fig. 5.12, and nearly all metals adopt one or other of these. Two are based on close-packing of spheres, i.e. the metal ions (assumed to be spherical) are arranged to fill the space as closely as possible.

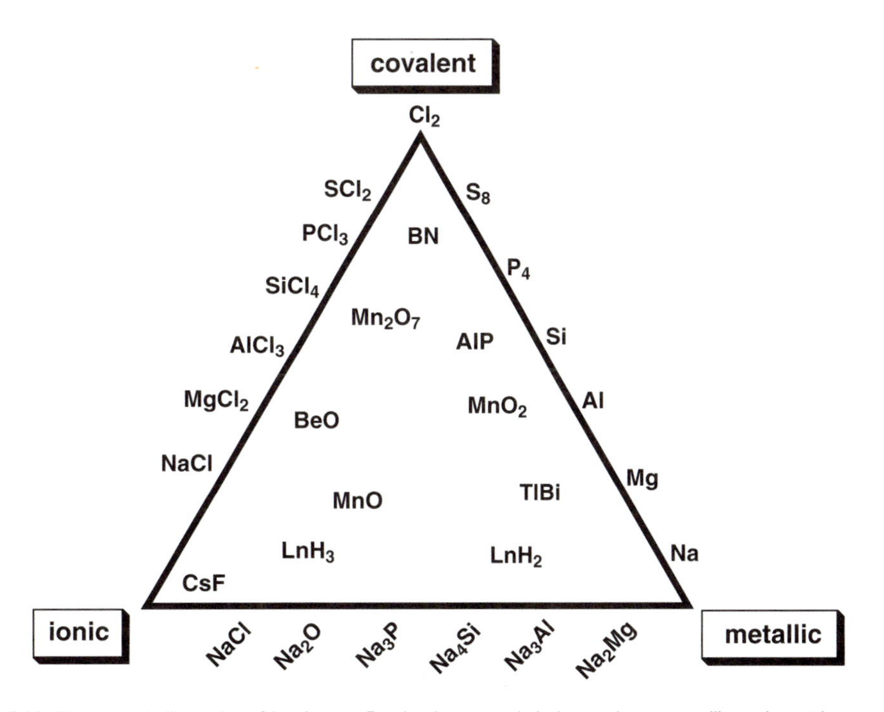

FIG. 5.11 Diagrammatic illustration of bond types. Few bonds are purely ionic, covalent or metallic, and most have some characteristics of all three types and would lie within a triangular plot of the type shown. This presentation also emphasizes that there is no sharp boundary between bonds of different types. Note that the electronegativity increases along the direction from metallic to covalent from the lowest to the highest values. Along the directions from points on this edge to the *ionic* apex there is an increase in the difference between the electronegativities of the two components

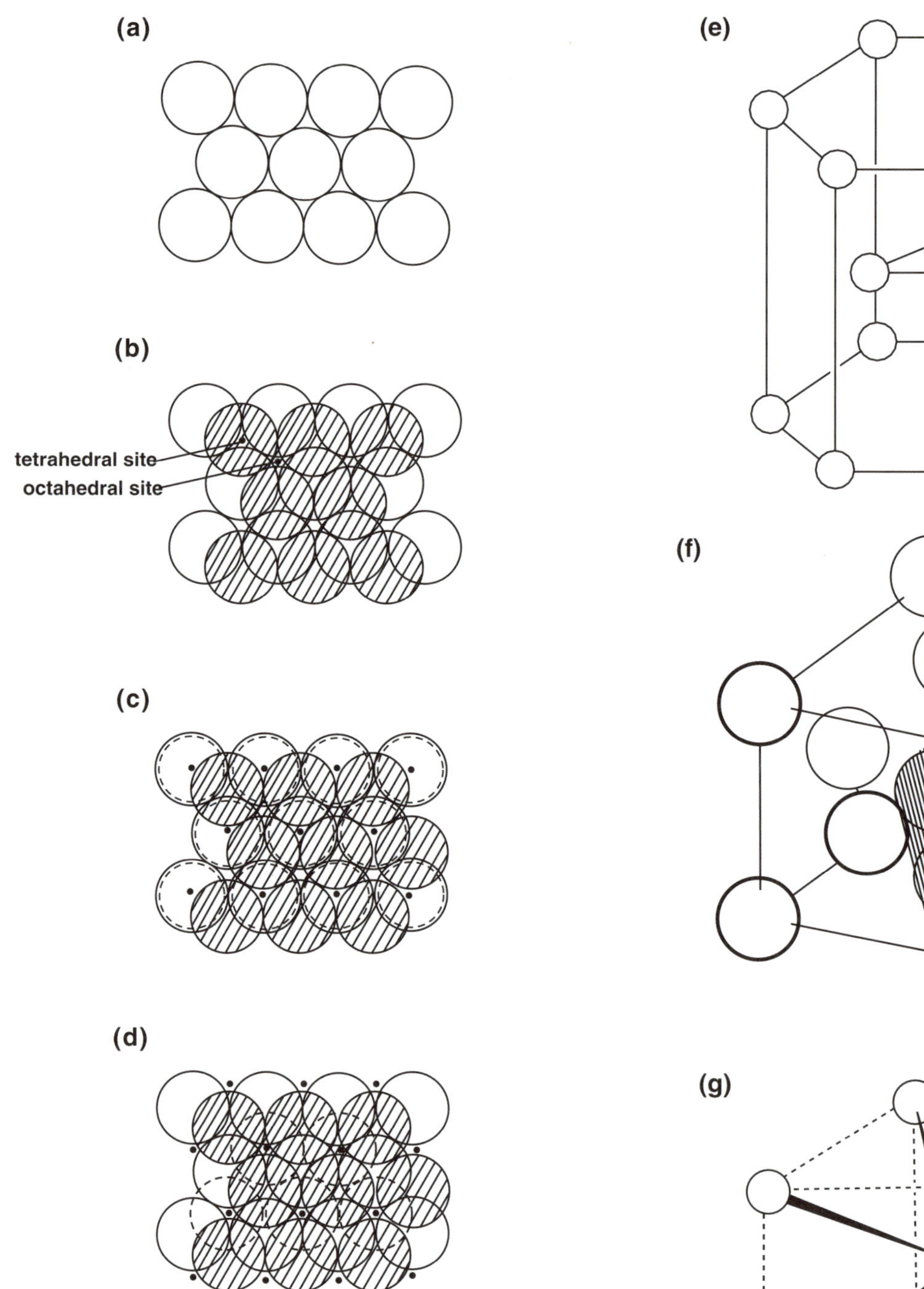

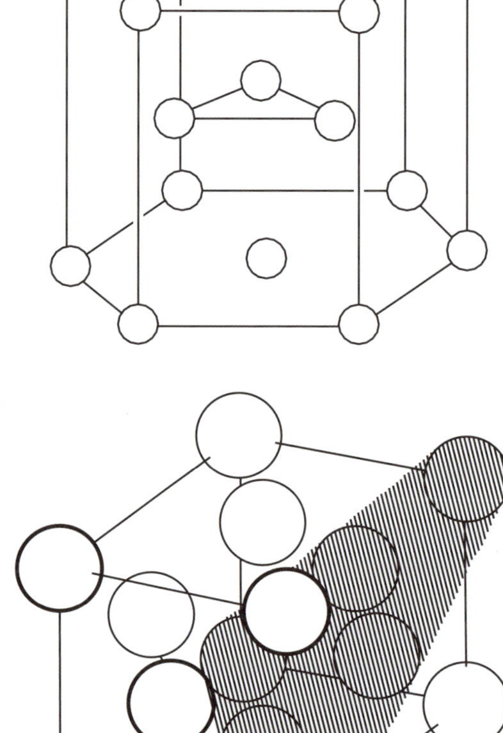

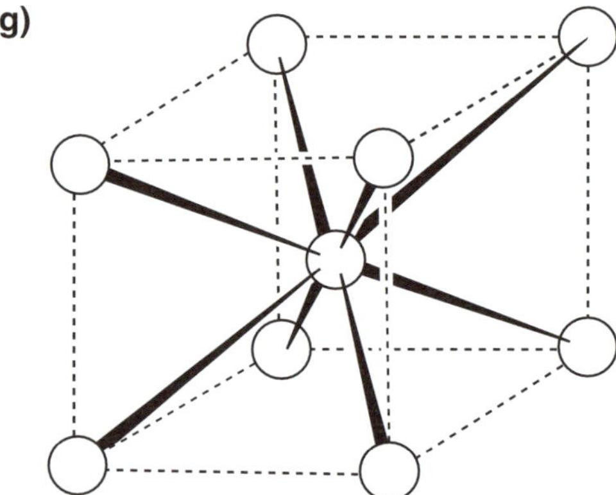

FIG. 5.12 Common metal structures and their construction: (a) a close-packed layer of spheres, (b) addition of a second layer to (a) in close-packing, (c) the addition of a third layer to (a) and (b) directly above (a), (d) alternative way of adding third layer, (e) hexagonal close-packing, (f) cubic close-packing (fcc), close-packed layer shaded, (g) the body-centred cube

For close packing, a layer of spheres can be arranged in only one way, as shown in Fig. 5.12a, in which each sphere is in contact with six neighbours. A second layer can be arranged on top of the first, again in only one way if close-packing is to be preserved, and this is shown in Fig. 5.12b. There are two possible ways of adding the third layer, both preserving close-packing. These are either directly above the first layer (Fig. 5.12c) or in a third position illustrated in Fig. 5.12d. The first case then gives a repetition of the second layer for the fourth one, the first and third layers for the fifth one, and so on in an ABABAB ··· arrangement. This gives rise to *hexagonal close-packing* (*hcp*) which is shown in Fig. 5.12e.

In the second case, the arrangement of the first three layers may be labelled ABC and the repeated pattern is then ABCABCABC ···, to give *cubic close-packing* (*ccp*). This is shown in Fig. 5.12f. In this structure, the close-packed layer is not parallel to the base of the cube, as it was in hcp, but lies along the body diagonal of the cubic array as indicated by the shaded plane in Fig. 5.12f. Put another way, the close-packed layers of Fig. 5.12d have to be turned through 45° to give the unit cube. An alternative name for ccp is *face-centred cube* (*fcc*) describing the orientation of Fig. 5.12f.

The third common metal structure is the *body-centred cube* (*bcc*), shown in Fig. 5.12g, which is not close-packed and in which the coordination number is eight. These high coordination numbers are typical of metal structures.

A number of common crystal structures are closely related to one or other of the close-packed structures above. In cubic close-packing, there are two different kinds of interstitial sites, see Fig. 5.12b. One of these is a tetrahedral site between four of the spheres, and the other is an octahedral site between six of the spheres. There are as many octahedral sites as there are spheres, and twice as many tetrahedral sites as spheres. Now suppose that a small atom was inserted into each octahedral site in a close-packed structure, leaving the large atoms in contact: the structure which results is of formula AB. The two kinds of atoms are 6-coordinated and this is the same as the sodium chloride structure. Thus the sodium chloride structure may be described as derived from cubic close-packing with all the octahedral sites occupied. Of course, in some compounds with the sodium chloride structure, the ions are of such relative sizes that the smaller ones force the larger ones out of contact with each other so that they are no longer close-packed, but their relative positions remain the same and the structure is often described in terms of the close-packed form.

Many of the other common structures may be described in similar terms, with the larger ions—usually the anions—in cubic or hexagonal close-packing, and the smaller ions occupying some fraction of the octahedral or tetrahedral sites in the close-packed structure. Table 5.3, summarizes the common structures in terms both of the coordination numbers and of their relation to the close-packed structures.

Apart from metals and alloys, metallic bonding is found in a number of other types of compound. One class consists of the so-called 'interstitial' hydrides, carbides, nitrides and borides of the transition metals. These compounds are formed with a variety of compositions—W_2C, TiN, ZrH_2, etc.—which do not commonly fit any normal ideas of valency. The compounds have metallic properties, such as conductivity and magnetic ordering, but have different structures from the parent metals. The bonding is still the subject of some controversy but most theories agree in leaving some valency electrons in conduction bands to give metallic properties. Thus these compounds may be pictured as,

METAL–METAL BONDS

There is a complete range of bond types involving metal atoms, from those involving only two metal centres, through polycentred clusters, giant clusters, microcrystallites, to metal crystals. There is much technological interest in the intermediate sizes—and an interesting question about what size is needed for characteristic metallic properties to appear (the current answer seems to be as low as a few hundred atoms). Bonds between two metal atoms are found in R_6M_2 for M = Pb, Sn, Ge and R a variety of organic groups; $(CO)_5Mn\text{-}Mn(CO)_5$; a number of transition metals bonded to Hg, Ge, Sn, as in $Pt(SnCl_3)ClL_2$ (L = π-bonding ligand) and so on.

Examples of clusters start with triangles, as in $Re_3Cl_{12}^{3-}$ (Fig. 15.25), include the $M_6X_8^{4+}$ ions of molybdenum and tungsten (Section 15.4), the $M_6X_{12}^{2+}$ ions of niobium and tantalum (Section 15.3), both of which contain octahedra of metal atoms, the Bi_9^{5+} cluster (Section 17.6.4), and the clusters of Sections 16.8 and 18.4.

The field of metal–metal bonding in clusters or in small molecules has attracted a lot of attention in the hope that these studies may throw light on the action of metallic catalysts.

TABLE 5.3 Summary of common structures and their relation to close-packing

Structure	*Figure*	*Coordination*	*Description in terms of close-packing*
		FORMULA TYPE AB	
Zinc blende ZnS	5.1c	4:4 Both tetrahedral	S atoms ccp with Zn in half the tetrahedral sites (every alternate site occupied)
Wurtzite ZnS	5.1d	4:4 Both tetrahedral	S atoms hcp with Zn in half the tetrahedral sites
Sodium chloride NaCl	5.1a	6:6 Both octahedral	Cl atoms ccp with Na in all the octahedral sites
Nickel arsenide NiAs	5.10c	6:6 Ni octahedral As trigonal prism	As atoms hcp with Ni in all the octahedral sites (note that the two types of position cannot be equivalent in hcp)
Caesium chloride CsCl	5.1b	8:8 Both cubic	Not close-packed. The AB_8 and A_8B arrangements are like bcc
		FORMULA TYPE AB_2	
β-cristobalite SiO_2	5.3c	4:2 Tetrahedral and linear	The Si atoms occupy both the Zn and S positions in zinc blende (this is equivalent to two interpenetrating ccp lattices) and the O atoms are midway between pairs of Si
Rutile TiO_2	5.3a	6:3 Octahedral and triangular	Not close-packed. The Ti atoms lie in a considerably distorted bcc
Fluorite, CaF_2 and anti-fluorite, Li_2O	5.3b	8:4 Cubic and tetrahedral	Ca atoms ccp with F in all the tetrahedral sites O atoms ccp with Li in all the tetrahedral sites
Cadmium iodide CdI_2	5.10a	6:3 layer lattice octahedral and the 3:coordination is irregular	I atoms are hcp and Cd atoms are in octahedral sites between every second layer The $CdCl_2$ structure is similar but the Cl atoms are ccp
		FORMULA TYPE AB_3	
Rhenium trioxide ReO_3	5.4a	6:2 Octahedral and linear	O atoms are in $\frac{3}{4}$ of the ccp sites and Re atoms are in $\frac{1}{4}$ of the octahedral sites
Bismuth triiodide BiI_3	5.10b	6:2 layer lattice octahedral and the 2:coordination is non-linear	I atoms are hcp and Bi atoms occupy $\frac{2}{3}$ of the octahedral sites between every second layer The $CrCl_3$ structure is similar but the Cl atoms are ccp
		FORMULA TYPE M_2O_3	
Corundum Al_2O_3	5.4b	6:4 Octahedral, and at four of the six corners of a trigonal prism	O atoms are hcp with Al atoms in $\frac{2}{3}$ of the octahedral sites (cf. NiAs with $\frac{1}{3}$Ni missing). Ilmenite ($FeTiO_3$) is the same structure with alternate layers of Fe and Ti atoms in the Al sites
Manganese(III) oxide Mn_2O_3	5.10d	6:4 Six of eight cube corners, and tetrahedral	Mn atoms ccp with O atoms in $\frac{3}{4}$ of the tetrahedral sites (cf. fluorite)
		OTHER TYPES	
Perovskite $CaTiO_3$	5.4c	Ca–12O (ccp) Ti–6O (octahedral) O–2Ti (linear) and 4Ca (square) giving distorted octahedron	Ca and O atoms together are ccp, with Ca in $\frac{1}{4}$ of the positions in a regular manner. The Ti atoms are in $\frac{1}{4}$ of the octahedral sites
Spinel $M_2^{III}M^{II}O_4$	5.4d	M^{II}–tetrahedral M^{III}–octahedral	O atoms are ccp and $\frac{1}{8}$ of the tetrahedral sites and $\frac{1}{2}$ the octahedral sites are occupied by metal atoms. In a *normal* spinel, the M^{II} ions are tetrahedral and M^{III} octahedral. In *inverse* spinels, M^{II} ions are octahedral and half M^{III} are octahedral and half are tetrahedral

Notes: (i) Cubic close-packed = ccp: hexagonal close-packed = hcp: body-centred cube = bcc. (ii) In both ccp and hcp, there are two tetrahedral sites and one octahedral site for each atom in the close-packed lattice.

for example, ionic structures but with additional electrons providing metallic properties (see Section 9.5 on the metallic hydrides). Many low oxidation state halides of the transition metals may belong to this class. These compounds, which exist only in the solid state, like ThI_2 or $NbCl_2$, may be regarded as containing $M^{\nu+}$ cations (ν = normal Group oxidation state of the metal) and ν electrons which form anions and provide conduction electrons. Thus, ThI_2 would contain Th^{4+} ions, two I^- ions and two conduction electrons per formula unit.

5.7 Complex ions

Although the discussion so far has been concerned with simple ions, most of the points made apply to complex ions as well. Many of the structures of compounds containing complex ions are simply related to those discussed for simple ions. For example, the relationship between the calcium carbide structure of Fig. 5.13, and the sodium chloride structure is obvious. Similarly, sodium nitrate, calcium carbonate and potassium bromate all have structures in which the anion occupies the chloride ion position of the sodium chloride structure. Sodium iodate has a caesium chloride structure while potassium nitrate and lead carbonate have nickel arsenide structures. In most cases of complex ion salts, as in all the above examples, the actual symmetry of the lattice is lower than that of the sodium chloride lattice as the complex ions are not spherical. The calcium carbide lattice, for example, is elongated in the direction parallel to the axes of the carbide ions to give a tetragonal rather than a cubic lattice. However, there are a number of cases where complex ionic lattices are of high symmetry. This occurs if the complex ion is able to rotate in its lattice position at room temperature. The resulting, averaged-out, configuration is spherical as in the alkali metal borohydrides, MBH_4. These compounds have actual sodium or caesium chloride configurations at room temperatures as the BH_4^- ions are freely rotating. An interesting intermediate case is provided by the alkali metal hydroxides. These have lattices of low symmetry at room temperature, in which the OH^-

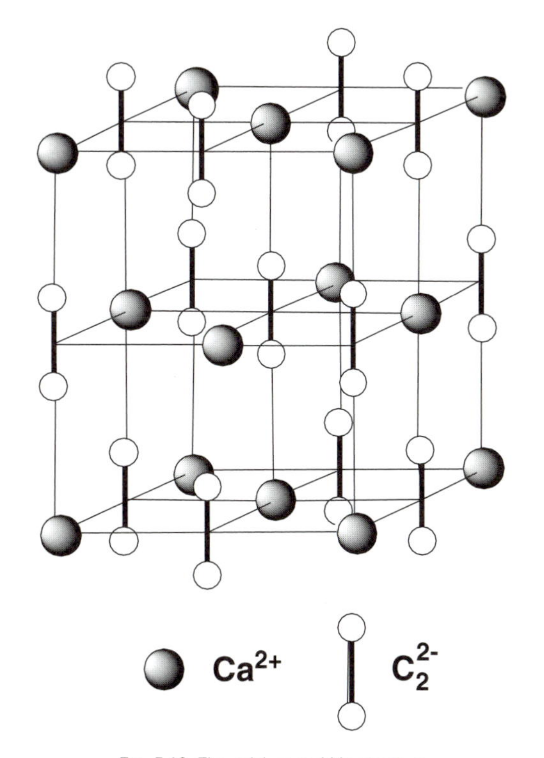

FIG. 5.13 The calcium carbide structure

groups are not rotating. When the temperature is raised, enough energy is present to allow the hydroxide ions to rotate and the alkali hydroxides undergo a transition to a high temperature form with the sodium chloride structure. Complex cations behave in exactly the same way; most ammonium salts, for example, have sodium chloride or caesium chloride structures as the NH_4^+ ion is freely rotating. Hydrated and other complex cations also typically form lattices of high symmetry with large anions. Thus $[Co(NH_3)_6][TlCl_6]$ has the sodium chloride structure while $[Ni(H_2O)_6][SnCl_6]$ crystallizes in the caesium chloride structure.

Compounds of more complicated formula type also mirror the structures of the simple ionic types. One example is provided by the many compounds which crystallize with the K_2PtCl_6 structure. This is related to the structure of calcium fluoride but with the cations in the fluoride positions and the large anion in the calcium site (called the *anti-fluorite* structure).

Compounds containing complex ions may also be found with the less symmetrical structures associated with significant nonionic contribution to the bonding. Lead carbonate and one form of calcium carbonate have the ions in the same positions as in nickel arsenide while a number of complex fluorides, such as K_2GeF_6, have layer structures related to that of cadmium iodide.

When complex ions are present, a permanent dipole often exists, and the interaction with this dipole has to be added to the interactions between ions, and between induced dipoles and ions, which have already been discussed. One example is provided by the hydroxide ion where there is a permanent dipole $O^{\delta-}$–$H^{\delta+}$ in addition to the negative charge. In the high temperature form where the hydroxyl group is freely rotating, the ion–dipole forces are equally directed, but when the hydroxyl groups become fixed in orientation, the existence of the dipole means that the forces between anion and cation differ in the direction of the dipole from those in directions perpendicular to the dipole. Salts containing small complex ions may thus be equivalent to those with simple ions when the complex ions can rotate freely, or the presence of permanent dipoles may introduce a directional element into the bonding in the crystal. Just as simple ions may occur in compounds which are ionic or which have nonionic contributions to the forces in the crystal, so do small complex ions occur in symmetrical crystals which are ionic to a high degree of approximation and also in less symmetrical compounds with layer or other 'nonionic' structures.

In addition to compounds with small, discrete, complex ions, there are very extensive series of compounds of large condensed ions, especially those containing condensed oxyanions such as polyphosphates or silicates (see Section 18.6). Such compounds form a vast topic of their own and the structural problems involved have often been very difficult to study.

5.8 The crystal structures of covalent compounds

In the largely ionic or metallic compounds discussed above there is a reasonable uniformity of bond strength throughout the crystal. Either all the bonds are identical, as in salts of simple ions and in elemental metals, or, even when there are strongly bonded units within the solid, as in the salts of complex ions, these units are bonded together by strong ionic forces. Such relative uniformity of bond strength is reflected in the hardness and fairly high melting points which are common among the compounds discussed above. When compounds in which the bonding is largely covalent are examined, these properties are no longer found except in relatively few examples, like diamond or silica, which are very hard and high melting. The vast majority of covalent compounds form soft, low-melting solids of low crystal symmetry, and the actual structure of the solid form of these compounds is usually of relatively minor importance for an understanding of their chemistry.

The hard, high-melting covalent compounds are those where covalent, electron pair bonding extends throughout the crystal so that the whole crystal is a 'giant molecule'.

STRUCTURES OF THE ELEMENTS OF THE CARBON GROUP

The structures of the elements in the carbon Group give a good illustration of this type of giant molecule and the effect of deviations from it. Carbon, silicon, germanium and tin (in the grey form) all occur with the diamond structure but the valency electrons become increasingly mobile with increasing atomic weight so that silicon, germanium and grey tin have conductivities increasing in that order. These conductivities are much higher than those of insulators, such as diamond, but many times less than the conductivities of true metals. Such compounds are called *semiconductors*. The valency electrons are largely fixed in the bonds but have a small mobility; in other words, these elements are at the start of the trend from covalent to metallic properties.

White tin and lead have different structures from the diamond one (Fig. 5.14) and tend much more towards metallic forms. White tin has the atoms in a distorted octahedral configuration with four nearest neighbours and two more a little further away, while lead has an approximately close-packed structure but with an abnormally large interatomic distance. These two elements represent further steps away from the covalent, low-coordinate structures towards the high-coordinate structures with close-packing which are typical of metals. Both have conductivities in the metallic range, and thus mobile electrons. However, even lead differs a little from the typical metal in its high interatomic spacing. These elements would all lie along the covalent-metal edge of the ionic–covalent–metallic triangle of bond types,

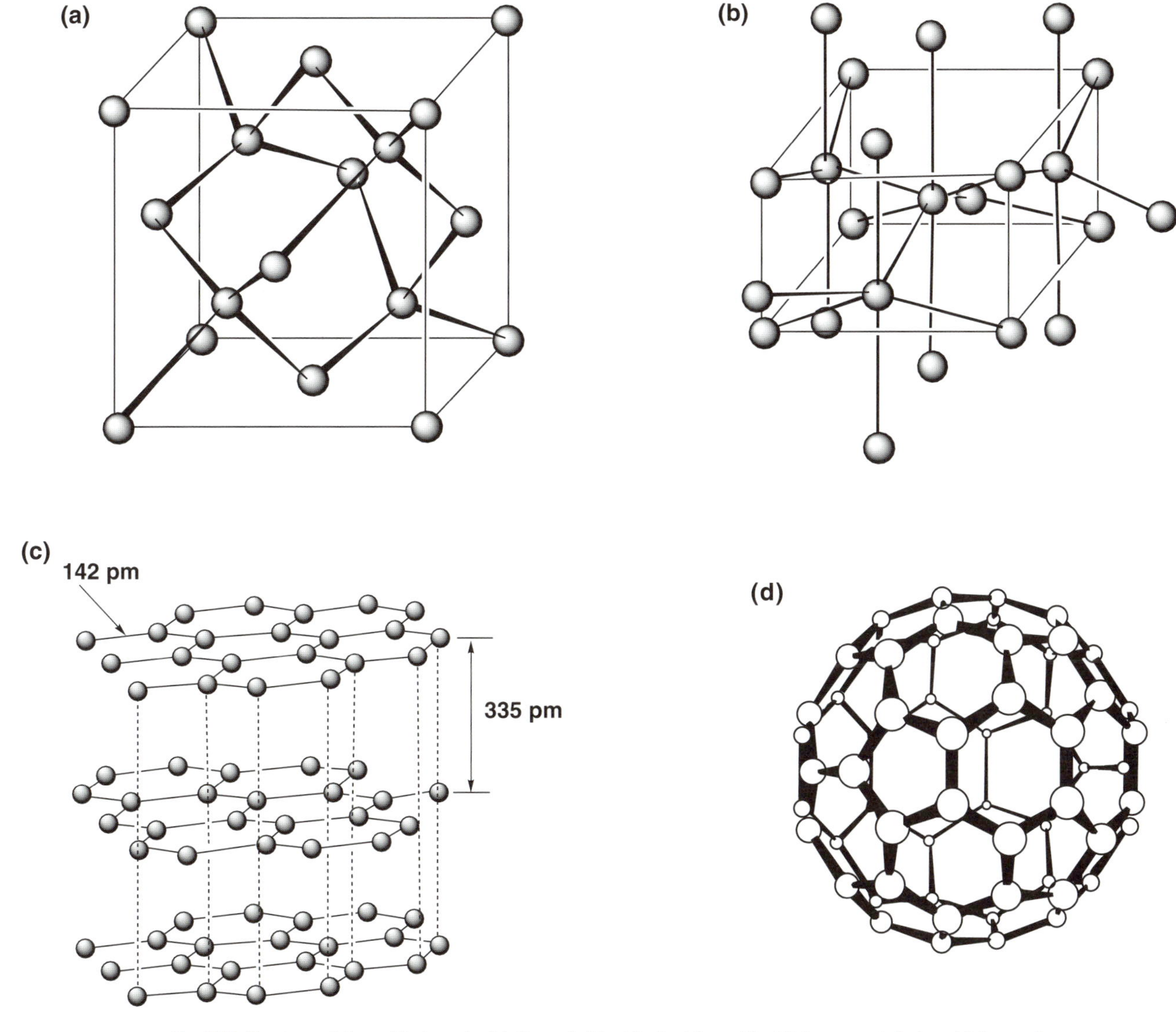

FIG. 5.14 Structures of Group 14 elements: (a) diamond, (b) white tin, (c) graphite, (d) the soccer ball-shaped fullerene, C_{60}

Fig. 5.11, with diamond at the covalent apex, silicon, germanium and grey tin near the covalent end of the edge, and white tin and lead nearer the metallic end of the edge.

As well as diamond, carbon is found in two other forms—so far unique to it—graphite and fullerene. Graphite is a much softer solid than diamond with a pronounced horizontal cleavage so that it readily flakes. These properties reflect the bonding as the carbons are 3-coordinated in a planar sheet, and the fourth electron and the fourth orbital form delocalized π bonds extending over the sheet. The sheets of carbon atoms are very strongly bound but the forces bonding the sheets together are relatively weak (shown, for example, by the C-C distances of 142 pm within the sheets and 335 pm between them), so the sheets readily slide over one another, giving graphite its lubricating properties. Thus the strength and external properties of the solid as a whole reflect the bonding, especially the weakest links in the solid.

Very recently, a new family of related structures has become established called fullerenes—these can be viewed as a new allotropic form of carbon. The simplest members are molecular structures, the type member being C_{60} which is a polyhedron consisting entirely of carbon atoms arranged in linked hexagons and pentagons. These molecules pack together efficiently and the molecule crystallizes in a face-centred cubic lattice. Such linked polyhedral structures were developed by the American architect Buckminster Fuller in the form of geodesic domes and many names have been coined such as buckminsterene, fullerene, 'buckyballs' and 'buckytubes'. The most-studied fullerene, C_{60}, has the shape of a soccer ball, see Fig. 5.14d, and contains twelve pentagons. This is the critical condition for forming closed polyhedra of this sort. Other members, such as C_{70}, C_{82}, etc., have the same 12 pentagons fused to an increasing number of hexagons. Larger extended clusters have also been identified, together with related polymeric forms including carbon 'nanotubes' made up of cylindrical graphite-like sheets with fullerene-like ends. These materials are currently the subject of intensive research around the world. In the few short years since their discovery, a whole new area of chemistry has opened up. A more detailed account is given in Section 19.3.

An obvious example is diamond in which each carbon atom is bonded tetrahedrally to four others and this structure continues throughout. In order to melt or fracture such a crystal, strong covalent bonds must be broken and this requires considerable energy. Similar examples are silica, with 4-coordinate silicon and 2-coordinate oxygen throughout, silicon and silicon carbide and the tetrahedral form of boron nitride—all with the diamond structure—and oxides of other elements of moderate electronegativity, such as aluminium.

When ordinary covalent compounds are considered, it is the weak intermolecular forces which are reflected in the crystal properties. In a simple molecule such as methane, the atoms are strongly linked together by directed bonds in the molecule, but the only forces between molecules are the very weak 'Van der Waals' interactions due to induced dipoles. Thus the compound melts readily and the solid is soft and readily fractured as only these weak interactions have to be overcome. Naturally, in such a case, properties of the solid give no information about the bonds in the compounds.

5.9 Defect structures and nonstoichiometric solids

Real crystals rarely show the ideal structures discussed in the earlier sections of this chapter. Some of the departures from ideality are trivial. For example, most solids are made up of a large number of small domains of ideal structure (called crystallites) which fit discontinuously at the edges and may have missing ions at these boundaries. Another case is found in many minerals where ions of similar size substitute one for another, for example Fe^{2+} for Mg^{2+}, so that there may be a continuum of compositions between the pure Fe and pure Mg species.

Note that the entropy effect means that normal solids will necessarily include defects.

Within a crystallite, or single crystal, departures from the ideal arrangement may occur with maintenance of the overall stoichiometry. In the *Frenkel defect*, Fig. 5.15a, an atom or ion is displaced from its regular position into a non-lattice site. This will usually be the cation, the smaller species. In the *Schottky defect*, Fig. 5.15b, atoms or ions are missing. These, and other, rarer defects are often marked by colour and may markedly affect

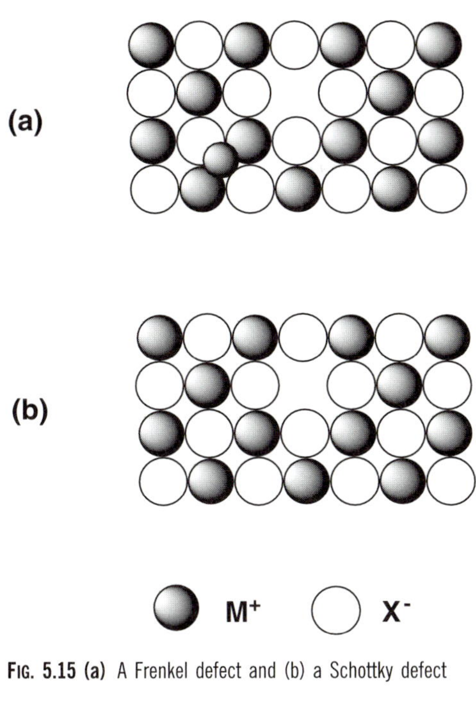

FIG. 5.15 (a) A Frenkel defect and (b) a Schottky defect

the conductivity of crystals. Solid state reactions, which involve atoms migrating through the lattice, will be markedly affected in rate. Clearly it requires much less energy for atoms or ions to move via vacant sites than by exchanging in a complete lattice.

Defects have two effects on the energy. Loss of an ion, or placing it in an irregular site, correspond to loss of energy compared to the ideal arrangement. There will also be an entropy contribution (arising from the number of ways the ions and defects may be interchanged). Both these contributions mean that the number of defects will be temperature dependent. While defects can be 'frozen in', at equilibrium the proportion of defects will be low at temperatures well below the melting point, but can be high within a few tens of degrees of the m.pt., and the formula may depart markedly from stoichiometric if, say, cations are more readily removed than anions.

This leads us to compounds which are markedly nonstoichiometric. In the 18th century, Dalton and Berthollet argued whether materials had fixed or variable composition. Dalton's view of fixed composition carried the day, but we now know of many systems where variable composition is normal and the name *berthollide* is often used as a descriptive term for these.

A good example is provided by the transition metal hydrides (Section 9.4). For example, titanium hydride can be regarded as forming from cubic close-packed Ti atoms with H atoms entering the tetrahedral sites. The hydrogen is taken up at elevated temperature and pressure. A whole range of uniform materials may be formed ranging from (say) $TiH_{0.1}$ to $TiH_{1.8}$ in overall composition (the limits depend on the temperature). A species of formula TiH has no special stability, but simply represents the stage where half the tetrahedral sites are occupied at random. At $TiH_{1.8}$, no metal phase remains, but 10% of the sites are empty. It requires prolonged treatment, or high H_2 pressures, to force the composition up to TiH_2. Since the defect structure has a higher entropy, if a lower enthalpy, the stable equilibrium composition may not be the stoichiometric one.

Similar nonstoichiometric phases are widely found among oxides, e.g. uranium and the actinides, ZnO, iron oxides (see Section 14.6); among sulfides and among the lower-valent halides.

The structural, electrical and optical properties of nonstoichiometric solids lead to a number of applications, but by far the most important is the controlled nonstoichiometry of semi-conductors. While silicon, or germanium, are intrinsically semiconductors (see above, Section 5.8), their properties may be vastly modified and tailored by adding impurities. If, say, indium is added in small amounts to silicon or germanium, it will occupy a site in the lattice but, as it contains one electron less than the Group 14 atom, there will be an electron vacancy (or positive hole). If a potential is applied an electron will move into the vacancy, leaving a hole and so on—i.e. the effect is of a positive charge migrating across the crystal. Similarly, addition of a Group 15 atom such as arsenic or antimony gives an excess electron and this negative charge will move under a potential. These p-type and n-type semiconductors are then united in various combinations to give all the components of modern electronics.

Problems

> In this chapter, we are concerned with quite complex three-dimensional arrays of atoms. Read the *note on crystal structure diagrams* on p. 109, and also look at as many models and structural texts as you can.
> You will need to use atomic parameters, especially from Sections 2.13 to 2.16, and these should be revised.

5.1 (a) From the values in Table 2.12, and following the discussion on p. 50, calculate a set of ionic radii consistent with the 'experimental' values.
(b) Calculate radius ratios from the data derived in (a) for the solids listed in Table 5.1. Comment on the predictions which result. Treat Shannon radii similarly.

5.2 Calculate correct to 2 significant figures the Madelung constant of a rectangular array of spacings r and $3r$.

5.3 From Table 2.12 and Section 5.3, calculate the approximate lattice energies of CaO, CaS and NaF based on the value in Fig. 5.5 for NaCl.

5.4 Calculate lattice energies, as in Question 5.3, for Na_2O, Na_2S, CaF_2, $CaCl_2$. (Assume the same Madelung constant as fluorite.)

5.5 The heats of dissociation of F_2 and O_2 are respectively 78.9 and 249.2 kJ mol^{-1}. The heat of formation of Ca atoms from the metal is 176 kJ mol^{-1}, and of S atoms from S_8 is 238 kJ mol^{-1}.
Work out the heats of formation of the solids in Questions 5.3 and 5.4 (see Tables 2.8 and 2.9). Discuss the order of stability of these solids. (See also Question 9.5.)

5.6 Decide what approximate lattice energy is appropriate for the hypothetical ionic solids CaF and CaCl. Hence calculate their heats of formation.
Why do CaF and CaCl not exist as stable species?

5.7 The experimental heat of formation of CuCl is 136 kJ mol^{-1}, the calculated lattice energy is 880 kJ mol^{-1}, and the heat of formation of copper atoms is 337 kJ mol^{-1}. Discuss whether CuCl is likely to be ionic.

CHAPTER 6

AQUEOUS SOLUTIONS 133
6.1 SOLUBILITY 133
6.1.1 Ionic Substances 133
6.1.2 Covalent Substances 135
6.2 ACIDS AND BASES 136
6.2.1 Strengths of Oxyacids 138
6.3 OXIDATION AND REDUCTION 138
NONAQUEOUS SOLVENTS 145
6.4 SOLUBILITY AND SOLVENT INTERACTION IN NONAQUEOUS SOLVENTS 146
6.5 ACID–BASE BEHAVIOUR IN NONAQUEOUS SOLVENTS 148
6.6 GENERAL USES OF NONAQUEOUS SOLVENTS 150
6.7 LIQUID AMMONIA 151
6.8 ANHYDROUS ACETIC ACID 153
6.9 'SUPERACID' MEDIA 155
6.10 BROMINE TRIFLUORIDE 156
6.11 SUPERCRITICAL FLUIDS 159
PROBLEMS 159

Solution Chemistry

Aqueous solutions

Most work in inorganic chemistry is carried out in solution and the observed results depend to a large extent on the properties of the solvent. These will be examined in this chapter. Water is still by far the commonest solvent, but an ever-increasing number of reactions are carried out in other solvent media. These range from solvents like liquid ammonia, which have much in common with water, through more exotic media like anhydrous hydrogen fluoride or liquid bromine trifluoride, to molten salts and even molten metals. The major features of aqueous chemistry are discussed first and then follows the extension of these principles to nonaqueous solvents, with detailed discussion of the properties of a small number of representative solvent systems.

The discussion of solution behaviour is divided into three sections;

(i) solubility and solvolysis, which depend directly on the interaction between solvent molecules and the solute,
(ii) acid–base behaviour,
(iii) oxidation–reduction behaviour.

None of these types of behaviour is independent of the others but it is convenient to make these broad divisions. They will be discussed in turn, first of all as they apply to solutions in water.

6.1 Solubility

6.1.1 Ionic substances

The energy changes which govern solubility are first considered for the process of dissolving an ionic solid in water. The principal enthalpy changes may be related by a simple cycle diagram such as Fig. 6.1. The process of solution is split into two steps:

(a) the ions in the solid are separated to infinity as gaseous ions, which requires the input of the lattice energy, U (see Section 5.3),
(b) the separated gaseous ions interact with the water molecules with the output of the heats of hydration H_{aq} of the cation and anion.

The heat of solution, H_s, is the difference between the lattice energy and the heats of hydration. This treatment should be compared with that used to analyse the formation of ionic solids (Section 5.2).

The factors affecting the lattice energy were discussed in Section 5.3. It will be recalled that the lattice energy is greatest for small ions of high charge and is increased if polarization effects are present to add the attractions between the ionic charges and induced dipoles to those between the ions themselves.

The heats of hydration of the gaseous ions arise from the electrostatic attractions between the ionic charges (and dipoles if any) and the dipole of the water molecules. These interactions are shown schematically in Fig. 6.2. In the water molecule, each O-H bond is polarized in the sense $O^{\delta-}$–$H^{\delta+}$ by the uneven sharing of the bonding electrons between the very electronegative oxygen and the less electronegative hydrogen. In addition, the effective negative charge on the oxygen atom is increased by the two unshared pairs of electrons. In the case of an anion, the relatively positive hydrogen atoms interact with the negative charge to give the anionic heat of hydration. The cations interact with

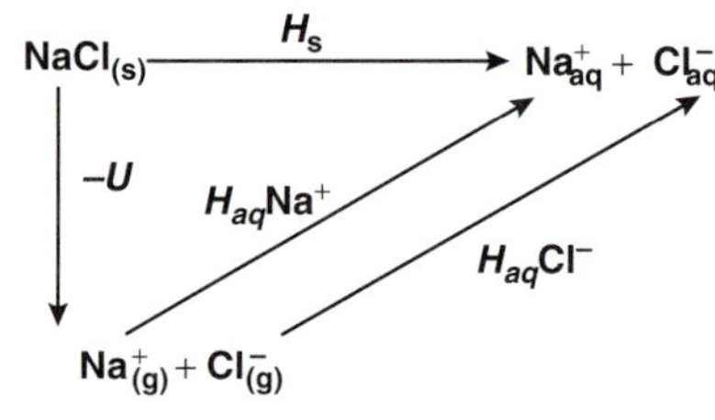

U = lattice energy of NaCl
H_s = heat of solution of NaCl
$H_{aq}Na^+$ = heat of hydration of Na^+
$H_{aq}Cl^-$ = heat of hydration of Cl^-

FIG. 6.1 Enthalpy changes in the solution of sodium chloride. The process of solution may be split up into simple steps, whose heat changes are known or measurable, in the same way as the formation of an ionic solid was treated in the Born–Haber cycle

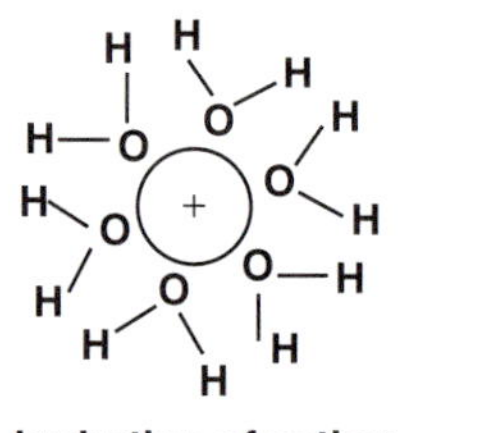

hydration of cation by water

hydration of anion by water

FIG. 6.2 Diagrammatic representation of the solvation of ions

the relatively negative oxygen atoms of the water molecules and the major effect is probably that due to the lone pairs. The energy of hydration of the cation is usually the most important exothermic term as the cation–lone pair interaction is strong and is enhanced by the general small size and consequent high charge density of cations. The strength of the anion and cation interactions with the water molecules increases with the charge on the ions and is inversely proportional to their sizes. The effect of ion size on the hydration energies is seen in Fig. 6.3 where the heats of hydration of the alkali halides are plotted against the anion radius.

It can be seen that the task of predicting the solubilities of ionic solids is very similar to that of predicting the formation of an ionic lattice which was discussed in the last chapter. The heat of solution, and therefore the probable solubility, is the result of the balance between the lattice energy of the solid and the heats of hydration of the gaseous ions. Both these factors are increased if the ions are small and of high charge, so that the difference between them may be expected to vary in a random manner and a general correlation of solubility with ion properties is impossible, each individual case having to be calculated separately. Solubility is yet another example of a phenomenon which depends on the small difference between large energies.

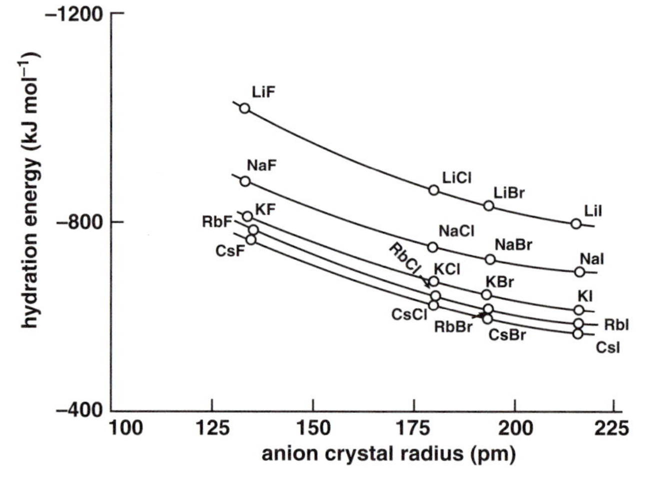

FIG. 6.3 Heats of hydration of the alkali halides. The heats of hydration are plotted against the anion radius to illustrate the effect of ion size on the hydration energy

There are, however, two factors which appear to have an over-riding effect on the balance of energies involved in solution. One is the effect of the ionic charge. When this is increased, the ionic sizes and the crystal structure remaining constant, the lattice energy increases more than the heats of solvation of the ions. Thus, A^+B^- solids, for example, are much more soluble than $A^{2+}B^{2-}$ solids of the same structure. Highly charged salts are generally of low solubility. A second effect which usually tends to decrease the solubility is the presence of polarization in the solid. A polarizing cation increases the lattice energy by the additional ion–dipole force which its presence introduces, and such a cation would also polarize the solvent molecules and increase the hydration energy in a similar manner. In this case, the effect of these forces appears to be largest in the solid and the presence of polarizing cations usually leads to low solubility, at least in cases like the silver and thallous salts which were discussed in Chapter 5.

The discussion above has been centred on heat changes whereas the true driving force is the free energy of solution which includes the entropy of solution as well as the heat. In general, entropy changes would be expected to favour the formation of solutions as the solute is much less ordered than the solid. (It will be recollected that a high degree of order corresponds to low entropy.) However, in a coordinating solvent such as water, the disorder of the solvent is reduced when ordered hydration sheaths are formed around the ions. If the ions are of high charge density, this increase in the ordering of the solvent may outweigh the decreased order of the solute. Thus, entropy changes should favour solution processes in general, but may oppose them for ions of high charge density which would be strongly solvated.

Another property of water, in addition to its power of hydrating ions, is important in determining its solvent power for ionic compounds. This is its very high dielectric constant $D = 81.1$. The attraction between opposite charges e_1 and e_2 at a distance r in a medium of dielectric constant D is $e_1e_2/4\pi\epsilon_0Dr^2$. Thus the interposition of a medium of high dielectric constant reduces the forces between ions in solution, hindering their precipitation, and assists the passage into solution of ions in the solid. Water is among the best of all solvents for ions combined with the high dielectric constant.

6.1.2 Covalent substances

By contrast, water is a poor solvent for covalent compounds, especially if these have no dipole or only a relatively weak one. To return to the heat cycle of Fig. 6.1, a covalent compound may go into solution as single molecules or as dimers or polymers, and the energy required to separate the appropriate species is analogous to the lattice energy U of an ionic compound.

In Section 5.8 we discussed the two extreme cases from the range of covalent compounds. For the 'giant molecule' type, with covalent bonding throughout, a very large energy would be required to split out any smaller particle and this is unlikely to be offset by the energy of hydration of the particle. Thus we find that solids such as diamond or silica are completely insoluble in water.

At the other extreme, the solid consists of individual molecules which are held in the crystal only by very weak Van der Waals forces. The heat required to overcome these attractions, and to dissolve the solid as individual molecules, is thus small. This should be readily available from hydration, which would arise from the interaction of the water dipoles with the dipoles induced by them in the covalent molecules. Yet we find that the solubilities of molecular covalent solids in water are usually very low. This apparent anomaly arises because our discussion so far has neglected a third term in the heat balance.

Just as the solid is prepared for solution by adding the lattice energy to separate the individual particles—ion, molecules, dimers, etc.—so we have to separate the solvent molecules to accommodate the solute particles. That is, heat is required to overcome the attractions between the water molecules and allow the solute molecules to enter between them.

This energy of the 'water structure' which arises from the attractions between the partial charges of the water dipoles, is negligible compared with the energies involved in the solution of an ionic solid and was neglected in the earlier discussion, but it becomes important when the much weaker interactions with covalent molecules are considered. Thus, before a covalent solid dissolves, enough energy must be available to separate the solute molecules and to separate the water molecules, and this can only be provided by the heat of hydration of the solute molecules. For nonpolar covalent molecules, the energy of the water structure is the dominant term and such compounds are of low solubility.

Between the extremes of giant molecule and weakly bonded molecular solid, we find the whole range of polar and polarizable covalent molecules. When polar molecules are involved, both the forces within the solid and the forces between the solute molecules and the water are increased, and the water structure energy no longer dominates the energy balance. Solubilities in such cases depend on the detailed balance of enthalpies but they tend to be higher than those of nonpolar compounds. For example, the solubility of the nonpolar molecule methane in water at room temperature is about 0.004 mole per litre, while the polar methyl iodide dissolves to the extent of 0.110 mole per litre.

With some solutes, interaction with the water molecule goes further than simple coordination and new chemical species are formed. Such a reaction, is termed *hydrolysis*. The distinction between hydration and hydrolysis is not completely clear-cut but hydrolysis implies a more extensive and less reversible interaction of the solute with the water molecule.

Some examples are given below:

$$SiCl_4 + 4H_2O \rightarrow Si(OH)_4 + 4HCl^*$$
$$O^{2-} + H_2O \rightarrow 2OH^-$$
$$HCl + H_2O \rightarrow H_3O^+ + Cl^-$$
$$BF_3 + 2H_2O \rightarrow H_3O^+ + (HOBF_3)^-$$

SOLUBILITY

This discussion may be summarized as follows:

(a) In the case of ionic solids strong forces are involved, those between ions in the solid and those between ions and dipoles in solution. Solubility depends on the detailed balance in each case, but salts with balanced numbers of ions with charges greater than one are of low solubility and compounds where there are appreciable polarization effects are also of low solubility.

(b) In the case of 'giant molecules' the strength of the forces binding the solid predominates and such compounds are insoluble in water.

(c) Covalent nonpolar solids are weakly bound and would be somewhat more strongly bonded to water molecules than to themselves, but there is insufficient energy to break down the 'water structure' and such compounds are insoluble in water.

(d) Polar covalent compounds present an intermediate range of interactions between those of (a) and (c). Solubilities vary but tend to be higher than those in class (c).

6.2 Acids and bases

The qualitative properties of acids—sharp taste, solvent power, effect on the colours of dyes, etc.—and the corresponding properties of bases were first listed by Boyle, and the succeeding centuries have seen a number of theories for acid–based properties developed and discarded. At present, three concepts are used to deal with acid–base phenomena in various solvents. These three theories overlap considerably although each has special uses and weaknesses. That of most value when dealing with aqueous solutions is the *protonic concept* of Brönsted and Lowry who characterized and related acids and bases by the equation:

$$\text{A (acid)} \rightleftharpoons \text{B (base)} + \text{H}^+ \text{ (proton)} \tag{6.1}$$

An acid is a proton donor and a base is a proton acceptor. Since the proton in this theory is the hydrogen nucleus which has such a high charge density that it is never obtained free in the condensed state (see Chapter 9), it follows that this is a 'half-equation' and the acidic properties of a molecule are observed only when it is in contact with the basic form of a second species. That is, the observed equation is always:

$$A_1 + B_2 \rightleftharpoons A_2 + B_1 \tag{6.2}$$

Such an equilibrium is displaced in the direction of the weaker acid and base, since the stronger acid is a stronger proton donor than the weaker acid and thus more of its molecules are in the basic form. The base corresponding to a given acid is called the conjugate base; clearly a strong acid has a weak conjugate base, and vice versa. The acid strength can be expressed in terms of the equilibrium constant of the above reaction:

$$K = \frac{[A_2][B_1]}{[A_1][B_2]} \tag{6.3}$$

This gives a method of expressing relative acid strengths. If one acid–base pair is taken as the standard, then the strengths of all other pairs may be expressed in terms of the equilibrium constants involving the standard acid. It is most convenient to choose as the

* These products undergo further reaction.

standard acid–base pair, the one involving the solvent. Thus in water, the acid strength is defined with respect to the pair H_3O^+-H_2O and the equilibrium used to define acid strengths is:

$$A + H_2O \rightleftharpoons B + H_3O^+ \text{ with } K' = \frac{[B][H_3O^+]}{[A][H_2O]} \tag{6.4a}$$

Providing dilute solutions are being used, the concentration of water $[H_2O]$ is essentially constant, so that the strength of an acid can be defined by:

$$K = [B][H_3O^+]/[A] \tag{6.4b}$$

The strengths of bases are quite naturally expressed by means of the K values and there is no need for a separate scale of base strengths. Since strong acids have weak conjugate bases and vice versa, the order of base strengths is the inverse of the order of acid strengths. Some typical K values for acids and bases in water are shown in Table 6.1. Also included are pK values which are the negative logarithms, analogous to the well-known pH values.

The relative strengths of the acids and bases which lie between the two acid pairs of water

$$\begin{aligned} H_3O^+ &\rightleftharpoons H_2O + H^+ \\ \text{and} \quad H_2O &\rightleftharpoons OH^- + H^+ \end{aligned} \tag{6.5}$$

are well defined, but it is impossible to measure the acid strengths of acids stronger than H_3O^+ (or of bases stronger than OH^-) as they are completely converted to the water acid (or base), that is, the equilibrium in Equation (6.4a) lies completely to the right. No acid stronger than the hydrated proton can exist in water. One way of determining the relative strengths of acids such as nitric acid or perchloric acid, which are completely dissociated in water, is to compare their catalytic powers in an acid-catalysed reaction. The catalysis of the inversion of sucrose gives an order of strengths as follows: $HClO_4 > HBr > HCl > HNO_3$. An alternative method is to measure the dissociation in a solvent which is more acidic than water.

An alternative definition of acids and bases was proposed by Cady and Elsey and is often termed the *solvent system* definition. They suggested that the acid and base in a particular solvent should be defined in terms of the ions formed in the self-dissociation of the solvent. For example, water dissociates slightly in the sense:

$$2H_2O \rightleftharpoons \underset{\text{acid}}{H_3O^+} + \underset{\text{base}}{OH^-} \tag{6.6}$$

TABLE 6.1 Strength of acids in water

Acid	*Conjugate base*	*K*	p*K*	
$HClO_4$	ClO_4^-	ca. 10^8	ca. −8	*K* very high
HCl	Cl^-	ca. 10^7	ca. −7	HBr and HI greater
HNO_3	NO_3^-	ca. 10^4	ca. −4	Uncertain *K* but high
H_2SO_4	HSO_4^-	ca. 10^3	ca. −3	Uncertain *K* but high
H_3O^+	H_2O	55.5	−1.74	
HSO_4^-	SO_4^{2-}	2×10^{-2}	1.70	Compare with H_2SO_4
H_3PO_4	$H_2PO_4^-$	7.5×10^{-3}	2.12	
HF	F^-	7.2×10^{-4}	3.14	Compare with HCl
$H_2PO_4^-$	HPO_4^{2-}	5.9×10^{-8}	7.23	
NH_4^+	NH_3	3.3×10^{-10}	9.24	Typical weak base
HPO_4^{2-}	PO_4^{3-}	3.6×10^{-13}	12.44	Compare with H_3PO_4 and $H_2PO_4^-$
H_2O	OH^-	1.07×10^{-16}	15.97	Strong base
OH^-	O^{2-}	Below 10^{-36}	Above 36	

TABLE 6.2 Strengths of oxyacids, H_yXO_x

$(x-y)=1$	Acid	HNO_2	H_2SO_3	H_3AsO_4	H_5IO_6
	pK	3.3	1.90	3.5	3.29
$(x-y)=0$	Acid	HClO	H_3BO_3	H_4GeO_4	H_6TeO_6
	pK	7.50	9.22	8.59	8.80

and an acid in water is any substance which enhances the concentration of the solvent cation, H_3O^+, while a base is any substance which enhances the concentration of the solvent anion, OH^-.

This definition is closely related to that of Lowry and Brönsted, the dissociation of Equation (6.6) corresponding to the two Brönsted acid pairs of water as given in Equation (6.5). A similar correlation between the solvent system and Lowry–Brönsted definitions holds for any protonic solvent, but the solvent system definition has the advantage that it is readily extended to solvents which do not contain dissociable hydrogen atoms.

The third definition of acids and bases is that due to Lewis, who defines an acid as a lone pair acceptor and a base as a lone pair donor. This definition is of little advantage in discussing reactions in water so an account of it is deferred to Section 6.5.

6.2.1 Strengths of oxyacids

Many compounds which show acidic properties fall into the class of oxyacids, that is, they contain the Group X-O-H. Two general observations may be made about the strengths of these acids.

First, where there are a number of OH groups attached to the central atom, the pK values for the removal of the successive ionizable hydrogens increase by about five each time (compare the values for phosphoric acid and the two phosphate ions in Table 6.1).

Second, the strength of the acid depends on the difference $(x-y)$ between the number of oxygen atoms and the number of hydrogen atoms in the molecule H_yXO_x. When x is equal to y, pK is about 8.5 ± 1. If $(x-y)=1$, pK is about 2.8 ± 1. If $(x-y)$ is two or more, the pK value is markedly less than zero. Examples of the last case are provided by the first three oxyacids in Table 6.1 while some examples of the first two cases are given in Table 6.2.

As oxygen is the second most electronegative element, when the central atom X has an oxygen atom attached to it as well as the OH group, that is in O=X-O-H, electron density will be withdrawn from X by the oxygen and this effect will be transmitted to the O-H bond, making it easier for the hydrogen to dissociate as the positive ion. The acid strength should therefore rise with the number of X=O groups, that is with $(x-y)$ as the tables show. Two cases are known where simple acids do not fit this pattern. One is provided by the lower acids of phosphorus, phosphorous acid H_3PO_3, where $(x-y)=0$ but p$K=1.8$, and hypophosphorous acid, H_3PO_2, with $(x-y)$ equal to -1 and p$K=2$. In both these molecules, the value of $(x-y)$ does not correspond to the number of X=O bonds as there are direct P-H bonds present, one in phosphorous acid and two in hypophosphorous acid. Allowing for these, both acids have one P=O bond, which should correspond to pK values in the range 2.8 ± 1, as is found.

The second exception is provided by carbonic acid, H_2CO_3, which is expected to have a pK value of about 2.8 and has an actual value of 6.4. In this case, as well as the acid dissociation equilibrium, there is a further nonprotonic equilibrium in solution:

$$H_2CO_3 \rightleftharpoons H_2O + CO_{2(aq)}$$

When allowance is made for this dissolved carbon dioxide, the effective pK value of carbonic acid is about 3.6, which just falls within the expected range.

6.3 Oxidation and reduction

NOTE: In this section particularly, use is made without proof of various thermodynamic relationships. The reader will find these explained in any standard textbook of physical chemistry.

Oxidation, originally defined in terms of combination with oxygen and later generalized to include combination with other electronegative elements, is nowadays commonly defined as the *removal of electrons* from the element or compound which is oxidized. *Reduction*, similarly, has been defined in terms of removal of oxygen and electronegative elements, addition of hydrogen or electropositive elements and now in terms of *gain of electrons* by the element or compound in question. In cases involving ions, for example in:

$$Na + \tfrac{1}{2}Cl_2 \rightarrow Na^+Cl^-,$$
$$\text{or} \quad 2Fe^{3+} + Sn \rightarrow 2Fe^{2+} + Sn^{2+}$$

the direction of electron transfer is clear from the equations. When only covalent species are involved, it is usually clear that some electron rearrangement has taken place but there is often no obvious electron transfer, as in:

$$\tfrac{1}{2}H_2 + \tfrac{1}{2}Cl_2 \rightarrow HCl$$

where the difference in electron density at the hydrogen atom, say, is the relatively small one due to the polarization of the H-Cl bond compared with the unpolarized H-H bond. It is in these cases that the definitions of oxidation and reduction in terms of oxidation numbers, using electronegativity values, discussed in Chapter 2 are so useful.

Some quantitative measure of oxidizing or reducing power is necessary and this is known as the *redox potential* of the reactant. The tendency to gain or lose an electron may be measured as an electrical potential under standard conditions and expressed relative to a suitable standard value (as only the relative values of potentials may be measured). The standard conditions used are a temperature of 25°C, unit activity (which is usually taken as unit concentration) of the ions concerned and, for gases, one atmosphere pressure. The standard potential is provided by the hydrogen electrode (Fig. 6.4), which consists of a platinum plate, coated with platinum black, partly dipping into a solution containing hydrogen ions at unit activity. Hydrogen gas at one atmosphere pressure is passed over the platinum plate and bubbled through the solution so that the platinum is in contact with both the gas and the hydrogen ions in solution and catalyses the attainment of equilibrium between them:

a is activity

$$H^+(aq, a = 1) + e^- \rightleftharpoons \tfrac{1}{2}H_2 \text{ (gas, } p = 1 \text{ atm)}$$

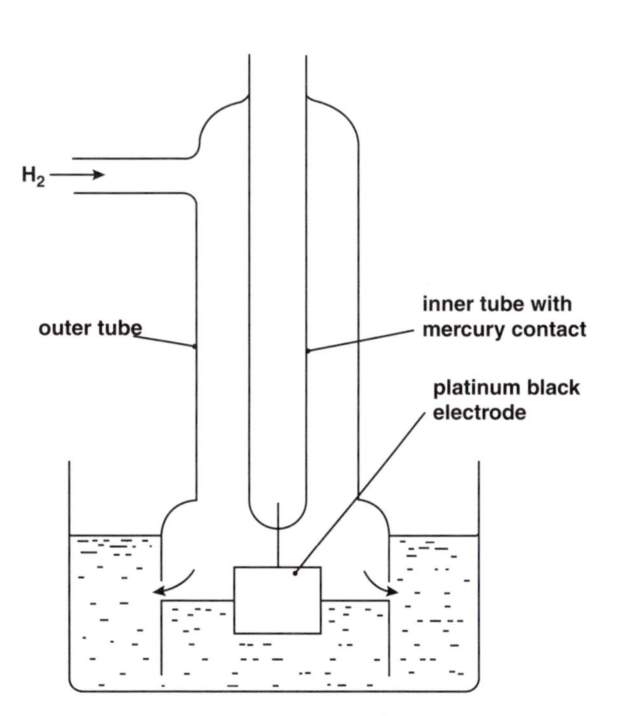

FIG. 6.4 The hydrogen electrode

The potential of this electrode is defined to be zero at 25°C (298 K) and this is written $E^0(298\text{ K}) = 0.000$ V.

The potential of any other oxidation-reduction or redox system is measured by immersing a platinum wire or other inert conductor in a solution which contains both the oxidized and the reduced form of the system at unit activity, and measuring the potential of this half-cell against the hydrogen electrode. For example, the standard potential of the ferric/ferrous system is measured from the cell:

It is usually more convenient to use some other electrode of accurately known potential in place of the hydrogen electrode which is awkward to handle.

$$\text{Pt}, \text{H}_2(p = 1\text{ atm})|\text{H}^+(a = 1)\left\|\begin{matrix}\text{Fe}^{3+}\ (a = 1)\\ \text{Fe}^{2+}\ (a = 1)\end{matrix}\right|\text{Pt}$$

If the potential of a gaseous element is to be determined, an electrode similar to the hydrogen electrode is used, while the half-cell used to determine the potential of a metallic element consists simply of a rod of the metal (which is defined to have unit activity) immersed in a solution of the metal ions at unit activity. For example, the oxidation potential of iron going to ferrous ions is measured as the standard potential of the cell:

$$\text{hydrogen electrode}\|\text{Fe}^{2+}(a = 1)|\text{Fe}$$

In the half-equation describing the change in the redox system, the oxidized form is always written on the left-hand side, that is the equations are in the form:

$$\text{Ox} + ne^- \rightleftharpoons \text{Red} \qquad (6.7)$$

where Ox stands for the oxidized form and Red for the reduced form,

Equation (6.7) is a reduction from left to right, which is the normal way of reading, but it is an oxidation from right to left. It has been suggested that these standard electrode potentials may be called reduction potentials. The term 'redox' is, however, recommended to emphasize the reversible nature of the reactions.

$$\text{e.g.}\quad \text{Fe}^{3+} + \text{e}^- \rightleftharpoons \text{Fe}^{2+}$$
$$\text{or}\quad \text{Fe}^{2+} + 2\text{e}^- \rightleftharpoons \text{Fe}$$

The short form for describing the electrode is written similarly: Ox|Red or, in the examples above, $Fe^{3+}|Fe^{2+}$ and $Fe^{2+}|Fe$.

The sign convention which applies for these potentials takes a negative sign for the potential Ox|Red to mean that the reduced form of the redox system is a better reducing agent than is hydrogen under standard conditions. This is tantamount to measuring the sign of the metal in the $M^{z+}|M$ electrode relative to the hydrogen electrode. Thus the standard potential for $Fe^{2+}|Fe$ is –0.41 V, showing that iron going to ferrous ions is a better reducing agent than hydrogen going to hydrogen ions. The potential for $Fe^{3+}|Fe^{2+}$ is +0.77 V so that ferrous ions going to ferric ions provide a much worse reducing agent than hydrogen. This is, of course, equivalent to saying that ferric going to ferrous is a useful oxidizing agent. The above sign convention is the one recommended by the International Union of Pure and Applied Chemistry (IUPAC) but the opposite convention (reversing the equation and the sign) is also found.

The full range of redox potentials extends over about six volts, from –3 V for the alkali metals which are the strongest reducing agents to +3 V for fluorine, the strongest oxidizing agent. The values at these extremes cannot be measured directly as these very reactive elements decompose water, but they can be derived by calculation. The range of useful redox reagents in an aqueous medium is about half of the total range, from about 1.7 V for strong oxidizing agents like ceric or permanganate to about −0.4 V for a strong reducing agent like chromous, or about –0.8 V for reduction by fairly active metals such as zinc. The values for many common oxidation–reduction couples are given in Table 6.3.

The potential of a redox system, where the oxidized and reduced forms are not at unit activities, is related to the standard potential, E^0, by the Nernst equation:

$$E = E^0 + \frac{RT}{nF}\ln\frac{[\text{Ox}]}{[\text{Red}]} \qquad (6.8)$$

where R is the gas constant, T the absolute temperature, F the Faraday constant, and n the number of electrons transferred in the oxidation process as in Equation (6.7). Transferring to logarithms to the base ten and introducing the values of the constants

$$E = E^0 + \frac{0.059}{n} \log \frac{[\text{Ox}]}{[\text{Red}]} \text{ at } 25^\circ\text{C} \quad (6.9)$$

This equation permits the extent of reaction between two redox systems to be calculated, and can be used, for example, to decide whether a particular reaction will go to completion or not. For the general reaction:

$$a \text{ Ox}_1 + b \text{ Red}_2 \rightleftharpoons b \text{ Ox}_2 + a \text{ Red}_1 \quad (6.10)$$

$$E_1 = E_1^0 + \frac{RT}{nF} \ln \frac{[\text{Ox}_1]^a}{[\text{Red}_1]^a} \quad \text{and} \quad E_2 = E_2^0 + \frac{RT}{nF} \ln \frac{[\text{Ox}_2]^b}{[\text{Red}_2]^b}$$

At equilibrium, $E_1 = E_2$ so that:

$$\log \frac{[\text{Ox}_2]^b [\text{Red}_1]^a}{[\text{Ox}_1]^a [\text{Red}_2]^b} = \log K = \frac{n}{0.059}(E_1^0 - E_2^0) \quad (6.11)$$

where K is the equilibrium constant for the reaction. Consider as examples the oxidations of iodide ion, first by ceric ion and secondly by ferric ion. In the first case:

$$Ce^{4+} + I^- \rightarrow Ce^{3+} + \tfrac{1}{2}I_2$$

E^0 for I_2/I^- is 0.54 V and E^0 for Ce^{4+}/Ce^{3+} is 1.61 V. Hence:

$$\log K = \frac{1}{0.059}(1.61 - 0.54) = 18.48$$

$$\text{i.e. } K \simeq 10^{18}$$

Thus iodide is completely oxidized to iodine by ceric ion.

In the second case, E^0 for Fe^{3+}/Fe^{2+} is 0.77 V so that for the reaction:

$$Fe^{3+} + I^- \rightarrow Fe^{2+} + \tfrac{1}{2}I_2,$$

$$\log K = (1/0.059)(0.76 - 0.52) = 4.068$$

Hence K is approximately 10^4 so that a small but significant proportion of iodide would remain in equilibrium with the iodine. For a one-electron change, an equilibrium constant of one million requires a difference in the standard potentials of the two redox systems of 0.354 V, so this is about the minimum difference necessary for a complete reaction.

The values of the standard potentials may also be used to determine the stability of different oxidation states in solution. If the potentials for the various oxidation states of iron in Table 6.3 are examined, it will be seen that iron going to ferrous ions is a much better reducing agent than ferrous ions going to ferric ions. Oxidation of metallic iron in aqueous solution therefore gives ferrous ions first and stronger oxidation is required to get to ferric ions. By contrast, the values for the different oxidation states of copper show that copper going to cuprous ions is a worse reducing agent than copper going to cupric ions (that is, an oxidizing agent strong enough to oxidize copper to cuprous, $E^0 = 0.52$ V, is more than strong enough to oxidize cuprous to cupric, $E^0 = 0.15$ V). The cuprous state is therefore avoided in aqueous solution. Put alternatively, cuprous ions in water disproportionate to cupric ions and metallic copper:

$$2Cu^+ \rightarrow Cu^{2+} + Cu$$

In this reaction, the Cu^+ ion and Cu may be taken as Ox_1 and Red_1 and then the Cu^{2+} ion and the Cu^+ ion become Ox_2 and Red_2: that is, the cuprous ion Cu^+ may be regarded as filling two roles, as both an oxidizing and a reducing agent, in the general Equation (6.10). Substituting the standard potentials in Equation (6.11) gives

$$K = \frac{[Cu^{2+}][Cu]}{[Cu^+]^2} = \text{approx. } 10^6$$

so that only one part in a million of the original cuprous copper remains in solution as cuprous copper at equilibrium. This disproportionation of the copper oxidation states may be reversed by adding some reagent which forms a very stable complex with the cuprous ion, so stable that

TABLE 6.3 Examples of standard redox potentials in acid solution

Couple	*Reaction equation*	$E^0(V)$
Li^+/Li	$Li^+ + e^- \rightleftharpoons Li$	−3.05
$M^+/M (M = K, Rb, Cs)$	$M^+ + e^- \rightleftharpoons M$	−2.93
Ba^{2+}/Ba	$Ba^{2+} + 2e^- \rightleftharpoons Ba$	−2.90
Sr^{2+}/Sr	$Sr^{2+} + 2e^- \rightleftharpoons Sr$	−2.89
Ca^{2+}/Ca	$Ca^{2+} + 2e^- \rightleftharpoons Ca$	−2.76
Na^+/Na	$Na^+ + e^- \rightleftharpoons Na$	−2.71
Mg^{2+}/Mg	$Mg^{2+} + 2e^- \rightleftharpoons Mg$	−2.38
La^{3+}/La	$La^{3+} + 3e^- \rightleftharpoons La$	−2.37
$\frac{1}{2}H_2/H^-$	$\frac{1}{2}H_2 + e^- \rightleftharpoons H^-$	−2.23
Be^{2+}/Be	$Be^{2+} + 2e^- \rightleftharpoons Be$	−1.70
Al^{3+}/Al	$Al^{3+} + 3e^- \rightleftharpoons Al$	−1.67
Zn^{2+}/Zn	$Zn^{2+} + 2e^- \rightleftharpoons Zn$	−0.76
Cr^{3+}/Cr	$Cr^{3+} + 3e^- \rightleftharpoons Cr$	−0.74
S/S^{2-}	$S + 2e^- \rightleftharpoons S^{2-}$	−0.51
H_3PO_3/H_3PO_2	$H_3PO_3 + 2H^+ + 2e^- \rightleftharpoons H_3PO_2 + H_2O$	−0.50
$CO_2/H_2C_2O_4$	$2CO_2 + 2H^+ + 2e^- \rightleftharpoons H_2C_2O_4$	−0.49
Fe^{2+}/Fe	$Fe^{2+} + 2e^- \rightleftharpoons Fe$	−0.41
Cr^{3+}/Cr^{2+}	$Cr^{3+} + e^- \rightleftharpoons Cr^{2+}$	−0.41
H_3PO_4/H_3PO_3	$H_3PO_4 + 2H^+ + 2e^- \rightleftharpoons H_3PO_3 + H_2O$	−0.28
Sn^{2+}/Sn	$Sn^{2+} + 2e^- \rightleftharpoons Sn$	−0.14
$H^+/\frac{1}{2}H_2$	$H^+ + e^- \rightleftharpoons \frac{1}{2}H_2$	0.00
$S_4O_6^{2-}/S_2O_3^{2-}$	$S_4O_6^{2-} + 2e^- \rightleftharpoons 2S_2O_3^{2-}$	0.09
Sn^{4+}/Sn^{2+}	$Sn^{4+} + 2e^- \rightleftharpoons Sn^{2+}$	0.15
Cu^{2+}/Cu^+	$Cu^{2+} + e^- \rightleftharpoons Cu^+$	0.15
Cu^+/Cu	$Cu^+ + e^- \rightleftharpoons Cu$	0.52
$\frac{1}{2}I_2/I^-$	$\frac{1}{2}I_2 + e^- \rightleftharpoons I^-$	0.54
H_3AsO_4/H_3AsO_3	$H_3AsO_4 + 2H^+ + 2e^- \rightleftharpoons H_3AsO_3 + H_2O$	0.56
O_2/H_2O_2	$O_2 + 2H^+ + 2e^- \rightleftharpoons H_2O_2$	0.68
Fe^{3+}/Fe^{2+}	$Fe^{3+} + e^- \rightleftharpoons Fe^{2+}$	0.77
Hg_2^{2+}/Hg	$\frac{1}{2}Hg_2^{2+} + e^- \rightleftharpoons Hg$	0.80
Hg^{2+}/Hg	$Hg^{2+} + 2e^- \rightleftharpoons Hg$	0.85
$\frac{1}{2}Br_2/Br^-$	$\frac{1}{2}Br_2 + e^- \rightleftharpoons Br^-$	1.09
IO_3^-/I^-	$IO_3^- + 6H^+ + 6e^- \rightleftharpoons I^- + 3H_2O$	1.09
ClO_4^-/ClO_3^-	$ClO_4^- + 2H^+ + 2e^- \rightleftharpoons ClO_3^- + H_2O$	1.19
$IO_3^-/\frac{1}{2}I_2$	$IO_3^- + 6H^+ + 5e^- \rightleftharpoons \frac{1}{2}I_2 + 3H_2O$	1.20
$\frac{1}{2}Cr_2O_7^{2-}/Cr^{3+}$	$\frac{1}{2}Cr_2O_7^{2-} + 7H^+ + 3e^- \rightleftharpoons Cr^{3+} + 7/2H_2O$	1.33
$\frac{1}{2}Cl_2/Cl^-$	$\frac{1}{2}Cl_2 + e^- \rightleftharpoons Cl^-$	1.36
$HIO/\frac{1}{2}I_2$	$HIO + H^+ + e^- \rightleftharpoons \frac{1}{2}I_2 + H_2O$	1.45
MnO_4^-/Mn^{2+}	$MnO_4^- + 8H^+ + 5e^- \rightleftharpoons Mn^{2+} + 4H_2O$	1.51
$BrO_3^-/\frac{1}{2}Br_2$	$BrO_3^- + 6H^+ + 5e \rightleftharpoons \frac{1}{2}Br_2 + 3H_2O$	1.52
$HBrO/Br_2$	$HBrO + H^+ + e^- \rightleftharpoons Br_2 + H_2O$	1.59
H_5IO_6/IO_3^-	$H_5IO_6 + H^+ + 2e^- \rightleftharpoons IO_3^- + 3H_2O$	1.60
Ce^{4+}/Ce^{3+}	$Ce^{4+} + e^- \rightleftharpoons Ce^{3+}$	1.61
$HClO/\frac{1}{2}Cl_2$	$HClO + 2H^+ + e^- \rightleftharpoons \frac{1}{2}Cl_2 + H_2O$	1.64
$HClO_2/HClO$	$HClO_2 + 2H^+ + 2e^- \rightleftharpoons HClO + H_2O$	1.64
H_2O_2/H_2O	$H_2O_2 + 2H^+ + 2e^- \rightleftharpoons 2H_2O$	1.77
$\frac{1}{2}S_2O_8^{2-}/SO_4^{2-}$	$\frac{1}{2}S_2O_8^{2-} + e^- \rightleftharpoons SO_4^{2-}$	2.01
O_3/O_2	$O_3 + 2H + 2e^- \rightleftharpoons O_2 + H_2O$	2.07
$\frac{1}{2}F_2/F^-$	$\frac{1}{2}F_2 + e^- \rightleftharpoons F^-$	2.87
$\frac{1}{2}F_2/HF$	$\frac{1}{2}F_2 + H^+ + e^- \rightleftharpoons HF$	3.06

there is significantly less than one part per million of cuprous ion in equilibrium with the cuprous complex. The cuprous ion is thus removed from the cuprous–cupric equilibrium by complexing and the disproportionation reaction reverses. One example of such a complexing agent is cyanide ion which gives the very stable $Cu(CN)_2^-$ complex ion.

The above examples give some indication of the ways in which the redox potentials may be used. Let us examine the potential in more detail and analyse its components.

The type of cell used for determining these potentials, as in Section 6.3

$$\mathrm{Pt, H_2|H^+||M^+|M}$$

implies the reaction:

$$\tfrac{1}{2}\mathrm{H_2} + \mathrm{M^+} \longrightarrow \mathrm{H^+} + \mathrm{M}$$

The electrode potentials of the left and right electrode are E^0_{H} and E^0_{M} respectively and the potential difference, the e.m.f., of the cell is $(E^0_{\mathrm{H}} - E^0_{\mathrm{N}})$. By convention we set $E^0_H = 0$.

The electrode potential E^0 of the cell is related to the standard free energy change ΔG^0 of the reaction by the equation:

$$\Delta G^0 = -nFE^0$$

and ΔG^0 is related to the heat of reaction ΔH^0 and entropy of reaction ΔS^0 at temperature T by

$$\Delta G^0 = \Delta H^0 - T\Delta S^0$$

Let us first examine ΔH^0 in detail. This means the enthalpy of the products minus the enthalpy of the reactants which we may write

$$\Delta H^0 = H^0(\mathrm{M}) + H^0(\mathrm{H^+}) - H^0(\mathrm{M^+}) - H^0(\tfrac{1}{2}\mathrm{H_2})$$

which can be regrouped as

$$\Delta H^0 = \{H^0(\mathrm{H^+}) - H^0(\tfrac{1}{2}\mathrm{H_2})\} - \{H^0(\mathrm{M^+}) - H^0(\mathrm{M})\}$$

and abbreviated as

$$\Delta H^0 = \Delta H^0(\mathrm{H}) - \Delta H^0(\mathrm{M}) \qquad (6.12)$$

The terms $\Delta H^0(\mathrm{H})$ and $\Delta H^0(\mathrm{M})$ may be analysed by means of a cycle, similar to those used to discuss lattice energies or solubilities (compare Figs 5.5 and 6.1). For the formation of a cation from the element in aqueous solution we may examine the contribution of various terms to $\Delta H^0(\mathrm{M})$:

$$\begin{array}{ccc} \text{element in standard state} & \xrightarrow{H_{\mathrm{A}}} & \text{gaseous atom} \\ \Delta H^0(\mathrm{M})\big\downarrow & & \big\downarrow I \\ \text{cation in solution} & \xleftarrow[H_{aq}]{} & \text{gaseous cation} \end{array}$$

where H_A is the heat of atomization of the element, I (see Table 2.8) is the ionization potential (or the sum of the first z ionization potentials if a M^{z+} cation is formed) and H_{aq} is the heat of hydration of the gaseous ion. An exactly similar cycle may be constructed for the formation of an anion in solution except that the ionization potential must be replaced by the electron affinity, E (see Table 2.9), when the gaseous atom goes to the gaseous anion.

Thus, for a cation,

$$\Delta H^0(M) = H_A + I + H_{aq}$$

and for an anion,

$$\Delta H^0(X) = H_A + E + H_{aq}$$

The values of H_A, I (or E) and H_{aq} are known from experiment and for hydrogen, equal respectively +218, +1310, and −1070 kJ mol^{-1}, giving $\Delta H^0(\mathrm{H}) = 452$ kJ mol^{-1}.

A similar treatment can be carried through for the entropy terms ΔS^0.

However, we can simplify the discussion by recognizing:

(a) That when similar systems are being compared (say Fe^{2+}/Fe and Zn^{2+}/Zn) the entropy changes will be very similar in each, so that the *difference* in entropy contributions to the systems may be neglected to a first approximation. Note that this approximation is not valid when one component is a gas, as for $O^{-2}/\frac{1}{2}O_2$.

(b) That each system is being compared to the hydrogen electrode so that ΔS^0 for hydrogen subtracts out in a comparison between two different metals M.

Thus, for an approximate treatment, and especially for comparative purposes, the entropy changes may be neglected and the approximate relation

$$-nFE^0 \approx \Delta H^0 \tag{6.13}$$

may be used.

We may now write from Equations (6.12) and (6.13):

$$-nFE_M^0 \approx 452 \text{ kJ mol}^{-1} - \Delta H^0(\text{M})$$

and E_M^0 will be negative if ΔH^0 (M) is less than 452 kJ mol^{-1}.

The alkali metals have ΔH^0 values of about 188 kJ mol^{-1} reflecting their low ionization potentials and sublimation energies, so that their standard potentials are strongly negative. The fall in standard redox potentials from caesium or barium to sodium or magnesium shown in Table 6.3 reflects the regular decrease in ionization potential and sublimation energy with increasing size, while the anomalous position of lithium is mainly the result of the greater hydration energy of the small cation. In the case of the halogens, the sum of the atomization energy (half the bond energy of X_2) and the electron affinity is approximately constant, so that the fall of the redox potential from fluorine to iodine again reflects the fall in hydration energy as the size of the ion increases.

This treatment need not be confined to the case of elements, for example the Fe^{3+}/Fe^{2+} system depends on the energy changes:

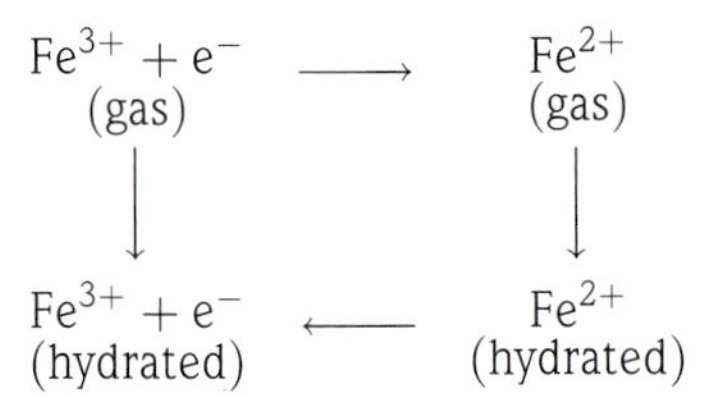

where the overall change depends on the hydration energies of the Fe(II) and Fe(III) ions and on the third ionization potential of iron.

These cycles provide one method of calculating potentials which are unobtainable experimentally, either because the species are not stable in water or because the attainment of equilibrium is too slow.

It must be noted that all the above treatment of standard redox potentials gives no indication of the rates of reaction. Although it may seem from the values of the potentials that a certain molecule should be readily oxidized by another, the rate of reaction might be so slow that nothing is observed. For example, the values given in Table 6.3 show that iodine should oxidize thiosulfate to tetrathionate (as in the standard volumetric method) and that oxidation of thiosulfate to sulfate should also be significant. However, the reaction to tetrathionate is quantitative, showing that the oxidation to sulfate must be extremely slow. In other words, the quantitative nature of oxidation to tetrathionate depends on a kinetic factor and is not revealed by the thermodynamic approach involved in redox potentials.

The potentials which appear in Table 6.3 as involving simple ions do in fact apply to the hydrated forms of these ions, since the potentials are determined in aqueous solution. In some of these cases, the water molecules are strongly held and the heat of hydration is an important factor in the potential. If these coordinated water molecules are replaced by other ligands, the potential changes. For example, the potential of 0.77 V for Fe^{3+}/Fe^{2+}

applies to the hydrated ions (approximately $Fe(H_2O)_6^{3+}/Fe(H_2O)_6^{2+}$). For the cyanide complexes, $Fe(CN)_6^{3-}/Fe(CN)_6^{4-}$ the potential drops to 0.36 V, and it rises as high as 1.2 V for the dipyridyl complexes, $Fe(dipy)_3^{3+}/Fe(dipy)_3^{2+}$. The potential also changes with the acidity when, for example, hydrated species in dilute acid change to hydroxy complexes in more alkaline media. The acidity is particularly important when oxyions are involved in the redox equation. Equations such as (6.8) or (6.9) include $[H^+]$ in equations from Table 6.3 which involve changing oxygen content by use of H^+. Thus the hydrogen ion concentration can enter in powers as high as 8 (for MnO_4^- going to Mn^{2+}). Standard conditions involve $[H^+] = 1$ i.e. $pH = 0$, and departures from this must be included in calculations of the redox potential.

If the redox potential in a solvent other than water is in question, it will clearly differ from that of the hydrated species in water. In the case of a simple ion, the difference lies in the heat of solvation compared with the heat of hydration, the contribution of the heat of atomization and the ionization potential (or electron affinity) remaining the same as in water. Although the solvation energy of any set of ions will neither be the same nor in the same order as in water, gross differences are less likely than small ones. In other words, ions which are strongly hydrated are likely to be strongly solvated by another ionizing solvent. This means that the general order of oxidizing powers should be similar in all solvents to that obtaining in water, although there may be marked differences in detail from one solvent to the next. Since, in general, thermodynamic measurements in nonaqueous solvents are much more sparse than those made in water, this generalization is usually the best that can be obtained. Similar remarks apply to oxidation–reduction reactions in systems which differ even more from that of aqueous solution, such as fused salts and other high temperature melts. In such systems, considerable changes take place in the relative stability and redox powers of chemical species, and, although the values derived from aqueous measurements may provide some guide, it is often better to go back to the fundamental parameters, such as ionization potentials, and base predictions on these.

For the use of redox potentials to display oxidation state stabilities, see the discussion of Ebsworth diagrams in Section 8.6.

Nonaqueous solvents

Water is so familiar and accessible a solvent that its properties are usually taken for granted and underlie most general statements about inorganic compounds. Such general ideas as 'stability' usually imply 'stability in an environment containing air and water', and many inaccessible compounds or highly reactive oxidation states in this type of environment become stable and manageable if handled in a nonaqueous medium. In fact, one of the advantages of studying nonaqueous solvents has been the increased attention to, and wider insight into, chemistry in water that has resulted.

The practical reasons for using solvents other than water lie partly in the extended range of experimental conditions which are available and partly in the exclusion of water as a reactant. Many anhydrous compounds, for example, cannot be prepared from the hydrates and were unknown until the use of other solvents was introduced. Such compounds often have unusual and unexpected properties. The nonaqueous solvents range widely in intrinsic reactivity, from solvents such as sulfur dioxide which commonly act only as inert media for the reaction, to solvents such as anhydrous hydrogen fluoride which react with nearly all nonfluoride solutes. A number of selected examples of such solvent systems are discussed individually later but first the discussion of the early part of this chapter on solubility and solvent interaction and on acid–base behaviour will be extended to include nonaqueous solvents. It has already been indicated how the treatment of redox behaviour may be extended and further illustrations occur in the course of the discussion of specific solvents.

ORGANIC SOLVENTS IN INORGANIC CHEMISTRY

Water can be an excellent solvent for many inorganic compounds, especially coordination complexes, and has many desirable features of a solvent (inexpensive, nonflammable, nontoxic and environmentally friendly). However, it is a poor solvent for compounds containing significant numbers of nonwater-soluble organic groups (e.g. triphenylphosphane, a commonly used tertiary phosphane ligand), and furthermore many compounds (especially organometallic ones) may react with water. There is therefore a need for alternative organic solvents, which can be obtained in an anhydrous state. Tetrahydrofuran (thf) is one such example.

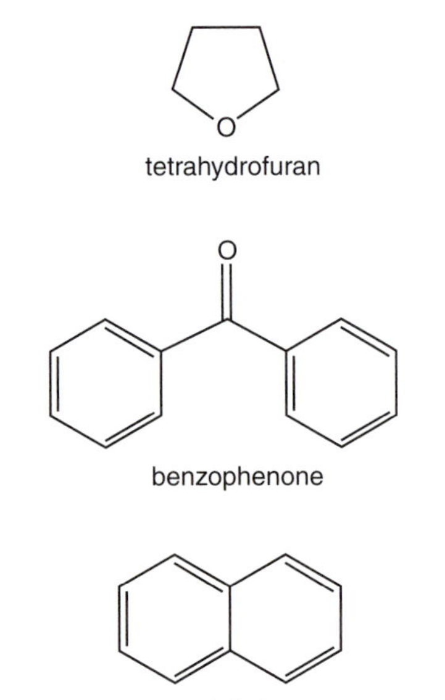

Tetrahydrofuran can be readily dried by distillation from a mixture of an alkali metal (typically sodium) with an added organic substance (R), such as benzophenone or naphthalene which can form a highly coloured charge-transfer compound Na^+R^-, which behaves as a source of 'soluble sodium'; the electron in the R^- anion occupies the π^* orbital of the aromatic hydrocarbon. This species reacts with the trace amounts of moisture by the usual reaction:

$$Na + H_2O \longrightarrow NaOH + H_2$$

Other ethers, such as the commonly used diethyl ether, can also be dried using this reagent. For solvents such as toluene, sodium metal itself is commonly used, since the melting point of sodium is less than the boiling point of toluene, so that the molten sodium constantly presents a fresh surface to any water present. For other solvents such as aliphatic hydrocarbons (often termed petroleum spirits), the sodium charge-transfer compounds described above have a low solubility, so instead compounds such as calcium hydride (CaH_2), which remove any traces of water (via the reaction $H_2O + H^- \rightarrow H_2 + OH^-$) can be used.

In many cases, compounds may be reactive towards air as well as moisture. The solvents are therefore distilled under a rigorously dried inert gas atmosphere (typically nitrogen or argon), and handled using appropriate techniques, such as vacuum or Schlenk lines using syringes. When materials are exceptionally sensitive to air and moisture (e.g. lanthanide organometallic compounds, Section 16.5), they are handled in an inert atmosphere drybox (often termed a glovebox). Further details on the manipulation of air- and moisture-sensitive compounds are given in the references below:

Synthesis of organometallic compounds. A practical guide, Ed. S. Komiya, John Wiley & Sons, 1997, Chapter 4.

The manipulation of air-sensitive compounds, D. F. Shriver and M. A. Drezdzon, Wiley Interscience, 1986.

For information on drying agents for solvents, see *Vogel's textbook of practical organic chemistry*, B. S. Furniss, A. J. Hannaford, P. W. G. Smith and A. R. Tatchell, 5th edition, Longman, 1989.

6.4 Solubility and solvent interaction in nonaqueous solvents

The extension of the discussion of solubility in Section 6.1 to nonaqueous solvents is fairly easy. A very generalized energy diagram is shown in Fig. 6.5. E_1 is the energy required to change the solute into the form in which it will exist in solution. In the case of an ionic solid, this means providing the lattice energy to separate the ions, while for a solute which dissolves as molecules it means providing the energy required to separate the molecules or to form dimers or other species. E_2 is the energy required to separate the solvent molecules, breaking up any solvent 'structure' to allow the admission of the solute particles. E_3 is the energy given out when the separate solute particles associate with the solvent molecules. Two general cases exist. One, the generalization of the 'strong forces' case of Section 6.1, is the case of ions or strongly polar molecules dissolving in a polar solvent. Here, the important balance of energies involves strong attractions between ions and permanent dipoles. The other extreme is the 'weak forces' case of weakly bound, nonpolar, covalent molecules dissolving in nonpolar solvents. Here, the balance of forces involves weak attractions between induced dipoles, and the order of magnitude of the effects is less than that in the

It may be noted that the energy of hydration as commonly measured is not E_3 but the difference $[E_3 - E_2]$.

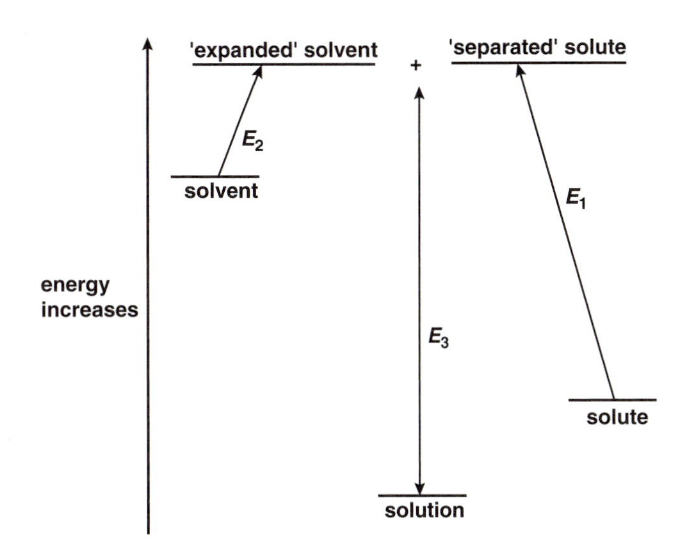

FIG. 6.5 Energy changes is the solution of a solid in a solvent. This very generalized version of Fig. 6.1 shows the energy changes for the solution of any solid in any solute, whether an ionic or a covalent system is involved

strong forces case by a factor of about ten. The major conclusion is that 'mixed cases' will all represent situations where solubility is low. Thus, ions will not dissolve in nonpolar solvents because, although E_2 is low, E_3 is also low and the energetics are dominated by E_1. Conversely, nonpolar molecules are insoluble in polar solvents as E_1 and E_3 are both low and there is insufficient energy to supply E_2. This analysis is at the base of the very old generalization about solubilities that 'like dissolves like'.

Although a clear answer can be given in the mixed cases, it must be noted that the detailed variation of solubilities within the 'strong forces' or within the 'weak forces' class depends on the detailed balance of energies of similar magnitudes, so that a knowledge of solubilities in one solvent gives only a very general guide to the solubilities of similar compounds in similar solvents. This is illustrated by the solubilities given in Table 6.4.

For the alkali halides, the values in ammonia and sulfur dioxide follow the same order as in water, which is, in turn, the inverse order of the lattice energies, and most of the solubilities fall going from water to ammonia to sulfur dioxide but the relative sizes of the solubilities change enormously. The much greater solubilities of the iodides and the silver salts in ammonia and in sulfur dioxide show that polarization effects producing induced dipoles are more important in these two solvents than in water. It is worth noting that the solubility of the silver halides in liquid ammonia is in the reverse order to their solubility in aqueous ammonia.

TABLE 6.4 Solubilities in liquid ammonia and sulfur dioxide

	Solubility (millimole/litre at 0°C)		
Compound	*Water*	*Ammonia*	*Sulfur dioxide*
NaCl	6100	2200	Insoluble
NaBr	7710	6210	1.4
NaI	10 720	8800	1000
KCI	3760	18	5.5
KBr	4490	2260	40
KI	7720	11 060	2490
AgCl	0.005	20	0.05
AgBr	0.003	135	0.16
AgI	10^{-6}	4000	0.68

The dielectric constants of nonaqueous solvents have an important effect on their power to dissolve ions. The general drop in solubilities of the alkali halides from H_2O to SO_2 reflects the drop in dielectric constant from 81 for water to 22 for ammonia and 12 for SO_2. Although important, the dielectric constant alone is a poor guide to solvent power for ions. The value for liquid hydrogen cyanide of 123 is one of the highest measured and yet this is a poor solvent for ions as the energy of solvation is low.

The general interaction between solvent molecules and ions in solution, which is analogous to hydration in water, is termed *solvation* and follows similar mechanisms. In liquid ammonia, for example, cations are ammonated by interaction with the lone pair of electrons on the nitrogen atom plus the interaction with the negative end of the $N^{\delta-} \cdot H^{\delta+}$ dipoles. The anions interact with the relatively positive hydrogens.

In addition to solvation, the solute may react with the solvent molecules to break up the solvent molecule, the solute species, or both, in the reaction called *solvolysis*—analogous to hydrolysis in water. The distinction between solvation and solvolysis is not completely clear-cut but solvolysis implies a more extensive and less reversible interaction with the solvent molecule. Some examples are given in Table 6.5.

TABLE 6.5 Examples of solvolysis and other interactions

Reaction	Type
$SnCl_4 + 4H_2O \rightleftharpoons Sn(OH)_4 + 4HCl$	Hydrolysis*
$+ 8NH_3 \rightleftharpoons Sn(NH_2)_4 + 4NH_4Cl$	Ammonolysis*
$+ 4HF \rightleftharpoons SnF_4 + 4HCl$	Solvolysis
$+ 2SeOCl_2 \rightleftharpoons SnCl_4.2SeOCl_2$ (or $2SeOCl^+ + SnCl_6^{2-}$)	Solvation (?)
$LiBr + NH_3 \rightleftharpoons$ dissolves : $Li(NH_3)_4Br$ recovered	Solvation
$+ SO_2 \rightleftharpoons Li_2SO_3 + SOBr_2$	Solvolysis*
$KNO_3 + 4HF \rightleftharpoons K^+ + H_2NO_3^+ + 2HF_2^-$	Solvolysis

*(=further reaction of the product takes place).

6.5 Acid–base behaviour in nonaqueous solvents

Of the two theories of acids and bases discussed in Section 6.2, the protonic theory of Lowry and Brönsted is restricted to solvents containing ionizable hydrogen atoms. The application of the theory is then exactly the same as for aqueous systems, except that the strengths of different acids are most conveniently measured relative to the acid–base pairs characteristic of the solvent. For example, K values in liquid ammonia may be measured relative to:

$$NH_4^+ \rightarrow NH_3 + H^+$$

Such values would fall in a similar order to those obtained in water or in any other solvent, although there might be minor inversions.

The solvent system definition was developed for use in nonaqueous solvents and provides a definition of acid and base in nonprotonic solvents. It is very convenient to use if work is being carried out on one particular solvent and it provides a guide to the properties to be expected for a new system, but it does suffer the disadvantage that the definition of acid or base changes from one solvent to another. A more important reservation which must be made is that it only suggests an acid–base system for a particular solvent and does not prove that the solvent ions do in fact behave as acid and base. This point must be proved experimentally. This is well illustrated by the case of sulfur dioxide—see the box on page 155.

As in the case of water (Section 6.2), the solvent's acid–base properties place a limit on the range of the acid strengths of solutes in it. The strongest acid existing in a solvent is the solvent acid and all solutes which are more acidic are converted into the solvent acid—and similarly for bases—so that the number of solutes which are strong acids (i.e. those which are completely converted to the solvent acid) varies with the intrinsic acid strength of the solvent. In an acidic solvent, such as glacial acetic acid, some solutes which are strong acids in water are not completely dissociated to the solvent acid, that is, they are weak acids in this medium. This is the case for HNO_3 or HCl in acetic acid. Indeed some molecules which are weakly acidic in water are basic in acetic acid. This latter effect is even more marked in strongly acidic solvents such as absolute sulfuric acid where even HNO_3 acts as a base and accepts a proton from the solvent. The relative strengths of those solutes which are strong acids in water may thus be differentiated by examining their properties in more acidic solvents. On the other hand, a basic solvent like liquid ammonia has a very weak solvent acid so that all solutes which are acids, strong or weak, in water are strong acids in ammonia. Base strengths in basic and acidic solvents are affected in an analogous manner.

An acid solute which *is* completely dissociated in an acidic solvent, such as perchloric acid in glacial acetic acid, is a much stronger acid than in water and finds valuable application in analysis for determining very weakly basic functions. In a similar way, the solution of a strong base in a basic solvent may be used for the estimation of very weakly acidic groups. Examples of these applications are given in Sections 6.7 and 6.8.

Both the proton-transfer and solvent system definitions of acids and bases are in general use for nonaqueous solvents and they are practically equivalent when applied to a protonic solvent. The solvent system definition—for example:

$$\begin{array}{rcll} & & \text{acid} & \text{base} \\ NH_3 & \rightleftharpoons & NH_4^+ & + NH_2^- \\ 2H_2SO_4 & \rightleftharpoons & H_3SO_4^+ & + HSO_4^- \\ 2HCN & \rightleftharpoons & H_2CN^+ & + CN^- \end{array}$$

is equivalent in each case to a pair of Brönsted acids.

$$\begin{array}{ccc} \text{Acid} & \rightleftharpoons & \text{conjugate base} + H^+ \\ NH_4^+ & & NH_3 \\ NH_3 & & NH_2^- \\ H_3SO_4^+ & & H_2SO_4 \\ H_2SO_4 & & HSO_4^- \\ H_2CN^+ & & HCN \\ HCN & & CN^- \end{array}$$

A third, much more extensive, theory of acids and bases was proposed by Lewis, who defined an acid as an electron pair acceptor and a base as an electron pair donor. This definition includes the Brönsted and solvent system definitions as special cases. For example, the proton is an electron pair acceptor and H_2O or NH_3 or H_2SO_4 or HCN, among our examples, act as donors. This definition emphasizes the similarity between all coordination reactions, for example:

Lewis acid	Lewis base	product
H^+	H_2O	H_3O^+
	OH^-	H_2O
	F^-	HF
BF_3	H_2O	$H_2O:BF_3$
	OH^-	$(HOBF_3)^-$
	F^-	BF_4^-

It includes solvation reactions:

Lewis acid	Lewis base	
M^{z+}	xH_2O	$[M(H_2O)_x]^{z+}$
	xNH_3	$[M(NH_3)_x]^{z+}$

and at least the initial steps of solvolysis reactions, for example:

$$SnCl_4 \quad + \quad NH_3 \quad \rightarrow \quad (SnCl_4:NH_3)$$

but also includes many other types of reaction, including oxidation–reduction reactions. The Lewis theory is more extensively applied in America than in Europe but use of the general term 'Lewis acid' for acceptor molecules like BF_3 is very common, even by authors who make no other use of the theory.

The concept of *hard* and *soft* acids and bases (acceptors and donors) may be seen as an amplification of the Lewis approach. This idea is introduced in Section 2.16 and its application to transition metal complexes is discussed in Section 13.7.

6.6 General uses of nonaqueous solvents

It is now possible to outline some general reasons for choosing to work in a solvent other than water and to suggest points which will guide the choice of solvent. These are best seen by comparison with the properties of water itself.

(a) *Solvation and solvolysis*. The choice of a solvent is clearly dependent on its solvent powers. Solvents of high polarity and coordinating power are good ionizing media, and the range runs from these down to solvents such as the hydrocarbons which take up only molecular solids. In preparation reactions, a solvent is often chosen because of the insolubility of a certain compound in it. For example, the reaction in water between KCN and $NiSO_4$ gives $Ni(CN)_2$ by direct precipitation but the product is very finely divided and difficult to filter and wash. The same reaction in liquid ammonia gives an easily handled precipitate of K_2SO_4, not $Ni(CN)_2$, and the potassium salt is readily removed and the nickel cyanide recovered by evaporation.

A solvent may also be chosen because of the way in which it reacts with solutes. Since water coordinates strongly, it is often impossible to remove it without decomposing the compound, and many anhydrous compounds are available only by syntheses which avoid water, or other strongly coordinating solvents. One example is anhydrous copper nitrate, $Cu(NO_3)_2$, which is a volatile solid prepared by reaction in N_2O_4. When solvolytic reactions are possible, the solvent may be chosen to avoid these, as in the use of liquid sulfur dioxide for reactions with nonmetal halides. On the other hand, a solvent is often used because it does react with the solute. An important example is anhydrous hydrogen fluoride which solvolyses almost every compound placed in it and provides a standard method for preparing fluorine-containing compounds.

(b) *Acidity*. The effect of the acid–base properties has already been discussed. In water, the strongest acid is H_3O^+ and the strongest base is OH^-, all solutes of higher acidity or basicity being completely converted to these. If the effect of more acidic or more basic media is to be investigated, clearly a suitable nonaqueous solvent must be chosen. The most fully investigated solvents from this point of view are glacial acetic acid, anhydrous hydrogen fluoride and anhydrous sulfuric acid for acidic media, while ammonia and the amines have been the most-studied basic media.

One hundred percent sulfuric acid is so strong a proton donor that many substances which act as acids in water accept a proton (i.e. are bases) in sulfuric acid. The few proton donors include $H_2S_2O_7$ ('pyro' or di-sulfuric acid), fluorosulfuric acid FSO_3H and its trifluoromethyl analogue CF_3SO_3H. The solutions of such species in sulfuric acid are extremely powerful protonating agents and allow the preparation of species such as the iodine cations I_n^+ (see Chapter 17), which are highly susceptible to attack by bases. An extreme case is the 'superacids' (Section 6.9), and see also Section 18.5.

At higher temperatures, fused salts provide suitable media. Acidic systems include ammonium salts of strong acids, acidic anion salts such as HSO_4^- or HF_2^-, and acidic oxides such as silica or phosphorus pentoxide. Basic systems include the oxides or hydroxides of the alkali metals.

(c) *Redox*. In a rather similar way, the oxidation–reduction properties of the solvent place limits on the reactions which may be carried out in it. Water is attacked rapidly by oxidizing agents with standard potentials greater than about 1.7 V, and by reducing agents with oxidation potentials below about −0.4 V. Outside these limits, use of water as the solvent is undesirable. Nonaqueous solvents which are useful for carrying out oxidizing reactions include sulfuric acid and higher valency oxides such as dinitrogen tetroxide. The most important solvent for strong reductions is liquid ammonia, which dissolves the alkali and alkaline earth metals to give a very strongly reducing system.

A fairly wide variety of compounds have been used as nonaqueous solvents. Examples are listed in Table 6.6 with notes on their special uses. The general points made above are illustrated by more detailed discussion of specific systems. Glacial acetic acid and liquid ammonia represent protonic acid and base solvents respectively, and we then cover the extremely acid media termed 'superacids'. Bromine trifluoride is characteristic of a non-protonic and extremely active solvent for which the solvent system definition appears to hold. Mention is also made of sulfur dioxide where the current consensus has reversed the older view that self-ionization occurs.

Table 6.7 shows the more important physical properties of these solvents.

6.7 Liquid ammonia

Liquid ammonia was one of the first nonaqueous ionizing solvents to be studied and is now the one which is most extensively used and best known. It is readily available in reasonable purity and is easily dried by the action of sodium. The liquid range is from –77°C to –33°C, and this presents some handling problems, but these are partly compensated by the relatively high latent heat of vaporization. Liquid ammonia is normally handled in cooled vessels—and it is often used as its own coolant—or at room temperature under pressure. A wide range of special techniques have been developed for handling ammonia so that few problems are now met.

The dielectric constant and the self-ionization are both lower than for water, indicating that liquid ammonia will be a poorer ionizing solvent than water. Soluble salts include most ammonium salts, nitrates, thiocyanates and iodides. Fluorides and most oxysalts are insoluble, and solubility among the halides increases $F^- < Cl^- < Br^- < I^-$. Calcium and zinc chlorides, which are extremely soluble in water, are quite insoluble in liquid ammonia, but they do take up a large amount of ammonia, forming solid ammoniates with eight and ten molecules of ammonia respectively. Calcium chloride is thus a useful absorbent for traces of ammonia. Since ammonia is less highly associated than water, it is a better solvent for organic compounds, especially for those with fairly small carbon radicals. Unsaturated hydrocarbons, alcohols, esters, ammonium salts of acids and most nitrogen compounds are soluble.

The self-ionization of ammonia is written:

$$2NH_3 \rightleftharpoons NH_4^+ + NH_2^-$$

and ammonium compounds behave as acids while amides are bases. Ammonium iodide, nitrate or thiocyanate are very soluble and concentrated solutions will slowly dissolve metals with the evolution of hydrogen:

$$Mg + 2NH_4^+ \rightarrow Mg^{2+} + H_2 + 2NH_3$$

This reaction is rapid with the active alkali and alkaline earth metals but slow with less active metals such as magnesium or iron. As ammonia is more basic than water, these acids are weaker, so the solvent power for metals does not extend to the less active metals. The commonest base is potassium amide, KNH_2, which is much more soluble than sodium amide. Acid–base titrations between potassium amide and ammonium salts may be carried out and followed conductiometrically or by using phenolphthalein. The ammonia system brings out very weakly acidic functions in molecules. Thus, urea, $CO(NH_2)_2$, which is a weak base in water, acts as a weak acid in liquid ammonia and may be neutralized by amide.

There are many examples of amphoteric behaviour in liquid ammonia. For example:

$$Zn^{2+} \underset{NH_4^+}{\overset{NH_2^-}{\rightleftharpoons}} Zn(NH_2)_2\downarrow \underset{NH_4^+}{\overset{NH_2^-}{\rightleftharpoons}} Zn(NH_2)_4^{2-}$$

TABLE 6.6 Typical nonaqueous solvents

Solvent	*Postulated self-ionization*	*Acids*	*Bases*	*Comments*
NH_3	$NH_4^+ + NH_2^-$	Ammonium salts	Amides, e.g. KNH_2	See Section 6.7 for discussion. A number of similar solvent systems have been studied including N_2H_4, NH_2OH and organic amines such as $C_2H_4(NH_2)_2$ and CH_3NH_2
$HCONH_2$ $HCON(CH_3)_2$ and CH_3CONH_2		Protonic and Lewis acids	Organic nitrogen bases, e.g. pyridine	H-bonding, high dielectric constant, make these good solvents for ionic compounds. Dimethylformamide is especially of current interest
HF	$H_2F^+ + HF_2^-$	F^- acceptors, e.g. BF_3 $HF + BF_3 = H_2F^+BF_4^-$	Alkali fluorides	Good ionizing solvent: many nonfluorine species react, e.g. $H_2SO_4 \rightarrow HSO_3F$. Used in the preparation of fluorine compounds, e.g. $AgNO_3 \rightarrow Ag^+ + H_2NO_3^+ + 2F^-$ $BF_3 + F^- \rightarrow BF_4^-$, then $Ag^+ + BF_4^- \rightarrow AgBF_4$
CH_3COOH	$CH_3COOH_2^+ + CH_3COO^-$	Protonic acids	Ionizable acetates	See Section 6.8. Other carboxylic acids behave similarly, especially HCOOH
H_2SO_4	$H_3SO_4^+ + HSO_4^-$	$H_2S_2O_7$ FSO_3H ($HClO_4$ very weak)	Soluble bisulfates H_2O, HNO_3, H_3PO_4	Most common acids are bases in H_2SO_4
N_2O_4	$NO^+ + NO_3^-$	NOCl	Alkali nitrates	Used to prepare anhydrous nitrates, nitrato- and nitro-complexes and nitrosyl compounds: commonly used in admixture with organic compounds such as ethyl acetate
SO_2	Conflicting evidence, see box on p.156			Useful as an inert reaction medium
$SeOCl_2$	$SeOCl^+ + Cl^-$ (or $SeOCl_3^-$)	Cl^- acceptors e.g. $SnCl_4$	Organic bases like pyridine giving $C_5H_5N.SeOCl^+Cl^-$	Dissolves many elements with reaction and dissolves a variety of metal chlorides. Other salts are converted to the chloride
Other oxyhalides, NOCl and $POCl_3$ for example, are similar. Some conductiometric and potentiometric titration data exist to support the self-ionization mode and acid–base behaviour which is postulated				
BrF_3	$BrF_2^+ + BrF_4^-$	F^- acceptors	Ionizable fluorides	Discussed in Section 6.10. IF_5 behaves similarly
$AsCl_3$	$AsCl_2^+ + AsCl_4^-$	Cl^- acceptors	Cl^-	$AsBr_3$, $SbCl_3$ and ICl are analogous

TABLE 6.7 Physical properties of nonaqueous solvents

	H_2O	*NH_3*	*CH_3COOH*	*SO_2*	*BrF_3*
m.p. (°C)	0	−77.7	16.6	−75.5	9.0
b.p. (°C)	100	−33.4	118.1	−10.2	126
Dielectric constant	78.5 (18°C)	23 (b.p.)	9.7 (18°C)	17.3 (−16°C)	–
Specific conductivity ($\Omega\ m^{-1}$)	6×10^{-8} (25°C)	5×10^{-9} (b.p.)	0.5–0.8 $\times 10^{-8}$ (25°C)	4×10^{-8} (b.p.)	8×10^{-3} (m.p.)
Heat of vaporization ($kJ\ mol^{-1}$)	40.7	23.6	24.3	25.0	42

ANALYTICAL USES

This enhancement of weakly acidic functions by the basic solvent finds application in analysis. Ammonia itself is rarely used because of its low boiling point, but simple organic derivatives such as ethylamine, $CH_3CH_2NH_2$ (which is related to ammonia as ethanol, CH_3CH_2OH, is to water), or ethylenediamine, $H_2NCH_2CH_2NH_2$, are used as solvents for the determination of weak acids. Titrants used include methoxides of the alkali metals and soluble hydroxides which are easier to purify and standardize than amides. Applications include the determination of phenols and related compounds in ethylenediamine, and the determination of carbon dioxide which may be separated from other gases by solution in acetone and determined as a weak acid with sodium methoxide, CH_3ONa. Suitable indicators or potentiometric methods are used to determine the end points.

compare with

$$Zn^{2+} \underset{H_3O^+}{\overset{OH^-}{\rightleftharpoons}} Zn(OH)_2\downarrow \underset{H_3O^+}{\overset{OH^-}{\rightleftharpoons}} Zn(OH)_4^{2-}$$

Amides and imides of the less-reactive metals dissolve in excess potassium amide:

$$\text{e.g.} \quad PbNH + NH_2^- + NH_3 = Pb(NH_2)_3^-$$

and even sodium amide, which is insoluble in ammonia, dissolves in potassium amide to give the amidosodiate:

$$NaNH_2 + 2NH_2^- = Na(NH_2)_3^{2-}$$

Hydrolysis of heavy metal salts to basic salts (oxy- and hydroxy-compounds) is paralleled by reactions between ammonia and such salts. The analogues of the oxy- and hydroxy-compounds in the aqueous system are compounds containing the groups, amide $-NH_2$, imide $=NH$, and nitride $\equiv N$. For example, $MoCl_5$ dissolves in liquid ammonia and the compound $Mo(NH)_2NH_2$ may be isolated from the solution. If an ammonium salt is added, this compound dissolves, while the addition of potassium amide precipitates it.

The most striking property of liquid ammonia is its ability to dissolve the active metals to give fairly stable, blue solutions. As ammonia is more resistant to reduction than is water, the reaction:

$$M + NH_3 = MNH_2 + \tfrac{1}{2}H_2$$

is much slower than the reaction:

$$M + H_2O = MOH + \tfrac{1}{2}H_2$$

If the reagents are pure and dry, sodium solutions in liquid ammonia may be preserved for several weeks and even the much more reactive caesium gives solutions which may be kept overnight. These solutions are formed by all the alkali metals, by the alkaline earth metals, and by the reducible lanthanide elements such as samarium. In addition, very dilute solutions may be formed from less reactive elements, such as magnesium, beryllium, aluminium and the other lanthanides, by electrolytic means. The alkali metals are also soluble in amines and dilute solutions are formed in certain ethers.

Dilute metal solutions are coloured a very deep blue and concentrated ones have a metallic, coppery appearance. The solutions are conducting and strongly reducing. In these solutions, the valency electrons are ionized and become solvated:

$$Na \rightarrow Na^+_{(ammoniated)} + e^-_{(ammoniated)}$$

The unique properties of the solutions are associated with the presence of these readily available electrons.

The major application of metal solutions is in reduction reactions. In organic chemistry, the skeletal single bonds, C-C, C-O, C-N, are stable in these solutions as are isolated double bonds and single benzene rings, but nearly all other functional groups and most unsaturated compounds are reduced. In inorganic chemistry, the most interesting reductions have been the formation of polyanions, the reductions of hydrides, and the formation of transition metal complexes in unusually low oxidation states. Some examples are given in Table 6.8.

6.8 Anhydrous acetic acid

To some extent, glacial acetic acid is the converse solvent to liquid ammonia. It is readily available, easily purified, and differs from water in undergoing less self-ionization, and in having a markedly lower dielectric constant. It is a moderately strong acid and is used to investigate weakly basic functions, just as ammonia and the amines are used for weakly acid functions. The lower dielectric constant makes acetic acid a poorer solvent for ions, and most compounds which are insoluble in water are insoluble in acetic acid. Soluble

TABLE 6.8 Reduction reactions by metal solutions in liquid ammonia

Reactant	*Reduction products*
O_2	Metal peroxides O_2^{2-}, and superoxides, O_2^-
S, Se, Te, As, Sb, Bi, Sn, Pb and their oxides or halides	White binary compounds such as M_2S or M_4Pb and highly coloured polyanions such as M_2S_x ($x = 2$ to 7) (deep red), M_3Bi_3 (violet) or M_4Pb_9 (deep green) (cf. Section 18.4.4)
Metal oxides, halides and dissociable complexes	Metal
Complexes which do not dissociate e.g. $Ni(CN)_4^{2-}$; $Co(CN)_6^{2-}$; $Ni(C\equiv CH)_4^{2-}$	$Ni_2(CN)_6^{4-}$; $Ni(CN)_4^{4-}$; $Co(CN)_4^{4-}$; $Ni(C\equiv CH)_4^{4-}$
Hydrides of Groups IV and V	Ions formed by removal of one or two hydrogens PH_2^- or SnH_2^{2-} and SnH_3^-
Ge_3H_8	$2GeH^{-3} + GeH_2^{2-}$ (i.e. each Ge-Ge bond is broken)

compounds include most acetates, nitrates, halides, cyanides and thiocyanates. The strong acids all dissolve in glacial acetic acid as do basic compounds like water and ammonia. A wide range of polar organic compounds is also soluble.

The self-ionization of acetic acid is:

$$2CH_3COOH \rightleftharpoons CH_3COOH_2^+ + CH_3COO^-$$

so that acetates of the reactive metals are bases. The normal protonic acids are acids in acetic acid by virtue of reactions such as:

$$HClO_4 + CH_3COOH \rightleftharpoons CH_3COOH_2^+ + ClO_4^-$$

(compare $HClO_4 + H_2O \rightleftharpoons H_3O^+ + ClO_4^-$), while ammonia, say, is a base as the acetate ion is produced by the reaction:

$$NH_3 + CH_3COOH \rightleftharpoons CH_3COO^- + NH_4^+$$

Of the mineral acids which are strong acids in water (that is, are completely dissociated to the H_3O^+ ion), only perchloric acid is strongly dissociated in glacial acetic acid. The relative strengths of the common acids in acetic acid are $HNO_3 = 1 < HCl = 9 < H_2SO_4 = 30 < HBr = 160 < HClO_4 = 400$. The acetates which are strongest bases are those of the alkali metals and ammonium, while bismuth, lead and mercuric acetates are ten to a hundred times weaker. Even the strongest bases and acids are poor electrolytes in acetic acid, largely as a result of the low dielectric constant of the solvent.

Acid–base titrations between the acetates and the mineral acids in acetic acid may be demonstrated by indicators or electrometrically. In addition, perchloric acid in glacial acetic acid is a widely used reagent for the estimation of the weakly basic functions of amines, amino acids, metal salts of organic acids, and the like.

Amphoteric behaviour has also been demonstrated in acetic acid. For example, the addition of sodium acetate to a solution of a zinc salt precipitates zinc acetate, which redissolves in excess acetate:

$$Zn^{2+} \underset{CH_3COOH_2^+}{\overset{CH_3COO^-}{\rightleftharpoons}} Zn(CH_3COO)_2\downarrow \underset{CH_3COOH_2^+}{\overset{CH_3COO^-}{\rightleftharpoons}} Zn(CH_3COO)_4^{2-}$$

Copper and lead(II) acetates also show amphoteric behaviour, while lead(IV) tetra-acetate decreases in solubility as sodium acetate is added to its solution, which corresponds to 'salting out' a poor electrolyte.

Acetic acid is a convenient reaction medium for the preparation of covalent hydrolysable compounds. For example, tin reacts smoothly and controllably with halogens in acetic acid:

$$Sn + 2X_2 \longrightarrow SnX_4$$

and the stannic halide is readily isolated by distillation or crystallization.

Other weak acids which have been studied as solvent systems include formic acid and hydrogen cyanide, but investigations in these systems have been largely confined to studies of solubility and acid–base relationships. Of the strong acids, attention has been concentrated on hydrogen fluoride and sulfuric acid, in which the principal type of reaction is solvolysis. This is also true of bromine trifluoride, Section 6.10.

6.9 'Superacid' media

We have mentioned above (Section 6.6) the use of H_2SO_4 as a strongly acidic medium. However, there are some systems which are even more powerfully proton-donating and these are of interest (a) in protonating species which are very weak H^+ acceptors and (b) in preparing species which are highly susceptible to base attack. (Clearly, the more strongly acidic the medium is, the more weakly basic it is.)

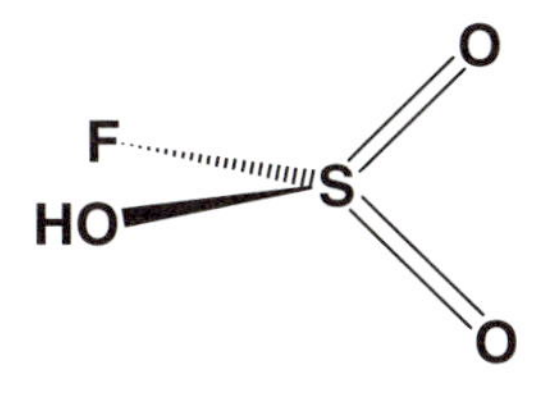

The structure of FSO_3H

Probably the strongest proton donor of all simple substances is disulfuric acid, $H_2S_2O_7$, formed from SO_3 and H_2SO_4. However, this is a very difficult medium to handle as it is very viscous and the equilibria between various components are complex. The next strongest proton donor is fluorosulfuric acid, related to sulfuric acid by replacing an OH group by an F.

This change (a) lowers the viscosity by reducing the amount of H-bonding and (b) increases the acidity as F withdraws more charge than OH. (Compare Section 6.2.) Handling properties are more convenient, with a m.pt. of −89°C and b.pt. of 163°C. The low m.pt. allows ^1H nmr studies at low temperatures (see Section 7.8) which have been extensively used to determine the site of protonation. At the higher m.pt. of H_2SO_4 (10.4°C), exchange processes and rearrangements are already too rapid for the simple species to be observed. When free from HF, fluorosulfuric acid can readily be handled in ordinary glassware.

Using the Lowry–Brönsted approach, the self-ionization is

$$2HSO_3F \rightleftharpoons H_2SO_3F^+ + SO_3F^-$$

From the conductivity, the concentration of the solvent ions at 25°C is about 2×10^{-4} mol kg^{-1}, much lower than in H_2SO_4 but higher than in H_2O or CH_3CO_2H.

Acids in HSO_3F are species which increase the $H_2SO_3F^+$ concentration, either by proton donation

$$HA + HSO_3F = H_2SO_3F^+ + A^- \quad \text{(A1)}$$

or by removing SO_3F^- ions as SbF_5 does:

$$SbF_5 + 2HSO_3F = H_2SO_3F^+ + [SbF_5(SO_3F)]^- \quad \text{(A2)}$$

Bases are alkali metal fluorosulfates or proton acceptors

$$KSO_3F = K^+ + SO_3F^- \quad \text{(B1)}$$

$$B + HSO_3F = BH^+ + SO_3F^- \quad \text{(B2)}$$

Because the medium is so acidic, there are very few proton donors: for example, H_2SO_4 does not follow (A1) but acts as a weak base according to (B2) forming $H_3SO_4^+$. Similarly, although SO_3 dissolves to give HS_2O_6F, this analogue of $H_2S_2O_7$ does not ionize in HSO_3F, and so the 'fluoro-oleum' system contrasts with SO_3/H_2SO_4. Similarly HF, $HClO_4$ and other acids which are strong in water do not donate H^+ in HSO_3F: e.g.

$$KClO_4 + HSO_3F \rightarrow K^+ + SO_3F^- + HClO_4$$

Thus the only acids are the strong acceptors which follow Equation (A2). Of these, two systems are important. SbF_5 behaves as a weak acid, i.e. ionization according to Equation (A2) is not complete and a titration of SbF_5 with the base KSO_3F shows a conductivity minimum about 0.4:1. However, if SO_3 is added, a range of species $SbF_{5-n}(SO_3F)_n$ is formed ($n = 1, 2, 3$) and such species act as strong acids. For example, $SbF_2(SO_3F)_3$ gives a 1:1 titration with KSO_3F.

Having established the very high acidity of HSO_3F itself, it follows that these even more acidic systems HSO_3F/SbF_5 and $HSO_3F/SbF_5/SO_3$ have astonishingly high proton donor powers. They have been colourfully termed *superacids,* while the term 'magic acid' has been applied to HSO_3F/SbF_5.

As examples, HSO_3F alone protonates organic acids to give $RC(OH)_2^+$, ketones to give $R_2C(OH)^+$ and haloaromatics to give for example $FC_6H_6^+$ from FC_6H_5. The superacid systems will completely protonate trinitrobenzene and even alkanes to give intermediates such as CH_5^+. Such ions readily lose H_2 to give carbonium ions, e.g. CH_3^+.

In inorganic chemistry, fluorosulfuric acid is an excellent agent for preparing fluorides and fluorosulfates as in

$$As_2O_5 \longrightarrow AsF_5 + AsF_2(SO_3F)_3$$
$$SiO_2.xH_2O \longrightarrow SiF_4$$
$$P_4O_{10} \longrightarrow POF_3$$
$$BaTeO_4 \longrightarrow TeF_5(SO_3F)$$
$$ClO_4^- \longrightarrow ClO_3F$$

Also of importance is the use of this medium to prepare polyatomic ions using $S_2O_6F_2$ as oxidant, as in

$$Se + S_2O_6F_2 \longrightarrow Se_4^{2+} + 2SO_3F^-$$

or

$$I_2 + \tfrac{1}{2}S_2O_6F_2 \longrightarrow I_2^+ + SO_3F^-$$

Such polyatomic species are susceptible to base attack and need an acid medium to survive. Often they were first seen in the sulfuric acid system but only characterized in the more manageable fluorosulfuric acid one. For instance, I_2 was reported to dissolve in oleum to give a blue solution 20 years before the blue species was identified by work in HSO_3F as I_2^+ and not I^+ as originally thought. Other examples are given in Chapter 17.

6.10 Bromine trifluoride

Bromine trifluoride is a much more restricted solvent than those discussed above but it is one to which the ideas of the solvent system theory appear to apply and it has been quite widely used as a preparative medium. It must be handled by special techniques which require experience but these are now highly developed and present few problems in a well-equipped laboratory. The main requirements are a rigorous exclusion of moisture and of materials such as tap greases which can be fluorinated or oxidized. (See also the comments on fluorine handling in Section 17.8.)

The alkali metal fluorides are soluble but most other ionic fluorides are rather insoluble. The more covalent fluorides of elements in higher oxidation states are also soluble. Most other compounds are either insoluble or converted to fluoro-compounds.

The self-ionization of bromine trifluoride is much more extensive than that of the other solvents in Table 6.6 and is presumed to follow the equation:

$$2BrF_3 = BrF_2^+ + BrF_4^-$$

By the solvent system definition, solutes which increase the conventration of BrF_2^+ are acids and those which increase the concentration of BrF_4^- are bases. The bromofluorides of the alkali metals and a number of other elements, e.g. $AgBrF_4$, are known and act as bases, while the alkali metal fluorides add on a molecule of the solvent and may also be regarded as bases in the system. No solute is known which produces the solvent cation by dissociation (as the mineral acids produce H_3O^+ in water) that is, there are no *donor acids* known, but a number of solutes are known which react with the solvent molecule by removing a fluoride ion and these are termed *acceptor acids*. This is true of most covalent fluorides, which form fluoro-complexes, for example:

$$VF_5 + BrF_3 = BrF_2^+ + VF_6^-$$
$$SnF_4 + 2BrF_3 = 2BrF_2^+ + SnF_6^{2-}$$

Neutralization reactions are represented by the interaction of these acids and bases and may be followed conductiometrically:

$$\underset{\text{acid}}{BrF_2^+VF_6^-} + \underset{\text{base}}{KBrF_4} = \underset{\text{salt}}{KVF_6} + \underset{\text{solvent}}{2BrF_3}$$

Solvolysis of some of these complex fluorides is also observed. For example, if hexafluorotitanate is treated with bromine trifluoride the reversible reaction

$$K_2TiF_6 + 4BrF_3 \rightleftharpoons (BrF_2)_2TiF_6 + 2KBrF_4$$

is observed leading to the formation of the solvent base in equilibrium.

The main use of bromine trifluoride is in the preparation of fluorides and fluorocomplexes of elements in the higher oxidation states. Among the transition metal compounds which have been prepared in bromine trifluoride are the hexafluoro-complexes of Ti(IV), V(V), Nb(V), Ta(V), Mn(IV), Rn(IV, V), Os(IV, V), Rh(IV), Ir(IV, V), Pd(IV), Pt(IV) and Au(III). The first preparation of gold trifluoride was also made in this solvent. Another interesting series of reactions involves the reactions of non-metal oxides with fluorocomplexes in bromine trifluoride to give fluorosulfonates, nitronium complexes and nitrosonium complexes, such as:

$$SO_3 + KF \longrightarrow KSO_3F$$
$$NO_2 + Au \longrightarrow (NO_2)AuF_4$$
$$NO_2 + As_2O_3 \longrightarrow (NO_2)AsF_6$$
$$NOCl + GeO_2 \longrightarrow (NO_2)GeF_6$$

(Note that several of these reactions involve oxidations as well as fluorination.)

SULFUR DIOXIDE AS A SOLVENT—A CLASSICAL EXAMPLE REASSESSED

Along with liquid ammonia, sulfur dioxide was one of the earliest nonaqueous solvents to be studied and such work extends back nearly a century. Early experiments showed a small conductivity, so that self-ionization into SO^{2+} and SO_3^{2-} was postulated and the Cady–Elsey solvent system model was applied.

This was supported by the observation of a 'neutralization' reaction between SO_3^{2-} and $SOCl_2$ taken as a source of SO^{2+} (i.e. an acid in the SO_2 solvent system). However, the physical properties raised doubts—especially that doubly charged ions had to be separated in a liquid of low dielectric constant. Later investigations, especially highly sensitive radioisotope exchange studies which showed no exchange of ^{18}O or ^{35}S between SO_2 and $SOCl_2$, led to the abandonment of the solvent system approach. The reaction was not one between solvent cation and anion, but must instead have been a two-step process, or indeed an artefact arising from traces of moisture. Self-ionization is negligible or absent and thoroughly dried SO_2 is now used as a noninteracting solvent for reactions which do not involve oxidizing agents.

Sulfur dioxide is often a useful reaction medium. It is convenient for low temperature reactions, and can be easily removed by allowing it to boil off at room temperature. Reaction temperatures up to around 50°C can be attained under moderate pressure. It supports reactions between Brönsted acids and bases, and is generally useful for reactions between weakly ionic or fairly polar molecules. Species showing high solubilities include the iodides (Table 6.4), thiocyanates and carboxylates of the heavier alkali metals or ammonium, together with the halides and pseudohalides of the heavier Main Group elements. A major use is as solvent for reactions of such readily hydrolysable compounds, as in

$$SOCl_2 + 2SCN^- = SO(SCN)_2 + 2Cl^-$$

or in the preparation of halo-complexes such as

$$NOCl + SbCl_5 = (NO)^+(SbCl_6)^-$$

Other fluorine compounds, such as IF_5 or SeF_4, also allow synthesis of fluorides. For elements with a range of oxidation states, HF is the medium for preparing fluorides of the lowest oxidation states, IF_5 gives intermediate states while BrF_3 gives high oxidation fluorides. To force an element to form fluorides of very strongly oxidizing states it may be necessary to use the even more vigorous ClF_3, as a medium. Of course, most fluorides result from direct reaction with F_2, but these halogen fluoride solvents may be very useful when it is necessary to discriminate among a range of oxidation states.

6.11 Supercritical fluids

Above a certain temperature and pressure, called the *critical point*, some substances cannot be classified as either gases or liquids; instead, these fluids are termed *supercritical fluids*, and they have properties intermediate between those of liquids and gases. While supercritical fluids have been known for around 180 years, it is only in recent years that significant practical applications of these fluids have been developed.

A number of substances have critical temperatures close to room temperature, and two of the most common substances are water and carbon dioxide. Both supercritical water (scH_2O) and carbon dioxide ($scCO_2$) are able to dissolve nonpolar substances such as organic compounds (scH_2O is considerably less polar than liquid water). $scCO_2$ is of particular interest, because it is inexpensive, and can be recycled from the reaction medium by simple evaporation followed by recompression. The main disadvantage in using supercritical fluids is the high pressures which are typically involved. However, the use of microbore metal pipework makes this technique much more accessible in the laboratory.

Supercritical CO_2 has been developed as an industrial solvent, and the best known application of $scCO_2$ is in the removal of caffeine from green coffee beans, termed the Hag process. However, applications in inorganic chemistry have been slower to develop. There are a number of desirable reasons for using supercritical fluids in inorganic chemistry, and a number of examples serve to illustrate this nicely. Firstly, it is known that H_2 is completely miscible with supercritical fluids, whereas it often shows low solubility in organic solvents (in which reactions such as catalysed hydrogenation reactions are commonly done). The use of H_2 in a supercritical fluid has allowed the isolation of the dihydrogen complex $(C_5H_5)Mn(CO_2)(\eta^2\text{-}H_2)$ (see Section 16.11 for dihydrogen complexes) which was previously believed to be too unstable to isolate. Secondly, supercritical xenon is readily accessible (with a critical temperature around room temperature), and is completely transparent in the ultraviolet, visible and infrared regions of the electromagnetic spectrum, and thus gives no background peaks. scXe has been used to detect the formation of coordinated xenon atoms in $M(CO)_5Xe$ ($M = Cr$, Mo, W), and has been used to study reactions of alkanes (such as methane and ethane) with highly reactive metal complexes; in this case, other solvents must be avoided because they too would undergo reaction.

Other applications of supercritical fluids include supercritical fluid chromatography (SFC). The low viscosity and good solvent properties of a supercritical fluid gives improved separation of mixtures of non-volatile compounds when compared to traditional chromatographic techniques such as high performance liquid chromatography (hplc).

Problems

This chapter is intended to extend your consideration of the conditions of chemical reactions from the familiar use of water as a medium to the use of other solvents which have many similarities to water.
Sections 6.1 to 6.3 are intended to bridge on to material normally covered in detail in physical chemistry courses. Only a few examples are given, but many others can be constructed, especially by applying the data of Table 6.3 to Equation (6.11).
The material of Sections 6.4 to 6.10 is best mastered by making your own correlations among the different solvents, and by relating to the chemistry discussion in later chapters. You should therefore look at the preparations of species of very low or high oxidation states.

6.1 Arrange the oxyacids of sulfur (Table 17.15) in their likely order of acid strengths from their formulae.

Does knowledge of the structures alter the position of any of the compounds?

6.2 Find four couples from Table 6.3 which would be completely converted to the reduced form by Cr (but not by Cr^{2+}) going to Cr^{3+}.

6.3 Find three examples in the systematic chemistry chapters of preparations which had to be carried out in a nonaqueous solvent.

CHAPTER 7

SEPARATION METHODS 160
7.1 ION EXCHANGE 160
7.2 CHROMATOGRAPHY 161
7.3 SOLVENT EXTRACTION 162
STRUCTURE DETERMINATION 164
7.4 DIFFRACTION METHODS 164
7.5 SPECTROSCOPIC METHODS AND THE ELECTROMAGNETIC SPECTRUM 166
7.6 ELECTRONIC SPECTRA 168
7.7 VIBRATIONAL SPECTRA 169
7.8 NUCLEAR MAGNETIC RESONANCE (NMR) 171
7.9 FURTHER METHODS OF MOLECULAR SPECTROSCOPY 175
7.9.1 Electron spin resonance (esr) spectroscopy 175
7.9.2 Mössbauer spectroscopy 175
7.9.3 Mass spectrometry 176
7.10 FOURIER TRANSFORM METHODS 177
7.11 OTHER METHODS 178
7.11.1 Magnetic measurements 178
7.11.2 Dipole moments 179
7.11.3 Scanning electron microscopy (SEM) 180
7.11.4 Scanning tunnelling microscopy (STM) 180
DETERMINATION OF ENERGY LEVELS 182
7.12 PHOTOELECTRON SPECTROSCOPY 182
7.12.1 The energy source 182
7.12.2 Vibrational structure 183
PROBLEMS 184

Experimental Methods

In the discussion of the chemistry of the elements which makes up the later part of this book, the structures of a number of compounds are described. It is the aim of this chapter to give a brief indication of the more important experimental methods of determining these structures, and a description of some of the more recently developed methods of separating compounds is also given. Fuller accounts of the individual methods are given in the references.

Separation methods

7.1 Ion exchange

The use of clay minerals and zeolites for base exchange and water treatment has been established for many years, but a major advance was made when the use of synthetic resins containing specific functional groups was introduced. The materials in present use are cross-linked polystyrene resins and similar types containing active groups. These are:

sulfonic groups	$-SO_2OH$	giving strongly acid cation exchangers
carboxylic groups	$-COOH$	giving weakly acid cation exchangers
quaternary ammonium groups	$-NR_3^+$	giving strongly basic anion exchangers
amine groups	$-NR_2$; $-NH_2$; $-NHR$	giving weakly basic anion exchangers,

where the bond is to the polymer skeleton. The polymer skeleton of the resin acts as an inert, unreactive framework to support the functional groups and has a relatively porous structure with about ten percent cross-linking.

The sulfonic acid resins* are strong acids and the hydrogen atoms ionize and are readily replaced by cations. If a solution of a salt, say NaX, is passed down a column containing the resin, the Na^+ is replaced:

$$c-\text{ResinH} + \text{NaX} \rightarrow c-\text{ResinNa} + \text{HX}$$

In a similar way, the strongly basic resins* containing quaternary ammonium groups have hydroxyl groups which ionize completely and are exchangeable with anions:

$$a-\text{ResinOH} + \text{NaX} \rightarrow a-\text{ResinX} + \text{NaOH}$$

$$\text{or } a-\text{ResinOH} + \text{HX} \rightarrow a-\text{ResinX} + H_2O$$

This illustrates one important use of these resins: a strong acid cation exchanger used in series with a strong base anion exchanger will remove all dissolved salts from water and is widely used for water treatment, especially in boilers to prevent scale formation.

Strongly acid cation exchangers in the sodium form will exchange the sodium ions for other cations:

$$c-\text{ResinNa} + M^{n+} = c-\text{ResinM} + n\text{Na}^+$$

These reactions are equilibrium reactions and the affinity of the resin for the cation in solution depends on the charge, the size of the ion and the concentration. In 0.1 mol l^{-1} solutions, the series $Th^{4+} > Fe^{3+} > Al^{3+} > Ba^{2+} > Pb^{2+} > Sr^{2+} > Ca^{2+} > Fe^{2+} >$

*Cation exchange resins are abbreviated as c-resins and anion exchange resins as a-resins.

$Co^{2+} > Mg^{2+} > Ag^{+} > Cs^{+} > Rb^{+} > NH_4^{+} = K^{+} > Na^{+} > H^{+} > Li^{+}$ has been established, but in concentrated solutions the effect of valency is reversed and univalent ions are favoured over multi-valent ones. Thus, the sodium form of the resin may be used to remove, say, calcium ions from a dilute solution and the calcium can be recovered by treating the resin with a concentrated solution of a sodium salt. Analogous use is made of strongly basic anion exchangers.

The weakly acid cation exchangers behave as insoluble weak acids. They may be buffered so that exchange takes place at a controlled pH. In acid solution they are undissociated and have little exchange capacity, but in neutral or alkaline media they behave similarly to the strong acid resins as cation exchangers, being rather more selective for divalent cations. The hydrogen form is useful to produce a weak acid from one of its salts:

$$\text{ResinH} + CH_3COONa \rightarrow \text{ResinNa} + CH_3COOH$$

Similar remarks apply to the weakly basic anion exchange resins.

The uses of the ion-exchange materials will be fairly obvious from the above outline of their properties. In preparations or analysis, any specific ion may be replaced by another one or by a hydrogen or hydroxyl ion. In the latter cases the liberated acid or alkali may be titrated to determine the quantity of cation or anion respectively in the original solution. An example of a preparative application is the use of a strongly basic anion exchanger to prepare carbonate-free alkali metal hydroxides. A solution of, say, sodium hydroxide contaminated with carbonate is run down a column containing a strong base anion exchanger. This has a higher affinity for the doubly charged carbonate ion which is then replaced by hydroxyl ion from the column. Pure sodium hydroxide solution is recovered.

An obvious extension of this technique is the removal of interfering ions in analysis, and the resins may also be used to concentrate trace constituents. The resins also find use as catalysts, especially as acid or base catalysts, and in the determination of dissociation constants and activity coefficients. As the exchange process is an equilibrium, the resins may be used in the separation of isotopes. For example, a separation of ^{14}N and ^{15}N has been effected by the exchange between a cation exchange resin and ammonium hydroxide:

$$\text{Resin}^{14}NH_4^{+} + {}^{15}NH_4OH \rightarrow \text{Resin}^{15}NH_4^{+} + {}^{14}NH_4OH$$

A band of NH_4^{+} travelling down a long column of the resin by means of a large number of absorption–desorption steps gradually becomes enriched with $^{15}NH_4^{+}$ at the trailing edge and with $^{14}NH_4^{+}$ at the leading edge. By the time the band is extended to about forty times its original width, the tail fraction contains 99% ^{15}N.

7.2 Chromatography

Chromatography is a process for separating a mixture which depends on the redistribution of the components between a stationary phase and a mobile one. The components may be adsorbed on the stationary phase or be held by more specific chemical bonds. The typical arrangement has the stationary phase in a column and the mixture is passed up or down the column in the gas or liquid phases. One form of chromatography involves the use of ion exchangers and follows on from the discussion in the previous section.

Suppose that a mixture of cations in solution have similar chemical properties, they will not be separated from each other by the simple ion exchange processes discussed above. If the cations are absorbed at the top of a cation exchange column and then treated with a weak complexing agent—one with an affinity for the cation similar to that between the cation and the column material—then the equilibrium,

$$\text{cation on ion exchanger} + \text{complexing agent} \rightleftharpoons \text{cation in complex in solution}$$

will depend sensitively on the nature of the cation. If the band of mixed cations on the column is washed down (*eluted*) by a solution of the complexing agent, those cations which

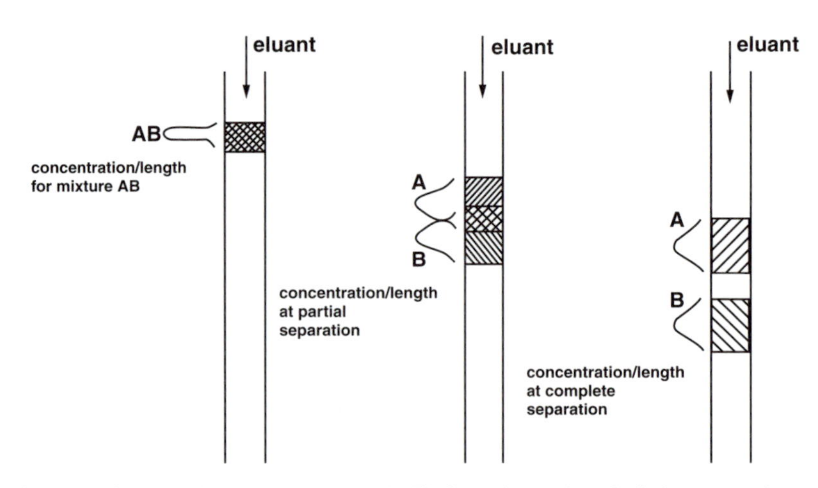

FIG. 7.1 Separation of a mixture by column chromatography. The figure shows schematically the concentration gradients of the two components, expressed as Gaussian distributions, at different stages in the separation

form the strongest complexes-will spend more time in solution than on the column, and will travel down the column faster than the cations whose complexes are weaker and which remain longer on the ion exchanger. As the cation band travels down the column, it starts to separate into its components and, if the column is long enough, the different components are recovered separately. The process is shown diagrammatically in Fig. 7.1. As there are also random processes of diffusion affecting the ions, the distribution of cation concentration down the column is Gaussian. The classic example of this method of separation was the separation of the lanthanide elements, using citrate as the weak complexing agent. A similar use of anion exchange resins allows a separation of anions. If the ions are not too closely similar, the separation may be effected without the use of a complexing agent. For example, the halides may be separated on an anion exchange resin by eluting with a sodium nitrate solution, and an additional element of control is introduced by varying the concentration of the eluant.

Many other systems have been devised for chromatographic separation. Materials such as silica gel or cellulose which act by adsorption, or by a mixture of adsorption and chemical interaction with adsorbed water, are widely used, and one common application is in paper chromatography. A wide variety of ions and molecules may be separated by using suitable solvents moving over paper. For example, the alkali metals may be separated by using a mixture of alcohols as the solvent, and the various phosphate anions may be separated using a two-dimensional method with a basic solvent flowing in one direction followed by an acidic one in a direction at right angles, Fig. 7.2. The analogous, but more controlled and sensitive, method of *thin layer chromatography* uses a uniform thin layer of silica spread on a glass plate as the separating medium.

Another modification of this method is in the separation of gas mixtures by passing them in a stream of nitrogen, or helium, through a column containing a high-boiling liquid supported on an inert material. For example, the silicon or germanium hydrides may be separated by this *vapour phase chromatography* by passing the hydride mixture through a column of silicone oil on powdered brick.

7.3 Solvent extraction

A mixture may be separated into its components by treating it with two or more immiscible liquids, most commonly with an aqueous phase and an organic phase. Ionic and strongly polar species remain in the aqueous phase while less polar species dissolve into the organic liquid. For example, germanium may be separated by forming the tetrachloride in contact with a solvent, such as carbon tetrachloride, in which it is soluble. As the hydrolysis of the tetrachloride is reversible in presence of hydrochloric acid, all the germanium ends up in the organic phase:

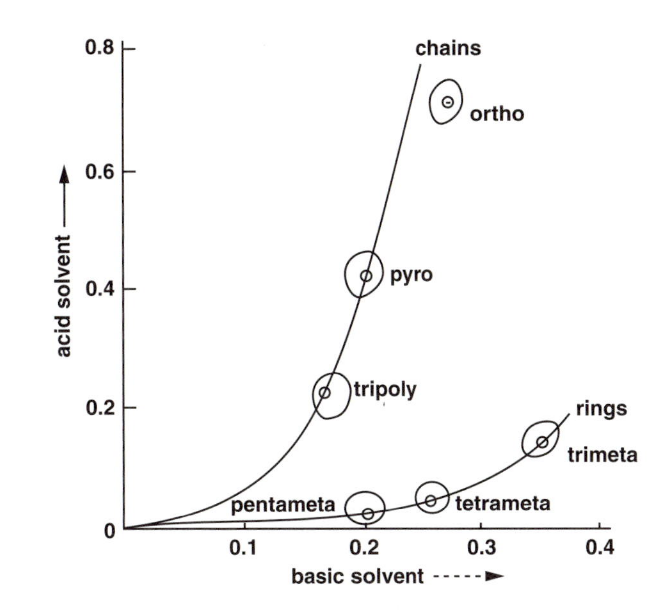

FIG. 7.2 Separation of phosphate anions by two-dimensional paper chromatography. The ring anions fall on one curve and the chain ones on another. Higher members of each type would lie on the prolongations of these curves towards the origin

$$\text{Ge(IV) in aqueous solution (as oxy- or hydroxy-species)} + HCl \rightleftharpoons GeCl_4 + H_2O$$
$$GeCl_4 \searrow \text{solution in } CCl_4$$

The nature of the solute may be altered by changing the pH, by adding complexing agents, or by adding common ions, while considerable variation in the organic phase is also possible. There is thus a fair chance that any given mixture may be separated by suitably varying the conditions.

Solvent extraction processes are very widely used in industry for the extraction of metals from ores or spent catalyst residues or the recovery of toxic metals from effluent streams. Perhaps the classic example of the application of this technique is the separation of uranium and plutonium from each other and from the fission products that accumulate in an atomic pile. One method involves the solution of the material in nitric acid, in which the uranium is in the VI state and is more covalent than the plutonium, which remains in the IV state. The uranium is extracted into ether and then the plutonium and fission products are precipitated and strongly oxidized to convert the plutonium to Pu(VI). This can then be extracted from the fission products in turn. A variation, which particularly lends itself to a cyclic process of repeated extraction, is to separate the plutonium and uranium in the VI state into an organic solvent which is then carried into a second vessel where it is subject to mild reduction, in contact with water, which leaves the U(VI) unchanged but converts the plutonium to Pu(III), which is washed back into the aqueous phase by dilute nitric acid.

Three general classes of extractants are recognized, according to their structures and mechanism of extraction. Inert extractants are simple noncoordinating solvents, such as CCl_4 in the case of $GeCl_4$ extraction just described.

Solvating extractants are reagents which have one or more donor atoms available for metal complexation. These complexes allow extraction of the metal into the organic phase, from where it may be recovered. The principles of coordination chemistry are of great importance in designing a new extractant for a given purpose and extractants are therefore basically just ligands which have been specifically designed to make them highly soluble in organic solvents. Thus, extractants containing soft, polarizable donor atoms will show a strong tendency to extract soft metal atoms. An example is the phosphane sulfides, $R_3P{=}S$, used for the recovery of silver from photographic wastes. By contrast, extractants containing hard donor atoms (such as nitrogen or oxygen) typically extract hard metals, and a good

See Section 13.7 for a detailed discussion on hard and soft acid/base behaviour.

example is tri-octyl phosphane oxide, $octyl_3P{=}O$, used in the separation of uranium from plutonium. Instead of solvating the metal itself, metals may often be extracted into organic solvents by solvating an associated proton, as in the case of the extraction of the H^+ $FeCl_4^-$ ion pair into ether. Ionic extractants, such as organic soluble ammonium salts R_3NH^+ or R_4N^+, or sulfonium salts R_3S^+, work by a similar mechanism of forming an organic soluble ion pair with the anionic metal complex, usually a halide, such as $FeCl_4^-$, or $CoCl_4^{2-}$.

The third major class of extractant is the acidic extractant, such as an organophosphorus acid or a carboxylic acid, which functions by replacing the acidic proton by a metal. These acids, designated H-A, typically exist as hydrogen-bonded dimers (see Section 9.7 on the hydrogen bond), and in the presence of an excess of the extractant these dimers can persist in the extracted metal complex which can be, for example, of the type $M(HA_2)_n$ (n is typically 2 or 3). In addition, the neutral unionized acid dimer H_2A_2 can also act as a solvating ligand in these complexes. Organophosphorus acids are widely used in processes to separate the individual lanthanide elements, as described in Section 11.3. Another type of extractant of this class are hydroxyoximes of the type shown which are highly selective for the extraction of copper and used on a very large scale worldwide. The extractant functions by deprotonation of the phenolic -OH group; copper then coordinates to the resulting O^- group and the oxime nitrogen, forming a six-membered chelate ring.

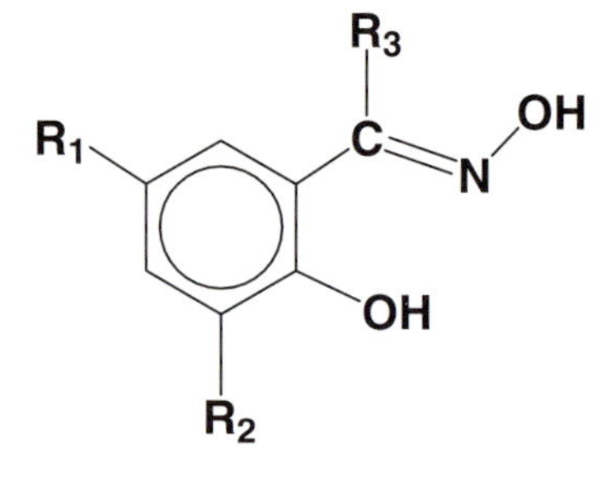

Hydroxyoxime used in copper extraction

Solvent extraction processes are not restricted to the extraction of metallic species and many other inorganic and organic species (such as phenolic compounds from effluents) can be recovered. Another of the classic examples of solvent extraction is the 'wet-acid' process for phosphoric acid purification. In this process crude H_3PO_4, produced by treating phosphate rock with sulfuric acid, is extracted into an organic solvent, such as methyl isobutyl ketone, from which the purified H_3PO_4 is recovered by extraction into water. Food-grade H_3PO_4 can be readily manufactured this way, resulting in considerable energy savings over the old 'thermal' process in which pure white phosphorus (P_4) is burned in air giving P_4O_{10} which is then dissolved in water giving H_3PO_4 (see Section 17.6).

All these solvent extraction processes may be adapted for the separation of very similar solutes by using a continuous flow method, when fresh organic phase is brought into contact with the partially extracted aqueous phase and vice versa. This 'counter-current' method corresponds basically to a chromatographic separation involving two mobile liquid phases instead of one mobile phase and one held stationary on an absorbing phase. The close similarity of all these methods means that they may often be used interchangeably and the problem of separating chemically similar elements or compounds can now be treated by very powerful and versatile methods.

Structure determination

7.4 Diffraction methods

X-ray diffraction. The most accurate and powerful method of determining the structure of solids is by X-ray diffraction. Just as light can be diffracted by a grating of suitably spaced lines, so can the much shorter wavelength X-rays be diffracted. The regular array of atoms or ions in a crystalline solid provides a suitably spaced three-dimensional grating for the diffraction of a beam of X-rays.

When a beam of X-rays passes through a crystal it meets various sets of parallel planes of atoms. The diffracted beams from atoms in successive planes cancel unless they are in phase, and the condition for this is given by the Bragg relationship:

$$n\lambda = 2d \sin\theta$$

where λ is the wavelength of the X-rays, d is the distance between successive planes and θ is the angle of incidence of the X-ray beam on the plane. From a set of distances d for

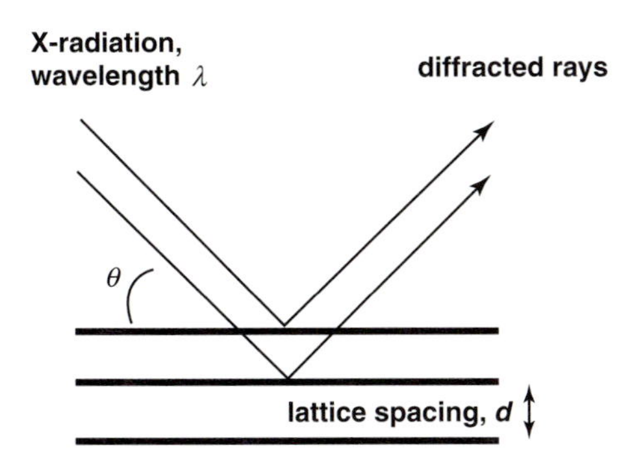

Definition of parameters in X-ray diffraction

different sets of planes in the crystal, the positions of the atoms can be derived. The angles θ can be measured and the values of d can be calculated if the wavelength of the X-rays is known. If the intensities of the diffracted beam in each direction can also be measured, complete structure determinations are possible. As each atom in the lattice acts as a scattering centre, the total intensity in a given direction of the diffracted beam depends on how far the contributions from individual atoms are in phase. The essential feature of complete X-ray structure determinations is a trial-and-error search for that arrangement of atoms which best accounts for the observed intensities of reflections. For large molecules, or structures of low symmetry, this process involves numerous calculations and is best carried out by computer.

For the first 50 years of X-ray work, the diffraction pattern was detected using a photographic film and analysis involved measuring the position of the diffraction spots and estimating their intensities. Processing involved extensive hand calculation and a single structure took several years to work out. By the 1970s, alternative detectors were available and a structure could be worked out in a few weeks using all the power of the institution's mainframe computer. Modern detectors are more sensitive, automated, and structures can be worked out on a desktop computer in a matter of hours.

There are a number of experimental methods for studying X-ray diffraction but two are widely used. One uses a single crystal of the compound which is mounted so that it can be rotated about a crystal axis. A monochromatic beam of X-rays (i.e. of a single wavelength λ) shines on the crystal. As it rotates, successive sets of planes are brought into reflecting positions and the reflected beams are recorded as a function of the crystal rotation. The positions and intensities of the reflected beams are measured and the complete structure can be derived from the resulting data. In modern equipment the crystal can be cooled to liquid nitrogen temperature which both reduces the thermal 'noise', giving more accurate atom positions, and allows the study of less stable compounds.

A second method is used for the many cases when substances are available only as a crystalline powder and no single crystal is obtainable. The powder is packed in a thin capillary tube and illuminated by a monochromatic beam inside a circular camera. All possible crystal orientations occur at random, so that some crystallites will be correctly orientated to fulfil the Bragg condition for each value of d. The reflections are recorded as lines on a film placed round the inside of the camera and these positions give the d values. A complete analysis is possible only for crystals of high symmetry but the method is invaluable as a qualitative technique, e.g. in mineral analysis.

Electron diffraction. Although X-ray methods give the fullest and most reliable data, they are restricted to solids and are not very suitable for determining the positions of very light atoms. Two other diffraction methods are available, electron diffraction and neutron diffraction. By virtue of the wave properties of electrons, a beam of electrons may be diffracted. Electron diffraction is used to determine the structures of gaseous molecules and gives very accurate values for bond lengths and angles. The process of interpretation depends on matching the observed diffraction pattern with ones calculated for different model structures, so that it is not fully reliable as a method of structure determination but does give accurate parameters for molecules whose overall structure is known. Electron diffraction is a function of the atomic number and hydrogen atoms are, as a rule, not detectable.

Neutron diffraction. A beam of neutrons behaves in a similar manner to a beam of electrons and gives information about the positions of light atoms and is thus complementary to the other two techniques. Neutron diffraction determines the positions of the nuclei, and the sensitivity does not vary much between elements. X-rays are diffracted by the electrons, and thus X-ray diffraction is dominated by the positions of largest electron density. In compounds which contain both light and heavy atoms, X-ray scattering from the low electron density around the light atom may not be detectable and only heavy atom positions are found. If a neutron diffraction is carried out as well, the heavy atom positions from the X-ray study allow a ready solution, and the light atom centres are easily determined. This is particularly so for hydrogen compounds, as H, and even more D, scatters neutrons efficiently. In addition, distinguishing between two adjacent atoms in the Periodic Table, such as iron ($Z = 26$) and cobalt ($Z = 27$), can be difficult by X-ray diffraction, but the different neutron scattering abilities of these two atoms makes distinguishing them by neutron diffraction a relatively straightforward matter. The disadvantages of the neutron diffraction method are the inaccessibility of neutron sources (commonly an atomic pile) and the requirement for large single crystals. Indeed, the difficulty of preparing crystals without twinning or other flaws is the main limitation on both X-ray and neutron diffraction which are otherwise the most powerful structural tools.

Since the diffraction pattern has contributions from every atom in the unit cell, it becomes rapidly more difficult to solve the structure as the molecular size increases. While there have been tremendous successes in recent work on large biomolecules containing metal atoms, analysis is very slow and many systems are known only to relatively low spatial resolution, say at the 5 Å level.

In such cases, a further X-ray method is valuable. This is the extended X-ray absorption fine structure or EXAFS experiment. When an X-ray photon is absorbed by an atom (the plot of X-ray absorption rises steeply at this point giving the *X-ray absorption edge*) an electron is expelled. This may be back-scattered by neighbouring atoms, giving a fine structure on the absorption edge which gives information about the nearest neighbours of the absorbing atom. The technique is particularly valuable in probing large molecules or noncrystalline materials.

To illustrate, a recent study involved $\{[(RO)_2P(S)_2]_4Zn_2\}$ which has a known structure containing $Zn\{(S)_2P_2\}_2$ units. This compound is typical of antioxidants for engine oils. Their mode of action is difficult to investigate as the oxidized products are not crystalline. An EXAFS study showed the presence of O with four Zn nearest neighbours followed by six PS_2 groups and an indication of three Zn third neighbours. This suggests a central O bonded tetrahedrally to 4 Zn, bridged on each edge by a S-P-S unit (presumably retaining some of the P-OR ligands) and probably linked in turn to further OZn groups (compare Fig. 10.9). Such information suggests how the antioxidant works and allows design of improvements.

7.5 Spectroscopic methods and the electromagnetic spectrum

If an atom, molecule or ion, with two energy states differing in energy by ΔE, is irradiated with continuous electromagnetic radiation, the radiation of frequency corresponding to ΔE will be absorbed and the species raised to the upper energy state. This absorption, or the consequent emission of radiation as the species returns to the ground state, may be detected and provides information about the energy states. The energy change is related to the frequency ν' of the radiation by the relation

$$\Delta E = h\nu' \tag{7.1}$$

where h is Planck's constant equal to 6.6256×10^{-34} J s. The frequency is related to the wavelength, λ, as their product equals the speed of light

$$\nu'\lambda = c = 2.998 \times 10^{10}\ \text{cm s}^{-1} \tag{7.2}$$

Thus the energy change per mole obeys the relations

$$\Delta E = Nh\nu' = Nhc/\lambda = Nhc\tilde{\nu} \tag{7.3}$$

where $\tilde{\nu}$ is called the *wave number* or Kayser and is the number of cycles per centimetre; hence the unit is cm^{-1}. Equation (7.3) leads to the relation between different units given on p. 4 and in Table 7.1.

The electromagnetic spectrum spans a very wide range, from gamma and cosmic rays with energies in excess of 10^6 kJ mol^{-1} down to radiowaves corresponding to small fractions of a joule per mole. In different regions of the spectrum, the energy corresponds to differences in two states of a system spanning many kinds of transition. Thus a nuclear transformation involves very high energies, in the gamma ray region, while the reversal of an electron spin in a magnetic field of 0.1 T would involve a tiny energy corresponding to the shorter radiowaves. Nevertheless, both these processes, and many others of intermediate energy, may be used to yield information of value to the chemist. The major regions of the electromagnetic spectrum, and the transitions corresponding to the interaction with

TABLE 7.1 Principal regions of the electromagnetic spectrum

Region	Approximate range in			Transition excited
	Wavelength (m)	Energy		
		(SI)	(Commonly-used)	
Gamma rays	$<10^{-10}$	$>10^6$ kJ mol^{-1}	$>10^4$ eV	Nuclear transformations
X-rays	10^{-8} to 10^{-10}	10^4 to 10^6 kJ mol^{-1}	100 to 10^4 eV	Transitions of inner shell electrons
Ultraviolet	4×10^{-7} to 10^{-8}	10^2 to 10^4 kJ mol^{-1}	1 to 100 eV or 10^4 to 10^6 cm^{-1}	Transitions of valence shell electrons including $d \rightarrow d$ and $f \rightarrow f$
Visible	8×10^{-7} to 4×10^{-7}			
Infrared	10^{-4} to 2.5×10^{-6}	1 to 50 kJ mol^{-1}	100 to 4000 cm^{-1}	Molecular vibrations
Microwave and far-infrared	10^{-2} to 10^{-4}	10 to 1000 J mol^{-1}	1 to 100 cm^{-1}	Molecular rotations
Radio frequency	$\sim10^{-2}$	~10 J mol^{-1}	3×10^4 MHz	Electron spin reversal in magnetic field of 1 A m^{-1}
	~10	~0.01 J mol^{-1}	10 to 100 MHz	Nuclear spin reversal in magnetic field of 1 A m^{-1}

1 eV = 96.49 kJ mol^{-1} = 23.06 kcal mol^{-1} and is equivalent to 8068 cm^{-1} and 2.419×10^8 MHz: 1 A m^{-1} = 10^4 G. Ranges are rounded-off and are not converted exactly from one unit to the next.

them, are listed in Table 7.1. There is no clear boundary between any two regions and the energy ranges and types of transition overlap: we give arbitrary round-figure ranges (which are thus slightly different for different units) whose boundaries are often set by arbitrary experimental factors. As there is such a wide range of energies, and so many different types of transition are involved, there have grown up a number of separate areas of study which have developed fairly independently and with their own conventions. In particular, this means that a variety of units have been used and some of the traditional ones in the important regions are indicated in Table 7.1. Notice that these include units of energy, wavelength and frequency.

The gamma and X-ray regions of the spectrum give information about the nucleus and the inner, closed shell, electrons and are therefore of less direct interest to the chemist although the Mössbauer effect, which involves gamma resonance absorption, and the derivation of atomic energy levels from X-ray spectra, as in Moseley's experiments, are important.

The average chemist is most likely to make direct measurements of *electronic spectra* in the ultraviolet and visible region, of *vibrational spectra* in the infrared region and in the Raman effect, and of *nuclear magnetic resonance* (*nmr*) *spectra* in the radiofrequency region. These methods are therefore discussed in detail in the following sections. Pure rotational spectra, studied in the microwave region, give accurate values for the moments of inertia and thus provide values for bond lengths and angles for sufficiently small or symmetric molecules which must also have a permanent dipole moment to interact with the radiation. By using isotopic substitution, a number of independent parameters may be determined, but even so, only a limited number of molecules are simple enough to be analysed. Thus microwave spectroscopy gives information complementary to that derived from electron diffraction for small molecules in the gas phase. Electron spin resonance, which is observed for species with unpaired electrons, is also a tool available for only a restricted number of species. Rotational changes often accompany vibrational changes and are seen as a fine structure to the vibrational absorptions, and, similarly, electronic transitions may have fine structures due to concomitant rotational and vibrational transitions.

Study of ultraviolet, visible and infrared spectra requires the same basic equipment. The sample must be placed in a beam of radiation which can be continuously varied in frequency and a detector is required to show absorption of energy. For example, visible spectra require a source of white light which is scanned in frequency by using a rotating prism, while the detector is a photocell whose output may be converted to a movement of a pen on a recorder. An infrared spectrum may be recorded using a heated element as source, an alkali halide prism or a grating to change the frequency, and a thermocouple as detector.

Vibrational transitions may also be detected in the Raman effect. In this, the sample is illuminated with strong monochromatic radiation, and some of the re-emitted quanta are found to have gained or lost a (smaller) quantum corresponding to the energy of one of the fundamental vibrational modes. The spectrum of the scattered radiation thus consists of a very strong line corresponding to the incident radiation and a number of other lines whose energy differences from the primary line give the energies of vibrational transitions of the molecule. Raman spectroscopy requires an intense radiation source with a very narrow frequency spread. The laser is ideal and Raman spectroscopy has greatly expanded in scope since the advent of lasers.

7.6 Electronic spectra

Transitions of outer shell electrons fall in the general wave number range of 100 000 cm^{-1} to 10 000 cm^{-1}, that is in the ultraviolet, visible and near-infrared regions of the electromagnetic spectrum. The transitions involved are those between σ, π or nonbonding, n, orbitals in the valence shell, such as those involved in Figs 4.25 or 4.30. Not all transitions are allowed, only

$$\sigma \longrightarrow \sigma^*$$
$$n \longrightarrow \sigma^*$$
$$\pi \longrightarrow \pi^*$$
$$n \longrightarrow \pi^*$$
$$\delta \longrightarrow \delta^*$$

while σ to π or π to σ transitions are forbidden. It is clear from energy level diagrams, such as those in Chapter 4, that this is the approximate order of decreasing energy. The largest energy is required for $\sigma \longrightarrow \sigma^*$ transitions and these are found in the far-ultraviolet regions in which the molecules of the atmosphere also absorb. Evacuated spectrometers have to be used and this region is less studied.

Most work is done in the 50 000 to 10 000 cm^{-1} region, 200 nm to 1 μm, where transitions involving nonbonding and π electrons are involved. For lighter atoms, such transitions are usually in the ultraviolet, giving colourless compounds, but extended conjugation, giving rise to a large number of closely spaced π orbitals, may lower the energy of the transition and give a coloured compound. In addition heavy atoms have their outer orbitals closer together in energy so that the transitions again fall into the visible region. Such transitions account for the colours of many iodides, for example. In addition, transition elements with partly occupied d or f orbitals show bands, usually in the visible region, due to d $\longrightarrow$ d or f $\longrightarrow$ f transitions. Such transitions are formally forbidden and give rise to weak bands: they are discussed in more detail in Chapters 11 and 13.

While the position of an absorption band corresponds to the energy of the transition, its intensity depends on the nature and quantity of the absorbing material. The relation between the *absorbance, A,* the path length l and the concentration C is given by the Beer–Lambert law

$$A = kCl \tag{7.4}$$

where k is a constant characteristic of the material. For a path length of 1 cm, and with the concentration expressed in moles per unit volume, the constant k becomes ε, the *molar extinction coefficient*. Allowed transitions generally have molar extinction coefficients in the range 10^3 to 10^5 l mol^{-1} cm^{-1} while d–d or f–f bands are much weaker with ε typically 10^{-1} to 10 l mol^{-1} cm^{-1}. The absorbance, formerly termed the optical density, is defined as

$$A = \log(I_0/I) \tag{7.5}$$

where I_0 is the intensity of the incident light and I the intensity after the light has passed through the sample. Most modern spectrometers record the absorbance directly.

Electronic spectra are used primarily to give information about the energy levels of the valency orbitals, and for inorganic chemistry this is particularly widely studied for transition elements. Use may also be made, through Beer's law, of absorbance measurements in quantitative analysis. Here, measurements are preferably made on strong bands and at, or near, maxima in the absorption. Thus, for example, the strong allowed bands of CrO_4^{2-} would be used for estimating Cr, rather than the weak d–d transitions of Cr^{3+} ions. (Note that species like chromate or permanganate, which are in the Group oxidation state, have no d electrons not involved in bonding so that these colours are not due to d–d transitions.) In a more qualitative way, electronic spectra may also be used to indicate the presence of particular groupings of atoms. For example, $\pi - \pi^*$ transitions in a benzene ring will be relatively constant in position and intensity as long as π bonding substituents are absent: thus the presence of a phenyl group would be indicated by such bands in the electronic spectrum. Such a use is more common in organic chemistry where such a chromophore is likely to occur in a fairly constant environment, but similar applications are of value, particularly in organometallic chemistry.

7.7 Vibrational spectra

Vibrational modes of a molecule are excited by the absorption of quanta whose energies lie in the infrared region of the spectrum, from about 4000 cm^{-1} downwards. Vibrational transitions are also detected in Raman scattering. As the selection rules for infrared absorption differ from those governing Raman scattering, the two techniques are complementary and both infrared and Raman spectra need to be measured to obtain the maximum amount of information.

The information obtainable from vibrational spectroscopy depends on the size and symmetry of the molecule. For a diatomic molecule, assuming simple harmonic motion, the wavenumber is given by

$$\tilde{\nu} = \frac{1}{2\pi c}\sqrt{\frac{k}{u}} \tag{7.6}$$

where k is the force constant (the proportionality between the extension of the bond and the restoring force) in N m^{-1} and u is the reduced mass ($1/u = 1/m_1 + 1/m_2$) of the two atoms. Thus the vibrational frequency is directly related to the force constant, which in turn is related to the bond strength. Absorptions of successive quanta of vibrational energy will continue until the molecule dissociates, and the frequency at which this occurs gives the bond energy. From the rotational fine structure, the moment of inertia may be derived and hence the bond length (if the atomic masses are known). The existence of isotopes of an element may be proved by observing different moments of inertia for the same compound. For example, hydrogen chloride is found to have a bond length of 12.1 pm and two moments of inertia corresponding to $H^{35}Cl$ ($I = 2{\cdot}649 \times 10^{-47}$ kg m^2) and $H^{37}Cl$ ($I = 2{\cdot}653 \times 10^{-47}$ kg m^2).

For polyatomic molecules, the position is more complicated. Vibrations involve not only bond stretching, but angle deformation and often twisting modes as well. An n-atom species has $3n$–6 degrees of vibrational freedom ($3n$–5 for a linear species) and a corresponding number of force constants are required to describe the vibrations. However, it is unlikely that all $3n$–6 vibrations will be observable in the majority of cases. This is partly due to practical difficulties of detecting weak bands and resolving closely overlapping ones, and partly due to degeneracy. For example, a tetrahedral molecule like GeH_4 has $3 \times 5 - 6 = 9$ modes of vibration but the maximum number of bands observable in the infrared is two and in the Raman four. This is because two of the modes are degenerate, giving only one fundamental, and two further groups of three are triply degenerate. Thus there are only four observable bands, two triply degenerate, one doubly degenerate and one nondegenerate and these are found in the Raman spectrum. Of these, only the triply degenerate modes involve a dipole change so that only these two are also observed in the infrared. Analysis in terms of force constants is thus difficult, though isotopic substitution may help.

For small and moderate-sized molecules, the number of bands expected in the vibrational spectrum is predictable from the symmetry of the molecule by the methods of *group theory* (this leads to the prediction above for GeH_4) and thus the spectrum may be used to determine which one out of a number of structures is the correct one. For example, a square planar AB_4 species has three infrared active bands (two of which are doubly degenerate) and three Raman bands which do not coincide with the infrared modes. The ninth vibration is inactive in both the infrared and Raman. If this is compared with the prediction for a tetrahedral AB_4 species given above, it will be seen that these two shapes could readily be distinguished unless the intrinsic intensities or spacings were very unfavourable. In a similar way, the two possible structures for a species like perchloryl fluoride could be distinguished, as the $FClO_3$ structure, based on a tetrahedron, would have six bands active in both infrared and Raman (as three are doubly degenerate) while the hypofluorite form, $FOClO_2$ has no degenerate bands and all nine vibrations would be seen in both infrared and Raman.

It should be noted that structural evidence of this sort may be used to support a structure but does not offer absolute proof. As some bands are inevitably difficult to observe, the assignment of six bands for perchloryl fluoride does not prove the first structure, as it is quite possible that the remaining three would be too weak or too close to others to be observed. The negative argument is much more definite—observation of seven or more fundamentals *would* disprove the first structure. Similarly, a species AB_4 showing three or more fundamentals in the infrared could not be a tetrahedron. These examples show the type of evidence that may be obtained by applying arguments based on the molecular symmetry, using group theory. The full discussion of these methods is beyond our scope, but the first step is to determine the molecular symmetry and this is discussed in Appendix C.

It will also be seen from these examples how valuable it is to be able to observe vibrational spectra both in the infrared and in the Raman. Often different modes are active in different effects. The extreme example is provided by species which have a centre of symmetry (such as square planar AB_4) where no modes are both Raman and infrared active. Even where all the expected modes are active in both effects, it is likely that bands which are weak in the infrared will be strong in the Raman, and vice versa. For example, stretching modes involving similar heavy atoms are often weak in the infrared, as the dipole change is small, but are strong in the Raman because a fairly extended electron cloud is moving. Thus the Si-Ge stretching mode in H_3SiGeH_3, which is allowed in both the infrared and the Raman, is too weak to be observed in the infrared but gives a strong Raman band at about 350 cm^{-1}. A similar effect is found for many metal–metal stretching modes.

In more complicated species this approach breaks down because the number of predicted bands becomes so large that detailed assignment is impossible. The vibrational spectrum is still useful, but in a more qualitative way. First, certain groups in a molecule may absorb in fairly constant regions of the spectrum and can thus be identified. For example, a CN group, with its triple bond, absorbs at about 2000 cm^{-1} a far higher wavenumber than modes involving single bonds and heavier atoms. Thus the presence of cyanide as a ligand may always be detected, and similarly of CO in metal carbonyls. Indeed, symmetry information may be derived from the number of bands in the 2000 cm^{-1} region, to distinguish *cis* and *trans* $ML_3(CN)_3$ for example, as these vibrations are little affected by the presence of other groups (compare also Fig. 7.3.). From the mass effect, modes involving hydrogen are also found at higher frequencies than those involving any other substituent and these also are readily distinguished. Such partial analyses are often valuable, and much information may be derived by comparing related species. Thus, the Cl-F and Cl-O-F alternatives for ClO_3F above might be further distinguished by comparison with chlorine fluorides and other

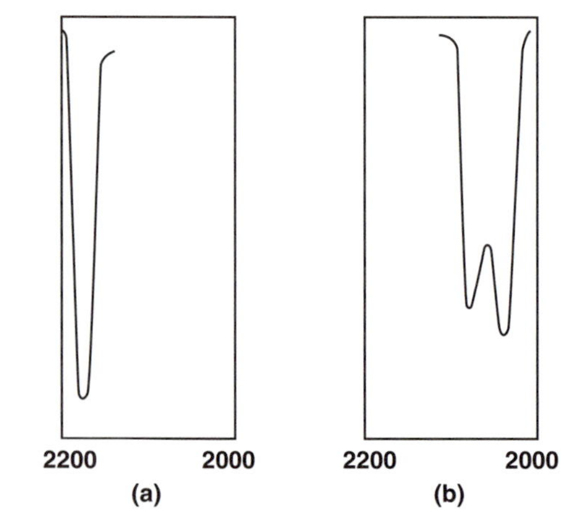

FIG. 7.3 C–N stretching bands in the infrared spectra of (a) $Ni(CN)_4^{2-}$, and (b) $Ni_2(CN)_6^{4-}$ ions. The CN absorption occurs in a similar region for both compounds (a) at 2124 cm^{-1} and (b) at 2045 cm^{-1} and 2075 cm^{-1}. The difference in symmetry between the two ions is reflected in the number of absorptions

hypofluorite species to see if bands appear in regions characteristic of Cl-F stretching or of O-F stretching.

Finally, a purely qualitative approach may be made in which a complex species may be identified with a known compound if their spectra are identical. In this case, the more complex the spectrum, the more definite would be the identification.

7.8 Nuclear magnetic resonance (nmr)

When an atom, such as hydrogen, with a nuclear spin of $\frac{1}{2}$ is placed in a magnetic field, the spin may take up one of two orientations, either parallel or antiparallel to the field vector. These two correspond to different energy states and, for hydrogen in a field of about one tesla (10^4 G) the energy difference is about 10^{-2} J mol^{-1}. This corresponds to a frequency of about 40 MHz. In modern instruments, higher fields are common and proton resonance is commonly run at 60, 90 or 100 MHz with iron core magnets or at fields from 220 to 750 MHz using superconducting magnets. In the experimental arrangement commonly used, the sample is placed in a cylindrical tube in the field and irradiated with a fixed frequency. The field is slightly modulated by passing a current through coils and energy absorption is detected at resonance when

$$h\nu = \Delta E = g_N \beta_N B \qquad (7.7)$$

Here B is the magnetic flux density, β_N, is the nuclear magneton $= eh/2\pi Mc = 5.0505 \times 10^{-27}$ J T^{-1} and g_N is a constant, called the nuclear g factor, which is a characteristic of the element.

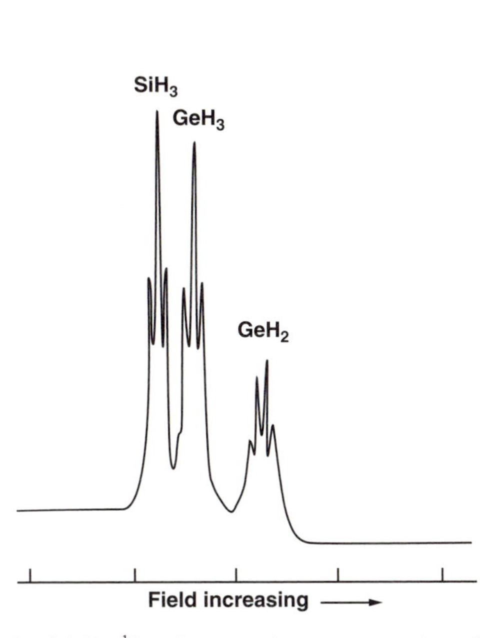

FIG. 7.4 The ^{1}H nuclear magnetic resonance spectrum of $SiH_3GeH_2GeH_3$. The resonances are, in order of increasing field, (1) the triplet due to the SiH_3 protons split by coupling to the two GeH_2 protons, (2) the triplet due to the GeH_3 protons, also split by the GeH_2 coupling, and (3) the multiplet due to the GeH_2 protons split by coupling both to the SiH_3 and to the GeH_3 protons. Notice that the coupling constant $\mathscr{J}$ (SiH_3GeH_2) is almost identical to the constant, $\mathscr{J}$ (GeH_3GeH_2), shown by the near-identity of the splittings in the two triplets

The great value of nmr to the chemist is that the magnetic field actually experienced by a particular nucleus is the sum of the applied field and fields induced in the electrons around the magnetic nucleus. Thus atoms of the same element which are in different chemical environments resonate at slightly different values of the external field and these differences in *chemical shift* may be detected and yield structural information. In other words, the position of the absorption depends on the atom to which the hydrogen is bonded and on the other bonds nearby. The classic case is that of ethanol, CH_3CH_2OH, where the absorption due to the hydrogen bonded to oxygen is in a different position to the absorptions of the carbon-bonded hydrogens. Furthermore, the hydrogens in the methyl group are in a different electronic environment to those in the methylene (CH_2) group, so the absorptions due to the two types of carbon-bonded hydrogens are separated. The three methyl protons are in identical environments, as are the two methylene ones and the resulting spectrum consists of three peaks, in the ratio of 3:2:1, for the three types of hydrogen atoms, and in positions typical of methyl, methylene and hydroxyl hydrogens. An example of the ^{1}H nmr spectrum of an inorganic hydride, $SiH_3GeH_2GeH_3$, is shown in Fig. 7.4. There are three main envelopes of absorption, each showing fine structure—see below—in intensity ratio 3:3:2 which arise from the SiH_3, GeH_3 and GeH_2 protons respectively.

Such detailed information about the environment of an atom may be sufficient to determine the structure, or at least goes a long way towards this. For example, diborane B_2H_6 (cf. Section 9.6), has had a number of structures proposed for it, including an ethane-like one (A) and a bridged structure (B).

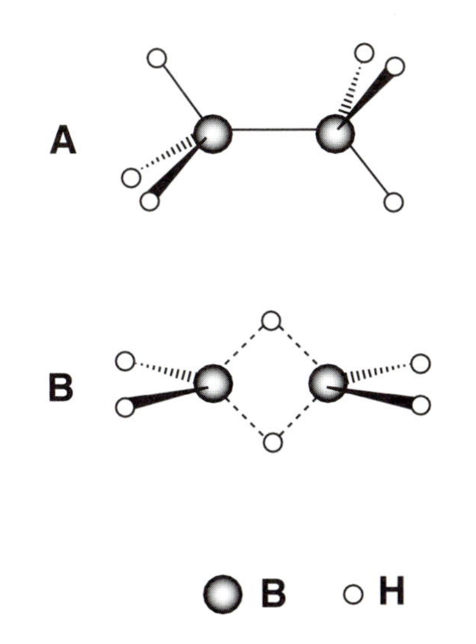

The hydrogen magnetic resonance spectrum of (A) would consist of only one line, as all the hydrogen atoms are equivalent, whereas the hydrogen spectrum of (B) would have two lines, in the ratio of 4:2, corresponding to the terminal and bridge hydrogens respectively. The latter spectrum is observed, supporting the bridge structure.

Nuclear resonance may be observed for atoms other than hydrogen which have a spin of one half, and also for atoms with higher nuclear spins, although the spectra are more complicated in the latter cases. No resonance is possible when atoms have zero spin and this includes some common atoms such as ^{12}C and ^{16}O. Hydrogen is the easiest nucleus to observe but nmr studies have now been extended to practically all the suitable isotopes in the Periodic Table. Modern techniques which involve observation by pulsed irradiation followed by Fourier transformation have greatly enhanced sensitivities. Wide-ranging studies on lighter nuclei include ^{2}H (deuterium), ^{11}B, ^{13}C, ^{14}N, ^{19}F, ^{29}Si and ^{31}P. Of these, carbon-13 is the most widely studied after hydrogen, and then phosphorus. One example of fluorine resonance is in the confirmation of the square pyramid structure of BrF_5, and IF_5 by the observation of two lines in the fluorine resonance, with intensities in the ratio 4:1, corresponding to the basal and apical fluorines respectively.

Further information may be obtained from the fine structure of the nmr bands which arises from the effects of spin–spin coupling. If an atom with a nuclear spin is bonded to a second one which also has a nuclear spin, the local magnetic field will be affected by the orientation of the spin of the second nucleus. For example, in PH_3 the phosphorus-31 nucleus, which has a spin of $\frac{1}{2}$, may be aligned with or against the field, giving two different local resultant fields. As the energy difference between the two orientations is so small, there will be essentially equal numbers of molecules with each P spin alignment. Thus, in the proton resonance signal there will be two components corresponding to hydrogens bonded to P atoms with spins parallel or antiparallel to the external field. Thus the proton resonance signal is a doublet with components of equal intensities. The phosphorus atom also shows a resonance signal, though at much lower frequency, and this is split by the spins of the three H atoms into four components. There are four possible arrangements of the proton spins: all parallel, which we can label $+\frac{1}{2}, +\frac{1}{2}, +\frac{1}{2}$, or with two, one or no spins parallel to the field and respectively one, two or three antiparallel spins, labelled $+\frac{1}{2}, +\frac{1}{2}, -\frac{1}{2} : +\frac{1}{2}, -\frac{1}{2}, -\frac{1}{2} : -\frac{1}{2}, -\frac{1}{2}, -\frac{1}{2}$. There are thus four different net fields and four components to the signal. In addition, while the $+\frac{1}{2}, +\frac{1}{2}, +\frac{1}{2}$ and $-\frac{1}{2}, -\frac{1}{2}, -\frac{1}{2}$ arrangements can only result in one way, any one of the three protons may be the antiparallel one in the other two combinations. Thus, a net spin of $+\frac{1}{2}$ results from any one of the three sets $+\frac{1}{2}, +\frac{1}{2}, -\frac{1}{2}$ or $+\frac{1}{2}, -\frac{1}{2}, +\frac{1}{2}$ or $-\frac{1}{2}, +\frac{1}{2}, +\frac{1}{2}$, and similarly for a net spin of $-\frac{1}{2}$. As all possible spin combinations are equally probable, there will be three times as many molecules in the sample with net spin $+\frac{1}{2}$ or $-\frac{1}{2}$ as with net spin $\frac{3}{2}$ or $-\frac{3}{2}$. Thus the phosphorus signal becomes a quartet with relative intensities 1:3:3:1. In a similar way, the signal of any atom bonded to n equivalent atoms of spin $\frac{1}{2}$ becomes an $(n+1)$ multiplet with intensities in the ratio of the binomial coefficients given by Pascal's triangle, Fig. 7.5. An example of the application of this is shown in Fig. 7.6 for the ^{31}P and ^{19}F nuclear magnetic resonance spectra of the octahedral PF_6^- ion. Thus, the ^{31}P nmr spectrum appears as a seven line multiplet, due to the phosphorus coupling to six equivalent fluorines, with the intensities of the lines 1:6:15:20:15:6:1. The ^{19}F spectrum shows the expected doublet.

It is worth noting that, for resonances of this type having a large number of lines, the outer lines are relatively weak and could lead to incorrect assignment of the spectrum if the outer lines are not identified (for example, due to poor signal-to-noise in the spectrum).

Such spin–spin coupling is not limited to atoms which are directly bonded, as it is transmitted via the bonding electrons, and it may be observed for atoms separated by several bonds. Thus the CH_3 and CH_2 protons in ethanol couple to make the methyl signal a

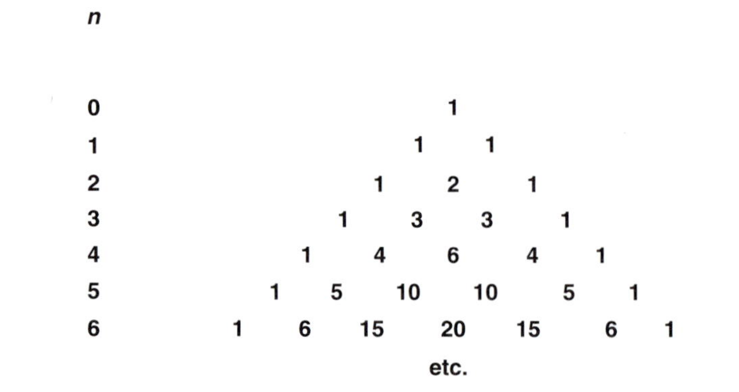

n
0 1
1 1 1
2 1 2 1
3 1 3 3 1
4 1 4 6 4 1
5 1 5 10 10 5 1
6 1 6 15 20 15 6 1
etc.

FIG. 7.5 Pascal's triangle for predicting the intensities of lines in a multiplet formed by spin-spin coupling of a spin $\frac{1}{2}$ nucleus with n identical spin $\frac{1}{2}$ nuclei

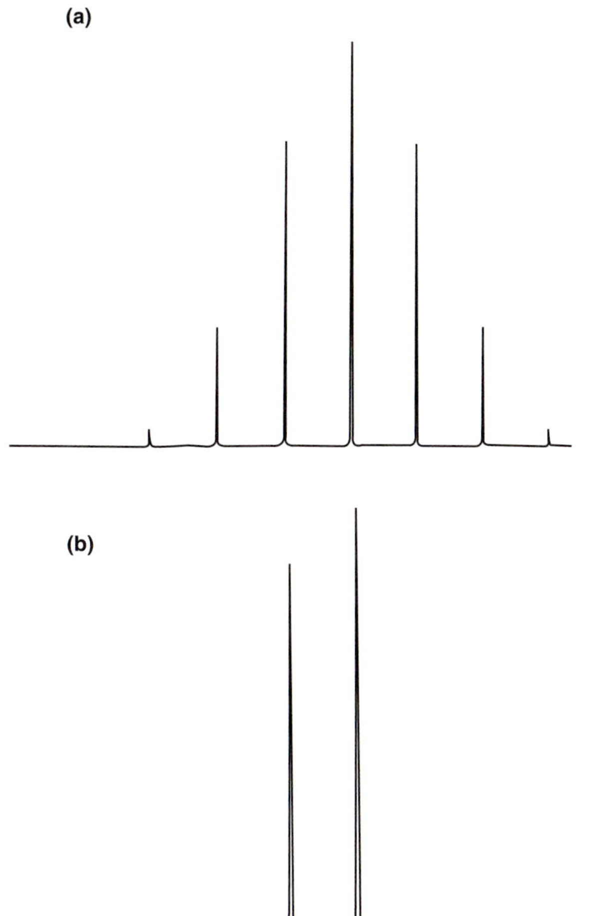

FIG. 7.6 (a) ^{31}P and (b) ^{19}F nmr spectra of the octahedral PF_6^- ion. The ^{31}P spectrum shows the expected 7 line multiplet with peak intensities in the ratio 1:6:15:20:15:6:1, see text. The splitting between the two lines in the ^{19}F spectrum is identical to the splitting between any adjacent pair of lines in the ^{31}P spectrum, this splitting being the one-bond ^{31}P–^{19}F coupling constant (708 Hz)

triplet (1:2:1 intensity ratio) and the methylene signal a quartet. This latter may be further split by coupling to the OH proton. Similarly, the SiH_3 signal in $SiH_3GeH_2GeH_3$ is split into a 1:2:1 triplet by coupling to the GeH_2 protons and the GeH_3 signal is also a triplet for the same reason. The GeH_2 signal is a more complex multiplet as these protons are coupled both to the SiH_3 ones and to the GeH_3 ones. The resultant fine structure is seen in Fig. 7.4.

The magnitude of the splitting of the lines caused by spin–spin coupling, i.e. the coupling constant $\mathcal{J}$, often holds a substantial amount of information for the chemist. A superscript is often used to denote the number of bonds over which the coupling is occurring, for example the one-bond phosphorus-31 to fluorine-19 coupling constant, $^1\mathcal{J}$ (^{31}P–^{19}F), is 708 Hz. Many factors can influence the magnitudes of coupling constants, such as the percentage of s-character in the bonding orbital, and three-bond couplings often obey a very strong torsion angle dependence. This is becoming of increasing importance in inorganic chemistry and while we do not have the space for a comprehensive discussion here, the reader is referred to the many nmr spectroscopy texts which cover this, in particular the book by Ebsworth, Rankin and Cradock (see Appendix A).

The magnitudes of spin–spin coupling constants prove to be very useful in the study of coordination complexes, and a good example comes from the study of square-planar platinum–phosphane complexes of the type $PtX_2(PR_3)_2$ (see Section 15.8 for a discussion of platinum and palladium chemistry). First of all, however, we need to say something about the spectra of platinum-containing complexes, which are slightly more complicated since not all of the platinum nuclei are spin-active and the resulting observed spectrum is

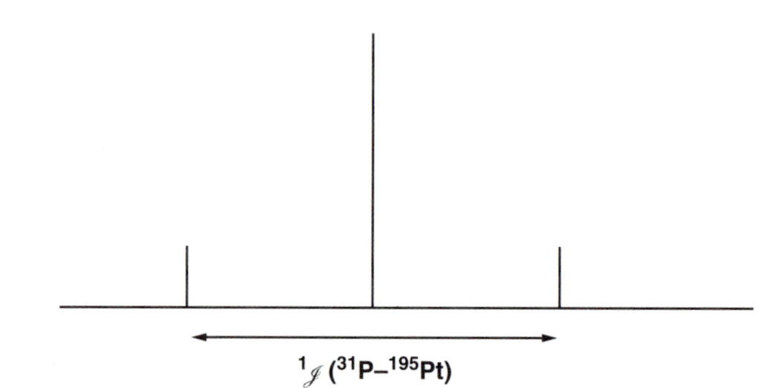

FIG. 7.7 A diagrammatic representation of the ^{31}P nmr spectrum of square-planar platinum(II) complexes of the type $PtX_2(PR_3)_2$. The spectrum is proton-decoupled, to simplify it, and is composed of two superimposed sub-spectra consisting of a doublet from molecules containing a ^{195}Pt atom and a singlet from molecules not containing ^{195}Pt. The separation of the two outer lines is the one-bond platinum phosphorus coupling constant, $^1\mathcal{J}$(PtP)

therefore composed of two sub-spectra. This is illustrated in Fig. 7.7 which shows the typical ^{31}P nmr spectrum of the complexes $PtX_2(PR_3)_2$. There are several isotopes of platinum but the only one which is nmr-active is ^{195}Pt, with spin $= \frac{1}{2}$ and making up approximately 33% of platinum. When ^{31}P is bonded to ^{195}Pt the nmr spectrum shows a doublet due to spin–spin coupling. For the remaining 67% of ^{31}P nuclei which are bonded to non-magnetic Pt, there is no splitting and there is a singlet at the chemical shift position. Thus, as the chemical shift lies at the mid-point of the doublet, the spectrum shows three lines of intensities 33/2, 67, 33/2, i.e. a non-binomial triplet with relative intensities 1:4:1, easily confused at first sight with a 1:2:1 binomial triplet. The separation of the two outer lines is the one-bond ^{31}P–^{195}Pt coupling constant.

Note that in the case of $SiH_3GeH_2GeH_3$, (Fig. 7.4), it is the separation of the successive pairs of lines of the SiH_3 or GeH_3 binomial triplets which gives the three-bond ^{1}H–^{1}H coupling constants.

When coupled with other nmr techniques such studies allow the characterization of a range of coordination complexes containing nmr-active nuclei. The ^{103}Rh isotope is a useful one since it is spin $\frac{1}{2}$ and 100% abundant. The direct observation of ^{103}Rh nmr spectra is rather difficult, since the nucleus resonates at a very low frequency, but the additional splitting from the ^{103}Rh nucleus in proton, carbon or phosphorus nmr spectra usually provides sufficient structural information. A good recent example of the utilization of ^{103}Rh coupling in nmr spectroscopy comes from the nmr identification of highly reactive rhodium–carbonyl intermediates in the rhodium-catalysed carbonylation of methanol. This process, developed by Monsanto, is used industrially to manufacture large quantities of acetic acid and is discussed further in Section 15.7 on rhodium and iridium chemistry.

If the nucleus has a spin of more than $\frac{1}{2}$, the coupling splittings follow different rules. A nucleus of spin $n/2$ has $n+1$ orientations each equally likely, and this gives an $(n+1)$ multiplet of equal intensities. Thus the proton signal of the terminal hydrogens in diborane, above, when the coupling to ^{11}B with spin $= \frac{8}{2}$ is taken into account, is a quartet with all components of equal intensity.

THE *TRANS* INFLUENCE IN PLATINUM COMPLEXES

The magnitude of the value of $^1\mathcal{J}$ (^{31}P–^{195}Pt) is highly dependent on the nature of the group *trans* to it, namely its *trans*-influence. The *trans*-influence of a ligand is its ability to weaken the bond *trans* to it. It is thermodynamic in origin and should not be confused with the *trans*-effect (see Section 13.9) which is a kinetic effect. A phosphane ligand *trans* to a high *trans*-influence ligand (such as another phosphane, a hydride ligand, or an alkyl or aryl group) will have its bond to platinum weakened, and will therefore show a smaller value of $^1\mathcal{J}$(^{31}P–^{195}Pt). Thus for a *cis*-$PtX_2(PR_3)_2$ complex, the PR_3 ligands are *trans* to X groups, and $^1\mathcal{J}$(PtP) values vary from around 2000 Hz, when X is a high *trans*-influence alkyl group, to around 3600 Hz when X is a much lower *trans*-influence chloride ligand. Correspondingly, phosphanes *trans* to low *trans*-influence ligands, such as halides or carboxylates, will show large values of $^1\mathcal{J}$(^{31}P–^{195}Pt). It is therefore a relatively simple matter to follow substitution reactions at platinum centres using phosphorus nmr spectroscopy.

Thus the nmr investigation gives information about the relative numbers of magnetic nuclei of each type in a molecule, from the position and intensities of the signals, and shows something of the structure of the molecule from the spin–spin coupling.

Apart from its use identifying compounds, information about reaction kinetics and exchange processes may be derived by studying the nmr signals over a range of temperatures.

While the nmr spectra of the majority of inorganic complexes, as well as organic compounds, such as ligands, are investigated in solution, there is a very large number of important materials which are highly insoluble. However, special methods are available to study nmr spectra of solid materials. In solution samples, molecules are rapidly tumbling such that the nmr experiment observes their average values of chemical shift. Solids, however, are rigid and since most solids consist of microcrystals in all possible orientations, this typically results in very broad peaks in the solid-state nmr spectrum. This is overcome by spinning the sample at an angle (called the 'magic angle') to the applied magnetic field and sharp spectra can thus be obtained. Good examples come from the study of ^{29}Si and ^{27}Al nmr spectra of silicate and aluminosilicate minerals and glasses, and ^{13}C nmr spectra of a wide variety of materials such as coal and even freshly cut celery!

7.9 Further methods of molecular spectroscopy

7.9.1 Electron spin resonance (esr) spectroscopy

While nmr is concerned with reversal of nuclear spin, *electron spin resonance* (esr) involves a very similar phenomenon, the reversal of the spin of an electron in a magnetic field. This involves a higher energy than the nuclear spin reversal, about 10 J mol^{-1} and thus a higher frequency of around 28 000 MHz. In an electron pair, the spins are already opposed and any reversal of one spin would be cancelled by that of the other (the energies involved are much lower than those required for excitation to the triplet state where the spins are parallel). Thus esr measurements can only be made on species with unpaired electrons like radicals and transition element compounds. For these, changes in the electron g factor (compare Equation 7.7), analogous to the nmr chemical shift, and spin–spin coupling to magnetic nuclei may be observed. One application is in the study of molecular orbitals. For example, if an electron is added to benzene to give the anion, $C_6H_6^-$, this electron will enter the lowest available orbital, which is one of the antibonding orbitals. The detailed structure of the electron resonance absorption then yields information about the interaction of the electron in this delocalized π orbital with the atomic nuclei, and hence the distribution of this π orbital over the atoms of the molecule may be determined and compared with the calculated one. In particular, an atom lying on a nodal plane should not interact with the electron, and this may be checked.

7.9.2 Mössbauer spectroscopy

If an isotope, such as ^{119}Sn, has a metastable excited state which transforms to the normal ground state of the nucleus by emitting a gamma ray, the excited state can be used as a source and the ground state as a target for the emission and resonant reabsorption of the gamma ray. This is the Mössbauer effect and is made use of in *Mössbauer spectroscopy.* The striking characteristic of the effect, which makes it so valuable, is the extremely well-defined energy of the transition. The line width is only about 10^{-12} of the energy, compared with about 1 cm^{-1} in 1000 cm^{-1} for a sharp infrared line for example. This means that the very small effect on the nuclear energy arising from the chemical environment may be detected in Mössbauer spectroscopy. Two interactions are important to us. First, as the nucleus has a slightly different radius in its ground and excited states, the resonance energy changes slightly with the electron density at the nucleus. This gives rise to a chemical shift whose magnitude reflects the s electron density as only electrons in s orbitals have any probability of being at the nucleus. A good example is provided by tin species where chemical shifts lie in the sequence Sn(IV) $<$ Sn(0) $<$ Sn(II). In metallic tin (α-form) the element has the diamond structure with four more tin atoms surrounding each atom tetrahedrally. The configuration is thus s^1p^3 giving one s electron per tin atom.

In tin(IV) compounds, all the valency electrons tend to be used in bonding so that the s electron population is less than one (and becomes 0 for the Sn^{4+} ion). On the other hand, in Sn(II) compounds there is one unshared pair of electrons whose configuration approximates to s^2. Thus the order of chemical shifts in the Mössbauer follows from the configurations $s^0 < s^1p^3 < s^2$. Clearly, intermediate shifts give information about the s electron density, for example shifts between those for Sn^{4+} and for the metal show the direction of s electron drift in bonds to covalent tin(IV).

In the second phenomenon, the Mössbauer reasonance may be split by interaction with the nuclear quadrupole moment. This arises where the nucleus has a spin of more than $\frac{1}{2}$, either in the ground state or in the metastable state. The nuclear quadrupole moment interacts with electric field gradients at the nucleus, and thus the quadrupole coupling indicates the degree of departure from spherical symmetry at the nucleus. That is, the quadrupole coupling gives information about the p and d electron densities. A further interaction which may be detected in the Mössbauer effect is the splitting of nuclear energy levels in a magnetic field. This may be imposed externally or arise internally from ferromagnetic or paramagnetic interactions and gives information about these.

The main limitation of Mössbauer spectroscopy is that only a limited number of elements have a suitable metastable nuclear state. All these have very short lifetimes and occur in the course of some decay sequence, so that the work requires an irradiating source and a suitable sequence of nuclear decay processes. The Mössbauer effect has been observed, or is predicted, for 49 elements but all with $Z \geqslant 26$ (iron), except for potassium. It is experimentally easiest to study iron and tin but a fair amount of work has been done on others including Te, I, Xe, Au and several of the lanthanides. It has the advantage that the sample need only be a powder, so that it provides a method of studying insoluble, poorly crystalline materials of the heavier elements which are difficult or impossible to study in any other way.

7.9.3 Mass spectrometry

One further spectroscopic technique which continues to make a significant impact on modern inorganic chemistry is mass spectrometry. Mass spectrometry has been traditionally applied to organic compounds. However, techniques for the analysis of metal carbonyls, other organometallic compounds and coordination compounds are now well established. The basic principle of mass spectrometry involves generating gas-phase ions from the substance of interest (a wide variety of ionization methods are available, and are discussed later). These ions are then passed through a magnetic field (and in some cases an electric field as well), and ions are separated according to their mass-to-charge (*m/z*) ratio. The mass spectrum then comprises signals (of differing intensities) for ions at different *m/z* values. If the parent ion $[M]^+$, or in many cases a pseudo-parent ion formed for example by protonation, $[M + H]^+$, can be detected, then the molecular weight of the molecule can be determined. The use of double focusing (magnetic and electric fields) yields masses accurate to about 1 ppm, and thus allows distinction of very similar masses, for example CO may be distinguished from $^{14}N_2$. For routine, relatively low-resolution spectra, the development of *quadrupole mass analysers* has resulted in a substantial decrease in size (and cost) of commercial instruments.

In traditional electron ionization mass spectrometry involving organic or organometallic compounds, the sample is heated under vacuum, which generates a vapour pressure of the compound. The vapour is then passed through a beam of electrons which have energy in the region of 70 eV. These electrons ionize the parent molecules M:

$$M + e^- \rightarrow M^+ + 2e^-$$

The resulting M^+ ions may then fragment further. Electron impact mass spectrometry has a relatively limited application in inorganic chemistry, because many samples are either too involatile, or are thermally sensitive, such that the heating regime required may result in decomposition of the sample.

Of more use to the inorganic chemist are the relatively recent soft ionization techniques. These involve a much gentler ionization process, such that molecular (or pseudomolecular ions) are more likely to be observed and the spectra are subsequently simpler. Perhaps the most widely accepted technique is electrospray ionization. In this, a solution of the sample is sprayed through a metal capillary held at a high potential, assisted by a gas (typically nitrogen). This produces a spray of charged droplets, from which the solvent is evaporated, leaving gas-phase ions, which are then analysed in the usual ways. The key advantages of electrospray ionization are that volatility of the sample is not necessary (it only needs to be soluble in a solvent to the extent of a few ppm), it allows the investigation of positive or negative ions equally well, and it is a soft ionization technique (so that minimal fragmentation can be selected). If fragmentation is desired (to obtain structural information), then this can be readily achieved, for example, by accelerating the ions through a potential difference (often termed the *cone voltage*), which causes collisions with solvent molecules, resulting in fragmentation. Electrospray mass spectrometers can also be readily interfaced with liquid chromatography (lc) systems, giving the so-called hyphenated technique of lc–ms. In this, a mixture of compounds can be separated, and a mass spectrum of each component of the mixture obtained as it exits the chromatography column.

As an example of the simplicity of an electrospray mass spectrum, Fig. 7.8 shows the negative ion electrospray ionization mass spectra of the perrhenate (ReO_4^-) ion, at two cone voltages, 20 V and 100 V. At 20 V, the spectrum consists solely of the ReO_4^- ion with major peaks at *m/z* 249 and 251, while at 100 V fragmentation occurs giving the ReO_3^- ion, with peaks at *m/z* 233 and 235.

Other relatively soft ionization techniques are also available. The oldest of these is *fast atom bombardment* (FAB) ionization, in which the sample is impacted by a beam of heavy atoms such as Cs^+ or Xe. Compared to electospray ionization, however, FAB spectra indicate a harsher ionization, and spectra are typically more complex. Another recent technique is *matrix assisted laser desorption ionization* (MALDI), in which a laser is used to generate ions from the sample dispersed in a liquid matrix.

7.10 Fourier transform methods

Many of the spectroscopic methods discussed above were developed using a sequential, point-by-point scanning of the wavelength region of interest. Such methods are slow, needing anything from several minutes to many hours for accurate infrared or nmr studies, for example. When methods are of low intrinsic sensitivity, such as nmr, the slowness limits the chance of improving sensitivity by repeated scanning. A solution to this problem is to replace the point-by-point scan with the simultaneous observation of the whole spectral region. This excites all the transitions simultaneously, and the resulting signals are found as beating patterns which can be analysed by the mathematical process of Fourier transformation. With the development of computers, the Fourier calculations are rapidly performed, and the technique is now standard. As the observation time is short, this allows (a) much enhanced sensitivity (as the results of up to many thousands of scans may be summed) and (b) experiments on a very short timescale.

Finally, we note that, although Fourier transform methods were a major motivation to add computers to spectrometers, computers have had other major impacts. Most intermediate and advanced level instruments are now operated more or less under computer control, and the data output is processed by computer. The possibility of accumulating a number of spectra to improve sensitivity is important even where point-by-point scanning is used, for example in Raman spectra see box titled 'The impact of enhanced sensitivity'. As well as allowing expansion, smoothing, determination of maxima and so on, the computer is useful in matching with library spectra, subtracting background or solvent absorptions, and further calculation of quantities such as concentrations or absorption coefficients.

The two main peaks, of different intensities, for the ReO_4^- ion arise because of the two isotopes of Re, ^{185}Re (60% relative abundance) and ^{187}Re (100% relative abundance). Oxygen is largely ^{16}O, but small amounts of the heavier isotopes ^{17}O and ^{18}O also contribute to the overall isotope pattern and intensities.

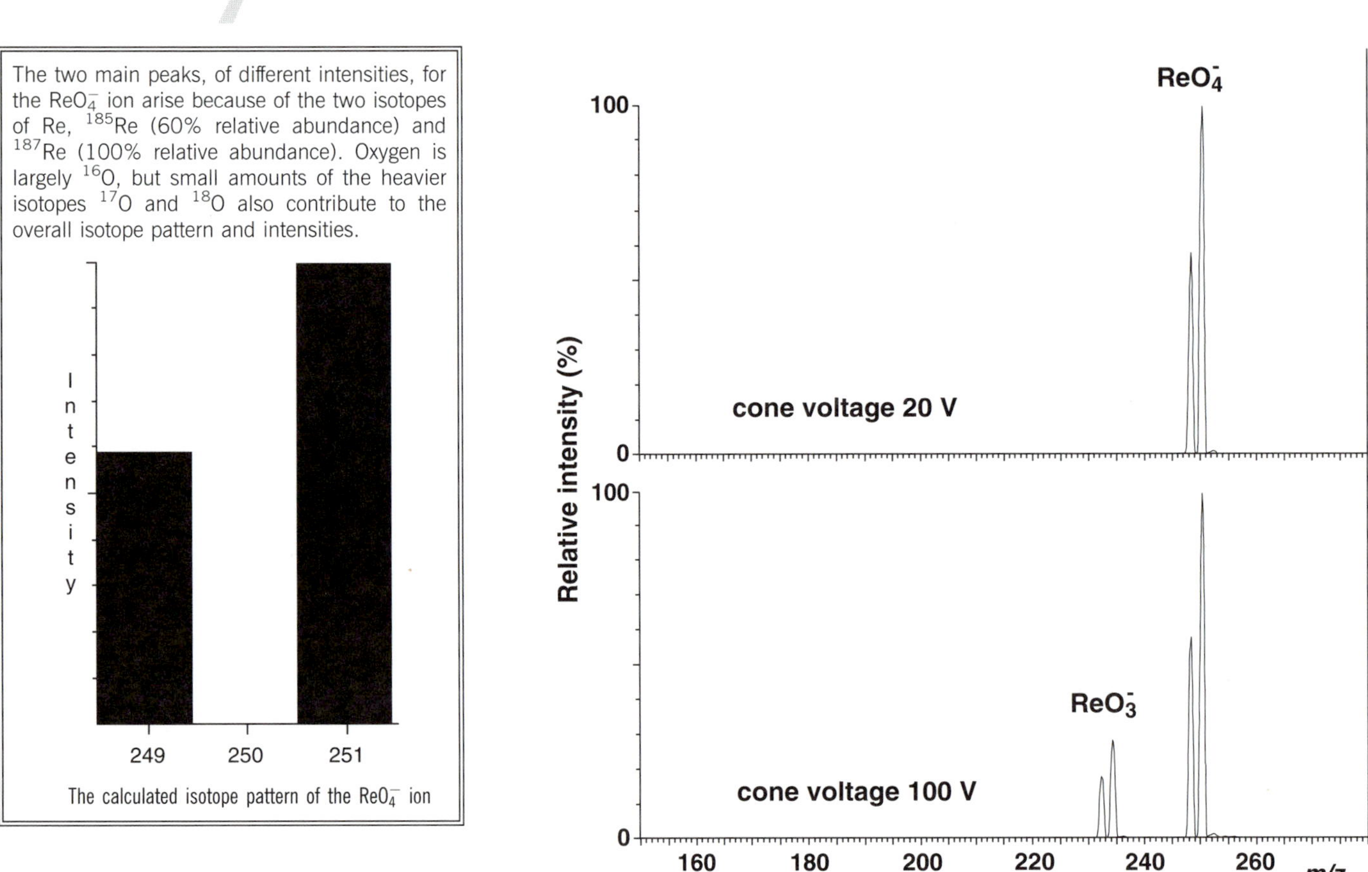

FIG. 7.8 Negative-ion electrospray mass spectra of ReO_4^- (as the NH_4^+ salt), at low (20 V) and high (100 V) cone voltages

7.11 Other methods

7.11.1 Magnetic measurements

If a sample of a compound is weighed in a magnetic field and then in absence of the field, a weight change will be observed. Most compounds are repelled by the field and show a decrease in weight; these are termed *diamagnetic.* The diamagnetism arises from the repulsion between the applied field and induced magnetic fields in the compound, and is a very small effect which occurs for all compounds. However, some compounds show a net attraction to a magnetic field and an increase in weight; these are termed *paramagnetic.* The paramagnetism arises where there are one or more unpaired electrons in the compound, and is a much larger effect than diamagnetism. An unpaired electron corresponds to an electric current, and hence to a magnetic field, by virtue of two effects, its spin, and its orbital motion. In most compounds, the effect of the orbital contribution is quenched out by the electric fields of surrounding atoms, and the spin-only magnetic moment is observed. This is given by:

$$\mu = 2\sqrt{[S(S+1)]}$$

where μ is the magnetic moment in units of Bohr magnetons, and $S = \frac{1}{2}n$ equals the number of unpaired spins multiplied by the spin quantum number. This formula holds, to within ten percent, for most compounds, allowing a direct determination of the number of unpaired electrons. In some cases, particularly when the unpaired electrons are in an f orbital, the orbital contribution is not quenched out and a more complex formula:

$$\mu = \sqrt{[4S(S+1) + L(L+1)]}$$

THE IMPACT OF ENHANCED SENSITIVITY

The principal technique where the enhanced sensitivity was important is nmr, as indicated above. Here, because the energy of the transition is so low, the population of the excited state very easily becomes equal to the ground state and the signal disappears (called 'saturation'). The Fourier transform method avoids this problem by using a short pulse of radiation, followed by a pause while the populations revert to thermal equilibrium. Measuring this relaxation time gives a further parameter not easily available from the older experiment. In modern studies every isotope with a nuclear spin is accessible, which means that practically every element in the Periodic Table yields at least some information. For the easier nuclei, the enhanced sensitivity allows work at lower concentrations, on metal carbonyls using ^{13}C at natural abundance, for example. Other major improvements arise because the computer may also be used to control and vary the experiment. For example, by irradiating with a sequence of pulses of different character, signals may be decoupled, enhanced in sensitivity, correlated or separated from noise. Thus, in a complex molecule such as a polytungstate (Section 15.4) two-dimensional nmr methods allow a structural determination in solution as it is possible to determine which atom bonds to which. In cases like this, where the species in solution may not be the same as the one isolated as a crystal, the method gives unique information. Other striking examples include studies on hydration and hydrolysis (a classic is of aluminium ions using ^{27}Al nmr), and the work on polyphosphorus compounds described in Section 18.3.1.

The advantages of enhanced sensitivity or shorter timescale apply to all techniques, and are of particular advantage where the sensitivity of the method is intrinsically low, as in Raman or far-infrared spectroscopy. Of equal importance, even in methods of reasonable natural sensitivity, has been the reduction in the timescale of the experiment. This allows the study of short-lived species under actual conditions. Thus, infrared studies of the catalytic oxidation of hydrocarbons has allowed the detection of intermediates like CH_3O-M as transient intermediates on metal or metal oxide catalyst surfaces.

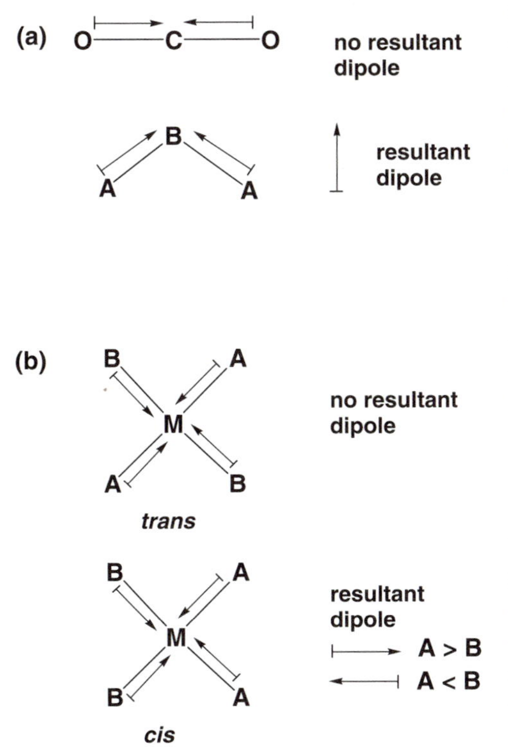

FIG. 7.9 The use of dipole moments to yield structural information. (a) shows how the two C-O moments cancel in linear O-C-O but would give rise to a resultant moment if the molecule was V-shaped. As CO_2 is observed to have no dipole moment, the linear structure is the correct one. (b) shows how a *cis* square planar complex MA_2B_2 will have a resultant moment while the *trans* form will not

which involves the orbital quantum number, L, holds (see Fig. 11.8). In some cases, the determination of the number of unpaired spins gives direct structural information. For example, consider a nickel(II) compound $NiL_4.2S$, where L is any ligand and S is a molecule of solvent. If the solvent molecules are not coordinated to the nickel, the NiL_4 species could well be square planar and have no unpaired electrons, while coordinated solvent would mean an octahedral NiL_4S_2 species with two unpaired electrons (compare Section 14.8).

It is possible to gain much more detailed information from magnetic measurements than is indicated above. Other effects such as ferromagnetism and antiferromagnetism are observed, and much valuable information results from studying the variation of magnetic moment with temperature, concentration, and field strength. However, the simple Gouy method of weighing the sample in a magnetic field is readily carried out and yields considerable information, especially in transition metal chemistry.

7.11.2 Dipole moments

The measurement of the dipole moment of a compound may also yield useful structural information. As any bond between atoms with different electronegativities is polarized, any molecule will have a dipole moment unless such bond dipoles are so arranged as to cancel out. As a simple example, if CO_2 is linear, the two C-O dipoles oppose each other and no resultant moment is observed, while if the molecule is bent a resultant dipole is observed (Fig. 7.9). The figure also illustrates, as a further example, how *cis* and *trans* isomers may be distinguished by dipole moment measurements. Care, however, must be exercised in interpreting dipole moments: thus, NF_3 has an almost zero dipole moment (whereas NH_3 has a marked moment), not because the molecule is planar as once thought, but because the bond and lone pair dipoles cancel. However, with care in interpretation, dipole moments have proved a very useful adjunct to structural determinations, and the method has the advantage that it does not destroy the sample, and adaptations are available which require only a small amount of material.

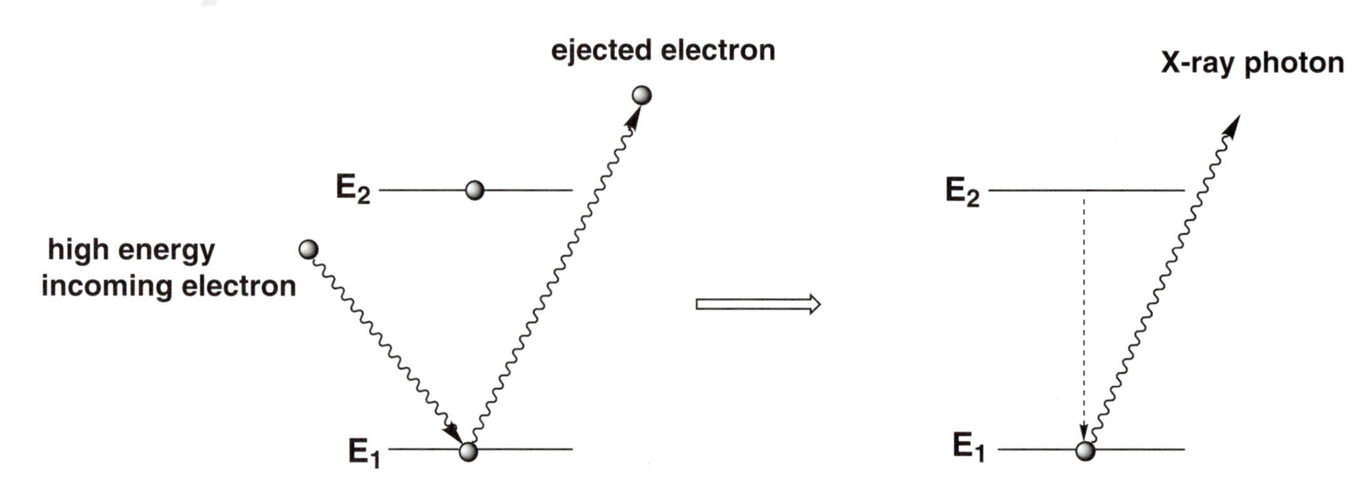

FIG. 7.10 The electronic process behind X-ray microanalysis. A highly energetic electron (from the scanning electron microscope) ejects one of the core electrons of the atom. An electron from a higher energy shell in the atom then fills the gap, emitting an X-ray photon in the process which is detected. The energy of the X-ray is equal to the separation of the two energy levels $E_2—E_1$

Remember that electrons have wave as well as particle characteristics, see Section 2.5.

7.11.3 Scanning electron microscopy (SEM)

The basic difference between a conventional light microscope and an electron microscope is in the type of radiation used. The electron microscope uses a beam of electrons as the 'light' source and the shorter wavelength therefore means that the resolution of an electron microscope is much greater, allowing smaller features to be resolved. While the applications of SEM to disciplines such as earth sciences and biology are immediately obvious, the technique is also becoming of increasing importance to chemists and especially materials scientists. For example, the degree of crystallinity of a solid sample, be it a synthetic zeolite (see Section 18.6), a copper oxide superconductor (see Section 16.1) or some other material, can readily be determined.

In addition to providing information on the morphology of the sample, information may also be obtained on its elemental composition, and this is of direct benefit to the chemist. Irradiation of a sample with an electron beam results in electrons being ejected from the inner electron shells of atoms. When electrons from outer valence shells drop down to fill this newly formed vacancy, they emit an X-ray photon, the energy of which is characteristic of the element concerned and the various energy levels involved. This is illustrated schematically in Fig. 7.10. The analysis of these X-rays then provides information on the relative numbers of each type of atom in the sample, though it can only be applied with any accuracy for elements heavier than about fluorine. A typical X-ray spectrum is shown in Fig. 7.11 for the potassium salt of a polyoxovanadate species (see Section 14.3.1). The spectrum consists of a rather broad background upon which are superimposed a number of sharp peaks, each being composed of a number of channels, or wavelengths, over which the X-rays are analysed. This analytical method, commonly called Energy-Dispersive X-ray Microanalysis, is particularly useful in qualitative analysis for the number of elements present in a sample though by the use of appropriate standards quantitative analyses may also be undertaken.

7.11.4 Scanning tunnelling microscopy (STM)

Scanning tunnelling microscopy is a very powerful and fairly recent (early 1980s) technique for studying surfaces of solids with atomic resolution. The principle behind this technique utilizes the fact that electron tunnelling occurs between a very sharp metal tip and a conducting or semiconducting surface—essentially the wave functions of the tip and the surface atoms overlap at very small distances. By scanning the metal tip over the surface of the solid, whilst keeping a constant flow of electrons (called the *tunnelling current*) between the tip and the solid, or vice versa, a 'map' of the surface can be built up. The fact that atomic resolution can be attained makes this an exceedingly powerful technique.

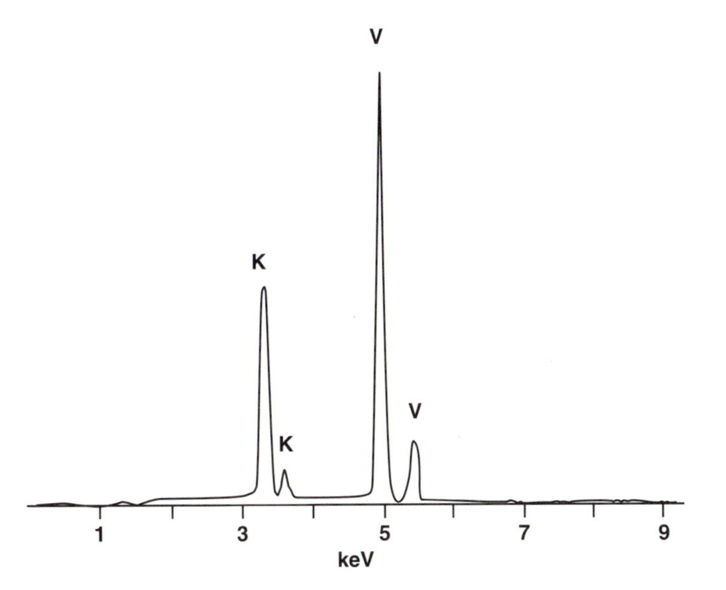

FIG. 7.11 A typical X-ray spectrum of a compound containing vanadium and potassium. Sharp, 'stepped', peaks due to X-rays emitted from vanadium and potassium atoms can be clearly seen superimposed on a broad baseline. The peaks due to oxygen atoms also present in the sample are very weak

Figure 7.12 shows an image obtained from a mica sample (see Section 18.6); the network of interlocking rings can be clearly distinguished.

The early studies in this area tended to concentrate on materials such as metals and semiconductors. However, recently chemists have begun to add this to their armoury of techniques. Applications include the investigations of silicon single crystals (of importance to the semiconductor industry), copper oxide superconductors and layered transition metal chalcogenides. This latter class of compounds includes molybdenum sulfide, MoS_2, a material which ordinarily has a layered structure but which has recently been found to exist in a range of polyhedral structures. Recent advances in this area are discussed in greater detail in Section 19.3.3. These layered chalcogenides have been found to undergo a very subtle periodic distortion at low temperatures, called a charge-density wave. (These distortions arise due to the unequal electron occupation of degenerate valence bands of the solid which results in a distortion occurring to relieve the degeneracy, similar in many respects to the Jahn–Teller distortion in certain coordination complexes, see Section 13.4.) STM provides a very powerful technique for the study of these charge-density wave distortions. Other areas of inorganic chemistry which have benefited from the STM technique include the study of zeolite morphology, metal colloids and metal carbonyl

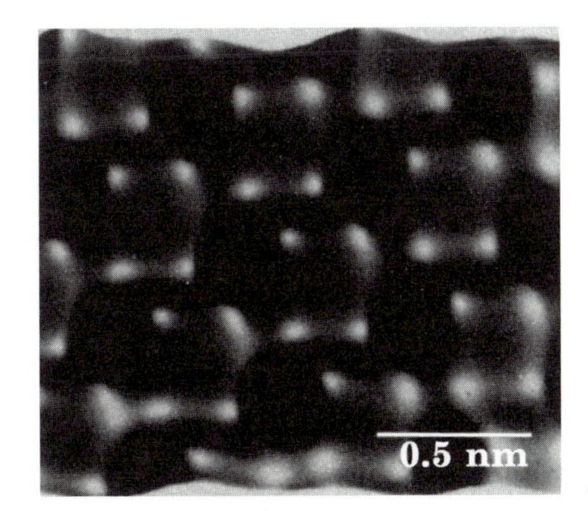

FIG. 7.12 Scanning tunnelling microscope image of a mica surface (see Section 18.6)

clusters. A recent article, given in Appendix A, summarizes the application of STM to the above areas.

Determination of energy levels

7.12 Photoelectron spectroscopy

When a molecule, M, interacts with a quantum of ultraviolet radiation, it is possible for all the energy of the quantum to be used in expelling an electron from a valence level orbital

$$M + h\nu \rightarrow M^{+} + e \qquad (7.8)$$

The quantum energy has first to provide the ionization energy, I, of the electron, and the remainder appears as the kinetic energy, E_{elect}, of the expelled electron. A small part of the energy, E_{vib}, may also be used in exciting vibrations of the molecular ion, M^{+}, and there are other small corrections which need not concern us. Thus the energy equation is

$$E = h\nu = I + E_{\mathrm{vib}} + E_{\mathrm{elect}} \qquad (7.9)$$

If we use monochromatic radiation, $h\nu$ is fixed and is known. The energy of the electron can be measured by finding the size of repelling electric field which just stops the electron reaching a detector. Thus we can determine the ionization energy (Equation 7.10).

$$I + E_{\mathrm{vib}} = h\nu - E_{\mathrm{elect}} \qquad (7.10)$$

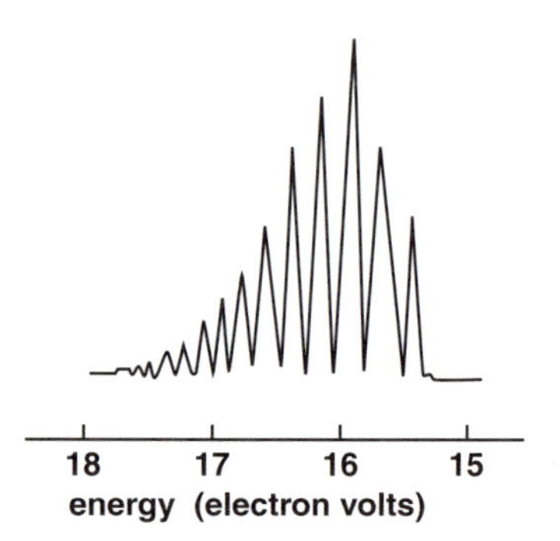

FIG. 7.13 The photoelectron spectrum of H_2

Since vibrational energy is quantized (and the quantum is much smaller than I) we observe a series of bands corresponding to $I + nh\nu_{\mathrm{vib}}$ where n is the change in the vibrational quantum number and may take the values $0, 1, 2, \ldots$ etc. Thus the band of lowest energy of the series represents the value of I, and the separations give the vibrational energy of M^{+}. This is clearly seen in the photoelectron spectrum of H_2 (Fig. 7.13). The lowest energy component about 15.5 eV gives the ionization energy and the separation, which corresponds to 2260 cm^{-1}, is the vibrational energy of H_2^{+}. Note that the vibrational spacing decreases, as more and more quanta are excited, as the ion moves towards its dissociation limit at about 18 eV.

If we compare with Fig. 3.6b (and making certain assumptions which are discussed in the references) we can identify the ionization energy I, as the stabilization energy of an electron in the orbital ϕ_{B}. That is, I is the difference in energy between the ϕ_{B} level and the energy zero which (compare Sections 2.6 and 2.7) corresponds to the complete separation of electron and remaining ion. Consequently, Figs 3.6b and 7.13 may be combined as in Fig. 7.14. In this way, we can use photoelectron spectra to determine by experiment the energies of occupied orbitals.

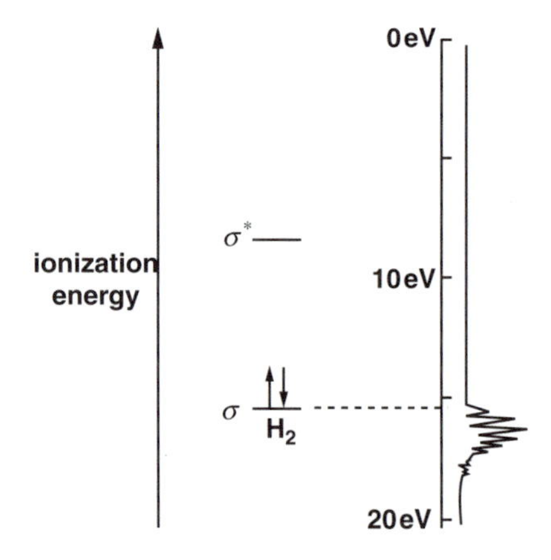

FIG. 7.14 Energy level diagram and photoelectron spectrum of H_2

Note that photoelectron spectroscopy gives no direct information about the unoccupied level ϕ_{A}. However, ordinary electronic spectroscopy (Section 7.6) does give the difference between ϕ_{B} and ϕ_{A} (the $\sigma \rightarrow \sigma^*$ transition, found in the ultraviolet for H_2). Thus these two techniques, in combination, can fix all the levels in molecular energy level diagrams.

In more complex molecules, where a number of valence level orbitals are occupied, the ionization from each is observed (in different molecules of the sample of course) and a spectrum with several bands is obtained as in Figs 3.22 or 4.21

7.12.1 The energy source

Monochromatic radiation (i.e., all quanta of same energy) is needed, and the most convenient experimental source is the helium (I) emission, with a quantum energy of 21.218 eV (2047.4 kJ mol^{-1}, wavelength 58.43 nm). This arises from the $1s \rightarrow 2p$

EXPERIMENTAL

Ultraviolet photoelectron spectroscopy is carried out on a gas sample, at relatively low pressure, which is irradiated with a collimated, monochromatic beam. The expelled electrons are sorted according to their kinetic energies by passing them through a retarding field. The electron count is plotted against the ionization energy (which is the photon energy less the electron kinetic energy). For X-ray photoelectron spectroscopy, solids may be used as well as gases. There is no convenient monochromatic source between the ultraviolet (40–50 eV) and the X-ray region (1000–1500 eV) and little work has been done on the intermediate region.

excitation. This energy is sufficient to ionize most of the valence level electrons. For more tightly bonded levels the helium (II) line (40.81 eV or 30.4 nm) arising from He^+, is used but is experimentally more difficult. Thus ionization energies of more than 21 eV are generally known with rather lower accuracy and are not always determined.

If a quantum of much higher energy is used, in the X-ray region of the electromagnetic spectrum, then the most tightly bound electrons from the inner closed shells are ionized. This *X-ray photoelectron spectroscopy (XPS)* is of less direct interest to us than *ultraviolet photoelectron spectroscopy (UPS)*, and it is the latter which is referred to when the general phrase *photoelectron spectroscopy* or *PES* is used. However, XPS does give chemical information since, although the inner shell electrons, such as the 1s electrons of C, N or O, are not involved in bonding, there are small changes in their energies with changes in their chemical environment. For example, the oxygen 1s binding energy in $RMn(CO)_5$ changed from 538.8 to 540.3 eV as R varied from C_5H_5 to $SiCl_3$ and this change correlated with other measures of the degree of electron back-donation from Mn to CO in these molecules.

XPS and UPS are usually carried out with quite different instruments, and there have not been many studies on the same molecule. Jolly (see reference) has emphasized the advantages of using the two techniques in conjunction. Basically, XPS reflects the effects of the nuclear charge and the overall electron shells, while UPS depends on these core effects plus the valency level ones. XPS can thus be used to allow isolation of the valency effects in UPS, and hence greatly expand the approach to valency energy levels. UPS results are used in other chapters: X-ray PS methods are further described in the references.

7.12.2 Vibrational structure

The spacing of the vibrational fine structure gives the vibrational frequency of the molecular ion, M^+, resulting from the ionization. If the electron was lost from a bonding orbital in M, then M^+ will be less tightly bound, it will require less energy to stretch the bond, and thus the vibrational frequency in M^+ will be lower. For example, the frequency of 2260 cm^{-1} in H_2^+ compares with 4280 cm^{-1} for H_2 (compare Section 3.3, especially Figs 3.7a and 3.7b).

Conversely, loss of an antibonding electron would lead to a higher stretching frequency for M^+. If a nonbonding electron is lost, there will be little change. Not all changes are as large as those for hydrogen, and the accuracy of measurement of the frequency from photoelectron spectra is much less than by other methods such as infrared spectroscopy. However, the vibrational frequencies are a very valuable guide to the assignment of ionizations to particular orbitals.

In addition to the frequencies, the number of vibrational components and their intensity pattern also vary with the bonding character of the lost electron. The detailed argument is given in the references, but their characteristics may be summarized as in Table 7.2.

TABLE 7.2 Vibrational structure of photoelectron bands

Orbital from which electron is ionized	*Vibrational frequency in M^+ compared with M*	*Number of vibrational components*	*Most intense vibrational components*
Bonding	Markedly smaller	Large	Near centre
Nonbonding	Similar	Small	First
Antibonding	Similar or larger	Medium	Near centre
Example NO: stretching frequency 1890 cm^{-1}			
Bonding π	1200	8	4th
Nonbonding σ	1610	3	1st
Antibonding π^*	2260	5	2nd

Polyatomic molecules show much more complex vibrational structure, since a number of different vibrational energies may be involved. Often the individual components are not resolved but only the overall envelope is seen. Even in this case, non-bonding ionizations tend to give sharp asymmetric envelopes, steepened towards the leading edge while bonding and antibonding envelopes are rounder and more symmetric.

Problems

7.1 Consider how each method reviewed in this chapter contributes to the basic scheme for identifying an unknown, given in Fig. 3.1. Note that some methods, like chromatography, are precursors to the scheme in providing the unknown as a single substance, and other methods, like photoelectron spectroscopy, extend the scheme to find bonding or other properties in terms of the structural formula.

7.2 If a compound occurs only (a) as a gas or (b) as a solid, discuss the limitations imposed on the experimental methods available for its study.

CHAPTER 8

8.1 VARIATION IN ENERGIES OF ATOMIC ORBITALS WITH ATOMIC NUMBER 185
8.2 EXCHANGE ENERGY 188
8.3 STABLE CONFIGURATIONS 190
8.4 ATOMIC AND IONIC SIZES 193
8.5 CHEMICAL BEHAVIOUR AND PERIODIC POSITION 194
8.6 METHODS OF SHOWING THE STABILITIES OF OXIDATION STATES 196
8.7 THE ABUNDANCE AND OCCURRENCE OF THE ELEMENTS 200
8.8 THE EXTRACTION OF THE ELEMENTS 203
PROBLEMS 207

General Properties of the Elements in Relation to the Periodic Table

8.1 Variation in energies of atomic orbitals with atomic number

In Chapter 2 we described how the existence, shapes and energies of atomic orbitals could be derived from the wave equation. Then the structure of the Periodic Table resulted from filling the atomic orbitals in order of increasing energy. A more detailed discussion of the properties of the elements demands a closer look at the variation in energy of the atomic orbitals as the atomic number increases.

Consider first the s orbitals. The energy of an electron in the 1s orbital of a hydrogen-like atom of atomic number Z, is $-Ze^2/8\pi\epsilon_0 a_0$; so that the energy decreases as Z increases (compare Equation 2.3). As the nuclear charge and the number of electrons in the atom increase, account must be taken of the repulsive effect of the extra electrons, as well as that of the increased nuclear charge. This is done by replacing the actual nuclear charge Z, by the effective nuclear charge, Z^*, which is the resultant of the nuclear charge and the electron charges as experienced by an electron in a particular orbital. For example, the 2s electron in lithium experiences an effective charge which is the resultant of the nuclear charge of +3 and the charges of the two 1s electrons. The effect of the inner electrons in reducing the effective charge experienced by the outer one is termed *shielding*. If the shielding effect of the two 1s electrons in lithium were perfect, the outer electron would experience an effective nuclear charge, Z^*, of 1.0 (3−2), but the 2s orbital has finite electron density at the nucleus (see Fig. 2.9b) so that an electron in the 2s orbital penetrates the 1s shell and thus experiences a greater nuclear charge than that calculated from perfect shielding. The result of shielding by inner electron shells is that the effective nuclear charge experienced by the outer electrons in an atom is always markedly less than the actual nuclear charge, Z, but, as the shielding is not perfect, the effective nuclear charge increases as Z increases, but more slowly. A useful indication of the shielding effects is given by Slater's rules which are summarized in Table 8.1.

Other, more detailed, methods of allowing for shielding have been proposed, but the Slater scheme represents the principles and needs no information outside the quantum numbers.

Application of these rules shows that the effective nuclear charge experienced by one 1s electron in, say, carbon is 5.7 and in nitrogen, 6.7. Similarly, a 2s electron in carbon experiences an effective charge of 3.25, while a nitrogen 2s electron experiences a charge of 3.9. Such calculations, or direct experimental determination of energy levels by electron spectroscopy in one of its forms, allows the construction of diagrams showing the variation in energy levels with atomic number, such as Fig. 8.1.

In this figure, the energies start at the calculated values for H, and fall off with Z (the scale is logarithmic) so that the electron is increasingly tightly held. One measure of the energy involved is given by the subshell binding energy measured by X-ray photoelectron spectroscopy. This energy is that required to remove one electron from a specific subshell in a free atom, say the energy to remove a 1s electron from a free C atom. For removal of the first electron, this is the same as the ionization potentials, but successive ionization potentials involve removal of electrons from ions (e.g. the 5th ionization potential of C is the energy to remove one 1s electron from a C^{4+} ion). Some binding energies are listed in Table 8.2.

TABLE 8.1 Slater's rules for shielding contributions

The effective nuclear charge, Z^*, is given by $Z-\sigma$, where σ is the sum of the shielding contributions of all the other electrons in the atom, as follows

Principal quantum number, n, of shielding electrons	*Shielding contribution, σ*
n higher than principal quantum number of the electron under consideration	Zero
n equal to principal quantum number of the electron under consideration	0.35 (for each electron) except that 0.30 is used for σ for a 1s electron acting on the second 1s electron
n is one less than the principal quantum number of the electron under consideration	
(a) for an s or p electron under consideration	0.85
(b) for a d or f electron under consideration	1.00
n is less by two, or more, than the principal quantum number of the electron under consideration	1.00

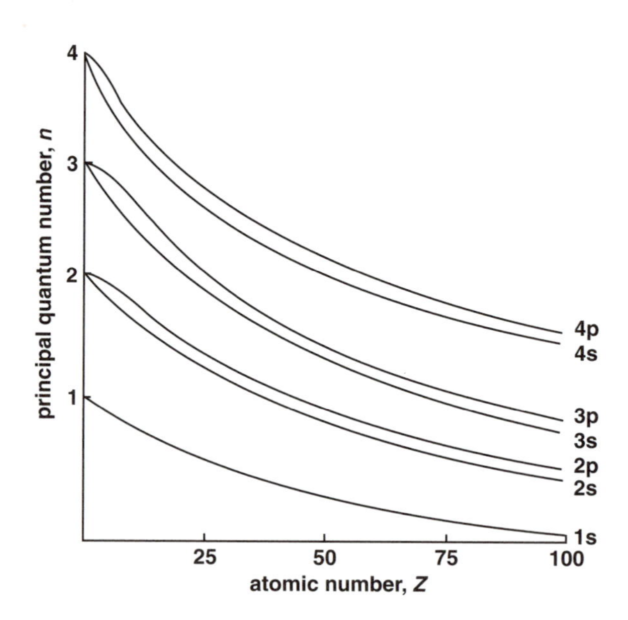

FIG. 8.1 Energies of the lower s and p orbitals as functions of Z

In hydrogen-like atoms, the p orbital has the same energy as the s orbital with the same value of n, but in all atoms with more than one electron, shielding effects come into play. The p orbital is more shielded than the corresponding s orbital as it does not penetrate so far towards the nucleus. It accordingly experiences a smaller effective nuclear charge and is of higher energy. The curves of p orbital energies versus atomic number run roughly parallel with the curves of the corresponding s orbitals as Z increases (Fig. 8.1). The gap in energy between the s and p orbital of a given n value is much smaller than that separating the p orbital and the s orbital with the next higher n value.

The case of the d orbitals is more complicated. Figure 8.2 shows the variations in energy with increasing Z of the 3d orbital with respect to the 3s, 3p, 4s and 4p orbitals. The 3d orbital has the same energy as the 3s and 3p orbitals in the hydrogen atom and, since it scarcely penetrates the first and second quantum shells at all, it is perfectly shielded from the increase of nuclear charge as these two atomic levels are filled. Thus the 3d level is subject to the same effective nuclear charge (about unity) for Z values up to $Z = 10$, and the plot of the 3d energy

TABLE 8.2 Free atom subshell binding energies (eV)

	H	He						
1s	13.6	24.6						
	Li	Be	B	C	N	O	F	Ne
1s	58	115	192	288	409	538	694	870
2s	5.4	9.3	12.9	16.6	20.3	28.5	37.9	48.5
2p			8.3	11.3	14.5	13.6	17.4	21.6
	Na	Mg	Al	Si	P	S	C1	Ar
1s	1075	1308	1564	1844	2148	2476	2828	3206
2s	66	92	121	154	191	232	277	326
2p	34	54	77	103	134	168	207	249
3s	5.2	7.6	10.6	13.5	16.2	20.2	24.5	29.2
3p			6.0	8.2	10.5	10.4	13.0	15.8

Note: p values averaged over two multiplicities.

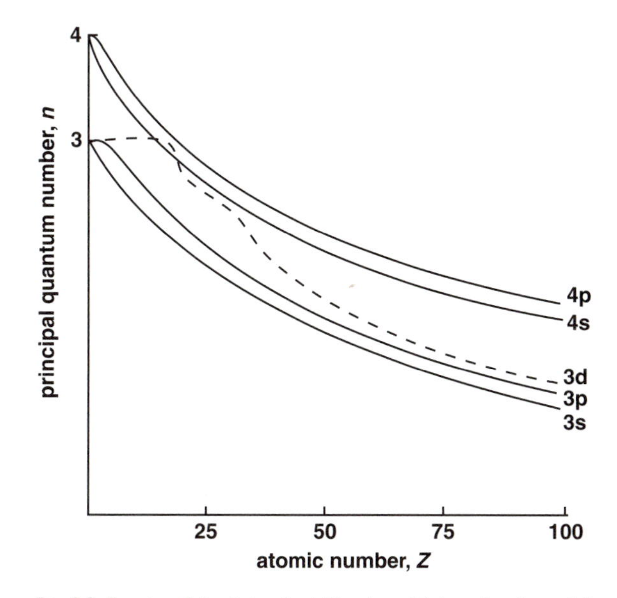

FIG. 8.2 Energies of the 3rd and neighbouring orbitals as functions of Z

against Z remains level. On the other hand, the 4s and 4p orbitals—which are of considerably higher energy than the 3d orbital in the lightest elements—do penetrate the inner electron shells significantly, are less shielded, and drop steeply in energy as Z increases. When the 3s and 3p levels are filling, the 3d level still remains almost unaffected and the energy of the 4s level falls below it at about $Z = 15$. As a result of this, when the 3p shell is filled at argon, the next lowest energy level is 4s and not 3d. The nineteenth and twentieth electrons therefore enter the 4s level, into which the 3d level does strongly penetrate. It experiences a marked increase in effective nuclear charge, and its energy falls from being nearly equal to that of the 4p level towards that of the 4s level. The twenty-first electron and the next nine enter the 3d level whose energy falls below that of the 4s level. As these two remain very close in energy, electrons readily switch between them; for example, copper, which might be $3d^9 4s^2$, is actually $3d^{10} 4s^1$, and gains the extra stability of the filled d shell by transferring an s electron. When the 3d shell is filled, the level next in energy is the 4p orbital. This filled d shell introduces an extra shielding effect on the higher orbitals, and the energy gap between the 4p and 5s levels at $Z = 36$, where the 4p level is filled, is larger than that expected by simple extrapolation from the 2p–3s and the 3p–4s differences.

A similar effect occurs for the 4d level relative to the 5s one, and the f levels show analogous relationships, Fig. 8.3. As the f orbitals are even less penetrating than the d levels, an extra

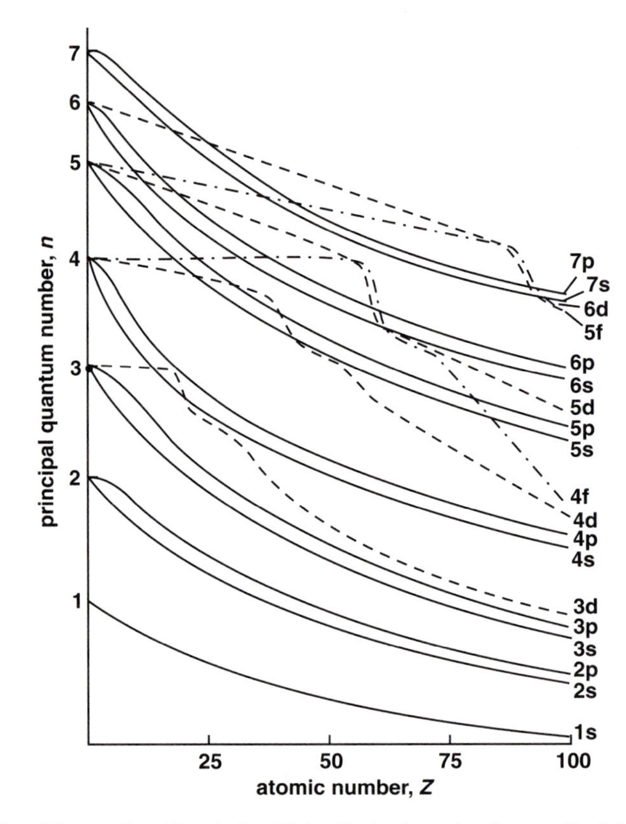

FIG. 8.3 The variation of the energies of the atomic orbitals with atomic number. Compare Fig. 2.4 which shows relative energies at the *Z* values where the orbitals start filling

quantum shell is filled before they become involved. The 5s, 5p, 6s, 6p and 5d levels all drop below the 4f level in energy, as Fig. 8.3 shows. The filling order is 5s, 4d, 5p, and then 6s. Both the 4f and 5d levels are strongly affected by the filling of the 6s level and drop very steeply in energy below the 6p level, but remain almost equal in energy to each other. Lanthanum, which comes after the 6s level is filled, has its outer electron in the 5d shell, but the following element cerium has the outer configuration $4f^2 5d^0 6s^2$. The next electrons fill 4f, with minor variations at the half-filled and filled levels (compare Table 11.1), then 5d and 6p. For the heaviest elements, where 5f and 6d are even closer, the configurations follow the same general principles with variation in detail (Table 12.1).

8.2 Exchange energy

The energy of an electron in an orbital depends on other factors besides the attraction of the nuclear charge and the electrostatic interaction between the electrons. There is a second, quantum-mechanical, interaction between electrons which is known as the *exchange energy*. There is no classical analogue to this energy which derives from the indistinguishability of electrons and the arrangement of their spins. The exchange energy is a function of the number of pairs of electrons with parallel spins, i.e. $E_{ex} = K \times P$, where K is a constant and P is the number of pairs of parallel electrons. P is equal to the combination ${}_nC_2$, where n is the number of parallel spins, i.e. P has the following values:

n	1	2	3	4	5	6	7
$P = {}_nC_2 = \dfrac{n(n-1)}{2}$	0	1	3	6	10	15	21

Hund's Rule, Section 2.8 (that electrons which enter orbitals of equal energy have parallel spins as far as possible), thus amounts to maximizing the exchange energy. For example,

In addition to the exchange energy from electrons in orbitals of the same kind—as in the p orbitals tabulated—exchange energy also results from interaction between electrons of the same spin in neighbouring orbitals of similar energy. Thus in the Cr atom, compare Section 14.4, the configuration d^5s^1 allows exchange between 6 parallel spins.

Number of electrons	*Exchange energy if Hund's rule is followed* ($\times K$)		*Exchange energy for maximum pairing* ($\times K$)		*Loss of energy in latter case* ($\times K$)
1	↑	0	↑	0	0
2	↑ ↑	1	↑↓	0+0	1
3	↑ ↑ ↑	3	↑↓ ↑	1+0	2
4	↑↓ ↑ ↑	3+0	↑↓ ↑↓	1+1	1
5	↑↓ ↑↓ ↑	3+1	↑↓ ↑↓ ↑	3+1	0
6	↑↓ ↑↓ ↑↓	3+3	↑↓ ↑↓ ↑↓	3+3	0
	due to spins	↑ ↓	due to spins	↑ ↓	

the exchange energies for the various possible configurations of the electrons in the three p orbitals are shown above (note that the parallel and antiparallel electrons act as independent sets).

Although the exchange energy is relatively small, it becomes significant when similar species are compared, as it changes with the number of electrons in a different way from the larger attractive and repulsive forces. An example of the stabilization due to exchange energy has already been noted in the case of the ground state of copper. The exchange energy of the actual configuration $d^{10}s^1$ is $20K$, from the two sets of five parallel electrons in the d shell. The energy of the alternative configuration, d^9s^2 is $16K$ (from a set of five plus a set of four in the d shell; the two s electrons are antiparallel of course). The exchange energy gain thus favours the d^{10} configuration, but against this must be set the loss in orbital energy in moving the electron from the 4s orbital to the 3d one. In the case of copper, the gain in exchange energy more than balances this loss, while in the case of nickel, which has the configuration d^8s^2 but could be $d^{10}s^0$, the balance appears to lie the other way and the former configuration is ground state. That the balance of energies is very close is shown by the configurations in the nickel and copper Groups:

nickel	$3d^84s^2$	palladium	$4d^{10}5s^0$	platinum	$5d^96s^1$
copper	$3d^{10}4s^1$	silver	$4d^{10}5s^1$	gold	$5d^{10}6s^1$

The exchange energy contributes to the stability of filled shells, and also shows the greatest relative change at the half-filled shell electron count. Illustrations are to be found in the ground state configurations of transition and inner transition atoms where the d^5 and f^7 half-filled shells are favoured. Examples can be found in Table 2.5, including the configurations of chromium and gadolinium and their neighbours. This preference for half-filled and filled shell configurations is general in the Periodic Table, although the nice balance of energies means that configurations are not readily predictable (compare the ground state electronic configurations of the second transition series from ytterbium to cadmium in Table 2.5). The examples quoted so far have been confined to the ground states of atoms, but the stability of these special arrangements also shows up in the general chemistry of the elements. Manganese, for example, is particularly stable in the + II state which is a d^5 configuration.

The interelectronic forces and the changes in nuclear charge play an important part in determining the stability and configurations of ions, but it must be noted that it is not possible to determine the detailed chemistry of an ion from the ground state configuration of its parent atom. For example, in most of the transition metals the $(n+1)$s shell is filled while the nd level is only partly occupied. That is, in the atom the s shell is more stable than the d shell. However, when any of the transition elements form ions, it is always the s electrons which are lost first. Further, once one or more electrons are lost from an atom, the order of orbital stabilities is not necessarily the same as in the undisturbed atom. Thus, europium has the configuration, $4f^75d^06s^2$, while the next two elements have the configurations $f^7d^1s^2$ and $f^9d^0s^2$, yet all three lose three electrons to give a stable trivalent cation of valency configuration f^6, f^7 and f^8 respectively, just as if each element had the outer configuration $f^nd^1s^2$.

8.3 Stable configurations

With the reservations expressed above, it is possible to generalize about stable electronic configurations by considering the interplay of the nuclear attraction, the repulsion by electrons already present, particularly one in the same orbital, and other effects such as exchange energy. If a line is drawn in Fig. 8.3 through the points at which the orbitals fill, it follows an irregular path. From H to He, it drops along the 1s curve so the energy is more negative and it takes nearly double the ionization potential to remove the first electron from He. The third electron leaps up to the 2s curve and the fifth to the 2p curve, and then the last electron becomes steadily more tightly bound until the second quantum shell is filled at Ne. Similar jumps occur as we continue through the Periodic Table, with the largest between the np and the $(n+1)$s levels. If we add to this the more detailed effects of the electron configuration, we can understand the variation across a Period as shown in Fig. 8.4. From the value of 24.6 eV to remove the first electron from He, the potential drops sharply to 5.4 eV for Li because the least tightly held electron is in a new quantum shell, further from the nucleus and quite effectively shielded by the inner 1s electrons. From Li to Ne, there is an overall strong increase in the ionization potential, as the effective nuclear attraction is increasing and the electrons are being added in the same shell. The 1s orbital binding energies show the unshielded nuclear charge effect.

The finer detail in Fig. 8.4 results from the differences between the s and p orbitals. As the electron in the p orbital has less probability of being close to the nucleus than an s electron, the first p electron, at B, is relatively less tightly bound than the Be s electron. The potential then rises for three successive elements as the nuclear charge increases and the electrons enter the three different p orbitals, maximizing exchange energy and minimizing interelectron repulsions. The fourth p electron has to be placed in an orbital already occupied so that there is no gain in exchange energy and a significantly increased repulsion. Thus it takes less energy to remove the first electron from O than from N. Alternatively expressed, the half-filled shell at N shows up relative to the configuration with one less electron, at C, or with one more electron, at O. Similar relative effects can be discerned at p^3 for other elements—e.g. compare the second and third potentials of F with its neighbours. Finally, filling the last three electrons into the 2p orbitals shows smoothly increasing ionization potentials.

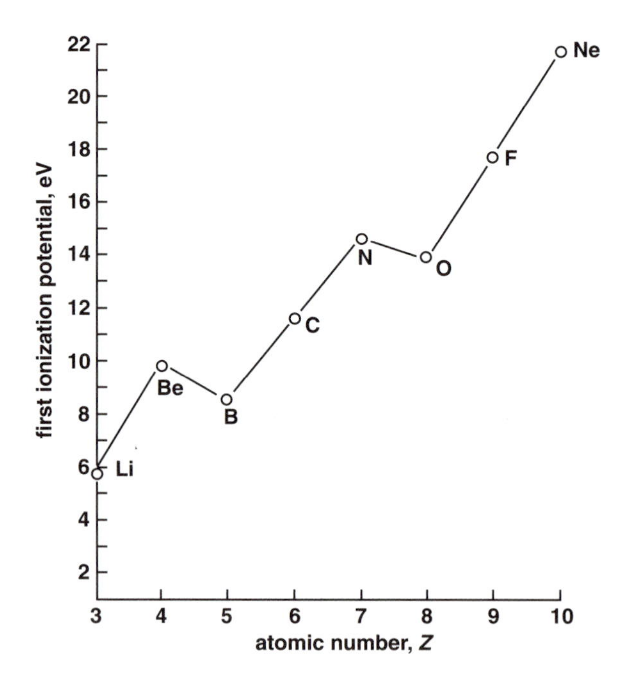

FIG. 8.4 Variations of first ionization potential across the first short Period

The energy gaps between successive levels with the same *l* value decrease as the *n* values increase, so that all the atomic orbitals get closer in energy as the atomic number increases. This trend is not completely regular, and larger than average energy gaps occur between the 4p and 5s levels where the first set of d orbitals has been filled, and between the 6s and 7s levels where the first of the f levels comes. These energy jumps reflect the poorer-than-average shielding powers of d and f electrons.

As the ionization potentials measure the energy required to remove the least tightly bound electron from an atom or ion, values reflect the stability of the configuration from which the electron is being removed. Table 2.8 gives the ionization potentials of the elements.

Apart from the major discontinuities at the rare gases, there is also a gap in energy wherever the outermost electron enters a new atomic orbital. These gaps correspond to stabilization of the filled shell configurations, s^2, $d^{10}s^2$ and $f^{14}d^{10}s^2$, before the p orbitals are occupied, and also suggest the possibility of transfer of s electrons into the d shell to give the d^{10} configuration, or of d electrons into the f shell to give the f^{14} arrangement, which was discussed in the previous section.

The stability of the rare gas configurations can be seen, both from the high energies required to remove an electron from the rare gases themselves, and from the leap in the values of the potential when the rare gas configuration has to be broken (i.e. when the second electron is removed from an alkali metal, the third electron is removed from an alkaline earth, the fourth electron from a boron Group element, etc.). The very low first ionization potentials of the alkali metals, and, to a lesser extent, the low first and second potentials of the alkaline earths, show how loosely held are the first one or two electrons outside the rare gas configuration.

The relative stabilities of filled and half-filled shells show up when the variation of the first ionization potentials across a Period is plotted. Figs 8.4 and 8.5 show such plots across a short and a long Period, respectively.

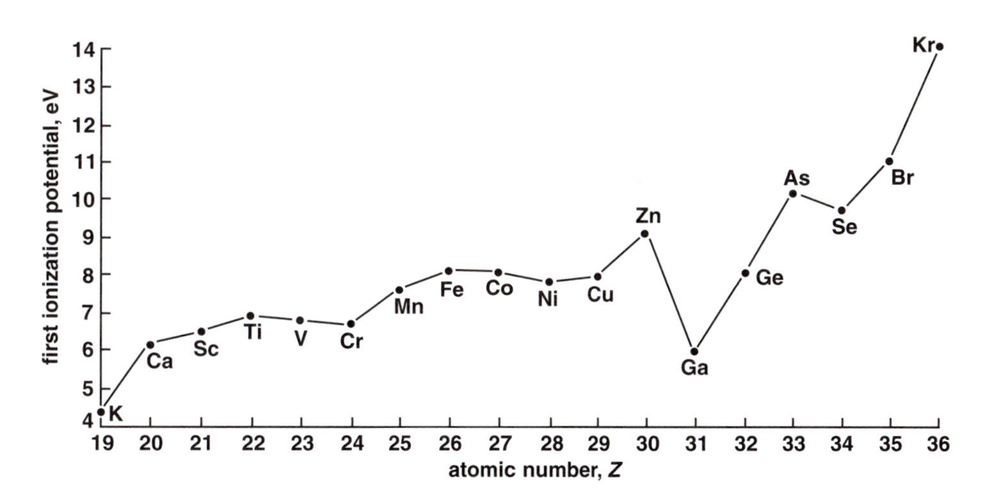

FIG. 8.5 Variation of the first ionization potential across the first long Period

As the atomic size increases within a Group, where the outer electronic configuration is the same, it becomes easier to remove the outermost electrons and this is shown by a decrease in the ionization potentials down a Group. Figure 8.6 shows this for the first ionization potential of the alkali metals and for the sum of the first three potentials of the boron Group. In the latter, the effect of the insertion of the first d shell is shown by the potentials of gallium, and that of the presence of the f series by the value for thallium. These effects are most marked in the boron Group, which follows immediately after the transition series, but they continue to show in the rest of the p block of elements, both in the ionization potentials and in small discontinuities in the chemistry of gallium, germanium, arsenic, selenium and bromine. The complete graph for all the elements is shown in Fig. 8.7. Ionization potentials increase from left to right across a Period and decrease with increasing atomic weight down a Group. The variation for elements filling an s or p level is much greater than that for the d or p elements.

Since the electron affinities measure the energy of the process converse to ionization, the energy of gaining an electron, the stable configurations are similarly reflected in the

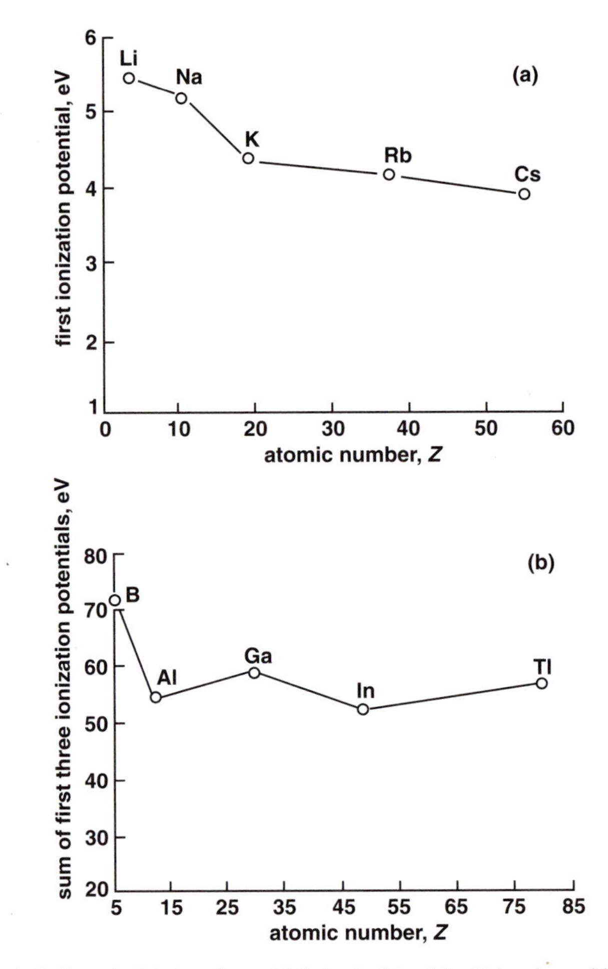

FIG. 8.6 Variation in ionization potentials in a Group: (a) first potentials of the lithium Group; (b) sum of the first three potentials of the boron Group

electron affinities, as the values in Table 2.9 show. Thus the addition of an electron to a halogen atom is exothermic, where the electron fills up to the rare gas configuration, while the addition of an extra electron to the rare gases is an endothermic process. Similarly, for

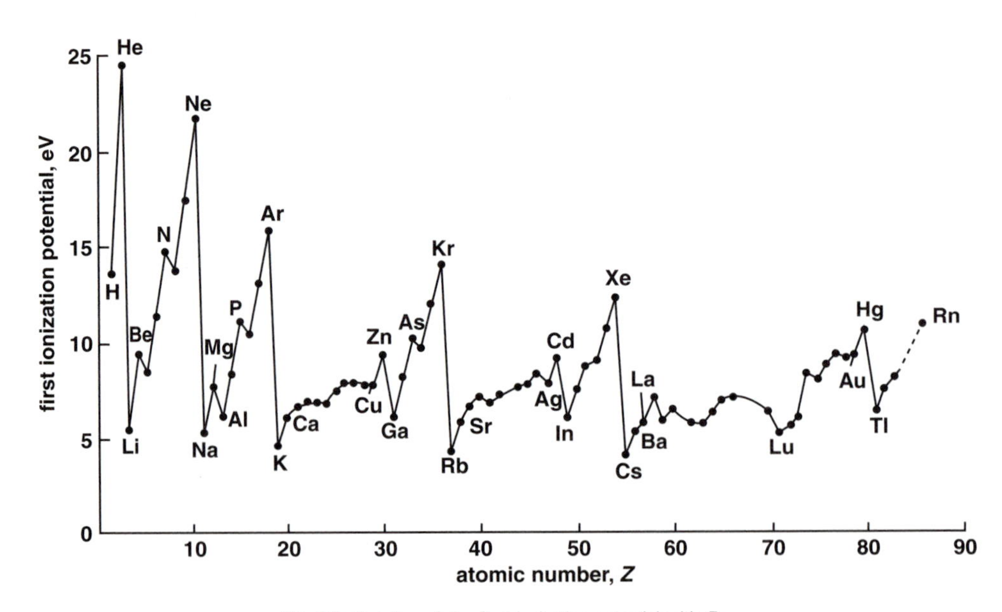

FIG. 8.7 Variation of the first ionization potential with Z

the copper Group a large negative electron affinity marks the tendency to complete the $d^{10}s^2$ configuration. The values for the alkali metals, the chromium Group and the carbon Group show the tendency to attain, respectively, s^2, d^5 and p^3 configurations, while the values for the succeeding groups are markedly lower, showing that the addition of a further electron to the filled or half-filled shell is a much less favoured process. However, less weight can be put on the trends in electron affinities as the values are more tentative than those for ionization potentials.

8.4 Atomic and ionic sizes

The definition and determination of the various sets of atomic and ionic radii is discussed in Section 2.15.

atomic radii in Table 2.10—for predicting covalent bond lengths,
Van der Waals radii in Table 2.11—for predicting non-bonded closest approach distances,
ionic radii in Table 2.12—for predicting distances between ions in solids,
metallic radii in Table 2.13—for predicting distances between atoms in metallic solids.

As discussed in Chapter 2, it is possible to derive an approximately self-consistent set of atomic radii which apply to all the elements and these are shown in Fig. 8.8, plotted against the atomic number. A similar set for real or hypothetical cations and anions with the rare gas structures are shown for the Main Group elements in Fig. 8.9. Figure 8.8 has many features in common with the ionization potential plot of Fig. 8.7. The main discontinuities in size come between the rare gases and the alkali metals where the outermost electron has to enter a completely new quantum shell. There is thus a marked increase in size and a marked decrease in ionization potential at this point. Due to the imperfect shielding of the valency shell electrons by each other, the effective nuclear charge increases, on average, across a Period. The outer electrons become more and more tightly bound and the atomic radius decreases while the ionization potential increases. The discontinuities at the filled shell configurations are particularly clear. The slow changes across the d and f series contrast markedly with the sharp changes in the s and p blocks, and the general decrease in size across the lanthanide series has a noticeable effect in reducing the sizes of the following elements.

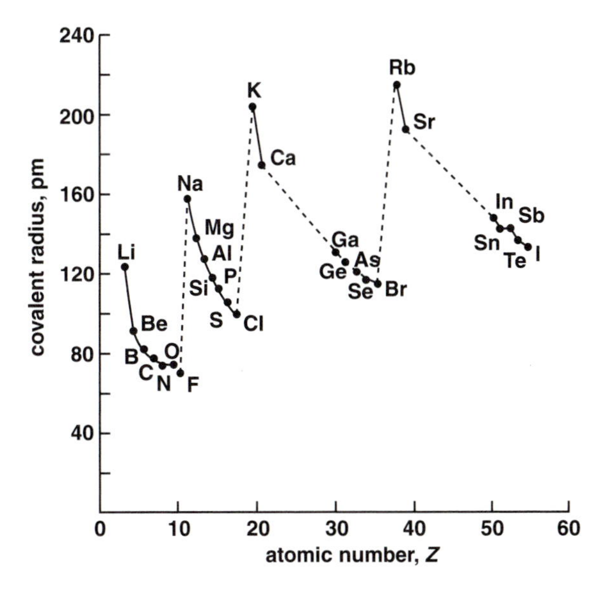

FIG. 8.8 Variation of 'covalent' atomic radii with Z

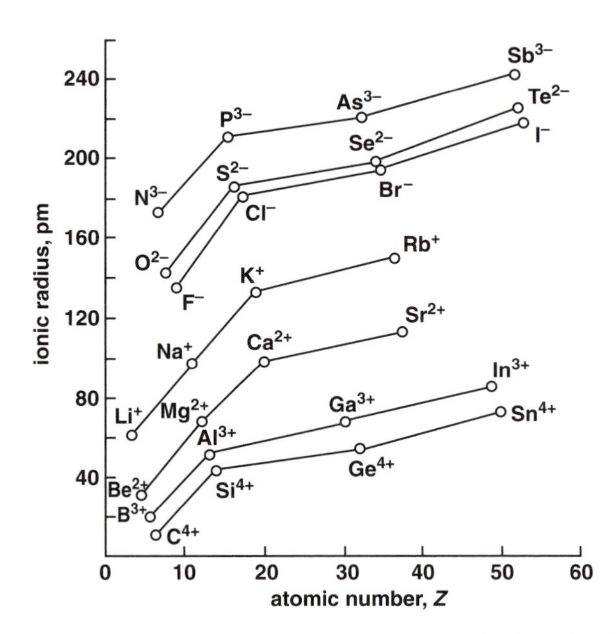

FIG. 8.9 Variation of ionic radii, corresponding to rare gas configurations, with Z, for Main Group elements

The variation in atomic size may be generalized as a decrease on going from left to right across a Period and an increase in going down a Group. These changes are the exact reverse of the ionization potential changes, as would be expected. The changes in ionic radii generally reflect these changes in the atomic sizes.

The parallelism between the changes in radius and ionization potential is not, of course, accidental but follows from the existence and arrangement of the atomic orbitals. Both the size and the ionization potential would be expected to change as the number of electrons in the atom increases, and this change would be discontinuous whenever a new orbital was occupied. The inter-electronic forces, including both the electrostatic repulsions and the exchange forces, modify this pattern of change but leave the main outlines. The change embodied in the effective nuclear charge clearly affects both the extension of the electron cloud and the energy required to remove an outer electron. The property of electronegativity discussed in Section 2.16 is a summarizing parameter which gives effect to the pattern of changes discussed above. It will be seen from the electronegativity values given in Table 2.14 that these increase towards the right of the Periods and decrease down the Groups. In addition, they reflect the other, smaller, variations which have been remarked; for example, the changes in the Main Groups are more pronounced than in the Transition Groups, and the discontinuity in properties of the elements from gallium to bromine when compared with the rest of their respective Groups is reflected in their electronegativity values.

8.5 Chemical behaviour and periodic position

The detailed chemistry of the elements is discussed in the succeeding chapters. In this section, the skeleton of the periodic properties is outlined to provide a framework for the more detailed account which follows.

Those elements where the outermost electrons are in a new quantum level, after a rare gas configuration, normally react by losing these loosely bound electrons and forming cations. This mode of behaviour is typical of the elements of the lithium, beryllium and scandium Groups together with the lanthanide elements, which have the respective valency shell configurations, s^1, s^2 and d^1s^2. All these elements, with the exception of beryllium itself, lose these outer electrons completely with the formation of cations; M^+ in the lithium Group, M^{2+} in the beryllium Group, and M^{3+} for scandium, yttrium and the lanthanides.

The elements of the boron, carbon, nitrogen, oxygen, and fluorine Groups, where the outermost electrons are in p orbitals, show more complicated behaviour.

(a) Elements with electron configurations close to the rare gases can acquire electrons to form anions with complete rare gas shells. Thus the elements of the halogen Group all form X^- ions, and we also find stable compounds containing O^{2-}, S^{2-} and N^{3-} anions.
(b) In covalent compounds, these p elements show a maximum oxidation state equal to the sum of the s and p electrons (the Group Oxidation State) and the other relatively stable oxidation states differ from the Group state by multiples of two. Thus the boron Group elements, with the configuration s^2p^1, show the Group oxidation state of III and the other stable state in this Group is I; the halogens, s^2p^5, show the Group state of VII and also the states V, III, I, and –I.

The Group oxidation state is the most stable one for the lighter elements, especially of the earlier Groups, and a state two less than the Group state becomes the most stable for the heavier elements. Thus boron and carbon are stable in the III and IV states respectively, while their heaviest congeners are stable in the I and II states. In the Groups to the right of the Periodic Table, where a larger variety of oxidation states is possible, the picture is more complex, compare Section 17.1.

The elements where the d orbitals are filling have a Group oxidation state equal to the sum of the s and d electrons. This state is shown in the earlier Groups but involves too many electrons in the later Groups of the transition series where only low states occur. The highest oxidation state shown in the Periodic Table is VIII, by ruthenium, osmium and xenon. The common oxidation states of the d elements vary in single steps, instead of the double ones shown by the p elements. Thus in the Group with the same number of valency electrons as the halogens, the manganese Group d^5s^2, the oxidation states found are VII, VI, V, IV, III, II, I and 0. Another distinction between the behaviour of a d Group and a p Group is that the heavier elements of a d Group are more stable in the higher oxidation states, in contrast to the trend in the p Group. Thus, the most stable state of manganese is II while its heavier congener, rhenium, is stable in the IV and VII states. The lower oxidation states of the transition metals, particularly the II state, often occur as cations, while the higher states are commonly bound to oxygen or to the halogens in covalent molecules or anions.

The f elements of the lanthanide series show only one stable oxidation state, the III state. This corresponds to the configuration $4f^1$ for cerium(III) and to $4f^n$ for the other elements up to $4f^{14}$ in lutetium(III). As already noted, three electrons are lost, as if the configuration was $4f^n5d^16s^2$, despite the fact that most of the elements do not have the d electron in the ground state. One or two of these elements do show fairly stable oxidation states other than the III state, and most of these correspond to the f^0, f^7 or f^{14} configurations. Thus, cerium shows a IV state corresponding to f^0, europium has a II state and terbium a IV state, corresponding to f^7; while ytterbium has a II state corresponding to f^{14}.

In the heaviest elements, where the 5f and 6d levels are filling, the pattern of oxidation states is less simple than with the lanthanides. As these two energy levels are very close, the earlier elements show a considerable variety of oxidation states with the maximum rising from III for actinium to VI at uranium and VII at neptunium and plutonium. The later actinide elements resemble the lanthanides more closely, and the III state becomes the most stable one at about curium (see Jørgensen in the references).

These patterns of behaviour lead to the division of the Periodic Table, Fig. 8.10, into four major blocks: the s elements, the p elements, the d elements and the f elements, together with a number of Groups which serve to bridge these divisions. The s elements are those of the lithium and beryllium Groups; the boron, carbon, nitrogen, oxygen and fluorine Groups make up the p block; the lanthanides and actinides form the f block; and the remaining transition elements, the d block.

As the typical behaviour of d elements depends on the presence of both d electrons and available d orbitals, the scandium Group (which always loses its solitary d electron and forms the M^{3+} ion) and the zinc Group (which always preserves the filled d^{10} configuration) are not characteristic d elements and are best regarded as bridging Groups. The scandium

Group links the s elements and the d block, while the zinc Group links the d block to the p elements, and both Groups show the appropriate intermediate properties. The chemistry of the scandium Group links strongly, also, with that of the f elements. Indeed, the general chemistry of the lanthanides in the III state is almost identical with that of yttrium and lanthanum. There are bigger differences between the chemistry of actinium and the actinide elements. It is convenient to treat scandium, yttrium, lanthanum, actinium and the lanthanides all together, and to treat the actinides independently.

The remaining Group in the Periodic Table is the helium Group. This Group forms the division between the p block and the s block, and the recently discovered chemistry of xenon shows strong links with that of iodine. Finally, there is the lightest element, hydrogen, which falls into no Group so far discussed and is best regarded as a unique introductory element to the Periodic Table.

SYSTEMATIC CHEMISTRY CHAPTERS AND THE PERIODIC TABLE

The arrangement of the following chapters reflects this division of the Periodic Table. The chemistry of hydrogen and its compounds is treated first in Chapter 9, giving a microcosm of the properties of the elements. Then follow Chapters 10 to 12 covering, respectively, the s elements, the scandium Group and the lanthanides, and the actinides. The d elements are treated systematically in Chapters 14 and 15 following an overview in Chapter 13. Finally, the chemistry of the p elements is covered in Chapter 17. The other Chapters treat selected topics of great current interest.

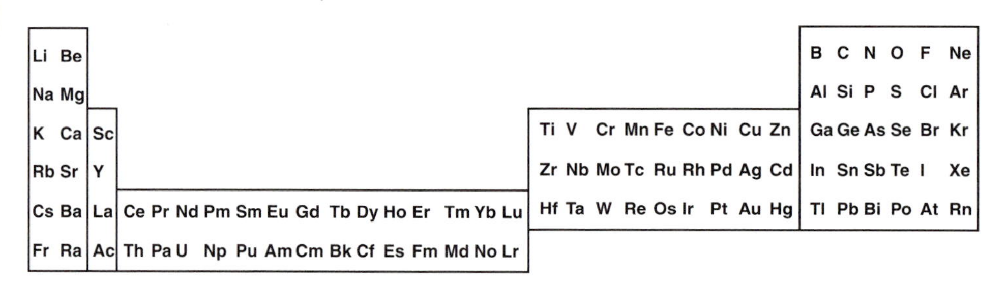

FIG. 8.10 Divisions of the Periodic Table

8.6 Methods of showing the stabilities of oxidation states

In many cases, as implied in the last section, elements show a number of oxidation states and some method of determining and portraying the relative stabilities of these states is necessary. Similarly, it is useful to be able to compare relative stabilities and show the trends in stabilities among related elements. Of course, in a full and complete description of the chemistry of an element, these stabilities are clear from the range of compounds of a given oxidation state and their ease of formation and decomposition. However, it is impossible to give a complete description of the known chemistry of any element within the space available in a general textbook so that methods of summarizing and illustrating the general behaviour are required. Stabilities vary a good deal with the chemical environment—the temperature, solid, liquid or gaseous state of the compound, solvent, presence of air or moisture, and so forth. However, there are two chemical states which are very common, as a solid and in solution in water.

The stability of an element in a particular oxidation state in the solid is qualitatively indicated by the variety of ions or ligands with which it reacts to form solid compounds. Thus, a strongly oxidizing state will form compounds with nonoxidizable ligands only, and *vice versa* for a reducing state, while a stable state will give compounds with a wide variety of ligands. For example, consider the relative stabilities of the II and III states of iron, cobalt and nickel as shown by the existence of the solid compounds of the II state with oxychloride anions.

Fe(II) $Fe(ClO_4)_2$: ClO_3^- and ClO_2^- oxidize to Fe(III)

Co(II) $Co(ClO_4)_2$ and $Co(ClO_3)_2$: ClO_2^- oxidizes to Co(III)

Ni(II) $Ni(ClO_4)_2$, $Ni(ClO_3)_2$ and $Ni(ClO_2)_2$

The oxidizing power of the oxychloride ions increases in the order, perchlorate < chlorate < chlorite, so the existence of the compounds shown above illustrate that the order of stabilities of the II state is Fe < Co < Ni.

A convenient set of ligands, for the purpose of demonstrating stabilities, is provided by the halides which form compounds with almost all oxidation states. An element in a stable oxidation state will form all four halides, a strongly reducing state will tend not to have a fluoride, while a strongly oxidizing state will tend not to have an iodide or bromide. Similarly, an oxidizing state will show an oxide but no sulfide, while a reducing state will form a sulfide but no oxide. Thus, the existence and stabilities of the oxides, sulfides and halides of the elements in their different states provides a useful general guide to the stabilities, in the solid state, of the various oxidation states. In the following chapters, Tables 13.3, 13.4 and 13.5, of transition element halides and oxides (sulfides are omitted here as their stoichiometry is often in doubt) and Table 17.2, of p block element oxides, sulfides and halides, are used to provide a general, overall view of the chemistry of these elements.

Stabilities in aqueous solution are expected to be broadly similar to stabilities in the solid state, but to differ in detail due to differences between lattice energies and hydration energies (compare Chapters 5 and 6). The relative stabilities of the various oxidation states of an element in solution are given by the free energy changes of the set of half-reactions connecting each pair of oxidation states. For example, the stabilities of the states of copper depend on the free energy changes of the half-reactions:

$$Cu^{2+} + e^- = Cu^+$$
$$Cu^+ + e^- = Cu$$
$$\text{and } Cu^{2+} + 2e^- = Cu$$

(These three free energies are not independent: any one may be derived from the other two.)

Such free energies are related to the corresponding redox potentials, since $-\Delta G = nFE$, see Section 6.3. A full list of redox potentials for the half-reactions of all the elements is available, but a method is required for displaying these values to the best advantage. It was suggested by Ebsworth (see references) that free energies may be usefully displayed graphically. In this, the oxidation states of the element are plotted against the free energy change, in one electron steps. The method is most readily discussed in terms of particular cases, for example uranium and americium whose potentials in acid solution have the values shown below:

	E^0 (volts)	
	$M = U$	$M = Am$
$MO_2^{2+} + e^- = MO_2^+$	0.05	1.64
$MO_2^{2+} + 4H^+ + 2e^- = M^{4+} + 2H_2O$	0.33	
$MO_2^+ + 4H^+ + e^- = M^{4+} + 2H_2O$	0.62	1.26
$M^{4+} + e^- = M^{3+}$	−0.61	2.18
$M^{3+} + 3e^- = M$	−1.80	−2.32
$MO_2^{2+} + 4H^+ + 3e^- = M^{3+} + 2H_2O$		1.69

The diagrams, Fig. 8.11, are plotted by taking the value for the element itself as zero and plotting the free energy changes for the half-reactions against the oxidation states of the element. The free energies are given as $-\Delta G^0/F = nE^0$ where n is the number of electrons involved in the change.

Using americium as our example the diagram is constructed as follows (see Fig. 8.11).

(1) Take the value for the element as zero.

(2) For the change

$$Am^{3+} + 3e^- = Am$$

E^0 is −2.32 V, therefore nE^0 is −6.96 V and is so plotted.

(3) For the change

$$Am^{4+} + e^- = Am^{3+}$$

E^0 is 2.18 V, therefore nE^0 is 2.18 V and this is added to −6.96 V to give −4.78 V, which is plotted for Am(IV).

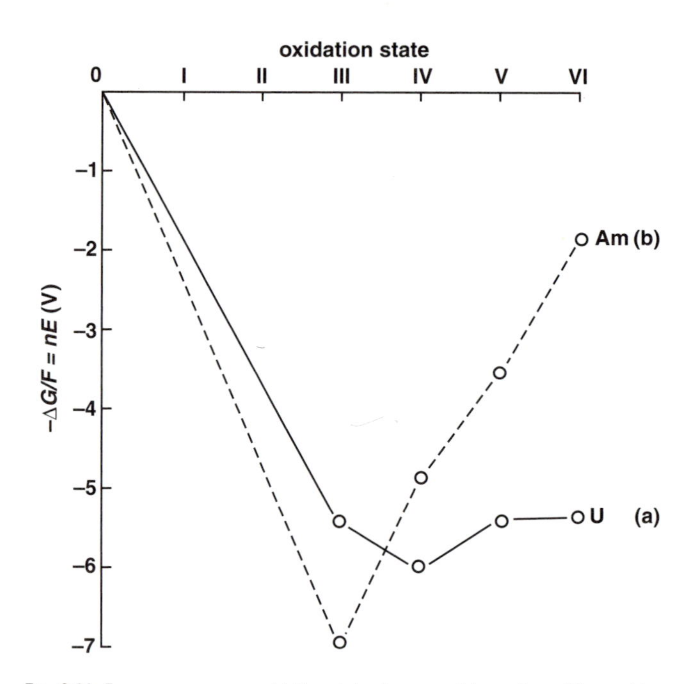

FIG. 8.11 Free energy versus oxidation state diagrams: (a) uranium; (b) americium

(4) For the change

$$\mathrm{Am(V)} \rightarrow \mathrm{Am(IV)} \quad (\mathrm{AmO_2^+} \rightarrow \mathrm{Am^{4+}})$$

E^0 is 1.26 V, and nE^0 is 1.26 V. Adding this to −4.78 V gives −3.52 V to be plotted for Am(V).
(5) For Am(VI), continuing in the same way one obtains −1.88 V.

Plots like these provide a kind of cross-section of the chemistry of the elements. The lower a state lies on the diagram, the more stable it is. Also, since a change from one state to another which involves moving down a slope involves a negative change of free energy, it is thermodynamically favourable. Similarly, a change which involves an 'uphill' move is unfavourable. These points about changes apply to those involving any two states, not only states which are nearest neighbours. Thus the changes from U(0) to any of the states U(III), U(IV), U(V) or U(VI) are favourable, and the changes from U(IV) to any of the states U(0), U(III), U(V) and U(VI) are unfavourable. Clearly, the greater the slope, the greater the driving force, so that Am(0) goes to Am(III) more readily than to Am(IV).

A particular oxidation state may be represented on such a diagram by a point which is (a) a minimum, (b) a maximum, (c) a concave point (i.e. one lying below the line joining the two neighbouring points), (d) a convex point (i.e. one lying above the line joining its neighbours), or (e) a 'linear' point (i.e. one lying on the line joining its neighbours). Table 8.3 lists the stability properties of the oxidation states represented by these different types.

The properties of the states represented by concave, convex and linear points are not restricted to being with respect to the two nearest neighbour states but may hold with respect to any two states. For example, in Fig. 8.11, Am(V) is linear, not only with respect to Am(IV) and Am(VI), but also with respect to Am(III) and Am(VI) and the disproportionation products will include Am(III). In a similar way, in the case of phosphorus (Figure 17.27) the points representing the 0, I and III states are all convex with respect to the −III and V states. That is, the element (0 state), hypophosphorous acid (I state) and phosphorous acid (III state) all disproportionate to phosphine, PH_3, (−III) and phosphoric acid (V state).

Applying these results to uranium and americium, Fig. 8.11 shows, at a glance, that uranium(V) is unstable in aqueous solution, uranium(IV) is stable, and uranium(III) is relatively stable. Uranium(VI) is also stable, with the U(VI) → U(IV) change being mildly oxidizing. By contrast, americium(VI) is unstable and the Am(VI) → Am(III) change is strongly oxidizing. The most stable state of americium is Am(III) and americium(IV) and (V) both disproportionate, with the V state a little more stable than the IV state.

TABLE 8.3 Types of oxidation state in free energy diagram

State represented by a point	*Example from Fig. 8.12*	*Further examples (figure numbers in brackets)*	*Properties of the oxidation state*
Minimum (a)	U(IV), Am(III)	Mn(II) (14.21) Cr(III) (14.16)	Stable relative to neighbouring states.
Maximum (b)	—	N(−I) (17.25)	Unstable relative to neighbouring states.
Concave (c)	U(III)	V(IV) (14.8) Re(IV) (15.20)	Relatively stable with respect to disproportionation into the neighbouring states.
Convex (d)	U(V)	Mn(III) (14.21) Re(III) (15.20)	Relatively unstable with respect to disproportionation.
Linear (e)	Am(V) and (IV)	All the intermediate states of Cl (17.51)	*Intermediate with respect to disproportionation.

*That is to say that, at equilibrium, there are approximately equal amounts of the original state and the states to which it has disproportionated.

The elements lying between uranium and americium—neptunium and plutonium—also show III, IV, V and VI states and the VI state becomes more oxidizing and less stable in order from uranium, through neptunium and plutonium, to americium. This is clear from Fig. 12.1, in which the curves for all four elements are plotted. The VI state of uranium lies lowest and the others in the order U<Np<Pu<Am. In a similar way, the relative stabilities of the II and III states of iron, cobalt and nickel, referred to above, are clear from Fig. 14.1.

These free energy diagrams will be extensively used in the following chapters but one or two reservations about them must be kept in mind. First, all the oxidation potential data are not of equal validity and some values may be in error so it is not wise to make too much of small effects—such as whether a point is slightly convex or slightly concave. It is unlikely, however, that there are any gross inaccuracies. The second point is that the properties listed in Table 8.3 are thermodynamic and there is no information about the rates of reactions. Thus a state may be thermodynamically unstable but persist in solution because its rate of reaction or disproportionation is slow. Similarly, the potential data apply only to systems in equilibrium, and the rate of attaining equilibrium may be slow. This applies particularly to systems involving solids—for example, an element. Many elements which are strongly reducing react only very slowly due to surface effects and the like. Throughout the later chapters, curves are plotted for potential data in acid solution, at a pH of 0. Similar data are available for alkaline solution, and a set of equivalent curves could be drawn for such media. These are less useful as many more states appear as solids (as hydroxides or hydrated oxides) in alkali.

The occurrence and stability of the solid halides, oxides and sulfides of the elements, taken together with the free energy diagrams linking the different oxidation states, gives an adequate guide to the general chemistry of an element in its common compounds. In general, a stable state may be assumed to form compounds with all the common anions and ligands—all the oxyanions, pseudohalides, organic acid anions, hydride, nitride, carbide, amide, sulfur anions and so on. Unstable states will form a much more limited set of compounds, down to those states represented by only one or two examples. A state which is unstable and oxidizing will form no compounds with oxidizable ligands such as nitrite or organic groups, and similarly, a reducing state will form no compounds with oxidizing ligands, as with the chlorite and chlorate compounds of iron mentioned above. By avoiding cataloguing such compounds of the elements, space is preserved for the mention of the more unusual compounds formed and as many of these as possible have been discussed in the later chapters which may be regarded as an introduction and supplement to the systematic chemistry given in the major inorganic textbooks which are listed in the reading lists.

8.7 The abundance and occurrence of the elements

One of the most satisfying scientific constructs of the second half of the 20th century was the picture of the genesis of the elements. Once 19th-century chemists had discovered that each element emitted a distinctive spectrum when heated, the way was open to identify elements in the stars by their spectral lines. In this way, helium (from the Greek *helios*, sun) was found in the solar spectrum about a decade before it was discovered on Earth. Such studies led to the evaluation of the abundance of the different elements in the Universe and a striking picture emerged. Hydrogen and helium together account for about 99% of the mass of the Universe and over 99.9% of the atoms. At the other extreme, the heaviest elements are present to the extent of about 10^{-12} of that of hydrogen. The distribution of abundances has the following features:

(a) The abundance of the elements declines steeply and exponentially from H to a mass about 100, then more linearly to U

(b) Within this broad pattern, Li, Be and B are markedly scarce relative to their neighbours, by a factor of 10^{10} for Li and B, and 10^{12} for Be.

(c) Less pronounced relative deficiencies, by factors between 10 and 1000, are found for elements around F, Sc, As, In and Ta

(d) There is a marked excess of around 10^3, over the trend line, for elements near Fe, and less marked excesses near Zr, Xe and Pt

(e) Finally there is a very pronounced alternation in abundances between successive elements

Table 8.4 gives some general values.

These abundance figures accumulated over more than a century of observation and analysis. They have been explained in considerable detail by a theory based on the genesis of the elements from hydrogen in the stars. This was refined by many workers over some 30 years up to the publication of the definitive review by Burbidge, Fowler and Hoyle in 1957. More recent developments may be found in Fowler's Nobel lecture in 1984. The picture is extraordinarily satisfying as the model not only accounts for the general pattern of abundances, but explains most of the fine detail. The main features are:

(1) An initial primitive universe with about 75% H and 25% He.

(2) Condensation by gravity into early stars where H nuclei fused into He, and further fusion processes built up heavier elements. Eventually stellar condensation led to

TABLE 8.4 Cosmic abundances of the elements

Element	*Relative abundances*			
	By number	*% of total*	*By weight*	*% of total*
Hydrogen	4.0×10^{10}	92.8	4.0×10^{10}	75.5
Helium	3.1×10^{9}	7.1	1.2×10^{10}	23.1
Li, Be, B	1.4×10^{2}	3.3×10^{-7}	1.3×10^{3}	2.4×10^{-6}
C, N, O, Ne	4.0×10^{7}	0.09	6.5×10^{8}	1.2
Na to Sc	2.7×10^{6}	0.006	7.3×10^{7}	0.13
Fe group (a)	6.4×10^{5}	0.001 5	3.6×10^{7}	0.07
Middle group (b)	1.1×10^{3}	2.6×10^{-6}	7.7×10^{4}	1.4×10^{-6}
Heavy group (c)	28	6.5×10^{-8}	4.6×10^{3}	8.6×10^{-8}

(a) A_r from 50 to 62
(b) A_r from 63 to 100
(c) A_r over 100

destruction of the star and scattering of the elements to be re-formed into further stars and the process continued.

(3) Eventually, the Sun and our planets condensed out of an accumulation of such cosmic dust. The Sun has proceeded along the H-fusion path while the cooling and evolution of the Earth has produced the crystal distribution of elements of the present day.

It is the details of the fusion processes which account for the present abundances. Briefly, the first major process is the formation of He by a number of steps from 4 H nuclei. This evolves huge amounts of energy and is the major source of stellar radiation. Then follows further fusion of He nuclei (mass 4) into heavier elements. Such a process bypasses Li (isotopes of masses 6 and 7) and B (10 and 11) but passes through the Be isotope of mass 8. This is extremely unstable (half-life about 10^{-16} s) but survives long enough to fuse with a further He to give C (mass 12). Similar processes with alpha particles add further mass in steps of 4 to give O (16), Ne (20), Mg (24), Si (28), S (32), and so on up to Ti (48). The actual abundance of these elements represents the balance between their ease of formation and their stability to further reaction. As the number of steps increases the abundance falls, since each element depends on the previous formation of its precursor. Thus the general exponential fall-off in abundances is to be expected. All the elements of intermediate masses form from intermediate and usually less favoured processes, and the abundances are now understood in detail in terms of the energies of formation, and the liability to further reaction.

Different processes occur in stars of different masses and temperatures, and at different times in the course of their evolution. Thus the Sun is in the basic H-burning stage, as are the majority of stars in the galaxy.

Nuclear configurations around Fe are the most stable, with lighter elements combining exothermically towards mass 56, and heavy elements breaking down, again by exothermic processes, towards this mass range. Under conditions of very high temperature and pressure, an equilibration process takes place to give the relatively high ratios of the elements around this stable minimum.

Building nuclei of greater mass than Fe involves neutron capture, and two processes occur. In the slow process, which occurs over a period of years, there is sufficient time after the capture for the nucleus to rearrange, by electron emission, to give the most stable proton/neutron ratio. In the fast process, typically in a supernova explosion, several neutrons are captured in a very short period before nuclear reorganization occurs. These two processes thus give rise to the heavier isotopes, and, with minor processes like proton capture, explain the distribution features in detail.

The above is only a brief outline of a very full theory. Further refinements continue—for example an explanation of an unexpectedly high ratio of ^{11}B to ^{10}B in meteorities in 1995. We have sketched the theory here as it is too little known that we can not only reach out into the Universe and assess the abundance of the elements in distant stars, but we can also account for the observed abundances in terms of quantitatively known processes. While such cosmic, nuclear, processes are not themselves included in inorganic chemistry, they do create our starting materials, the elements.

The elements are now known to form simple combinations in interstellar space. In dark regions of the Universe, matter is present which absorbs starlight, though the concentration is so minute—of the order of atoms per cubic metre—that it is a million million million times less than in an ultrahigh vacuum on Earth. Atoms do link into units which may be detected spectroscopically. Many of the species are radicals—it takes a long time for an OH unit to pick up another H atom, for example. Diatomic units involving the more abundant elements (Table 8.4) have all been identified, and multiatom units with as many as ten atoms are suggested. Entities involve H with C and O, and also the less abundant N and even S. We cannot further pursue this fascinating area of cosmochemistry here, but as more information is collected we can foresee the development of an interstellar inorganic chemistry!

The element abundances in the universe have a broad relation to the pattern of abundances in the Earth's crust, but further processes of loss (especially of H and He) and of concentration have taken place so that there are major differences in detail. The distribution of the elements on Earth is now reasonably well evaluated (and we have preliminary figures for the Moon and Mars), though further refinement will occur, especially as we penetrate deeper into the crust. A good start has been made in understanding the geochemical processes which have collected and redistributed the elements into the rocks, minerals and ores which we find today. A start has even been made in reproducing on a laboratory scale the enormous pressures and temperatures of the mantle processes which form the igneous rocks. Table 8.5 shows estimates of the abundances of the commoner elements in the surface crust and in the whole Earth. The latter figures, especially the high values for Fe, Mg and S, broadly reflect the cosmic abundances given that substantial proportions of the lighter elements would be lost as gases. The surface figures are broadly rationalized by seeing the rocks as SiO_2 with some Si replaced by Al and the Na, K, Mg and Ca incorporated to balance the charge—compare Section 18.6. Abundances of the remaining elements range from a few hundred parts per million down to three parts in 10^{-16} for actinium.

It is worth remarking on another interface zone—that between the atmosphere and space. Here the abundance of molecules is extremely low, the radiation level from the Sun is very high, and the component elements are those of the atmosphere. Thus we find relatively high concentrations of species which are too reactive to accumulate near the surface, such as radicals like OH and OOH or molecules such as ozone and hydrogen peroxide, O_3 and H_2O_2. Ozone shields against hard ultraviolet radiation from the Sun and its loss due to pollution is a major problem. This and other pollution problems, such as the greenhouse effect, have focused attention on the chemistry of the upper and lower atmosphere. A more detailed discussion is given in Section 20.3.

The abundant elements are readily accessible, as are those elements which, though rare overall, occur in localized concentrated deposits. One example is boron, which occurs to the extent of only three parts per million in the crust but is found in concentrated deposits as borax. Also accessible are those elements which are found native or are readily recovered from their ores, for example the precious metals silver and gold. All the chemistry of common, or readily accessible, elements is well explored.

A second group includes all those relatively rare elements which occur only in small proportions in the crust and are found only as trace constituents in the ores of more important minerals. We should also include the more expensive of the precious metals such as gold and platinum. Such elements form an intermediate group where intensive study has been more recent. Although all these elements are now well understood, interest often reflects their possible applications. For example, germanium, gallium and indium have attracted attention for their semiconductor applications, while the catalytic properties of their organometallic compounds has focused attention on the rarer heavy transition metals. Such studies have been greatly helped by the development of large scale chromatographic and other methods to separate very similar elements, such as hafnium from zirconium or the lanthanides from each other.

THE COMPOSITION OF THE ATMOSPHERE

The gases of the atmosphere make up a relatively small proportion of the total earth mass, but are important in climate and in determining the form of minerals at the surface (e.g. oxides, hydroxides, carbonates). Of a total mass of 5.14×10^{15} tonnes, about 1.7×10^{13} tonnes is water. As a percentage of dry air, the significant components are

N_2	78.08
O_2	20.95
Ar	0.93
CO_2	0.032
Ne	1.82×10^{-3}

He, Kr, CH_4 are each about 10^{-4}
Xe, H_2, N_2O and CO are each about 10^{-5}
NH_3, NO_2, SO_2, H_2S and O_3 occur in smaller amounts down to 10^{-9}.

TABLE 8.5 The most abundant Earth elements (atoms percent)

Element	*In the crust*	*Whole Earth*
Oxygen	60	50
Silicon	20	14
Aluminium	6	1
Hydrogen	3	0.1
Sodium	2.5	0.5
Calcium	2	0.5
Iron	2	17
Magnesium	1.7	14
Potassium	1.3	0.05
Titanium	0.2	0.04
Sulfur	0.04	1.7

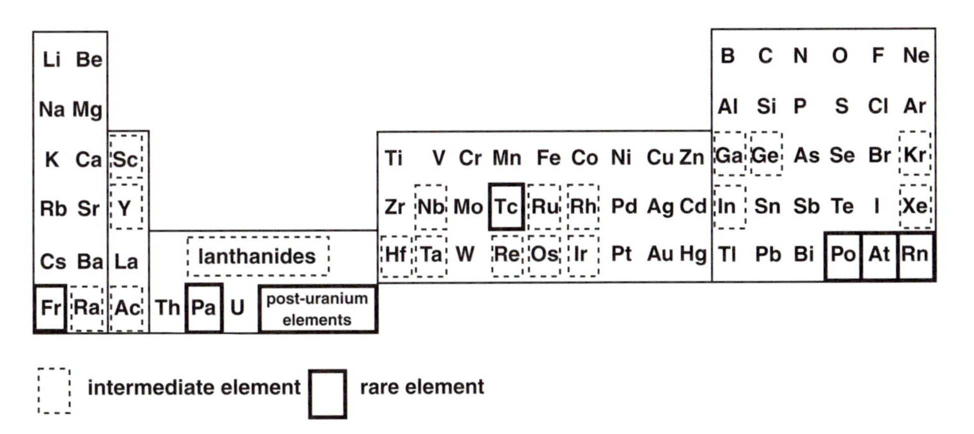

FIG. 8.12 Distribution of the elements. The elements are divided into three classes, common, intermediate and rare, on the basis of their natural abundance combined with their accessibility. The boundaries between types are necessarily somewhat arbitrary

The rarest elements are the artificial elements which have no naturally occurring isotopes. These include all the post-uranium elements and a few lighter ones such as promethium and technetium. Supplies of many of these elements have become available recently from the fission products or synthesis products of nuclear reactors, while the attempt to synthesize ever heavier elements continues. Elements up to 114 have been announced recently, with the identification based on only a few atoms. Figure 8.12 gives a broad picture of element accessbility.

8.8 The extraction of the elements

The extraction processes for producing elements from their ores may be divided into three classes in order of increasing power:

(i) mechanical separation and simple heat treatment,
(ii) separations involving chemical reduction,
(iii) separations by electrolytic reduction.

Although the days of the gold rush may be gone, mechanical separation on a huge scale is still the basic process of gold and diamond production, the final recovery of gold being by chemical reduction with zinc of a cyanide complex in solution. Also included in the first class of mild treatments are the recoveries, by distillation or thermal decomposition, of elements such as zinc or mercury. A further reaction of wide application in this class is the Van Arkel and De Boer process which is used to produce very pure metals on a small scale. In this process, elements which form volatile iodides are purified by a cyclic process in which the iodide is formed at a low temperature and decomposed on a heated wire, at a higher temperature, to the element and iodine. The iodine is recycled to form more iodide. For example, zirconium gives ZrI_4 when heated at 600°C with iodine and this may be decomposed at 1800°C, on a heated tungsten or zirconium filament, to zirconium and iodine.

Most commercial separations fall into the second class, commonly involving reduction by carbon, as in iron production. Reduction by other elements is also found on a small scale: examples include the preparation of pure molybdenum by hydrogen reduction and the reduction of titanium tetrachloride by magnesium in the Kroll process.

Electrolytic reduction represents the most powerful method available, but it is expensive compared to the chemical methods and is only used either for very reactive metals, such as magnesium or aluminium, or for the production of samples of high purity as in the electrolytic refining of copper (which has the additional advantage of allowing the recovery of valuable minor contaminants such as silver and gold). The main commercial application of electrolysis is, of course, in aluminium manufacture. Here, the ore bauxite, which is impure Al_2O_3, is purified by alkaline treatment, then dissolved in molten cryolite (Na_3AlF_6) and reduced electrolytically in this fused-salt system. The alkali metals and calcium and magnesium are also produced by electrolysis in a fused salt melt, while copper and zinc are among those elements recovered by electrolysis in an aqueous medium.

This is also true of their occurrence. Elements of similar chemistry would have behaved similarly when the rocks crystallized out of the primitive magma, and later processes, such as leaching out of soluble salts, would further tend to concentrate similar elements in similar forms.

The type of process chosen for any one element is a complex function of the chemical properties, nature of the ore and relative economics, and many elements are processed in different ways in different parts of the world. Nevertheless, Fig. 8.13 does give a valid broad picture of the occurrence and method of recovery of the elements. As the extraction of the elements reflects their general chemistry, elements which are close in the Periodic Table are treated similarly.

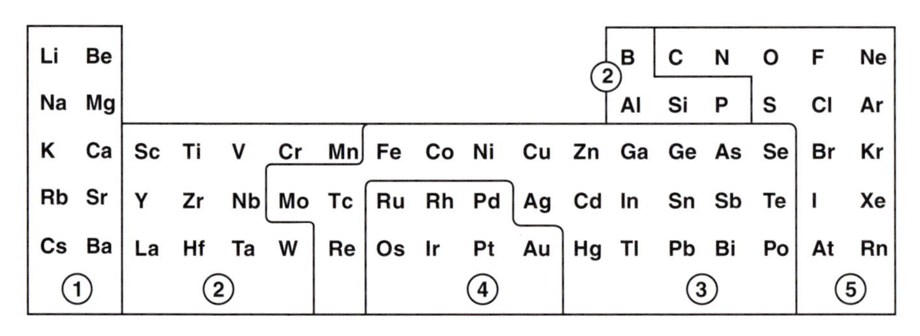

FIG. 8.13 Methods used for the extraction of the elements. (1) Reactive metals extracted electrolytically in a nonaqueous system. Main sources of alkaline earth minerals are insoluble sulfates and carbonates, while alkali metals often form deposits of soluble salts, e.g. NaCl or KCl; (2) Reactive metals of high charge with strong affinity for oxygen and occurring as oxyanions or double oxides. Separation is by electrolytic or chemical reduction, especially by active metal replacement; (3) Elements occurring in sulfide ores or otherwise associated with sulfur. Extracted usually by roasting to oxide and then reducing or treating thermally; (4) Elements occur native or in easily decomposed compounds yielding to thermal treatment; (5) Nonmetals which occur free, in the atmosphere, or as anions. Ag, Hg and also Zn, can be classified equally well as class 3 or class 4

The process of extraction by carbon reduction illustrates several interesting general points in the chemistry of metal recovery and is worth discussing in fuller detail. The extent of the reduction of one element from its oxide by a second element depends on the difference in free energy of the two oxidation reactions of the type:

$$M + \frac{x}{2}O_2 \rightarrow MO_x \tag{8.1}$$

It will be recalled that reactions which evolve free energy tend to occur spontaneously, so that the equilibrium between two elements and their oxides:

$$M'O + M'' \rightleftharpoons M''O + M' \tag{8.2}$$

(or the corresponding equations for different oxide stoichiometries), will favour that oxide whose free energy of formation (Equation 8.1) is most negative. The free energy change, ΔG, is separable into two components, the heat change ΔH, and the energy involved in the change of entropy, $T\Delta S$, where T is the absolute temperature:

$$\Delta G = \Delta H - T\Delta S$$

In the formation of a metal oxide according to Equation (8.1), the heat change is usually favourable but, as the reaction uses up a gaseous component (the oxygen) which has a relatively large entropy, the entropy term is unfavourable and this energy increases with increasing T. As a result, the free energy change for metal oxide formation in Equation (8.1) falls off with rising temperature in a broadly similar way for any metal, as in the examples shown in Fig. 8.14. It will be seen that the metal oxides can be divided into two classes: first those which are intrinsically unstable at normal temperatures, such as gold oxide, or at accessible temperatures such as silver oxide or mercury oxide, and the second class which contains the elements with a favourable free energy of oxide formation at any accessible temperature. Gold, silver and mercury, fall into the class (i) (Page 203) of elements which can be extracted by simple heat treatment alone, while the second class of elements contains those whose extraction process falls into class (ii) or (iii), requiring reduction. Any metal will reduce the oxide of any second metal which lies above it in Fig. 8.14, according to Equation (8.2), as the net change in the free energy will be negative (i.e. favourable) by an amount equal to the difference between the two curves at the appropriate temperature. For example, magnesium will reduce all the other oxides shown in Fig. 8.14.

The formation of a metal oxide thus involves a free energy change which becomes less negative with increasing temperature, because the oxidation proceeds with the consumption of a gaseous component with a corresponding loss of entropy. The formation of carbon oxides is quite different. The main reaction between carbon and oxygen at higher temperatures is:

$$2C + O_2 \rightarrow 2CO$$

in which an excess of one mole of gas is produced for each mole of oxygen used. This reaction therefore involves a positive entropy change and its free energy becomes more negative as the temperature rises. At temperatures below 500°C, the main reaction is:

$$C + O_2 \rightarrow CO_2$$

where there is no overall change in the amount of gaseous reactant. The entropy change is thus small and the free energy of this reaction is almost independent of temperature. Figure 8.15 shows the overall change in free energy as carbon is oxidized at increasing temperatures. The total free energy curve falls as T rises and will eventually cross every curve, of the type shown in Fig. 8.14, for the free energy of metal oxide formation. It follows

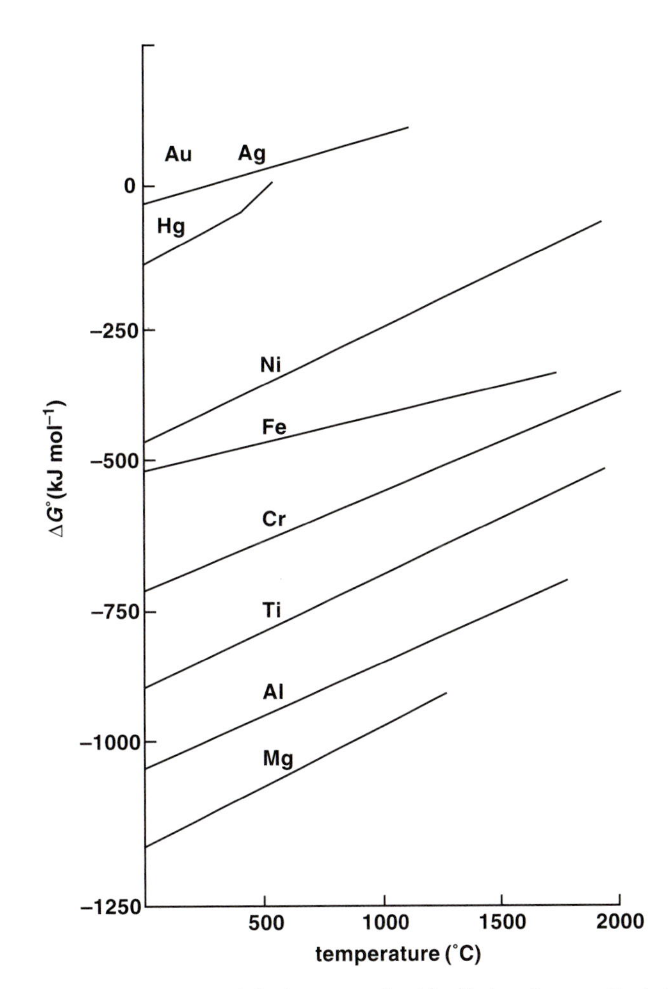

FIG. 8.14 The variation with temperature of the free energy of metal oxide formation according to Equation (8.1)

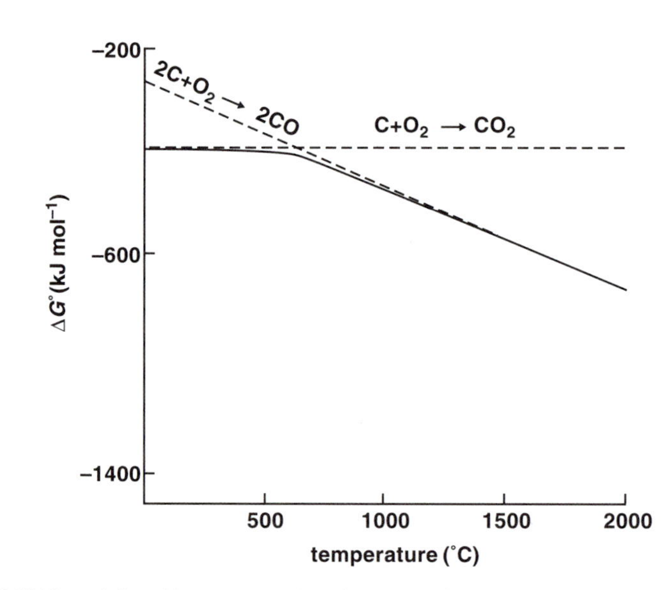

FIG. 8.15 The variation with temperature of the free energy of the reaction of carbon with oxygen

that, in principle, carbon may be used to reduce any metal oxide if a high enough temperature can be attained. In practice, of course, temperatures which would be high enough to allow carbon to reduce the more stable oxides such as TiO_2 or Al_2O_3 are not

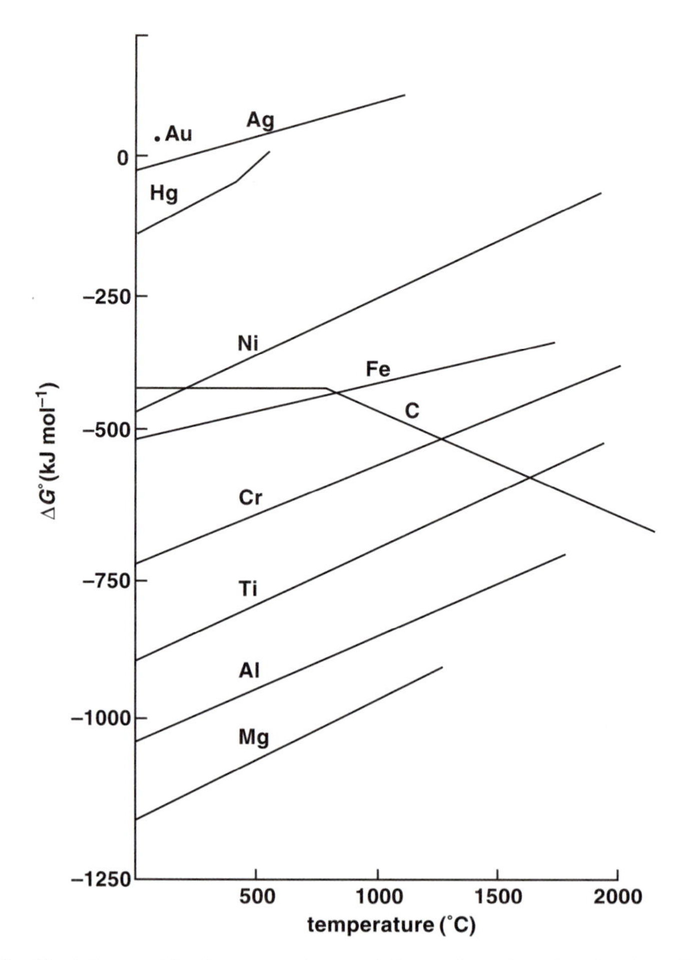

FIG. 8.16 Combined diagram of the free energy changes of the reactions of metals and carbon with oxygen

accessible economically on a large scale. The usefulness of carbon as a reducing agent may be seen in Fig. 8.16, in which the curves of Figs. 8.14 and 8.15 are superimposed. Of course, other chemical factors come into play in choosing reduction conditions, such as the formation by many elements of unwanted carbides, which limit the use of carbon in practice.

Once again there are important reservations to be kept in mind when applying such a thermodynamic analysis as that sketched above. Free energy changes are calculated on the assumption that the system is in equilibrium—which is far from the case in practice—and they give no indication of the kinetics of the reaction so that a particular reduction may have a favourable free energy change but be too slow. However, the thermodynamic analysis does distinguish reactions which will occur from those which will not, and it gives some indication of the conditions, for example of temperature, which are suitable. For a fuller account of this topic, and curves for sulfide and halide systems, the monograph by Ives (see references) should be consulted.

LOWER ENERGY EXTRACTION METHODS

While thermal and electrolytic processes are still important for recovering the pure elements, hydrometallurgical processes using solvent extraction principles are becoming increasingly popular for the separation of aqueous solutions of metals into their components, from which the pure metals can be recovered if necessary. The basic principles behind such processes have already been briefly outlined in Section 7.3 and space unfortunately does not permit a more detailed account. Specific examples used industrially include uranium extraction and nuclear fuel reprocessing (Section 7.3) separation of the various lanthanide elements (Section 11.3), separation of platinum from palladium, and the recovery of copper from ores.

Another method used industrially for recovering metals from ores uses bacteria. This method is particularly suited to the recovery of metals from low-grade ores, and is therefore likely to grow in importance as reserves of the various metals become depleted. Acidophilic, or 'acid-loving', bacteria such as *Thiobacillus ferrooxidans* oxidize metal sulfide ores to the metal sulfate solution, obtaining energy by the oxidation of sulfide and Fe(II). The leach solution is then collected and can be subjected to the procedures described above for the recovery of its various metal components. This technique is well placed for the recovery of gold from low-grade ores. A review of this inter-disciplinary extraction method is given in Appendix A for the interested reader.

Problems

This chapter should be related back to Chapter 2 and forward to Chapters 10 to 12 and 14 to 17. The emphasis is on *patterns* of behaviour in the Periodic Table which are used as a framework to correlate the detailed chemistry of the elements.

8.1 Replot diagrams such as Figs 8.4 to 8.9 against the effective nuclear charge, Z^*, rather than against Z. Discuss the features which disappear and the ones which remain.

8.2 Plot the variation of the free atom binding energies from Table 8.2 against Z. Compare how each level varies, and compare also with the ionization potential plots for the same elements. How far do the changes reflect the changes in Z, Z^* or the electron configurations?

8.3 Calculate the difference in exchange energy between the element E and the ion E^+ for the first Short Period in units of K, plot against Z, and compare with Fig. 8.4.

8.4 Plot (a) the first ionization potential and (b) the 2s and 2p core binding energies for Na to Ar against the covalent radii. Discuss any relationships which emerge.

8.5 Choose a number of elements and look up their chemistry in the systematic chemistry chapters. Determine how valid are the generalizations of Section 8.5, and where their properties fit into the patterns of this chapter.

Try jotting down the expected oxides and halides—for example, of Fe, Sc, In and As (which all show a III state perhaps with others) from Section 8.5 before looking up the systematic chemistry.

8.6 The redox potentials of the heavier actinides are

M =	Cm	Bk	Cf	Es	Fm	Md	No
$M^{3+} + 3e = M$ (eV)	−2.7	−2.4	−2.1	−2.0	−2.1	−2.2	−2.5

Use these together with the values in Table 12.2 to plot an Ebsworth diagram and discuss the relative stabilities of the oxidation states.

8.7 The redox data for xenon are reported as

in acid

$$H_4XeO_6 + 2H^+ + 2e = XeO_3 + 3H_2O \quad 3.0\,V$$
$$XeO_3 + 6H^+ + 6e = Xe + 3H_2O \quad 1.8$$
$$XeO_3 + 2HF + 4H^+ + 4e = XeF_2 + 3H_2O \quad 1.6$$
$$XeF_2 + 2H^+ + 2e = Xe + 2HF \quad 2.2$$

in base

$$HXeO_6^{3-} + 2H_2O + 2e = HXeO_4^- + 4OH^- \quad 0.9$$
$$HXeO_4^- + 2H_2O + 4e = XeO + 5OH^- \quad 0.7$$
$$HXeO_4^- + 3H_2O + 6e = Xe + 7OH^- \quad 0.9$$
$$XeO + H_2O + 2e = Xe + 2OH^- \quad 1.3$$

Plot Ebsworth diagrams for species (a) in acid and (b) in base. Compare with the halogens. (Treat XeF_2 as you would Xe^{2+}.)

8.8. From one of the major textbooks, find data on the abundance of the elements in the Earth's crust and in the ocean. Plot selected values, representing the full range of abundances, of the crustal figures against the ocean ones. Discuss how far they are in parallel, and discuss major differences. Carry out the same exercise versus the abundances in the Cosmos.

8.9 Compare the isolation methods detailed in the systematic chemistry chapters with the discussion in Section 8.8. How far do they fit, and how far are alternative methods used when very pure samples are needed?

8.10 Plot the pattern of the historical discovery of the elements (Chapter 1) on the Periodic Table. Discuss how far it reflects the distributions of Figs 8.12 and 8.13.

8.11 Extend the information of Fig. 8.16 to other elements and take the two major prehistoric metals, copper and iron, as your markers. Which other elements were potentially accessible by the technology used for (a) copper and (b) iron? Which of these were actually known? Discuss reasons for selected omissions—why not Ni which is less demanding than Fe, for example. (If you become interested in primitive metallurgy, you may wish to read further and discuss whether the other ferrous metals were completely unknown.)

CHAPTER 9

9.1 GENERAL AND PHYSICAL PROPERTIES OF HYDROGEN 209
9.2 CHEMICAL PROPERTIES OF HYDROGEN 211
9.3 IONIC HYDRIDES 213
9.3.1 Complex hydride anions 214
9.4 METALLIC HYDRIDES 215
9.5 COVALENT HYDRIDES 218
9.5.1 Preparation 218
9.5.2 Properties 220
9.6 ELECTRON DEFICIENT HYDRIDES 222
9.6.1 Boron hydrides 222
9.6.2 Wade's rules 224
9.6.3 Other electron deficient hydrides 226
9.7 THE HYDROGEN BOND 228
PROBLEMS 233

Hydrogen

9.1 General and physical properties of hydrogen

Hydrogen is the simplest of the elements with its one valency orbital and single electron. It can react only by gaining or sharing another electron and its behaviour is therefore fairly uncomplicated. Hydrogen combines with almost all the other elements, and, as its electronegativity value comes in the middle of the range, its bonds have a wide range of polarity from the strongly positive hydrogen of the hydrogen halides to the negative, anionic hydrogen in the active metal hydrides. The properties of the hydrides thus illustrate the variation in chemical properties with Periodic Table position. In addition, the small size of the hydrogen atom presents no steric barriers and the hydrides have the shapes expected from the electronic structure of the central atom. This small size does permit the close approach of other, nonbonded, atoms and special properties result, especially the *hydrogen bond* between atoms of high electronegativity values which is discussed later.

Hydrogen does not fit into any of the Groups of the Periodic Table and is best regarded as an introductory element to the Periodic classification. It does show significant analogies to three of the other Groups which are worth noting as a guide to its chemistry:

(1) in common with the halogens, it has a tendency to gain an electron and forms the hydride ion H^-.

(2) in common with the alkali metals, it may lose an electron to form a cation, though this statement must be treated with considerable reserve as will be seen.

(3) in common with the carbon Group elements, hydrogen has a half-filled valency shell and forms covalent bonds with a wide range of polarities.

This last comparison is especially apt, particularly if hydrogen is compared with carbon with one remaining free valency as in methyl, H_3C-. There is a very close relationship between the stability and properties of the hydrides and organometallic compounds of many elements, as Table 9.1 illustrates, and it is often useful to regard the hydride as the parent compound of a homologous series of organometallic compounds, e.g. $HSnCl_3$, CH_3SnCl_3, $C_2H_5SnCl_3$, etc. Organometallic chemistry, which cuts across the boundary between inorganic and organic chemistry, has become a major field of research in its own right in the last three decades (compare Chapters 16 and 20 in particular). Thus this third relationship—between hydrogen chemistry and carbon Group chemistry—is of considerable interest and importance.

Hydrogen is the most abundant element in the Universe (Section 8.7). On Earth, hydrogen is found almost entirely in combination—especially in water and organic compounds, though H_2 is emitted by volcanoes. Hydrogen is widely used, especially in the manufacture of ammonia and in hydrogenation reactions in petrochemical production. Most hydrogen is formed as part of the overall process, either from water, from oil fractions or from methane (natural gas or, on a smaller scale, from anaerobic fermentation).

Where electricity is cheap, electrolysis of water provides hydrogen—either as the direct product or as a byproduct in processes like chlorine manufacture by the electrolysis of brine.

An alternative production of H_2 from both water and hydrocarbons is by the steam-reforming process over a nickel catalyst at around 850°C:

$$C_nH_{2n+2} + H_2O \rightarrow CO + H_2$$

TABLE 9.1 Comparison of some metal-hydrogen and metal-alkyl compounds

Type	*R = H*	*R = alkyl, e.g. methyl*
NaR	Ionic, Na^+H^-	Ionic, $Na^+CH_3^-$
AlR_3	Polymeric solid linked by electron-deficient Al····H····Al bridges	Dimer linked by electron-deficient alkyl bridges, Al····CH_3····Al
SiR_4 to PbR_4	Covalent gaseous molecules, M-H decreases in stability Si > Ge > Sn > Pb	Volatile covalent compounds, M-R decreases in stability from Si to Pb though more stable than M-H
GeR_2	Polymeric oxidizable solids of obscure structure	Rings $(GeR_2)_n (n = 4, 5, 6)$ or polymeric solids
$RCo(CO)_4$	Unstable complex hydride with σ Co-H bond	Unstable organometallic complexes with σ Co-R bonds
$RPtX(PPh_3)_2$	Hydride complex with Pt-H bond stabilized by Pt–phosphane interaction	Stable organometallic complex with σ Pt-C bond stabilized by Pt-phosphane interaction

THE 'HYDROGEN ECONOMY'

At present-day costs, electrolysis is relatively expensive but this will change as accessible oil and gas stocks run out. A scenario has been suggested where future sources of cheap electricity, for example from improved solar cells, is converted into hydrogen which is storable and could readily replace most current uses of oil products, with major advantages in reducing pollution. This 'hydrogen economy' would steadily replace the 'oil economy'. The energy derived by burning a unit weight of hydrogen is very high, but an important drawback is that the energy per unit volume is very low. For use as a transport fuel, storage of hydrogen must not carry a weight penalty—see Section 9.4 for a possible solution by storing hydrogen as a metal hydride.

HEAVY HYDROGEN ISOTOPES

Pure deuterium is normally separated from light hydrogen by the electrolysis of water. The lighter isotope is evolved preferentially and almost pure deuterium oxide remains by the time the bulk is reduced a millionfold. Tritium is most conveniently prepared by the irradiation of lithium with slow neutrons in a reactor:

$$^{6}_{3}Li + ^{1}_{0}n = ^{3}_{1}H + ^{4}_{2}He$$

and the tritium is separated by oxidation to T_2O. Most deuterated or tritiated compounds are made from isotopically substituted water. For example, deuterated acids may be made simply by solution:

$$P_2O_5 + 3D_2O \rightarrow 2D_3PO_4$$

$$SO_3 + D_2O \rightarrow D_2SO_4$$

or tritiated ammonia by the reaction of a nitride:

$$Mg_3N_2 + 3T_2O \rightarrow 2NT_3 + 3MgO$$

Since the chemical behaviour of all the isotopes of an element is identical (although rates of reaction may differ), deuterium or tritium substituted hydrides are widely used in studying the mechanisms of reactions involving hydrogen (see Section 2.3).

Further reaction with steam can convert CO to CO_2 which is removed chemically or separated by molecular sieve. An alternative source of process hydrogen is from hydrocarbons by thermal cracking.

Pure hydrogen may be prepared by diffusion through palladium tubes (which pass only H_2) or, on a small scale, by hydrolysis of active metal hydrides like CaH_2.

There are three well established isotopes of hydrogen:

$^{1}_{1}H$ normal or light hydrogen, mass = 1.008, natural abundance = 99.98%
$^{2}_{1}H$ deuterium (D) or heavy hydrogen, mass = 2.015, natural abundance = 0.02%
$^{3}_{1}H$ tritium (T), mass = 3.017, natural abundance = 10^{-17}%, radioactive, with $t_{\frac{1}{2}} = 12.4$ years, decay process $^{3}_{1}H \rightarrow ^{3}_{2}He + ^{0}_{-1}e$

Some of the important properties of hydrogen are listed in Table 9.2.

It has already been noted in Section 2.15 that atomic and ionic radii show some variation with the chemical environment of the species. This variability is particularly

TABLE 9.2 Properties of hydrogen

Heat of dissociation of H_2	$H_{2(gas)} = 2H_{(gas)}$, $\Delta H = 435.9$ kJ mol^{-1} (443.3 kJ mol^{-1} for D_2)
Ionization potential	$H_{(gas)} = H^+_{(gas)} + e^-$, $I = 13.595$ eV $= 1309$ kJ mol^{-1}
Electron affinity	$H_{(gas)} + e^- = H^-_{(gas)}$, $E = -68.99$ kJ mol^{-1} (exothermic)
Radius, anionic H^-	= 112 to 154 pm (measured in ionic hydrides)
	= 208 pm (calculated for free H^-)
cationic H^+	= 10^{-3} pm
covalent H	= 37.07 pm (from H_2 bond length)
	28 pm (from bond lengths of hydrogen halides)
	32 pm (from MH_4 distances in carbon Group hydrides)

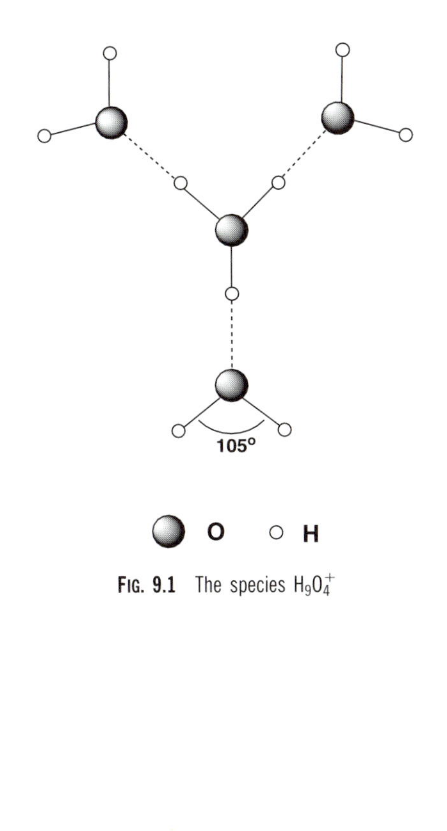

FIG. 9.1 The species $H_9O_4^+$

marked in the ions and covalent molecules involving hydrogen, as Table 9.2 shows, since there is only a single nuclear charge on hydrogen and the 1s orbital is unshielded by any inner electron shells. The hydride ion, H^-, is especially sensitive to change of electric field intensity in its environment as it consists of two electrons in the field of the single nuclear charge and therefore has a very diffuse electron cloud. The free hydride ion has been calculated to have a radius of 208 pm, twice as large as that of helium, which has two electrons in twice the nuclear field. The measured values of the hydride ion radius in ionic lattices are much smaller than this and show considerable variation, as the values in the alkali metal hydrides illustrate:

MH	Li	Na	K	Rb	Cs
H^- radius (pm)	126	146	152	153	154

(Compare these values with 133 pm for F^- and 145 pm for O^{2-}, where the larger numbers of electrons are offset by the increase in the nuclear charges and consequent greater density of the electronic cloud.) The covalent radius of hydrogen also shows considerable variability, although the effect is less pronounced than with the anionic radius, apart from the high value for the hydrogen molecule itself.

The most striking change in dimension would come if the hydrogen atom were to lose its electron to become the positive ion, H^+. This is the bare proton with a radius of about 10^{-3} pm and is a hundred thousand times smaller than any other ion (compare Li^+, radius 60 pm). The charge density on the proton is thus enormously higher than on any other chemical species and it would have a powerful polarizing effect on any other molecule in its neighbourhood. As a result, the free proton has no independent existence in any chemical environment and always occurs in association. Thus, if an acid dissociates in water, H_3O^+ and not (except as a shorthand) H^+ is formed, and this 'hydrogen ion' or 'hydroxonium' ion is then further solvated just as any other cation. A similar situation holds for any other protonic solvent, as discussed in Chapter 6 (although it must be noted that in the Brönsted definition of an acid it is the proton which is transferred). In water, the proton is in a rapidly changing environment and interacting with varying numbers of water molecules. The type of species present in solution is illustrated by the various hydrated protons which have been isolated and characterized. Examples are known where the proton interacts with 2, 3, 4 or even more H_2O molecules, including $H_9O_4^+$, shown in Fig. 9.1, which was extracted from an aqueous acid solution into an immiscible organic base. A simpler species, where the proton is linked to two water molecules, was isolated in a solid complex, see Fig. 9.2.

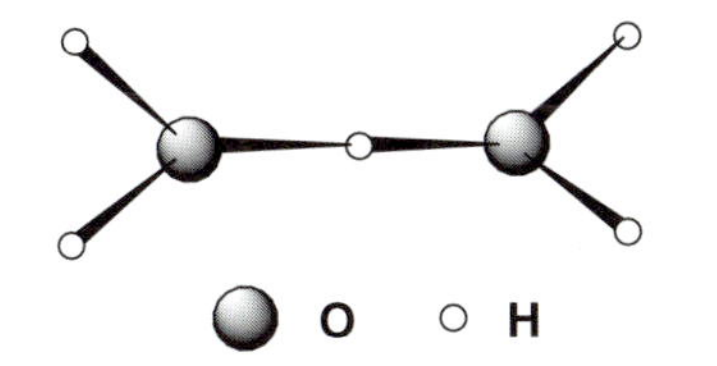

FIG. 9.2 The structure of $H_5O_2^+$ in $[Co(en)_2Cl_2]Cl^-H_5O_2^+Cl^-$. The (H_2O)-H-(OH_2) link is symmetrical and the hydrogens of the water molecule are *trans* to each other. The distance O-H-O is 243.1 pm, making the OH bridge length 121.6 pm. H-O in the H_2O molecule is 99.5 pm. Angles at oxygen are 109° to terminal H atoms and 114° for H (bridge)-O-H (terminal)

9.2 Chemical properties of hydrogen

As the hydrogen molecule bond energy, of 436 kJ mol^{-1}, is high, molecular hydrogen is fairly unreactive at ordinary temperatures. At higher temperatures it combines, directly or with aid of a catalyst, with most elements. Some of the more important reactions are shown in Fig. 9.3.

ORTHO- AND *PARA-*HYDROGEN

One further physical modification of the hydrogen molecule should be briefly mentioned: *ortho-* and *para-*hydrogen. These forms arise from the different ways in which the nuclear spins may be lined up. If the nuclear spins are parallel, the form is *ortho-*hydrogen, while if they are antiparallel, the form is *para-*hydrogen. These two forms of molecular hydrogen have different physical properties, such as thermal conductivities, and the two coexist at ordinary temperatures. (Other symmetrical molecules with nuclear spins, such as N_2 or Cl_2, have *ortho-* and *para-* forms but only H_2 and D_2 show significant differences in physical properties.) The more stable form at low temperatures is *para-*hydrogen and this makes up 100% of hydrogen at absolute zero. At higher temperatures, the equilibrium proportion of *ortho-*hydrogen rises, and reaches its maximum of 75% at room temperature. The equilibrium proportions at any given temperature may be calculated theoretically, and the interconversion may easily be followed experimentally. The interconversion is slow but subject to catalysis by a number of materials, especially by paramagnetic compounds, and this *ortho–para* conversion is frequently used in studies of catalysis.

Atomic hydrogen may be produced in high-intensity electric arcs and is very short-lived and reactive. It finds uses in welding where the recombination of the atoms takes place on the metal surface, yielding up the heat of dissociation and at the same time providing a protective atmosphere against oxidation.

Hydrogen forms a molecular ion H_3^+ which has an equilateral triangular structure at equilibrium. This molecule, the simplest of all polyatomic molecules, has been of interest from a bonding point of view and it has also been proposed as a chain intitiator in reactions occurring in interstellar clouds. H_3^+ has also been recently detected for the first time outside of the laboratory—in the atmospheres of Jupiter, Uranus and Supernova 1987 A.

Binary compounds of hydrogen (i.e. those containing hydrogen and one other element) are termed hydrides whether they contain the hydride ion, H^-, or are covalent, and this term is commonly extended to less simple hydrogen compounds, as in 'transition metal hydride complexes' and 'complex hydride ions'. It is convenient to discuss the chemistry of the hydrides in three groups, reflecting the three ways in which the hydrogen electron enters into bonding. These are (i) gain of an electron to form ionic compounds containing H^-, (ii) sharing the electron in covalent hydrides, (iii) forming metallic bonds with the electron delocalized in the so-called interstitial or metallic hydrides.

We first note how these types distribute in the Periodic Table. As the electron affinity of hydrogen is low compared with that of the halogens, the distribution of ionic hydrides in the Periodic Table is much more restricted than the distribution of ionic halides. Ionic hydrides are formed only by the elements of the alkali, alkaline earth and, possibly, the

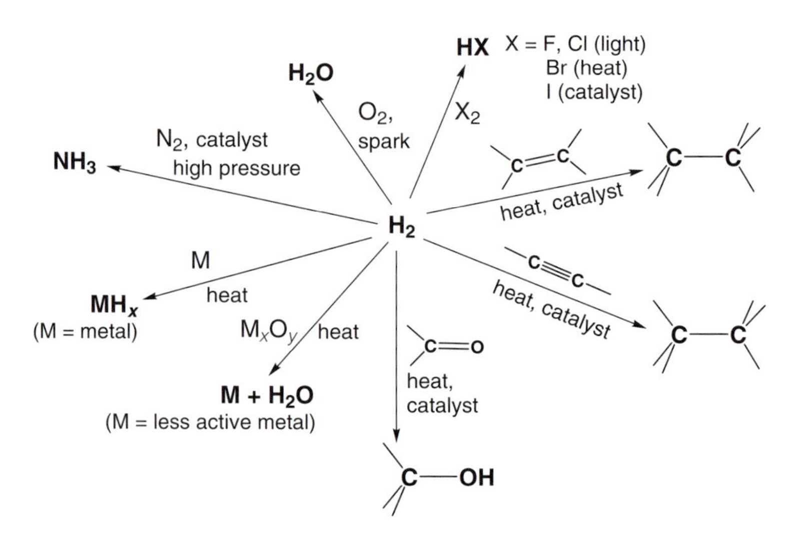

FIG. 9.3 Reactions of molecular hydrogen

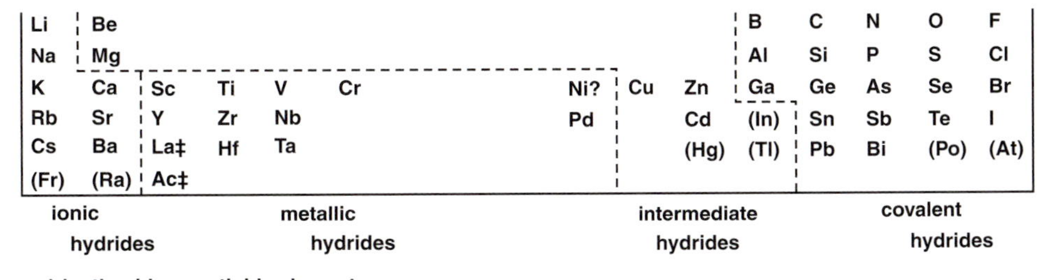

FIG. 9.4 Types of hydride in the Periodic Table

lanthanide Groups. Most other elements of the main Groups form covalent hydrides, while the transition elements give metallic hydrides. Figure 9.4 shows the approximate distribution of the various types of binary hydride within the Periodic Table. The boundaries between classes are by no means sharp, and a number of intermediate types are observed. In addition, there is some doubt whether the hydrides of a number of elements, especially of the heaviest elements, exist at all.

In passing across a Period, the type of hydride changes from ionic compounds at one extreme to volatile covalent molecules at the other. In the middle of the short Periods, the transition between the two types is marked by solid hydrides, say of magnesium and aluminium, of polymeric structure with bonding of intermediate character. Among the transition elements in the long Periods, the type changes through interstitial hydrides to hydrides of dubious existence at the right of the transition block, before coming to the covalent hydrides of the p elements. In any Main Group of the Periodic Table, the stability of the hydrides tends to fall with increasing atomic weight.

Finally we note that the transition metals also form hydrido-complexes such as $[ReH_9]^{2-}$ (Fig. 15.23) or $RuH(CO)(PPh_3)_3$. In addition, the dihydrogen molecule can coordinate to metals. Metal hydride complexes are included in Chapter 15 and the M-H_2 compounds are discussed in Section 16.11.

9.3 Ionic hydrides

The gain of an electron by a hydrogen atom gives the helium configuration, $1s^2$, and is analogous to halide ion formation. However, the formation of the hydride ion is much less favourable than the formation of a halide ion, see Fig. 9.5, as the electron affinity of hydrogen is much less exothermic and because more energy is needed to break the H-H bond. As a result only the most active elements, whose ionization potentials are low, form ionic hydrides. The alkali metal hydrides and CaH_2, SrH_2 and BaH_2 are the only compounds which are clearly ionic. Magnesium hydride, MgH_2, is intermediate between the ionic hydrides and the solid covalent hydrides like AlH_3. The dihydrides of the lanthanide elements are metallic and only approach the ionic type as the hydrogen content rises towards the MH_3 stoichiometry. However, the two lanthanide elements which show a relatively stable II state in their general chemistry (compare Chapter 11)—europium and ytterbium—do form ionic hydrides, EuH_2 and YbH_2. These two compounds are isomorphous with CaH_2 and differ in structure from the other lanthanide dihydrides which have the fluorite structure.

The ionic hydrides are formed by direct reaction between hydrogen and the heated metal. They are all reactive and reactivity increases with atomic weight in a Group. The alkali metal hydrides are more reactive than those of the corresponding alkaline earth elements. The alkali metal hydrides have the sodium chloride structure (the radius of H^-, about 150 pm, is comparable with the halide radii, $F^- = 133$ pm, $Cl^- = 181$ pm). The alkaline earth hydrides, and EuH_2 and YbH_2, have the same structure as $CaCl_2$. The metal atoms are in approximately hexagonal close packing and each is surrounded by nine hydride ions in a slightly distorted lead dichloride structure. In the regular $PbCl_2$ structure, the metal atom is coordinated by six metal ions at the corners of a trigonal prism and three more beyond the rectangular faces. In the dihydrides, the metal atom

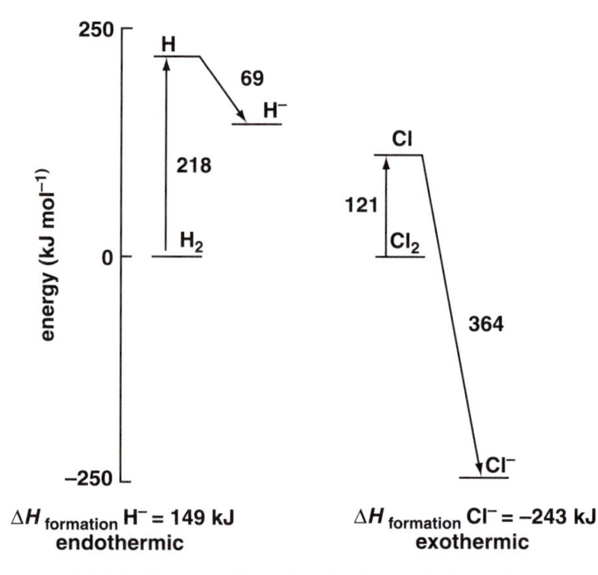

FIG. 9.5 Energies of formation of hydride and chloride ions

has seven hydride ions at equal distances and two more distant ions (e.g. CaH_2 has seven Ca-H distances of 232 pm and two of 285 pm). In all these ionic hydrides, the metal–metal distances are less than they are in the elements so that the hydrides are denser than the metals.

Evidence for the ionic nature of these hydrides is:

(i) Molten LiH shows ionic conductance and the melt gives hydrogen at the *anode* on electrolysis. The other hydrides decompose before melting, but they may be dissolved in alkali halide melts without decomposition and, on electrolysis, give hydrogen at the anode.

(ii) It has been possible, by a combination of X-ray and neutron diffraction, to construct an electron density map for LiH. This shows that 0.8 to 1.0 of an electron has been transferred to hydrogen from each lithium atom (i.e. to give Li^+H^-). Thus lithium hydride is almost completely ionic. As polarization effects are greatest in lithium hydride (see Chapter 4) it follows that the other alkali hydrides are also ionic, with full transfer of an electron from each metal atom.

(iii) The crystal structures of the hydrides show no indication of directional bonding (chains, sheets or discrete molecules) and are reasonable for ionic compounds with the radius ratios of the hydrides.

(iv) Observed and calculated lattice energies are in good agreement (compare Section 5.2).

USES

The ionic hydrides find use in the laboratory for drying solvents, and as reducing agents, though they have been largely superseded in the latter application by the advent of the complex hydrides. On the industrial scale, NaH and CaH_2, which are relatively inexpensive and easily handled, are commonly used as condensing agents in organic syntheses and as reducing agents, e.g.:

$$CaH_2 + MO \rightarrow CaO + M + H_2 \quad \text{(at 500–1000°C)}$$

$$CaH_2 + 2NaCl \rightarrow 2Na\uparrow + CaCl_2 + H_2$$

The ionic hydrides react readily, and often violently, with water or any other source of acidic hydrogen and with oxidizing agents. The reaction with acidic hydrogen may be represented as:

$$X^{\delta-}\text{–}H^{\delta+} + H^- \rightarrow H_2 + X^-$$

Examples of this and other common reactions of the hydride ion are shown in Fig. 9.6.

9.3.1 Complex hydride anions

One important reaction of the alkali metal hydrides is in the preparation of the complex hydride anions. If lithium hydride is reacted in ether with aluminium trichloride, the tetrahydroaluminate, $LiAlH_4$, results.

$$4LiH + AlCl_3 \rightarrow LiAlH_4 + 3LiCl$$

$LiAlH_4$ is usually called by its nonsystematic name of lithium aluminium hydride.

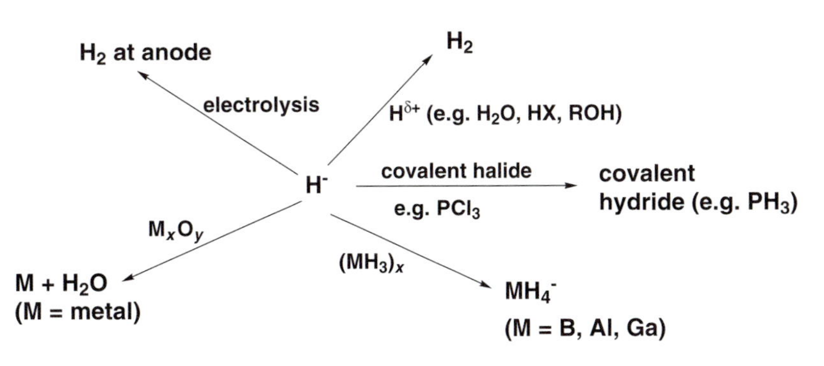

FIG. 9.6 Reactions of the hydride ion

Recent work has shown that the complex aluminium hydrides may also be made by direct reaction of the elements at high pressures, giving a much cheaper route as all the expensive lithium ends up in the product

$$Li + Al + 2H_2 \xrightarrow[250\ atm, 120-150^\circ C]{ether\ solvent} LiAlH_4 \quad \text{(also for Na or K).}$$

$NaBH_4$, sodium tetrahydroborate (borohydride), Li_2BeH_4, $LiGaH_4$, and similar compounds may be prepared by one or other of these routes, as can the aluminium hexahydrides like Na_3AlH_6.

The alkali metal compounds MBH_4, $MAlH_4$ and $MGaH_4$ all contain tetrahedral EH_4^- anions (i.e. E = B, Al or Ga) although there may be a weak interaction with the cation, especially for M = Li, of the multicentred electron deficient type discussed in Section 9.6. The M_3AlH_6 compounds contain octahedral AlH_6^{3-} anions.

$NaBH_4$, $LiAlH_4$ and, to a lesser extent, the other complex hydrides are extremely useful reducing agents, especially the lithium compounds which are appreciably soluble in ether. Examples of their uses in the preparation of covalent hydrides are given in Section 9.5 and organic applications include the reduction of aldehydes or ketones to alcohols and of nitriles to amines. The different complex hydrides vary in reactivity thus BH_4^- is a milder reducing agent than AlH_4^-. The introduction of organic groups, as in $Na^+[HB(OR)_3]^-$ or in $Na^+[H_3B(OR)]^-$ for $R = CH_3$, etc., gives reagents of varying reactivity which can be highly selective. The introduction of such reagents in the last few decades has revolutionized reductive preparations in inorganic, organometallic and organic chemistry. As simple examples, yields in the preparation of the hydrides of the heavier elements of the carbon and nitrogen Groups have been raised to 80–90% as compared with the 10–20% which was common with the older methods using ionic hydrides.

Less electropositive elements also form borohydrides or aluminohydrides, but these are more closely related to the parent boron and aluminium hydrides, which are electron deficient covalent species and are discussed in Section 9.6.

9.4 Metallic hydrides

Nonstoichiometric compounds between hydrogen and many transition metals have been known for a long time, but their nature and bonding have not been well understood. These compounds are brittle solids with a metallic appearance and with metallic conductivity and magnetic properties. Typical formulae are $TiH_{1.8}$, $NbH_{0.7}$ or $PdH_{0.6}$. These compounds were called 'interstitial' hydrides as it was originally conjectured that the hydrogen atoms were held in the interstices of the metal lattice. The problem was further complicated by the fact that most transition metals physically adsorb hydrogen so that many spurious compounds were reported.

It is now known that all the lanthanide and actinide elements, the elements of the titanium and vanadium Groups, chromium and palladium combine exothermically and reversibly with hydrogen to form metallic hydrides in which the metal atoms are in a different structure from those of the element. These metallic hydrides have the idealized formulae shown in Table 9.3, but, as usually prepared, are commonly nonstoichiometric and hydrogen deficient.

There is thus a 'hydride gap' in the middle of the Periodic Table where binary hydrides are not formed. The exothermic heats of formation decrease from left to right towards this gap, and it has been calculated that hydrides of the ferrous metals would have heats of formation of around zero.

Many of these hydrides have structures where the hydrogens occupy the tetrahedral sites in cubic close-packed metal lattices. With all these sites occupied, this corresponds to the formula MH_2 and the fluorite structure (compare Table 5.3). It is thought that VH, NbH and TaH have the hydrogens in tetrahedral sites and these compounds have slightly distorted body-centred cubic metal lattices which are closely related to cubic close-packed arrangements. In the lanthanide hydrides, and those of yttrium and the heavier actinides, the MH_2 phase can take up further hydrogen in octahedral sites. The lighter lanthanides form MH_3 phases in this way which remain cubic, but the heavier elements undergo a structural change around the composition $MH_{2.5}$ to give a hexagonal lattice. UH_3, PaH_3 and Th_4H_{15} have more complex structures. In two of the hydrides, CrH and PdH, the hydrogens are in octahedral sites only. Palladium hydride (which has been prepared only up to the $PdH_{0.7}$ composition) forms a sodium chloride lattice but chromium hydride has a hexagonal, *anti*-nickel arsenide structure.

There has been much discussion of the bonding in these metallic hydrides, and in similar compounds such as some low-valency halides of the transition elements which also show metallic properties. One theory, which accounts for many of the facts, regards the metallic hydrides as modified metals. A metal with n valency electrons is regarded as forming M^{n+} cations and having the n electrons per metal ion in completely delocalized orbitals. Then, as hydrogen atoms enter the lattice, each acquires one of these delocalized electrons to form a hydride ion. Thus, a metal hydride, MH_x contains M^{n+} ions, xH^- ions and $(n–x)$ delocalized

TABLE 9.3 Metallic hydrides (idealized formulae and structures)

ScH_2 (fluorite)	TiH_2 (fluorite)	VH (bcc)	CrH (anti-NiAs)	–	–	–	(NiH)(c) (NaCl)	CuH (wurtzite)
		$VH_{1.6}$ (fluorite?)	CrH_2? (fluorite?)					
YH_2 (fluorite)	ZrH_2 (fluorite)	NbH (bcc?)	–	–	–	–	$PdH_{0.7}$ (NaCl)	–
YH_3 (hexagonal)		NbH_2 (fluorite?)						
LaH_2 (fluorite)	HfH_2 (fluorite)	TaH (bcc)	–	–	–	–	–	–
LaH_3 (cubic)								
	Lanthanide elements	MH_2 (fluorite) formed by Ce, Pr, Nd, Sm, (a), Gd, Tb, Dy, Ho, Er, Tm, (a), Lu MH_3 (cubic) formed by Ce, Pr, Nd, Yb MH_3 (hexagonal) formed by Sm, Gd, Tb, Dy, Ho, Er, Tm, Lu						
AcH_2 (fluorite)								
	Actinide elements	MH_2 (fluorite) formed by Th, (b), Np, Pu, Am MH_3 (hexagonal) formed by Np, Pu, Am MH_3 (cubic, complex structure) formed by Pa, U						

Notes: (a) EuH_2 and YbH_2 have orthorhombic, ionic hydrides discussed in Section 9.3.
(b) Thorium dihydride has a distorted fluorite structure: a second hydride, Th_4H_{15} also exists.
(c) NiH exists only under hydrogen pressures of 10^4 atm.

electrons. The relative numbers and sizes of the metal cation and hydrogen anions govern the structure of the hydride, while the remaining (*n*–*x*) electrons give the hydride its metallic properties. For example, TiH_2 would be regarded as consisting of Ti^{4+}, two H^- and two conduction electrons per formula unit.

The transition metal hydrides are usually prepared by direct combination between the metal and hydrogen at moderate temperatures and, often, high pressures. They may be decomposed by raising the temperature. This reversible hydrogenation is made use of in two ways. One is to provide a convenient source of very pure hydrogen. The metal hydride is formed leaving any impurities in the hydrogen behind. Then, by heating the hydride to a higher temperature, pure hydrogen is evolved. The second use is to provide the metal in a finely divided and highly reactive form. Many of the transition metal hydrides differ sufficiently from the metal in lattice parameters so that when the hydride is formed from the bulk metal, it is produced as a fine powder. The other hydrides are brittle and may be much more readily powdered than the metal. The powdered hydride is then heated to remove the hydrogen, leaving the metal in a suitable form and free of surface oxide for further reactions. In addition, the metal hydrides themselves often provide suitable starting materials for synthesizing other compounds of the metal.

Related to the simple hydrides are a wide range of mixed metal hydrides. These often involve a more electropositive metal which, alone, forms an ionic hydride, but the mixed species retains metallic properties. Of particular interest are mixed metal hydrides formed by those metals which do not have a stable binary hydride phase such as Mg_2NiH_4. Similar mixed metal hydrides are known for most of the metals in the 'hydride gap'.

These compounds are intermediate between the binary metallic hydrides, and the hydrido-complexes of the transition metals which have covalent M-H bonds (compare the ReH_9^{2-} ion and related species, Section 15.5.1). Mg_2FeH_6 also approaches the limit of an FeH_6^{4-} ion, but there is a significant H···Mg interaction remaining (Section 14.6.3). One further, somewhat different, intermediate group is the metallic hydride-halides and similar species represented by ZrHCl. This shares the metallic properties of the binary hydride and of the 'subhalide' (Section 15.2).

The hydrides of the remaining transition metals are quite different. Copper hydride, CuH, is formed endothermically by the reduction of copper salts with hypophosphorous acid and it decomposes irreversibly. In the zinc Group, ZnH_2, CdH_2 and HgH_2 are formed by the reaction of the halides with $LiAlH_4$ but stability decreases rapidly from Zn to Hg. A mixed hydride-halide, H_3Zn_2X, is formed similarly for X = Cl or Br, while cadmium gives CdHX. All these species probably resemble $(AlH_3)_x$ and belong to the electron deficient class.

INDUSTRIAL APPLICATIONS OF METAL HYDRIDES

A number of the hydrides find industrial application, particularly in powder metallurgy where the hydrogen evolved during fabrication gives a protective atmosphere. Another application is as a moderating material in atomic piles. A metal hydride, such as zirconium hydride, provides a higher density of hydrogen than conventional moderators, such as water, and they may be used to higher temperatures.

A potential use of metal hydrides of high significance is as a portable, relatively safe, hydrogen store. Hydrogen is an excellent fuel for urban transport and may be produced directly from renewable energy sources. It would be a highly desirable substitute for petrol if its advantages were not offset by the need for weighty storage cylinders. The metal hydrides offer the prospect of better hydrogen/weight storage, exploiting the readily reversible formation and decomposition of the metal hydrides. The most promising single metal is probably titanium which is relatively light and can be charged close to the 2H:1Ti ratio. The mixed-metal hydrides may be even more attractive, such as the Mg_2NiH_4/Mg_2Ni system which gives a better H_2/weight

COLD FUSION

In 1989, the 'cold fusion' phenomenon burst onto the scene amidst a blaze of publicity: interest in palladium hydrides multiplied many times overnight. It was reported that electrolysis in D_2O using a palladium electrode produced excess heat which the investigators could not account for by any chemical reaction. They therefore ascribed it to a nuclear process: that D atoms were present at high local concentration in the Pd lattice and a few underwent nuclear fusion, liberating the extra energy. As the only established nuclear fusion is that at high temperatures in the H-bomb or the Sun, the palladium process was termed 'cold fusion'. Such claims aroused enormous interest and soon enormous controversy. Many research groups, including the original authors, attempted to reproduce or extend the initial results and most failed. Many other, less innovative, explanations were put forward to account for the observations and the weight of opinion remains sceptical. Unexpected effects have been observed, but by the accepted standards of science the claimed phenomena have not yet been established. For such a striking departure from current understanding and expectation, it is essential that observations must be reproduced consistently by a number of independent investigators. This has not so far happened, though the topic is not yet completely dead.

9.5 Covalent hydrides

The hydrogen atom may attain the inert gas structure by sharing an electron pair in a covalent bond. All the remaining binary hydrides fall into this group. MgH_2 has properties intermediate between ionic and covalent hydrides while CuH, ZnH_2 and CdH_2 are intermediate between metallic and covalent species. Covalent hydrides are formed by all the elements with an electronegativity down to about 1.5 (e.g. aluminium) which are just on the border for forming ionic hydrides. As hydrogen has an electronegativity of 2.1 it follows that bond polarities range from those, as in the hydrogen halides, where the hydrogen end is strongly positive, i.e. $H^{\delta+}—X^{\delta-}$, to those cases where the hydrogen end of the dipole is negative, as in $B^{\delta+}—H^{\delta-}$ or $Ga^{\delta+}—H^{\delta-}$. If the second element has an electronegativity less than about 1.2, the hydride becomes definitely ionic.

These are Pauling electronegativity values, see Table 2.14b.

The covalent hydrides fall into two distinct classes. First there are the compounds of the carbon, nitrogen, oxygen and fluorine Groups which have normal electron pair bonds between the element and hydrogen. Secondly, there are compounds exemplified by the simplest boron hydride, B_2H_6, which do not have enough valency electrons to form electron pair bonds to all the hydrogens, and these are termed *electron deficient*. Into this class fall the hydrides of beryllium, boron, aluminium and gallium. MgH_2, ZnH_2 and perhaps CdH_2 and CuH have some affinities with this class. Other members are the borohydrides and aluminohydrides of elements which are not sufficiently electropositive to give the complex hydride ions: examples are $Be(BH_4)_2$ and $Al(BH_4)_3$. The electron deficient hydrides are treated in Section 9.6.

Covalent electron pair bonds are also formed between hydrogen and the transition elements, in compounds where the metal is also bonded to ligands capable of forming π bonds, such as CO, phosphanes, arsanes, sulfanes or NO. Such compounds as $(R_3P)_2PtH_2$—where R stands for a variety of aliphatic and aromatic substituents—or $(CO)_5MnH$, have metal to hydrogen covalent bonds of sufficient stability to allow for their isolation. Such covalent bonds to hydrogen seem to be most readily formed by those transition metals in the 'hydride gap' which do not form binary metallic hydrides. Some examples are given in Chapters 13 to 16.

9.5.1 Preparation

There are three general methods available for the preparation of covalent hydrides, although many others are available in specific cases. These general methods are:

(i) simple direct combination, especially with the more reactive elements:

$$H_2 + Cl_2 \xrightarrow{\text{light}} 2HCl$$

(ii) hydrolysis of a binary compound of the element with an active metal by any non-oxidizing dilute acid:

$$Mg_3B_2 \longrightarrow B_2H_6$$
$$Al_4C_3 \longrightarrow CH_4$$
$$Ca_3P_2 \longrightarrow PH_3$$

(iii) reduction of a halide or oxide by an ionic hydride or by a complex hydride:

$$GeO_2 + BH_4^- \longrightarrow GeH_4$$
$$SiCl_4 + LiH \longrightarrow SiH_4$$
$$AsCl_3 + LiAlH_4 \longrightarrow AsH_3$$
$$R_2SbBr + BH_4^- \longrightarrow R_2SbH.$$

The reactions of all these hydrides may be carried out in ether solution and BH_4^- may also be used in an aqueous system.

(iv) A fourth method of synthesis, in increasing use, involves the interconversion of hydrides in a suitable discharge. This is particularly valuable for forming longer chains from simple hydrides or for forming long chain halides which may then be reduced to the hydride. The discharge may be of radio or microwave frequency or may be of the ozonizer type.

(v)
$$GeH_4 \longrightarrow Ge_2H_6 + Ge_3H_8 + \text{hydrides up to } Ge_9H_{20}$$
$$SiH_4 + GeH_4 \longrightarrow GeH_3SiH_3 + Ge_2H_6 + Si_2H_6$$

$$SiCl_4 \longrightarrow Si_2Cl_6 \xrightarrow{LiAlH_4} Si_2H_6$$
$$SiH_4 + PH_3 \longrightarrow SiH_3PH_2 + (SiH_3)_2PH + Si_2H_5PH_2, \text{ etc.}$$

The p element hydrides are listed in Table 9.4.

Of these four methods of preparation, (i) is of limited applicability but is being developed for the less reactive elements under high temperature and pressure, (ii) gives low yields but is the most direct way of obtaining the higher members of homologous series, (iii) is usually the best and most convenient method on a laboratory scale, while (iv) gives better yields than (ii) of the higher hydrides but requires a supply of the simple material. Higher

TABLE 9.4 Hydrides of the p elements

B_2H_6 (and many higher hydrides: see Table 9.5)	C_nH_{2n+2}, etc. (no limit to n is known)	NH_3 N_2H_4	H_2O H_2O_2	HF
$(AlH_3)_x$ (solid polymer)	$Si_nH_{2n+2}^{(1),(2)}$ (characterized up to $n=8$; straight-and branched-chain isomers occur)	$PH_3^{(2)}$ $[P_xH_y]^{(3)}$	H_2S H_2S_n (characterized up to $n=6$)	HCl
$[(GaH_3)_x]$	$Ge_nH_{2n+2}^{(1),(2)}$ (characterized up to $n=9$; straight- and branched-chain isomers occur)	$AsH_3^{(2)}$ $[As_xH_y]^{(3)}$	H_2Se	HBr
	$[SnH_4]$ $[Sn_2H_6]$	$[SbH_3]$	$[H_2Te]$	HI
	$[PbH_4]$?	$[BiH_3]$?	$[H_2Po]$?	[HAt]?

[ ] = unstable at room temperature
[ ]? = existence unconfirmed or transient
1. Mixed hydrides $Si_xGe_yH_{2(x+y)+2}$ are also known.
2. Solids MH_x of uncertain composition are widely reported but few properties are established.
3. An extensive class of such hydrides is characterized: see Section 18.3

hydrides, containing a chain of atoms of the central element, are known for many p block elements but, for most, chain lengths are short and stabilities are low. The exceptions are silicon and germanium, where hydrides M_nH_{2n+2} are characterized up to about $n = 10$. Both straight and branched chains are known, and hydrides containing mixed chains of silicon and germanium atoms are found. For example, pentagermane is found in all three isomeric forms $(GeH_3)_2GeHGeH_2GeH_3$, $GeH_3GeH_2GeH_2GeH_2GeH_3$ and $Ge(GeH_3)_4$, and Si_2GeH_8 occurs as $SiH_3SiH_2GeH_3$ and $SiH_3GeH_2SiH_3$. Mixed hydrides containing silicon or germanium and certain other atoms are also fairly stable. Examples include the silicon–phosphorus hydrides SiH_3PH_2, both isomers $(SiH_3)_2PH$ and $SiH_3SiH_2PH_2$, and $(SiH_3)_3P$.

9.5.2 Properties

The thermal stabilities of the hydrides decrease in each Group as the atomic weight of the central element increases. Except in the case of carbon compounds (and, to some extent, boron ones), the thermal stability decreases fairly rapidly with increasing molecular weight for the members of a homologous series. The variation of stability across a Period is irregular, with the carbon Group and the halogens forming the most stable hydrides, e.g. the hydrides of Ga $<$ Ge $>$ As $<$ Se $<$ Br in stability.

The structures of the hydrides are as predicted from the number of electron pairs on the central element, cf. Chapter 4. All the carbon Group hydrides are tetrahedral MH_4 molecules and the higher homologues are also based on tetrahedra. The nitrogen and oxygen Group molecules are based on tetrahedra with, respectively, one and two lone pairs. The bond angles decrease towards 90° with increasing atomic weight of the central element. The boron Group hydrides are discussed later. Hydrazine, N_2H_4, and hydrogen peroxide, H_2O_2, adopt the structures shown in Fig. 9.7 which separate the lone pairs as much as possible.

The reactions of the hydrides are varied and many are familiar: for example, the reactions of hydrogen sulfide, hydrogen halides, water and ammonia. In general terms, all the hydrides are reducing agents and react strongly with oxygen and halogens. Stability to oxygen varies from the relatively stable germane, GeH_4, and hydrogen halides, to the cases such as silane, SiH_4, and phosphane, PH_3, which explode or inflame in air. With the halogens, reactions may also be violent, although iodine often reacts smoothly to cleave only one bond as in:

$$GeH_4 + I_2 \rightarrow GeH_3I + HI$$

The hydrogen halides also react with many hydrides to give partial substitution:

$$SiH_4 + HX \xrightarrow{AlX_3} SiH_3X + H_2$$

The hydrides of most elements are strong reducing agents which reduce many heavy metal salts to the element:

$$PH_3 + AgNO_3 \rightarrow Ag$$

$$SnH_4 + CuSO_4 \rightarrow Cu, \text{ etc.}$$

Exceptions are provided by the hydrides of the first short Period which are relatively unreactive. One reason for this probably lies in the absence of an easy reaction path. Thus silane, SiH_4, reacts readily as the silicon 3d orbitals can provide a means of coordinating an attacking reagent in a reaction intermediate; in methane, CH_4, there is no such pathway. Reaction depends on the breaking of a C-H bond which requires much more energy and is correspondingly slower.

One important reaction of many covalent hydrides is ionization to give a positive hydrogen species. The hydrides of the most electronegative elements are already polarized, with the hydrogen positive. Although the formation of the free proton is impossible, as shown above, the hydrogen bonded to an electronegative element can ionize to give a proton stabilized by bonding to one or more neutral molecules. In the hydrides of the most electronegative elements, F, O and N, this is accomplished by self-ionization in the liquid phase when the proton is associated with a molecule of the hydride (compare Section 6.5 and Figs 9.1 and 9.2):

(a)

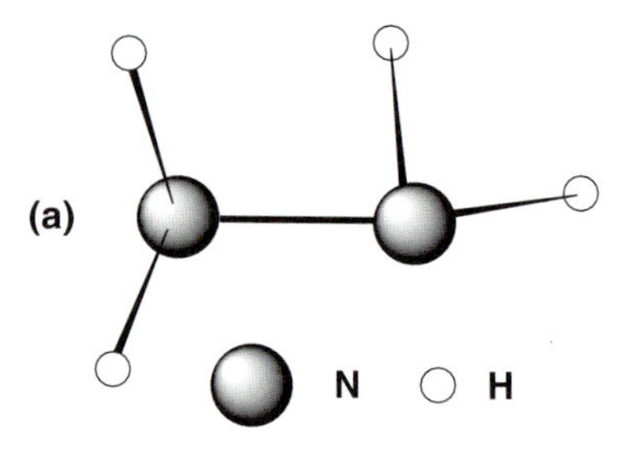

(b)

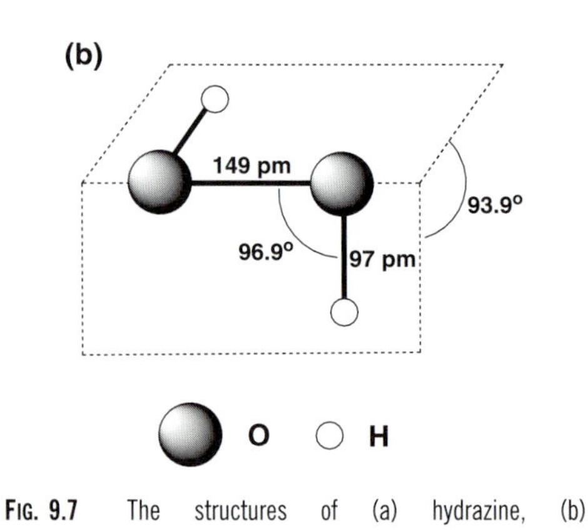

FIG. 9.7 The structures of (a) hydrazine, (b)

$$2NH_3 \rightarrow NH_4^+ + NH_2^-$$
$$2H_2O \rightarrow H_3O^+ + OH^-$$
$$2HF \rightarrow H_2F^+ + F^-$$

The hydrides of less electronegative elements do not self-dissociate in the liquid phase (largely because the liquids are less efficient ion supporters), but they do dissociate when dissolved in an ionizing solvent such as water. Such hydrides are those of the other halogens, and of the other chalcogens, which dissociate in water or ammonia. The energies involved in such ionizations may be illustrated by the case of hydrogen chloride. The formation of the free proton in the gas phase, $H \rightarrow H^+$, requires 1340 kJ mol^{-1}, and the hydration energy of the gaseous proton is about –1090 kJ mol^{-1}. For hydrogen chloride:

$$HCl_{(gas)} \rightarrow H^+_{(gas)} + Cl^-_{(gas)} \qquad \Delta H = 1380 \text{ kJ mol}^{-1}$$

$$H^+_{(gas)} + Cl^-_{(gas)} \rightarrow H^+_{(aq)} + Cl^-_{(aq)} \qquad \Delta H = -1460 \text{ kJ mol}^{-1}$$

so ionization of hydrogen chloride in aqueous solution is exothermic by about 80 kJ mol^{-1}. The hydrogen halides are fully dissociated in water but most other hydrides which dissociate in water do so only very weakly. For instance, the dissociation constant for

$$H_2S + H_2O \rightleftharpoons H_3O^+ + HS^-$$

is only about 10^{-7}. The hydrides of phosphorus and the other members of the nitrogen Group do not dissociate measurably in water. This rapidly decreasing tendency to dissociate in solution clearly parallels the fall in the polarity of the element–hydrogen bond.

The formation of H_3O^+ or NH_4^+ may be considered as the result of the donation of an electron pair to the proton by the oxygen or nitrogen atoms, $H_3N: \rightarrow H^+$ or $H_2O: \rightarrow H^+$ This is one case of the general donor–acceptor (Lewis base–acid) behaviour of the hydrides. The hydrides with one or more lone pairs of electrons may donate them to suitable acceptor molecules to form coordination complexes. This is, of course, the reaction involved in the solvation of cations by water or ammonia, but it is a general reaction possible for all the hydrides of the nitrogen, oxygen or halogen Groups. In general, donor power falls as the number of lone pairs on the central atom increases and as the size of the central atom increases. Thus the hydrogen halides are weak donors as there are three lone pairs on the halogen atoms, and the heavier analogues of ammonia and water, such as SbH_3 or H_2Se, are also weak. However, in super-acidic systems such as HF/SbF_5 mixtures, the sulfonium and selenonium-containing compounds $H_3S^+SbF_6^-$ and $H_3Se^+SbF_6^-$ can be isolated and characterized.

Acceptor molecules among the hydrides are largely confined to those of the boron Group, although there is some evidence of weak d orbital acceptor power in the carbon Group tetrahydrides. The complex hydride anions, BH_4^-, AlH_4^- and GaH_4^-, may be regarded as being formed by the acceptance of the electron pair on the hydride ion by the MH_3 species, $H^-: \rightarrow MH_3$. Donor–acceptor complexes are also formed between the boron Group hydrides and those of the nitrogen and oxygen Groups. Compounds such as H_3BNH_3 are formed at low temperatures but, on warming towards room temperatures, these compounds lose hydrogen and polymerize (in this case to the ring compound $B_3N_3H_6$ discussed in Section 9.6). If organic derivatives are used, the complexes are more stable; both H_3BNMe_3 and Me_3BNH_3 are stable at room temperature. The much less-stable polymeric hydrides of aluminium and gallium also form such complexes and considerable stabilization of the M-H bonds results. Complexes such as R_3NAlH_3 and R_3NGaH_3 are well characterized. The gallane complex (where $R = CH_3$) is monomeric in the crystalline state, whereas the corresponding alane complex is a dimer, with two hydride bridges. However, the chemistry of this system is clearly more complex than initially thought, since changing the amine to dimethylamine (Me_2NH) results in both the alane and gallane derivatives $(Me_2NH)EH_3$ (E = Al, Ga) being dimeric in the solid state, with highly unsymmetrical E-H· · · ·E bridges. These elements also make use of their d orbitals to accept more than one donor molecule. Thus, the compound $AlH_3.2NMe_3$ has the trigonal bipyramidal structure

ORGANIC-SUBSTITUTED DERIVATIVES AS ELECTRON-PAIR DONOR LIGANDS

Organic-substituted hydrides of the nitrogen and oxygen Group elements, especially the phosphanes, R_3P, and the sulfanes, R_2S (where R includes aliphatic and aromatic groups), do act as donors to a wide variety of species. Acceptor molecules with suitable empty orbitals available include the p acceptors of the boron Group (and beryllium), the *n*d acceptors such as the tetrafluorides of the carbon Group, and especially the $(n-1)$ d acceptors among the transition metals. Phosphane ligands have a very important role in the chemistry of the transition metal elements.

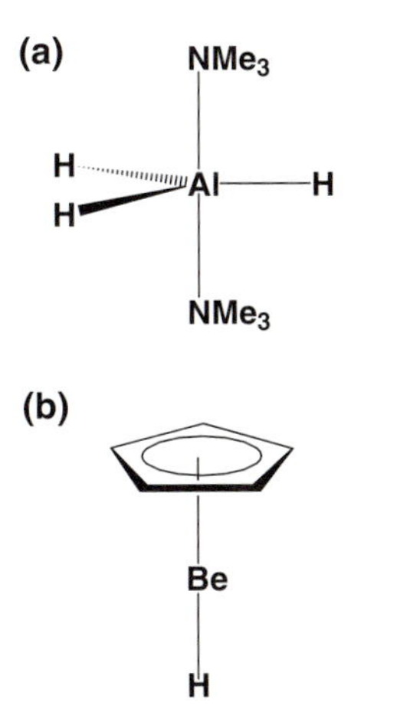

FIG. 9.8 (a) The structure of $AlH_3.2NMe_3$. (b) The structure of $(C_5H_5)BeH$

shown in Fig. 9.8a. The interesting species $(C_5H_5)BeH$, Fig. 9.8b, has a planar C_5H_5 ring π-bonded to Be with Be-H on the five-fold axis.

Many of these ions and hydrides form isostructural sets, all based on four electron pairs:

four bond pairs	tetrahedron	BH_4^-, CH_4, NH_4^+
three bond pairs and one lone pair	pyramid	NH_3, H_3O^+
two bond pairs and two lone pairs	V-shape	H_2O, H_2F^+

9.6 Electron deficient hydrides

9.6.1 Boron hydrides

Boron and related elements form a range of hydrogen compounds which cannot be accounted for by classical ideas of electron pair bonds between two atoms. It is found that the simple monomeric species, such as BeH_2 or BH_3, do not occur but that more complicated compounds are formed. Thus the simplest boron hydride is B_2H_6 while beryllium and aluminium form high molecular weight polymers, $(BeH_2)_x$ and $(AlH_3)_x$. The most fully studied compounds of this class are the hydrides of boron and these are treated first.

The work of Stock, starting in 1909, on the boron hydrides, was one of the sources of the renaissance in inorganic chemistry in the 20th century. He characterized the compounds listed in the first column of Table 9.5. It took 25 years, until 1958, before the next compound, B_9H_{15}, was added to Stock's list and only in the last decade has the chemistry of compounds larger than his $B_{10}H_{14}$ been much explored. The full extent of Stock's genius and experimental skill becomes apparent when it is realized that all these hydrides are inflammable, many are very unstable, and all the products identified by Stock and his team appeared as mixtures either from the hydrolysis of magnesium boride or from interconversions of these.

The modern development of this field started in the 1960s and has continued rapidly ever since, expanding into further classes of compounds, like the carboranes which include C atoms in the skeleton. Table 9.5 lists the established binary boron hydrides. Clearly the field is extremely complex and can only be introduced here, and we shall focus on the simpler compounds from the first column.

The boron hydrides form part of an exceptional group of compounds. On simple valency grounds, the lowest hydride of boron is expected to be BH_3, but no such molecule has ever been discovered—despite considerable search. The simplest boron hydride is diborane, B_2H_6, which many studies have shown to have the bridge structure of Fig. 9.9 (see Section 7.8). There are only twelve valency electrons in B_2H_6 (3 per B and 1 per H), so that the molecule cannot be made up of electron pair bonds, which would require

An interesting account of the development of the structure of diborane is given by P. LASZLO, *Angewandte Chemie, International Edition* **39**, 2000, 2071.

TABLE 9.5 Some boron hydrides

B_2H_6	B_6H_{14}	$B_{11}H_{15}$
B_4H_{10}	B_7H_{13}	$B_{12}H_{16}$
B_5H_9	B_8H_{12}	$B_{13}H_{19}$
B_5H_{11}	B_8H_{14}	$B_{14}H_{18}$
B_6H_{10}	B_8H_{16}	$B_{14}H_{20}$
B_6H_{12}	B_8H_{18}	$B_{14}H_{22}$
$B_{10}H_{14}$	*n*- and iso-B_9H_{15}	$B_{15}H_{23}$
	$B_{10}H_{16}$	$B_{16}H_{22}$
	$B_{10}H_{18}$	*syn*- and *anti*-$B_{18}H_{22}$
	$B_{10}H_{20}$	$B_{20}H_{16}$
		9 isomers of $B_{20}H_{26}$
		$B_{30}H_{38}$

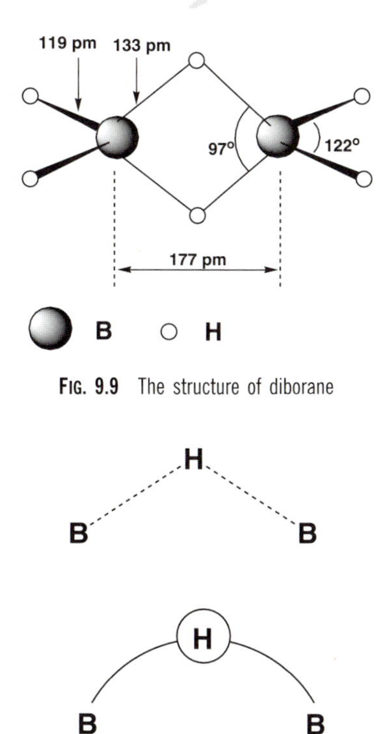

FIG. 9.9 The structure of diborane

Representations of a three-centre B···H···B bond

sixteen electrons for the structure of Fig. 9.9. Such molecules, with insufficient electrons to form two-centre electron pair bonds between all the atoms, are termed *electron deficient*. The evidence of bond lengths and angles, and of the infrared stretching frequencies, suggests that the terminal B-H bonds are normal single bonds. This leaves the two bridging H atoms, together with four electrons, to be fitted into the picture. The most satisfactory description of the bridge was that proposed by Longuet-Higgins. He suggested that, in place of the normal electron pair bond centred on two nuclei, a three-centre bond should be considered, made up of the hydrogen 1s orbital and appropriate hybrids on the two boron atoms.

If the boron atoms form sp^3 hybrid orbitals, two of which form the bonds to the terminal hydrogens, then the others may overlap with the hydrogen 1s orbitals as shown in Fig. 9.10. Figure 9.11 gives the energy level diagram for such bonds. The three atomic orbitals from B_1, H_A and B_2 combine to form a bonding, a nonbonding and an antibonding three-centred molecular orbital centred on these atoms. (Note that three atomic orbitals give three molecular orbitals.) The three atomic orbitals on B_1, H_B and B_2 give an identical set of three molecular orbitals. If two of the four electrons, left over after the terminal B-H bonds were formed, are placed in the $B_1H_AB_2$ bonding orbital and the other two in the $B_1H_BB_2$ bonding orbital, as in Fig. 9.11b, the result is to use all the valency electrons and to fill only the three-centred molecular orbitals which are bonding. Diborane is thus described as having four two-centre B-H bonds and two three-centre B····H····B bonds, all σ, all holding two electrons, and accounting for the twelve valency electrons.

The description of these three-centre bonds is one example of the polycentred σ bonding discussed in Chapter 4. Here the three-centre description is by far the most satisfying of the alternatives possible.

Diborane serves as a model compound for the other boron hydrides and other hydrogen compounds which are formulated with similar polycentre bonding. For example, tetraborane, B_4H_{10}, is shown in Fig. 9.12a. This has six two-electron B-H bonds, one B-B bond, and four two-electron three-centre B····H····B bonds. The structures of higher boron hydrides may be built up similarly, but other types of polycentre bonds may also be present including three-centre B····B····B bonds and even five-centre bonds involving five boron atoms, as found in B_5H_9, Fig. 9.12.

In addition to the boron hydrides listed in Table 9.5, there is a wide range of boron hydride ions, of similar formulae and a variety of structures, and bonding in these species also involves multi-centred bonding. Examples include $B_6H_6^{2-}$ and $B_{12}H_{12}^{2-}$ which have all

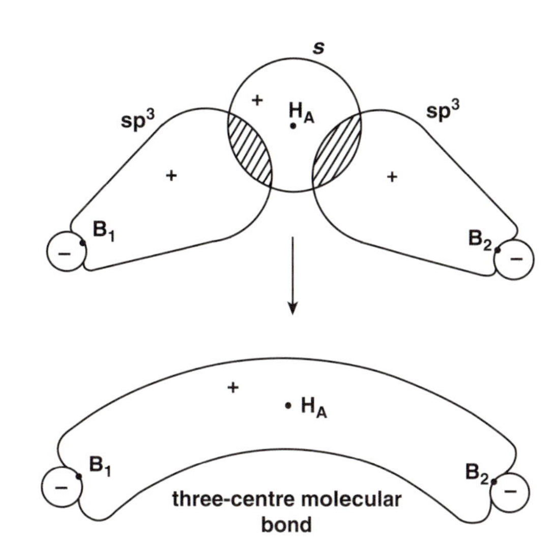

FIG. 9.10 Three-centre bonds in diborane. A tetrahedral hybrid orbital on B_1, an s orbital on H_A and a tetrahedral orbital on B_2 combine to form the bonding orbital shown, an antibonding orbital with nodes between each B and the H, and a nonbonding orbital which places the electron density mainly on the two B atoms. A similar set of three three-centred orbitals is formed by H_B with two other tetrahedral orbitals on B_1 and B_2

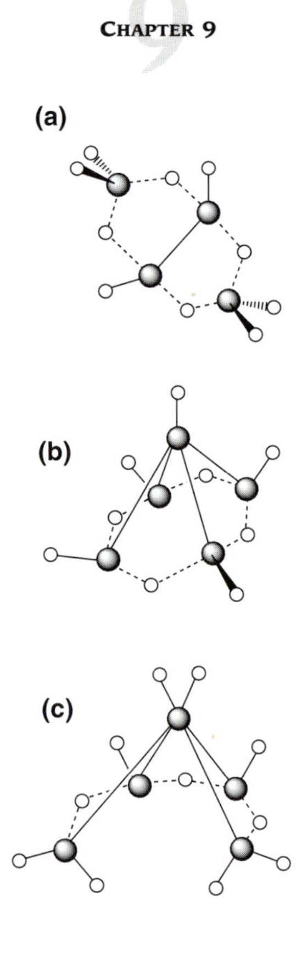

FIG. 9.12 Structures of (a) tetraborane, (b) pentaborane-9, (c) pentaborane-11 (a) Tetraborane forms a boat-shaped molecule with four B····H····B three-centre bonds and a direct B-B single bond. (b) Here the boron atoms form a square pyramid. The edges of the square base are bonded by three-centre B····H····B bonds while the apical B atom is bonded to the four base B atoms by five-centre bonds using boron orbitals directed into the body of the pyramid. (c) Pentaborane-11 is an unsymmetrical square pyramid where, in addition to B····H····B bonds, there are three-centre bonds between three boron atoms in the triangular faces—the so-called 'central' bond B B / B

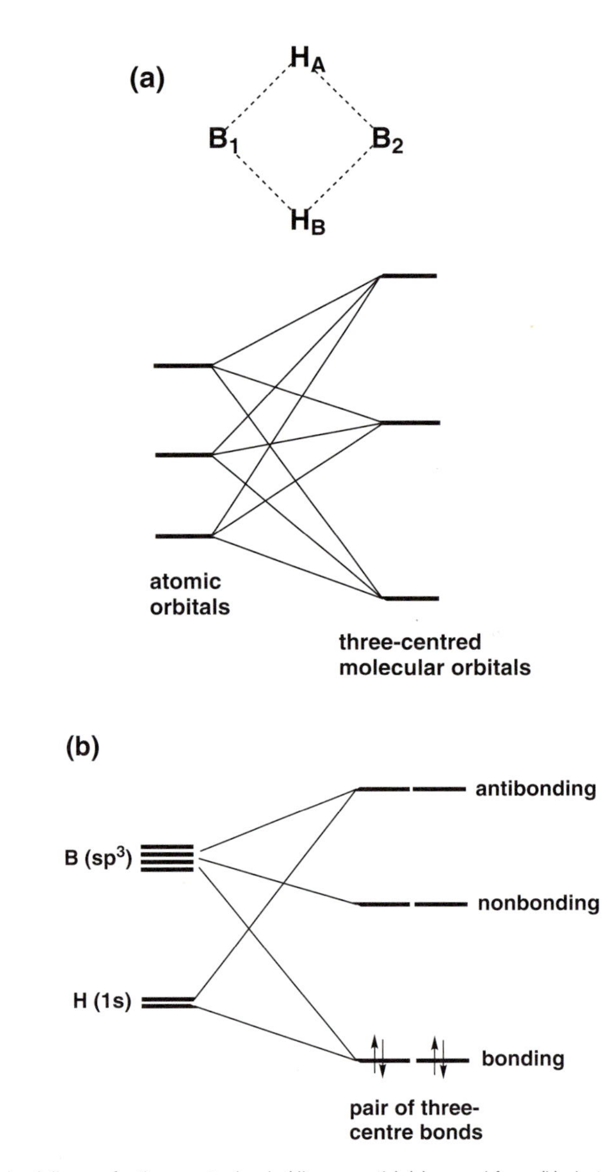

FIG. 9.11 Energy level diagram for three-centre bonds (diagrammatic) (a) general form, (b) electronic structure for the pair of three-centre bonds in diborane. Figure (a) shows the generalized energy level diagram for three atomic orbitals combining to form three three-centred molecular orbitals. The energy level diagram (b) corresponds to the two sets of three-centred orbitals formed by the s orbitals on H_A and H_B and two suitably directed orbitals on each of B_1 and B_2 in the diborane bridge. The six atomic orbitals give six three-centred molecular orbitals and, in diborane, the two bonding orbitals are filled

the hydrogens present as terminal B-H bonds. The B_6 skeleton is a regular octahedron and the B_{12} one a regular icosahedron held together by multi-centred B_n bonds. Some of these ions contain atoms other than boron. Thus there is an ion $B_9C_2H_{11}^{2-}$ which can be regarded as derived from $B_{12}H_{12}^{2-}$ by removing one apex of the icosahedron (taking away one BH unit) and substituting two other B atoms with carbon atoms. The chemistry of such *carboranes* has expanded very rapidly in the last decade or so. Compounds range from relatively small molecules such as $C_2B_3H_7$ (isoelectronic with B_5H_9) and $C_2B_3H_8$ up to icosahedral species such as $C_2B_{10}H_{12}$. A very rich and varied chemistry has developed which is introduced in the references.

9.6.2 Wade's rules

The wide variety of structures of the boron hydrides and their analogues have been rationalized by several different approaches. One of the simplest, which applies also to other clusters, is the Skeletal Electron Pair (SEP) approach which was presented in a convenient form by Wade, and is thus often known as Wade's rules. It is useful to introduce the ideas here in the context of the simpler boron hydrides—see Section 18.4.2 for a fuller discussion.

This approach starts by building a boron hydride from n B atoms, each with one outward-pointing B-H bond external to the cluster. (Any further hydrogens are attached in other ways.) That is, any hydride B_nH_m is formulated as $(BH)_nH_{m-n}$. The external B-H is a normal two-electron, two-centre bond and thus uses one orbital (conveniently regarded as an sp^3 hybrid) from the B.

The skeletal electron pairs are then calculated by:

(a) finding the total number of valency electrons (i.e. 3 per B, 1 per H, plus or minus electrons for any charge)
(b) subtracting 2 electrons for each BH
(c) leaving the skeletal electron pairs, which bind the cluster.

Then the key predictions are:

Regular solids are those whose faces are all equilateral triangles.

(1) The structure observed for a molecule with $(n+1)$ skeletal electron pairs is the regular solid with n vertices, provided there are n B atoms. Such structures are closed clusters and given the descriptor *closo*.

(2) If there are only $(n-1)$ B atoms in a cluster with $(n+1)$ SEP, the structure is that formed by removing one vertex from the *closo* cluster (called *nido*).

nido means nest and *arachno* means spider and hence web.

(3) Similarly, for $(n-2)$ B atoms, two vertices are removed (called *arachno*). Any greater deficiency is rare.

(4) If there are more than n B atoms with $(n+1)$ SEP, the additional B go into capping positions over the triangular faces.

For mixed clusters, such as carboranes, the SEP approach also applies. Thus C is assumed to form an external C-H bond, and it contributes four valency electrons.

The theory has been considerably refined but this basic level suffices to account for simple species. Thus the two pentaboranes in Fig. 9.12 are treated as follows:

Pentaborane-9 (B_5H_9)

valency electrons for 5B =	15
for 9H =	9
less a pair of electrons for 5BH =	−10

This leaves 14 electrons or 7 SEP and the structure is predicted to be related to the 6-vertex regular solid which is the octahedron. As there are only 5 B, one vertex is removed. Thus the predicted structure is the *nido*-octahedron or square pyramid as observed.

Pentaborane-11 (B_5H_{11})

5B =	15 electrons
11H =	11 electrons
less 5BH =	−10 electrons

This gives 8 SEP, which is a structure based on the 7-vertex regular solid, the pentagonal bipyramid. If one vertex and one of the pentagonal plane atoms are removed, the result is the structure of Fig. 9.12c.

In these two molecules the extra hydrogens bridge the edges in B_5H_9 and are also added as terminal bonds to the boron atoms in the open face for B_5H_{11}.

It is readily seen that the $[B_6H_6]^{2-}$ and $[B_{12}H_{12}]^{2-}$ ions count to 7 and 13 SEP respectively, giving the octahedron and the 12-vertex figure which is the icosahedron. B_4H_{10} counts to 7 SEP, so the structure should be the *arachno*-octahedron. Removing two adjacent vertices leaves two triangles joined by an edge as observed. A summary of the commoner shapes is given in Table 9.6.

It can be seen from Table 9.6 that, while the simple theory rationalizes observed shapes, it is not predictive in detail. Thus the alternatives for 7 SEP and 4 atoms depend on whether the two removed vertices are adjacent or opposite, and the simple theory gives no guide. Likewise, placing two capping atoms often offers a choice. As the trigonal bipyramid is equivalent to the capped tetrahedron, we find the tetrahedron (and the triangle) occurring for different SEP counts.

TABLE 9.6 Shapes predicted by Wade's rules for simple clusters

SEP (n+1)	*Closo n vertices, n atoms*	*nido (n−1) atoms*	*arachno (n−2) atoms*
4	Triangle		
5	Tetrahedron	Triangle	
6	Trigonal bipyramid	Tetrahedron	Triangle
7	Octahedron	Square pyramid	Square or linked triangles
8	Pentagonal bipyramid	Pentagonal Pyramid	Irregular (compare B_5H_{11})

Despite these ambiguities, the theory has the advantage of being extremely simple and of applying to the vast majority of the large number of boranes, carboranes and related compounds. It accounts simply for the finding that some 6-atom molecules are octahedra and some are pentagonal pyramids, or that 5-atom molecules can be trigonal bipyramids or square pyramids, etc. It may be extended to transition metal clusters (see Section 18.4.2).

9.6.3 Other electron deficient hydrides

Another example of Al····H····Al bridging is found in the molecule $Al_2H_5NMe_2$ which has the structure.

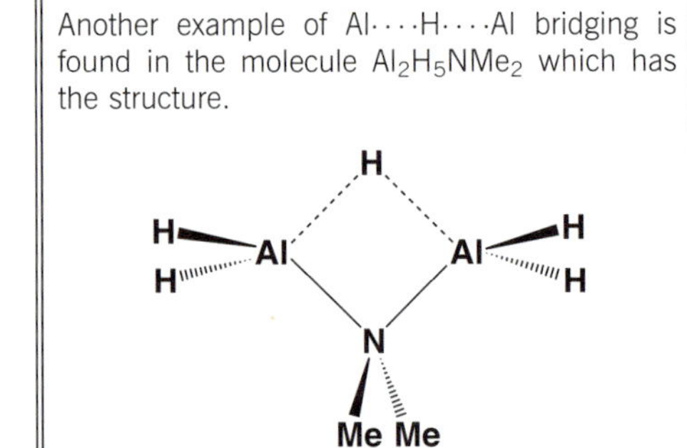

Electron deficient bonding is by no means confined to boron compounds. Aluminium hydride is an insoluble polymer $(AlH_3)_x$, and a recent structural determination by X-ray and neutron diffraction shows that it contains Al atoms surrounded octahedrally by six H atoms. The structure is a giant molecule of AlH_2Al bridges and is analogous to the structure of AlF_3. This is an example of electron deficient bonding analogous to that in diborane. The simple AlH_3 molecule has been prepared as a highly reactive molecular species in a noble gas matrix, by the reaction of Al atoms and hydrogen.

The somewhat elusive binary gallium hydride $[GaH_3]_n$ has recently been synthesized from dimeric $[H_2GaCl]_2$ and $LiGaH_4$ at low temperature. The principal product is the diborane-like dimer $[GaH_3]_2$, but there is evidence for higher molecular weight species in the solid state containing four or more Ga atoms, but still showing terminal Ga-H bonds. The mixed galloborane $H_2Ga(\mu\text{–}H)_2BH_2$ has been synthesized by the corresponding reaction of $[H_2GaCl]_2$ with $LiBH_4$. The compound again has the diborane-type structure, but aggregation occurs in the solid state, with an X-ray diffraction study showing a polymeric network with helical chains made up of alternating tetrahedral GaH_4 and BH_4 units linked through single hydrogen bridges. Beryllium hydride, $(BeH_2)_x$, is also polymeric and insoluble. The structure is reasonably postulated to be a chain polymer with bridging hydrogens.

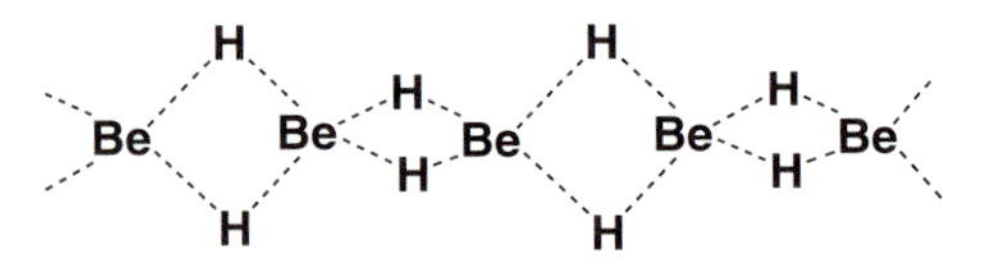

Hydrogen bridging to beryllium is established in a number of other species, for example, the ion $Et_2BeH_2BeEt_2^{2-}$ with the structure

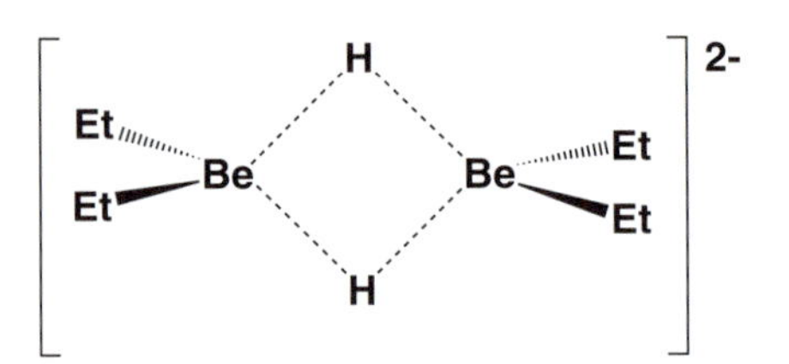

Electron deficient bonding is also found in some of the organic analogues of the hydrides. For example, trimethyl-aluminium exists as a dimer, $Al_2(CH_3)_6$, with a *methyl-bridged* structure very similar to that of diborane. This is based on three-centre bonds formed by an sp^3 hybrid on each aluminium together with an sp^3 hybrid on the carbon atom, as shown in Fig. 9.13. The mixed gallium alkyl/halide/hydride compounds $Me_2Ga(\mu\text{-}H)_2GaMe_2$ and $Me_2Ga(\mu\text{-}Cl)_2GaMe_2$ have hydride or halide bridges in preference to alkyl ones. For the aluminium analogue $[Me_2AlH]_n$ (prepared by reaction of Me_3Al with $LiAlH_4$) the predominant form at low pressure and a temperature of 470 K is the dimer ($n = 2$) but at low temperature (330 K) some tetramer ($n = 4$) occurs. The related compound $Me_2Ga(\mu\text{-}H)_2BH_2$ and $[Et_2Be(\mu\text{-}H)_2BeEt_2]^{2-}$ have similar structures, with two hydride bridges.

These electron deficient bond systems are of reasonable strength. Thus, the boron hydrides resemble the silicon hydrides in thermal stability. The hydrides share the strong reducing properties of all hydrides and the boron hydrides react violently with oxygen. The electron deficiency shows up mainly in the susceptibility of the hydrides to attack by electron pair donors, which provide the extra electrons to allow the formation of electron pair two–centre bonds. Diborane, for example, readily reacts to give borine adducts, $H_3B \leftarrow D$, with suitable donors, D, such as the amines.

The boron Group of elements also makes use of the empty p orbital in the formation of the complex hydrides. The BH_4, AlH_4 and GaH_4 groups have been prepared and they decrease in stability in that order. Many elements form these complex hydrides and they range from the ionic compounds of the active metals, such as $Na^+BH_4^-$ or $Li^+AlH_4^-$, to the compounds of the less active metals which appear to be covalent and electron deficient. A structure with unsymmetric three-centre B····H····Be bonds has been reported for beryllium borohydride, $Be(BH_4)_2$.

For aluminium borohydride, $Al(BH_4)_3$, the Al atom is surrounded by six hydrogen atoms, forming three AlH_2B bridges while the low temperature form of $Zr(BH_4)_4$ probably contains zirconium surrounded tetrahedrally by four boron atoms linked by three-hydrogen bridges. In the related anionic complex $[Al(BH_4)_4]^-$, the four BH_4 units are arranged tetrahedrally around the aluminium, with each BH_4 contributing two hydrogen bridges to the aluminium, giving 8-coordination. Complex aluminium hydrides of the less electropositive metals probably have similar bridging structures. Even in $LiAlH_4$, the Li-H distance is shorter than in LiH, suggesting some Li····H interaction which probably accounts for its solubility in ethers. The complex beryllium hydrides, Li_2BeH_4 and Na_2BeH_4, also behave more like electron deficient polymers than as species containing BeH_4^{2-} ions.

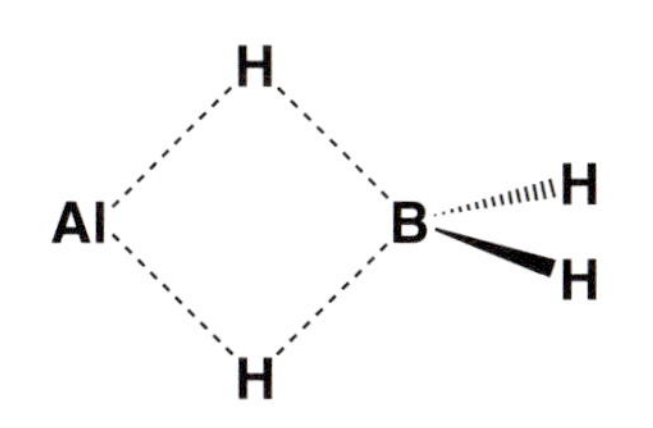

A two-hydrogen bridge in Al(BH₄)₃

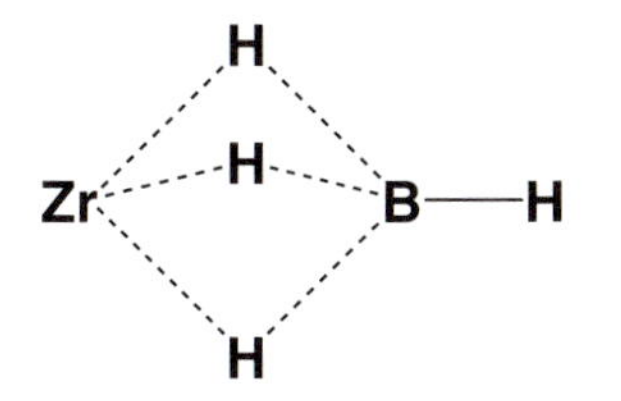

A three-hydrogen bridge in Zr(BH₄)₄

It has been seen that hydrogen-bridged, electron deficient species, where polycentred σ bonding has to be postulated, occur quite widely. Although best known for boron, it is also established for compounds of beryllium, aluminium and zinc together with the B····H····M links to many transition metals in their borohydrides. M····H····M bridges are also established in metal-hydrido complexes such as $HCr_2(CO)_{10}^-$. It is also probable that a weaker interaction, of the form Al-H····Li, occurs in complex beryllium, zinc and aluminium hydrides. Thus electron deficient polycentred bonding is typical of hydrides of elements which have fewer valence shell electrons than valency orbitals. Its presence allows use of orbitals which would otherwise remain empty, and the interaction is sufficiently strong to give aluminium, and probably other elements, a coordination number greater than four. Similar conditions hold for the formation of methyl and similar bridges.

The boron group hydrides react readily with electron pair donors to revert to two-centre two-electron bonding and adducts such as $H_3B.NMe_3$ are readily formed. The heavier elements also give 5- and 6-coordinate adducts (compare Fig. 9.8a). While the simple gallium hydrides decompose readily, coordinated hydrides like Me_3NGaH_3 are much more stable and readily characterized. As expected the coordination at Ga is tetrahedral.

If the donor group is itself a hydride, further interaction may occur with elimination of H_2. For example, the product of the reaction between ammonia and diborane at low temperatures is the expected adduct, H_3BNH_3. On warming to room temperature, this compound loses hydrogen and gives a product of empirical formula, HBNH. The latter is actually the trimer, called borazole or borazine, whose structure is shown in Fig. 9.14. The molecule, $B_3N_3H_6$, is isoelectronic with benzene, C_6H_6, and has a similar, planar structure with all the B-N bonds the same and a delocalized π-bonding system. A wide variety of substituted borazines have been made, either by substitution reactions on borazine or, more commonly, by altering the composition of the starting adduct. Thus, $(CH_3)H_2BNH_3$ gives, on heating, $(CH_3)_3B_3N_3H_3$—called B-trimethylborazine (where the methyl groups are on the boron). The corresponding *N*-trimethylborazine is formed from the adduct $H_3BNH_2(CH_3)$. Many similar compounds can be made such as the chloro-boron derivative of Fig. 9.14c. Later work showed the existence of an aluminium analogue $(MeAlNAr)_3$.

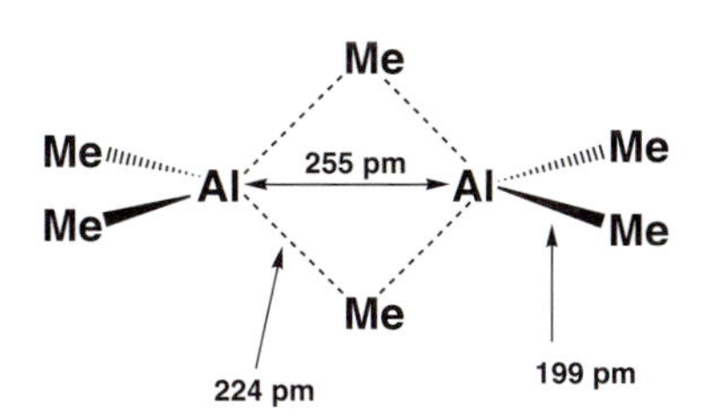

FIG. 9.13 The structure of Al_2Me_6. This diagram gives the form of aluminium trimethyl dimer. The methyl group is bonded to two aluminium atoms in three-centre bonds formed by overlap of tetrahedral hybrid orbitals on each aluminium and on the carbon atom. This bonding resembles the B····H····B bond in diborane, and also the central B–B–B bonds found in pentaborane-11

In contrast, if the ammonia is replaced by phosphane, PH_3, the adduct H_3BPH_3 loses less hydrogen to give $(H_2BPH_2)_3$ which has a different ring system. This is the analogue, not of

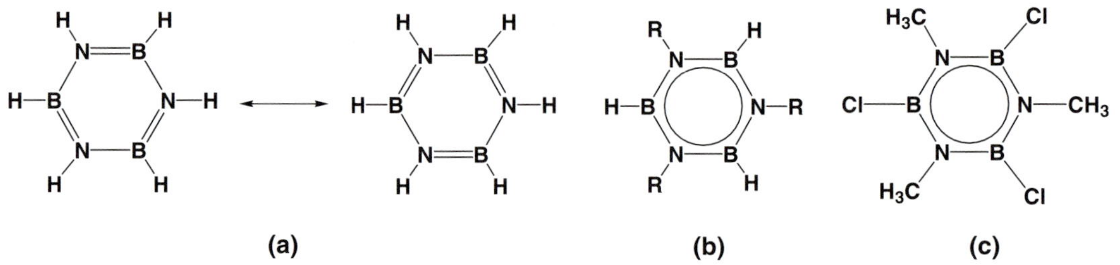

(a) (b) (c)

FIG. 9.14 (a) The structure of borazine ($H_3B_3N_3H_3$), (b) *N*-trialkyl borazine, (c) B-trichloro-*N*-trimethylborazine

benzene but of cyclohexane, C_6H_{12}, and the ring is no longer planar but chair-shaped. Similar compounds may be made by starting from substituted phosphanes.

9.7 The hydrogen bond

Compounds containing hydrogen bonded to a very electronegative element, especially F, O or N, show properties which are consistent with an interaction between the hydrogen bonded to one electronegative atom and a second atom. This secondary bond is relatively weak and is termed the hydrogen bond. It may be written X-H····Y, where X is the atom to which the hydrogen is bonded by a normal bond (compare also Figs 9.1 and 9.2). The importance of hydrogen bonding cannot be overstressed since substances such as water and DNA, both essential to life, are extensively hydrogen-bonded. While the double-helix structure of the DNA molecule is now well understood, the structure of liquid water still poses many challenges to the chemist, due to the formation of 'dynamic' hydrogen bonds. Protons are able to be transferred very rapidly from the oxygen atom of one water molecule to another, with a frequency of around $10^{12}\,s^{-1}$. The migration of a proton in water is accomplished via a cooperative cascade of proton transfers between adjacent water molecules, called the *Grotthus mechanism*. In contrast to liquid water, the various crystalline polymorphs of ice are better understood since experimental methods give much more direct information on the structure of the solid form.

Evidence for hydrogen bonding is widespread. Some of the important facts are:

(i) *Evidence of molecular association* from melting points, boiling points, heats of evaporation and, in some cases, molecular weight determinations. For example, the boiling points of the hydrides of the carbon, nitrogen, oxygen and fluorine Group elements are shown in Fig. 9.15. For the carbon Group hydrides, the boiling point

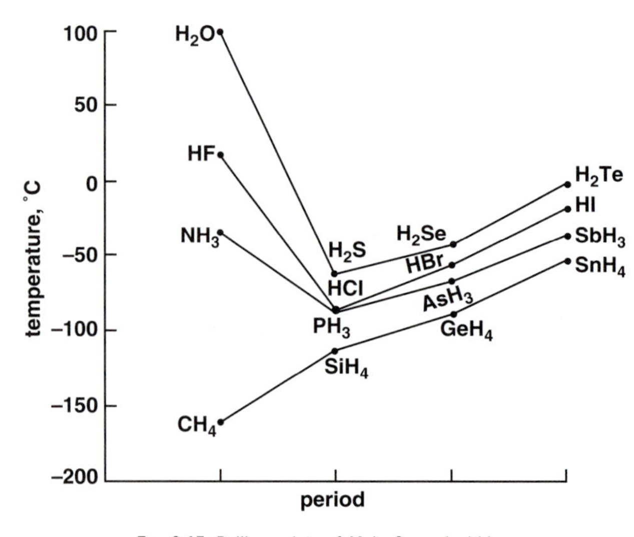

FIG. 9.15 Boiling points of Main Group hydrides

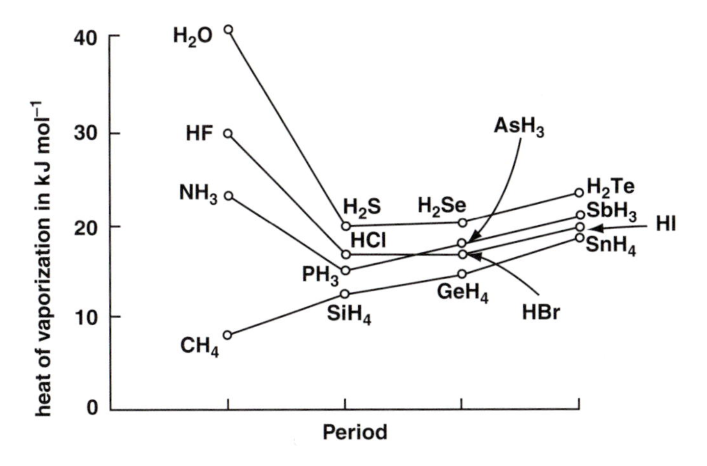

FIG. 9.16 Latent heats of vaporization of Main Group hydrides

is a linear function of atomic weight. However, the positions of ammonia, water and hydrogen fluoride are clearly anomalous. These unexpectedly high boiling points indicate an extra interaction between the molecules in the liquid, which is not found for the heavier hydrides. A similar anomaly appears in the latent heats of vaporization shown in Fig. 9.16, and also in the melting points and the latent heats of fusion, showing that the interaction is also present in the solid. Fig. 9.16 serves to give some idea of the size of the effect; the increases in the heats of evaporation, compared with those found by extrapolating from the heavier hydrides, is of the order of 20 ± 10 kJ mol^{-1}. In general, the hydrogen bond energies may be taken to lie in the range from 4 to 40 kJ mol^{-1}, about one tenth of normal bond energies but several times greater than normal weak interactions between molecules.

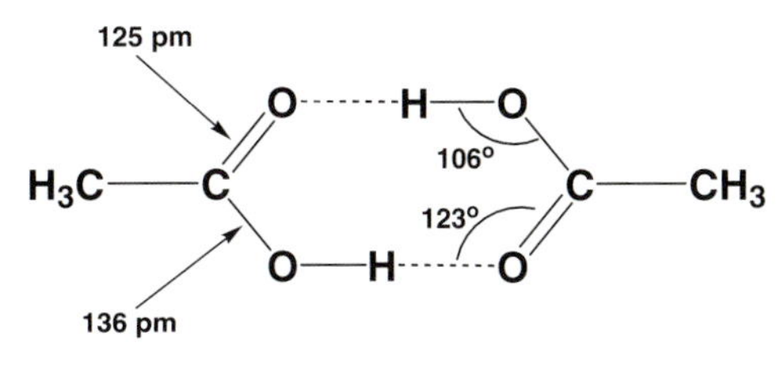

FIG. 9.17 The structure of the acetic acid dimer

Molecular weight studies show clear evidence of association in many cases: perhaps the most striking case is the dimerization of the lower carboxylic acids. The heat of association of acetic acid:

$$CH_3COOH = \tfrac{1}{2}[CH_3COOH]_2$$

is found to be 28.9 kJ mol^{-1} (of monomer). Electron diffraction studies have shown that the structure of the acetic acid dimer is as given in Fig. 9.17.

Other carboxylic and related oxy-acids behave similarly, for example dialkyl phosphoric acids, $(RO)_2P(O)OH$. Organic solvent-soluble long- and branched-chain analogues of these compounds are employed industrially in hydrometallurgical extraction and separation of various metals (e.g. the separation of cobalt from nickel).

(ii) *X-ray studies.* The evidence from boiling points, etc., above gives no indication of the atomic arrangements in the hydrogen bond and this information might be expected from X-ray studies. However, as hydrogen is so light compared with other nuclei, it is not an efficient scatterer of X-rays and it is not usually possible to determine the positions of hydrogen atoms in this way. One exception is boric acid, H_3BO_3, which contains only light nuclei and whose structure consists of loosely bound sheets with the configuration shown in Fig. 9.18.

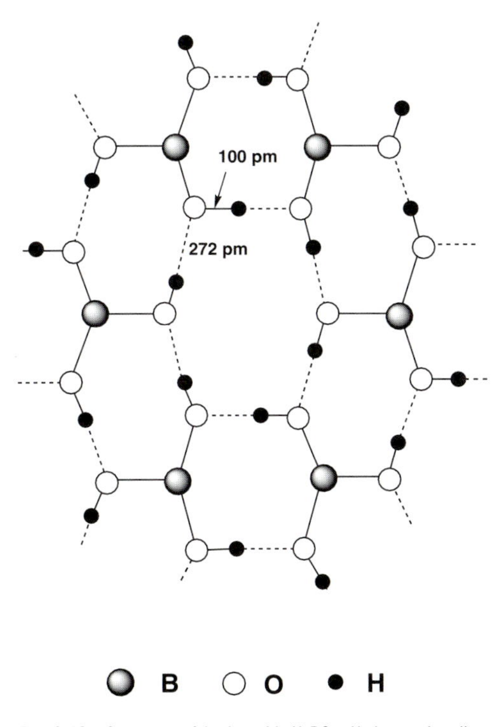

FIG. 9.18 Structure of boric acid, H_3BO_3. Hydrogen bonding holds the molecules in sheets

In general, X-ray evidence gives only the X-Y distance, and it is found that this is less than the sum of the van der Waals' radii, which would determine the distance in the absence of hydrogen bonding. See examples in Table 9.7.

(iii) *Neutron diffraction*. The scattering of neutrons does allow the location of hydrogen atom positions and a number of studies have appeared. Neutron diffraction, electron diffraction, and refined X-ray methods, now available for crystal studies, have contributed to the structures in Fig. 9.19.

(iv) *Infrared studies*. When an X-H grouping is involved in hydrogen bonding, the characteristic infrared stretching of the X-H bond is shifted and it is found that this is

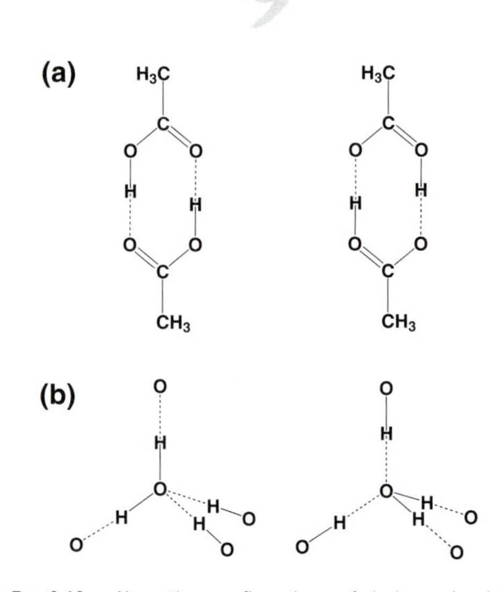

FIG. 9.19 Alternative configurations of hydrogen-bonded species leading to non-zero entropy at absolute zero: (a) acetic acid dimer, (b) ice

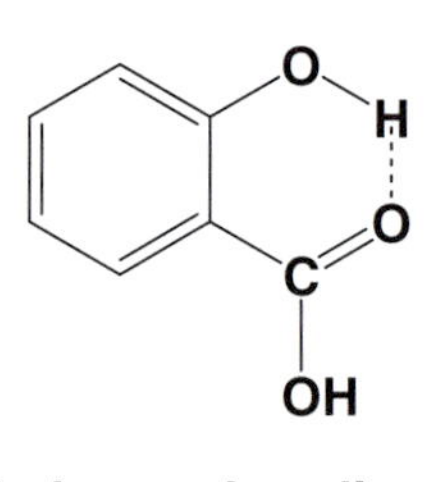

Hydrogen bonding in salicylic acid

always to lower frequencies. There is a general correlation between the extent of this frequency shift and the hydrogen bond energy. Some examples are given in Table 9.7. As infrared spectra are readily measured, this frequency shift is usually the most convenient diagnostic test for hydrogen bonding.

(v) *Entropy data*. Evidence of hydrogen bonding may also be given by entropy data determined at low temperatures. As entropy is linked with disorder, a definition in terms of the number of possible arrangements of the system is possible. This is $S=-R\ln W$, where R is the gas constant and W is the probability of the state of the system. Now, a perfect crystal, when cooled to absolute zero (to eliminate thermal motions of the atoms), has only one possible arrangement of the atoms and the entropy at absolute zero will be $-R\ln 1=0$ (remember that certainty = a probability of one). However, if hydrogen bonding occurs in a crystal, more than one configuration may be possible at absolute zero. For example, in the acetic acid dimer, these are the two configurations of Fig. 9.19a: similarly, ice has the oxygen atoms surrounded by four hydrogen atoms and a number of configurations are possible, such as those shown in Fig. 9.19b. As a result, such hydrogen-bonded molecules have nonzero values of the entropy at 0 K. In the case of ice, it has been calculated that the entropy should be 3.39 J K^{-1} mol^{-1} and it has been measured as 3.39 + 0.21 J K^{-1} mol^{-1}.

The hydrogen bond can be shown to exist by the methods given above, and the next question is about its shape. The bond may be linear or bent and the hydrogen atom may lie symmetrically or unsymmetrically between X and Y. As far as present evidence goes, it appears that XHY are usually arranged linearly unless there is a good steric reason opposing this—as in the intramolecular hydrogen bonds in molecules like salicylic acid or nickel dimethylglyoxime (Fig 14.30).

The symmetry of the position of the H atom varies. In the bifluoride ion, FHF^-, the hydrogen is symmetrically placed between the two fluorine atoms, as is the hydrogen in the OHO links in the nickel dimethylglyoxime complex. In most other cases of known structure, however, the unsymmetrical position of the hydrogen is adopted, as in water or acetic acid. Apart from structure determinations, it should be noted that the entropy determination will distinguish symmetrical structures as these will have no residual

TABLE 9.7 Properties of hydrogen bonds

Bond	*Compound*	*Bond length X-Y distance* (pm)	*Van der Waals' distance X-Y* (pm)	*Depression of stretching frequency* (cm^{-1})†	*Bond energy* (kJ mol^{-1})
F-H-F	$K^+HF_2^-$	226	270	2700	113
(symmetrical)	$NH_4^+HF_2^-$	232	270		
F-H···F	$(HF)_n$	255	270	700	28.0
(unsymmetrical)					
O-H-O	$Ni(DMG)_2$*	240	280	1200	
(symmetrical, bent?)					
O-H···O	KH_2PO_4	248	280	900	
(unsymmetrical)	$(HCOOH)_2$	267	280	600	29.7
	Ice	276	280	400	18.8
N-H···N	NH_4N_3	294 299	300		
	Melamine	300	300	120	ca. 25
N-H···F	NH_4F	263	285		
N-H···O	$NH_4H_2PO_4$	291	290		
(slightly bent)					
N-H···Cl	$N_2H_4.HCl$	313	330	500	

*DMG = dimethylglyoxime.
†The stretching frequency figures are rounded off to the nearest hundred wave numbers and represent averages of two frequencies in some cases.

TABLE 9.8 Hydrogen bonds in hydrated protons

Ion	*Structure*	*O···O length* (pm)	*Compound*	*Comments*
$H_5O_2^+$	$(H_2O\text{-}H\text{-}OH_2)^+$ probably symmetrical cf. Fig. 9.2	242.4	$HClO_4.2H_2O$	$HCl.xH_2O$ ($x = 2, 3$) contain similar ions with short hydrogen bonds (242–250 pm)
	$(H_2O\text{-}H\cdots OH_2)^+$	257	$HAuCl_4.4H_2O$	Unsymmetric, O-H = 99 pm and H···O = 148 pm. $H_5O_2^+$ units are linked through H_2O by further hydrogen bonds of 274 pm
$H_7O_3^+$	$(H_2O\cdots H\text{-}O\text{-}H\cdots OH_2)^+$ with H on central O	247 and 250	$HBr.4H_2O$	Central H_3O surrounded pyramidally by two outer H_2O and one Br
$H_9O_4^+$	Similar to Fig. 9.1	250 and two of 259	$HBr.4H_2O$	Central H_3O surrounded pyramidally by three outer H_2O

entropy. In fact it is found that there is no residual entropy for HF_2^- in potassium hydrogen fluoride. It appears that, in general, the stronger hydrogen bonds are the more symmetrical.

Table 9.7 summarizes the data available for some cases of hydrogen bonds involving F, O and N. One or two cases are worth further mention.

The solvated proton. As discussed above, the free H^+ ion cannot exist in a chemical environment, and there are a number of structural studies which demonstrate its occurrence as H_3O^+. Thus the solid monohydrates of perchloric, sulfuric and hydrochloric acids have been shown to exist as $H_3O^+ClO_4^-$, $H_3O^+HSO_4^-$ and $H_3O^+Cl^-$ respectively.

Studies of higher hydrates of protonic acids have shown that more complex forms of the hydrated proton exist, whose structures involve hydrogen bonding. For example, $HCl.2H_2O$ is best formulated as $H_5O_2^+Cl^-$, while $HBr.4H_2O$ is $(H_7O_3)^+(H_9O_4)^+ 2Br^-.H_2O$. The structures of these forms of the hydrated proton are summarized in Table 9.8. The $HAuCl_4$ hydrate was studied by neutron diffraction so that the hydrogens could be located.

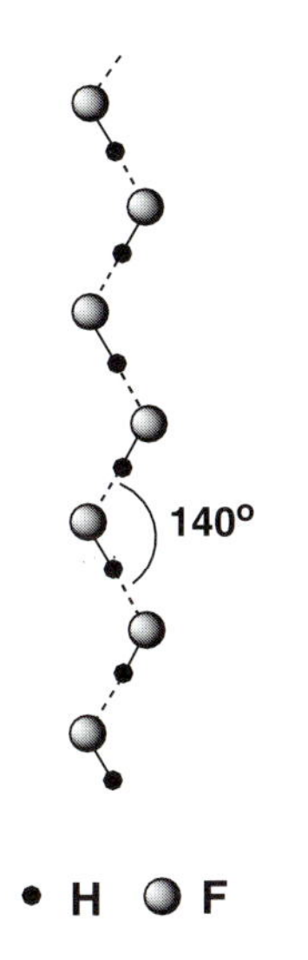

FIG. 9.20 The structure of hydrogen fluoride

The symmetric $H_5O_2^+$ ion has a short hydrogen bond length, similar to that in the nickel dimethylglyoxime molecule (Table 9.7). The unsymmetric $H_5O_2^+$ ion, $H_7O_3^+$ and $H_9O_4^+$ are structurally more like a H_3O^+ ion hydrogen bonded to one, two or three water molecules but with many of the hydrogen bond lengths distinctly short.

Hydrogen fluoride. This has been shown to have the zig-zag structure shown in Fig. 9.20. There seems to be no convincing explanation of the wide H-F-H angle. The hydrogen bonding in hydrogen fluoride persists into the gas phase, where small polymers are found.

Ammonium fluoride. NH_4F shows a different structure from the other ammonium halides. The latter have the CsCl or NaCl structure (the transition to NaCl taking place below 200°C), but the fluoride has the wurtzite structure (Fig. 5.1d), in which each N atom forms four N-H-F bonds of length 263 pm to its four neighbouring N atoms which are arranged tetrahedrally around it. The structure resembles that around the O atoms in ice (Fig. 9.19).

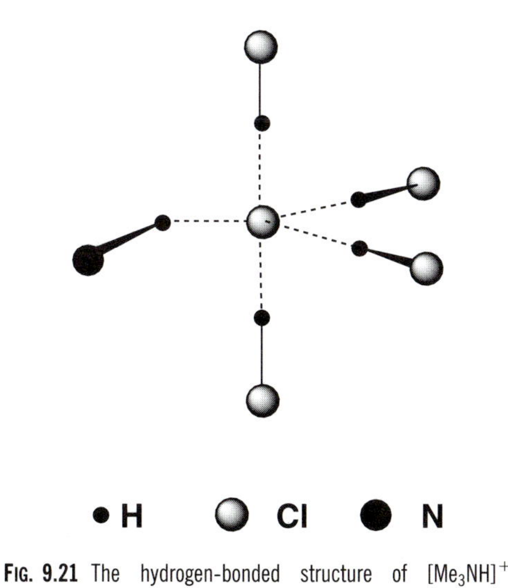

FIG. 9.21 The hydrogen-bonded structure of $[Me_3NH]^+[Cl(HCl)_4]^-$

Apart from F, O and N, there are few atoms which form hydrogen bonds. Chlorine occasionally enters into hydrogen bonding and the HCl_2^- ion is formed in the presence of large cations such as Cs^+ or NR_4^+. Indeed, recent work has shown that HBr_2^- and HI_2^- also exist under such conditions. There is no evidence as to their degree of symmetry. One phase formed by trimethylamine and hydrogen chloride is a low-melting solid, $Me_3N.5HCl$, which provides examples of two of the weaker hydrogen bond systems, N-H···Cl and Cl-H···Cl, Fig. 9.21. The compound is best formulated as $[Me_3.NH]^+[Cl(HCl)_4]^-$ where Cl^- is at the centre of a trigonal bipyramid with two axial and two equatorial HCl groups and with $[Me_3NH]^+$ occupying the third equatorial position. The C-H···Cl bonds are nearly linear and vary in length from 343 to 353 pm, substantially longer than those in Table 9.6. The N-H···Cl bond is bent, with an angle of 141° at H, and the hydrogen bond length is 327 pm.

The reader is also referred to Section 16.11 for a discussion of transition-metal dihydrogen complexes.

There is also some evidence for hydrogen bonding involving carbon, C-H···X, if the carbon is bonded to electronegative groups as in HCN. One case has even been established of an N-H···C bond of length 258 pm. All these other hydrogen bonds are much weaker than those involving F, O and N.

The fact that hydrogen bonds occur only between strongly electronegative elements, and that small elements form the strongest bonds, suggests an electrostatic origin for the interaction. The relatively positive hydrogen in O-H, N-H or F-H bonds interacts with the dipole in neighbouring atoms to form the bond. This happens only with hydrogen, probably because of its small size and lack of inner shielding electron shells.

HYDROGEN – HYDROGEN BONDING

While the importance of hydrogen bonding involving electronegative elements cannot be overemphasized, there are increasing numbers of examples of systems where a metal–hydride bond (where the metal is less electronegative than hydrogen) acts as the hydrogen donor

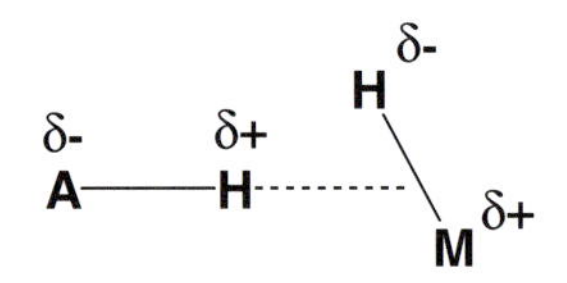

This constitutes a 'hydridic to protonic' interaction, $H^{\delta-} \rightarrow H^{\delta+}$, and the terms dihydrogen bonding and hydrogen–hydrogen bonding have been coined to describe this effect. The strength and directionality characteristic of such hydrogen–hydrogen bonds can be comparable to those of 'classical' hydrogen bonds involving electronegative atoms, and it is believed that the interaction is primarily electrostatic. X-ray and neutron diffraction studies are typically the means by which such H···H interactions are detected. As an example, the solid state structure of $NaBH_4.2H_2O$ contains O-H···H-B hydrogen–hydrogen bonding, with the OH vectors pointing towards the middle of the B-H bonds

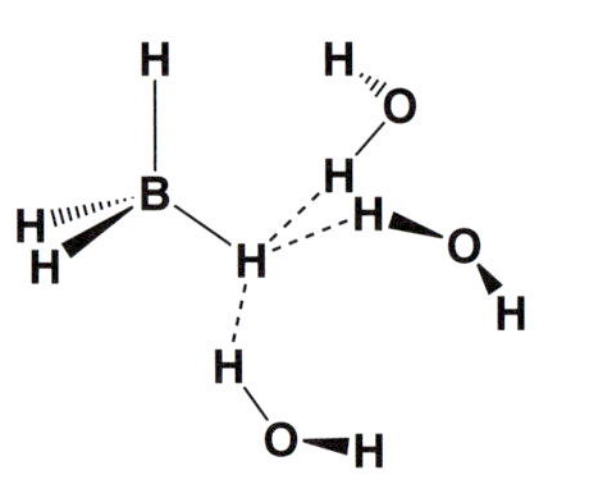

Transition metals, particularly those in the centre of the 4d and 5d series, where metal hydride complexes are readily formed, also form hydrogen–hydrogen bonds. One example of this involves the tungsten–hydride complex

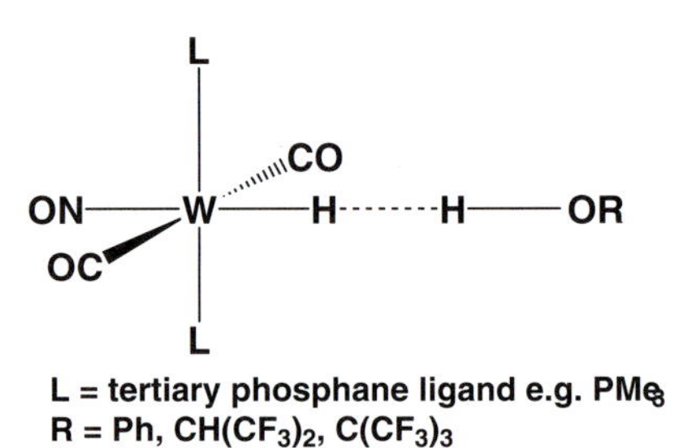

The strength of the hydrogen–hydrogen bonding increases with the donor ability of the ligand L (which makes the hydride atom more electron-rich) and with the acidities of the alcohol ROH. Many other examples of hydrogen–hydrogen bonding are given in the reference below. It is worth noting that the interaction between a hydridic hydrogen, $H^{\delta-}$, to a protonic hydrogen, $H^{\delta+}$, may result in the formation of a coordinated H_2 ligand, and ultimately to the loss of H_2 from the metal centre. Given that classical hydrogen bonding is already of great importance in many solid state structures, the conversion of solids with H···H bonds that can be 'upgraded' through loss of H_2 into strong covalent bonds might lead to novel ways of constructing new types of covalently bonded inorganic lattices.

Reference

Dihydrogen bonding: structures, energetics, and dynamics, R. CUSTELCEAN and J. E. JACKSON, *Chemical Reviews* **101**, 2001, 1963–1980.

Problems

It is important, in learning systematic chemistry, to correlate properties and highlight similarities and differences. The chemistry of the hydrides of the various elements should be compared with that of their other simple compounds.

9.1 Compare the shapes of the covalent hydrides with those predicted by the VSEPR method (Chapter 4).

9.2 (a) Assuming bond energies are additive, compare the calculated energy of formation from the elements of NH_3, PH_3 and AsH_3 with those of the analogous fluorides, using data in Table 17.3, p.430. (The heat of atomization of P = 334 and As = 289 kJ mol^{-1}.)
(b) Comment on the fact that the thermal stability of fluorides is generally higher than that of the corresponding hydrides.

9.3 Formulate the bonding in $Al(BH_4)_3$ (see p.225). How does this compare with aluminium hydride?

9.4 Calculate the heats of formation of LiH, NaH and KH, using the experimental lattice energies in Table 5.2, p.118.

9.5 The heat of formation, under standard conditions, of CaH_2 is -173 kJ mol^{-1}. Assuming the Madelung constant value is 5.5, calculate the lattice energy as in Section 5.3 and compare with the Born–Haber cycle value (see also Question 5.5).

9.6 You should also formulate answers to questions of the type:
Compare and contrast the chemistry of the hydrides with that of the fluorides/chlorides/iodides of the Main Group elements/transition elements etc.

CHAPTER 10

10.1 GENERAL AND PHYSICAL PROPERTIES, OCCURRENCE AND USES 234
10.2 COMPOUNDS WITH OXYGEN AND OZONE 237
10.3 CARBON COMPOUNDS 238
10.4 COMPLEXES OF THE HEAVIER ELEMENTS 239
10.5 CROWN ETHERS, CRYPTATES AND ALKALI METAL ANIONS 240
10.6 SPECIAL FEATURES IN THE CHEMISTRY OF LITHIUM AND MAGNESIUM 241
10.7 BERYLLIUM CHEMISTRY 243
PROBLEMS 245

The 's' Elements

10.1 General and physical properties, occurrence and uses

Elements with their outermost electrons in an s level are those of the lithium Group (alkali metals) and of the beryllium Group (beryllium, magnesium and the alkaline earth metals).

Some properties of the elements are given in Table 10.1 and important parameters in Figs 10.1 and 10.2

Of the lithium Group elements, compounds of the first three are produced commercially. The common source is salt deposits from ancient seas, or brines from seawater evaporation. Lithium is also mined as *spodumene*, an aluminosilicate. The commonest form of production for all three is as the chloride, with Li and Na also being converted to the carbonate.

The cheaper metals, especially sodium, are utilized industrially in some processes requiring powerful reduction. One example is the formation of titanium or zirconium from their tetrachlorides by reaction with sodium or calcium. Alloys of sodium and potassium (which are liquid at ambient temperatures for compositions containing 40 to 90% K) are used as the primary coolant in the experimental 'breeder' reactor.

Uses of the elements in the laboratory, and in fine chemical production, are based on their high reactivity. As a result of its light weight, lithium is now widely incorporated in high energy density batteries. Lithium compounds are also used in glasses, ceramics, and for specialized lubricants. Lithium therapy has made significant contributions to the treatment of mental illnesses. Sodium compounds in major use are the industrial alkalis, Na_2CO_3 and NaOH, and NaCl, especially for preparation of Cl_2 by electrolysis (and also further conversion to NaOCl or $NaClO_3$). The main use of potassium is as a primary fertilizer, where application is usually as KCl. The carbonates are the main vehicle for incorporating Na_2O and K_2O into glasses, and in cement manufacture.

The beryllium Group elements, with the exception of beryllium itself, occur mainly as carbonates or in brines and salt deposits. The principal source of beryllium is the aluminosilicate mineral beryl, $Be_3Al_2Si_6O_{18}$ (see Section 18.6 for further discussion on silicate minerals). Beryl is found in a number of coloured gem forms including aquamarine (sea blue) and emerald (green), the most highly prized. Beryllium is an element which presents an interesting dichotomy. On the one hand it is the most toxic nonradioactive element known (second in the Periodic Table to plutonium) and great care must be exercised in handling beryllium compounds. The reader may be interested to learn that more research papers have been published on theoretical aspects of beryllium chemistry than on practical aspects. On the other hand, however, the metal itself has a number of highly specialized uses. As a result of its transparency to X-rays, it is used in windows in X-ray apparatus, and its strength and lightweight nature have even led to its use as a component of an alloy from which golf clubs are made!

Calcium carbonate (lime) and magnesium carbonate are important in agriculture, as components of cements, and as the source of the oxides. MgO, prepared at high temperature, is inert and refractory and is extensively used as a furnace lining, especially in the steel industry. Barium carbonate finds a significant application in high-density concrete. Organomagnesium compounds are extensively used in laboratory syntheses and in the fine chemical industry.

The elements are all prepared by electrolysis of the fused halides. The relatively volatile rubidium and caesium are also conveniently prepared in the laboratory by heating the

TABLE 10.1 The s elements

Element	*Symbol*	*Electronic configuration*	*Abundance (ppm of the crust)*	*Accessibility**	*Common coordination numbers*
Lithium	Li	$[He]2s^1$	65	Common	4,6
Sodium	Na	$[Ne]3s^1$	28 300	Common	6
Potassium	K	$[Ar]4s^1$	25 900	Common	6
Rubidium	Rb	$[Kr]5s^1$	310	Rare	6
Caesium†	Cs	$[Xe]6s^1$	7	Rare	6,8
Francium	Fr	$[Rn]7s^1$	–	Very rare	?
Beryllium	Be	$[He]2s^2$	6	Becoming common	2,4
Magnesium	Mg	$[Ne]3s^2$	20 900	Common	6
Calcium	Ca	$[Ar]4s^2$	36 300	Common	6
Strontium	Sr	$[Kr]5s^2$	300	Common	6
Barium	Ba	$[Xe]6s^2$	250	Common	6
Radium	Ra	$[Rn]7s^2$	Trace	Rare	

* See Section 8.7 for a discussion of this point.
† The name 'cesium' is also widely found.

chlorides with calcium metal and distilling out the alkali metal. The alkali metals have melting points ranging from 180°C–29°C and boiling points in the range 1340°C–670°C, in both cases falling with increased atomic weight so that caesium has the second-lowest melting point of any metal. The beryllium Group elements are much less volatile, with melting points ranging from 1300°C–700°C and boiling points six or seven hundred degrees higher. Again there is a general tendency for volatility to increase with atomic weight.

The chemistry of the elements of the s block is dominated by their tendency to lose the s electrons and attain a stable, rare gas configuration. (Beryllium is an exception to this, see Section 10.7). This tendency is shown by the low ionization potentials and strongly negative redox potentials. The metals are among the most powerful of chemical reducing

PREVIOUS REFERENCES TO PROPERTIES OF THE s ELEMENTS

Many properties have been listed in the earlier chapters, including atomic masses and numbers (Table 2.5), ionic radii (Table 2.12), ionization energies (Table 2.8), electronegativities (Table 2.14) and structural details of halides, etc. (Table 5.1). Structures of s element compounds include NaCl (Fig. 5.1a), CsCl (Fig. 5.1b) and CaC_2 (Fig. 5.13).

The energetics of formation of alkali halides and alkaline earth chalcogenides are discussed in Section 5.3.

ELECTROLYSIS: NEW TECHNIQUE GAVE FIRST ISOLATION OF s ELEMENTS

Early in the 19th century, Humphrey Davy developed electrolysis and promptly isolated all the common s elements by electrolysis of their fused chlorides—K (the first) and Na in 1807 and Mg, Ca, Sr and Ba in 1808. In 1808, he also used his potassium to isolate B from its oxide, and identified chlorine as an element in 1810 (previously thought to contain oxygen). Eight new elements in three years is an all-time record!

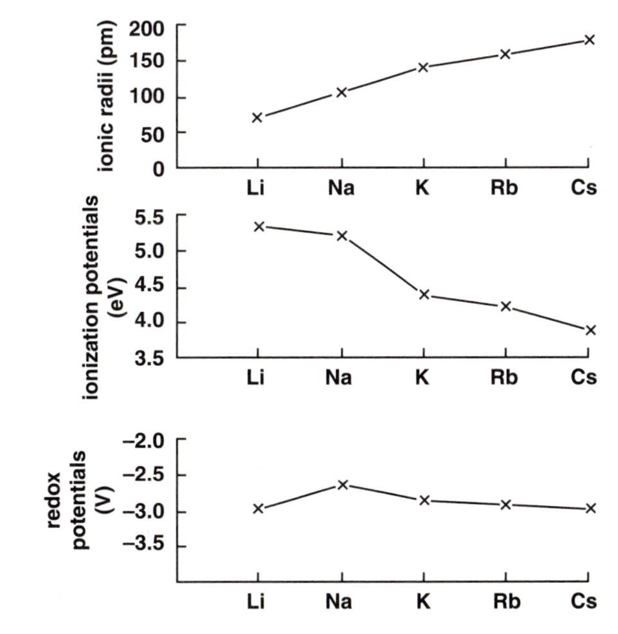

FIG. 10.1 The ionic radii, ionization potentials, and redox potentials of the alkali metals. The relatively large changes between Li and Na, and to a lesser extent between Na and K are reflected in the general chemistry

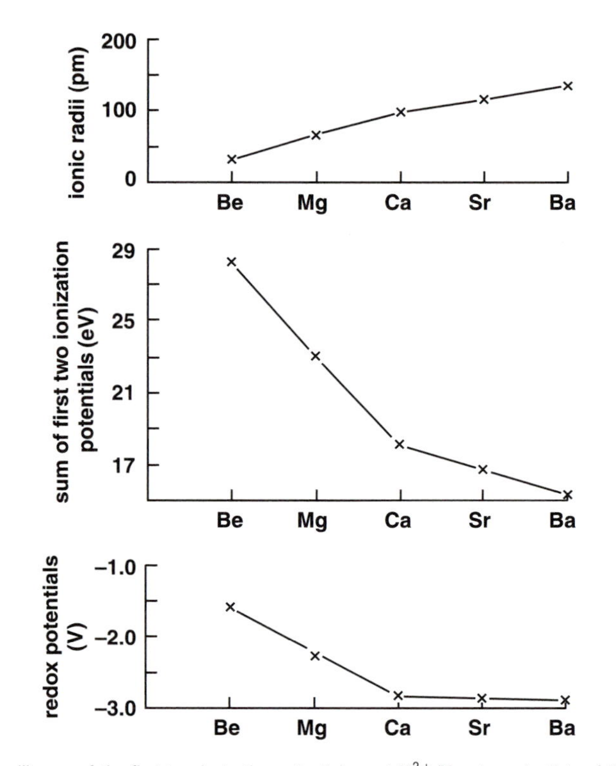

FIG. 10.2 The ionic radii, sum of the first two ionization potentials, and M^{2+}/M redox potentials of the beryllium Group of elements. As in Fig. 10.1 the major changes in properties come between the first and second members. The similarity between Ca, Sr, Ba is clear

HIGHER OXIDATION STATES OF CAESIUM

The chemistry of the Group 1 metals is dominated by the +1 oxidation state. However, there have been indications that caesium might form higher oxidation species. Thus, electrochemical oxidation of $[CsL]PF_6$ (L = 18-crown-6 or cryptand [222]—compare Fig. 10.6) gives evidence for the Cs^{2+} and Cs^{3+} ions. Compounds containing caesium in higher oxidation states are yet to be isolated.

agents and combine directly, and usually violently, with most nonmetals to yield ionic compounds. The cations formed by these elements (M^+ by the lithium Group, M^{2+} by Mg and the alkaline earths) are very stable and form salts with the most strongly oxidizing or reducing anions. Caesium is particularly useful to stabilize large anions, as in the formation of the bihalide ions, HX_2^-, mentioned in Section 9.7.

The reactivity increases down each Group and the lithium Group elements are more reactive than the corresponding members of the beryllium Group. The reactions of the elements with a number of typical reagents are given in Table 10.2. Although metallic lithium is the most reducing of the metals as judged by redox potentials, see Table 6.3, the heavier metals are considerably more reactive, caesium most of all. This poses much greater difficulties in handling, partly on account of its low melting point (28.5°C) when compared to lithium (180°C) or sodium (98°C).

The lightest elements are atypical, as their small size leads to a high charge density on the ions with resulting strong polarizing effects and high heats of solution. Lithium and beryllium show marked differences from their heavier congeners and sodium and magnesium also show distinct, though smaller, differences. There are many similarities between lithium and

THE RADIOACTIVE HEAVIEST s ELEMENTS

Francium and radium both occur only as radioactive isotopes. Francium is inaccessible, with the longest-lived isotope, ^{223}Fr, having a half-life of only 21 minutes. Its chemistry is little known, but, in the few reactions which have been studied, it resembles the other heavy alkali metals. For example, it has an insoluble perchlorate and hexachloroplatinate, like rubidium and caesium. The longest-lived isotope of radium, ^{226}Ra, $t_{\frac{1}{2}} = 1600$ years, is much more stable and radium chemistry is well-established. Radium was used as a 'medicinal tonic' called Radithor in the late 1920s though, thankfully, the fad of consuming radioactive products is now long gone.

TABLE 10.2 Reactions of the s elements

M is used for one mole of an alkali metal or a half-mole of a beryllium Group element

Reaction	*Notes*
$2M+X_2=2MX$	X = halogen: alkali metals also form polyhalides, e.g. CsI_3, $KICl_4$ or KIF_6, see Chapter 15
$2M+Y=M_2Y$	Y = O, S, Se, Te: higher oxides are also formed (Section 10.2) and also polysulfides $M_2S_{2\text{–}6}$
$3M+Z=M_3Z$	Z = N, P: Li reacts even at room temperature, Mg to Ra at red heat. The alkali metals also react with As, Sb and Bi. Metal solutions in ammonia give polyanions such as $(Bi_9)^{4-}$ or $(Sb_5)^{3-}$
$2M+2C=M_2C_2$	Most rapidly with Li of the alkali metals. Ca, Sr, Ba and Ra at high temperatures. Be forms Be_2C
$2M+H_2=2MH$	At high temperatures. Ionic hydrides. Not with Be, Mg.
$M+H_2O=MOH+\frac{1}{2}H_2$	By Be and Mg only slowly at room temperatures. $Be(OH)_2$ is amphoteric, all others are strong bases. $Mg(OH)_2$ is insoluble, others dissolve readily. All except beryllium hydroxide absorb CO_2 to give M_2CO_3, and alkali metals give $MHCO_3$ also
$M+NH_3=MNH_2+\frac{1}{2}H_2$	With gaseous ammonia at high temperatures or liquid ammonia plus catalyst. Mg and Be only by reaction with amide of a more reactive metal, $Be+2NaNH_2 \rightarrow Be(NH_2)_2 \rightarrow NaBe(NH_2)_3$

its diagonal neighbour in the beryllium Group, magnesium. Beryllium (Section 10.7) differs markedly from all the other elements.

Salts of the s elements tend to be among the most soluble of their kind in water and other ionizing solvents, though solubilities vary widely with the anion. This is especially the case with the compounds of the alkaline earths where the double charge results in high lattice energies and high heats of solvation. AB_2 compounds tend to be soluble (e.g. the chlorides), while the insolubility of such AB compounds of the alkaline earths as the sulfates, oxalates and carbonates is well known in qualitative and quantitative analysis.

All the s elements dissolve in liquid ammonia to give intensely blue, conducting solutions which contain the metal ions and electrons which are free of the metal and appear to be associated with the solvent. (Beryllium and magnesium give only dilute solutions by electrolysis.) These 'solvated electrons' are very reactive and the metal–ammonia solutions are powerful reducing agents which can be used at low temperatures. (Compare Section 6.7) The heavier alkali metals may be recovered unchanged on evaporating off the ammonia, while lithium gives $Li(NH_3)_4$ and the alkaline earths yield hexammoniates, $M(NH_3)_6$. On standing, or in the presence of catalysts such as transition metal oxides, the metals react with the ammonia to form the metal amide and hydrogen e.g.

$$M + NH_3 = MNH_2 + \tfrac{1}{2}H_2$$

$$\text{(more accurately)}\quad e^-_{(\text{solvated})} + NH_3 = NH_2^- + \tfrac{1}{2}H_2$$

Similar, but more dilute, solutions are formed in amines and methyl ethers. With aromatic hydrocarbons, such as benzophenone PhC(O)Ph and naphthalene, alkali metals form highly coloured charge-transfer compounds, M^+Ar^-. These are used as 'soluble sodium' in reduction reactions and are also invaluable for rigorous drying of solvents like diethyl ether or tetrahydrofuran. The electron occupies the π^* orbital of the aromatic hydrocarbon and is highly reactive, though less so than in liquid ammonia.

10.2 Compounds with oxygen and ozone

The reaction between the s elements and oxygen may go further than to the simple oxide, and a number of higher oxides are formed when the metals are burned in air or are oxidized by O_2 in liquid ammonia. Peroxides, O_2^{2-}, are formed by all the elements except beryllium. In addition, sodium, potassium, rubidium, caesium and calcium superoxides, O_2^-, have been prepared. The normal products of the combustion of the metals in an adequate supply of air are:

oxide	Li, Be, Mg, Ca, Sr
peroxide	Na, Ba (and Ra?)
superoxide	K, Rb, Cs

The peroxides contain the ion $^-$O-O$^-$ and are salts of hydrogen peroxide (compare the relation of the hydroxides and oxides to water). The superoxides contain the ion, O-O$^-$. It will be recalled that oxygen, O_2, has its two outermost electrons unpaired in π^* orbitals. The superoxide ion and the peroxide ion have, respectively, one and two electrons more than in O_2. The superoxide ion has thus the configuration $(\pi^*)^3$ and has paramagnetism corresponding to one unpaired electron and a bond order of one and a half. The peroxide ion has the configuration $(\pi^*)^4$ with no unpaired electrons and a bond order of one. It is isoelectronic with F_2. The MO_2 solids have the tetragonal lattice of calcium carbide. The increasing stability of the peroxides and superoxides with increasing cation size is noteworthy, and provides another example of the stabilization of large anions by large cations through lattice energy effects.

See Section 3.5 for further discussion and a list of the bond lengths of dioxygen ions.

When the metals potassium, rubidium and caesium are treated with ozonized oxygen, or when ozone is passed into their solutions in liquid ammonia, the ozonides $M^+O_3^-$ are formed. These are yellow or orange and contain the group (O-O-O)$^-$ which is paramagnetic with one unpaired electron. For example, rubidium ozonide has an overall caesium chloride structure with a bent anion. The O-O distance is 134 pm and the bond angle is 114°. LiO_3 and NaO_3 can be obtained in liquid ammonia solution by an ion exchange reaction starting from CsO_3. However, solid LiO_3 and NaO_3 cannot be obtained. Instead, removing the solvent results in oxidation of the ammonia, giving LiOH or NaOH together with N_2 and O_2. The properties of the lighter alkali metal ozonides thus differ from their heavier congeners. Red, crystalline complexes are formed between all of the alkali metal ozonides and crown ethers or cryptands (compare Section 10.5) and in these adducts the geometry of the ozonide anion is very similar to that in the parent MO_3 compounds.

The heavier elements, rubidium and caesium, also form oxides which are enriched in the metal (called suboxides). Established formulae include $Cs_{11}O_3$, Cs_7O, Rb_9O_2 and Rb_6O. The structures involve M_6O octahedra (which may be linked by shared faces) and may also include metallic regions. Thus $Cs_{11}O_3$ is three Cs_6O units sharing faces and Cs_7O is $[Cs_{11}O_3]$ Cs_{10}. These compounds often form on surfaces and are usually metallic or highly coloured.

10.3 Carbon compounds

If acetylene is passed through a solution of an alkali metal in liquid ammonia, or is reacted with the heated metal, the following reactions take place:

$$M + C_2H_2 = MHC_2 + \tfrac{1}{2}H_2$$
$$MHC_2 + M = M_2C_2 + \tfrac{1}{2}H_2$$

The carbon compounds M_2C_2 and MHC_2 are termed acetylides and contain the discrete anions, $(C{\equiv}C)^{2-}$ and $(C{\equiv}CH)^-$, arising from the displacement of both or one of the relatively acidic hydrogens in the acetylene molecule. Acetylides also result from the direct reaction between carbon and heated lithium, sodium, magnesium and alkaline earth metals. All form acetylene on hydrolysis. The structure of calcium acetylide (commonly called calcium carbide) has been determined (Fig. 5.13) and is related to that of sodium chloride (Fig. 5.la).

These acetylides are the principal carbides (binary compounds of element and carbon) formed by the s elements, but two others of interest exist. Magnesium also forms a carbide of formula Mg_2C_3, which yields propyne, $HC{\equiv}C\text{-}CH_3$, as the main product on hydrolysis.

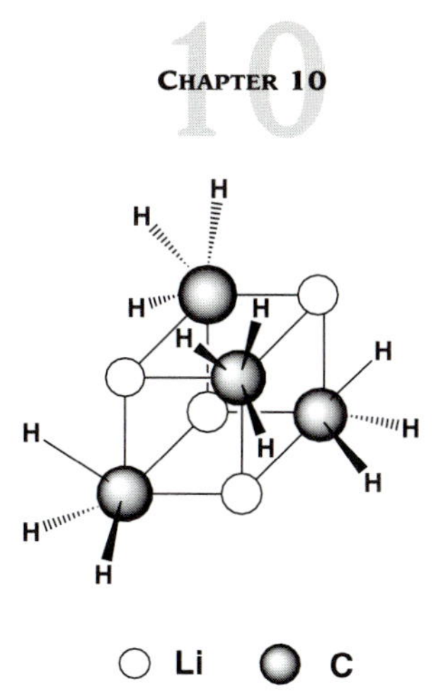

FIG. 10.3 The tetrameric structure adopted by methyl-lithium $(CH_3Li)_4$

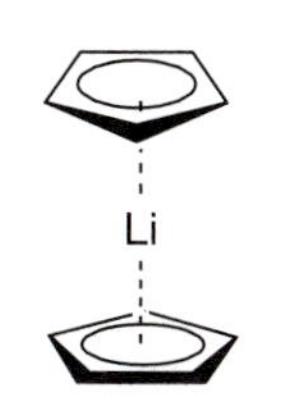

The structure of the cyclopentadienyl lithium ion $[Li(C_5H_5)_2]^-$

COMPLETELY METALLATED METHANE

It is interesting that calculations first suggested that CLi_4, i.e. methane entirely substituted by Li, would be stable and even that CLi_6 species containing octahedral carbon should exist. These have both since been detected in the gas phase at 1000 K. It has even proved possible to synthesize such a startling compound on a bench scale by displacing HgX from $C(HgX)_4$ (X = halogen or alkyl) with an organolithium, RLi. The CLi_4 is a grey, extremely pyrophoric solid.

This suggests the presence of a C_3 unit in this carbide. The product of direct combination between beryllium and carbon is the carbide Be_2C. This cardide probably contains single carbon atoms as the main hydrolysis product is methane. It has the anti-fluorite structure. All these carbides have many of the properties of ionic solids, with colourless crystals which are nonconducting at ordinary temperatures.

A second group of ionic compounds containing carbon is formed by the more reactive s elements. These are the metal alkyls and aryls such as ethyl-sodium, $C_2H_5^-Na^+$, or phenylpotassium, $C_6H_5^-K^+$. Such compounds are extremely reactive solids which inflame in air and react violently with almost all compounds apart from nitrogen and saturated hydrocarbons. They are involatile solids which decompose before melting and the evidence available indicates that they are ionized, R^-M^+. Crystal structures, supported by neutron diffraction, are reported for CD_3M species (D is preferred to H for neutron scattering). These show that methylpotassium is purely ionic, $K^+CD_3^-$, and the CD_3^- ion is pyramidal with CD = 109 pm and DCD = 106° (compare isoelectronic NH_3). The structure of methyl-sodium is more complex. Half the units exist as separate ions and half as $(CD_3Na)_4$ tetramers, like methyl-lithium (see below). The tetramers are arranged in chains in the crystal and these are linked together via Na···C contacts with the free ions.

Anions like $C_5H_5^-$ are more stable as the charge can delocalize in π orbitals, though they are still very air- and water-sensitive. They are widely used in the synthesis of sandwich compounds: thus $Na^+C_5R_5^-$ (R = H or Me) is used to form the sandwich compounds called *metallocenes*, such as ferrocene, $Fe(C_5H_5)_2$ and all its many analogues (see Section 16.4 especially Fig. 16.7).

The corresponding lithium and magnesium compounds are covalent and much less reactive. The alkyl-lithiums, for example, are liquids or low-melting solids which are soluble in ethers or hydrocarbons. They are relatively involatile and exist as associated molecules with a highly polar C-Li bond. Butyl-lithium, for example, is hexameric in hydrocarbon solvents and dimeric in ether. Methyl-lithium has been isolated as the tetramer, $(CH_3Li)_4$, which contains a tetrahedron of lithium atoms with a methyl group bridging each triangular face (Fig. 10.3). When LiC_5H_5 is dissolved in tetrahydrofuran at low temperature, an equilibrium is set up with $[Li(C_5H_5)_2]^-$ (the lightest possible metallocene) and the solvated Li^+ cation. The anion, $[Li(C_5H_5)_2]^-$, can be isolated using a large cation such as $(Ph_4P)^+$ and has the structure shown in the margin diagram, with the Li atom lying between parallel cyclopentadiene rings, compare Fig. 16.7a. The organolithium compounds find extensive uses in organic and organometallic syntheses.

The principal class of organomagnesium compounds are the halides, or Grignard reagents, RMgX. These resemble the organolithiums in reactivity and these two types of reagent complement each other usefully. The more reactive ionic alkyls are much less useful as they are so difficult to handle, but they find some application where particularly vigorous conditions are required. Unsubstituted organomagnesiums, R_2Mg, are much less fully studied. They are bridged polymers with covalent properties. For example, $(C_5H_5)_2Mg$ is a low-melting, volatile solid which has a sandwich structure like the isoelectronic lithium anion $(C_5H_5)_2Li^-$.

10.4 Complexes of the heavier elements

The cations of the heavier s elements are very poor electron pair acceptors as their positive charge density is low. Solvates, such as hydrates or ammoniates, are not found for the heavier alkali metals although sodium gives a moderately stable tetrammoniate in the iodide, $Na(NH_3)_4I$. The alkaline earths are of higher charge density and more strongly hydrated. With large anions, octahedral $Mg(H_2O)_6^{2+}$ is found, while a 7-coordinated capped trigonal prism configuration is found around Ca in $2[Ca(H_2O)_7][Cd_6Cl_6(H_2O)_2]$. In the complex double azide, $CaCs(N_3)_3.H_2O$, the Ca is 7-coordinate to H_2O plus 6N from the azides, while the Cs is 9-coordinate.

In complexing agents for these elements, the best donor atom is oxygen, sometimes together with nitrogen. Chelating agents with donor oxygen give a few complexes, of which

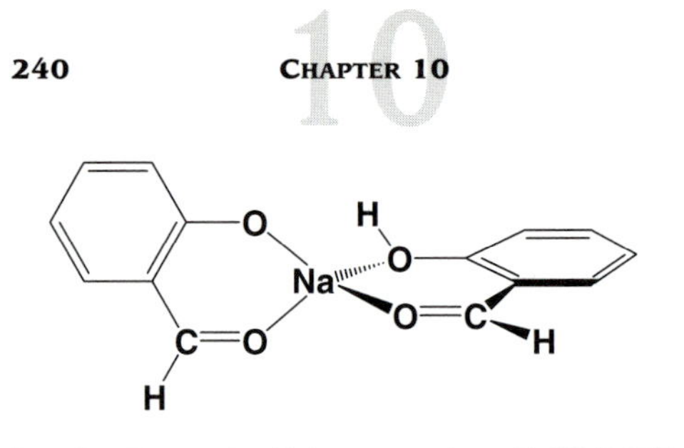

FIG. 10.4 The salicylaldehyde complex, $Na(OC_6H_4CHO)(HOC_6H_4CHO)$

the 4-coordinated salicylaldehyde complex of the alkali metals shown in Fig. 10.4 is typical. Potassium, rubidium and caesium also give 6-coordinated complexes $M(OC_6H_4CHO)(HOC_6H_4CHO)_2$.

Because of their higher charge, the alkaline earth elements form a rather wider variety of complexes with compounds containing donor oxygen or nitrogen atoms. One complexing agent of considerable value in the quantitative analysis of these elements is ethylenediamine-tetra-acetic acid (EDTA). This forms 6-coordinated complexes of the type shown in Fig. 10.5 and is especially useful in the determination of calcium and magnesium.

Calcium is also complexed by polyphosphates and this reaction is the basis of methods of removing hardness from water.

10.5 Crown ethers, cryptates and alkali metal anions

An important class of strongly complexing reagents has been developed since the mid-1960s. A typical example is the cyclic ether shown in Fig. 10.6. This is called a *crown ether*, from the shape of the metal complex. Other similar reagents, called *cryptates* (because the metal is *hidden* in the structure), are also used. These poly-functional ligands provide a variety of O and/or N donor atoms which strongly complex the M^+ or M^{2+} ions of the s elements. They provide unusual solubilities in organic solvents, stabilize large anions, and give model systems for the movement of ions through biological membranes.

One example is the greatly enhanced solubility of alkali metals in ethers. In absence of cryptate ligands, sodium and potassium form extremely dilute blue solutions in ethers, analogous to their solutions in liquid ammonia (Section 6.7) but at 10^{-4} of the concentration. When a crown ether or cryptate is added, the concentrations of alkali metal can be increased by a factor of 10^3 to 10^4. From such solutions have been isolated a golden compound, 2Na(Crypt 222) with interesting properties. The ^{23}Na nmr spectrum showed two signals, one in the position of $Na(crypt)^+$ as seen in ordinary salts such as $Na(crypt)^+ Br^-$. The other was in an unusual position 60 ppm upfield and this suggested the signal came from the Na^- ion where the two s electrons would strongly shield the nucleus. Calculations on the gaseous Na^- ion confirmed the position. The other alkali metal solutions, in the presence of cryptates or crown ethers, contained similar M^- ions with upfield shifts as follows: K^- (90 ppm), Rb^- (190 ppm) and Cs^- (250 ppm). The metal anions also show a distinctive optical spectrum, and crystal structure studies have confirmed the formulation. Mixed cation–anion systems may be prepared, such as $K(crypt)^+Na^-$.

The stability of these anions is ascribed to the shielding effect of the cryptate on the cation. The large ligand completely isolates the cation so that electron transfer from Na^- to Na^+ is prevented (otherwise sodium metal forms). By using an analogue of the cryptand ligand in Fig. 10.6b in which all the O and X groups are replaced by NCH_3 units, the stability of crystalline salts containing Na^- or K^- can be significantly improved. The M^- anions have the s^2 configuration, isoelectronic with the elements of the beryllium Group. A similar metallide anion is found for gold (the $d^{10}s^2$ auride anion, Au^-) which is discussed in Section 15.9.2.

Perhaps the most spectacular species is illustrated by the dark blue Li(crypt). This is formulated as $[Li(crypt)]^+e^-$ and contains the stabilized electron acting as an anion. Such *electrides* are formed by all the alkali metals with suitable cryptates or crown ethers, for example, $[Cs(15\text{-crown-}5)_2]^+e^-$.

The large, well-shielded cation allows the isolation of other anions which would normally be unstable to reaction with the positive charge. One example is $K(crypt)HgTe_2$ containing the linear Te-Hg-Te$^-$ ion. The metal crypt solutions provide a major route to cluster anions of the p-block elements (compare Section 18.4), where the large cation is an important contributor to the stabilization.

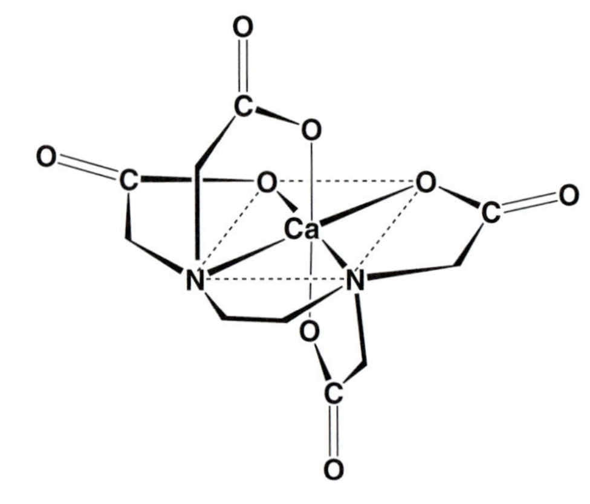

FIG. 10.5 EDTA complex of calcium

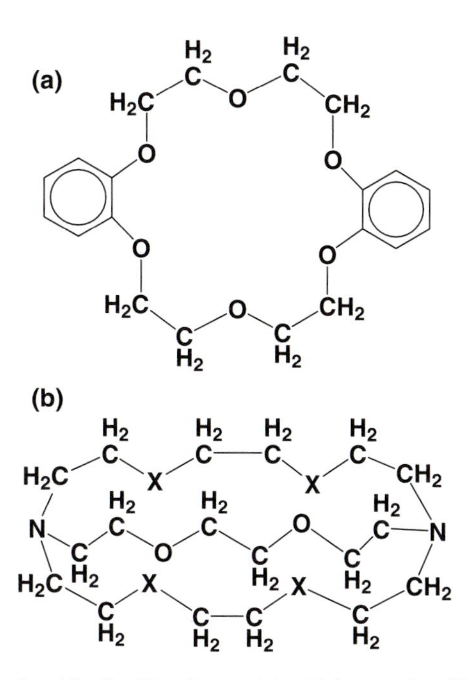

FIG. 10.6 Examples of macrocyclic polyfunctional ligands or cryptates. (a) A crown ether, (b) a mixed-donor cryptate: X can be O or S in various combinations. Such ligands complex strongly to the s elements through the oxygen atoms, forming complexes of the type (polyether)M^+ or (polyether)M^{2+} containing a large number of 5-membered rings. A variety of ring sizes, with various combinations of N and O donor atoms, is available

FURTHER NOTES ON CROWN ETHERS AND CRYPTATES

(a) *Nomenclature*. A rather full nomenclature has grown up to describe these complex molecules. We shall only indicate the approach.

(b) The ether of Fig. 10.6a is (18-crown-6), showing an 18-membered ring with 6 O atoms in every third position. A widely used relation is (15-crown-5) with a similar 15-membered ring and 5 O atoms. Terms like Crypt 222 are used to designate particular forms of Figure 10.6—in this case where all the X atoms are O.

(c) *Structure variations*. Since these species are seen as models for the ion channels in biological membranes, there has been much interest in their detailed structure. Thus $[Na(18\text{-crown-}6)]^+X^-$ have structures which depend in detail on weak interactions—even hydrogen bonds as weak as C-H $\cdots$ X (compare Section 9.7).

For example, where $X = ClO_4^-$, NO_3^- or ReO_4^- the complex is isolated as $[Na(crown)H_2O(X)]$ containing hydrogen bonds between the water molecule and the crown ether. In contrast, if $X = I_3^-$ or N_3^-, we find $[Na(crown)(H_2O)_2]^+X^-$ with weak interactions with the Na. Such subtle changes depending on very weak interactions are typical of biomolecules and the cryptate compounds are valuable models.

10.6 Special features in the chemistry of lithium and magnesium

Lithium differs from the other alkali metals in a number of ways. These variations stem from the small size of the lithium cation (radius 68 pm, c.f. $Na^+ = 98$ pm) and its resulting higher polarizing power. This has already been seen in the more covalent nature of its alkyls. Magnesium resembles lithium in much of its chemistry, as the higher charge is offset by the greater size, and these elements are one example of the diagonal similarity which holds in parts of the first two short Periods. We have already seen differences in their organic compounds in Section 10.3.

Lithium and magnesium resemble each other in the direct formation of the nitride and carbide (Table 10.2), in their combustion in air to the normal oxide, and in the properties of their organic compounds, as already remarked in the previous sections. They are also similar to each other, and different from the heavier elements, in the thermal stability of their oxysalts and in the mode of decomposition of these. For example, lithium and magnesium nitrates decompose on heating to give the oxide and dinitrogen tetroxide, e.g.

$$2LiNO_3 = Li_2O + N_2O_4 + \tfrac{1}{2}O_2$$

while the other alkali metal nitrates give the nitrite on heating ($NaNO_2$ decomposing further to a mixture of oxides):

$$MNO_3 = MNO_2 + \tfrac{1}{2}O_2$$

and the alkaline earth nitrates give nitrite and oxide mixtures.

In the case of the carbonates, lithium and magnesium (and also calcium and strontium) give the oxide on heating:

$$MgCO_3 = MgO + CO_2$$

while the carbonates of the other alkali metals and of barium and radium are stable to heat.

Lithium and magnesium are more strongly hydrated than their heavier congeners and their salts are similar in solubility. Most salts are soluble but the carbonate and phosphate are insoluble in each case. The high solubility of lithium perchlorate in diethyl ether has led to its use in organic chemistry as a reagent for accelerating Diels–Alder cyclization reactions. Magnesium fluoride and magnesium hydroxide are rather insoluble while lithium fluoride and hydroxide are both markedly less soluble than the sodium salts. Lithium and magnesium halides are also similar in being soluble in organic solvents such as alcohol. With larger anions, solubilities tend to fall with cation size in both Groups. Thus the alkali metal perchlorates are relatively insoluble for K, Rb, Cs and Fr and the solubilities of carbonates and sulfates are in the order $Mg > Ca > Sr > Ba > Ra$. The fluorides and hydroxides present an exception to this order—for example, the solubilities of the fluorides are in the order $Mg < Ca < Sr < Ba$ —and this must result from the effect of the relatively small OH^- and F^- anions on the lattice energies.

Lithium and magnesium are also more strongly complexed by nitrogen donors, in ammonia and amines, than are the heavier elements. The ammoniated salts $[Li(NH_3)_4]I$ and $[Mg(NH_3)_6]Cl_2$, for example, can be precipitated from liquid ammonia solution by double decomposition reactions.

The high hydration energy of lithium more than compensates for its relatively high ionization potential in the case of reduction reactions which are carried out in water. The result is that the redox potential in water of lithium is as strongly reducing as that of caesium, although the latter is much more reactive under anhydrous conditions.

Mg and Li form a wider range of complexes than the other s elements, and show a variety of coordination behaviour in addition to the monomers with 4-coordinate Li or 6-coordinate Mg. Thus, the dimer $[R_2NLi.OEt_2]_2$ contains the unit

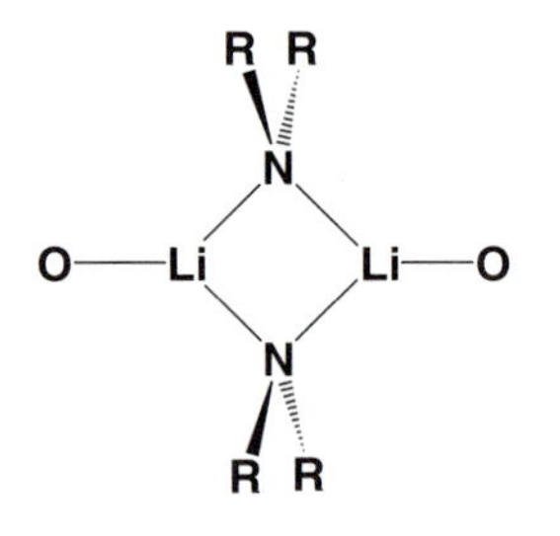

where R is a bulky group. Similarly, DMg(OR)Br (for D a donor such as Et_2O) is a dimer of the form

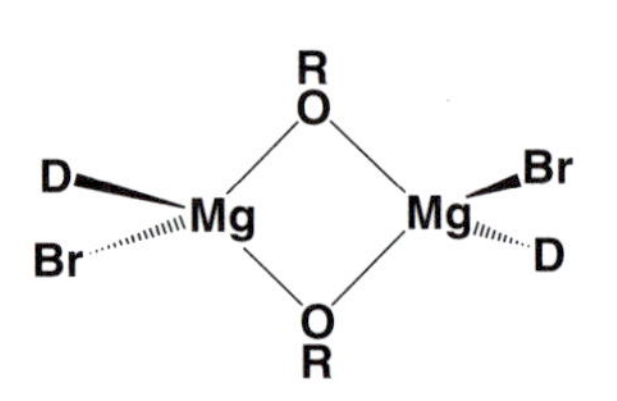

In the complex cation, $[Mg_2Cl_3.6THF]^+$ the Mg atoms are 6-coordinated to three THF molecules and three bridging Cl:—$[(THF)_3Mg(Cl)_3Mg(THF)_3]^+$.

Magnesium and lithium can activate molecular N_2, which gives one route to transition metal complexes containing $M(N_2)$ units. One interesting example where the product contains magnesium is the species $[(Me_3P)_3Co(N_2)Mg(THF)_4]_2$. This has the central structural unit Co-N≡N-Mg-N≡N-Co with an angle of 150° for N-N-Mg. The synthesis depends on the activation of N_2 by a fresh magnesium surface.

10.7 Beryllium chemistry

The size effects which produce the differences between the chemistry of lithium and the other alkali metals, have a much more marked effect on the chemistry of beryllium as compared with its heavier congeners. On passing from lithium to beryllium, the size of the atom grows less and the charge on the possible ion doubles. The result is to give such a high charge density on the hypothetical Be^{2+} ion that it is too polarizing to exist and all beryllium compounds are either covalent or contain solvated beryllium ions, such as $Be(H_2O)_4^{2+}$. Even the anhydrous halides are only feebly ionic; beryllium fluoride is one of the few metal fluorides which is not completely ionized in solution, while the conductivity of fused beryllium chloride is only one thousandth of that of sodium chloride under the same conditions.

The small size and strong hydration effects have a marked influence on the solubility of beryllium compounds as compared with those of the other elements of the Group. In water, the size effects increase the solvation energies more than the lattice energies, and the solubilities of compounds such as the sulfate, selenate or oxalate are markedly greater than those of the corresponding calcium compounds. An extreme case is provided by beryllium fluoride which is a thousand times more soluble than magnesium fluoride.

In water, the hydrated beryllium ion differs from the other s element ions in being hydrolysed. Beryllium therefore more closely resembles aluminium, which also undergoes extensive hydrolysis in solution. $Be(H_2O)_4^{2+}$ exists in strongly acid solutions but in neutral or weakly acid solutions this is hydrolysed and polymerized, for example:

$$2Be(H_2O)_4^{2+} \rightarrow (H_2O)_3Be - O - Be(H_2O)_3^{2+} + 2H_3O^+$$

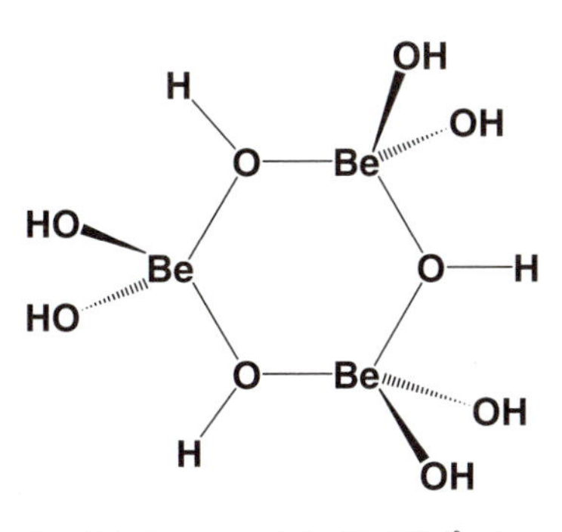

FIG. 10.7 Structure of the $[Be(OH)_3]_3^{3-}$ ion

and more complex species are formed. One stable form in dilute solutions is $[Be(OH)_3]_3^{3-}$ which is thought to have the ring structure of Fig. 10.7. Beryllium hydroxide is eventually precipitated on going to more alkaline conditions, and this differs from the other s element hydroxides in being amphoteric and dissolving in excess base, probably with the formation of the species $Be(OH)_4^{2-}$.

As would be expected from the above, beryllium forms more, and more stable, complexes than the other s elements. There is a strong tendency to assume a coordination number of four—in which use is made of all the valency orbitals. Thus the halides are monomeric and linear in the gas phase (Section 4.2), but they dissolve in ether and a dietherate, $BeX_2.2Et_2O$, is recovered. This contains beryllium in an sp^3 configuration with the ether oxygen atoms acting as lone pair donors (Fig. 10.8a). A large number of similar 4-coordinated complexes are formed with ligands such as ethers, aldehydes and ketones. In the liquid and solid states, the beryllium halides polymerize in order to achieve 4-coordination. In the solid, the unshared pair on a halogen atom of one molecule donates to the beryllium atom of the next and a long chain structure containing

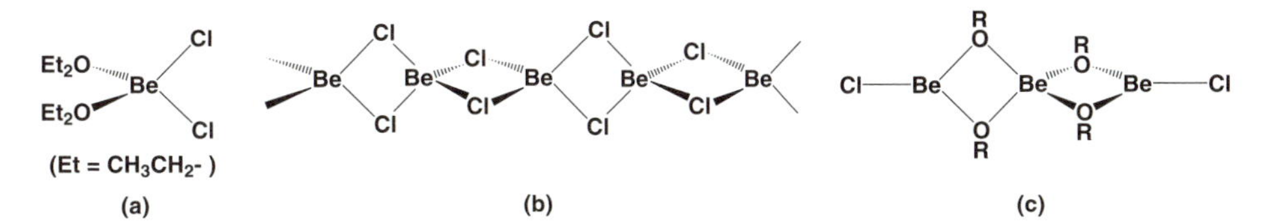

FIG. 10.8 Beryllium chloride compounds: (a) etherate $BeCl_2.Et_2O$, (b) $BeCl_2$ polymer in the solid, (c) $Be[Be(OR)_2Cl]_2$

tetrahedral beryllium results (Fig. 10.8b). Planar 3-coordinate Be is found together with tetrahedral Be in the dimer $[Be(OR)_2Cl]_2$ when R is the bulky tert-butyl group, Fig. 10.8c. Solid beryllium fluoride is a glassy material which also appears to contain chains but these are disordered and the material does not crystallize. The fluoride reacts with fluoride ion to form the tetrafluoroberyllate ion, BeF_4^{2-}, which is tetrahedral and usually isostructural with sulfate.

A more complex example, which illustrates the tendency of beryllium compounds to hydrolyse, and also to become 4-coordinated, is provided by the compound called 'basic beryllium acetate'. This is a very stable compound formed by the partial hydrolysis of beryllium acetate:

$$4Be(OOCCH_3)_2 + H_2O \rightarrow Be_4O(OOCCH_3)_6 + 2CH_3COOH$$

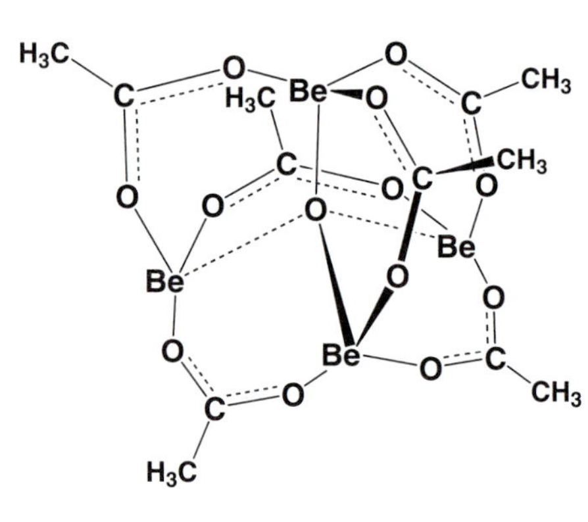

FIG. 10.9 Basic beryllium acetate, $Be_4O(OOCCH_3)_6$

It is a volatile solid which may be purified by sublimation. It is soluble in organic solvents such as chloroform, insoluble in water, and stable to heat and moderate oxidation. The structure is shown in Fig. 10.9. The four beryllium atoms are placed tetrahedrally around the central oxygen and they are linked in pairs by the six acetate groups, each of which spans one edge of the tetrahedron. Similar compounds of other carboxylic acids have been prepared. In basic beryllium nitrate, $Be_4O(NO_3)_6$, the nitrate groups are linked to two beryllium atoms, Be-O-N(O)-O-Be, in the same way as the acetate groups in the basic acetate.

Complexes also exist in which nitrogen is the donor atom. For example, beryllium chloride readily takes up ammonia to give the ammine, $[Be(NH_3)_4]Cl_2$. This compound is very stable to thermal decomposition but the ammonia groups are readily displaced by water.

Beryllium also achieves 4-coordination in the hydride and in the beryllium alkyls by polymerization through electron deficient bridges. Thus beryllium dimethyl, $Be(CH_3)_2$, has a chain structure very similar to that of beryllium chloride, and the bridge bonding is the same as that in the aluminium trimethyl dimer of Fig. 9.13.

Just as there is some resemblance in the properties of lithium and magnesium, so beryllium and its diagonal neighbour, aluminium, have similar properties. In this case, the resemblance in general chemistry is very close and beryllium is traditionally difficult to distinguish from aluminium in qualitative analysis. The basic resemblance is due to the high charge density on each element in the hypothetical cations, and to the existence of an empty p orbital or orbitals giving the elements acceptor properties. Among the detailed resemblances there may be noted:

(i) Similar redox potentials, $Be^{2+}/Be = -1.70$ V, $A1^{3+}/Al = -1.67$ V.

(ii) Both metals dissolve in alkali with the evolution of hydrogen.

(iii) The hydroxides are amphoteric, and the salts are readily hydrolysed.

(iv) The halides form polymeric solids, beryllium halides forming chains and aluminium halides, dimers. The halides are also similar in their solubilities in organic solvents, their electron pair acceptor properties (e.g. $AlCl_3.OEt_2$), and their catalytic effects.

(v) Both metals form carbides by direct combination and these yield mainly methane on hydrolysis.

(vi) Both elements form hydrides and alkyls which polymerize through electron deficient bridges.

Problems

When consulting alternative sources on systematic chemistry, you should check out 'Notes to the reader' at the beginning of Appendix A. Further Reading, together with the comments on the different sections of this list.

10.1 The 'diagonal relationship' is a long-standing generalization about the similarity in chemical properties of elements with neighbours 'one down to the right' in the Periodic Table.

Discuss similarities and differences between the following sets:

(a) lithium and magnesium
(b) beryllium and aluminium
(c) potassium, strontium and lanthanum.

10.2 The melting points of the alkali metals are (°C): Li = 453; Na = 371; K = 337; Rb = 313; and Cs = 302. Discuss these values and find out whether (a) the boiling points and (b) the values for the Be Group show similar trends.

10.3 The alkali metals form the commonest group of M^+ cations. Which other M^+ cations exist? Compare and contrast the properties of their compounds with those of the alkali metals. Extend this discussion to NH_4^+ species.

10.4 'The chemistry of the s elements is dominated by their tendency to lose the s electrons.'
Discuss this statement, and comment on

(a) the parameters which reflect the tendency
(b) compounds formed only by the more reactive s elements
(c) 'exceptions that prove the rule'.

10.5 Compare and contrast the chemistry of the s^1 and s^2 elements.

10.6 Research the topic: 'The biological activity of the alkali and alkaline earth elements'.

10.7 Magnesium and calcium are often found in different environments in silicate and aluminosilicate minerals. Find from the references three examples of this.

CHAPTER 11

11.1 GENERAL AND PHYSICAL PROPERTIES 246
11.2 CHEMISTRY OF THE TRIVALENT STATE 248
11.3 THE SEPARATION OF THE ELEMENTS 250
11.4 OXIDATION STATES OTHER THAN III 252
11.4.1 Cerium(IV) 252
11.4.2 Europium(II) 252
11.4.3 Other IV states 253
11.4.4 Other II states 253
11.5 PROPERTIES ASSOCIATED WITH THE PRESENCE OF f ELECTRONS 255
PROBLEMS 256

The Scandium Group and the Lanthanides

11.1 General and physical properties

The scandium Group of elements has the outer electronic configuration d^1s^2 and is formally part of the d block of the Periodic Table. However, as the chemistry is dominated by the III oxidation state, which involves the loss of the d electron, this Group is best regarded as forming a transitional region between the s elements and the main d block. Following lanthanum, the third member of the scandium Group, the 4f shell fills for the next fourteen elements. As the general chemical behaviour in these *lanthanide elements* (or rare earths) is that of the III state and is very similar to that of the scandium Group proper, it is convenient to include these elements here. Properties are listed in Table 11.1.

REFERENCES TO LANTHANIDES IN OTHER CHAPTERS:

Table 2.5 atomic numbers and masses
Table 2.8 ionization potentials
Table 2.14 electronegativities
Table 6.3 redox potentials
The organometallic chemistry of the lanthanide elements is the topic of Section 16.5

When the 5f shell fills following actinium, which is the fourth member of the scandium group, the very small energy gap between the 5f level and the 6d level gives rise to a considerable variability in the chemistry of the actinide elements which are therefore treated separately in Chapter 12.

Of these elements, scandium and actinium are very rare and are not completely investigated. The lanthanide elements and lanthanum and yttrium all occur together, and there is a certain amount of segregation into larger and smaller ions. Thus the lighter (and larger) lanthanides make up ca. 90% of the two commercial ores *bastnasite* (a carbonate–fluoride), and *monazite* (a phosphate), with Ce ca. 45%, La ca. 25% and Nd ca. 15% as the major components. The best source of the smaller elements (yttrium and the heavier lanthanides) is the relatively rare phosphate ore *xenotime* (60% Y, 9% Dy, 6% Yb, 5% Er and other heavy elements 10%).

Separation of the lanthanides was difficult by classical methods, and chromatography or solvent extraction are now the methods of choice (see Section 7.2). Cerium may be removed chemically by oxidation to the IV state, and promethium does not occur naturally. It is thus much easier to separate the lighter lanthanides with these two gaps in the series and this, coupled with the abundances, makes the earlier members much cheaper to obtain than the later ones.

The largest scale use of lanthanides is in the fabrication of special steels. For this, the elements are not separated, but the naturally occurring mix is converted to the chlorides and then reduced electrolytically to *mischmetal* which is added to steels, and also used as lighter flints. A second, rapidly growing, use of mixed lanthanides is their addition to zeolites (see Section 18.6) to increase their catalytic activity, especially in petroleum crackers. Gd, Sm, Eu and Dy have large cross-sections for neutron capture and are used in control rods in nuclear reactors.

TABLE 11.1 Properties of the scandium elements and the lanthanides

Element	*Symbol*	*Electronic configuration*	*Abundance (ppm of crust)*	*Radius of ions (pm)*						*Oxidation states*	*Redox potentials $M^{3+}+3e=M$ (V)*
				M^{2+}		M^{3+}		M^{4+}			
				CN 6	8	6	8	6	8		
Scandium	Sc	$[Ar]3d^14s^2$	5			88.5	101.0			III	1.88
Yttrium	Y	$[Kr]4d^15s^2$	28			104.0	115.9			III	2.37
Lanthanum	La	$[Xe]5d^16s^2$	18			117.2	130.0			III	2.522
Actinium	Ac	$[Rn]6d^17s^2$	trace			126				III	2.13
Cerium	Ce	$[Xe]4f^25d^06s^2$	46			115	128.3	101	111	III, IV	2.483
Praseodymium	Pr	$[Xe]4f^35d^06s^2$	5			113	126.6	99	110	III, (IV)	2.462
Neodymium	Nd	$[Xe]4f^45d^06s^2$	24		143	112.3	124.9			III, (IV?) (II)	2.431
Promethium	Pm	$[Xe]4f^55d^06s^2$	unstable			111	123.3			III	2.423
Samarium	Sm	$[Xe]4f^65d^06s^2$	6	136	141	109.8	121.9			III, (II)	2.414
Europium	Eu	$[Xe]4f^75d^06s^2$	1	131	139	108.7	120.6			III, II	2.407
Gadolinium	Gd	$[Xe]4f^75d^16s^2$	6			107.8	119.3			III, (II)	2.397
Terbium	Tb	$[Xe]4f^95d^06s^2$	1			106.3	118.0	90	102	III, (IV)	2.391
Dysprosium	Dy	$[Xe]4f^{10}5d^06s^2$	4	121	133	105.2	116.7			III, (IV?) (II)	2.353
Holmium	Ho	$[Xe]4f^{11}5d^06s^2$	1			104.1	115.5			III	2.319
Erbium	Er	$[Xe]4f^{12}5d^06s^2$	2			103.0	114.4			III	2.296
Thulium	Tm	$[Xe]4f^{13}5d^06s^2$	0.2	117		102.0	113.4			III, (II)	2.278
Ytterbium	Yb	$[Xe]4f^{14}5d^06s^2$	3	116	128	100.8	112.5			III, (II)	2.267
Lutetium	Lu	$[Xe]4f^{14}5d^16s^2$	0.8			100.1	111.7			III	2.255

*Bracketed states are unstable: states marked (?) are very unstable.

> Consult the reading list, Appendix A, for the history of the discovery and separation of the lanthanides which gives an interesting insight into the experimental skill and persistence required to solve the problems by classical methods.

The elements are electropositive and reactive, with the heavier elements resembling calcium in reactivity while scandium is similar to aluminium. Two of the elements, promethium and actinium, occur only as radioactive isotopes. Actinium is found associated with uranium and the most readily available isotope, ^{227}Ac, has a half-life of 22 years. It is, however, very difficult to handle as the decay products are intensely active and build up in the samples. Its chemistry fits in as that of the heaviest element of the scandium Group. The missing lanthanide, promethium, occurs only in radioactive forms with the longest-lived isotope, ^{145}Pm having $t_{\frac{1}{2}} = 30$ years. Its chemistry fits in with its place in the series. The analysis of samarium and neodymium isotope ratios provides a geochronological method for dating very old rocks.

Reactivity increases with increasing atomic weight in the scandium Group, just as in the s Groups. The elements are prepared by the reduction of the chlorides or fluorides with calcium metal. Some reactions of the metals are shown in Fig. 11.1. These direct reactions with elements broadly parallel those of the s elements given in Table 10.2.

The hydrides, formed by direct combination, illustrate the transitional character of this Group (compare Section 9.4). They form stable MH_2 and MH_3 phases, which usually occur in nonstoichiometric form, and the MH_3 formula for the most highly hydrogenated species is never fully attained. The hydrides have some salt-like properties and appear to contain the H^- ion. There are also available extra delocalized electrons giving metallic properties, so that the overall properties of the hydrides are a mixture of the ionic character shown by the s element hydrides and the metallic character of the hydrides of the d elements. The hydrides also resemble the ionic hydrides in their reactivity to oxygen and water. Mixed element hydrides involving lanthanides and transition elements show promise as hydrogen storage materials. The alloy $LnNi_5$ reacts reversibly with hydrogen gas to form $LnNi_5H_6$ at

APPLICATIONS BASED ON THE OPTICAL OR MAGNETIC PROPERTIES OF THE LANTHANIDES

Because of their specific magnetic and optical properties, due to the f electrons (Compare Section 11.5), specific elements find magnetic, electronic and optical applications. 'Didymium', the natural mixture of Pr and Nd, is used in protective glasses to shield from UV and high intensity light. Uses as activators in phosphors include Eu^{3+} for the red colour in TV, and with Tb^{3+} (green), in fluorescent tubes. Europium(III) complexes with certain cryptate, aryl-nitrogen or aminopolycarboxylate ligands show strong luminescence and high efficiency in converting light energy. These *light-conversion molecular devices* have interesting potential applications as luminescent materials. In medical X-ray intensifiers, Eu^{2+}, Tm^{3+} and Tb^{3+} act as activators and La or Gd oxide species as support media. Neodymium lasers are very important. The neodymium may be held in a glass or in yttrium aluminium garnet, and there is also a liquid laser using neodymium oxide dissolved in selenium oxychloride. A liquid system has many technical advantages, but this was only the second liquid system to be announced and it was by far the most powerful. The key aspect seems to be the use of a solvent which does not contain light atoms, such as hydrogen, so that most of the input energy is emitted in the laser beam and not transferred to heat the solution.

Yttrium is used in the 90 K $YBa_2Cu_3O_{7-x}$ superconductor, and other lanthanides have been tried in similar formulations (see Section 16.1). Yttrium and gadolinium in *garnet* oxides, $Ln_3M_5O_{12}$ (M = a trivalent element, especially Fe or Ga) are used in bubble devices for memory storage, microwave components and other magnetic applications.

In the laboratory several lanthanide complexes are used as shift reagents for spreading out the signals in proton nmr. The advent of high-field nmr spectrometers, together with powerful multi-dimensional nmr techniques, has resulted in a decline in the use of these reagents. However, lanthanide complexes, particularly those of gadolinium, are finding ever-increasing use as magnetic resonance imaging (MRI) agents. In this technique, gram quantities of gadolinium complexes are injected into the body and the proton nmr signals, largely of tissue water molecules, are spatially mapped; the role of the strongly paramagnetic (f^7) gadolinium ion is to change the relaxation time of nearby water molecules, thereby improving the contrast of the MRI technique. MRI is just one of a rapidly growing array of inorganic-based diagnostic and therapeutic medical techniques, discussed in greater detail in Chapter 20.

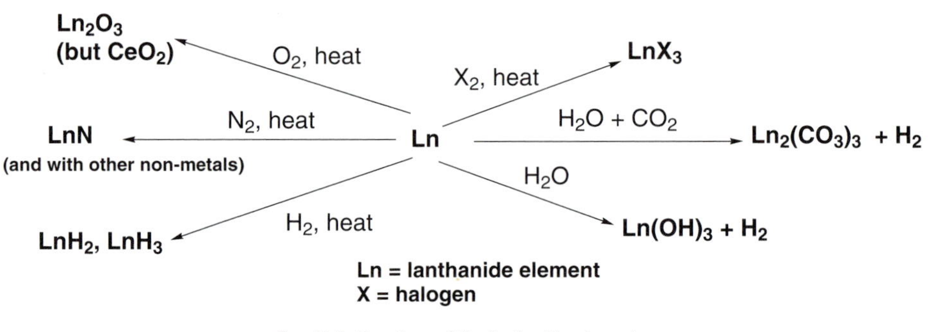

FIG. 11.1 Reactions of the lanthanide elements

readily accessible temperatures and pressures, and the properties can be 'tailored' by varying Ln = La, Ce, Pr or Nd, or mixtures of these.

11.2 Chemistry of the trivalent state

As Table 11.1 shows, the oxidation state of +III is shown by all these elements and is the most stable state. Other states are found only where f electrons are present. For the group Sc, Y, La and Ac, the oxides, M_2O_3, and hydroxides, $M(OH)_3$, increase in basicity with increasing atomic mass. Scandium, because of its smaller size, is more easily hydrolysed in solution than the other ions, and its oxide is amphoteric. Hydrated scandium triflate $[Sc(O_3SCF_3)_3.9H_2O]$ is isomorphous with the hydrated lanthanide triflates, and contains a tricapped trigonal prismatic $Sc(H_2O)_9^{3+}$ ion. There does not appear to be a definite scandium hydroxide, although the species ScO(OH) is well established (compare the existence of AlO(OH)). The oxide, with water, forms the hydrous oxide, $Sc_2O_3.nH_2O$, which dissolves in excess alkali to form anionic species such as $Sc(OH)_6^{3-}$. The other elements form oxides and hydroxides which are basic only. These hydroxides are precipitated from solution by the addition of dilute alkalis and do not dissolve in excess alkali. Yttrium oxide and hydroxide are strong enough bases to absorb carbon dioxide from the atmosphere, while lanthanum oxide 'slakes', with evolution of much heat like calcium oxide, and rapidly absorbs water and carbon dioxide. The lanthanide element basicities decrease towards lutetium, which is similar to yttrium in the properties of its oxide. Actinium compounds are more basic than the lanthanum ones.

Among the trihalides, the fluorides are insoluble and their precipitation, even from strongly acidic solution, is a characteristic test for these elements. Scandium fluoride dissolves in excess fluoride with the formation of the complex anion, ScF_6^{3-}. The fluorides of the heavier lanthanide elements are also slightly soluble in hydrogen fluoride, probably because of complex formation. All the elements form oxide-fluorides MOF from the reaction of MF_3 with M_2O_3. More complex phases such as $Y_7O_6F_9$ are also reported.

The other trihalides are very deliquescent and soluble (compare $CaCl_2$). $ScCl_3$ is much more volatile than the other trichlorides. It resembles $AlCl_3$ in this, but it is monomeric in the vapour (aluminium trichloride is dimeric) and it has no activity as a Friedel–Crafts catalyst. The chlorides are recovered from solution as the hydrated salts and these, on heating, give the oxychlorides, MOCl (with the exception of scandium which goes to the oxide). Actinium also forms oxyhalides but only by reaction with steam at 1000°C, a treatment which produces oxide from the lower members of the Group. This is a further example of the increase in basicity from scandium to actinium. Bromides and iodides resemble the chlorides in general behaviour.

Among the oxysalts, most anions are to be found including strongly oxidizing ones. The carbonates, sulfates, nitrates and perchlorates, for example, all resemble the calcium compounds. The carbonates, phosphates and oxalates are insoluble while most of the others are rather more soluble than the calcium salts. Scandium carbonate differs from the others in dissolving in hot ammonium carbonate, with double salt formation, and this affords a method of separating scandium from yttrium and lanthanum. Double salts are very common and include double nitrates, $M(NO_3)_3.2NH_4NO_3.4H_2O$, and sulfates such as $M_2(SO_4)_3.3Na_2SO_4.12H_2O$. Such salts were used for separation of the lanthanides by fractional crystallization methods.

These salts are fully ionic and lanthanum is useful as one of the few available stable ions with a charge higher than +2. The scandium ion is more readily hydrolysed than the others, and polymeric species of the type $[Sc\text{-}(OH)_2\text{-}Sc\text{-}(OH)_2\text{-}]_n$ have been identified with the chain length increasing as the acidity falls. The other ions are only slightly hydrolysed in the sense:

$$M(H_2O)_6^{3+} + H_2O \rightleftharpoons M(H_2O)_5(OH)^{2+} + H_3O^+$$

with the tendency to hydrolyse increasing as the size decreases.

Although these ions have a charge of +3, the tendency to complex formation is relatively slight. When compared with transition metal ions, such as Fe^{3+} or Cr^{3+}, which readily form complexes, this reluctance to complex may be ascribed to the greater size of the scandium Group ions, and to their low electronegativity which decreases any possible covalent contribution to the bonding. The best donor atom is oxygen, and insoluble complexes are formed by β-diketones such as acetylacetone (Fig. 11.2). Of some importance are the water-soluble complexes formed by chelate ligands such as EDTA, and especially the complexes formed by hydroxycarboxylic acids such as citric acid, $HOOCCH_2C(OH)(COOH)CH_2COOH$, which are used in the separation of these elements by ion exchange methods (see Section 11.3). The lanthanide elements are often 6-coordinated in complexes, but higher coordination numbers are known. Eight-coordination is found in $La(acac)_3(H_2O)_2$, where acac = acetylacetonato, and in $[Y(CF_3COCHCHCF_3)_4]^-$. Shapes include antiprismatic (compare Fig. 15.5b) and dodecahedral (compare Fig. 13.12). An interesting example is found in the $[HoAl_3Cl_{12}]_n$ polymer which contains square antiprismatic $HoCl_8$ units formed by four $Ho(\mu^2\text{-}Cl)_2Al$ bridges (compare Fig. 17.12) and these four Al atoms bridge further in a polymeric structure.

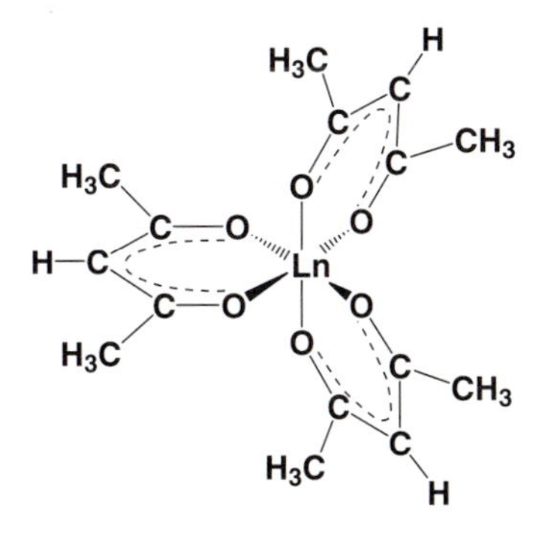

FIG. 11.2 Acetylacetonato complex of the lanthanide elements

$Nd(BrO_3)_3.9H_2O$ contains the 9-coordinate $[Nd(H_2O)_9]^{3+}$ ion and similar ions exist for other lanthanides. $Yb(NO_3)_3(H_2O)_3$ is also 9-coordinate with bidentate nitrates (as in the cerium compound below). The nine O atoms form a tricapped trigonal prism (compare Fig. 15.23). EuN_9 coordination is also found in Eu_2L_3 (L = a pyridine/benzimidole long-chain ligand) which self-assembles into a triple helix. Ten-coordination is found in LaEDTA-$(H_2O)_4.3H_2O$. The EDTA occupies six positions and there are four water molecules attached to the lanthanum. Three water molecules lie on one side of the lanthanum, the two nitrogen atoms of the EDTA lie opposite them, while the four EDTA oxygens and the final water molecule lie in a rather distorted medial plane. The complex nitrate, $Ce(NO_3)_5^{2-}$ contains 10-coordinate Ce. The nitrates are present at the apices of a trigonal bipyramid and each nitrate is coordinated by two oxygens.

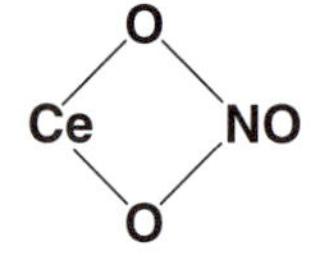

Coordination of the nitrate (NO_3^-) ligand to cerium

Yttrium, and the other lanthanides investigated, also form 10-coordinate $M(NO_3)_5^{2-}$ complexes but Sc in $Sc(NO_3)_5^{2-}$ is only 9-coordinate with one of the nitrate groups bonded through only one oxygen. In $Ce(NO_3)_6^{3-}$, the coordination number is twelve. In the ion $La(C_6H_9N_3)_4^{3+}$, the twelve nitrogen atoms form an almost regular icosahedron (compare Fig. 17.7d) around the La atom.

Scandium, because of its small size, forms more stable complexes than the other elements. For example, the scandium acetylacetonate may be sublimed at about 200°C, while all the others decompose on heating. Continuing this trend to the heaviest element, actinium is less ready to form complexes than the others. Thus, the lanthanides can be extracted into an organic solvent by means of tributylphosphate, $OP(OC_4H_9)_3$, which forms a complex, but actinium extracts much less readily under these conditions.

11.3 The separation of the elements

One separation of scandium is mentioned above, and actinium occurs separately, but yttrium, lanthanum and the lanthanides are commonly found together in minerals. From the radii given in Table 11.1 it will be seen that the lanthanides are very close in size, with a small but regular decrease from lanthanum to lutetium, and that yttrium is close to dysprosium and holmium. A similar gradation is found in the redox potentials of the elements, M_{aq}^{3+}/M, which range from −2.52 V for La to −2.25 V for Lu, in steps of about 0.01–0.02 V between each element and the next. Again yttrium, −2.37 V, fits in near dysprosium, −2.35 V. These resemblances in size and behaviour are much closer than those between elements in the same Group (compare Sr^{2+} and Ba^{2+} which differ by 19 pm) and mean that the chemistry of all the lanthanide elements is practically identical.

Compare Section 8.1 for effective nuclear charge.

The slow decrease in size from La to Lu just about balances the normal increase in size between the elements in one period and the next. This is shown by the similarity between yttrium and the heavier lanthanides. The decrease is termed the *lanthanide contraction* and arises from the slow increase in effective nuclear charge as the f electrons are added. This accounts for the decrease in size and the increase in oxidation potential. As will be seen, the lanthanide contraction also affects the chemistry of the heavier transition metals.

As the elements and ions are so similar in size and properties, the separation of the individual lanthanide elements is extremely difficult. In the classical studies of the elements, fractional separations had to be adopted. These included fractional crystallization of double salts, such as the nitrates, fractional precipitation of the hydroxides and fractional decomposition of the oxalates. These processes were very slow, and as many as 20 000 operations are reported in some cases before pure samples were obtained. The separation of the lighter elements, up to gadolinium, was relatively easy as cerium could be removed by oxidation to the relatively stable IV state, promethium was missing from the natural sources, and samarium and europium could be reduced to the II state. The heavier elements and yttrium were much more difficult to separate as such chemical aids were not available. Despite this, all the lanthanide elements had been separated and correctly characterized before the advent of more powerful methods, in what was one of the most painstaking series of studies in chemistry.

The separation problem is greatly simplified by the use of ion exchange or solvent extraction techniques (see Chapter 7). One common ion exchange method uses the soluble citrate complexes. To a cation exchange resin, which will be written resin-H, a solution of the lanthanides is applied, and the acid formed washed out:

$$3\text{Resin-H} + \text{Ln}^{3+} \rightarrow \text{Resin}_3\text{-Ln} + 3\text{H}_3\text{O}^+$$

Then citric acid, buffered with ammonia to constant pH, is added, and the equilibrium:

$$\text{Ln-Resin}_3 + 3\text{HCit} \rightleftharpoons 3\text{H-resin} + \text{LnCit}_3$$

is set up (Ln is used as a general symbol for any lanthanide element). As the buffered citrate flows down the column, the concentration of lanthanide ions changes and the equilibrium reverses many times. As the heavier ions are smaller, they will be more strongly complexed by the citrate and so will tend to spend more time in solution and less on the resin. As a result, the heavier lanthanide elements are washed down the column first and will eventually be eluted. If the conditions are correct, the different elements will be separated into pure components. Figure 11.3 gives an example of such an elution curve. The whole process is analogous to the classical fractionations but with the numerous operations taking place *in situ* on the column. The process leads to considerable dilution: in one example 0.4 g of mixed oxides per litre was used and collection of about fifty litres of eluate gave each element in about 80% purity. Each fraction would then be concentrated by precipitation of the oxalate and the exchange repeated to give pure samples.

Similar separations are possible using solvent extraction methods (also called hydrometallurgy), typically in a counter-current extraction technique. For example, by extraction of lanthanides from a strong nitric acid solution into tributylphosphate, 95% pure gadolinium has been prepared on a kilogram scale. Similar separation processes involving organophosphorus acids such as R(RO)P(O)OH, where R is, for example, 2-ethylhexyl, have also been developed. There is a gradual change in the pH at which the trivalent lanthanide extracts into the organic solvent, as shown in Figure 11.4, and this leads to a separation of the elements. Such processes are currently operated industrially. Solvent extraction processes are of considerable importance in the nuclear industry, as discussed in the next chapter, and have contributed to the characterization of the synthetic post-plutonium elements.

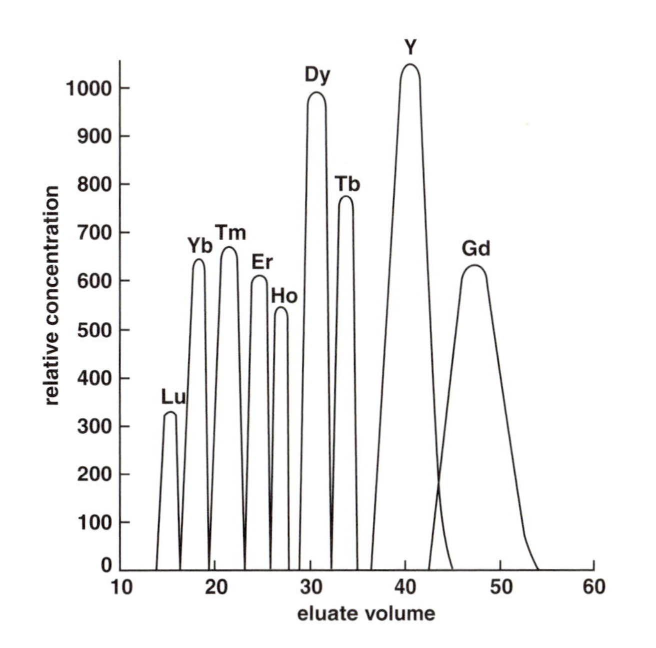

FIG. 11.3 Elution curve of the lanthanide elements from an ion exchange resin column

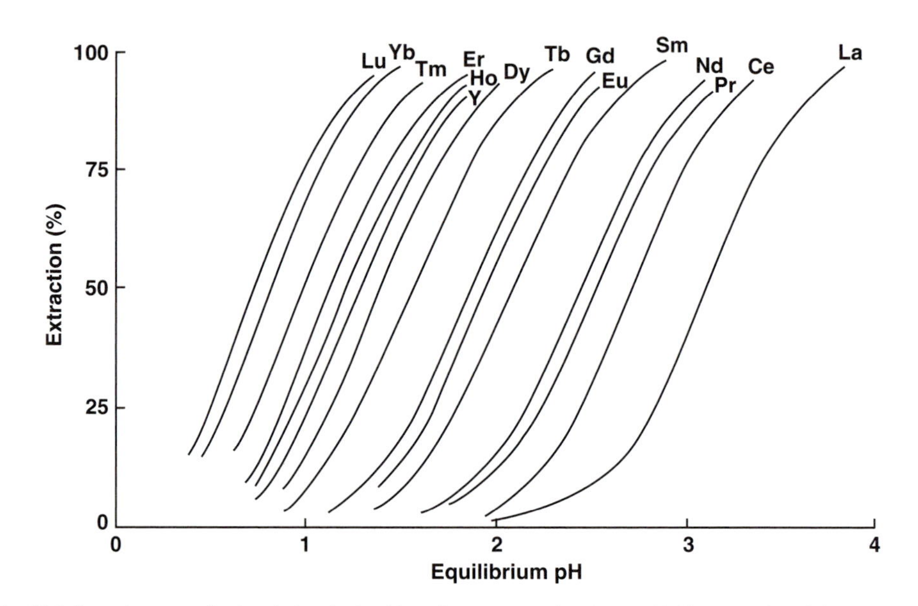

FIG. 11.4 Extraction curves for the trivalent lanthanides using an organophosphorus acid. The separation of the curves is a measure of the selectivity of the extractant—two lanthanides with curves close together will require many repeat extraction stages to separate them completely

11.4 Oxidation states other than III

A number of the lanthanide elements exist in oxidation states other than III and the most stable of these are Ce(IV) and Eu(II). The cerium(IV) state corresponds to the loss of the four outer electrons to give an $f^0d^0s^0$ rare gas configuration, while the europium(II) state corresponds to the loss of only the two s electrons to retain the half-filled f^7 shell.

11.4.1 Cerium(IV)

Cerium is the only lanthanide which exists in the IV state in solution. Ceric oxide, CeO_2, which is colourless when pure, is the product resulting from heating the metal, or decomposable cerium(III) oxysalts such as the oxalate, in air or oxygen. It is inert and insoluble in strong acids or alkalis. It does dissolve in acids in the presence of reducing agents to give cerium(III) solutions, and these, in turn, give cerium(IV) in solution on treatment with strong oxidizing agents such as persulfate. A yellow, hydrated, form of ceric oxide, $CeO_2.nH_2O$, is precipitated from such cerium(IV) solutions by the action of bases. Another solid cerium(IV) compound known is the tetrafluoride, prepared by the action of fluorine on the trichloride or trifluoride. The aqueous chemistry of cerium(IV) resembles that of zirconium and hafnium, or of the four-valent actinides such as thorium.

In more complex compounds, the oxidizing power of cerium(IV) is modified and a wider range of reactions can be studied. A range of water-soluble cerium(IV) salts has been prepared to act as models for the active photosynthetic centre (compare Chapter 20) in bacteria. In contrast to chlorophyll (see Fig. 20.1), the bacterial centre has a double four-nitrogen ring. The cerium(IV) compounds are of the form $Ce(P)_2$ where P^{2-} stands for a porphin ring (see also Fig. 20.6) with various substituents on the bridging carbons. The structure has these rings staggered above and below the 8-coordinated central cerium atom so that the coordination is square antiprismatic CeN_8, as shown in Fig. 11.5. Such complexes undergo a variety of redox and aggregation reactions.

The very high charge on the Ce^{4+} ion leads to its being strongly hydrated. The hydrated ion is acidic and hydrolyses to give polymeric species and hydrogen ions, except in strongly acidic solution. The solution of cerium(IV) in acid is widely used as an oxidizing agent, and its redox potential depends quite strongly on the acid used, ranging from 1.44 V in molar sulfuric acid to 1.70 V in perchloric acid. This variation probably arises from the formation of complex ions by association with the acid anion in nitric or sulfuric acids. As perchlorate shows no tendency to complex in this way, the potential in perchloric acid probably characterizes the plain hydrated ion, $Ce^{4+}.nH_2O$.

The high charge density means that cerium(IV) forms stronger complexes than the tripositive lanthanides. It is much more readily extracted by tributylphosphate, and the hexachloro-complex $CeCl_6^{2-}$ has been prepared as the pyridinium salt.

11.4.2 Europium(II)

Europium has the most stable divalent state of all the lanthanides. Europium(II) chloride is prepared as a solid by the action of hydrogen on the trichloride:

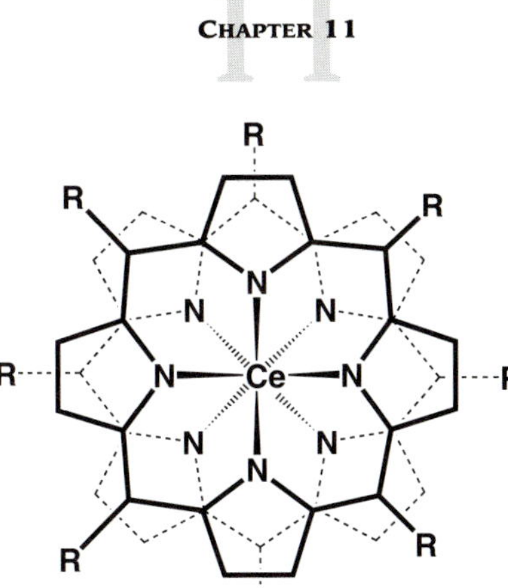

FIG. 11.5 The structure of an 8-coordinate cerium bisporphyrinate complex. The dotted lines represent the bottom porphyrin ligand.

$$EuCl_3 + H_2 \longrightarrow EuCl_2 + HCl$$

Its dihydrate is very insoluble in concentrated hydrochloric acid (like $BaCl_2.2H_2O$) and this is used as a means of purification. Europium in solution is readily reduced to the II state, for example by magnesium, zinc or alkali metal amalgams, and it resembles calcium in this state. Thus the sulfate or carbonate may be precipitated from solution. The oxide does not exist, but EuS, EuSe or EuTe can all be prepared. EuH_2 is ionic and isomorphous with CaH_2.

The redox potential of $Eu^{3+} + e^- = Eu^{2+}$ is -0.43 V so that europium(II) is a reducing agent of similar power to Cr^{2+}. A careful magnetic investigation has shown that the magnetic properties of Eu(II) are identical with those of Gd(III), over a wide range of temperature, confirming the $f^7d^0s^0$ arrangement in the ion. The dichloride, dibromide and diiodide all have moments of 7.9 Bohr magnetons, corresponding to the seven unpaired electrons.

Europium(II), although the most stable of the lower oxidation states, is a strong reducing agent as the potential shows, and its solutions are readily oxidized in air. The solids are rather more stable. Europium, together with ytterbium which also has a II state, dissolves in liquid ammonia to give a concentrated blue solution. The other lanthanides are either insoluble or give only weak solutions on electrolysis.

11.4.3 Other IV states

Other elements which form the IV state are praseodymium, neodymium, terbium and dysprosium. Of these, only Tb(IV) can be accounted for by the tendency to equally occupied f orbitals, in this case f^7. All the states are very unstable and have only been prepared as solid compounds.

Ignition of praseodymium compounds in air gives a complex oxidation product of approximate composition Pr_6O_{11}. Heating finely divided Pr_2O_3 in oxygen at 500°C and 100 atmospheres gives the stoichiometric oxide, PrO_2. No binary fluoride, PrF_4, has been prepared but solid solutions of PrF_3 in CeF_3, containing less than 90% PrF_3, do react completely with fluorine to the composition PrF_4/CeF_4.

There is no firm evidence for the existence of Nd(IV) in oxide systems, but the fluorination of NdF_3 in presence of CsF gives compounds containing 10–20% Nd(IV) in the form of a double salt.

A higher oxide of terbium, of approximate composition Tb_4O_7, has long been known as a product of ignition. A careful study has yielded three oxide phases, in the range TbO_x ($x = 1.5-1.8$), in ignition products of oxalate or nitrate. The TbO_2 composition results from the reaction of atomic oxygen on Tb_2O_3. This, like PrO_2, has the fluorite structure. Structures containing linked $Tb^{IV}O_6$ octahedra are found in the oxyanion species M_2TbO_3, $M_{16}Tb_3O_{14}$ and $M_6Tb_2O_7$ where M is an alkali metal, especially Li. Terbium(IV) is also formed as fluoride by fluorination of the trifluoride. TbF_4 is isostructural with CeF_4. Terbium(IV) is probably the most stable of the IV states after cerium(IV), but it is an extremely powerful oxidizing agent and there is no question of its existence in an aqueous medium.

Dysprosium(IV) resembles neodymium(IV) in being found only in a fluorine system. Fluorination of DyF_3, in presence of CsF, gives materials containing up to 50% Dy(IV).

11.4.4 Other II states

Elements found in the II state are neodymium, dysprosium, samarium, gadolinium, thulium and ytterbium. Ytterbium(II) corresponds to the completed f^{14} level. All these elements are much less stable in the II state than is europium. Yb(II) and Sm(II) may be prepared in water but are oxidized by water on standing; the others are found only in the solid state, or in nonaqueous solvents. The order of stability is Nd(II) $=$ Gd(II) $<$ Tm(II) $<$ Sm(II) $<$ Yb(II).

Divalent neodymium and gadolinium are prepared, as the dichloride or diiodide, by the reaction of the metals with the fused trihalides, and the diiodides can be prepared from the metal and iodine. $NdCl_2$ is isostructural with $EuCl_2$. Controlled reduction of $NdCl_3$ gives the mixed (II)/(III) KNd_2Cl_5. Thulium diiodide may be prepared in a similar manner by the action of Tm on TmI_3 at 600°C. It is isostructural with YbI_2. These low valency halides tend to be nonstoichiometric and they have metallic conduction and other properties.

Samarium(II) iodide, largely as a result of its rather oxophilic nature, is beginning to find quite widespread application in the synthetic organic chemistry laboratory. This, together with the use of cerium(IV) salts as powerful oxidizing agents, are examples of the current upsurge in the use of lanthanide reagents in chemical syntheses.

Low formal oxidation states are also found in a number of 'subhalides' with metallic properties. Among the best established are the monochlorides LnCl formed by Se, Y and all the lanthanides. The structure consists of a four-layer repeat unit in the order Cl–Ln–Ln–Cl in cubic close-packing, the same structure as ZrCl (Section 15.2). Two electrons per metal atom are delocalized, giving metallic conductivity. Similar MBr phases are found for M = Y, La and Pr. Small atoms such as C, O or H may intercalate between the metal layers, and cations may be held between the Cl sheets of successive four-layer units. Thus phases such as $ScCCl_{0.56}$ or $K_{0.26}YClC_{0.4}$ are found.

Phases of composition approximating M_2Cl_3 are also known. Sc_7Cl_{10} consists of Sc octahedra linked by shared Cl atoms. Similar phases, also with M-M bonding, are known for M = Y, La and Gd.

Samarium(II) occurs in a number of compounds, including the halides, sulfate, carbonate, phosphate and hydroxide. It may be extracted from a lanthanide mixture, along with Eu(II), by reduction of the trichlorides with alcoholic sodium amalgam. The mixture of Eu(II) and Sm(II) chlorides is readily separated by controlled oxidation, which produces Sm(III) only.

A variety of ytterbium(II) compounds exist, including all those found for samarium, and also possibly the monoxide. The dihalides may be prepared by hydrogen reduction of the trihalides and, in the case of the diiodide, by thermal decomposition. Yb(II) is more stable than Sm(II) and has been estimated to have an oxidation potential of -1.15 V with respect to the III state. YbH_2, like EuH_2, is ionic and isomorphous with CaH_2.

Work on the complex chemistry of Ln(II) species is increasing. For example, complexes of the diiodides with coordinated solvent molecules (ethers) are now well known, such as $LnI_2(MeOCH_2CH_2OMe)_3$ (Ln = Nd, Dy, Tm) and $LnI_2(THF)_5$ (Ln = Nd, Dy). $NdI_2(THF)_5$ is prepared as black crystals by crystallization of NdI_2 from THF, and the X-ray crystal structure shows a trigonal bipyramidal structure with the two iodine atoms in axial positions.

One interest in lanthanides is their possible incorporation into III–V or II–VI semiconductors. For Yb(II), this has led to the synthesis of various $Yb(ER)_2L_n$ (n = 4 for E = S, Se or 5 for E = Te) species where R is usually an aromatic group and L is a donor ligand such as pyridine. Such molecules are synthesized from the blue solution of Yb metal in liquid ammonia (see Section 6.7 and 10.1) or from $YbCl_3$ with reduction. The structures (Fig. 11.6a) have the two ER groups *trans* to each other and 4L groups for E = S or Se giving an octahedral configuration. The larger Tc–Yb distances allow 5L groups to coordinate to Yb, forming a pentagonal bipyramid (Fig. 11.6b). The Yb–E distances are 283 pm for E = S, 296 pm for E = Se and 328 pm for E = Te, reflecting the changes in the ionic radii of E^{2-} (see Table 2.12).

Very gentle oxidation of lanthanide amalgams (solutions of the metal in mercury) in the presence of oxygen donors such as diglyme, DIME, (see Appendix B), gives complex ions, e.g. $[(DIME)_3Ln]^{2+}$ for Ln = Sm or Eu. There are three donor O in each DIME molecule so the coordination is nine-fold and approximately tricapped trigonal prismatic, as in Fig. 15.23. Similar preparations using Yb in acetonitrile gave the complexes $[(DIME)_2Yb(CH_3CN)_2]^{2+}$ and $[(DIME)Yb(CH_3CN)_5]^{2+}$. In the first of these, three O from one DIME and the N from one CH_3CN form a square on each side of the Yb, giving an overall square antiprismatic arrangement, while the second has a similar configuration except that one face is made up of four acetonitrile nitrogens.

It will be noted that, although there are a significant number of examples of other oxidation states, the III state is by far the most stable among the simple compounds of the lanthanide elements. The most stable of the other states, Ce(IV) and Eu(II), are very reactive and most other examples are found only as solid state species. The exceptions

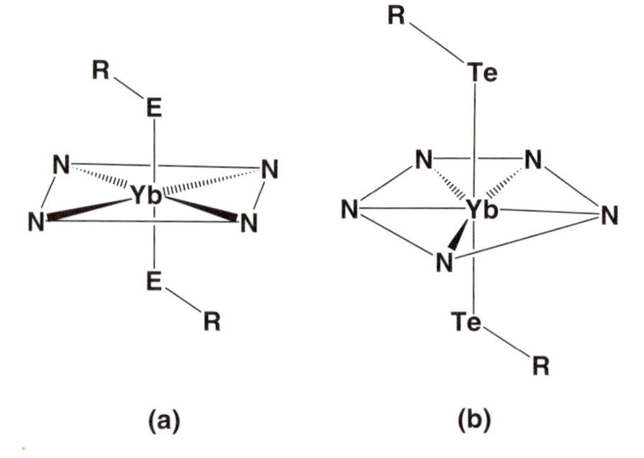

FIG. 11.6 (a) Structure of *trans*-$Yb(EPh)_2(N)_4$, where E=S or Se and N=C_5H_5N (pyridine). The octahedral Yb(II) ion is bound to *trans* phenylthiolato or phenylselenolato ligands. (b) Structure of $Yb(TePh)_2(N)_5$. The 7-coordinate Yb(II) ion contains *trans* phenyltellurolato ligands separated by a ring of pyridine ligands in a pentagonal bipyramidal geometry

are complexes of large multidentate ligands which do modify the normal high reactivity of Ln(IV) or (II). This dominance of the III state in the lanthanides presents a marked contrast with the behaviour of the actinide elements and is probably a result of the larger energy gap between the 4f and 5d levels than that existing between the 5f and 6d levels.

11.5 Properties associated with the presence of f electrons

As is implied in the discussion in the earlier sections, the electrons in the 4f level are too strongly bound to be involved in the chemistry of the elements except under unusual conditions. In addition, the f orbitals appear to be too diffuse to enter into bonding generally, so that there are few chemical effects from the presence of f electrons or unfilled f orbitals. There are, however, electronic effects which show up in the spectra and magnetic properties of the lanthanides.

The lanthanide ions show absorptions in the visible or near-ultraviolet regions of the spectrum, except La^{3+} with no f electrons and Lu^{3+} with no empty f orbitals. These colours are due to transitions between f levels, f–f transitions, and, as the f levels lie deep enough in the atom to be shielded from much perturbation by the environment, these transitions appear in the visible and near-ultraviolet spectra as sharp bands. This is in contrast to the d–d transitions found for the transition elements, which usually appear as broad bands due to environmental effects. Figure 11.7 illustrates a typical lanthanide ion spectrum. As these bands are so sharp, they are very useful for characterizing the lanthanides and for quantitative estimations. The positions of the absorptions shift with the f configuration, giving rise to the visible colours of the different ions as shown in Table 11.2.

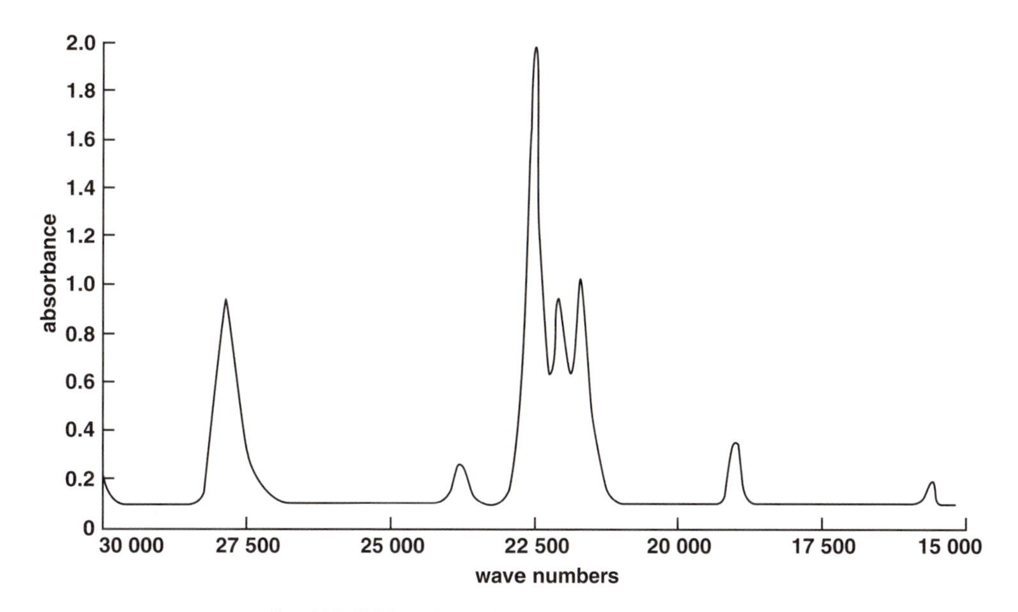

FIG. 11.7 Visible and near-ultraviolet spectrum of holmium

TABLE 11.2 Typical colours of lanthanide compounds

f^1 or f^{13}	Ce(III), Yb(III)	UV absorption
f^2 or f^{12}	Pr(III), Tm(III)	Green
f^3	Nd(III)	Blue–violet
f^4 or f^{10}	Pm(III), Ho(III)	Pink or yellow
f^5 or f^9	Sm(III), Dy(III)	Cream
f^6, f^7 or f^8	Eu(III), Eu(II), Gd(III), Tb(III)	UV absorption
f^{11}	Er(III)	Pink

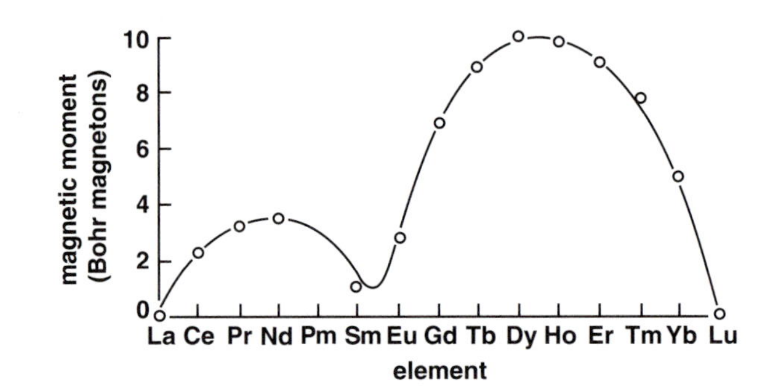

FIG.. 11.8 Calculated (—) and experimental (○) values of the magnetic moments of the lanthanides

Solutions of samarium(II) are red, and of ytterbium(II) are green. Note that the much more intense colours of cerium(IV) are not due to f–f transitions, but to a different mechanism involving charge transfer between ion and coordinated ligand.

In the transition metals the orbital contribution is usually quenched out by interaction with electric fields of the environment—at least to a first approximation—but the f levels lie too deep in the atom for such quenching to occur.

All the f states, except f^0 and f^{14}, contain unpaired electrons and are therefore paramagnetic. These elements differ from the transition elements in that their magnetic moments do not obey the simple 'spin-only' formula (Section 7.11). In the f elements, the magnetic effect arising from the motion of the electron in its orbital contributes to the paramagnetism, as well as that arising from the electron's spinning on its axis. When the moments are calculated on the basis of spin and orbital contributions, there is excellent agreement between experimental and calculated values, as Figure 11.8 shows.

The one case in which contributions to the bonding from the f orbitals is possible is in complexes of the heavier elements in which the coordination number is high. Use of the s orbital, together with all the p and d orbitals of one valency shell, permits a maximum coordination number of nine in a covalent species. Thus, higher coordination numbers imply either bond orders less than unity or else use of the f orbitals. In addition, certain shapes (such as a regular cube) of lower coordination number also demand use of f orbitals on symmetry grounds. These higher coordination numbers have only become clearly established recently, but their occurrence in lanthanide or actinide element complexes suggest the possibility of f orbital participation. Examples include the 10-coordinate complexes mentioned above, $LaEDTA(H_2O)_4$ and $Ce(NO_3)_5^{2-}$ or 10-coordinate $La_2(CO_3)_3.8H_2O$; 11-coordinate $Th(NO_3)_4.5H_2O$ (coordination by four bidentate nitrate groups and three of the water molecules); and the 12-coordinate lanthanum atoms in $La_2(SO_4)_3.9H_2O$—with twelve sulfate O atoms around one type of La atom position.

Problems

When consulting alternative sources on systematic chemistry, you should check out 'Notes to the reader' at the beginning of Appendix A, Further Reading, together with the comments on the different sections of this list.

11.1 Find out the details of one method which was used classically to separate the lanthanide elements and one method currently used, other than the examples given in Section 11.3.

11.2 Investigate the names of the lanthanide elements: how far do these reflect their chemical similarities?

11.3 Compare and contrast the chemistry of lanthanum (a) with calcium (see also Question 10.1), and (b) with thallium.

11.4 How far does the chemistry of their hydrides reflect the general characteristics of the lanthanide elements?

11.5 Find out from the literature

(1) further examples of coordination numbers of 8 or more.

(2) the use of lanthanide 'shift reagents' in nmr.

(3) examples of the application of the narrow-line f–f spectrum (e.g. as wavelength standards, in estimation, in lasers).

11.6 To what extent does the chemistry of scandium parallel that of aluminium. Discuss why there should be similarities and differences.

CHAPTER 12

12.1 Sources and physical properties 257
12.2 General chemical behaviour of the actinides 258
12.3 Thorium 261
12.4 Protactinium 263
12.5 Uranium 264
12.6 Neptunium, plutonium and americium 268
12.7 The heavier actinide elements 270
Problems 272

The Actinide Elements

12.1 Sources and physical properties

All the elements lying beyond actinium in the Periodic Table are radioactive, and many of them do not occur naturally. Uranium and thorium are available as ores. Although actinium, protactinium, neptunium and plutonium are available in small amounts in these ores their isolation is difficult and expensive and it is more convenient to isolate them from the fuel materials of nuclear reactors.

There are two main nuclear reactions which are used in the synthesis of elements beyond plutonium in the actinide series. One is the capture of neutrons, followed by β emission, which increases the atomic number by one unit. The second method is by the capture of the nuclei of light elements, ranging from helium to neon, which increases Z by several units in one step.

Atomic piles provide intense neutron sources and samples can be inserted into piles for irradiation, so the first method is readily carried out, but it is a process of diminishing returns as each element has to be made from the one before. For example, in the irradiation of plutonium-239, less than one percent of the original sample appears as californium-252, after capture of thirteen neutrons. The main stages are

	$^{239}Pu \rightarrow$	$^{241}Pu \rightarrow$	$^{245}Cm \rightarrow$	$^{247}Cm \rightarrow$	$^{251}Cf \rightarrow$	^{252}Cf
% of original sample	100	30	10	1.5	0.7	0.3

Neutron capture by a nucleus increases its neutron/proton ratio until this becomes too high for stability. Then a neutron is converted to a proton, with emission of a β-particle, and an increase of one in the atomic number, for example:

$$^{242}_{94}Pu + ^{1}_{0}n \rightarrow ^{243}_{94}Pu \rightarrow ^{0}_{-1}e + ^{243}_{95}Am$$

In the synthesis of heavier elements by successive neutron capture, not only does the yield of a heavier nucleus fall off sharply as the number of neutron addition steps from the starting material increases, but the process is made even less favourable by the general decrease in nuclear stability with increasing atomic mass. The heaviest elements may thus be obtained only by means which short-circuit the step-by-step addition of neutrons. One means is provided by an atomic explosion, where there is a vast flux of fast neutrons. A number of neutrons are added to the target nucleus simultaneously before the intermediate nuclei can decay. Thus, einsteinium and fermium were first discovered in the fall-out products of the first atomic bomb explosion.

A second, and more amenable, method for jumping a number of places in one operation is to bombard the starting material with species containing several nuclear particles. Bombardment by α-particles is the easiest way of achieving this, and many of the actinides, such as ^{248}Cm, ^{249}Bk, ^{249}Cf and ^{256}Md, were first made in this way, e.g.

$$^{253}_{99}Es + ^{4}_{2}He \rightarrow ^{256}_{101}Md + ^{1}_{0}n$$

α-particle bombardment requires the target to be the element with an atomic number of two less than the desired element, and such target elements will themselves be scarce and only obtained in small amounts where the desired elements is one of very high atomic mass. To make the very heaviest elements, therefore, bombarding nuclei heavier than the α-particle are required. The last two actinide elements to be discovered were obtained by heavy nucleus bombardment in this way:

$$^{246}_{96}\mathrm{Cm} + ^{12}_{6}\mathrm{C} \rightarrow ^{254}_{102}\mathrm{No} + 4^{1}_{0}\mathrm{n}$$
$$^{252}_{98}\mathrm{Cf} + ^{11}_{5}\mathrm{B} \rightarrow ^{257}_{103}\mathrm{Lr} + 6^{1}_{0}\mathrm{n}$$

Table 12.1 indicates the current sources of each element.

The intense activity of the samples of these elements, and of their sources, demands special handling techniques which involve manipulating microgram amounts of the sample by remote control. Such work demands specialized facilities and training and is limited to a few laboratories in the world, the most famous being at the University of California where most of the transuranium elements were discovered.

The ease with which a radioactive element may be handled depends on the type and intensity of the radiation from the isotope and from its decay products which accumulate in the sample. This activity is indicated by the half-life, which is a measure of the rate of the decay process. The degree of activity of an isotope governs the extent to which its chemistry may be studied. In decay processes, the extremely energetic emitted particles break bonds and disrupt crystal structures. The energy given out appears largely as heat in the sample; for example, the heat produced in a millimolar solution of curium-242 salts in water would be sufficient to evaporate the solution to dryness in a short time. The breaking of bonds by emitted particles in a sample of a radioactive element is equivalent to a continuous process of self-reduction. As a result, it may be impossible to prove the existence of an oxidizing state for very reactive isotopes. For example, the evidence for the oxidizing IV states of a number of the heavy elements had to await the production of isotopes which were more stable than the very short-lived ones originally available.

In a number of cases, neutron bombardment, which is readily carried out, does not give rise to the isotope of longest half-life. An example is provided by berkelium where the most accessible isotope, Bk-249, has a much shorter half-life than Bk-247, which is only available in tiny amounts from ion bombardment.

The electronic configurations of these elements gave rise to considerable controversy in the early days of work in this field. The elements which were available before the advent of the atomic pile, especially thorium and uranium, strongly resemble the transition metals (hafnium and tungsten) in their chemistry, and the heaviest elements were accordingly placed in the d block. As new elements were studied, it became increasingly obvious that an f shell was being filled but it was not clear whether this started after actinium, paralleling the lanthanides, or later on in the middle of a d series. It became clear, however, that curium corresponded to gadolinium in properties and was thus the $f^7d^1s^2$ element, implying that the series are genuinely *actinides*. Later, it was possible to interpret the very complicated atomic spectra and determine the electronic configurations as given in the table. Magnetic studies have also confirmed these. Element 103, lawrencium, completes the actinide series. Elements 104 and 105 are discussed in Section 16.12.

12.2 General chemical behaviour of the actinides

Table 12.2 lists some important chemical properties of the actinides, with actinium included for comparison. The M^{4+}/M^{3+} redox potentials clearly show the increasing stability of the III state for the heaviest elements. Diagrams of the free energy changes per electron in oxidation–reduction reactions are shown in Fig. 12.1, illustrating the stable oxidation states for the elements. The stabilization of the III state to the right of the actinide series is again shown here.

The most stable oxidation state of the elements up to uranium is the one involving all the valency electrons. Neptunium forms the VII state, using all its valency electrons, but this is oxidizing and the most stable state is Np(V). Plutonium also shows states up to VII and americium up to VI but the most stable states are Pu(IV) and Am(III). Later elements also tend to be most stable in the III state. This pattern of higher oxidation state stabilities has more in common with that of a d series where all the electrons are used in the Group oxidation state, until the middle of the series—Mn(VII) or Ru and Os(VIII)— and this state becomes more oxidizing across the series. Only for the later actinide elements does the III

TABLE 12.1 Some properties of the actinide elements

Z	*Element*	*Symbol*	*Mass of most accessible isotope (ii)*		*Half-lives*	*Source (i)*	*Electronic configuration* [Rn] 5f 6d 7s
89	Actinium	Ac	227		21.77 yr	Natural	0 1 2
90	Thorium	Th	231.2 (natural mixture)	232	1.40×10^{10} yr	Natural	0 2 2
91	Protactinium	Pa	231	231	32 500 yr	Natural and fuel elements	2 1 2
92	Uranium	U	238.1 (natural mixture)	235	7.04×10^{8} yr	Natural	3 1 2
				238	4.47×10^{9} yr		
93	Neptunium	Np	237	237	2.14×10^{6} yr	Fuel elements	4 1 2
94	Plutonium(iii)	Pu	239	239	24 100 yr	Fuel elements	6 0 2
				242	3.75×10^{5} yr		
				244	8.00×10^{7} yr	Neutron bombardment	
95	Americium	Am	241	241	433 yr	Fuel elements	7 0 2
				243	7370 yr	Neutron bombardment	
96	Curium	Cm	242	242	163 d	Neutron bombardment	7 1 2
			244	244	18.1 yr		
				247	1.56×10^{7} yr		
				248	3.48×10^{5} yr		
97	Berkelium	Bk	249	247	1400 yr	Ion bombardment	9 0 2
				249	326 d	Neutron bombardment	
98	Californium	Cf	252	249	351 yr	Neutron bombardment	10 0 2
				252	2.64 yr		
99	Einsteinium	Es	254	252	472 d		11 0 2
				253	20.5 d	Ion bombardment	
				254	276 d	Neutron bombardment	
100	Fermium	Fm	253	253	4.5 d	Neutron bombardment	12 0 2
				257	100 d		
101	Mendelevium	Md	256	256	77 min	Ion bombardment	13 0 2
				258	51.5 d		
				260	27.8 d		
102	Nobelium	No	254	254	3 s	Ion bombardment	14 0 2
				255	3 min		
				259	58 min		
103	Lawrencium(iv)	Lr	257	256	35 s	Ion bombardment	14 1 2
				260	3 min		
				261	39 min		
				262	3.6 h		

Notes: (i) The source given is the most recent for manageable quantities of the isotope in question, except that elements 100 onwards are not available in weighable amounts at present. The most accessible isotope of the heavier elements is usually the one produced by fewest steps and is not necessarily the longest-lived. Half-lives are given only for the most accessible and for the longest-lived isotopes. (ii) The isotopes Th-232, U-238, Np-237, Pu-239 and Am-241 are available in kilogram quantities; other isotopes from fuel elements in gram amounts, as are Am-243 and Cm-244; Ac-227, Bk-249 and Cf-249 are available at the 1-10 mg level and Es-253 somewhat less. From fermium onwards, quantities range from micrograms down to atoms. (iii) For plutonium, the abundance of isotopes in fuel elements is $239 > 240$ (half-life 7000 yr) > 241 (14 yr) $> 242 > 238$ (88 yr). The choice of isotope for a particular study is a balance of abundance, half-life, and type of radiation. (iv) Alternative configuration $5f^{14}6d^{0}7s^{2}7p^{1}$.

state become dominant and the resemblance to the lanthanides appears. Although the IV state of berkelium is strongly oxidizing, it is more stable than the IV states of curium and americium. In this, it is showing a parallel to the properties of terbium, where the IV state, corresponding to the f^7 configuration, has some stability. Americium does not form the II state in aqueous media, but it has been reported in a chloride melt, so that there is some slight resemblance to europium which attains the f^7 configuration in its moderately stable II state. An oxide phase, MO, exists for Pu and Am. The heavier actinides, though studied mainly by carrier methods, show some evidence for a II state in addition to the stable III state but attempts to oxidize Md^{3+}, No^{3+} or Lr^{3+} to the IV state were unsuccessful. These II states are distinctly more stable than for the corresponding lanthanides, and a significant difference between the two f series is evident. In particular, Md(II) is moderately stable and No(II) is markedly more stable than No(III), with No^{2+} requiring an oxidizing agent comparable with permanganate or ceric to form No^{3+}. The f^{14} configuration would probably be attained by No^{2+} but this state is relatively much more stable than the analogous Yb^{2+} in the lanthanides.

The regular trend in ionic radii resembles that shown by the lanthanide elements, and it is possible to talk of an *actinide contraction* similar to the lanthanide contraction and arising from a similar increase in effective nuclear charge due to poor screening by the f electrons. The actinide contraction means that the actinide elements should show similar ion exchange behaviour to the lanthanides, and this has been made use of in a striking way in the identification of the newer heavy elements. The elements beyond curium have all very similar properties chemically, and the methods of synthesis mean that they are formed in only small amounts in the presence of excess target material. The identification of the heavier elements depends upon detecting their characteristic radiation (which can be predicted theoretically).

The method which was successfully adopted was to dissolve the irradiated targets and pass the actinides in solution through an ion exchange column and count the radiation from each fraction. Due to the tiny scale of the experiments, the 'column' was a few beads of

TABLE 12.2 Chemical properties of the actinides and actinium

Element	*Oxidation states*	*Crystal radii (pm)*			*Potential (V)*		*1st ionization energy** (eV: 1 eV = 96.48 kJ mol^{-1})
		M^{3+}	M^{4+}		$M^{4+} + e = M^{3+}$	$M^{3+} + 3e = M$	
		CN*6*	*6*	*8*			
Ac	III	126				−2.13	5.17
Th	(III), IV		108	119	−3.0	−1.17	6.31
Pa	(III), IV, V	(118)	101	115		−1.49	5.89
U	III, IV, V, VI	116.5	103	114	−0.631	−1.66	6.05
Np	III, IV, V, VI, VII	115	101	112	+0.155	−1.79	6.26
Pu	III, IV, V, VI, VII	114	100	110	+0.982	−2.00	6.03
Am	(II), III, IV, V, VI, (VII)†	111.5	99	109	+2.0	−2.07	5.97
Cm	(II)†, III, (IV)	111	99	109	+3.2	−2.06	5.99
Bk	(II), III, IV	110	97	107	+1.6	−1.97	6.20
Cf	(II)†, III, (IV)	109	96	106		−2.01	6.28
Es	(II), III	(108)				−1.98	6.42
Fm	(II), III					−1.95	6.50
Md	II, III					−1.66	6.58
No	II, III					−1.78	6.65
Lr	III					−2.06	

*Data for elements Es to No are based on extrapolation; for elements where only a tiny amount was available (10^{12} atoms, 0.4 ng), such as Bk or Cf, a mass spectrometric method was used.
†Transient only. Stable oxidation states are underlined: unstable states are in brackets.

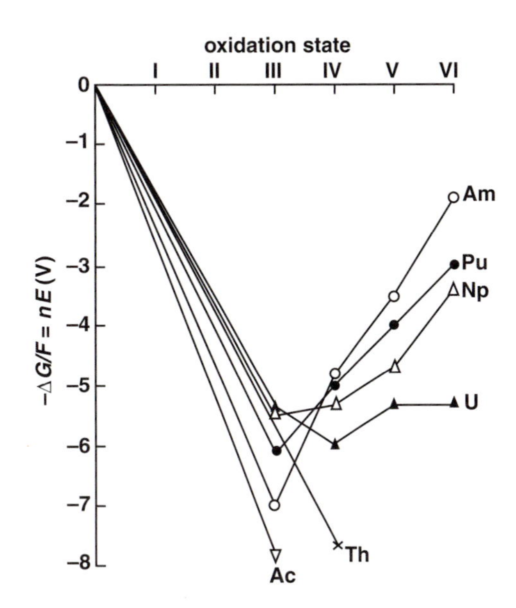

FIG. 12.1 Free energy changes per electron for actinide element oxidation–reduction couples in acid solution. Here the free energy change (in volts) is plotted against the oxidation state for all the actinide elements, as far as data are available. The uranium and americium diagrams are discussed in Section 8.6, Fig. 8.11

resin and the fractions were single drops. The order of elution, and the elution positions, of the tripositive actinide ions and the tripositive lanthanide ions, are the same on the same resin, and this was used in the first identification of the elements from americium up to mendelevium on a weighable scale (i.e. apart from the use of carrier methods). A composite elution diagram of these elements is shown in Fig. 12.2 along with a similar diagram for the heavier lanthanides. The one-to-one correspondence in positions is clear. The scale of the operations is made evident by the fact that the first identification of element 101, mendelevium, was based on the count of five decompositions, i.e. of the fission of five individual atoms.

The actinide metals resemble the lanthanides, are of low electronegativity, and are very reactive. The metals are produced by electrolytic reduction of fused salts or by treating the halides with calcium at high temperatures. They are all extremely dense, with densities ranging from 12 to 20 g cm^{-3}. The direct reactions of the metals (for example, with oxygen, halogens and acids) are similar to those of the scandium Group elements. The metals also react directly with hydrogen with the formation of nonstoichiometric hydrides, such as Th_4H_{15}. Phases with idealized compositions MH_2 and MH_3 are most common. These hydrides are reactive and often form suitable starting materials for the preparation of other compounds. On heating, they decompose, leaving the metal in a very finely divided and reactive form.

12.3 Thorium

Thorium has been known since 1828. Its principal source is the mineral monazite, which is a complex phosphate of thorium, uranium, cerium and lanthanides. Thorium is extracted by precipitation as the hydroxide, along with cerium and uranium, and then separated by extraction with tributyl phosphate from acid solution. The metal is made by calcium reduction of the oxide or fluoride, and pure samples can be prepared in the Van Arkel process by decomposing the iodide, ThI_4, on a hot filament.

The only stable oxidation state is the IV state in which thorium resembles hafnium. This is very stable and, because of the large size, the Th^{4+} ion has a low enough charge density to be capable of existing without excessive polarization effects. This is the highest-charged monatomic ion known. The hydroxide is precipitated from thorium solutions and gives the

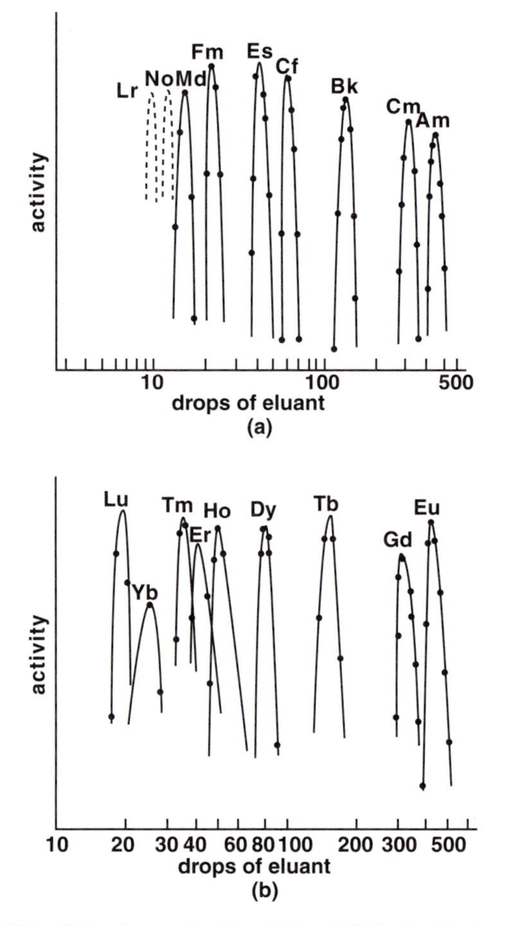

FIG. 12.2 Elution diagrams for (a) actinide and (b) lanthanide elements

oxide, ThO_2, on ignition. This is also formed directly from the metal and oxygen, and on ignition of oxy-salts. It is a stable and refractory material (m.p. 3050°C) and is soluble only in hydrofluoric/nitric acid mixtures. The anhydrous halides, ThX_4, are prepared by dry reactions such as metal plus halogen or oxide plus hydrogen halide at 600°C. The tetrafluoride is involatile but the others sublime in vacuum above 500°C. Treatment of the halides with water vapour gives the oxyhalides, $ThOX_2$.

Dilute thorium solutions in strong aqueous acid contain the hydrated thorium ion, $Th^{4+}.nH_2O$, but hydrolysis takes place on concentration or when the pH is raised. At a pH of about 6, the hydroxide, $Th(OH)_4$, is precipitated. This has a crystal structure containing chains of thorium atoms linked by oxy- and hydroxy-bridges and 8-coordinated Th.

The commonest salt is the nitrate, $Th(NO_3)_4.5H_2O$ which is very soluble in water, alcohols and similar solvents. The fluoride, oxalate and phosphate are very insoluble and may be precipitated even in strong acid solution [compare hafnium and cerium(IV)]. Thorium also gives a borohydride, $Th(BH_4)_4$ which sublimes in vacuum at about 40°C.

The coordination of thorium(IV) is variable and tends to be high. $ThCl_4$ and $ThBr_4$ have the distorted 8-coordination of UCl_4 while 7-, 8- and 9-coordinate Th atoms are all found in $ThOCl_2$ (and in the isomorphous Pa, U and Np analogues). Eight-coordination to sulfur is found in the complex $Th(S_2CNEt_2)_4$ and to oxygen in $[Th(NO_3)_6]^{4-}$. Nine-coordination, by sharing fluorines, is found in $(NH_4)_4ThF_8$ and in Na_2ThF_6 and the ThF_9 arrangement is similar to that shown in Fig. 15.23.

See Appendix B for oda.

An alternative form of 9-coordination is found in the oxydiacetato (oda) complex $[Th(oda)SO_4(H_2O)_2]$. Here the Th atom is coordinated by 9 O atoms in a capped square antiprism arrangement (compare Fig. 12.5) with a ninth O atom above one of the square faces.

The amber Th^{3+} ion, which has an f^1 electronic configuration and is strongly reducing, has been generated in aqueous solution by reaction of Th^{4+} with hydrazoic acid (HN_3, Section 17.6). There is also evidence for the triiodide, ThI_3, formed by heating the metal with the stoichiometric amount of iodine in vacuum at 555°C. Using a deficiency of iodine gives the diiodide which can also be prepared electrochemically. It has been shown that ThI_3, and two different crystal modifications of ThI_2, can be also prepared by heating ThI_4 with thorium metal. The triiodide converts to the diiodide on further heating. Both ThI_2 and ThI_3 react with water with the evolution of hydrogen and formation of thorium(IV). These compounds are metallic with bonding which can be explained on the same model as that used for the metallic hydrides, Section 9.4. One nonmetallic thorium(III) compound is the white ThOF, prepared by reducing a ThF_4/ThO_2 mixture with thorium metal at 1500 K. Thorium forms a range of alkoxide compounds $[Th(OR)_4]_n$, with the isopropoxide being dimeric in isopropanol and tetrameric in benzene solutions.

12.4 Protactinium

Protactinium was first identified in uranium in 1917. It is a product of uranium-235 decay and, in turn, gives actinium by α-particle emission:

$$^{235}_{92}U \rightarrow ^{4}_{2}He + ^{231}_{90}Th \rightarrow ^{0}_{-1}e + ^{231}_{91}Pa \rightarrow ^{4}_{2}He + ^{227}_{89}Ac$$

A further isotope occurs in the decay of neptunium-227, but both these naturally occurring isotopes have low concentrations in equilibrium. The element is most readily obtained by synthesis:

$$^{232}_{90}Th + ^{1}_{0}n \rightarrow ^{233}_{90}Th \rightarrow ^{233}_{91}Pa + ^{0}_{-1}e$$

This isotope has a half-life of 27.4 days but is a β-emitter and more readily handled than the α-emitting Pa-231. The latter is now available from the fuel elements of atomic piles and is commonly used. Because of its relative scarcity until recently, and because of the strong tendency of its compounds to hydrolyse and form polymeric colloid particles which are adsorbed on reaction vessels, protactinium chemistry is comparatively less well known.

The oxide system is complex and compounds range in composition from PaO_2 to Pa_2O_5. The pentoxide is obtained on igniting protactinium compounds in air and is a white solid with weakly acidic properties, being attacked by fused alkali. On reduction with hydrogen at 1500°C, the black dioxide PaO_2 is formed.

Among the halides, two fluorides are known. The pentafluoride PaF_5 results from the reaction of bromine trifluoride on the pentoxide. It is a very reactive and volatile compound. The complex anion, PaF_7^{2-}, is known and was used in the classical isolation of the element. In this ion, the protactinium is 9-coordinate with Pa linked by two fluorine bridges to a neighbour on either side, giving a chain structure. The structure of the PaF_9 units is the same as that of the ReH_9^{2-} ion shown in Fig. 15.23. In the complex Na_3PaF_8 the PaF_8^{3-} ion is a slightly distorted cube and the sodium ions are also 8-coordinated. The Na_3MF_8 compounds of uranium(V) and neptunium(V) are isostructural. The second fluoride, PaF_4, is a red, high-melting solid which results from the reaction of hydrogen and hydrogen fluoride on the oxide. The oxyfluoride, PaO_2F, and complexes $MPaF_5$ and M_4PaF_8 are known.

In recent studies, all the Pa^{IV} and Pa^{V} chlorides, bromides and iodides of the types PaX_4, PaX_5, $PaOX_2$, $PaOX_3$ and PaO_2X have been prepared, together with the complexes M_2PaX_6 and $MPaX_6$. $PaCl_5$ and PaF_5 are polymeric structures with pentagonal bipyramidal coordination, like β-UF_5. $PaBr_5$ consists of dimeric units with two bridging bromines giving 6-coordinate Pa, while $PaBr_4$ is an infinite polymer with all the bromine atoms bridging pairs of Pa atoms giving $PaBr_8$ coordination. PaI_3 exists in the solid state. More complex oxyhalides of Pa(V) include Pa_2OX_8 (X = F, Cl) and Pa_3O_7F. There is only one reported alkoxide complex of protactinium, $[Pa(OEt)_5]_n$.

The solution chemistry is obscure because of the formation of colloids, but anionic complexes like $(PaOCl_6)^{3-}$ have been claimed. Lower oxidation states may be obtained in solution by reduction with zinc amalgam. The tetravalent state is stable in absence of air but evidence for the III state is slight and based on polarographic results. The absorption spectrum of $PaCl_4$ in water shows three maxima and is similar to that of Ce^{3+}, providing some evidence for the presence of a single f electron in Pa(IV).

12.5 Uranium

Uranium is the longest known of the actinide elements, having been discovered in 1789, but it attracted little interest until the discovery of uranium fission in 1939. It is now of importance as a fuel, and its chemistry has been very fully explored in the course of the atomic energy investigations.

Natural uranium contains two main isotopes, ^{238}U 99.3% and ^{235}U 0.7%, and also traces of a third, ^{233}U. The vital isotope from the nuclear energy point of view is ^{235}U because this reacts with a neutron, not by building up heavier elements as in the examples discussed in Section 12.1, but by fission to form lighter nuclei. This fission process releases considerable energy and more neutrons, which, in turn, fission uranium-235 nuclei and allow the build-up of a chain of fissions. The energy of such nuclear processes is about a million times the energy released in chemical reactions, such as the burning of a fuel or the detonation of a high explosive. This is the reason for the value of atomic fission as an energy source, and for the horror of fission as a source of explosive energy in a weapon. A typical fission process is:

$$^{235}_{92}U + ^{1}_{0}n \rightarrow ^{92}_{36}Kr + ^{140}_{56}Ba + 3^{1}_{0}n + \text{about } 8 \times 10^{9} \text{ kJ mol}^{-1}$$

The nuclei formed in fission fall into two main groups, a lighter set with masses from about 70 to 110 and a heavier one with masses from 125 to 160. Splitting into approximately equal nuclei is about a thousand times rarer than splitting to an unequal pair such as that shown in the equation. The neutrons evolved in the fission are either used in other fissions, absorbed by nonfissionable nuclei such as uranium-238, or escape through the surface of the uranium mass. The essence of running an atomic pile is to ensure that one neutron per fission is available to cause another fission. More than one leads to a rapidly increasing chain reaction and explosion, while less than one means that the process dies out. The absorption of neutrons by uranium-238 leads to the formation of heavier elements, of which the most important is plutonium which is itself a nuclear fuel. In appropriate conditions, more plutonium can be produced from the uranium-238 than the amount of uranium-235 consumed, and such an arrangement 'breeds' nuclear fuel in the 'breeder reactor'.

In its chemistry, uranium resembles the three succeeding elements, neptunium, plutonium and americium, in having four main oxidation states, III, IV, V and VI. The most stable state drops from VI for uranium through V for neptunium, IV for plutonium, to III for americium. This is illustrated by the oxides and halides found for these elements, shown in Table 12.3, and also by the free energy diagrams for the changes in oxidation state in acid solution, Fig. 12.1. In solution, the VI state is present as the uranyl ion, UO_2^{2+}, and the V state also occurs as an oxycation, UO_2^{+}. The IV and III states are present as simple cations, U^{4+} and U^{3+}. Since the change from U^{3+} to U^{4+}, and that from UO_2^{+} to UO_2^{2+} (and the reverse changes) involve only transfer of an electron, these two pairs of redox reactions occur rapidly. On the other hand, oxidations such as U^{4+} to UO_2^{2+} involve oxygen transfer as well and are slow and often irreversible. In solution, the III, IV and VI states all exist, but UO_2^{+} has only a transitory existence. However, in a nonoxide medium, such as anhydrous hydrogen fluoride or a chloride melt, uranium(V), although still unstable, is well represented.

Uranium metal, produced by reduction of the tetrafluoride with calcium or magnesium:

$$UF_4 + 2Ca \rightarrow U + 2CaF_2$$

is reactive and combines directly with most elements. It dissolves in acids but not in alkalis. Direct reaction with hydrogen at about 250°C gives the hydride, UH_3, usually in a form

NUCLEAR POWER AND THE PROBLEM OF THE FISSION PRODUCTS

The major problem of using nuclear power arises from the fission products, arising like ^{140}Ba above. These are neutron-rich isotopes which are themselves radioactive. The witches' brew which results is highly radioactive and contains isotopes with a wide range of chemical and radiochemical properties. At present, the fuel rods are allowed to stand in a strongly shielded store for a period to allow the short-lived component of the radioactivity to disappear. Then they are treated—all by remote control—to recover unchanged uranium, plutonium and other useful elements. Increasingly, uses are being found for some components, see the discussion of technetium (Chapter 20) for example.

No method which is acceptable to general public opinion has yet been formulated to deal with the remaining mixture of unwanted radioactive material. Most of the spent fuel residues from current commercial power production remain in temporary storage awaiting a long-term solution to the storage problem. It is quite feasible to safely store the mixture until the short-lived isotopes have decayed to negligible proportions. Twenty half-lives reduce the amount of isotope by a factor of about one million, so storage for a few years deals with everything with half-lives of the order of days, or less. Likewise isotopes with extremely long half-lives have very low activities and thus do not create a major problem. The great difficulties arise from isotopes of intermediate half-lives, especially those of elements which are readily taken up by living organisms. Thus the major problems in Britain and Western Europe from the Chernobyl disaster were caused by ^{137}Cs which has a half-life of 30.2 years, and which mimics potassium in its metabolism. This continues to cause concern, particularly for sheep farming.

Most current work is directed towards immobilizing such residues in glasses or concrete, with a view to permanent storage deep underground in a geologically stable formation. This has often seemed a good idea—until a proposal arises to locate such a store near a particular community! There are general fears that radioactive material will eventually leach out and get into the environment. There are also justifiable doubts whether such stores could possibly be maintained in isolation for the thousands of years required. It may be that the ultimate solution will come from quite different directions—either an alternative source of power (possibly fusion power) which will remove the need for atomic reactors, or a radically different method of dealing with radioactive materials. It may be possible to convert the difficult isotopes by further irradiation, for example.

In the meantime, however, methods are still needed to 'manage' the intermediate half-life nuclides. A recent process using solvent extraction, the Truex or transuranium extraction process, has been developed in order to significantly reduce the volume of waste requiring deep burial underground. In this process a carbamoylmethylphosphane oxide is used to extract, and thereby concentrate, the transuranic elements. Importantly, the process extracts all transuranic elements (in a range of oxidation states) including americium-241, which is responsible for much of the radiation in nuclear wastes.

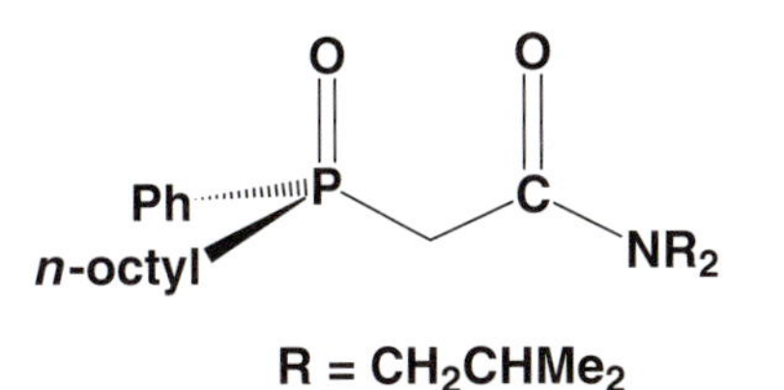

Carbamoylmethylphosphane oxide used in actinide extraction

with a small deficiency of hydrogen. This is a very reactive compound which is a useful starting material for the preparation of uranium compounds of the III and IV states. The normal product is the IV compound, for example:

$$UH_3 + H_2Y \rightarrow UY_2 \quad (Y = O, S \text{ at about } 400°C)$$

$$UH_3 + HF \rightarrow UF_4$$

$$\text{but } UH_3 + HCl \rightarrow UCl_3 \quad (Cl_2 \text{ gives } UCl_4)$$

The uranium–oxygen system is very complex as the oxidation states are of comparable stability and nonstoichiometric phases are common. The main uranium ore, *pitchblende*, is an oxide approximating to UO_2 in composition. The other stoichiometric oxides are U_3O_8, which is the ultimate product of ignition, and UO_3 which is obtained by the decomposition of uranyl nitrate, $UO_2(NO_3)_2$, at about 350°C. The trioxide can be reduced to the dioxide by the action of carbon monoxide at 350°C, and it goes to U_3O_8 on heating to 700°C. All three oxides dissolve in nitric acid to give uranyl, UO_2^{2+}, salts. More recent studies have established U_3O_7 and U_4O_9 as definite compounds.

TABLE 12.3 Halides and oxides of uranium, neptunium, plutonium and americium

Uranium	UF_6		UF_5	U_2F_9	UF_4	UF_3	
				U_4F_{17}			
	UCl_6		UCl_5		UCl_4	UCl_3	
					UBr_4	UBr_3	
					UI_4	UI_3	
	UO_3	U_3O_8	U_3O_7	U_4O_9	UO_2		
Neptunium	NpF_6		NpF_5		NpF_4	NpF_3	
					$NpCl_4$	$NpCl_3$	
					$NpBr_4$	$NpBr_3$	
						NpI_3	
		Np_3O_8	Np_2O_5		NpO_2		
Plutonium	PuF_6			Pu_4F_{17}	PuF_4	PuF_3	
					($PuCl_4$)	$PuCl_3$	
						$PuBr_3$	
						PuI_3	
					PuO_2	Pu_2O_3	
Americium					AmF_4	AmF_3	
						$AmCl_3$	$AmCl_2$
						$AmBr_3$	$AmBr_2$
						AmI_3	AmI_2
					AmO_2	Am_2O_3	

The known halides of uranium are listed in Table 12.3 and the interconversions of the fluorides and chlorides are shown in Figs 12.3 and 12.4, respectively. The hexahalides are octahedral. UF_6 is volatile, sublimes at 56°C, and has been used in the separation of uranium isotopes by gaseous diffusion. The compound is a powerful fluorinating agent and is rapidly hydrolysed. Two complex fluoride ions, UF_7^- and UF_8^{2-}, contain uranium(VI). Uranium pentafluoride and pentachloride both readily disproportionate to give the hexahalide and the tetrahalide. Uranium(V) also occurs in fluoride complexes, UF_8^{3-} and UF_6^-. The latter is formed in HF solution and may be the reason for the relative stability of the V state in this medium. UCl_6^- and UBr_6^- have also been prepared. The pentahalides all have polymeric structures. The pentafluoride has two forms. In α-UF_5, an octahedron of fluorines around the uranium is completed by the sharing of fluorine atoms and the formation of long chains of linked octahedra. In the β-form of UF_5, the uranium atom is 7-coordinated with three fluorines attached to only one uranium and the other four shared with four different uraniums in a polymeric structure. In U_2F_9, all the uranium atoms have nine fluorines at equal distances and this (black) compound probably contains crystallographically equivalent uranium(IV) and uranium(V) atoms. UCl_5 is a dimer of two UCl_6 octahedra sharing an edge, with 2Cl bridges. This is the simplest form: polymeric forms also occur.

The most stable halides of uranium are the tetrahalides, to which the higher halides are readily reduced and the trihalides oxidized. The exception is UI_4 which slowly converts to UI_3 and iodine at room temperature. UCl_4 is molecular both in the gas phase and in the solid state, where the UCl_4 molecules have a distorted tetrahedral shape, with Cl-U-Cl angles of 66.4°. In the solid, UCl_3 (like many other lanthanide and actinide trihalides), crystallizes with 9-coordinated U^{3+} ions, in a structure where the U atom has three chloride ions coplanar with it, three above this plane, and three below. UI_3, and the tribromides and triiodides of Np, Pu and Am, have an 8-coordinated layer lattice. All the halides form $U^{IV}X_6^{2-}$. The uranium III ion, UCl_4^-, is also known, though this is very readily oxidized.

The standard syntheses of UX_3 and other AnX_3 involve high temperature routes, producing unreactive polymeric solids which are not very useful starting materials for the

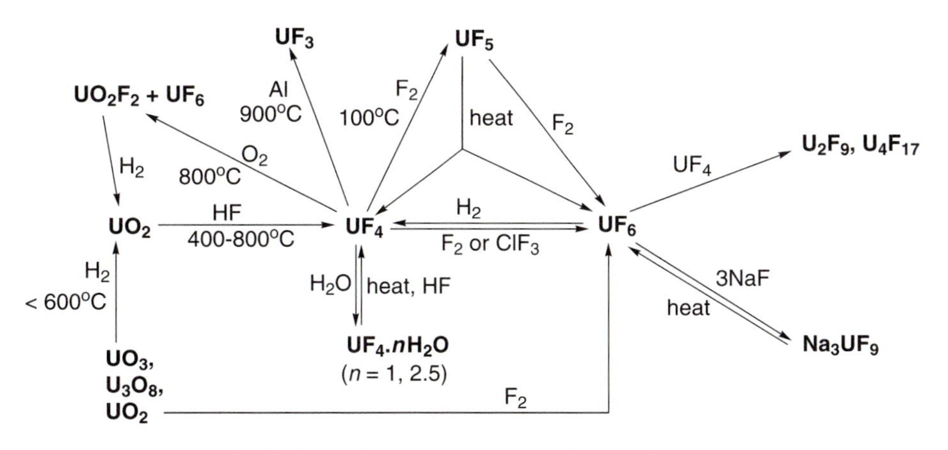

FIG. 12.3 Reactions and interconversions of uranium fluorides

preparation of other An(III) compounds. A much lower temperature route is provided by the direct reaction of a slight excess of the metal in THF at room temperature:

$$An + 1.5X_2 \longrightarrow AnX_3(THF)_4$$
$$(An = U,\ Np\ or\ Pu;\ X = Br\ or\ I)$$

The products are pentagonal bipyramidal in shape, with two X atoms at the apices and the third X and the four THF molecules in the equatorial positions.

One interesting product accessible through such species is the 3-coordinate $An[N(SiMe_3)_2]_3$ which is soluble in hydrocarbons and sublimes at 60°C. The low coordination is stabilized by the presence of the very bulky $N(SiMe_3)_2$ ligands:

$$AnI_3(THF)_4 + 3NaN(SiMe_3)_2 \longrightarrow An[N(SiMe_3)_2]_3 + 3NaI$$

[An=U, red-purple; An=Np, dark blue; An=Pu, orange]

Oxyhalides of uranium(VI), UO_2X_2 and UOF_4 are also known. These are made from the oxides or halides by partial substitution:

$$UO_3 + 2HF \xrightarrow{400^\circ C} UO_2F_2 + H_2O$$

Oxyhalides of uranium(V) are UO_2X (X = F, Cl, Br), UOX_3 (X = Cl, Br) and for uranium(IV), UOX_2 (X = F, Cl, Br).

A number of alkoxides of uranium have been described. The uranium(IV) alkoxides, $U(OR)_4$, are green sublimable compounds which, like their Th and Pa counterparts, are oligomeric in solution. Alkoxides in the V and VI states, such as $U(OR)_5$ and $UO_2(OR)_2$ are also known, as are the analogous amides, $U(NR_2)_4$.

In solution, hydrolysis occurs for all oxidation states, being least for the III state. Uranium(III) and (IV) exist as ions in strong acid, and uranium(IV) hydrolyses in more dilute solution in a similar manner to Th^{4+}. The U^{4+} ion also gives insoluble precipitates with

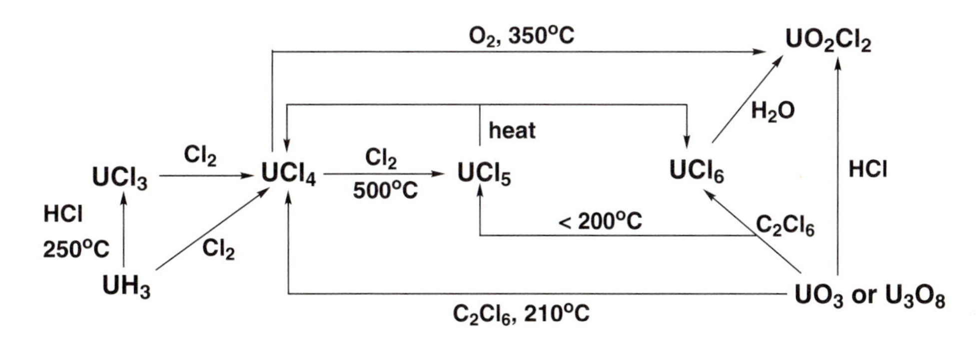

FIG. 12.4 Reactions and interconversions of uranium chlorides

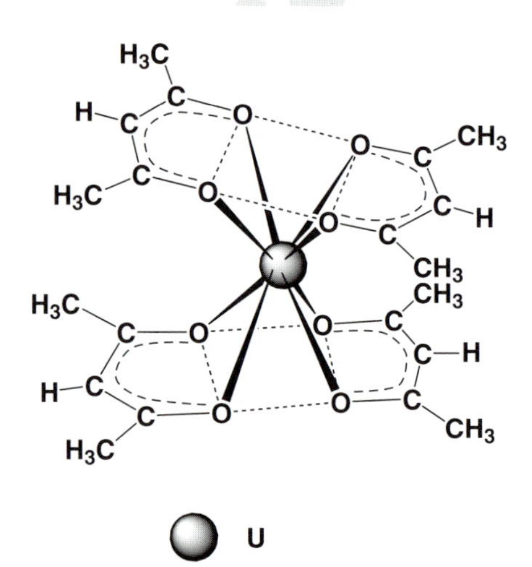

FIG. 12.5 The structure of uranium (or thorium) acetylacetonate, $U(acac)_4$. The uranium (or thorium) atom is coordinated to eight oxygen atoms which are arranged in a square antiprism around it

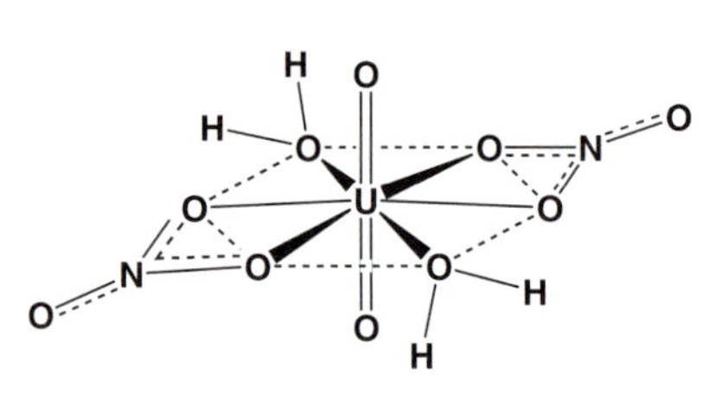

FIG. 12.6 The structure of uranyl nitrate hydrate, $UO_2(NO_3)_2.2H_2O$. The UO_2 group is linear and two water molecules and the two nitrate groups are coordinated in the central plane to the uranium. As the nitrate groups here are bidentate, the uranium has a coordination number of eight

For a discussion of $U(C_8H_8)_2$ (uranocene) and other $M(C_8H_8)_2$ species see Section 16.6.

similar reagents—F^-, PO_4^{3-}, etc. — as does Th^{4+}. Uranium(V) has a strong tendency to disproportionate to U^{4+} and UO_2^{2+}. It is most stable in fairly acid solution at a pH of about 3. Uranium(VI) also hydrolyses in solution, this time giving double hydroxy bridges so that polymers of the type $\cdots UO_2(OH)_2UO_2(OH)_2 \cdots$ are formed. Uranium(VI), as uranyl, forms the only common uranium salts, and the most usual starting material is uranyl nitrate, $UO_2(NO_3)_2.nH_2O$, ($n=2$, 3 or 6). This is soluble in water and in a variety of organic solvents.

The stereochemistry of uranium, and of the other actinides, shows a tendency to high coordination numbers, as illustrated by the halide structures above.

In $U(BH_4)_4$, 14-coordination of uranium occurs. The structure contains two of the four BH_4 groups bonded through three hydrogens to one U atom while the other two bridge two U atoms, sharing two hydrogens with each.

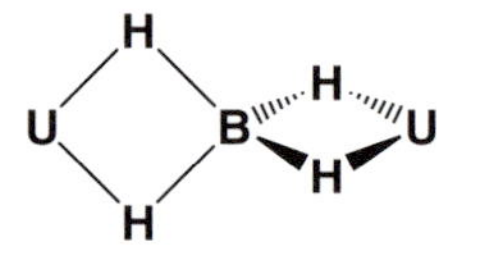

A bridging BH_4 group in $U(BH_4)_4$

Thus the total structure is *cis* $(HBH_3)_2U(H_2BH_2)_{4/2}$ with 14 U-H-B links. Borohydrides also form with U(III). In $U(BH_4)_3.3THF$, the BH_4 groups bonded through three bridging hydrogens, together with the three THF groups, make up 12-coordination around U.

A tendency to 8-coordination for M^{4+} actinide ions seems to be general. For example, uranium and thorium form an acetylacetonate, $M(C_5H_7O_2)_4$, which has the 8-coordinating oxygen atoms at the corners of a square antiprism, as shown in Fig. 12.5. (The square antiprism is most readily visualized as a cube with the top face twisted 45° relative to the bottom one.) Another interesting 8-coordinate species is $M(NCS)_8^{4-}$. When the cation is Cs^+, the anion structure is a square antiprism (M = U, Pu). When a bulkier cation, in $(NEt_4)_4M(NCS)_8$, is used, the anion structure becomes a cube (M = Th, Pa, U, Np and Pu). $(NEt_4)_4$ $M(NCSe)_8$ (M = Pa, U) also contains a cubic anion.

In uranyl compounds, the UO_2 group is linear and complexes exist in which four, five or six donor atoms lie coplanar with the uranium atom, giving 6-, 7- or 8-coordination overall. An example is shown in Fig. 12.6, where one form of hydrated uranyl nitrate has two water molecules and the two nitrate groups coordinated to uranium through both oxygens, all in the central plane, to give the 8-coordinated structure shown. Nitrogen analogues of the uranyl ion, containing the linear R-N=U=N-R group, or the mono-substituted O=U=N-R group (where R = e.g. PPh_3), have also been recently synthesized.

12.6 Neptunium, plutonium and americium

The three elements, neptunium, plutonium and americium, which follow uranium, also show the four oxidation states, III, IV, V and VI. In addition, strong oxidation in alkaline media by reagents like peroxide, ozone, periodate or XeO_3 produce Pu(VII) or Np(VII). Attempts to prepare plutonium(VIII) have so far been unsuccessful. Compounds isolated include Li_5MO_6 and M'_3MO_5 (M = Np or Pu; M′ = K, Rb or Cs) and also $Ba_3(NpO_5)_2$. On heating, these compounds lose O_2 and revert to M(VI) ternary oxides. The compounds are isostructural with their heptavalent iodine or rhenium analogues. The VII states also exist in strong alkaline solution but reduce rapidly to the VI state on acidification. The standard reduction potential for the Np(VII)/Np(VI) couple in strongly alkaline solution is 0.38 V. There is evidence that Np(VII) in alkaline solution contains a tetraoxo first coordination sphere, slightly distorted from a square-planar geometry. This has a Np-O bond length of 1.87 Å, with two additional oxygens, possibly hydroxides, bonded at a greater distance, giving a species of the type $[NpO_4(OH)_2]^{3-}$.

The III, IV, V and VI states are present in solution as the M^{3+}, M^{4+}, MO_2^+ and MO_2^{2+} ions, as in the case of uranium, and they have a similar tendency to hydrolyse. As

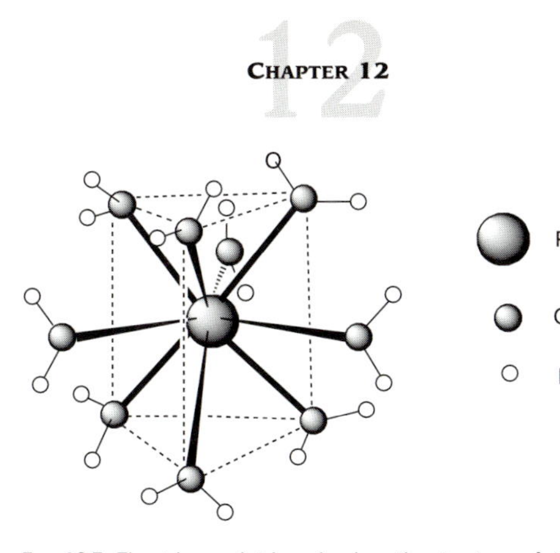

FIG. 12.7 The tricapped trigonal prismatic structure of the $[Pu(H_2O)_9]^{3+}$ cation. The trigonal prism is defined by the dotted lines

expected, the ions are heavily hydrated to make up the high coordination number, and as an example, the structure of the ion $[Pu(H_2O)_9]^{3+}$ (as its $CF_3SO_3^-$ salt) has been determined by X-ray crystallography, and found to have a tricapped trigonal prismatic geometry, as shown in Fig. 12.7. The speciation of plutonium(IV) in nitric acid solutions is of significant practical importance because of the industrial extraction of Pu(IV) nitrate species in the processing of nuclear wastes. It has been found that, as the concentration of nitric acid increases, the number of coordinated nitrates increases at the expense of coordinated water molecules. Thus, in dilute (3 mol l^{-1}) HNO_3 the principal Pu(IV) species is $Pu(NO_3)_2^{2+}$, while in 13 mol l^{-1} HNO_3 it is $[Pu(NO_3)_6]^{2-}$, analogous to the related hexanitrato species of other lanthanides and actinides such as Ce and Th. The hydrolysis of Pu(IV) is of environmental importance, since solid $Pu(OH)_4$ has an exceedingly low solubility product (log $K_{sp} = -54$). This, together with the rather complex aqueous chemistry, can limit the amount of plutonium in aqueous solution, decreasing its bioavailability and potential to contaminate groundwaters, etc. However, it is currently thought that *siderophores* (low molecular weight iron chelating agents produced by microorganisms), which are present in soils, could mobilize plutonium and thereby increase its bioavailability. In sea water, the soluble PuO_2^+ ion is the principal form of dissolved plutonium, at a concentration of around 10^{-14} mol l^{-1}.

The relative stability of the four oxidation states varies among the elements. The VI state becomes increasingly unstable and oxidizing from uranium to americium, and AmO_2^{2+} solutions are as strongly oxidizing as permanganate. The V state in solution, as MO_2^+, is the most stable state for neptunium, while PuO_2^+ and AmO_2^+, while less stable than NpO_2^+, are more stable than UO_2^+. Thus, UO_2^+ disproportionates in aqueous solution to uranium(VI) and uranium(IV) (see Fig. 12.1) while NpO_2^+ is stable. In the compound, $Cs_7(NpO_2)(NpO_2)_2Cl_{12}$, the balance of charges shows that the Np(V) and Np(VI) oxycations coexist. When hydrofluoric acid is used instead of water as solvent, the relative stabilities of the V states of U and Np reverse and $CsUF_6$ dissolves unchanged while $CsNpF_6$ disproportionates to give $NpF_4 + NpF_6$. This example is a useful reminder that stabilities differ markedly in different media, and our generalizations apply only to stabilities in aqueous solutions in air. The IV state is the most stable one for plutonium, while americium resembles the following elements in being most stable in the III state.

Plutonium(IV) forms complexes with the carbonate ion in solution, and these are important in the mobilization of Pu in groundwater. The ion $Pu(CO_3)_5^{6-}$ has been isolated and shows 10-coordination of Pu to O.

The relative stabilities of the oxidation states of Np, Pu and Am in solution is shown by the free energy diagrams, Fig. 12.1, and is paralleled in the solid state, as Table 12.3 shows. The plutonium tetrahalides do not parallel the stability of plutonium(IV) in solution, with no bromide or iodide, and $PuCl_4$ existing only in a Cl_2 atmosphere. Thus the stability in solution probably reflects the high solvation energy of the Pu^{4+} ion. Similarly, although neptunium(V) is the most stable state in solution, the synthesis of NpF_5 caused great difficulty. It was finally achieved by using very reactive KrF_2 (compare Section 17.9.5) to react with NpF_4 in anhydrous HF. Thus the general stability of Np(V) must depend on the formation of NpO_2^+.

As found for the transition elements (compare Chapters 14 and 15), oxyhalides are often found for oxidation states where the simple halides are missing, perhaps because of the lower coordination demand. We find in the VI state, MO_2F_2 for M = Np, Pu, Am and MOF_4 for M = Np and Pu: in the V state, NpO_2F occurs as well as MOF_3 for Np and probably Pu. Plutonium also forms PuOF.

The first preparations of all the hexafluorides were by direct reaction between metal and F_2. Later, it was found that O_2F_2 will react with any plutonium oxide, oxyfluoride or lower fluoride to form PuF_6 at room temperature. Plutonium hexafluoride is a liquid at room temperature and is somewhat different in properties to uranium hexafluoride, the volatility of which permits separation of the U-235 and U-238 isotopes. PuF_6 is also a more powerful fluorinating agent than UF_6, and reacts with SF_4 producing SF_6, and also converts iodine to IF_7.

As elements, Np, Pu and Am closely resemble uranium. They are reactive metals which combine with most elements. Plutonium metal displays some rather interesting characteristics. At the melting point of liquid plutonium (640°C), the solid is 2.4% less dense than the liquid and therefore floats on the top; plutonium is one of only a handful of materials showing this property, the best known of which is water. The properties of metallic plutonium also differ somewhat from those of its neighbour, uranium. The difference in boiling points between Pu (3327°C) and U (3818°C) can be used as a means of efficiently separating these two metals. The metallic characteristics of plutonium (*viz.* metallic and electrical conductivities, and malleability) are also low, with the metallic conductivity of the pure metal being only one tenth that of silver. These properties have been explained by the way in which the 5f electrons are distributed. For the pre-plutonium metals, the electrons are delocalized, resulting in more metallic behaviour, whereas the transplutonium metals show properties which are better explained by a localized, or nonmetallic model. Plutonium is considered to show intermediate behaviour, which can be altered by changing the pressure.

All three elements, Np, Pu and Am, form hydrides, but these resemble the lanthanide hydrides rather than uranium hydride. Stoichiometric hydrides, MH_2, are formed and also nonstoichiometric systems up to a composition, $MH_{2.7}$. The properties of the metals show increasing similarities to those of the lanthanide elements; the similarity is greatest for americium. Americium is the analogue of europium and americium(II) does exist, at least in a halide melt as AmX_2 (X = Cl, Br, I) and as AmO.

These three elements provide the most striking evidence, in their chemical properties, that the heaviest elements are, in fact, filling an f shell and are not a fourth d series. While thorium and uranium show similarities to hafnium and tungsten, neptunium and plutonium are clearly not the analogues of rhenium and osmium. Neptunium(VII) is strongly oxidizing while rhenium(VII) is stable and there is no indication as yet of an VIII state for plutonium analogous to osmium(VIII). Even the VI state of these two elements is relatively unstable and americium, with its stable III state, conforms to the expected pattern and links up with the chemistry of the succeeding elements.

AMERICIUM AND SMOKE DETECTORS

Americium, as a result of its essentially mono-energetic radiation (it emits 5.4 and 5.5 MeV alpha and 0.6 MeV gamma radiation), has found a variety of industrial and even household uses. Many smoke detectors contain a small amount of Am-241 which serves to ionize the air in the detector. Industrial applications include the measurement of air humidity and the determination of the uniformity of thin films, by the passage of (poorly penetrating) alpha radiation through the material.

12.7 The heavier actinide elements

Although the first characterization was by tracer methods, the next four actinides are now available in sufficient quantity for their macroscopic chemistry to be studied. Curium is available in gram quantities, berkelium and californium in tens of milligrams, and einsteinium in tenths of a milligram. The oxygen and halogen compounds of these four elements are listed in Table 12.4.

The III state, the most stable for americium, is also the most stable state for these four actinides. Magnetic measurements have confirmed the f^7 configuration in curium(III) compounds. Curium(IV) does not exist in solution and, for example, attempts to oxidize Cm^{3+} with Na_4XeO_6 (which produces Pu(VII)) were unsuccessful. Thus curium, the f^7 actinide, shows a resemblance to gadolinium, the f^7 lanthanide, in its solution behaviour. Curium(IV) is found in the solid state in the compounds listed in Table 12.4, while a good range of curium(III) solid species are reported. For the next element, berkelium, the IV state, although strongly oxidizing, is rather more stable and is found in solution and in the solid state. In solution, green berkelium(III) species may be oxidized by bromate to yellow or orange berkelium(IV) compounds whose oxidizing power is similar to that of cerium(IV). Californium(IV) is relatively unstable and the compounds listed in Table 12.4 have been prepared only as solids.

Both Bk(III) and Cf(III) are well established in solution and in solid compounds. Einsteinium does not form the IV state, the III state is stable, and the II state is well established in the dihalides. Cf(II) compounds also exist but are less stable. These II states are strongly reducing (compare the potentials in Table 12.5).

As for the earlier actinides, coordination numbers above six are found. For example, the $[M_6F_{31}]^{7-}$ species has a complex structure of linked antiprisms.

TABLE 12.4 Oxygen and halogen compounds of the heavier actinides

Curium	CmF_4	CmX_3 (X = F, Cl, Br, I)	
	$LiCmF_5$		
	$Na_7Cm_6F_{31}$		
	CmO_2	Cm_2O_3	
		$Cm(OH)_3$	
		CmOX (X = F, Cl, Br, I)	
Berkelium	BkF_4	BkX_3 (X = F, Cl, Br, I)	
	$[BkF_6]^{2-}$		
	$Na_7Bk_6F_{31}$		
	$[BkCl_6]^{2-}$	$Cs_2NaBkCl_6$	
	BkO_2	Bk_2O_3	
		$Bk(OH)_3$	
		BkOX (X = Cl, Br, I)	
Californium	CfF_4	CfX_3 (X = F, Cl, Br, I)	$[CfBr_2]$
	$Na_7Cf_6F_{31}$		CfI_2
	CfO_2	Cf_2O_3	
	[stable only under	$Cf(OH)_3$	
	high O_2 pressure]		
		CfOX (X = F, Cl, Br, I)	
Einsteinium	EsF_4(?)	EsX_3 (X = F, Cl, Br, I)	EsX_2
	EsO_2	Es_2O_3	(X = Cl, Br, I)
		EsOX (X = Cl, Br, I)	

^{249}Bk, while the half-life is only 320 days, is now readily available from spent fuels, allowing a more complete study of berkelium chemistry. It forms BkE (E = N, P, As, Sb), Bk_2E_3 (E = S, Se as well as O), both the cubic hydride BkH_{2+x} and the hexagonal BkH_3, and also the organometallic compounds $BkCp_2$ and $[BkCp_2Cl]_2$. In all this chemistry, there is a close parallel with the behaviour of terbium, the corresponding lanthanide.

The remaining four actinide elements have only been available in much smaller amounts and have been studied by carrier methods. In this way the behaviour of these elements has been elucidated by findng which one of a mixture of metal ions is accompanied by their characteristic radiation in the course of a chemical reaction. For example, Md accompanies Eu but not La when solutions of the M^{3+} ions are reacted with sodium amalgam and extracted. As this reaction forms Eu^{2+} but leaves La^{3+} unreduced, the formation of Md^{2+} is indicated. The scale of the experiments decreases very rapidly across these four elements. By the mid-1980s, it was possible to work on fermium with about 10^{11} atoms per experiment, for mendelevium using 10^6 atoms, for nobelium 10^3 and for lawrencium only about 10 atoms. Lawrencium chemistry, in particular, is so far based on only a handful of experiments each involving the detection (by their characteristic disintegrations) of a few atoms.

For Fm and Md, the III states are stable in solution and in solids, with a range of fermium solution chemistry already well explored. There is also a well-attested, strongly reducing Fm(II) state. The reduction potential ranges from about −1.6 V to −1.1 V, depending on the medium. This Fm(II) state is more stable than Es(II) and this trend continues to Md(II) and No(II). Mendelevium(II) is mildly reducing while nobelium(II) is the stable state and No(III) is strongly oxidizing. Potentials are judged by reaction with known oxidizing and reducing agents, and also by polarography. The current best estimates are shown in Table 12.5. While the stability of No(II) could mark the completion of the f shell with 14 electrons in No^{2+}, these II states in the later actinides are markedly more stable than is found for the lanthanides. Thus the actinides differ from the lanthanides both in the range of higher oxidation states found at the beginning of the series and in the stability of lower states at the end. Evidence for Md^+ has been reported, but is subject to dispute.

Lawrencium behaves as expected for the $f^{14}d^1s^2$ configuration by forming only Lr(III)—$f^{14}d^0s^0$—and resisting oxidation or reduction. New lawrencium isotopes 260, 261 and 262, formed by bombarding ^{254}Es with neon ions, have half-lives up to 400 times longer than ^{256}Lr and ^{257}Lr, which were the first ones prepared. This should allow Lr chemistry to be

TABLE 12.5 Standard redox potentials for the later actinides (V)

	Es	*Fm*	*Md*	*No*	*Lr*
$M^{3+} + e = M^{2+}$	−1.55	−1.2	−0.2	+1.4	
$M^{2+} + 2e = M$	−2.20	−2.4	−2.4	−2.4	
$M^{3+} + 3e = M$	−1.98	−2.0	−1.7	−1.1	−2.1

more fully established—particularly the intriguing possibility that the electron configuration is $5f^{14}7s^27p^1$ rather than $5f^{14}6d^17s^2$ because of relativistic effects (see Section 16.13). However, recent studies using the ^{262}Lr isotope have suggested that any relativistic effects are insufficient to stabilize the Lr(I) state, since Lr(III) was resistant to reduction by the strong reducing agents Sm(II), Cr(II) and V(II). This places an upper limit of −1.56 V for the Lr(III)/Lr(I) couple in aqueous solution.

Recent work, particularly on the heavier elements, has thus filled in the picture of the actinide elements as a whole. The total view indicates that, while there are useful analogies with the chemistry of the lanthanides and of the 5d series transition metals, the chemistry of the actinides presents an individual pattern reflecting the relatively small difference between the 5f and 6d energies for all these elements.

Post-actinide elements are covered in Section 16.12.

Problems

When consulting alternative sources on systematic chemistry, you should check out "Notes to the reader" at the beginning of Appendix A, Further Reading, together with the comments on the different sections of this list.

12.1 Reconsider Question 8.5

12.2 Review the structures of compounds of the actinides where the coordination number is more than 6.

12.3 It used to be thought that the elements from actinium onwards formed the heaviest transition series. How far does the *chemical* evidence support this view

(a) in matching Th and U with Hf and W, or
(b) for the elements Th to Pu matching Hf to Os?

12.4 The other expected relationship is between lanthanides and actinides—which obviously fails for the early elements. Discuss critically the parallels between Am to Lr and Eu to Lu including both chemical properties and numerical parameters.

12.5 As an alternative to Questions 12.3 and 12.4, discuss the extent to which the chemistry of the elements actinium to lawrencium is unique in the Periodic Table. How far are parallels with the transition elements and lanthanide elements justified?

12.6 Give an account of the halides, oxyhalides and oxides of all elements in the periodic Table with oxidation states of VI and above.

12.7 Outline the main oxidation states expected for elements 104 and 105, Rf and Ha. Speculate on the chemistry that might be found for element 126.

CHAPTER 13

13.1 INTRODUCTION TO THE TRANSITION ELEMENTS 273
13.2 THE TRANSITION ION AND ITS ENVIRONMENT: LIGAND FIELD THEORY 277
13.3 LIGAND FIELD THEORY AND OCTAHEDRAL COMPLEXES 279
13.4 COORDINATION NUMBER FOUR 285
13.5 STABLE CONFIGURATIONS 288
13.6 COORDINATION NUMBERS OTHER THAN FOUR OR SIX 289
13.7 EFFECT OF LIGAND ON STABILITY OF COMPLEXES 292
13.8 ISOMERISM 296
13.9 MECHANISMS OF TRANSITION METAL REACTIONS 297
13.10 STRUCTURAL ASPECTS OF LIGAND FIELD EFFECTS 302
13.11 SPECTRA OF TRANSITION ELEMENT COMPLEXES 304
13.12 π BONDING BETWEEN METAL AND LIGANDS 307
PROBLEMS 310

The Transition Metals: General Properties and Complexes

13.1 Introduction to the transition elements

The elements of the transition block are those with d electrons and incompletely filled d orbitals. The zinc Group, with a filled d^{10} configuration in all its compounds, is transitional between the d block and the p elements and is discussed later.

The Groups of the d block contain only three elements and correspond to the filling of the 3d, 4d and 5d shells respectively. In between the 4d and 5d levels is interposed the first f level, the 4f shell, which fills after lanthanum. It has already been seen (Chapter 11) that the occupation of this level is accompanied by a gradual decrease in atomic and ionic radius from La to Lu and the total lanthanide contraction is approximately equal to the normal increase in size between one Period and the next. The result is that in the transition Groups there is the normal increase of about 20 pm in radius between the first and second members (filling the 3d and 4d shells), but the expected increase between the second and third members is just balanced by the lanthanide contraction so that these two elements are almost identical in size. This effect is illustrated by the radii given in Table 13.1, where the normal increases in the alkaline earths and in the scandium Group contrast sharply with the figures for the succeeding Groups.

As the pair of heavy elements have almost identical radii, and therefore very similar characteristics in other ways (e.g. ionization potentials, solvation energies, redox potentials, lattice energies), their chemistry is very similar. Thus each transition Group typically divides into two parts—the lightest element with its individual chemistry, and the pair of heavy elements with almost identical chemistries.

The three elements within each Group have a number of properties in common, of course. They show the same range of oxidation states in general, though these differ in relative stabilities. All the d and s electrons are involved in the chemistry of the earlier elements, so that the Group oxidation state is the maximum state shown. Once the d^5 configuration is exceeded, there is less tendency for all the d electrons to react, and the Group oxidation state is not shown by iron (though Os(VIII) and Ru(VIII) exist), nor by any elements of the cobalt, nickel or copper Groups. Since, in the Group oxidation state, all the valency electrons are involved and since the properties of the elements then depend on valency and size only, there are similarities between the properties of Main Groups and Transition Groups of the same Group oxidation state. Thus sulfates and chromates, both MO_4^{2-}, are isostructural, while molybdenum and tungsten show higher coordination numbers with oxygen (especially six), just as does tellurium. The principal differences between the first and the heavier elements in a transition Group are those of size, and stability of oxidation states. The larger elements commonly show higher coordination numbers, and the higher oxidation states are more stable for the heaviest elements. Thus, chromium(VI) is strongly oxidizing while molybdenum and tungsten are stable in the VI state.

The effects of the lanthanide contraction die out towards the right of the d block. In the titanium and vanadium Groups, which immediately follow the lanthanides, the heavier

TABLE 13.1 Radii showing the effect of the lanthanide contraction (Pauling)

M^{2+} (pm)	M^{3+} (pm)	M^{4+} (calc) (pm)	atomic radii (pm)		
Ca = 99	Sc = 70	Ti = 68	Ti = 132	V = 122	Cr = 117
Sr = 113	Y = 90	Zr = 74	Zr = 145	Nb = 134	Mo = 129
Ba = 135	La = 106	Hf = 75	Hf = 144	Ta = 134	W = 130

elements are practically identical and their separation is more difficult than the separation of a pair of lanthanides. The next two Groups show clear differences between the two last elements, though these are still slight. In the platinum metals, the differences are increasing until, in the copper Group, there are few points of resemblance between silver and gold. Finally, in the zinc Group, the pattern approaches that in a p Group and zinc and cadmium resemble each other with mercury as the singular member.

TABLE 13.2 Oxidation states of the transition elements

	Ti	V	Cr	Mn	Fe	Co	Ni	Cu	
Group	$\underline{\text{IV}}$	V	VI (ox)	VII (ox)					d^0
O.S.	III (red)	$\underline{\text{IV}}$	(V) (d)	(VI) (d)					d^1
	(II) (red)	III (red)	(IV) (d)	(V) (d)	(VI) (ox)				d^2
		(II) (red)	$\underline{\text{III}}$	IV (ox)	(V) (ox)				d^3
	(0)	(I)	II (red)	(III) (ox)	(IV) (ox)				d^4
	(−I)	(0)	(I)	$\underline{\text{II}}$	$\underline{\text{III}}$	(IV) (ox)			d^5
		(−I)	0	(I)	$\underline{\text{II}}$	$\underline{\text{III}}$	(IV) (ox)		d^6
			(−I)	0		II	(III) (ox)		d^7
			(−II)	(−I)	0	(I)	$\underline{\text{II}}$	(III) (ox)	d^8
						0	(I)	$\underline{\text{II}}$	d^9
							0	I	d^{10}
	Zr	**Nb**	**Mo**	**Tc**	**Ru**	**Rh**	**Pd**	**Ag**	
Group	$\underline{\text{IV}}$	$\underline{\text{V}}$	$\underline{\text{VI}}$	$\underline{\text{VII}}$	VIII (ox)				d^0
O.S.	(III) (red)	(IV) (d)	V	VI (d)	(VII) (d)				d^1
	(II) (red)	III (red)	IV	(V) (d)	VI				d^2
		(II) (red)	III	$\underline{\text{IV}}$	(V)	(VI) (ox)			d^3
			(II)	(III)	IV				d^4
				?	$\underline{\text{III}}$	IV			d^5
			0		II	$\underline{\text{III}}$	IV		d^6
				0		II			d^7
					0	(I)	$\underline{\text{II}}$	(III) (ox)	d^8
						0	(I?)	II (ox)	d^9
							0	$\underline{\text{I}}$	d^{10}
	Hf	**Ta**	**W**	**Re**	**Os**	**Ir**	**Pt**	**Au**	
Group	$\underline{\text{IV}}$	$\underline{\text{V}}$	$\underline{\text{VI}}$	$\underline{\text{VII}}$	VIII (ox)				d^0
O.S.	(III) (red)	(IV) (d)	V	VI	(VII)				d^1
	(II) (red)	(III) (red)	IV	(V)	VI				d^2
		(II)	(III)	$\underline{\text{IV}}$	(V)	(VI) (ox)			d^3
			(II)	$\underline{\text{III}}$	$\underline{\text{IV}}$	(V) (ox)	(VI) (ox)		d^4
				(II)	(III)	$\underline{\text{IV}}$	(V) (ox)		d^5
			0	(I)	(II)	$\underline{\text{III}}$	$\underline{\text{IV}}$	V (ox)	d^6
				0	(I)	(II)	?		d^7
					0	(I)	$\underline{\text{II}}$	$\underline{\text{III}}$	d^8
						0	(I)		d^9
						(−I)	0	I	d^{10}

Notes: ox = oxidizing, red = reducing, unstable states bracketed, d = disproportionates, most stable state(s) for any given element underlined. State 0 usually in carbonyls and related complexes: the element itself is not counted as a 0 state here.

TABLE 13.3 Transition element oxides

Oxidation state							
+II	*+III*	*+IV*	*+V*	*+VI*	*+VII*	*+VIII*	*Other compounds*
TiO	Ti_2O_3	TiO_2					
		ZrO_2					
		HfO_2					
VO	V_2O_3	VO_2	V_2O_5				
		NbO_2	Nb_2O_5				
		(TaO_2?)	Ta_2O_5				
CrO	Cr_2O_3	CrO_2		CrO_3			
		MoO_2	Mo_2O_5	MoO_3			
		WO_2	(W_2O_5?)	WO_3			
MnO	Mn_2O_3	MnO_2			Mn_2O_7		Mn_3O_4
		TcO_2		TcO_3	Tc_2O_7		Also Tc_2S_7
	$Re_2O_3^*$	ReO_2	(Re_2O_5)	ReO_3	Re_2O_7		Re_2S_7
FeO	Fe_2O_3						Fe_3O_4
	$Ru_2O_3^*$	RuO_2		(RuO_3)*		RuO_4	
		OsO_2		(OsO_3)*		OsO_4	
CoO	(Co_2O_3)*	(CoO_2)*					Co_3O_4
RhO	Rh_2O_3	RhO_2					
	Ir_2O_3	IrO_2		(IrO_3)			
NiO	(Ni_2O_3)*	(NiO_2)*					
PdO		(PdO_2)*					
(PtO)*	(Pt_2O_3)*	PtO_2		(PtO_3)*			Pt_3O_4
CuO							Cu_2O
AgO	(Ag_2O_3?)						Ag_2O
	Au_2O_3						Au_2O

Most stable compounds underlined.
*Hydrous oxides of these states are reported.

Table 13.2 summarizes this discussion in terms of the oxidation states shown by the d elements, and the stabilities of these. The general behaviour is also illustrated by Tables 13.3 to 13.5 which give the oxides, fluorides and other halides of the transition elements. A stable state will show all these compounds while a strongly oxidizing state will be more likely to have a fluoride than an iodide; similarly, a reducing state will be more likely to show a heavier halide than a fluoride. In fact, the highest oxidation state of a transition metal is typically obtained in an anionic complex, particularly oxides and fluorides. As a good illustration of this, the AgF_4^- and NiF_6^{2-} ions have been known for some time, but the parent binary fluorides AgF_3 and NiF_4 have only recently been synthesized as highly reactive, polymeric solids. A promising route for the synthesis of these high oxidation state compounds involves fluoride-ion removal from the anionic fluoro-complex by a powerful fluoride ion acceptor such as AsF_5.

This division between the lighter transition elements and the two heavier Periods is quite marked and is reinforced in practice by the relative inaccessibility of most of the heavier elements. The latter have therefore been less fully studied, especially the less available member of the pair (for example Hf, Nb, Tc). In addition, there are strong horizontal resemblances, especially among ions of the same charge, and horizontal trends in

TABLE 13.4 Transition element fluorides

Oxidation state						
+II	*+III*	*+IV*	*+V*	*+VI*	*+VII*	*Notes and other compounds*
	TiF_3	$\underline{TiF_4}$				
	(ZrF_3)	$\underline{ZrF_4}$				
		$\underline{HfF_4}$				
VF_2	VF_3	$\underline{VF_4}$	VF_5			Nb_6F_{15}
	NbF_3	(NbF_4)	$\underline{NbF_5}$			
			$\underline{TaF_5}$			
CrF_2	$\underline{CrF_3}$	CrF_4	CrF_5	(CrF_6)		(CrF)
	MoF_3	MoF_4	MoF_5	$\underline{MoF_6}$		
		WF_4	WF_5 (d)	$\underline{WF_6}$		
$\underline{MnF_2}$	MnF_3	MnF_4				
			TcF_5	$\underline{TcF_6}$		
		ReF_4	ReF_5	ReF_6	$\underline{ReF_7}$	
FeF_2	$\underline{FeF_3}$					
	$\underline{RuF_3}$	RuF_4	$\underline{RuF_5}$	RuF_6		
		$\underline{OsF_4}$	OsF_5	$\underline{OsF_6}$	(OsF_7)	*Note* No OsF_8
$\underline{CoF_2}$	CoF_3					
	$\underline{RhF_3}$	RhF_4	(RhF_5)	RhF_6		
	IrF_3	$\underline{IrF_4}$	(IrF_5)	IrF_6		
$\underline{NiF_2}$	(NiF_3)	(NiF_4)				
$\underline{PdF_2}$	$[PdF_3]$	PdF_4				
		$\underline{PtF_4}$	PtF_5	PtF_6		$PdF_3 = Pd^{2+}(PdF_6)^{2-}$
$\underline{CuF_2}$						
AgF_2						Ag_2F, $\underline{AgF}$
	$\underline{AuF_3}$		(AuF_5)			

Most stable compounds underlined. d = disproportionates. ( ) = compound well-established but unstable at room temperature.

Structures: MF_2 rutile or distorted rutile.
MF_3 ReO_3, i.e. octahedra linked through all corners; M = Au, linked planar AuF_4 units with longer bonds to next layers completing a distorted octahedron.
MF_4 octahedra sharing edges; M = Zr, square antiprisms linked through all F.
MF_5 octahedra sharing corners; chains or closed rings.
MF_6 octahedron.
MF_7 pentagonal bipyramid.

properties, with increasing number of d electrons in a given oxidation state, that make it convenient to divide the discussion of the transition block into two sections, one on the first row elements (Chapter 14) and one (Chapter 15) on the heavier elements of the second and third rows. Selected transition metal topics of active current interest are reviewed in Chapter 16.

The pattern of oxidation state stabilities outlined above is complex and there are exceptions to most of the generalizations which can be made about it. The picture is complicated by the use of the term 'stability' in a number of different senses. In the most general sense, it is used to mean that a compound exists in air at around room temperatures: that is, that it is thermally stable at room temperature, that it is not oxidized by air, and that it is not hydrolysed, oxidized or reduced by water vapour. In turn, terms such as thermal stability may cover a number of processes. Thus a higher oxide such as MO_2 might decompose thermally to $M + O_2$ or to $MO + \frac{1}{2}O_2$ and the free

TABLE 13.5 Transition element halides*

Oxidation state					
+II	*+III*	*+IV*	*+V*	*+VI*	*Notes*
TiX_2	TiX_3	TiX_4			
(ZrX_2)	ZrX_3	$\underline{ZrX_4}$			ZrCl
$HfCl_2$, Br_2	$HfCl_3$, Br_3	$\underline{HfX_4}$			Also HfCl?
VX_2	$\underline{VX_3}$	VCl_4, Br_4			VBr_4 very unstable
$(NbBr_2)$	NbX_3	NbX_4	$\underline{NbX_5}$		Nb_6X_{14}, Ta_6X_{14}
$TaCl_2$	$TaCl_3$, Br_3	TaX_4	$\underline{TaX_5}$		All $=(M_6X_{12})^{2+}(X^-)_2$ (a)
					Nb_3Br_8, Nb_3I_8, Nb_6I_{11}
CrX_2	$\underline{CrX_3}$				
MoX_2	MoX_3	MoX_4	$MoCl_5$		MoX_2 and WX_2
WX_2	WX_3	WX_4	WCl_5, Br_5	WCl_6, Br_6	$=(M_6X_8)^{4+}\ X_4^-$
$\underline{MnX_2}$					
		$TcCl_4$		$(TcCl_6)$?	
(ReX_2)	ReX_3	$\underline{ReX_4}$	$ReCl_5$, Br_5	$(ReCl_6)$?	$ReCl_3$, $ReCl_4$ are trimers
					ReX_2 in complexes only
$\underline{FeX_2}$	$FeCl_3$, Br_3				
	$\underline{RuX_3}$	$RuCl_4$			
		OsX_4	$OsCl_5$		$OsX_{3.5}$
$\underline{CoX_2}$					
	$\underline{RhX_3}$				
$(IrCl_2)$?	$\underline{IrX_3}$	$(IrCl_4)$			
$\underline{NiX_2}$					Platinum trihalides may be
$\underline{PdX_2}$					mixtures of Pt(II)+Pt(IV)
$\underline{PtX_2}$	PtX_3?	$\underline{PtX_4}$			$PtCl_2$ structure consists of
					Pt_6Cl_{12} units
$CuCl_2$, Br_2					Also $\underline{CuX}$
					$\underline{AgX}$
	$AuCl_3$, Br_3				AuCl, I

Most stable compounds underlined.
*The symbol X is used when the chloride, bromide and iodide all occur.
(a) All species $M_6X_{12}^{n+}(X^-)_n$ for $n=2$, 3 and 4 occur for M = Nb, X = Cl; M = Ta, X = Cl, Br. $Nb_6I_{11}=(Nb_6I_8)^{3+}(I^-)_3$.

energy change of each process would have to be evaluated before conclusions could be drawn about the stability. Similarly, a compound may exist for a long time at room temperature, not because it is thermodynamically stable with respect to decomposition, but because the decomposition process occurred at a negligible rate. Thus, whether a compound can be kept 'in a bottle' depends on a wide variety of thermodynamic and kinetic factors.

13.2 The transition ion and its environment: ligand field theory

In the discussion of the energy levels of an atom given in the earlier chapters, the levels of a given p, d or f set were treated as of equal energy. This is true of isolated atoms, or of those in an electric field which is spherically symmetrical around the atom, but is not true when the atom lies in an unsymmetrical field. This may be readily seen by considering an atom which is

STABILITIES OF TRANSITION ELEMENT OXIDATION STATES

Attempts have been made to examine, predict and rationalize the stabilities of the transition element oxidation states although most treatments to date are either limited in scope or are empirical. One example of a general approach which may be quoted is that of Sheldon who proposes that the preferred oxidation state of a transition element (defined as that which, in simple binary compounds such as the oxides or halides, is the most stable under normal laboratory conditions) is related to the quantity $rH/40$. Here, r is half the interatomic distance and H the heat of atomization of the metal—both well-known quantities. This expression leads to the following predictions for the most stable oxidation states of the transition elements:

VI for W, Re, Os, Ir
V for Nb, Mo, Tc, Ru, Ta
IV for Ti, V, Zr, Rh, Hf, Pt
III for Cr, Fe, Co, Ni, Pd, Au
II for Mn, Cu, Ag

If these predictions are compared with Tables 13.2 to 13.5, it will be seen that they are surprisingly accurate for such a simple formula. The only really poor predictions are those for nickel, palladium and silver where the states predicted to be stable are nonexistent or very unstable.

A much more fundamental and searching analysis is that, discussed in the next chapter, on the stability of trihalides of the first row of transition elements, but this is limited to one particular decomposition reaction. Further work on rigorous thermodynamic analysis should lead to a greatly increased understanding of stabilities and periodic trends.

strongly coordinated to two other groups in a linear configuration. If these groups lie in the $\pm z$ directions, the orbitals which point along the z axis will lie in the field of these ligands and be perturbed by them more than orbitals lying in other directions. As ligands are regions of negative charge (they coordinate through lone electron pairs and also have negative charges or the negative end of a dipole directed towards the central atom or ion) the z-axis directed orbitals on the central atom will be in a region of higher negative field than the non-z orbitals and electrons will avoid entering them as far as possible. This means that such orbitals as p_z, d_{z^2} and, to a lesser extent, d_{xz} and d_{yz}, are of higher energy than the remaining ones and the degeneracy of the p and d set is split in such a z-directed field in the manner shown in Fig. 13.1. The size of such a splitting will depend on the size of the ligand field and this, in turn, depends on the distance of the ligand and thus on the intensity of the attraction between the central atom and the ligand. Such ligand fields occur in all chemical environments. Their effects are generally negligible, except for d orbitals, either because the fields are small (as in the case of f orbitals) or because the orbitals are equally populated, as is

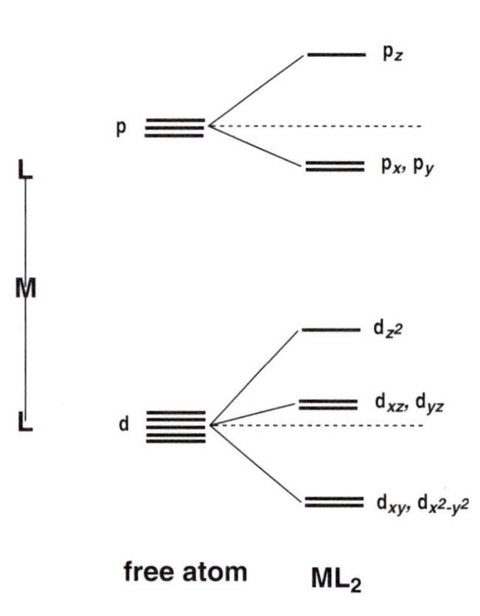

FIG. 13.1 Energy level diagram for a linear field. The z axis is taken as the direction of the coordinated groups. The ligands are regions of high negative charge density, so orbitals on the central atom with components in the $\pm z$ direction are less stable than those with no such component. Among the d orbitals, the d_{z^2} orbital is most strongly destabilized as it has the greatest density in the z direction. The atom levels are those of an atom in a spherically symmetrical field of the average effect of the ligands

usually the case for p orbitals. Thus our discussion of ligand fields is confined to d element chemistry.

The strength of the ligand field effect is marked in the case of the transition elements. The ions are small so the M^{2+} and M^{3+} ions are centres of high charge density which coordinate strongly with lone pair donors such as water or ammonia. The case of the first row of the transition block is particularly interesting as the energy differences introduced by ligand field splittings are of the same magnitude as the various energy changes involved in electron pairing. The effects of the fields of different ligands on ions of differing numbers of d electrons, and in different environments, show up in the numbers of unpaired d electrons. These are readily determined by magnetic measurements. Ligand field effects are also seen in a number of other properties such as ionic radii, lattice energies, reaction mechanisms and electronic spectra, but it was the magnetic effects which first attracted attention and gave rise to the interest in *ligand field theory*. The application of the theory may be examined in more detail in the case of octahedral complexes of the first row transition elements. This is the commonest geometry shown by these elements, in solution, in solvated or coordinated individual ions, and also in solids such as the oxides or fluorides. In later sections, the extension of the theory to other coordination numbers and to the heavier elements will be discussed.

13.3 Ligand field theory and octahedral complexes

Regular 6-coordination is most readily pictured by placing the ligands at the plus and minus ends of the three coordinate axes. In the *xy* plane, the positions of such ligands relative to the d orbitals are shown in Fig. 13.2a, while the corresponding diagrams for the *xy* and *yz* planes are shown in Figs 13.2b and c. In the *xy* plane, the orbital d_{xy} lies between the ligands while $d_{x^2-y^2}$ points directly at the ligands. An electron in the $d_{x^2-y^2}$ orbital is therefore most affected by the field of the ligands and is raised in energy relative to an electron in the d_{xy} orbital. Similarly, electrons in d_{z^2} are less stable than ones in d_{xz} or d_{yz} (Figs 13.2b and c). If the alignments of the three orbitals, d_{xy}, d_{xz} and d_{yz}, relative to the ligands are compared, it will be seen that these are identical. It follows that, in the full three-dimensional case, electrons in these three orbitals are identical in energy and are stabilized relative to the other two. Electrons in orbitals d_{z^2} and $d_{x^2-y^2}$ are also identical in energy and are destabilized. The combined energy level diagram is therefore composed of two upper orbitals, of equal energy, and three lower orbitals, which are also degenerate (Fig. 13.3). The energy zero is conveniently taken as the weighted mean of the energies of these two sets of orbitals; the lower trio are thus stabilized by $-2/5\Delta E$ while the upper pair are destabilized by $3/5\Delta E$, where ΔE is the total energy separation.

It is easier to accept this equivalence if it is recalled that d_{z^2} is compounded of two orbitals similar to $d_{x^2-y^2}$.

The e_g and t_{2g} symbols are symmetry labels arising from group theory and are now the most commonly used symbols. It may help in remembering them that e signifies a doubly degenerate state and t a triply degenerate one.

Consider the case of a d^1 system in an octahedral field, for example the hydrated titanium (III) ion, $Ti(H_2O)_6^{3+}$. The orbitals are split as in Fig. 13.3, and the single d electron naturally enters the lowest available one, here one of the t_{2g} set. In doing so it gains energy, equal to $-2/5\Delta E$, relative to the energy it would have had if the octahedral splitting had not occurred. In this case the energy gain is equal to about 90 kJ mol^{-1}. This energy gain, relative to the case of five equal d orbitals, is termed the *ligand field stabilization energy*, or, since it was first observed in crystals, the crystal field stabilization energy or CFSE. This was first observed when it was found that calculations of the lattice energy of simple transition element oxides and fluorides, by the electrostatic method which was so successful for s element salts, gave answers which did not agree with the experimental values. Including the effect of the crystal field on the d orbital energies (Section 13.7) led to full agreement. The CFSE is an additional energy increment to the system which has to be added to the other attractive and repulsive energies, both in solids and in calculation of solvation energies and the like.

The size of ΔE is most readily measured spectroscopically by observing the energy of the electronic transitions between the t_{2g} and e_g orbitals. The energy usually lies in the visible or near-ultraviolet region of the spectrum and it is such d-d transitions which are responsible for the colours of most transition metal compounds. The magnitude of ΔE depends on the ligand and on the nature of the transition metal ion. One of the simplest examples is that of titanium(III) complexes, where the configuration is d^1.

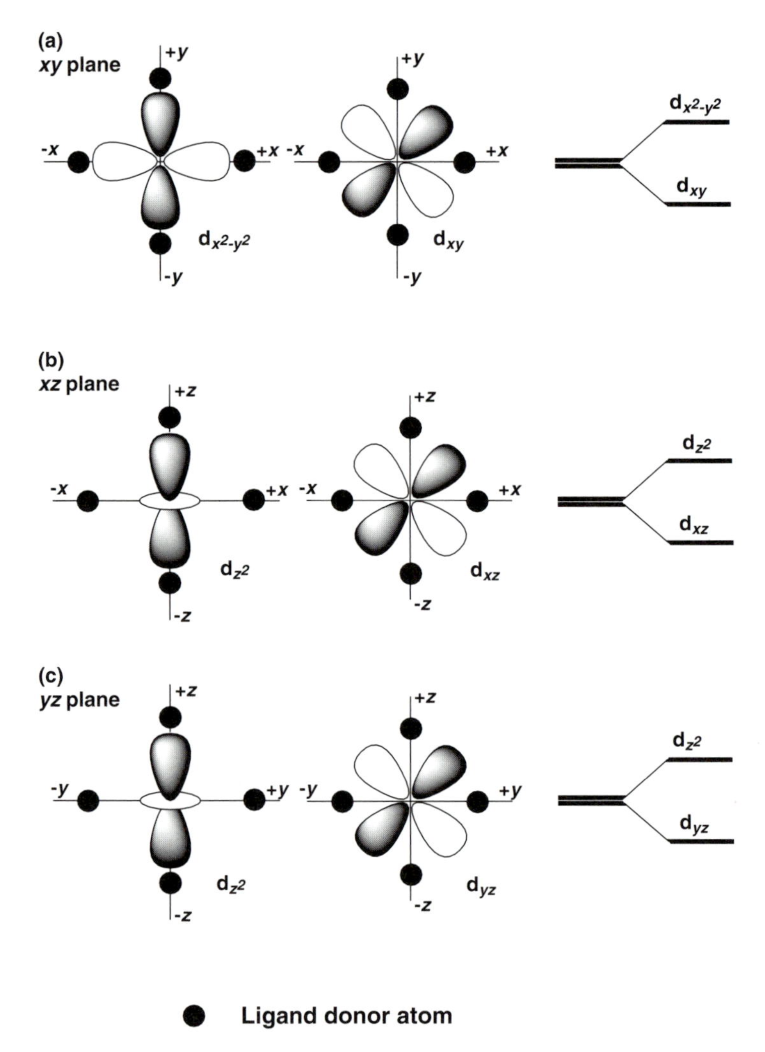

FIG. 13.2 Positions of ligands and d orbitals in an octahedral complex: (a) the xy plane, (b) the xz plane, (C) the yz plane

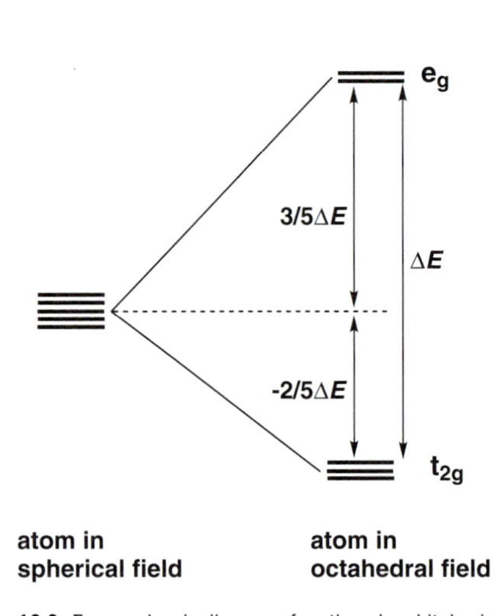

FIG. 13.3 Energy level diagram for the d orbitals in an octahedral field. Note that electrons in the d levels of the free atom would be more stable than when the atom is in a spherical field. The energy gap ΔE is often labelled 10 Dq or Δ_{oct}

The transition from the t_{2g} to e_g level of the single electron gives rise to a single absorption band in the visible region (Fig. 13.4). The position of this band gives the size of ΔE (compare the discussion in Sections 7.5 and 7.6). When the ligand in $Ti^{III}L_6$ changes from $L = H_2O$ to $L = Cl^-$, the position of the absorption band moves to lower energy, thus ΔE for Cl^- is smaller than for H_2O. By examining a whole series of complexes with different ligands, L, the size of ΔE for each ligand may be determined, both for titanium(III) and for other metals in various oxidation states. It is found that the order of increasing ligand effects is approximately constant from one transition ion to the next and increases in the order:

$$I^- < Br^- < Cl^- < F^- < H_2O < NH_3 < en < NO_2^- < CN^-$$
$$(en = ethylenediamine)$$

The ligand field increases by a factor of approximately two from halide to cyanide. A large number of other ligands are to be fitted into this series, of course, but these are representative ligands, and cyanide has the strongest ligand field of all common ligands. The series is termed the *spectrochemical series*. The main effects of the transition metal ion are those due to charge and to Period. The splitting increases by about 30% between members of the same Group in successive Periods, and there is an increase of roughly 50% on going from the divalent to the trivalent ion of any element. These trends are illustrated by the values shown in Table 13.6. The d-d transitions for configurations other than d^1 are rather more complicated and are discussed in Section 13.11.

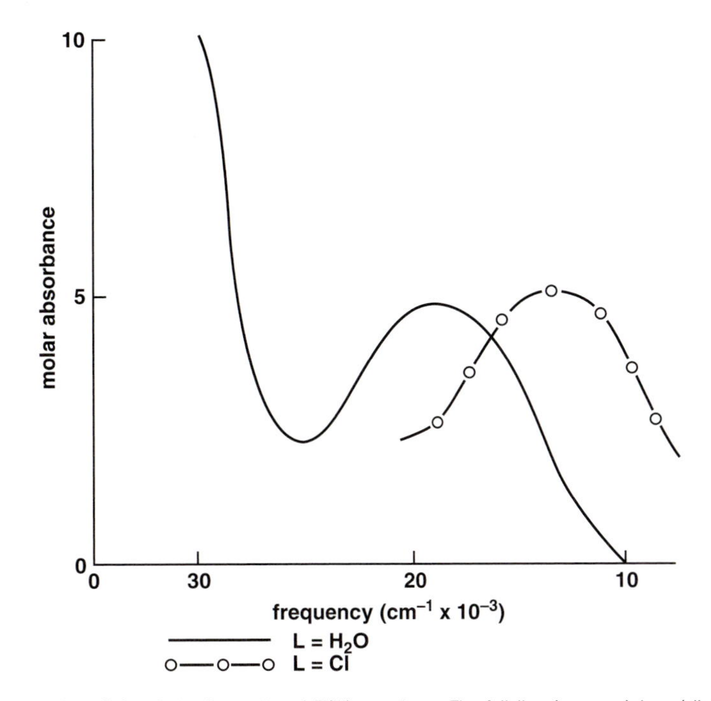

FIG. 13.4 Representation of the electronic spectra of Ti(III) complexes. The full line (———) is a (slightly simplified) representation of the spectrum of $Ti(H_2O)_6^{3+}$ with a maximum at 20 300 cm^{-1}. The broken line (–o–o–o–) represents the spectrum of $TiCl_6^{3-}$, with a maximum at 13 000 cm^{-1}. The steeply rising portion of the curve for the hydrate is the edge of the strong, allowed, charge-transfer transition in the ultraviolet

In the case of titanium(III), there is no ambiguity as to the location of the d electron, and this is so for the d^2 and d^3 configurations also. The d electrons enter the t_{2g} set of orbitals with parallel spins to give CFSE values of $-4/5\Delta E$ for d^2 and $-6/5\Delta E$ for d^3. Similarly, for d^8,d^9 and d^{10} only one configuration is possible—think of these as having respectively 2,1 and 0 holes in the shell.

In the remaining configurations alternative allocations between t_{2g} and e_g levels are possible. In the case of d^4, two configurations are possible. The first three electrons enter the

TABLE 13.6 Values of the ligand field splitting ΔE in octahedral complexes

Ion		*6Cl⁻*	*6H₂O*	*6NH₃*	*6CN⁻*
		ΔE (*kJ mol⁻¹*)			
Ti^{3+}	$3d^1$		243		
V^{3+}	$3d^2$		226		
Cr^{3+}	$3d^3$	163	213	259	314
Mn^{3+}	$3d^4$		(250)		
Fe^{3+}	$3d^5$		162		
Co^{3+}	$3d^6$		222	296	406
Mo(III)	$4d^3$	230			
Rh(III)	$4d^6$	243	322	406	
Ir(III)	$5d^6$	297			
Pt(IV)	$5d^6$	347			
V^{2+}	$3d^3$		151		
Cr^{2+}	$3d^4$		(170)		
Mn^{2+}	$3d^5$		92	approx. 100	
Fe^{2+}	$3d^6$		126		393
Co^{2+}	$3d^7$		113	121	
Ni^{2+}	$3d^8$	88	100	130	
Cu^{2+}	$3d^9$		(150)	(180)	

Values for d^4 and d^9 configurations are approximate because of distortion in these octahedral complexes.

three t_{2g} orbitals while the fourth may either remain parallel to the first three, thus producing maximum exchange energy, and enter the higher energy e_g level, or it may pair up with one of the electrons already present in the t_{2g} level and produce maximum crystal field stabilization energy. The first configuration is termed the *high-spin* or *weak field* configuration, while the arrangement with the paired electrons is the *low-spin* or *strong field* case. In the case of d^4, the CFSE for the $t_{2g}^3e_g^1$ configuration is $-6/5\Delta E+3/5\Delta E=-3/5\Delta E$, while the CFSE for the low-spin t_{2g}^4 configuration is $-8/5\Delta E$, so that the adoption of the low-spin configuration means the gain of ΔE in excess of the CFSE in the high-spin configuration. On the other hand, the exchange energy of four parallel electrons is $6K$ (see Section 8.2) while that of the three parallel electrons in the low-spin configuration is only $3K$. Which configuration is actually adopted therefore depends on the relative sizes of ΔE and K.

The K values are difficult to determine but remain approximately constant for atoms of the same quantum shell. In the case of the first transition series the loss of exchange energy usually lies within the range of values found for ΔE. Thus, for any configuration where alternative electronic arrangements are possible, there will be a particular value of ΔE where the change from high-spin to low-spin values takes place. A large value of ΔE obviously favours the low-spin arrangement, hence the alternative name of strong field configuration. Alternative electronic configurations are possible for d^4, d^5, d^6 and d^7 ions in octahedral complexes. Table 13.7 lists the values of the CFSE and exchange energy for all the d configurations, while Table 13.8 shows the differences for high- and low-spin configurations in the states d^4 to d^7. Notice that there are no examples known of intermediate configurations such as $t_{2g}^4e_g^1$. In all cases the electrons are either paired as far as possible or parallel as far as possible. Table 13.9 gives the approximate magnetic moments, based on the 'spin-only' formula for each configuration. Experimental magnetic moments usually

TABLE 13.7 Crystal field stabilization and exchange energies in octahedral configuration

Number of d electrons	*Electron configuration*					*CFSE ΔE*	*Exchange energy K*
	t_{2g}			*e_g*			
1	↑					−2/5	0
2	↑	↑				−4/5	1
3	↑	↑	↑			−6/5	3
4 high-spin	↑	↑	↑	↑		−6/5+3/5	6
4 low-spin	↑↓	↑	↑			−8/5	3
5 high-spin	↑	↑	↑	↑	↑	−6/5+6/5	10
5 low-spin	↑↓	↑↓	↑			−10/5	3+1
6 high-spin	↑↓	↑	↑	↑	↑	−8/5+6/5	10
6 low-spin	↑↓	↑↓	↑↓			−12/5	3+3
7 high-spin	↑↓	↑↓	↑	↑	↑	−10/5+6/5	10+1
7 low-spin	↑↓	↑↓	↑↓	↑		−12/5+3/5	6+3
8	↑↓	↑↓	↑↓	↑	↑	−12/5+6/5	10+3
9	↑↓	↑↓	↑↓	↑↓	↑	−12/5+9/5	10+6
10	↑↓	↑↓	↑↓	↑↓	↑↓	−12/5+12/5	10+10

Exchange energies shown separately for the parallel and antiparallel sets.

TABLE 13.8 Balance of exchange and crystal fields energies for states of alternative configurations in the octahedral field

Number of d electrons	*Gain in CFSE of low-spin relative to high-spin configuration*	*Loss in exchange energy of low-spin relative to high-spin configuration*
4	ΔE	$3K$
5	$2\Delta E$	$6K$
6	$2\Delta E$	$4K$
7	ΔE	$2K$

TABLE 13.9 'Spin only' magnetic moments for octahedral arrangements

Number of d electrons	Magnetic moment, Bohr magnetons		
	High-spin		Low-spin
1		1.73	
2		2.83	
3		3.87	
4	4.90		2.83
5	5.92		1.73
6	4.90		0.00
7	3.87		1.73
8		2.83	
9		1.73	
10		0.00	

'Spin only' moment equals $2\sqrt{[S(S+1)]}$, where $S = \frac{1}{2}n$ = number of unpaired spins × spin quantum number.

The 'spin-only' formula is usually a good approximation for transition metal ions although orbital coupling occurs to a small extent in most cases and is marked in the cases of Co^{2+} and Co^{3+}. (Compare section 7.11.1).

agree with those calculated by the 'spin-only' formula (Section 7.11) to within 10%, quite close enough to distinguish high-spin from low-spin configurations.

Theories. Any theory of bonding in d element compounds has to treat crystal field stabilization energy and also the possibility of alternative electronic configurations. Two independent approaches have developed—as an electrostatic theory or as a molecular orbital theory. The results of these two approaches are very similar so that either may be used as seems most appropriate and the combined theory is termed the *ligand field theory*. The term *crystal field theory* is sometimes reserved for specific application to the electrostatic version, but usages vary among different authors.

These approaches will be examined briefly in turn. In the electrostatic approach, the bond energy is held to arise purely from the electrostatic attractions between the central ion and the charges, dipoles and induced dipoles on the ligands, with the repulsions between dipoles, induced dipoles and charges on different ligands taken into account. To these forces is to be added the CFSE arising from the d electron arrangement in d orbitals split by the ligand field. The situation is similar to the electrostatic treatment of ionic crystals with the addition of the crystal field stabilization energy. As no covalent bonds enter into this treatment, the energy level diagram is simply that of the atomic energy levels in the transition element, of which the vital part is the d electron diagram as shown in Fig. 13.3. This approach has the major advantage that the electrostatic calculations can actually be carried out, without drastic approximations, so that energies of formation and reaction mechanisms or stabilities can be predicted. The disadvantage of the theory is that it neglects the clear evidence that some covalent bonding does occur in transition metal compounds and it can throw no light on those cases, such as nickel carbonyl, where the central element is in a low, zero, or even negative oxidation state when the electrostatic forces would be weak or nonexistent.

Covalent bonding is incorporated in the molecular orbital theory. This constructs seven-centred (in an octahedral complex) molecular orbitals from the six ligand orbitals holding the lone pairs which are to be donated to the central atom, together with six orbitals of suitable energy and symmetry, on the central atom. In a transition metal of the first series, these six atomic orbitals are the $3d_{x^2-y^2}$, $3d_{z^2}$, 4s and the three 4p orbitals. These are the metal orbitals directed towards the ligands and of the right energy. The bonding molecular orbitals from these combinations of atomic orbitals with ligand orbitals are shown in Fig. 13.5; the numerical constants are chosen to weight the contributions of the ligand orbitals so that each adds up to unity. The antibonding orbitals corresponding to these six are those with the sign reversed between the central orbitals and the ligand combinations. Some of these are shown in Fig. 13.6, which also illustrates that orbitals such as d_{xy} cannot form σ bonds with any ligand combination. It will be noticed that these are delocalized polycentric σ orbitals, similar to those discussed in Section 4.7. The six atomic orbitals combine with the six ligand orbitals to form six bonding molecular orbitals and six

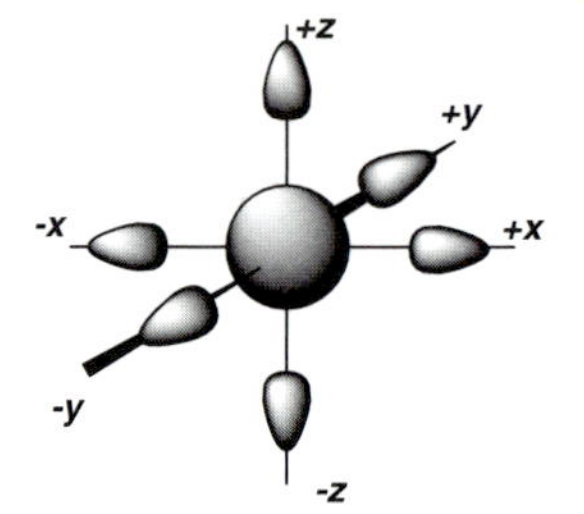

$\psi_{a_{1g}} = \phi_s + 1/\sqrt{(6)}(l_x + l_{-x} + l_y + l_{-y} + l_z + l_{-z})$

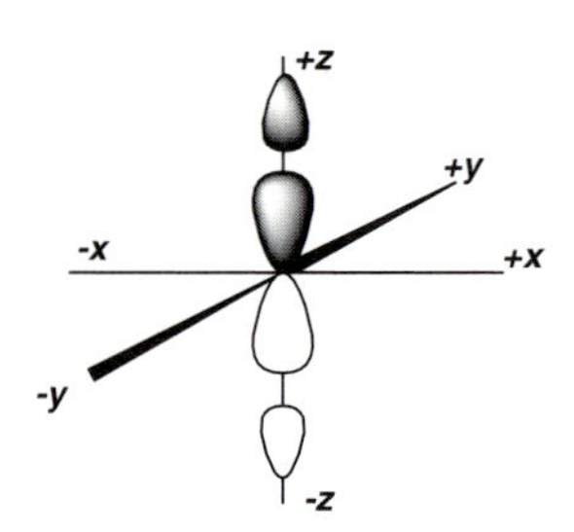

$\psi_{t_{1u}} = \phi_{p_z} + 1/\sqrt{(2)}(l_z - l_{-z})$

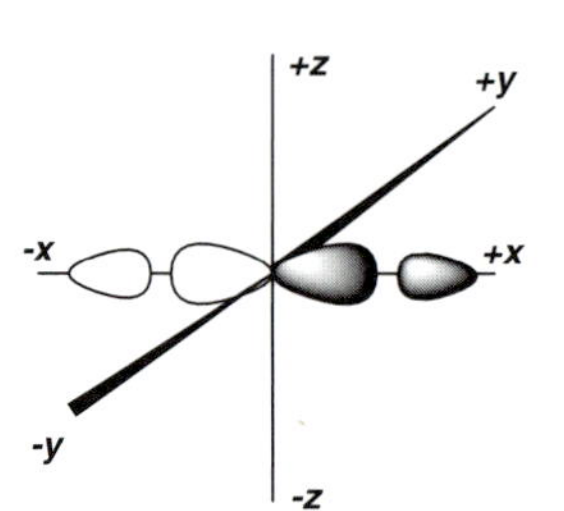

$\psi_{t_{1u}} = \phi_{p_z} + 1/\sqrt{(2)}(l_x - l_{-x})$

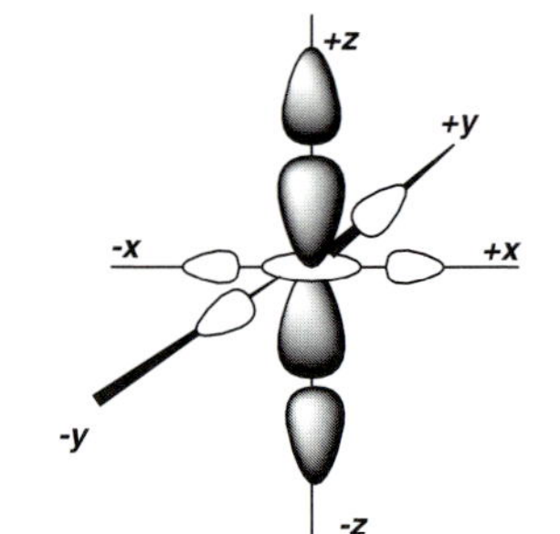

$\psi_{e_g} = \phi_{d_{z^2}} + 1/(2\sqrt{3})(2l_z + 2l_{-z} - l_x - l_{-x} - l_y - l_{-y})$

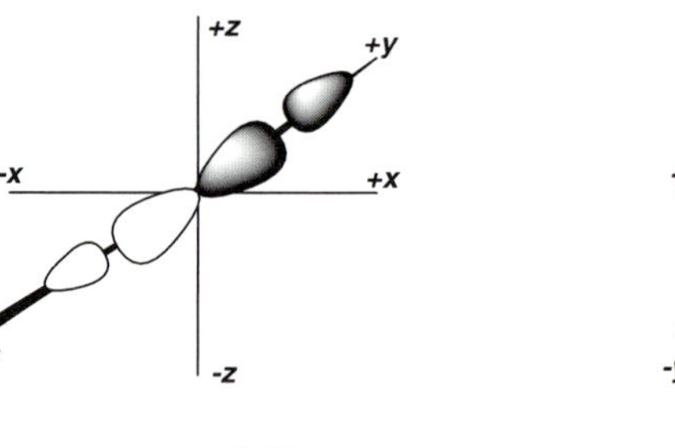

$\psi_{t_{1u}} = \phi_{p_y} + 1/\sqrt{(2)}(l_y - l_{-y})$

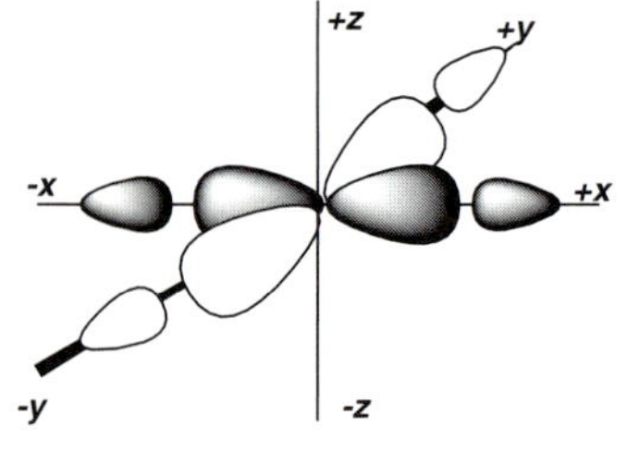

$\psi_{e_g} = \phi_{d_{x^2-y^2}} + \frac{1}{2}(l_x + l_{-x} - l_y - l_{-y})$

FIG. 13.5 The six bonding molecular orbitals formed by the six ligand orbitals and the s, p and d_{z^2} *and* $d_{x^2-y^2}$ orbitals on the central atom in an octahedral complex. The symbols a_{1g}, t_{1u} etc., are symmetry-indicating labels. Here they are used as a convenient way of distinguishing the orbitals. Note that there are three levels in sets labelled t and two in sets labelled e

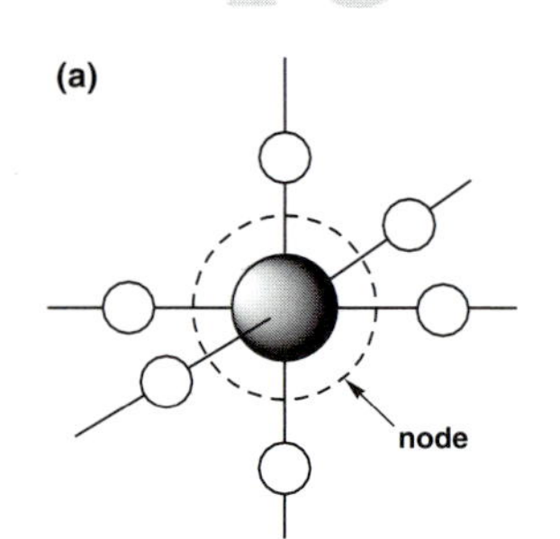

$\psi^*_{a_{1g}} = \phi_s - 1/\sqrt{(6)}(l_x + l_{-x} + l_y + l_{-y} + l_z + l_{-z})$

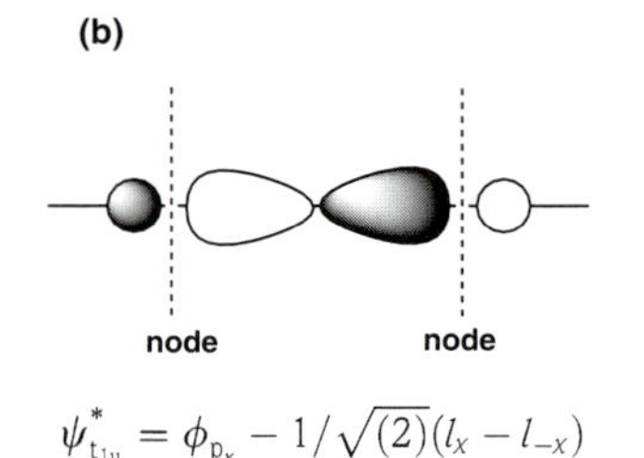

$\psi^*_{t_{1u}} = \phi_{p_x} - 1/\sqrt{(2)}(l_x - l_{-x})$

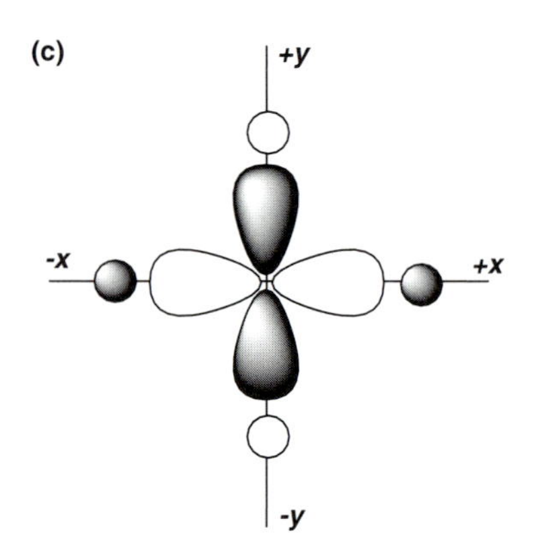

$\psi^*_{e_g} = \phi_{d_{x^2-y^2}} - \frac{1}{2}(l_x + l_{-x} - l_y - l_{-y})$

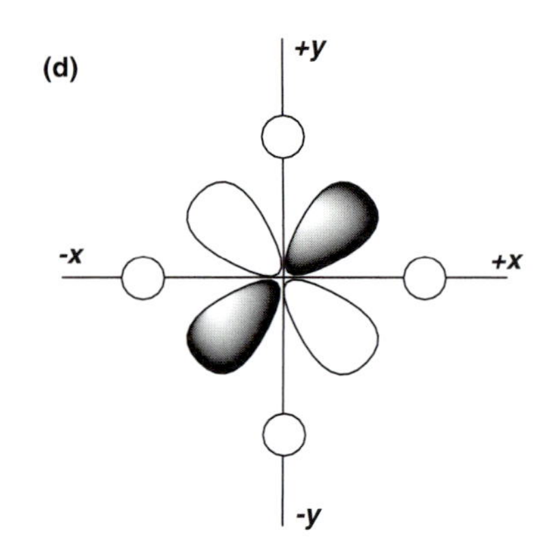

$\phi_{d_{xy}}$ — no σ-bonding combination possible

FIG. 13.6 Some nonbonding and antibonding combinations in an octahedral complex. Figs (a) to (c) are antibonding combinations of the s, one p and one d orbital, with ligand orbitals, corresponding to some of the bonding orbitals in Fig 13.5. Fig. (d) shows how the d_{xy} orbital is wrongly aligned to form any σ bond; the other two t_{2g} orbitals are similarly nonbonding. The labels a^*_{1g} etc., are used to distinguish the antibonding orbitals corresponding to the bonding orbitals of Fig. 13.5

antibonding molecular orbitals, while the remaining three 3d orbitals are nonbonding. The energy level diagram of the molecular orbitals in the octahedral complex is shown in Fig. 13.7.

Let us place the 12 electrons from the ligand lone pairs in the six bonding orbitals. Then the nonbonding t_{2g} set of 3d orbitals and the antibonding e^*_g pair are the next five orbitals available to accommodate the d electrons on the metal. Thus the descriptions in the electrostatic and molecular orbital theories are very similar. Both theories produce five orbitals in a lower set of three and an upper set of two, separated by ΔE, to accommodate the d electrons. In the electrostatic theory these sets are, respectively, the atomic d_{xy}, d_{yz} and d_{zx} orbitals and the d_{z^2} and $d_{x^2-y^2}$ atomic orbitals, while, in the molecular orbital theory, the lower set are the same atomic orbitals and the upper set are antibonding molecular orbitals composed of the d_{z^2} and $d_{x^2-y^2}$ atomic orbitals with ligand orbital contributions.

Thus in their essential description of the varying magnetic properties and d–d transitions, these two theories are identical. The molecular orbital theory has the advantage that it is easily extended to include π-bonding and it gives a prediction of the alteration of energy levels in such a case. It is also more useful for the interpretation of spectra as it provides information, not only about the d levels, but also about the higher energy antibonding orbitals to which excitations occur when higher energy quanta are absorbed. On the other hand, the molecular orbital theory suffers from the disadvantage of all wave

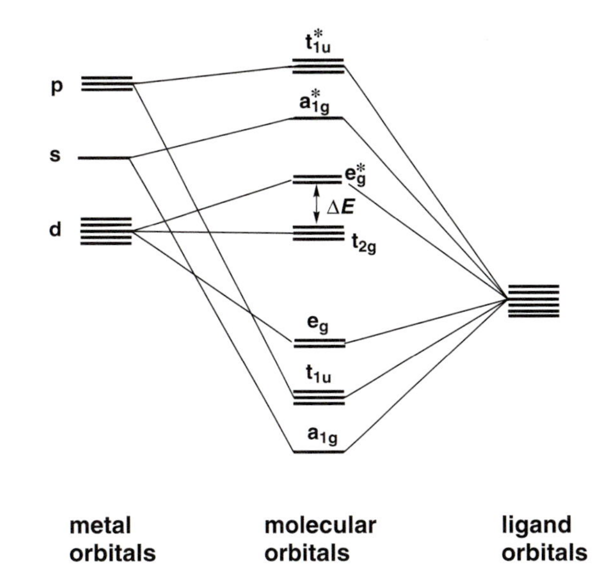

FIG. 13.7 Energy level diagram for the bonding, nonbonding, and antibonding molecular and atomic orbitals in a σ-bonded octahedral complex. The levels are labelled as in Figs 13.5 and 13.6.

mechanics, that it is impossible to calculate bond energies, heats of formation and the like directly.

In practice, these two theories may be used interchangeably, as most convenient. Both rely on experimental data to fix the energy levels; for example, ΔE is usually determined spectroscopically.

13.4 Coordination number four

Two configurations other than octahedral are common among the elements of the first transition series. Both involve 4-coordination and are the tetrahedral and square-planar configurations. The square-planar configuration may be considered as derived from the octahedral one by removing the ligands on the z axis. An elongation of the M–L distances on the z-axis leads to a decrease of the interaction between the ligand field and those metal orbitals with components in the z direction. The energy levels therefore split as indicated in Fig. 13.8a with the d_{z^2} level and the d_{xz} and d_{yz} levels falling below the others. The $d_{x^2-y^2}$ and d_{xy} levels rise slightly as the metal–ligand distances in the xy plane shorten a little because of the decrease in repulsion from the z ligands. Such an intermediate case corresponds to elongation of the metal–ligand distances on the z axis, to unsymmetric substitution on the z axis in a complex such as MX_4Z_2, (both of which are tetragonal distortions) or to the case of the 5-coordinated square pyramidal configuration. If the z ligands are removed completely to give the square-planar configuration, the energy level diagram of Fig. 13.8b results. Here, the d_{z^2} and d_{xy} levels have crossed over. Notice that, as the configuration in the xy plane is similar to that in the octahedron, the energy separation between $d_{x^2-y^2}$ and d_{xy} remains practically the same as the octahedral ΔE.

In the tetrahedral arrangement, the d orbitals are split into a lower set of two (called e in this symmetry) and an upper set of three (t_2). No orbital points directly at a ligand in the tetrahedral case, but the d_{xy} type lies closer to the ligand than $d_{x^2-y^2}$ or d_{z^2}. Fig. 13.9 shows that the distances are as half the side of a cube compared with half the face diagonal. This lack of direct interaction between orbitals and ligands in the tetrahedral configuration reduces the magnitude of the crystal field splitting by a geometrical factor equal to 2/3, and the fact that there are only four ligands instead of six reduces the ligand field by another 2/3. The total splitting in the tetrahedral case is thus approximately $2/3 \times 2/3 = 4/9$ of the octahedral splitting. In theory, alternative electron configurations are possible for the cases d^3, d^4, d^5 and d^6 in the tetrahedral field but the gain in CFSE is reduced by the smaller size of the splitting, and the loss of exchange energy is never counterbalanced. Thus

> The value of 4/9 assumes the bond lengths remain the same. For the smaller coordination number, the bonds should be shorter as there is less ligand–ligand repulsion.

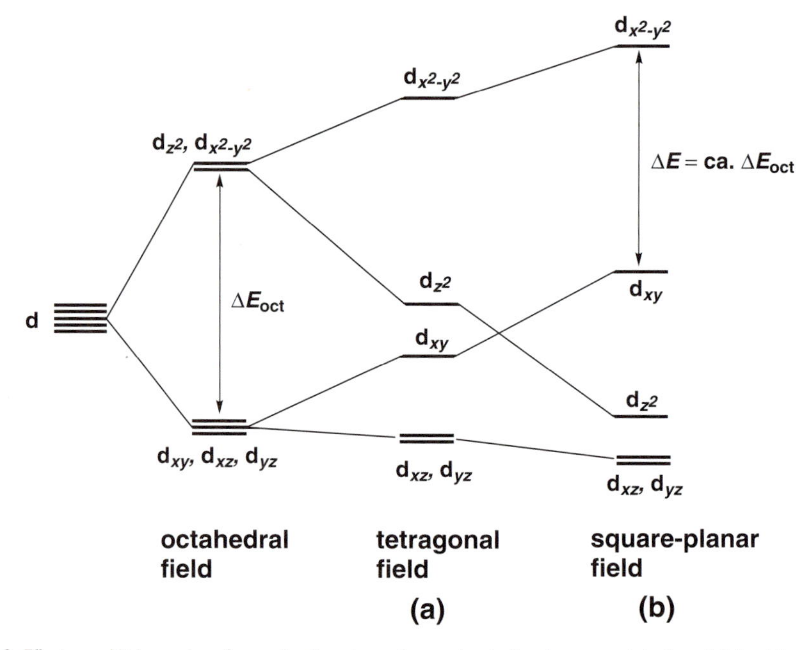

FIG. 13.8 Effect on orbital energies of removing ligands on the z axis, starting from an octahedron: (a) the tetragonal field with nonequivalent z ligands, (b) the square-planar field with no z ligands

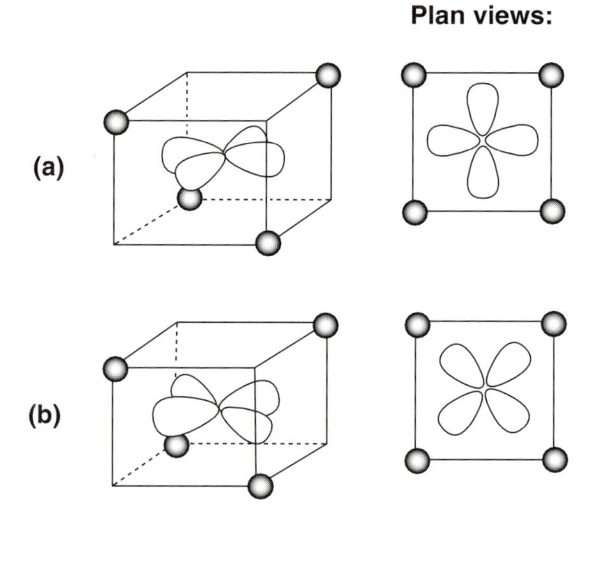

FIG. 13.9 Alignment of the d orbitals and the ligands in a tetrahedral complex: (a) the $d_{x^2-y^2}$ orbital, (b) the d_{xy} orbital

all tetrahedral complexes are high spin. The CFSE values for these high-spin configurations are given in Table 13.10.

If Table 13.10 is compared with Table 13.7, it will be seen that the CFSE in an octahedral configuration is always greater than that in the corresponding tetrahedral configuration, except in the cases of d^0, d^5 and d^{10} where both values are zero. The configurations with the next smallest CFSE loss in the tetrahedral field compared to the octahedral field are d^1 and d^6 if the octahedral state is high spin. The adoption of 4-coordination rather than 6-coordination is expected, in general, to be accompanied by loss of energy of formation as only four interactions occur instead of six. In addition, there is commonly a loss of CFSE as well.

TABLE 13.10 CFSE values in a tetrahedral field

Number of d electrons	Electronic configuration					CFSE	
	e		t_2			(as ΔE_{tetr})	(as ΔE_{oct}) (assuming $\Delta E_{tetr} = 4/9\ \Delta E_{oct}$)
1	↑					−3/5	−0.27
2	↑	↑				−6/5	−0.53
3	↑	↑	↑			−6/5+2/5	−0.36
4	↑	↑	↑	↑		−6/5+4/5	−0.18
5	↑	↑	↑	↑	↑	−6/5+6/5	−0.00
6	↑↓	↑	↑	↑	↑	−9/5+6/5	−0.27
7	↑↓	↑↓	↑	↑	↑	−12/5+6/5	−0.53
8	↑↓	↑↓	↑↓	↑	↑	−12/5+8/5	−0.36
9	↑↓	↑↓	↑↓	↑↓	↑	−12/5+10/5	−0.18
10	↑↓	↑↓	↑↓	↑↓	↑↓	−12/5+12/5	−0.00

These considerations, and the data for the CFSE values, allow a prediction of the most probable cases in which 4-coordination will be found for transition elements of the first series. The tetrahedral configuration is expected to be unfavourable compared with the octahedral one, except in the case of large ligands with low positions in the spectrochemical series, or in the ions with 0, (1), 5, (6) or 10 d electrons. The larger ligands will experience steric hindrance to formation of 6-coordinated complexes and the interactions will also be reduced due to the increased metal–ligand distances. The low position in the spectrochemical series ensures that any loss of CFSE is not too serious due to the low intrinsic value of ΔE (similarly, a low charge on the transition metal ion affects the ΔE value so divalent ions should form tetrahedral complexes more readily than trivalent ones, and *a fortiori* for lower charges). Finally, the d configurations listed are those with lowest CFSE differences. In practice, tetrahedral complexes are typically formed by the halides (except fluoride) and related ligands.

The case of the square-planar configuration is rather different. Because the bond distances in the *xy* plane are essentially the same in octahedral and square-planar configurations, steric effects are negligible and the increase in attractions due to forming six bonds rather than four will normally overwhelmingly favour the regular octahedral complex. Changes in CFSE for low numbers of d electrons either favour the octahedral case or are small. Consider, however, the case of d^8. In an octahedral complex, there are two electrons in the e_g level, while the square-planar configuration allows these to be paired in the d_{xy} orbital (Fig. 13.10) with a gain in CFSE of about $2\Delta E_{oct}$. (This is only approximate as the lower levels in the square-planar case do not match the octahedral t_{2g} levels, but they are quite close.) This is offset by a reduction in exchange energy from the $13K$ of the octahedral case to $12K$ for two sets of four electrons. However, if ΔE is sufficiently large, it is possible for the gain in CFSE to overbalance the loss in bonding interactions and the small loss in exchange energy. It is found that square-planar complexes are indeed formed by d^8 ions with ligands to the right of the spectrochemical series: for example Ni^{2+} forms a square-planar cyanide complex, $Ni(CN)_4^{2-}$, while its hydrate or ammine are octahedral, e.g. $Ni(NH_3)_6^{2+}$. The larger ΔE values of heavier elements or of more highly charged elements extend the scope of formation of square complexes. Thus all complexes, even halides, of platinum(II) and gold(III) (both d^8) are square-planar.

This extra CFSE found in square-planar complexes of d^8 elements also occurs for the d^7 and d^9 configurations. However, the CFSE gain is only ΔE. Table 13.11 lists the species which typically form square-planar complexes. While such square complexes of the first row elements are comparatively rare, those formed by heavy transition elements are the majority of the representatives of these oxidation states, especially for the d^8 configurations.

In configurations other than d^8, the CFSE gain relative to octahedral is small, and most of these configurations are distorted in the octahedral case, so it is often difficult to decide what

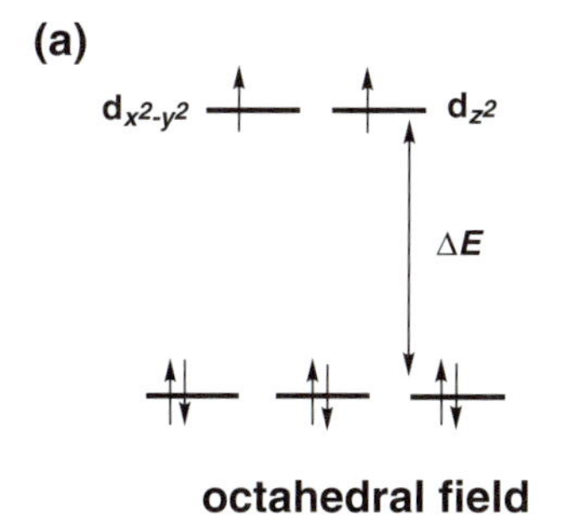

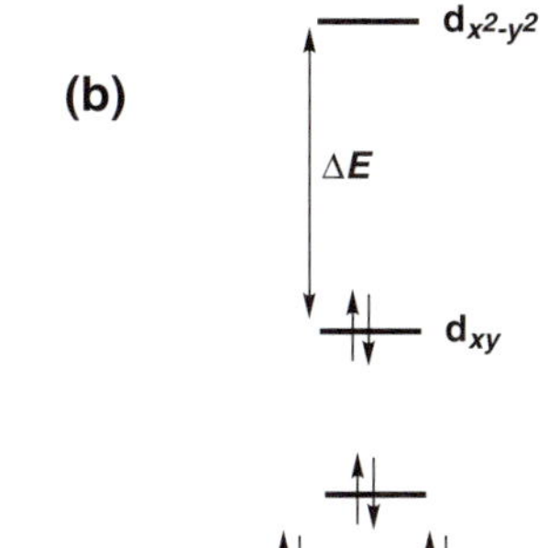

FIG. 13.10 The configuration of electrons in a d complex: (a) octahedral and (b) square-planar.

TABLE 13.11 Species forming square-planar complexes

d electron configuration	*Species*			*Approx. CFSE*	*Unpaired electrons*
d^8	Ni(II)	Pd(II) Rh(I)	Pt(II) Ir(I) Au(III)	$2\Delta E_{oct}$	0
d^9	Cu(II)	Ag(II)		ΔE_{oct}	1 (in $d_{x^2-y^2}$)
d^7	Co(II)			ΔE_{oct}	1 (in d_{xy})
d^6	Fe(II)			$\frac{1}{2}\Delta E_{oct}$	2
d^4	Cr(II)			$\frac{1}{2}\Delta E_{oct}$	4

The Cr(II) spin corresponds to one electron in each of the four stable orbitals, and the Fe(II) value indicates two filled and two half-filled orbitals. The CFSE are with respect to the octahedral configuration, those for d^4 and d^6 being for the weak field configuration.

has happened. For example, copper(II) compounds often show four short bonds in a square plane with two longer ones, or even three sets of pairs of bonds with different lengths.

The distortion mentioned in the last paragraph arises whenever the d_{z^2} and $d_{x^2-y^2}$ orbitals are unequally occupied. If, for example, there is one electron in the d_{z^2} orbital, ligands on the z axis are more shielded from the nuclear field than are ligands on the x and y axes. The ligand–metal distances in the z direction are therefore shorter than those in the xy plane. If the electron is, instead, in the $d_{x^2-y^2}$ orbital the four distances in the xy plane are shorter. Such distortions, which are less simple than described here, are one manifestation of the Jahn–Teller theorem, which states that if a system has unequally occupied, degenerate energy levels it will so distort as to raise the degeneracy. Cases where distortions are expected are d^4 high spin, d^7 low spin and d^9. Distortions involving t_{2g} levels are normally too small to be detected.

13.5 Stable configurations

With all the factors discussed in the last three sections in mind, it is possible to make some general remarks about the stabilities of various d configurations. These apply particularly to the first row elements which only occasionally show coordination numbers other than four or six (see Section 13.6). The extent to which these apply in practice will become clear when the chemistry of the individual elements is discussed.

The traditionally stable configurations of the empty, half-filled and filled shells should still be stable for d elements, the stability in the last two cases stemming from the high exchange energy as well as from the general symmetry of the electron clouds. In octahedral environments, d^5 will be unusually stable only in the high-spin arrangement. Hence, it should be more stable in the divalent ion Mn^{2+} than in the trivalent d^5 element, Fe^{3+}, and also more stable with ligands of relatively low field in each case. The configurations d^0 and d^{10} may well be found in stable tetrahedral complexes.

In the octahedral configuration, the states d^3 and low-spin d^6 are also expected to have special stability as they correspond to half-filled and filled t_{2g} orbitals. In this case, stability increases with increasing ΔE and should be most marked for trivalent ions, here Cr^{3+} (d^3) and Co^{3+} (d^6).

It will be noted that the d^4 configuration in an octahedral environment lies between two others that are especially stable, so that d^4 species are expected to be unstable, both to oxidation to d^3 and to reduction to d^5. In particular, the M^{2+} d^4 state should readily oxidize to the M^{3+} d^3 ion where the increase in charge increases ΔE and the CFSE, while the M^{3+} d^4 ion should be readily reduced to the M^{2+} d^5 ion which is favoured by the reduced ΔE. Such behaviour is indeed found, the ions in question being respectively Cr^{2+} and Mn^{3+}.

At the d^8 configuration, there is competition between the square-planar and the octahedral arrangements. In nickel(II) the latter is more common but the former is dominant for Pd(II), Pt(II) and Au(III). In both configurations, d^8 tends to be more stable than either d^7 or d^9 which have neither the CFSE of square-planar d^8 nor the equally occupied e_g arrangement of octahedral d^8. There is some tendency for d^9 to go to d^{10} but a number of factors come into play here as the oxidation states of the copper Group elements show. It might be noted that d^{10} species have some tendency to linear configurations as Cu(I) and Au(I), as well as Hg(II), show. This is in addition to their ready adoption of tetrahedral structures, and the 2-coordinate species are found for relatively low charge densities on the ions, either in large ions or in M^+ species.

The discussion is summarized in Table 13.12.

TABLE 13.12 General stabilities of d configurations

Number of d electrons	*Comments*
0	Stable in earlier Groups as the Group oxidation state
1, 2	Tend to oxidize to d^0, e.g. Ti^{3+}, Ti^{2+}, V^{3+}
3	Stable, especially for trivalent ion, e.g. Cr^{3+}
4	Unstable, e.g. Cr^{2+} oxidizes to Cr^{3+}; Mn^{3+} reduces to Mn^{2+}
5	Stable, especially Mn^{2+}; Fe^{3+} is also relatively stable
6	Stable in spin-paired state, e.g. Co^{2+} (except in hydrate); Fe^{2+} is also relatively stable. Pt(IV) very stable
7	Relatively unstable, Co^{3+} oxidizes except as hydrate or halide complex; Ni^{3+} is scarcely known
8	Stable, Ni^{2+} as octahedral and square-planar complexes; Pd(II), Pt(II), Au(III) very stable square-planar species
9	Relatively unstable except in case of Cu(II)
10	Stable, e.g. Ag^+; zinc Group never show states above II

13.6 Coordination numbers other than four or six

While the majority of compounds of the first row elements exhibit octahedral, tetrahedral or square-planar coordination, there are now well-established examples of a range of other coordinations. For the heavier transition elements, octahedral and square-planar configurations are common, but higher coordination numbers are well known. An alternative form of

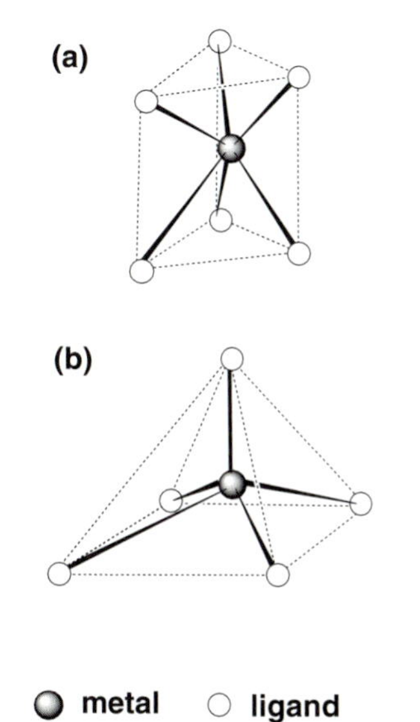

FIG. 13.11 Lower-symmetry coordinations: (a) trigonal prism form of 6-coordination, (b) square pyramid form of 5-coordination. Trigonal prismatic coordination is often found when the donor atoms are sulfurs, as in MoS_2 or in the dithiolene complexes $M(S_2C_2R_2)_3$ where M=V, Mo, Re. In the square pyramid, note that the metal atom lies above the base plane, making the base-L/metal/apex-L angle greater than 90°

Coordination number 2. This is found in d^{10} configurations such as copper (I), silver(I), gold(I) and mercury(II). The arrangement is always linear, as in $Ag(NH_3)_2^+$ or $AuCl_2^-$, and is simply that derived from the electron pair repulsion theory.

Coordination numbers greater than six are most usually found among the compounds of the second or third row transition elements, and also for the lanthanides and actinides as illustrated in Chapters 11 and 12. For these higher coordination numbers, it is usual to find several alternative shapes, and the energy differences between them are commonly very small. There is at present no generally accepted, unambiguous, method available for predicting which shape will be adopted, either for the higher coordination numbers or for 5-coordination.

6-coordination, the trigonal prism of Fig. 13.11, is also found but is rare. It is the configuration of arsenic in NiAs (Fig. 5.10c) and is found for molybdenum in MoS_2.

Coordination number 5. Five-coordination is known for most of the first row elements, but is uncommon for the heavier transition metals. As well as the trigonal bipyramid (Fig. 4.4), square-pyramidal 5-coordination is found for transition metal compounds. As the discussion of five electron pairs in Chapter 4 showed, there is little energy difference between the trigonal bipyramid and square pyramid, and the balance of stability is readily tipped one way or the other by additional stabilizations arising from the d electron configuration, the shape and properties of the ligands, or from interactions in the solids. The square pyramid configuration has the central metal atom raised *above* the plane of the four base atoms (Fig. 13.11). Notice that this increases the bond angles at the central atom, and is the opposite distortion from that found in AB_5L species, like BrF_5, derived from the octahedron, where the repulsion of the lone pair decreases BAB angles so that the central atom lies *below* the base plane. (Compare BrF_5 with Fig. 14.13.)

The small differences in energy between the two shapes is strikingly illustrated by the compound

$$(Cr(en)_3)[Ni(CN)_5)]\tfrac{3}{2}H_2O.$$

whose crystal structure contains both square pyramidal and distorted trigonal bipyramidal $Ni(CN)_5^{3-}$ ions. Table 13.13 gives some illustrative examples of 5-coordinate complexes, chosen either for their relative simplicity or because a family of compounds of one type exist. In some of the structures, distortions from the ideal shapes occur, while many other cases are known where the structure is intermediate between the two geometries.

Coordination number 7. Seven-coordinate complexes are commonly found in one of three shapes. The most symmetric is the regular pentagonal bipyramid adopted by ReF_7, OsF_7, by the ions MF_7^{3-} (M = Zr, Hf) in their sodium salts, and also by the main group compound IF_7. Another example is the ytterbium complex $Yb(TePh)_2(pyridine)_5$ whose structure is given in Fig. 11.6b. As the in-plane angle of a regular pentagonal bipyramid is only 72°, some distortion of the five equatorial ligands is likely to reduce steric interactions,

TABLE 13.13 Examples of 5-coordination

	Trigonal bipyramid	*Square pyramid*
Titanium(IV)	$TiCl_5^-$, $TiOCl_2(NMe_3)_2$	Y_2TiO_5
(III)	A	
Vanadium(IV)	$VOCl_2(NMe_3)_2$	$VO(acac)_2$ (Fig. 14.13)
(III)	A	
Niobium(V)	$NbCl_5$ (in gas phase only)	
Tantalum(V)	$TaCl_5$ (in gas phase only)	
Chromium(III)	A	
(II)	B	
Molybdenum(V)	$MoCl_5$ (in gas phase only)	$Mo_3O_{10}^{5-}$ includes Mo^VO_5 units
Manganese(II)	B	C
Rhenium(V)	$ReOX_4^-$ (X = Cl, Br, I)	
Iron(0)	$Fe(CO)_5$	
(II)	B	C
(III)		$Fe(S_2CNEt_2)_2Cl$
Cobalt(II)	B	C
Nickel(II)	$Ni(CN)_5^{3-}$ (a), B	$Ni(CN)_5^{3-}$ (a)
Platinum(II)	$Pt(SnCl_3)_5^{3-}$	
Copper(II)	B, $Cu(bipy)_2X$	$[Cu(NO_3)_2(C_5H_5NO)_2]_2$, $Cu(acac)_2$ quinoline
Zinc(II)	B	C, see also Fig. 13.11

A = $M^{III}X_3L_2$ (X = halogen, L = ligand like NMe_3).
B = $[M^{II}(\text{tetra N})Br]^+$ where tetra N = $(Me_2NCH_2CH_2)_3N$, compare Appendix B.
C = $(terpy)M^{II}Cl_2$ (terpy = terpyridyl, compare Appendix B).
(a) Pentacyanonickel(II) is found in both shapes in the $Cr(en)_3^{3+}$ compound.

either by buckling out of the plane or by variation of the M-F bond lengths in the equatorial plane. Some uranyl compounds, such as $UO_2F_5^{3-}$, also show this shape as the UO_2 unit has to be linear and the other five substituents lie in a plane at right angles to this, around the uranium atom (compare Fig. 12.6 for the corresponding UO_2L_6 situation).

A second 7-coordinate arrangement found is that formed by inserting an extra substituent into the triangular face of an octahedron, and spreading out the three ligands forming this face. Such a positioning is termed *capping* the face. One example is $NbOF_6^{3-}$ illustrated in Fig. 15.5a (which emphasizes that the seventh ligand is on the three-fold axis).

The third shape found for 7-coordination is that obtained by inserting a substituent above one of the rectangular faces of a trigonal prism—a capped triangular prism. This structure is shown in the ammonium salt of ZrF_7^{3-} (Fig. 15.1) or by MF_7^{2-} (M = Nb, Ta). These structures are all very similar in energy and change from one to another is readily induced, for example by the change of counterion in the heptafluorozirconium complexes.

Coordination number 8. While the 8-coordinate shape of highest symmetry is the cube, this is common only in extended structures like the ionic CsCl or CaF_2 structures or in metallic body-centred cube forms. A smaller repulsion between ligands results when one face of the cube is twisted by 45° relative to the opposite one to form the square anti-prism. This structure is shown by TaF_8^{3-} (Fig. 15.5b) and ReF_8^{2-}, or by $Zr(acac)_4$ and $U(acac)_4$ (Fig. 12.5).

The other common 8-coordinate structure, the dodecahedron or bis-bisphenoid (bisdisphenoid), may also be derived from the cube. As every second corner of a cube defines a tetrahedron (as can be seen in Fig. 13.9), the cube may be regarded as two interpenetrating tetrahedra. If one of these tetrahedra is flattened and the other elongated, the 8-coordinate structure of Figs 15.2 or 15.18 results. A tetrahedron distorted in this way is called a bisphenoid, hence the name bisbisphenoid. If the distortions are equal and produce equilateral triangular faces, the structure is a regular dodecahedron, shown in Fig. 13.12 where the vertices labelled A define one bisphenoid, and those labelled B, the other. A number of structures approximate to a regular dodecahedron, for example the MF_8^{4-} species of Fig. 15.2, but quite distorted forms also occur. These result particularly from the presence of bidentate ligands (see Section 13.7) where the two donor atoms are close together. Thus $Co(NO_3)_4^{2-}$, with two oxygens bridging Co to each nitrate, forms a distorted dodecahedron with the Co-O bond lengths equal to 204 pm in the elongated bisphenoid and averaging 255 pm in the flattened one. A similar distorted structure is found in the chromium(V) peroxy compound $K_3Cr(OO)_4$ where both oxygens of the O-O group are bonded to Cr.

Higher coordination numbers. As a transition element of the d block has only nine valence shell orbitals (five d plus one s plus three p) the maximum coordination number is limited to nine unless metallic bonding (compare close-packed structures with coordination number 12, Chapter 5), purely electrostatic bonding, or electron deficient bonding (compare Section 9.6) occurs. One example is $Zr(BH_4)_4$ which forms two hydrogen bridges to each BH_4 group at room temperature, but transforms to a three-bridge structure at −160°C, $Zr(H)_3BH$. Thus the Zr is 12-coordinate in the low temperature form. If f orbitals are available, this restriction no longer holds as illustrated by the 10-coordinate $La(H_2O)_4$ EDTA compound mentioned in Chapter 11.

The 8-coordination of $UO_2(NO_3)_2.2H_2O$, Fig. 12.6 shows a fourth shape—the hexagonal bipyramid. This is only found for such MO_2 species where the O-M-O group has to be linear.

Coordination number 9. The one configuration found for 9-coordination is that formed by placing one ligand above each of the three rectangular faces of a trigonal prism. This structure is illustrated in Fig. 15.23 for ReH_9^{2-} and is also shown by the lanthanide hydrates like $Nd(H_2O)_9^{3+}$ mentioned in Chapter 11.

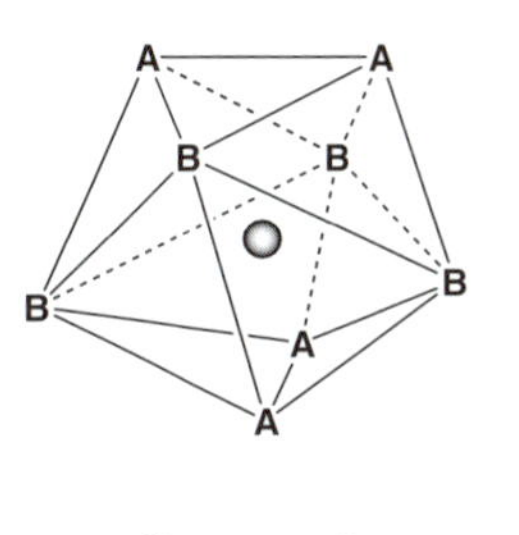

FIG. 13.12 The dodecahedron. This 8-coordinate shape is related to that of the cube by elongating one of the interpenetrating tetrahedra (A vertices) and flattening the other (B vertices). For the regular figure, all the faces are equilateral triangles

13.7 Effect of ligand on stability of complexes

The type of ligand has a distinct influence on the stabilities of complexes and this must be discussed in more detail. The spectrochemical series gives one classification of the ligands and some points about this should be noticed. In an extended form of the series, it is seen that position depends largely on the donor atom in the order:

$$\text{Halogen or S} < \text{O} < \text{N} < \text{P} < \text{C}$$

Thus, nitrate which donates through O is much weaker than nitrite which donates through N. An even more striking example is the thiocyanate group, SCN, which is sometimes found coordinated through S, when it has a weak ligand field, and sometimes coordinated through N, when it shows a much stronger field. A similar example is shown by the isomeric form of nitrite, nitrito, which coordinates as ONO, donating through O, instead of the NO_2 form of nitrite which donates through N. The nitrito form has a weaker ligand field than the nitrite form. These features govern the effect of the ligands by way of their contribution to the crystal field stabilization energy, but this term may be only a minor contribution to the overall stabilization, and it is necessary to take care not to equate strong crystal field unthinkingly with stability.

In terms of the Lewis nomenclature where a lone pair donor is a base and an acceptor is an acid, the metal ions are classified as hard or soft acids, and the ligands as hard or soft bases; see Section 2.16.

A number of attempts have been made to find general relationships which indicate overall stability. One of the most extensive is Pearson's classification into *hard and soft* ligands and metal ions. In this, a ligand is a hard base if it is nonpolarizable as for most ligands with a first row donor atom, and the ligand is a soft base if it is polarizable, as with sulfur or phosphorus ligands. Similarly, a metal ion is a soft acid if it has easily polarizable electrons, or is large, or has a low charge, while a hard acid is a metal ion of high charge, small size and with valence electrons which are not polarizable. The classification extends beyond transition elements and some examples of hard and soft ligands and metal ions are given in Table 13.14.

The most important generalization about stabilities is then that soft ligands form stable complexes with soft metal ions, and hard ligand–hard ion complexes are also stable. Mixtures of hard ion–soft ligand or soft ion–hard ligand are less stable. A second general trend is that the substitution of a ligand of a particular type tends to make the metal ion behave more as that type of acid. Thus a borderline ion like Fe^{2+} becomes harder when a number of hard ligands are coordinated so that the intermediate $Fe(H_2O)_3^{2+}$ (say) is more likely to add further water molecules than would a ferrous ion coordinated by soft ligands.

This hard–soft classification has been greatly extended and ramified, and some features are still subject to controversy. However, the basic predictions derived from the simple version are most useful. One example is provided by thiocyanate, SCN^-, which can bond through S or through N (in isothiocyanates). The S-bonded form is a soft ligand, and is preferred in complexes of soft ions such as Hg^{2+}. The N-bonded form is harder and is most usually found in first row complexes. A neat example is $HgCo(NCS)_4$ which is an extended lattice formed of Hg-SCN-Co links.

Chelate comes from the Greek word for claw.

The next effect is the *chelate effect*. If a ligand contains more than one donor atom it may coordinate to more than one position on the cation giving ring formations. The existence of such chelation in a complex is accompanied by increased stability and this is shown by increased heats of formation and resistance to substitution, and also by the higher position in the spectrochemical series of the chelating ligand as compared to an analogous nonchelating ligand. An example is provided by the case of ethylenediamine, $H_2NCH_2CH_2NH_2$, where both nitrogen atoms can donate, forming a five-membered ring. Such a reagent with two donor atoms is termed *bidentate*. It is observed that the treatment of ammonia complexes with ethylenediamine results in the displacement of the ammonia:

(en = ethylenediamine)

$$Co(NH_3)_6^{3+} + 3\,en \rightarrow Co(en)_3^{3+} + 6NH_3$$

The driving force of such a replacement is probably the entropy change. The equation above has four particles on the left hand side and seven on the right, so that reaction proceeds with increase of entropy. For example, the change in free energy in the copper complexes is:

TABLE 13.14 Classification of ligands and ions under hard/soft formalism

Ligands or bases	
Hard	H_2O, ROH, R_2O, OH^-, OR^-, NO_3^-, RCO_2^-, SO_4^{2-}, CO_3^{2-}, $C_2O_4^{2-}$, PO_4^{3-} (donors through O)
	NH_3, NR_3, NHR_2, NH_2R, Cl^-
Soft	R_2S, RSH, RS^-, SCN^-, $S_2O_3^{2-}$ (S donors) R_3P, R_3As, I^-, CN^-, H^-, R^-
Borderline	py, Br^-, N_3^-, SO_3^{2-}, NO_2^-
Metal ions or acids (charges are formal only)	
Hard	Mn^{2+}, Cr^{3+}, Fe^{3+}, Co^{3+}, Ti^{4+}, VO_2^+, VO^{2+}, Zr^{4+}, MoO^{3+}
	H^+, all s element ions, M^{3+} for M = Al, Ga, In, Sc, Y, Ln
Soft	Cu^+, Ag^+, Au^+, Hg_2^{2+}, Hg^{2+}, Pd^{2+}, Pt^{2+}, Pt^{4+}
	Tl^+, Tl^{3+}
Borderline	Fe^{2+}, Co^{2+}, Ni^{2+}, Cu^{2+}, Zn^{2+}, Ru^{2+}, Os^{2+}, Rh^{3+}, Ir^{3+}
	Sn^{2+}, Pb^{2+}, Sb^{3+}, Bi^{3+}

(Recall that $\Delta G = \Delta H - T\Delta S$)

$$Cu(en)_2^{2+} - Cu(NH_3)_4^{2+}$$

$$\text{difference in } \Delta G = 18.0\,\text{kJ mol}^{-1}$$

$$\text{difference in } \Delta H = 10.9\,\text{kJ mol}^{-1}$$

$$\text{difference in } T\Delta S = -7.1\,\text{kJ mol}^{-1}$$

This entropy term thus provides about 40% of the free energy change, whereas the substitution of one monodentate ligand by another, for example the exchange of ammonia for water, usually has only a small entropy effect which is commonly neglected in calculations. The optimum ring size for elements of the transition series appears to be a five-membered ring. Thus the substitution of 1,3-propane-diamine for ethylenediamine, increasing the ring size to six atoms (Fig. 13.13b), results in a loss of free energy of formation of 16.3 kJ mol^{-1} in the case of the copper complex above and 35.6 kJ mol^{-1} in the case of the nickel complex Ni(diamine)$_3^{2+}$.

The chelating effect is increased still further if the chelating molecule has more donor atoms. The triamine (trien) $H_2NCH_2CH_2NHCH_2CH_2NH_2$ forms an even more stable complex than ethylenediamine. Chelating agents with up to six donor atoms are available, an example of the latter being ethylenediaminetetra-acetic acid, EDTA, shown in Fig. 10.5. The common complexing and precipitating agents of analysis are nearly all chelating agents which form five- or six-membered rings with the metal atom which is being determined. Some examples are shown in Fig. 13.13. Polydentate chelating agents may give rise to complexes of unusual coordination, for example, the less common 5-coordination in the zinc complex shown in Fig. 13.14.

Quantitative indication of the process of forming a complex comes from the evaluation of the stability constants which characterize the equilibria corresponding to the successive addition of ligands. That is, we consider the steps

$$M + L \rightleftharpoons ML$$

$$ML + L \rightleftharpoons ML_2$$

and so on down to

$$ML_{n-1} + L \rightleftharpoons ML_n$$

These are characterized by equilibrium constants $K_1, K_2 \ldots, K_n$ such that

$$K_1 = [ML]/[M][L]$$

$$K_2 = [ML_2]/[ML][L]$$

and,

$$K_n = [ML_n]/[ML_{n-1}][L]$$

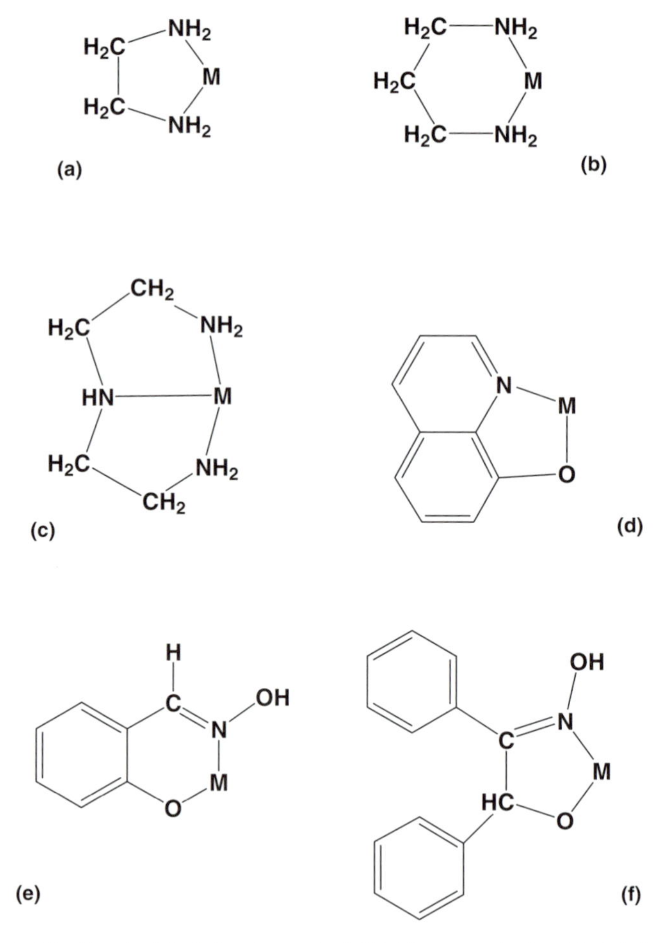

FIG. 13.13 Some chelating ligands: (a) ethylenediamine (en), (b) 1,3-propanediamine, (c) diethylenetriamine (trien), (d) 8-hydroxyquinoline (oxine), (e) salicylaldoxime, (f) α-benzoinoxime (cupron)

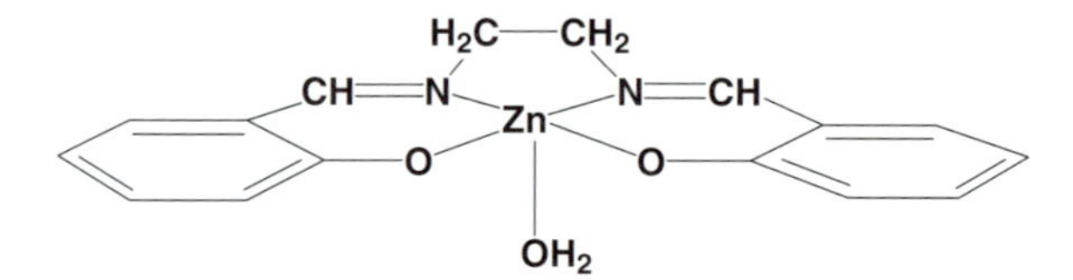

FIG. 13.14 A 5-coordinated zinc complex, *N,N'*-disalicylidene-ethylenediamine zinc hydrate. The quadridentate ligand coordinates through its two nitrogen atoms and two oxygen atoms to the zinc, giving a structure of a shallow square pyramid, with the zinc atom 34 pm above the NNOO plane. The water molecule is coordinated on the opposite side of the zinc with Zn-OH_2 bond in the direction of the axis of the pyramid

These constants K, are termed *stepwise formation constants*. An alternative formulation is to consider the overall formation reaction

$$\mathrm{M} + n\mathrm{L} \rightleftharpoons \mathrm{ML}_n$$

characterized by the nth *overall formation constant* β_n.

$$\beta_n = [\mathrm{ML}_n]/[\mathrm{M}][\mathrm{L}]^n = K_1K_2 \ldots K_n$$

In most preparations, the complex is formed in aqueous solution, and the stability constants refer to steps where L replaces coordinated water. The evaluation of formation constants usually calls for a good deal of experimental ingenuity and will not be discussed here. We may simply note some values and their interpretation. For example, the logarithms of the successive formation constants of various nitrogen complexes formed in aqueous solution are shown in Table 13.15.

TABLE 13.15 Stepwise formation constants of some nitrogen complexes

Ligand		Co^{2+}	Ni^{2+}	Cu^{2+}	Zn^{2+}
NH_3	$\log K_1$	2.1	2.8	4.2	2.4
	$\text{Log } K_2$	1.6	2.2	3.5	2.4
	$\log K_3$	1.1	1.7	2.9	2.5
	$\log K_4$	0.8	1.2	2.1	2.2
	$\log K_5$	0.2	0.8	−0.5	
	$\log K_6$	−0.6	0.03		
en	$\log K_1$	6.0	7.5	10.6	5.7
	$\log K_2$	4.8	6.3	9.1	4.7
	$\log K_3$	3.1	4.3	−1.0	1.7
trien	$\log K_1$	8.1	10.7	16.0	8.9
6-en	$\log K_1$	15.8	19.3	22.4	16.2

en = ethylenediamine (Fig 13.13a): trien = diethylenetriamine (Fig 13.13c).
6-en = pentaethylenehexamine, the analogous 6-nitrogen molecule (compare Appendix B)

The values in Table 13.15 illustrate a number of features which give a quantitative indication of the stability properties discussed above. There is first a general tendency for K values to fall as the number of ligands increases. This is probably due, at least in part, to the statistical effect that the number of sites for substitutions is reduced as substitution proceeds.

The values for nickel show a steady progression to $Ni(NH_3)_6^{2+}$ and zinc, similarly, goes steadily to $Zn(NH_3)_4^{2+}$. The cobalt values show that the sixth ligand is unstable, probably reflecting the effect of the e_g electron, and this is even more marked for copper where the fifth NH_3 is only added in presence of a large excess of ammonia and the sixth is not taken up at all. This reflects the general tendency of the d^9 ions to form four strong bonds and two, much longer, weaker ones.

It will also be noted, comparing K_1 values for example, that the stability order of M^{2+} ions increases to copper. This is part of a more extensive sequence, the Irving–Williams order, which shows that stability constants vary

$$\mathrm{Mn(II)} < \mathrm{Fe(II)} < \mathrm{Co(II)} < \mathrm{Ni(II)} < \mathrm{Cu(II)} > \mathrm{Zn(II)}$$

towards a particular ligand.

The values also illustrate the effect of chelation on the stabilities. If we compare $\log \beta_2$ for NH_3 with $\log K_1$ for $H_2NCH_2CH_2NH_2$, we are comparing values where two N atoms are coordinated in each case. As $\beta_2 = K_1K_2$, $\log \beta_2 = \log K_1 + \log K_2$. It will be seen that $\log K_1$ for the ethylenediamine complexes is uniformly greater than $\log \beta_2$ for ammonia, and likewise $\log K_2$ and $\log K_3$ for en are greater than the corresponding $\log \beta_4$ and $\log \beta_6$ values for NH_3. Similarly, the tridentate chelate, trien, gives a value of $\log K_1$ which exceeds $\log \beta_3$ for the analogous NH_3 species. Finally, the hexadentate ligand pentaethylenehexamine has a value of $\log K_1$ (coordination of all six nitrogens) which is greater than $\log \beta_3$ for the $M(en)_3^{2+}$ species and than $\log \beta_6$ for $M(NH_3)_6^{2+}$. It is also noteworthy that, while copper and zinc fail to become 6-coordinated to ammonia, they do form 6-coordination to nitrogen when the ligand is a chelating amine.

The effect of chelate ring size is illustrated by the values for 3-propanediamine (Fig. 13.13b) which has $\log K_1 = 0.0$ and $\log K_2 = 7.2$ for the copper compounds.

Anomalous changes in stability constants may often indicate electronic changes. One example is provided by the Cr(II) dipyridyl values where $\log K_1 = 4.5$, $\log K_2$ is 6.0, instead of the expected decrease, and $\log K_3$ is 3.5. In this case, the d^4 Cr^{2+} species is high spin in its hydrate, and low spin in $Cr(dipy)_3^{2+}$. The anomalous value for K_2 suggests that spin pairing occurs as the second dipyridyl is added.

A full set of stability constants is to be found in *SI Chemical Data* (see references).

The properties of the ligands affect the replacement reactions which they undergo. It has already been seen that a chelating ligand will replace a monodentate ligand with the same donor atom. Such chelating ligands also commonly come higher in the spectrochemical series than the corresponding simple one: en lies above ammonia, for example. However, care must be taken, in general, to avoid equating high ligand field strength with high energy of formation or high stability of a complex. Obviously, in all configurations where there is some ligand field stabilization energy, the size of the field of a given ligand is an important factor in its reactions, but the CFSE varies with the metal atom, as has been seen. It is safe to expect a general correlation between stability and ligand field strength, for example cyanide is expected to replace water in general, but the detailed behaviour in any one case is a balance of different effects.

It is all too easy to unthinkingly equate high ligand field with stability so it is important to remember that ligand field energies are only one contribution to the total energy change but the hydride complexes (see box) give warning. Kinetic as well as thermodynamic effects occur. Thus $Co(NH_3)_6^{3+}$ is prepared in water and is stable indefinitely to exchange of the coordinated ammonia for water, while $Cr(NH_3)_6^{3+}$ can only be prepared in liquid ammonia and rapidly exchanges the ammonia groups for water when dissolved in water.

ELIMINATION OF STABLE ENTITIES

A most marked example is the case of transition metal complexes containing metal–hydrogen bonds, such as $(R_3P)_2PtH_2$. Hydrogen lies high in the spectrochemical series with ΔE values approaching those of cyanide, yet many of its compounds with the transition metals are unstable, and isolable compounds containing σ-M-H bonds (as opposed to metallic bonds in nonstoichiometric hydrides) are found only in the presence of a few stabilizing ligands. This instability may arise from the fact that, when metal–hydrogen bonds dissociate, hydrogen atoms are produced. These are very reactive and combine readily with each other to form H_2 with a high heat of formation. Similar effects are found with other complexes, such as those containing alkyls as ligands. By contrast, when the normal metal–ligand bond dissociates, stable lone pair entities like water, ammonia or halide ions are produced and the metal–ligand bond can reform in equilibrium with small quantities of ligand.

13.8 Isomerism

The use of polydentate chelating agents aids the study of *stereoisomerism* in complexes. In octahedral complexes, both optical and geometrical isomerism is possible, as illustrated by the ethylenediamine complexes of cobalt in Fig. 13.15. The existence of optical isomers may be shown by optical rotation measurements and by the resolution of isomers in favourable cases. The scheme for the resolution of the tris-ethylenediamine complex of cobalt(III) is shown in Fig. 13.16, making use of naturally occurring *d*-tartaric acid to separate the isomers. The existence of geometrical isomers is generally noticed when two compounds of different properties are found to have identical molecular formulae. In favourable cases the identity of the isomers may be determined by experiment, as in the separation of the optically active forms of the *cis*-$Co(en)_2X_2^+$ compound shown in Fig. 13.15. It is also possible to distinguish geometrical isomers in some cases by substitution reactions with a bidentate ligand. Such a ligand can replace two monodentate ligands lying *cis* but not two lying *trans*, as they are too far apart. Fig. 13.17 shows an example of such a reaction scheme. It must be noted that such proofs of configuration by reaction are not fully definite, as change of configuration may occur during reaction. The final proofs depend on structural evidence, usually from crystallography, although other techniques such as nmr may be useful on occasion.

Stereoisomerism is not confined to octahedral coordination, of course. Fig. 13.18 illustrates a classic case of isomerism among square-planar platinum complexes.

Other types of isomerism occur among complexes. The case of linkage isomerism in which a ligand can coordinate through one or other of alternative atoms has been exemplified above by the cases of nitro–nitrito and thiocyanate–isothiocyanate. The nitro–nitrito isomerization occurs in the same compound, for example both forms of the cobalt-pentammine are known, Fig. 13.19. Another example is provided by the NCS group, which is found bonded through either N or S in certain compounds of palladium, platinum and copper. A good example is provided by the series $LCu(NCS)_2$ which is brown, LCu(NCS)(SCN) which is yellow-green, and $LCu(SCN)_2$ which is deep green. Here, L may be one of a number of bidentate nitrogen ligands.

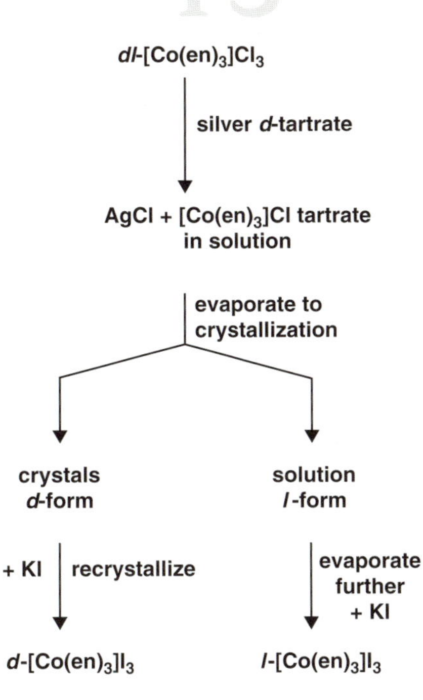

FIG. 13.16 Resolution into optical isomers of $Co(en)_3I_3$

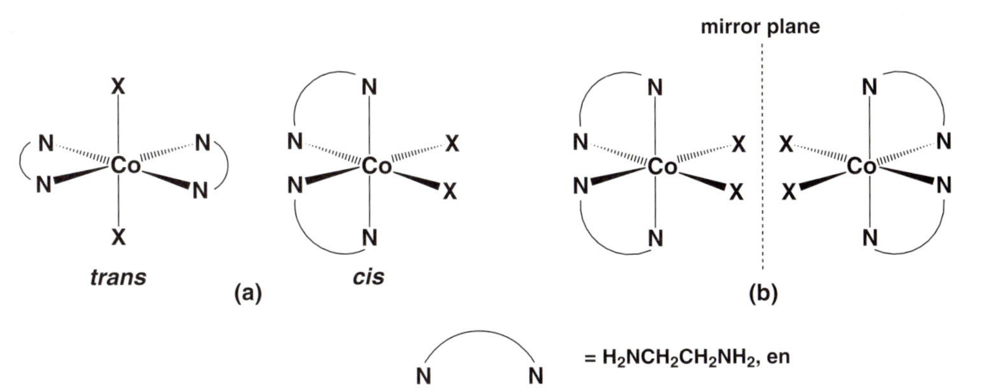

FIG. 13.15 Optical and geometrical isomerism in octahedral complexes: the example of $Co(en)_2X_2^+$. Fig. (a) shows the geometrical isomers, *cis* and *trans* forms, while Fig. (b) shows the non superimposable optical isomers of the *cis* form. The *trans* form has a plane of symmetry and is inactive

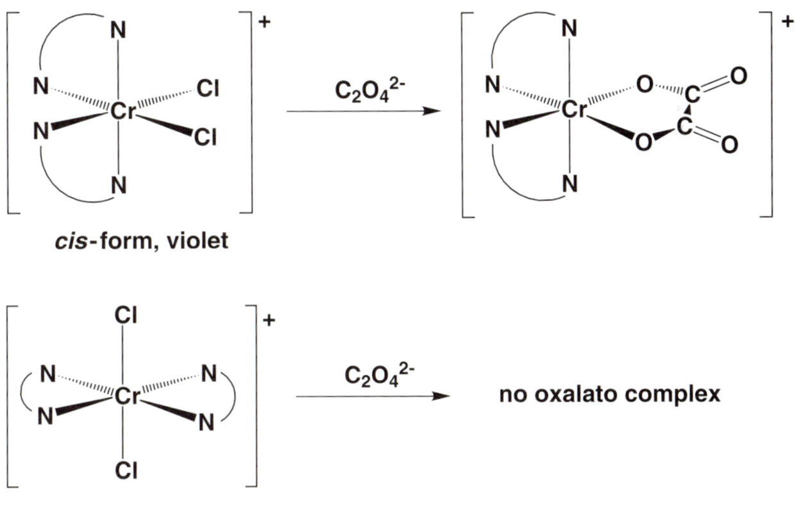

FIG. 13.17 Differentiating between geometrical isomers by a replacement reaction with a bidentate ligand

FURTHER TYPES OF ISOMERISM

A number of other types of isomer have been defined, though these are of less significance. Examples are given in Table 13.16. In type (a) a group present as a ligand in one isomer becomes the counter-ion in the other. In (b) the ligands are exchanged between the complex cation and complex anion while (c) involves dimerization—in the general case, polymerization—as well as ligand rearrangement.

TABLE 13.16 Further types of isomerism in complexes

(a) $[Co(NH_3)_4Cl_2]NO_2$ and $[Co(NH_3)_4Cl(NO_2)]Cl$
$[Pt(NH_3)_4Cl_2]Br_2$ and $[Pt(NH_3)_4Br_2]Cl_2$

(b) $[Co(NH_3)_6][Cr(CN)_6]$ and $[Cr(NH_3)_6][Co(CN)_6]$
$[Pt^{II}(NH_3)_4][Pt^{IV}Cl_6]$ and $[Pt^{IV}(NH_3)_4Cl_2][Pt^{II}Cl_4]$

(c) $[Pt(NH_3)_2Cl_2]$ and $[Pt(NH_3)_4][PtCl_4]$

13.9 Mechanisms of transition metal reactions

A number of different classes of reaction of transition element complexes exist and we review some of the better-established cases.

The simplest type of reaction is that of oxidation–reduction by *electron transfer*. For example, if a mixture of ferricyanide, $Fe(CN)_6^{3-}$, and ferrocyanide, $Fe(CN)_6^{4-}$ is made, and if an electron is lost from the $Fe(CN)_6^{4-}$ ion and gained by the $Fe(CN)_6^{3-}$ ion, an oxidation–reduction reaction has taken place, even though there is no change in the overall

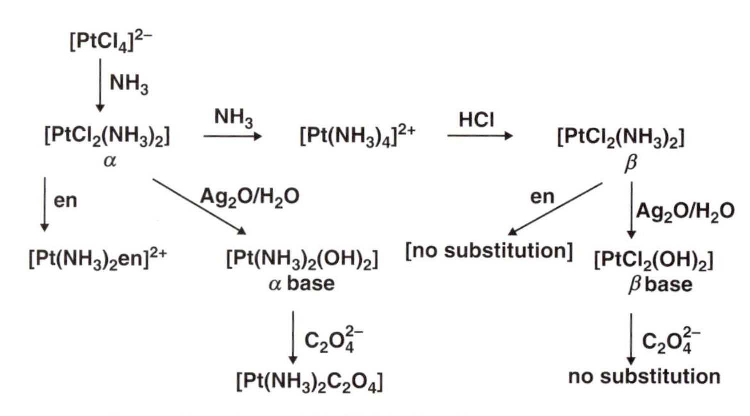

FIG. 13.18 Stereoisomerism in square-planar complexes of platinum

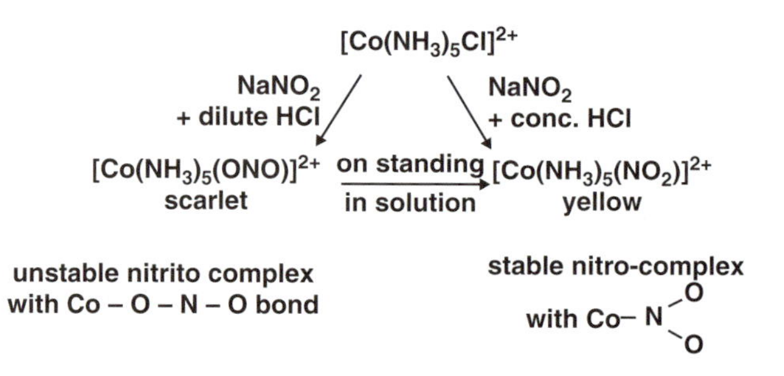

FIG. 13.19 Linkage isomerism: nitro- and nitrito-pentammine-cobalt(III)

composition of the mixture. The presence of such reactions is most readily demonstrated by making one of the components with a radioactive isotope. If the radioactivity is found, at a later time, to be spread between both components the reaction is indicated (star indicates the radioactive isotope)

$$^{*}\mathrm{Fe(CN)_6^{4-}} + \mathrm{Fe(CN)_6^{3-}} \rightleftharpoons {}^{*}\mathrm{Fe(CN)_6^{3-}} + \mathrm{Fe(CN)_6^{4-}}$$

As there is no net change in the reaction mixture, there is no heat change in such a reaction. There is, however, a requirement for activation energy. This is because the Fe-CN bond length in ferricyanide is shorter than that in ferroyanide. Thus the simple electron transfer between ions in their equilibrium configurations would produce ferrocyanide ions with compressed bonds, and ferricyanide ions with extended ones, that is the product ions would be vibrationally excited. The electron transfer takes place between matched ions, thus there is an activation energy required to stretch the bonds in ferricyanide and compress those in ferrocyanide to the intermediate matching configuration, by vibrational excitation. The matched anions must approach closely, but they do not have to be in actual contact for the electron transfer to occur.

Similar electron transfer reactions occur for a variety of matched ions, such as MnO_4^-/MnO_4^{2-} (permanganate/manganate), $IrCl_6^{2-}$/$IrCl_6^{3-}$ or $Mo(CN)_8^{3-}$/$Mo(CN)_8^{4-}$. The indications of the electron transfer mechanism are, first, that the reaction is second order (that is, the rate is a function of the concentration of each ion) and, second, the reactions are much faster than ones involving ligand exchange which are discussed below. Electron transfer reactions are favoured by the presence of ligands like cyanide or phenanthroline which allow electron delocalization from the metal through a conjugated system.

A slower group of electron transfer reactions is that exemplified by the $Co(H_2O)_6^{3+}$/$Co(H_2O)_6^{2+}$ system. Here, not only are the bond lengths different in the two oxidation states, but cobalt(III) is spin-paired with configuration t_{2g}^6 while cobalt(II) is high spin in the hydrate, configuration $t_{2g}^5e_g^2$. Thus electron transfer between the ground state configurations would yield excited electronic states in the products as shown by the first row of configurations below the equation:

$$^{*}Co(H_2O)_6^{3+} + Co(H_2O)_6^{2+} \rightleftharpoons {}^{*}Co(H_2O)_6^{2+} + Co(H_2O)_6^{3+}$$

ground state initial configurations

$$t_{2g}^6 \qquad t_{2g}^5e_g^2 \qquad t_{2g}^6e_g^1 \qquad t_{2g}^5e_g^1$$

pre-excitation needed

$$t_{2g}^5e_g^1 \qquad t_{2g}^6e_g^1 \qquad t_{2g}^5e_g^2 \qquad t_{2g}^6$$

To overcome this, a further excitation energy is needed to produce an excited electronic configuration, such as that shown in the second line below the equation, in addition to the vibrational excitation needed to adjust the bond lengths. Thus, electron transfer reactions which involve changes in spin pairing need a higher excitation energy, and are slower than those which require only the matching of bond lengths.

A related class of reactions is that where oxidation–reduction takes place by *electron transfer accompanied by transfer of a ligand*. In the transition state, the transferred ligand (sometimes in conjunction with other species) forms a link between the two metals, e.g. $L_5Cr\text{-}Cl\text{-}CoL'_5$. This type of mechanism is often termed *inner sphere* or *bridged*. Such reaction mechanisms are most readily established when the complex to which the ligand is transferred is *substitution-inert*. The rates of substitution processes vary enormously. For most d configurations substitutions take place very rapidly, but one or two configurations are substitution-inert, that is to say, reaction times are measurable in hours or days rather than in fractions of a second. Note that this inertness has nothing to do with the thermodynamic stability of the reactants or products but is a question of the reaction rate. The most common inert configurations are d^3 and d^6, for example Cr(III), Co(III), Mo(III), W(III), Re(IV), Rh(III), Ir(III), Ru(II), Os(II), Pd(IV) and Pt(IV). One striking example, used by Taube, is the study of the oxidation mechanism by using chromium(II) oxidized to chromium(III). As the latter is substitution-inert, any transfer of an atom or group during the oxidation process will be detected by its appearance in the chromium(III) complex. For example, in the reaction

$$\underset{Cr(II)}{Cr(H_2O)_6^{2+}} + \underset{Co(III)}{Co(NH_3)_5X^{2+}} \rightarrow \underset{Cr(III)}{Cr(H_2O)_5X^{2+}} + \underset{Co(II)}{Co(NH_3)_5(H_2O)^{2+}}$$

the transfer, in the oxidation process, of the group X has been demonstrated for $X^- =$ halide$^-$, NCS^-, N_3^-, SO_4^{2-} and PO_4^{3-}. Although the equation has been balanced by giving the cobalt(II) product as the pentammine hydrate, exchange in the labile cobalt(II) complex means that the hexahydrate would be recovered if the reaction was carried out in water.

The electron transfer reactions make up one group of transition metal reactions, the other major class are *ligand substitution reactions*. The mechanisms of ligand replacement reactions may be discussed in the light of ligand field theory. Two limiting modes of reaction for an octahedral complex are conceivable: either a ligand may be removed, leaving a 5-coordinated intermediate which then picks up the substitution ligand, or else the incoming ligand may become coordinated to the original complex, giving a 7-coordinated intermediate, which subsequently expels one of the original ligands. That is, either:

$$MX_6 \xrightarrow{\text{slow}} MX_5 \xrightarrow{Y} MX_5Y \tag{13.1}$$

or,

$$MX_6 + Y \rightarrow MX_6Y \rightarrow MX_5Y \tag{13.2}$$

The first is labelled S_N1 (*s*ubstitution, *n*ucleophilic, *first* order) and the second S_N2, as it is second order. That is, the rate of S_N1 reactions is governed by the first, dissociation, step which is far slower than the subsequent uptake of Y, so that the rate law is simply

$$\frac{d[MX_5Y]}{dt} = k[MX_6] \tag{13.3}$$

It should be noted that reaction (13.1) is almost indistinguishable from that in which the solvent, S, enters the sixth position and is then displaced by the incoming ligand

$$MX_6{+}S \xrightarrow{\text{slow}} MX_5S \xrightarrow{\text{Y, fast}} MX_5Y \qquad (13.4)$$

The rate law for (13.4) would be

$$\frac{d[MX_5Y]}{dt} = k'[MX_6][S] \qquad (13.5)$$

and, as the concentration of the solvent is effectively constant, $k'[S]$ may be replaced by a constant k'', giving Equation (13.5) the same form as (13.3).

In the S_N2 reaction, again making the reasonable assumption that the first step is slow and the expulsion of a ligand from the 7-coordinate intermediate is fast, the rate law would be

$$\frac{d[MX_5Y]}{dt} = k[MX_6][Y] \qquad (13.6)$$

Whether a given Y will replace any particular ligand is governed by the usual balance of energies with the additional factor of the change in CFSE. If the latter is an important factor, it will allow not only a prediction of whether the substitution will occur but also a prediction of the path. The CFSE of the 5- and 7-coordinated intermediates may be calculated and the most likely path determined. Consider the case of substitution in a low-spin cobalt(III) complex (which has six d electrons and a strong CFSE contribution). It may be shown that the possible intermediates have the CFSE values shown (all given in terms of the octahedral splitting of ΔE_{oct}):

Shape:	*Octahedral*	*Pentagonal bipyramid* (7)	*Trigonal bipyramid* (5)	*Square pyramid* (5)
CFSE:	2.4	1.55	1.25	2.0

The square pyramid therefore corresponds to the lowest loss of CFSE of all the possible intermediates in this case, the loss being equal to about 92 $kJ\,mol^{-1}$ if the ammine complex is the case in point. It can be calculated that the total activation energy for substitution in the cobalt(III) hexammine complex ion ranges from 521 $kJ\,mol^{-1}$ for the trigonal bipyramidal intermediate, through 431 $kJ\,mol^{-1}$ for the pentagonal bipyramid, to a range between 25 and 395 $kJ\,mol^{-1}$ for the square pyramid, depending on whether the empty site in the last case is occupied by a solvent molecule or not. Thus the CFSE variation accounts for about 50% of the total activation energy in each case. It would be predicted that the mechanism involving the 5-coordinated square-pyramidal configuration for the intermediate is the most likely and the activation energy found by experiment, about 140 $kJ\,mol^{-1}$, appears to bear this out.

It must be borne in mind that, while the kinetic study of reactions is a valuable guide to mechanisms, the order of the reaction is not necessarily the same as the molecularity of the critical step in the reaction path. One example is provided by systems where a rapid equilibrium is first set up

$$MX_6{+}Y \rightleftharpoons MX_6Y \xrightarrow{\text{slow}} MX_5Y + X \qquad (13.7)$$

and the slow step is the elimination of X from the intermediate, associated, species $MX_6.Y$. The actual rate-determining step is the slow one, which is unimolecular but the rate law will involve the constants for the forward and reverse steps of the equilibrium and the concentration of both MX_6 and Y: that is, the rate law will have the overall form of a second order reaction. In this case, the MX_6. Y does not involve coordination of Y to M but denotes some association with a definite lifetime, such as the formation of an ion pair. Such species become especially important when reactions are carried out in less polar solvents than water, such as acetone or methanol.

A further important class of reactions exists which are apparently S_N2, but where the basic step is unimolecular, and the overall kinetics result from the combination of this step with a pre-equilibrium. This is the class of reactions where the attacking ligand is OH^- and there is a replaceable hydrogen present in the ligands on the metal. Thus the reaction

$$Co(NH_3)_5Cl^{2+} + OH^- \rightleftharpoons Co(NH_3)_5OH^{2+} + Cl^- \quad (13.8)$$

obeys the rate law

Note that in the rate equation the square brackets indicate concentrations.

$$\frac{d[Co(NH_3)_5Cl^{2+}]}{dt} = k[Co(NH_3)_5Cl^{2+}][OH^-] \quad (13.9)$$

Such reactions are about 10^6 more rapid than most ligand replacements. It is believed that this reaction is not S_N2, but instead involves the initial abstraction of a proton from one of the ammonia ligands in a fast pre-equilibrium.

$$Co(NH_3)_5Cl^{2+} + OH^- \rightleftharpoons Co(NH_3)_4(NH_2)Cl^+ + H_2O \quad (13.10)$$

and this is followed by a slower, rate-determining step involving the expulsion of chloride from this amide complex

$$Co(NH_3)_4(NH_2)Cl^+ \xrightarrow{slow} Co(NH_3)_4(NH_2)^{2+} + Cl^- \quad (13.11)$$

and the 5-coordinate amido-intermediate rapidly reacts with water to give the hydroxy complex

$$Co(NH_3)_5(NH_2)^{2+} + H_2O \xrightarrow{fast} Co(NH_3)_5(OH)^{2+} \quad (13.12)$$

Thus the rate-determining step is unimolecular, but the rate law is second order, with a constant which is a combination of the forward and backward constants of reaction (13.10) and the constant for the slow step (13.11). Thus, the overall reaction (13.8) is the sum of (13.10), (13.11) and (13.12) obeying the overall rate law (13.9) but the critical step is unimolecular. As the amido-complex $Co(NH_3)_4(NH_2)Cl^+$ is the conjugate base (see Chapter 6) of the starting complex $Co(NH_3)_5Cl^{2+}$, this mechanism is usually known as the S_N1 *CB* (*sub*stitution, *nu*cleophilic, *u*nimolecular, conjugate *b*ase) mechanism. Such reactions are found whenever OH^- attacks a complex ion which contains ionizable hydrogen atoms.

These remarks about mechanisms apply, of course, to coordinations other than the octahedral one. In square-planar complexes, an additional effect, the *trans effect*, becomes important. In reactions of square-planar complexes of the general form:

$$(MLX_3) + Y = (MLX_2Y) + X$$

the group X which is displaced may be *cis* or *trans* to the ligand L and two isomeric products are possible, one or both of which may be observed. The common ligands may be arranged in an order of increasing ability to direct incoming substituents to a position *trans* to themselves:

$$H_2O < NH_3 < Cl^- < Br^- < NO_2^- < CN^-$$

The examples given serve only to illustrate the type of work which is being carried out in this field and are by no means even a complete summary of the known results. The field is obviously of importance for the understanding of transition metal chemistry as all reactions of transition elements in solution are reactions between complexes. Thus, when simple reactions involving transition metal ions in solution are discussed, the species actually involved are the complexes formed between ion and solvent molecule.

For octahedral complexes, S_N1 mechanisms and electron transfers with or without ligand transfer are well established, together with those involving ion-pairing or conjugate base pre-equilibria which are unimolecular in their rate-determining step, although following a second order rate law, such as Equation (13.6), for the overall sequence of reactions. There is, however, little firm evidence of true S_N2 mechanisms, going through a

USE OF THE *TRANS* EFFECT IN THE SYNTHESIS OF THE ANTICANCER DRUG CISPLATIN, *CIS*-$[PtCl_2(NH_3)_2]$

The knowledge of *trans* directing powers opens up methods of synthesizing desired compounds. Consider the method of synthesizing the *cis* and *trans* isomers of $[PtCl_2(NH_3)_2]$. If the biologically active *cis* isomer is required, the starting material must be the tetrachloroplatinate ion:

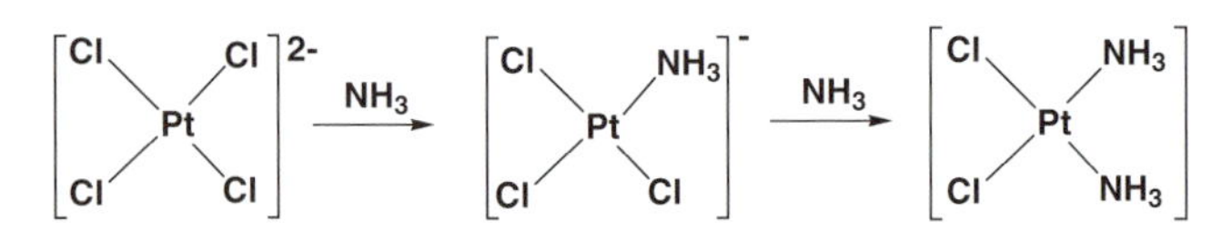

The first ammonia enters to give $Pt(NH_3)Cl_3^-$ and then, as the *trans* effect of the chloride ion is greater than that of ammonia, the second ammonia molecule enters *trans* to chlorine giving the *cis* isomer. To prepare the *trans* isomer, the starting material must be the tetrammine. In the intermediate $Pt(NH_3)_3Cl^+$, the *trans* effect of the chloride is greater than that of the ammonia, and the second chloride enters *trans* to the first to give the *trans* isomer:

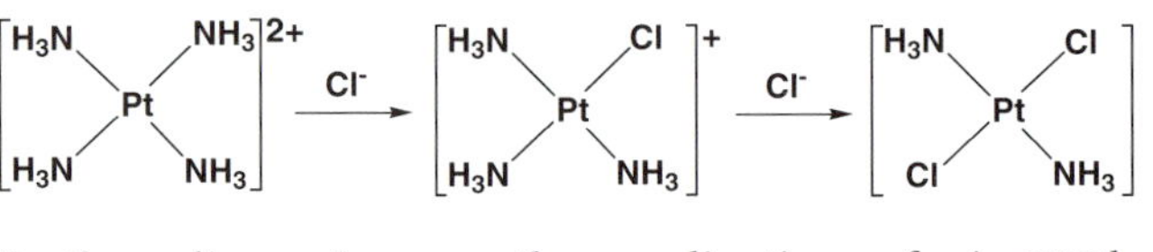

Further discussion on the application of *cis*-$[PtCl_2(NH_3)_2]$ as an anticancer drug is given in Chapter 20.

7-coordinate intermediate, for ligand substitution in octahedral complexes. In square-planar complexes, and also in 5-coordinate species, there is little steric hindrance to an increase of coordination number, and both S_N1 and S_N2 mechanisms are available for these lower coordination numbers.

One mechanism which is related to this question is that of *oxidative addition* in which an increase in oxidation state is accompanied by an increase in coordination number. This occurs most readily where a low coordination number is stable in a low oxidation state of an element while a higher one is the preferred one for a higher oxidation state. One striking example is provided by platinum where the +II state is d^8 and found in square-planar configurations while the +IV state is d^6 and most stable in the octahedral configuration (Table 13.12). Thus a change

$$Pt^{II}L_4+X_2 \rightarrow Pt^{IV}L_4X_2$$

is particularly favoured, compare Section 15.8. Such a reaction may lead to oxidations by relatively unreactive species under very mild conditions, as in the reaction at room temperature and normal pressure

$$Ir(CO)Cl(PPh_3)_2+H_2 \rightarrow Ir(CO)(H)_2Cl(PPh_3)_2$$

13.10 Structural aspects of ligand field effects

This is similar to the lanthanide contraction discussed in Chapter 11.

If the radii of divalent ions in a given Period are considered, it is found that the change of radius across a transition series follows a curve with two minima, such as that shown for the first transition series in Fig. 13.20. A similar effect is observed for trivalent ions but the minima come one element later. A smooth curve may be drawn through the values for the radii of the d^0, d^5 and d^{10} ions representing the fall in radius with increasing number of d electrons as the increase in nuclear charge is imperfectly balanced by the shielding of d electrons. The actual radii for the ions with empty, half-filled and filled shells lie on this curve, as the d electron clouds are spherically symmetrical. The radii of ions corresponding to other d configurations lie below this curve, and the largest differences are found for those ions with the largest excess of t_{2g} over e_g electrons. (The discussion and figure refer here to high-spin configurations in the octahedral field—extensions to other cases are obvious, though less well documented.) Consider first the case of Ti^{2+} (divalent scandium is unknown). In this ion, there are two d electrons and these lie in the t_{2g} level, that is, in orbitals which are directed away from the ligands. Electrons in such orbitals have only a slight effect in screening the nuclear charge of the transition ion from the ligands—a much smaller screening effect than if they were disposed in a spherically symmetric shell. It

It will be recalled from Chapter 2 that ionic radii are derived from measured ion–ion or ion–molecule distances.

follows that the metal–ligand distance is shorter than if these two electrons were spherically distributed, thus the octahedral radius is shorter, when measured, than that found by interpolation between the spherical d^0 and d^5 configurations. The difference between measured and interpolated radii is even greater for V^{2+} where there are three electrons in the t_{2g} orbitals, and then the difference decreases for Cr^{2+} with one electron in the e_g set which is directed at the ligands. In the d^5 ion, Mn^{2+}, the configuration is $t_{2g}^3e_g^2$ which is symmetrical, and the radius lies on the d^0-d^{10} curve. A similar pattern is repeated from d^6 to d^{10}.

Such a variation in radius will contribute a corresponding variation to the lattice energy of isostructural compounds of a given formula involving transition metal ions. Such a lattice energy curve is shown in Fig. 13.21. The lattice energy rises as the ionic radii fall, as the lattice energy is inversely proportional to the interionic distance. Thus, the points for the d^0, d^5 and d^{10} ions lie on a smoothly rising curve, and those for intermediate configurations lie above this curve. If the crystal field stabilization energies are subtracted from the experimental values, the resulting points lie on the curve. Notice that the CFSE values and the effects on radii both result from the ligand field splitting and the resulting electron configuration. Whether the lattice energy values are regarded entirely as an effect of CFSE or the radial effects are included depends on whether calculated or experimental values are used (calculated values already take account of the variation in radius).

A similar curve is found if other thermodynamic values are considered. Fig. 13.22 shows the hydration energies of the divalent and trivalent ions of the first row, that is the energy of:

$$M^{n+}_{(gas)} + 6H_2O_{(gas)} = M(H_2O)^{n+}_{6(gas)}$$

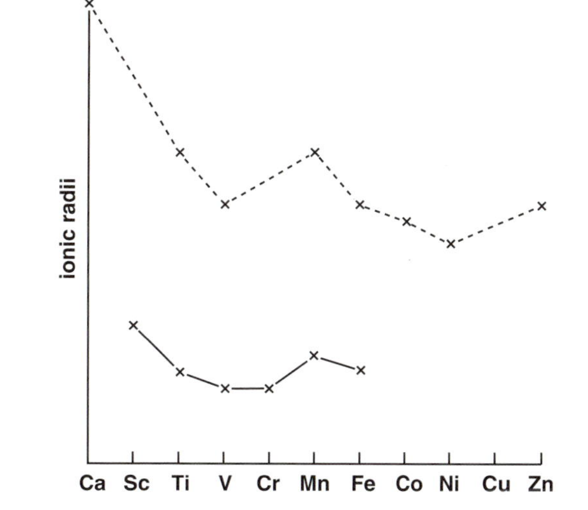

FIG. 13.20 Radii of the divalent (- - -) and trivalent (—) ions of the first transition series

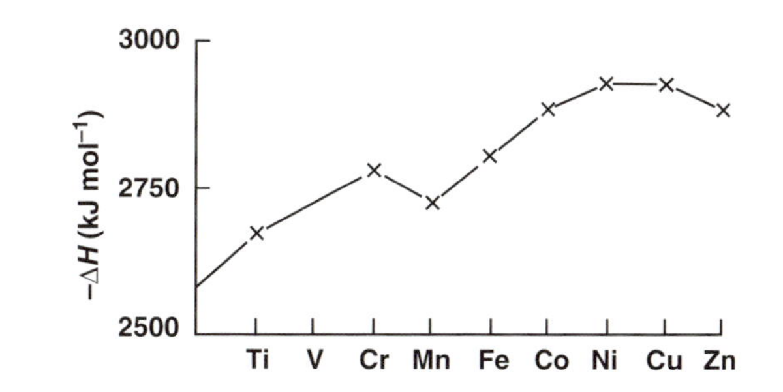

FIG. 13.21 The lattice energies of the difluorides of the elements of the first transition series

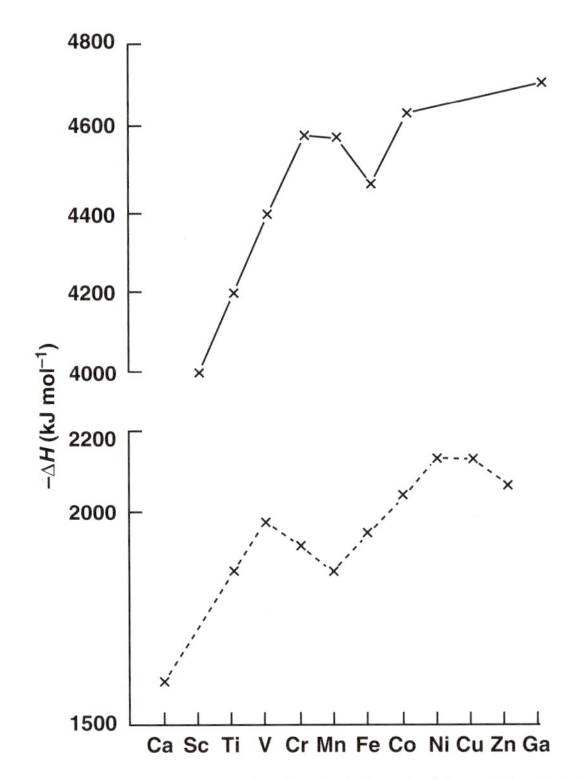

FIG. 13.22 The hydration heats of the divalent (- - -) and trivalent (—) ions of the first transition series

In each case, a curve with two maxima and a minimum at the d^5 configuration is observed, and the values fall on a smooth curve rising from d^0 to d^{10} when the CFSE values are subtracted from the experimental values. Notice that such a curve may be used in reverse to find the crystal field stabilization energies for different configurations, and hence the value of ΔE. Such thermodynamically determined values of the ligand field splitting agree closely with those derived from spectroscopic data.

13.11 Spectra of transition element complexes

When only one d electron is present, a simple spectrum consisting of a single band is observed (Fig. 13.4). The band is relatively broad both because the excited and ground states interact to some extent with their environment (usually with molecules of the solvent) and because the equilibrium bond lengths in the excited state would be greater than those in the ground state so that the upper state is produced with vibrational modes excited. These vibrational levels are too close to be resolved into individual bands but they give a general broadening of the d–d absorption.

The effect of an octahedral ligand field on the d^1 ion may be shown as in Fig. 13.23a. The t_{2g} and e_g levels are separated more widely as the ligand field increases. The symbol ^2D at the left is a symmetry label which describes the d^1 configuration in the absence of the ligand field splitting. We shall use such symbols here simply as labels and readers who study this subject further will find out how they are derived and their full significance.

Consider now the d^9 configuration in the octahedral field. We can describe this in a way which is related to the description of d^1 and use this as a method of determining the d^9 splitting diagram. In the ground state, d^9 can be regarded as derived from the filled d^{10} shell by forming a single hole in the e_g levels. Thus d^1 has a single electron in the t_{2g} level, outside a filled shell configuration, and d^9 has a single hole in the e_g level. In d^9, when a t_{2g} electron is excited to the e_g level, the latter level becomes filled and the hole appears in the t_{2g} level. Thus, the excitation may be described as the transition of a hole, that is, an electron vacancy, from the e_g to the t_{2g} level, while the d^1 transition is described as the

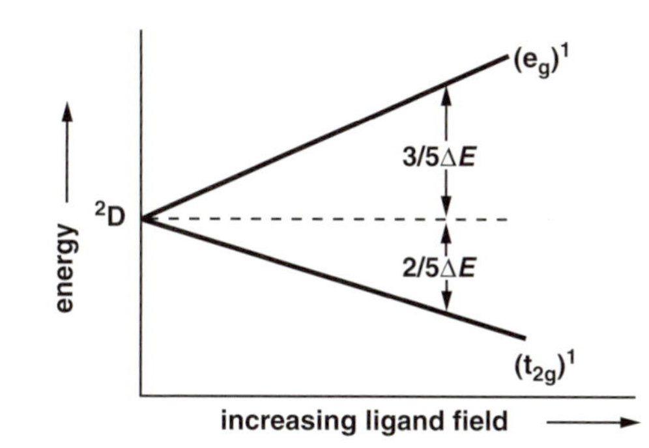

FIG. 13.23a Energy level diagram for the d^1 configuration in an octahedral field

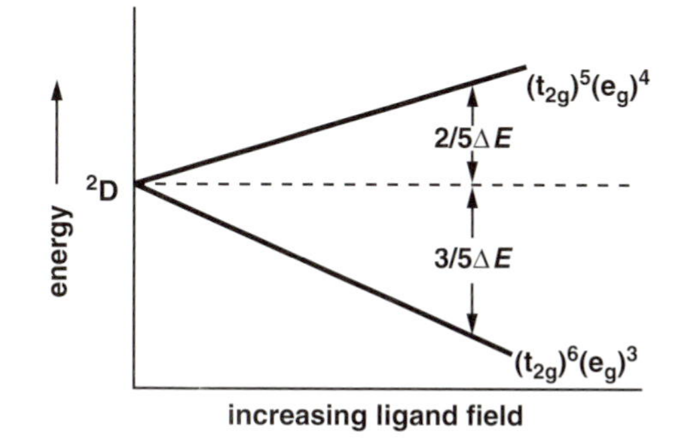

FIG. 13.23b Energy level diagram for the d^9 configuration in an octahedral field

transition of an electron from t_{2g} to e_g. Thus the transition in d^9 is shown on an energy level diagram which is the *reciprocal* of that for d^1, as shown in Fig. 13.23b.

Consider now a tetrahedral field. Here, the lowest level is the doublet e level and the upper state is the triplet t_2 level. Thus the energy level diagram for d^1 configuration in a tetrahedral field is the reverse of d^1 octahedral, and therefore qualitatively the same (though the ΔE value is only about 4/9) as that for d^9 octahedral. Similarly, d^9 tetrahedral is qualitatively the same as d^1 octahedral.

Yet another related configuration is that of d^6 high spin. Here, each d orbital is singly occupied and the sixth electron, in the octahedral field, is in one of the t_{2g} orbitals with an opposed spin. Now, transitions which require spin reversal are formally forbidden, and give rise to very weak bands if they can be observed at all, so that we may neglect any transition of one of the set of five electrons with parallel spins. Thus the only band we are likely to observe is that due to the excitation of the single antiparallel electron from the t_{2g} level to the e_g level. Thus the energy level diagram for d^6 high spin in the octahedral field is the same as that for d^1 in an octahedral field.

In a similar manner, the high spin d^4 configuration in the octahedral field may be described, by the hole formalism, in exactly the same way as d^9 octahedral. Furthermore, d^6 tetrahedral and d^4 tetrahedral are qualitatively described by the same diagrams as, respectively, d^4 octahedral and d^6 octahedral.

Thus, to summarize, by combining Figs 13.23a and 13.23b into one, as shown in Fig. 13.24 we can describe all those configurations with one electron in excess of an empty or half-filled d shell and, by the hole formalism, all the configurations with one electron less than a half-filled or filled d shell. This description applies both to the octahedral field, and to the tetrahedral field which causes the reciprocal splitting. Thus Fig. 13.24 describes qualitatively the effect of the ligand field on d^1 octahedral, d^6 octahedral, d^4 tetrahedral and d^9 tetrahedral, by the portion to the right of the origin, and on d^4 octahedral, d^9 octahedral, d^1 tetrahedral and d^6 tetrahedral by the portion to the left of the origin. For all these configurations, there is only one state above the ground state, separated by ΔE, so that there is only one transition and it occurs at a position in the spectrum corresponding to ΔE.

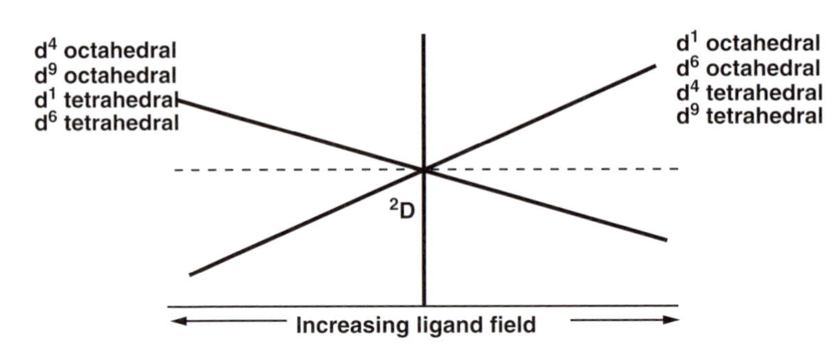

FIG. 13.24 Combined energy level diagram for single-electron or single-hole configurations in octahedral and tetrahedral fields. The zero crystal field level is shown at the origin and the left portion is the reciprocal of the right hand one. Such diagrams are termed *Orgel diagrams* after their originator

If the d^2 configuration is now considered, the energy level diagram is found to be more complicated. This is because, while the spin and orbital quantum numbers of a single d electron give rise to only one state, the ^2D state above, the combinations of these quantum numbers for two d electrons give rise to four different states. Some of these have the two electron spins antiparallel, so that transition to them would involve spin reversal. These states may therefore be neglected and we are concerned with two states for d^2, which are labelled ^3F (the ground state) and ^3P and are shown in the centre of Fig. 13.25. In an octahedral field, the ^3F state splits up, in a similar manner to the ^2D state in Fig. 13.23, but into three components while the ^3P state remains unsplit. Furthermore, interaction between the levels derived from ^3F and ^3P causes some of these levels to curve away from each other, to give the dependence of energy level on ligand field which is shown in Fig. 13.25. This interaction depends on a single parameter, B—the Racah parameter—and this may be evaluated in the course of the assignment. Thus, the d^2 diagram, for an octahedral field, has three states above the ground state to which an electron may be excited without changing its spin. This means that the spectrum of a d^2 complex contains three bands, in place of the single band of d^1, and the positions of these three bands together give the value of the ligand field for the complex. By the same arguments that applied to d^1, the energy level diagram is qualitatively the same for d^2 octahedral and for d^7 octahedral (high spin) and also for d^8 tetrahedral and d^3 tetrahedral, while the reciprocal diagram applies to d^8 and d^3 octahedral and also to d^2 and d^7 tetrahedral. All these levels are shown in Fig. 13.25. A d^8 spectrum is illustrated in Fig. 13.26.

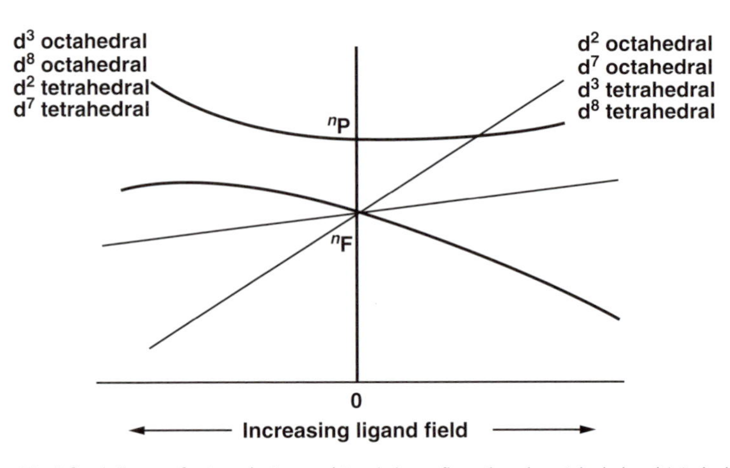

FIG. 13.25 Combined Orgel diagram for two-electron and two-hole configurations in octahedral and tetrahedral fields. The multiplicities are, for d^2 and d^8, $n = 3$ and for d^3 and d^7, $n = 4$. States of lower multiplicity are omitted. On the d^3 octahedral side of the diagram, the upper state which is derived from the ^4P interacts with the highest level derived from the ^4F and these levels curve away from each other. The other two levels derived from ^4F do not interact with any other and vary linearly with the ligand field. On the d^2 octahedral side of the diagram, the same two states also interact (derived now from ^3P and ^3F) but, as the ^3F state is now the lowest one, the interaction is less marked

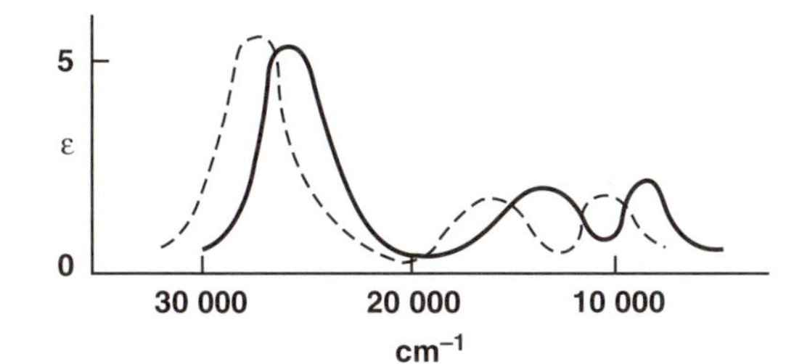

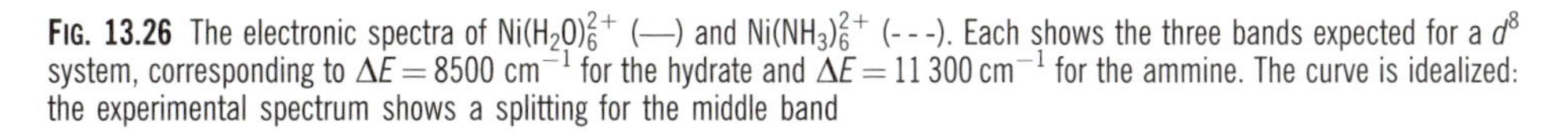
FIG. 13.26 The electronic spectra of $Ni(H_2O)_6^{2+}$ (—) and $Ni(NH_3)_6^{2+}$ (- - -). Each shows the three bands expected for a d^8 system, corresponding to $\Delta E = 8500\ cm^{-1}$ for the hydrate and $\Delta E = 11\,300\ cm^{-1}$ for the ammine. The curve is idealized: the experimental spectrum shows a splitting for the middle band

Finally, the d^5 high spin configuration has all d orbitals half-filled with parallel spins in the ^6S state. Any electron transition must involve spin reversal so that the d–d bands in a d^5 complex, such as a manganese(II) compound, are very weak. There are four excited states which involve only one spin reversal, and these split up into a total of ten states in the ligand field. Thus, a d^5 complex might show anything up to ten very weak absorptions in the visible and ultraviolet, but since these transitions are formally forbidden, their intensities are about 100 times weaker than normal d–d transitions in an octahedral field. Hence most d^5 compounds look almost colourless; witness the very pale pink of the majority of manganese(II) species.

A further point about intensities is the difference between octahedral and tetrahedral complexes. A further selection rule forbids transitions between states with the same symmetry with respect to a centre of inversion (see Appendix C) and this applies to octahedral complexes but not to tetrahedral ones as the tetrahedron has no inversion centre. Thus tetrahedral complexes are commonly more strongly coloured than octahedral ones, as in the case of the cobalt(II) complexes where the blue of the $CoCl_4^{2-}$ species is much more intense than the pink of the octahedral aquo complex, $Co(H_2O)_6^{2+}$. All these intensities are much less than those of fully allowed transitions, see Section 7.6, so that d–d transitions are only observed where allowed bands of the ligands or of the total complex are sufficiently far into the ultraviolet that they do not swamp the weaker d–d bands.

Low-spin complexes may be analysed in a similar way to the above, although the detailed arguments are more difficult and will not be treated here. Further effects found in the spectra must also be considered in a full discussion. For example, Jahn–Teller effects may cause absorption bands to split, as also will interaction between orbital and spin angular momentum. However, all these effects may be taken into account and it is found that experimental spectra agree well with those predicted by application of ligand theory.

The spectra of square-planar and other configurations may be analysed similarly.

The major effects in the spectra of transition complexes may be summarized as follows. For configurations d^n where $n = 1, 4, 6$ and 9, one band is expected and its position gives ΔE directly. Configurations with $n = 2, 3, 7$ or 8 give rise to three bands whose positions depend on two parameters, ΔE and B, the latter measuring the interaction between energy levels derived from F and P states. As there are three bands depending on only two parameters, both ΔE and B may be evaluated unambiguously. Finally, when $n = 5$, a much larger number of transitions may be observed which may be analysed to give the value of ΔE. Low-spin configurations for $n = 4, 5, 6$ and 7 may be similarly analysed to give ΔE values, though the treatment is more complicated. Apart from allowed transitions, the most intense absorptions are found in tetrahedral complexes, those in octahedral species are weak, while transitions in any d^5 high-spin complex are extremely weak indeed.

13.12 π Bonding between metal and ligands

The molecular orbital discussion of Section 13.3 is readily extended to include π bonding between metal and ligands. Such π bonding helps to account for a number of the properties we have discussed, such as the positions of some of the members of the spectrochemical series and the existence of complexes of metals in low oxidation states where the charge would not provide the attractive forces required in the crystal field theory.

Let us consider an octahedral complex: similar considerations will apply for other shapes.

π bonds may be formed between ligand orbitals of suitable symmetry and metal d orbitals of the t_{2g} set or metal p orbitals. The most important interaction is with the t_{2g} set, and Fig. 13.27 shows typical examples of the in-phase bonding components formed with ligand p or d orbitals. Clearly such orbitals will concentrate electron density between the metal and the ligands and form π bonding orbitals. The corresponding out-of-phase interactions will be antibonding.

Other π interactions are possible, such as those of t_{1u} symmetry involving the metal p orbitals (Fig. 13.31), but these are less significant and can be neglected in a simple treatment. There are also combinations of the ligand π orbitals (labelled t_{1g} and t_{2u}) which are the wrong symmetry to combine with any metal orbitals. We neglect these also, and discuss only the three ligand combinations of the t_{2g} set.

There are two cases to consider: (a) where the ligand π orbitals are empty and of relatively high energy and (b) where the ligand π orbitals are filled and relatively stable. Examples of (a) include phosphanes, PR_3, or sulfur ligands like SR_2, where the ligand π orbitals are the empty 3d orbitals on P or S. Examples of (b) are fluoride or oxide where π bonds can potentially be formed using the filled p orbitals not involved in σ bonding.

(a) *Acceptor π ligands* When the ligand π orbitals are empty, they will commonly be of higher energy than the metal d orbitals. The appropriate energy level diagram is shown in Fig. 13.28. The effect of the π interaction is to produce the orbitals, labelled π, which are more stable than the originally nonbonding metal t_{2g} levels together with the corresponding π^* orbitals which lie among the other antibonding levels. The precise relationships among these orbitals will, of course, vary from one compound to the next.

Since the ligand π orbitals are empty, there are no more electrons to place in this scheme than there were in that of Fig. 13.7, where only σ bonds were formed.

Thus the effect of the additional π interaction in this case is to increase the separation, ΔE, between π and e_g^* compared with that between t_{2g} and e_g^* in the simple case. Thus ligands with empty orbitals capable of forming π bonds will have a large value of ΔE, and lie to the right of the spectrochemical series. This explains the positions of ligands such as the phosphanes, or of CN^- which accepts into the empty π^* orbitals (Fig. 13.29). Note in this latter case that, although the π^* orbitals are antibonding as far as the interaction between C and N is concerned, the total M-C-N π interaction is bonding overall. The increase in M-C stabilization exceeds the loss of C-N stabilization.

This type of interaction involves the transfer of charge from the metal to the ligands as electrons which would have been entirely on the metal in t_{2g} nonbonding orbitals are now shared in the π orbitals. Thus the metal can be regarded as a donor and the ligands as charge acceptors. This interaction clearly serves to offset the build-up of charge on the metal which occurs in the σ bonding. We come back to this in the discussion of carbonyl compounds in Chapter 16.

(b) *Donor π ligands* When the ligand π orbitals are filled, they will commonly be of very similar energy to the σ levels, and will be more stable than the metal t_{2g} levels. The appropriate change from Fig. 13.7 is thus that shown in Fig. 13.30. The ligand π orbitals interact with the t_{2g} level to form π orbitals in the complex which are more stable than the ligand level and π^* orbitals in the complex which are less stable than t_{2g}. The energy gap, ΔE, between π^* and e_g^* is thus less than that between t_{2g} and e_g^*.

Let us place electrons in this energy level diagram. The σ bonding levels are filled as before (Section 13.3). Since the ligand π levels are filled, we have these six electrons to accommodate and, if we regard them as entering the molecular π level, their stabilization compared with the ligand π levels will be the main driving force in forming the complex. However, the remaining nonbonding metal d electrons must enter π^* and e_g^* , corresponding to the situation of weak ligand field complexes.

This type of reaction involves transfer of charge from the ligand to the metal both in the σ and in the π interactions. It is therefore favoured by a high formal charge on the metal, i.e. by a high metal oxidation state. It is not always clear whether π donor behaviour occurs,

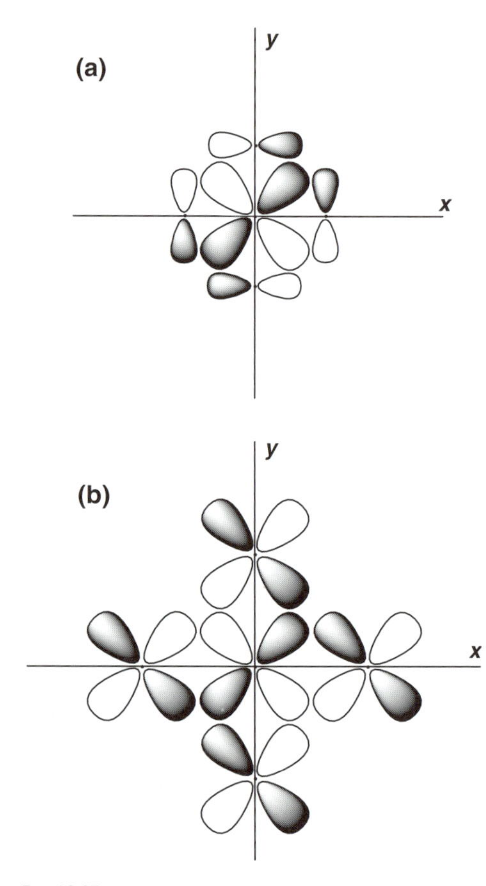

FIG. 13.27 π bonding involving t_{2g} orbitals: (a) with ligand p orbitals, (b) with ligand d orbitals. Similar components occur in the *xz* and *yz* planes.

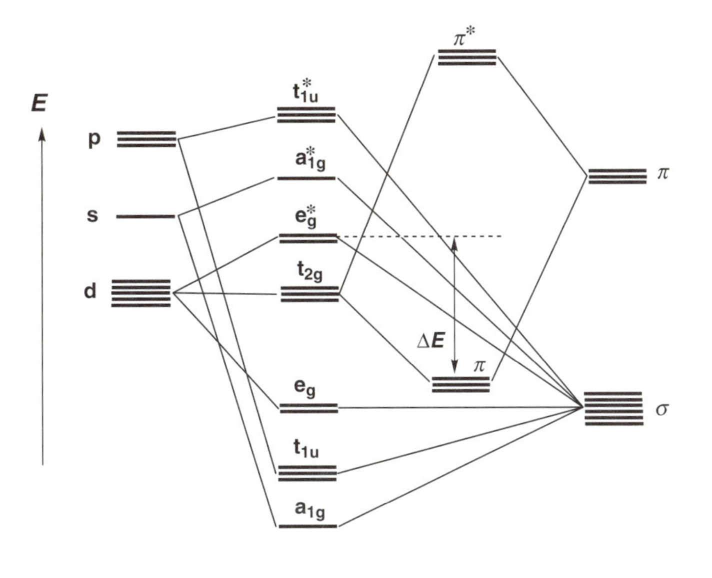

FIG. 13.28 Energy level diagram showing the interaction between the d orbitals in an octahedral complex with ligand orbitals of π symmetry. The significant effect of such π bonding is to create three more stable orbitals using the t_{2g} set and thus increase ΔE. Note: the diagram indicates only the three ligand π orbitals most directly involved

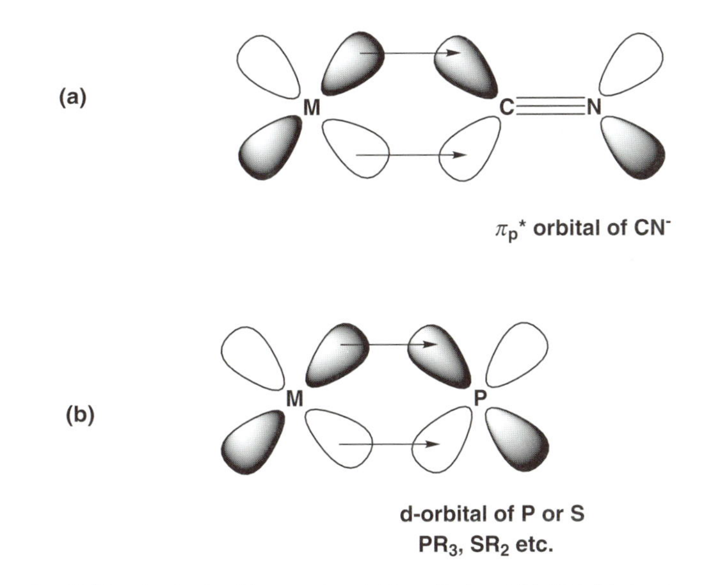

FIG. 13.29 π bonding involving acceptor orbitals on the ligand (a) CN^-, (b) phosphane

but it is more likely to contribute in, say the nickel(IV) species NiF_6^{2-} than in the nickel(II) one, NiF_6^{4-}.

The discussion above applies to other shapes, such as square-planar or tetrahedral, with differences in detail but with the same general effect on ΔE: (a) empty π orbitals on the ligand increase ΔE, correspond to ligands to the right of the spectrochemical series, and are commonly involved with the lower oxidation states from III downwards, (b) if full π orbitals on the ligand are delocalized onto the metal ΔE is decreased corresponding to ligands on the left of the spectrochemical series, and such π bonding commonly occurs for high oxidation states especially in oxyanions such as CrO_4^{2-} or MnO_4^-.

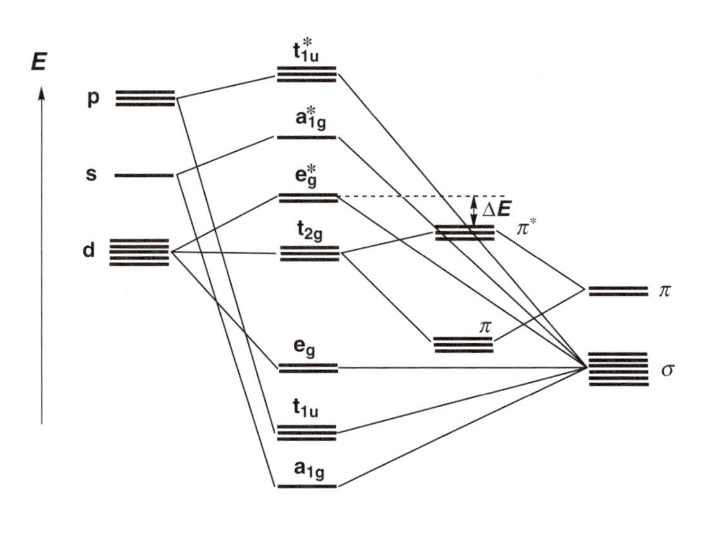

FIG. 13.30 Energy level diagram for molecular orbitals in an octahedral complex where π bonding occurs from donor ligand orbitals. Here (contrast Fig. 13.28) the π interaction lowers the filled ligand π levels to the molecular π orbitals, leaving the molecular π^* orbitals to hold electrons originally in the t_{2g} level. Thus ΔE is decreased

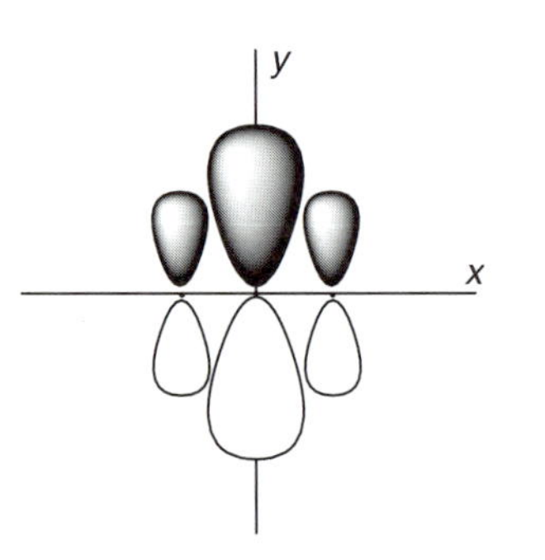

FIG. 13.31 Possible π bonding involving p orbitals, One component of the t_{1u} set

In all shapes, π bonding can also involve the metal p orbitals, or molecular orbitals derived from them like the t_{1u} levels of an octahedral complex. An example is shown in Fig. 13.31. Such interactions are less significant as they involve less stable orbitals and so contribute less to the stabilization. They are included in more detailed treatments, particularly of excited states.

Problems

This chapter is mainly concerned with the properties of the d^n electrons of a transition element while Chapters 14 and 15 discuss their detailed chemistry. A great deal of reference back and forward between these chapters will be needed.
A first reading should concentrate on Sections 13.1 to 13.5.

13.1 *General:* for each oxidation state of each element (as discussed in Chapters 14 and 15), decide the d configuration and hence use Chapter 13 to determine the expected configuration, stability, magnetism, reactivity and spectroscopic properties. Conversely, from Chapter 13 decide which configurations are expected to be stable or otherwise and determine from 14 and 15 how far these predictions are reasonable.

13.2 Draw a diagram analogous to Fig. 13.1 for the case ML_3, where the ligands form an equilateral triangle around M (choose z as the axis perpendicular to the ML_3 plane).

13.3 Would Fig. 13.3 be altered by choosing new x and y axes at 45° to the ones used in Fig. 13.2?

13.4 Calculate CFSE values for $M(H_2O)_6^{3+}$ and $M(CN)_6^{3-}$ ions of the first series elements (compare Tables 13.6 and 13.7). Estimate the corresponding values for $M(H_2O)_4^{2+}$ (tetrahedral).

13.5 Calculate the change in CFSE on oxidizing M^{2+} to M^{3+} in water for V, Cr, Mn, Fe and Co.

13.6 The experimental magnetic moments (Bohr magnetons) of a number of complex ions are listed. Comment on (a) the d electron configuration and (b) the validity of the 'spin only' formula.

Moment	*Complex*
1.8	$Ti(H_2O)_6^{3+}$
1.8	$Co(NO_2)_6^{4-}$
2.2	$Mn(CN)_6^{4-}$
3.8	$Cr(H_2O)_6^{3+}$
4.8	$Cr(H_2O)_6^{2+}$
5.1	$Co(H_2O)_6^{2+}$
5.9	$Mn(H_2O)_6^{2+}$

13.7 Draw out each of the orbital combinations of a square-planar complex ML_4, as was done for octahedral ML_6

in Figs 13.5 and 13.6. Thus draw up a full qualitative orbital diagram for ML_4 corresponding to Fig. 13.7.

13.8 It is thought that the crossover point between high- and low-spin Co(III) comes just at $Co(H_2O)_6^{3+}$. Use this to evaluate K (Table 13.8). Assuming K is constant for the elements of the first transition series, decide which of the complexes in Table 13.6 are likely to be low spin.

13.9 It is valuable to see the relations across the transition series of the major energy parameters. For the elements Ca to Zn, plot first, second and third ionization energies (from Table 2.8) against atomic number. Compare the curve for I_3 with Figs 13.20 to 22.

> *Note to questions 13.9 to 13.11. After* you have completed these exercises, consult the valuable article by P. F. Lang and B. C. Smith in *Education in Chemistry*, March 1986, 50–53. (Note: these authors use the ionization *number* derived by dividing the ionization energy in wavenumbers by the Rydberg constant to get a dimensionless number, which is often convenient.)

13.10 The curves for I_1 and I_2 in Question 13.9 are less regular than the discussion of this chapter would indicate. One reason arises from the variation in electron configuration between the 4s and 3d orbitals. While I_3 is for the loss of a d electron for each element, I_2 arises from $3d^n4s^1$ to $3d^n4s^0$ for Ca, Sc, Ti, Mn, Fe and Zn but for the remaining elements, there is no electron in the 4s orbital in the M^+ ion and the ionization refers to the loss of a d electron. By plotting the $4s^1$ loss for the above six elements, and assuming a smooth curve, estimate the energy differences between the $3d^n4s^1$ and $3d^{n+1}4s^0$ configurations for the other elements.

Plot the sum of $I_1 + I_2$ against atomic number. Which elements are anomalous? What is your estimate of the relevant excitation energies?

13.11 Carry out the same exercise as in Questions 13.9 and 13.10 for the second transition period, and discuss the results.

13.12 Values for ionic radii (Shannon, compare Table 2.12) for six-coordination are (pm):

	Sc	Ti	V	Cr	Mn	Fe	Co	Ni	Cu	Zn
M^{2+}		100	93	94	97	92	79	83	87	88
M^{3+}	89	81	78	76	79	79	75	74		
M—F				179.5	181.1	176.9	175.4	172.9	171.3	174.5

(a) Use these values to plot out Fig. 13.20 more precisely.

(b) The bond lengths in certain difluorides in the gas phase, i.e. linear MF_2 molecules, are also shown above. Construct a similar figure to Fig. 13.20 and discuss differences from the curve for 6-coordination in terms of the energy level diagram of Fig. 13.1.

CHAPTER 14

14.1 GENERAL PROPERTIES 312
14.2 TITANIUM, $3d^2 4s^2$ 314
14.2.1 Titanium(IV) 315
14.2.2 Titanium(III) and lower oxidation states 317
14.3 VANADIUM, $3d^3 4s^2$ 318
14.3.1 Vanadium(V) 319
14.3.2 Vanadium(IV) 322
14.3.3 Vanadium(III) 323
14.4 CHROMIUM, $3d^5 4s^1$ 324
14.4.1 Chromium(VI) 325
14.4.2 Chromium(V) and chromium(IV) 325
14.4.3 Chromium(III) 327
14.4.4 Chromium(II) and lower oxidation states 328
14.5 MANGANESE, $3d^5 4s^2$ 329
14.5.1 The high oxidation states, manganese(VII), (VI) and (V) 329
14.5.2 Manganese(IV) and manganese(III) 330
14.5.3 Manganese(II) and lower oxidation states 332
14.6 IRON, $3d^6 4s^2$ 333
14.6.1 The iron–oxygen system 334
14.6.2 The higher oxidation states of iron 334
14.6.3 The stable states, iron(III) and iron(II) 335
14.7 COBALT, $3d^7 4s^2$ 338
14.7.1 Cobalt oxidation states greater than (III) 338
14.7.2 Cobalt(III) 339
14.7.3 Cobalt(II) 340
14.7.4 Lower oxidation states of cobalt 341
14.8 NICKEL, $3d^8 4s^2$ 342
14.8.1 Higher oxidation states of nickel 342
14.8.2 Nickel(II) 343
14.9 COPPER, $3d^{10} 4s^1$ 344
14.9.1 Copper(IV) and copper(III) 345
14.9.2 Copper(II) 345
14.9.3 Copper(I) 347
14.10 THE RELATIVE STABILITIES OF THE DIHALIDES AND TRIHALIDES OF THE ELEMENTS OF THE FIRST TRANSITION SERIES 348
PROBLEMS 350

The Transition Elements of the First Series

14.1 General properties

The transition elements of the first series, dealt with here, are those elements, from titanium to copper, where the 3d level is filling. The free energy diagrams of all the elements are shown on a small scale in Fig. 14.1 so that a general comparison may be made. The individual diagrams for important systems are given on a larger scale in later sections.

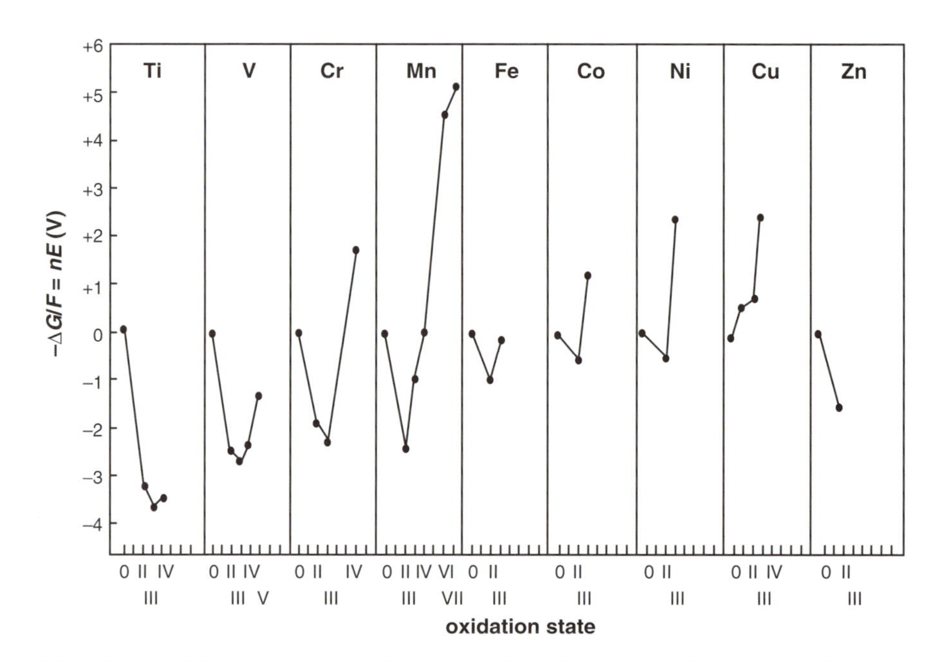

FIG. 14.1 Oxidation state free energy diagrams of the first transition series elements. It is clear that the Group oxidation state becomes increasingly unstable towards manganese. Thereafter, all states above III are unstable, and the variation in the relative stability of the II and III states (and I for copper) is of most importance. (Compare Section 8.6.)

The lists of the oxides and halides and the free energy diagrams give largely complementary pictures of the stabilities of the various oxidation states, in the solid state and in aqueous solution respectively. At titanium, the Group oxidation state of IV is the most stable and the lower states become increasingly reducing. Moving along the series, the Group oxidation state becomes more unstable and more oxidizing so that, at manganese, only a few compounds of the VII state are known and all are strongly oxidizing. Beyond manganese, the Group state disappears and only a few, unstable, strongly oxidizing states greater than III exist. Among the lower states, either the II or the III state is the most stable state from chromium onwards, and the relative stability of these two states varies with the number of d electrons as was indicated in the last chapter, with the II state finally becoming the most stable at nickel and copper. This variation in stability is shown up in the free energy diagrams by the increasing height above zero of the Group state up to manganese, and by the increasing instability of the III state at the right of the figure. (See also Section 14.10.)

PROPERTIES OF FIRST SERIES TRANSITION ELEMENTS COVERED IN OTHER CHAPTERS

Table 13.2, in the last chapter, shows their oxidation states, while Tables 13.3 to 13.5 give their oxides, fluorides and other halides. Other properties which are listed earlier include the electronic configurations, atomic weights and numbers (Table 2.5) and the ionization potentials (Table 2.8). Some values of the redox potentials, especially of species used in quantitative analysis, are given in Table 6.3.

Specific groups of compounds are covered in later sections—carbonyls in Section 16.2, organic derivative including metallocenes in Sections 16.3 and 16.4, higher order M-M bonding in Section 16.7, dioxygen, dinitrogen and dihydrogen species in Sections 16.9 to 16.11, complexes with polysulfur and related ligands in Section 19.2, and some metal species of major biological importance in Section 20.1.4

Oxidation states which lie between the Group state and the II or III states have a tendency to disproportionate. Some of them, such as Cr(IV) and (V), and Mn(V) and (VI), are very rare and poorly represented. The one state of moderate stability among these intermediate ones is manganese(IV), which owes part of its importance in chemistry to the insolubility of MnO_2 in neutral or basic solution.

All the elements of the first transition series are common and all are important commercially, titanium, iron and copper in their own right, and the others largely as constituents or coatings of iron and steel. Most are produced in the blast furnace, and very pure nickel is produced in the Mond process (Section 14.8). High grade copper for electrical applications is prepared by electrolysis.

The metals are reactive and electropositive, with both properties decreasing towards the right of the series. However, a number form a thin layer of coherent and impermeable oxide on the surface and so resist attack. In particular, this is the action of chromium, hence its use as chromium plating to protect iron.

The 3d cations are relatively small, so that the charge density on an ion like Fe^{3+} is high. In water, the ion is strongly hydrated as $Fe(H_2O)_6^{3+}$, and the cation withdraws electron density from the water molecules to allow ionization:

$$[Fe(H_2O)_6]^{3+} + H_2O \rightarrow [(H_2O)_5Fe(OH)]^{2+} + H_3O^+$$

As a result, hydrates act as acids and their existence depends on the charge density. In particular, no transition element forms a simple M^{4+} ion and $[M(H_2O)_6]^{3+}$ species require strong acid media to prevent deprotonation. The divalent hydrates do exist in more neutral media. Thus, while it is often convenient to write M^{3+} or even M^{4+} species, this should be seen only as a shorthand for all the hydrated/hydroxy species in solution. When more than one OH group is present, a water molecule may be eliminated, forming an oxy species. For example, the di-hydroxy-vanadium(IV) ion, $[V(OH)_2(H_2O)_4]^{2+}$, goes to $[VO(H_2O)_5]^{2+}$, often written simply as VO^{2+} (called vanadyl(IV)), and the +(V) species $[V(OH)_4(H_2O)_4]^+$ gives the vanadyl(V) species $[VO_2(H_2O)_4]^+$ or VO_2^+. The OH groups, or the O group formed by water elimination, may bridge between two metal atoms leading to progressively more complex and polymeric species until eventually the hydroxide, or hydrated oxide, precipitates. Thus in moderately basic media, chromium(III) exists as hydrated $[Cr(OH)]^{2+}$, $[Cr_2(OH)_2]^{4+}$ or $[Cr_3(OH)_4]^{5+}$.

The transition elements form nonstoichiometric hydrides, carbides, oxides and similar compounds—often termed *interstitial*—where the small atoms are incorporated into the metallic lattice. It is now known that, although the metal atom arrangements in such compounds are commonly in one of the forms characteristic of metals (cubic or hexagonal close-packed or body-centred cube), the arrangement of the metal atoms in the interstitial compound is rarely the same as that in the metal itself. The best theory of these compounds regards the included atoms as 'metallic'. They show larger coordination numbers than normal—e.g. six for carbon in TiC—and contribute valency electrons to the common pool of delocalized electrons in the metallic bonding. Interstitial compounds show many metallic properties such as hardness, metallic appearance, high conductivity with negative temperature coefficient and high melting point which support this view. The interstitial compounds are among the hardest known and show extremely high melting points, e.g.:

See also the discussion of metallic hydrides, Section 9.4.

	m.p. (°C)	Hardness, Moh's scale
TiC	3140	8–9
HfC	4160	9
W_2C	3130	9–10

The hardnesses are on Moh's scale on which diamond = 10.

The binary compound 4TiC + ZrC has the highest recorded melting point of 4215°C. The earlier metals, of the titanium, vanadium, and chromium Groups, form interstitial compounds with small metal–nonmetal ratios, such as MC, M_2C, MN, or MH_2 (all the formulae are idealized; nonstoichiometry marks most of these phases), and these have regular structures. Thus the MN and MC compounds have the sodium chloride structure and the M_2C ones have defect NaCl structures. These compounds are very unreactive.

The elements of the later Groups form less well-defined compounds with complex metal–nonmetal ratios such as Fe_3C or Cr_7C_3. Such compounds are similar in general physical properties but are much more reactive chemically. They are attacked by water and dilute acids to give hydrogen and mixtures of hydrocarbons.

Borides, and also silicides, are similar to the carbides and nitrides in structures and properties. These provide a link between the interstitial compounds and metal alloys.

14.2 Titanium, $3d^24s^2$

The free energy diagram for titanium is shown in Fig. 14.2. The titanium(IV) state is the most stable one and is well characterized with a wide variety of compounds. The III state is reducing but reasonably stable; in water, it is a reducing agent somewhat stronger than tin(II). The II state is very strongly reducing. The only solid compounds are unstable polymerized solids and it rapidly decomposes in water.

This is the Kroll process.

Titanium metal is used in high-performance situations where a high strength/weight ratio is needed. It is also resistant to corrosion in harsh environments. Titanium is prepared by calcium or magnesium reduction of $TiCl_4$. Since Ti reacts readily with air to form the nitride, carbide or hydride, which are all brittle compounds which ruin the metal quality, the final step in the commercial reduction is carried out in an atmosphere of argon. This was the first large-scale use of a rare gas in industrial metallurgy.

A new production method, developed to the pilot plant stage, has considerable promise. In this, titanium dioxide is reduced to the metal electolytically in a calcium chloride melt. The process is cheaper than the Kroll process, and avoids the hazard of chlorine. If successfully scaled up, we should see a much wider use of titanium in place of stainless and other special steels. It will also make titanium alloys, such as superconducting titanium–niobium or shape-memory titanium–nickel, more accessible.

TITANIUM DIOXIDE—THE BRIGHTEST WHITE PIGMENT

Apart from the use of the metal, the other major industrial production of titanium compounds is that of TiO_2 which is an excellent white pigment. The ore *rutile* is the dioxide but this is much less abundant than another prolific source, the iron–titanium oxide, *ilmenite*. Finely divided, pure, TiO_2 is much the most effective commercial white pigment with great opacity and excellent hiding power. Any trace of iron would impart a yellowish or reddish tinge so the preparation must exclude iron. One commercial method converts the titanium into the tetrachloride by treating the ore with coke and reacting with chlorine at 1000°C. $TiCl_4$ is a volatile liquid (iron chlorides are not) which can be purified by distillation. The chlorine is recovered and recycled, making the process environmentally friendly, if relatively expensive and requiring more energy. The alternative route uses sulfuric acid to dissolve the ore and the TiO_2 is recovered by precipitation and purified. This process gives more waste product for disposal.

Both processes produce needle-like rutile crystals which give the best colourant properties, especially for exposed surfaces. In addition, the sulfate process can yield a second form of TiO_2, called *anatase*, which forms tabular crystals which are preferred for some applications such as paper or fibre.

Annual production of TiO_2 is several million tonnes and it has almost completely displaced the old toxic white lead pigments.

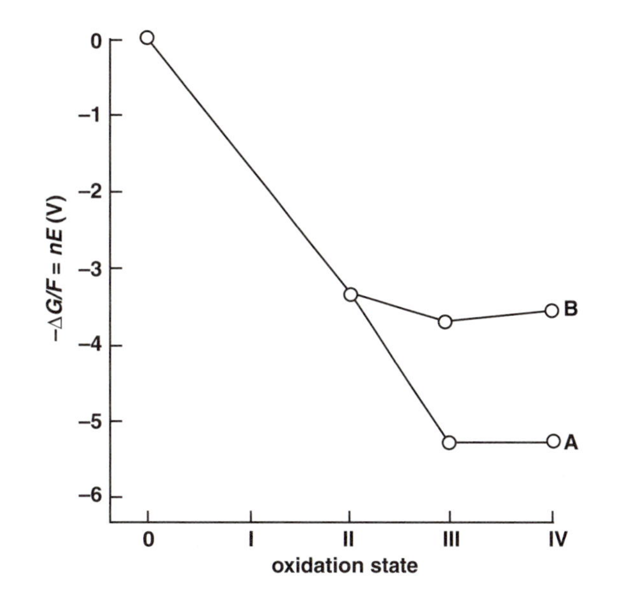

FIG. 14.2 The oxidation state free energy diagram of titanium. The IV state is stable and the lower states reducing. Titanium redox potentials are uncertain: the Ti(IV) potential (A) is for $Ti(OH)^{3+}$ and the value (B) is for TiO_2

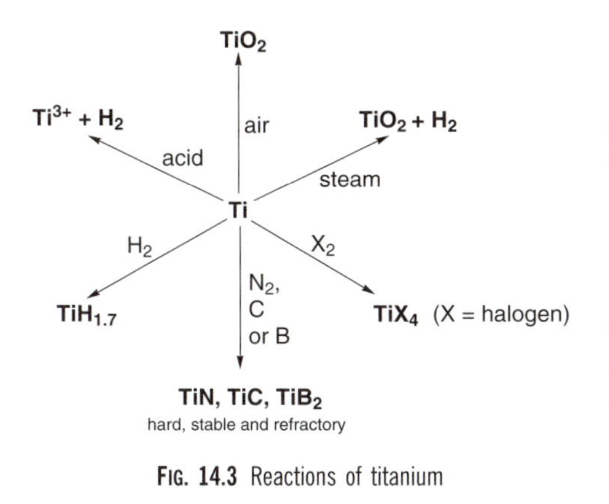

FIG. 14.3 Reactions of titanium

The bonding and chemistry of C_5H_5–metal compounds are discussed in Section 16.4.

While titanium is inert at ordinary temperatures, it does combine with a variety of reagents on heating, see Fig. 14.3. Elements such as carbon, silicon or iron increase the strength of titanium but also make it more brittle. They are often added to titanium in controlled amounts to modify its properties.

Titanium compounds play an important role in Ziegler–Natta catalysis (see Section 17.4.3). As well as the long-standing use of $TiCl_4$ in the original catalysts, 'second generation' catalysts based on $TiCl_3$, often supported by $MgCl_2$, give better molecular weight distributions and easier work-up, while homogeneous catalysts have been developed using $(C_5H_5)_2TiR_2$ (for R = Cl or alkyl) plus AlR_3. Such C_5H_5 compounds of titanium have also been tested as anti-tumour agents.

14.2.1 Titanium(IV)

Titanium dioxide is weakly acidic and weakly basic, dissolving in concentrated base or acid but easily hydrolysing on dilution:

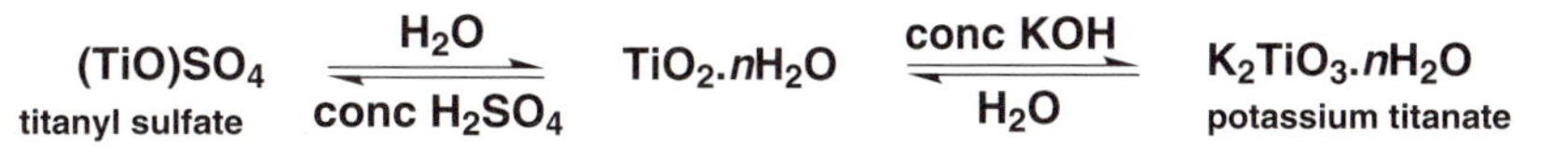

No definite hydroxide, such as $Ti(OH)_4$, exists, the compound produced on hydrolysis being a hydrated form of the oxide. The titanyl group, TiO^{2+}, does not exist as the simple monomeric cation. It is present in acidic solutions, but is in equilibrium with oligomeric species such as hydrated $Ti_3O_4^{4+}$. The TiO^{2+} moiety may be distinguished only in solids, as in $TiOCl_4^{2-}$, and in the [TiO(ring)] species where ring = porphyrin or phthalocyanin (see Appendix B). In this behaviour, titanium(IV) contrasts with V(IV) and Cr(V), where the MO^{n+} unit does exist on its own. The structure of titanyl sulfate, for example, consists of $(TiO)_n^{2n+}$ chains held together in the crystal by the sulfate groups. The alkali metal titanates do not contain discrete anions but consist of TiO_6 octahedra linked into layers or three-dimensional arrays and holding the alkali metals in sites of high coordination (7 up to 10). Thus they are mixed oxides, not oxytitanium anions.

Titanium(IV) forms peroxy compounds which are the formula analogues of the oxy-compounds as shown in Fig. 14.4. The yellow colour in acid is extremely intense and is the basis of the colorimetric determination of titanium. A more detailed study of the acid solutions shows that the principal species at pH < 1 is $Ti(O\text{-}O)(OH)^+$, and at pH = 1–2 are the ions

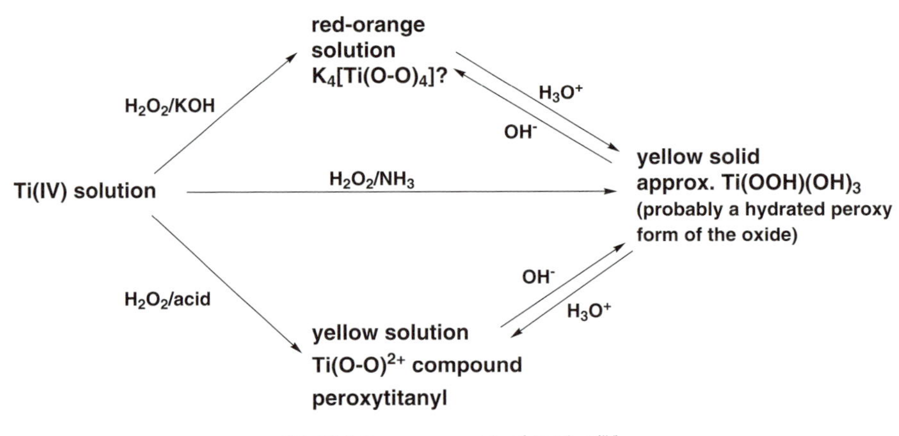

FIG. 14.4 Peroxy compounds of titanium(IV)

$$\left[(O_2)Ti-O-Ti(O_2)\right](OH)_x^{2-x} \quad (x = 1 \text{ to } 6)$$

which then precipitate $TiO_3.nH_2O$ ($n = 1$ or 2).

The titanium halides may all be made by direct reaction of the metal and halogen, and $TiCl_4$ may be converted to the others by the reaction of the hydrogen halide. All the tetrahalides are readily hydrolysed, although TiF_4 is more stable than the others, and the intermediate oxyhalides may be isolated under careful conditions:

$$TiX_4 \xrightarrow{H_2O} TiOX_2 \xrightarrow{H_2O} TiO_2.nH_2O$$

Titanium tetrachloride forms the hexachloride in concentrated hydrochloric acid, and salts, M_2TiCl_6, are known but are very unstable. The hexafluoride, TiF_6^{2-}, is readily formed and is very stable. The shape is regular octahedral.

An interesting series of compounds, illustrating the three ways of linking octahedra by bridging groups, is provided by the species formed by the action of TiF_4 on TiF_6^{2-} in liquid SO_2 as solvent. As the ratio varies from 1:3 through 1:1 to 3:1, the ions $Ti_2F_{11}^{3-}$, $Ti_2F_{10}^{2-}$ and $Ti_2F_9^{-}$ are formed, with the structures shown in Fig. 14.5 suggested by fluorine nmr.

The oxyfluoride may also accept F^- ions, giving complexes like $TiOF_3^-$, and, just as there are peroxy analogues of oxyanions, so peroxyfluoroanions such as $[Ti(O_2)_2F_2]^{2-}$ and $[Ti(O_2)F_5]^{3-}$ exist. If the O-O unit is seen as occupying an O position, then these are pseudo-tetrahedral and -octahedral respectively. Alternatively, since the O-O unit is bonded edge-on (compare, for example Fig. 14.12), then these peroxyanions may be described respectively as 6- or 7-coordinated.

The tetrahalides act as acceptors to a wide variety of donor ligands such as R_3P, R_2O or py, to give complexes TiX_4L_2. These are usually in the *cis* configuration unless the ligands are bulky. $Ti_2Cl_{10}^{2-}$ has the same structure as the fluoride (Fig. 14.5b). Among 5-coordinate complexes are the unusual hydride derivative $TiCl_4.AsH_3$, where the arsane molecule is

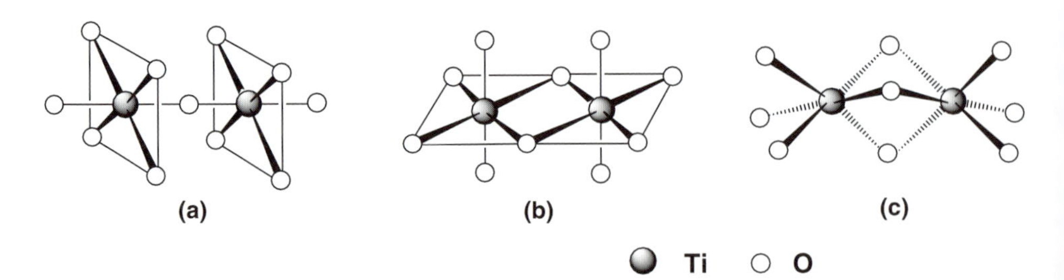

FIG. 14.5 The dinuclear fluorotitanium anions; (a) $Ti_2F_{11}^{2-}$, (b) $Ti_2F_{10}^{2-}$, (c) $Ti_2F_9^{2-}$

thought to occupy an equatorial position in a trigonal bipyramid, and the compound $Et_4N^+(TiCl_5)^-$ where the $TiCl_5^-$ ion may be trigonal bipyramidal like the isoelectronic $SnCl_5^-$ ion.

Two further simple titanium(IV) compounds are of interest. In the anhydrous nitrate, $Ti(NO_3)_4$, the nitrate groups form a tetrahedron around the titanium, but are each bonded through two oxygen atoms so that the titanium is 8-coordinate. The coordination is nearly regular dodecahedral. Although the perchlorate ion, ClO_4^-, does not usually coordinate, it is found that $TiCl_4$ reacts with anhydrous $HClO_4$ to form $Ti(ClO_4)_4$ which is a volatile solid. It has an 8-coordinate structure analogous to $Ti(NO_3)_4$. The only other volatile perchlorate is $Cu(ClO_4)_2$.

14.2.2 Titanium(III) and lower oxidation states

Titanium(III) compounds are readily formed by reduction and, as they contain a d electron, are coloured. Some preparations are shown in Fig. 14.6. The existence of differently coloured hydrates of titanium trihalides arises as both water and halide may be directly coordinated to the titanium ion. The violet trichloride is $[Ti(H_2O)_6]^{3+}Cl_3^-$ while the green form is $[Ti(H_2O)_5Cl]^{2+}Cl_2^-$.

Titanium(III) is much more basic than titanium(IV) and the purple, hydrated oxide, $Ti_2O_3.nH_2O$, which is precipitated from titanium(III) solutions by base, is insoluble in excess alkali. Anhydrous TiOCl is formed by heating TiO_2 with $TiCl_2$ at 700°C. The pure compound is stable in air.

The titanium(III) ion is a d^1 system with the electron in a t_{2g} orbital. Only one d–d transition is possible, Fig. 13.23a, and a simple spectrum with one band in the visible region is observed, Fig. 13.4. Magnetic measurements on titanium(III) compounds give values close to the spin-only value of 1.73 Bohr magnetons for one unpaired electron.

An interesting structure is found for $Ti(BH_4)_3$, which is volatile at room temperature. Each BH_4 group is linked to Ti by three bridging hydrogens, giving 9-coordinate Ti(III). Gas phase electron diffraction shows that the Ti and the 3 B atoms are coplanar.

Titanium(II) is a very unstable state represented only by solid compounds. The dihalides may be made by heating the trihalides, as indicated in Fig. 14.6, but the dihalide disproportionates to metal plus tetrahalide at a temperature below that of its formation so that it is never obtained uncontaminated by metal. The oxide, TiO, is made by heating the dioxide with titanium.

FIXING NITROGEN WITH TITANIUM IN A LOW OXIDATION STATE

Titanium(II) is thought to be the active intermediate in one of the few reversible systems of *nitrogen fixation* so far announced. A number of halides (including $TiCl_4$) yield a species which fixes N_2 gas when reduced by organometallic compounds but these produce only NH_3 on hydrolysis with the concomitant destruction of the active intermediate (see under nitrogen, Section 17.6). In a recent study, it was shown that nitrogen could be fixed and converted to ammonia by a cycle, involving titanium alkoxides, which also regenerated the starting reagent so that an overall catalytic system is possible. The main steps of the reaction are

$$Ti(OR)_4 + 2Na \rightarrow Ti(OR)_2 + 2RONa \quad (14.1)$$

$$Ti(OR)_2 + N_2 \rightarrow [Ti(OR)_2N_2]_n \quad (14.2)$$

$$[Ti(OR)_2N_2]_n + 4Na \rightarrow [\text{intermediate}] + 6ROH \rightarrow 2NH_3 + Ti(OR)_4 + 4RONa \quad (14.3)$$

Thus, the overall reaction is

$$N_2 + 6e^- + 6ROH \rightarrow 2NH_3 + 6RO^-$$

The sodium source in Equations (14.1) and (14.2) is the naphthalide complex, $Na^+C_{10}H_8$. In this, the valency electron of the sodium is transferred to the lowest antibonding π level of the aromatic hydrocarbon, where it is readily available for reductions but is in a more manageable form than if the metal was used alone (compare Section 6.1). The critical step is the uptake of N_2 from the gas phase at normal temperature and pressure by the titanium(II) alkoxide. The formulation of the reduced intermediate is still uncertain.

This cycle does not consume titanium and is thought to model the biological nitrogen-fixation process which also probably acts through the reversible formation of a low valency transition metal intermediate to which the nitrogen becomes coordinated. A model for the coordinated nitrogen is a sideways-bound N_2 (compare Section 16.10), and such a unit is involved in nitrogen fixation by the organotitanium system $(C_5H_5)_2TiCl_2$-Mg which gives N^{3-} as the initial fixed form.

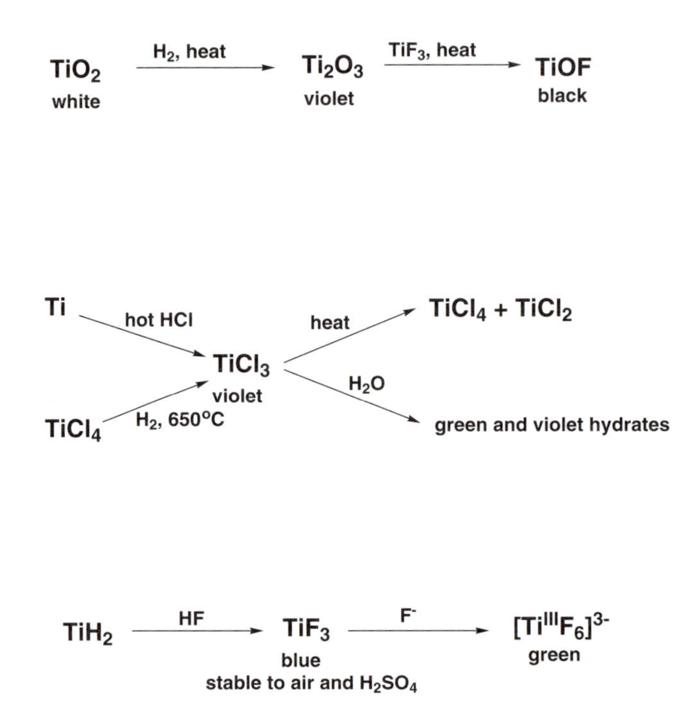

FIG. 14.6 Preparations and reactions of titanium(III) compounds

2,2'-dipyridyl (dipy)

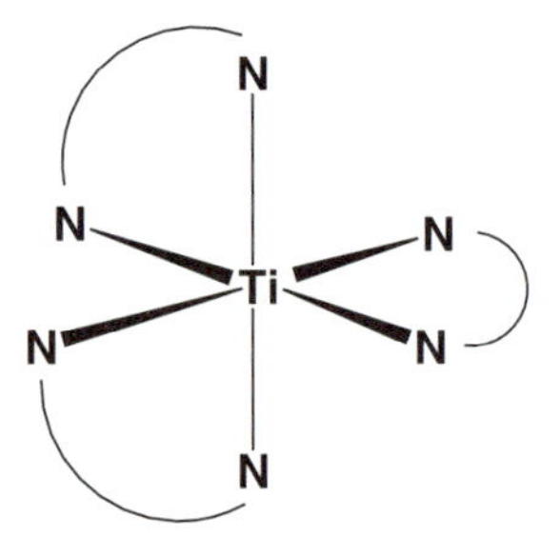

FIG. 14.7 2,2'-dipyridyl, and its titanium(0) compound, $Ti(dipy)_3$

The titanium–oxygen system is, in fact, an extremely complex one with at least twenty phases with different structures recognized. Some of the main ones are as follows:

Up to $TiO_{0.5}$	Ti atoms hcp with O in trigonal prismatic sites so $TiO_{0.33}$ is anti-BiI_3 and $TiO_{0.5}$ is anti-CdI_2 (layer structures with O in the Bi or Cd sites—see Table 5.3).
Around TiO	NaCl structures but defective with equal numbers of cation and anion sites empty (Schottky defects, see Section 5.10). At $TiO_{1.0}$, at room temperature, one-sixth of the sites are vacant.
Ti_2O_3	Corundum structure (Fig. 5.4b)
Ti_3O_5, Ti_4O_7 etc to $Ti_{10}O_{19}$	TiO_6 octahedra sharing edges forming sections of rutile structure joined by shared faces.
TiO_2	Rutile (Fig. 5.3b) together with two other forms.

Thus TiO_2 and Ti_2O_3 are stoichiometric phases but distinct species exist between these Ti(III) and Ti(IV) oxides. TiO is part of a range of structures while the 'lower' oxides formally Ti in oxidation states of I or less—are better understood in terms of hcp Ti metal taking up oxygens and filling sites in a regular manner.

One compound of titanium in the oxidation state 0 has been reported. This is the compound $Ti(dipy)_3$, where dipy = 2,2′-dipyridyl (Fig. 14.7). The oxidation states –I and –II are reported in the same system, in the compounds Li(Ti $dipy_3$)3.5THF and Li_2(Ti $dipy_3$).5THF respectively. These compounds result when $TiCl_4$ is reduced in the presence of dipyridyl by lithium in tetrahydrofuran (THF).

14.3 Vanadium, $3d^34s^2$

The free energy diagram for vanadium is shown in Fig. 14.8. There are five valency electrons and oxidation states from V to –I are known, with the ones from II to V of importance. Vanadium(V) and (IV) are both stable, with the former mildly oxidizing and represented mainly by oxy-species. Vanadium(IV) is stable, but shows disproportionation reactions in the solid when volatile species, like VF_5, can be driven off. Vanadium(III) is reducing but less so than Ti(III); it is stable to water and is only slowly oxidized in air.

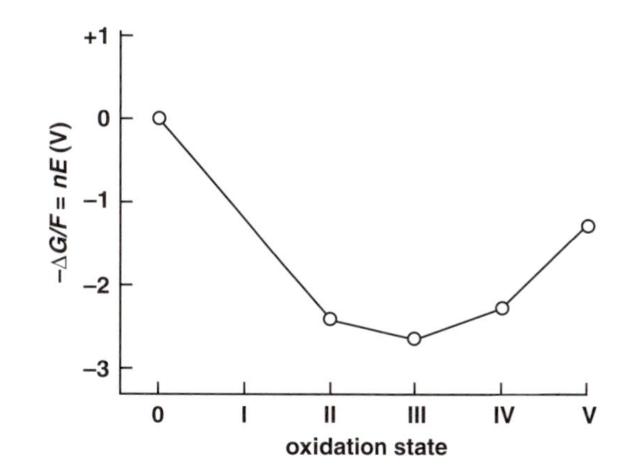

FIG. 14.8 The oxidation state free energy diagram of vanadium (Compare Section 8.6)

Particular interest in aqueous vanadium chemistry since the early 1980s has stemmed from the discovery that vanadium occurs in the nitrogen-fixing system of some bacteria. This is a second fixing system, independent of the better known one which involves molybdenum. Vanadium is also involved in peroxidases of some marine organisms and vanadate, in combination with hydrogen peroxide, mimics the effect of insulin.

Vanadium(II) is strongly reducing; it attacks water and is rapidly oxidized in air. Again, in the solid state, its compounds tend to disproportionate, yielding vanadium(III) and the element. The oxidation state free energy diagram shows that none of the intermediate states disproportionates in solution.

The metal resembles titanium in readily forming carbides and nitrides, but as it is seldom used alone this is less of a disadvantage. Its main use is in steels and it is usually produced by reducing a mixture of V_2O_5 and Fe_2O_3 with aluminium to form ferrovanadium which is added directly in the steel manufacture. Vanadium metal is prepared by reduction with calcium of the pentoxide, V_2O_5. Pure samples of the metal are best prepared by the van Arkel process in which the iodide is decomposed on a hot filament under vacuum. The pure metal resembles titanium in being relatively unreactive at low temperatures. Although it is thermodynamically a strong reducing agent, vanadium easily becomes passive and reacts readily only with oxidizing agents such as nitric acid. When heated, vanadium does react, as shown in Fig. 14.9. Note that only O_2 or F_2 give the V state.

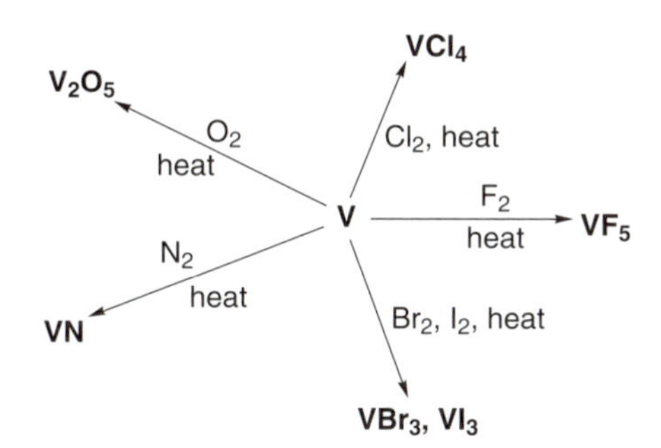

FIG. 14.9 Reactions of vanadium

14.3.1 Vanadium(V)

Simple compounds found in the V state are V_2O_5, VF_5, VO_2X (X=F, Cl) and all the oxyhalides, VOX_3. The pentoxide is prepared, and reacts, as shown in Fig. 14.10. It is amphoteric, dissolving in acid and base to give a variety of species. In strong base, the mononuclear vanadate ion, VO_4^{3-}, is present and, as the pH is reduced, these units link up into binuclear species, and then into polynuclear ions until, at pH about 6, $V_2O_5.nH_2O$ is precipitated. In more acidic solutions, this dissolves to form cationic vanadyl species. The approximate species present at different pH values are shown in Table 14.1, together with the colour changes.

While the lower vanadium oxide systems are similar to the titanium species, being based on VO_6 units, vanadium(V) oxide and the vanadium(V) oxyanions are based on VO_4 tetrahedral units.

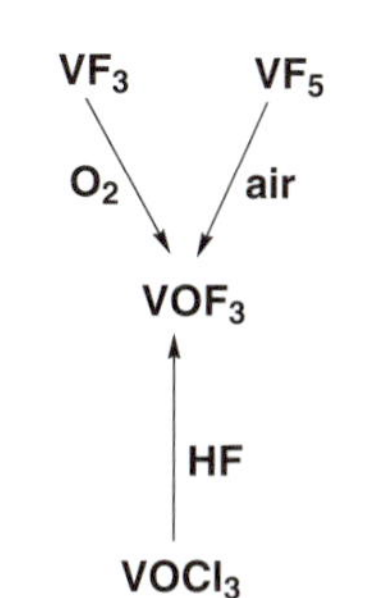

Preparation of VOF_3

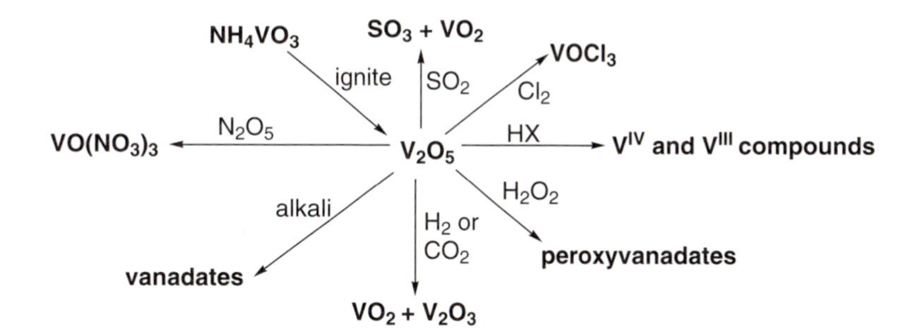

FIG. 14.10 Preparation and reactions of vanadium pentoxide

TABLE 14.1 Vanadium(V) species in basic and acidic solutions

pH	Above 12	12–9	9–7	7–6.5
Number of V atoms	1	2	3 → 4	5 → ∞
Approximate formula	$(VO_4)^{3-}$	$[V_2O_6(OH)]^{3-}$	$(V_3O_9)^{3-}$	$V_5O_{14}^{3-}$*, $V_5O_{16}^{5-}$*
Colour	←	Colourless	→	Red → red-brown

pH	6.5–2.2	Below 2.2
Number of V atoms	∞	10 ⇌ 1 (V_1 being favoured as the pH is lowered)
Approximate formula	$V_2O_5.nH_2O$*	$V_{10}O_{28}^{6-} \rightleftharpoons VO_2^+$ (or VO^{3+})
Colour	Brown	Yellow

*These species are representative of solids precipitated at the appropriate pH; they do not necessarily correlate with the formulae of species in equilibrium with them in solution. The other formulae apply in solution; the V_{10} species is rather well established and, for example, V_9 or V_{11} give a much poorer fit to the data.

Vanadium forms a red peroxy cation in acid peroxide solution and this is probably the peroxy analogue of the vanadyl cation, $V(O\text{-}O)^{3+}$. In alkaline solution, a yellow pervanadate is formed with two peroxy groups per vanadium; the simplest formulation is $[V(O\text{-}O)_2O_2]^{3-}$. The use of stronger base and 30% H_2O_2 gives the blue, fully substituted $[V(O_2)_4]^{3-}$ and also blue $[V(O_2)_3]^-$.

Other ligands are found in peroxy complexes, as in $[VO(O_2)_2(C_2O_4)]^{3-}$ (Fig. 14.12) and the related complex with bidentate CO_3^{2-} replacing the oxalate. In these complexes, the VOO triangles are unsymmetric with the shorter V-O bonds *trans* to the oxalate. Similar ions are known with F or Cl as ligands in $[VO(O_2)_2X]^{2-}$. Treatment of the pentoxide with

POLYVANADATES

An extensive series of polyvanadates is known, not all with determined structures. One species which has been characterized is $HV_4O_{12}^{3-}$. This is formed of four VO_4 tetrahedra linked into a ring by sharing oxygen. The V-O-V bridges have relatively open angles at oxygen of 132–146° The H protonates one of the nonbridging oxygens. The parent tetravanadate $[V_4O_{12}]^{4-}$ has recently been found to have a very similar structure. Anhydrous MVO_3 species (M = alkali metal) have infinite chains of linked VO_4 tetrahedra while $KVO_3.H_2O$ is made up of edge-linked VO_5 units. Complex oxyanions of vanadium(V) are becoming established whose building blocks are VO_6 units and which parallel the features of the polymolybdates and polytungstates (Section 15.4.1, especially Table 15.1). In particular, vanadium species $[XV_{12}O_{40}]^{n-}$, known as Keggin structures, are under study where an XO_4 tetrahedron is encapsulated in the structure of Fig. 15.14a, formed by 12 linked MO_6 units sharing edges. In a recent determination the anion $[AsV_{12}O_{40}(VO)_2]^{9-}$ has been characterized which has the classical Keggin structure of a central AsO_4 tetrahedron within 12 VO_6 octahedra. A special feature is the presence of two VO^{3+} ions in cavities of the structure with a square pyramid $O{=}VO_4$ configuration. The presence of these additional cations avoids the very high charge which may well prevent the formation of the simple $[AsV_{12}O_{40}]^{15-}$ cluster.

Such compounds illustrate the strong tendency towards forming polymeric anions in this part of the periodic table. The most explored and striking exemplars are vanadium, molybdenum and tungsten, but their neighbours behave similarly though with a less extensive range of compounds.

In the early 1990s, attention was focused on a further striking class of polyvanadates where neutral molecules and anions are enclosed within basket-like or shell structures. In the $[Ph_4P]^+$ salt, the anion $[V_{12}O_{32}]^{4-}$ may incorporate an acetonitrile molecule with the MeCN unit held 'head downwards' in an open basket structure (Fig. 14.11a) and the interaction was shown to persist in solution by nmr spectroscopy. Even more surprising was the incorporation, despite the like charges, of a halide or carbonate anion within $[V_{15}O_{36}]^{5-}$, which could be described as a basket with a lid.

With the incorporation of a range of anions within polyvanadate structures (Figs 14.11b,c) it is becoming clear that the guest anion is acting as a template on which the polyvanadate builds up. An alternative view is that the polyvanadate unit is acting as a macroligand analogous to the cryptands and crown ethers (compare Section 10.5).

Similar clusters may contain vanadium(IV) as well as (V), and there is even an example of one in which two ammonium ions and two chloride anions are encapsulated (Fig. 14.11d). Related fully reduced species, all vanadium(IV), are well established, with interesting magnetic and redox properties.

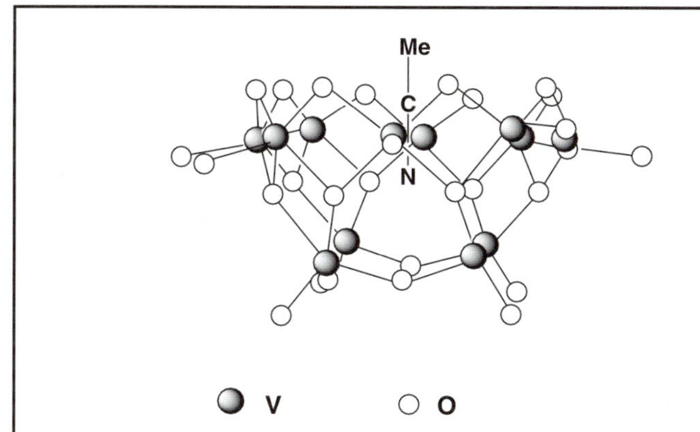

FIG. 14.11a Structure of the $[V_{12}O_{32}]^{4-}$ ion showing the incorporated acetonitrile (CH_3CN) molecule

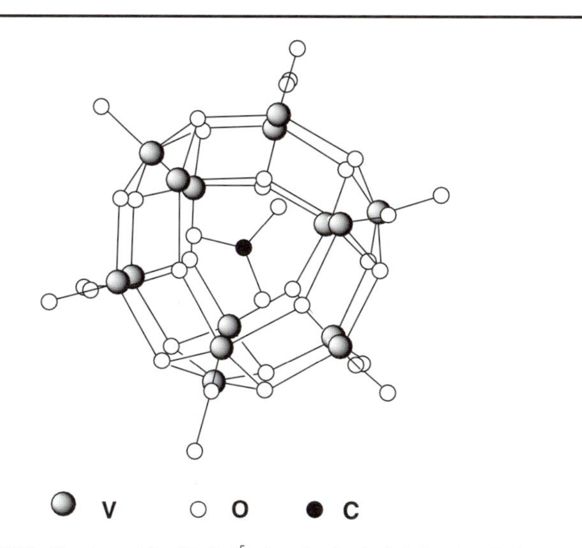

FIG. 14.11b Structure of the $[V_{15}O_{36}]^{5-}$ ion showing included carbonate ion

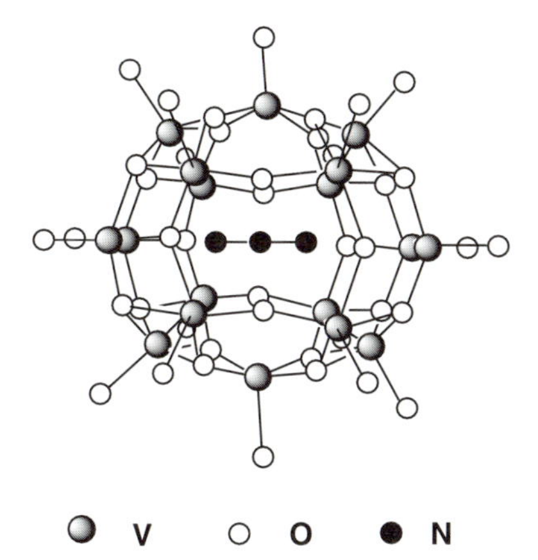

FIG. 14.11c Structure of the $[H_2V_{18}O_{44}(N_3)]^{5-}$ ion showing the encapsulated azide (N_3^-) ion

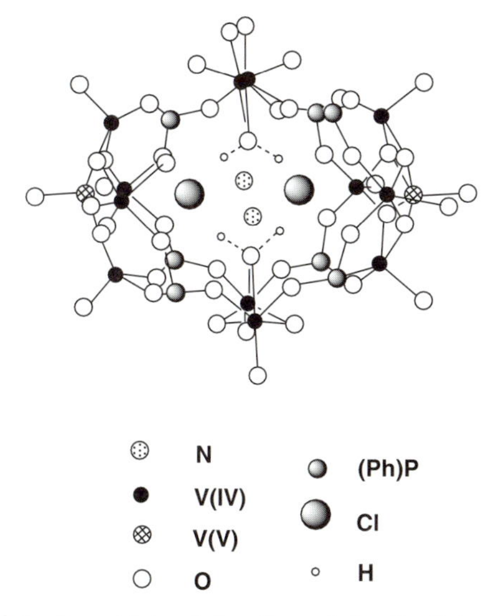

FIG. 14.11d A polyvanadate ion cluster with a central cavity which contains two NH_4^+ ions and two Cl^- ions. The cluster can be viewed as two $[V_5O_9]^{3+}$ units which form the left-hand and right-hand sections, and two $V(IV)_2(\mu\text{-}O)_2$ units at the top and bottom. These are linked up to form the cavity by eight $PhPO_3$ units (two at each corner). This cavity contains the Cl^- to the left and right and the NH_4^+ in front and behind. The ammonium ions are linked by H-bonding, part of which is indicated. Note all the H were located by crystallography. The overall formula is $[2NH_4^+, 2Cl^- \subset V_{14}O_{22}(OH)_4(H_2O)_2(PhPO_3)_8]^{6-}$ where $\subset$ indicates that the group is encapsulated within a cavity

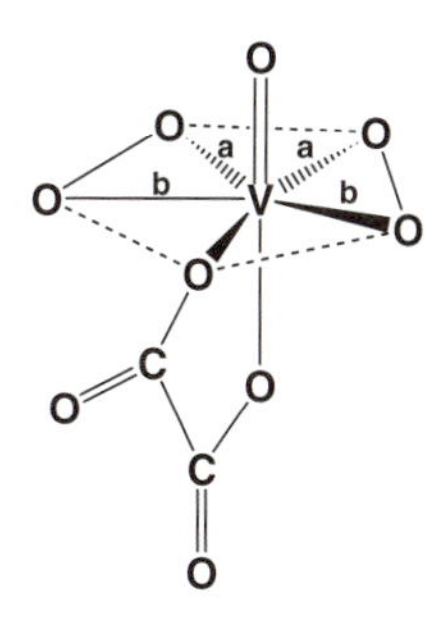

FIG. 14.12 The structure of the peroxy (oxalate) vanadate(V), $K_3VO(O_2)_2(C_2O_4)$. Bond lengths to peroxide oxygens (a) are 186 pm, (b) are 192 pm. The four peroxy O atoms and one oxalate O lie approximately in a pentagon, with the second oxalate O and the V=O completing a pentagonal bipyramid

H_2O_2 in presence of F^- in strong base gave deep blue $[V(O_2)_3F]^{2-}$, and the corresponding chloride is known. Thus from one to four O atoms in a vanadium–oxygen species may be replaced by an O-O group.

The only pentahalide of vanadium is VF_5. This results from fluorine plus the metal at 300°C or by disproportionation:

$$VF_4 \xrightarrow[N_2]{600^\circ C} VF_5 \uparrow + VF_3$$

The pentafluoride is a volatile white solid melting at 20°C. The crystal structure is composed of chains of VF_6 octahedra with two bridging fluorines in *cis* positions to each other. Such a structure is indicated by the useful nomenclature: $VF_4(F_{2/2})$ to show the four F atoms bonded only to one vanadium and two F atoms shared between two vanadiums, and also the overall 6-coordination. A related structure is shown by the oxyfluoride ion, $VO_2F_3^{2-}$, which also contains 6-coordinated vanadium with *cis* fluorine bridges in the structure $VO_2F_2(F_{2/2})$. The O atoms are *trans* to the bridging fluorines. The hexafluoro-complex ion also exists and is made by reacting VCl_3 and an alkali halide in liquid BrF_3 which acts as a fluorinating and oxidizing agent:

$$Cl^- + VCl_3 + BrF_3 \rightarrow VF_6^- + ClF + BrF$$

Vanadium pentafluoride reacts with air to form the oxyfluoride, which also may be made by oxidizing or fluorinating suitable compounds.

The oxychloride is made by the reaction of chlorine on any of the vanadium oxides, and the oxybromide similarly. The oxyhalides are relatively volatile. An electron diffraction study shows $VOCl_3$ is a trigonal pyramid in the gas phase, and the monomeric structure probably persists in the solid. A second oxyfluoride, VO_2F, is now established, together with the Cl analogue which adds Cl^- to give the tetrahedral ion $VO_2Cl_2^-$.

$VOCl_3$ is violently hydrolysed by water and is inert to metals, even to sodium. It shows no reaction with salts and dissolves many non-metals which suggests possible uses as a non-aqueous solvent.

14.3.2 Vanadium(IV)

Vanadium(IV) is readily prepared by mild reduction of vanadium(V), for example by ferrous salts. The oxide, VO_2, is formed from V_2O_5 in this way and is dark blue. It reverts to the pentoxide on heating in air. It is also mildly amphoteric, but is much more basic than acidic. It dissolves in acid to give blue solutions of the vanadyl(IV) ion VO^{2+}, and a variety of salts containing this cation are known. It dissolves in alkali to form vanadites which are readily hydrolysed and relatively unstable. A number of vanadate(IV) anions are known, such as VO_3^{2-}, VO_4^{4-} and $V_4O_9^{2-}$. These are generally prepared by reaction of VO_2 with alkali, or alkaline earth oxides in the fused state. The best known vanadyl(IV) salts are the sulfate, $VOSO_4$, and the halides, VOX_2. A number of complexes of vanadium(IV) are known, all derived from the vanadyl ion. Examples include the halides, VOX_4^{2-}, and corresponding oxalate and sulfate compounds. Neutral complexes are formed by the enol forms of β-diketones as in vanadylacetylacetonate, $VO(acac)_2$, which has the square pyramidal coordination shown in Fig. 14.13. Lone pair donors can coordinate weakly in the sixth position to complete the octahedron as in $VO(acac)_2C_5H_5N$.

The tetrachloride is formed by direct reaction and gives the tetrafluoride with HF. Alternatively, VF_4 may be prepared directly in crystalline form from vanadium powder

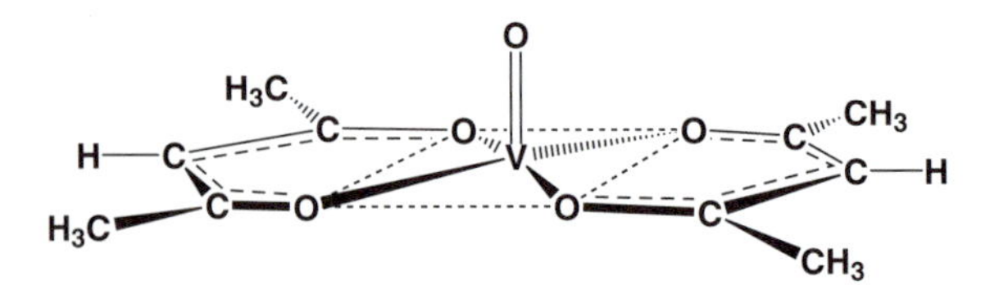

FIG. 14.13 Structure of vanadylacetylacetonate, $VO(acac)_2$

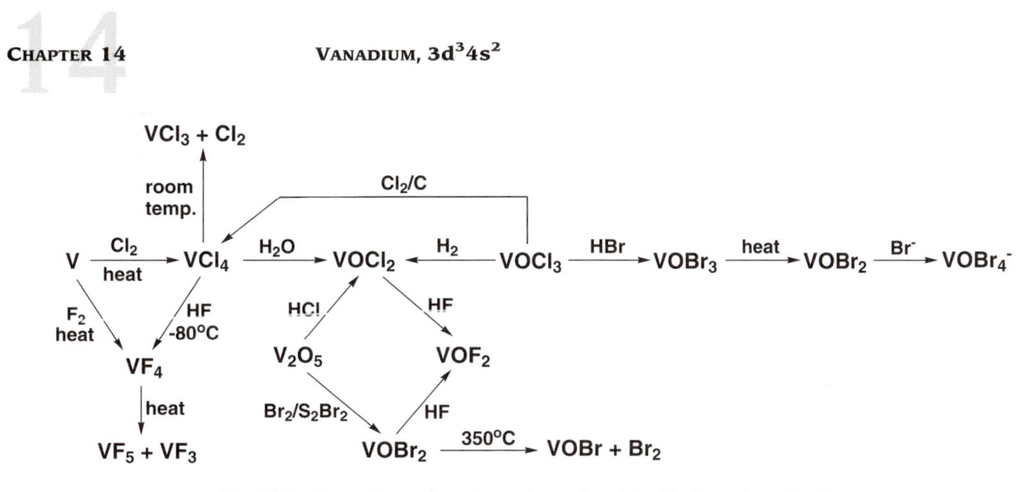

FIG. 14.14 Preparation and reactions of vanadium tetrachloride and oxychlorides

and elemental fluorine in an autoclave. The structure of VF_4 contains layers of corner-sharing VF_6 octahedra, similar to the structure of SnF_4.

This can be alternatively described as two octahedra sharing a triangular face—compare Figs 14.5c and 16.3d.

In an unusual reaction, a product with two different complex chloride ions of vanadium(IV) was formed when VCl_4 was treated with $Ph_4P^+Cl^-$ to give $(Ph_4P)_2^+(V_2Cl_9)^-(VCl_5)^-$. The VCl_5^- ion is a trigonal bipyramid. The $(V_2Cl_9)^-$ ion has a structure often found for M_2R_9 species where three R groups bridge the two M, completing 6-coordination at each. In $(V_2Cl_9)^-$ there is also a long weak V···V interaction. Hydrolysis of the tetrahalides leads to VOX_2. The interrelationships are summarized in Fig. 14.14. The tetrabromide is very unstable and VI_4 is known only in the vapour above VI_3. VCl_4 contains one d electron, which might be expected to lead to some distortion in the shape, but the molecule is tetrahedral in the gas. The liquid is dimeric. Notice, in Fig. 14.14, the tendency for these halogen compounds of the IV state to disproportionate to give the III and V states.

14.3.3 Vanadium(III)

Vanadium(III) is produced by fairly strong reduction by hydrogen or red-hot carbon. It resembles other trivalent ions such as chromium or ferric in size and general properties, apart from reduction behaviour. The solid oxide and trihalides are known and the state exists in aqueous solution as the green $V(H_2O)_6^{3+}$ ion.

This state has no acidic properties: the action of alkali on the solution of V^{3+} precipitates hydrated $V(OH)_3$ which has no tendency to redissolve. The green hydroxide oxidizes rapidly in air, and the solutions are also oxidized in air.

All four trihalides exist. VBr_3 and VI_3 are the products of direct combination with the metal while VF_3 and VCl_3 result from the disproportionation of the tetrahalides, as shown in Fig. 14.14. The trichloride itself disproportionates on heating and some of the vanadium chloride relationships are given in Fig. 14.15. It will be seen that heating with removal of chlorine gives VCl_2, while heating in excess of chlorine gives the tetrachloride. VCl_3 forms a 2:1 adduct with trimethylamine, $VCl_3.2NMe_3$, which is a trigonal bipyramid with axial NMe_3 groups. The triiodide also disproportionates

$$VI_3 \rightleftharpoons VI_2 + VI_4 \quad \text{(decomposes)}$$

The structure of VI_3 is a layer lattice like BiI_3 (Fig. 5.10).

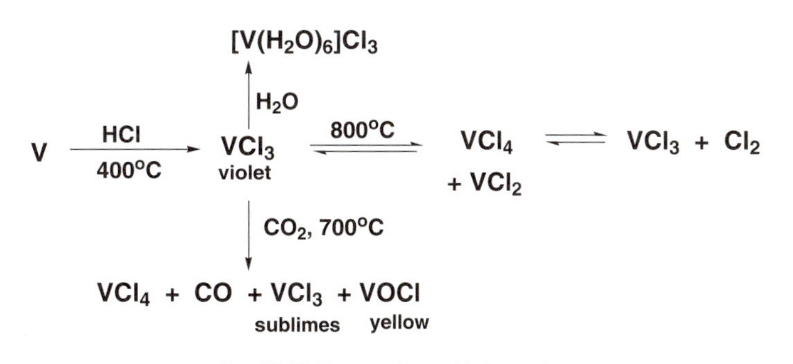

FIG. 14.15 The vanadium–chlorine system

VANADIUM(II) AND LOWER OXIDATION STATES

The vanadium(II) state is strongly reducing and evolves hydrogen from water, although it is more stable in acid solution. The oxide, VO, is made by reaction of the pentoxide with vanadium and is commonly nonstoichiometric. All four halides, VX_2, exist. The chloride and iodide result from disproportionation of the trihalides, the dibromide from reduction of VBr_3 with hydrogen. VF_2 is made by the action of HF on VCl_2 or by the reduction of VF_3 by H_2/HF at 1200°C. VF_2 forms blue crystals with the rutile structure while VI_2 exists in the CdI_2 layer lattice (see Figs 5.3 and 5.10). The dihalides dissolve in water to give violet solutions, from which $V(OH)_2$ is precipitated by alkali. The solutions soon turn green due to formation of $V(H_2O)_6^{3+}$. All these compounds are unstable and readily oxidized. The divalent ion in solution is probably octahedral and a few complexes are isolable, including the cyanide, $K_4V(CN)_6.7H_2O$. $K_4[V(CN)_7]$ is quantitatively reduced by potassium in liquid ammonia, first to $K_4[V(CN)_6]$ and then to the vanadium (I) species $K_5[V(CN)_6]$.

Low oxidation states are represented by a few compounds including dipyridyl complexes of vanadium I, 0, and –I, $V(dipy)_3^+$, $V(dipy)_3$, and $V(dipy)_3^-$, the latter being isolated as the etherate of the lithium salt, as in the case of titanium. The carbonyl, $V(CO)_6$, is known and differs from the sequence of first row carbonyls in being monomeric in the gas and therefore in not having 18 electrons on the vanadium. The anion, $V(CO)_6^-$, is also known and is a further case of V(−I). In all the low oxidation state compounds, the geometry is octahedral.

Vanadium(III) also forms a triacetate which exists as a dimer. Its structure is a sandwich one, with four bridging acetates and one terminal one on each vanadium. $AcOV(OAc)_4VOAc$. This is similar to the chromium(II) acetate (Fig. 14.20) or copper acetate structure (Fig. 14.33) with terminal acetates in place of the two water molecules.

Oxyhalides, VOX, are reported for X = F, Cl and Br. VOF has the rutile structure with both F and O in the oxygen positions of TiO_2. The corresponding TiOF is isostructural.

The V^{3+} ion forms a variety of complexes, typical of transition ions. All the complexes are labile as expected of a d^2 ion where the empty t_{2g} orbital presents a low-energy path for attack. The commonest shape is octahedral and most complexes contain vanadium coordinated to oxygen, nitrogen or halogen. The hexafluoride and related complexes are stable, e.g. VF_6^{3-} and $VF_4(H_2O)_2^-$, but other halogen complexes are more readily oxidized. A cyanide, $V(CN)_6^{3-}$, can be made in alcohol but precipitates $V(CN)_3$ on addition of water. A second cyano-complex is $V(CN)_7^{4-}$, an example of the relatively rare coordination number seven. It has a pentagonal bipyramid structure with the five equatorial ligands close to a pentagon with bond angles averaging 72°. The axial C-V-C angle is slightly bent at 171°. While the pentagonal bipyramid structure is found for molecular IF_7, 7-coordinate complexes commonly show one of the alternative structures of Fig. 15.1 or Fig. 15.5.

Among the complexes formed by the heavier vanadium(III) halides is $(V_2Cl_9)^{3-}$ and the corresponding bromide. This has the face-sharing octahedral structure, which we met above for the analogous vanadium(IV) complex $(V_2Cl_9)^-$, except that the structure is fully symmetrical with no V···V interaction. Analogous M(III) dimeric chloro-complexes are now known for Ti, Cr and a number of heavy transition metals.

14.4 Chromium, $3d^54s^1$

The oxidation state free energy diagram for chromium, Fig. 14.16, illustrates the increased oxidizing power of the Group oxidation state of VI, as compared with the cases of titanium and vanadium, and the stability of the III state. This is further shown in the solid state by the existence of CrO_3 and the unstable CrF_6 only (Tables 13.3 to 13.5). The III state corresponds to the d^3 configuration and is very stable. Other states shown range to chromium(−II) in carbonyl anions but only the II state is well represented, the IV and V states disproportionating very readily.

Chromium metal is formed by aluminium reduction of Cr_2O_3. Like titanium and vanadium it is unreactive at low temperatures, due to the formation of a passive surface coating of oxide. The main use of chromium is as a protective plating, and as an additive to steels. It dissolves in hot hydrochloric or sulfuric acids and reacts with oxygen or the

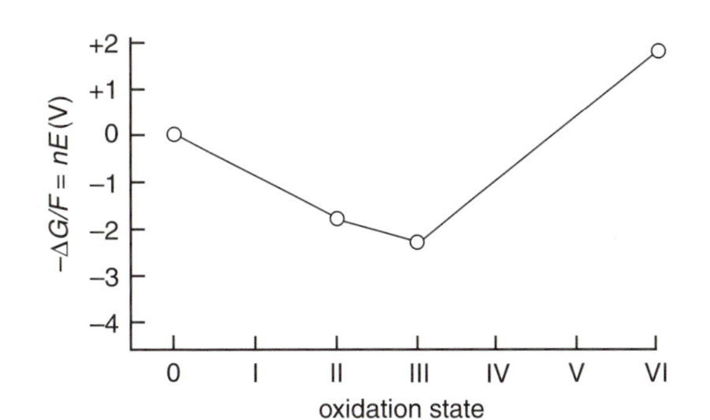

FIG. 14.16 The oxidation state free energy diagram for chromium. It is clear that Cr(III) is stable, Cr(VI) is strongly oxidizing, and Cr(II) is reducing. The other states disproportionate, but data are not very reliable

halogens on heating to give chromium(III) compounds, and with hydrogen chloride to give $CrCl_2$ and hydrogen.

14.4.1 Chromium(VI)

The limited number of chromium(VI) compounds includes CrO_3, the chromium oxyions $Cr_nO_{3n+1}^{2-}$ ($n=1,2,3,4$), $CrOX_4$ (X = F, Cl), CrO_2X_2 and CrO_3X^- (X = F, Cl, Br, I) and peroxy compounds. CrO_3 reacts with F_2 to yield a volatile lemon-yellow solid when the reaction is carried out at 170°C and 25 atmospheres. At normal pressures, the reaction gives CrO_2F_2 at 150°C and $CrOF_4$ at 220°C. The lemon solid was identified as CrF_6 but this has been a matter of controversy and the solid has also been interpreted as CrF_5, sometimes contaminated with $CrOF_4$. The second product, CrO_2F_2, is a monomer in the liquid and an F-bridged polymer, with 6-coordinate Cr, in the solid. In the monomer the OCrO angle is 108° and FCrF is 112°. Very similar values of 108.5° and 113° are found for CrO_2Cl_2. An electron diffraction study on the third product shows $CrOF_4$ has C_{4v} symmetry with a square plane of four fluorine atoms, the Cr=O group perpendicular to this and the Cr raised out of the F_4 plane.

CrO_3 is acidic and dissolves in alkali to give solutions of the chromate ion. On acidification, these give dichromate solution, which exists in strongly acid solutions down to a pH of 0. In very concentrated acid the trioxide is precipitated and there are no cationic forms of chromium(VI) in solution. This behaviour is similar to that of vanadium(V) but the chromium(VI) is more acidic and polymerization does not go so far. The equilibrium between chromate and dichromate is rapid and the two forms coexist over a wide range of pH. Many chromates of heavy metals, such as Pb^{2+}, Ag^+ or Ba^{2+}, are insoluble and these may be precipitated from chromium(VI) solutions, even at a pH where the major part exists as dichromate, as the equilibrium is so rapidly established. The CrO_4^{2-} ion is tetrahedral while dichromate has two tetrahedral CrO_4 groups joined through an oxygen with the Cr-O-Cr angle about 115°. The polychromate series is much more limited than polyvanadates or polymolybdates, but the next two members, $Cr_3O_{10}^{2-}$ and $Cr_4O_{13}^{2-}$ have been characterized. They are chains of 3 and 4 CrO_4 tetrahedra linked by corners.

> In a well-known laboratory demonstration, ammonium dichromate $(NH_4)_2Cr_2O_7$ deflagrates briskly when a small mound is ignited. However, very serious explosions have been reported when the material decomposes under confined conditions in a closed vessel.

If a dichromate solution is heated in the presence of chloride ion and concentrated sulfuric acid, the red, hydrolysable chromyl chloride, CrO_2Cl_2, distils out (b.p. 117°C). The chromyl bromide and iodide exist only in low-temperature matrices, and the fluoride is made only by the action of fluorine on chromyl chloride, so the latter compound distinguishes chlorine from the rest of the halogens and is used in this way in qualitative analysis. The chromyl halides are covalent compounds which hydrolyse in water to chromate, and there is no evidence for any CrO_2^{2+} cation. The intermediate halochromate ions, CrO_3X^-, are known for X = F, Cl, Br, with the chlorochromate the most stable. $CrOCl_4$ is also reported.

14.4.2 Chromium(V) and chromium(IV)

There are few chromium(V) compounds and they readily convert to chromium(VI) or (III). They include the peroxide (see above), CrF_5, $CrOF_3$, some oxyhalide ions such as $CrOF_4^-$

PEROXY COMPOUNDS OF CHROMIUM

A very complicated series of reactions occur between Cr(VI) species and hydrogen peroxide, which depend on pH and on Cr concentration. The ultimate product is Cr(III) and complete decomposition of the H_2O_2. When hydrogen peroxide is added to acidified chromate solution, a deep blue, transient colour appears. This is unstable in water but extracts into ether (the colour test for chromium) and a stable solid pyridine adduct, $CrO_5.C_5H_5N$, may be prepared. This is a peroxy-analogue of the trioxide. The structure of the pyridine species is a pentagonal pyramid, Fig. 14.17a, with the two peroxy groups and the donor atom forming the pentagonal base. The dipyridyl analogue, CrO_5(dipy) has the related pentagonal bipyramid structure, Fig. 14.17b.

In less acidic media, a violet species which is probably $CrO(O_2)_2(OH)^-$, i.e. the anion of $CrO_5.H_2O$, is found, and has been isolated as the violet, explosive, potassium salt $KHCrO_6$.

In alkaline solution, a red-brown species forms which is more stable, and well characterized as K_3CrO_8. This contains the tetraperoxy ion $Cr(O_2)_4^{3-}$ of chromium(V). The paramagnetism corresponds to the d^1 configuration of Cr(V). The structure is dodecahedral, like $Mo(CN)_8^{4-}$ in Fig. 15.18. Unlike the symmetrical CrO_2 triangle of the blue species, this has one Cr-O distance of 194 pm and one of 185 pm in each peroxy-chromium triangle. K_3CrO_8 is isostructural with K_3MO_8 (M = Nb, Ta), reinforcing the Cr(V) assignment. Two further Cr(V) peroxy species have been characterized, forming a family with 8-coordinate $[Cr(O\text{-}O)_4]^{3-}$. These are the 7-coordinate $[Cr(OH)(O\text{-}O)_3]^{2-}$ and the octahedral $[CrO(OH_2)(O\text{-}O)_2]^-$ which has the water molecule *trans* to the Cr=O group. While the tetraperoxy complex forms in alkali at high concentrations of hydrogen peroxide, the lower peroxides form at progressively decreasing H_2O_2 concentration and in increasingly acidic solution. It is this variation in behaviour with changes in conditions which has made the understanding of peroxide species so difficult for all the transition metal compounds.

In the presence of ammonia, a dark red-brown species $(NH_3)_3CrO_4$ may be isolated. This contains two peroxy groups and is a chromium(IV) compound, again with a pentagonal bipyramid structure as shown in Fig. 14.17c.

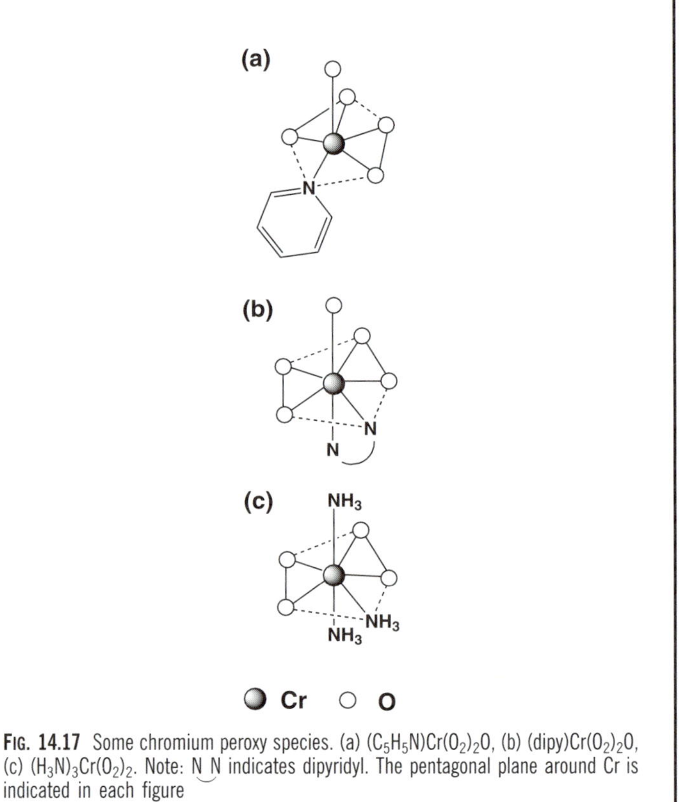

FIG. 14.17 Some chromium peroxy species. (a) $(C_5H_5N)Cr(O_2)_2O$, (b) $(dipy)Cr(O_2)_2O$, (c) $(H_3N)_3Cr(O_2)_2$. Note: N N indicates dipyridyl. The pentagonal plane around Cr is indicated in each figure

and $CrOCl_5^{2-}$, and the oxyanion, CrO_4^{3-}, which results when potassium chromate is heated in molten KOH (compare $K_2Mn^{VI}O_4$). A recent structural study indicates that the CrO_4^{3-} ion is of distorted tetrahedral structure, as might be expected for a d^1 species. Red CrF_5 is a *cis*-bridged polymer and is formed by heating CrO_3 with F_2 at 200°C or by treating CrO_2F_2 with XeF_2. In purple $CrOF_3$, square-pyramidal $CrOF_4$ units are linked into a three-dimensional array by sharing F corners. $CrOF_3$ disproportionates into Cr(VI) and Cr(III) in water, gives CrF_3 and O_2 on heating, and adds fluoride to form $CrOF_4^-$ which is probably also a fluorine bridged polymer.

An unusually stable class of Cr(V) complexes is the $CrOL_2^-$ group where L = OCR_2COO, formed by loss of two protons from α-hydroxybutyric acid and its relatives. The Cr=O is axial and four O from the two ligands form the base of a square pyramid. This compound is stable as a solid in air, and stable in solution in presence of excess ligand. It finds important uses as a specific oxidizing agent.

Chromium(IV) compounds are also rare. CrO_2, CrF_4 and $CrOF_2$ exist and there are reports that $CrCl_4$ and $CrBr_4$ exist in the vapour phase at high temperatures in mixtures of the trihalides and halogen. The complex ion, CrF_6^{2-}, has been isolated and mixed oxides containing chromium(IV) are known. The structure of one of these, $Ba_2(CrO_4)$, has been partially determined and appears to contain discrete CrO_4^{4-} ions. Chromium(IV) is also found in some mixed oxidation state species. The sulfide Cr_5S_8 is formulated as $Cr_4^{III}Cr^{IV}S_8$, and the selenium analogue is similar. In a structural study on the oxygen species $M_2^ICr_3O_9$, linked chains of $Cr^{VI}O_6$ octahedra and $Cr^{IV}O_4$ tetrahedra were indicated.

14.4.3 Chromium(III)

By far the most stable and important oxidation state of chromium is chromium(III). It is the most stable of all the trivalent transition metal cations in water and a wide variety of compounds and complexes are known. The complexes are octahedral, inert to substitution, and have a half-filled t_{2g} set of orbitals.

The oxide is green Cr_2O_3, with the corundum structure, and is used as a pigment. It, and the hydrated form precipitated from Cr^{3+} solution by OH^- ions, dissolve readily in acid to give $Cr(H_2O)_6^{3+}$ ions and also in concentrated alkali to give the chromites. The species in the latter solutions are not identified but may be $Cr(OH)_6^{3-}$ and $Cr(OH)_5(H_2O)^{2-}$. They are readily hydrolysed and precipitate hydrated oxide on dilution. Cr_2S_3 is a black stable solid made by direct combination and is inert to nonoxidizing acids.

CrOF and all four anhydrous trihalides are known and may be prepared by the standard methods. The trichloride gives the dichloride and chlorine when heated to 600°C, and sublimes in the presence of chlorine at this temperature. It is a red-violet solid which is rather insoluble in water, except in the presence of Cr^{2+} ions. A similar effect is found in the case of chromium(III) complexes in solution. These are inert to substitution except in the presence of chromium(II) which may behave similarly in abstracting a ligand *via* a bridging and oxidation process.

> It is thought that these assist the solution process by attaching through a Cl bridge to Cr^{3+} in the crystal—Cr^{3+}_{solid}-Cl-Cr^{2+}_{soln}—which then transfers an electron to give divalent chromium in the solid. This Cr^{2+} does not fit the crystal lattice and dissolves to repeat the process, Cr^{2+}_{solid}-Cl-$Cr^{3+}_{soln} \rightarrow Cr^{2+}_{soln}$.

The trihalides in aqueous solution give rise to violet $Cr(H_2O)_6^{3+}$ ions and to a number of aquo-halogeno ions, some of which are indicated in Fig. 14.18.

A wide variety of chromium(III) complexes exist, all octahedral. The hexammine, $Cr(NH_3)_6^{3+}$, and similar complexes with variety of substituted amines and related molecules are found, and all possible aquo-ammine mixed complexes have been prepared. There are also a wide variety of aquo-X and ammino-X mixed complexes (X = acid radical like halide, thiocyanate, oxalate, etc.) and even aquo-ammino-X species. A commonly used example is Reinecke's salt, $NH_4[Cr(NH_3)_2(NCS)_4].H_2O$, where the large monovalent anion is widely

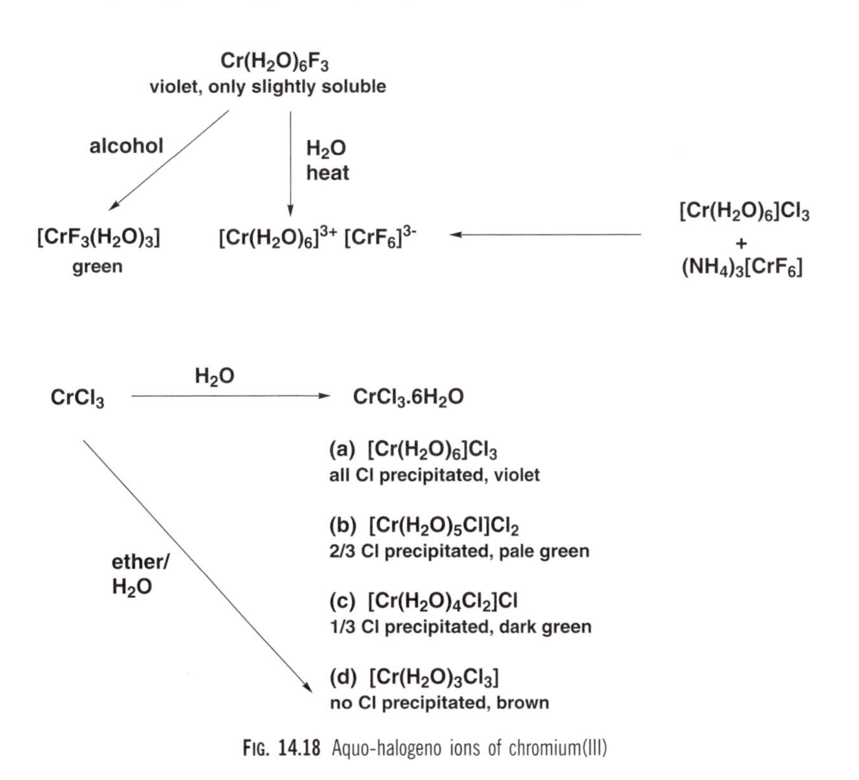

FIG. 14.18 Aquo-halogeno ions of chromium(III)

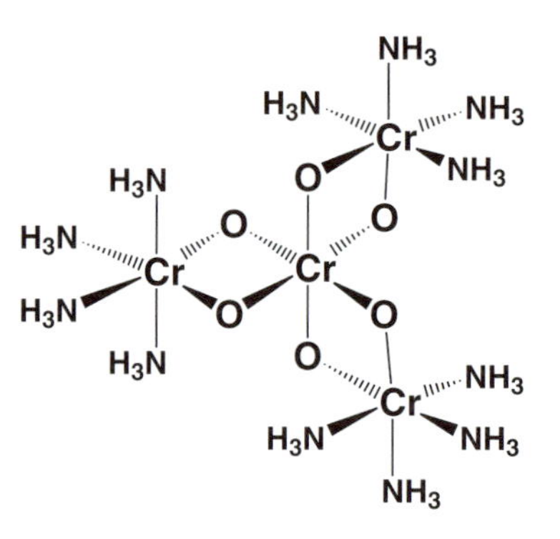

FIG. 14.19 The structure of the ion $[Cr\{Cr(OH)_2NH_3)_4\}_3]_6^+$. A central octahedron, $Cr(OH)_6$ is linked to three outer $Cr(OH)_2(NH_3)_4$ octahedra by bridging OH groups. [O = OH]

used as a precipitant for large cations. Apart from the simple hydrated cation, hydrated chromium also occurs in an extensive series of alums, $M^ICr(SO_4)_2.12H_2O$, where Cr^{3+} replaces Al^{3+} which is of similar size.

The unusual 3-coordination appears to occur in the chromium(III) species $Cr(NR_2)_3$ which have been prepared for R = isopropyl and R = $SiMe_3$. The configuration may be planar and its adoption is probably aided by the presence of the bulky ligands.

In anhydrous $KCrF_4$ the structure is based on distorted octahedral CrF_6 units. First, three of these are linked into a ring by sharing *cis* fluorine corners, and then these are linked further by the axial fluorines (perpendicular to the ring) into an infinite array.

The ion $[Cr\{Cr(OH)_2(NH_3)_4\}_3]^{6+}$ is linked through a central CrO_6 octahedron as shown in Fig. 14.19.

An interesting problem is presented by the complex $[Cr(H_2O)_5NO]^{2+}$ which can be formulated either as a Cr(III) complex with an NO^- ligand or as a Cr(I) compound with NO^+. The NO stretching frequency in the chromium complex is 1747 cm^{-1}, compared with 1890 cm^{-1} for NO, and this may be interpreted as evidence for a weaker N-O bond as expected for NO^- (compare Section 3.5). However, other evidence led to the opposite view but a recent crystal structure shows bond distances appropriate to the Cr(III)/NO^- formulation—in particular the Cr-OH_2 values are typical of Cr(III). The Cr-NO value of 168 pm suggests multiple bonding between Cr and N, and the Cr-OH_2 distance *trans* to NO is longer than the four *cis* values. All these observations are compatible with the Cr(III) formulation. A similar problem arose with $[Fe(H_2O)_5NO]^{2+}$, the material formed in the 'brown ring' test for iron, which is also now accepted as the Fe(III)/NO^- species.

14.4.4 Chromium(II) and lower oxidation states

The mixed Cr(II)–Cr(III) fluoride, Cr_2F_5, is found. The structure consists of chains of CrF_6 octahedra linked by apices with the Cr(II) and Cr(III) chains alternating. The Cr(II)F_6 unit is distorted with four short and two long CrF bonds while the Cr(III) environment is regular. This is one example of a mixed valence compound where the two oxidation states are found in distinct environments (compare Ga_2Cl_4, Section 17.4.3, p. 443 and Pb_3O_4, Fig. 17.20).

A fair number of chromium(II) compounds are known, including all the dihalides and CrO. A considerable variety of complexes are also found, although all are unstable to oxidation. In water, the sky-blue $Cr(H_2O)_6^{2+}$ ion is formed. This is a strong reducing agent which is just too weak to reduce water. It has a number of uses as a reducing agent including the removal of traces of oxygen from nitrogen. The solutions are less stable when not neutral, and hydrogen is evolved. They react rapidly with the air. Chromium(II) complexes include the hexammine, $Cr(NH_3)_6^{2+}$, and related ions. The dipyridyl complex is found to disproportionate to give a chromium(I) species:

$$2Cr(dipy)_3^{2+} = Cr(dipy)_3^+ + Cr(dipy)_3^{3+}$$

The halides can be prepared by reaction of the appropriate HX on the metal at 600°C or by the reduction of the trihalides with hydrogen at a similar temperature. The iodide, CrI_2, and CrS, may also be made by direct combination.

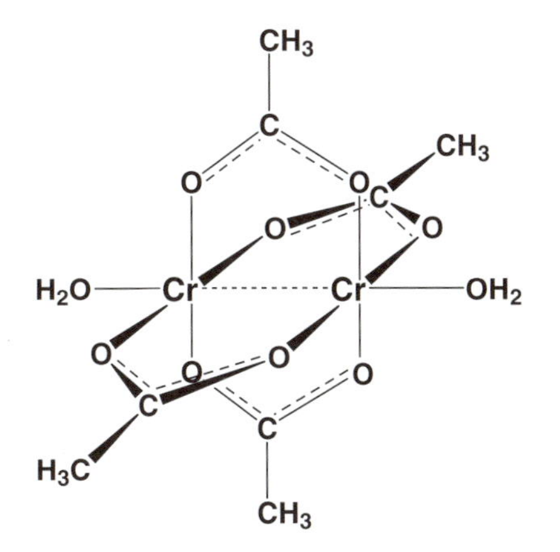

FIG. 14.20 Structure of chromium(II) acetate, $Cr_2(CH_3COO)_4.2H_2O$. The oxygen atoms of the acetate groups lie in a square plane around each Cr and the acetates link the edges together. The Cr-Cr distance is short, implying direct metal-metal bonding. Compare with copper acetate, Fig. 14.33

Among the chromium(II) salts, the hydrated acetate, $Cr_2(CH_3COO)_4.2H_2O$, is the commonest. It is readily prepared by adding chromium(II) solution to sodium acetate, when it is precipitated as a red crystalline material. It has the unusual bridged structure shown in Fig. 14.20, with a very short Cr-Cr distance suggesting metal–metal interaction. An interesting structure is found in $CrSO_4.5H_2O$. Four of the water molecules lie in a square plane around the Cr and the last two positions of the octahedral array are filled by O atoms from two different sulfate groups. These bridge two Cr atoms giving an infinite array of $\cdots$O-Cr$(H_2O)_4OSO_2O$-Cr$\cdots$ units with further hydrogen bonding linking SO_2 oxygens to coordinated water molecules on Cr atoms in parallel chains.

Lower oxidation states are represented by a few compounds including the chromium(I) dipyridyl complex above. The 0 state is shown in the stable, octahedral carbonyl, $Cr(CO)_6$, in carbonyl ions, $Cr(CO)_5X^-$, and in dibenzene chromium (Fig. 16.7b). Carbonyl anions of chromium(–I) and (–II) are the compounds $Na_2[Cr_2(CO)_{10}]$ and $Na_2[Cr(CO)_6]$ respectively, prepared by sodium reduction of the hexacarbonyl. Chromium(I) is also found in the cyanide, $K_3Cr(CN)_4$, and the zero state in $K_6Cr(CN)_6$ which may be used to prepare $Cr(diphos)_3$ and $Cr(triphos)_2$ (for di- and tri-phosphane ligands see Appendix B).

14.5 Manganese, $3d^54s^2$

Manganese, with seven valency electrons, shows the widest variety of oxidation states in the first transition series. The oxidation state energy diagram, Fig. 14.21, shows that Mn(II) is the most stable state, and comparison with the diagrams for previous elements shows that manganese(VII) is the most strongly oxidizing of all the high oxidation states known in aqueous solution. Many of the intermediate states are rare, with a strong tendency to disproportionate.

Manganese is abundant in the earth's crust and its principle ore, pyrolusite, is a crude form of the dioxide, MnO_2. The metal is obtained by reduction with aluminium, or in the blast furnace. The metal resembles iron in being moderately reactive and dissolving in cold, dilute non-oxidizing acids. It combines directly with most nonmetals at higher temperatures, sometimes quite vigorously. Thus it burns in N_2 at 1200°C to give Mn_3N_2 and in Cl_2 to give $MnCl_2$. The product of high temperature combination with oxygen is Mn_3O_4. The main use of manganese is in steel, for which ferromanganese is formed by reducing the mixed ores, as in the case of vanadium. It also finds limited uses in other alloys.

14.5.1 The high oxidation states, manganese(VII), (VI) and (V)

Note! This heptoxide requires great care in handling, as it can decompose or react with explosive violence.

Manganese(VII) compounds are strongly oxidizing. Green Mn_2O_7 is produced by treating $KMnO_4$ with sulfuric acid. Both in the gas phase and in the solid, the structure contains the linked tetrahedra of Fig. 14.22. On treatment with chlorosulfonic acid, the explosively unstable oxychlorides result:

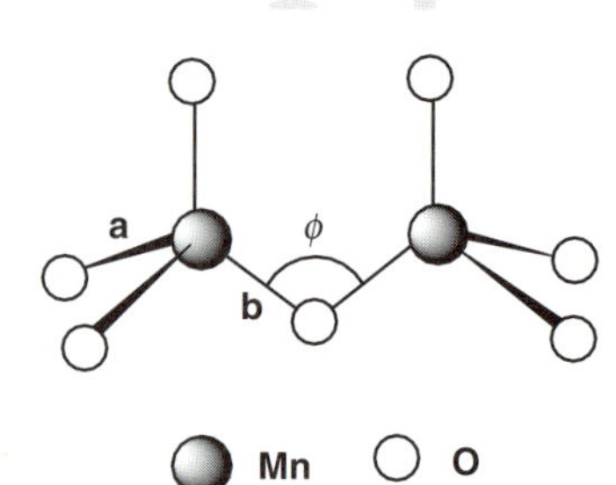

FIG. 14.22 The structure of dimanganese heptoxide: a = 159 pm, b = 17 pm, $\phi = 121°$

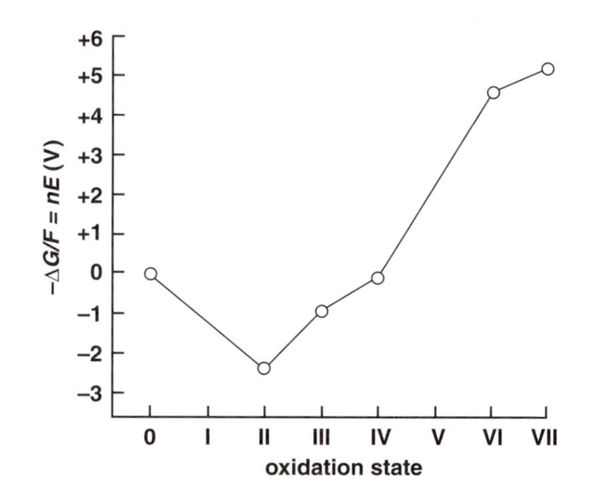

FIG. 14.21 Oxidation state free energy diagram for manganese. The stability of Mn(II) is striking and it is clear that the III and VI states disproportionate. Compare Fig. 14.1 and Section 8.6

$$Mn_2O_7 + ClSO_3H \rightarrow \underset{\text{black}}{MnO_3Cl} + \underset{\text{brown}}{MnO_2Cl_2} + \underset{\text{green}}{MnOCl_3}$$

corresponding to the VII, VI and V states. They are more stable when handled in solution in carbon tetrachloride. The oxyfluoride, MnO_3F, has been reported, but the only stable representative of the VII state is the permanganate ion, MnO_4^-, which is most common as the potassium salt. This is a strong oxidizing agent in acid solution:

$$MnO_4^- + 8H_3O^+ + 5e^- \rightarrow Mn^{2+} + 12H_2O, 1.51V$$

and is also strong in basic media:

$$MnO_4^- + 2H_2O + 3e^- \rightarrow MnO_{2(\text{solid})} + 4OH^-, 1.23V$$

when manganese dioxide is precipitated. However, in concentrated alkali the anion of the dibasic acid, manganate, is formed in preference, and this reverts to permanganate plus dioxide on acidification:

$$MnO_4^- + OH^- \rightarrow MnO_4^{2-} \underset{OH^-}{\overset{H^+}{\rightleftharpoons}} MnO_4^- + MnO_2$$

The permanganates are, of course, purple, while the manganate ion is an intense green, and is the only stable representative of the manganese(VI) state. Permanganate is about the strongest accessible oxidizing agent which is compatible with water and even it undergoes slow decomposition in acid solution to give MnO_2 and oxygen.

In an interesting disproportionation of the manganese(VI) oxyion, $Mn_7O_{24}^{2-}$ has been isolated from the action of $BaMnO_4$ with sulfuric acid. The oxidation equivalent is 6.57 per manganese atom, and the compound is formulated as the mixed manganese(IV)–manganese(VII) species $[Mn(MnO_4)_6]^{2-}$. The structure shows a central Mn(IV) coordinated by six Mn(VII)O_4 tetrahedra, each sharing one oxygen with the Mn(IV). The MnO_6 unit is octahedral and the distances are Mn(IV)-O = 190 pm, Mn(VII)-O (bridge) = 174 pm and Mn(VII)-O (terminal) = 160 pm.

The manganese(V) state is represented only by $MnOCl_3$ and the blue MnO_4^{3-} ion, formed by the action of alkaline formate on MnO_4^{2-}. Manganese(V) is relatively stable in alkaline melts, e.g. it is the major form of manganese in molten NaOH or KOH at about 400°C. It has also been stabilized in the apatite-like structure of $Ba_5(MnO_4)_3Cl$ and similar blue or green species. Here the Mn(V) takes the place of P(V) in normal apatites. The MnO_4^{3-} ion is tetrahedral with Mn-O = 170 pm.

14.5.2 Manganese(IV) and manganese(III)

Manganese(IV) does not have a very extensive chemistry although black MnO_2 is well known and is a common precipitate from manganese compounds in an oxidizing medium. It is very insoluble and usually only dissolves with reduction, as in its reaction with HCl:

$$MnO_2 + 4HCl \rightarrow MnCl_2 + Cl_2 + 2H_2O$$

A few complex salts are known including the hexachloride and fluoride, MnX_6^{2-} and also $Mn(CN)_6^{2-}$. The only simple halide reported is the fluoride, MnF_4.

The oxide Mn_5O_8 contains Mn(IV) and Mn(II). The structure is based on the CdI_2 layer type (Fig. 5.10a) and contains Mn(IV) in a distorted octahedron of oxygens together with $Mn(II)O_6$ trigonal prisms.

The III state is represented by the oxide, Mn_2O_3, and the fluoride, MnF_3. The oxide stable at high temperatures, Mn_3O_4, is correctly formulated as a mixed oxide of the II and III states, $Mn^{II}Mn_2^{III}O_4$. The structure of MnF_3 is a polymeric one with all the F atoms bridging pairs of Mn atoms. There are two different MnF_6 sites, each a distorted octahedron. In one, there are three pairs of *trans* Mn-F bonds each of different length increasing from 183 pm,

MANGANESE-OXYGEN COMPOUNDS AND THEIR OCCURRENCE IN BIOLOGICAL SYSTEMS

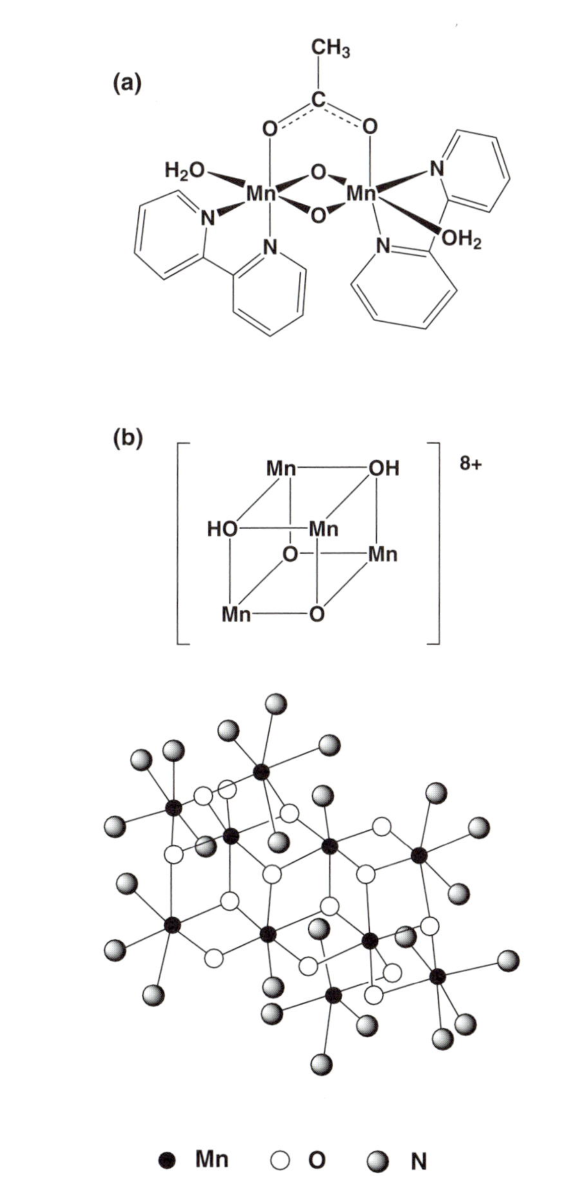

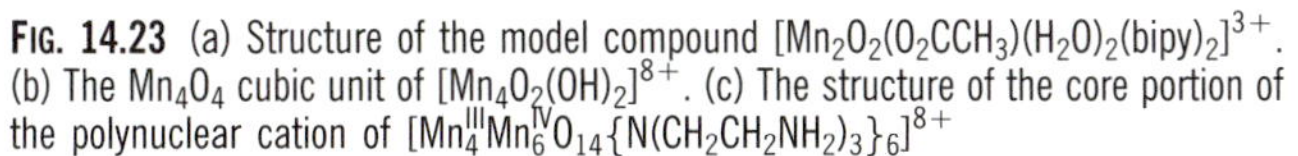

FIG. 14.23 (a) Structure of the model compound $[Mn_2O_2(O_2CCH_3)(H_2O)_2(bipy)_2]^{3+}$. (b) The Mn_4O_4 cubic unit of $[Mn_4O_2(OH)_2]^{8+}$. (c) The structure of the core portion of the polynuclear cation of $[Mn_4^{III}Mn_6^{IV}O_{14}\{N(CH_2CH_2NH_2)_3\}_6]^{8+}$

While the chemistry of simple Mn(IV) and Mn(III) compounds is limited, more complex species containing these oxidation states have been of increasing interest in biological chemistry. A number of metalloproteins containing manganese are known and their active sites have been increasingly identified. Parallel to this is the synthesis of new manganese compounds to act as models for the biological species. Compounds containing from one to four Mn are well known and larger molecules with up to 12 Mn have been studied. The established oxidation states are Mn(II), (III) and (IV) with some indication of Mn(V). Many of the manganese–protein compounds are enzymes dealing with the control of dioxygen species in the organism—especially in protecting against damage by peroxide or superoxide radicals. Human superoxide dismutase has been shown to contain four separate Mn atoms, each in an identical protein subunit. These are linked in pairs through coordination to the Mn, and the pairs are further linked by four-helix interfaces. The characteristic Mn site in a superoxide dismutase is 5-coordinated Mn(III) with a trigonal bipyramid formed from 3N and an O from amino acids in the protein, with the fifth site occupied by OH^- or OH_2. Other enzymes contain two Mn(III) atoms in octahedral sites linked by three bridges which are different combinations of oxide and carboxylate bridges, Mn-O-Mn and Mn-OC(R)O-Mn. The structure of a similar Mn(IV) dimanganese compound, prepared as a model species, is shown in Fig. 14.23a. Among the larger units is the active site in 'photosystem II' which is involved in the oxidation of H_2O to O_2 driven by visible light. One proposal for this site is the Mn_4O_4 cubane core, shown in Fig. 14.23b, which may function by opening up and closing again in a four-step cycle, transferring an electron at each step. An example of a larger structure is given in Fig. 14.23c, which is a simplified representation of $[Mn_4^{III}Mn_6^{IV}O_{14}\{N(CH_2CH_2NH_2)_3\}_6]^{8+}$. The structure can be seen as a series of Mn_3O_4 cubanes lacking a Mn corner and sharing Mn_2O_2 faces, terminated by N atoms from the ligand. The Mn atoms are all 6-coordinated to O or to mixtures of O and N.

through 192 pm, to 210 pm. In the second type of site all six Mn-F distances differ and range from 180 to 208 pm. These distortions probably reflect the Jahn–Teller effect expected for the high-spin d^4 configuration of Mn(III) (compare Section 13.4).

No other simple trihalides exist but the red $MnCl_5^{2-}$ complex ion is known, as is the corresponding fluoride. $(NH_4)_2MnF_5$ contains chains of MnF_6 octahedra linked by *trans* bridging fluorides, in the structure $MnF_4(F_{2/2})$. Large cations stabilize MnF_6^{3-}. Complex ions of manganese(III) include $MnCl_3L$ and $MnCl_3L_3$ types, where L is a nitrogen ligand, and the acetylacetonates $MnCl_2(acac)$, $MnCl(acac)_2$ and $Mn(acac)_3$ are also reported.

In solution manganese(III) is unstable both to reduction and to disproportionation:

$$2Mn^{3+} + 6H_2O \rightarrow Mn^{2+} + MnO_{2(solid)} + 4H_3O^+$$

If manganese(II) in sulfuric acid solution is oxidized with permanganate in the stoichiometric ratio, an intensely red solution results which contains all the manganese as manganese(III), presumably as a sulfate complex. This solution is as strongly oxidizing as permanganate and was once used as an alternative oxidizing agent for sulfate media. A rather similar oxidation of the acetate gives $Mn(OAc)_3.2H_2O$ as a solid. This manganese(III) acetate is readily prepared and used as a starting material for most manganese(III) studies. By heating the oxides at 500°C for 14 days, red Na_5MnO_4 crystals were produced which contained compressed $Mn(III)O_4$ groups matching the MO_4 units found for M = Fe, Co or Ni in the (III) state.

14.5.3 Manganese(II) and lower oxidation states

In contrast to all the above unstable or poorly represented states, the manganese(II) state is very stable and widely represented. It is the d^5 state and all the compounds contain five unpaired electrons, except for the cyanide and related complexes, for example $Mn(CN)_6^{4-}$ and $Mn(CN)_5(NO)^{3-}$, which are low spin with only one unpaired electron. All the high-spin manganese(II) compounds are very stable and resist attack by all but the most powerful oxidizing or reducing agents. They have all very pale colours, for example the hydrate $Mn(H_2O)_6^{2+}$ is pale pink, as the absorptions due to the d electron transitions are very weak. This is because a d–d transition which involves the reversal of an electron spin (as it must be in a high-spin d^5 system) is an event of low probability, compared with one which is 'spin-allowed'. The absorptions in manganese(II) compounds are therefore about a hundred times weaker than the general run of transition compound absorptions.

The stability of the high-spin manganese(II) state is illustrated by the wide variety of stable compounds formed. Some of these are listed in Table 14.2.

A variety of complex ions exist, including $Mn(NH_3)_6^{2+}$ and octahedral complex ions with EDTA, oxalate, ethylenediamine, and thiocyanate but the hexahalide ions are unknown. The equilibrium constants for the formation of such ions in solution are low, as the Mn^{2+} ion is relatively large (Fig. 13.20) and there is no CFSE. Thus one important source of the energy required to displace the coordinated water molecules is lacking.

By contrast to the high-spin compounds, the low-spin complexes are much more reactive and oxidize readily, for example:

$$Mn(CN)_6^{4-} \xrightarrow{air} Mn(CN)_6^{3-}$$

TABLE 14.2 Examples of high-spin manganese(II) compounds

MnX_2	X = F, Cl, Br, I	Isomorphous with the Mg halides.	Stable at red heat
MnY	Y = O, S, Se, Te	Sodium chloride structure.	Very stable when dry but the hydrated forms slowly oxidize to MnO_2 in air
$Mn(OH)_2$		This is a true hydroxide, not a hydrated oxide, isomorphous with $Mg(OH)_2$	
$MnSO_4$		Very stable even at red heat. The hydrate is isomorphous with copper sulfate	
$Mn(ClO_4)_2$		Very soluble. Stable to 150°C, then the perchlorate oxidizes the Mn^{2+} to the dioxide	
$MnCO_3$		Insoluble. Very stable for a transition metal carbonate. It goes to MnO and CO_2 at about 100°C	
$Mn(OOCCH_3)_2$ and other organic acid salts		Stable: prepared by heating $Mn(NO_3)_2$ with the acid anhydride	

Manganese(II) is unstable in these low-spin compounds probably because the crystal field stabilization energy—though large enough to cause spin pairing—is only a little greater than the loss of exchange energy due to spin pairing. In the trivalent state, the CFSE is increased because of the greater charge. A somewhat similar situation arises with cobalt(III) in water.

A few examples exist of manganese in a square-planar environment. In $Mn(acac)_2.2H_2O$, the bidentate acetylacetonate groups form a square plane around the manganese and the water molecules complete a distorted octahedron: when the compound is dehydrated, it is probable that the $Mn(acac)_2$ remaining is truly square-planar. The sulfate, $MnSO_4.5H_2O$, is isostructural with $CuSO_4.5H_2O$ and therefore contains square-planar $Mn(H_2O)_4^{2+}$ units.

Some tetrahedral manganese units are found, especially the halogen anions, MnX_4^{2-}, which may be prepared as the salts of large cations. These tetrahedral complexes are unstable in water, or other donor solvents, and go to octahedral complexes of the solvent. It is interesting that, in the presence of a large stabilizing cation, it is even possible to observe only 4-coordination in the tetrahedral $[Mn(II)(CN)_4]^{2-}$ ion isolated as the PPN^+ salt. Four-coordination by cyanide is rare and otherwise found only in the square-planar complexes of the nickel group and the tetrahedral cyanides of the zinc group.

$PPN^+ = Ph_3PNPPh_3^+$, a widely used large cation

One or two examples of 3-coordinate Mn(II) are known, where the low coordination is stabilized by bulky groups R. One example is the amide $(R_2N)_4Mn_2$, where $R = Me_3Si$. The structure is based on the unit

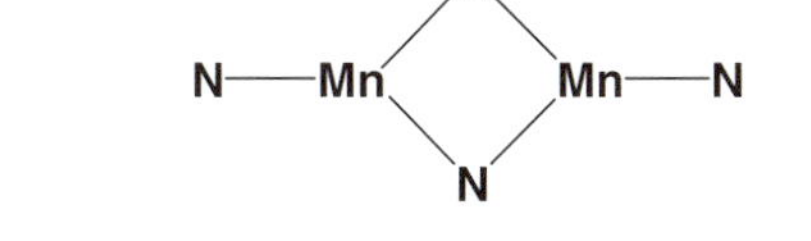

A similar species is found for Co(II), another stable +II state.

This very wide range of compounds and coordination behaviour underlines the stability of Mn(II) and its dominance in manganese chemistry.

In its low oxidation states, manganese forms the cyanide, $K_5Mn(CN)_6$, and the carbonyl halides, like $Mn(CO)_5Cl$, in the I state, the carbonyl, $Mn_2(CO)_{10}$, in the 0 state, and the anion, $Mn(CO)_5^-$, in the –I state. The latter is a trigonal bipyramid, as is the related $Mn(NO)(CO)_4$ with NO in an equatorial position. In its carbonyl chemistry, manganese forms an extensive range of polynuclear and substituted products.

14.6 Iron, $3d^6 4s^2$

When iron is reached in the first transition series, the elements cease to use all the valency electrons in bonding, and the Group oxidation state of VIII is not found. The highest state of iron is VI and the main ones are II and III. The oxidation energy diagram, Fig. 14.24, shows that the III state is only slightly oxidizing while the II state lies at a minimum and is stable in water. By comparison with the II and III states of other transition elements,

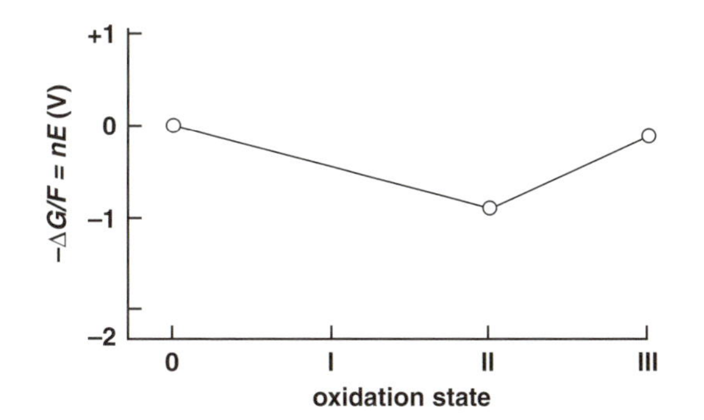

FIG. 14.24 Oxidation state free energy diagram for iron. No potential values for higher states than III are known: the few Fe(IV) compounds known are strongly oxidizing. The II state is the more stable but (compare Fig. 14.27) the III state is also relatively stable

the iron(II) and (III) states lie much closer together in stability, and this accords with the well-known properties of ferrous and ferric solutions which are readily interconverted by the use of only mild oxidizing or reducing agents.

Iron is the most abundant of the fairly heavy elements in the Earth's crust and is used on the largest scale of any metal. Its production in the blast furnace is well known, and basically involves the reduction of the oxide by carbon. The resulting metal contains a small proportion of carbon, and the various types of iron and steel result from varying carbon contents, and from other metal additions. In particular, three forms of iron exist at different temperatures, the carbide phase is Fe_3C, called *cementite*. Thus, a steel cooled slowly from above 720°C and containing about 1% C, consists of layers of α-iron interleaved with cementite to give the steel called *pearlite*, a soft malleable material. Alternatively rapid cooling of the same mixture prevents separation of the layers and *martensite* is formed which is hard but brittle. Technology has evolved to a point where a huge variety of special steels of varying properties is available. Iron is reactive in air and the process of rusting involves the formation of a coat of hydrated oxide on the surface in moist air. This is noncoherent and flakes away revealing fresh surfaces for attack. Iron combines at moderate temperatures with most nonmetals and it readily dissolves in dilute acids to give iron(II) in solution, except with oxidizing acids which yield iron(III) solutions. Very strongly oxidizing agents, such as dichromate or concentrated nitric acid, produce a passive form of the metal, probably by forming a coherent surface film of oxide.

14.6.1 The iron–oxygen system

Iron forms three oxides, FeO, Fe_2O_3 and Fe_3O_4, which all commonly occur in nonstoichiometric forms. Indeed, FeO is thermodynamically unstable with respect to the structures with a deficiency of iron. The three oxides show a number of structures, some based on a cubic close-packed array of oxide ions. Properties of the oxides are summarized in Table 14.3.

It will be seen that the structures of the cubic forms of all these oxides are related. If the cubic array of oxide ions is taken, then all the structures result from different dispositions of ferrous and ferric ions in the octahedral and tetrahedral sites. This explains the tendency to nonstoichiometry and the interconversion of oxides. If the oxide lattice has its octahedral sites filled up with Fe^{II} ions, the FeO structure builds up as the Fe/O ratio approaches one. If a small portion of the Fe^{II} is missing or replaced by Fe^{III} in the ratio of two Fe^{III} for every three Fe^{II}, defect FeO forms result, with a stability maximum at $Fe_{0.95}O$. If the process is continued until there are two Fe^{III} atoms for every Fe^{II} and if half these Fe^{III} ions enter tetrahedral sites, the structure becomes that of Fe_3O_4. Conversion of the remaining Fe^{II} to Fe^{III} gives the cubic form of Fe_2O_3. Compare Section 5.9.

14.6.2 The higher oxidation states of iron

If hydrated ferric oxide in alkali is treated with Cl_2, a red-purple solution of iron(VI) is obtained containing the ferrate ion, FeO_4^{2-}. The sodium and potassium salts are very

TABLE 14.3 The iron oxides

Oxide	Preparation	Structure
FeO (black)	Prepared by thermal decomposition of ferrous oxalate at a high temperature, followed by rapid quenching to prevent disproportionation to $Fe+Fe_3O_4$	Structure is sodium chloride, i.e. O^{2-} ions ccp and Fe^{2+} ions in all the octahedral sites
Fe_2O_3 (brown)	Occurs naturally. Otherwise by ignition of hydrated ferric oxide, precipitated from a ferric solution by ammonia	The structure has the oxide ions ccp with Fe^{3+} ions randomly distributed over the octahedral and tetrahedral sites. A second form is hexagonal
Fe_3O_4 (black)	Occurs naturally as *magnetite*. It is the ultimate product of strong ignition in air of the other two oxides	The structure is inverse spinel*: that is the O^{2-} ions are ccp, the Fe^{2+} ions are in octahedral sites, and the Fe^{3+} ions are half in octahedral sites and half in tetrahedral sites

*It will be recalled from Chapter 5 that a normal spinel is an oxide AB_2O_4 with A a divalent metal ion and B a trivalent one. The oxide ions are cubic close-packed and the A atoms are in tetrahedral sites while the B atoms are in octahedral sites. The inverse structure of magnetite may be ascribed to the much greater CFSE of d^6 (low-spin) ferrous ions in octahedral sites instead of tetrahedral ones. The d^5 ferric ions have no CFSE in either type of coordination.

IRON(III) IODIDE—THE SUCCESSFUL ISOLATION OF A 'NON-EXISTENT' COMPOUND

For a long time it was thought that the compound FeI_3 (ferric iodide) could not be obtained, due to the fact that Fe(III) is mildly oxidizing, while iodide is a reducing anion. In fact, this is exactly the case in aqueous solution—addition of I^- to an Fe(III) solution immediately produces elemental iodine:

$$Fe^{3+} + I^- \rightarrow Fe^{2+} + \tfrac{1}{2}I_2$$

However, very recent studies have shown that by carrying out the synthesis of FeI_3 in nonaqueous media it can be isolated as an unstable black compound, of structure yet to be defined. The reaction involves oxidative decarbonylation of the iron(II) carbonyl iodide $(OC)_4FeI_2$ in hexane:

$$(OC)_4FeI_2 + \tfrac{1}{2}I_2 \rightarrow FeI_3 + 4CO$$

Several derivatives of the FeI_4^- ion, together with diphosphane and diarsane derivatives of FeI_3, have also been prepared and characterized. These results were all obtained in nonaqueous media and show that Fe(III) iodine compounds are not intrinsically unstable, although they are too reactive to prepare in water.

soluble but the barium compound may be precipitated. The ferrate ion is stable only in a strongly alkaline medium; in water or acid it evolves oxygen:

$$2FeO_4^{2-} + 10H_3O^+ \rightarrow 2Fe^{3+} + \tfrac{3}{2}O_2 + 15H_2O$$

Ferrate is a stronger oxidizing agent than permanganate. It is a tetrahedral ion and the potassium salt is isomorphous with K_2CrO_4 and with K_2SO_4. The ferrate ion shows an unusual luminescence in the near-infrared region of the spectrum, suggesting a possible application in solid-state infrared laser materials.

A pentavalent oxyanion, FeO_4^{3-}, is also reported. Iron(V) also occurs 4-coordinated to oxygen in the complex phase, $SrLa_3LiFeO_8$.

Iron(IV) is also rare. Alkali metal iron oxides such as Sr_2FeO_4, $Sr_3Fe_2O_7$, $MFeO_3$ (M = Ca, Sr) are all known, and contain distorted FeO_6 octahedra which share vertices. However, the $MFeO_3$ only exist under a pressure of O_2 and disproportionate to Fe(V) plus Fe(III). Na_4FeO_4 is also known, and has a structure consisting of discrete FeO_4^{4-} ions plus Na^+ cations. The FeO_4^{4-} ions are strongly distorted, as predicted, since the Fe(IV) oxidation state has a high-spin d^4 electronic configuration and would be expected to undergo a Jahn–Teller distortion, as does the isoelectronic Mn(III). Fe(IV) also occurs in the coordination complex $Fe(diars)_2Cl_2^{2+}$, where diars = the *ortho*-diarsane derivative of benzene, o-$C_6H_4(AsMe_2)_2$, which acts as a bidentate ligand through the lone pair on each arsenic atom. The complex is produced as the salt with a large anion, $FeCl_4^-$ or ReO_4^-, by oxidation of the iron(III) complex with 15 mol l^{-1} nitric acid. Analogous diphosphane and phosphane–arsane complexes of iron(IV) are reported, some with Br in place of the Cl.

14.6.3 The stable states, iron(III) and iron(II)

While a distinct chemistry of Fe(VI), (V) and (IV) has developed, it is a very minor part of iron chemistry and all the compounds are unstable outside a strongly oxidizing regime. The dominant oxidation states are Fe(II) and Fe(III).

Apart from the oxides, solid compounds of the II and III oxidation states are now represented by all of the halides, and salts of Fe(II) are known with nearly all stable anions.

In solution, the relative stabilities of the III and II states vary widely with the nature of the ligand. As Fe^{II} is d^6 and Fe^{III} is d^5, changes in CFSE have an important effect on these relative stabilities. The effect is illustrated by the potentials below:

phen = o-phenanthroline, $C_{12}H_8N_2$, a bidentate aromatic nitrogen ligand. See Appendix B.

$$Fe(H_2O)_6^{3+} + e^- \rightarrow Fe(H_2O)_6^{2+} \quad 0.77\ V$$

$$Fe(CN)_6^{3-} + e^- \rightarrow Fe(CN)_6^{4-} \quad 0.36\ V$$

$$Fe(phen)_3^{3+} + e^- \rightarrow Fe(phen)_3^{2+} \quad 1.12\ V$$

$$Fe(C_2O_4)_3^{3-} + e^- \rightarrow Fe(C_2O_4)_2^{2-} + C_2O_4^{2-} \quad 0.02\ V$$

The cyanide and phenanthroline complexes are low spin while the other two are high spin, in all cases both in the II and III states. Since the ΔE value for the trivalent d^5 Fe^{III} ion is larger than that of the divalent d^5 ion, Mn^{II}, the cyanide of ferric iron is less unstable, from the CFSE versus exchange energy point of view, than the manganese (II) hexacyanide. There is a gain in CFSE on going from ferricyanide to the d^6 low-spin ferrocyanide, due to the additional t_{2g} electron, but this is relatively small. Ferricyanide acts as a mild oxidizing agent while ferrocyanide is stable. In addition, the d^5 ferricyanide is relatively labile to substitution and the cyanide may be replaced by water and other ligands, as in $Fe(CN)_5(H_2O)^{2-}$. Thus ferricyanide in solution is much more poisonous than is ferrocyanide.

In strong base, a number of Fe(III) oxyanions have been identified including FeO_4^{5-} (tetrahedron), $Fe_2O_6^{6-}$ (two tetrahedra sharing an edge), and $Fe_6O_{16}^{14-}$ which is a ring of six corner-linked tetrahedra.

In aqueous solution, ferric iron shows a strong tendency to hydrolyse. The hydrated ion, $Fe(H_2O)_6^{3+}$, which is pale purple, exists only in strongly acid solutions at a pH of about 0. In less acidic media, hydroxy complexes are formed:

$$Fe(H_2O)_6^{3+} + H_2O \rightarrow Fe(H_2O)_5(OH)^{2+} + H_3O^+$$
$$Fe(H_2O)_5(OH)^{2+} + H_2O \rightarrow Fe(H_2O)_4(OH)_2^+ + H_3O^+$$

These occur up to pH values of 2 to 3 and are yellow in colour, the typical colour of ferric salts in solution in acid. At lower acidities, above a pH of 3, bridged species are formed and the solutions soon form colloidal gels. As the pH is raised hydrated ferric oxide is precipitated as a reddish-brown gelatinous solid. This precipitate probably does not contain any of the hydroxide, $Fe(OH)_3$, and part of it is probably in the form FeO(OH) and part as the hydrated oxide. The hydrated oxide readily dissolves in acid and is also slightly soluble in strong bases, so the ferric state is weakly acidic as well as moderately basic. The basic solutions in strong alkali probably contain the $Fe(OH)_6^{3-}$ ion which has been isolated as the strontium and barium salts.

Two interesting Fe(III) sulfur species are the ion FeS_4^{5-} which contains discrete tetrahedral ions, and the complex ion $[(Et_3PFe)_6(\mu^3\text{-}S)_8]^+$. The structure, Fig. 14.25a, consists of an octahedral cluster of 6 Fe atoms, bridged on each of the triangular faces by the 8 S atoms, with a phosphane terminal on each Fe—compare the structure of $Mo_6Cl_8^{4+}$ shown in Fig. 15.19. A full series of oxidation/reduction steps have been demonstrated which vary the charge on the complex from 0 to 4^+.

Ferric iron forms many complexes with ligands which coordinate through oxygen, especially phosphate anions and polyhydroxy-organic compounds such as sugars. It also forms intensely red thiocyanate complexes, used to detect and estimate trace quantities of iron. These colours are destroyed by fluoride due to the formation of FeF_6^{3-}. In contrast, Fe^{III} forms no ammonia complex (ammonia precipitates the oxide from aqueous solutions) and is only weakly coordinated by other amine ligands. If the ligand field is sufficiently strong to produce spin pairing, much more stable complexes are formed and this occurs with dipyridyl and phenanthroline. In aqueous HCl media, the ion $FeCl_4^-$ dominates the chemistry; however, other species such as $FeCl_6^{3-}$ have been isolated using a large tripositive countercation, and $Fe_2Cl_9^{3-}$ is known as its caesium salt. From aqueous solutions $FeCl_5(H_2O)^{2-}$ has been isolated, but using a nonaqueous solvent mixture gave the unsolvated $FeCl_5^{2-}$ ion, which has the expected trigonal bipyramidal structure.

Iron(III) forms the 3-coordinate compound $Fe[N(SiMe_3)_2]_3$ like the chromium analogue. Structural studies show the presence of planar 3-coordinate FeN_3 and $FeNSi_2$ groups indicating extended π bonding. The nitrate complex, $Fe(NO_3)_4^-$, is 8-coordinated in a dodecahedral configuration. The mixed valent Fe(II)–Fe(III) fluoride, $Fe_2F_5.7H_2O$, is known. On dehydration, it yields FeF_2 and FeF_3.

Ferrous iron also forms a variety of complexes. In aqueous solution it exists as the $Fe(H_2O)_6^{2+}$ ion, which is pale sea-green in colour. This is slowly oxidized by air in acid, and is very readily oxidized when the hydrated oxide is precipitated in alkali. The anhydrous halides combine with ammonia gas to give the hexammine, $Fe(NH_3)_6^{2+}$, but this is unstable and loses ammonia when brought into contact with water. Stable complexes are formed, however, by chelating amines such as ethylenediamine. All these examples are octahedral, and we note the most famous Fe(II) octahedral complex of all, haem, which exists in haemoglobin as discussed in Chapter 20.

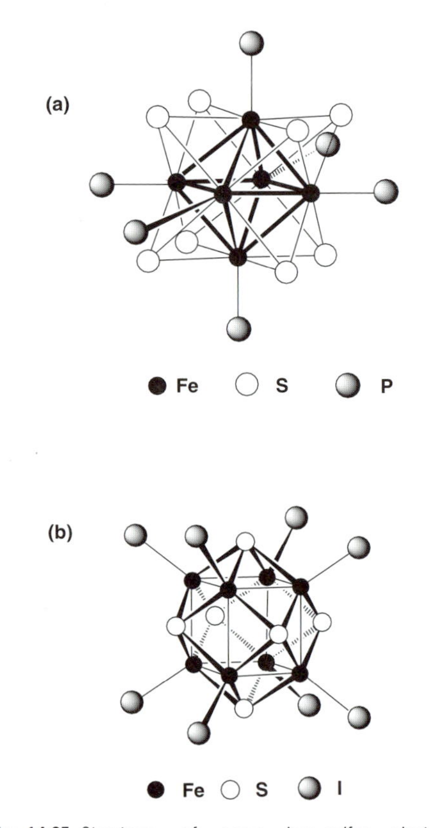

FIG. 14.25 Structures of some iron–sulfur clusters: (a) $(Et_3PFe)_6S_8|^+$ and (b) $[Fe_8S_6I_8]^{4-}$

Tetrahedral complexes are rare but the anions of the heavier halogens, $[FeX_4]^{2-}$, can be precipitated by large cations. The dimeric Fe(II) species $[Fe_2Cl_6]^{2-}$ has recently been prepared as a phosphonium salt and found to have a chloride-bridged structure very similar to that of Al_2Br_6 (Fig. 17.12) with tetrahedral geometry about the Fe atoms.

An A_6B_8 unit can be described either as a cube with six caps on the square faces or as an octahedron with eight caps on the triangular faces. The two descriptions are equivalent and whether the cube or the octahedron is seen as the basic structure depends on the point of view.

Many Fe(II)–sulfur compounds have been synthesized as inorganic models for biological systems and these are discussed in Chapter 20. A more complex Fe(II) sulfur species is the ion $[(IFe)_8(\mu^4\text{-}S)_6]^{4-}$ where a cube of 8 Fe atoms is bridged on each square face by a sulfur, and each Fe carries a terminal Fe-I bond (Fig. 14.25b). The iodine can be replaced by other ligands such as phosphane. All the iron atoms are Fe(II), and the structure has been suggested as a further model for the active centre of iron–sulfur proteins. This structure is the 'inverse' of the $Fe_6^{III}S_8$ ion of Fig. 14.25a (with Fe and S interchanged).

In the II state, iron forms a polyhydride complex, $[FeH_6]^{4-}$, whose structure has been fully determined by X-ray crystallography supported by neutron diffraction to determine the position of the hydrogens (compare Section 7.4). When $FeCl_3$ was treated with a Grignard reagent (RMgBr) and hydrogen in THF, a complex salt was isolated of formula $Mg_4(FeH_6)X_4 \cdot 8THF$ where X was Br+Cl in ratio 7:1. The structure contains a regular octahedral

PRUSSIAN BLUE AND RELATED COMPOUNDS

A well-known reaction among iron complexes is the formation of *Prussian blue* by the reaction of ferrocyanide with ferric solution, and of *Turnbull's blue* by the reaction of ferricyanide with ferrous. In addition, ferrous plus ferrocyanide gives a white potassium salt of ferrousferrocyanide, and ferric plus ferricyanide gives brown-green ferricferricyanide. It now appears that all these compounds are related structurally and that Prussian and Turnbull's blues are identical. The basic structure is that shown in Fig. 14.26. Ferric ferricyanide, $Fe[Fe(CN)_6]$, contains a cubic array of iron(III) atoms of unit cell length equal to 510 pm. This is the structure in the figure when the atoms are identical and the figure contains eight unit cells. If every second atom becomes Fe^{II} and one potassium ion is placed in the centre of every alternate small cube, a structure of unit cell length equal to 1020 pm results. This corresponds to the so-called soluble Prussian blue, $K[Fe(Fe(CN)_6)]$ where one iron is Fe^{II} and one is Fe^{III}, the alternative cases being indistinguishable from the X-ray data. Insoluble, or true, Prussian blue is $Fe_4^{III}[Fe^{II}(CN)_6]_3.14H_2O$ and is identical to Turnbull's blue. The cyanide coordinates strongly to the ferrous iron, giving an $Fe^{II}C_6$ octahedral unit. Both the nitrogen of the cyanides and some of the water molecules interact with the ferric atoms which are found in $Fe^{III}N_6$ and $Fe^{III}N_4O_2$ environments. When Prussian blue is heated in a vacuum at 400°C, the water is lost and the compound isomerizes to ferrous ferricyanide which is stable when dry, but reverts to the ferric ferrocyanide when it takes up water. The isomerism appears to involve the reversal of the CN groups, and intermediate environments such as $Fe(CN)_3(NC)_3$ occur.

When all the iron is in the Fe^{II} state, the unit cell reverts to 510 pm in length. A potassium ion placed at the centre of each small cube gives the formula, $K_2FeFe(CN)_6$, of the ferrousferrocyanide ion. The insoluble blues correspond to the formation of alkali-free complexes by replacing the potassium by ferrous or ferric ions as follows. If the cyanides are regarded as lying in the cube edges and coordinated through carbon to one iron atom and through nitrogen to the next iron atom, the framework of Fig. 14.26 corresponds to a superlattice of formula $Fe^{II}Fe^{III}(CN)_6^-$ in the case of the blue compounds. The compounds with ferric iron and ferrous iron then become $Fe^{2+}[FeFe(CN)_6]_2^-$ or $Fe^{3+}[FeFe(CN)_6]_3^-$. Other complex ferricyanides and ferrocyanides are related to these structures. Thus the cupriferricyanide ion, $CuFe(CN)_6^-$, is the same structure with the Fe^{II} ions replaced by Cu^{II} ions. Such structures probably hold for all ferrocyanide or ferricyanide complex ions of heavy metals apart from the alkali and alkaline earth metals.

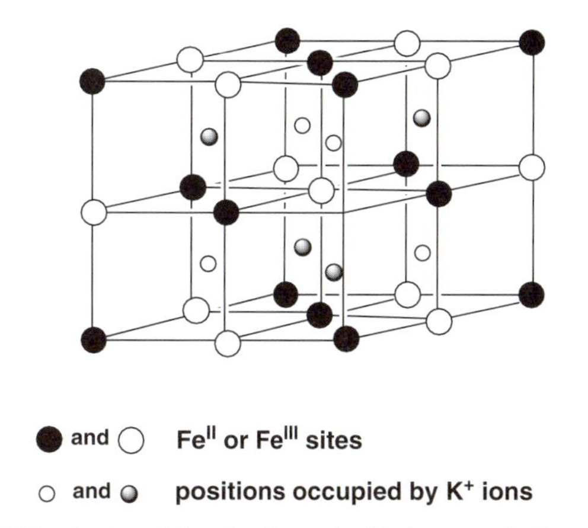

FIG. 14.26 The structure of Prussian blue and related compounds. If none of the cube centres are occupied, the structure is that of ferric ferricyanide (both black and white large circles occupied by Fe^{III}): if every second cube centre site (identified by the small grey circle) is occupied by K^+, the structure is that of soluble Prussian blue (black=Fe^{II}, white=Fe^{III}): if all the centre sites are occupied by K^+ (small white circles as well as grey circles) the structure is that of potassium ferrousferrocyanide (black and white sites=Fe^{II})

LOW OXIDATION STATES OF IRON

Iron(I) chemistry is limited to only a few compounds, many of them substituted carbonyls. In contrast, there is a well developed chemistry of iron(0), mainly of the carbonyls and their derivatives and analogues which are discussed further in Section 16.2. Three carbonyl compounds are known, $Fe_2(CO)_9$ and $Fe_3(CO)_{12}$ as well as the pentacarbonyl, and a fair number of compounds exist where the carbonyls are replaced by other π-bonding ligands, as in $(Ph_3P)_2Fe(CO)_3$. Trifluorophosphane behaves in a very similar manner to carbonyl and analogues of most carbonyl compounds exist. Thus, $Fe(PF_3)_5$ and $Fe_2(PF_3)_9$ are found, as are all the mixed carbonyl-phosphane analogues of the pentacarbonyl such as $(PF_3)Fe(CO)_4$ and $(PF_3)_3Fe(CO)_2$.

Ferrocene, $(C_5H_5)_2Fe$, is discussed in Section 16.4. It contains Fe(II) and is oxidizable to the ferricinium ion, $(C_5H_5)_2Fe^+$, which contains Fe(III).

$[FeH_6]^{4-}$ ion with an Fe-H distance of 160.9 pm, a reasonable value for a covalent bond. One other polyhydride complex, ReH_9^{2-} (see Section 15.5), has been known for a long time, and a few others—$Li_4RhH_{4\text{ and }5}$, Sr_2RuH_6 (probably also octahedral), Mg_2NiH_4, and perhaps Cu and Ir species—have been studied to varying extents. The iron compound is interesting, both as adding a well-established member to this small class, and in showing the dual interaction of H with Fe and Mg.

14.7 Cobalt, $3d^74s^2$

Fig. 14.27 shows the oxidation state energy diagram for cobalt. Only the II and III states have any stability in water and the II state is far more stable than Co^{III}, except in the presence of complexing ligands. Hydrated Co^{3+} decomposes water. In the solid state, the only trihalide is CoF_3, which is a strong oxidizing and fluorinating agent.

Cobalt metal is usually found in association with nickel in arsenical ores. It is relatively unreactive and dissolves only slowly in acids. It does not combine with hydrogen or nitrogen but it does react with carbon, oxygen and steam at elevated temperatures, giving CoO in the latter cases. Cobalt is mainly used in high-performance alloys, in magnets and in ceramics and paints where it provides a blue colour or, often, balances out yellow tinges arising from iron compounds.

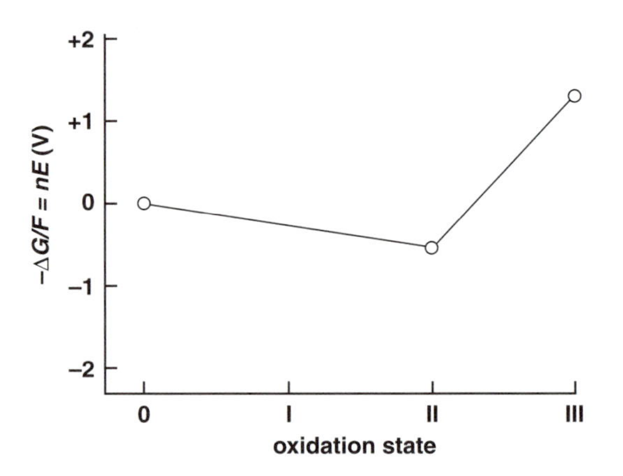

FIG. 14.27 Oxidation state free energy diagram for cobalt. In aqueous acid conditions, cobalt(II) is much more stable than cobalt(III)

14.7.1 Cobalt oxidation states greater than (III)

The highest oxidation state shown by cobalt is IV. An ill-defined, hydrated CoO_2 is formed when cobalt(II) solutions are oxidized in alkaline media. The species M_2CoO_3 (M=K, Rb, Cs), Li_8CoO_6 and Ba_2CoO_4 all contain cobalt(IV). The last has the same structure as K_2SO_4. Cobalt(IV) occurs in the well-defined dithiocarbamate complexes, $[Co(S_2NR_2)_3]^+$, formed by oxidation of the Co(III) species.

The green $(H_3N)_5Co\text{-}O\text{-}O\text{-}Co(NH_3)_5^{5+}$ was originally thought to be a peroxide (i.e. an O_2^{2-} derivative) with one Co(III) and one Co(IV) atom. It is now accepted as a superoxide (i.e. an O_2^- derivative) with two Co(III) atoms. It is prepared by oxidizing the brown cobalt(III) peroxide, $(H_3N)_5Co\text{-}O\text{-}O\text{-}Co(NH_3)_5^{4+}$.

14.7.2 Cobalt(III)

Cobalt(III) is strongly oxidizing in its simple compounds, and as the hydrated ion, but it forms a wide variety of stable octahedral complexes. It is the trivalent d^6 ion; therefore ΔE is large and spin pairing is expected, to take advantage of the large CFSE of this configuration in the low-spin state, t_{2g}^6. Spin pairing appears to occur in most complexes, although CoF_6^{3-} is high spin. The situation in the hydrate, $Co(H_2O)_6^{3+}$, is interesting. This is low spin and diamagnetic but the CFSE gain appears to be only slightly greater than the loss of exchange energy, and the hydrated ion readily reduces to $Co(H_2O)_6^{2+}$, which is high-spin d^7. As the CFSE gain in d^6 is so high, it only takes a small increment in the ligand field to make the cobalt(III) state the more stable. Thus the hexammine, $Co(NH_3)_6^{3+}$, is prepared by air oxidation of cobalt(II) in aqueous ammonia. A still higher ligand field gives complexes in which the cobalt(II) state becomes strongly reducing. This is shown by the redox potentials for Co^{III}/Co^{II} complexes of the ligands shown below (all for octahedral complexes in both states):

Ligand	H_2O	NH_3	CN^-
Redox potential (V)	+1.84	+0.1	−0.8

In the hydrates, cobalt(III) decomposes water by oxidizing it; in the ammines, cobalt(III) is stable to water; and in the cyanides, cobalt(II) decomposes water by reducing it. This range of activity resembles that already quoted for iron, but is more extreme because of the large ligand field effects.

By distinguishing different types of isomers and by proving the constancy through a series of chemical changes of certain groupings of atoms, Werner was able to formulate the idea of definite complex species and to determine the coordination numbers and shapes—all this before the development of any of the powerful modern techniques of structural determination.

Cobalt(III) complexes are extremely numerous and they undergo substitution only slowly so that a large variety of them are readily prepared and handled. It was the study of cobalt(III) and platinum(IV) complexes (both d^6), together with chromium(III) and square-planar platinum(II) compounds which first led to the development of the ideas of complex chemistry at the start of this century by Werner and his school. The variety and types of cobalt(III) complexes are best illustrated by some of these classical sequences of preparations and interactions in the field of cobalt–ammonia compounds, Fig. 14.28. A number of other examples, including ethylenediamine complexes, are discussed in Section 13.7.

Although nitrogen is probably the commonest donor atom in cobalt complexes, a variety of oxygen complexes exist with ligands of the type oxalate and acetylacetone, $Co(C_2O_4)_3^{3-}$ or

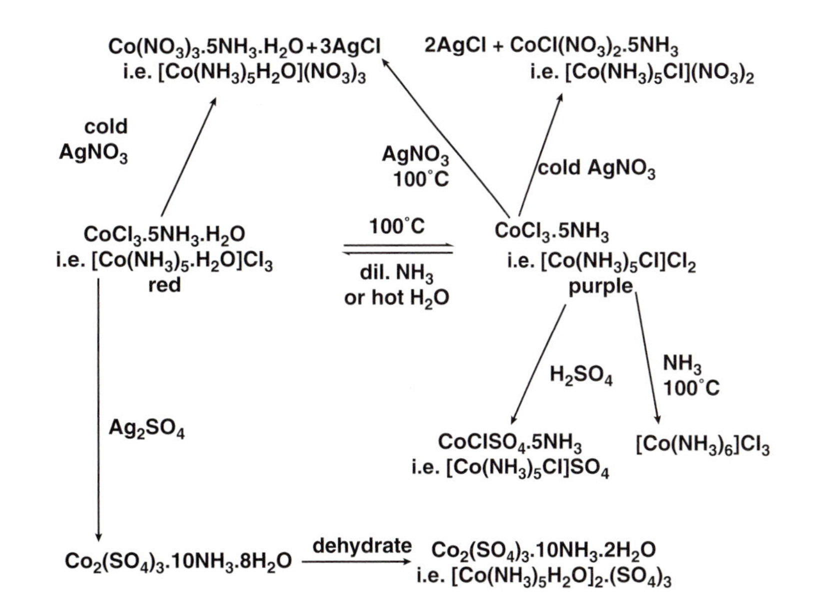

FIG. 14.28 Interconversions of cobalt(III) ammonia complexes. This is an example of one of the series of interconversions which led Werner to the formulation of the concept of a complex

$Co(acac)_3$. Cobaltinitrite, $Co(NO_2)_6^{3-}$, is coordinated through nitrogen of course, but there is some evidence for the existence of the isomeric nitrito-form, Co-O-N-O, in solution in equilibrium with the nitro-compound. Essentially all cobalt(III) compounds are octahedral.

The cobalt(III) nitrosyl species of empirical formula $[Co(NH_3)_5NO]^{2+}$ has presented an interesting and long-standing problem. At the time of its discovery, in 1903, it was found to form two different isomers, one red and the other black. The crystal structure of the black form has now shown it to contain the mononuclear ion $[(H_3N)_5Co(NO)]^{2+}$ with a non-linear Co-N-O group, angle about 120°. The red isomer was recently shown to be a dimer, containing the hyponitrite group, N_2O_2, as a bridge. The structure is nonsymmetric with both Co-N and Co-O bonds to the hyponitrite:

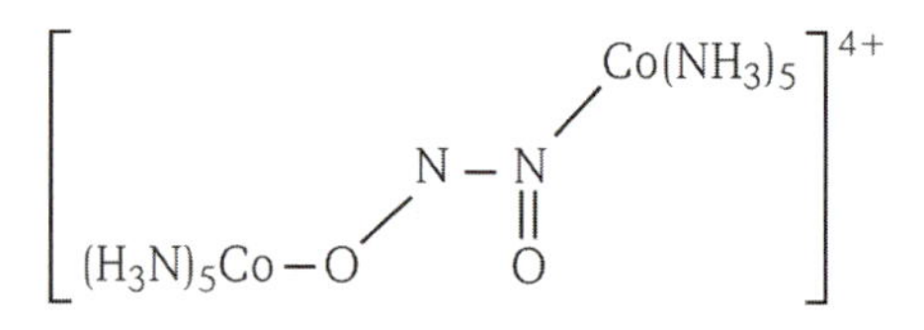

Addition of CN^- to the black isomer gives two more hyponitrite species in the orange and yellow isomers of $(NC)_5Co-N_2O_2-Co(CN)_5^{6-}$. These have the bridges

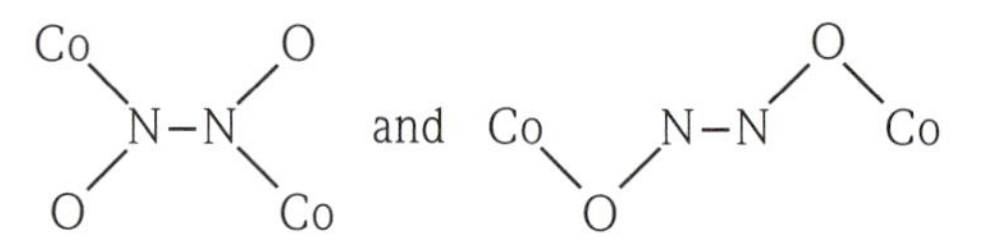

CoF_2 reacts readily with F_2 to give CoF_3 and the latter is a strong fluorinating agent, though less reactive than fluorine. It provides a suitable way of moderating fluorination reactions. The compound to be fluorinated is streamed over CoF_3, giving the desired product and CoF_2. CoF_3 can then be regenerated by passing fluorine over the cobalt difluoride and the process may be continued in a cyclic manner.

One example of a simple cobalt(III) derivative is the anhydrous nitrate, $Co(NO_3)_3$. This is prepared in nonaqueous solvents and has bidentate nitrate groups giving octahedral CoO_6 coordination.

14.7.3 Cobalt(II)

Although cobalt(III) exists largely in complexes and has unstable simple compounds, cobalt(II) is just the reverse. It is perfectly stable in simple compounds and salts and forms a number of complexes with ligands of relatively weak ligand field. However, with ligands further along the spectrochemical series than water, the CFSE gain on achieving the low-spin d^6 configuration is sufficiently great to make cobalt(III) the preferred state.

Cobalt(II) oxide, halides and sulfide are well known and may be made by normal methods. Red or pink hydrated cobalt salts of all the common anions are known. On addition of base to Co^{II} solutions, the pink (occasionally blue) hydroxide is precipitated and this dissolves in concentrated alkali to give the deep blue $Co(OH)_4^{2-}$ anion. The latter may be precipitated as the sodium or barium salt.

Cobalt(II) hydrate is octahedral, as is $Co(NH_3)_6^{2+}$. This, and a number of related complexes, may be prepared as long as an inert atmosphere is maintained to stop oxidation. $Co(CN)_5^{3-}$, and related species, are found in green and yellow forms, depending on the cation. In the green one, the coordination of the Co is square pyramidal with the sixth position occupied by a weakly coordinated cation or solvent molecule. The yellow form is a pure square pyramid. Similar species are found for Ni analogues, which also form distorted trigonal bipyramids in some cases.

The deep blue $CoCl_4^{2-}$, formed by addition of excess Cl^- to the pink hydrated solutions, is tetrahedral. The other halides form similar, blue anions as does thiocyanate, $Co(SCN)_4^{2-}$. These tetrahedral ions generally have to be precipitated from solution as salts of large cations. A related compound is the mercury complex, $CoHg(SCN)_4$, which is used as a calibrant in magnetic measurements. This contains cobalt(II) tetrahedrally coordinated by the nitrogen atoms of the thiocyanate groups while the sulfur atoms tetrahedrally coordinate mercury(II) ions to give a polymeric solid. Square-planar cobalt(II) complexes are found for some chelating ligands, such as dimethylglyoxime and salicylaldehyde-ethylenedi-imine, which form stable planar complexes in general (compare nickel dimethylglyoxime Fig. 14.30).

The tetrahedral complexes of cobalt(II) have three unpaired electrons and the square-planar ones have only one, both as expected for d^7. The octahedral complexes include both high-spin and low-spin cases, the former with three and the latter with one unpaired electron. The changeover appears to come between ammonia and the nitro anion, NO_2^-, in the spectrochemical series.

An interesting cluster structure is found in $[(PhSCo)_8(\mu^4\text{-}S)_6]$ which is the inverse of the $[(Et_3PFe)_6(\mu^3\text{-}S)_8]^+$ structure of the last section. In the cobalt(II) case, the 8 Co atoms form a cube with the six S atoms above the square faces—alternatively described as forming an octahedron enclosing the cube. The PhS groups lie terminally and uniformly bent on each Co (compare Fig. 14.25b).

14.7.4 Lower oxidation states of cobalt

Cobalt(I) is found in the hydride complex ion, CoH_5^{4-} which has a square-pyramidal structure. Most compounds in the I, 0 and −I states form with π-bonding ligands. The commonest are the carbonyls (see Section 16.2) and related species. The simplest carbonyl, $Co_2(CO)_8$, readily gives Co(−I) in the anion $Co(CO)_4^-$ and the hydride $HCo(CO)_4$. It also reacts with organic isonitriles to give Co(I) in the cation, $Co(CNR)_5^+$ which has a square-pyramidal structure in the case R = Ph. The cation can also be prepared by reduction of the cobalt(II) compounds, $Co(CNR)_4X_2$.

Interesting cluster compounds are found among the more complex cobalt carbonyl anions. For example, in $Na_4Co_6(CO)_{14}$, the $Co_6(CO)_{14}^{4-}$ anion consists of an octahedron of Co atoms, with each of the eight triangular faces bridged by a CO. The remaining six CO units are normal terminal groups, one on each cobalt. In other words, the structure is like that of $Mo_6Cl_8^{4+}$ shown in Fig. 15.19, with Co in place of Mo, carbonyl in place of the chlorines together with six terminal CO groups.

Cobalt carbonyl hydride, $HCo(CO)_4$, has been shown to be the active species in the OXO process for the conversion of alkenes to alcohols in presence of a cobalt catalyst

$$RCH{=}CH_2 + CO + 2H_2 \rightarrow RCH_2CH_2CH_2OH$$

The reaction proceeds by initial insertion of H-Co into the double bond (an example of a *hydrometallation* reaction) followed by CO insertion into the reactive Co-C bond

$$RCH{=}CH_2 + HCo(CO)_4 \rightarrow$$

$$RCH_2CH_2Co(CO)_4 \xrightarrow{+\,CO} RCH_2CH_2COCo(CO)_4$$

It is thought that the acyl-cobalt compound then loses CO and the resulting tricarbonyl is cleaved by hydrogen and the product aldehyde is reduced to the alcohol:

$$H_2 + RCH_2CH_2COCo + (CO)_n \rightarrow$$

$$(n = 3?)$$

$$RCH_2CH_2CHO + HCo(CO)_n$$

$$\searrow CO \searrow HCo(CO)_4$$

Another interesting cobalt(I) species is the compound $CoH(N_2)(PPh_3)_3$, hydridodinitrogentris(triphenylphosphane)cobalt(I). This is a representative of a growing class of compounds which have N_2 as a ligand (compare Section 16.10). The structure was one of the first N_2 complexes to be determined, and is shown in Fig. 14.29, with trigonal bipyramidal coordination at the cobalt. The end-on, linear, Co-N-N bonding is analogous to the long-known Co-C-O unit, with which it is isoelectronic.

The ligand $L{=}CH_3C(CH_2PPh_2)_3$, with three donor phosphorus atoms, forms halide complexes of cobalt(I), CoLX, which are tetrahedral.

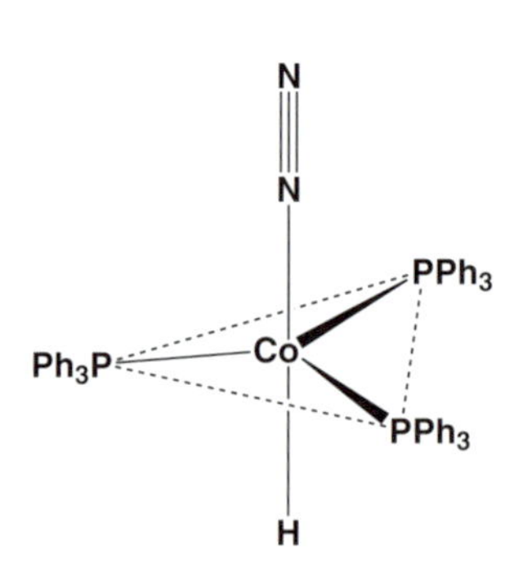

FIG. 14.29 The structure of the cobalt(I) nitrogen compound, $CoH(N_2)(PPh_3)_3$. The Co-N_2 unit is linear with bond length 180 ± 3 pm, the Co-H distance is 160 pm and the equatorial P atoms are bent slightly towards the hydrogen position (as is usual in hydrido-complexes) with N-Co-P angles of 97 ± 2°

Cobalt(0) also occurs in $K_4Co(CN)_4$ which is the reported product of the reduction of the cobalt(III) hexacyanide with potassium in ammonia. Cobalt carbonyls are reviewed in Section 16.2, and cobalt–carbon compounds in Sections 16.3 and 16.4.

14.8 Nickel, $3d^84s^2$

The chemistry of nickel is much simpler than that of the other first row elements. The only oxidation state of importance is nickel(II) and these compounds are stable. Nickel(II) is the d^8 ion and is able to form stable square-planar complexes as well as octahedral ones. Ligands with large crystal field favour square planar coordination, because of the more favourable CFSE. The heavier elements in the nickel Group are exclusively square planar in the II state.

The element occurs largely in sulfide or arsenic ores and is extracted by roasting to NiO and then reducing with carbon. Pure nickel is made by the Mond process in which CO reacts with impure Ni at 50°C to give $Ni(CO)_4$. This is decomposed to Ni and CO at 200°C to yield metal of 99.99% purity. The alternative large-scale preparation involves reduction with carbon followed by electrolytic refining. Nickel is used in steels, especially stainless steel, and a number of alloys are widely familiar. Cupro-nickel, used in coins, is about 4Cu to 1Ni; addition of zinc gives nickel silver which is the base of EPNS tableware when silver-plated. Nickel is an important hydrogenation catalyst, and is used in Ni-Fe and Ni-Cd batteries. The metal resists attack by water or air and is used as a protective coating. It has good electrical conductivity. The metal dissolves readily in dilute acids.

14.8.1 Higher oxidation states of nickel

Oxidation states above II are represented by some obscure oxides, some complexes and the binary fluorides NiF_3 and NiF_4. The NiF_6^{2-} ion is known in several salts such as K_2NiF_6. Treatment with a fluoride-ion acceptor (e.g. BF_3, AsF_5) gives NiF_4 as a tan-coloured solid. When NiF_4 is made and decomposed in the presence of some K^+ ions, NiF_3 is formed. M_3NiF_6($M = Na$, K) are also reported for Ni(III).

Oxidation of $Ni(OH)_2$, suspended in alkali, by a moderately strong oxidizing agent like Br_2 gives a black solid which can be dried to the composition $Ni_2O_3.2H_2O$. Attempts at further dehydration lead to decomposition to NiO. The action of powerful oxidizing agents in an alkaline medium give impure, hydrated products of approximate composition NiO_2. This is a powerful oxidant, forming permanganate from Mn^{2+} in acid, and decomposing water. If nickel is heated in oxygen in fused sodium hydroxide, sodium nickelate(III), $NaNiO_2$, results.

High oxidation states of nickel can also be stabilized in complexes. Oxidation by chlorine in strong base allowed the formation of Ni(IV) which was isolated as the dimethylglyoxime complex $[(DMG)_2Ni]^{2+}$ ion (compare Fig. 14.30 for the Ni(II) DMG complex).

Also using ligands with oxime groups and other donor atoms, relatively stable nickel(III) complexes, NiL_3, have been made. Here L contains one oxime and one donor N in a

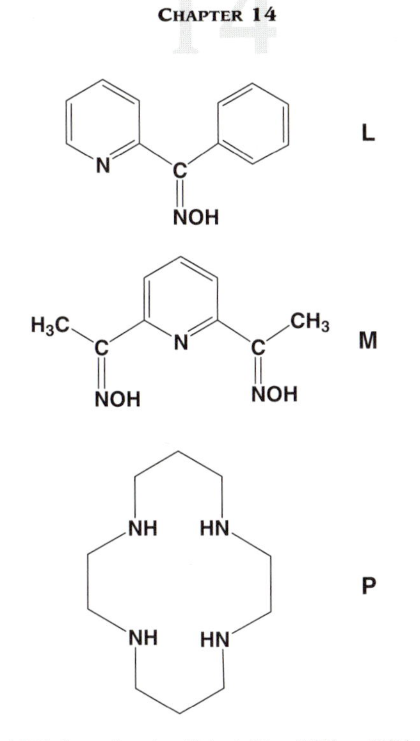

FIG. 14.31 Some ligands which stabilize Ni(III) or Ni(IV). In each case, hydrogen ionizes, giving L^-, M^{2-} or P^{4-}

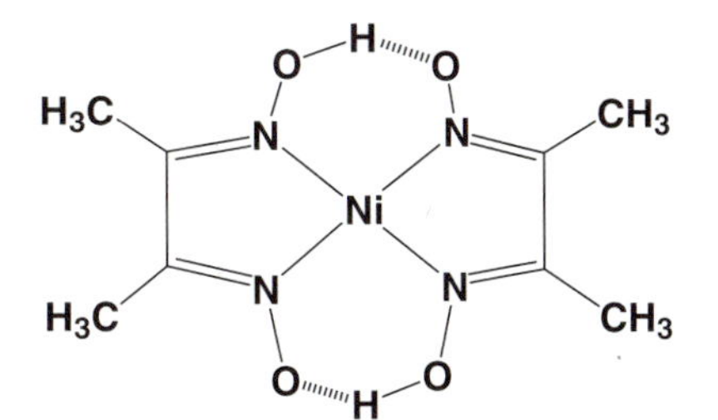

FIG. 14.30 Nickel(II) dimethylglyoxime. The dimethylglyoxime loses a proton and coordinates to Ni giving a five-membered ring. The complex is further stabilized by hydrogen bonding

pyridine group, and the complex is neutral and 6-coordinated. With a ligand M containing two oximes and a pyridine, NiM_2, containing 6-coordinated Ni(IV) was prepared. Nickel(III) is also stabilized by large cyclic ligands (Fig. 14.31), as in $[NiPX_2]^+$ or $NiL(SO_4)^+$. The products are formed by electrochemical oxidation of the Ni(II) analogue and are stable in dilute acid solutions: the oxidation potential is about 1 V.

Many other reported complexes have since been shown to contain Ni(II) and oxidized ligand, but trigonal bipyramidal $NiBr_3(PR_3)_2$ is well established. Ni(III) is also found in the green $(NiCl_2en_2)^+Cl^-$ complex, which has a magnetic moment corresponding to one unpaired electron, and also in the organometallic complex $Ni(C_6Cl_5)_4^-$, which has a square-planar structure.

14.8.2 Nickel(II)

A wide variety of simple compounds of nickel(II) exist, including all the halides and all the oxygen Group compounds. Ni^{2+} forms salts with even strongly oxidizing anions such as chlorite, and with relatively unstable ions like carbonate. The relative stabilities of the II and III states of iron, cobalt and nickel, as shown by the compounds formed with the oxychlorine anions, are discussed in Section 8.6. As chlorine in alkali oxidizes nickel(II) to nickel(IV), it is to be assumed that the extremely powerfully oxidizing ClO^- anion will not form even a nickel(II) salt. The complex ions with ligands such as water and ammonia are octahedral, the green $Ni(H_2O)_6^{2+}$ ions being responsible for the typical colour of hydrated nickel salts. The hexammines and related complexes such as $Ni(en)_3^{2+}$ are generally blue. All possible mixed forms occur such as $Ni(H_2O)_2(NH_3)_4^{2+}$.

With ligands of high field strength, square-planar, diamagnetic complexes are formed, such as the cyanide, $Ni(CN)_4^{2-}$ and the dimethylglyoxime complex used in quantitative analysis for nickel (Fig. 14.30). In the latter, there may be interaction between nickel atoms in the crystal, where the flat molecules are stacked vertically above each other, so that the compound could be considered as a very distorted octahedron.

There are several examples of tetrahedral nickel(II) complexes. These involve halogen compounds, either as anions with large cations, as in $(Ph_4As)_2^+(NiCl_4)^{2-}$, or in neutral complexes with phosphane or phosphane oxide and related ligands, as in $(PPh_3)_2NiI_2$ or $(Ph_3AsO)_2NiBr_2$. Such complexes are characteristically intensely blue due to a relatively intense absorption in the red end of the spectrum. These intense spectral lines distinguish tetrahedral nickel(II) from octahedral complexes, where the absorptions are relatively weak.

Since nickel(II) is found in octahedral, square-planar and tetrahedral environments, and there is no large energy difference between these, there are many cases where more than one form of a complex occurs, differing in stereochemistry, or where an equilibrium exists between two forms in solution. The salicylaldimato complexes (compare Appendix B) of the type shown in Fig. 14.32, provide examples of all three stereochemistries. When R = methyl or isobutyl, the complex of Fig. 14.32 is square planar, but when R = isopropyl, the coordination about the nickel becomes tetrahedral. Further, if the diamagnetic square-planar complex (e.g. R = Me) is dissolved in pyridine, it becomes paramagnetic with the moment expected for an octahedral complex, and an octahedral dipyridine derivative may be isolated.

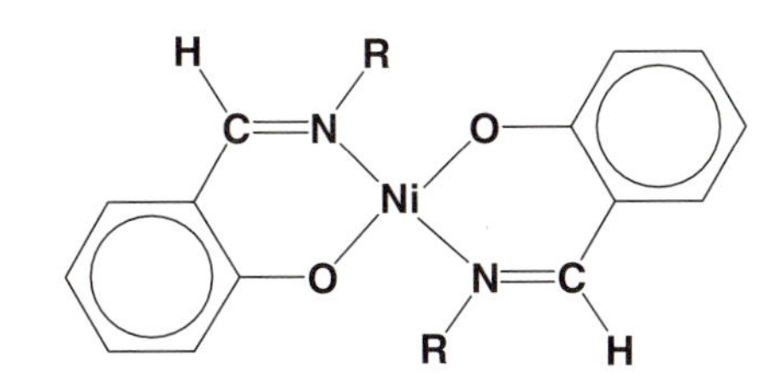

FIG. 14.32 Bis-(*N*-alkylsalicylaldimato)nickel(II) complexes

LOWER OXIDATION STATES OF NICKEL

Low oxidation states are represented by the cyanide complexes, $K_4Ni^I_2(CN)_6$ and $K_4Ni^0(CN)_4$, which result from the reduction of the $Ni^{II}(CN)^2_4$ species with potassium in liquid ammonia. The nickel(I) compound may also be made by reduction with hydrazine in an aqueous medium. It is relatively stable but the nickel(0) complex is extremely reactive and not well characterized. Both are oxidized in air or water. Nickel(I) is also represented by the phosphane complexes $NiX(PPh_3)_3$ and NiLX, where $L = CH_3C(CH_2PPh_2)_3$. Nickel(0) is found in the carbonyl, $Ni(CO)_4$, in $Ni(PF_3)_4$, and in all the mixed trifluoro-phosphane-carbonyls. The carbonyl anion, $Ni_2(CO)_6^{2-}$ contains nickel(−I). Although nickel carbonyl is one of the best-known carbonyls, the stability of such compounds is relatively low at this end of the d block and nickel has a much less rich carbonyl chemistry than the earlier elements.

When a ligand atom is a moderately weak donor, it is possible to isolate complexes with the same formula but different coordinations. For example, NiL_2Cl_2

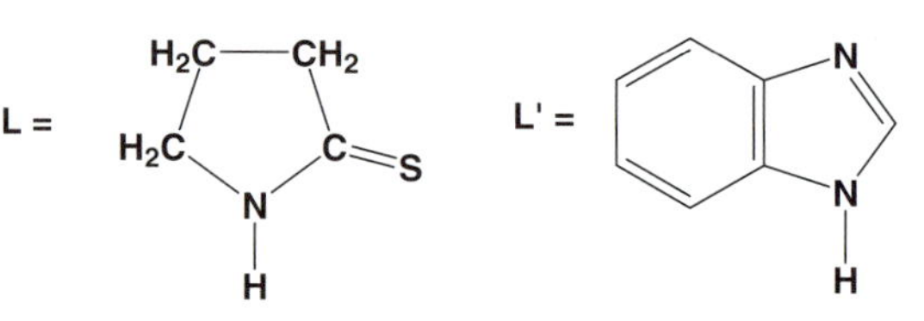

may be found in a blue tetrahedral form or in a yellow octahedral one. Similarly, NiL'_4Br_2 where L' = benzimidazole, is found in an orange square-planar form and a yellow octahedral one.

In the above cases, solids may be isolated and characterized, but often the two forms coexist in solution, even though only one is isolable. This is true of a wide range of complexes NiL_2X_2 which show tetrahedral–octahedral equilibrium (usually as blue-yellow changes) in solution, with the tetrahedral form most favoured when X = I. Such equilibria may be studied spectroscopically, and when L = phosphane with bulky groups, the exchange is slow enough to be followed by nmr at low temperatures.

One particularly striking example of the case of inter-conversion of these stereochemistries is provided by the compound $Ni(PRR'_2)_2Br_2$ where the substituents on phosphorus in the phosphane are R = benzyl ($C_6H_5CH_2$) and R′ = phenyl (C_6H_5). In the crystal of this compound, both square-planar and tetrahedral nickel environments are found. Thus any small energy difference between the two coordinations is compensated for by the readier packing of the two different sorts of molecule.

Nickel(II) also forms a range of 5-coordinate complexes (compare Section 13.6), both in the square-pyramidal configuration like NiX_2terpy (terpyridyl) and trigonal bipyramidal, as in many Schiff's base complexes (compare Appendix B). There is even the case, quoted in Section 13.6, of the $Ni(CN)_5^{3-}$ ion adopting both coordinations in the same compound.

An interesting complex structure is found in $Ni_9Te_6(PR_3)_8$ which was isolated in the course of studies on metal–tellurium compounds as possible precursors to interesting electronic materials. The structure is related to that of Fig. 14.25b, with an octahedron of Te atoms face-bridging a cube of Ni atoms, each carrying a terminal phosphane ligand. The ninth Ni atom lies at the centre of the whole structure, so there is a $Ni(Ni)_8$ cube having a common centre with a $Ni(Te)_6$ octahedron and the cube lies within the octahedron.

14.9 Copper, $3d^{10}4s^1$

There are three possible oxidation states in the copper Group, the I state corresponding to d^{10}, the II state which is common to the whole transition series, and the III state corresponding to d^8 which would be stable in a square-planar environment. The elements in this Group show these states, with II the stable state of copper, I the stable state of silver, and III the stable state of gold. Copper also exists in the I state, which is quite stable in solids, and one or two copper (III) compounds are reported. This wide variation in stabilities in this Group is probably a resultant of size, exchange energy and ligand field effects, and is discussed in Section 14.10.

Copper is the major component of two important alloys, making up 80% or more. *Bronze* is the alloy with tin, while *brass* is the alloy with zinc. Each of these alloys will contain a variety of minor components to modify their properties for particular uses.

Copper is found in sulfide ores and as the carbonate, arsenide or chloride. Extraction involves roasting in air to the oxide, reduction and purification by electrolysis. In a clever application of biotechnology, processes have been developed to recover copper from low-grade ores and mine tailings by using sulfide bacteria which obtain their energy from the oxidation of sulfide to sulfate, which goes into solution for further recovery. Solvent extraction processes employing hydroxyoximes are also used industrially to recover and purify copper (see Section 7.3).

Pure copper has an electrical conductivity second only to that of silver and its major application is in the electrical industry. The element is inert to nonoxidizing acids but reacts with oxidizing agents. With oxygen it combines on heating to give CuO at red heat and Cu_2O at higher temperatures. It also reacts with halogens and dissolves in hot nitric or sulfuric acid.

14.9.1 Copper(IV) and copper(III)

Cu(III) is also important in the new ceramic superconductors of formula $YBa_2Cu_3O_{7-x}$ (where x is about 0.1), which are discussed in detail in Section 16.1.

One example of copper(IV) is established. When the copper(II) complex, $CsCuCl_3$ is heated with F_2 at 250°C, the product is $Cs_2Cu^{IV}F_6$. The ion is a slightly distorted octahedron with two opposite Cu-F distances of 177 pm and the other four Cu-F a little shorter at 175 pm. Such a structure is appropriate for a d^7 ion, compare cobalt(II).

The III state of copper may be obtained by oxidation in alkali, yielding $MCuO_3$ for M = alkali metal. The structure is linked square-planar CuO_4 units. Reaction of the Cu(II) complex $CsCuCl_3$ with F_2 gives $CsCuF_4$ containing the square-planar $Cu^{III}F_4^-$ ion, while similar treatment of mixed MCl plus $CuCl_2$ gives alkali metal salts of octahedral CuF_6^{3-}, which have a very short Cu-F distance of 183 pm. A further example of copper(III) is the periodate, $K_7Cu(IO_6)_2.7H_2O$. It will be noticed, if the higher oxidation states of iron, cobalt, nickel and copper are compared, that the methods of preparing such compounds tend to be similar, for example, by oxidation in alkaline media and the use of oxidizing anions.

14.9.2 Copper(II)

Copper(II) is the main state in aqueous solution and the compounds are paramagnetic with one unpaired electron. The hydrated ion may be written $Cu(H_2O)_6^{2+}$ although the structure is not regular. There are four near ligands in a square plane and the other two are further away as a result of the unequal occupation of the two e_g orbitals. This distorted shape is common for copper(II) compounds as the examples of bond lengths below show:

The $[CuCl_6]^{4-}$ ion was originally thought, from an X-ray structure, to show the opposite distortion with only two short bonds. It is now clear that the ions are disordered in the crystal, so that the four 'longer' bonds are actually an average of two long and two short bonds. Several experimental techniques now agree that the configuration is actually the normal one with four Cu-Cl distances in the range 228–238 pm and two at 283 pm, comparable with the values tabulated.

Compound	*Distances* (pm)	
	Shorter	*Longer*
CuF_2	4 of 193	2 of 227
$CuCl_2$	4 of 230	2 of 295
$CsCuCl_3$	4 of 230	2 of 265
$CuCl_2.2H_2O$	2 of 231 (Cl)	2 of 298
	2 of 201 (O)	
K_2CuF_4	4 of 192	2 of 222

In contrast, $KAlCuF_6$ does contain $[CuF_6]^{4-}$ ions with the unusual compressed geometry, unlike the K_2CuF_4 compound listed above. In $KAlCuF_6$, there are four F ions in the equatorial plane at a distance of 212 pm and two much shorter bonds of 188 pm. The four equatorial F have Al^{3+} ions as neighbours and the plane is quite distorted with the Cu slightly above the plane, one FCuF angle of 120° and three of 80°.

Mixed complexes of water and ammonia are found up to $Cu(H_2O)_2(NH_3)_4^{2+}$, but replacement of the last two water molecules is impossible in aqueous solution and $Cu(NH_3)_6^{2+}$ can only be prepared in liquid ammonia. It is similarly possible to form $Cu(H_2O)_4en^{2+}$ and $Cu(H_2O)_2en_2^{2+}$ but formation of $Cuen_3^{2+}$ is difficult. These, and similar amine complexes, are all a much deeper blue than the hydrated ion.

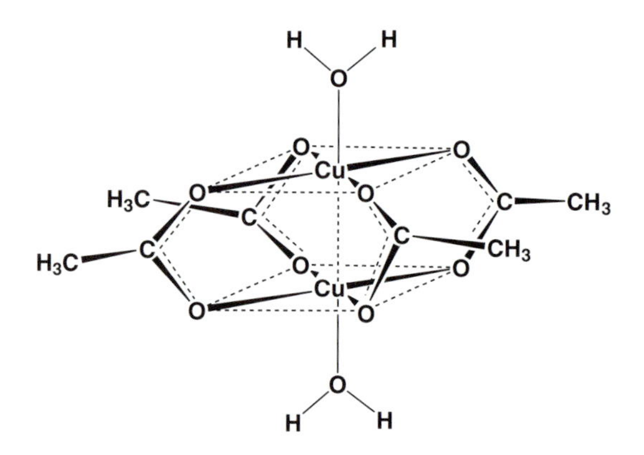

FIG. 14.33 Structure of hydrated copper(II) acetate. The two square-planar CuO_4 units are linked by the acetate residues and also by Cu-Cu bonding

Halide complexes are also distorted but are tetrahedral with a flattened structure, for example, $CuCl_4^{2-}$, which can be precipitated from a chloride medium by large cations. Such tetrahedral complexes are generally green or brown.

Among the salts, mention must be made of the unusual structures of copper(II) acetate and of anhydrous copper nitrate. The acetate is dimeric and hydrated, $Cu_2(CH_3COO)_4.2H_2O$, and has the structure of Fig. 14.33. The copper atoms are surrounded by a square plane of oxygen atoms and the acetate groups bridge the planes together. Similar structures are found for copper derivatives of other carboxylic acids.

Anhydrous copper nitrate cannot be made by dehydrating the hydrated salt as this decomposes to the oxide. However, copper metal dissolves in liquid N_2O_4/ethyl acetate mixture to give $Cu(NO_3)_2.N_2O_4$, and $Cu(NO_3)_2$ results when the solvating molecule is pumped off. The structure in the solid is shown in Fig. 14.34. The structure consists of chains of copper atoms bridged by NO_3 groups and cross-linked by other nitrate groups. Anhydrous copper nitrate is slightly volatile as single molecules.

There are some apparently 3-coordinated copper compounds but all are of more complex structure. For example, $KCuCl_3$ contains dimeric planar $Cu_2Cl_6^{2-}$ ions. In the lithium salt, these dimeric ions are joined into longer chains by long chloride bridges. A bromide

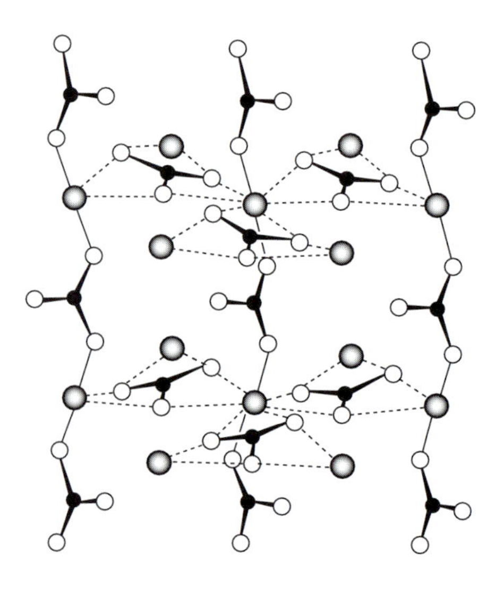

FIG. 14.34 Solid anhydrous copper nitrate, $Cu(NO_3)_2$. This structure consists of cross-linked chains of nitrate groups bonded to two copper atoms. A second crystal form of copper nitrate has been discovered recently

complex contains the $Cu_3Br_8^{2-}$ ion which is completely planar with three square-planar units linked through their edges. A very distorted octahedron around the Cu is completed by weak interactions with two Br from parallel ions above and below in an extended stack.

As with nickel, copper(II) appears in a variety of geometries and examples exist of the same compound occurring in two different structural modifications. One example is provided by complexes $CuCl_2L_2$, where L is an N-oxide such as pyridine-N-oxide, C_5H_5NO. These occur in a green form, which is thought to be *trans*-square planar and also in a yellow form in which the coordination is tetrahedral.

Copper(II) is found in 5-coordination which is usually square pyramidal although $CuCl_5^{3-}$ $CuBr_5^{3-}$ and $CuCl_2Br_3^{3-}$ are trigonal bipyramids (with axial Cl in the latter).

As many of these structures are distorted, and often show a variety of bond lengths to the same ligand atom, copper stereochemistry presents an extremely difficult field. Normal spectroscopic methods are inadequate to distinguish distorted shapes from those of lower coordination number, and even X-ray crystallography may give inconclusive results because of the problems of correlating several different interatomic distances with short, long or nonexistent bonds. Three-coordinate copper(II) is found in the halides CuX_3^- with large cations.

14.9.3 Copper(I)

A mixed Cu(I)- Cu(II) species is found in the intensely blue $Cu_2Cl_4^-$ ion, which consists of an infinite chain of distorted tetrahedral units sharing edges. One type of Cu has Cu-Cl = 234.2 pm and two ClCuCl angles of 92.8° and four of 118°, values appropriate to Cu(I). Its neighbour has Cu-Cl = 225.5 pm, and angles in pairs of 98°, 100° and 134°, all features indicating Cu(II). The blue colour should not be confused with that of Cu(II) in solution which is due to the hydrated ion. Cu(II) chloride species are generally yellow or red. The intense colour is due to electron transfer between the two oxidation states, and intense colours are characteristic of such systems (compare the discussion on Prussian blue, Section 14.6).

The I state of copper is represented largely by solid compounds which are insoluble in water. For example, if iodide is added to a copper(II) solution, the cupric iodide initially formed rapidly decomposes to give a precipitate of CuI. Similarly, one method of determining copper quantitatively is by precipitation of CuSCN. The stability of copper(I) in solution is low and the redox potentials

$$Cu^+ + e^- \rightarrow Cu \quad 0.52\ V$$
$$Cu^{2+} + e^- \rightarrow Cu^+ \quad 0.15\ V$$

show that Cu^+ unstable to disproportionation:

$$2Cu^+ \rightarrow Cu + Cu^{2+}, \quad E = 0.37\ V, \quad K = \frac{[Cu^{2+}]}{[Cu^+]^2} = 10^6$$

The reason for the marked instability of Cu^+ in water is not completely clear, as the d^{10} configuration, with its very high exchange energy, would be expected to be reasonably favoured. One possible explanation lies in the low hydration energy which is likely for the Cu^+ hydrated ion, as compared with that for the Cu^{2+} ion. The cuprous ion is larger and has only half the charge so that its charge density is markedly lower, and hence the energy of interaction with the water dipole is less. In addition, in known complexes of Cu^+, especially the ammine, Cu^+ is only 2-coordinated as in $Cu(NH_3)_2^+$. If the hydrate is reasonably supposed to have the same formula, then there would be only two interactions instead of the four strong and two weaker interactions in the Cu^{2+} hydrate. Thus Cu^+ has about half the interaction energy with half the number of water molecules compared with Cu^{2+}. When this case is compared with that of Ag^+, where the behaviour is quite the reverse and Ag^{2+} is quite rare and unstable, it will be seen that the greater size of silver tends to decrease these differences between the two oxidation states and the exchange energy probably then becomes the dominant term. Passing to gold, this trend again alters with the size, and the large CFSE term for a trivalent species of the third transition series helps to make square-planar Au(III), with a d^8 system, the preferred state, although Au(I) also occurs. It is probable that the marked variation in chemistry in this Group, where the three elements differ markedly more than in any other Group, is a function of the increase in size coupled with the existence of stable electronic configurations on either side of the divalent d^9 state.

In the complex copper(I) ion, $Cu_4I_6^{2-}$, there is no Cu-Cu bonding. Cu lies almost symmetrically in the triangular faces of the I_6 octahedron.

Apart from the insoluble CuCl, CuBr, CuI, and the cyanide and thiocyanate, copper(I) gives soluble complexes with these groups as ligands. Two cyanide complexes exist, soluble $Cu(CN)_4^{3-}$, and the compound $KCu(CN)_2$ which has a chain structure containing three-coordinate Cu^I:

$$- - - - \text{Cu-C-N-Cu(CN)-C-N-Cu} - - - -$$

For zinc, $3d^{10}4s^2$, see Section 15.10, the zinc Group.

The red cuprous oxide, Cu_2O, is well known. This is the compound precipitated in Fehling's test for sugars; it is produced by the reduction of the blue cupric tartrate complex by glucose or related molecules. The presence of direct Cu-Cu bonding in Cu_2O has recently been established.

It is still uncertain whether copper can exist in the 0 state as a number of reported compounds have been reformulated. One possible example is the diamagnetic species

$$Cu(NC\text{-}(CH_2)_4\text{-}CN)_2.$$

14.10 The relative stabilities of the dihalides and trihalides of the elements of the first transition series

Although, at present, the factors affecting the stabilities of oxidation states are incompletely understood, answers can be obtained in a rigorous manner from the thermodynamic parameters involved, providing that the problem is precisely formulated. One case discussed is the thermal stabilities of the trihalides (excluding fluorides) of the first row transition elements to decomposition according to the equation

$$MX_{3(\text{solid})} \rightarrow MX_{2(\text{solid})} + \tfrac{1}{2}X_2$$

at 25°C. This problem is only a small part of the question of the relative stabilities of the II and III states of the first row elements but its solution points the way for further work. Other compounds, other conditions, and even other decomposition routes of the trihalides, would have to be considered to extend our understanding of the general problem.

The observed stabilities show a fairly regular trend from scandium to zinc, with the trihalide stable relative to the dihalide at the left hand end of the transition series, and unstable with respect to the dihalide for the elements to the right, especially nickel, copper and zinc. There is an anomalous order of stability in the middle of the series with Mn $<$ Fe $>$ Co in the stability of the trihalide relative to the dihalide. This order holds for X = Cl, Br and I. In the detailed analysis, it turns out that the change of entropy in the decomposition reaction is small and nearly the same for all the elements. Thus the free energy change in the reaction depends on the changes in enthalpy. These are best analyzed by breaking the reaction down into a number of simpler steps in a Haber cycle:

Recall that exothermic processes are negative in sign.

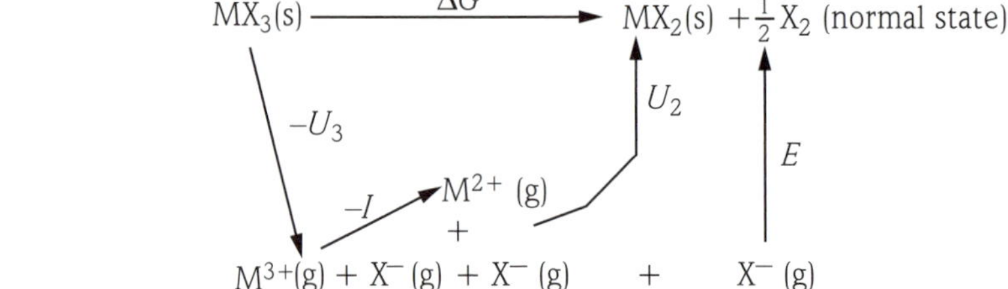

Thus $\Delta G = -U_3 - I + U_2 + E - T\Delta S$, where ΔG is the free energy of the decomposition reaction, U_2 and U_3 are the lattice energies of the di- and trihalide, $T\Delta S$ is the entropy energy, and E and I are the appropriate electron affinity and the third ionization potential of the metal. As the changes in the chlorides, bromides and iodides are parallel, attention may be focused on the chlorides.

As we are comparing the chlorides of one element with the next, the value of E is a constant for the series, and as $-T\Delta S$ is found to vary only slightly, we can take the sum of E and $-T\Delta S$ as a constant along the series. Thus the *variation* in ΔG along the series is a consequence of the variations in the terms $-I$, $-U_3$ and U_2. The values of the ionization energies, I, are well established (Table 2.8) and the lattice energies, U, may be evaluated by Born–Haber cycles, as discussed in Chapter 5, in terms of atomic properties.

The variations across the series of $-I$, U_2 and $-U_3$ are plotted in Fig. 14.35, together with the resultant value of ΔG which is the combination of the constant terms E and $-T\Delta S$ with (U_2-U_3-I). The lattice energy curves, U_2 and U_3, are similar to those shown in Figs 13.21 and 13.22 and clearly reflect ligand field stabilization energy changes. In particular, the d^5 species—$MnCl_2$ and $FeCl_3$—are less stable than their neighbours. (Note that U_2 and U_3 are plotted with opposite signs.) However, for the difference $(U_2 - U_3)$, these variations partly cancel (as shown in the top curve of Fig. 14.35) as they occur one element later for the trihalides than for the dihalides. It is clear that the most important single factor is the third ionization potential of the metal, which is exothermic for the cycle step shown

$$M^{3+}(g) + e^- \longrightarrow M^{2+}(g)$$

The free energy curve ΔG is clearly seen in the figure to be dominated by the value of $-I$, moderated to some extent by the variation in (U_2-U_3). The change in free energy of the reaction

$$MCl_{3(solid)} \longrightarrow MCl_{2(solid)} + \tfrac{1}{2}Cl_{2(gas)}$$

is positive for the elements at the left of the transition series and varies through near-zero values to negative at the right hand end. That is, the trichlorides (and similarly for the tribromides and triiodides) are stable with respect to decomposition at the left of the series while the decomposition to the dihalide is favoured towards the right. The variation of I, in turn, may be analysed as a general increase across the transition series as the nuclear charge increases (that is, $-I$ becomes more negative) as expected from the incomplete

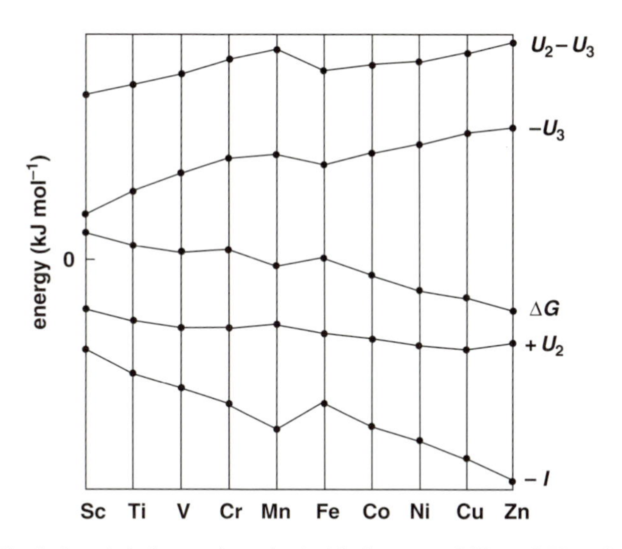

FIG. 14.35 The variation in the principal energy terms involved in the process $MX_3 \longrightarrow MX_2 + \frac{1}{2}X_2$. The resultant ΔG follows the $-I$ curve, moderated by the lattice energy difference U_2-U_3. For clarity, the zeros of the different curves have been shifted: the $-I$ curve ranges from -2430 to -3850 kJ mol^{-1}, the U_2 curve from -2340 to -2720 kJ mol^{-1}, the ΔG curve from $+375$ to -630 kJ mol^{-1}, and the $-U_3$ curve from $+5270$ to $+5860$ kJ mol^{-1}. (Recall that lattice energies are exothermic as defined so that $+U_2$ is exothermic and therefore has negative values while $-U_3$, the energy for the reverse process to forming the trihalide lattice, is endothermic and has positive values.)

shielding effect of the d electrons in the same shell. To this is added the major break between Mn($d^5 \rightarrow d^4$) and Fe($d^6 \rightarrow d^5$) which again reflects the exchange energy effect in the d^5 configuration.

Thus, although this decomposition reaction depends on the total effect of a large number of factors, it is seen that the pattern is set by exchange energy effects in the d^5 state as reflected in the value of I moderated by CFSE effects on the lattice energies of the halides.

Problems

The best way to build up a clear picture of systematic chemistry is to correlate one body of facts in as many ways as possible. Look at the chemistry of neighbouring elements, of other oxidation states of the same element, of the same d^n configuration, of ions of the same charge, etc. The problems below are a guide to many similar ones which you can devise.

Many such questions can be devised. Surveys should start with the first transition series in Chapter 14 and include material from Chapters 9 and 13. Many should be broadened to the heavy transition elements in Chapter 15, and often to the rest of the Periodic Table (see Chapters 10 to 12 and 17). The special topics of Chapters 16 and 20 will also often be relevant.

When consulting alternative sources on systematic chemistry, you should check out 'Notes to the reader' at the beginning of Appendix A, Further Reading, together with the comments on the different sections of this list.

14.1 From Fig. 14.1 (compare Section 8.6) determine the most stable oxidation state in solution for each of the transition elements of the first series. Compare Tables 13.2 to 13.5 and decide how far stability in the solid state matches stability in solution. Discuss any difference.

14.2 For each element of the first transition series, examine how far the detailed chemistry (as given in the appropriate Sections 14.2 to 14.9) matches the general deductions drawn in Question 14.1 (see also Chapter 13, Question 1).

14.3 The oxidation states of the first transition series are demonstrated by the oxides and fluorides in Tables 13.3 and 13.4. Collect together descriptions of mixed species (the oxyfluorides) from this chapter and compare the picture of oxidation states which results with that given by the tables. Do the same exercise for the heavier halides. Such an approach may be extended to the complex ions: compare the fluoro-complexes with the oxyfluoro- ones.

14.4 Compare and contrast the chemistry of every third element of the first transition series with that of its neighbour on either side.

14.5 (a) A potential of −1.9 V has been measured for $Fe^{3+} + 4H_2O = FeO_4^{2-} + 8H^+ + 3e$. Extend the Ebsworth diagram for Fe and discuss the properties of Fe(VI) compared with Mn(VI) and Cr(VI).

(b) Survey all oxidation states of VI or above throughout the Periodic Table. Discuss their occurrence, stability and properties.

14.6 Survey a general topic throughout the transition elements. Examples of such topics include

(a) octahedral complexes (compare particular Groups of elements)
(b) complexes of ligands with alternative donor atoms (e.g. −NCS and −SCN)
(c) coordination numbers greater than six
(d) shapes adopted for 5-coordination
(e) the relative stabilities of octahedral, tetrahedral and square-planar configurations
(f) particular classes of compound such as 'bridged oxygen species', or 'complexes with metal–sulfur bonds', or 'peroxides'.
(g) complexes of bidentate oxyanions (e.g. acetates, nitrates)
(h) low oxidation states.

14.7 Look up the paper 'Thermochemistry of the Potassium Hexafluorometallates(III) of the Elements from Scandium to Gallium' by P. G. Nelson and R. V. Rearse, in *J. Chem. Soc., Dalton Transactions*, 1983, pages 1977–1982. Compare the results with Fig. 14.35. Discuss how far the interpretations are compatible and whether there is an effect of the halogen.

CHAPTER 15

15.1 GENERAL PROPERTIES 351
15.2 ZIRCONIUM, $4d^25s^2$, AND HAFNIUM, $5d^26s^2$ 352
15.3 NIOBIUM, $4d^45s^1$, AND TANTALUM, $5d^36s^2$ 354
15.3.1 The V state 355
15.3.2 The IV state 357
15.3.3 Lower oxidation states of niobium and tantalum 357
15.4 MOLYBDENUM, $4d^55s^1$, AND TUNGSTEN, $5d^46s^2$ 359
15.4.1 Molybdenum(VI) and tungsten(VI) 360
15.4.2 The V state 363
15.4.3 The IV state 364
15.4.4 The lower oxidation states 365
15.5 TECHNETIUM, $4d^65s^1$, AND RHENIUM, $5d^56s^2$ 366
15.5.1 The VII oxidation state 367
15.5.2 The VI state 369
15.5.3 The V state 370
15.5.4 The IV state 370
15.5.5 The lower oxidation states 371
15.6 RUTHENIUM, $4d^75s^1$, AND OSMIUM, $5d^66s^2$ 372
15.6.1 The higher oxidation states 372
15.6.2 The lower oxidation states 374
15.7 RHODIUM, $4d^85s^1$, AND IRIDIUM, $5d^96s^0$ 376
15.8 PALLADIUM, $4d^{10}5s^0$, AND PLATINUM, $5d^96s^1$ 379
15.8.1 The VI, V and IV oxidation states 379
15.8.2 The III, II and lower oxidation states 380
15.9 SILVER, $4d^{10}5s^1$, AND GOLD, $5d^{10}6s^1$ 382
15.9.1 Silver 382
15.9.2 Gold 384
15.10 THE ZINC GROUP 385
15.10.1 The I state and subvalent compounds 386
15.10.2 The II state 388
PROBLEMS 390

The Elements of the Second and Third Transition Series

15.1 General properties

This chapter deals with the remaining elements of the transition block, those in which the 4d or 5d level is filling. The effect of the lanthanide contraction is to make the chemistry of the heavier pair of elements in the same Group very similar.

In these elements, higher oxidation states are more stable, in general, than in the first series. This is shown both by the relatively nonoxidizing behaviour of states like Re(VII) and Mo(VI) or W(VI), and by the existence of high oxidation states to the right of the series, which are not found for the first row elements: examples include Os(VIII) and Ir(VI). Complementary to this increased stability of high oxidation states is a decrease in the stability of many lower oxidation states, as shown by such examples as the strong reducing properties of Zr(III) or Nb(IV) and the virtual nonexistence of compounds of W(II). Some similarities do, of course, exist between the lighter and heavier elements, especially in the properties of the higher oxidation states to the left of the d block and in those of the lower oxidation states at the right hand end. The maximum range of oxidation states comes in the middle of the block, rising to the VIII state of osmium and ruthenium.

Some of these elements are rather rare or inaccessible, especially those in the Groups immediately following the lanthanides, which are difficult to separate. Both hafnium and niobium, which occur along with their more abundant congeners, zirconium and tantalum, are not well studied and there are gaps in their chemistries where they are presumed to be similar to their congeners but without full proof. Other rare elements are technetium, which has only unstable isotopes and has to be made artificially, and the platinum metals and gold which, though accessible, are expensive to work with. The elements become less reactive towards the right of the transition block, and the tendency to reduction to the metal is an important characteristic, especially of the platinum metals and gold, and also, to a lesser extent, of rhenium and silver. Few of these elements find large scale applications, though molybdenum and tungsten are components of highly resistant steels. The precious and semi-precious metals, apart from their uses in coinage and jewellery, find some application in precision instruments, electrical apparatus, and in surgery. Platinum, palladium and rhodium are used as catalysts in a number of industrial processes. Zirconium and niobium, the former especially, find use as 'canning' materials for the fuel in nuclear reactors. For this purpose, they must be separated from their congeners, hafnium and tantalum, which 'poison' the reactors by capturing neutrons. Most elements are extracted by carbon or metal reduction of the oxides or chlorides.

Many of these elements form nonstoichiometric carbides, nitrides and hydrides, similar to those already discussed for the first series elements.

As these elements are larger than the first row members higher coordination numbers are found. Although 6-coordination to singly bonded ligands is still common, 7- and 8-coordination are found, as in ZrF_7^{3-} or $Mo(CN)_8^{2-}$. To π-bonding ligands such as oxygen, six becomes a common coordination number as well as four; compare the polymeric oxyanions of Mo and W, which are based on MO_6 groups, with the vanadates and chromates, which contain MO_4 units.

PROPERTIES OF SECOND AND THIRD SERIES TRANSITION ELEMENTS COVERED IN OTHER CHAPTERS

A summary of the oxidation states of these elements, in relation to those of the first series, is given in Table 13.2, while Tables 13.3, 13.4 and 13.5 give the known oxides and halides. Other properties already listed include electronic configurations, atomic weights and numbers (Table 2.5) and ionization potentials (Table 2.8). Specific groups of compounds are covered in later chapters—carbonyls in Section 16.2, organic derivatives including metallocenes in Sections 16.3 and 16.4, higher order M-M bonding in Section 16.7, metal cluster compounds in Section 16.8, dioxygen, dinitrogen and dihydrogen species in Section 16.9 to 16.11, complexes with polysulfur and related ligands in Section 19.2, some metal species of major biological importance in Section 20.1.4, and some uses of technetium, platinum or gold in medicine in Section 20.2.

See Section 16.12 for elements of the fourth transition series (the 'superheavy' elements).

The increase in ligand field splittings which is shown in these larger atoms means the CFSE values increase markedly in all configurations, and the normal electronic configurations are the low-spin ones. Evidence for high-spin complexes is very rare.

15.2 Zirconium, $4d^25s^2$, and hafnium, $5d^26s^2$

Only the IV oxidation state is stable for these elements and potential data are not available for the lower states. The M/M^{IV} values for the two elements are similar and differ from that of titanium by a factor of about two, the elements being better reducing agents than titanium. $Ti^{IV}/Ti = -0.89$ V, $Zr^{IV}/Zr = -1.56$ V, and $Hf^{IV}/Hf = -1.70$ V. The IV state occurs in solution and in a variety of solid compounds, but the III and II states decompose water and are only found in solid products.

Zirconium ores occur, including the oxide and zircon, $ZrSiO_4$, but there are no discrete sources of hafnium, which is always found as a minor component in zirconium minerals. Zirconium was known for nearly a hundred and fifty years before hafnium was discovered in its ores and compounds. The elements are extracted as tetrahalides and reduced with magnesium in a process similar to that for titanium. Here also, argon has to be used to provide an inert atmosphere, as the metals combine with nitrogen.

The covalent radii, 145 pm and 144 pm respectively for Zr and Hf, and also the radii of the hypothetical ions Zr^{4+} (74 pm) and Hf^{4+} (75 pm), are so close that the chemistry of this pair of elements is virtually identical. Separation is even more difficult than for the lanthanides but ion exchange and solvent extraction procedures are now available. One example of these is the separation of the tetrachlorides dissolved in methanol on a silica gel column. Elution is by an anhydrous HCl/methanol solution, with the zirconium coming off the column first.

In the IV state, these elements show a general resemblance to titanium(IV) but differ in the acidity of the dioxides and the solvolytic behaviour of the oxycation. The dioxides are soluble in acid solution and addition of base precipitates gelatinous hydrated oxide, $ZrO_2.nH_2O$. No true hydroxide exists. The hydrated oxide gives ZrO_2 (or HfO_2) on heating and these are hard, white, insoluble, unreactive materials with very high melting points (above 2500°C). The more abundant zirconium dioxide is used for high temperature equipment, such as crucibles, because of these properties. ZrO_2, doped with Y_2O_3, finds important application as a solid state electrolyte in oxygen sensors. These are used, for example, in motor vehicle exhausts, to measure the oxygen content of the exhaust gases in a feedback loop which allows optimized combustion and minimum emissions. The hydrated oxide is quite insoluble in alkali and the dioxides therefore have no acidic properties. The solids M_2ZrO_3 contain ZrO_5 square pyramids linked into chains by sharing edges.

In solution, zirconium(IV) and hafnium(IV) hydrolyse less than titanium(IV). The main species in solution is frequently written as the oxyion ZrO^{2+} (zirconyl) or HfO^{2+} (hafnyl) but it is doubtful whether these simple species exist. The species in solution are probably $M(OH)_n^{(4-n)}$ and trimeric or tetrameric hydroxy species. Tracer experiments indicate the presence of Zr^{4+} ions in very dilute solution in perchloric acid. Compounds containing the MO group are common, for example $ZrO(NO_3)_2.2H_2O$ or $HfO(OOCCH_3)_2$, but their structures may not be simple. Thus, $ZrOCl_2.\ 8H_2O$ contains the cation $[Zr_4(OH)_8(H_2O)_{16}]^{8+}$

which has dodecahedral coordination around the zirconium. There is also evidence for the possibly polymeric, ion $Zr_2O_3^{2+}$ in solution. The complexity of these hydrolysis products is typical of the solution chemistry of this part of the Periodic Table, although the larger atoms mean that there is less extensive hydrolysis than for titanium(IV). However, the elements are not large enough to allow the formation of the M^{4+} ions. Simple zirconium or hafnium compounds, such as the tetrahalides or the tetraacetate, $Zr(OOCCH_3)_4$, are covalent compounds, not salts.

The halides may be made by standard methods and show the expected reactions:

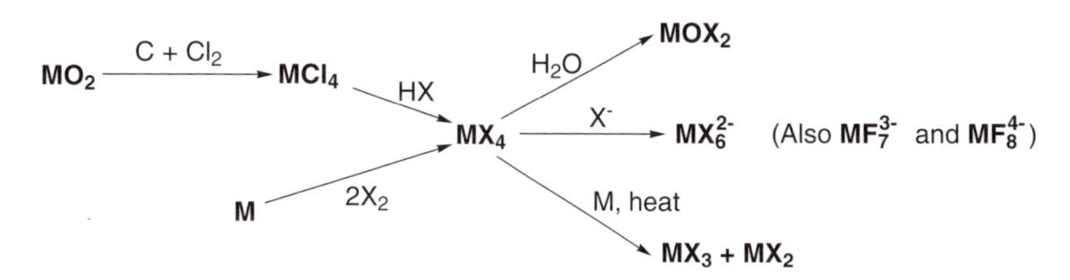

ZrF_4 forms a solid with 8:2 coordination where each Zr atom is surrounded by 8F atoms in a square antiprism configuration (compare Fig. 15.5b). The other tetrahalides are volatile solids. In the vapour phase, the structure is monomeric and tetrahedral, while in the solid the Zr or Hf atoms are in octahedral coordination. The structure of one form of $ZrCl_4$ is a chain of linked octahedra ($ZrCl_2Cl_{4/2}$) like the NbI_4 structure shown in Fig. 15.8. The halides hydrolyse vigorously to the oxyhalide, which is stable to further hydrolysis. The most stable complex halides are the fluorides, and the hexa-, hepta- and octafluorides are known. The octahedral ZrF_6^{2-} ion is found for example in Li_2ZrF_6. In the formally similar K_2ZrF_6, the structure involves bridging fluorines and ZrF_8 coordination. In the mixed potassium-cupric hexafluorozirconate, the structure is even more complex and is formulated as $K_2Cu(H_2O)_6[Zr_2F_{12}]$. The $(Zr_2F_{12})^{4-}$ anion has the unusual pentagonal bipyramidal coordination around each zirconium, and the two bipyramids share an edge—$F_5ZrF_{2/2}ZrF_5$. The pentagonal bipyramid is also found in the sodium salts Na_3ZrF_7 and Na_3HfF_7. These contain pentagonal bipyramid MF_7 groups, as in IF_7, but the ammonium salt of the heptafluorides has the same structure as the isoelectric niobium and tantalum heptafluorides. This is shown in Fig. 15.1 and may be described as a trigonal prism with the seventh fluoride added beyond the centre of one of the rectangular faces. In the MF_8 groups, the eight fluorines adopt the bisdisphenoid configuration shown in Fig. 15.2, and described in Section 13.6. If the cation is $Cu(H_2O)_6^{2+}$, in place of the alkali metals, the antiprismatic form is found for ZrF_8^{4-}, like the octafluorotantalate (Fig. 15.5b).

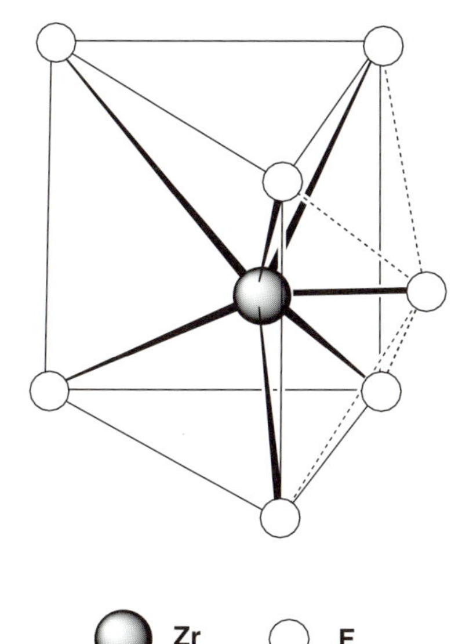

FIG. 15.1 The structure of the ZrF_7^{3-} ion in $(NH_4)_3ZrF_7$. This species is basically a trigonal prism with an extra ligand attached to one face. The NbF_7^{2-} and TaF_7^{2-} ions adopt the same structure

The chloro complex ion $[Hf_2Cl_{10}]^{2-}$ contains two octahedral $HfCl_6$ units, linked in a dimer by sharing an edge $[Cl_4Hf\text{-}(\mu\text{-}Cl)_2\text{-}HfCl_4]^{2-}$ (compare Fig. 14.5b). The zirconium analogue was later found, together with $[ZrCl_6]^{2-}$. Single $ZrCl_4$ molecules have been identified in the vapour phase and the dimer and small polymers in the melt. All these are probably linked through chloride bridges.

Zr and Hf, as well as U, form volatile borohydrides $M(BH_4)_4$, with 12-coordination where each BH_4 group is bonded to M through three shared H atoms. In contrast, the larger uranium allows even higher coordination (see p.267). The ion $M(BH_4)_5^-$ may be stabilized by large cations and appears to be 10-coordinate with M(–H–)₂B bridges.

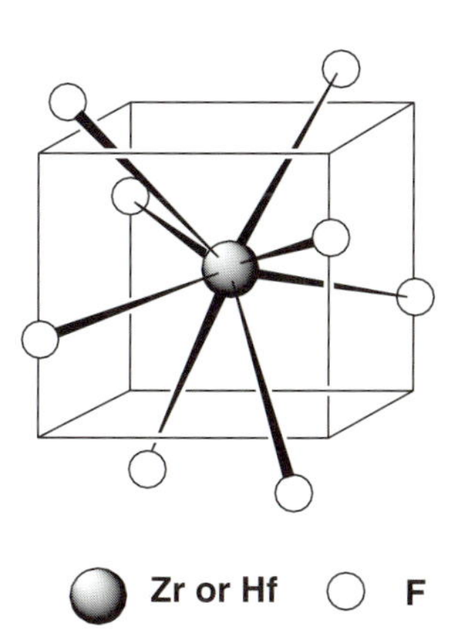

FIG. 15.2 The structure of ZrF_8^{4-} and HfF_8^{4-}. This form of 8-coordination is similar to that found in the octacyanomolybdate ion (Fig. 15.18). Contrast this with the structure of the octafluoride of tantalum (Fig. 15.5b)

The lower oxidation states of zirconium and hafnium are very strongly reducing. Mixtures of di- and trihalides result from reduction by the element, or by reduction with H_2 at 400–500°C. It may be noted that treatment of a mixture of $ZrCl_4$ and $HfCl_4$ with zirconium metal gives $ZrCl_3$, which is involatile, and leaves $HfCl_4$ unreacted. The latter may then be sublimed out of the reaction mixture, providing a method of separating the elements. The III and II states do not exist in solution. One striking example of the reducing power of the III state is provided by the production of the blue potassium solution in liquid ammonia, in the following reaction:

MIXED VALENCE ZIRCONIUM CLUSTERS

Zirconium also forms halogen-bridged clusters containing 6 Zr atoms which are of intermediate formal oxidation state. These are analogous to the long-established niobium and tantalum clusters, compare Section 15.3.3. If we start from the $[Nb_6Cl_{12}]^{2+}$ cluster of Fig. 15.9a and replace each Nb by Zr, we remove six electrons. In the known Zr_6 clusters, these missing electrons are partly offset by an encapsulated atom lying at the centre of the Zr_6 octahedron. Thus we find complexes like $[Zr_6Cl_{12}N]^{3+}(Cl^-)_3$, $[Zr_6Br_{12}C]^{2+}(Br^-)_2$ or $Cs^+[Zr_6I_{14}B]^-$. Where charges are balanced by anions, these coordinate in external apical positions to the Zr, so the fully symmetric complex would have 18 halogens, 12 bridging the edges and one coordinated to each Zr apex (Fig. 15.3, compare Fig. 15.9b). The encapsulated or interstitial atom is either a light element like H, Be, B, C or N, or a transition metal such as Mn, Fe or Co.

It was initially thought that the encapsulated atom was essential to the stability of the Zr_6 clusters but more recent work has demonstrated the existence of 'empty' clusters such as $[(R_3PZr)_6Cl_{12}]$ with an apical phosphane on each Zr. Further examples have various combinations of coordinated phosphane and halogens.

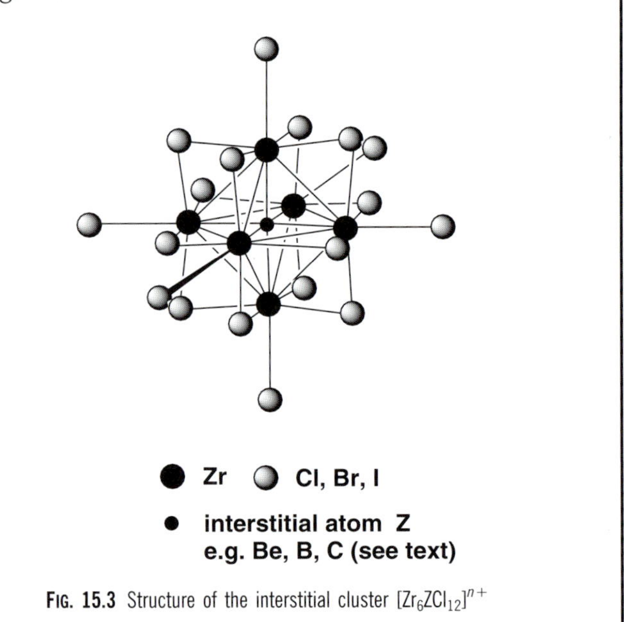

FIG. 15.3 Structure of the interstitial cluster $[Zr_6ZCl_{12}]^{n+}$

$$ZrCl_3 + 4KNH_2 \xrightarrow{NH_3} Zr(NH_2)_4 + K + 3KCl$$

$HfCl_4$ gives $HfCl_3$ when heated with hafnium metal, and the trichloride is stable to 350°C in the presence of $HfCl_4$. Hafnium dichloride is said to disproportionate to HfCl and $HfCl_4$ when heated to 627°C. In contrast there is no lower hafnium iodide with iodine content less than $HfI_{3.2}$.

The MX_3 species have either the BiI_3 layer structure (Fig. 5.10b) or the ZrI_3 structure which consists of a chain of octahedra sharing faces.

Treatment of Zr with $ZrCl_4$ in an inert container yields a new phase, ZrCl, which has metallic properties. The structure is a novel layer form, of nearly cubic symmetry, which can be regarded as the $CdCl_2$ structure (see Table 5.3) with an extra layer of metal atoms inserted giving a four-layer Cl-Zr-Zr-Cl arrangement. A Zr atom is approximately ccp with 6 Zr neighbours in the same layer, three Zr in the neighbouring layer on one side and 3 Cl atoms as neighbours on the other side (compare Section 5.6). ZrBr is similar.

The zero-valent state is rare but the violet dipyridyl, $[Zr(dipy)_3]$, is quite well established. This is formed, like the corresponding Ti compound, by reduction of $ZrCl_4$, with Li in ether in the presence of dipyridyl. A further possibility is a cyanocomplex of Zr(0), formed by reduction in liquid ammonia (compare Section 6.7), which is formulated as $[Zr(CN)_5]^{5-}$. However, this may be the dimer.

Compounds such as $Zr(benzyl)_4$ and $Zr(C_5H_5)_4$ show promise as homogeneous catalysts in the Ziegler–Natta process (see Section 17.4.3).

15.3 Niobium, $4d^45s^1$, and tantalum, $5d^36s^2$

These two elements resemble zirconium and hafnium in the very close similarity of their chemistries, although here it is the lighter niobium which is the rarer element. The different ground state electronic configurations appear to have no effect on the chemistry in the valency states. Rather more is known of the lower oxidation states of these elements than was the case with zirconium and hafnium, but the V state is by far the most stable and well known. The element/M^{5+} potentials show the same trend as in the preceding Group:

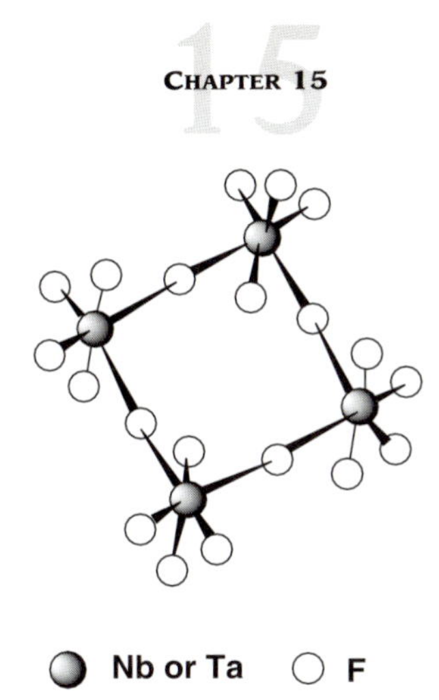

FIG. 15.4 The structure of NbF_5 or TaF_5. Most other pentafluorides of the transition metal adopt similar, tetrameric, structures

$$V^V/V = -0.25\ V,\ Nb^V/Nb = -0.65\ V,\ Ta^V/Ta = -0.85\ V$$

A potential of about −1.1 V has been estimated for Nb^{III}/Nb and one of about −0.1 V for Nb^V/Nb^{III}, both in sulfuric acid solution where complex species are probably formed in both states.

Niobium and tantalum generally occur together and are separated by fractional crystallization of fluoro-complexes:

$$\text{ore} \longrightarrow M_2O_5 + KHF_2/HF \longrightarrow \underset{\text{less soluble}}{K_2TaF_7} \xrightarrow{\text{electrolysis}} Ta$$

$$\searrow \underset{\text{more soluble}}{K_2NbOF_5} \xrightarrow{Al} Nb$$

The metals are very resistant to acid attack but will react slowly with fused alkalis and with a variety of nonmetals at high temperatures. They have very high melting points (Ta above 3000°C) and find some use in high temperature chemistry. Tantalum is also used in surgery as it can be inserted in the body, as in fracture repair, without causing a 'foreign body' reaction.

15.3.1 The V state

In the V state, the oxides, Nb_2O_5 and Ta_2O_5, may be prepared by igniting the metals, their carbides, sulfides or nitrides, or any compound with a decomposable anion. The oxides are inert substances which are generally brought into solution by alkali fusion, or treatment with concentrated HF. The oxides are therefore amphoteric but the acidity is very slight and the niobates are decomposed, even by as weak an acid as CO_2. The product of alkali fusion contains one metal atom and is written as NbO_4^{3-} (or TaO_4^{3-}) and termed orthoniobate (or orthotantalate). A monatomic metaniobate, $NaNbO_3$, is also known which has the perovskite structure. A number of more complex species are also known which contain 2, 5 or 6 metal atoms, for example, $M^I_4Nb_2O_7$ or $M^I_8Ta_6O_{19}$. The latter $Ta_6O_{19}^{8-}$ ion has also been shown to be present in solution. The structure of none of these species is known, nor are the formulae unambiguous but may include water molecules and hydroxyl ions. It does seem clear, though, that niobium and tantalum share with vanadium the tendency to form polymeric oxyanions. The neutralization of these niobate or tantalate solutions with acid leads to the precipitation of white gelatinous precipitates of the hydrated pentoxides. These dissolve in hydrofluoric acid, probably as fluoro-complexes, but there is no evidence of cationic forms of niobium or tantalum in solution analogous to the vanadium oxycations.

The pentafluorides may be prepared by the reaction of fluorine on the metal, pentoxide or pentachloride, or by HF on the pentachloride. Both are volatile white solids melting below 100°C and boiling near 230°C. The structures are tetrameric in the solid with MF_6 units and one bridging fluoride between each pair of metal atoms (Fig. 15.4), isostructural with MoF_5. In the liquid phase, there is evidence for similar *cis*-bridging, but forming polymers rather than tetramers. In the vapour above NbF_5, 98% is the trimer Nb_3F_{15} and about 2% is monomeric NbF_5.

The pentoxide dissolves in HF with formation of fluoro-complexes and these are also formed by the pentafluorides and F^-. Crystallization from solutions of moderate F^- concentration gives the salts $M^I Nb$(or Ta)F_6 containing octahedral MF_6^- ions. In the presence of excess F^-, TaF_7^{2-} and $NbOF_5^{2-}$ are formed. A larger excess of F^- leads to the formation of NbF_7^{2-} and, at very high F^- concentrations, $NbOF_6^{3-}$ and TaF_8^{3-} are formed. The octafluoroniobate has not been reported. The structures of the MF_7^{2-} ions are those based on the trigonal prism shown in Fig. 15.1. $NbOF_5^{2-}$ has an octahedral structure, while $NbOF_6^{3-}$ illustrates another form of 7-coordination. This, Fig. 15.5a, is based on the octahedron with the seventh ligand placed at the centre of one triangular face. The TaF_8^{3-} ion is the square antiprism (Fig. 15.5b) which reduces interactions between ligands to a minimum in 8-coordination. This corner of the d block gives a variety of 7- and 8-coordinated structures which is not found elsewhere in the Periodic Table.

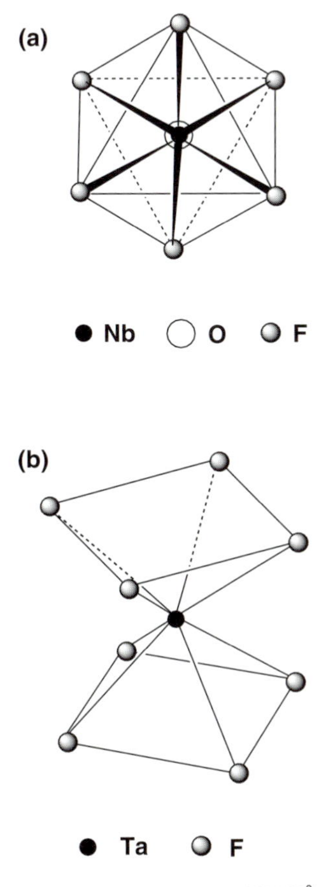

FIG. 15.5 Higher coordination numbers: (a) $NbOF_6^{3-}$ and (b) TaF_8^{3-}

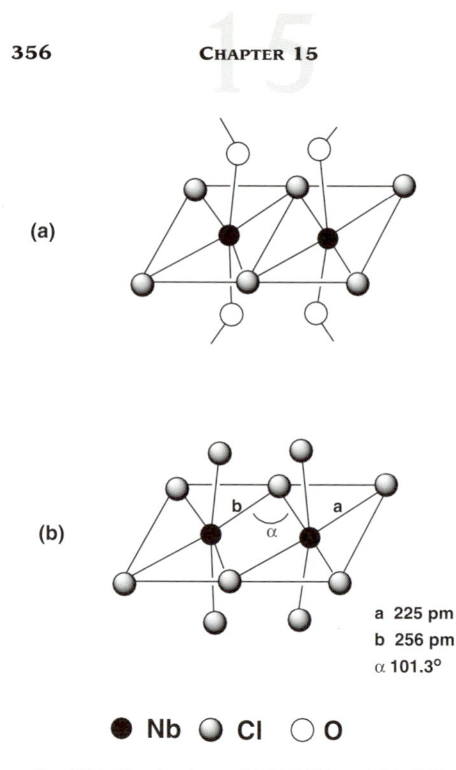

FIG. 15.6 The structures of (a) $NbOCl_3$ and (b) $NbCl_5$

The other pentahalides all exist and can be prepared by standard methods. All six compounds are volatile covalent solids with boiling points below 300°C. The vapours are all monomeric and electron diffraction results indicate the structures are probably the expected trigonal bipyramids. In the solid, an X-ray study shows that $NbCl_5$, is a dimer, $Cl_4Nb(\mu\text{-}Cl)_2NbCl_4$, with two bridging chlorides giving an octahedral configuration around each niobium. In solution in nondonor solvents, such as CCl_4, the dimeric form is retained. Tantalum pentachloride and both bromides have the same solid structure but the pentiodides are of a different, and unknown, structure. Much less is known about complexes of these heavier halides but there is good evidence for the existence of MCl_6^- species.

The halides all hydrolyse readily to the hydrated pentoxides and MOX_3 for X = Cl, Br, I have been isolated as intermediate products for both metals. A better preparation is by reaction between the pentahalide and oxygen. Treatment of M_2O_5 with aqueous HF gives MO_2F which yields progressively on heating MOF_3 then M_3O_7F. The corresponding heavy halide compounds were also established so that the complete series MOX_3, MO_2X and M_3O_7X is known for all the halides and both metals.

The mononuclear oxyhalides are covalent but less volatile than the pentahalides. The vapours appear to be monomeric and tetrahedral but the solids are polymeric. The structure of $NbOCl_3$ is shown in Fig. 15.6, with the pentahalide for comparison. The oxyhalide contains planar, binuclear $Cl_2Nb(\mu\text{-}Cl)_2NbCl_2$ groups which are linked into long chains by oxygen bridges between the niobium atoms.

Apart from the oxides and halides, there are few important compounds of these elements in the V state. One interesting one is the cyclopentadienyl hydride, $(\pi\text{-}C_5H_5)_2TaH_3$, which provides another example of the stabilization of transition metal–hydrogen bonds by the presence of a π-bonding ligand. The structure, shown in Fig. 15.7, is derived from the ferrocene one (Fig. 16.7a) by bending the rings towards each other and inserting the three hydrogens in the central plane pointing away from the rings. The ring-M-ring angle is 139° (Ta) and 142° (Nb) while the H_BMH_B angles are 126° in each case. The bond to the central hydrogen H_A bisects this angle and all three M-H distances are equal. Nmr studies confirm two different types of H atom environment occupied in the ratio 2:1.

A similar compound with angled cyclopentadiene rings is $[Nb(Cl)(O)(C_5Me_5)_2]$ where the Cl and the O occupy the H_B positions of Fig. 15.7, implying an Nb=O interaction.

The gas phase structures of two volatile 5-coordinate tantalum compounds, $TaMe_5$ and $Ta(NMe_2)_5$, have been determined by electron diffraction and show, unexpectedly, a square-pyramidal geometry. With the reservation that electron diffraction is not an absolute method (compare Section 7.4) and the amide particularly is very demanding for the method, the similar conclusions are mutually reinforcing. In the study on the amide the expected trigonal bipyramid structure was also tested and showed a poorer fit. Further, $SbMe_5$ was also measured for comparison with the pentamethyl and did show the expected trigonal bipyramid. Thus the TaR_5 square pyramid structures seem firmly based. In each case the Ta is above the plane of the four base atoms and the Ta–apex bond is shorter than the Ta–base one (Ta-Me values of 211 pm and 218 pm; Ta-N distances of 194 pm and 204 pm). It is suggested that, because the metal d orbitals are involved in the bonding, the simple VSEPR analysis fails. In particular, the least stable bonding orbital in the trigonal bipyramid form (which involves largely metal p_z overlap with the apical ligand orbitals) is significantly stabilized by the change to square pyramid without offsetting losses, if the ligands are of relatively low electronegativity and there is no significant π bonding.

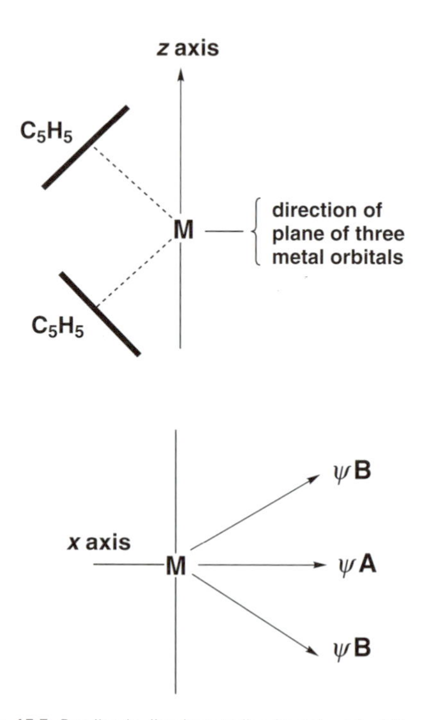

FIG. 15.7 Bonding in dicyclopentadienyltantalumtrihydride, $\pi\text{-}(C_5H_5)_2TaH_3$ This is a diagrammatic representation of the three orbitals which become available on the central atom when the C_5H_5-Ta-C_5H_5 angle is reduced from the linear arrangement in ferrocene. The two H atoms bonded to the B orbitals are identical and differ from the central A hydrogen

The electron diffraction structure shows that Me_3TaF_2 is a trigonal bipyramid with the three methyl groups equatorial (Ta-C = 213 pm; Ta-E = 186 pm). While TaF_5 is mainly tetramer in the vapour, it has been possible to determine the monomer which is also a trigonal bipyramid. Clearly, the balance between the two geometries is a very delicate one.

15.3.2 The IV state

Compounds in the IV state include NbO_2, all the tetrahalides, MX_4, and some oxyhalides, MOX_2, for both metals. In addition there are a number of niobium oxide phases in between Nb_2O_5 and NbO_2, such as Nb_8O_{19}, and $Nb_{11}O_{27}$. There is more uncertainty but TaO_2 probably does exist. The state is unstable, reducing, and shows some tendency to disproportionate. The dioxides are prepared by high temperature reduction of the pentoxides. They dissolve only in hot alkali, with reduction of the solvent.

All the tetrahalides exist except the tetrafluorides: the iodides are best known. NbI_4 results from prolonged heating of NbI_5 at 270°C, when iodine sublimes off, leaving the tetraiodide. TaI_4 is most easily made by heating TaI_5 with the metal. These iodides are diamagnetic solids, volatile at 300°C. The diamagnetism arises because the metal atoms are linked by long bonds in dimers, as shown in Fig. 15.8, thus pairing the single electron on each. The structure consists of chains of MI_6 octahedra, each joined by the edges to its neighbours, and with the metal atoms placed unsymmetrically in the octahedra and linked in pairs. $NbCl_4$ has a similar structure with a short Nb-Nb distance of 303 pm and the long (nonbonded) one equal to 379 pm. The tetrachlorides are made by reduction of the pentachlorides and both $NbCl_4$ and $TaCl_4$ disproportionate,

$$\text{e.g.}\quad TaCl_4 \xrightarrow{400\,^\circ C} TaCl_3 + TaCl_5 \uparrow$$

Similarly, on hydrolysis, $TaCl_4$, gives a precipitate of tantalum(V) oxide and a green solution of the trichloride, which is fairly stable unless heated. The IV state is also reported from the electrolytic reduction of niobium(V) in 13M hydrochloric acid. An orange Nb^{IV} solution results which probably contains the oxyhalide ion $NbOCl_4^{2-}$. This solution disproportionates to niobium(III) + (V). MOX_2 oxyhalides are known for X = Cl, Br and I. A significant complex is the chloro-species, $[Me_3P)_4M_2(Cl)_4(\mu\text{-}Cl)_4$. Here each M atom is bonded to two phosphanes and two terminal Cl atoms, and the two are linked together by four Cl bridges. The two M atoms are close enough to be bonded, thus the complex shows 9-coordination and parallels the NbI_4 interaction above.

15.3.3 Lower oxidation states of niobium and tantalum

In the III state, MX_3 is known for X = F, Cl and Br. Heating NbI_5 at 430°C forms the mixed oxidation state, Nb_3I_8, (and further heating gives Nb_2I_{11}). TaI_3 is not reported. The other niobium trihalides are formed by reducing the pentahalide with hydrogen at about 500°C. Tantalum trichloride is obtained from the tetrachloride as above, while the tribromide is made by hydrogen reduction. Most of the trihalides are brown or black and strongly

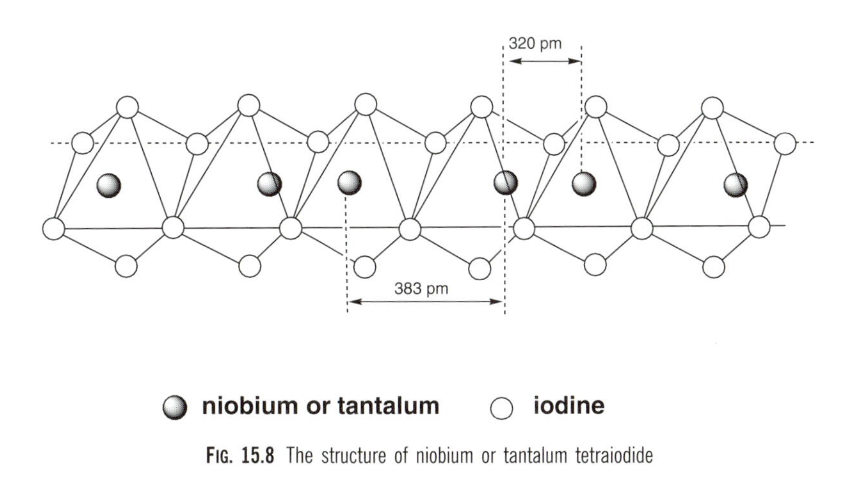

FIG. 15.8 The structure of niobium or tantalum tetraiodide

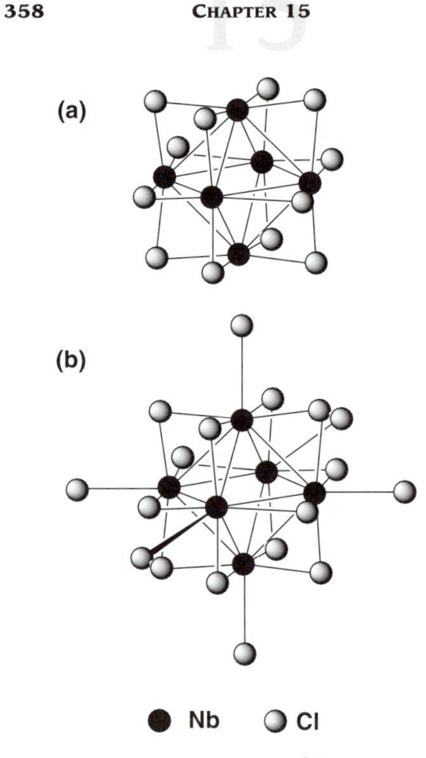

FIG. 15.9 The structures of (a) $[Nb_6Cl_{12}]^{2+}$ and (b) $[Nb_6Cl_{18}]^{4-}$ The octahedra of niobium atoms are bonded, not only by the bridging chlorines but by metal–metal bonding within the cluster

reducing, although the reactivity depends greatly on the thermal history of the sample. In strong hydrochloric acid, electrolytic reduction of niobium(V) to niobium(III) is reported. The III state solutions are yellow or blue, depending on the conditions used.

Nb(IV) and (III) are found in the octacyano complexes $Nb(CN)_8^{n-}$, where $n = 4$ or 5. Like the octacyanides of Mo and W (Section 15.4 and Fig. 15.18), both dodecahedral and square antiprismatic structures are found.

$TaCl_2$ is the only known tantalum dihalide. It is a nonstoichiometric, green-black solid resulting from the disproportionation of $TaCl_3$ at 600°C. It is much more reactive than the trichloride, and attacks water under all conditions. Small quantities of $NbBr_2$ have been produced from the reaction of the pentabromide with hydrogen in an electric discharge. Nothing is known of its properties. Electrolytic reduction of niobium in 10M HCl gives a violet solution colour attributed to niobium(II) but this state is not otherwise reported in solution.

Apart from the tetraiodides, nothing is known of the structures or degrees of polymerization of these other halides. They are all relatively involatile and the volatility drops from the tetrahalide to the trihalide to the dihalide; the structures are thus probably polymeric, although the ability of $TaCl_3$ to dissolve in water must be recalled.

In addition to these halides of simple stoichiometry, niobium and tantalum form compounds of formula M_6X_{14} for X = Cl, Br and I. Related compounds Nb_6F_{15}, Ta_6Cl_{15}, Ta_6Br_{15} and Ta_6Br_{17} are also found. These result from sodium amalgam reduction of the pentahalides (a better preparation is by reduction with cadmium metal at red heat followed by precipitation of CdS). The compounds are soluble in water and alcohol and the action of Ag^+ on the compounds M_6X_{14} precipitates only one-seventh of the halide as AgX. The compounds are therefore salts of the complex cation $(M_6X_{12})^{2+}$. The structure of this ion has been determined by X-rays. The metal atoms (Fig. 15.9a) form an octahedron and are bridged in pairs along the octahedron edges by halogen atoms. (Compare with the structure of the dihalides of molybdenum and tungsten.) The octahedron of metal atoms is an example of the *metal cluster* compounds which are attracting current attention. It is held together by polycentred metal–metal bonding as well as by the bridging halogen atoms.

A variety of similar clusters is reported which are related to the $[M_6X_{12}]^{2+}$ structure, either by having charges of +3 or +4, or by containing terminal M-X or M-L, groups in addition. The nicely symmetric structure of $[Nb_6Cl_{18}]^{4-}$, Fig. 15.9b, is the result of complete substitution. Such species are also formed by Ta, may be partly substituted, or have different halogens in the bridging and terminal positions. In the hydrate $Ta_6Cl_{14}.4H_2O$, the $[Ta_6Cl_{12}]^{2+}$ ion is elongated along one axis, apparently to optimize hydrogen bonding. More complex species have been characterized, such as clusters linked together through halogen bridges.

A related structure is seen in $K_4Al_2[Nb_{11}O_{20}F]$ which has two Nb_6 octahedra sharing a common Nb apex. There are 12 edge-bridging O atoms and 8 terminal Nb-O bonds on the equatorial Nb atoms. The remaining two apices, which are *trans* to the common Nb, form Nb-F-Nb bonds which then link up the paired octahedra into polymeric units.

The oxide of formula NbO is gray and has metallic properties. The structure is cubic with 8 Nb vacancies at the corners and an O vacancy at the centre of the NaCl structure. This gives square-planar NbO_4 coordination.

The lowest oxidation states are limited to a few carbonyl and organometallic species. For example, the (–I) oxidation state is found in the carbonyl anion $[M(CO_6)]^-$ which is formed by Nb and Ta (also V). These elements do not, however, form a neutral carbonyl like vanadium.

Multiple M-M bonded compounds are discussed in Section 16.7

It is thus seen that this Group is quite similar to the titanium one. The heavier elements are broadly similar to the lighter ones in the Group oxidation states of IV and V respectively, but the oxides are less hydrolysed in solution and less amphoteric. The heavier elements form complexes with fluorine of high coordination numbers. The vanadium Group has a wider range of oxidation states and the lower ones are not quite so unstable for niobium and tantalum as they are zirconium and hafnium, but they are much less stable than the Group state and are strongly reducing.

15.4 Molybdenum, $4d^5 5s^1$, and tungsten, $5d^4 6s^2$

These two elements, although there is still a close resemblance, show much more distinct differences in their chemistries than the two earlier pairs in these series. This is illustrated by the ready separation of the two elements in qualitative analysis, where tungsten appears in Group I of the conventional scheme and molybdenum in Group II. This is because tungsten(VI) which is soluble in neutral and alkaline solution precipitates an insoluble hydrated oxide, $WO_3.nH_2O$, in acid solution. Molybdenum(VI) oxide, on the other hand, dissolves again due to formation of chloro-complexes and molybdenum is first reduced by H_2S and then precipitates as MoS_2.

Examples of all oxidation states from VI to –II are found for these elements. The Group state of VI is the most stable, but the V and IV states are well represented and occur in aqueous solution. Strong reducing properties are not shown until the III and II states are met. This behaviour is illustrated by the occurrence of the various halides and oxides (Table 13.3 to 13.5) and by the free energy changes in solution shown in Fig. 15.10. This diagram shows the differences in behaviour of the two elements, in particular the greater relative stability of Mo^{V} and the similar effect at W^{IV}, both of which lie at shallow minima. These two elements show little resemblance to chromium: the stability of the VI state contrasts with the strong oxidizing nature of chromium(VI) while the III state, which is so stable for chromium, is very much less stable for molybdenum and tungsten.

The elements occur as oxyanions and molybdenum is also found as the sulfide, MoS_2. The ores are converted to the trioxides via the oxyanions and these are reduced to the metals by reduction with hydrogen (carbon cannot be used as the very stable carbides would result). Both metals are relatively inert, with very high melting points and with fairly high electrical conductivity. The metals are most readily attacked by alkaline peroxide or other fused, oxidizing alkaline medium. Attack by aqueous alkali and by acids is only slight.

> MoS_2 has found recent popularity as a solid lubricant, as it has a layer structure which allows the planes of atoms to slide over each other easily, see also Section 19.3.3.

Molybdenum and tungsten find extensive uses. Molybdenum is used in stainless steels, as a catalyst, and as an electrode material. One of the most important catalytic applications is in dehydrosulfurization of petroleum products. A Mo catalyst usually with added Co, supported on Al_2O_3, is used to treat a stream of crude petroleum with hydrogen to remove sulfur-containing organics by converting them to H_2S, and to hydrogenate olefinic bonds. The active species is probably MoS_2, formed *in situ*. Tungsten is used as the filament in light bulbs, in alloys, and, as WC, in applications where its hardness and wear resistance are important. Molybdenum oxide is used in the blast furnace or reduced with iron oxide by Al to give ferromolybdenum. Pure molybdenum is prepared by hydrogen reduction of ammonium molybdate (either the $Mo_2O_7^{2-}$ or $Mo_7O_{24}^{6-}$ species, see below) which is crystallized from a solution of the oxide in ammonia. Tungsten is recovered by dissolving the oxide in fused NaOH and extracting the tungstate. Acidification yields 'tungstic acid' a hydrated oxide which gives the metal on hydrogen reduction.

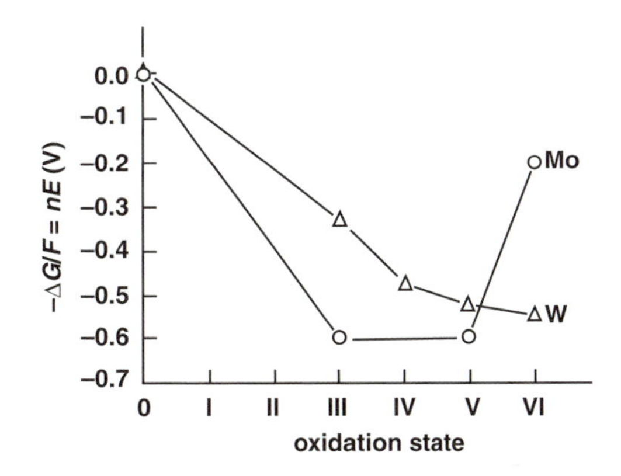

FIG. 15.10 Oxidation state free energy diagrams for molybdenum and tungsten. This illustrates the greater relative stability of the lower oxidation states of molybdenum compared with those of tungsten

15.4.1 Molybdenum(VI) and tungsten(VI)

In the VI state, the oxides, the hexafluorides, the hexachlorides and the hexamethoxides $[M(OMe)_6]$ all occur and tungsten also forms WBr_6. Uranium is the only other element to form a hexachloride, and hexabromides are unknown apart from the tungsten compound. There is also an extensive aqueous chemistry of Mo^{VI} and W^{VI} and a vast collection of polymeric oxyanions.

The oxides, MoO_3 and WO_3, are the final products of igniting molybdenum or tungsten compounds. Both oxides are insoluble in water but dissolve in alkali to give oxyanions. MoO_3 also dissolves in acids to give oxy-cations or related hydrolysed species but WO_3 is insoluble in acid. The simplest product of the solution in alkali is the MO_4^{2-} ion (molybdate or tungstate) which is tetrahedral. The imido analogue $[W(NBu^t)_4]^{2-}$ has been prepared. This ion is isoelectronic with $[WO_4]^{2-}$ and has a similar structure. The molybdates and tungstates of most metals except the alkalis, NH_4^+, Mg^{2+} and Tl^+, are insoluble. The neutralization of the alkaline solutions leads to the precipitation of the hydrated oxides, $MoO_3.2H_2O$ which is yellow, and $WO_3.2H_2O$ which is nearly white. These are definitely hydrates, as written above, not the hydrated acids, $H_2MO_4.H_2O$. As with the anhydrous oxides, which are derived from the hydrates by ignition, the hydrated oxides differ in their behaviour to acid, the molybdenum compound dissolving and the tungsten compound being relatively insoluble.

These simple, mononuclear oxyanions are found only in strong alkali. In less basic solutions, condensation occurs, as we have seen for other elements (compare Table 14.1). These condensed polyoxo-anions formed at lower pH values are called *isopoly-molybdates* or *tungstates* if only Mo or W is present, and *heteropoly* oxyanions when they contain other oxyanions as well, such as silicate or phosphate. This gives a rich and varied field of condensed structures, probably only exceeded in complexity by the silicates. It is not possible to deal with this group of compounds in anything like full detail but Table 15.1 lists some typical formulae. The polyacids are built up from MO_6 octahedra, Fig. 15.11, and most of the structures so far determined are rings, double rings or clusters. On the basis of older experimental results, it was thought that Mo and W behaved similarly and passed, on increasing condensation, through stages which were termed ortho-, para-, and meta-acids (ortho being the normal MO_4^{2-}). It is now known that some of these stages correspond to a number of different compounds which are not represented by the same formulae for Mo as for W. The ions or other species are most conveniently named on the basis of the number of Mo or W atoms. The structures of a number of ions are known, and ions with the same number of metal atoms appear to exist in solution, but there is no final proof that the solution equilibria do correspond to the solids found, and there is certainly evidence for a number of intermediate species in solution. Problems arise as it is difficult to obtain unambiguous analyses for alkali metals and Mo or W in the heavier species, and because mixed crystals are readily formed.

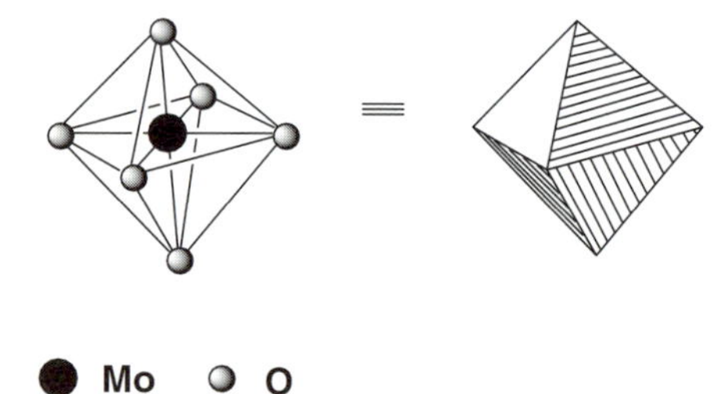

FIG. 15.11 The MO_6 octahedron, the unit from which the polymolybdates and tungstates are built up

The polyacids, and especially the heteropolyacids, form a variety of unusual environments and are at present attracting attention as a means of studying unusual coordinations or unusual oxidation states such as Ni^{IV}.

Recently these polyoxoanions have been studied as 'ligands' towards organometallic species and a number of hybrid complexes have been prepared. The polyoxoanions in these cases are considered as small and, most importantly, *soluble* pieces of metal oxide, which allow the nature of the bonding interactions between the metal oxide species and the organometallic fragment to be thoroughly studied by such techniques as X-ray crystallography and solution nmr spectroscopy. These complexes are models for the interactions between organometallic complexes supported on a metal oxide support, such materials being important catalysts. Polyoxoanions themselves are also used as industrial catalysts, for example in the oxidation of methacrolein to methacrylic acid.

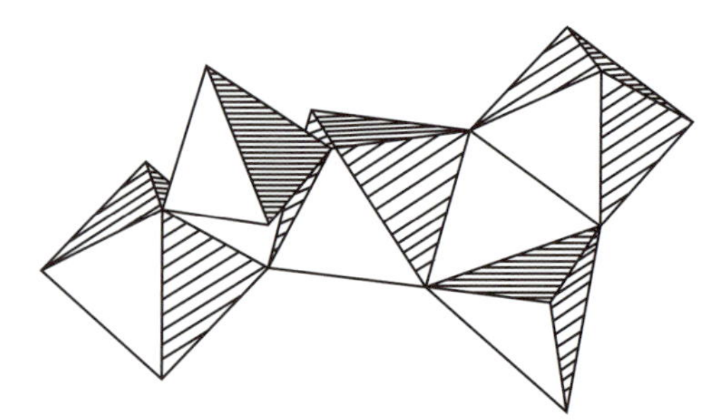

FIG. 15.12 Structure of the $[Mo_2O_7]^{2-}$ polymeric anion

Apart from these oxygen compounds, the VI state is found in halides, oxyhalides and complexes related to these. Both molybdenum and tungsten form MF_6, MOF_4, and MO_2F_2 compounds but molybdenum has a much lower affinity for the heavier halogens than has tungsten, as Table 15.2 shows.

The hexahalides are made by direct reaction. They are volatile and unstable to oxygen and moisture, with which they react readily to give oxyhalides. Tungsten hexabromide is thermally unstable and decomposes on gentle warming. The oxyhalides are also volatile, covalent compounds which hydrolyse to the trioxides, the MOX_4 type more rapidly than

TABLE 15.1 Iso- and hetero-poly molybdates and tungstates

Iso-compounds

(a) Molybdenum in solution

$MoO_4^{2-} \xrightarrow{pH=6} Mo_7O_{24}^{6-} \xrightarrow{\text{lower pH}} Mo_8O_{26}^{6-} \xrightarrow{\text{strong acid}} Mo_n$ species?

(7– and 8– ions may be protonated and hydrated.)

(b) Solid molybdenum compounds (all except the 1– and 2– species are heavily hydrated).

MoO_4^{2-}	Tetrahedron.
$(Mo_2O_7^{2-})_n$	Chain of MoO_6 octahedra sharing opposite corners and linked in pairs through adjacent corners by MoO_4 tetrahedra (Fig. 15.12). The NH_4^+ salt has two octahedra sharing an edge and successive pairs are linked by two tetrahedra sharing corners. $Ag_2Mo_2O_7$ is formed of octahedra only.
$Mo_7O_{24}^{6-}$	Linked octahedra (Fig. 15.13).
$Mo_8O_{26}^{6-}$	Linked octahedra.
$Mo_{10}O_{34}^{8-}$	The ammonium salt is a Mo_8O_{28} unit built up of linked octahedra with two MoO_4 tetrahedra linked to opposite corners.

(c) Tungsten in solution.

$WO_4^{2-} \rightleftharpoons HW_6O_{21}^{5-}$ (pH = 6); $HW_6O_{21}^{5-} \xrightarrow[\text{intermediate stages}]{\text{slowly, through}} H_{10}W_{12}O_{46}^{10-}$

$WO_4^{2-} \rightleftharpoons H_3W_6O_{21}^{3-}$ (pH = 4); $H_3W_6O_{21}^{3-} \rightleftharpoons\ ?\ \rightleftharpoons H_2W_{12}O_{40}^{6-}$

(d) Solid tungsten compounds (all are heavily hydrated as salts except the 1– and 2– species).

WO_4^{2-}	Tetrahedron.
$(W_2O_7^{2-})_n$	Linked tetrahedra and octahedra: see Mo compound.
W_6 species	Structures not known.
$W_{12}O_{46}^{20-}$	See Fig. 15.14a, linked octahedra.
$W_{12}O_{40}^{8-}$	Linked octahedra; isomorphous with the 12-hetero-acids (q.v.).

Hetero-compounds (In these formulae x = formal positive charge on hetero-atom M′)

(a) MO_6 (M = Mo, W) octahedra surrounding a central (hetero) $M'O_6$ octahedron (all compounds are heavily hydrated).

$M'M_6O_{24}^{(12-x)-}$	$M' = I^{7+}$, Te^{6+}, or a number of trivalent ions such as Co^{3+}, Al^{3+} or Rh^{3+}. Structure is a ring of six linked MO_6 octahedra with a central octahedral site for M′ (Fig. 15.14b).
$M'M_9O_{32}^{(10-x)-}$	$M' = Mn^{4+}$, Ni^{4+}. Structure consists of three sets of clusters of three MO_6 groups giving a central octahedral hole for M′.

(b) MO_6 octahedra surrounding a central (hetero) $M'O_4$ tetrahedron.

$M'M_{12}O_{40}^{(8-x)-}$	$M' = P^{5+}$, As^{5+}, Si^{4+}, Ge^{4+}, Ti^{4+}, Zr^{4+}, Sn^{4+}. Structure contains four sets of three MO_6 octahedra joined by edges and defining a central tetrahedral site. The structure of $W_{12}O_{40}^{8-}$ is the same as these with the central site empty.

Other heteroacids with tetrahedral $M'O_4$ groups occur with M′/M ratios of 1/11, 1/10, 2/18 and 2/17 and containing mainly P^{5+} or As^{5+}. For example:

$P_2Mo_{18}O_{62}^{6-}$	This is the 12-acid structure as above with the three MO_6 octahedra at the base removed to give the 'half-unit' PM_9O_{34} and two of these are linked, sharing six oxygens, to give the P_2M_{18} unit.

(c) More complex hetero-compounds.

Much greater complication is possible as illustrated by the phosphorus–molybdenum compounds:

$(MoP_2O_{11})_n$	Chains of MoO_6 octahedra formed by linking corners. The chains are cross-connected by P_2O_7 groups into a three-dimensional structure.
$(MoP_2O_8)_n$	Layers formed by MoO_6 octahedra sharing oxygens and these layers linked up by $(PO_3)_n$ chains which run perpendicular to the planes of the layers.

TABLE 15.2 Halides and oxyhalides of molybdenum(VI) and tungsten(VI)

MX_6		MOX_4		MO_2X_2	
MoF_6	WF_6	$MoOF_4$	WOF_4	MoO_2F_2	WO_2F_2
$MoCl_6$	WCl_6	$MoOCl_4$	$WOCl_4$	MoO_2Cl_2	WO_2Cl_2
	WBr_6		$WOBr_4$	MoO_2Br_2	WO_2Br_2
					WO_2I_2

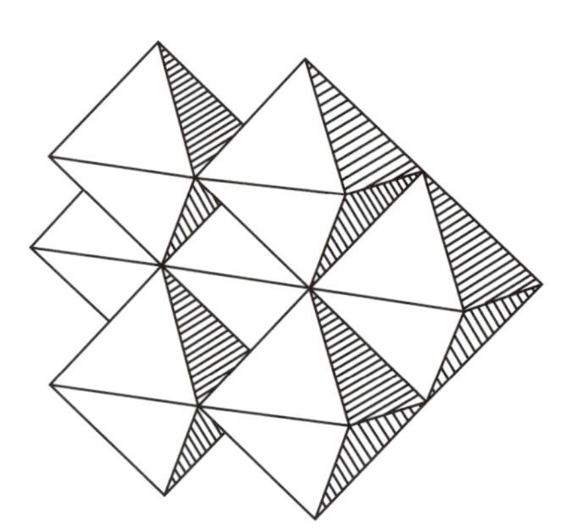

FIG. 15.13 The $Mo_7O_{24}^{6-}$ structure. This structure is more compact than that found in the similar heteropolyion (Fig. 15.14b)

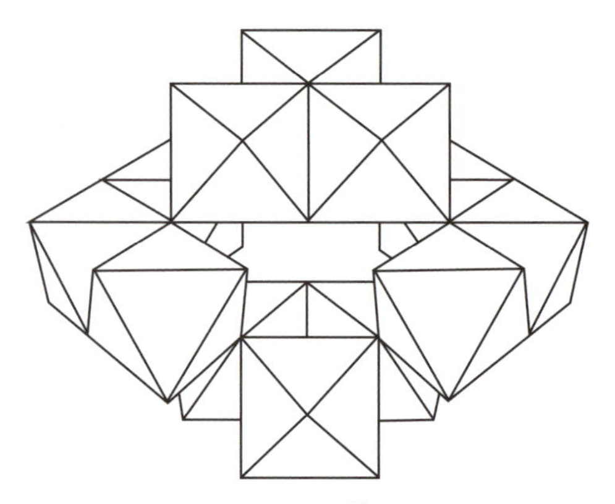

FIG. 15.14a The $W_{12}O_{46}^{20-}$ structure

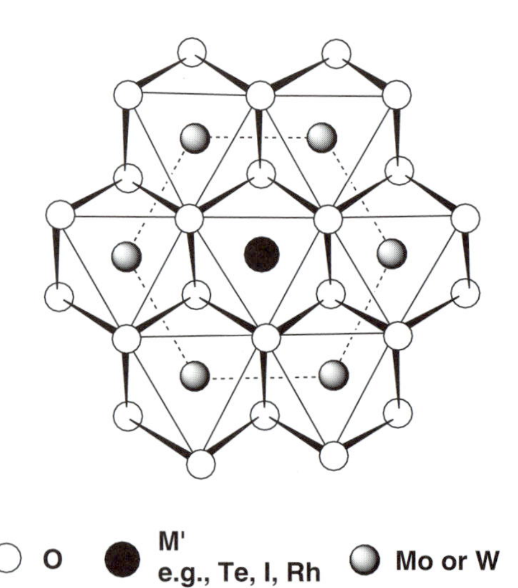

FIG. 15.14b The $M'M_6O_{24}^{(12-x)-}$ structure. This consists of a ring of six linked MO_6 octahedra forming a central octahedral site for the M′ metal ion

the MO_2X_2 compounds. Tungsten forms analogous compounds with the heavier elements of the oxygen group; WSF_4, $WSCl_4$ and $WSeCl_4$ are all relatively volatile solids with a square-pyramidal structure in the gas phase, where the W lies above the X_4 plane, just as in the oxygen compounds. WO_2Cl_2, which is yellow, disproportionates above 200°C to WO_3 and red $WOCl_4$.

The hexahalides are octahedral, and WOF_4 has the tetrameric structure of NbF_5 with F atoms in the bridging positions (Fig. 15.4). In contrast, $MoOF_4$ is formed into chains of octahedral units, linked by shared fluorines in both structures. In the gas phase, both MOF_4 molecules are square-pyramidal monomers.

Molybdenum and tungsten form a variety of complex halides in the VI state, of which the fluorine compounds are the most widely represented. Examples include $M^I_2WF_8$ and M^IMF_7 (for both M = Mo and W). The octafluorotungstate has a square antiprismatic structure like the Ta analogue, see Fig. 15.5b. The heptafluorides provide further examples of 7-coordination and the structures of a number of salts have been determined. Because all three established 7-coordination shapes (compare Figs 15.1, 15.5a and 15.22) are very similar in energy, different counterions may give different structures (compare the discussion on tetrahedral or square-planar Ni, Section 14.8.2). By varying the cation $(M^I)^+$ it was expected that the intrinsic shape could be decided. For several $(M^I)^+$, including Cs^+, NO_2^+ and Me_4N^+, the shapes of $[MoF_7]^-$ and $[WF_7]^-$ were all found to be capped octahedral, as found for $[NbOF_6]^{3-}$ (see Fig. 15.5a). This contrasts with the 7-coordinate Main Group structures so far determined which are all pentagonal bipyramids.

Mononuclear oxyhalide anions of both elements are also known. Of these $[MO_2F_4]^{2-}$, $[MO_2Cl_4]^{2-}$, and $[MO_3F_3]^{3-}$, and probably $[MOF_5]^-$, are octahedral monomers. More complex oxyfluoro compounds are built up of 6-coordinate units linked together. A dinuclear example is the anion $[O_2F_3W\text{-}F\text{-}WF_3O_2]^{3-}$ with one fluorine bridge. $[WO_2F_3]^-$ is a chain polymer with two terminal *cis* F, two similar O and two *trans* O atoms bridging to the next W atoms while the Mo analogue is similar but with F bridges. The chlorine analogue, $[WO_2Cl_3]^-$, is a dimer linked through two bridging oxygens. One well-established tetrahedral species is MoO_3Cl^-. Other oxy-species are known, including molybdenyl sulfate, MoO_2SO_4, formed from molybdenum trioxide and sulfuric acid. Such compounds are probably molecular, rather than salts of the oxycation.

Molybdenum and tungsten are sufficiently stable in the VI state to form sulfides, MS_3. These are precipitated as hydrated compounds when H_2S is passed through slightly acid M^{VI} solutions. In stronger acid, the H_2S reduces the VI state to the IV state, giving the disulfides MS_2. Sulfur-containing polyanions are found including $W_3S_9^{2-}$ which has two tetrahedral WS_4 units sharing edges with a central WS_5 unit which is a distorted square pyramid. The $W_3OS_8^{2-}$ analogue has the O at the apex position of the square pyramid, and Mo analogues occur. In the interesting mixed oxidation state sulfur anion, $W_3S_8^{2-}$, the structure shows a central planar WS_4 unit linked by sharing edges to two outer WS_4 tetrahedra. These outer units are W(VI), and the simple WS_4^{2-} ion is now well established, while the unique, central, square-planar unit is W(II).

Considerable interest in molybdenum-sulfur compounds derives from the work on nitrogen-fixing organisms (see Sections 16.10 and 20.1.4), where the active site contains Mo and S. This interest has led to the intense study of molybdenum and tungsten sulfur chemistry. Starting from the simple thiomolybdate and thiotungstate ions MS_4^{2-} ions, an extensive array of more complex molecules has been built up, with emphasis on those containing the cubane M_4S_4 unit (Fig. 15.15a), or a fragment of this such as that of Fig. 15.15b.

MOLYBDENUM BLUE IN ANALYSIS

Another significant application for molybdenum polyoxoanions is the use of 'molybdenum blue' in analysis. The reduction of a molybdate solution gives a deep blue cloudy solution. This rapid, simple test has been known for over 200 years and is widely used as a qualitative test for molybdenum. However, until recently, the structure of this 'molybdenum blue' was not known. Using a range of structural techniques, including infrared spectroscopy and X-ray powder diffraction, the molybdenum blue has been found to be a cyclic aggregate of 154 molybdenum atoms, with associated ligands including large numbers of oxo (O), hydroxo (HO) and water (H_2O) groups. The molybdenum atoms occur in two different oxidation states (V and VI) and this causes the deep blue colour, due to charge transfer processes, analogous to the mixed-valence iron(II,III) material Prussian blue (see Section 14.6).

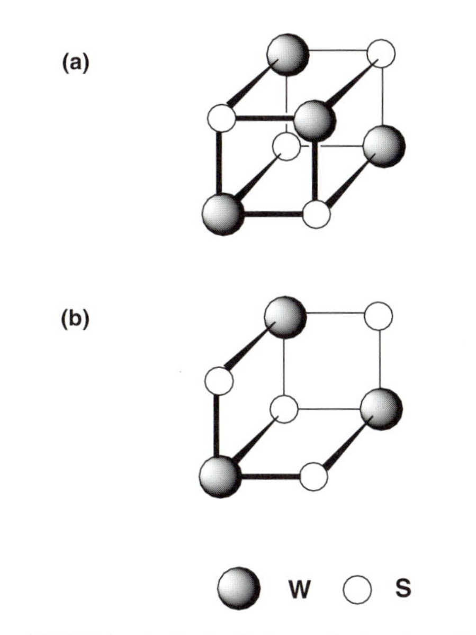

FIG. 15.15 Major structural units in complex tungsten sulfur clusters: (a) cubane (b) cubane missing one corner. (b) may alternatively be described as W_3S trigonal pyramid with each W-W edge bridged by S

The VI state is also sufficiently stable to allow the preparation of a polyhydrido-complex, $WH_6(PR_3)_3$. This shows 9-coordination and multiple substitution by hydrogen as in the similar rhenium hydrides discussed in Section 15.5. The structure is based on the tricapped trigonal prism of Fig. 15.23, with two of the phosphane groups replacing two H in one of the long edges, and the third phosphane taking the place of the opposite face-bridging H.

Two interesting types of compound are found which fall between the VI and V states. Mild reduction of the trioxides gives intensely blue oxides whose composition is intermediate between M_2O_5 and MO_3. These blue oxides appear to contain both M^V and M^{VI} in an oxide lattice, and the intense colour arises from the existence of two oxidation states in the same compound. Similarly intense colours are observed in other cases like this, for example in magnetite, Fe_3O_4, and in Prussian blue. The second compound is the product of reduction of sodium tungstate, of formula Na_nWO_3 (n lying between 0 and 1), called tungsten bronze. The colour varies from yellow to blue-violet as n varies from about 0.9 to 0.3. The structure of these bronzes is based on the ReO_3 structure shown in Fig. 15.16. Here, the metal atoms lie at the corners of a cube and the oxygens are at the mid-points of the edges. If a M^I ion lies at the centre of each cube, the structure is that of perovskite, M^IMO_3. The sodium bronzes are compounds where sodium ions appear at random in the cube centres. The metallic appearance and conductivity of the bronzes arises as the sodium valency electron is delocalized over the structure. It may be noted that the structure of WO_3 itself is a distorted form of the ReO_3 structure.

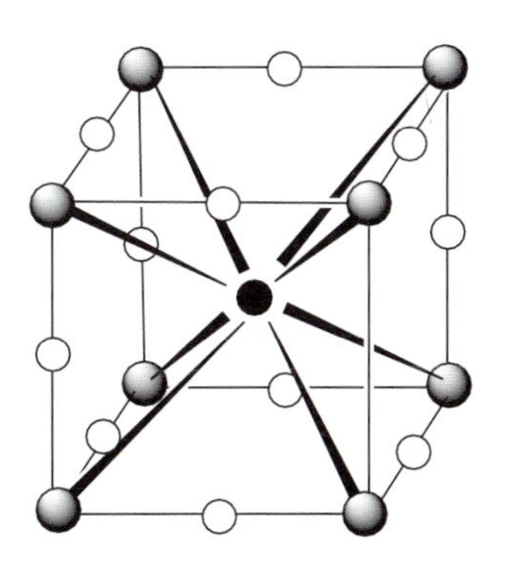

FIG. 15.16 Tungsten bronze: the relation between the structures of rhenium trioxide, tungsten bronze and perovskite. The figure shows the ReO_3 structure, with the black circle indicating the site occupied, at random, by sodium ions in the bronze. If all the black circles are occupied, the structure is that of perovskite

15.4.2 The V state

The V state is prepared by mild reduction of the VI compounds and is coloured, usually green or red. The oxides are not acidic. Although W^V does exist, there are fewer tungsten than molybdenum compounds in this state.

Mo_2O_5 may be made by reacting MoO_3 with Mo at 750°C. It is violet in colour and insoluble in water and dilute acids. It now seems likely that W_2O_5 does not exist and reports of its preparation apply to oxygen-deficient WO_3. Mo_2S_5 is also known from the reaction of H_2S on Mo^V solutions.

Among the halides, MF_5, MCl_5, (M = Mo, W) and WBr_5 are found. The existence of WBr_5 parallels the behaviour in the VI state. WF_5 is unstable and disproportionates to $WF_6 + WF_4$ above 50°C. Molybdenum pentafluoride is prepared by reduction with the carbonyl:

$$5MoF_6 + Mo(CO)_6 \rightarrow 6MoF_5 + 6CO$$

The pentachloride is the product of direct reaction between Mo and Cl_2. The tungsten pentahalides are formed by mild reduction of the hexahalides. All the pentahalides are covalent, relatively volatile solids. Solid MoF_5 has a tetrameric structure like NbF_5 (Fig. 15.4) while $MoCl_5$ and WCl_5, in the solid, are dimeric like $NbCl_5$. Among the oxyhalides of the V state are $MoOF_3$, $MOCl_3$(M = Mo, W), $WOBr_3$ and WO_2I.

Most complexes of the V state are halogen or oxy-halogen compounds or contain pseudohalogen groups like CN or SCN. Reaction of $M(CO)_6$ in liquid IF_5 in the presence of alkali halides gives the two formula types $M^{I}Mo(W)F_6$ and $M^{I}_3Mo(W)F_8$. The former are precipitated as salts of large cations and contain MF_6^- units. It is not known if the latter contain MF_8 units. Oxyhalide complexes result from the reduction of solutions of M^{VI}. The commonest types are MOX_5^{2-} and MOX_4^-, where X includes Cl, Br, SCN, and the Mo compounds are found in greater variety than the W ones. An interesting analogue is the $MoNCl_4^-$ ion which has a square-pyramidal structure and a molybdenum–nitrogen triple bond.

The M^{V} octacyanides, $Mo(CN)_8^{3-}$ and $W(CN)_8^{3-}$, are obtained from strong oxidation of the M^{IV} octacyanides and have similar structures. These are discussed below.

15.4.3 The IV state

The two elements are quite similar in the IV state. This is usually formed by stronger reduction from the VI state than is used to reach the V state. The two dioxides, the disulfides and all the tetrahalides are known. Complexes are not very numerous and are similar in type to the complexes of the V state.

The dioxides, MO_2, are prepared by reduction of the trioxides by hydrogen or by careful oxidation of the metals. They are readily oxidized by halogens or oxygen and are reduced to the metal in hydrogen at temperatures above 500°C. The dioxides are insoluble in nonoxidizing acids, but dissolve in nitric acid with oxidation to M^{VI}.

Both elements form the disulfide, MS_2. Molybdenum disulfide is an important naturally occurring molybdenum source. It has a layer lattice in which the Mo atoms are surrounded by six S atoms at the corners of a trigonal prism, see Section 19.3. The outer S planes of neighbouring layers are only weakly cross-linked and the material is a solid lubricant. Among the sulfur complex ions of the (IV) state are the trinuclear species $[M_3S_4]^{4+}$ together with oxygen analogues. Indeed, for Mo the whole series $Mo_3S_nO_{(4-n)}^{4+}$ is known for $n = 0, 1, 2, 3, 4$. The structures are the cornerless cubane of Fig. 15.15b. In $[W_3S_4(NCS)_9]^{5-}$, each W is approximately octahedrally coordinated to 3 NCS groups, two μ-S and the central μ^3-S.

All the tetrahalides are known, although the bromide and iodide are poorly characterized. The IV halides tend to disproportionate, thus:

$$MoO_2 + Cl_2 \longrightarrow MoCl_4 \longrightarrow MoCl_3 + MoCl_5$$

$$WCl_6 + H_2 \longrightarrow WCl_4 \xrightarrow{\text{heat}} W_6Cl_{12} + WCl_5$$

The WX_4 compounds are slightly more stable than the MoX_4 ones. All are coloured, involatile solids which are readily oxidized. The tetrachlorides have polymeric structures in which the metal coordination is octahedral and these are linked into chains by sharing Cl atoms. The $M\cdots M$ distances are alternately long and short, compare the niobium and tantalum tetraiodides, Fig. 15.8. Indeed, for $(WCl_4)_x$ the short W-W distance is only 268.8 pm, which suggests a bond order greater than 1 (compare Table 16.3) and this contrasts with the long $W\cdots W$ distance of 378.7 pm. $MoCl_4$ exists in a second form where six octahedra link into a hexameric ring by sharing opposite edges. The bridges show variable Mo-Cl distances between 243 and 251 pm, the terminal distance is 220 pm, while the Mo-Mo distance of 367 pm is too long for significant metal–metal bonding. The chlorocomplex, $Mo_2Cl_{10}^{2-}$, also has two octahedra sharing an edge.

Tungsten(IV) can be produced in solution by reduction of tungstate by tin and HCl. From the dark green solution the salt $K_2[W(OH)Cl_5]$ can be crystallized. This was formulated more recently as a dimer with bridging oxygen: $K_4[Cl_5W(O)WCl_5]$ and presumably hydrated. Some octahedral MX_6^{2-} complexes are also found, including the fluorides, chlorides and thiocyanates. Both elements form 7-coordinate complexes $MCl_4(PR_3)_3$ with the face-capped octahedral structure of Fig. 15.5a. Three Cl form one face, three P the opposite face and this is capped by the last Cl.

The complex ion $[W_4F_{18}]^{2-}$ has a closed cluster structure based on WF_6 octahedra. Each is linked to the other three by W-F-W bridges. The structure can alternatively be described as a tetrahedron of WF_3 units linked along each of the six edges by a W-F-W bridge.

Also in the IV state is the hydride-cyclopentadienyl class of compounds, $(C_5H_5)_2$-$Mo(W)H_2$ and $(C_5H_5)_2WH_3^+$. The latter is isoelectronic with the tantalum trihydride and

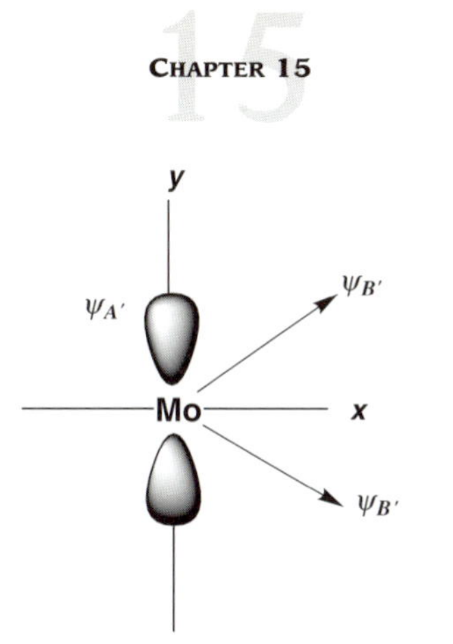

FIG. 15.17 Central plane orbitals in $(C_5H_5)_2MoH_2$. Contrast with Fig. 15.7. ψ_A now lies along the y axis and is filled with a nonbonding electron pair. As a result, the two $\psi_{B'}$ orbitals make a much smaller angle than the two ψ_B orbitals of Fig. 15.7

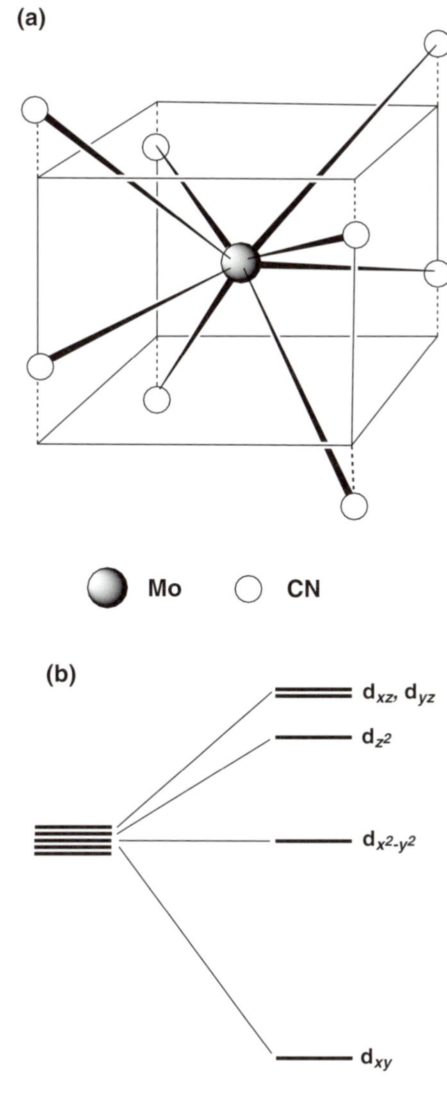

FIG. 15.18 (a) The structure of the octacyanomolybdate(IV) ion, $Mo(CN)_8^{4-}$, (b) the energy level diagram for this structure. (b) shows how this structure stabilizes one d orbital relative to the rest

appears to have the same kind of structure. The structure of $(C_5H_5)_2MH_2$ shows similar ring-M-ring angles of 146°, but the HMH angle is only 76°. It is suggested that this, and related $(C_5H_5)_2ML_2$ species, better fit an alternative combination of orbitals than that of Fig. 15.7, namely that of Fig. 15.17. Contrast with Fig. 15.7. $\psi_{A'}$ now lies along the y axis and is filled with a nonbonding electron pair. As a result, the two $\psi_{B'}$ orbitals make a much smaller angle than the two ψ_B orbitals of Fig. 15.7.

The best-known complex ion of the IV state is the octacyanomolybdate or tungstate, $M(CN)_8^{4-}$. This is an extremely stable grouping and is attacked only by permanganate or ceric which oxidize it only as far as the corresponding V octacyanide ion. The structure in the solid is dodecahedral and is shown in Fig. 15.18. A similar arrangement is shown in Fig. 15.2. This arrangement of ligands stabilizes the d_{xy} orbital relative to all the rest, as the energy level diagram (Fig. 15.18b) shows. The configuration is thus suitable for d^1 and d^2 arrangements with large ligand fields and the CFSE must be a major factor in the stability of these octacyanides.

The $M(CN)_8^{3-}$ ion, which results from oxidation of $M(CN)_8^{4-}$, is also found in the dodecahedral configuration. However, as we have seen in other cases of high coordination, the energy difference between alternative structures is small, and the octacyanides are also found in square antiprismatic coordination (as in Fig. 15.5b) when the cation is large. Examples include $Mo(CN)_8^{4-}$ in the $Cd(N_2H_4)_2^{2+}$ salt, and both $M(CN)_8^{3-}$ ions when the cation is $Co(NH_3)_6^{3+}$.

In the tungsten(IV) ion $(W_3O_4F_9)^{5-}$ the structure is a tetrahedron formed of the three W atoms and one O. Each W-W edge is bridged by an oxygen and there are three terminal F atoms per W.

15.4.4 The lower oxidation states

The lower oxidation states are unstable and strongly reducing. There is no evidence for simple cations in the III or II state. Tungsten(III) is not stable and compounds are limited to WCl_3 and some chloro-complexes. WCl_3 consists of Cl^- ions and the cluster $W_6Cl_{12}^{6+}$ which is isostructural with $Nb_6Cl_{12}^{2+}$ (Fig. 15.9). The complex ion $W_2Cl_9^{3+}$ has two octahedra sharing a face, together with a W-W bond. WBr_3 does not contain W(III) but is a polybromide $(Br_4)^{2-}$ containing the $W_6Br_8^{6+}$ cluster related to the $Mo_6Cl_8^{4+}$ structure (Fig. 15.19). By contrast, Mo^{III} is relatively stable and well represented. The oxide Mo_2O_3 is not known, but the sulfide exists. All the trihalides are reported, being derived from the higher oxidation states by strong reduction, or disproportionation. $MoCl_3$ is fairly stable, being only slowly oxidized in air and slowly hydrolysed by water. In solution, in presence of excess chloride, the complex anion $MoCl_6^{3-}$ may be prepared. A considerable number of other representatives of the type MoX_6^{3-} are known including the fluoride and thiocyanate. These are all octahedral complexes with three unpaired electrons, as expected. Some complex cations of Mo^{III} are known, including $Mo(dipy)_3^{3+}$ and $Mo(phen)_3^{3+}$, and neutral compounds like $Mo(acac)_3$ also occur. All these compounds are probably octahedral.

The II state is represented by the complex halides M_6X_{12}(M = Mo, W: X = Cl, Br, I) and by a few complexes with π-bonding ligands. The dihalides are relatively inert and insoluble materials. It was soon found that only a third of the halogen could be replaced, for example with $AgNO_3$ or by OH^-, and structural evidence has led to the formulation of the dihalides as hexamers:

$$\text{'}MX_2\text{'} = M_6X_{12} = (M_6X_8)^{4+}X_4^- \xrightarrow{Y^-} M_6X_8Y_4$$

The structure of the $Mo_6Cl_8^{4+}$ group is shown in Fig. 15.19; the other compounds are isomorphous. The 6 Mo form an octahedron whose faces are bridged by the 8Cl atoms (which themselves thus lie approximately in a cube circumscribing the octahedron). Compare this structure with that of Fig. 15.9 where the octahedron of metal atoms is bridged along each edge, giving the $M_6X_{12}^{n+}$ cluster unit. In both these types of compound, there is polycentred metal–metal bonding as well as the bonding involving the bridging chlorines.

$(W_6Cl_8)^{4+}$ oxidizes to $(W_6Cl_{12})^{6+}$, the W(III) cluster, but oxidation of molybdenum(II) chloride stops at the species $[Mo_6Cl_{12}]^{3+}$, corresponding to a formal oxidation state of $2\frac{1}{2}$.

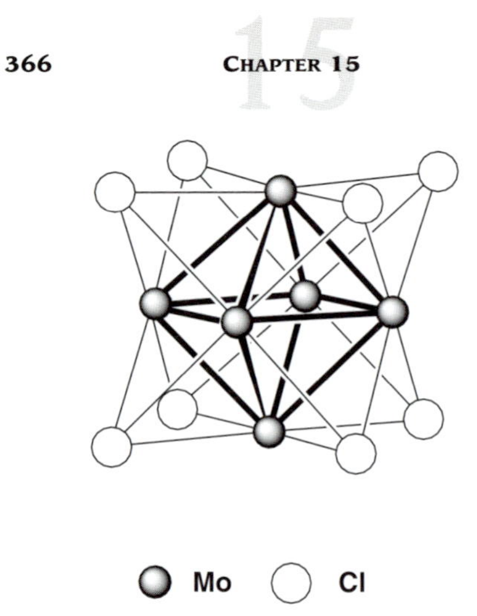

FIG. 15.19 The structure of the ion, $Mo_6Cl_8^{4+}$. This structure also contains metal–metal bonding in the Mo_6 cluster

See also Section 16.3 for low oxidation state alkyls.

These species are reminiscent of the $(Ta_6Cl_{12})^{n+}$ clusters where a range of charges is found. Addition of 6 Cl atoms in terminal positions on the metals gives the $M_6Cl_{14}^{2-}$ cluster. Removal of one MoCl unit from this gave the $Mo_5Cl_{13}^-$ species where the Mo atoms now form a square pyramid with the open square face edge bridged by Cl. These examples illustrate the extensive chemistry which is developing for such clusters. Linked clusters, compounds containing atoms in the octahedral site in the centre, and other variations are being explored.

Most other compounds of the divalent elements are carbonyl complexes such as $(C_6H_5)W(CO)_3Cl$ and $Mo(diars)_2(CO)_2X^+$. There is also an involatile acetate $Mo(CH_3COO)_2$, formed from $Mo(CO)_6$ and glacial acetic acid. This is a dimer and, like the red Mo(II) chloro-complex $K_4Mo_2Cl_8.2H_2O$, involves multiple Mo-Mo bonding. This is discussed in Section 16.7. Reduction of Mo(IV) cyano-complexes yields $Mo(CN)_7^{5-}$ which has a regular pentagonal bipyramidal structure (compare $V(CN)_7^{4-}$, p.323).

The I state occurs in π-bond compounds such as $Mo(C_6H_6)_2^+$ (compare the analogous Cr compound) or $[C_5H_5Mo(CO)_3]_2$.

The 0 state is represented by the octahedral and rather stable carbonyls, $Mo(CO)_6$ and $W(CO)_6$, which are white solids. Related compounds are also quite common, for example, $Mo(CO)_5I^-$, $W(PF_3)_6$, or $py_3Mo(CO)_3$. The −II state is shown by the carbonyl anion $M(CO)_5^{2-}$.

15.5 Technetium, $4d^65s^1$, and rhenium, $5d^56s^2$

The comparative chemistry of these two elements has substantially developed since the 1970s when technetium found extensive use in medicine. Since all Tc isotopes are radioactive, thorough study is quite recent. Tc and Re, in sharp contrast to Mn, are particularly stable in the VII state which has only weak oxidizing properties. The lower states are also quite stable, especially Tc(IV) and Re(III) and (IV). The V and VI states are relatively unknown and tend to disproportionate. Low oxidation states are strongly reducing and not well known, especially the II state which was so stable for Mn. These characteristics are illustrated by the oxides and halides in Tables 13.3–13.5 and by the oxidation state free energy diagrams in Fig. 15.20.

The longest-lived isotopes of technetium, ^{97}Tc and ^{98}Tc, have half-lives of the order of two million years. These are prepared by neutron bombardment of molybdenum isotopes. However, the most accessible source is nowadays the isotope ^{99}Tc which is one of the fission products of uranium and thus occurs to a considerable extent in spent fuel elements. This has a half-life of 212 000 years and is a weak β-emitter. It is therefore only mildly radioactive and relatively easy to handle. The metastable isotope, technetium-99m has a very useful application in medical imaging, discussed in Chapter 20.

Work on the usual isotopes of technetium involves their separation from the uranium and fission products in the fuel rods by oxidation, followed by distilling out the volatile

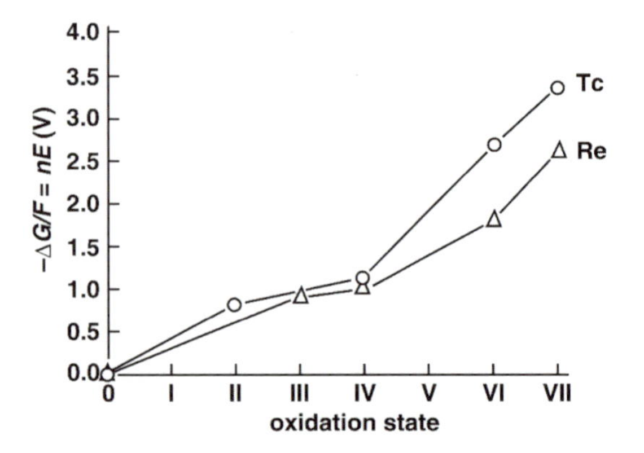

FIG. 15.20 The oxidation state free energy diagram for technetium and rhenium

Tc_2O_7. This may be separated from Re_2O_7, which is also volatile, by fractional precipitation of the sulfides. At an acid concentration above 8M HCl, Re_2S_7 precipitates but Te_2S_7 does not.

Rhenium is a very rare element in the Earth's crust and does not occur in quantity in any ore. However, it is found in molybdenum ores and can be quite readily recovered from these so that it is not too inaccessible and its chemistry is quite well known. It is left in oxidized solution as the perrhenate ion, ReO_4^-, whence it may be precipitated as the insoluble potassium salt. The metal is obtained by hydrogen reduction.

15.5.1 The VII oxidation state

The Group oxidation state is represented by the heptoxides, the acid and salts of the MO_4^- ion, by the heptasulfides, by oxyhalides and by the MH_9^{2-} complexes discussed below. Rhenium alone forms the heptafluoride, ReF_7. The structure of this compound has been the subject of much recent discussion as it is one of only two stable binary compounds of AB_7 stoichiometry (the other is IF_7). A recent low-temperature neutron diffraction study confirms that the configuration is a distorted pentagonal bipyramid (see Table 13.4).

The heptoxides result from heating the metals in oxygen or air. Tc_2O_7 is a yellow solid melting at 120°C and boiling at 310°C. It is stable up to the boiling point. Re_2O_7 is also yellow and the solid sublimes; the calculated boiling point is 360°C. In the vapour, the structure is $O_3M\text{-}O\text{-}MO_3$, with tetrahedral coordination. The crystal structures of Tc_2O_7 and Rc_2O_7 differ, with the technetium oxide having the same structure as the vapour, with a linear Tc-O-Tc bridge. The Re_2O_7 solid contains regular ReO_4 tetrahedra and ReO_6 octahedra with three short and three long bonds; the overall arrangement indicates that $ReO_3^+ReO_4^-$ may be a reasonable formulation. Technetium heptoxide is somewhat more oxidizing than the rhenium compound and these two heptoxides differ strikingly in stability from Mn_2O_7 which is also volatile but which rapidly decomposes at room temperature.

Compare N_2O_5 in the solid.

The heptoxides dissolve in water to give colourless solutions of the acids. Pertechnic acid, $HTcO_4$, is produced as dark red crystals on evaporation, but perrhenic acid, $HReO_4$, cannot be isolated although the colour of the solution changes to yellow-green on concentration and lines due to the acid appear in the Raman spectrum of the concentrated solution. These colours are due to the lowering of the symmetry on passing from the tetrahedral anion ReO_4^- to the acid $(HO)ReO_3$ on concentration. Perrhenic acid resembles periodic acid in having a second form H_3ReO_5, and salts derived from this are readily prepared. The normal perrhenates are formed in dilute solution while salts of the tribasic form of the acid are formed in media of higher basicity:

Compare HIO_4 and HIO_6.

$$ReO_4^- + K^+ \rightarrow KReO_4(\text{yellow})$$
$$ReO_4^- + K^+OH^- \rightarrow K_3ReO_5\ (\text{red})$$

or

$$Ba(ReO_4)_2 \underset{H_2O}{\overset{Ba(OH)_2}{\rightleftarrows}} Ba_3(ReO_5)_2$$

The peracids are strong acids, with perrhenic acid lying between perchloric acid and periodic acid in strength. Permanganic acid is stronger than perrhenic acid, so it is likely that pertechnic acid is also stronger. The structure of Na_3ReO_5 shows a square-pyramid ReO_5 unit with the axial Re-O at 184 pm a little longer than the four Re-O base values of 179 pm. A third anion is found in $Ba_5(ReO_6)_2$ which probably contains isolated ReO_6^{5-} ions. The technetium analogue of the tribasic anion has not been isolated but TeO_5^{3-} may exist in fused sodium hydroxide. Some reactions of perrhenic acid are shown in Fig. 15.21. Reactions of $HTcO_4$ are similar, as far as is known, apart from the lack of H_3TcO_5 and some differences in solubilities. Thus $KTcO_4$ is twice as soluble as $KReO_4$ and does not precipitate so readily. Both these potassium salts are very stable and can be distilled at temperatures of over 1000°C without decomposition. $KMnO_4$, on the other hand, loses oxygen above 200°C. The perrhenates, even of organic bases such as strychnine, may be isolated while permanganate readily oxidizes such compounds.

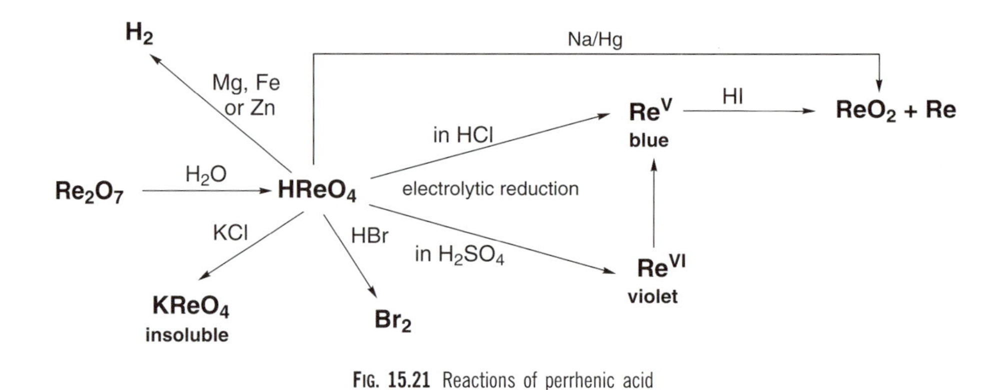

FIG. 15.21 Reactions of perrhenic acid

The imido analogue of the perrhenate ion, $[Re(NBu^t)_4]^-$, has been synthesized as the $Li(tmen)^+$ salt. It is an air- and moisture-sensitive solid, reflecting the greater reactivity resulting from the less electronegative nitrogen substituents compared with ReO_4^-.

The VII state is found in halides and oxyhalides as shown in Table 15.3.

TABLE 15.3 Halides and oxyhalides of the VII state

Re	*Tc*	*Mn*
ReF_7		
$ReOF_5$	$TcOF_5$	
ReO_2F_3	TcO_2F_3	
ReO_3F	TcO_3F	MnO_3F
ReO_3Cl	TcO_3Cl	MnO_3Cl?
ReO_3Br		

Technetium heptafluoride is not formed by direct combination at 400°C, the reaction which gives ReF_7, but TcF_6 is found instead. The halides and oxyhalides are generally colourless or pale yellow compounds which are either liquids or low-melting solids. The oxyhalides result from halogenation of the oxide or oxyion, or from the action of oxygen or water on the fluoride in the case of the rhenium oxyfluorides. The MO_2F_3 compounds have structures with a chain of octahedral units linked by M-F-M bridges. The two O atoms on each M are *cis* and the bridging fluorines are those *trans* to the O. Addition of F^- in acetonitrile to ReO_2F_3 gives the mononuclear $[ReO_2F_4]^-$ ion together with the polynuclear species $[Re_2O_4F_7]^-$ and $[Re_3O_6F_{10}]^-$ which also have F bridges and *cis* oxygens.

The structure of the $[ReOF_6]^-$ ion in both the Cs^+ and the NO_2^+ salts is a pentagonal bipyramid, Fig. 15.22, with the Re-O bond axial and shorter, at 167 pm, than the axial Re-F (192 pm). The equatorial Re-F bonds vary slightly from 187–191 pm and the central pentagon is slightly puckered. One sulfur analogue of the oxyhalides is reported. Treating ReF_7 with Sb_2S_3 gives maroon $ReSF_5$ which is very sensitive to water.

Few complexes of the VII state are known, but oxidation of the octacyanide of Re(VI) gives salts of what appears to be the $Re(CN)_8(OH)_2^{3-}$ ion, which may contain Re(VII) in 10-coordination.

A particularly striking example of the stability of the VII state is provided by the hydrogen complexes, K_2ReH_9 and K_2TcH_9. It has long been known that treatment of perrhenate solutions with potassium/ethylenediamine/water gave a water-soluble reactive rhenium species. This was originally identified as a 'rhenide' anion, analogous to the halide ions in the corresponding Main Group but it has now been shown to contain Re(VII) coordinated to nine hydrogen atoms. The compound is colourless and diamagnetic, which accords with Re(VII). It was soon afterwards shown that a similar Tc compound existed.

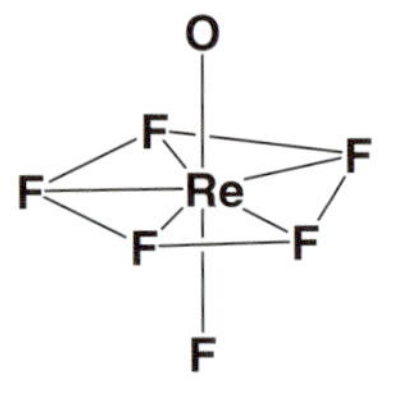

FIG. 15.22 The structure of the $[ReOF_6]^-$ ion. The structure is based on a pentagonal bipyramid with a slightly puckered central pentagon

The structure of the ion is shown in Fig. 15.23. The positions of the metal atoms were determined by X-rays and then the hydrogen positions were found from neutron diffraction. The Re-Re distances rule out the possibility of any metal–metal bonding. The Re atoms are at the centre of trigonal prisms of six H atoms with the other three H atoms beyond the rectangular faces. This lies inside a similar prism of K atoms. The Tc compound is

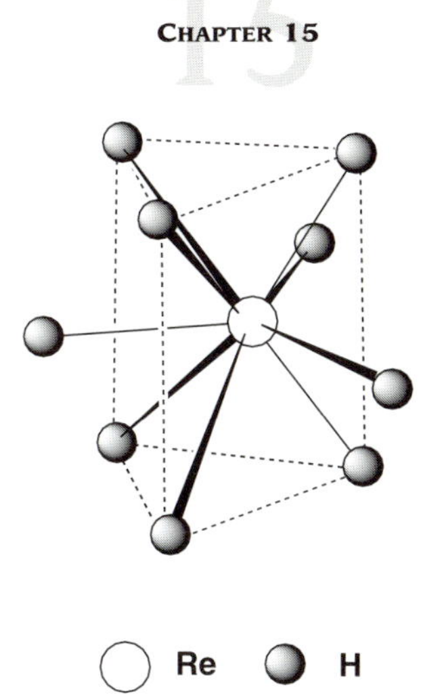

FIG. 15.23 The structure of the manohydridorhenium(VII) ion, ReH_9^{2-}

isostructural. The existence of such compounds, like the $WH_6(PR_3)_3$ species indicated in the last section, highlights the lack of oxidizing tendency in the high oxidation states of these heavy elements.

There has been a recent surge in interest in the chemistry of organometallics in high oxidation states, and that of technetium and rhenium in the VII oxidation state serves to illustrate this very well. High oxidation state organometallic compounds were once thought to be rare, but are actually turning out to be quite common. These compounds are of interest as catalysts, particularly in alkene epoxidation reactions which are of industrial importance. The reaction of the heptoxides M_2O_7 (M = Tc or Re) with $SnMe_4$ gives the very interesting tetrahedral M(VII) compounds $MeMO_3$; the rhenium compound is air-stable and water-soluble, whilst the technetium analogue is moisture-sensitive. The methyltrioxy-rhenium catalyses many reactions including olefin oxidations and metatheses. The pentamethyl cyclopentadienyl analogue of the rhenium complex $(\pi\text{-}C_5Me_5)ReO_3$ can also be prepared by oxidation of $(\pi\text{-}C_5Me_5)Re(CO)_3$. A wide range of Re(VII) derivatives containing different alkyl groups, oxo ligands and halides have also been prepared. The compound Me_3ReO_2 is another example of a rhenium(VII) oxo alkyl compound (compare Fig. 15.24).

15.5.2 The VI state

The VI state is represented by a number of compounds but shows a marked tendency to disproportionate, especially in aqueous solution. In contrast, in concentrated sulfuric acid, Re(VII) may be reduced to pink Re(VI) and then to blue Re(V) in distinguishable steps. ReO_3 is well known and there is one report of an oxide of composition $TcO_{3.05}$. The rhenium trioxide is made by reaction of rhenium on the heptoxide and is red. In vacuum at 300°C it disproportionates to ReO_2 plus Re_2O_7. It is inert to acids and bases in a nonoxidizing medium, and rhenates, ReO_4^{2-}, have to be made by fusing mixtures of perrhenate and rhenium dioxide. Technates exist in alkaline solution and pink $BaTcO_4$ may be precipitated, indicating less tendency to disproportionate than for rhenium(VI). Solutions of TcO_4^{2-} (violet) and ReO_4^{2-} (olive-green) result from controlled cathodic reduction of the MO_4^- ions. The analogous, highly reactive, tetrathiorhenate(VI) anion ReS_4^{2-} has been investigated by spectroelectrochemistry.

TABLE 15.4 Halogen compounds of the lower oxidation states of technetium and rhenium

VI		*V*		*IV*		*III*	
ReF_6	TcF_6	ReF_5	TcF_5	ReF_4		$[ReF_3]$	
$(ReCl_6)$	$(TcCl_6)$	$ReCl_5$		$ReCl_4$	$TcCl_4$	Re_3Cl_9	
		$ReBr_5$		$ReBr_4$	$(TcBr_4)$	Re_3Br_9	
				ReI_4		Re_3I_9	
Also ReX_2 in complexes, and possible compound ReI							
$ReOF_4$	$TcOF_4$	$ReOF_3$					
$ReOCl_4$	$(TcOCl_4)$		$TcOCl_3$				
$(ReOBr_4)$			$TcOBr_3$				
ReO_2Br_2?							
ReF_8^{2-}	TcF_8^{2}	ReF_6	TcF_6^-	ReF_6^{2-}	TcF_6^{2-}	$ReCl_6^{3-}$	
	TcF_7^-			$ReCl_6^{2-}$	$TcCl_6^{2-}$	$Re_3X_{12}^{3-}$	
				$ReBr_6^{2-}$	$TcBr_6^{2-}$	$Re_3X_{11}^{2-}$	
				ReI_6^{2-}	TcI_6^{2-}	$Re_3X_{10}^-$	
				Re_2X_9		$Re_2X_8^{2-}$	$Tc_2Cl_8^{2-}$
				(X = Cl, Br)		(X = Cl, Br)	
$ReO_2F_4^{2-}$?							
$ReOCl_6^{2-}$		$ReOCl_5^{2-}$	$TcOCl_5^{2-}$	$Re(OH)Cl_5^{2-}$	$Tc(OH)Cl_5^{2-}$		
$ReOCl_5^-$		$ReOBr_5^{2-}$	$TcOBr_5^{2}$	$Re_2OCl_{10}^{4-}$			
		$ReOX_4^-$	$TcOX_4^-$				
		(X = Cl, Br, I)	(X = Cl, Br)				

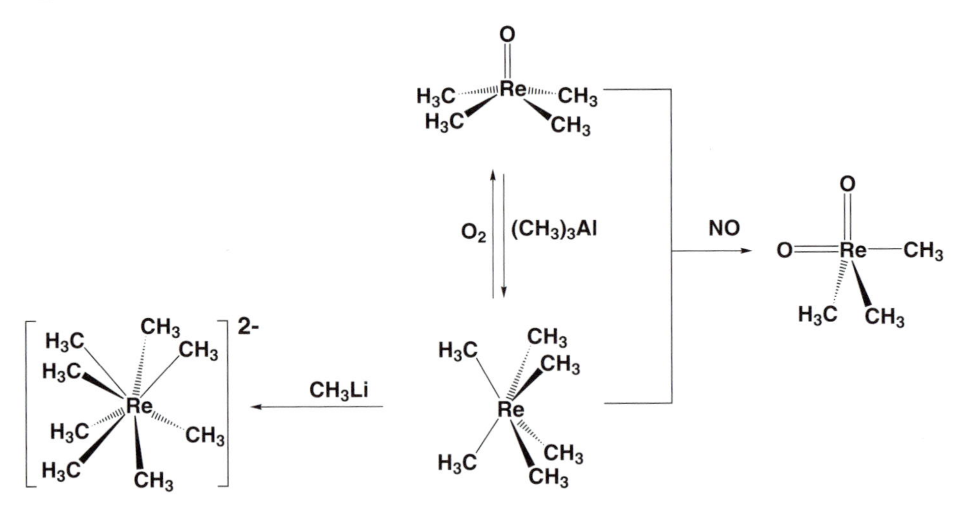

FIG. 15.24 Some rhenium VI and VII methyls

The halides, oxyhalides and related complexes, of the VI state are given in Table 15.4. The VI state of Re is well represented although $ReCl_6$ (and $TcCl_6$) readily lose Cl_2. Tc(VI) is much less represented, with only the oxide and oxyfluoride of reasonable stability. The fluorides are prepared by direct combination: ReF_6 may be purified from ReF_7 by heating with Re metal. $ReOF_4$ results from the reaction of ReF_6 with rhenium carbonyl. $TcOF_4$ exists in two forms: the blue variety contains infinite chains of octahedra while the green form contains trimers of octahedra. In both cases the octahedra are formed by sharing fluorines, $(TcOF_3F_{2/2})_n$. The other oxyhalides result from the reaction of the halides with air. $ReOCl_4$ is a square-pyramidal monomer in the solid, as is the isoelectronic $TcNCl_4^-$, formed from the reaction of azide on TcO_4^-. The sulfur analogue, $ReSF_4$, is also known. The $[ReF_8]^{2-}$ ion has been isolated as the NO^+ and $(NO_2)^{2+}$ salts and has the same square antiprismatic structure, Fig. 15.5b, as the W and Ta analogues.

Rhenium(VI) methyl compounds are also well established, and some of the inter-relationships between Re(VI) and Re(VII) methyls are summarized in Fig. 15.24. $ReMe_6$ is a distorted octahedron, as expected for a d^1 species, and $ReMe_8^{2-}$ is a square antiprism.

15.5.3 The V state

The V state is also relatively unstable and disproportionates:

$$3M^V = 2M^{IV} + M^{VII}$$

For example, $ReCl_5$ reacts with HCl to give ReO_2, perrhenic acid, and $ReCl_6^{2-}$. $ReCl_5$ exists as a dimer with octahedral rhenium, like other pentahalides such as $NbCl_5$ (Fig. 15.6). Where known, the MOX_4^- complexes are square pyramidal.

The complex of the V state which is of most interest is the cyanide, $Re(CN)_8^{3-}$. This is the d^2 octacyanide, isoelectronic with the Mo^{IV} compound, and it appears to have the same dodecahedral structure. On oxidation it gives the d^1 compound of Re^{VI}, $Re(CN)_8^{2-}$. Related compounds are produced from the dioxides in alkaline cyanide solutions. The complex ions $Tc(OH)_3(CN)_4^{3-}$ and $Re(OH)_4(CN)_4^{3-}$ are reported to be formed under similar conditions. It is interesting that Re is oxidized in this system while Tc remains in the IV state.

The diarsane ligand already encountered (p.334) appears to stabilize the relatively unstable II, III and V states of Tc and Re. Both metals form the $M(diars)_2Cl_4^+$ ion in which the element is in the V oxidation state and 8-coordinate. The $M^{II}(diars)_2Cl_2$ and $M^{III}(diars)_2Cl_2^+$ compounds are also known. In $ReH_5(PR_3)_3$ the skeleton forms a dodecahedron (compare Fig. 15.18a) with two of the PR_3 groups in the same edge.

> The related $ReH_7(PR_3)_2$ has been the subject of recent investigation. With the discovery of coordinated dihydrogen compounds (see Section 16.11) the question arises whether a polyhydride might contain $M\text{-}H_2$ units in place of 2MH bonds. Initially it was thought that the properties of $ReH_7(PR_3)_2$ indicated formulation as a $Re(V)\text{-}H_2$ species, but further work, including a neutron diffraction structure, has confirmed the original formulation as a Re(VII) heptahydride.

15.5.4 The IV state

The IV state is the second most stable oxidation state for both elements (cf. Table 15.4). The dioxides, MO_2, disulfides, MS_2, and most of the halides are known. A number of complexes exist including the very important class of hexahalides, MX_6^{2-}. These are formed by dissolving the dioxides in the hydrohalic acid and are a most useful starting point for preparations. The halide complexes $Re_2X_9^-$ are ReX_6 octahedra linked through a shared face: $X_3Re(\mu\text{-}X)_3ReX_3^-$.

The dioxides are formed by reducing the heptoxides, or from the metal by controlled oxidation. Apart from dissolving in the hydrogen halides, ReO_2 dissolves in fused alkali to give the oxyanion, rhenite, ReO_3^{2-}. This precipitates the dioxide when treated with water. It does not appear that the technetium dioxide is soluble in alkali.

A further example of the smaller tendency of Tc to enter the V state is provided by the reaction of the MI_6^{2-} complexes with KCN. TcI_6^{2-} reacts in methanol with KCN to give the Tc^{IV} cyanide complex, $Tc(CN)_6^{2-}$, but ReI_6^{2-} undergoes oxidation in the course of the reaction to the Re^V octacyanide $Re(CN)_8^{3-}$. It is easy to see that, if the octacyanide is to be formed, the M^V (d^2) compound will be more stable than the M^{IV}(d^3) compound, as the third electron is in an unstable orbital in the dodecahedral field (Fig. 15.18b). It is therefore likely that this difference in behaviour is to be ascribed to a steric effect favouring 8-coordination for Re rather more than for Tc.

15.5.5 The lower oxidation states

In the III state, technetium is quite unstable but rhenium forms the oxide, $Re_2O_3.nH_2O$, and the heavier halides. The oxide is formed by hydrolysis of the rhenium(III) chloride.

Rhenium trichloride and tribromide, formed by heating the pentahalides in an inert gas, are dark red solids which have been shown to be trimers, Re_3X_9. The structure is similar to that of the $Re_3Cl_{12}^{3-}$ ion shown in Fig. 15.25. In these species, a triangle of Re atoms is bridged along each edge by a halogen atom, and there are two more halogen atoms on each rhenium situated above and below the Re_3 plane giving a basic Re_3Cl_9 unit. In the ion $Re_3Cl_{12}^{3-}$, the structure is completed by adding a terminal halogen, in the plane of the triangle, to each rhenium. The related ions (see Table 15.4) $Re_3Cl_{11}^{2-}$ and $Re_3Cl_{10}^-$ lack one and two, respectively, of these in-plane terminal chlorines. In $[Re_3Cl_{10}(H_2O)_2]^-$ two of the in-plane terminal positions are occupied by water molecules and other donors will behave similarly. The corresponding bromides have similar structures. In the parent trihalide, the basic Re_3Cl_9 units are joined together by forming $Re(Cl)_2Re$ bridges, using the in-plane positions and one of the Cl atoms which are perpendicular to the plane from each Re. A range of related complexes are also found, of type $Re_3X_9L_3$ where neutral ligands occupy the in-plane terminal positions. Earlier reports of dimeric forms of rhenium trihalides have been shown to be unfounded. ReI_3 is a black solid, which results from heating the tetraiodide, and further heating leads to another black compound of formula ReI. The structure is also trimeric and similar to that of the trichloride. The overall linking of Re_3 units gives long chains in the iodide, while the chloride and bromide have layer structures.

A further type of complex halide ion of the III state of rhenium is formed by reduction of perrhenate in acid solution and contains dimeric units, $(Re_2Cl_8)^{2-}$. Similar species are $Mo_2Cl_8^{4-}$, $Re_2Br_8^{2-}$, a number of derivatives such as $Re_2Cl_6[PEt_3]_2$, and dinuclear carboxylic acid derivatives. Technetium forms $Tc_2Cl_8^{n-}$ for $n=2$ or 3. Such compounds contain M-M bonds of high order which are discussed in more detail in Section 16.4. Other complexes of the trichlorides with donor ligands are found, for example $(Ph_3PO)_2ReCl_3$. Rhenium(III) also gives hydrides of formula $(C_5H_5)_2ReH$ and $(C_5H_5)_2ReH_2^+$.

Representatives of lower oxidation states include complexes of MX_2 with donor ligands such as py_2ReI_2 and $(diars)_2TcCl_2$. Rhenium(I) is found in the stable dinitrogen complexes, $ReCl(N_2)L_4$, where the ligands L are various phosphanes, CO or PF_3. Oxidation with Ag^+ or Fe^{3+}, gives the rhenium(II) species, $ReCl(N_2)L_4^+$.

Both elements form a cyanide in the I state corresponding to the manganese compound, $K_5M(CN)_6$. The I state is also found in carbonyl halides and similar species, for example in $Re(PF_3)_4Cl$.

The 0 state appears in the carbonyls, $Tc_2(CO)_{10}$ and $Re_2(CO)_{10}$, which have the same structure as the manganese carbonyl. The anion, $Re(CO)_5^-$, is formed with the alkali metals and contains Re(−I).

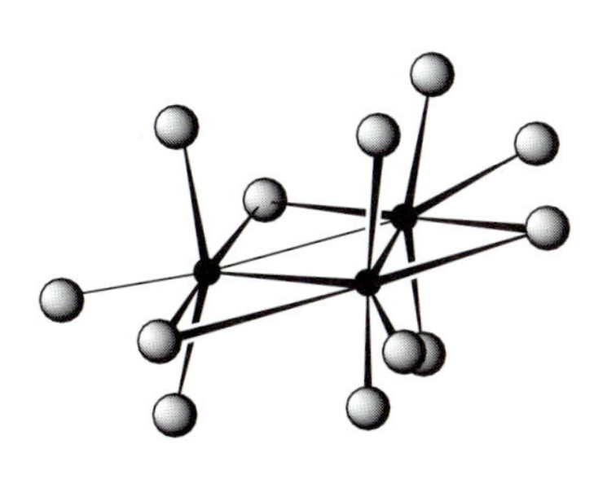

FIG. 15.25 The structure of the $[Re_3Cl_{12}]^{3-}$ ion

15.6 Ruthenium, $4d^7 5s^1$, and osmium, $5d^6 6s^2$

AQUA REGIA AND NOBLE METALS

A powerful acid, unrelated to the highly-acidic superacids, is aqua regia which is a 3/1 mixture of concentrated hydrochloric and nitric acids. This was known to the alchemists as the only solvent for gold (hence its name) and is still widely used to get gold and neighbouring elements into solution. The nitric acid oxidizes and the HCl gives chlorocomplexes.

The alchemists called gold 'noble' because it remained unchanged by known materials—e.g. it did not corrode when buried. The term has been expanded to include the platinum metals and neighbours, which are relatively inert, and whose simple compounds convert to the metals under mild conditions.

These are the first two of the six platinum metals, i.e. the six heavier members of the iron, cobalt and nickel Groups. These elements, together with rhenium and gold, are broadly similar in that the element is fairly unreactive—'noble'—and decomposition of compounds to the element is fairly ready. The platinum metals, gold and silver are commonly found together and a number of schemes are in current use for their separation. One method involves extracting the mixed metals with aqua regia and then treating the soluble and insoluble portions as in Fig. 15.26. Osmium may occur in either fraction and is removed as the volatile tetroxide, while ruthenium ends up in the VI state in fused alkali.

These two elements share with xenon the highest observed oxidation state of VIII, and their oxidation states range downwards to −II. The VI and IV states are stable while the VII and V states are poorly represented and tend to disproportionate. Osmium is most stable in the IV state and ruthenium in the III state.

Some inter-relations among the oxides and oxyions of the different oxidation states are shown in the reaction diagrams of Table 15.5, and for halogen compounds in Table 15.6.

15.6.1 The higher oxidation states

Note from the tables that osmium is more stable in the VIII state than ruthenium, the metal being oxidized directly to the tetroxide. The tetroxides are volatile (b.p. about 100°C), tetrahedral, covalent molecules which are strongly oxidizing. Osmium(VIII) also occurs bonded to N in $Os(NBu^t)_4$, with the bulky *tert*-butyl substituents contributing to the stability. The structure is a distorted tetrahedron with OsNC angles of 156°, all in accord with an Os=N formulation. The mixed oxygen–nitrogen species $Os(NBu^t)_2O_2$ and $Os(NBu^t)O_3$ have similar tetrahedral structures. Further osmium(VIII)–oxygen species include the hydroxide complexes $OsO_4(OH)_2^{2-}$, which is a *cis* octahedron, and $OsO_4(OH)^-$, which contains a distorted trigonal bipyramidal group with OsO_4 units bridged axially by hydroxides.

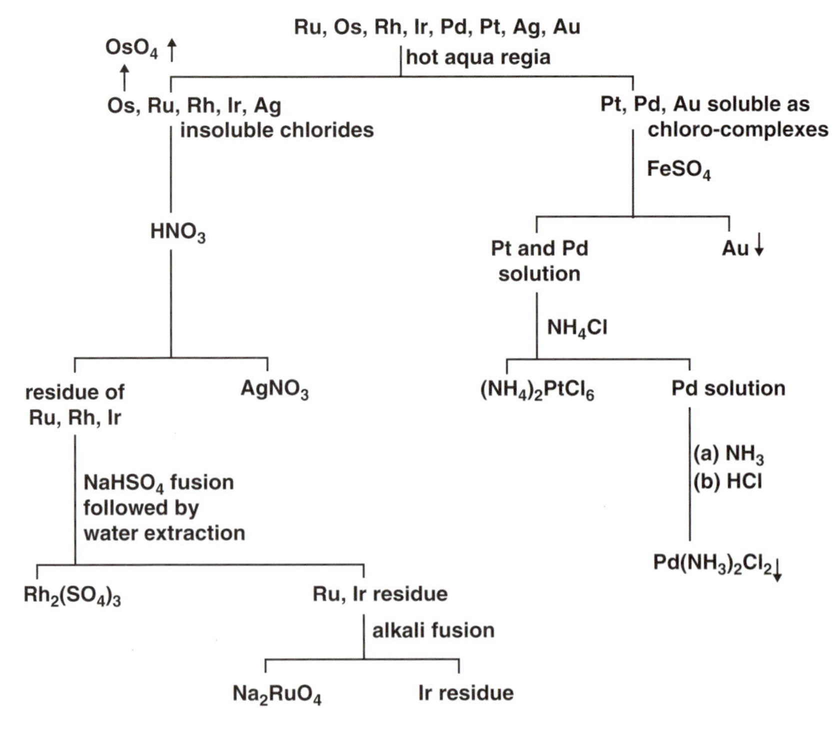

FIG. 15.26 A reaction scheme for the separation of the platinum metals, gold and silver

TABLE 15.5 Ruthenium and osmium oxygen compounds

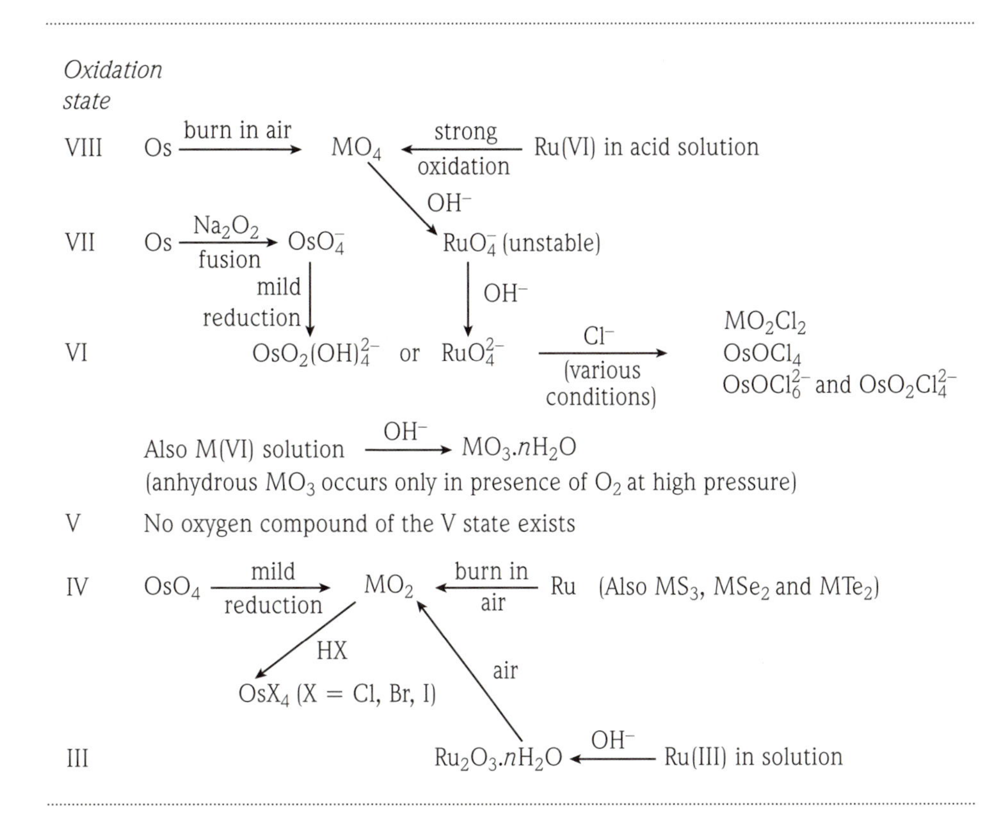

Osmium(VIII) and (VII) are also represented among halogen compounds although OsF_8 does not exist and OsF_7 survives only under a high pressure of F_2. The VIII state is found in the oxyfluorides OsO_2F_4 and OsO_3F_2. OsO_2F_4 (previously identified incorrectly as $OsOF_6$) is formed by the action of KrF_2 on OsO_4 and has a *cis* octahedral structure. KrF_2 is a powerful oxidizing and fluorinating agent, and oxygen in this reaction is oxidized from O(–II) to O(0):

$$OsO_4 + KrF_2 \rightarrow OsO_2F_4 + 2Kr + O_2$$

TABLE 15.6 Ruthenium and osmium halogen compounds

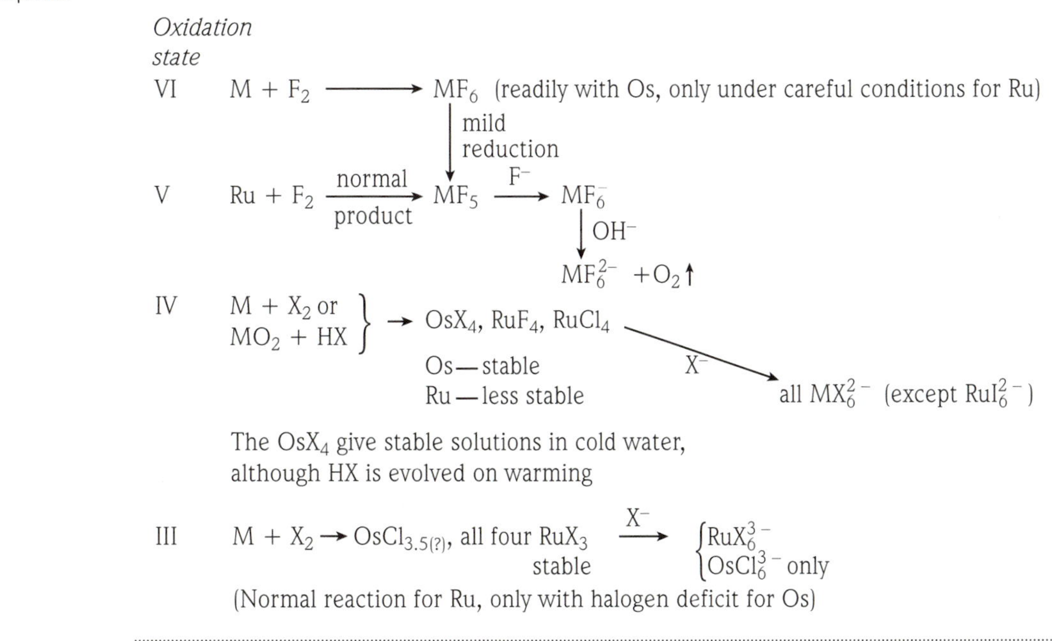

OsO_3F_2 is formed by the reaction of OsO_4 with liquid ClF_3 and has a structure of linked octahedra. The Os bears three O and one terminal F and is linked through Os-F-Os bridges to the octahedra on either side, with the two bridging F occupying *cis* positions.

This compound undergoes an interesting reaction where it is *reduced* by reaction with fluorine to Os(VII) and Os(VI) complexes, and oxygen is again oxidized, this time from O(–II) to O(II):

$$OsO_3F_2 + F_2 \longrightarrow OsOF_5 + OsF_6 + OF_2$$

Halo-complexes of Os(VIII) include $[OsO_4F_2]^{2-}$, which is a *cis* octahedron, and $[OsO_4Cl]^-$, which is a trigonal bipyramid with the Cl axial with a very long bond.

As well as the MO_4^- ions, osmium forms a second oxyanion in the VII state, OsO_6^{5-}. These two oxyanions compare with those found for rhenium. A second oxyfluoride of osmium(VII), OsO_2F_3, forms from OsO_3F_2 and $OsOF_4$, and disproportionates back to these VIII plus VI species at 60°C.

In the VI state, $OsOX_4$ (X = F, Cl or CH_3), are square pyramids in the gas phase with the oxygen in the axial position, like the rhenium and tungsten analogues. In contrast, the MOX_4 species formed by the Main Group elements are trigonal bypyramids with the oxygen in an equatorial position. RuF_6 is more reactive than OsF_6. It is thermally stable at room temperature but reacts vigorously with most materials. The structure is a fully symmetric octahedron in a low-temperature matrix. The osmium(VI) polyhydride, $OsH_6(PR_3)$, can be made by the reaction of osmium(IV) chloro-complexes with $LiAlH_4$. The thiosulfato complex $[OsO_2(S_2O_3)_2]^{2-}$, prepared by reduction of OsO_4 with aqueous sodium thiosulfate, has a distorted tetrahedral geometry about osmium, with the two thiosulfate ligands bonded through S. The oxo-carbonyl cation *trans*-$OsO_2(CO)_4^{2+}$, containing Os(VI), can be isolated as a highly air- and moisture-sensitive $Sb_2F_{11}^-$ salt from the reaction of OsO_4 and SbF_5 under CO. Cl and Br form the dimeric complex anions $[Os_2X_{10}]^{2-}$ which have octahedral units sharing an edge, compare Fig. 14.5b.

This species is a rare example of a high oxidation state metal carbonyl complex; CO typically forms complexes with metals in low oxidation states, but increasingly the range of metal oxidation states is being extended.

In the unstable V state both metals form the pentafluoride and its anion. Both MF_5 compounds are tetramers with nonlinear M-F-M bridges, similar to NbF_5 (Fig. 15.4). This tetrameric form is adopted, with minor variations, by most of the pentafluorides of the heavier transition metals. In the gas phase the major form is a similar fluorine-bridged trimer, again with four terminal F and two M-F-M bridges formed by *cis* fluorines. Dimers form the minor gas phase species. $OsCl_5$ is isomorphous with rhenium pentachloride, having a dimeric structure with two chloride bridges. The complex ions $OsCl_6^-$ and $OsBr_6^-$ have slightly distorted octahedral structures.

The bulky aromatic group mesityl helps to stabilize less-accessible species and allows the formation of Ru-C bonds in both the V and IV states in the compounds $[Ru(mes)_4]^+(PF_6)^-$ and $Ru(mes)_4$ respectively. The structures are tetrahedral.

Mesityl is 2,4,6-trimethyl benzene.

15.6.2 The lower oxidation states

The lower oxidation states are very accessible. Only mild reduction of osmium tetroxide is required to give the dioxide, and ruthenium burns in air to give RuO_2 directly. The dioxides are stable compounds with the rutile structure. The stability of the IV state is also illustrated by the existence of the disulfides and of the heavier chalcogenides.

In Table 15.6, the stable states of Ru(III) and Os(IV) show clearly. One dimeric anion in the IV state is $[Os_2Br_{10}]^{2-}$ which has a structure with two octahedra sharing an edge. It reacts with a range of ligands to yield Os(III) compounds. A fuller study has thrown doubt on the compounds originally reported as osmium trihalides. These have now been characterized as the oxyhalides, Os_2OX_6, which are Os(IV) compounds. The Os(III) halide complexes OsX_6^{3-} X = Cl, Br, I can be prepared in presence of the large cation $[Co(en)_3]^{3+}$.

Deep pink RuF_4 has a structure built up of apex-sharing octahedra forming a puckered sheet array with the bond length to the bridging F equal to 200 pm. The nonbridging F (Ru-F = 182 pm) are *trans* to each other, lying above and below the sheet. This fluorine bridging is very similar to that found in $(RuF_5)_4$ and also in RuF_3 where all the F bridge, giving a three-dimensional array.

Both elements form the mixed oxyhalide, $M_2OX_{10}^{4-}$ (M = Ru, Os(IV): X = Cl, Br). The oxygen bridges the M atoms, completing the octahedra, e.g. $[Cl_5Ru\text{-}O\text{-}RuCl_5]^{4-}$.

Apart from those given above, these elements form a variety of complexes in the IV, III and II states. In the IV state, osmium complexes are more extensive and stable than the ruthenium ones. The commonest are the halide, hydroxy-halide, amine and diarsine complexes. In the III state, the situation is reversed and there are more ruthenium species, mainly octahedral. Both elements give hexammines, $M(NH_3)_6^{3+}$, and ruthenium gives the whole range of mixed halogen–ammonia complexes down to $Ru(NH_3)_3X_3$. A variety of other ruthenium complexes occurs, including those with substituted amines, and complex chlorides with 4, 5, 6 and 7 chlorine atoms in the anion. A very interesting series is provided by the complexes $M(NCS)_6^{3-}$ for M = Ru or Os. The thiocyanate group can bond M-NCS as a 'hard' ligand or M-SCN as a 'soft' ligand (Section 13.7) and these metals are clearly on the hard–soft borderline in the +III state since a number of members of the series $M(NCS)_n(SCN)_{6-n}^{3-}$ are formed. For Ru, $n = 1$ to 4 are indicated while Os gives $n = 2$ to 4.

(See also Section 16.10)

Although there are few simple compounds of the II state, there are a variety of complexes of both ruthenium and osmium. All are formed by reduction of metal solutions, in the IV or III states, in presence of the ligands. Examples include $M(dipy)_3^{2+}$, the very

NITROGEN COMPLEXES OF RUTHENIUM AND OSMIUM

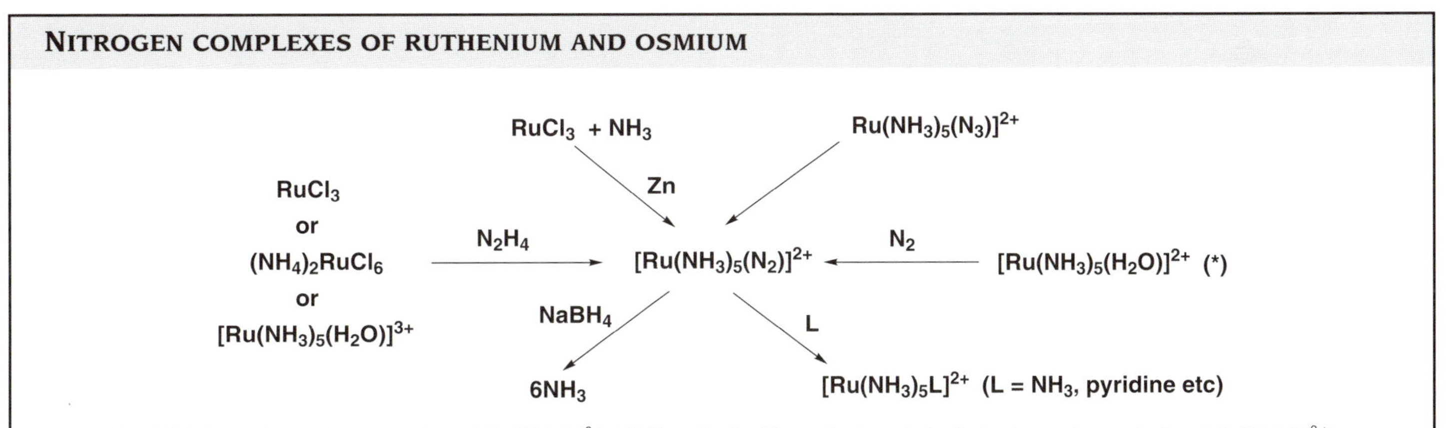

FIG. 15.27 Preparations and some reactions of $[Ru(NH_3)_5N_2]^{2+}$. (*) The ruthenium(II) complex is made *in situ* by zinc amalgam reduction of $[Ru(NH_3)_5Cl]^{2+}$

A little after the first reports of nitrogen fixation in the systems metal halide + organometallic reducing agent which were discussed under *Titanium* (Section 14.2), there was reported the first compound in which it was clearly shown that nitrogen was coordinated as a ligand. This was the ruthenium(II) complex $[Ru(NH_3)_5N_2]^{2+}$. This compound was first made by the action of hydrazine (which was the source of the coordinated N_2) on the trichloride, but a number of other routes have since been reported including the reaction of a ruthenium(II) complex with gaseous nitrogen. Some of these preparations and some reactions are summarized in Fig. 15.27.

This series of reactions shows that the ligand nitrogen may be derived from a number of sources, including the direct fixation of atmospheric nitrogen. The coordinated nitrogen is replaceable by a range of ligands, presenting a useful route to otherwise unobtainable ruthenium(II) complexes. In these cases, the nitrogen comes off as N_2 gas, so that the cycle of reactions does not lead to fixation of N_2 as a compound. The only indication of the latter process is in the reaction with $NaBH_4$ which yields six volumes of ammonia, that is, one N atom out of the two in the coordinated group is converted to NH_3.

A number of other metals in this part of the Periodic Table have been shown to form nitrogen complexes (Table 16.6). Osmium forms the analogue of the ruthenium complex, $[Os(NH_3)_5(N)_2]^{2+}$, by similar routes. The osmium complex is much more stable and resists replacement of the nitrogen. In $(NH_3)_5Ru(N_2)Ru(NH_3)_5^{4+}$, the NN unit bridges two metal atoms. Structural studies on these compounds indicate a linear M-N-N-M bridge, in contrast to bridging carbonyl which bonds to both metals through carbon in a nonlinear unit.

An N_2 ligand bridging two metal atoms

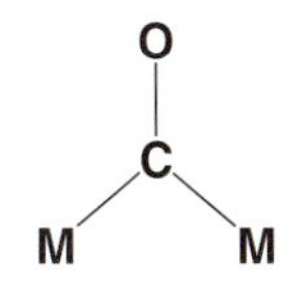

A CO ligand bridging two metal atoms

stable $M(CN)_6^{4-}$, and a variety of ammine and arsane complexes. As the II state is the d^6 configuration, there will be a large CFSE contribution to octahedral complexes of these heavy elements. An important group of compounds are those containing NO bonded to ruthenium(II), such as $[Ru(NO)X_5]^{2-}$, where X = halogen, OH, CN and many more. Another interesting group of Os(II) compounds is derived from the carbonyls, see below and Section 16.2. Oxidative fluorination gives a range of species in solution identified spectroscopically as *cis*-$[Os(CO)_4F_2]$, $[Os(CO)_5F]^+$ and $[Os_2(CO)_8F_3]^+$—the latter being two octahedra linked by Os-F-Os. In the solid CO is lost, giving the tetramer $[Os(CO)_3F_2]_4$.

In lower oxidation states, there are a number of compounds including $Os(NH_3)_6Br$ and $Os(NH_3)_6$—Os(I) and Os(0) respectively—formed by reduction with potassium in liquid ammonia. Other representatives of the 0 state include the carbonyls $M(CO)_5$ and $M_3(CO)_{12}$. Among the more complex carbonyls, there is the very interesting derivative $Ru_6C(CO)_{17}$. This contains an isolated carbon atom which is situated at the centre of a distorted octahedron of Ru atoms giving a CRu_6 environment which is also found in most interstitial metallic carbides (Section 17.5). Four of the Ru atoms carry three terminal CO groups and the other two have two terminal carbonyls and are bridged by the seventeenth one. The Ru_6C cluster involves delocalized polycentred bonding. There is a reported anion, $Ru(CO)_4^{2-}$, which would contain ruthenium(–II) and be isoelectronic with nickel carbonyl.

15.7 Rhodium, $4d^85s^1$, and iridium, $5d^96s^0$

In this Group, the Group oxidation state of IX is not shown and a strong tendency for lower oxidation states to become stable is seen, as compared with the previous Groups of heavy transition elements. The highest state found is VI, while the most stable states are iridium(IV) and (III) and rhodium(III). In the separation from the rest of the platinum metals, Fig. 15.26, both elements are found in the fraction insoluble in aqua regia. Iridium remains as the insoluble residue, after alkaline fusion, while rhodium is extracted in aqueous solution as sulfate after fusion with sodium bisulfate.

The highest oxidation state is represented by the hexafluorides, MF_6, formed by direct reaction. As in the previous Group, the heavier element is more stable in the high oxidation states and IrF_6 is much more stable that the rhodium fluoride. However, the highly reactive RhF_6, formed by high temperature and pressure fluorination of RhF_5, is thermally stable at room temperature in passivized nickel vessels. No oxygen species exist in the VI state: the reported IrO_3 is possibly a peroxide rather than an iridium(VI) compound.

The V state is also unstable and is found only in the IrF_6^- ion, whose salts are prepared by fluorinating mixtures of iridium trihalides and alkali halides in bromine trifluoride solution. Mixed fluoride–chloride Ir(V) complexes are made similarly with up to 3Cl, all *cis*. The first Ir(V) oxygen species was found in $KIrO_3$ and related salts. In strong base, the higher coordinated IrO_6^{7-} ion is formed. As with Re, Os and other elements, we see the formation of oxyanions in the same oxidation state with two different oxygen coordination numbers. As for ruthenium, stable organometallic compounds of iridium are available using mesityl as ligand. The iridium(V) species is the tetrahedral cation $[Ir(mes)_4]^+$, while the neutral species $[Ir(mes)_4]$ and $[Ir(mes)_3]$ form in the IV and III states. The latter, like $[Rh(mes)_3]$, is pyramidal with CMC angles around 108°. The hydride $IrH_5(PR_3)_2$ is formally Ir(V) and extends the heavy metal polyhydride series. It readily transfers H, e.g. to $Pt_2Cl_4L_2$.

The IV state is fairly stable for iridium but represented by only a few rhodium compounds. The products formed on igniting the metals in air are IrO_2 and Rh_2O_3, respectively, and RhO_2 can be made only by strong oxidation of rhodium(III) solutions. It can be dehydrated without reverting to rhodium(III) only by heating under a high pressure of oxygen. IrO_2 with alkali metal bases gives black Cs_4IrO_4 or red Na_4IrO_4. The crystal structures show the presence of square-planar IrO_4^{4-} ions. Both elements form the tetrafluoride, MF_4, and $IrCl_4$ is also reported. Rhodium gives only the halogen complexes, RhF_6^{2-} and $RhCl_6^{2-}$, but a wider variety of iridium(IV) complexes occur. Fairly stable hexahalides, IrX_6^{2-}, are formed for X = F, Cl, and Br, and other complexes include the oxalate, $Ir(C_2O_4)_3^{2-}$, which can be resolved into optical isomers.

In the III state both elements form all four trihalides and the oxides, M_2O_3. Ir_2O_3 can only be formed, by addition of alkali to iridium(III) solutions, as a hydrated precipitate in an inert atmosphere. Attempts to dehydrate it lead to decomposition, and it oxidizes in air to IrO_2. Rhodium trifluoride results, along with some tetrafluoride, from fluorination of $RhCl_3$, but IrF_3 can only be prepared by reduction of IrF_6 with Ir, as fluorination of iridium(III) compounds gives the tetra- or hexafluoride. The other trihalides are made by direct reaction with halogen. They are insoluble in water and probably polymeric.

A considerable number of rhodium(III) complexes are known, and these are generally octahedral. The hydrate, $Rh(H_2O)_6^{3+}$, the ammine, $Rh(NH_3)_6^{3+}$, and a large variety of partially substituted hydrates and ammines are found. For example, hydrates containing 1, 2, 3, 4 and 5 chloride ions in place of water molecules are formed, as well as $RhCl_6^{3-}$ and other hexahalides. Other halide complexes include RhX_5^{2-}, and RhX_7^{4-}, especially for the bromides. The dinuclear complex halides, $Rh_2X_9^{3-}$, have the face-sharing octahedral structure $[Cl_3Rh(\mu\text{-}Cl)_3RhCl_3]^{3-}$. A similar structure is found for the iridium chloro- and bromo-analogues.

The complexes of rhodium(III) are relatively inert to substitution so that isomers may be isolated, as in the resolution into optical isomers of the *cis*-$Rh(en)_2Cl_2$ ion. There is also an extensive series of iridium complexes which are generally similar to the rhodium compounds. Iridium also gives the hydrides, $(R_3P)_3Ir(H)_n(Cl)_{3-n}$ for $n = 1$, 2 and 3. Rhodium forms the analogues with $n = 1$ and 2. The related carbonyls, $(R_3P)_2Rh(H)(Cl)_2(CO)$ and $(R_3P)_2Ir(H)_3(CO)$ are known.

These elements form only a few compounds in the II state. Simple compounds are restricted to polymeric $IrCl_2$ and the reported oxide, RhO, which is not well established. There is a slowly growing number of simple complexes such as $[Ir(CN)_6]^{4-}$ and $Ir(NH_3)_4Cl_2$. An important binuclear type is the rhodium(II) carboxylate $[Rh_2(OOCR)_4]$ which acts as a catalyst in a number of applications. The molecules have the same structure as the Cr or Cu analogues, see Figs 14.20 or 14.33, but without the extra water molecules. As well as the acetate (R = Me), the fluoroacetate ($R = CF_3$) is particularly stable. Compounds where the carboxylate is optically active are particularly interesting as these act as enantioselective catalysts.

Low oxidation state compounds include the carbonyl monohalides of the I state, and the carbonyl anions $Rh(CO)_4^-$ and $Ir(CO)_4^-$, in the −I state. These are paralleled by trifluorophosphane analogues like $KIr(PF_3)_4$ which is oxidized from the −I state to the +I state by iodine to yield $Ir(PF_3)_4I$. Compounds in the 0 state include $Ir(NH_3)_5$ and $Ir(en)_3$ which are formed by reduction in liquid ammonia, like the similar osmium compounds. The 0 state is also represented in the carbonyls which include $M_2(CO)_8$ and the interesting hexanuclear cluster $Rh_6(CO)_{16}$. This has a structure with an octahedron of rhodium atoms, each with two terminal carbonyl groups, and the other four carbonyl groups are found in the middle of opposite faces. The structure is held together by delocalized metal–metal bonding in the Rh_6 cluster. A formal −III oxidation state is found for iridium in the anion $[Ir(CO)_3]^{3-}$, formed by reduction of $Ir(CO)_4^-$ by sodium metal.

OXIDATIVE ADDITION AND REDUCTIVE ELIMINATION REACTIONS, AND THEIR IMPORTANCE IN CATALYSIS

Among the most fully explored reactions are those of the square-planar compounds of the I state, which add neutral molecules to give octahedral compounds of the III state in the process which has been called oxidative addition. The reverse process, the loss of a neutral molecule from a complex in the III state and formation of a complex in the I state, is called reductive elimination. Rhodium and iridium complexes undergo such reactions, which are not restricted to these metals, and many other examples can be found, e.g. in the chemistry of platinum and palladium. An example is:

$$IrCl(CO)(PPh_3)_2 + XY \rightarrow XIrY(Cl)(CO)(PPh_3)_2$$

where XY = HCl, CH_3I, Cl-HgCl, etc. Molecules such as H_2, O_2 or SO_2 are also taken up and may be lost again in a reversible reaction. While H_2 gives two M-H bonds, the O_2 molecule is not split and the two O atoms are *cis* to each other at a distance O-O = 130 pm, suggesting coordination as superoxide, O_2^-.

A very important application of oxidative addition is in catalysis. The Rh(I) complex $[Rh(CO)_2I_2]^-$ catalyses the carbonylation of CH_3OH to acetic acid in the Monsanto process, shown in Fig. 15.28, via an oxidative addition (step 1) of CH_3I, forming $[Rh(CO)_2(CH_3)I_3]^-$ as an intermediate. One of the CO groups then inserts into the Rh-CH_3 bond in step 2, to give an Rh-C(O)CH_3 (acetyl) group. After the addition of another CO ligand in step 3 to regenerate the 6-coordinate complex $[Rh(COCH_3)(CO)_2I_3]^-$, this then undergoes a reductive elimination of $CH_3C(O)I$ in step 4, regenerating the original $[Rh(CO)_2I_2]^-$ catalyst. The cycle is completed by the $CH_3C(O)I$ reacting with water in step 5, producing the product CH_3COOH, together with HI. Reaction of the HI with methanol feedstock produces the methyl iodide required for the oxidative addition reaction in step 1, thus continuing the catalytic cycle. Development of this work has led to similar Rh catalysts for other reactions and modification of the catalyst, for example by bonding to an insoluble polymer backbone, to attempt to improve the control and activity. Square planar Rh(I) species with at least one CO group, usually a halide, and other ligands like phosphanes, are the compounds most studied. Recent work using low-temperature nmr spectroscopy has been successful in detecting the intermediate complexes, such as $[Rh(CO)_2(CH_3)_3I_3]^-$, which are produced transiently in very small amounts in the reaction mixture. In 1996, the new and more efficient Cativa process, involving an iridium catalyst, was introduced by BP Chemicals, and is now in operation in a number of manufacturing plants.

In a different type of reaction, of potential catalytic interest, hydride complexes of Ir and Rh, of the type $(Me_5C_5)ML_2$ (where L = CO, 2H, phosphane or similar ligands), activate the usually inert C-H bond. An organic group R-H adds to the metal forming R-M-H with elimination of L. This is seen as the first step in a potentially very important controlled conversion of hydrocarbons to functional organic compounds, and it is particularly important that such activation has been established for CH_4, as natural gas is a premium starting material for chemical synthesis.

References

Methanol carbonylation: *Journal of the Chemical Society, Dalton Transactions*, 1996, 2187–2196.
Cativa process: *Catalysis Today* **58**, 2000, 293–307; *Platinum Metals Review* **44**, 2000, 94–105.

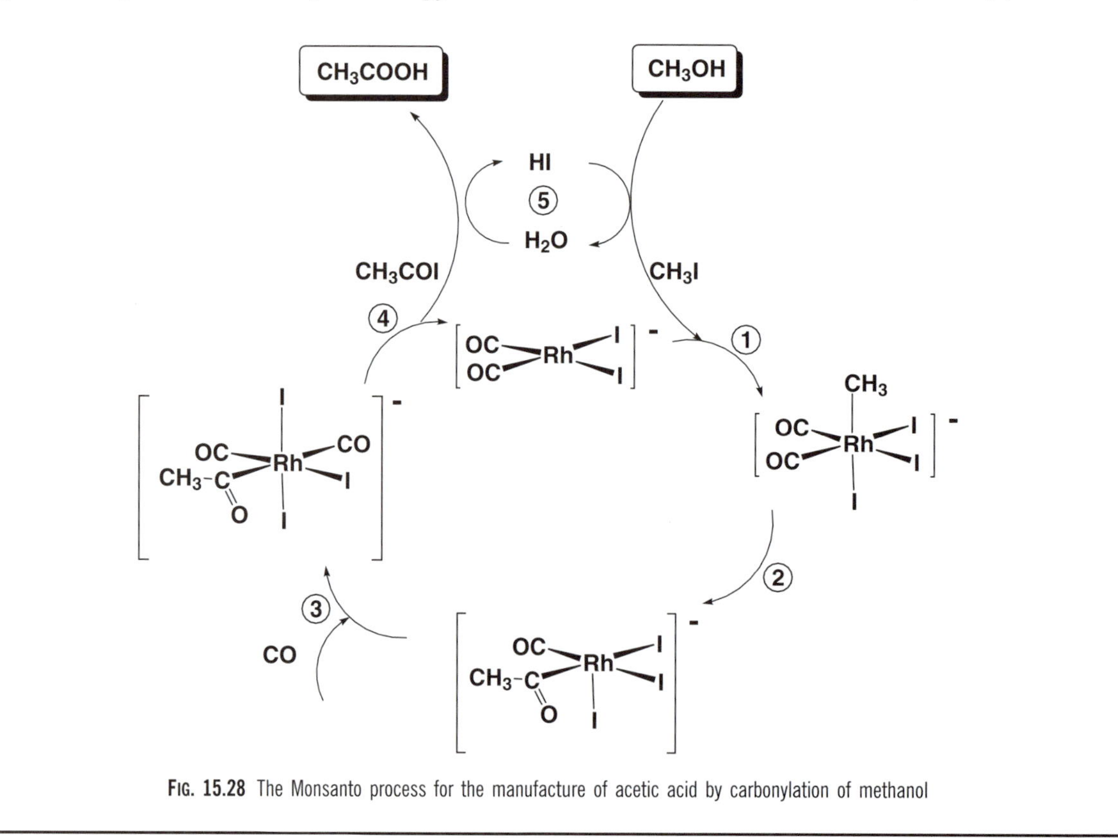

FIG. 15.28 The Monsanto process for the manufacture of acetic acid by carbonylation of methanol

The I state is also found in 5-coordinate compounds, usually involving phosphane ligands. The structure of $HRhCO(PPh_3)_3$ is trigonal bipyramidal, with the P atoms in the equatorial plane.

15.8 Palladium, $4d^{10}5s^0$, and platinum, $5d^96s^1$

This pair of elements continues the trends already observed. The higher oxidation states are unstable and the VI and V states are represented only by a few platinum compounds. The IV state is stable for platinum and well represented for palladium. This corresponds to d^6 and many octahedral Pt^{IV} complexes occur with high CFSE. The II state is the other common oxidation state. In this palladium and platinum are almost invariably 4-coordinated and square planar. Palladium continues the trend, which probably starts at technetium or molybdenum and is clearly seen for ruthenium and rhodium, in being less stable than the heaviest element of the Group in the higher oxidation states and in having a well-developed lower oxidation state.

Platinum metal is particularly important as a catalyst, both in the laboratory and in general. One of the best-known uses is that of Pt in car exhaust converters to reduce the proportion of pollutants (such as nitrogen oxides) emitted. Major applications in industry include the hydrogenation of benzene to cyclohexane, the dehydrogenation of petroleum hydrocarbons to aromatics and, in conjunction with rhodium, the conversion of ammonia to nitric acid. This last is one case of increasing use of 'bimetallic' catalysts which are usually superior to the individual metal. Usually the metal is dispersed as fine particles on a support such as α-alumina with a high surface area. Another important application of platinum complexes is in cancer chemotherapy where the complex *cis*-$PtCl_2(NH_3)_2$ is very widely used for treating certain types of cancers and this is discussed in greater detail in Chapter 20.

15.8.1 The VI, V and IV oxidation states

Only platinum forms a hexafluoride, PtF_6, and attempts to isolate the palladium compound have failed. The VI state of platinum may also occur in the reported oxide, PtO_3. There is also a compound of unknown structure which may contain Pt(VIII). This is $PtF_8(CO)_2$ which is reported to be formed by the reaction of CO under pressure on PtF_4. It is difficult to see why such a system should be oxidizing, but spectroscopic evidence shows no bridging carbonyl groups and no adduct molecules such as F_2CO.

The V state is found in PtF_5 and in the PtF_6^- ion. The latter was first found as a product of the reaction of O_2 and PtF_6 which yielded the unexpected oxygen cation in $O_2^+PtF_6^-$. The now famous first report of a rare gas compound was of the $Xe^+PtF_6^-$ complex and a number of other compounds of the anion have since been made.

The highest state for palladium, and the first stable state for platinum, is the IV state. This is represented by all four PtX_4 halides and by PtO_2. Palladium forms PdF_4 and PdO_2. The latter is found as the poorly characterized hydrated oxide but PtO_2 is the most stable oxide of platinum. It is obtained as a hydrated precipitate from the action of carbonate on Pt^{IV} solution and it is soluble in acid and alkali in this condition. It can be dehydrated by careful heating when it becomes insoluble. On heating to 200°C, it decomposes giving platinum metal and O_2.

PtF_4 is the major product of fluorination of platinum, although PtF_5 and PtF_6 also result from the direct reaction. PdF_4 is also formed by direct fluorination though here the main product is PdF_3. The other platinum tetrahalides are formed by direct halogenation and $PtCl_4$ also results when H_2PtCl_6, the product from the aqua regia solution of the metal, is heated. These heavier halides of platinum are quite stable, even PtI_4 does not decompose until about 180°C, when it goes to PtI_2 and iodine.

Palladium(IV) complexes are only a little more stable than the simple Pd^{IV} compounds. The common examples are the halides, PdX_6^{2-}, all of which are known except the iodide, and the tetrahalide amine, $Pd(amm)_2X_4^+$, where amm = ammonia, pyridine or related ligands.

By contrast, platinum(IV) forms a large number of complexes which are always octahedral, and are stable and inert in substitution reactions. As platinum(II) also gives a wide range of complexes, platinum is probably the most prolific complex-forming element of all. All the common types of complex are found, for example all members of the set

between $Pt(NH_3)_6^{4+}$ and PtX_6^{2-} are known for a variety of amines as well as ammonia, and for X = halogen, OH, NCS, NO_2, etc. However, fluoride is found only in PtF_6^{2-}.

One very interesting Pt(IV) complex is the sulfide $Pt(S_5)_3^{2-}$ which contains three S_5^{2-} units forming PtS_5 six-membered rings by linking into *cis* positions in the octahedron. As this species with three bidentate ligands has nonsuperimposable mirror images, the optical enantiomers may be resolved.

(Compare $Co(en)_3$ in Section 13.8: also Section 19.2 for metal–sulfur ring compounds).

The Pt^{IV} octahedral complexes may readily be obtained from the Pt^{II} square-planar complexes if the attacking ligand is also an oxidizing agent. Thus $Pt^{II}L_4 + Br_2 \rightarrow$ *trans*-$Br_2Pt^{IV}L_4$ for a variety of ligands (L). The halogen atoms simply add on opposite sides of the square plane.

An interesting class of complexes is the deeply coloured, usually green, type of compound which apparently contains trivalent platinum, such as $Pt(en)Br_3$ or $Pt(NH_3)_2Br_3$. These are not Pt^{III} compounds at all but chains made up of alternate $Pt^{II}(NH_3)_2Br_2$ and $Pt^{IV}(NH_3)_2Br_4$ units as shown in Fig. 15.29. The Pt-Br distances on the vertical axis are 250 pm for Pt^{IV}-Br and 310 pm for the weak Pt^{II}- Br interaction.

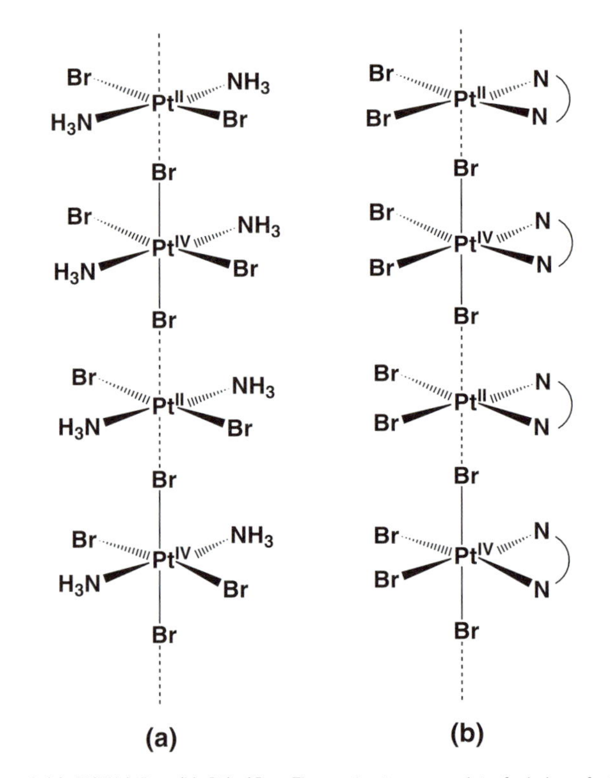

FIG. 15.29 Structures of (a) $Pt(NH_3)_2Br_3$, (b) $Pt(en)Br_3$. These structures consist of chains of alternate square-planar platinum(II) units, PtN_2Br_2, and octahedral platinum(IV) units, PtN_2Br_4 (N = NH_3 or half an ethylenediamine molecule)

15.8.2 The III, II and lower oxidation states

As in the example above, many compounds which appear to be in the III state are instead mixed valency ones. Thus PdF_3 is $Pd^{2+}PdF_6^{2-}$ and similarly for PtX_3. The interesting Pt_3I_8 consists of two octahedral $Pt(IV)I_6$ units linked to a central square-planar $Pt(II)I_4$ by shared edges. Hydrated oxides, M_2O_3, are reported but their identity is not proven.

However, Pd(III) was confirmed in the complex ions $Na^+[PdF_4]^-$ and Ag_3PdF_6. In NaK_2PdF_6, a structural study shows that the PdF_6^{3-} ion has four shorter Pd-F bonds of 195 pm, and two longer ones of 214 pm, just as expected for a low-spin d^7 state. Pt(III) has been observed by electron spin resonance after X-irradiation of a Pt(II) complex, $[Pt(en)_2]^{2+}$, in the presence of a large anion.

The structure of $PdCl_2$

In a second class of formally + III compounds, a Pd-Pd or Pt-Pt bond exists in a dimer in compounds of the type X-Pt(bident)$_n$Pt-X. Here bident = a bidentate ligand which bridges the two Pt atoms, holding them close enough together for a Pt-Pt bond to form. The M(III) compounds are made when the bridged Pt(II) complexes are treated with X_2 (compare the analogous gold compounds, Fig. 15.32). There may be between $n = 2$ and $n = 4$ bidentate groups present. Such Pt(III) compounds are well established and a few Pd analogues are known. More recently a Pt(III) species with an unsupported Pt-Pt bond has been found in the complex ion $[(CN)_5Pt\text{-}Pt(CN)_5]^{4-}$.

Like nickel, these elements have a very stable II state, but occur only in the square-planar configuration in this state. All the dihalides except PtF_2 are found. Palladium forms a stable PdO while platinum gives a hydrated PtO which readily oxidizes to PtO_2. Palladium dichloride has a chain structure of linked square-planar $PdCl_4$ units.

A second form of palladium dichloride, and the only form of the platinum compound, consists of M_6Cl_{12} units with the chlorine atoms placed above the edges of an octahedron of platinum atoms. This structure is reminiscent of the tantalum(II) and molybdenum(II) halide complex ions and is a further example of a metal cluster compound.

The complexes of palladium(II) and platinum(II) are abundant and include all common ligands. Some examples have been given in Chapter 13 (see Fig. 13.18). Palladium(II) complexes are a little weaker in bonding and react rather more rapidly than the platinum(II) complexes, but are otherwise very similar. The commonest donor atoms are nitrogen (in amines, NO_2), cyanide, the heavier halogens and phosphorus, arsenic and sulfur. The affinity for F and O donors is much lower, but the stability of Pt(II) is underlined by the isolation from nitric acid of $Pt(NO_3)_4^{2-}$ where the nitrate groups are bonded through a single O forming a square-planar PtO_4 coordination shell. Another example is provided by the planar PtH_4^{2-} ion, a red-violet compound synthesized by the action of NaH on Pt under hydrogen. This is a further case of the polyhydride ions remarked on in earlier sections.

While $Pt(NH_3)_4^{2+}$ ions are colourless in solution and $PtCl_4^{2-}$ is red, the salt $[Pt(NH_3)_4][PtCl_4]$—named *Magnus's green salt*—is strongly coloured. The square-planar units stack, alternate cation and anion, directly above each other in the crystal with a Pt-Pt distance of 325 pm. Similar species with small ligands which allow short Pt-Pt distances are also abnormally coloured, betokening a metal–metal interaction in the crystal. Such species have aroused interest as possible 'one-dimensional conductors'. If some electrons are removed, as in the partly oxidized compound $K_2Pt(CN)_4.0.3Br$, the crystals become strongly conducting but only in the direction parallel to the Pt chains.

A further interesting square-planar platinum(II) compound is Zeise's salt, one of the first organometallic compounds (see Fig. 16.5a). Similar ethylene complexes are formed with a variety of other ligands and for Pd as well. The dimeric species $[C_2H_4PdCl_2]_2$ rapidly gives CH_3CHO and Pd in water. When linked with $CuCl_2$ to reoxidize the Pd, this becomes the basis of the *Wacker process* for converting ethylene to acetaldehyde:

$$C_2H_4 + PdCl_2 + 3H_2O \rightarrow Pd + CH_3CHO + 2H_3O^+ + 2Cl^-$$

$$Pd + 2CuCl_2 \rightarrow PdCl_2 + 2CuCl$$

$$2CuCl + 2H_3O^+ + 2Cl^- + \tfrac{1}{2}O_2 \rightarrow 2CuCl_2 + 3H_2O$$

which adds up to $C_2H_4 + \frac{1}{2}O_2 \rightarrow CH_3CHO$. The process is viable as the reaction between Pd and $CuCl_2$ is quantitative and efficient, so only low Pd concentrations are required.

One or two examples do exist of coordination numbers other than four in the II complexes: $Pd(diars)_2Cl^+$ for example is a trigonal pyramid and $Pt(NO)Cl_5^{2-}$ is octahedral (though the latter might be formulated as an NO^- compound of Pt^{IV} rather than as the NO^+ compound of Pt^{II}). The $SnCl_3^-$ complexes also provide examples. Platinum(II) forms a square-planar ion $PtCl_2(SnCl_3)_2^{2-}$ (as does ruthenium) but it also gives the ion $Pt(SnCl_3)_5^{3-}$ which has a trigonal bipyramidal $PtSn_5$ skeleton. This species reacts with hydrogen under pressure to form the hydride ion $HPt(SnCl_3)_4^{3-}$. Palladium forms the ion $PdCl(SnCl_3)_2^-$, but it is not yet known if this is a monomer or forms a chloride-bridged dimer which would be square-planar around the palladium. Such compounds are commonly precipitated as salts of

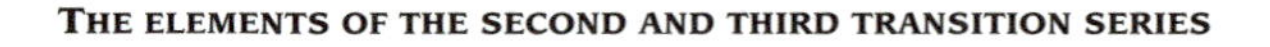

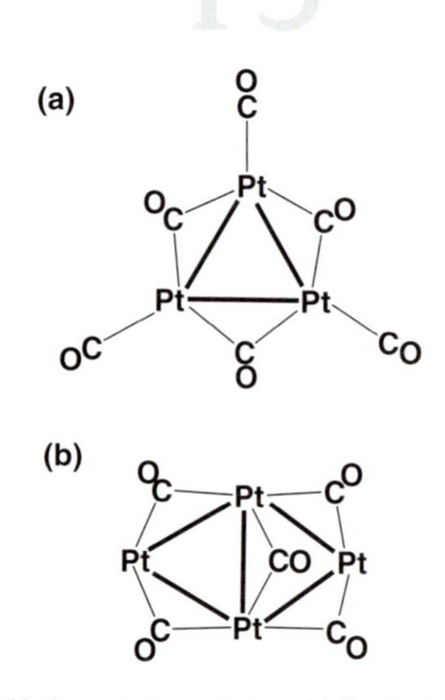

FIG. 15.30 Some platinum clusters. (a) The basic building unit of $[Pt_3(CO)_6]_n$ (known for n up to 12), (b) the $Pt_4(CO)_5$ skeleton of $[(R_3P)Pt]_4(CO)_5$

very large cations (like Ph_4As^+) from acid chloride solutions containing palladium, or platinum, and tin. Compounds with analogous M-Ge bonds, such as $(R_3P)_2M(GePh_3)_2$, are also well known.

The I state is uncommon and the established examples are M-M bonded dimers. This probably occurs in the reported $[Pd_2(CN)_6]^{4-}$ and has been found crystallographically in $[(Me_3P)_3Pd\text{-}Pd(PMe_3)_3]^{2+}$. The structure shows a square-planar arrangement around each Pd, with the two planes mutually perpendicular. On heating to 100°C, a Me group cleaves from a phosphane and the Pd(II) complex $[(Me_3P)_3PdMe]^+$ is formed. Platinum(I) is found in the Pt-Pt bonded carbonyl anion dimer, $[PtCl_2(CO)]_2^{2-}$, again with square-planar coordination.

In the 0 state, there are no stable carbonyls analogous to $Ni(CO)_4$, though $M(CO)_n$ species are formed in low-temperature matrix studies. The isoelectronic $M(CN)_4^{4-}$ ions, and $M(PF_3)_4$ species, do exist and CO-containing molecules occur such as $Pt(CO)_n(PR_3)_{4-n}$ ($n = 1$, 2). There are also reports of $Pt(NH_3)_5$ and $Pt(en)_2$ analogous to the iridium compounds. It has also been shown that the phosphane complexes, $(Ph_3P)_4Pt$ and $(Ph_3P)_3Pt$, are derivatives of Pt^0 rather than hydrides of higher states as originally reported. The structure of $Pt(PPh_3)_3$ has been shown to contain the planar PtP_3 group.

A plane trigonal unit, PdH_3, is found in the unusual species $NaBaPdH_3$ formulated as a Pd(0) complex hydride. The $Na^+\cdots H$ distance is short, suggesting further interactions (compare Section 9.4 for the unusual metallic hydride formed by palladium).

Platinum forms a number of irregular clusters containing carbonyls. The family of polymetallic 2-carbonyl anions of the general type, $[Pt_3(CO)_6]_n^{2-}$, are built up of the basic unit shown in Fig. 15.30a. The Pt_3 triangles stack up above each other in twisted trigonal prisms, and species with n up to 12 have been characterized. The cluster of Fig. 15.30b is found in $Pt_4(PR_3)_4(CO)_5$. The four Pt atoms, each with a terminal PR_3, form a tetrahedron which has five out of the six edges bridged by CO groups. See also Section 16.8.

15.9 Silver, $4d^{10}5s^1$, and gold, $5d^{10}6s^1$

The major part of silver chemistry is that of the d^{10} state, Ag(I). Gold is most often found in Au(III) square-planar complexes, but there are also a variety of Au(I) compounds, mostly complexes. Simple compounds of gold readily give the element, and silver is also reduced to the element fairly readily. These elements differ quite widely in their chemistries, and also differ from copper which occurs in the II state.

> Reasons for this were discussed in the copper section (Section 14.9).

The elements are found uncombined and in sulfide and arsenide ores. They may be recovered as cyanide complexes which are reduced to the metal, in aqueous solution, by the use of zinc. Gold is inert to oxygen and most reagents but dissolves in HCl/HNO_3 mixture (aqua regia) and reacts with halogens. Silver is less reactive than copper and is similar to gold, except that it is also attacked by sulfur and hydrogen sulfide.

It is easiest to treat the two elements separately as there is little resemblance in their chemistries.

15.9.1 Silver

There are a limited number of Ag^{III} compounds; however, these are of potential interest because silver(III) is isoelectronic with copper(III) (d^8), and because of the occurrence of both copper(II) and copper(III) in superconducting oxides (Section 16.10) it has been proposed that silver(II)–silver(III) fluorides might also show superconductivity.

Oxidation in basic solutions gives $[Ag(IO_6)_2]^{7-}$ or $[Ag(TeO_6)_2]^{9-}$, and copper(III) compounds are obtained similarly. Fluorination of a mixture of alkali and silver halides gives M^IAgF_4, and red, diamagnetic AgF_3 may be precipitated by addition of a fluoro Lewis acid to AgF_4^- solutions in anhydrous HF:

$$AgF_4^- + L \longrightarrow AgF_3 + LF^-$$
$$\text{(examples of L include } BF_3, PF_5 \text{ or } AsF_5)$$

The square-planar AgF_4^- also reacts with the silver(II) ion AgF^+, to give the maroon mixed-oxidation state compounds $Ag^{II}Ag^{III}F_5$ which, with AgF_3 gives $Ag^{II}Ag_2^{III}F_8$. This latter compound is also formed by the thermal decomposition of AgF_3. Solvated silver(III) is an extremely powerful oxidizing agent, even more powerful than krypton(II).

Electrolytic oxidation gives Ag_2O_3 which has the same structure as Au_2O_3, i.e. square-planar MO_4 units linked by shared corners into a network. A similar oxidation in base yielded square-planar $[Ag(OH)_4]^-$. The mixed Ag(II)–Ag(III) oxide, Ag_3O_4, has also been isolated from anodic oxidation work. All silver(III) compounds are strongly oxidizing.

When Ag_2O is oxidized by persulfate, black AgO is obtained. This is a well-defined compound which is strongly oxidizing. It is diamagnetic, which excludes its being an Ag^{II} compound which would contain the paramagnetic d^9 configuration. A neutron diffraction study has shown the existence of two units in the lattice, linear O-Ag-O and square-planar AgO_4 groups. This strongly suggests a formulation as $Ag^IAg^{III}O_2$. The square-planar configuration is expected for the d^8 Ag^{III} while the linear one is common for d^{10} as in Ag^I (compare Ag^I with Cu^I or Au^I which both show linear configurations and compare the III state with Au^{III}). AgO is stable to heat up to 100°C and it dissolves in acids, giving a mixture of Ag^+ and Ag^{2+} in solution and evolving oxygen. Reaction of Cl_2 with $CsAgCl_2$ gives the analogous mixed-oxidation chloride $Cs_2Ag^IAg^{III}Cl_6$. The structure shows square-planar Ag^{III} (Ag-Cl = 221 pm) and linear Ag^I (Ag-Cl = 227 pm).

The only simple compound of silver(II) is the fluoride, AgF_2, which is formed from the action of fluorine on silver or AgF. It is a strong oxidizing or fluorinating agent and can be used in reversible fluorinating systems in a similar way to CoF_3. AgF_2 occurs in two forms: $Ag^IAg^{III}F_4$ is a disproportionated, high-temperature form, while $Ag^{II}F_2$ is a nondisproportionated low-temperature form, which has a blue colour. The square-planar AgF_4^{2-} ion is also known. Salts containing the AgF^+ ion are well known, such as $AgF[BF_4]$, which has an infinite linear chain -Ag-F-Ag-F-Ag-F-.

Silver ions of the II state can exist in solution but only transiently. They are produced by ozone on Ag^+ in perchloric acid. The potential for $Ag^{2+} + e \rightarrow Ag^+$ has been measured as 2.00 V in 4M perchloric acid. This makes Ag^{2+} a much more powerful oxidizing agent than permanganate or ceric. Silver also gives a number of complexes in the II state. These are usually square-planar and paramagnetic corresponding to the d^9 state, and known structures, for example, of $Agpy_4^{2+}$, are isomorphous with the Cu^{II} analogues. Other examples include the cations $Ag(dipy)_2^{2+}$ and $Ag(phen)_2^{2+}$. These cations are stable in the presence of nonreducing anions such as nitrate, perchlorate or persulfate. The interesting species $Ag(MF_6)_2$ (M = Nb or Ta) consists of a central AgF_6 octahedron sharing three *cis* F with MF_6 octahedra. The AgF_6 octahedron has two long and four short Ag-F bonds consistent with d^9 Ag(II).

The normal oxidation state of silver is Ag^I and the chemistry of this state is already familiar, for example from the precipitation reactions of Ag^+ in qualitative analysis. Salts are colourless (except for anion effects) and generally insoluble apart from the nitrate, perchlorate and fluoride Structures are usually similar to those of the alkali metal equivalents, for example the silver halides have the sodium chloride structure. As discussed in Chapter 5, their lattice energies are higher because the Ag^+ ion is more polarizable, hence the lower solubilities.

A highly important application of Ag(I) salts is in photography. The basic process is the activation by light of Ag centres in the AgBr film, and these nucleate the formation of Ag metal particles when reduction (development) occurs. The unreacted salt is removed (fixing) by forming soluble Ag^+ complexes, typically sulfur species. Colour photography is based on the same general process with the interposition of dyes to filter out the three primary colours.

Silver(I) forms a wide variety of complexes. With ligands which do not π-bond, the most common coordination is linear, 2-coordination as in $Ag(NH_3)_2^+$, while π-bonding ligands give both 2- and 4-coordinate complexes and 3-coordination is found for some strongly π-bonding ligands such as phosphanes.

A number of very interesting complexes of silver(I) (and other late transition metals) with alkyl halides have been described recently, as detailed in the review given in Appendix A. Solvents such as dichloromethane and 1,2-dichloroethane are often used in synthetic inorganic chemistry because of their noncoordinating nature. However, the recent isolation of complexes of the type $[Ag(CH_2Cl_2)(OTeF_5)]$ and $[AgNO_3(CH_2I_2)]$, containing *coordinated* alkyl halides, clearly points to the term 'noncoordinating' now being of limited relevance. The isolation of such complexes is even more remarkable given that the silver(I) ion is a very powerful halide-abstracting group and the isolated halocarbon complexes may therefore be thought of, in some respects, as 'frozen intermediates' in the abstraction process.

The reader is referred to Sections 17.7.3 and 17.9.4 for further chemistry involving the $OTeF_5$ group.

It is noteworthy that ions containing the poorly coordinating $OTeF_5$ group have been successfully employed in the isolation of all of these unusual halocarbon silver species, and this suggests that a range of other novel complexes should be accessible using this type of group.

A formal 1/2 oxidation state is found in Ag_2F. This has a layer lattice of the anti-CdI_2 type and the Ag-Ag distance is similar to that in silver metal, which accounts for the low formal oxidation state. The properties are metallic.

15.9.2 Gold

The highest oxidation state found for gold is the V state, which has the d^6 configuration. Thus Au(V) corresponds to Pt(IV). Powerful oxidation of AuF_3 with F_2 and XeF_2 gives AuF_6^- as the $Xe_2F_{11}^+$ salt (compare Section 17.9). The AuF_6^- ion is a slightly distorted octahedron linked by long weak Au-F···Xe bonds to the two Xe atoms of the $Xe_2F_{11}^+$ ion. Other cations, including the larger alkali metals and O_2^+, stabilize the AuF_6^- ion. The KrF_2AuF_5 species has a stronger bridge and is best seen as $FKr\text{-}FAuF_5$ with a more covalent interaction. AuF_5 is formed by gentle heating of the krypton compound. The pentafluoride is extremely reactive and a violent fluorinating agent. It decomposes to AuF_3 and F_2 at 200°C and forms $HAuF_6$ which melts at 88°C. Among unusual adducts are $BrF_5.AuF_5$ which is oxidized by KrF_2 to $BrF_7.AuF_5$ although free BrF_7 does not exist (Section 17.8).

Gold(III) is the most common oxidation state of gold in compounds. The interrelations are illustrated in Fig. 15.31 starting off from the solution in aqua regia. The simple compounds readily revert to the element but the complexes are more stable. Square-planar 4-coordination is the most common shape. $AuCl_3$ and $AuBr_3$ form planar bridged dimers (as R_2AuX below with R = X) while AuF_3 forms a polymer made up of *cis*-bridged AuF_4 units i.e. $(AuF_2F_{2/2})_n$. There is evidence for a few 6-coordinate complexes such as $AuBr_6^{3-}$. Examples of the tendency to 4-coordination are provided by the alkyls such as R_2AuX. These contain σ Au-C bonds which are among the most stable metal–carbon bonds in the transition block. The halides have dimeric structures:

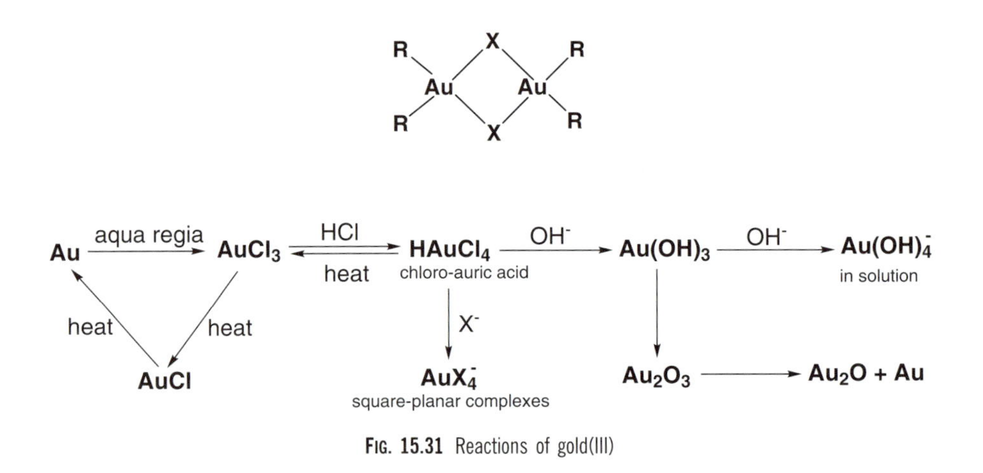

FIG. 15.31 Reactions of gold(III)

While, when X = CN, tetrameric structures are found:

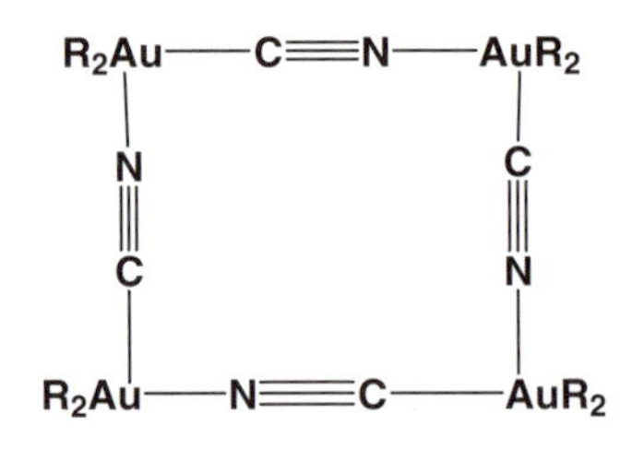

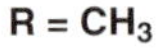

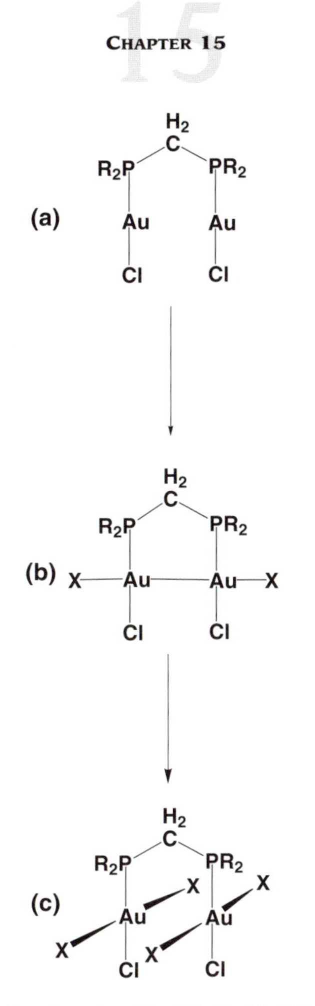

FIG. 15.32 Transformation of the R_2P-CH_2-PR_2 bridged digold compound. The short 'bite' of the bidentate phosphane holds the two Au atoms close together in a rigid conformation. (a) Two gold(I) centres linked through the diphosphane. (b) Oxidation forms the Au-Au bond giving gold(II). (c) Further oxidation gives two gold(III) centers linked through the diphosphane

Formal Au(II) results from Au-Au bonding in the same way as Pt(III) in the last section (see Fig. 15.32). Ligands with two donor atoms close together, such as $Ph_2PCH_2PPh_2$ bridge two Au atoms holding them close together. Then stepwise oxidation by successive additions of halogen converts the Au from (I) through (II) with the Au-Au bond to (III). However, relatively few examples of gold(II) complexes have been prepared, and good σ donors and π acceptors (such as dithiolate ligands) are required to stabilize the mononuclear, paramagnetic (d^9) gold(II) centre. The Au^{2+} ion has recently been generated in a solid matrix by heating gold(III) fluorosulfate, $Au(OSO_2F)_3$, or in solution by the reduction of the same compound with gold metal in $HOSO_2F$ solution.

Gold(I) is less stable but forms the simple halides and Au_2O. AuI and AuCl have similar chain structures

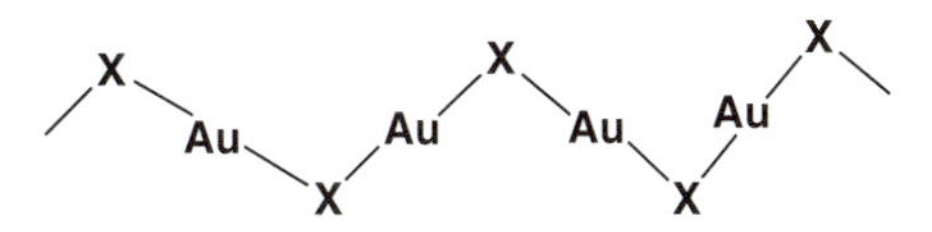

Complexes are usually linear, e.g. $[AuX_2]^-$ (X = halogen, CN, etc.) or R_3PAuX. X includes CH_3 as well as halides, etc.

The R_3PAu^- group itself forms a range of complexes, often of high coordination number, for example $R_3PAuV(CO)_6$ or $(R_3PAu)_3Re(CO)_4$. The long-known carbonyl Au(CO)Cl can now be readily made and is a valuable source of Au(I) complexes since the CO is very weakly bonded. Planar 3-coordinate Au(I) is found in $(R_3P)_2AuCl$ and $[(R_3P)_3Au]^+$.

Gold(I)–sulfur compounds are of interest as photosensitizers of silver halide layers in photography. The hydrogensulfido complex $[Au(SH)_2]^-$ has recently been prepared. It is suggested that this species is responsible for the transport of considerable quantities of gold in hydrothermal ore solutions.

Au has a relatively high electron affinity (compare Table 2.9) and Au(-I) has the $d^{10}s^2$ configuration which is fairly stable as found in Hg(0) (compare relativistic effects, Section 16.13). A very striking consequence is to have evidence for the Au^- ion. The first indications appeared half a century ago, particularly the observation that CsAu has relatively little metallic character and is best formulated as a salt, Cs^+Au^-, containing the *auride* Au^- anion, rather than as an alloy. In line with this the compound crystallizes in the caesium chloride structure, rather than the more densely close-packed structures which are usually adopted by metals. However, there has been recent discussion as to the relative contributions of ionic and covalent character in the bonding in CsAu and an equal contribution from both of these bonding types is currently the most favoured opinion. A range of other derivatives containing anionic gold has been prepared in liquid ammonia or ethylenediamine (compare Section 6.7), and isolated in complexes with various large cations, especially (crypt)M^+. The solid-state mixed-metal oxide derivative Cs_3AuO has been prepared by reaction of Cs_2O with CsAu and this compound also contains anionic gold. A related nitride auride, Ca_3AuN, has also been described.

15.10 The zinc Group

The zinc Group does not fit the general picture of the transition Groups as developed in the last two chapters. It shares with beryllium the property of belonging to one block of the Periodic Table and having many of the properties characteristic of another. In this case, the three elements of this Group resemble the three heavy elements of the boron Group.

The elements, zinc, cadmium and mercury, in this Group have the outer electronic configuration $d^{10}s^2$ and have the common oxidation state of II, corresponding to the loss of the two s electrons. In addition, mercury shows a well-established I state and cadmium and zinc form analogous but very unstable I compounds. Thus the heaviest element is more stable than the lighter ones in a low oxidation state, a characteristic of the main Groups. However, a recent theoretical study has suggested that HgF_4 should be thermodynamically stable or, at worst, only slightly endothermic with respect to gas-phase HgF_2 and fluorine,

The radii also reflect this transitional character. Shannon crystal radii for M(II) are (pm)

Coordination no.	4	6	8
Zinc	74	88	104
Cadmium	92	109	124
Mercury	110	116	128

The significant increases between successive Group members for the 4-coordinate radii is like that found for a main Group while the pattern of changes in 8-coordination is that typical of a transition Group. (See also relativistic effects, Section 16.13).

Mercury and many of its compounds are very poisonous and, as mercury has a relatively high vapour pressure at room temperature, mercury surfaces should always be kept covered to avoid vaporization.

Though the existence of the dimeric form of the mercury(I) state is without doubt, there is now also some evidence for the monomeric ion, Hg^+, both in solution and in solids.

whereas CdF_4 and particularly ZnF_4 are much less likely to be stable entities. It was suggested that a relativistic destabilization of the Hg(II)-F bonds, rather than a stabilization of HgF_4, is the reason for this and that the use of powerful fluorinating agents, such as KrF_2, might be suitable for the oxidative fluorination of HgF_2 to HgF_4.

The M^{II}/M redox potentials show a large difference between Cd and Hg which can be ascribed in part to the higher solvation energy of Cd^{2+}, which is an effect of smaller size. The potential values for $M^{2+} + 2e \rightarrow M$ are:

$$Zn = -0.762\ V,\ Cd = -0.402\ V,\ Hg = 0.854\ V$$

These values show the relatively high electronegativity of zinc and cadmium and reflect the reducing power of these elements. By contrast, mercury is unreactive and 'noble'.

The elements are all readily accessible as they occur in concentrated ores and are easily extracted. Zinc and cadmium are formed by heating the oxides with carbon and distilling out the metal (boiling points are 90°C for zinc and 767°C for cadmium). Mercury(II) oxide is decomposed by heating alone, without any reducing agent, at about 500°C, and the mercury distils out (b.p. = 357°C). Mercury is the lowest melting metal with a melting point of –39°C, while zinc and cadmium melt at about 420°C and 320°C respectively. Mercury is monatomic in the vapour, like the rare gases.

15.10.1. The I state and subvalent compounds

In the I state, mercury exists as the dimeric ion, Hg_2^{2+}. This has been demonstrated by a number of independent lines of evidence:

(i) The Raman spectrum of aqueous mercurous nitrate shows an absorption attributed to the Hg-Hg stretching vibration.

(ii) The crystal structures of a large number of mercury(I) salts show the existence of discrete Hg_2 units. The structure of Hg_2Cl_2 (Fig. 15.33) shows the presence of linear Cl-Hg-Hg-Cl units, but there are also longer Hg···Cl interactions which make the structure more complex. The stable organo-derivative RHgHgR [$R = (Me_3SiMe_2Si)_3Si$], however, has a discrete molecular structure, and confirms the existence of a linear Si-Hg-Hg-Si group.

(iii) All mercury(I) compounds are diamagnetic, whereas an Hg^+ ion would have one unpaired electron.

(iv) E.m.f. measurements on concentration cells of mercurous salts show that two electrons are associated with the mercurous ion. For example, the e.m.f. of the cell:

Hg	mercury(I) nitrate in (0.05 mol l^{-1}) 0.1 mol l^{-1} HNO_3	mercury(I) nitrate in (0.005 mol l^{-1}) 0.1 mol l^{-1} HNO_3	Hg

was found to be 0.029 V at 25°C. Now $E = \frac{RT}{nF}\ln(c_1)/(c_2)$, where the concentrations are c_1 and c_2, n is the number of electrons, and the other constants have their usual significance. If values are put in for the constants and conversion to logarithms to the base ten is carried out, the equation becomes $E = (0.059/n)\log(0.05/0.005)$: hence $n = 2$.

(v) Conductivities and equilibrium constants also fit for a dimeric ion of double charge and not for Hg^+.

The redox potentials involving mercury(I) are

$$Hg_2^{2+} + 2e^- \rightarrow 2Hg_{(liquid)} \quad E = 0.789\ V$$
$$2Hg^{2+} + 2e^- \rightarrow Hg_2^{2+} \quad E = 0.920\ V$$

It follows that, for the disproportionation reaction:

$$Hg + Hg^{2+} \rightarrow Hg_2^{2+}$$

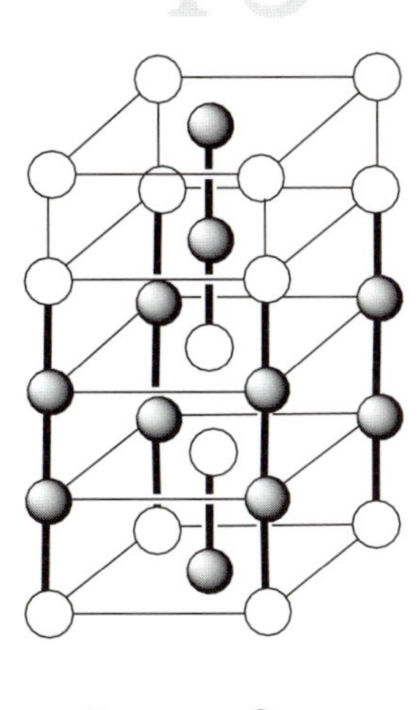

FIG. 15.33 The crystal structure of mercurous chloride, Hg_2Cl_2

AMALGAMS

The solution of a metal in mercury is given the specific name *amalgam* (an old alchemical term). For most metals, relatively dilute solutions are liquid, and thus useful for reactions by providing a fresh surface. For example, many reductions are conveniently carried out by shaking the reagents in solution with 5% sodium amalgam. Amalgams are also used for depositing a surface film, e.g. in gilding, when the mercury is removed by evaporation - though the toxicity of mercury vapour makes this procedure dangerous.

Amalgam is thus a specific term for alloys containing mercury.

the potential is $E = 0.131$ V. Then, as $E = (RT/nF) \ln K$, where K is the equilibrium constant for the reaction:

$$K = [(Hg_2^{2+}]/[(Hg^{2+})] = \text{about } 170$$

In other words, in a solution of a mercury(I) compound there is rather more than $\frac{1}{2}\%$ mercury(II) in equilibrium. Thus, if another reactant is present which either forms an insoluble mercury(II) salt or a stable complex, the mercury(II) ions in equilibrium are removed and the disproportionation goes to completion. If OH is added to a solution of Hg_2^{2+}, a grey precipitate of HgO mixed with Hg is formed. Similarly, sulfide precipitates HgS, and CN^- ions give a precipitate of mercury and the undissociated cyanide of mercury(II), $Hg(CN)_2$.

The existence of this disproportionation reaction means that a number of mercury(I) compounds, such as the sulfide, cyanide or oxide, do not exist. The main compounds of mercury(I) are the halides and a number of salts. Hg_2F_2 is unstable to water giving HF and HgO plus Hg. The other halides are very insoluble. Mercurous nitrate and perchlorate are soluble and give the insoluble halides, sulfate, acetate and other salts by double decomposition reactions. The nitrate and perchlorate are isolated as hydrates which contain the hydrated ion $[H_2O\text{-}Hg\text{-}Hg\text{-}H_2O]^{2+}$.

A three-atom chain exists in $Hg_3(AlCl_4)_2$ where the basic structure is a Z-shaped molecule with Cl bridging between approximately tetrahedral $AlCl_4$ units and approximately linear Hg-Hg-Hg chains.

In the reaction with fluoro-arsenic or antimony species, a number of poly-mercury ions result. For example,

$$3Hg + 3AsF_5 \longrightarrow Hg_3(AsF_6)_2 + AsF_3$$
$$4Hg + 3AsF_5 \longrightarrow Hg_4(AsF_6)_2 + AsF_3$$
$$6Hg + 3AsF_5 \longrightarrow [Hg_{2.85}AsF_6]_n + AsF_3 \qquad \text{(reaction in } SO_2\text{).}$$

The crystal structures show a linear symmetric Hg_3^{2+} ion with Hg-Hg = 255.2 pm. In Hg_4^{2+}, the structure is centro-symmetric and nearly linear with angle 176°:

$$Hg \overset{257}{\text{---}} Hg \overset{270}{\text{---}} Hg \text{---} Hg$$

Infinite chains are found in $[Hg_{2.85}AsF_6]_n$ which contains chains of Hg atoms passing in two directions at right angles through a lattice of octahedral AsF_6^- ions. The average Hg-Hg distance is 264 pm. Similar mercury chains are found where the As is replaced by Nb or Hf, but such compounds change on standing to give golden crystals which contain planes of close-packed Hg atoms separated by layers of MF_6^- ions.

Alkali metals dissolve in mercury to form *amalgams* and clusters of mercury atoms have been identified in the solids isolated from such amalgams. All these clusters involve formal oxidation states for mercury of around I. CsHg, KHg and Na_3Hg_2 phases have all been shown to contain rectangular Hg_4 clusters. Even more striking is $Rb_{15}Hg_{16}$ which contains slightly distorted Hg_4 squares and Hg_8 cubes (angles, 86–94°; Hg-Hg distances, 294–298 pm, except for two edges of the square at 304 pm). A parallel is to be drawn with the gold clusters discussed in Section 16.8.

Mercury(II) and (I) carbonyl cations have been isolated in liquid antimony fluoride from Hg^{2+} and CO at 100°C. The $[Hg(CO)_2]^{2+}$ and $[Hg_2(CO)_2]^{2+}$ ions were isolated as the $[Sb_2F_{11}]^-$ salts, following a similar preparation of the analogous $[Au(CO)_2]^+$ cation. While neutral binary carbonyls are not known for the heavy elements to the right of the d block (compare Section 16.2), reasonably stable M-CO species have now been isolated as ions, or in presence of additional ligands, for most of these elements.

The existence of a cadmium(I) ion, Cd_2^{2+}, has been conclusively proved in a molten halide system, $CdAlCl_4$ is obtained as a yellow solid and shown to be $Cd_2^{2+}(AlCl_4)_2^-$. This reacts violently with water to give cadmium metal and Cd^{2+} in solution. Cd^I may also exist in the deep red melts of Cd in cadmium(II) halides. A recent report indicates that zinc(I), Zn_2^{2+}, exists under similar conditions.

15.10.2 The II state

The most stable state for all three elements is the II state. In this, zinc and cadmium resemble magnesium and many of the compounds are isomorphous. Mercury(II) compounds are less ionic and its complexes are markedly more stable than those of zinc and cadmium. All three elements resemble the transition elements more than the main Group elements in forming a large variety of complexes.

The halides of all three elements are known. All the fluorides, MF_2, are ionic with melting points above 640°C. HgF_2 crystallizes in the fluorite lattice and is decomposed in contact with water. The structures of ZnF_2 and CdF_2 are unknown: these compounds are stable to water and are poorly soluble, due both to the high lattice energy and to the small tendency of the fluorides to form complex ions in solution. In this the zinc and cadmium fluorides resemble the alkaline earth fluorides. The chlorides, bromides and iodides of zinc and cadmium are also ionic, although polarization effects are apparent and they crystallize in layer lattices. The structures are approximately close-packed arrays of the anions with Zn^{2+} in tetrahedral sites in the zinc halides while, in the cadmium halides, Cd^{2+} ions occupy octahedral sites. The zinc and cadmium halides have lower melting points than the fluorides and are ten to thirty times more soluble in water. This is due, not only to lower lattice energies, but also to the ready formation of complex ions in solution. A variety of species result, especially in the case of cadmium halides. Thus a 0.5 mol l^{-1} solution of $CdBr_2$ contains Cd^{2+} and Br^- ions and, in addition, $CdBr^+$, $CdBr_2$, $CdBr_3^-$ and $CdBr_4^{2-}$ species, the most abundant being $CdBr^+$, $CdBr_2$ and Br^- (these species are probably hydrated). Hydrolysis also occurs and species such as Cd(OH)X are observed. The tetrahalides, ZnX_4^{2-} and CdX_4^{2-}, may be precipitated from solutions of the halides in excess halide by large cations. These are tetrahedral ions, as are all 4-coordinated species in this group.

By contrast, $HgCl_2$, $HgBr_2$ and HgI_2 are covalent solids melting and boiling in the range 250°C–350°C. $HgCl_2$ is a molecular solid with two Hg-Cl bonds of 225 pm and the next shortest Hg-Cl distance equal to 334 pm, so that there is little interaction between the mercury atom and these external chlorines. In the bromide and iodide, layer lattices are formed, but in the bromide, two Hg-Br distances are much shorter (248 pm) than the rest (323 pm) so that this is a distorted molecular lattice. HgI_2 forms a layer lattice in which there are HgI_4 tetrahedra with the Hg-I distance equal to 278 pm. In the gas phase, the mercury halogen distances in the isolated HgX_2 molecules are, for X = Cl, 228 pm; X = Br, 240 pm; X = I, 275 pm. Thus the Hg-Cl distances are the same in the solid and gas, underlining the molecular form of the solid. The Hg-Br distance is a little longer in the solid, while the Hg-I distance is markedly longer in the solid, showing the increasing departure from a purely molecular solid on passing from the chloride to the iodide. Mercury also forms halogen complexes, and the same species are found for mercury as for cadmium. The stability constants for the mercury complexes are much higher than those for the zinc and cadmium species. Halo-mercury ions are found, e.g. $(HgI)^+$ which exists as infinite chains, with I Hg I angles near linear and HgIHg angles around 90°. In presence of a large cation, the $HgCl_5^{3-}$ ion may be isolated, whose shape is a trigonal bipyramid with short axial Hg-Cl distances of 233 pm and long Hg-Cl equatorial distances of 303 pm. This can be seen as a linear Cl-Hg-Cl unit weakly coordinated by three further Cl^- ions.

THE ACTION OF CRYPTAND 222 ON MERCURIC HALIDES

The strongly coordinating ligand, cryptand 222 (see Fig. 10.6b and box in Section 10.6) has an interesting effect on mercury halides. Part of the mercury is coordinated as the very large cation, $[Hg(crypt\ 222)]^{2+}$ and this stabilizes complex anions $[Hg_2X_6]^{2-}$ and even $[Hg_3I_8]^{3-}$. The anion structures are based on tetrahedral HgX_4 units with halogen bridges.

Such an experiment could be extensively developed. For example, a cryptate with S donor atoms in place of the O would bond even more strongly to Hg (compare Table 13.14), possibly leading to sequestration of a higher proportion of Hg as cations and opening up possibilities of more complex iodo-mercury anions.

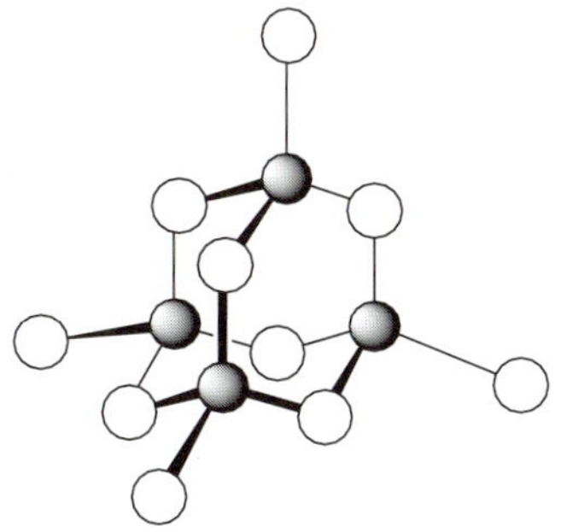

The oxides are formed by direct combination. ZnO is white and turns yellow on heating. CdO is variable in colour from yellow to black. The colours in both cases are due to the formation of defect lattices, where ions are displaced from their equilibrium positions in the crystal lattices to leave vacancies. These may trap electrons whose transitions give rise to colours in the visible region. HgO is red or yellow, depending on the particle size. Zinc oxide and the hydroxide are amphoteric. $Zn(OH)_2$ is precipitated by the addition of OH^- to zinc solutions and dissolves in excess alkali. $Cd(OH)_2$ is precipitated similarly but is not amphoteric and remains insoluble in alkali. Mercury(II) hydroxide does not exist; the addition of alkali to mercuric solutions gives a precipitate of the yellow form of HgO. The elements all form insoluble sulfides and these are well known in qualitative analysis. ZnS is somewhat more soluble than CdS and HgS, and has to be precipitated in alkali rather than the acid conditions under which yellow CdS and black HgS precipitate.

Most of the oxygen Group compounds of Zn, Cd and Hg have the metal in tetrahedral coordination in the zinc blende structure (see Fig. 5.1c) or in the related wurtzite structure (Fig. 5.4a). Both these are found for ZnS, with the wurtzite form the stable one at high temperatures. ZnO, ZnS, ZnTe, CdS and CdSe all occur in both the wurtzite and the zinc blende forms. ZnSe, CdTe, HgO, HgSe and HgTe are found in the zinc blende form only. HgS occurs in two forms, one is zinc blende and the other is a distorted NaCl lattice. The only other example of 6-coordination is CdO which forms a sodium chloride lattice. These structures again illustrate the strong tendency for these elements to form tetrahedral coordination. A growing range of sulfides and selenides of cadmium and zinc have been prepared which are polynuclear species containing segments of the tetrahedral ZnS structure. A small unit is seen in the Cd_4Se_{10} skeleton of $[(RSe)_6(CdX)_4]^{2-}$. Figure 15.34 shows the structure for X = SeR (note the SeR groups occupy both terminal and bridging positions). Other terminal groups, such as X = halogen, are found, SR may replace SeR, and Zn may replace Cd. The basic skeleton is the same structure as P_4S_{10} (Fig. 17.32b).

Cd Se

FIG. 15.34 The structure of the Cd_4Se_{10} skeleton of the $[Cd_4(SeR)_{10}]^{2-}$ ion

In $[S_4Cd_{10}(SR)_{16}]^{4-}$, four of these units are fused together, see Fig. 15.35. Here, four of the S atoms are bonded only to Cd, six edges are bridged by SR, and the remaining SR are terminal on Cd. Again, Zn and Se analogues are known. Even larger polytetrahedral clusters are known, and the ultimate product of successive fusion is the sulfide. Such compounds may reflect the coordination of Cd in biomolecules.

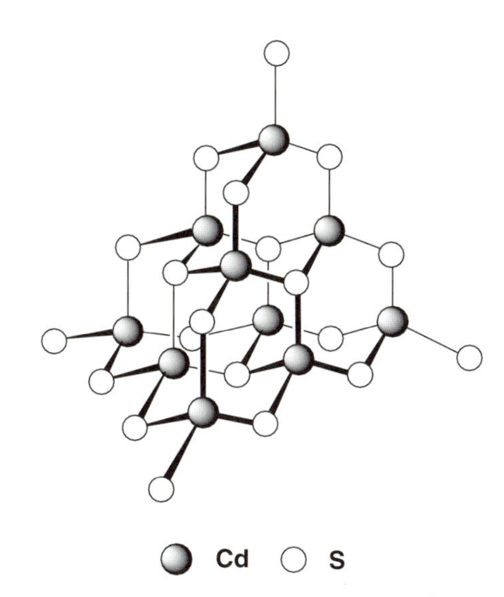

FIG. 15.35 The $Cd_{10}S_{20}$ skeleton of $[S_4Cd_{10}(SR)_{16}]^{4-}$

This formation of 4- and 6-coordinated species is also found in the complexes of these elements. Zinc occurs largely in 4-coordination in complexes like $Zn(CN)_4^{2-}$ or $Zn(NH_3)_2Cl_2$ and is also found in rather unstable 6-coordination, as in the hexahydrate and the hexammine, $Zn(NH_3)_6^{2+}$. Cadmium forms similar 4-coordinated complexes but is rather more stable than zinc in 6-coordination, due to the larger size. Mercury is commonly found in 4-coordination, though a few octahedral complexes such as $Hg(en)_3^{2+}$ are also found.

All three elements are also found in linear 2-coordination especially in their organometallic compounds, and in the halides and similar compounds. The organic compounds R_2M (M = Zn, Cd or Hg) are well known and mercury also forms RHgX compounds wih halides. The so-called RZnX and RCdX species are, like the Grignard reagents (RMgX), more complex and their structures are not fully understood. They are polymeric with some evidence for MX_2 and MR_2 groups and are usually coordinated by the ether used in their preparation.

Problems

The best way to build up a clear picture of systematic chemistry is to correlate one body of facts in as many ways as possible. Look at the chemistry of neighbouring elements, of other oxidation states of the same element, of the same d^n configuration, of ions of the same charge, etc. The problems below are a guide to many similar ones which you can devise.

Many such questions can be derived. Surveys should start with the first transition series in Chapter 14 and include material from Chapters 9 and 13. Many should be broadened to the heavy transition elements in Chapter 15, and often to the rest of the Periodic Table (see Chapters 10 to 12 and 17). The special topics of Chapters 16 and 20 will also often be relevant.

When consulting alternative sources on systematic chemistry, you should check out 'Notes to the reader' at the beginning of Appendix A, Further Reading, together with the comments on the different sections of this list.

15.1 Carry out, for the transition elements of the second and third series, the exercises given the Questions 14.1 and 14.2.

15.2 For each element of the second transition series in turn, compare and contrast the chemistry with that of the other two elements in the same Group.

15.3 Discuss the lanthanide contraction and its effects on transition metal chemistry.

15.4 Continue general surveys along the lines suggested for Chapter 14 Problems, e.g. Question 14.5.

CHAPTER 16

16.1 COPPER OXIDE CERAMIC SUPERCONDUCTORS 391
16.2 CARBONYL COMPOUNDS OF THE TRANSITION ELEMENTS 395
16.2.1 Formulae 395
16.2.2 Bonding 396
16.2.3 Structures 397
16.2.4 Related species 398
16.3 METAL–ORGANIC COMPOUNDS 399
16.3.1 Metal–Carbon σ Bonding 399
16.3.2 Metal–Carbon Multiple Bonding 401
16.4 π-BONDED CYCLOPENTADIENYLS AND RELATED SPECIES 402
16.5 THE ORGANOMETALLIC CHEMISTRY OF THE LANTHANIDES 403
16.6 ACTINIDE ORGANOMETALLIC CHEMISTRY 405
16.7 MULTIPLE METAL–METAL BONDS 406
16.7.1 Bonding 408
16.8 TRANSITION METAL CLUSTERS 409
16.8.1 Halide clusters 409
16.8.2 Carbonyl and related clusters 409
16.8.3 Gold clusters 411
16.8.4 Very large clusters 412
16.9 METAL–DIOXYGEN SPECIES 413
16.10 COMPOUNDS CONTAINING M-N_2 UNITS AND THEIR RELATIONSHIP TO NITROGEN FIXATION 415
16.10.1 Metal compounds with Coordinated dinitrogen species 416
16.10.2 Bonding between N_2 and M 418
16.11 METAL–DIHYDROGEN COMPLEXES 419
16.11.1 Discovery 419
16.11.2 Properties and bonding 420
16.12 POST-ACTINIDE 'SUPERHEAVY' ELEMENTS 421
16.12.1 Background 421
16.12.2 Synthesis and properties 421
16.13 RELATIVISTIC EFFECTS 424
PROBLEMS 426

Transition Metals: Selected Topics

The preceding three chapters have covered the basic themes of transition metal chemistry. In this chapter we select a few topics—each where work is proceeding actively—to give a further indication of the transition chemist's current interest. This selection is necessarily arbitrary and many areas of interest have been omitted. The treatment of each is limited to an introductory review.

Readers may wish to omit this chapter at first reading.

16.1 Copper oxide ceramic superconductors

The headlong development of superconductors with critical temperatures above the boiling point of liquid nitrogen (see Box titled 'Warm' superconductors) produced many variants of the original formulation, and much initial confusion over reported properties. Judged by two reasonably strict criteria that the resistance should drop to zero sharply over a range of only a degree or so, and that the sample should show the Meissner effect in which it expels magnetic field lines (most spectacularly, a small sample floats in a magnetic field)—there are two well-established classes of superconductors. One is based on the original species, reformulated as $La_{2-x}Ba_xCuO_{4-y}$, where x is 0.15 to 0.2, and y is undefined but small. These are cases of the established La_2CuO_4 layer perovskite with the K_2NiF_4 structure where up to 10% of the La^{3+} is replaced by Ba^{2+}, and there is a small oxygen deficiency. Of course, the changes were immediately rung on the constituents, especially replacing Ba by Ca or Sr, or substituting lanthanides, and T_c was raised to around 50 K very quickly.

The second system was of general formula $YBa_2Cu_3O_{7-x}$, called the 1-2-3 type from the numbers of metal atoms. This compound also has a perovskite structure. It is superconducting when x is around 0.1, but semiconducting when x is 0.5 or more, and the value of T_c is 93 K. Analogues included substitution of Y by Ho, the lanthanide with the same radius but paramagnetic. Despite theories to the contrary, the presence of paramagnetic ions did not destroy the superconductivity and the holmium species has T_c of 91 K. Sm, Eu, Nd, Dy and Yb have also been substituted, so the M^{3+} radius is not critical. Values of T_c around 90 K are well established for this series, and the Meissner effect is exhibited. The great advantage is that T_c is above the temperature of liquid nitrogen, which boils at 77 K. The next discovery was of systems with Bi or Tl in place of the lanthanides which brought T_c up to around 120 K. The structure is a perovskite, related to those of Fig. 16.1, but with CuO and BiO or TlO layers as the major feature. As an example, the thallium compound $Tl(Ba,Ca)_2Ca_3Cu_4O_{10.5+\delta}$ has four consecutive copper layers and adding more layers suggests that this will increase T_c further. By growing an oriented thin film of a cuprate compound belonging to the BiSrCaCuO family onto a $SrTiO_3$ single crystal, a material having eight adjacent cuprate layers has been produced and studies suggest this material may have a T_c as high as 250 K. If such materials can be prepared in bulk form, then very interesting applications may arise from them.

The key to the superconducting properties is thought to lie in the copper chemistry, and to understand this we have to look at the structures in more detail. If we start with the perovskite structure of Fig. 5.5c, we need to insert a number of different metal atoms in the 12-coordinate site. Thus the unit cell will be stretched from the cubic one of the simple perovskite along one axis to accommodate several units with different metal atoms. By crystallographic convention, this long axis is the c axis, parallel to the z coordinate, so we turn the Fig. 5.5c structure through 90° and extend it by a number of units.

THE CHALLENGE TO PREPARE COPPER OXIDE SUPERCONDUCTORS

The 90 K superconductors are perovskite phases which are brittle solids. Most have been made only as powders, and much effort is going into the problems of finding forms which can be fabricated on a large scale. Because the rewards are large, the problems are likely to be solved faster than the pessimists' estimate of twenty years.

Out of all the syntheses, the main route has been to grind together the metal oxides or carbonates in stoichiometric amounts and then heat in air, cool, regrind, heat in oxygen, anneal in oxygen and finally cool. In a modification which allows easier control of composition, the carbonates or citrates are precipitated from a stoichiometric mixture of soluble salts such as the nitrates. One preparation then continues: 'dry at 140°C overnight, heat in air at 950°C for 24 h, cool in air, grind and form into pellets and reheat in a flow of oxygen at 950°C for 16 h, cool in oxygen to 200°C over 1 h and finally cool in air'. These details show how critical the oxygen content is. Another synthetic route which is attracting an increased amount of interest uses molecular metal–organic precursors containing a precise ratio of the different desired metals which can then be converted to the desired mixed-metal oxide. Metal alkoxide compounds are of particular interest in this area.

Recall that the atoms on the edges of the unit cell are shared between four cells, so each adds 1/4 to the contents of the cell.

The formula La_2CuO_4, on which the first class of superconductors is based, is created by adding an extra layer of O atoms to the perovskite formula, $CaTiO_3$. Some 10% of the La ions are replaced with Ba ions, and to compensate for this loss of charge, a corresponding number of Cu atoms change from Cu(II) to Cu(III). It is thought that the superconductivity is due to electron transfer between the two copper states, mediated by the oxygens, perhaps also involving copper(I), and assisted by the small oxygen deficiency. From general copper chemistry, we recall that Cu(II) has a tetragonally distorted coordination sphere.

For the 1-2-3 structure, we need a tripled perovskite structure. To arrive at a formula $YBa_2Cu_3O_7$ we insert a Y^{3+} ion in the centre cell, with Ba^{2+} ions in the two outer cells, and also remove the oxygen atoms in the plane of the yttrium (which reduces the composition to $YBa_2Cu_3O_8$ from the $(ABO_3)_3$ formula of the tripled perovskite). If we further remove two of the four oxygens in the top and bottom faces of the unit cell, we arrive at the stoichiometry $YBa_2Cu_3O_7$ which is that shown in Fig. 16.1a. The final superconductor stoichiometry $YBa_2Cu_3O_{7-x}$ results from removing between 10% and 50% of the remaining oxygens in the end faces, with 50% at the limit of the superconducting composition range. If we remove all the oxygens in these two faces, the structure of Fig. 16.1b results, for formula $YBa_2Cu_3O_6$, and this phase is only a semiconductor.

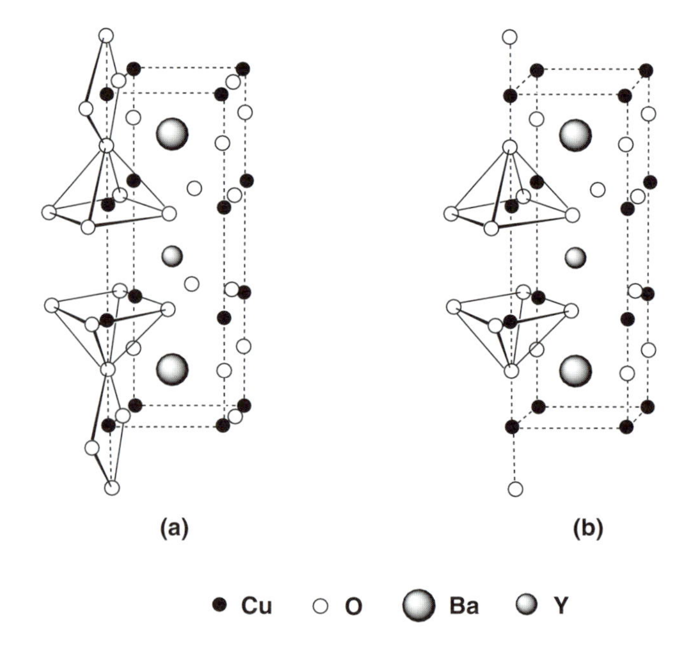

FIG. 16.1 Barium–yttrium–copper-oxide perovskite phases: (a) the structure of the superconductor $YBa_2Cu_3O_{7-x}$, (b) the structure of the semiconductor phase $YBa_2Cu_3O_6$

In the structure of $YBa_2Cu_3O_{7-x}$, there are two different copper positions. In those copper layers which lie between the Ba and the Y, the removal of the oxygens from the Y layer leaves the Cu 5-coordinated in a square pyramid. (Remember that the Ti atoms, which are at this position in the basic perovskite structure, are octahedrally coordinated to O.) The second copper site is in the end layers (i.e. between two layers of Ba in successive unit cells) and these Cu atoms are in square-planar 4-coordination in $YBa_2Cu_3O_7$, linear 2-coordination in $YBa_2Cu_3O_6$, and somewhere between these two in $YBa_2Cu_3O_{7-x}$.

From the stoichiometry, as neither Ba nor Y are likely to have variable charges, seven charges are shared among three Cu atoms in $YBa_2Cu_3O_7$. The square-planar positions are allocated to Cu(III), reasonable for the d^8 configuration, while the Cu(II) are allocated to the square-pyramidal positions, again reasonable in view of the usual distortions found for Cu(II). As x increases and O is removed, some of the square-planar Cu(II) sites are converted into trigonal, and eventually linear Cu(I) configurations (completely at the $YBa_2Cu_3O_6$ stoichiometry). Again, the superconductivity is thought to result from the ready transfer of electrons between the different copper oxidation states, mediated by the oxygens.

Since the current theory of superconductivity in the classical metal and alloy case involves electrons moving as pairs, the coexistence of Cu(I) and Cu(III) is thought to be important. However, there are many current suggestions and no generally accepted theory for the high-temperature superconductors at present.

These observations suggest that this type of superconductivity is particularly associated with copper. The combination of three accessible oxidation states, with different preferred coordination in each, and perhaps the tetragonal distortion associated with Cu(II) are important. If the rest of the transition elements are reviewed, this combination is unique. For example, while d^4 configurations are also distorted, choices such as Mn(III) or Cr(II) may be less good because the neighbouring oxidation states vary substantially in stability and do not show the same variation in preferred coordination. On the other hand, variations in the M^{2+} and M^{3+} species have minor effects, presumably as long as they are reasonably big. No doubt many other changes will be tried; S for O, doping of some M(I) sites with Ag or M(III) sites with Au.

Since the early discoveries of superconducting metal oxides, a steady increase in T_c has been observed in different systems. The maximum value now observed has exceeded 150 K. One interesting observation made recently is that an increase in T_c can be effected with an increase in pressure. When the compound $HgBa_2Ca_2Cu_3O_{8+\delta}$ is placed under a pressure of 150 000 atmospheres it shows a T_c of 153 K. It has been predicted that even higher pressures might increase T_c further. One way of mimicking the effects of increased pressure might be to replace the large barium atom by a somewhat smaller strontium atom. Of course, pressures such as these would rule out any direct applications within the forseeable future but, nevertheless, these studies serve to indicate that higher superconducting temperatures should be possible, given the right system.

'WARM' SUPERCONDUCTORS

Superconductivity was first discovered in mercury by Onnes in 1911, gaining him a Nobel Prize. When a metal such as Hg or Pb was cooled nearly to absolute zero, it was found to offer no resistance to electric current. This happens at a *critical temperature*, T_c, and work for the next 50 years led to a slow rise in the highest critical temperature found, up to around 20 K for alloys. By 1973, the T_c of Nb_3Ge at 23.7 K was the best that had been achieved, and similar but cheaper niobium–tin alloys, T_c about 22 K, were being used in magnet windings. The great advantage of superconductors is that, once a current has been created in a coil, it continues with no loss. Thus the first major application was in producing high-field electromagnets, used in particle accelerators, in nmr instruments (see Section 7.8), and in the related magnetic resonance imaging (MRI) instruments used in medicine. Unfortunately, the low temperatures needed to achieve criticality needed expensive liquid helium as a coolant. Large-scale commercial applications, such as power transmission, were out of the question.

In 1957, a theory explaining superconductivity was formulated by Bardeen, Cooper and Schrieffer (BCS). One feature of the BCS theory was the calculation that the supercurrent carriers (Cooper pairs of electrons) could not exist above about 30 K, thus placing an upper limit on the critical temperature. The situation in the 1980s was that a widespread search among metals, alloys, and also some mixed oxide phases, had failed to raise the critical temperature and the theory indicated an upper limit not far above that achieved.

In this situation, we can understand the great excitement which followed when Bednorz and Mueller, working in Switzerland in 1986, announced a mixed metal oxide, $La_{5-x}Ba_xCu_5O_{5(3-y)}$ with a T_c of 35 K. The species was a member of the perovskites, and turned attention away from metals and alloys in the quest for superconductors. The discovery created enormous interest, and reports of new records in T_c poured out from the USA, Japan, China and all round the world. Extra sessions on superconductivity were slipped into conferences, and short-notice seminars were filled to overflowing and continued into the night. The huge interest culminated in the award of the 1987 Nobel Prize in Physics to Bednorz and Mueller—the most rapid award ever.

Of course, in all the excitement and pressure, confusions arose. The preparation of the materials demands a specific regime to control the oxygen content, and lots of private 'recipes' flew about. Many samples were mixtures where only one phase was superconducting. In such systems, the resistance drops over a temperature range before true superconductivity sets in, and many of these high temperatures appeared as new 'records' in the popular press. One major landmark was passed—many of the new materials had critical temperatures above 77 K (−1196°C), which is the temperature of liquification of nitrogen. This is a much cheaper coolant than liquid helium and allows larger-scale applications. Continuing study of cuprate ceramic systems eventually gave an oxide of HgTlBaCaCu with the highest critical temperature to date, near 140 K.

Excitement died down as progress stalled and the difficulties became clearer. The behaviour of the copper oxide systems was critically dependent on the oxygen content, which was difficult to control, and could change even in samples held at room temperature. Thus different teams evolved different 'recipes' for preparations, and groups found it difficult to reproduce each other's results. Furthermore, the compounds were brittle ceramics and much ingenuity was demanded to develop methods of fabrication.

As critical temperatures rose, it was clear that the BCS theory did not apply to these cuprate materials, and a period of active theoretical speculation followed with several competing proposals. The experimental difficulties in preparing materials reproducibly meant slow progress in making critical tests of conflicting theoretical predictions. Matters are slowly clarifying and consensus is approaching. It seems generally agreed that superconductivity occurs primarily in the copper oxide layers in the structure (see Fig. 16.1) and valid theories reflect the fourfold symmetry of the crystal structure—as opposed to the spherical symmetry at the heart of BCS.

Thus, at the end of the 1990s, the field was settling down with a choice between classical metallic superconductors and the recent copper/metal oxide ceramics. The metals and alloys were well understood, relatively easy to fabricate but with critical temperatures below 30 K, demanding expensive liquid helium cooling. The cuprates were becoming understood, could use cheap liquid nitrogen coolant but were difficult to fabricate and thus extremely expensive materials.

At the same time, work on large-scale applications was actively pursued in several countries, particularly the USA and Japan. In 2001 several extended pilot scale projects were coming to fruition. For example, power was being sent via superconducting cables to 150 000 homes in Copenhagen, while, in a suburb of Detroit, three liquid nitrogen cooled superconducting power cables replaced 9 copper ones carrying power from the generating station to the locality. The primary advantage is avoiding the 10% loss of power in conventional transmission cables, but the higher capacity also has major benefits when the old system is up for replacement. Progress is continuing in other applications, such as the development of powerful superconducting motors.

Further classes of superconducting materials have been found. Superconductivity was discovered in another recent new family of compounds, the metal doped fullerenes (see Section 19.3), with critical temperatures up to 40 K. Then, at the turn of the millennium, a new observation was reported which started another wave of excitement. In 2000, Akimitsu in Japan found superconduction in MgB_2 (Fig. 17.7b) with a critical temperature of 39 K. Within six months, at an American Physical Society meeting, nearly 80 teams were reporting results on this material and related systems. As magnesium diboride is metallic, it reopens the potential for higher critical temperatures in easily fabricated materials. It is close to the classical superconductors, and most physicists suggest it can be fitted within the BCS theory. Later work developed methods of fabricating dense MgB_2 superconducting wire clad with iron. Other work has shown that bombarding with protons or allowing slight adsorption of oxygen increased the current carried.

Industrial scale superconducting cables with liquid nitrogen coolant are a major development, but there are still major difficulties of fabrication with copper oxide systems, while readily fabricated materials, like metal alloys or magnesium boride, require much more expensive liquid helium cooling. Progress is to be expected in raising the critical temperature in the more readily fabricated metallic systems, and in overcoming the fabrication problems in the higher critical temperature copper ceramics. As such materials come into wider use, they will make a large contribution to energy efficiency, and thus to alleviating global warming. The prospect of ambient temperature superconductors is distinctly more distant but not ruled out by present understanding.

16.2 Carbonyl compounds of the transition elements

A brief indication of the carbonyls of the individual elements is given in Chapters 14 and 15 in the discussions of the low oxidation states.

16.2.1 Formulae

Over a century ago, Mond found that nickel reacted with carbon monoxide to form an unusual volatile compound, $Ni(CO)_4$, which we call nickel carbonyl. $Ni(CO)_4$ is readily decomposed to Ni + 4CO by gentle heating. The formation and decomposition of $Ni(CO)_4$ became the *Mond process* for nickel refining which was used for many years (compare Section 14.8).

It has since been found that most transition metals form one or more carbonyl compounds. Although a number of specific syntheses are available, the preparation from metal or readily reducible compound and CO applies generally

$$M + nCO \rightarrow M(CO)_n \qquad (16.1)$$

though usually requiring raised temperatures and pressures. Table 16.1 lists the simpler carbonyls formed by each element. A very large number of more complicated species are reported, particularly for Re, Ru, Os, Co, Rh and Pt. The first M-CO compound to be reported was the mixed ligand $Pt(CO)_2Cl_2$.

The carbonyls, on the basis of Equation (16.1), are best regarded as forming from neutral CO and the metal in an oxidation state of zero. A convenient guide to the formulae is provided by the *18-electron rule*. Each CO provides 2 electrons and these, with the metal valence shell electrons, add up to the rare gas configuration of eighteen electrons in nearly every case. Thus nickel(0) has 10 valency electrons $[d^8s^2]$ and the 8 from the four CO groups give the rare gas configuration. Similarly, iron with 8 electrons and chromium with

Table 16.1 Properties of the simpler carbonyls

$V(CO)_6$ Dark green d. 70° Octahedron Paramagnetic	$Cr(CO)_6$ Colourless d. 130°, s Octahedron	$Mn_2(CO)_{10}$ Yellow m 154°, s Figure 16.3a	$Fe(CO)_5$ Yellow m−20°, b 103° Trigonal Bipyramid $Fe_2(CO)_9$ Orange d. 100° Figure 16.3d $Fe_3(CO)_{12}$ Green d. 140°, s Figure 16.4a	$Co_2(CO)_8$ Orange m 51°, s Figure 16.3b, c $Co_4(CO)_{12}$ Black d.60°	$Ni(CO)_4$ Colourless m−19.3°, b 42° Tetrahedron
	$Mo(CO)_6$ Colourless d. 180°, s Octahedron	$Tc_2(CO)_{10}$ Colourless m 160°, s As Figure 16.3a	$Ru(CO)_5$ Colourless m−16°, Trig. bipyramid $Ru_3(CO)_{12}$ Orange m 150° Figure 16.4b	$Rh_4(CO)_{12}$ Red d. 150°	
	$W(CO)_6$ Colourless d. 180°, s Octahedron	$Re_2(CO)_{10}$ Colourless m 177°, s As Figure 16.3a	$Os(CO)_5$ Colourless m 2° Trig. bipyramid $Os_3(CO)_{12}$ Yellow m 224° Figure 16.4b	$Ir_4(CO)_{12}$ Yellow d. 210° Tetrahedron of $Ir(CO)_3$ units	$Pt(CO)_4$ Transient existence

6 require, respectively, 5 and 6 CO groups to make up the 18. For manganese, with 7 electrons of the d^5s^2 configuration, we see that $Mn(CO)_5$ would have only 17 electrons while $Mn(CO)_6$ would have 19 electrons. In this case we find that two $Mn(CO)_5$ units join together forming a Mn-Mn bond and the two shared electrons of the bond count for each Mn. Thus there are 10 CO electrons, 7 Mn electrons plus 1 shared from the second Mn making 18. Similarly cobalt forms $Co_2(CO)_8$ where the count at each Co atom is 8 from 4 CO groups, 9 valency electrons and one from the Co-Co bond. The only exception to the 18-electron rule is provided by vanadium which forms the $V(CO)_6$ monomer with only 17 electrons. Possibly this is due to the steric problems of 7-coordination in a dimer.

The carbonyls are readily reduced, by such reagents as sodium amalgam or hydrides, to form anions. These all obey the 18-electron rule and we find species such as $Co(CO)_4^-$ or $Fe(CO)_4^{2-}$. Vanadium carbonyl very readily forms $V(CO)_6^-$ where the extra electron completes the eighteen. Addition of acid to the anions forms the carbonyl hydrides, such as $HMn(CO)_5$ or $H_2Fe(CO)_4$. These also all obey the 18-electron rule. The M-H bond is relatively weak and the simple hydrides readily lose H_2, forming the parent carbonyl.

Cationic species are rarer, but the large majority obey the 18-electron rule, for example $[Mn(CO)_6]^+$. The recently reported $[Pt(CO)_4]^{2+}$ is square-planar d^8 and a rare exception to the 18-electron rule (as is often found for this configuration).

As the known carbonyls, carbonyl ions, and also a range of other derivatives such as hydrides and halides, almost all follow the 18-electron rule, we can turn the argument round and use the 18-electron rule as a guide to suggest the formulation of more complex species. Thus, iron forms two more carbonyls in addition to $Fe(CO)_5$. In $Fe_2(CO)_9$, there are 18 electrons from the CO groups, 16 from the two Fe atoms, and so we need one Fe-Fe bond to complete 18 electrons at each iron. Similarly, $Fe_3(CO)_{12}$ with 24 CO electrons and 24 Fe electrons needs 3 Fe-Fe bonds (i.e. 2 per Fe atom) to complete 18 electrons per iron.

16.2.2 Bonding

Since the metal is zero-valent, and the CO group is a very poor lone pair donor, it is clear that a simple electrostatic approach cannot account for the carbonyls. We find that the carbonyls are readily accommodated by the molecular orbital approach with π bonding between metal and ligand, as in case (a) of Section 13.12, where the ligand has empty orbitals of π symmetry which accept electron density from the metal.

The basic M-CO bond is illustrated schematically in Fig. 16.2. The CO molecule has three occupied σ orbitals (see Table 3.4 and Fig. 3.23b) of which σ_1 is strongly bonding between C and O. σ_2 and σ_3 are weakly bonding or nonbonding between C and O, and are directed outwards on O and C respectively. Thus σ_3 is suitable for donation to the metal as in Fig. 16.2a. As the M accepts electrons from σ interactions with a number of CO groups, the electron density on M builds up. However, the π^* orbitals on CO are empty and suitably aligned to form π bonds as in Fig. 16.2b. This removes electron density from the metal (making it a better acceptor) and increases electron density on the CO (making it a

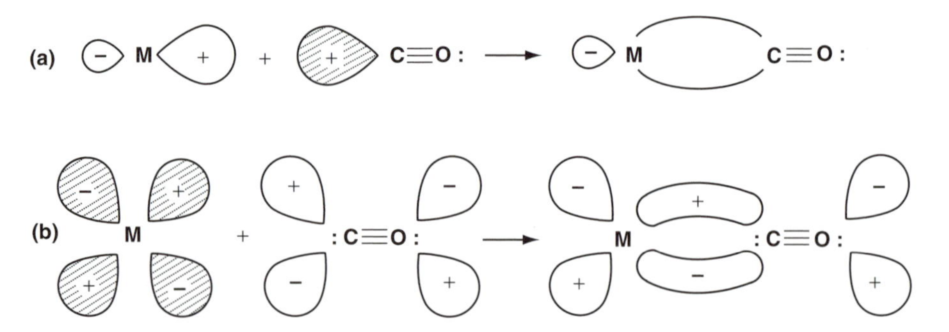

FIG. 16.2 The σ and π bonds between a metal and CO. (a) The σ bond formed when the lone pair on the CO donates into a suitable metal orbital; (b) the π bond formed when electrons from a filled metal orbital donate back into the empty π^* antibonding orbital on the CO. Cross-hatching indicates the orbitals which originally held the donated electrons

CARBONYL COMPLEXES OF METALS IN HIGHER OXIDATION STATES, AND RELATED COMPLEXES

In addition to the 'classical' metal carbonyl complexes, described in Section 16.2, a significant number of complexes are now known where CO is a ligand in complexes where the metal is either in a relatively high oxidation state, or where the metal has previously demonstrated a reluctance to form CO complexes. The most important criterion for identifying 'nonclassical' metal carbonyls is a C-O infrared stretching frequency which is higher than the value for free CO (2143 cm^{-1}). This indicates that the usual backbonding into the CO π^* orbital is of lesser importance for the nonclassical CO complexes. The metals at the right hand side of the d block have now been shown to form nonclassical CO complexes, illustrated by $M(CO)_4^{2+}$ (M = Pd, Pt), $M'(CO)_n^+$ (M'=Cu, Ag; $n = 1$–4), $Hg(CO)_2^{2+}$ and $Hg_2(CO)_2^{2+}$. As an example of the CO stretching frequencies, those for the $Ag(CO)_n^+$ cations lie in the range 2189 to 2220 cm^{-1}. Studies have been extended to carbonyl cations of other transition metals, e.g. $M(CO)_6^{2+}$ (M=Fe, Ru, Os) and $Ir(CO)_6^{3+}$. The $Fe(CO)_6^{2+}$ cation is formally isoelectronic with, but less stable than, the well-known $Fe(CN)_6^{2-}$ (ferrocyanide) ion, as a result of the cyanide ligand being a much better σ donor. Finally we note the synthesis of the dinuclear platinum(I) carbonyl complex $\{Pt(CO)_3\}_2^{2+}$, formed by reaction of PtO_2 with CO in concentrated sulfuric acid. These recent results clearly suggest that the chemistry of metal carbonyl complexes remains an active field of research, with many surprises yet in store.

References

Nonclassical Metal Carbonyls, A. J. LUPINETTI, S. H. STRAUSS AND G. FRENKING, *Progress in Inorganic Chemistry* **49**, 2001, 1–112.

Copper(I) and Silver(I) Carbonyls – To Be or Not To Be Nonclassical, S. H. STRAUSS, *Journal of the Chemical Society, Dalton Transactions* 2000, 1–6.

Homoleptic Metal Carbonyl Cations of the Electron-rich Metals: Their Generation in Superacid Media Together With Their Spectroscopic and Structural Characterization. H. WILLNER AND F. AUBKE, *Angewandte Chemie, International Edition* **36**, 1997, 2402–2425.

Homoleptic Noble Metal Carbonyl Cations, L. WEBER, *Angewandte Chemie, International Edition* **33**, 1994, 1077–1078.

better donor) so that the σ and π processes reinforce each other. The term *synergic* has been applied to describe this mutual reinforcement. Notice that this process weakens the CO π bonds, as it populates the CO π^* orbital, but this is more than compensated by the enhanced M-C bonding so the overall process is favourable.

This CO weakening is reflected by a drop in the CO stretching frequency (compare Section 7.7). As the CO frequencies are usually very strong bands in the infrared spectrum, this gives a conveniently observed way of studying carbonyl bonding. In this way we find the infrared absorption due to the carbonyl stretching at 2037 cm^{-1} in $Ni(CO)_4$, distinctly less than the frequency of 2143 cm^{-1} in free CO. If we increase the negative charge on the metal, we expect more π donation from M to π^* (as in Fig. 16.2b), a weaker CO bond and a reduced carbonyl frequency. Thus for $Co(CO)_4^-$ and $Fe(CO)_4^{2-}$ (which are isostructural with $Ni(CO)_4$) the frequencies drop to 1918 cm^{-1} and 1788 cm^{-1} respectively, reflecting the increased negative charge.

The measurements of other parameters, such as M-C bond lengths, all support this picture. Therefore the bonding model (a) of Section 13.12, as applied in Fig. 16.2, is an adequate way of describing the metal carbonyls. More detailed treatments, taking account of all the CO groups and the symmetry of the complex and extending to other substituents, are very successful, but all are based on the core concepts described here.

We note that the model helps to account for the distribution of the simple carbonyls: to the left of the d block the metals have insufficient electrons to form the π bonds while to the right the elements lack empty orbitals to accept the σ electrons.

16.2.3 Structures

The simple carbonyls have the symmetric shapes expected (Table 16.1). In $M_2(CO)_{10}$ (M = Mn, Tc, Re) the second metal atom completes the octahedron, Fig. 16.3a.

The other dimeric carbonyls of the first transition series, $Co_2(CO)_8$ and $Fe_2(CO)_9$, introduce a new type of carbonyl bonding, the bridging carbonyl group. The structure of $Co_2(CO)_8$, in the solid state, is shown in Fig. 16.3b. Each Co is bonded to three ordinary terminal CO groups and to the second Co by a Co-Co bond. The two remaining CO groups bond to both Co atoms, bridging the Co-Co bond.

Although sometimes written in a keto form,

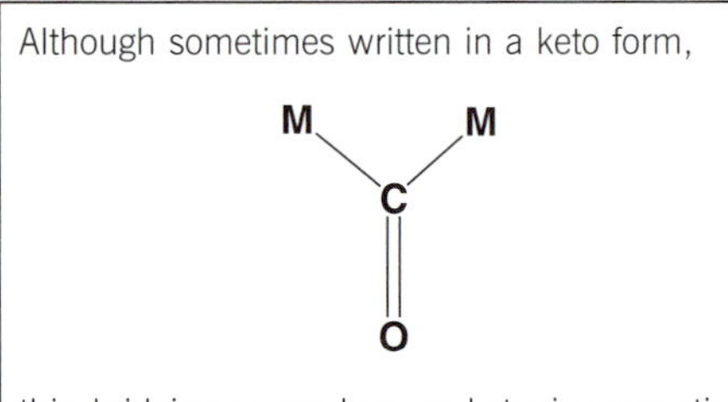

this bridging group has no ketonic properties. The bonding is best regarded as a three-centre overlap, similar to that of the B_3 faces in B_5H_{11} (Fig. 9.12c).

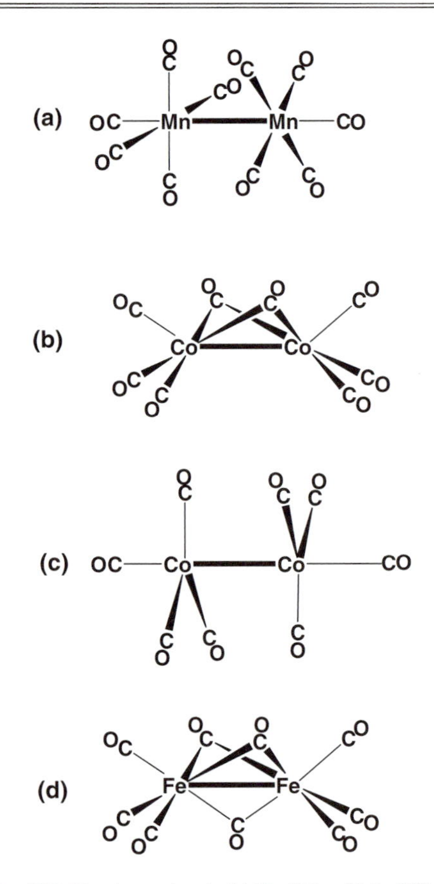

FIG. 16.3 Binuclear carbonyls: (a) $Mn_2(CO)_{10}$, (b) $Co_2(CO)_8$ in solid, (c) $Co_2(CO)_8$ in solution, (d) $Fe_2(CO)_9$

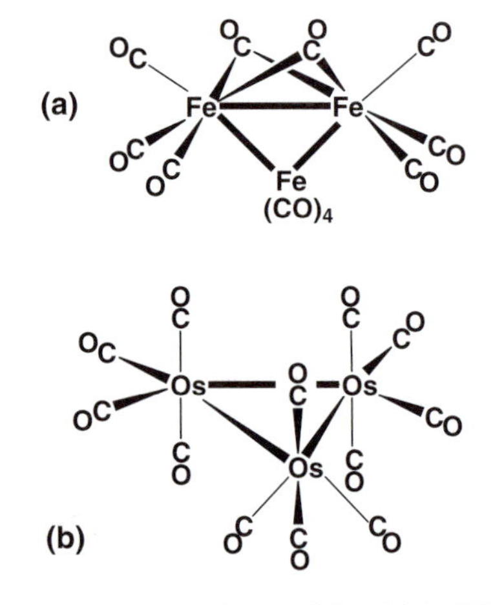

FIG. 16.4 Structures of (a) $Fe_3(CO)_{12}$, (b) $Os_3(CO)_{12}$

In terms of electron counting, the bridging group contributes one electron to each metal, so two groups bridging two metal atoms have the same effect as if each was terminal on one metal. Thus the 18-electron rule does *not* allow us to distinguish bridging and non-bridging structures. This is well illustrated by the existence of the second isomer of cobalt carbonyl, Fig. 16.3c, which is in equilibrium with the bridged form in solution.

$Fe_2(CO)_9$ has the structure shown in Fig. 16.3d, with three bridging CO groups placed symmetrically around the Fe-Fe bond. The $Co_2(CO)_8$ form of Fig. 16.3b is similar to this with one bridging CO removed. For $Fe_2(CO)_9$, the electron count is 8 valency electrons plus 6 from three terminal CO groups plus 3 from three bridging CO groups plus 1 from the Fe-Fe bond making 18 at each iron atom.

The presence of bridging carbonyls is clear from the infrared spectrum where the stretching frequencies of bridging groups are found at lower energies. Thus the terminal CO groups of $Co_2(CO)_8$ give vibrations between 2028 and 2104 cm^{-1} while the bridging modes are at 1898 and 1867 cm^{-1}.

The structures of the more complex carbonyls are built up in similar ways. For example, the $M_3(CO)_{12}$ species of the iron group have the structures of Fig. 16.4. $Fe_3(CO)_{12}$ involves bridging carbonyls and can be envisaged as $Fe_2(CO)_9$ with one bridging CO replaced by a $Fe(CO)_4$ group. The Fe_3 skeleton is an isosceles triangle with the three Fe-Fe bonds indicated by the 18-electron rule. $Ru_3(CO)_{12}$ and $Os_3(CO)_{12}$ form equilateral triangles with no bridging carbonyls. A similar pattern is seen for the $M_4(CO)_{12}$ species of the cobalt group, where the metal atoms form a tetrahedral cluster. For M = Co and Rh, the base triangle face consists of terminal $(CO)_2M$ units with a bridging CO along each edge and the apex is occupied by an $M(CO)_3$ group. For M = Ir, the structure is symmetric with four $Ir(CO)_3$ terminal units in a regular tetrahedron. It is found generally that bridging CO groups are commoner for the lighter elements.

Hydrides with more complex structures are well known. Figure 16.5 shows some of the simpler examples. The edge-bridging hydrogens (as in Fig. 16.5c) are analogous to the B-H-B bridges in the boron hydrides (Fig. 9.10). In $HFe_3(CO)_{11}^-$, the H replaces one of the bridging CO groups of Fig. 16.4a.

The simplest stable carbonyl of iridium is the tetranuclear species $Ir_4(CO)_{12}$ which has all the CO in terminal positions and the Ir_4 core bonded together into a regular tetrahedron. The CO are oriented equally about the threefold axes so the overall symmetry is T_d. The rhodium analogue has an alternative configuration where one $Rh(CO)_3$ unit is bonded to a triangle of three $Rh(CO)_2$ units with the last three CO edge-bridging this triangle (overall C_{3v}). $Co_4\ (CO)_{12}$ has the same sructure as the Rh carbonyl.

Larger carbonyl clusters are a well-established and rapidly growing field which is included in Section 16.8.

16.2.4 Related species

There are three other groups of compound which are closely related to the carbonyls. First the nitrosyls, where the NO group behaves similarly to CO but acts as a 3-electron donor. Thus $Cr(NO)_4$ is an 18-electron species and we can construct the isoelectronic series from this through $Mn(CO)_3(NO)$, $Fe(CO)_2(NO)_2$ and $Co(CO)_3(NO)$ to $Ni(CO)_4$, all of which are known. Further isoelectronic species of this type are $Mn(CO)_5^-$ (obtained by Na/Hg reduction of $Mn_2(CO)_{10}$) and $Cr(CO)_4(NO)^-$.

Another ligand closely related to CO is PF_3. Unlike organic phosphanes, such as $P(CH_3)_3$, PF_3 is a poor σ donor, as the electron density of the phosphorus lone pair is attracted by the electronegative fluorines. However, this same attraction makes the P a good π acceptor, using its empty 3d orbitals. As a result of these effects, PF_3 turns out to be a ligand very similar to CO, and PF_3 can replace terminal CO in most formulae, so we find $Ni(PF_3)_4$, $Fe(PF_3)_5$ or $Cr(PF_3)_6$ for example. The properties of such species are very similar to those of the carbonyl analogues. It has been very striking to find that $V(PF_3)_6$ can be made and parallels the carbonyl. It is a volatile paramagnetic molecule, corresponding to the same 17-electron count as the exceptional $V(CO)_6$, and it very readily forms the $[V(PF_3)_6]^-$ anion, completing the 18 electrons.

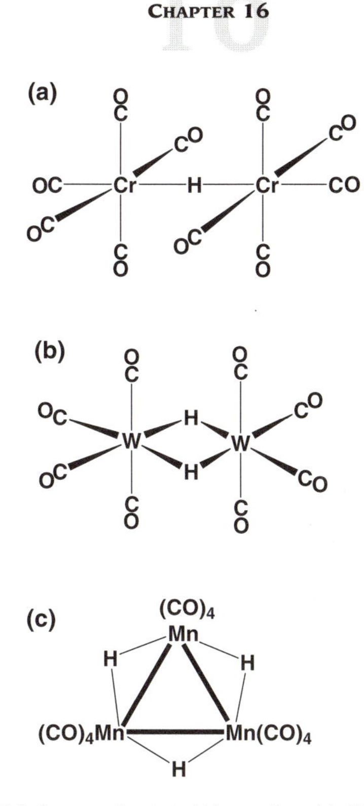

FIG. 16.5 Some carbonyl hydride species: (a) $HCr(CO)_{10}$, (b) $H_2W_2(CO)_8^{2-}$, (c) $H_3Mn_3(CO)_{12}$

There is also a huge range of mixed-ligand compounds containing the carbonyl group. For example, we find the reaction

$$(CH_3)_3SnW(CO)_5^- + NO^+ \rightarrow (CH_3)_3SnW(CO)_4NO$$

or the reaction of $[(C_5H_5)Fe(CO)(PR_3)I]$ to form other $[(C_5H_5)Fe(CO)(PR_3)X]$ species with halogens.

Another class of related ligand, isoelectronic with carbon monoxide, is isocyanide (also known as isonitrile), RNC. One advantage of isocyanides over carbon monoxide as a ligand is that the steric and electronic properties of the former may be readily varied by changing the R group. Nevertheless, the number of isocyanide compounds is substantially less than that of carbonyls. Analogues of carbonyl species such as $Ni(CNR)_4$ and $Cr(CNR)_6$ have been prepared, and in dinuclear and cluster compounds bridging isocyanides akin to bridging carbonyl ligands are also known. The isocyanide group is, in general, a better σ donor than CO, and stabilizes complexes in higher oxidation states than does CO, for example the complex $[Mn(CNR)_6]^{2+}$, which has no known carbonyl analogue.

The isoelectronic cyanide ligand CN^-, while invariably a carbon donor ligand, is a much poorer π-acceptor than CO or RNC and is best considered as a pseudohalide ligand. However, it is still useful to mention some related CO and CN^- complexes in the iron(II) system. Ferrocyanide (Section 14.6) is a very well-known and highly stable species, but the corresponding CO analogue, $Fe(CO)_6^{2+}$, has only recently been prepared. Mixed CO-CN species are also known; $Fe(CN)_5(CO)^{3-}$ was reported in the 19th century, while *trans*-$Fe(CN)_4(CO)_2^{2-}$ was only reported in 2001. The synthesis of the latter complex is surprisingly simple, involving the addition of 4 equivalents of NaCN to an aqueous solution of $FeCl_2$ under a CO atmosphere.

16.3 Metal–organic compounds

Many of the early attempts to make metal–organic compounds were unsuccessful, and molecules which were isolated, such as $CH_3CH_2Mn(CO)_5$, decomposed easily (in this case, at −30°C). The discovery in 1951 of the very stable organometallic compound, ferrocene $(C_5H_5)_2Fe$, was thus of considerable moment. This was quickly found to have an unusual structure and the interest created led to a very rapid expansion of the organometallic field. The two chemists most involved in the organometallic revival, Fischer and Wilkinson, were jointly awarded the 1973 Nobel prize for this work.

16.3.1 Metal–carbon σ bonding

The simplest system to consider is that of a metal bonded to a methyl group. Here, the only bonding interaction of any significance is a σ bond using appropriate orbitals on M and CH_3. Such bonds are very similar to the transition metal, M-H, bond and also to the relatively stable bonds to Main Group metals as in $Sn(CH_3)_4$. One the whole, σ-bonded metal–organics are unstable. A number of simple compounds exist, notably $Ti(CH_3)_4$, $Nb(CH_3)_5$ and $W(CH_3)_6$ with a few related species, which are well characterized but which decompose at or below normal temperatures, often violently. In the presence of other ligands, stability increases. Thus $Ti(CH_3)_4$ starts to decompose above −78°C but $(CH_3)_2TiCl_2$ can be prepared at −20°C and diamine complexes are stable at 0°C. The higher alkyls are generally less stable, but aromatic derivatives are more stable than the methyls.

It is instructive to examine the factors affecting that metal–carbon bond stability. A number are of importance:

(1) *the metal–carbon bond is probably relatively weak*
There is little quantitative data but we note the thermal instability and the fairly low force constants for the metal–carbon stretching vibration.
(2) *organic products of M-C cleavage are highly reactive*
Whether $M\text{-}CH_3$ cleaves to give methyl radicals or organic ions, these will react rapidly with the solvent or with each other. Contrast this with the cleavage of most metal–ligand bonds, e.g. $M\text{-}OH_2$ or M-Cl, which give unreactive species such as H_2O or Cl^-.
(3) *bonds formed by the organic cleavage products are relatively strong*
We expect to find bonds such as C-C, C-O, C-N, C-halogen, depending on the system.

Thus we expect that, in a model reaction such as

$$2M\text{–}CH_3 \rightarrow H_3C\text{–}CH_3 + M \text{ products} \qquad (16.2)$$

the forward reaction will be exothermic and thermodynamically preferred. Further, any equilibrium involving the first cleavage step, e.g.

$$M\text{–}CH_3 \rightarrow M\cdot + \cdot CH_3 \quad (16.3)$$

will be driven to the right as the $\cdot CH_3$ will rapidly be removed by further reaction.

However, even if compounds are thermodynamically unstable with respect to their decomposition products, their lifetime depends on the rate of the reaction, i.e. on their *kinetic stability*. Lifetimes range from the extremely short to the indefinitely long. The main factor affecting kinetic stability is the size of the energy input—the activation energy—required to get the molecule into a state where reactions such as Equations (16.2) or (16.3) occur. Such an input may involve bond weakening or breaking, formation of an intermediate, population of antibonding orbitals and so forth. If the required activation energy is large, the compound may be stable indefinitely, even if the overall decomposition reaction has a strongly favourable free energy change. Similar comments apply to stability to other reactions such as oxidation or hydrolysis.

With these comments in mind we can list some factors which will generally be expected to lead to relatively stable metal–carbon σ bonds.

(1) If we make the reasonable assumption that population of the antibonding orbitals leads to bond breaking, any factor that increase the energy gap between the highest filled orbitals and the antibonding ones, will improve stability. Such factors are:

(a) a d^0 configuration—no electrons in the relatively high nonbonding orbitals such as t_{2g} in an octahedral complex.

(b) in any other configuration, all factors increasing the ligand field ΔE (compare Sections 13.3 to 13.5, and 13.12).

(2) All factors which decrease the kinetic contribution such as:

(a) steric hindrance from large ligands in higher coordination numbers.

(b) substitution-inert configurations such as d^6 octahedral (see Section 13.9).

(c) absence of low-energy pathways such as β-elimination (in the interaction with β-hydrogen below).

(3) Factors which directly increase the metal–carbon bond strength. These include π contributions to be M-C bond as discussed below.

Of course, such contributions to stability do not necessarily have only one effect. Thus, π bonding ligands L in $L_nM\text{-}(CH_3)_x$ will increase ΔE (see 1b) but will probably increase the metal–carbon bond energy as well. This energy, in turn, affects the stability through both its thermodynamic effect (decreasing the free energy change for the forward reaction of Equation (16.1)), and through its kinetic effect increasing the activation energy input for M-C bond cleavages as in Equation (16.2).

Table 16.2 lists some examples of metal–carbon σ bonds, and it will be seen that the factors outlined above apply in most cases. Note that there are many further examples with π-bonding ligands such as C_5H_5, CO, phosphanes, etc.

The stability order $C_6H_5 > CH_3 > C_2H_5$ is accounted for by two features:

Table 16.2 Some examples of metal–carbon σ bonds

$(CH_3)_4Ti$	$(CH_3)_5M$	$(CH_3)_6W$	$RM(CO)_5$
CH_3TiCl_3	M=Nb, Ta	$(CH_3)_8W^{2-}$	M=Mn, Tc, Rc
Diamine $Ti(CH_3)_4$	$(C_5H_5)_2VR$	$(C_5H_5)M(CO)_3R$	
	$V(C_6H_5)_6^{4-}$	M=Cr, Mo, W	
		$RCr(H_2O)_5^{2+}$	
$(C_5H_5)M(CO)_2R$	$RCo(CN)_5^{3-}$	$(PR_3)_2MR_2$	$(PR_3)AuR$
M=Fe, Ru, Os	$RM(X)_2(CO)(PR_3)_2$	(M=Ni, Pd, Pt)	$(PR_3)AuR_3$
(diphos)Ru(R)(H)	(M=Rh, Ir)	$[(CH_3)_3PtX]_4$	

Notes (a) R = alkyl or aryl: stability usually decreases R = aryl > CH_3 > C_2H_5.
(b) Stability increases down group e.g. M = Ni < Pd < Pt.
(c) Note the common configurations are d^0 or d^{10} or the substitution-inert ones.

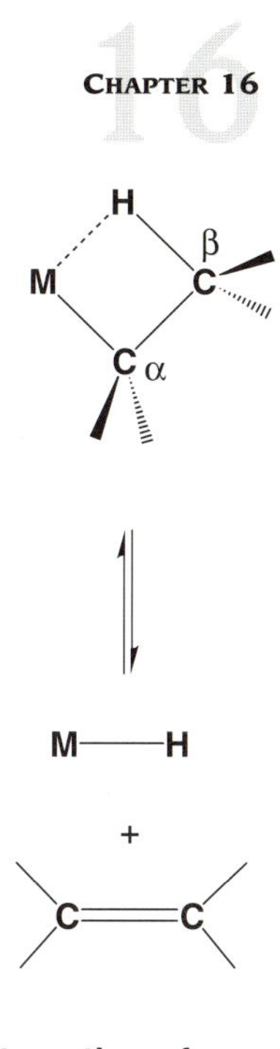

Interaction of a metal with a β-H atom

(a) aromatic ligands (and unsaturated ones generally) allow for additional M-R π interactions between metal d orbitals and ligand π^* ones. Note, for example, that a metal–acetylide, M-C≡CH, is isoelectronic with the carbonyls, M-C≡O, discussed in the last section.

Further, since the charge distribution is usually $M^{\delta+}-C^{\delta-}$, any organic ligand able to delocalize the negative charge should help stability. This applies to aromatic ligands, halogen-substituted ones, etc., culminating in the generally high stability of M-CF_3 groups. These, and related ligands, can be used to stabilize metal centres in high oxidation states. Thus, the copper(III) species, $Cu(CF_3)_4^-$, has been isolated as a stable colourless solid, which has a nearly square-planar structure, and the electron-withdrawing aryl group C_6Cl_5 forms the square-planar, d^7, 15-electron nickel(III) complex $Ni(C_6Cl_5)_4^-$.

(b) It is commonly found that systems which do not contain H on the carbon β to the metal atom are distinctly the more stable. Thus, in equivalent species, M-CH_3, M-CH_2-$C(R)_3$, M-CH_2-Ar, M-CH_2-SiR_3 or M-CH_2-NR_2 are more stable than M-CH_2CH_3 or M-CH_2-CHR_2 in general. This effect arises because the metal may interact with a β-H atom. This provides an additional pathway for reaction (and the reverse reaction provides a synthesis).

16.3.2 Metal–carbon multiple bonding

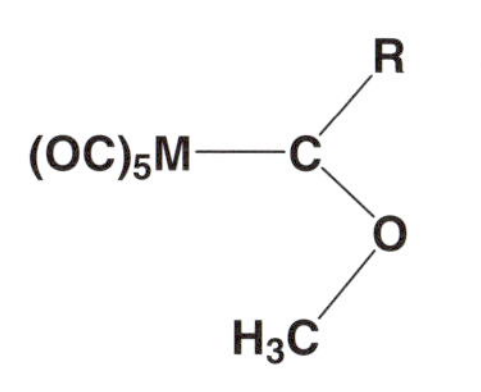

R = CH_3, C_6H_5
M = Cr, C-Cr = 204 pm
M = W, C-W = 205 pm

An interesting development has been the evidence that metal–carbon double and triple bonds may occur in the *carbenes* M = CR_2, and *carbynes* M≡CR respectively. One of the first examples of a carbene was $(CO)_5WC(R)OMe$ whose crystal structure shows the CR(OMe) group completing an octahedron at W. The W-C bond length of 205 pm is greater than W-CO (190 pm). The bonding is formulated as (a) a W-C σ bond plus (b) a π bond formed by a carbon p orbital overlapping with, say, d_{xz} on W (z is W-CR_2 axis). As the d_{yz} overlap would be identical, there is cylindrical symmetry around the z axis and the CR_2 group is free to rotate. A range of metal–carbenes is now known.

More recently a further family of compounds has been reported, the carbynes, with an M-CR unit. An example is $I(CO)_4WC(C_6H_5)$. For this and related compounds, the structure is octahedral with the halogen *trans* to the CR and with the dimensions.

For R = CH_3 or C_6H_5 and X = Br or I
$X(CO)_4M$-CR Cr-C = 169 pm
W-C = 190 pm.

The M-C-R angle is near 180° for R = CH_3 but about 170° for R = C_6H_5.

The bonding is expressed as an overlap of p_x and p_y orbitals on C with metal d_{xz} and d_{yz} orbitals, in addition to the σ component. Metal p_x and p_y may also contribute.

Thus we have three classes of metal–carbon compound with the bonding along the M-C axis. These are analogues of C-C, C=C and C≡C bonding respectively although the M-C, M=C and M≡C systems are much weaker. A further, very widespread, class involving metal–carbon π bonding is that with unsaturated organic ligands bonding 'sideways-on' to the metal, and this is discussed in Section 16.4.

The carbene or carbyne metal complexes essentially consist of a highly reactive organic fragment stabilized by coordination to the transition metal centre. They are examples of a general effect which is now very important in organometallic chemistry—that highly reactive species can be stabilized on coordination. A wide and diverse range of such stabilized reactive entities is now established, including highly strained cyclic alkynes, such as cyclobutyne and cyclohexyne, cyclobutadiene, benzyne and trimethylenemethane $[C(CH_2)_3]$. Inorganic entities are also stabilized, such as silylenes $[R_2Si]$ and the corresponding germylenes and stannylenes.

A major reason for interest in metal–carbon bonded species comes from this stabilization, which applies to intermediates in many metal-catalysed organic reactions. Examples are the oxo-process (Section 14.7) the Monsanto acetic acid process (Section 15.7) or the Wacker process (Section 15.9).

16.4 π-bonded cyclopentadienyls and related species

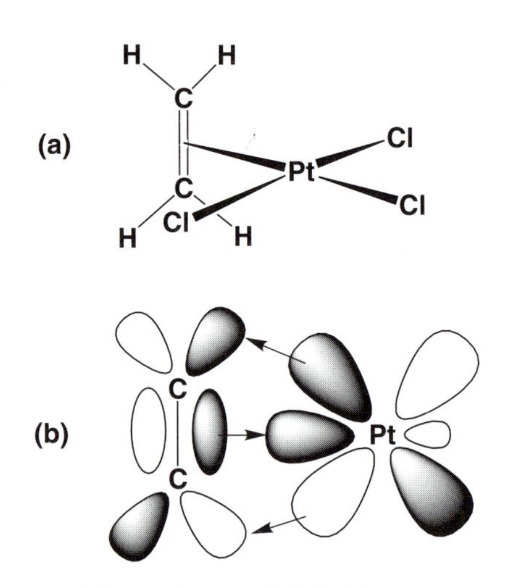

Fig. 16.6 Zeise's salt anion, $[PtCl_3(C_2H_4)]^-$: (a) structure, (b) Pt ethylene bonding

Substituted cyclopentadienyl ligands

Related substituted cyclopentadienyls are commonly used as π ligands, especially C_5Me_5 (often labelled cp*) where steric effects might be significant. One of the most substantial is C_5Ph_5, 'supercp', in which the steric demand often forces alternative geometries. The first perfluorinated analogue, $(C_5F_5)Ru(cp^*)$ was recently discovered.

A further large class of π-bonded transition metal complexes, which we can refer to only briefly here, is the group containing unsaturated organic molecules as ligands. The two classical examples are *Zeise's salt*, $K[PtCl_3(C_2H_4)].H_2O$, and *ferrocene*, $Fe(C_5H_5)_2$.

In Zeise's salt, the three Cl atoms and the mid-point of the C=C bond form a square plane around the Pt atom, and the ethylene molecule lies perpendicular to this plane (Fig. 16.6a). The bonding is illustrated in Fig. 16.6b. As in the carbonyls, sigma donation is postulated from the ligand, this time from the filled C-C π orbital, and π donation from the metal into the ligand π* orbital.

In the ferrocene molecule the metal atom is sandwiched between the organic parts, C_5H_5-Fe-C_5H_5, and the planes of the organic rings are parallel (Fig. 16.7a). A very similar situation is found for other aromatic systems, as in dibenzene chromium (Fig. 16.7b). An interesting extension of this class is the triple sandwich (Fig. 16.7c). Many more layers in such sandwich complexes have been constructed of late and even a hexadecker sandwich has been described. This consists of five cobalt atoms and six cyclopentadienyl or boron-substituted cyclopentadienyl ligands.

Cyclopen dienyl compounds are formed quite generally. Most transition metals form $(C_5H_5)_2M$ species. However, some, such as $[(C_5H_5)_2Ti]$, are less simple than first thought, and all structurally characterized titanocenes have required the coordination of a neutral ligand such as CO or phosphanes to stabilize the d^2 metal centre. However, use of a substituted cyclopentadienyl ligand gives $(C_5Me_4SiMe_2^tBu)_2Ti$, which has a ferrocene-like mononuclear structure with parallel cyclopentadienyl rings.

The related ions, such as $(C_5H_5)_2Fe^+$, are also readily formed and redox relationships between ions and neutral species reflect the 18-electron rule. One interesting case is cp^*_2Fe where the neutral compound is staggered, like the H analogue, but the cation is in the eclipsed configuration.

Molecules are found with one cp substituent together with other ligands, such as $C_5H_5Mo(CO)_3Cl$, and the reader is referred to the review in Appendix A for a detailed account of compounds of this type.

Dibenzene compounds are less widespread, but are typified by $(C_6H_6)_2Cr$, Fig. 16.7b, which has the Cr sandwiched between two parallel benzene rings. Other aromatic systems appear in similar compounds including the cation of cycloheptatriene $(C_7H_7)^+$ and the dianion of cyclobutadiene $(C_4H_4)^{2-}$. The latter system has long appeared as a hypothetical aromatic system with no evidence of its actual occurrence. The parent hydride is still unknown, but the substituted molecule with four phenyl or methyl groups in place of the hydrogens has been attached to a nickel atom, as in the compound shown in Fig. 16.8. The four-membered ring is planar and the aromatic electrons appear to be fully delocalized. A very wide range of other aromatic systems containing heteroatoms has also been complexed to transition metals in a π-manner. Examples include pyridine (C_5H_5N), phosphabenzene (C_5H_5P) and the cyclic P_5 ligand analogous to cyclopentadienyl which is described in more detail in Section 18.3.1.

The bonding in these aromatic sandwich compounds may be described briefly. The basic bond in ferrocene is a single bond of π symmetry between the iron atom and each ring. This bond is formed by overlap of the d_{xz} and d_{yz} orbitals on the iron (the z axis is the molecular axis, and these two d orbitals are of equal energy) with that aromatic orbital on each ring which has one node passing across the ring. These two metal d orbitals and the two ring orbitals combine to give two bonding π orbitals, of equal energy, and two antibonding orbitals. Each of these orbitals is 'three-centred' on each ring and the iron atom. There are four electrons in ferrocene which fill the two bonding orbitals. In the corresponding cobalt and nickel compounds, the extra electrons enter the antibonding orbitals, making these compounds less stable and giving cobaltocene one unpaired electron and nickelocene two unpaired electrons (as there are two degenerate π* orbitals). One component of the π bond is shown in Fig. 16.9. Other overlaps add smaller contributions to the bonding, but this is the main interaction. The basic feature is that there is a single metal–ring bond and the rings are aromatic, undergoing aromatic substitution and similar reactions. Analogous orbitals can be constructed for the species with other C_nH_n rings.

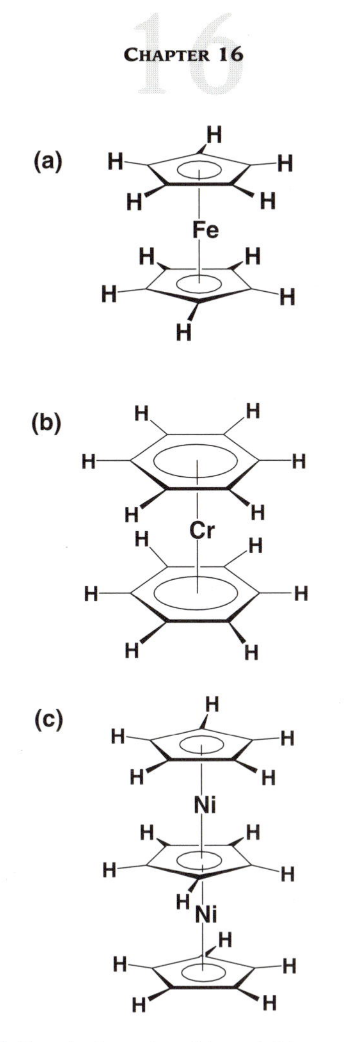

FIG. 16.7 The structure of metal sandwich compounds: (a) ferrocene, $(C_5H_5)_2Fe$, (b) dibenzene chromium, $(C_6H_6)_2Cr$, (c) tris(cyclopentadienyl)dinickel cation $(C_5H_5)Ni(C_5H_5)Ni(C_5H_5)^+$

Many compounds are known which contain both carbonyl and π-bonded organic ligands. The wide variety of types is illustrated by the species π-$C_5H_5M(CO)_x$ (M = V, Nb, Ta, $x = 4$; M = Mn, Tc, Re, $x = 3$; M = Co, Rh, Ir, $x = 2$). The dimers $[\pi\text{-}C_5H_5M(CO)_x]_2$ are found with M = Cr, Mo, W and $x = 3$ or with M = Fe, Ru, Os and $x = 2$, or with M = Ni, Pt and $x = 1$.

Mixed-metal species include

$$\pi\text{-}C_5H_5(CO)_2FeMo(CO)_3(\pi\text{-}C_5H_5) \text{ and}$$
$$\pi\text{-}C_5H_5(CO)_3MoMn(CO)_5,$$

and other π ligands include, in addition to cyclopentadienyl, C_5H_5, the groups C_4H_4, C_6H_6 or C_7H_7 or derivatives of these. Just as C_5H_5 may be regarded as contributing five electrons to the central atom, these latter groups contribute respectively four, six or seven electrons. Also well known are allyl derivatives like π-$C_3H_5Mn(CO)_4$ where the C_3H_5 group contributes three electrons, and diene derivatives where the substituent contributes four electrons, as in π-$C_6H_8Re(CO)_3H$.

Boron hydride analogues of ferrocene exist. For example, the $B_9C_2H_{11}^{2-}$ ion, which presents an open face consisting of a pentagon of three boron and two carbon atoms, can replace the $C_5H_5^-$ ions. Compounds $(B_9C_2H_{11})Fe(C_5H_5)^-$ and $(B_9C_2H_{11})_2Fe^{2-}$ are formed which are ferrocene analogues, and these are oxidizable to Fe(III) compounds, such as $(B_9C_2H_{11})_2Fe^-$, which are analogues of the ferricinium ion. This and many other *carborane* ions have been shown to replace the cyclopentadienyl ion in a variety of other compounds such as the cobalticinium ion and $(C_5H_5)Mn(CO)_3$.

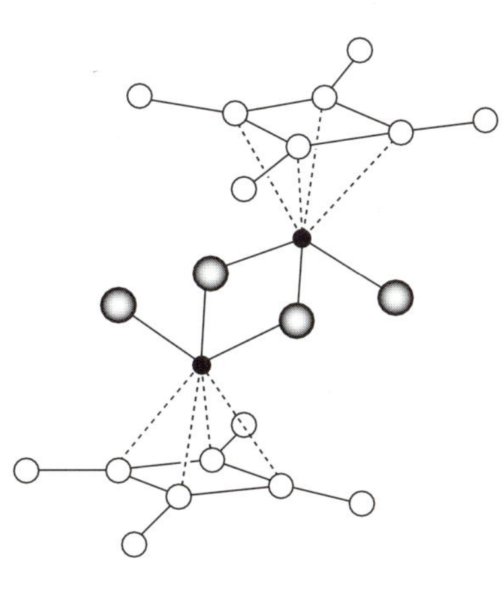

FIG. 16.8 Di(tetramethylcyclobutadienylnickeldichloride), $(C_4Me_4NiCl_2)_2$

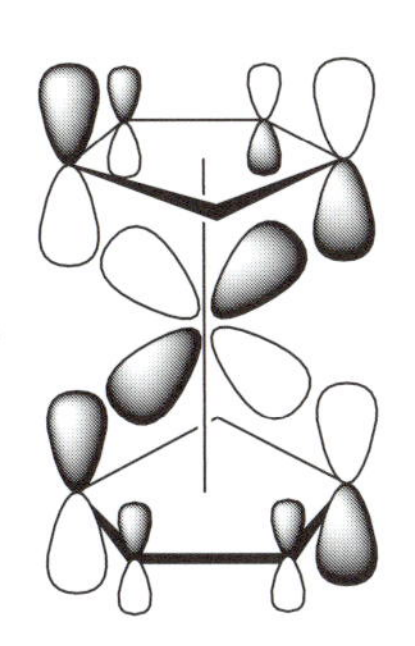

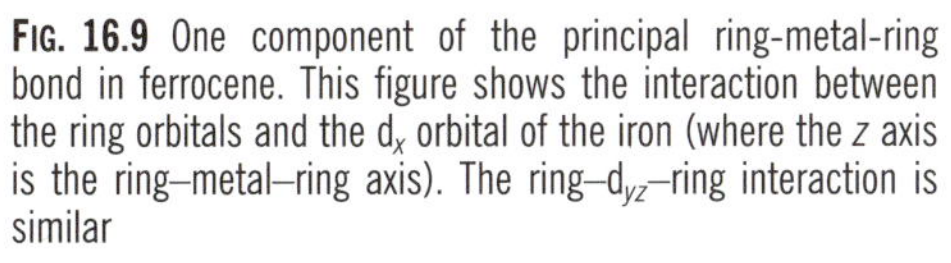

FIG. 16.9 One component of the principal ring-metal-ring bond in ferrocene. This figure shows the interaction between the ring orbitals and the d_x orbital of the iron (where the z axis is the ring–metal–ring axis). The ring–d_{yz}–ring interaction is similar

16.5 The organometallic chemistry of the lanthanides

An organic chemistry of the lanthanides has developed steadily since the establishment in the 1950s of the cyclopentadienyl $(C_5H_5)^-$ compounds formed by the reaction

$$LnCl_3 + 3NaC_5H_5 \rightarrow Ln(C_5H_5)_3 + 3NaCl$$

Such species are known for all lanthanides, though the route to $Eu(C_5H_5)_3$ had to be indirect to avoid reduction to the europium(II) species $Eu(C_5H_5)_2$. The compounds are air and moisture sensitive and all except $Sc(C_5H_5)_3$ readily form $Ln(C_5H_5)_3D$ species with a range of donor molecules including D = NH_3, THF, or R_3P. The properties are basically ionic. The commonest structures have the C_5H_5 rings and are planar with the Ln^{3+} ion lying above the centre with all five Ln-C distances equal. Further weaker interactions occur leading to polymeric units. Thus in $Sc(C_5H_5)_3$, one ring bridges two Sc atoms (in the 1, 3 positions) giving a structure with long chains $\{-(C_5H_5)_2Sc\text{-}C_5H_5\text{-}Sc(C_5H_5)_2\text{-}C_5H_5\}_\infty$. In a similar way, the related $Nd(C_5H_4Me)_3$ is a tetramer in the solid.

Closely related halides are also well known, e.g.

$$LnX_3 + 2NaC_5H_5 \rightarrow Ln(C_5H_5)_2X$$

or

$$LnX_3 + Ln(C_5H_5)_3 \rightarrow Ln(C_5H_5)_2X$$

The structures are dimers with halogen bridges

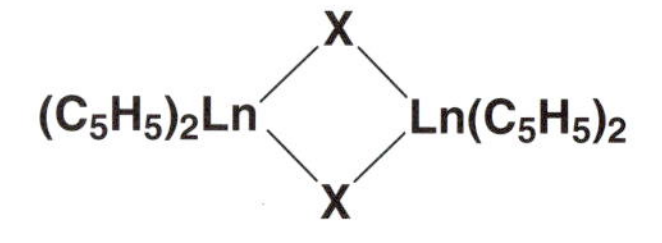

and similar bridges form to other elements as:

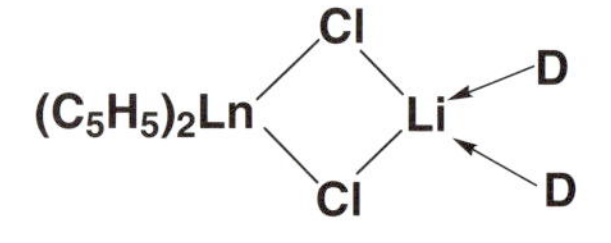

These halides may be converted to other organometallics such as $[(C_5H_5)_2LnR]_x$ where R = H or an organic group. These compounds are usually dimers with electron deficient bridges:

(compare Figs 9.9 and 9.13). If R is a bulky group, such as $C(CH_3)_3$ or $CH(SiMe_3)_2$, monomers are formed and these are also stabilized by donor molecules in $(C_5H_5)_2$ LnR.D species. One interesting reaction is the abstraction of H from a $C(CH_3)_3$ substituent:

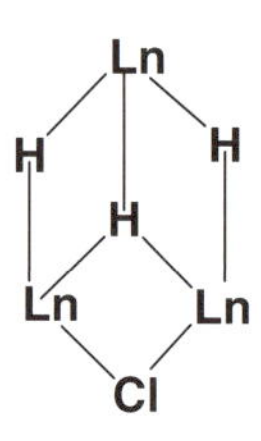

$$3(C_5H_5)_2Ln[C(CH_3)_3].THF + LiCl \rightarrow [Li(THF)_4][\{(C_5H_5)_2Ln\}_3(H)_3Cl]$$

where the anion has a core structure containing two types of bridging H.

Bulky groups allow the isolation of species with direct Ln-C bonds of formulae LnR_3 or $[LnR_4]^-$. The latter is stabilized by bulky cations such as $(Li(THF)_4]^+$. Such compounds are found for ligands R = $C_6H_3Me_2$, $C(CH_3)_3$, CH_2SiMe_3, etc.

These organolanthanide compounds are broadly parallel to those formed by the s elements, but they do allow a unique opportunity to assess steric effects. The Ln^{3+} ions are large and compounds are clearly most stable when the ligands are very bulky. As the size of Ln^{3+} varies in small steps from La^{3+} to Lu^{3+}, effects sensitive to size can be seen. Thus $[Ln(C_6H_3Me_2)_4]$ $[Li(THF)_4]$ could be isolated only for Ln = Yb or Lu. Similarly, $[Ln\{C(CH_3)_3\}_4][Li(THF)_4]$ was formed for Ln = Sm, Er, Yb and La.

Two interesting types of compound resulted when the ligands were made more bulky. First, the cyclooctatetrane ($C_8H_8^{2-}$) sandwich compounds first found for the actinides (see $U(C_8H_8)_2$, Fig. 16.11a) were paralleled in the species $K(diglyme)^+Ln(C_8H)_2^-$. The crystal structure of the cerium compound shows Ce in the centre of a sandwich formed by parallel planar C_8H_8 rings. Divalent lanthanide species $M(C_8H_8)$ and $M(C_8H_8)_2^{2-}$ are known for M = Yb, Eu and Sm.

A second bulky ligand of interest is the $C_5Me_5^-$ ring. The stabilizes the Ln-R bond in a range of compounds $(C_5Me_5)_2LnR$, which are often dimeric (e.g. when R = H). In addition, the C_5Me_5 group stabilizes divalent lanthanide compounds, such as in $Sm(C_5Me_5)_2$ and its solvated analogues:

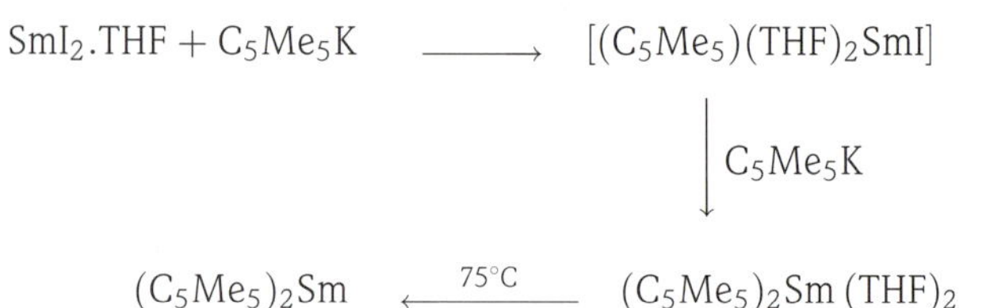

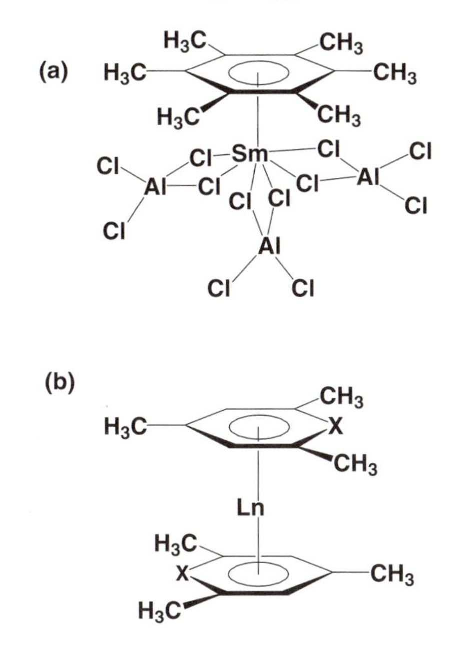

FIG. 16.10 π-arene complexes of the lanthanide elements: (a) $(\pi\text{-}C_6Me_6)Sm(AlCl_4)_3$, (b) $(\pi\text{-arene})_2Ln$ (arene = trimethyl or tri-*t*-butyl benzene or tri-*t*-butyl pyridine)

In $(C_5Me_5)_2Sm(THF)_2$, the coordination around Sm is roughly tetrahedral. When the THF molecules are removed, the structure remains bent with the ring–Sm–ring angle = 140°. Similarly, in $(C_5Me_5)_2Yb$, the angle = 158°. The unsubstituted compound $Sm(C_5H_5)_2$ is also known. $(C_5Me_5)_2Sm$ reacts with $Ag^+BPh_4^-$ in toluene to give the unsolvated cation $[(C_5Me_5)_2Sm]^+BPh_4^-$, and the analogous Nd and Tm complexes have been prepared by reaction of the allyl complexes $(C_5Me_5)_2Ln(CH_2CHCH_2)$ with $[Et_3NH]^+BPh_4^-$. $[(C_5Me_5)_2Ln]^+BPh_4^-$ (M = Sm, Nd) react with C_5Me_5K to give $(C_5Me_5)_3Sm$ and $(C_5Me_5)_2Nd$.

Complexes of the type $Ln(C_5Me_5)_3$ were thought to be too sterically crowded to exist, but in 1991 the first complex of this type, $Sm(C_5Me_5)_3$ was prepared by reaction of $Sm(C_5Me_5)_2$ with cyclooctatetraene (C_8H_8). Other routes to $Sm(C_5Me_5)_3$ have since been reported, and $Nd(C_5Me_5)_3$ is also known.

Whereas the predominant oxidation state for organic derivatives of the lanthanides is the trivalent state (with the exceptions caused by the stability of the f^0, f^7 or f^{14} configurations, typified by Eu(II), Ce(IV), etc.), there is a developing chemistry of zero-valent organometallic compounds of the lanthanides. In an attempt to prepare the first example of a zero-valent sandwich complex $Sm(p\text{-}C_6Me_6)_2$, analogous to dibenzene chromium (Fig. 16.7b), the reaction of $SmCl_3$, Al, $AlCl_3$ and C_6Me_6 was found to give the Sm(III) complex $(\pi\text{-}C_6Me_6)Sm(AlCl_4)_3$, Fig. 16.10a. The synthesis of zero-valent π-arene complexes was eventually achieved by employing the technique of metal vapour synthesis in which metal atoms (from a heated filament) are co-condensed with the organic molecule of interest at low temperature. By this method a number of zero-valent π-sandwich complexes have been synthesized from 1,3,5-tri-*t*-butyl benzene, and 2,4,6,-tri-*t*-butyl pyridine, such as the complexes shown in Fig. 16.10b.

Organolanthanide chemistry presents interesting parallels with the chemistry of the s elements and B or Al species, a way of undertaking detailed study of steric effects, and a number of unusual species not found elsewhere in the Periodic Table.

16.6 Actinide organometallic chemistry

Interest in the organometallic chemistry of the actinides is relatively recent. The larger size gives a preference for higher coordination and the main examples are formed by the IV and III states.

The cp_4E molecules are formed by Th, Pa, U and Np and the centroids of the four rings are arranged tetrahedrally around E. As well as cp and cp*, a variety of cyclopentadienyls form similar compounds. Th and U also form cp_3ER and cp_3EX compounds in the IV state. In these molecules electron transfer from E into the ring π orbitals is relatively unimportant in the bonding and the main interaction is donation from ring π into the 6d and 5f orbitals. Both are significant, but the f contribution is weaker as the orbitals are large and dispersed. There are similar cp_2EX_2 and cp_2ER_2 species but these are less common.

In oxidation state III all the actinides from Th to Cf have been shown to form cp_3E, again with a variety of cp analogues. The centroids are arranged in a plane triangle around E. Cp_3Cm and cp_3Cf are the only well-characterized organometallic compounds of these elements.

In a benzene π-complex, $(C_6Me_6)U(BH_4)_3$, the ring is planar and the centre, together with the three borohydrides, gives a tetrahedral array at U. The U-B distance indicates $U(\mu\text{-}H)_3BH$ coordination through three bridging hydrogens.

While the carbonyl chemistry of the actinides is not well developed, we note the isolation and characterization (by single-crystal X-ray diffraction) of the uranium carbonyl complex, $cp_3^{**}U(CO)$ (where cp** is the bulky tetramethyl-substituted ring, C_5Me_4H). The CO and the three ring centres lie tetrahedrally around U. The U-CO bonding (compare Section 16.2.2) consists of σ donation by CO into the $6d_{z^2}$ orbital on the U and back-donation from U 5f orbitals into the CO π^* orbitals.

A further effect of the large size is the ability to stabilize a larger ring system. There was great interest in the discovery, in 1968, of *uranocene,* $U(C_8H_8)_2$, which has two planar 8-membered rings forming a sandwich structure analogous to ferrocene (Fig. 16.11a). Similar $E(C_8H_8)_2$ compounds were soon reported for Sc and other f elements including Th,

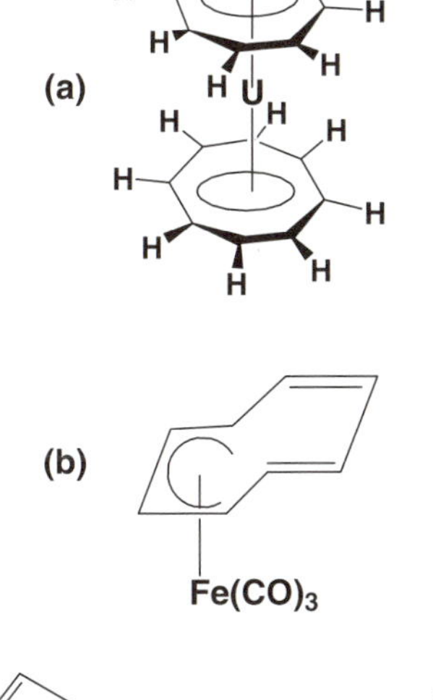

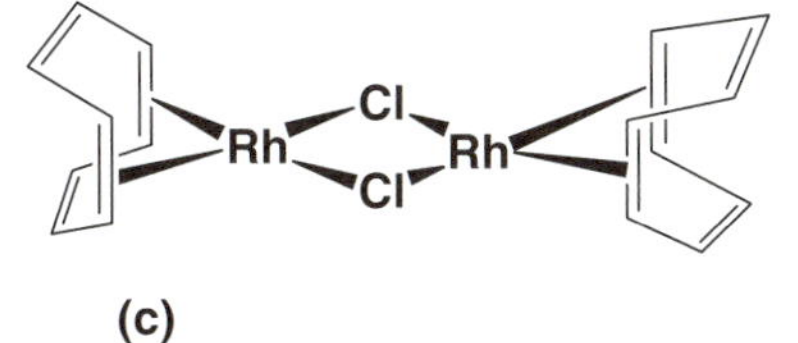

FIG. 16.11 The coordination modes of cyclooctatetraene: (a) as a planar dianion in uranocene, $U(C_8H_8)_2$ (i.e. octahapto), (b) as a delocalized diene (tetrahapto) in $(C_8H_8)Fe(CO)_3$, and (c) as four alkenes (dihapto) in the $[(C_8H_8)RhCl]_2$ dimeric complex

Pa, Np and Pu. In contrast, the cyclooctatetraene dianion prefers alternative coordination modes to transition metals, as illustrated in Figs 16.11b and c.

The $E(C_8H_8)_2$ species are very sensitive to oxygen and water but have now been made with a wide range of substituted rings. Half-sandwiches, such as $Th(C_8H_8)Cl_2(THF)_2$, are also found.

Simple alkyls and aryls are rare, though the anion $[ThMe_7]^{3-}$ has been stabilized with a complex cation and $Th(CH_2Ph)_4$ is also relatively stable. E-alkyl bonds are more stable in the presence of π ligands as noted above.

16.7 Multiple metal–metal bonds

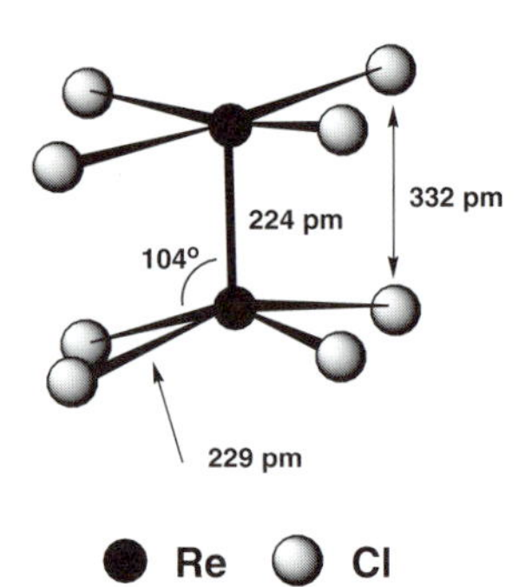

FIG. 16.12 Structure of the $Re_2Cl_8^{2-}$ ion

The discovery in 1964 of the ion $Re_2Cl_8^{2-}$ brought into the focus of chemists' attention the existence of metal–metal bonds of high bond order. The $Re_2Cl_8^{2-}$ ion has the eclipsed structure shown in Fig. 16.12. There are two extraordinary features of this structure: the extremely short Re-Re distance of 224 pm (compare 274 pm in Re metal) and the eclipsed configuration which makes the Cl-Cl distance between opposite halves of the molecule markedly less than twice the Cl van der Waals' radius (compare Table 2.11).

It was soon demonstrated that a wide range of similar rhenium(III) species existed, of general type $Re_2X_8^{2-}$ where X was a univalent anion, and with Re-Re distances in a very narrow range, around 222 pm. Such species often occurred with further ligands, such as H_2O, on the Re atoms in the axial position *trans* to the Re-Re bond but this substitution had only a minor effect on the Re-Re distance. A further series of compounds with very short M-M bond lengths were those with carboxylic acid groups bridging the two metals, of the type $Re_2(O_2CR)_4^{2+}$. Mixed species like $Re_2Cl_4(O_2CCH_3)_2$ are also found. The dinuclear species bridged by four acetate groups, $M_2(O_2CCH_3)_4$, was already well known e.g. for M = Cr (Fig. 14.20) or M = Cu (Fig. 14.33). The molybdenum acetate, $Mo_2(O_2CCH_3)_4$, has a short Mo-Mo distance of 211 pm, thus extending the class of multiple metal–metal bonds to Mo(II). The chromium compound has Cr-Cr of 238 pm (compare the metal, 258 pm) so it also belongs to this class. (Note that not all dimeric acetates show multiple bonds—the copper compound has a long Cu-Cu distance of 265 pm compared with 256 pm in the metal).

The Cr_2 molecule has been identified in the gas phase from photolysis of $Cr(CO)_6$ and has an extremely short Cr-Cr bond length of 168 pm, which was discussed as a sextuple bond.

A further group of short metal–metal bonded species was found for neutral compounds M_2X_6 where M = Mo(III) or W(III) and X = halide, NR_2^-, OR^-, etc. Other formulae include the rather rare M-C σ bond in $W_2(CH_3)_8^{4-}$, bridging carbonate or sulfate analogs to the acetates, e.g. $Re_2(SO_4)_4^-$, and the cyclooctatetraene derivatives $M_2(C_8H_8)_3$ (M = Cr or W) with one bridging C_8H_8 group (Fig. 16.13a).

The first stable technetium species was $Tc_2(O_2CCMe_3)_4Cl_2$ (Fig. 16.13b) which shows axial ligands and seems to require them for stability as $Tc_2Cl_8^{4-}$ is unstable.

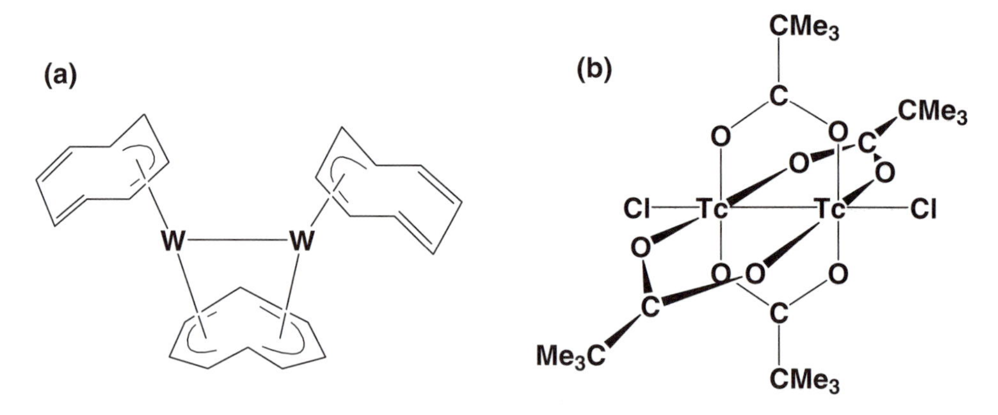

FIG. 16.13 Unusual metal–metal bonding: (a) $W_2(C_8H_8)_3$, (b) $Te_2(O_2CCMe_3)_4Cl_2$

TABLE 16.3 Species with multiple metal–metal bonds

	M-M *distance (pm)*	*Bond order*	*Comments*
$V_2[C_6H_3(OMe)_2]_4$	220	3	Few examples
$Cr_2[O_2CR]_4$	185–254	4	Many examples: very variable
$Cr_2[C_8H_8]_3$			bond length
$Cr_2[C_5Me_5(CO)_2]_2$	228	3	
$Mo_2[O_2CR]_4$ } $Mo_2X_8^{4-}$ }	209–214	4	Many examples including mixed species and extra D
$Mo_2[SO_4]_4^{3-}$	216	3.5	Compare 211 pm in $Mo_2[SO_4]_4^{4-}$
Mo_2X_6	220–224	3	Many examples including extra D
$W_2[Cl_3Me_5]^{4-}$	226	4	Also $W_2Me_8^{4-}$
$W_2[C_8H_8]_3$	238	4	
W_2X_6	227–230	3	Many examples including extra D
$Te_2[O_2CCMe_3]_4Cl_2$	219	4	$Tc_2Cl_8^{4-}$ very unstable
$Te_2Cl_8^{3-}$	212	3.5	
$Re_2X_8^{2-}$	220–224	4	Many examples including mixed species and extra D
$Re_2[O_2CR]_4^{2+}$	223	3	
$Re_2O_8^{8-}$	226	3	In $La_4Re_2O_{10}$
$Ru_2(O_2CR)_4Cl$	228	4 (3 unpaired electrons)	Chain structure linked through anion Cl^-
$Rh_2(O_2CCH_3)_4$	239	3?	Compare 246–255 for Rh=Rh and 280 ± 15 for Rh-Rh

Notes (1) Where a range is given, this shows the commonest bond lengths. Unusual species greatly widen the range, e.g. 218 pm in $Re_2Me_8^{2-}$ and 226 pm in $Re_2Cl_5(diphos)_2$.
(2) X = halide, OR^-, NR_2^- and other monovalent monodentate anions. O_2CR = carboxylic anion or other bridging bidentate species. D = neutral donor such as H_2O, R_3N, R_3P.
(3) Multiple bonds are found in rather different types of compound,
(a) For Fe in $[(C_4R_4)_2Fe](\mu\text{-}CO)_3$ and $[(R_3P)_3Fe]_2(\mu\text{-}H)_3$ which have Fe-Fe = 218 pm and 234 pm respectively (bond order 3 postulated).
(b) For Ir, in $[(R_3P)_2(H)Ir]_2(\mu\text{-}H)_3^+$ a triple bond is also postulated.
(c) Long triple bonds are found for Cr, Mo, W in species of the type $(C_5H_5)_2M_2(CO)_4$.

While a multitude of homonuclear multiply bonded complexes are known, Table 16.3, compounds with two different metals are rare. Until recently, the only examples of such heteronuclear compounds contained two metals from the same Group of the Periodic Table, such as Cr-Mo or Mo-W dinuclear complexes. Only very recently, the dinuclear quadruply bonded complex $[(TPP)MoRe(OEP)]^+$, containing two metals from different Groups (Mo and Re), has been synthesized. In this complex, both metals are coordinated by macrocyclic porphyrin ligands; the porphyrin ligand on Mo is a tetraphenylporphyrin (TPP)

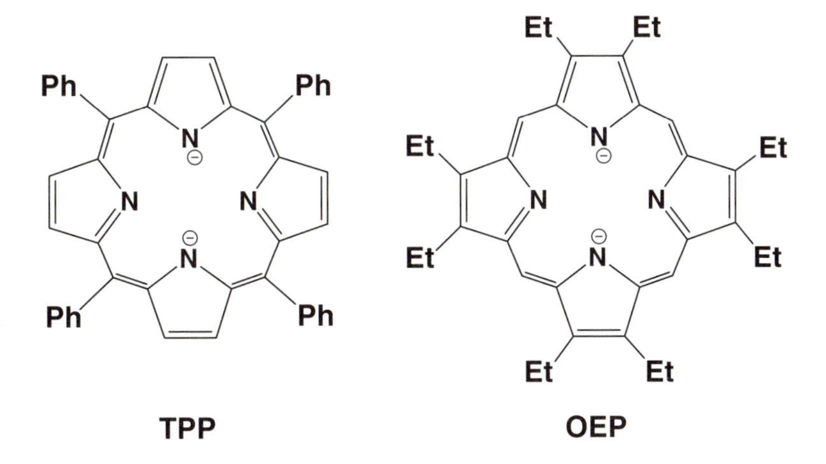

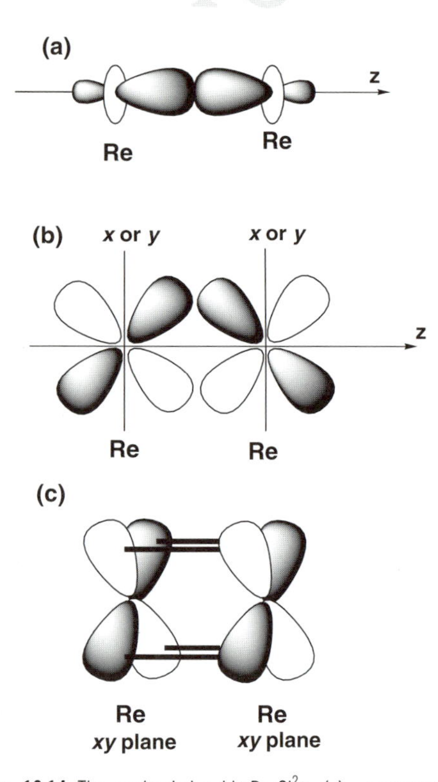

FIG. 16.14 The quadruple bond in $Re_2Cl_8^{2-}$: (a) σ component, (b) π (one of the two components), (c) δ component

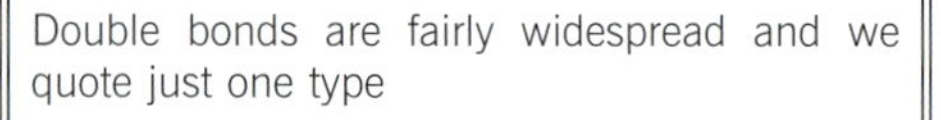
Double bonds are fairly widespread and we quote just one type

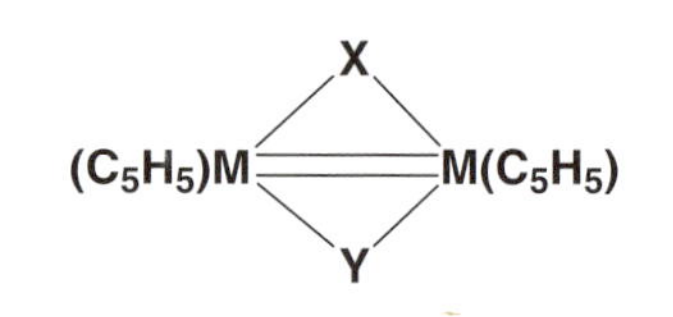

found for a number of metals M (e.g. Fe, Rh) and a range of bridging groups X and Y including CO, NO, NR_2 and organic groups.

while that on Re is an octaethylporphyrin (OEP); the two porphyrins differ in the positions and type of substituents on the macrocycle.

16.7.1 Bonding

The unusual properties of $Re_2Cl_8^{2-}$ were explained by Cotton by postulating a quadruple Re-Re bond, Fig. 16.14. If we consider the Re-Re axis as z, and x and y to lie along Re-Cl directions, then the square-planar $ReCl_4$ unit will be bonded using the Re s, p_x, p_y and $d_{x^2-y^2}$ orbitals. The Re-Re σ bond is then formed using the d_{z^2} orbital, on each Re. (If we include p_z contributions as well, a second orbital on each Re, pointing outwards along the z axis, is suitable for bonding the additional axial ligands.) The d_{xz} and d_{yz} orbitals on each Re may then overlap to give two Re-Re π bonds. This leaves the d_{xy} orbital on each rhenium to form a δ bond.

Thus a bond of order 4 may be formulated with the occupancy $\sigma^2\pi^4\delta^2$. We see that the δ component explains the eclipsed configuration (and indeed was proposed because of this configuration) whereas a *pair* of π bonds allows any configuration. The δ component makes a relatively weak contribution to the total bond strength since the two nodal planes reduce the electron density (compare π bonds which are in turn weaker than σ bonds—see Fig. 3.13). It has been estimated that the δ contribution is less than 15% of the total bond energy. Calculated stabilization energies in $Mo_2Cl_8^{4-}$ were approximately -7.3, -6.1 and -4.9 eV for σ, π and δ electrons, respectively.

If we take the $Re(III)_2X_8^{2-}$ and the carboxylate species as bases, it will be clear that the corresponding Cr(II), Mo(II) or W(II) species $M_2X_8^{4-}$ and $M_2(O_2CR)_4$ are isoelectronic and will also be formulated with a quadruple bond. If we then move to Mo(III) or W(III), and at the same time take two X^- ligands away, we leave enough electrons to form $\sigma^2\pi^4$.

Thus we obtain the well-represented M_2X_6 class. As the triple bond does not impose barriers to rotation, these species are staggered in configuration like ethane. We can add a donor group (retaining the triple bond), as in $Mo_2(OR)_6.2NHMe_2$, and find that the M_2X_8 configuration is adopted with approximately planar $M(OR)_3(NHMe_2)$ units, but the two ends of the molecule are again staggered.

A triple bond also results if 2 electrons are added to the quadruple bond since the lowest empty orbital is δ^*. Thus we get the $\sigma^2\pi^4\delta^2\delta^{*2}$ configuration for $Re_2Cl_4(PEt_3)_4$ (regard the 4P lone pairs as replacing $2Cl^-$ lone pairs and 2Cl single electrons—hence the two additional electrons).

Finally, if we remove one electron from the quadruple bond we get $\sigma^2\pi^2\delta^1$, corresponding to a bond order of 3.5. Two pairs related in this way are $Mo_2(SO_4)_4^{4-}$ with $Mo_2(SO_4)_4^{3-}$ and the unstable $Tc_2Cl_8^{2-}$ with $Tc_2Cl_8^{3-}$. In most cases, oxidation causes structure change as in the reaction of $Re_2X_8^{2-}$ with halogens to form the face-sharing bis-octahedral structure of $Re_2X_9^-$.

Quite often, reactions of these multiply bonded species occur with little change to the M_2 unit. A striking case is the addition of axial donor groups: for example the Mo-Mo distance in $Mo_2[O_2CCF_3]_4$ is 208 pm and in $(C_6H_5NMo)_2[O_2CCF_3]_4$ it is 213 pm. Some M_2X_6 species undergo insertion reactions:

$$W_2(NMe_2)_6 + 6CO_2 \rightarrow W_2[O_2CNMe_2]_6$$
$$Mo_2(NMe_2)_6 + 4CO_2 \rightarrow Mo_2[(NMe_2)_2(O_2CNMe_2)_4]$$

The substituted species have very similar M-M bond lengths but the M coordination number increases. For example, $W_2(O_2CNMe_2)_6$ has two bridging $W\text{-}OC(NMe_2)\text{-}OW$ units together, one bidentate and one monodentate ligand on each W, making W 5-coordinate to oxygens.

These metal–metal bonds of high order complete a natural progression which starts from very weak interactions—e.g. Cr-Cr of 391 pm in $Cr_2Cl_9^{3-}$ and the $Cu_2[O_2CCH_3]_4$ case mentioned above—through single bonds (compare Figs 16.3 and 16.4), and double bonds to these cases of order 3 to 4.

The steps in bond order are not nearly as sharp as they are in Main Group chemistry (e.g. between C-N, C=N, and C≡N) but the whole range of species represents a continuum of metal–metal interactions.

16.8 Transition metal clusters

Many areas of interesting chemistry involve compounds where a number of metal atoms are bonded together directly and the M-M bonds are often supported by bridging ligands. If we use a broad definition of a cluster as any species with more than two M atoms linked together, then we would include the compounds of Figs 15.25, 15.30, 16.4 and 16.5 as clusters. Such compounds represent only the tip of the iceberg of a large field of substantial current interest. Clusters are known under ambient conditions ranging in size from three to over a hundred metal atoms and an even larger variety has been detected by spectroscopic and other means. The development has stemmed partly from improved synthesis and characterization—in particular in the greatly increased accessibility and improved performance of X-ray crystallography. In addition, the work has been driven by the possibility of using soluble metal clusters in place of heterogeneous catalysis on metal surfaces. With the smaller clusters this idea largely failed, mainly because a group of a few metal atoms does not reproduce the properties of bulk metal or even of metal surfaces. However, the larger clusters which have become available more recently do begin to act as bulk metal and interesting developments can be expected. Thus, in a Pt_{309} cluster, it was found that 147 core atoms do behave like bulk metal whereas the 13 core atoms in Au_{55} still differ in charge density from the bulk. Work with clusters of this size is right at the limit of current techniques, falling within the limits in which X-ray crystallography and scanning electron microscopy are most effective (Section 7.11). Much work has been reported on smaller clusters, with intensive study of species containing from 4 to about 20 metal atoms.

We will look briefly at three classes of these metal clusters.

16.8.1 Halide clusters

We have already met a few examples of transition metal clusters in the M_6X_n species formed by Zr (Fig. 15.3), Nb and Ta (Fig. 15.9), or Mo and W (Fig. 15.19). Further halogen-bridged clusters are reported, including $[Ti_6C](\mu\text{-}Cl_{12})Cl_6$ which resembles Fig. 15.9b but also contains a central C atom in an octahedral site.

Extended halogen-bridged structures are known, including chains and sheets of linked octahedra, and this area extends into the layer and three-dimensional structures of the lower halides and oxides. An exotic example is $[Y_{16}Ru_4I_{20}]_n$ where a central tetrahedron of Ru atoms is sheathed by 16 Y which form fused octahedra containing Ru at the centre. The whole metal cluster is sheathed by Y-I bonds and the I atoms bridge to other clusters giving a three-dimensional network.

However, a major focus in cluster work has been on species like the carbonyls which have no or few bridging groups and are clearly held together by M-M cluster bonding.

16.8.2 Carbonyl and related clusters

To sample this extensive field of chemistry we look first at some moderate-sized clusters and then at a small selection of larger species.

In Table 16.1 we have noted the formation of $M_4(CO)_{12}$ carbonyls by Co, Rh and Ir and there are a variety of related compounds such as $[H_2Ir_4(CO)_{10}]^{2-}$, $Ru_2Co_2(CO)_{13}$ or $[Fe_4(CO)_{13}]^{2-}$ which all contain a tetrahedron of metal atoms with various combinations of terminal and bridging CO. In contrast, $[HFe_4C(CO)_{12}]^-$ and $[Os_4N(CO)_{12}]^-$ have a butterfly structure with two triangles sharing an edge (Fig. 16.15a). Similarly, $Os_5(CO)_{16}$ contains a trigonal bipyramidal Os_5 skeleton (Fig. 16.15b) while $Os_5C(CO)_{15}$ has a square pyramid of Os atoms (Fig. 16.15c). Many examples of octahedral clusters of metal atoms are known including $[M_6(CO)_{16}]$ [M = Co, Rh, Ir (Fig. 16.15d) with 2 terminal CO on each M and the remaining CO bridging], and $[M_6(CO)_{18}]^{2-}$ [M = Fe, Ru, Os (Fig. 16.15d) with 3 terminal CO on each M] and the similar $[M_6(CO)_{17}]$ [with C at the centre of the octahedron]. Two octahedra joined by a Rh-Rh(μ^2-CO)$_2$ bridge occurs in $[Rh_{12}(CO)_{30}]^{2-}$. Interstitial atoms are now well established with B, C and N found in smaller clusters while larger units, such as C_2 or even C_3, and heavier atoms like Si, Ge, Sn, P, As or Sb, are found interstitially in larger clusters.

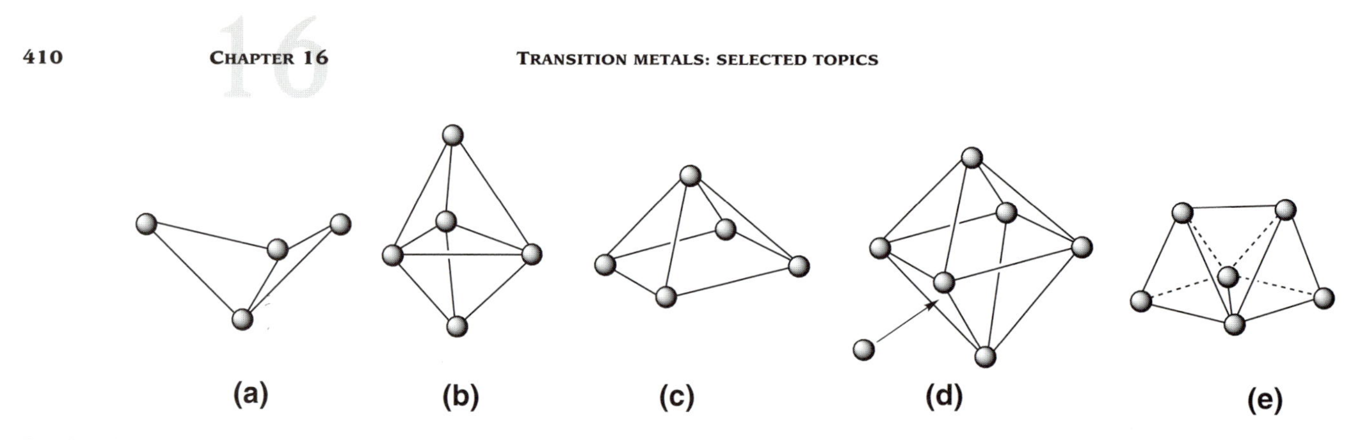

FIG. 16.15 Some small cluster skeletons: (a) M_4 butterfly, (b) M_5 trigonal bipyramid, (c) M_5 square pyramid, (d) M_6 octahedron, also indicating a capping site for M_7, (e) M_6 capped trigonal bipyramid

Even interstitial H is now well established, usually in an octahedral site as in $[HRu_6(CO)_{18}]^-$.

Rationalization of such variations has been approached by use of the 18-electron rule (see Section 16.2.1) or by an extension of SEP theory (compare Section 9.6). You can determine that the 18-electron rule works for the four- and five-metal examples above but fails for the six-metal ones.

For example, the two Fe_4 species give electron counts of 60 for $[Fe_4(CO)_{13}]^{2-}$ and 62 for $[HFe_4C(CO)_{12}]^-$. To achieve $4 \times 18 = 72$ electrons thus requires six Fe-Fe bonds in the former case and five in the latter. The tetrahedron allows 6 M-M bonds along the six edges, whereas the pair of triangles sharing one edge need only 5 M-M bonds. The cluster $[Fe_6(CO)_{18}]^{2-}$, for example, has 86 electrons, compared with 108 required by the 18-electron rule which would require 11 M-M bonds. However, the octahedron has 12 edges (and the structure is fully symmetrical) so the idea of each metal being joined by a normal bond breaks down.

To apply Wade's rules we note that the modification required is to count 12 electrons per metal atom for external bonding, in place of the 2 electrons per B atom in the boranes. The calculation of the number of skeletal electron pairs for $[Fe_6(CO)_{18}]^{2-}$, for example, then proceeds: from the 86 electrons subtract $6 \times 12 = 72$ electrons for external bonding, leaving 7 SEP. Thus (compare Table 9.6) the expected structure is an octahedron. Note that in electron counting of these species atoms like C or N which are encapsulated are taken to add all their valency electrons to the count and to be incorporated in the skeletal bonding without requiring any specific allocation of electrons.

A further example is the two Os_5 cases quoted above:

$Os_5(CO)_{16}$

5 Os valency electrons $= 5 \times 8$	$= 40$
16 CO contributing 2 electrons each	$= 32$
less 12 electrons per Os	$= -60$
giving 12 electrons or 6 SEP	

Thus the structure is the 5-vertex cluster, a trigonal bipyramid [Fig. 16.15b: all vertices are $Os(CO)_3$ apart from one equatorial $Os(CO)_4$].

$Os_5C(CO)_{15}$

5 Os	40 electrons
C	4 electrons
15 CO	30 electrons
less 12 per Os	− 60 electrons
giving 7 SEP	

Thus the structure is an octahedron less one vertex, a square pyramid (Fig. 16.15c: all vertices are $Os(CO)_3$ and the C lying just below the base).

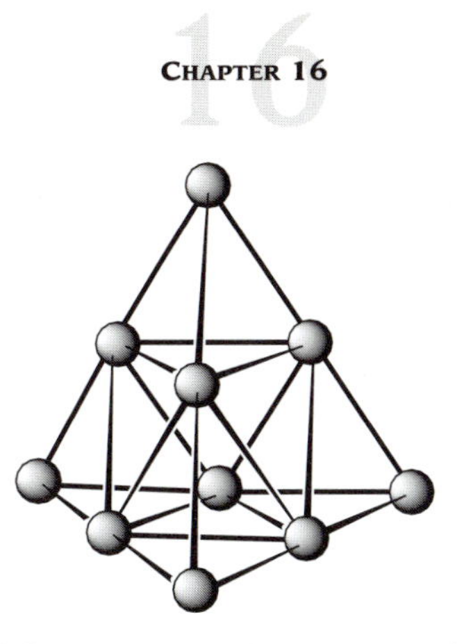

FIG. 16.16 The cubic close-packed metal core of $[H_4Os_{10}(CO)_{24}]$

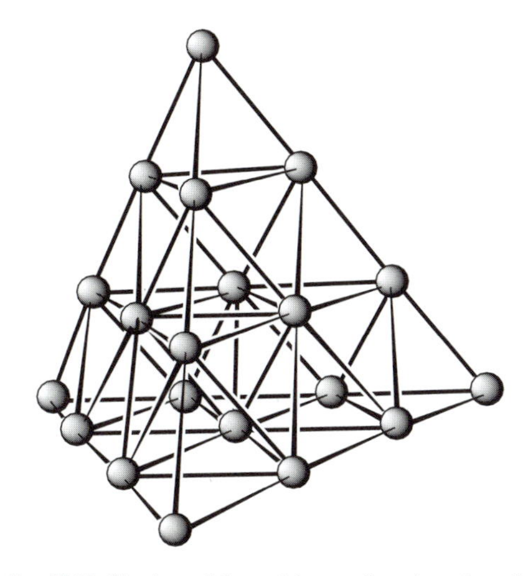

FIG. 16.17 Structure of the metal core of osmium atoms in the giant cluster $[Os_{20}(CO)_{40}]^{2-}$

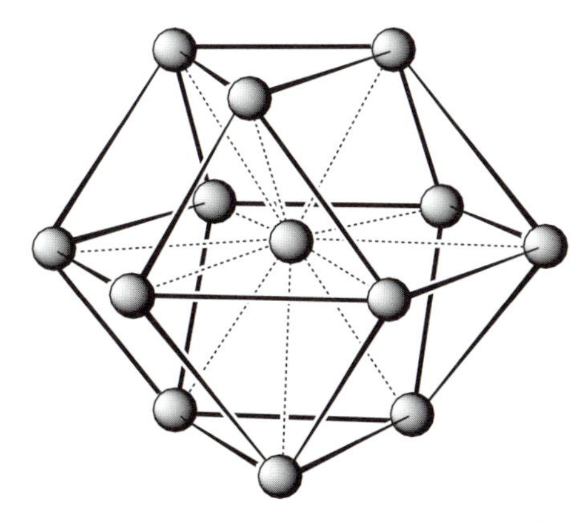

FIG. 16.18 The metal core of the cluster $[Rh_{13}(CO)_{24}H_x]^{(5-x)-}$

The series $[Os_6(CO)_{18}]^{2-}$, $Os_7(CO)_{21}$ and $[Os_8(CO)_{22}]^{2-}$ is interesting as the counts are each 7 SEP. Thus the expected structures are octahedra and the extra metal atoms in the two latter cases are in capping positions (compare Fig. 16.15d).

A significant test case is $Os_6(CO)_{18}$. While there are a large number of octahedral M_6 complexes, this molecule is one of a few examples of the capped trigonal bipyramid (bicapped tetrahedron) shown in Fig. 16.15e with all vertices $Os(CO)_3$. The electron count $(48 + 36$ less $6 \times 12)$ gives only 6 SEP, thus predicting a trigonal bipyramid with the extra atom taking up a capping position.

Out of the large number of carbonyl clusters with 10 or more metal atoms we can look briefly at only one or two structural themes. Figure 16.16 shows the skeleton of the $[H_5Os_{10}(CO)_{24}]^-$ ion and related clusters. The electron count gives 7 SEP so this is an octahedron with four caps. These are the outer apices in the figure and each is an $Os(CO)_3$ unit. All the rest are $Os(CO)_2$ units defining the octahedron. An alternative way of looking at this structure is as a large tetrahedron with atoms placed at the mid-points of each edge or, alternatively, built up of layers containing 1, 3 and 6 atoms. A big brother is known, $[Os_{20}(CO)_{40}]^{2-}$ (Fig. 16.17), which is a giant tetrahedron containing layers of, successively, 1, 3, 6 and 10 atoms. In a rather similar way $[Ni_{38}Pt_6(CO)_{48}]^{6-}$ is close to a giant octahedron with 10 atoms in each face (or layers of 1, 4, 9, 16, 9, 4 and 1 atoms).

Such layers are in the positions corresponding to close-packing. A further family of structures can also be regarded as being built up from close-packed layers. The simplest example is $[Rh_{13}(CO)_{24}]^{5-}$, Fig. 16.18, which has a central Rh inside a regular cluster—alternatively seen as close-packed layers of 3, 7 and 3 atoms. Related species also contain encapsulated H atoms. Larger clusters of this type are for example, a Rh_{22} species with layers arranged 3, 6, 7, 3, $[Pt_{26}(CO)_{32}]^{2-}$ with 7, 12, 7 layers and a Pt_{38} species with 7, 12, 12, 7 layers.

A further theme is stacked polygons. Compounds with up to five stacked triangles have been established and stacked squares or pentagons have also been found, along with a hexagonal prism. A particularly symmetric member of this family is $[Pt_{19}(CO)_{22}]^{4-}$ which has $\mathbf{D_{5h}}$ symmetry and consists of three pentagonal bipyramids sharing apices (alternatively described as a 1, 5, 1, 5, 1, 5, 1 system).

16.8.3 Gold clusters

Different structural features are seen in the rapidly developing field of gold clusters which often incorporate silver atoms also.

Gold(I)–sulfur chemistry has yielded the complex ion $[Au_{12}S_8]^{4-}$ which has an unusual cubic structure in which the S atoms are at all the corners and the 12 Au atoms are in the centres of the edges. The structure is slightly distorted with angles at S in the range 87–93° and the Au atoms in essentially linear configuration, as preferred. The Au···Au distances are 318–335 pm which suggests a weak interaction (see Section 16.13 for discussions of Au···Au interactions).

In lower formal oxidation states gold forms an extensive range of cluster compounds. In $[Au_4(PR_3)_4]^{2+}$ gold atoms, each with one terminal phosphane, form a regular tetrahedron with very short Au-Au edges. Short distances are also found, for example, in $[Au_6(PR_3)_6]^{2+}$ which has formal oxidation state 0.33. The gold atoms, again each with one terminal phosphane, form an axially compressed octahedron. This form is paralleled in $[Au_7(PPh_3)_7]^+$ in which the 7 Au form a pentagonal bipyramid with a short axial Au-Au distance.

Further examples of the poly-gold clusters include $[Au_9L_8]^{3+}$, $[Au_{11}L_7]^{3+}$, $[Au_{13}L_{12}]^{3+}$ and related species where L includes phosphanes and halides. The Au_{13} species has one Au atom at the centre of an Au_{12} icosahedron. The other two species also have a centred Au with the remaining atoms forming an incomplete icosahedron. Evidence has been presented for the much larger $[Au_{55}(PPh_3)_{12}Cl_6]$ cluster containing an inner Au_{13} icosahedron surrounded by an outer shell of Au, Au-P and Au-Cl units. Very large mixed silver–gold clusters have also been found. A closely related pair are $[Au_{13}Ag_{12}(PR_3)_{12}Cl_7]^{2+}$ and $[Au_{13}Ag_{12}(PR_3)_{10}Br_8]^+$ (Fig. 16.19) whose structures consist of two icosahedra sharing

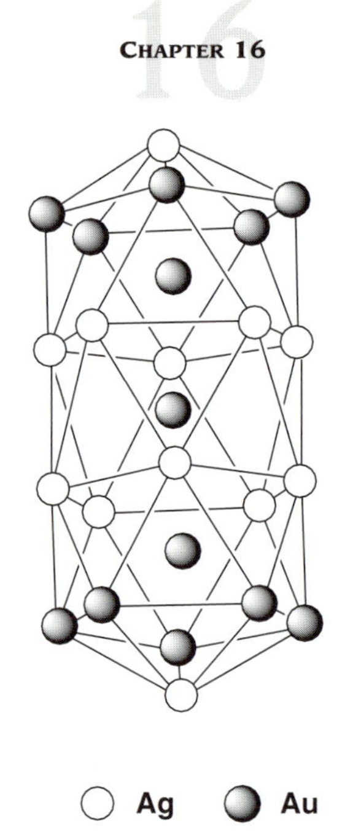

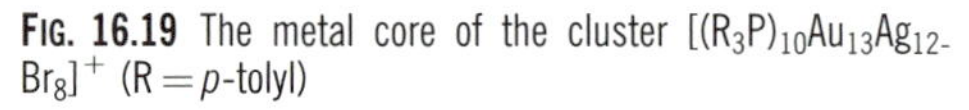

FIG. 16.19 The metal core of the cluster $[(R_3P)_{10}Au_{13}Ag_{12}Br_8]^+$ (R = *p*-tolyl)

a vertex with a gold atom at the centre of each. An alternative description of this structure is as a succession of alternate pentagons and single atoms in an array: Ag, 5Au, Au, 5Ag, Au, 5Ag, Au, 5Au, Ag. The two structures differ in the relative orientations of the pentagons and in the distribution of the ligands. In the chloride ten of the phosphane ligands (R = *para*-Tol) bond to the Au atoms and two more to the two apical Ag. In the bromide the halogens bridge the two central Ag_5 pentagons and also bond to the terminal Ag.

A number of larger clusters belong to this family, including $[Au_{18}Ag_{17}(PPh_3)_{12}Br_{11}]^{2-}$ and $[Au_{22}Ag_{24}(PPh_3)_{12}Cl_{12}]$ whose structures are also based on icosahedra sharing vertices. These large and magnificent structures show how giant molecules can be assembled using clusters (in these cases the centred M_{13} icosahedron) as the building blocks to form clusters of clusters.

The L-Au(I) unit readily attaches to other metal clusters so that gold is incorporated in clusters with many metals. Examples are VAu_3L_9, containing a VAu_3 trigonal pyramid, and $Fe_4Au_2(C)L_{14}$ in which the 2 Au atoms occupy neighbouring positions in an octahedron enclosing the C. In such compounds L is generally CO or a phosphane, or similar ligands in combination.

16.8.4 Very large clusters

The larger metal clusters quoted so far have reached a size where experimental problems arise in isolation and characterization. In particular, X-ray crystallography meets difficulties as the strong scattering from the metal core starts to swamp the picture. Thus, there are sometimes uncertainties about how many CO or other lighter groups are present. Despite this, compounds with around 40 metal atoms have been identified, such as $[Pt_{38}(CO)_{44}]^{2-}$ which contains five layers with, respectively 4, 9, 12, 9 and 4 Pt atoms in approximately square arrays. A mixed-metal example is $[Ni_{36}Pt_4(CO)_{45}]^{6-}$ where a tetrahedron of Pt atoms is encapsulated in a truncated giant tetrahedron of Ni atoms (one size up from Fig. 16.17). One approach to systematic synthesis of larger clusters notes that the ratio of ligand to metal decreases as complexes get larger—e.g. 3 or 4 CO per M in M_3, M_4 or M_6 species down to about 2 for Figs 16.17 and 16.18. Schmid used even lower M/ligand ratios, and also bulkier ligands such as PPh_3 in place of CO, and isolated even larger clusters termed *ligand-stabilized clusters*. In this way, $[Au_{55}(PPh_3)_{12}Cl_6]$ was prepared which has been used as the model compound for investigating the onset of metallic properties (see Box). Larger, though less completely defined, clusters formed by this route include mixed-metal clusters containing silver and gold, a Pt_{309} species, and palladium complexes with 500 to 2000 metal atoms.

Working in the other direction, downwards in size from colloids, it is possible to probe the region between large clusters and metallic particles. Colloids such as palladium absorb CO readily and surface Pd–CO groups have been identified by infrared and ^{13}C nmr spectroscopy. From the cluster side, organic groups bonded to a metal atom in a cluster like Fig. 16.17 do interact with other metal atoms in the face and undergo unusual reactions. Such studies serve to clarify reactions on metal catalysts leading to novel or improved applications.

> The smallest colloid particles are some 10 times the diameter of these large metal clusters and contain 5000 to 20 000 atoms.

Such studies have built up structures where some of the metal atoms have only other metal atoms as their nearest neighbours and the question arises, at what point do they start showing properties of bulk metals? Thus, in Fig. 16.18, the central Rh is surrounded by 12 Rh atoms in a (slightly expanded) close-packing array (compare Section 5.6). A similar configuration is found for each of the three central gold atoms of Fig. 16.19. In larger clusters, a group of metal atoms may be similarly surrounded such as the Pt_4 tetrahedron described above. Schmid's Au_{55} cluster has an inner Au_{13} icosahedron surrounded by an outer shell of Au, Au-P and Au-Cl units so the Au atom at the centre of the icosahedron is surrounded by a double shell of metal atoms.

The largest synthesized clusters start to resemble metals, and it is expected that such materials will display interesting electronic properties. The diameters of these clusters are on the scale of nanometres (1.4 nm for the Au_{55} core of the Schmid cluster) and are being studied as part of the current interest in 'nanomachines' and nanoscale electronics (see Box).

LIGAND-STABILIZED CLUSTERS, QUANTUM DOTS AND MOLECULAR WIRES

When does a metal cluster act like a metal? Studies in the last two decades clearly show that metal particles in the 1–4 nm size range are electronically characterized by quantum size effects—that is, instead of electrons being in conduction bands which are delocalized over the whole cluster in three dimensions (compare Section 5.6) they fall into discrete energy levels. This is a transitional region between the fully delocalized electrons in a metal, and the electrons in a molecule which occupy discrete energy levels. Particles showing such quantum size effects have been termed *quantum dots* and they show novel electronic properties which are of great interest in microelectronics.

The ligand-stabilized clusters fall into this size range and the Au_{55} cluster has been studied in detail. It turns out that not only are its two uppermost electrons trapped in the quantum box of the 55Au core, but that the diameter of the core plus ligands, equal to 2.1 nm, allows electrons in the first excited state to tunnel between clusters when they are held in contact.

Exploitation of these properties depends on being able to order the clusters. A one-dimension ordering or 'molecular wire' can be achieved by adsorbing the clusters onto an atomic step on a crystal plane. An alternative is to use nanopores in alumina as tubes and fill them with clusters—creating bundles of molecular wires. The formation of parallel tubes in Al_2O_3 formed by anodizing aluminium is a well-understood process and experiments to fill 1–200 nm lengths show promise.

Ordering in two or three dimensions is more difficult as the complexes precipitate as powders. Studies have explored bonding the complexes to metal or silica surfaces by substituting the phenyl groups of the PPh_3 with active groups, as in $P\text{-}C_6H_4SO_3^-$. Work in two- and three-dimensional ordering has also explored the use of 'spacers' created by using chains of different lengths and rigidities bonded to the phenyl groups to create bridges between clusters or to substrates.

The chemistry, by providing flexibility in the size of the metal cluster and in the properties of the stabilizing ligands, creates a series of compounds which are valuable in the study of materials on the borderline between metals and molecules.

Reference

G. SCHMID, *Journal of the Chemical Society, Dalton Transaction* 1998, 1077–82.

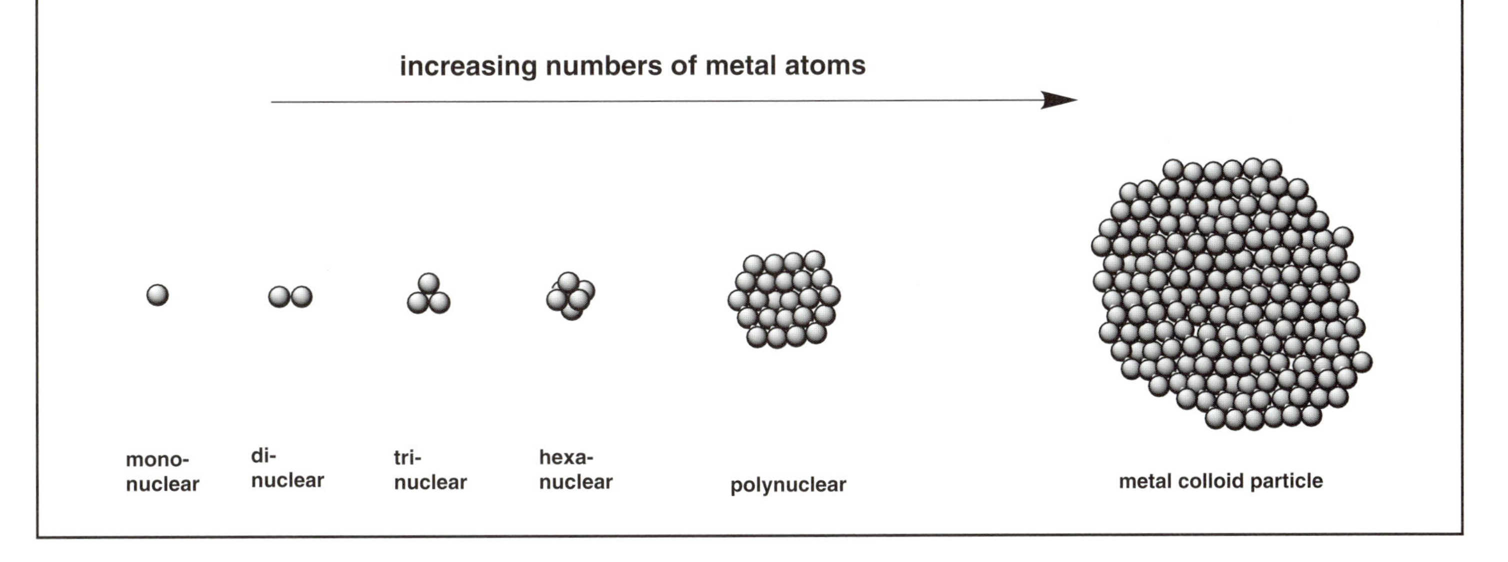

16.9 Metal–dioxygen species

Several areas of work have converged to create the current interest in species containing O_2 groups bonded to metals. Compounds resulting from the action of hydrogen peroxide have long been of interest, e.g. in analysis— see Sections 14.2 to 14.4 for the Ti, V and Cr species as examples. Secondly, the reversible uptake of oxygen gas by haemoglobin (see Fig. 20.4) and other oxygen-carrying proteins has been shown to depend on dioxygen–metal coordination. Much work has gone into the study of such systems and into the syntheses of simpler model compounds which might aid understanding. A third area of interest lies in catalysis of oxidation by O_2: here the basic species is again likely to be an O_2 M unit formed on the metal surface in the case of a heterogeneous catalyst or formed by homogeneous catalysts such as metallo-enzymes.

Table 16.4 M-O and O-O bond lengths (pm) in peroxo MO_2 units

Ti	*V*	*Cr*	*Mn*	*Fe*	*Co*	*Ni*
145–46	144–47	140–46			142–45	obs.
185–89	187–88	181–92			187–90	
Zr	*Nb*	*Mo*	*Tc*	*Ru*	*Rh*	*Pd*
obs.	148–51	138–55	obs.	obs.	142–47	obs.
	197–204	191–97			202–03	
Hf	*Ta*	*W*	*Re*	*Os*	*Ir*	*Pt*
obs.	obs.	150	obs.	obs.	130–52	145–51
		193			200–07	201

(a) Upper value is range of O-O distances in the reported compounds and the lower value is the M-O range.
(b) Obs. = peroxo species observed but no structural data.

There has been considerable debate about the formulation of metal–dioxygen species. First we recognize that it may be difficult to distinguish, let us say, M(I) plus neutral O_2 from M(II) plus O_2^- (superoxide) from M(III) plus O_2^{2-} (peroxide), since it may not be possible to determine the degree of electron transfer from M to the O_2 species. An example is given by the cobalt species in Section 14.7.1, where green $[(NH_3)_5Co\text{-}O\text{-}O\text{-}Co(NH_3)_5]^{5+}$ was originally formulated as a superoxo species containing one Co(IV) atom. Secondly, the method of synthesis does not, as was thought, give any guidance. For example, direct addition of molecular O_2 is found (Equation 16.4).

$$\underset{A}{[IrCl(CO)(PR_3)_2]} + O_2 \rightarrow \underset{B}{[(O_2)IrCl(CO)(PR_3)_2]} \quad (16.4)$$

but A is iridium(I) and B is best formulated as a peroxo complex of iridium(III). Indeed, it has now been demonstrated that the same species results either by the action of O_2 or by the traditional reaction of hydrogen peroxide in a number of cases such as Equation (16.5)

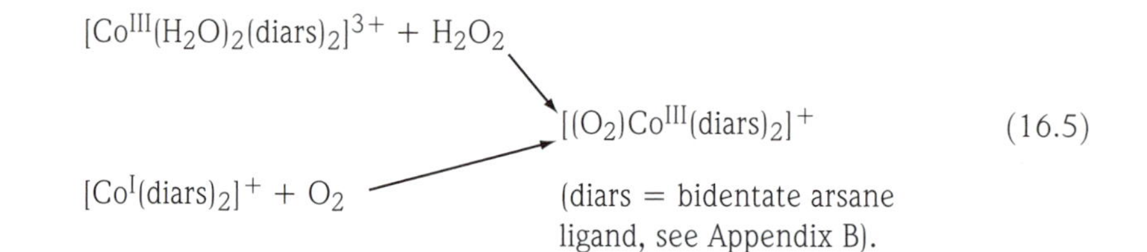

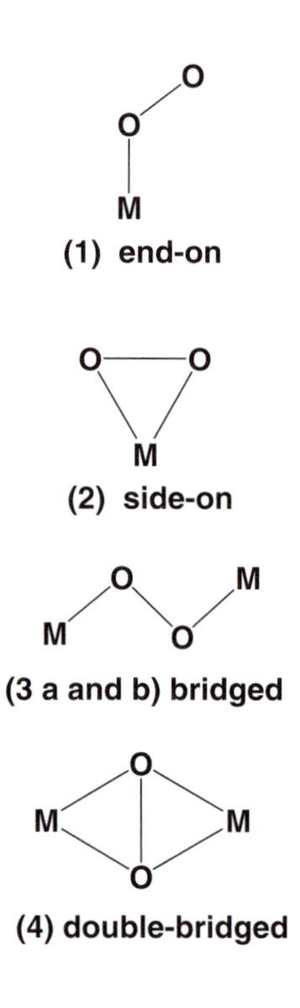

However, there is now sufficient evidence from molecular structure determinations to show that $M\text{-}O_2$ units exist as one or other of the forms (1) or (2) (see margin).

There are also two bridged forms: 3(a) where the M-O-O-M unit is all in one plane and 3(b) which is non-planar with a twist angle generally about 145° between the two MOO planes. Finally a double-bridged species has recently been reported.

The *side-on* form (2) is by far the commonest, and the structures of about fifty species have been determined. The two M-O distances are usually equal within experimental limits, and M-O increases with metal size while O-O distances are relatively constant. Table 16.4 lists the ranges found. The O-O distances fall in the range 130–155 pm and 90% are between 140 and 152 pm, while the overall average is 146 pm (compare 149 pm in O_2^{2-} and 147 pm in H_2O_2).

The number of established end-on structures of type (1) is much smaller but these are characterized by much shorter O-O bond lengths (compare $O_2^- = 128$ to 134 pm). This bond length evidence is supplemented by the O-O stretching frequencies, which have been assigned for a much larger number of complexes than have had crystal structures determined. Furthermore, when the M-O-O-M bridged structures are examined it is found that the values for the planar species (3a) fall into the same range as the end-on MO_2 species (1), while those for the non-planar species (3b) agree with the side-on MO_2 species (2). The values are summarized in Table 16.5.

Table 16.5 Parameters in metal–dioxygen species

Bond type	*Metal(a)*	*Average O-O distance* (pm)	*Average O-O frequency* (cm^{-1}) (*b*)
End-on (1)	Fe(III), Co(III), Rh(III)	125	1134
Side-on (2)	Table 16.4	145	881 (3d elements) 872 (4d elements) 850 (5d elements)
Planar bridge (3a)	Co(III)	131	1110
Twisted bridge (3b)	Mn(III), Fe(III), Co(III), Rh(III), Mo(VI), Co(II)?	144	807
Double bridge (4)	V(IV)	149	

(a) Metal species forming this structure type.
(b) Stretching frequencies (cm^{-1}) of $O_2 = 1556$ cm^{-1}, $O_2^- = 1145$ cm^{-1}, $O_2^{2-} = 770$ cm^{-1}.

Thus, apart from the double-bridge example, there are only two basic types of metal–dioxygen species:

(A) those with a shorter O-O distance and stretching frequency above 1075 cm^{-1}.

(B) those with a longer O-O distance and stretching frequency below 950 cm^{-1}. Type A shows the end-on structure or the planar bridge while type B shows the side-on structure or twisted bridge. While there are a few marginal cases, most known species fall firmly into one or other class.

The species of type A are formulated as *superoxo* compounds, formally derived from the superoxide ion O_2^-. The species of type B are *peroxo-*, based on the peroxide ion O_2^{2-}. Bonding to the metal mainly involves the filled π^* orbitals (compare Section 3.4) of the dioxygen unit. These electrons are markedly stabilized by their delocalization into metal orbitals and the reduced π^* density in the O_2 unit leads also to some stabilization of the O-O bond (seen especially in the slight shortening of the O-O bond and increase in stretching frequency for the peroxo-compounds). These descriptions for A and B are supported by a number of other physical properties and by theoretical analysis.

16.10 Compounds containing M-N_2 units and their relationship to nitrogen fixation

There is considerable chemical interest in the activation of a molecule as stable and strongly bonded as N_2, and in comparisons between complexes of N_2 and the isoelectronic species, especially CO and C_2^{2-}. By studying metal–dinitrogen complexes we can begin to obtain an understanding of the fundamental chemical steps which occur during nitrogen fixation by the so-called *nitrogenase* enzymes. This is of great technological importance—much energy is utilized annually in synthesizing ammonia fertilizer from dinitrogen at high temperatures and pressures. If we can begin to understand the fundamental chemistry occurring in these enzymes we can design catalysts to synthesize ammonia under very mild (and hence inexpensive) conditions.

A discussion on the nature of the structures of the active sites of the nitrogenases is given in the section on bioinorganic chemistry in Chapter 20, together with some of the attempts at directly mimicking the complexes responsible for the nitrogen fixation process. In this section we look more closely at the interaction of the dinitrogen molecule with transition metal centres and what reactions the coordinated N_2 molecule undergoes, which is of relevance to the problem of nitrogen fixation. Of particular importance is the reduction of a coordinated N_2 molecule to reduced, but still metal-bonded, nitrogen-containing species and ultimately to the desired, fully reduced product—ammonia.

16.10.1 Metal compounds with coordinated dinitrogen species

There are two distinct classes of reaction which fix dinitrogen into other chemical compounds:

(a) *Under non-aqueous conditions and with powerful reducing agents,* N_2 can be fixed in the form of air-sensitive and often poorly characterized materials which react with H_2O or other proton sources to yield NH_3, or partly reduced species such as N_2H_4. Of course, the simplest species of this type, such as Li which forms Li_3N with N_2, have long been known (compare Section 10.5). More complex systems are illustrated by the titanium species discussed in Section 14.2. A number of similar systems are well established but all depend on active organometallics as the reducing agents.

[Me_5C_5 is the fully substituted analogue of C_5H_5: the structures are probably 'bent sandwiches' of the type found in $(C_5H_5)_2TaH_3$, see Fig. 15.7).]

Thus these systems do not appear to provide a suitable route for industrial N_2 fixation, but they model the take-up of N_2 under very mild conditions and allow study of the steps of the subsequent conversion. Thus $TiCl_3$ and Mg react with N_2 at 25°C forming a species postulated as $TiNMg_2Cl_2$, via Ti^{II} and subsequent reaction with Mg. Under even milder conditions, at −78°C, $(Me_5C_5)_2Ti$ gives $(Me_5C_5)TiN_2$ which changes structure at −62°C and evolves N_2 quantitatively at 20°C. The two structures suggested contain N_2 respectively end-bonded and side-bonded to the $(Me_5C_5)_2Ti$ fragment. Other model reactions include the formation of N_2H_4 from N_2 via $Ti(OR)_2$ polymer [R = $(CH_3)_2CH$], and the ready formation of a substituted diimine complex $(C_5H_5)_2Ti(N_2Ph_2)$ perhaps with a sideways link as illustrated.

The compound $[Me_5C_5ZrN_2]_2N_2$ contains both linear Zr-N≡N units (N≡N of 112 pm) and bridging Zr-N≡N-Zr (N≡N of 118 pm). On addition of HCl, both N_2 and N_2H_4 are evolved and tracer studies showed half the hydrazine nitrogen was from the terminal N_2 and half from the bridge. The intermediate $(Me_5C_5)_2Zr(N_2H)_2$ was postulated.

One recent development in this field is the 3-coordinate molybdenum complex $Mo(NRAr)_3$ (Ar = 3,5-dimethylphenyl; R = *t*-butyl), Fig. 16.20 which cleaves dinitrogen. The first and slowest step (which takes several days) involves the uptake of molecular N_2 and formation of a bridging N_2 intermediate. The second stage of the reaction, which is much more rapid (half-life about 30 minutes) involves cleavage of the N_2 group to give two molybdenum-nitrido complexes, containing Mo≡N triple bonds. The $Mo(NRAr)_3$ complex also reacts with white phosphorus (P_4) to give the corresponding Mo≡P complex.

We may add to this class of reaction the work in aqueous systems which are a long way away from biological conditions of moderate pH and solution reactions. An example is provided by the observation that a coprecipitate in base of Mo(III-V) with 10% Ti(III) converts N_2 via N_2H_2 to hydrazine, with the active species probably Mo(IV). This type of system may become important in larger-scale synthesis, as it is cheap and accessible, even though it does not model the natural fixation route. We note again the formation of hydrogenated N-N compounds.

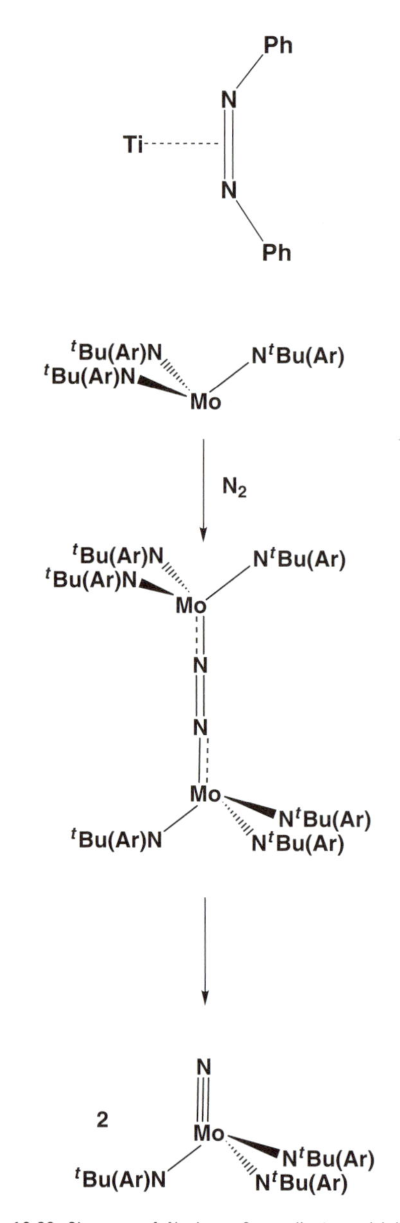

FIG. 16.20 Cleavage of N_2 by a 3-coordinate molybdenum complex (Ar = 3,5-dimethylphenyl)

(b) *In aqueous media,* relatively stable complexes are formed, mainly by elements in the middle of the transition series especially in d^6 states. These contain N_2 as a ligand but the complexes are unreactive or release unchanged N_2 when attempts are made to reduce them under mild or moderate conditions. Thus these compounds allowed a detailed study of the modes of M-(N_2) bonding.

The first of these compounds to be made was $[Ru(NH_3)_5N_2]^{2+}$ (see Fig. 15.27). Many others are now known (see Table 16.6 for some examples). Common preparations involve strong reduction in presence of N_2 and other ligands, as in

$$WCl_4(PR_3)_2 + Na/Hg + N_2 \rightarrow W(N_2)_2(PR_3)_4$$

or replacement of labile ligands, which often occurs reversibly under mild conditions.

$$Ru(NH_3)_5(H_2O)^{2+} + N_2 \rightleftharpoons Ru(NH_3)_5(N_2)^{2+} + H_2O$$
$$CoH_3(PR_3)_3 + N_2 \rightleftharpoons CoH(N_2)(PR_3)_3 + H_2$$

Table 16.6 Some representative stable N_2 complexes

Terminal N_2	Bridging N_2	
	$[(C_5H_5)_2Ti]_2N_2$	
$(C_6H_6)Cr(CO)_2N_2$		
$P_4M(N_2)_2$	$[(C_6H_6)PM]_2N_2$	M = Mo, W
$(C_5H_5)M(CO)_2N_2$		M = Mn, Re
$(C_5H_5)Re(CO)(N_2)_2$		
$P_4ReCl(N_2)$		
$P_3M(H)_2N_2$		M = Fe, Ru, Os
$[(NH_3)_5MN_2]^{2+}$	$[(NH_3)_5M]_2N_2$	M = Ru, Os
$P_3(H)CoN_2$	$(P_3Co)_2N_2$	
$P_2MCl(N_2)$		M = Rh, Ir
$P_3NiH(N_2)$	$[P_2Ni]_2N_2$	

Notes (1) PR_3 (phosphane) or $\frac{1}{2}\times$(diphosphane) = P.
(2) All are d^6 except the compounds of Ti(d^3), Co(d^8 and d^9), Rh and Ir(d^8) and Ni(d^9 and d^{10}).

A wide range of such coordinated dinitrogen compounds has been made, with various types of NN-M bonding, and differing stabilities. The next step in modelling fixation is to consider that it is unlikely that the naturally occurring reduction

$$N_2 \longrightarrow 2NH_3$$

is a direct one-step process. Instead, it seems probable that a number of intermediates follow the formation of the first complex, *S*-N_2 (*S* = substrate: here the metal atom site of nitrogenase).

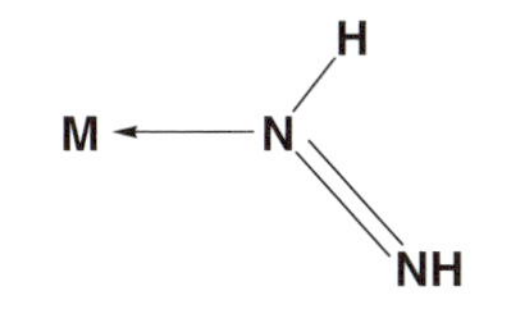

The coordinated diimine in (A)

A scheme which is consistent with most of the currently accepted observations is shown in Fig. 16.21. Here each addition step involves the supply of both H^+ and an electron, and the paths branch at some points. Thus, in addition to complexes containing bound N_2 in all its modes, work has also been carried out on diazenido (NNH), diazene or diimine (NHNH), hydrazido ($NHNH_2$), and hydrazine complexes, as well as single-nitrogen species containing M-N, M-NH and M-NH_2 groups.

To illustrate this, we cite one series of experiments.

The addition of H to N_2 bound to W has been established with reactions such as

$$trans\ [W(N_2)_2(diphos)_2] + 2HCl \longrightarrow [W(N_2H_2)Cl_2(diphos)_2] + N_2 \qquad (A)$$

M=N—NH_2

The coordinated hydrazide ligand in (B)

The Mo complexes behave similarly, as does HBr (diphos is a bidentate ligand such as $R_2PCH_2CH_2PR_2$). The complex (A) is 7-coordinate but will lose a halide ion

$$(A) + BF_4^- \longrightarrow [W(N_2H_2)Cl(diphos)_2]^+[BF_4^-] + Cl^- \qquad (B)$$

Structural studies show that the N_2H_2 group differs between (A) and (B). In (A), it behaves as a 2-electron donor—the unsymmetrically coordinated diimine HN = NH.

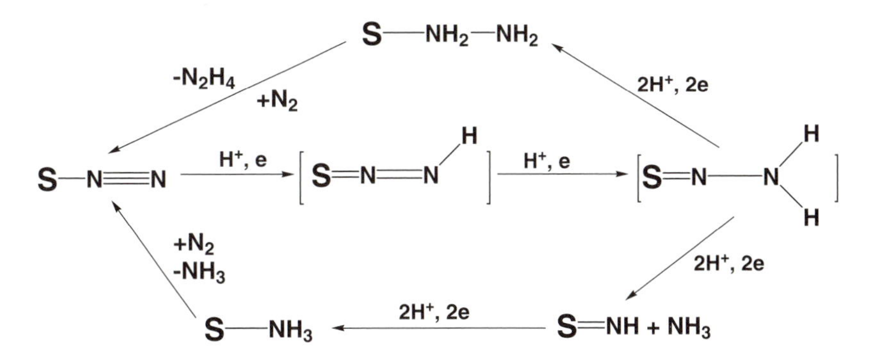

FIG. 16.21 The transformations involved in the conversion of substrate (S) bound dinitrogen into NH_3 or N_2H_4. Each step involves transfer of a proton and the corresponding electron

However, the loss of the halide ligand in (B) would give the metal a 16-electron configuration if the N_2H_2 group remained a 2-electron donor. Instead it forms the isomeric di-anion of hydrazine, $H_2N\text{-}N^{2-}$ and acts as a 4-electron donor.

In both (A) and (B), the N_2H_2 group is stable to further reduction. However, if the diphosphane is replaced by two mono-phosphanes, such as PMe_2Ph or $PMePh_2$, then such Mo or W dinitrogen complexes are reduced completely to NH_3 plus some N_2H_4 as in

$$trans\text{-}[(N_2)_2W(PMePh_2)_4] + H_2SO_4 \rightarrow 1.9NH_3 + \text{trace } N_2H_4$$

The Mo analogue yields $0.7NH_3$ per mole of complex. The overall scheme postulated, via successive additions of H^+, is

$$M{-}N{\equiv}N \rightarrow M{-}N{=}NH \rightarrow M{=}N{-}NH_2 \rightarrow M{-}NHNH_2 \rightarrow M{=}NH + NH_3 :$$
$$M{=}NH \rightarrow M{-}NH_2 \rightarrow M(VI) + NH_3$$

The metal goes from (0) to (VI), and a number of the intermediates have been isolated.

Experiments such as this have been carried out to clarify all the steps in the general scheme of Fig. 16.21, and several other minor paths have been found. The main thrust of the work, in addition to synthesis, has been the study of the protonation–deprotonation reactions of these nitrogen complexes leading to overall nitrogen fixation reactions such as

$$L_xM + N_2 \rightarrow L_xMN_2 + 6H^+ \rightarrow L_xM + 2NH_3$$

There has also been considerable exploration of parallels between these nitrogen complexes and C_2H_4 complexes or the various dioxygen species (compare previous section).

The work on model fixation reactions may be summarized by saying that separate laboratory reactions are now known which reproduce many of the observed or postulated steps in the reaction of N_2 with *nitrogenase*, such as those in Fig. 16.21, but these have not been combined into one catalytic cycle. Different steps have been demonstrated with different nitrogen complexes, and S represents a variety of transition metal+ligand combinations, but not yet the natural substrate in *nitrogenase*.

The whole fixation problem has turned out to be a very complex one. Three decades of work have led to an enormous increase in understanding and development of quite new classes of compound.

16.10.2 Bonding between N_2 and M

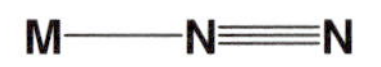

M—C≡O

Comparison of N_2 and CO bonded to a metal, M

Although many fewer N_2 complexes are found, their general properties are similar to the carbonyls (Section 16.1) and a similar linear bond is expected. This is indeed found in all structural studies of stable terminal nitrogen complexes (compare Fig. 14.29). The M-N-N angle is usually very close to 180° and the NN bond, in accurately known structures, is in the range 112 ± 2 pm, compared with 109.8 pm in the N_2 molecule. Again, as with carbonyls, the stretching frequency is found in the triple bond region, usually in the range 1990–2160 cm^{-1} compared with 2345 cm^{-1} in N_2.

This all points to a bonding model for M-N≡N very similar to M-C≡O (see Fig. 16.2). However, there are significant differences. First (see Table 3.4), the highest filled orbital in N_2 is at −15.57 eV compared with −14.01 eV in CO. This is the σ orbital involved in the σ $M \leftarrow NN$ (or $M \leftarrow CO$) bond. That is, the electrons donated *to* the metal are more tightly held in N_2 than in CO, and the σ bond to the metal provides less energy in the nitrogen complexes. This is partly compensated by the lower energy of the empty π^* orbital (−7 eV for N_2, −6 eV for CO) involved in the π donation *from* the metal $M \rightarrow NN$ or $M \rightarrow CO$, but the difference of 0.6 eV (ca. 58 kJ mol^{-1}) is an indication of the expected increased stability of the carbonyl over the dinitrogen system. This is reflected by the much smaller number of N_2 complexes known and by the fact that most contain only one N_2 group. In the M-N-N unit, the σ donation is more important for the M-N bond formation than is the π back donation, and the end-on coordination should be appreciably stronger than the side-on configuration. The main effect of the π back donation in both configurations is to weaken the N-N bond, providing the required activation. In the side-on configuration, the N_2 unit

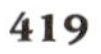

donates both σ and π electrons to the metal, considerably weakening the N-N bond, and it is suggested that this may be the active intermediate in the reduction reaction.

Although we have not discussed them in detail, the terminal metal acetylides, M-C≡CH, are also very similar to the nitrogen complexes.

When we come to bridging groups, we find a linear M-N≡N-M system quite different from bridging carbonyls (compare Fig. 16.3b and d) but found for acetylides M-C≡C-M. Here each M-N bond is similar to that in the terminal complexes, and a fully linear system is expected. As charge transfers are relatively small, and the total effect of the upper energy levels in M-N≡N-M is approximately nonbonding for NN, we find the bond lengths and stretching frequencies in the bridged nitrogen complexes are generally similar to those in the terminal species, again comparable with the acetylides and in contrast to the bridging carbonyls.

Although 'sideways-on' nitrogen, $M\cdots\overset{N}{\underset{N}{|||}}$, was postulated in reactions, stable complexes with this bonding are rare. One report is of *trans*-$(R_3P)_2Rh(Cl)N_2$ and the analogues $(R_3P)_2Rh(Cl)O_2$ and $(R_3P)_2RhCl(C_2H_4)$ [R = $CH(CH_3)_2$]. Each contains a $Rh\cdots\overset{X}{\underset{X}{|}}$ unit (X = N, O or CH_2) with the X_2 perpendicular to the $RhClP_2$ plane and its mid-point completing the square-planar configuration expected for Rh(I). The reaction with Rh is relatively weak.

A new mode of N_2 binding has recently been determined in a complex containing two tantalum atoms; the N_2 unit is end-on bonded to one tantalum, and side-on bonded to the other tantalum.

The carbonyl type of bridging *is* found when the N-N bond order is reduced as in the imine complex $[Pt(PR_3)_2N_2H]_2^{2+}$ which contains the bridges

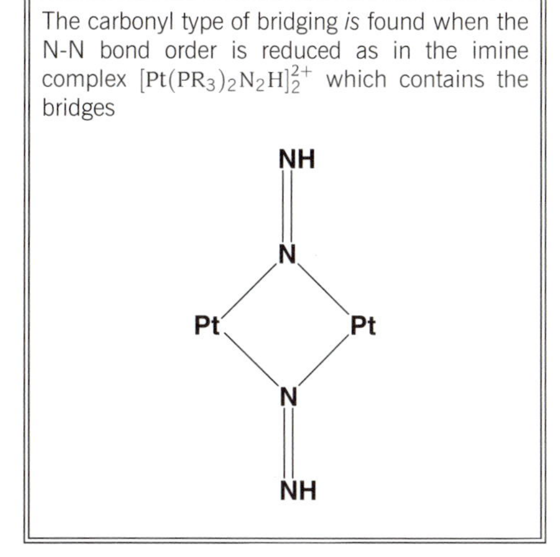

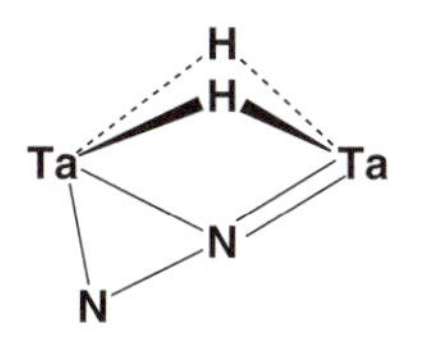

The core of the tantalum complex containing an end-on, side-on bonded N_2 ligand. Each Ta is also bonded to a tridentate $(PhNSiMe_2CH_2)_2PPh$ ligand (not shown).

16.11 Metal–dihydrogen complexes

16.11.1 Discovery

In Section 15.5.1 we discuss the formation of the remarkable $[ReH_9]^{2-}$ complex which was unambiguously proved by X-ray and neutron diffraction (compare Section 7.4) to contain 9 M-H bonds. Following this and other determinations of complexes L_xMH, it became accepted that H could act as a ligand even in complexes of metals in states which are relatively strongly oxidizing. Polyhydrides of the more stable high oxidation states such as Ta(V), W(VI) or Os(VI) are noted in Chapter 15. It was also widely accepted that metal–dihydrogen units, M-H_2, were intermediates in the activation of H_2 by metal catalysts, and low-temperature matrix studies gave evidence for species such as Pd(H_2) and $Cr(CO)_5(H_2)$.

In 1983 startling evidence appeared that the H_2 molecule could coordinate as an entity in complexes which were stable at room temperature. During a study of SO_2 complexes a purple compound turned yellow under N_2 and yielded the dinitrogen complex $M(CO)_3L_2(N_2)$, from which the N_2 could be displaced to form the desired SO_2 complex. The purple compound was later identified as the 16-electron intermediate $M(CO)_3L_2$ (M = Mo, W; L=a bulky phosphane such as P(*iso*-Pr)$_3$). The significant discovery was that H_2 behaved in exactly the same way as N_2 to give a yellow complex containing 2 H on addition to the purple intermediate. Furthermore, addition of N_2 resulted in replacement of H_2 by nitrogen to give the dinitrogen complex. Two interpretations were possible: (i) that two normal M-H bonds were formed giving $M(CO)_3L_2(H)_2$ (Fig. 16.22b) and involving oxidation to M(II) which happens to be yellow, or (ii) that a new type of molecule is formed, $M(CO)_3L_2(H_2)$ (Fig. 16.22a), in which dihydrogen is bound as a molecule as N_2 is in dinitrogen complexes.

The experimental resolution of this question is interesting as it highlights the need that often arises to use a combination of techniques. First, the infrared spectrum showed no band in the regions 2300–1700 cm^{-1} and 900–700 cm^{-1} where M-H stretching or bending occur, but there were bands at 1570, 960 and 465 cm^{-1} which shifted in the D_2 analogue, as expected from the mass effect (see Section 7.7) for vibrations involving hydrogen. An X-ray crystal structure study can often not detect H in presence of heavy

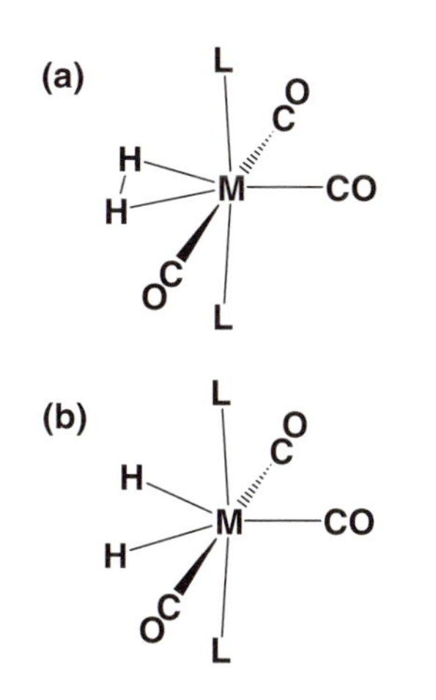

FIG. 16.22 Complexes formed between a metal entity and H_2: (a) coordinated dihydrogen, (b) metal dihydride with M-H bonds

atoms, and there were additional experimental problems involving disorder. The incomplete structure that emerged did show the three CO and the two phosphane ligands in octahedral positions, suggesting that the hydrogen occupied the sixth site. The hydrogen was located using neutron diffraction on the W complex, which showed an H-H unit coordinated sideways on to the metal, parallel to the P-W-P axis (Fig.16.22a). However, the disorder left some remaining doubt and since this totally new mode of bonding by H_2 required further proof, nmr studies were undertaken.

Metal hydrides have a characteristic resonance position and the H signal is split by coupling to the P atoms and to the magnetic ^{183}W isotope (Section 7.8). Any coupling between two M-H bonds would be fairly small and give a doublet. For W complexes sharp lines would be expected. The yellow hydrogen complex showed a resonance close to the M-H range but the signal was unexpectedly broad. The HD analogue was therefore made for nmr study. D has a spin $I = 1$ and therefore gives a 1:1:1 triplet on coupling. This was indeed found with $\mathcal{J}$(H-D) = 33.5 Hz whereas for coupling between separate M-H and M-D bonds, $\mathcal{J}$ is 1–2 Hz. In addition, the HD complex now showed a band in the infrared at 2360 cm^{-1}, which was assigned as the HD stretch (the H-H stretch was later detected in the region 2700–3000 cm^{-1} in $M(H_2)$ complexes). This adds up to strong supporting evidence for the complex and also shows that the W-H_2 structure persists in solution. Additional nmr effects, especially the relaxation times, were also found to be characteristic of dihydrogen coordination.

Other dihydrogen complexes were quickly identified including hydrogen–dihydrogen species such as $[MH(H_2)L_4]^+$ (for M = Fe, Ru, Os) and even $[IrH_2(H_2)_2L_2]^+$. In most examples L_x = phosphane or a combination of phosphane with ligands like CO or cp. The series $M(CO)_3L_2(H)_2$ was completed by the synthesis of the M = Cr species, which was stable only under an atmosphere of H_2.

Re-examination of some of the established polyhydrides led to their re-formulation as dihydrogen complexes. Nmr and neutron diffraction established the structure $MH_2(H_2)(PR_3)_3$ (M = Fe, Ru) a much more comfortable status than the original formulation as a 7-coordinate Fe(IV) tetrahydride. However, many polyhydrides remain established as solely M-H compounds. While there was some question about Re(VII) species, the current conclusion is that the ReH_7L_2 analogues of $[ReH_9]^{2-}$ are indeed Re(VII) heptahydrides and not dihydrogen complexes of Re(V).

See Section 9.7 for a discussion of hydrogen–hydrogen bonding involving metal–hydride compounds

16.11.2 Properties and bonding

The H-H bond distance is best determined by neutron diffraction or nmr methods. The observed values range from 85 pm to 102 pm (compare H_2 = 74 pm, Chapter 9). The values of the HD coupling in the nmr vary from 35 to 22 Hz, and the coupling decreases as the bonds get longer—as expected, since the coupling is mediated by the bonded electrons ($\mathcal{J}$(HD) = 43.2 Hz for the free molecule). The H-H stretching frequency ranges from 3080 to 2690 cm^{-1}, with the corresponding H-D values around 2300 cm^{-1} (compare free H_2 at 4390 cm^{-1} and HD at 3820 cm^{-1}). The other infrared bands are the $M(H)_2$ stretches in the 1500 cm^{-1} and 900 cm^{-1} regions and lower frequency bending modes.

The heat of formation of the dihydrogen W complex from the purple intermediate is measured as −42 kJ mol^{-1} compared with −57 kJ mol^{-1} for the corresponding N_2 compound.

All these properties indicate an interaction which weakens the H-H bond compared with the H_2 molecule. As H_2 is extremely simple, with no accessible orbitals other than σ and σ^* (Fig. 3.6b), the interaction with the metal must involve only these. The proposed model is shown in Fig. 16.23 and involves a side-on orientation with donation from the H_2 σ bonding orbital into an empty metal d orbital, supported by back-donation from a filled metal orbital into σ^*. The model is closely analogous to the long-standing one for ethylene–M coordination (Fig. 16.6b), except that there the π orbitals were involved. The effect of such an electron transfer is to weaken the H-H interaction, especially by back-donation, in order to gain the M-H interaction. The system is very delicately balanced but, as the H-H bond is very strong, some weakening is tolerable before dissociation occurs. The back-donation is thought to be important for the formation of $M(H_2)$ complexes since acceptors which lack filled d orbitals do not form them. More detailed calculations show that the σ interaction is

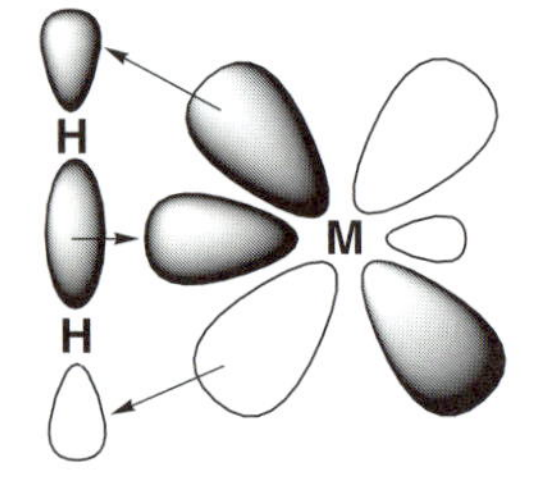

FIG. 16.23 σ donation and π back-donation proposed for the M-H_2 interaction

the major one, and give values for the parameters which are a good match to the observed ones.

For many complexes there is nmr evidence in solution for dissociation of the dihydrogen complex into 2MH bonds, i.e. an equilibrium between forms corresponding to Fig. 16.22a and b. The dihydrogen complexes thus capture the initial stage of hydrogen activation and support the model proposed for the mechanism of metal catalysis. It is reasonable to apply a similar model to other cases of activation of strong single bonds, such as C-H. It has been observed that C-H bonds in ligands such as phosphanes sometimes lie in an abnormal position close to the transition metal in what is termed an *agostic* interaction, suggesting bond formation of a type very similar to that proposed for dihydrogen. One example is found in the 16-electron purple intermediate above. This clearly models the first stage of C-H activation. Similar activation is seen in Wilkinson's compound, $RhCl(PPh_3)_3$, where an H of one of the phenyl groups of the phosphane ligand coordinates agostically to Rh, explaining the good hydrogenation catalyst properties of the complex.

16.12 Post-actinide 'superheavy' elements

16.12.1 Background

For the elements up to about $Z = 100$, the mode of decay is principally by α- or β-particle emission and it is the likelihood of these events which governs the half-lives. It will be seen from Table 12.1 that the half-lives tend to decrease as Z increases. However, at about the atomic number of the heaviest elements, another mode of decomposition, spontaneous fission, becomes the dominant one. In this mode half-lives again decrease with increasing atomic number. Estimates of isotope stability require an understanding of the binding forces in the nucleus, and this has only slowly developed. Earlier approaches were too pessimistic and the experimenters continued to find isotopes which were longer lived than expected. For example, initial estimates suggested that the half-life for spontaneous fission decay would be a few seconds for element 103 and would have dropped to about 10^{-4} seconds by element 110, making it unlikely that elements in this range could be synthesized. A decade or so later, several isotopes of lawrencium, $Z = 103$, had been reported and the longest half-life, Lr-262 = 214 minutes, was about 500 times longer than predicted. While this suggested that it would be somewhat easier to reach heavier elements, the expectation remained of ever-decreasing half-lives as Z increased. At the same time, the experimental difficulties of synthesizing new nuclei remained enormous and required skilled and very specialized teams.

Much of the work for several decades was done by groups of experimenters in the United States and in Russia, and these were joined more recently by a team in Germany. (See the box titled 'The race to superheavy elements'.)

In the last decade or so, the pace of new discoveries has picked up, with element 112 indicated in 1995 and 114 plus new isotopes of earlier elements by 1999. In this atomic number zone, a new feature of nuclear structure comes into play. Theorists predict an 'island of stability' based on 'magic numbers' of protons for $Z = 114$, 120 or 126. These magic numbers reflect the existence of 'closed shells' of protons and neutrons in the nucleus (broadly analogous to the stable electron shells of the rare gas configurations), which provide a further barrier to decay, offsetting the overall decline in stability with increasing Z. While the latest discoveries are still too neutron-poor to reach the point of highest predicted stability, they do show much longer half-lives than would be extrapolated from elements 104 to 110.

16.12.2 Synthesis and properties

The actinide series is complete at lawrencium, $Z = 103$ (see Section 12.7).

The same techniques of bombarding actinides with accelerated nuclei of light atoms produced the next two elements, 104 and 105. Thus, bombarding californium with ^{15}N nuclei gave an isotope of 105, Unp-260, which has a half-life of 1.6 s and decays to Lr-256 with emission of an α-particle.

The routes to the heavier elements have used preferred targets whose nuclei are close to the 'magic numbers', corresponding to high stability. Thus elements 106 to 109 have been approached through lead or bismuth targets corresponding to $Z = 82$ or $N = 126$, and the energy of the bombarding particle is kept relatively low in the 'cold fusion' approach. Thus an announcement of the synthesis of element 109 from Darmstadt involved bombarding Bi-209 with accelerated Fe-58 nuclei. Only one in 10^{14} interactions gave element 109, which was formed at the rate of one atom in a week. Identification was purely by use of nuclear properties (using a new velocity filter technique for separation), and four independent properties matched calculation. Decay was by successive α-emissions to Bh-262, then to Db-258 and then by electron capture to Rf-258. Similar experiments yielded a few atoms of element 111 and 112 in the period 1994–6. Overall, the Darmstadt methodology allowed a significant jump forward, yielding elements 107 to 112.

This approach has two advantages:

(1) the α-decays are well defined and eventually lead to a known daughter product, allowing identification through the chain;

(2) as the initial products contain more neutrons than protons and the α loss removes two of each, the daughter products may be more neutron-rich and of longer half-lives than the isotopes of such daughter elements which were directly synthesized.

This latter effect is one reason why an initial discovery of an element is often followed by the report of further isotopes with longer half-lives.

Despite these advantages, the Darmstadt technique was reaching its limits and much higher accelerating energies would be needed to go to Z values above 112.

In a further leapfrog the USA team at Berkeley and the Russian group at Dubna have each achieved elements above 112. The Berkeley experiments used the same approach as the German one but followed a prediction that a particular magic number pair—Kr-86 plus Pb-208—would produce element 118. A positive claim was made in 1999 that three atoms were seen of the 293 isotope of element 118 which decayed by a series of α emissions through $Z = 116$, 114, 112, 110, 108 to $^{269}_{106}Sg$. Unfortunately, the work could not be repeated and the claim was withdrawn in 2001. This incident underlines the extreme difficulty of such experiments, but also gives confidence in the checking procedures in place.

Also in 1999, the Dubna team reported experiments using much higher bombarding energies to react the rare, but very neutron-rich, isotope ^{48}Ca with targets of U or Pu. The high energy increases the probability of fusing the two nuclei, but also increases the probability of fission, and gives a messier system. Bombarding ^{238}U gave two events assigned to the isotope of element 112 with mass of 283 with a half-life of 1.5 minutes (much longer than the 277 isotope prepared earlier). Similar experiments using Pu-244 gave a single atom of element 114 of mass 289 which decayed by three α emissions through elements 112, 110 to 108 which then fissioned. A third experiment showed that starting from the lighter Pu isotope of mass 242 gave the isotope of element 114 of mass 287. This decayed to the same isotope $^{283}112$ which was made from uranium-238, as expected. These three experiments were self-consistent. While the product mixes from the Dubna method were more difficult to analyse, it complements the Berkeley approach in giving more neutron-rich isotopes, which have longer half-lives. It highlights the difficulties, and also the immense sophistication of the experimental techniques, to note that in all these experiments only about one in a trillion of the bombarding atoms produces one of the new nuclei.

Table 16.7 summarizes the properties of the post-actinide elements reported by 2001.

For element 104, rutherfordium, tracer studies show properties expected for the 6d congener of hafnium. Thus it forms $RfCl_4$, comparable to $HfCl_4$ and unlike the involatile trichlorides of the actinides and lanthanides. Rutherfordium was manipulated in solution as the hydroxybutyrate anion and its tracer behaviour followed hafnium. Thus rutherfordium behaves as the first post-actinide element, as the heaviest element of the titanium Group.

Dubnium, element 105, formed a chloride and a bromide, which were more volatile than the hafnium tetrahalide but less volatile than the niobium pentahalides. $DbBr_5$ has an estimated boiling point of 430°C. Tracer experiments showed that dubnium adsorbs to

THE RACE TO SUPERHEAVY ELEMENTS

Because of the experimental difficulties, there have been conflicting claims to be the first to synthesize several of the new 'superheavy' elements (including the later actinides). The Russian team used heavy ion bombardment which allowed leaps forward in Z but gave isotopes which decayed by spontaneous fission. These were more difficult to identify unequivocally than the American experiments which used only stepwise advances in Z. This approach had the advantage that it gave isotopes decaying by alpha-particle emission into directly identifiable daughter products. Cold War suspicions precluded any collaboration in a field as sensitive as nuclear chemistry.

The result was that competing claims arose between the Russians and the Americans for priority in discovery and for the right to propose names for the new elements. As a result, different names for particular elements got into the literature.

By international agreement, the final decision on names is made by the International Union of Pure and Applied Chemistry, IUPAC. It was agreed that a new element would be accepted only after it was produced by two independent experiments. Often, a long time was required to sort out competing claims and provide the verifying observations. Some of this confirmation came from the German team with a new methodology.

A systematic terminology was proposed by IUPAC for use until formal names were finally decided. The ten numbers are shown by

0	nil	1	un	2	bi	3	tri
4	quad	5	pent	6	hex	7	sept
8	oct	9	en				

Names are formed from these number elements in order with rules about elision. For example, element 104 is named *unnilquadium* and the symbol is Unq. While this scheme gives rather unlovely names, it is unfortunate that it was not generally adopted, as much confusion would have been saved.

An interim determination of names was published by IUPAC about 1995, but this led to further heavy lobbying and a number of changes were made in the final determination of the official names. In this edition, the official names are used throughout—for example, in Tables 2.5 and 16.7, and elsewhere in this section.

However, as many of the alternative names for the superheavy elements have had some currency in the literature, these are recorded for reference below

Element	*Proposed names*	*Proposed symbol*	*Proposer*
104	Dubnium	Db	Interim IUPAC
	Rutherfordium	Rf	USA
	Kurchatovium	Ku	Russia
	Unnilquadium	Unq	IUPAC systematic
105	Joliotium	Jl	Interim IUPAC
	Neilsbohrium		Russia
	Hahnium	Ha	USA
	Unnilpentium	Unp	IUPAC systematic
106	Rutherfordium	Rf	Interim IUPAC
	Seaborgium	Sg	USA
	Unnilhexium	Unh	IUPAC systematic
107	Bohrium	Bh	Interim IUPAC
	Neilsbohrium	Ns	Germany
	Unnilseptium	Uns	IUPAC systematic
108	Hahnium	Hn	Interim IUPAC
	Hassium	Hs	Germany
	Unniloctium	Uno	IUPAC systematic
109	Meitnerium	Mt	Interim IUPAC, Germany
	Unnilennium	Une	IUPAC systematic

glass like niobium and tantalum, but in contrast to zirconium and hafnium. Thus dubnium is assigned as the heaviest element of the vanadium Group. However, in solvent extraction experiments, dubnium remained with niobium in the aqueous phase while tantalum extracted into the organic phase, showing that properties of the heaviest element cannot always be predicted by simple extrapolation from the 4d and 5d congeners.

The chemistry of seaborgium, element 106, has also been studied thanks to the relatively long half-lives of the 265 and 266 isotopes. In several tracer experiments, Sg followed Mo and W(VI) and did not follow U(VI). The species formed were probably oxy- or oxyhalide complexes such as $[SgO_4]^{2-}$, SgO_2F_2 or $[SgO_2F_3]^-$. Seaborgium therefore fits as the 6d series member of the chromium Group.

For bohrium, the preparation of the longer-lived Bh-267 isotope allowed the identification of the very volatile BhO_3Cl behaving as the heavier homologue of Tc and Re. Similarly, hassium behaves as a reasonable extrapolation from Ru and Os.

To date, there has been insufficient material to study the heavier elements from meitnerium onwards. Theoretical calculations of properties such as gas chromatographic or

TABLE 16.7 Superheavy element isotopes

Element	*Isotope*	*Half-life*	*Other isotopes*
104 rutherfordium, Rf	257	0.8 s	All from 253 to 262
	259	1.7 s	
	261	65 s	
105 dubnium, Db	258	4 s	255, 257, 260, 261
	262	34 s	
	263	27 s	
106 seaborgium, Sg	265	7 s	259, 260, 261, 263
	266	21 s	
107 bohrium, Bh	261	12 ms	262
	264	0.45 s	
	267	ca. 20 s	
108 hassium, Hs	265	1.8 ms	264, 269, 277
	267	74 ms	
109 meitnerium, Mt	266	3.4 ms	
	268	72 ms	
110	269	0.17 ms	267, 273, 281
	271	1.1 ms	
111	272	1.5 ms	271
112	277	0.28 ms	285
	283	1.5 min	
114	287	5.5 s	
	289	34 min?	

Predicted centre of island of stability: $Z = 114$, mass $= 298$.

solution chemistry match reasonably with observations for the earlier elements and thus give a reasonable indication of the probable behaviour of the heavier ones.

All this work was carried out on a tiny scale, often using only a few tens of atoms. Earlier experience with the post-uranium elements was that interpretations based on tracer studies were often amended when larger quantities became available. With such reservations in mind, the observations to date are in accord with the view that the superheavy elements from rutherfordium onwards are filling the 6d post-actinide period. If this is the case, meitnerium, 110, 111 and 112 complete the 6d series.

The next elements are expected to be members of the Main Groups, with the outermost electrons in the 7p shell. If it is possible to reach element 123, we may find that the first g level, the 5g shell, will start to fill. A little further on, the $Z = 126$ magic number presents a variety of interesting possibilities, as we would expect the 5g, 6f and 7d levels to be very close in energy.

Recently discovered element 114 is expected to be 'ekalead' while 118 would be the heaviest member of the rare gas Group. If the half-lives in this area of the Periodic Table are as long as predicted, and if the developing new methods give sufficient yields, then there are exciting possibilities for a relatively well-developed chemistry. Already predictions are being made—for example, that 114 will be even more stable than lead in the +II state and that 115 will have a stable +1 ion. Perhaps the next edition will see a more extended discussion—transferred to its correct position in a Main Group chapter!

The estimated 'centre of the island of stability' comes at even higher neutron numbers around a mass of 298 for element 114. Thus these latest experiments arouse great interest, as the higher Z values and the lengthening half-lives seem to show we are approaching the stable region. New or improved methods are also being developed.

16.13 Relativistic effects

When the nuclear charge becomes large, the radial velocity of the inner electrons rises to become a significant fraction of the speed of light. This can be envisaged on a simple planetary model—as the positive charge on the nucleus rises, the negative electron has to move faster to remain in a particular orbit. Calculations show that, for elements around Hg with $Z = 80$, the average radial velocity of the 1s electron is about 60% of the speed of light. At such speeds, the special theory of relativity shows that the mass increases, by about 20%, and, as a result, the average radius of the orbital contracts by about 20%. The effect is most prominent for orbitals with electron density close to the nucleus, so the main effect is on s orbitals. There is a lesser effect on p orbitals with $m_l = 0$ but not on the +1 or −1 values, causing a splitting of the p set into groups of 1 and 2 orbitals.

There are two principal consequences. First, the s orbitals of the heavier elements become more stable than otherwise expected, and therefore have higher ionization potentials for electron loss or more exothermic electron affinities for electron gain. The second consequence is that the more contracted s orbitals shield the outer orbitals of the d and f sets more effectively from the nuclear charge, so that these orbitals expand and their energies are less.

The chemical consequences are seen mainly in the heaviest elements. While the lanthanide contraction accounts for the similarity between the 4d and 5d series, it has been calculated that relativistic effects contribute about 20% to the contraction of Hf, so that the extremely close similarity between the earlier members of the two series depends in part on the relativistic effect. As we move along the 5d series, the resemblance to the 4d congener decreases (compare Sections 15.6 to 15.10) and this is a combination of both relativistic effects. The reduced binding of the 5d electrons allows them to participate more fully, which is seen in the increasing number and stability of the higher oxidation states. Examples are Pt(IV) or Au(III) versus the much less stable Pd(IV) or Ag(III), and the existence of Pt(VI) or Au(V) with no Pd or Ag counterparts. Other examples are clear in Chapter 15.

A second effect is the stabilization of the $6s^2$ configuration. This can be clearly seen on comparing the properties of mercury with its neighbour, gold (and also to a lesser extent with thallium). Metallic mercury has anomalously low melting and boiling points, is monoatomic in the gas phase and the density of mercury (13.53 g cm^{-3}) is markedly different from that of gold (19.32 g cm^{-3}). Mercury thus appears to be behaving as if it has a rare-gas electronic configuration and this can be nicely accounted for by the stabilization of the $6s^2$ configuration by relativistic effects.

A similar effect, which has been encountered previously (Section 15.9.2), is the striking tendency of gold to adopt the same $6s^2$ configuration in the Au^- ion. Thus, the intermetallic

GOLD IS SMALLER THAN SILVER

Gold is often described as having a covalent radius which is the same, or slightly larger, than that of its lighter Group member, silver. However, as a result of a relativistic contraction, described in this section, it has been predicted that gold should be slightly smaller than silver. Until 1996, no experiment had been carried out which could give unambiguous results. The experiment involved synthesizing analogous complexes of gold(I) and silver(I), with the same ligands and counteranions, the same coordination geometry, isomorphous crystal structures, and equal experimental conditions. Bis(trimesitylphosphane) (= L) complexes were prepared as their BF_4^- salts, i.e. $[L_2M]^+BF_4^-$, and characterized by X-ray diffraction. The two complexes had identical structures, except that the Au-P bond length was 0.09(1) Å (9 pm) smaller for gold than for silver. The covalent radii of gold(I) and silver(I) were thus calculated to be 125 and 133 pm respectively, which corresponds to a reduction in covalent radius of 6% on going from silver to gold.

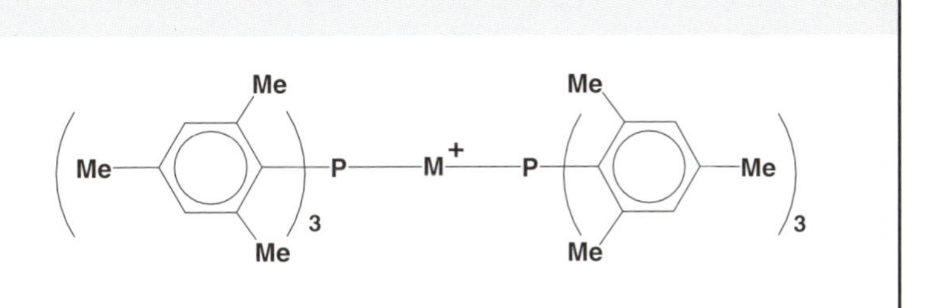

The isostructural gold and silver trimesitylphosphane complexes used in the study.

Reference

J. Amer. Chem. Soc. **118**, 1996, 7006–7007.

compound CsAu is a semiconductor rather than a metal and has a significant amount of ionic character, Cs^+Au^-. Consistent with this, gold has the highest electron affinity and electronegativity outside of the 'typical' electronegative elements, higher than sulfur and almost as high as iodine. Again, gold is showing a strong tendency to form a stable $6s^2$ configuration. In addition, the surprisingly stable gas-phase Au_2 molecule has a stronger bond dissociation energy than I_2 (221 vs 151 kJ mol^{-1}), and gold, in many respects, can be thought of as a pseudo-halide analogous to iodine, both elements being one electron short of an inert-gas electronic configuration. Another phenomenon in gold chemistry arising from relativistic effects is the strong tendency for gold to form Au···Au contacts in the solid state, similar to halogen···halogen interactions in organic compounds. The energy of this interaction has been estimated to be as strong as 30 kJ mol^{-1}, comparable with a hydrogen bond, and this is therefore a significant factor in determining the solid-state structures of gold complexes.

A further manifestation of relativistic effects in chemistry is the 'inert pair' effect in the heaviest Main Group elements Tl, Pb and Bi, where the most stable oxidation state is two less than the Group oxidation state (compare Sections 17.1, 17.3 and the chemistries of these elements). This is again nicely accounted for by the relativistic stabilization of the $6s^2$ configuration.

The relativistic contraction does not change smoothly with Z, but increases markedly while the 5d shell is filling, with the maximum effect at Au. The effect diminishes to Bi and then changes only slowly through the 5f shell so that effects comparable to Au are seen only around Fm. It appears that the chemistry of the superheavy elements will eventually develop, though probably rather slowly, and the influence of relativistic effects on the chemistry of these elements should prove very interesting indeed. Recent relativistic calculations on the superheavy elements has suggested that these effects may be quite subtle, particularly with regard to p orbital occupation. One interesting pointer is the possible configuration of Lr where the last electron may be in the 7p level rather than the 6d one, as a consequence of the relativistic effect on the $m = 0$ level. It has also been recently suggested that the ground state electronic configuration for element 104 (rutherfordium) may be [Rn] $5f^{14}$ $6d^1$ $7s^2$ $7p^1$, while on the other hand that of element 105 (hahnium) may be [Rn] $5f^{14}$ $6d^3$ $7s^2$, by analogy with the configuration of tantalum. Around ekalead (element 114) there exists the possibility of an 'inert quartet' effect.

The relativistic effect is not the only contribution to the unusual chemistry of the heavier elements, and other rationalizations of the inert pair effect have been proposed, but it is a substantial contribution. It is clear that calculated changes in the relative energies of the s and d levels are in accord with observation and account for many of the unusual features.

Problems

16.1 Discuss the following organometallic compounds in terms of the 18-electron rule:

$(\eta^6\text{-}C_6H_6)Cr(CO)_3$ $Mn_2(CO)_{10}$ $(\eta^5\text{-}C_5H_5)Fe(CO)_2Br$
$(\eta^5\text{-}C_5H_5)Ni(NO)$ $Fe_3(CO)_{12}$ $Mn(CO)_6^+$

16.2 Why is the average C-O stretching frequency in the infrared spectrum of $(Ph_3P)_3Cr(CO)_3$ lower in energy than that of $Cr(CO)_6$?

16.3 Explain the following infrared spectroscopic data:

Compound	C-O stretching frequency (cm^{-1})	M-C stretching frequency (cm^{-1})
$V(CO)_6^-$	1860	460
$Mn(CO)_6^+$	2090	416

CHAPTER 17

17.1 INTRODUCTION AND GENERAL PROPERTIES 427

17.2 THE FIRST ELEMENT IN A p GROUP 431

17.3 THE REMAINING ELEMENTS OF THE p GROUP 433

17.4 THE BORON GROUP ns^2np^1 435
- 17.4.1 The elements, general properties and uses 435
- 17.4.2 The III state 439
- 17.4.3 The I oxidation state and mixed oxidation state compounds 441

17.5 THE CARBON GROUP, ns^2np^2 445
- 17.5.1 General properties of the elements, uses 445
- 17.5.2 The IV state 449
- 17.5.3 Hydride and organic derivatives 450
- 17.5.4 The II state 452
- 17.5.5 Reaction mechanisms of silicon 454

17.6 THE NITROGEN GROUP, ns^2np^3 456
- 17.6.1 General properties 456
- 17.6.2 The V state 461
- 17.6.3 The III state 464
- 17.6.4 Other oxidation states 468

17.7 THE OXYGEN GROUP, ns^2np^4 470
- 17.7.1 General properties 470
- 17.7.2 Oxygen 473
- 17.7.3 The other elements: the (VI) state 476
- 17.7.4 The IV state 478
- 17.7.5 The II state: the −II state 479
- 17.7.6 Compounds with an S-S or Se-Se bond 481
- 17.7.7 Sulfur–nitrogen ring compounds 482

17.8 THE FLUORINE GROUP, ns^2np^5 (THE HALOGENS) 483
- 17.8.1 General properties 483
- 17.8.2 The positive oxidation states 485
- 17.8.3 The −I oxidation state 493

17.9 THE HELIUM GROUP 494
- 17.9.1 Xenon compounds 495
- 17.9.2 Preparation and properties of simple compounds 496
- 17.9.3 Structures 499
- 17.9.4 Reactions with fluorides 500
- 17.9.5 Other rare gas species 502

17.10 BONDING IN MAIN GROUP COMPOUNDS: THE USE OF d ORBITALS 502

PROBLEMS 504

The Elements of the 'p' Block

17.1 Introduction and general properties

Those elements which have their least tightly bound electron in a p orbital lie in the Main Groups of the Periodic Table headed by B, C, N, O, F and He. Some aspects of the chemistry of these elements have been discussed in the earlier chapters. The hydrides of the p elements are included in Chapter 9, the structures are discussed in Chapter 4, solid state structures are treated in Chapter 5, while aspects of aqueous and nonaqueous chemistry, including properties of oxyacids are covered in Chapter 6.

This chapter aims to give a broad picture of Main Group chemistry, illustrated largely by the simpler compounds. A very limited selection from the huge range of topics of current interest is surveyed in the next chapter.

Many of the applications of p block elements are listed under the individual elements. One general theme is polymers with inorganic backbones which include the silicones, the polyphosphazenes and the sulfur nitrides. These are studied to find polymers with specific advantages over organic ones in properties such as electrical or thermal conductivity, or resistance to heat and oxidation. Another application which ranges across various groups is that of high performance ceramics which include B_4C, SiC, Si_3N_4, BN, AlN, Al_2O_3, MgO and mixed oxides with transition elements such as Al_2TiO_5 or $PbZrTiO_3$. In the important field of electronics, the semiconductors Ge and Si are now supplemented by the isoelectronic mixed species like GaAs, InP, ZnSe and ternary analogues. The use of chemical vapour deposition, as a means of building up precisely oriented layers of specific composition and thickness by pyrolysis of volatile compounds of the contributing elements, has demanded the preparation and handling of very high purity hydrides like SiH_4 or GeH_4, and alkyls like Me_3As, Me_3Ga or $InEt_3$. Solar cells involve deposit of Si, or oxides, by similar processes.

The maximum oxidation state shown by a p element (the 'Group oxidation state') is equal to the total of the valency electrons, i.e. to the sum of the s and p electrons, and is the same as the Group Number in the Periodic Table (or the Group Number less 10 in the 1 to 18 form of the table). In addition to this oxidation state, p elements may show other oxidation states which differ from the Group state by steps of two. Clearly, the number of possible oxidation states increases towards the right of the Periodic Table. The most important oxidation states in the various Groups of p elements are shown in Table 17.1. where oxidation states other than these occur, they usually arise either from multiple bonding, as in the nitrogen oxides, or from the existence of element–element bonds as in hydrazine, $H_2N\text{-}NH_2$ (N = −II) or disilane, $H_3Si\text{-}SiH_3$ [Si-Si bond makes this formally Si(III)].

As fluorine is the most electronegative element, it can show only negative oxidation states and always exists in the −I state. Oxygen is also always negative, except in its fluorides, and is found in the −I state in peroxides (due to the O-O link) and in the −II state in its general chemistry. The other highly electronegative elements, nitrogen, sulfur and the halogens also show stable negative oxidation states in which they form anions, hydrides and organic derivatives.

In the boron, carbon and nitrogen Groups, the Group oxidation state is the most stable state for the lighter elements in the Group while the state two less than the Group oxidation state is the most stable one for the heaviest element in each Group. The relative stabilities of these two states varies down the Group; thus, in the carbon Group, lead(II) is stable and lead(IV) is strongly oxidizing, tin(II) and (IV) are about equal in stability, germanium(II) is represented by a handful of compounds only, and germanium(IV) is the stable state, while silicon shows only the IV state.

TABLE 17.1 Oxidation states among the p elements

Group headed by	B	C	N	O	F	He
Group oxidation state	III	IV	V	VI	VII	VIII
Other states	I	II	III	IV	V	VI
		(−IV)*	I	II	III	IV
			−III	−II	I	II
					−I	

*As the electronegativity of C lies between that of H and those of O, N, or halogen, carbon in CH_4 is −IV while carbon in CF_4 is IV, and all intermediate cases occur. The other carbon Group elements are formally in the IV state in their hydrides, and show II and IV in their general chemistry.

In the oxygen and fluorine Groups the position is more complicated because of the wider ranges of oxidation states. Oxygen and the halogens are most stable in the negative states. The VI state is stable for sulfur and falls in stability for the heavier elements of the oxygen Group, but both the IV and the II states are relatively stable for the heavy elements. Among the halogens, chlorine and iodine both show the Group state of VII in oxyions; bromine(VII) was only recently found. Iodine is also fairly stable in the V and III states. Thus, the general trend for the lower oxidation states to be more stable for the heavier elements is shown by elements such as iodine or tellurium, but more than one state is involved and no simple rule may be given. Of the rare gas compounds, the difluorides are known to exist for radon and krypton as well as for xenon. The other oxidation states in Table 17.1 are shown by xenon.

The general pattern of oxidation state stabilities in the Main Group elements may be summarized: the common oxidation states vary in steps of two, with states lower than the Group state being the most stable for the heavier elements. This trend is opposite to that found among the transition elements.

This pattern of stabilities is reflected in the oxides, sulfides and halides formed by the Main Group elements, as given in Tables 17.2a and b. A stable oxidation state will form

TABLE 17.2a Oxides and sulfides of p elements

Elements in oxidation state = number of valency electrons							
B_2O_3	B_2S_3	CO_2	CS_2	N_2O_5			
Al_2O_3	Al_2S_3	SiO_2	SiS_2	P_4O_{10}	P_4S_{10}	SO_3	
Ga_2O_3	Ga_2S_3	GeO_2	GeS_2	As_2O_5	As_2S_5	SeO_3	
In_2O_3	In_2S_3	SnO_2	SnS_2	Sb_2O_5	Sb_2S_5	TeO_3	
Tl_2O_3		PbO_2		(Bi_2O_5?)		(PoO_3?)	
Elements in an oxidation state lower by two							
		CO		N_2O_3			
				P_4O_6		SO_2	
Ga_2O	Ga_2S	GeO	GeS	As_4O_6	As_4S_6	SeO_2	
In_2O?		SnO	SnS	Sb_4O_6*	Sb_2S_3	TeO_2	
Tl_2O	Tl_2S	PbO	PbS	Bi_2O_3	Bi_2S_3	PoO_2	
Elements in an oxidation state lower by four							
				N_2O			
						TeO	
						PoO	PoS
Other compounds							
$(BO)_x$	GaS	C_3O_2		NO, NO_2, N_2O_4,			
	Ga_4S_5			(NO_3 or N_2O_6)			
	InS			N_4S_4			
	In_6S_7			(PO_2)		S_2O	
				P_4S_3, P_4S_5, P_4S_7			
				As_4S_3, As_4S_4			
				(SbO_2)			
		Pb_3O_4					

*Antimony trioxide also occurs as $(Sb_2O_3)_n$ chains.

TABLE 17.2b Halides of the elements

Fluorides of the Groups headed by				*Chlorides of the Groups headed by*				*Bromides of the Groups headed by*				*Iodides of the Groups headed by*			
B	C	N	O	B	C	N	O	B	C	N	O	B	C	N	O
Elements in oxidation state = Number of valency electrons															
MF_3	MF_4	MF_5	MF_6	MCl_3	MCl_4	MCl_5	MCl_6	MBr_3	MBr_4	MBr_5	MBr_6	MI_3	MI_4	MI_5	MI_6
B	C			B	C			B	C			B	C		
Al	Si	P	S	Al	Si	P		Al	Si	P		Al	Si		
Ga	Ge	As	Se	Ga	Ge			Ga	Ge			Ga	Ge		
In	Sn	Sb	Te	In	Sn	Sb		In	Sn			In	Sn		
Tl	Pb	Bi	Po?	(Tl)	(Pb)			(Tl)							
Elements in oxidation state = two less than the number of valency electrons															
MF	MF_2	MF_3	MF_4	MCl	MCl_2	MCl_3	MCl_4	MBr	MBr_2	MBr_3	MBr_4	MI	MI_2	MI_3	MI_4
		N				(N)				(N)				(N)	
		P	S			P	(S)			P				P	
	Ge	As	Se	Ga?	Ge	As	Se	Ga?	Ge	As	Se	Ga?	Ge	As	
	Sn	Sb	Te	In	Sn	Sb	Te	In	Sn	Sb	Te	In	Sn	Sb	Te
Tl	Pb	Bi	Po	Tl	Pb	Bi	Po	Tl	Pb	Bi	Po	Tl	Pb	Bi	Po
Other compounds															
B_2F_4		N_2F_2 N_2F_4 NF_2	OF_2 O_2F_2 O_3F_2 O_4F_2	B_2Cl_4 B_nCl_n ($n = 4$, 8–12) $(B_9Cl_8)_2$											
	Si_nF_{2n+2} (up to $n = 14$)		S_2F_2 S_2F_{10}		Si_nCl_{2n+2} (up to $n = 10$)	P_2Cl_4	S_nCl_2 (up to n ca. 100)		Si_2Br_6 Si_3Br_8 Si_4Br_{10}		S_2Br_2		Si_2I_6	P_2I_4	
Ga_2F_4			Se_2F_{10}	Ga_2Cl_4	Ge_2Cl_6		$SeCl_2$ Se_2Cl_2	Ga_2Br_4			$SeBr_2$ Se_2Br_2	Ga_2I_4			
InF_2?			Te_2F_{10}	In_2Cl_3			$TeCl_2$	In_2Br_4			$TeBr_2$	In_2I_4		(Sb_2I_4)	
				In_2Cl_4 Tl_2Cl_3 Tl_2Cl_4		BiCl*	$PoCl_2$	Tl_2Br_3 Tl_2Br_4			$PoBr_4$	$Tl^{I}(I_3)$			

the oxide, sulfide and all four halides, while a state which is unstable and oxidizing, such as the Group states of the heavy elements, thallium, lead and bismuth, will not form compounds of the readily oxidizable sulfide or heavy halide ions. Similarly, an oxidation state which is unstable and reducing, such as gallium(I), will form the sulfide and heavier halides, but not the oxide or fluoride.

A similar picture of the stabilities of the oxidation states, but this time in solution, is given by the oxidation state free energy diagrams of Figs 17.5, 17.18, 17.27, 17.43 and 17.55. These diagrams show how the Group oxidation states become unstable with respect to lower oxidation states for the heavier elements in each Group, and they also indicate the instability, with respect to disproportionation, of the intermediate oxidation states in Groups, such as the halogens, which show a number of states.

The effects shown in Tables 17.2 and 17.3 and in the oxidation state free energies are also reflected in the ionization potentials (Table 2.8), the electron affinities (Table 2.9) and the electronegativities (Table 2.14) of the p elements. It will be seen that the five elements in a p Group, although they have many properties in common, split into three sets when their detailed chemistry is examined. This division is:

(i) the lightest element
(ii) the three middle elements
(iii) the heaviest element.

The lightest element shows the most marked differences from the rest of the Group, in properties which are discussed in detail in the next section. The heavier elements are discussed in the following section; the division between the heaviest element and the rest is not so marked as that between the first element and the Group, but it is quite distinctive and has been noted above in the stability of the lower oxidation states. It might also be noted that the middle element in the Group does not always fit between the second and fourth elements (compare electronegativities and chemical evidence such as the non-existence of $AsCl_5$), but these deviations are minor and there is a fairly regular trend of properties among the four heavy elements in each Group.

TABLE 17.3 Some bond energies (kJ mol^{-1})

(a) From diatomic molecules

H-H	436.0	H-F	566	F-F	158	F-Cl	257
O=O	497.3	H-Cl	431	Cl-Cl	242	F-Br	234
N≡N	945.6	H-Br	366	Br-Br	194	F-I	197
C≡O	1075	H-I	299	I-I	153	Cl-Br	222
NO	626.3					Cl-I	211
						Br-I	179

(b) *E* values from polyatomic molecules

C-C	348	N-N	160	O-O	146	C-H	416	C-F	485
Si-Si	297	P-P	215	S-S	265	Si-H	326	Si-F	582
Ge-Ge	260	As-As	134	Se-Se	172	Ge-H	289	C-Cl	327
Sn-Sn	240	Sb-Sb	126	Te-Te	138	Sn-H	251	Si-Cl	391
Si-C	301	Bi-Bi	105	O-F	190	N-H	391	Ge-Cl	342
Ge-C	270	N-F	272	O-Cl	205	P-H	322	Sn-Cl	320
Sn-C	226	N-Cl	193	S-F	326	As-H	247	Pb-Cl	244
Pb-C	130	P-F	490	S-Cl	250	O-H	467	C-Br	285
C-N	292	P-Cl	319	S-Br	212	S-H	374	Si-Br	310
C-P	264	P-Br	264	Se-F	285	Se-H	277	Ge-Br	276
C-O	336	P-I	184	Se-Cl	243	Te-H	238	Sn-Br	272
C-S	272	As-F	464	Te-F	335			C-I	213
		As-Cl	317					Si-I	234
		As-Br	243	O-Si	369			Ge-I	213
		As-I	178					Sn-I	272
		Sb-Cl	212	S-Si	227				
		Bi-Cl	278						
Multiple bonds		C=C	682	C=N	640	C≡C	962	C≡O	1075
		N=N	450	C=O	732	N≡N	946	C≡N	937
		O=O	402	C=S	477				
				N=O	481				

(c) Some bond dissociation energies, *D*

O-ClO	243	O=CO	536	F_3P=O	544	H-CH_3	435
HO-Br	239	H_2C=CH_2	699	Cl_3P=O	511	H-NH_2	448
O-NO	306	HC≡CH	963	Br_3P=O	498	H-OH	498
O-NN	167	H-CN	540			H-CF_3	444
HO-OH	213	NC-CN	607			CH_3-F	452
F_2N-NF_2	83						

It is useful here to summarize some bond energies of the p elements and hydrogen. Two different types of bond energy are encountered. First, if the bond between each pair of atoms in most molecules is independent of the rest of the molecule, we might hope to compile a list of the energies of individual bonds which would allow us to predict the energy of formation of the molecule. For example, the energy of the process of forming SiH_3F from its atoms

$$Si + F + 3H \rightarrow SiH_3F$$

should be given by three times the Si-H energy plus the Si-F energy. Bond energies determined in this way are given the symbol E and are written E(Si-H), etc. Experimentally, such energies are obtained from heats of formation and heats of atomization.

In general, D and E values differ, and D values for successive steps also differ; thus $D(SiH_2F\text{-}H)$ is not the same as E(Si-H), D(SiHF-H) or D(SiF-H).

It is also possible to determine the energy required to break a particular bond, leaving two fragments in their ground state and this is called the bond dissociation energy and given the symbol D. Thus D(HO-OH) is the energy required to break the O-O bond in hydrogen peroxide to give two OH radicals. We can also define the average value of the Si-H dissociation energy in SiH_3F, written $\bar{D}$, and this is usually fairly close in value to the corresponding E value. Clearly, for a diatomic molecule, D and E values are equal.

In Table 17.3 are listed bond energies for some diatomic molecules, which may also be used as E values for these bonds, together with a set of other E values chosen to reasonably reproduce known heats of formation. Also included are a few D values to emphasize the difference between the two. For prediction of heats of formation, the E values should be used, but the D values give a better indication of relative bond strengths. Note that values for first row elements forming single bonds, for example the X_2 energies for the halogens, are often unusually small compared with the heavier members. This is another effect distinguishing the lightest elements.

17.2 The first element in a p Group

In a p Group, the first element differs sharply from the remaining members. The valency shell configuration of the first row elements of the p Groups is $2s^22p^n$, and the orbital next in energy is the 3s level. This is separated from the 2p level by a considerable energy gap and is not used in bonding by these elements. The first row elements are accordingly restricted to a maximum coordination number of four (using the 2s and the three 2p orbitals). In contrast, the second element of a p Group, with the configuration $3s^23p^n$, has the 3d orbitals lying between the 3p level and the 4s level in energy (compare Fig. 8.2). The second row elements make use of these d orbitals in bonding to a substantial extent. For this reason, and because of the larger size, coordination numbers greater than four occur. Compare, for example, the fluorine complexes of boron and aluminium where boron can form only BF_4^- while aluminium gives the AlF_6^{3-} ion. Similar behaviour is shown by the third and subsequent members of each p Group, which also have d orbitals available in the valence shell. A further effect reflects the fact that the first element is unique in the Group in having only the underlying $1s^2$ shell. As a result, the 2p orbitals are less diffuse relative to the 2s ones than is the case for the higher quantum shells. Thus, the contributions of the 2s and 2p orbitals to the bonding and lone pairs in a compound are more nearly matched than for the later members. Related to this is the size effect which leads to larger repulsion effects: for example, for an angle of 90°, the separation of the H atoms in an EH_2 unit is about 14 pm if E is the first member of a Group, and around 25 pm if a second member. In general, see for example Table 4.3, bond angles at second or later members are narrower than at the first Group member. Other, more general, effects of size also contribute to the differences: as with the lightest s elements, Li and Be, the biggest relative change of size is between the first and second Group members.

A second, major, difference between the first and subsequent members of a p Group is the ability of the first row elements to form strong double bonds using only p orbitals (named p_π–p_π bonds). One striking example is provided by carbon and silicon dioxides. Carbon dioxide is a volatile, monomeric molecule in which all the valency electrons are used in forming σ and π bonds between carbon and oxygen (Section 4.8). In silicon dioxide, π-bonding between the silicon and oxygen does not occur. Instead, each silicon forms a single bond to four oxygen atoms, and each oxygen links two silicon atoms, to form a giant molecule with a three-dimensional structure of single Si-O bonds.

Other examples are provided by the oxides of other first row elements, given in Table 17.2a. In each case (compare, for example, the electronic structure of NO in Fig. 3.18), the oxide of the first row element, is a small molecule with p_π–p_π bonds e.g. CO, CO_2, NO, NO_2, N_2O_5, etc. The elements of the second, and subsequent, rows in the Main Groups rarely form p_π–p_π bonds, probably because their greater size leads to a much weaker π overlap between p orbitals. The situation is illustrated diagrammatically in Fig. 17.1 which shows that the sideways extension of the 3p orbital is insufficient to give good π overlap in association with the longer σ bond formed by the larger second row atom.

A second factor in this contrast between the tendency of first and second row elements to form p_π–p_π bonds is the strength of the homonuclear *single* bonds of many of the first row elements. For example, in Table 17.3 it may be seen that the bond energies of F-F, O-O or N-N are distinctly smaller than, respectively, Cl-Cl, S-S or P-P. This reflects repulsions between the nonbonding electrons on these atoms when they are brought close enough together to form the single bond. That is, there are two factors to be balanced, first that the p_π–p_π bond is stronger for the first row elements, and second that the sigma bond may be weaker. Thus these elements are found to form a sigma and pi bond between them, rather than using the second orbital and electron to form a second sigma bond to a third atom. For the heavier element the converse case holds, and it is more stable forming two different σ bonds rather than one σ and one π bond between the same pair of atoms. Strong π bonding using only p orbitals is confined to the first element of a p Group and often leads to small, volatile molecules, as with the carbon and nitrogen oxides.

> Although the doubly bonded atoms are even closer, some of the electrons are now used in the π bond.

The ability to form p_π–p_π bonds, and the weakness of simple σ bonds, are effects of the small size of the first row elements. Other properties, such as ionization potential and electron affinity, which depend in part on size, also change sharply between the first and second elements of a p Group. This shows up very clearly in the electronegativity values in Table 2.14, where there is a major drop in value between the first and second elements in each p Group, and then a much slower fall in values down the rest of the Group.

The size effect also shows up in the bond strengths, as long as π bonding plays no part. The larger atoms usually form weaker σ bonds, so that bond strengths fall on passing down a p Group, and the largest relative changes come between the first two elements of the Group. This point is illustrated by the stabilities of the hydrides, discussed in Chapter 9, and by a similar fall in the stability of element–carbon bonds in the organic analogues

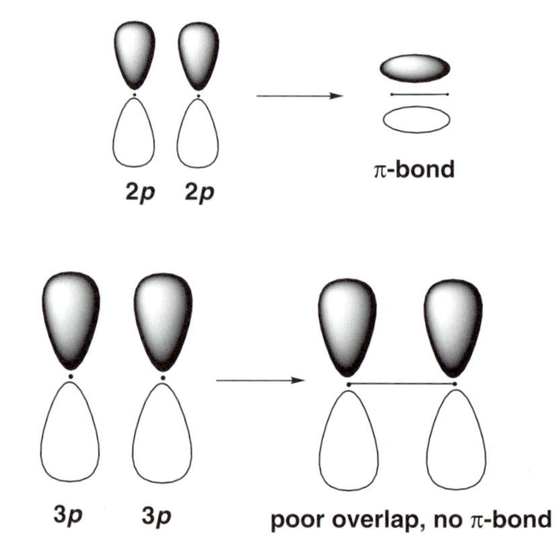

FIG. 17.1 The poor overlap of 3p and higher p orbitals in π-bonding. As the second row elements are larger than those of the first row, the σ bond formed between them is longer. Thus the sideways extension of the 3p orbital is not sufficiently large to give good π overlap. Consequently, a double bond is less favourable than two single bonds

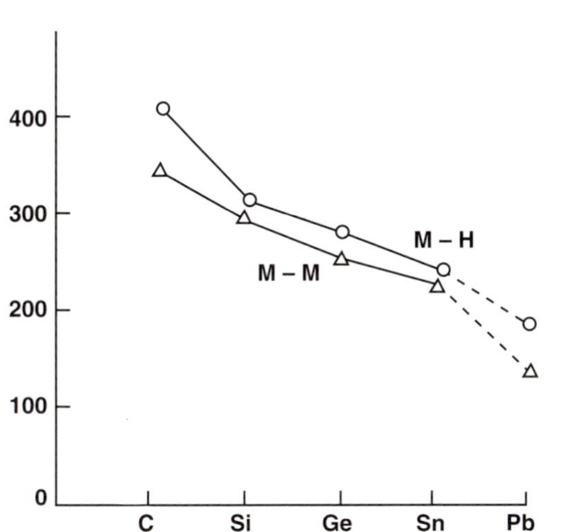

FIG. 17.2 The element–hydrogen and element–element bond energies in the carbon Group. Similar trends are found for the other Main Groups

of the hydrides. These changes are illustrated by the values of the element–hydrogen and element–element bond energies shown in Fig. 17.2.

THE CHARACTERISTIC PROPERTIES OF THE FIRST ELEMENT IN EACH p GROUP

These may be summarized as:

(i) small size, with the greatest relative size change occurring between the first and second members of a Group

(ii) other properties which result from the size effect change in a similar way: for example, electronegativity and the tendency to form anions and negative oxidation states

(iii) the ability to form strong π bonds using the p orbitals is restricted to the first row elements, with the result that many simple compounds such as the oxides are small molecules instead of polymers

(iv) The first element has no low-lying d orbital and is limited to a maximum coordination number of four.

The effects sketched above are not independent of each other, but they add up to a unique character for the first member of each p Group.

17.3 The remaining elements of the p Group

The last member of each p Group shows distinct differences from the other elements in the Group, though these differences are less pronounced than those which mark off the first member. The most obvious property of the last row of p elements is the stability of an oxidation state two less than the Group state in thallium, lead and bismuth. The other two heavy elements, polonium and astatine, occur only as radioactive isotopes and their chemistry is less well known but, in polonium at least, the trend appears to continue and the IV, and also the II, states of polonium are more stable than the Group state of VI. This effect has been termed the *inert pair effect* as it corresponds to the oxidation state which arises if two of the valency electrons are inactive. This phenomenon is due, at least in large part, to the relativistic effect (see Section 16.13). The stability of the lower state, and the oxidizing properties of the Group state, in these heavy elements is clearly shown in Tables 17.2 and 17.3 and in the oxidation state free energy diagrams. The heavy elements show the Group oxidation state only in their oxides and fluorides, while sulfides and other halides of the elements exist only for the lower oxidation states.

The heaviest element in each p Group is also the most metallic and least electronegative member of its Group (as the larger size means that the outer electrons are less tightly held). Most of the heavy elements are metallic except polonium, which is metalloid, and astatine, whose character is not clear. They all form basic oxides and appear in solution in cationic forms. The single bond strengths of these elements with hydrogen, organic groups and halogens are lower than for the lighter elements in the p block and the very existence of some of the hydrides is dubious.

The chemistry of the middle three elements in a p Group follows a graded transition in properties from the small electronegative element at the head of the Group to the large, much more basic element at the foot. This is shown by the increasing stability of the oxidation state two less than the Group state, the increasing basicity of the oxides, and by the lower stability of bonds with hydrogen and organic groups as the Group is descended. This discontinuity between first and second element is most marked in the middle Groups of the p block. Boron and aluminium each have an empty p orbital in the valence shell in their trivalent compounds and thus have acceptor properties, forming 4-coordinate species. The fact that Al enters into further bonding to form 5- or 6-coordinate species makes a difference in degree, but not in kind of behaviour so that boron and aluminium resemble each other reasonably closely. In sharp contrast, the ability of silicon, phosphorus and sulfur to use the 3d orbitals, and the ability of carbon, nitrogen and oxygen to form p_π bonds combine to create marked differences in properties between the first and second members of these Groups. The ability to increase the coordination number above four has important effects on reactivity, and thus on the stability of compounds to air and water. The most striking manifestation is in the carbon Group, where there is neither an empty low-lying orbital nor unshared electron pairs to provide a low-energy reaction intermediate. Among the hydrides and halides, compounds such a CH_4 and CCl_4 are stable in air and react only under vigorous conditions. In contrast, silicon hydrides inflame or explode in air and the silicon tetrahalides are hydrolysed violently on contact with water. Some part of this difference, particularly in the case of the hydrides, may be ascribed to the high Si-O bond energy, but this cannot be the sole explanation. For example, the heats of the reaction:

$$MCl_4 + 2H_2O \rightarrow MO_{2(aq)} + 4HCl$$

are exothermic and very similar in the two cases, M = C and M = Si. A simple explanation of the difference arises if the mechanism of the reaction is examined. Carbon tetrachloride can react with water only if the strong C-Cl bond is first broken, or considerably weakened, as there is no other way in which the water molecule can coordinate to the carbon. In contrast, silicon tetrachloride may be readily attacked by water if the 3d orbitals are used to expand the silicon coordination number above four, to give a reaction intermediate such as $Si(OH_2)Cl_4$ which then loses HCl. In support of this theory, silicon tetrachloride is known to form complexes, $SiCl_4.D$ or $SiCl_4.D_2$, with donor ligands, such as amines or pyridine (see Fig. 17.3). Thus, the great difference in reactivity between carbon and silicon compounds may be ascribed, in considerable part, to the availability of a low-energy reaction path which makes use of the silicon d orbitals. Similar effects are probably present in the chemistry of the other p Groups, but, as boron has an empty p level, and nitrogen, oxygen and fluorine have lone pairs in their compounds, other means for providing low-energy reaction intermediates are present, so differences are less marked in these Groups.

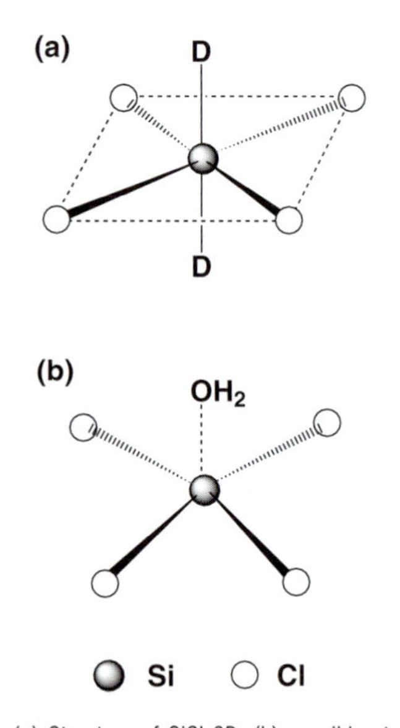

FIG. 17.3 (a) Structure of $SiCl_42D$, (b) possible structure for intermediate in hydrolysis of $SiCl_4$

At oxygen, and still more at fluorine, the tendency to enter the negative oxidation states as ions or covalent molecules becomes more important and reintroduces some resemblance in the general chemistry, especially between fluorine and chlorine. These trends may be linked with those in the s block elements so that, in the Main Groups as a whole, the first row element differs from the rest of its Group. This difference is least at the ends of the Periods in the lithium and fluorine Groups, and is most marked in the centre, in the carbon and nitrogen Groups.

Size effects in the chemistry of the heavier elements are much less distinctive than the changes observed at the head of the Group. There is a general tendency for the heavier elements to show higher coordination numbers but this is offset to some extent by the weakening of element–ligand single bonds already noticed. In oxyacids, antimony, tellurium and iodine give compounds in the Group oxidation state where the coordination to oxygen is six-fold in place of the 4-coordinated oxyions of the lighter elements. In halogen compounds there is evidence for TeF_7^- species and IF_7 exists. It is not known whether this trend continues for polonium and astatine.

One further feature of the chemistry of the p Groups is the slight discontinuity in properties observed at the middle element in each Group (i.e. in the Period from gallium to bromine) which was mentioned above. These middle elements do not always have properties which interpolate between those of the second and fourth element. These effects are ascribed to the insertion of the first d shell immediately preceding this row of Main Group elements—the effect is reminiscent of the lanthanide contraction following the appearance of the first f shell and has even been termed the 'scandide contraction'. The effects are much less striking, however, than in the case of the lanthanide contraction. In the chemistry of the elements, the effect shows up in minor anomalies such as the low stability of $AsCl_5$ compared to phosphorus and antimony pentachlorides. Similarly, GeH_4 is found to be stable to dilute alkali while SiH_4 and SnH_4 are rapidly attacked. This is not a major effect, and most of the chemistry of the second, third and fourth elements in any p Group follows a smooth trend, but the second order anomaly clearly exists.

The extent of d orbital occupation appears, from modern calculations, to be less than implied by the classical hybridization ideas like sp^3d or sp^3d^2. Substantial energy contributions can arise for d populations of only a fraction of an electron. We shall use the more familiar terminology in this chapter, but see the discussion of this in Chapter 18, Section 9. See also the comment on relativistic effects in Section 16.13.

THE CHARACTERISTIC PROPERTIES OF THE HEAVIER ELEMENTS IN EACH p GROUP

These points may be summarized:

(i) The heaviest element in a p Group differs from the rest in the stability of oxidation states lower than the Group state, in its more metallic character with more basic oxides, and in its weak bonding to hydrogen and related ligands.

(ii) The properties of the middle three elements in a p Group form a transition to those of the heaviest element, with the lower oxidation states becoming more stable and the oxides more basic, etc.

(iii) There is a sharp discontinuity in properties with the first element, especially in the carbon, nitrogen and oxygen Groups.

(iv) The heavier elements tend to show higher coordination numbers.

(v) Minor anomalies are observed in the chemistry of the elements of the middle row of the p block as compared with the properties of the rows above and below.

17.4 The boron Group ns^2np^1

17.4.1 The elements, general properties and uses

Table 17.4 lists some properties of the elements and Figs 17.4 and 17.5 give the variations of certain properties with Group position. It will be seen from the free energy diagram, Fig. 17.5, that the III state is the most stable one, except for thallium. Figure 17.6 shows some typical reactions of the boron Group elements.

Boron reacts directly with most metals to give hard, inert, binary compounds of various formulae. These borides somewhat resemble the interstitial carbides and nitrides. Table 17.5 lists some typical formulae, and some structures, which all involve chains, sheets or clusters of boron atoms, are shown in Fig. 17.7. The binary compound, boron nitride, BN, is interesting as it is isoelectronic with carbon and occurs in two structural modifications.

OTHER REFERENCES TO THE PROPERTIES OF BORON GROUP ELEMENTS

References to the properties of the boron Group elements given in earlier chapters include:

Ionization potentials	Table 2.8 and Fig. 8.6
Electronegativities	Table 2.14
Hydrides	Chapter 9, especially electron-deficient hydrides, Section 9.6

Boron clusters are treated in Sections 18.4.1 and 18.4.2. For an example of aluminium in biochemistry see Section 20.1.3, and for boron neutron capture therapy, Section 20.2.5.

See Section 16.1 for MgB_2 as a 'warm' superconductor.

OCCURRENCE AND USES OF THE BORON GROUP ELEMENTS

Aluminium, the most common metallic element in the earth's crust, is extracted from the hydrated oxide, bauxite, by electrolysis of the oxide (after purification by alkaline treatment) dissolved in molten cryolite, sodium hexafluoroaluminate. Boron is found in concentrated deposits as borax, $Na_2B_4O_5(OH)_4.8H_2O$, and similar tri-, tetra-, and pentaborates of Na and Ca. The element is formed by magnesium reduction of the oxide. The other three elements are found only in the form of minor components of various minerals, and the elements are produced by electrolytic reduction in aqueous solution. Gallium, indium and thallium are relatively soft and reactive metals which readily dissolve in acids. Aluminium is also a reactive metal but is usually found with a protective, coherent oxide layer which renders it inert to acids, although it is attacked by alkalies. Boron is nonmetallic and the crystalline form is very hard, inert and nonconducting. However, boron often shows considerable analogies with metals, specifically the transition metals, in its chemistry. This is particularly apparent when clusters containing these elements are considered (see Sections 16.8 and 18.4). The amorphous form of boron, which is more common than the crystalline variety, is much more reactive.

Boric oxide and borates find extensive application. Borax and other borates find uses in water treatment, and in preserving timber from insects. Larger amounts of sodium or calcium borates, and of boric acid or oxide, are used in glass manufacture. Borosilicate glasses have a lower coefficient of thermal expansion than the more conventional ones, and are therefore more robust under heating. Sodium perborate, approximately $NaBO_3.4H_2O$ in composition, is widely used as a bleach. This material was once formulated as borate with H_2O_2 of crystallization, but it is now established as a true peroxyborate with B-O-O links.

Aluminium oxide, alumina and related complex oxides include a number of widely used species. Recall also that the clay minerals, the feldspars, zeolites and other important groups are aluminium–silicon–oxygen species (see Section 18.6). Aluminium phosphates, which are isoelectronic with the silicates, have a similar range of structures and show promise in similar uses, for example in catalysis like the zeolites. The compact, inert α-alumina occurs as the minerals corundum (Fig. 5.5b) and emery, both very hard, which find extensive uses in abrasives and in refractory and ceramic materials. The lower temperature form, γ-alumina, has a much more open structure and can be prepared in forms with very high surface areas which are widely used as catalysts and as supports for catalysts (compare dehydrosulfurization, for example, Section 15.4). Such 'activated' alumina is also used for chromatography. Alumina, with zirconium dioxide, has been produced in the form of very fine, strong fibres which have important uses as reinforcing in light-weight materials. Thus, aluminium metal reinforced by alumina fibres can have up to five times the strength of the metal alone. As well as the clay minerals, which in turn give ceramics, two other important aluminium–metal ternary oxides are spinel (Fig. 5.5d) and the species called β-alumina which actually contains sodium ions and has the approximate formula $NaAl_{11}O_{17}$. This is used as the electrolyte in solid state electrical cells. The structure is blocks of spinel separated by layers of composition NaO. These sodium oxide layers are very open, with a number of equivalent Na^+ positions, so that sodium ions move readily, giving rise to the very low resistivity. Finally, we note that a major constituent of Portland cement is $Ca_9Al_6O_{18}$ which contains a ring formed by six AlO_4 tetrahedra linked by sharing corner oxygens (compare the similar 3-tetrahedron ring in silicates, Section 18.6).

TABLE 17.4 Properties of the elements of the boron Group

Element	*Symbol*	*Oxidation states*	*Common coordination numbers*	*Availability*
Boron	B	III	3, 4	common
Aluminium	Al	III	3, 4, 6	common
Gallium	Ga	(I), III	3, 6	rare
Indium	In	I, III	3, 6	very rare
Thallium	Tl	I, (III)	3, 6	rare

All the elements have high boiling points (above 2000°C), but gallium has an unusually low melting point at 29.8°C which gives it the longest liquid range of any element.

One has a layer structure, like graphite (but is light in colour) and is soft and lubricating, while the other, formed under high pressure, has a very hard, stable, tetrahedral structure as in diamond.

In Li_3Ga_{14} there are Ga_{12} icosahedra linked into a three-dimensional network.

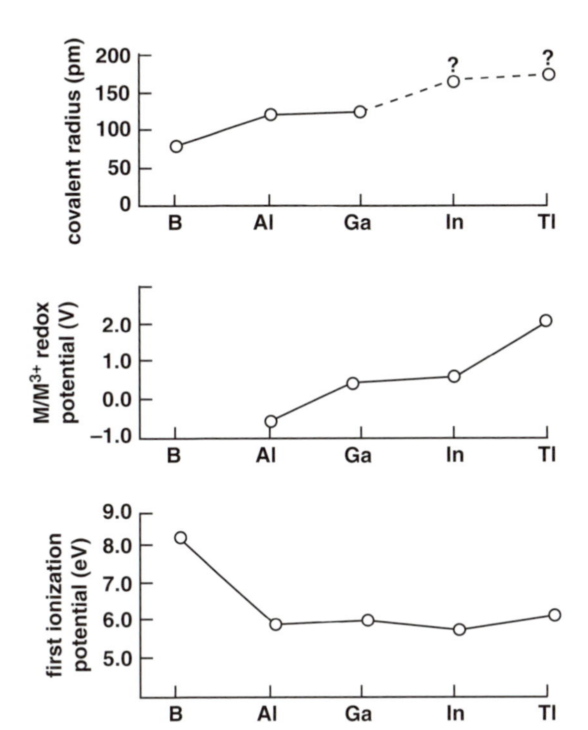

FIG. 17.4 Some properties of the boron Group elements. The figure shows the covalent radii, the oxidation potentials and the first ionization potentials as functions of Group position

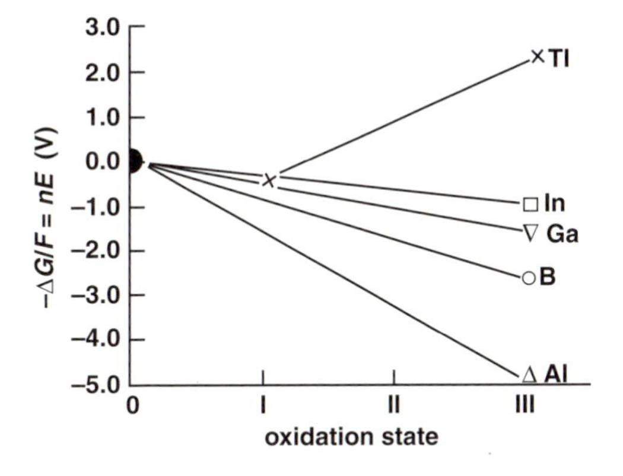

FIG. 17.5 The oxidation state free energy diagram for the boron Group elements

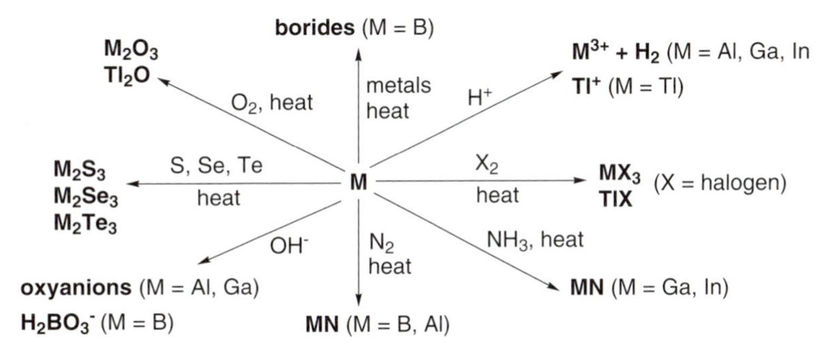

FIG. 17.6 Reactions of the elements of the boron Group

TABLE 17.5 Some typical borides

Formula	*Boron atom structure*	*Boride examples*
M_2B	Single atoms	Be, Cr, Mn, Fe, Co, Ni, Mo, Ta, W
$M_2M'B$	Pairs	M = Al, Ti, Mo; M′ = Cr, Fe, Ni
MB	Single chain (Fig. 17.7a)	V, Cr, Mn, Fe, Co, Ni, Nb, Ta, Mo, W
M_3B_4	Double chains	V, Cr, Mn, Ni, Nb, Ta
MB_2	Sheets (Fig. 17.7b)	Be, Mg, Al, Sc, Ti, V, Cr, Mn, Y, Lu, U, Pu, Zr, Hf, Nb, Ta, Tc, Re, Ru, Os, Ag, Au
MB_4	Sheets linking B_6 octahedra	Mg, Ca, Mn, Y, Ln, Th, U, Pu, Mo, W
MB_6	B_6 octahedra (which occupy Cl^- positions in a CsCl structure with the metals (Fig. 17.7c))	Na, K, Be, Mg, Ca, Sr, Ba, Sc, Y, La, Th, Pu
MB_{12}	Three-dimensional lattice consisting of linked B_{12} clusters and with metal atoms in the middle of each cluster	Be, Mg, Al, Sc, Y, Ln from Tb to Lu, U, Zr

Other more unusual species include CuB_{22}, $B_{13}M_2$(M = P, As) and $Ru_{11}B_8$.

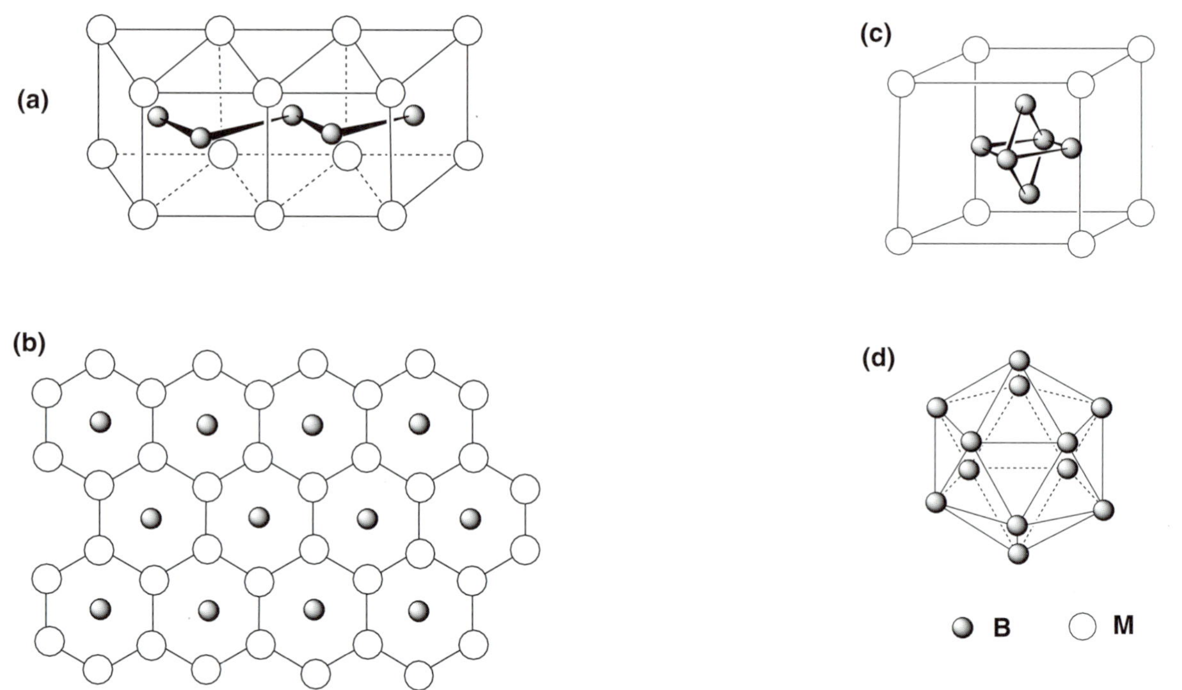

FIG. 17.7 Examples of the structures of boron units in borides: (a) shows the configuration of the boron chain in MB compounds, (b) the sheet structure found in MB_2 and (c) the B_6 cluster in MB_6 compounds. The B_{12} icosahedron shown in (d) is found in boron itself, in some of the boron hydrides, and in BeB_{12} and AlB_{12}. The other MB_{12} species have a related structure. In the tetraborides, hexaborides and dodecaborides, the boron clusters (such as the B_6 unit in (c)) are themselves linked so that a bonded boron framework extends completely through the compound

The elements form sulfides, selenides and tellurides, of similar formulae to the oxides, whose structures are based on 4- or 6-coordination. Several have interesting electronic properties. Two unusual structures are those of $Ga_4S_{10}^{8-}$, which is the same as P_4O_{10} (Fig. 17.32b) while $Ga_6Sc_{14}^{10-}$ consists of six edge-sharing tetrahedra (Fig. 17.8), and an extended form related to the many M_2X_6 examples (compare Fig. 17.13). The other important compounds formed by the heavier elements of this Group are 'III/V' binaries, especially GaAs and InP, which are very important semi-conductors. Structures are ZnS type where each element is tetrahedral, and related to Si with which they are isoelectronic.

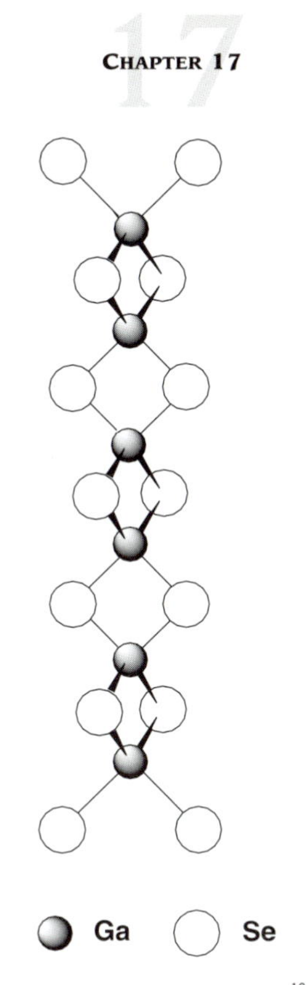

FIG. 17.8 The structure of the $Ga_6Se_{14}^{10-}$ ion

17.4.2 The III state

The Group state of III is shown by all the elements, and is the most stable state for all except thallium. It is represented by a wide variety of compounds, of which the oxygen and halogen compounds are typical. There is no evidence for a free M^{3+} ion, either in the solid or in solution. A number of solids, especially fluorides and oxides, are high melting and strongly bonded, but the bonds are intermediate between ionic and covalent and the stabilities of the solids are due to the formation of giant molecules with uniform bonding. For example, aluminium chloride, bromide and iodide are volatile, covalent solids, while aluminium trifluoride is high melting and a giant molecule. Similar effects are seen for the other trifluorides, except for BF_3, and for the oxides. In solution, extensive hydration and hydrolysis occur and ionic species (though often written as cations for convenience) are actually much more complex, e.g. $Al(OH)(H_2O)_5^{2+}$ has been shown to occur in 'Al^{3+}' solutions.

All the elements form the trioxides, usually as hydrated species by precipitation from solution or by hydrolysis of the trihalides. Chemical and structural properties are given in Table 17.6.

Oxides of the I state are treated later.

Hydration of the oxides gives a variety of hydrates and hydroxy-species. Boric oxide gives boric acid, $B(OH)_3$, on hydration which forms crystals in which the $B(OH)_3$ units are linked together by hydrogen bonding (Fig. 9.18). When boric acid is heated, it dehydrates first to metaboric acid, HBO_2, and ultimately to boric oxide:

$$B(OH)_3 \underset{+H_2O}{\overset{-H_2O}{\rightleftharpoons}} HBO_2 \underset{+H_2O}{\overset{-H_2O}{\rightleftharpoons}} B_2O_3$$

Metaboric acid exists in three crystalline forms, one of which contains the cyclic unit shown in Fig. 17.9. The structures of the other two are not known with certainty but appear to contain chains of BO_3 and BO_4 units. The cyclic anion is also found in sodium and potassium metaborates. A wide variety of other oxyanions of boron exists with very varied structural types. Not only are discrete ions, rings, chains, sheets and three-dimensional structures found—as with the silicates—but boron occurs both in planar BO_3 units and in

TABLE 17.6 Oxides of the III state of the boron Group elements

Oxide	Properties	Structure
B_2O_3	Weakly acidic Many metal oxides give glasses with B_2O_3 as in the 'borax bead' test.	Glassy form—random array of planar BO_3 units with each O linking two B atoms. Crystalline form—BO_4 tetrahedra linked in chains.
Al_2O_3 and Ga_2O_3	Amphoteric	α-form—inactive, high-temperature form. Oxide ions ccp with metal ions distributed regularly in octahedral sites. γ-form—low-temperature form, more reactive. Metal ions arranged randomly over the octahedral and tetrahedral sites of a spinel structure.
In_2O_3 and Tl_2O_3	Weakly basic Tl_2O_3 gives O_2 and Tl_2O on heating to 100°C	The structure has the metal ions in irregular 6-coordination, and 4-coordinated oxygens. The same structure is adopted by most oxides of the lanthanide elem€ nts, Ln_2O_3.

Other Oxides: $(BO)_x$ formed by heating $B + B_2O_3$ at 1050°C. This probably contains both B-O-B and B-B links as it reacts with BCl_3 to give B_2Cl_4.
gem forms of alumina : ruby—Al_2O_3 + traces of Cr^{3+}
blue sapphire—Al_2O_3 + traces of Fe^{2+}, Fe^{3+} or Ti^{4+}
white sapphire—this is the gem form of alumina itself

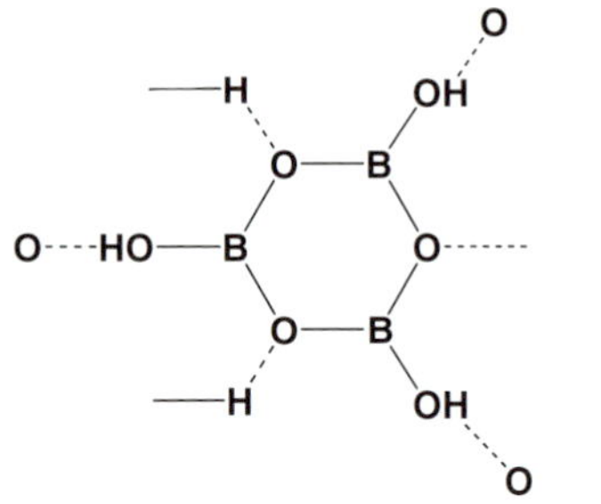

FIG. 17.9 The cyclic form of metaboric acid

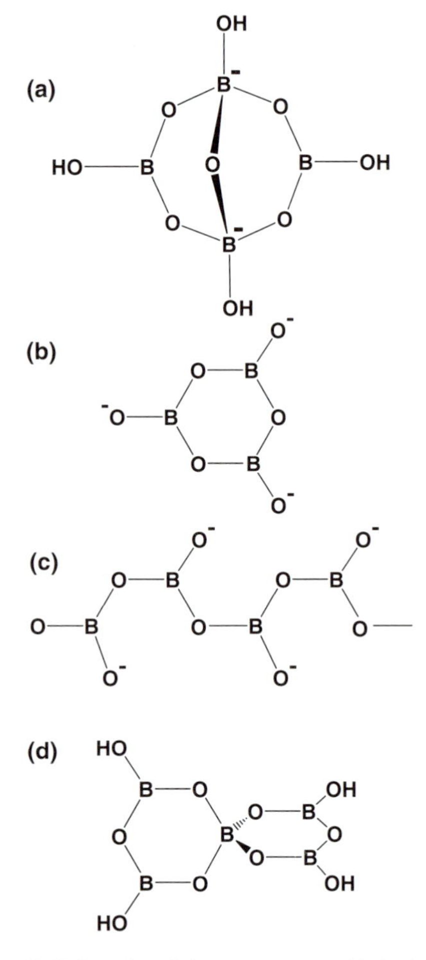

FIG. 17.10 Examples of borate structures: (a) in borax, $Na_2B_4O_7.10H_2O$, (b) in metaborates, $M_3B_3O_6$ (cyclic anion), (c) in linear metaborates, CaB_2O_4, (d) $B_5O_{10}H_4^-$

tetrahedral BO_4 units, and many borates contain OH groups. It is impossible to discuss all the borates, but Fig. 17.10 gives a few representative borate structures.

Aluminium and gallium form hydrated oxides of two types—MO(OH) and $M(OH)_3$. These are precipitated from solution by, respectively, ammonia and carbon dioxide. Indium gives a hydrated oxide, $In(OH)_3$. In these compounds the metal is 6-coordinated to oxygen.

Hydroxy species also occur in solution. Thus boric acid accepts an OH^- group in dilute solution and polymerizes in more concentrated solutions:

$$B(OH)_3 + 2H_2O \rightarrow B(OH)_4^- + H_3O^+$$
$$3B(OH)_3 \rightarrow B_3O_3(OH)_4^- + H_3O^+ + H_2O$$

The hydrates of the other elements of the Group behave similarly. For example:

$$M(H_2O)_6^{3+} \rightleftharpoons M(H_2O)_5(OH)^{2+} \rightleftharpoons \text{intermediate stages} \rightleftharpoons M(OH)_6^{3-}$$

and a compound, $Ca_3[Al(OH)_6]_2$ has been isolated. Among species identified in solution, by a range of methods including ^{27}Al nmr, are $Al(H_2O)_6^{3+}$, $Al(H_2O)_5(OH)^{2+}$, $Al_2(OH)_2(H_2O)_8^{4+}$ and more highly polymerized entities such as $Al_8(OH)_{20}(H_2O)_n^{4+}$ and $Al_{13}O_4(OH)_{24}(H_2O)_{12}^{7+}$. The Al environments are probably all based on AlO_6 octahedra.

All the trihalides of all the boron Group elements exist and all correspond to the III state, except TlI_3 which is the triiodide, I_3^-, of Tl^+. The normal trihalides are planar molecules which have an empty p orbital in the valence shell. Most of the trihalides make use of this empty orbital, both in the structure of the trihalide and in the formation of complexes of the form $MX_3.D$, where D is a lone pair donor. Table 17.7 lists these applications for the halides and for the halide complexes. Aluminium, and the heavier elements, also use their d orbitals to become 6-coordinate. It will be seen that the formation of a p_π–p_π bond in BF_3, and the use of d orbitals, especially in the fluorides, mirrors the discussion of these effects in Section 17.2. An interesting example of a mixed dimer is provided by $NbAlCl_8$ which has octahedral $NbCl_6$ linked to tetrahedral $AlCl_4$ by sharing an edge (compare Figs 15.6b and 17.12).

Most of the trihalides react with water to give the hydrated oxides, but boron trifluoride gives 1:1 and 1:2 adducts, $BF_3.H_2O$ and $BF_3.2H_2O$, which are not ionized in the solid state. The 1:1 adduct has the expected donor structure, $F_3B.OH_2$, but the structure of the second is unknown. When these adducts are melted, they each ionize:

$$2BF_3.H_2O \rightleftharpoons (H_3O.BF_3)^+ + (HO.BF_3)^-$$
$$\text{and} \quad BF_3.2H_2O \rightleftharpoons H_3O^+ + (HOBF_3)^-$$

Among the complex halides, all the MX_4^- species are tetrahedral while the MX_6^{3-} ones are octahedral. All the boron Group trihalides act as catalysts in the Friedel–Crafts reaction, where their function is to abstract a halide ion (giving MX_4^-) from the organic molecule, leaving a carbonium ion. As well as the MX_4 and MX_6^{3-} ions shown in the table, the isolation of MCl_5^{2-} ions has been reported for indium and thallium. A crystal study shows that, in its tetraethylammonium salt, $InCl_5^{2-}$ is a square pyramid, similar to the structures described in Section 13.6. This is the first report of this structure for a Main Group element (not to be confused with species like IF_5 with a lone pair in addition). The form probably results from the way the ions pack with the large cation into the crystal. There are only a few polynuclear ions, including the binuclear Tl(III) complex $[Tl_2Cl_9]^{3-}$. The structure of this ion, shown in Fig. 17.13, consists of two $TlCl_6$ octahedra which share a common face.

Two different aluminium fluoride anions are found in the unusual compound $Sm(II)AlF_5$:

(a) $[Al_2F_{10}]^{4-}$ ions with two octahedra sharing an edge
(b) a long chain ion formed by linking AlF_6 octahedra through *trans* apices.

The trihalides, and other trivalent MX_3 species, readily form tetrahedral complexes such as BH_4^-, $AlX_3.NR_3$, or $GaH_3.NMe_3$. A wide selection of 1:1 $BX_3.D$ complexes exist where D is a lone pair donor such as ammonia, amine, water or ether, phosphane, sulfide, etc., and X is halogen, hydrogen or an organic group. The organic compounds $R_3B.D$ have been much studied to find the factors, such as electron attracting power and steric effects, which most

TABLE 17.7 Acceptor and structural properties of the trihalides of the boron Group elements

Halide	*Structural use of empty* p *orbital*	*Halide complex*
BF_3	Internal p_π–p_π bonding: see note (1) and Fig. 17.11	BF_4^-
BX_3	Possibly slight π bonding in BCl_3, otherwise none.	BX_4^-
AlF_3	Accepts lone pair from fluorine (as do two d orbitals) to give AlF_6 units in a highly polymerized solid. M.p. above 1000°C	AlF_6^{3-}, AlF_4^-
AlX_3 (note 2)	Accepts one lone pair from a halide to give Al_2X_6 dimer (Fig. 17.12). m.p. 100–200°C.	AlX_4^-
MF_3 (M = Ga, In, Tl)	As AlF_3 (m.p. about 1000°C).	MCl_6^{3-}, MBr_6^{3-}
GaX_3, InX_3	As AlX_3 (m.p. 100–600°C).	MCl_5^{2-}, MCl_4^-
TlX_3	As AlX_3 (note 3)	MBr_4^- (M = Ga, In)
	In all cases above, X = Cl, Br and I.	$TlCl_5^{2-}$

Note (1) The evidence for internal π bonding in BF_3 derives from two sources. First, the B-F bond length is shortened compared with that in the BF_4^- ion, 130 pm compared with 142 pm. Second, the order of acceptor strengths for the boron trihalides (forming BX_3.D) is $BBr_3 > BCl_3 > BF_3$. As ability to accept an electron pair depends on the electron density at the boron, the strongly electronegative fluoride would be the strongest acceptor unless other effects intervene.
Note (2) For X = Cl, the dimer is found in the vapour but solid $AlCl_3$ exists as a slightly deformed $CrCl_3$ layer lattice structure (compare Table 5.3) with 6-coordination of aluminium.
Note (3) $TlBr_3$ decomposes to TlBr and Br_2 at room temperature, and $TlCl_3$ loses chlorine similarly at 40°C.

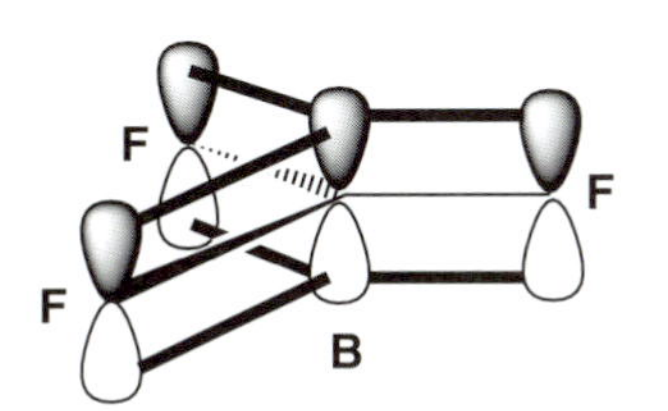

FIG. 17.11 Internal p_π-p_π bonding in boron trifluoride. The p orbital on the boron which is not used in the σ bonds, accepts electrons from the three corresponding fluorine orbitals to give internal π-bonding in BF_3

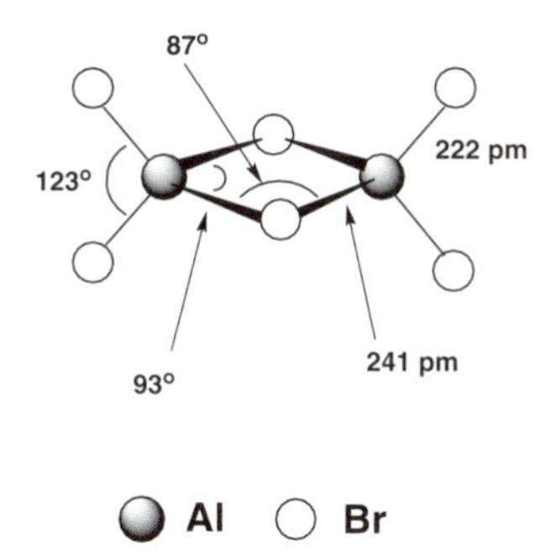

FIG. 17.12 The structure of the aluminium tribromide dimer. The aluminium makes use of its empty p orbital to accept a lone pair from a bromine atom in a second $AlBr_3$ molecule, giving Al_2Br_6

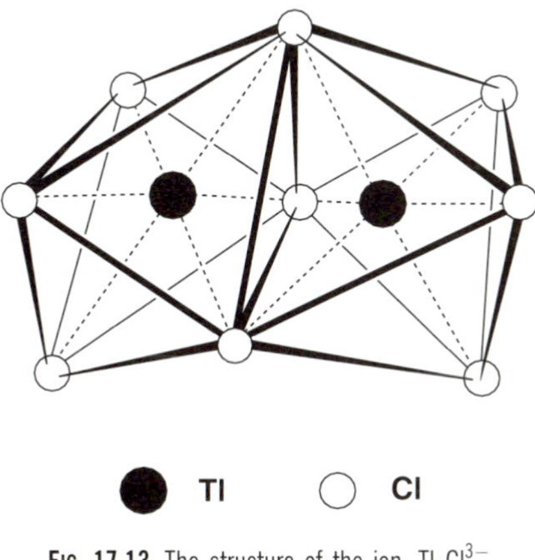

FIG. 17.13 The structure of the ion, $Tl_2Cl_9^{3-}$

influence Lewis acid–base behaviour. One anion of analytical importance is the tetraphenylboronate ion $B(C_6H_5)_4^-$, which forms insoluble salts with potassium and the heavier alkali metals and is used in their gravimetric determination.

Complexes of the elements other than boron include both tetrahedral types as above and also octahedral complexes, of which important examples are the oxygen chelate complexes shown in Fig. 17.14 and the 8-hydroxyquinoline complex of Fig. 17.15 which is used in the gravimetric determination of aluminium. These elements form hexahydrates, $M(H_2O)_6^{3+}$, which hydrolyse in solution. Also, hydrated salts with this cation and a variety of oxyanions are known. Aluminium also forms the well-known series of double salts, $MAl(SO_4)_2.12H_2O$, called alums. M is any univalent cation and the aluminium may be replaced by a variety of trivalent ions such as Cr^{3+}, Fe^{3+}, Co^{3+}, Ga^{3+} or Ti^{3+}. The crystals contain $M(H_2O)_6^+$, $Al(H_2O)_6^{3+}$—or other $M(H_2O)_6^{3+}$ ions—and sulfate ions.

17.4.3 The I oxidation state and mixed oxidation state compounds

The I oxidation state is most important in thallium chemistry where it is the most stable state. The few common thallium(III) compounds, such as the oxide and halides, are strongly oxidizing, and the potential Tl^{3+}/Tl^+ of 1.3 V in acid solution makes thallium(III) in solution as oxidizing as chlorate or MnO_2. Thallium(I) compounds are stable and show some

THE GALLEX EXPERIMENT

One interesting application of $GaCl_3$, involving about 100 tons of a concentrated aqueous solution, is in the experiment (called GALLEX) to detect low energy solar neutrinos produced in the Sun as a result of fusion reactions. In this experiment, Ga nuclei are converted, in *extremely* low yield, by reaction with a neutrino and an electron into ^{71}Ge atoms, which are then separated and counted after conversion to germane, GeH_4. As an indication of the difficulties which had to be overcome, less than one ^{71}Ge atom was produced per day consequently it was no mean technological feat to separate the germanium atoms from the $GaCl_3$!

ORGANIC COMPOUNDS AND ZIEGLER–NATTA CATALYSTS

All the boron Group elements form organic compounds, R_3M, and also mixed types, R_2MX and RMX_2, with halogens and related groups. The boron compounds are monomers with planar BC_3 skeletons, but the later members of the Group give dimeric or polymeric compounds. The halides dimerize through halogen bridges similar to those in Al_2X_6. The purely organic compounds polymerize through electron deficient carbon bridges, as in $Al_2(CH_3)_6$, shown in Fig. 9.13, similar to the bonding in diborane.

Organo-aluminium compounds are involved in a number of related, and commercially important, catalytic processes which arose from the work of Ziegler. The basic discovery was that Al-H bonds add across the alkene double bond coupled with the fact that aluminium metal reacts under fairly easy conditions with hydrogen in the presence of aluminium alkyl. Thus, Al-H bonds may be formed and converted into Al-C in the overall process shown in equation (1)

$$Al + \tfrac{3}{2}H_2 + 3RCH{=}CH_2 \xrightarrow{AlR_3} Al(CH_2CH_2R)_3 \quad (1)$$

The addition is to terminal double bonds. The Al-C bond in turn will add across a terminal double bond, in a series of steps which leads to growth of the alkyl chain

$$R_2Al - R + R'CH{=}CH_2 \rightarrow R_2AlCH_2CH(R)R' \text{ etc.} \quad (2)$$

Finally, the process may be terminated by the reverse step to (l), yielding a long-chain α-alkene, or hydrolysis and oxidation is used to yield an alcohol

$$R_2Al - CH_2CHRR' \xrightarrow{O_2} R_2AlOCH_2CHRR' \xrightarrow{H_2O} R_2AlOH + RR'CHCH_2OH. \quad (3)$$

All the Al-R links are eventually broken in step (3).

There are several applications of this process: firstly, using ethylene and high pressures at 160°C, steps (1), (2) and (3) can be arranged to produce alcohols with chain length about C_{14} which are used in the production of bio-degradable detergents. Secondly, using ethylene at about 100°C, steps (1), many repetitions of (2), terminated by the reverse of (1), produces polythenes with average chain-length about C_{200}. Thirdly, longer chain olefins may be dimerized, as in the formation of isoprene via the dimerization of propene

$$2CH_3CH{=}CH_2 \xrightarrow{(1)\text{ and }(2)} CH_3CH_2CH_2C(CH_3){=}CH_2 \xrightarrow{heat} CH_4 + CH_2{=}CHC(CH_3){=}CH_2$$

Isoprene polymerization itself takes place in the presence of catalysts including aluminium alkyls.

A further process, developed in part by Ziegler and partly by Natta, is an extension of the olefin polymerization. The process above, steps (1), (2) and the reverse of (1), gives a wide spectrum of chain lengths and disordered polymers which are soft and low melting. In the Ziegler–Natta process, a transition metal halide, such as $TiCl_4$, is added to an aluminium trialkyl and the resulting reaction mixture is found to catalyse the polymerization of alkenes, giving a stereo-regular product (one where, for example, all the sidechains lie the same way) and these regular polymers are much higher melting and more crystalline. For example, while ordinary polythene softens below 100°C, polythene from the Ziegler–Natta process melts at 130–135°C. Later improvements include 'second generation' catalysts involving Ti(III) or Cr, often on $MgCl_2$ supports, and also the use of homogeneous catalysts functioning in solution, involving organometallic Al and metal compounds. Such improvements aim at narrower ranges of molecular weights in the polymers, particular geometrical order, or greater ease of processing.

The exact nature of the catalytic process is still under study. A variety of transition metal halides may be used together with other active organometallic species in place of the AlR_3. For $TiCl_4$, it is established that the titanium is reduced to the III (or lower) state and one theory is that the catalysis takes place on the crystal surface of the reduced species. A chain growth process like (2) occurs but, as it is on a surface, the approach of the incoming olefin is oriented, giving a regular polymer. The aluminium alkyl acts as the reducing agent and also forms Ti-R groups on the surface to provide growth sites. An alternative theory suggests some Ti···X···Al bridge which provides the active site, where X may be an organic group or a halogen. Here again, the orientation of substituents is postulated to restrict the attacking alkene into a regular and repeatable orientation, giving a regular orientation of the product.

This is supported by the observation that optical isomers may be separated during the polymerization. The active site is regarded as an ordering matrix analogous to, though simpler than, the active sites on enzymes in biochemical catalyses.

resemblances to both lead(II) compounds and to those of alkali metals. The oxide, Tl_2O, and the hydroxide, TlOH, are strongly basic, like the alkali metal compounds, and absorb carbon dioxide from the atmosphere. The halides resemble lead halides in being more soluble in hot water than in cold and behave in analysis like the lead or silver compounds. Tl^I forms a number of stable salts which are generally isomorphous with the alkali metal ones; examples include the cyanide, perchlorate, carbonate, sulfate and phosphates. TlF has a deformed

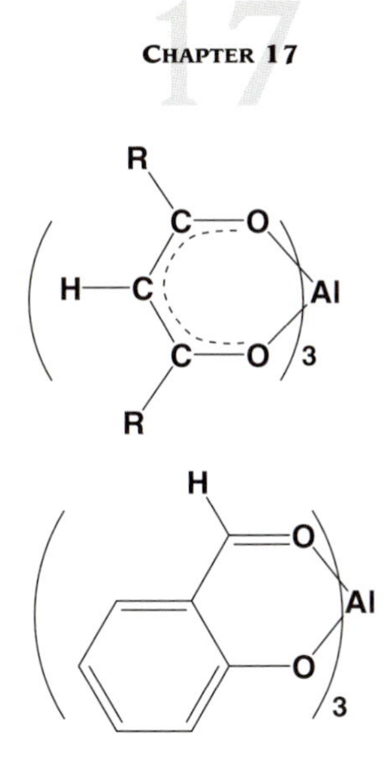

FIG. 17.14 Oxygen chelate complexes of aluminium

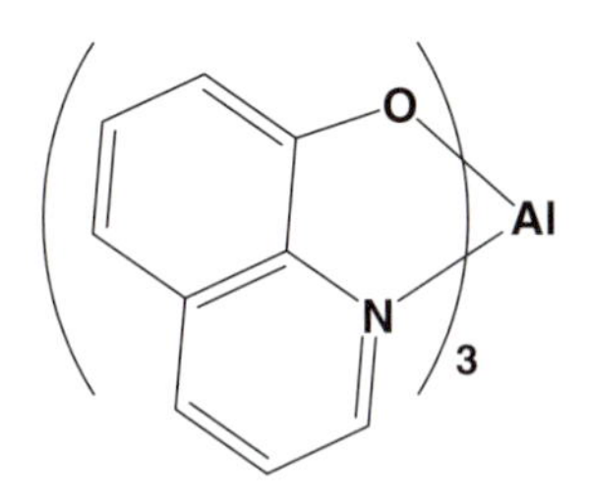

FIG. 17.15 The 8-hydroxyquinoline complex of aluminium. This is commonly used to determine aluminium gravimetrically

sodium chloride structure in the solid, while the other thallous halides crystallize with the caesium chloride structure. The chemistry of thallium is indicated in Fig. 17.16 which shows the interrelation between the two oxidation states.

The I oxidation state becomes rarer and less stable as Group 13 is ascended. Among other low-valent gallium compounds, the oxide Ga_2O is formed by heating Ga_2O_3 with Ga, and disproportionates at high temperatures. Indium(I) halides are more stable than their gallium counterparts, but In_2O needs further study, while both gallium and indium form various sulfide phases as well as M_2S_3. The monohalides of indium and thallium are interesting. The yellow form of InCl has the distorted NaCl structure like TlF. In its red modification, together with InBr, InI and TlI, a 7:7 coordinated structure is found, where the coordination shell is 1:4:2—that is, like an octahedron but with two M atoms spanning the site of one apex. This structure may be seen as intermediate between the 6:6 NaCl one, and the 8:8 CsCl structure of TlCl, TlBr and the second form of TlI.

Gallium, indium, and also thallium, form halides of formula MX_2 which are mixed valency species containing the I state cation and the III state complex anion, $M^+[MX_4]^-$. A similar mixed valency formulation is found for In_2Cl_3, Tl_2Cl_3 and Tl_2Br_3 which have octahedral M(III) complex anions, $M_3^+[MX_6]^{3-}$. Related compounds with these anions, such as $Ga^+[AlCl_4]^-$ or $Tl_3^+[InCl_6]^{3-}$, are known. In contrast, M_2Br_3, for M = Ga or In, contain the diatomic $[M\text{-}M]^{2+}$ cation, formally M(I), and the $M_2Br_6^{2-}$ anion mentioned below.

Crystal structures of these mixed halides have allowed the determination of the M^+ radii, which show less variation than the alkali metal ions. Thus Goldschmidt radii are $Ga^+ = 141$ pm, $In^+ = 143$ pm, and $Tl^+ = 149$ pm, compare Table 2.12.

Gallium(II) is found in the $Ga_2X_6^{2-}$ species for X = Cl, Br, I, prepared electrolytically from gallium in strong acid. The structure appears to be $X_3Ga\text{-}GaX_3^{2-}$, like the isoelectronic Ge_2Cl_6. A related solvated species $[Ga_2Cl_4(dioxane)_2]_x$ is a polymeric adduct of the $Cl_2Ga\text{-}GaCl_2$ unit with 1,4-dioxane, $O(CH_2CH_2)_2O$. In these compounds the formal +II state arises because of the gallium–gallium bond. The halides Ga_4X_6 are mixed oxidation state species containing the ions $(Ga^+)_2(Ga_2X_6)^{2-}$.

Formal (II) states are shown in the ME species when M = Ga, In or Tl and E = S, Se or Te. Two structures are common. The lighter combinations have the GaS structure where there are successive layers in the sequence –S-Ga-Ga-S-S-Ga where the (II) state arises because of the Ga-Ga bonds, and the elements are 4-coordinated. For InTe and the Tl species, the apparent II state arises from mixed (I) plus (III) as in Tl(I)[Tl(III)S_2] where the TlS_2 formula represents an infinite chain of TlS_4 tetrahedra linked by shared edges.

An extremely interesting group of compounds involves benzene and methyl-substituted benzenes. It was reported in 1881 that the compounds of formula GaX_2 were unexpectedly soluble in benzene and could be isolated as benzene-containing solids. Studies a hundred years later isolated Ar_2Ga^+ complexes, where Ar = C_6H_6 or $C_6H_3Me_3$ and the Ga was equidistant from the six carbons (compare dibenzene chromium, Section 16.3). A similar Tl(I) species is also found. When a more hindered benzene is used, at a lower ratio, the compound $ArGa^+$ was isolated as the $GaBr_4^-$ salt for Ar = C_6Me_6. The ring is almost

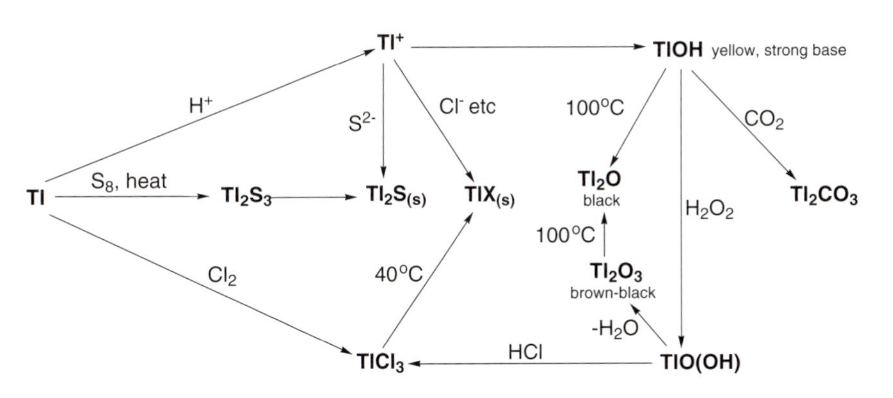

FIG. 17.16 Reactions of thallium compounds

LOW OXIDATION STATE ALUMINIUM AND GALLIUM COMPOUNDS AND CLUSTERS

Metastable compounds MX, such as AlCl, AlBr, AlI, GaBr, etc., can be obtained by heating the metal and HX in the gas phase at high temperature and low pressure, followed by trapping the reactive MX at low temperature in liquid nitrogen. Upon warming, the expected disproportionation reaction occurs, e.g.

$$3AlCl \rightarrow AlCl_3 + 2Al$$

If a suitable neutral co-ligand is added, the aluminium(I) halide can be trapped. Thus, triethylamine-stabilized AlBr can be obtained as a tetramer by addition of NEt_3. An X-ray structural study shows that the compound contains a square Al_4 ring, with Al-Al bonds.

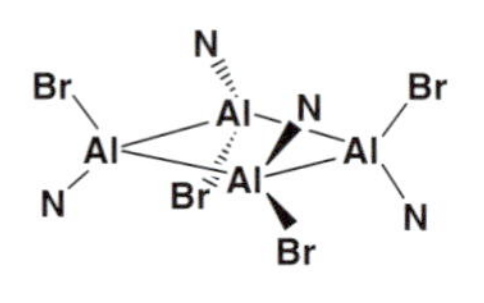

Structure of $[AlBr(NEt_3)]_4$

Each Al atom is bonded to one terminal Br and one triethylamine ligand. When AlBr is co-condensed with toluene and THF, the cluster $Al_{22}Br_{20}.12THF$ is obtained. This is based on an Al_{12} icosahedron, with $AlBr_2$ units bonded to 10 of the 12 Al vertices, with the remaining two vertices, in opposite positions, bonded to THF molecules only.

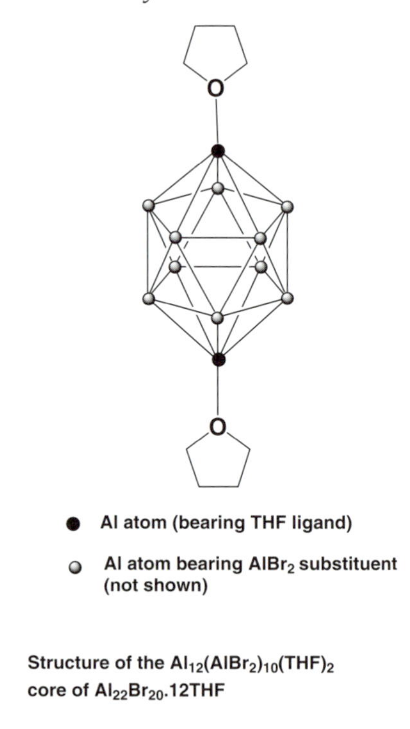

Structure of the $Al_{12}(AlBr_2)_{10}(THF)_2$ core of $Al_{22}Br_{20}.12THF$

The remaining THF molecules are not bonded to the core. The icosahedral Al_{12} unit is highly reminiscent of the B_{12} unit present in one form of elemental boron, in some metal borides (Fig. 17.7) and some boron hydrides.

By replacement of the halide ligands of the above, other substituted Al clusters can be obtained; these correspond to trapping of the disproportionation reaction shown above, before complete formation of Al metal. When AlCl is reacted with the amide ligand $LiN(SiMe_3)_2$(LiR) the cluster $Al_{12}R_8^-$ is obtained. The structure of this anion resembles a part of the metallic aluminium lattice. Similarly, reaction of AlI with LiR gave $Al_{77}R_{20}^{2-}$, which has a core consisting of a central Al atom surrounded by three concentric polyhedral shells of 12, 44 and 20 Al atoms, with the outer layer protected by the $N(SiMe_3)_2$ groups. An even larger cluster is $Ga_{84}R_{20}^{4-}$, likewise formed from GaBr and LiR.

In related chemistry starting from the III oxidation state, the reaction of $GaCl_3$ with ${}^tBuLi$ gives trivalent Ga^tBu_3 as well as the cluster $Ga_9{}^tBu_9$ formed by a reduction reaction, which has been shown to have a tricapped trigonal prismatic arrangement of Ga atoms.

Thus, the lighter Group 13 elements clearly have a more extensive chemistry in their lower oxidation states than previously thought. In the formation of clusters, the heavier Group 13 elements are paralleling the chemistry of boron, and also the transition metals which also form extensive series of metal clusters, (Section 16.8.2). Transition metal clusters invariably have their surface protected by stabilizing ligands, such as CO or phosphanes, which prevent aggregation of the clusters to give the bulk metal. Likewise, in the new aluminium and gallium clusters, ligands such as amides or alkyls also prevent decomposition to the metal plus a trivalent amide or alkyl derivative (formed through disproportionation). Many new developments are expected in this area.

References

Aluminium(I) and Gallium(I) Compounds: Syntheses, Structures and Reactions, C. DOHMEIER, D. LOOS AND H. SCHNÖCKEL, *Angewandte Chemie, International Edition*, **35**, 1996, 129–149.
From AlX/GaX Monohalide Molecules To Metalloid Aluminium and Gallium Clusters, H. SCHNÖCKEL AND A. SCHNEPF, Advances in Organometallic Chemistry, **47**, 2001, 235–281.

planar and the Ga lies 255 pm above its centre completing a hexagonal pyramid. The interaction is stronger than in the Ar_2Ga^+ ions (Ga to ring = 267 pm) and there is a weak interaction (distances 320 to 360 pm) with 5 Br atoms from $GaBr_4^-$ ions (the Ga(III)-Br distance is 232 pm). Analogous, isoelectronic Ar-Sn(II) complexes are also known.

Although Al_2O and AlO have been identified in the vapour phase above 1000°C, aluminium chemistry at ordinary temperatures is almost entirely of the III state. The formal II state, arising from an Al-Al bond is found in species formed by

$$R_2AlX + 2K \rightarrow R_2Al{-}AlR_2 + 2KX$$

where R is a bulky organic group like $(CH_3)_2CH$. With smaller R groups, or on heating, R_3Al and aluminium metal are formed. $AlCl_2$ has also been reported.

Boron is found in low formal oxidation states in the hydrides and in a variety of halides. The latter include the dihalides, B_2X_4, and lower-valent boron fluorides which are related to BF_3 by replacing F by a BF_2 group: $(F_2B)_nBF_{3-n}$ for $n = 0, 1, 2$ and 3. These are prepared by treating B_2Cl_4 with SbF_3 or by strongly heating BF_3 with boron. $FB(BF_2)_2$ disproportionates at –30°C, to give BF_3 and B_8F_{12}, while the $n = 3$ member is not stable alone but does form adducts with L = CO or PF_3, $(BF_2)_3BL$ analogous to H_3BCO or H_3BPF_3. It has been suggested that the rather unstable B_8F_{12} is an analogue of diborane. Like diborane, it gives monomer adducts

$$B_8F_{12} + CO \rightarrow (BF_2)_3BCO \text{ (compare } B_2H_6 + CO \rightarrow H_3BCO).$$

F F
B
F_2B B B BF_2
F_2B BF_2
B
F F

Proposed structure of B_8F_{12}

17.5 The carbon Group, ns^2np^2

17.5.1 General properties of the elements, uses

Table 17.8 summarizes some properties of the elements, and the variation with Group position of ionization potentials, radii and oxidation state free energy is indicated in Figs 17.17 and 17.18

Carbon reacts, when heated, with many elements to give binary carbides. Numerous silicides also exist and these are similar to the borides in forming chains, rings, sheets, and three-dimensional structures. Table 17.9 summarizes the various carbide types and Figs 17.19 and 5.13 give some of the structures.

The carbon Group shows the same trend down the Group towards metallic properties as in the boron Group. The II state becomes more stable and the IV state less stable from

TABLE 17.8 Properties of the elements of the carbon Group

Element	*Symbol*	*Structures of elements*	*Oxidation states*	*Coordination numbers*	*Availability*
Carbon	C	G, D	IV	4, 3, 2	Common
Silicon	Si	D	IV	4, (6)	Common
Germanium	Ge	D	(II), IV	4, 6	Rare
Tin	Sn	D, M	II, IV	4, 6	Common
Lead	Pb	M	II, (IV)	4, 6	Common

G = graphite, D = diamond and M = metallic forms

OTHER REFERENCES TO THE PROPERTIES OF CARBON GROUP ELEMENTS

References to the properties of the carbon Group elements include:

Ionization potentials	Table 2.8
Atomic properties and electron configuration	Table 2.5
Radii	Table 2.10
Electronegativities	Table 2.14
Redox potentials	Table 6.3
Hydrides	Chapter 9
Structures of elements	Sections 5.9, 19.3
Structures of silicates	Section 18.6

The use of carbon in metal extraction is discussed in Section 8.7.

Fullerenes and related forms of carbon are covered in Section 19.3, silicates in Section 18.6 and cluster ions of the heavier elements in Sections 18.4.4 and 18.4.5. See also Section 20.1.3 for the role of silicon in biochemistry and Section 20.3 for the environmental impact of these elements, especially the role of carbon compounds in the ozone hole and the greenhouse effect.

OCCURRENCE AND USES OF THE CARBON GROUP ELEMENTS

All the elements are common except germanium, which occurs as a minor component in some ores, and also in trace amounts in some coals. Carbon, of course, occurs in all living things, and in deposits derived from them such as coals, oils and tars. Hard coals like anthracite have high carbon contents. Heating coals to form coke removes hydrogen components, leaving carbon containing a low percentage of metal compounds. Pure carbon is formed by pyrolysis of hydrocarbons and, as graphite, finds substantial industrial and electrical uses. While production of artificial diamonds is feasible, large pressures and temperatures are needed. More recently diamond films have been produced on metals by deposition from vapour-phase decomposition of methane, by plasma methods or by ion-beam deposition. Such interest in diamonds stems from their optical and semiconductor properties, and the ability to deposit films would be of great advantage where wear resistance is required.

Silicon, with oxygen, is the major component of the Earth's crust and the vast majority of rocks, minerals and their breakdown products the sands and clays, are silicates. Tin occurs in concentrated deposits as the oxide *cassiterite*, SnO_2. Lead also occurs in concentrated form as the sulfide, *galena,* PbS. As both Sn and Pb are readily formed from their ores by heating with carbon in the form of wood fires, these two metals were among the earliest to be produced and used by humans. Lead was particularly widely used by the Romans for water pipes ('plumbing' comes from Latin *Plumbum*, lead) while tin was the vital additive to copper (also an ancient metal) to form the much harder alloy, bronze. Current uses still reflect the ancient patterns, with much of tin consumption being in alloys, and lead being phased out from water supply uses only in the second half of the 20th century as awareness of its toxicity increased. About half of modern tin production is used in tin plate, where its inertness protects the underlying steel. Another use is in glass manufacture where the absolutely smooth surface required for the formation of sheet glass is provided by a bath of molten tin (making use of the inertness and nontoxicity of tin). A major outlet for both metals is in alloys—*solders* which are basically Sn/Pb, *type metals* which are Sn/Pb/Sb, *pewters* which are now mainly Sn/Sb but formerly contained some Pb, *bearing alloys* which are around 9 Sn to 1 Pb, and the tin–copper alloys *bronze* and *brass* (with Zn). All such alloys have a range of compositions optimized to different uses, and contain a number of other elements as minor components to refine the properties. The toxicity of lead is starting to limit some of its traditional uses, such as ceramic glazes. (It is thought that some of the less rational behaviour of the Roman emperors was due to lead poisoning from glazes used on wine jars.) It is widely used in batteries, in priming pigments, in sheathing for heavy-duty cables, as well as in alloys. The structures of the elements are discussed in Section 5.8 and illustrated in Fig. 5.14.

First germanium, and then silicon, became extensively used in very pure forms in semiconductor devices, which are at the basis of the whole electronic industry, including computer hardware. Germanium now accounts for only a few percent of the electronic uses, and other materials, like the binary GaAs, make significant contributions, but the major use is of silicon. Since the detailed tailoring of a semiconductor demands the controlled addition of different elements at less than the parts per million level, silicon or germanium are first produced from purified oxide, via a cycle of MCl_4 distillation, and reduction to the element. They are then produced in very high degrees of purity (better than $1:10^9$) by zone refining. In this process, the element is formed into a rod which is heated near one end to produce a narrow molten zone. The heater is then moved slowly along the rod so that the molten zone travels from one end to the other. Impurities are more soluble in the molten metal than in the solid and thus concentrate in the liquid zone which carries them to the end of the rod.

Uses of compounds of the carbon Group elements include all the well-known industrial organic chemicals like the oil and plastics industries (see below for uses of organometallic compounds of the other elements of the Group). Silicon dioxide and silicates are the major components of glasses, ceramics, pottery and related products, while pure SiO_2 is important in finely divided high surface area forms for adsorption uses in industry, medicine and the laboratory. Fused silica is a very high melting, inert, glass with very low expansion coefficient. Silicon carbide, SiC, is the abrasive carborundum.

Tin dioxide is a component of heterogeneous catalysts, is used widely in ceramics and enamels, and—as a very thin film—in electroluminescent devices. Treatment of glass with $SnCl_4$ deposits a thin film of SnO_2 on the surface which adds markedly to the toughness. All the carbon Group elements are fairly unreactive, with reactivity greatest for tin and lead. They are attacked by halogens, alkalis and acids. Silicon is attacked only by hydrofluoric acid, germanium by sulfuric and nitric acids, and tin and lead by a number of both oxidizing and non oxidizing acids.

carbon to lead. Carbon is a nonmetal and occurs in the tetravalent state. Silicon is metalloidal, but nearer nonmetal than metal, and forms compounds only in the IV state, apart from the occurrence of catenation. Germanium is a metalloid with a definite, though readily oxidizable, II state. Tin is a metal and its II and IV states are both reasonably stable and interconverted by moderately active reagents. The Sn^{4+}/Sn^{2+} potential is –0.15 V and

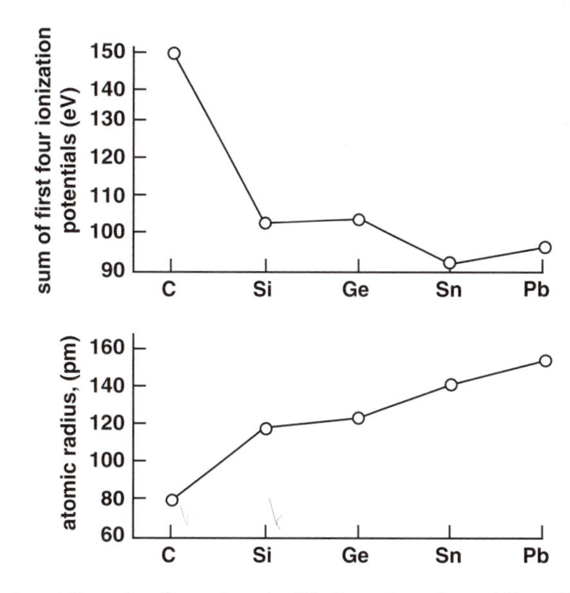

FIG. 17.17 Some properties of the carbon Group elements. This figure shows the variation, with Group position, of the atomic radii and the sum of the first four ionization potentials. The characteristic differences between first and second elements, and the similarity between second and third elements, are noticeable here

tin(II) in acid is well known as a mild reducing agent. Lead is a metal with a stable II state. Lead(IV) is unstable and strongly oxidizing.

The elements of the carbon Group are particularly characterized by their tendency to *catenation*, i.e. to form chains with links between like atoms. Carbon, of course, has this property in an exceptional degree. In the hydrides, chains of up to ten atoms are established for silicon and germanium, as in $Si_{10}H_{22}$, and distannane, Sn_2H_6 is known. Silicon also forms long-chain halides but germanium is limited to $GeCl_4$ and Ge_2Cl_6 as far as present studies go. However, when the chain is fully substituted by organic groups, as in $M_n(CH_3)_{2n+2}$, there is no apparent limit to n for M = Si, Ge and Sn. In these compounds, as with silanes and germanes, the restriction is the experimental difficulty of handling high molecular-weight compounds. While the hydrides readily oxidize, and the halides

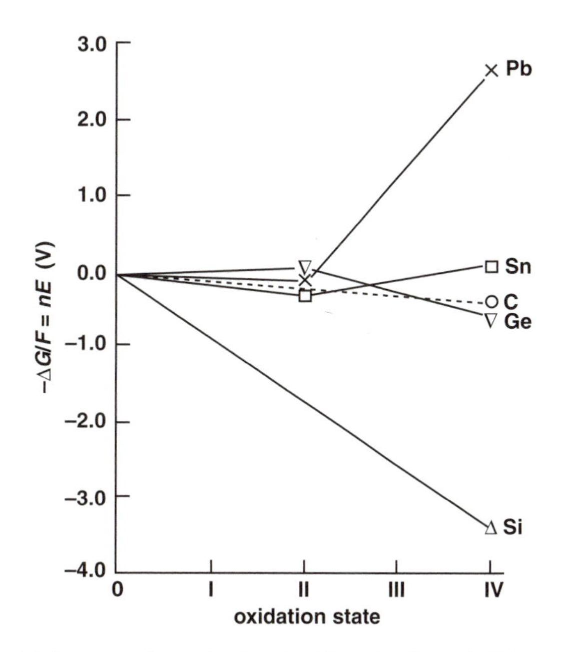

FIG. 17.18 The oxidation state free energy diagram for elements of the carbon Group. It will be seen that Ge(II) is unstable, Sn(II) of nearly the same stability as Sn(IV),and Pb(IV) is very unstable relative to Pb(II)

TABLE 17.9 Types of binary carbide

State of aggregation of the carbon atoms	*Properties*	*Examples and structures*
Single atoms		
(a) Salt-like carbides	Yield mainly CH_4 on hydrolysis	Be_2C (antifluorite) Al_3C_4
(b) Transition element carbides	(i) Conducting, hard, high melting, chemically inert	MC, M = Ti, Zr, Hf, Ta, W, Mo (sodium chloride) W_2C, Mo_2C
	(ii) Conducting, hard, high melting, but chemically active: give C, H_2, and mixed hydrocarbons on hydrolysis	Compounds of the elements of the later transition Groups, e.g. M_3C where M = Fe, Mn, Ni
Linked carbon atoms		
(a) C_2 units 'acetylides'	(i) CaC_2 type—ionic, give only acetylene on hydrolysis	MC_2, M = Ca, Sr, Ba: structure related to NaCl (Fig. 5.13)
	(ii) ThC_2 type—apparently ionic, give a mixture of hydrocarbons on hydrolysis	ThC_2 and MC_2 for M = lanthanide element. Structure (Fig. 17.19) also related to sodium chloride.
(b) C_3 chains	Gives propyne, $H_3C–C{\equiv}sCH$, on hydrolysis	Li_4C_3, Mg_2C_3 and $Ca_3Cl_2C_3$ all contain C_3^{4-} ions. C=C bond length = 134.4 pm, angle at C = 169–176°.
(c) C_n chains	C-C spacing in chain is similar to that in hydrocarbons	Cr_3C_2.-C-C-C- chains running through a metal lattice, compare FeB
Carbon sheets (lamellar structures derived from graphite)		
(a) Buckled sheets	Nonconducting, carbon atoms are 4-coordinated	(i) 'Graphite oxide' from the action of strong oxidizing agents on graphite. C:O ratio is 2:1 or larger and the compounds contain hydrogen. C=O, C-OH and C-O-C groups have been identified. (ii) 'Graphite fluoride' from the reaction with F_2. White, idealized formula is $(CF)_n$, with n about 1.1. The C atoms within the sheets are bonded to one F, while the sheet edges are CF_2 units. These species have advantageous properties as high temperature lubricants.
(b) Planar sheets	Conducting π system is preserved	(i) Large alkali metal compounds—of K, Rb, or Cs, e.g. C_8K. The metal is ionized and the electron enters the π system, while the metal ions are held between the sheets. (ii) Halogen compounds. X^- ions are held between the sheets and positive holes are left in the π system which increase the conductivity

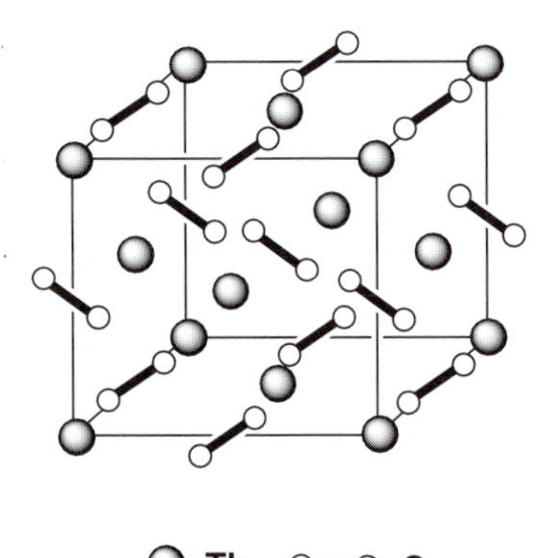

FIG. 17.19 Structure of thorium carbide, ThC_2. As in CaC_2, the C_2 units occupy halide ion positions in a NaCl-type structure but differ in being oppositely aligned in successive layers (compare Fig. 5.13)

hydrolyse rapidly in air, the organo-derivatives are moderately stable to attack by air. Lead compounds are more restricted, but Pb_2R_6 species are well known. In addition, these elements form compounds $(MR_2)_n$ which are ring compounds. Rings with $n = 4$, 5 and 6 are well known for M = Si, Ge and Sn, for R = Ph or related aryls, and some alkyl species are reported as well. For example, $Me_{10}Si_5$ has been made and converted to Me_9Si_5X (X = Cl, Br) which allows further reaction. A few three-membered rings occur but only with bulky ligands. Larger rings are established for Si and Sn. There is also definite evidence for the formation of branched chains. The compounds $(Ph_3Ge)_3GeH$ and a number of mixed silylgermanes, $(H_3Si)_nGeH_{4-n}$ ($n = 2$, 3, 4), and methyl derivatives, $(H_3Si)_n(CH_3)GeH_{3-n}$ ($n = 1$, 2, 3), have been reported. Branched chain hydrides of M_4 and M_5 forms (M = Si, Ge) are also indicated by chromatographic experiments. For tetrasilanes and tetragermanes, the *n*- and *iso*- forms have been separated on the macroscale, as have two of the three Ge_5H_{12} isomers and related silicon–germanium species (see Section 9.5). Recently, a number of compounds $(Ph_3M)_4M'$ (M = Ge, Sn, Pb and M′ = Sn, Pb) have been reported. The pentaplumbane of this *neo*-form, $Pb(PbPh_3)_4$, is the only other species with Pb-Pb bonds.

17.5.2 The IV state

The IV state is found for all the elements of the Group, and is stable for all but lead. Its properties are well illustrated by the oxygen and halogen compounds. The oxygen compounds are listed in Table 17.10. All the dioxides are prepared by direct reaction between the elements and oxygen. They are also precipitated in hydrated form (except CO_2 of course) by addition of base to their solutions in acid. No true hydroxide, $M(OH)_4$, exists for any of the elements. The very marked effect of p_π–p_π bonding on the structures of the carbon, as compared with the silicon, compounds is obvious, as is the tendency towards a higher co-ordination number to oxygen for the heavier elements.

Carbonic acid, H_2CO_3, was long thought to be a 'nonexistent' compound, but was recently synthesized at low temperatures by high energy irradiation of CO_2/H_2O mixtures. Free carbonic acid is surprisingly stable, but in the presence of water, its stability dramatically decreases. This is consistent with the observation that carbon dioxide dissolved in water is largely weakly hydrated CO_2, with only a small proportion (*ca.* 0.2%) existing as carbonic acid. However, because of the well-known equilibrium

$$H_2CO_3 \rightleftharpoons HCO_3^- + H^+$$

solutions of CO_2 in water are acidic. The surprising stability of carbonic acid when prepared under special conditions suggests that other 'nonexistent' acids such as H_2SO_3 should also be reinvestigated. The protonated carbonic acid molecule, $C(OH)_3^+$ has recently been prepared as its AsF_6^- and SbF_6^- salts, but these are only stable at low temperatures, decomposing to CO_2 and $H^+MF_6^-$ at about $-10°C$.

The oxyanions are also listed in Table 17.10 and comprise a wide array of compounds because of the strong tendency to form condensed polynuclear species based on EO_4 coordination for $E = Si$, Ge (compare silicates, Section 18.6) or EO_6 units for the heavier elements. While the naturally occurring silicates tend strongly towards ring, sheet, and three dimensional structures, use of different counterions may give alternative structures. It is perhaps significant that the first short chain silicate $(Si_4O_{13})^{10-}$ was recently isolated as the Ag^+ salt.

The formation of the heavier chalcogenides also reflects the relative stabilities of the II and IV states. While Si, Ge and Sn form such compounds as GeS_2 or $SnSe_2$, there are no such compounds of lead. Among the interesting complex species (again underlining the IV state stability) are $Sn_2Te_6^{4-}$ (two tetrahedra sharing an edge) and $Si_2Se_8^{2-}$ where the apparently anomalous oxidation state arises from a Se-Se bridge which links two $SiSe_3$ units, completing a tetrahedron at each Si.

All the tetrahalides, MX_4, are found except for lead(IV) which is too oxidizing to form the tetrabromide or iodide. All may be made from the elements, from the action of hydrogen halide on the oxide, or by halogen replacement. All the carbon tetrahalides, all the chlorides, bromides and iodides, and also SiF_4 and GeF_4, are covalent, volatile molecules. The volatility and stability fall in a regular manner with increasing molecular weight of the tetrahalide. By contrast, SnF_4 and PbF_4, are involatile solids with melting or sublimation points at 705°C and 600°C respectively. They have polymeric structures based on MF_6 octahedra, with partially ionic bonding, as has aluminium trifluoride. Thus the tetrafluorides of the carbon Group parallel the trifluorides of the boron Group in changing from volatile to involatile and polymeric, but the changeover comes further down the Group. Bond lengths are given in Table 2.10b.

Carbon tetrafluoride (and all the fluorinated hydrocarbons) and carbon tetrachloride are very stable and unreactive, though CCl_4 will act as an oxidizing and chlorinating agent at higher temperatures. Carbon tetrabromide and iodide are stable under mild conditions, but act as halogenating agents on warming, and are also decomposed by light.

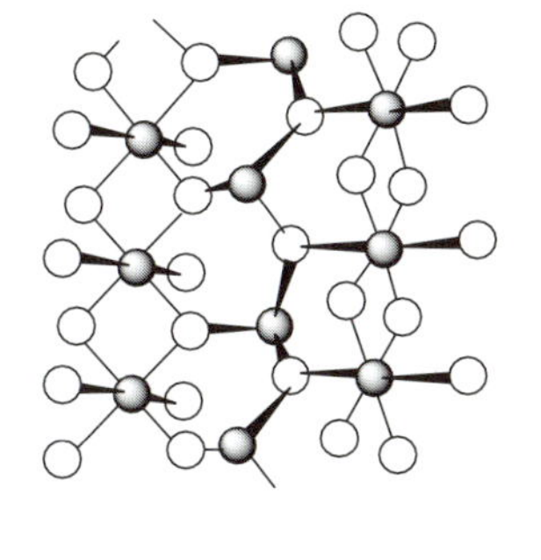

FIG. 17.20 The structure of Pb_3O_4. The structure contains $Pb^{IV}O_6$ octahedra linked together into chains by sharing edges. These chains are, in turn, linked by $Pb^{II}O_3$ pyramidal units which both link two of the Pb^{IV} chains and form a chain of $Pb^{II}O_3$ units

The silicon tetrahalides, except the fluoride, are hydrolysed rapidly to 'silicic acid' which is hydrated silicon dioxide. The heavier element tetrahalides also hydrolyse readily, but the hydrolysis is reversible and, for example, $GeCl_4$ can be distilled from a solution of germanium(IV) in strong hydrochloric acid.

A very wide range of 6-coordinate complexes $MX_4.2D$ is formed by the tetrahalides, and 5-coordinate species such as $R_nSnX_{5-n}^-$ ($n = 1, 2, 3$) are also well known. The latter have a trigonal bipyramidal structure. Seven-coordination is found in $Me_2Sn(NCS)_2$(terpy) where

TABLE 17.10 Oxygen compounds of carbon Group elements

Compound	Properties	Structure	Notes
Dioxides			
CO_2	Monomer, weak acid	Linear, O=C=O	p_π–p_π bonding between first row elements giving π-bonded monomer
SiO_2	Involatile, weak acid	4:2 coordination with SiO_4 tetrahedra (see Fig. 5.4b)	There is some weak p_π–d_π bonding in some Si-O-Si systems, though not in the oxide.
GeO_2	Amphoteric	Two forms: one with a 4:2 silica structure, and one with the rutile (6:3) structure	The Ge/O radius ratio is on the borderline between 6- and 4-coordination
SnO_2	Amphoteric	Rutile structure	
PbO_2	Inert to acids and bases	Rutile structure	Strongly oxidizing
Other oxides (excluding monoxides)			
C_3O_2	Carbon suboxide, prepared by dehydrating malonic acid $CH_2(COOH)_2 \xrightarrow[300°C]{P_2O_5} O=C=C=C=O$	Linear, C-C and C-O distances are intermediate between those expected for single and for double bonds	The molecule contains extended π bonding of the same type as in CO_2.
Pb_3O_4	Red lead: oxidizing, contains both Pb(IV) and Pb(II)	Figure 17.20. The structure consists of $Pb^{IV}O_6$ octahedra linked in chains; the chains are joined by pyramidal $Pb^{II}O_3$ groups	
Oxyanions			
Carbonate	CO_3^{2-}	Planar with π bonding	Again C and O give p_π bonds
Silicates	Wide variety (Section 18.6)	Formed from SiO_4 units	
Germanates	Variety of species	Contain both GeO_4 and GeO_6 units	Compare the two forms of germanium dioxide
Stannates, Plumbates } e.g. $M(OH)_6^{2-}$		Contain octahedral MO_6 units	
Oxyhalides			
Carbonyl halides			Rapidly hydrolysed
COF_2	b.p. −83°C	All are planar $X_2C=O$	Very poisonous: has been used as nonaqueous solvent
$COCl_2$ (phosgene)	Stable		
$COBr_2$	Fumes in air		

Other mixed oxyhalides such as ClC(O)Br and FC(O)Cl are known. The $ClCO^+$ cation can be stabilized in the solid state by the $Sb_3F_{16}^-$ anion.

Silicon oxyhalides

These are all single-bonded species containing -Si-O-Si-O- chains (for example, $Cl_3Si\text{-}(OSiCl_2\text{-})_nOSiCl_3$ with $n=4$, 3, 2, 1 or 0) or rings (for example $(SiOX_2)_4$ where X = Cl or Br)

the 3 nitrogens of the terpy (see Appendix B) and the two NCS groups form an almost regular pentagon with the methyls on the axis completing a pentagonal bipyramid. Eight-coordinate lead is found in $(C_6H_5)_2(CH_3COO)_3Pb^-$. The structure is a hexagonal bipyramid with the phenyl groups on the axis and the three bidentate acetates lying in the central plane.

17.5.3 Hydride and organic derivatives

Hydride halides of the types MH_3X, MH_2X_2 and MHX_3 are also formed. Most representatives of these formulae (for X = F, Cl, Br, I) are found for silicon and germanium, but a few tin compounds, such as SnH_3Cl, are also known. Such compounds are key members of synthetic routes to organic and other derivatives, as in reactions such as:

$$SiH_3Br + RMgX \rightarrow RSiH_3 \quad (R = \text{organic radical})$$

$$\text{or} \quad GeH_3I + AgCN \rightarrow GeH_3CN + AgI$$

It has recently been discovered that the higher hydrides of silicon and germanium behave similarly, and all the compounds M_2H_5X, for M = Si or Ge, and X = F, Cl, Br and I, have been prepared.

Organotin and organolead compounds have been studied for their pharmaceutical and biocidal properties and organogermanium compounds may have similar properties. Butyl–tin compounds find widespread use in marine antifouling paints but concerns about the accumulation of organo-tin in the environment is leading to a search for alternatives. A number of organotin complexes show promising anti-tumour activity. Tetraethyl lead is manufactured on a large scale as an anti-knock agent for petrols, though it is being reduced in usage.

The elements from silicon to lead have an extensive organometallic chemistry and a wide variety of MR_4 and M_2R_6 compounds exist, with σ metal–carbon bonds of considerable stability. All the tetraalkyl and tetraaryl compounds are stable, although stability falls from silicon to lead and the aryls are more stable than the alkyls. For example, tetraphenylsilicon, Ph_4Si, boils at 530°C without decomposition, tetraphenyllead, Ph_4Pb, decomposes at 270°C, while tetraethyllead, Et_4Pb, decomposes at 110°C. A wide variety of organocompounds, with halogen, hydrogen, oxygen or nitrogen linked to the metal, is also known and this class includes the *silicone polymers*. These are prepared by the hydrolysis of organosilicon halides:

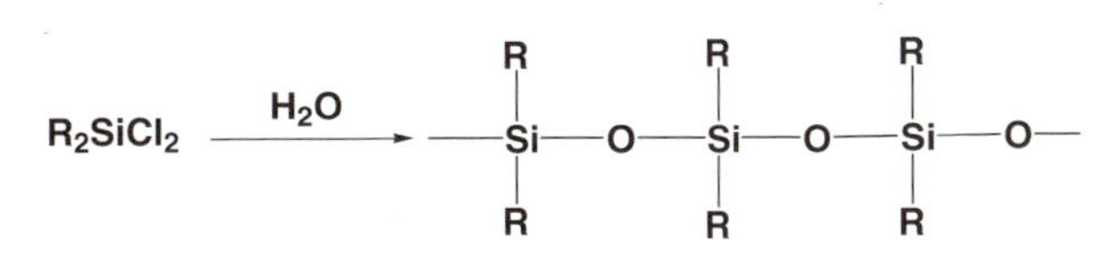

This long-chain polymer is linked by the very stable silicon–oxygen skeleton and the organic groups are also linked by strong bonds so the polymer has high thermal stability. The organic groups also confer water-repellent properties. The chain length is controlled by adding a proportion of R_3SiCl to the hydrolysing mixture to give chain-stopping -$OSiR_3$ groups, while the properties of the polymer may also be varied by introducing cross-links with $RSiCl_3$:

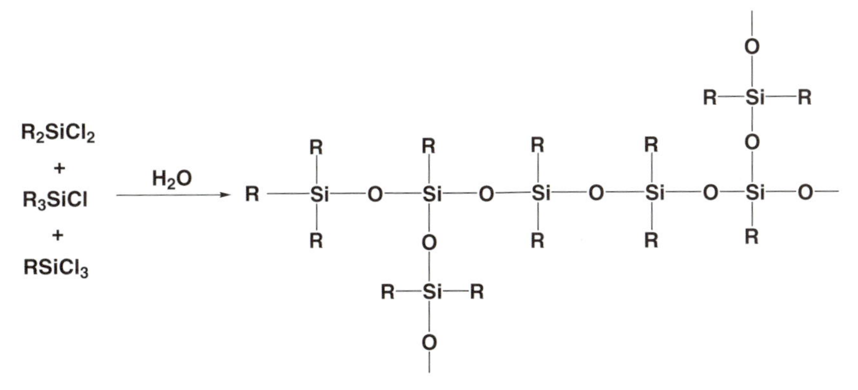

The elements from silicon to lead use their d orbitals to form 6-coordinated complexes which are octahedral. All four elements give stable MF_6^{2-} complexes with a wide variety of cations. The MCl_6^{2-} ion is formed for M = Ge, Sn and Pb, and tin also gives $SnBr_6^{2-}$ and SnI_6^{2-}. In addition, a variety of MX_4L_2 complexes are formed by the tetrahalides with lone pair donors such as amines, ethers or phosphanes. The chemistry of the tin compounds is particularly well explored, and both *cis* and *trans* compounds are known. When stannic chloride was reacted with phosphorus pentachloride, Cl^- transfer took place (compare PCl_5 itself, Section 17.6.2) to yield $(PCl_4^+)_2SnCl_6^{2-}$, and the mixture also yielded $Sn_2Cl_{10}^{2}$ (with a structure involving two octahedra sharing an edge) and $SnCl_5^-$ (trigonal bipyramid). Six-coordinate complexes may also be formed by the organic derivatives of Ge, Sn and Pb, as long as there are enough electronegative substituents to give reasonable acceptor power. Thus we find $Me_2SnCl_4^{2-}$, and $Me_4Sn_2Cl_6^{4-}$, each with octahedral Sn and trans Me groups, and the latter with two bridging Cl.

There are a much more limited number of 5-coordinated species including MF_5^- for M = Si, Ge or Sn, and MCl_5^- for Ge and Sn. The chlorides are trigonal bipyramids, as expected, as are the fluorides in the presence of large cations. However, with smaller cations the fluorides form a *cis* fluorine-bridged polymeric structure.

The d orbitals are also used in internal π bonding, especially in silicon compounds. The classic case is trisilylamine, $(SiH_3)_3N$, Fig. 17.21. This has a quite different structure from the carbon analogue, trimethylamine, $(CH_3)_3N$, shown in Fig. 17.22. The pyramidal structure of trimethylamine is similar to that of NH_3 and reflects the steric effect of the unshared pair on the nitrogen. The NSi_3 skeleton, by contrast, is flat and the nitrogen in trisilylamine shows no donor properties. This is due to the formation of a π bond involving the nitrogen p orbital and d orbitals on the silicon, as shown in Fig. 17.23. The lone pair electrons donate into the empty silicon d orbitals and become delocalized over the NSi_3 group, and hence there is no donor property at the nitrogen. Structural evidence comes from the infrared and Raman spectra and from the zero dipole moment. Although trisilylamine is a gas at room temperature, it has recently been possible to carry out an X-ray study of the solid at low temperatures and this has confirmed the planar Si_3N skeleton. It is noteworthy that $N(CF_3)_3$ is nearly planar ($CNC = 118°$) and $N(C_2F_5)_3$ even closer ($CNC = 119.3°$) showing the effect when strongly electron-withdrawing ligands remove much of the lone pair electron density.

FIG. 17.21 The structure of trisilylamine, $(SiH_3)_3N$

Tetrasilylhydrazine, $(SiH_3)_2NN(SiH_3)_2$, also shows differences in symmetry compared with the methyl analogue, arising from similar π bonding. Another clear case is the isothiocyanate, MH_3NCS. Where M = C, the C-N-C-S skeleton is bent at the nitrogen atom due to the steric effect of the nitrogen lone pair while, when M = Si, the Si-N-C-S skeleton is linear and there is d_π–p_π bonding between the Si and N atoms. Similar effects are observed in M-O bonds: dimethyl ether, CH_3OCH_3, has a C-O-C angle of 110° which is close to the tetrahedral value, while the bond angle in disilyl ether, SiH_3OSiH_3, is much greater—140° to 150° indicating delocalization of the nonbonding pairs on the oxygen.

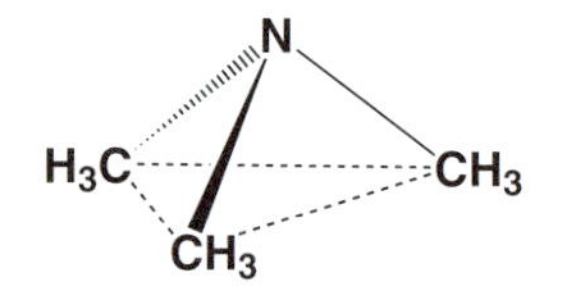

FIG. 17.22 The structure of trimethylamine, $(CH_3)_3N$

This formation of π bonds using metal d orbitals and a p orbital on a first row element is most marked in the case of silicon, but there is some evidence that it occurs in germanium compounds as well. For example, the Ge-F bond in GeH_3F is very short, which may indicate d_π–p_π bonding but GeH_3NCO and GeH_3NCS have bent Ge-N-C skeletons in contrast to their silicon analogues. Other evidence for π bonding by germanium is indirect and derived from acidities and reaction rates in substituted phenylgermanes.

> An alternative, and increasingly favoured, view of the bonding in compounds such as trisilylamine is that d orbitals play only a relatively minor role. Instead, interaction between the filled nitrogen p orbital and a σ antibonding Si-H orbital occurs, i.e. $n \rightarrow \sigma^*(SiH)$.

17.5.4 The II state

The II state is the stable oxidation state for lead, and the IV state of lead, like thallium(III) in the last Group, is strongly oxidizing. Pb^{2+} ions exist in a number of salts, though hydrolysis occurs readily in solution:

$$Pb^{2+} + 2H_2O \rightleftharpoons PbOH^+ + H_3O^+$$

and a further equilibrium is found:

$$4Pb^{2+} + 12H_2O \rightleftharpoons Pb_4(OH)_6^{2+} + 6H_3O^+$$

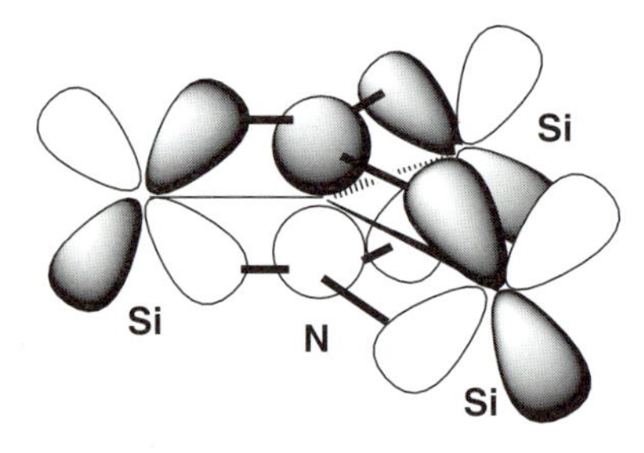

FIG. 17.23 π bonding in trisilylamine

A considerable variety of lead(II) salts is known, and these generally resemble the coresponding alkaline earth compounds in solubility, e.g. the carbonate and sulfate are very insoluble. The halides are less similar and $PbCl_2$ is, like TlCl, insoluble in cold water though more soluble in hot water.

Addition of alkali to lead(II) solutions gives a precipitate of the hydrated oxide, which dissolves in excess alkali to give plumbites. The hydrated oxide may be dehydrated to PbO, called litharge, which is yellow-brown in colour. The structure of PbO is an irregular one in which the lead is coordinated to four oxygen atoms at corners of a trigonal bipyramid, with the fifth position occupied by an unshared pair of electrons. Just as lead forms the mixed oxidation state oxide (see Table 17.10), a mixed oxidation state oxyanion species is formed in compounds like $KPbO_2$. In a related example, $KNa_7[PbO_4][PbO_3]$, a crystal structure shows an approximately tetrahedral $Pb(IV)O_4^{4-}$ ion and a pyramidal $Pb(II)O_3^{4-}$ ion. Another interesting structure is found in the $Pb_8O_4^{8+}$ cluster ion whose core is a cubane Pb_4O_4 unit with a further Pb bonded to each O corner, so that these 4 Pb atoms transcribe a tetrahedron around the cubane.

Lead forms all the heavier chalcogenides, PbE, as do Ge and Sn (E = S, Se, Te). Where known, the structures are layer lattices.

The II state of tin is mildly reducing but otherwise resembles lead(II). In solution, Sn^{2+} ions hydrolyse as in the first equation for Pb^{2+} above. Addition of alkali to stannous solutions precipitates $SnO.xH_2O$ (neither $Sn(OH)_2$ nor $Pb(OH)_2$ exist), and the hydrated oxide dissolves in excess alkali to give stannites. Dehydration gives SnO, which has a similar

> Stannite is formally named oxostannate(II)

structure to PbO. One stannite structure is known. $K_2Sn_2O_3$ is built of trigonal pyramidal SnO_3 units with Sn-O-Sn bridges. Tin(II) gives all four dihalides and a number of oxysalts. In the vapour phase, $SnCl_2$ exists as monomeric molecules with the V-shape characteristic of species with two bonds and one lone pair. One molecule of water adds to this molecule to give the pyramidal hydrate, $SnCl_2.H_2O$. SnF_2 has a tetrameric structure based on an Sn_4F_4 puckered 8-membered ring. The Sn-F-Sn bridge angles are 135°, the SnF distances average 215 pm, and each Sn also carries external F with a 207 pm bond length. The Sn atom is thus at the apex of a trigonal pyramid with FSnF angles of 84°. Much longer Sn-F bridges of about 290 pm link these 8-membered rings together.

Germanium in the II state gives a number of compounds; GeO, GeS and all four dihalides are well established. These compounds all appear to have polymeric structures and are not too unstable, probably because attack on the polymeric molecules is relatively slow. The known structures are those of GeI_2, which has the CdI_2 layer structure, and of GeF_2 which has a long-chain structure similar to that of SeO_2. The II state is readily oxidized to the IV state: thus the dihalides all react rapidly with halogen to give the tetrahalides, while the corresponding reaction of tin(II) compounds is slow. Yellow GeI_2 disproportionates to red GeI_4 and germanium on heating. The divalent compounds are involatile and insoluble, in keeping with a polymeric formulation. One complex ion of the II state is known, $GeCl_3^-$, in the well-known salt $CsGeCl_3$ and adducts $R_3P.GeI_2$ are also reported. The $GeBr_3^-$ ion has been characterized as the Rb^+ salt. The structure shows a pyramidal ion (Ge—Br = 253 pm, BrGeBr = 95.5°) which is more weakly linked to three Br of other ions (Ge···Br = 324 pm) giving distorted octahedral geometry at Ge. Germanium(II) may also be obtained in acid solution, in the absence of air, and addition of alkali precipitates the yellow hydrated oxide, $GeO.xH_2O$.

The divalent Si(II) halides, SiX_2(X = F, Cl, Br and I), can be prepared as reactive species and have angular molecules. For example, $SiBr_2$ shows a Br-Si-Br angle of 103°, and Sr-Br bond distances of 224 pm. Liquid $SiCl_2$ is a viscous mixture of polymers in which $[SiCl_2]_n$ rings have been identified for $n = 4$, 5 and 6. Sublimation of the tetramer gives an extremely air-sensitive linear polymer with Si-Si = 241 pm and a SiSiSi angle of 114°.

Tin and lead form more complexes in the II state than does Ge, though fewer than in the IV state. Halogen complexes, MX_3^-, are pyramidal monomers for the heavier halides while polymeric forms are found for X = F. Thus SnF_3^- is an infinite chain of SnF_4 square pyramids with Sn at the apices and linked by sharing F atoms at *trans* corners. Similarly, MOX or $M(OH)X_2$ compounds are polymers, all with sterically active lone pairs. Cation complexes MX^+ are also polymers.

Formally divalent organometallic compounds R_2M are mostly ring or long-chain species with M-M bonds, where M = Si, Ge or Sn (see above). Rings with 4, 5 and 7 Si, Ge or Sn atoms are relatively more stable than their carbon analogues. However, simple molecular R_2M species are known to be highly reactive species which can be trapped in matrices and studied. These divalent organic derivatives, especially of Si, Ge or Sn, are also proven as reaction intermediates. A few examples have been more fully characterized. Large ligands are necessary to allow stable solids, as in the trifluoromethylphenyl compound $[(CF_3)_3.C_6H_2]_2Sn$, where the CSnC bond angle is 98°. The corresponding lead compound is also known and has CPbC = 94°. Similarly, carbenes such as CCl_2 are well known in organic chemistry and provide a route for the synthesis of cyclopropanes, by the addition to alkenes. The use of large ligands allows stable compounds to be isolated and compounds like $Ge[CH(SiMe_3)]_2$ or $Sn[N(CMe_3)]_2$ are monomers. In a series of recent reports, a number of stabilized carbenes, silylenes and germylenes (shown in Fig. 17.24) have been described which show extraordinary thermal stability, a property not expected for such reactive species. The silylene is stable for at least four months at 150°C. This stability has opened up the debate on the electronic structure of such species, and whether or not they are true carbenes, silylenes or germylenes. It is likely that the lone pairs on the nitrogen atoms of such species are donating electron density to the electron deficient carbon, silicon or germanium atom. It has even been suggested that these species are aromatic in nature, due to the possibility of having six π-electrons (two from the C=C double bond, and two from each of the nitrogens), similar to benzene.

These R_2M species are termed silylenes, getmylenes, stannylenes and plumbylenes for Si, Ge, Sn and Pb, respectively.

$(C_5H_5)_2M$ (M = Ge, Sn, Pb) are monomers in the gas phase with ring-M-ring angles about 145° and the lone pairs pointing away from the rings. In the solid, they are polymeric. With the very bulky substituent in $[Ph_5C_5]_2Sn$ the rings are planar and parallel (compare Fig. 16.7). A few mixed species, RMX, RM(OH) or $(RM)_2O$ are found. All are polymeric structures with sterically active lone pairs.

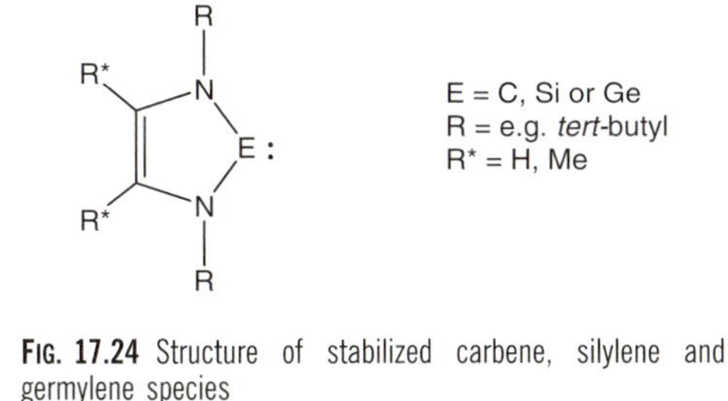

FIG. 17.24 Structure of stabilized carbene, silylene and germylene species

Polymetal clusters such as Sn_5^{2-} are discussed in Section 18.4.

Although the monomers in the II state have a lone pair, and thus could act as donors (Lewis bases), such behaviour is much rarer than the acceptor modes above. One case is that of $(RNM)_4.2AlCl_3$ (for M = Ge or Sn and R is the bulky CMe_3 group) where the M_4N_4 skeleton is a cubane (with R on N) and two of the four M atoms form donor bonds to $AlCl_3$ groups.

17.5.5 Reaction mechanisms of silicon

Work on inorganic reaction mechanisms is less developed than the corresponding area in organic chemistry, mainly because of the large variety of systems and of experimental difficulties. In particular, many nonorganic reactions are extremely fast. Of all the p elements (for mechanisms at d elements, see Section 13.9), silicon presents one of the most favourable cases for study, and we illustrate something of what is known about Main Group mechanisms by this outline of mechanisms at silicon.

Mechanisms are postulated (and remember that all reaction mechanisms are only hypotheses) on the basis of reaction kinetics, and study of silicon has the major advantage that kinetic work may be independently supported by evidence from optically active compounds. The isolation and resolution of active silicon species has given a powerful tool which has been used, particularly by Sommer and his colleagues, to study mechanisms of substitution.

A number of optically active silicon species have been reported, one of the first being Ph(α-Nt)(Me)SiX, where α-Nt = α-naphthyl. This was resolved using X = (−)menthoxide (menthol being a naturally occurring optically active species) by recrystallization from pentane at −78°C. We shall abbreviate the optically active species as R_3Si^*X. This isolation was greatly aided by the presence of bulky aromatic groups which reduce the rate of reaction. Even so, R_3Si^*X commonly reacts about a thousand times faster than similar carbon compounds.

It was first shown that stereospecific substitutions did occur by cycles of changes analogous to the Walden cycle, e.g.

$$(+)R_3Si^*H + Cl_2 \rightarrow (-)R_3Si^*Cl \xrightarrow{LiAlH_4} (-)R_3Si^*H$$

$[\alpha]_D$ values: $+34°$ $-6°$ $-34°$

Thus one of these steps must occur with inversion, and one with retention, and both must be highly stereospecific. Later work showed that the same relative configuration occurred in the following species R_3Si^*X, shown with the rotations

X	H	Cl	OH	OMe	Br	F
$[\alpha]_D$	+34°	−6°	+20°	+17°	−22°	+47°.

Thus, the chlorination above is retention, while the reduction involves inversion of configuration.

These observations establish that stereospecific substitutions do take place. Extensive further work has led to the postulation of four main mechanisms at silicon. These are briefly outlined.

S_N2. This is similar to the mechanism at carbon, but is much faster. It is found for R_3Si^*X in polar, but poorly ionizing, solvents and particularly when X is a halogen. The reaction takes place with inversion of configuration, and is postulated to proceed through a trigonal bipyramidal intermediate conformation, in which the organic groups are in the central plane

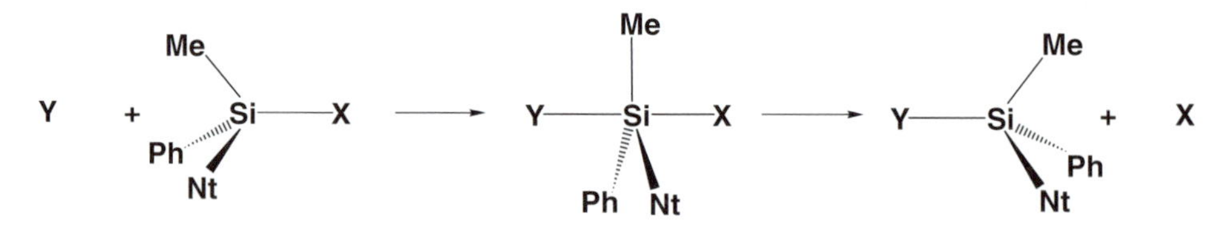

Typical examples are hydrolyses, or other replacements of Si-X by Si-OR, and the formation of the hydride above. The addition product between SiH_3Cl and dimethyl ether can be considered to be a model for the intermediate formed in hydrolysis reactions; the unreactive

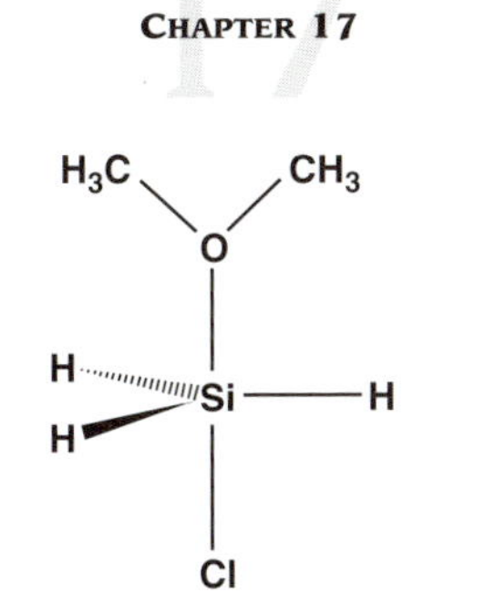

FIG. 17.25 The structure of the adduct $SiH_3Cl[O(CH_3)_2]$ formed from SiH_3Cl and dimethyl ether

O-CH_3 groups prevent any further reaction in the case of the reactive O-H groups. The structure, Fig. 17.25, has been found to be the expected trigonal bipyramid.

While such a process is assisted by using one of the silicon d orbitals to achieve 5-coordination (which probably accounts for the speed of reaction) it does not necessarily follow that a stable intermediate forms. This could happen, or the effect of the d orbital may simply be to lower the activation energy compared with the carbon analogue.

S_Ni. When X = OR, and hydride or organometallic reagents are used in nonpolar solvents, a slow reaction is found which proceeds with retention of configuration. This cannot be S_N2, as the intermediate would undergo fast loss of H or R (leading to racemization) rather than undergo cleavage of the very strong Si-O bond. It is therefore postulated that the reaction proceeds via a four-centre intermediate, and it is termed internal nucleophilic substitution. The intermediate may be represented as shown, where E is the electrophilic and N the nucleophilic part of the reagent. Thus, for a Grignard reagent, N = R, and E = MgX; or for AlH_4^-, N = H and E = AlH_3. The process may be understood as a nucleophilic attack assisted by the electrophilic coordination to oxygen which helps to overcome the strong Si-O binding energy. As an example,

$$R_3Si^*OMe + LiAlH_4 \xrightarrow{\text{ether, 16 h}} R_3Si^*H$$

$$[\alpha]_D = 16^\circ \qquad\qquad [\alpha]_D = 30^\circ (90\% \text{ retention})$$

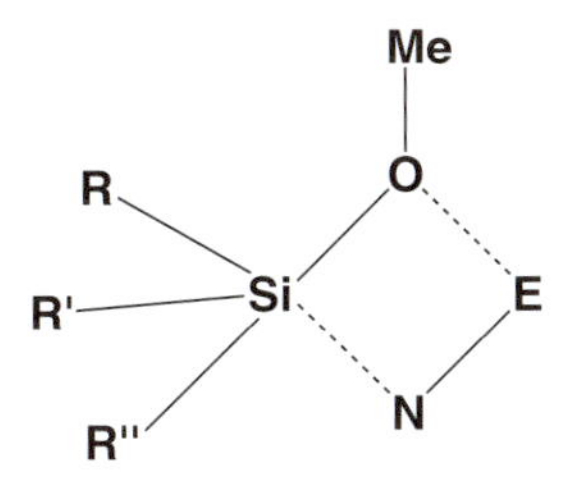

A similar four-centred mechanism is postulated for the very wide range of reactions called *hydrometallations* in which an M-H bond adds across a double bond. These are found for many metals, M, of which the most important are for M = B, Al (see last section), Si or Sn.

S_N1. This is less common than in carbon chemistry, and is found typically for halides in polar solvents of high dielectric constant. Thus, while R_3Si^*Cl is recovered unchanged from solution in CCl_4 or an ether, when it is dissolved in acetonitrile or nitromethane (CH_3CN or CH_3NO_2, both with high dielectric constants) racemization takes place rapidly. This has been postulated to proceed through a solvent-stabilized silyl cation, $RR'R''Si^+$(solv). The silyl cation is analogous to the carbonium ion $RR'R''C^+$ known in organic chemistry.

In fact, the search for the tricoordinated and unsolvated silyl cation is an ongoing area of activity. Since the silyl cation $RR'R''Si^+$ is a highly reactive species, it tends to coordinate a 'ligand', either the solvent or the anion itself. Accordingly, the search for the free silyl cation has been paralleled by the search for the least coordinating anion. Classical non-coordinating anions, such as ClO_4^-, BF_4^-, PF_6^- and $CF_3SO_3^-$, for example, have all been found to coordinate to metals in all parts of the Periodic Table. The use of the even more poorly coordinating anions $B(C_6F_5)_4^-$ and carboranes, e.g. $CB_9H_{10}^-$, which have their negative charges spread over a large number of atoms, has allowed the isolation of compounds which contain an even weaker, but still persistent, interaction between the R_3Si^+ ion and the anion or the solvent. One example is the compound $[(toluene)SiEt_3]^+[B(C_6F_5)_4]^-$ which contains an interaction between the toluene solvent molecule and the silicon atom. Other poorly coordinating anions, such as $[B(OTeF_5)_4]^-$ and $[Nb(OTeF_5)_6]^-$ which have their negative charge delocalized over a large number of electronegative fluorine atoms, also reduce the potential for these anions to act as 'ligands' towards Si^+ and other reactive species. The use of anions of this type in the isolation of silver carbonyls is described in Section 15.9.1.

EO (expanded octet). One special mechanism is sometimes involved when X = F. For most reactions, fluorides behave as other halides and give the above mechanisms. However, the Si-F bond is much stronger than, for example, Si-Cl and this allows a further mechanism. An example is the reaction in which R_3Si^*F is racemized in dry pentane solution by the addition of MeOH. This reaction has the following characteristics which exclude any of the three mechanisms outlined above:

(a) the rate is retarded in formic acid, a solvent of high dielectric constant, hence the reaction is not S_N1. Further, addition of HF retards the reaction so that the reaction does not proceed by loss of F^- as this would be stabilized as HF_2^-.

(b) A mixture of R_3Si^*F and R_3Si^*OMe plus MeOH gives unchanged R_3Si^*OMe and racemic R_3SiF. Thus the racemization is not via R_3SiOMe or any species which could give rise to it, excluding S_N2 and S_Ni.

These features led to the postulate of an expanded-octet mechanism with a 5- or 6-coordinate intermediate formed by addition of OMe, and which subsequently loses OMe again:

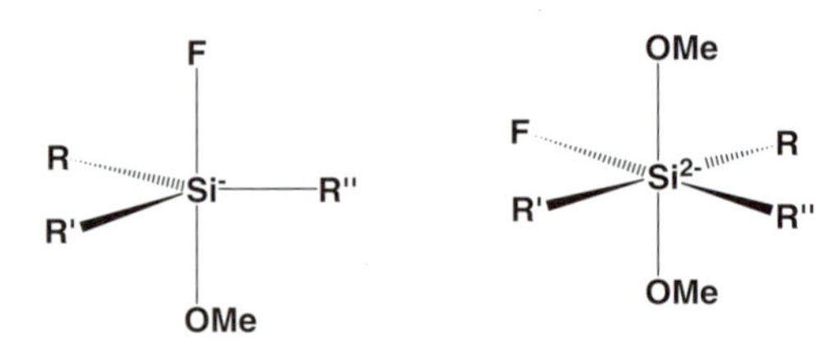

As the intermediates are labile or inactive, racemization occurs. There is no breaking of the very strong Si-F bond. Formation of an expanded octet would be assisted by the presence of the fluorine substituent. Note that this mechanism must be rare or there could be no isolation of optically active silicon compounds at all.

These conclusions from optical studies may be supported by kinetic studies in favourable cases. Thus, the formation in a fast step of a relatively stable intermediate, followed by a slow dissociation to products

$$A + X{-}Y \underset{\leftarrow}{\overset{\text{fast}, k_1}{\longrightarrow}} A{-}X{-}Y \underset{\leftarrow}{\overset{\text{slow}, k_2}{\longrightarrow}} A{-}X + Y$$

would be characterized by a dependence on k_2 alone, and by the fact that the rate of consumption of A was not equal to the rate of appearance of Y. Thus, in S_Ni, EO and some S_N2 reactions (if the intermediate was relatively long-lived) the above difference in rates would be detected.

Conversely, if the intermediate was unstable and immediately gave the product (i.e. if the second step above was very fast) the rate of appearance of Y would equal the rate of loss of A, and k_1 would be rate-determining. This reaction would thus be second order. Such kinetics would characterize the normal S_N2 reaction.

Finally, the S_N1 reaction is first order and the determining step is the dissociation into cation and anion.

While many mechanisms give rise to an intermediate kinetic picture (there may be a wide range of lifetimes for the A-X-Y intermediate, for example) if kinetic and optical studies agree, the postulated mechanism is quite strongly supported. As far as the silicon mechanisms outlined above are concerned, such kinetic studies as are reported do validate the proposed mechanisms.

17.6 The nitrogen Group, ns^2np^3

17.6.1 General properties

Of these elements, nitrogen and bismuth are found in only one form while the others occur in a number of allotropic forms. Nitrogen exists only as the triply bonded N_2 molecule, and bismuth forms a metallic layer structure shown in Fig. 17.28. There are a number of

OTHER REFERENCES TO THE PROPERTIES OF NITROGEN GROUP ELEMENTS INCLUDING

References to the properties of the nitrogen Group elements which have occurred in the earlier part of the book include:

Ionization potentials	Table 2.8
Atomic properties and electron configurations	Table 2.5
Radii	Table 2.10, Table 2.11
Electronegativities	Table 2.14
Redox potentials	Table 6.3
Structures	Chapter 5

Fixation of nitrogen is discussed under titanium (Section 14.2), and nitrogen complexes in Sections 15.6 and 16.10. Polynuclear P and As species are covered in Section 18.3, the biological role of NO in Section 20.1.3 and phosphates in detergents in Section 20.3.1.

Table 17.11 lists some of the properties of the elements and the variation with Group position of important parameters is shown in Fig. 17.26. The oxidation free energy diagram is shown in Fig. 17.27.

TABLE 17.11 Properties of the nitrogen Group elements

Element	*Symbol*	*Oxidation states*	*Coordination number*	*Availability*
Nitrogen	N	-III, III, V	3, 4	Common
Phosphorus	P	(-III), (I), III, V	3, 4, 5, 6	Common
Arsenic	As	III, V	3, 4, (5), 6	Common
Antimony	Sb	III, V	3, 4, (5), 6	Common
Bismuth	Bi	III, (V)	3, 6	Common

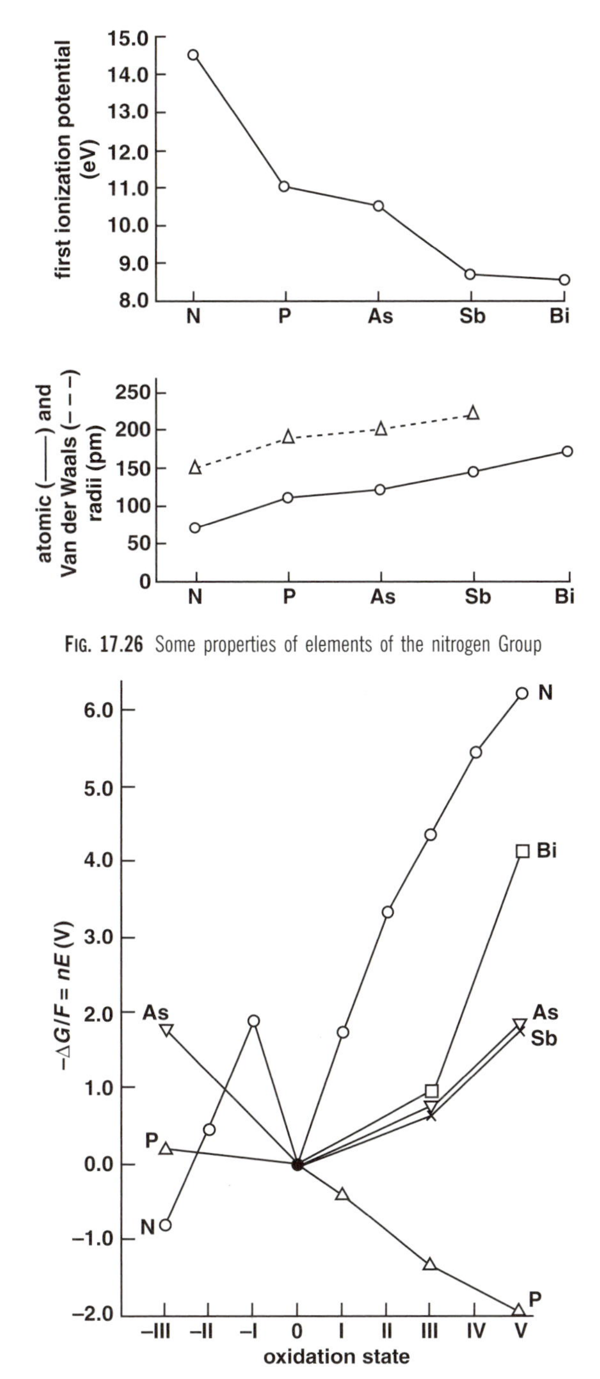

FIG. 17.26 Some properties of elements of the nitrogen Group

FIG. 17.27 Oxidation state free energy diagrams of elements of the nitrogen Group. This is the most complex oxidation state diagram of all the Main Groups. The properties of nitrogen are the most individual, with the element and the -III states as the most stable. All the positive states between 0 and V tend to disproportionate in acid solution (though many form gaseous species in equilibrium with the species in solution) and the -I state is markedly unstable. The curves for P, As, Sb and Bi form a family in which the -III state becomes increasingly unstable (values for Sb and Bi in this state are uncertain) and the V state becomes less stable with respect to the III state from P to Bi. All intermediate states of phosphorus tend to disproportionate to PH_3 plus phosphorus(V). The diagram also illustrates the very close similarity between As and Sb and the strongly oxidizing nature of bismuth(V)

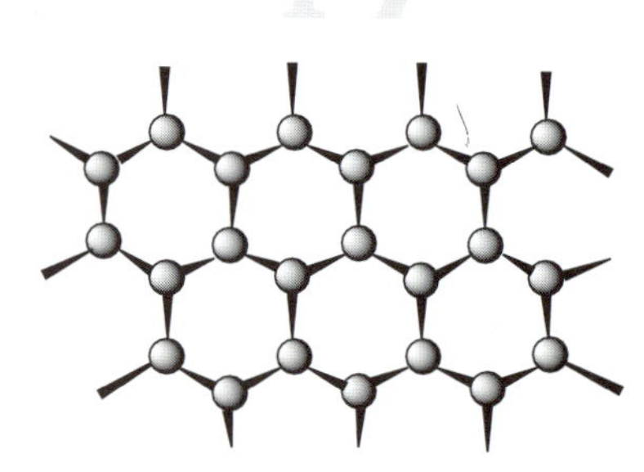

FIG. 17.28 The structure of bismuth

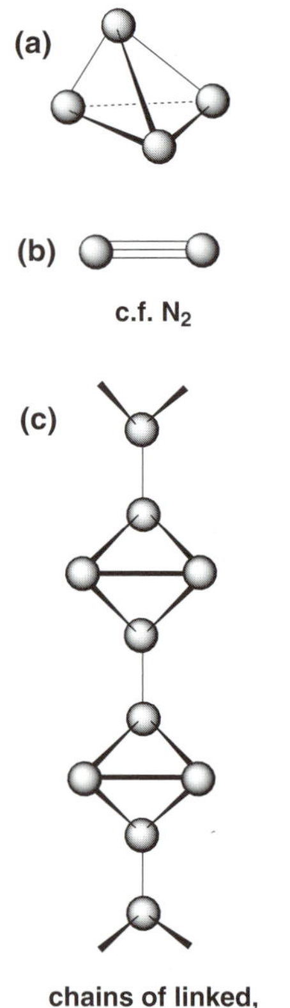

FIG. 17.29 Structures of phosphorus allotropes: (a) white phosphorus, (b) brown phosphorus, (c) red phosphorus (postulated)

allotropes of phosphorus. In white phosphorus, and also in the liquid and vapour states, the unit is the P_4 molecule where the four phosphorus atoms form a tetrahedron. If the vapour is heated above 800°C, dissociation to P_2 units starts and rapid cooling of the vapour from 1000°C gives an unstable brown form of phosphorus which probably contains these P_2 units. When white phosphorus is heated for some time above 250°C, the less reactive red form is produced. The structure of this form is not yet established and it exists as a number of modifications with various colours—violet, crimson, etc. These might be different structures or due to different crystal sizes, but they may also be due to the incorporation of part of the catalysts used in the transformation. When white phosphorus is heated under high pressure, or treated at a lower temperature with mercury as a catalyst, a dense black form results which has a layer structure like bismuth. A further vitreous form is also reported which results from heating and pressure. The structures of some of these allotropes are shown in Fig. 17.29 and the interconversion of the allotropes in Fig. 17.30. White phosphorus is the most reactive of the common allotropes, red phosphorus is much less reactive, and black phosphorus is inert. Substantial further insight is provided by the recently established structures of the polyphosphorus hydrides and anions outlined in Section 18.3.

Arsenic and antimony each occur in two forms. The most reactive is a yellow form which contains M_4 tetrahedral units and resembles white phosphorus. These yellow allotropes readily convert to the much less reactive metallic forms which have the same layer structure as bismuth.

Binary compounds of these elements are similar to those of previous Groups. Nitrides range from those of the active metals, which are definitely ionic with the N^{3-} ion, through the transition metal nitrides which resemble the carbides, to covalent nitrides like BN and S_4N_4. Heated Ba metal reacts with N_2 giving a product which hydrolyses to give 30% N_2H_4 and 70% NH_3. This suggests that the N_2^{4-} ion is formed as well as N^{3-}. The phosphides are similar. The interesting polyphosphide ion, P_3^{4-}, is isoelectronic with ClO_2 and contains an unpaired electron. The structure is V-shaped with an angle of 118° and a P-P length of 218.3 pm. The heavier elements form compounds with metals which become more alloy-like as one passes from phosphorus to bismuth. One important feature is the appearance of ionic nitrides, as compared with monatomic carbides which are not ionic. Ionic nitrides include the Li, Mg, Ca, Sr, Ba and Th compounds. They are prepared by direct combination or by deammonation of the amides:

$$3Ba(NH_2)_2 \longrightarrow Ba_3N_2 + 4NH_3$$

By contrast, the corresponding carbides either contain polyanions, like the acetylides, or are intermediate between ionic and giant covalent molecules.

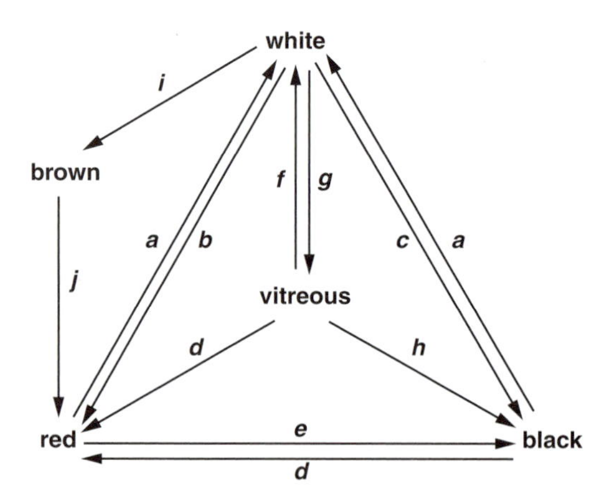

FIG. 17.30 The interconversion of the allotropes of phosphorus. The reaction conditions are indicated by letters as follows: (a) melting followed by quenching, or vacuum sublimation, (b) heating above 250°C, (c) heating to 220°C under pressure, (d) heating above 450°C, or on prolonged standing at room temperature, (e) at 25°C under high pressure, (f) vacuum sublimation, (g) heating above 250°C under pressure, (h) heating to 400°C under pressure, (i) rapid cooling of vapour with liquid nitrogen, (j) warming above liquid nitrogen temperature (-196°C). Many of these interconversions may also be brought about by catalysis, especially by mercury

OCCURRENCE AND USES OF THE NITROGEN GROUP ELEMENTS

Nitrogen gas is now produced on a substantial scale, largely as a byproduct of the isolation of oxygen from air for steelmaking. It is used to provide a cover for processes which are sensitive to air, and as liquid, it is in ever-increasing use as a coolant. One of the largest manufacturing applications using nitrogen is the Haber process for producing NH_3 from N_2 and H_2 under pressure over an iron catalyst. Recent work on catalysts, such as ruthenium on carbon, allows the reaction to occur at atmospheric pressure and ordinary temperatures, foreshadowing a cheaper Haber process. The ammonia is used mainly as a fertilizer, as salts, and part is converted to nitrate by catalytic oxidation, again mainly for use as a fertilizer. Hydrazine, N_2H_4 is used in agricultural chemicals and herbicides and as an intermediate in pharmaceuticals. Elementary phosphorus, made by carbon reduction of phosphate minerals, is extensively used in matches. The old formulation, based on poisonous white phosphorus, has now been entirely superseded by red phosphorus mixtures which also include sulfides of P and Sb. The oxidant is chlorate, and safety matches have separate oxidant in the match, with the phosphorus species on the striking surface. The major production of phosphorus compounds is that of phosphate fertilizer, mostly using the family of *apatite* ores, $Ca_5(PO_4)_3X$ ($X = OH$, Cl or F). These are of low solubility and superphosphate fertilizer, of higher solubility, is produced by treatment of the crude phosphate rock with sulfuric acid. Phosphoric acid is produced from the phosphate rock by treatment with sulfuric acid followed by purification by a solvent extraction process (see Section 7.3). Phosphate, especially as a component of DNA and RNA, is essential to all life processes. The major component of teeth and the main bone mineral is hydroxyapatite, accompanied in the bone by amorphous calcium phosphate. In teeth F replacement, giving fluoroapatite, increases toughness and resistance to caries. Phosphates are widely used as food additives (such as phosphoric acid in cola drinks and sodium phosphates in processed foods), as flame retardants and in the prevention of scale and corrosion in water-cooling systems. Phosphorus sulfides are also manufactured on a large scale as starting materials for the synthesis of organophosphorus compounds which are widely used as pesticides, for example malathion. Many of the military nerve gases are organophosphorus compounds, e.g. sarin $CH_3P(O)(OPr^i)F$.

Traditional uses of arsenic compounds in insecticides and fungicides, e.g. as wood preservatives, are being phased out as less toxic alternatives become available. All three heavier elements find uses in alloys (see under tin in the last section). As and Sb are becoming increasingly important in the III/V semiconductor materials mentioned under gallium above. Sb_2O_3 finds substantial use, along with various phosphorus compounds, as flame retardants in plastics.

Nitrogen, because of the high strength of the triple bond (heat of dissociation = 962 kJ mol^{-1}), is inert at low temperatures and its only reaction is with lithium to form the nitride. At higher temperatures it undergoes a number of important reactions including the combination with hydrogen to form ammonia (Haber process), with oxygen to give NO, with magnesium and other elements to give nitrides and with calcium carbide to give cyanamide:

$$N_2 + CaC_2 \rightarrow CaNCN + C$$

The other elements react directly with halogens, oxygen and oxidizing acids.

This Group shows a richer chemistry than the boron and carbon Groups, as there are more than two stable oxidation states and there is a wider variety of shapes and coordination numbers. Nitrogen shows a stable –III state in ammonia and its compounds, as well as in a wide variety of organo-nitrogen compounds. Phosphorus has an unstable –III state in the hydride and also forms acids and salts, which contain direct P-H bonds, in the III and I states. In the normal states of V and III a variety of coordination numbers is found. The MX_3 compounds, where M = any element in the Group and X = H, halogen or pseudohalogen, or organic group, are pyramidal with a lone pair on M. The MX_5 compounds are trigonal bipyramids in the gas phase and adopt a variety of structures in the solid. It is interesting that, while PPh_5 and $AsPh_5$ are trigonal bipyramids, $SbPh_5$ is a square pyramid both in the solid and in solution. This is a second (with $InCl_5^{2-}$) Main Group example of the square pyramid as the alternative five electron pair structure.

A number of MX_4^+ species are known, as well as $MX_3.A$ (where A = any acceptor molecule such as BR_3) and these are tetrahedral. A few $M^{III}X_5^{2-}$ complexes also exist and these are square pyramids with a lone pair in the sixth position. Finally, a variety of $M^VX_6^-$ and $M^VX_5.A$

TABLE 17.12 Coordination numbers and stereochemistry in the nitrogen Group

Number of electron pairs	*Number of π bonds*	*Number of non-bonding pairs*	*Shape*	*Examples*
4	0	0	Tetrahedron	NH_4^+, MR_4^+, PX_4^+ (X = all halogens)
4	0	1	Pyramid	MH_3, MR_3, MX_3 (X = all halogens)
4	1	1	V	NO_2^-
5	0	0	Bipyramid	MF_5, PCl_5, PBr_5, PPh_5 ($SbPh_5$—see text)
5	1	0	Tetrahedron	MOX_3, MO_4^{3-}, HPO_3^{2-}, $H_2PO_2^-$
5	2	0	Plane	NO_3^-
6	0	0	Octahedron	MF_6^-, PCl_6^-, $SbPh_6^-$
6	0	1	Square pyramid	SbF_5^{2-}

(M = P, As, Sb, and Bi, R = simple alkyl radical, Ph = phenyl)

compounds are found which are octahedral. Many of these shapes are repeated in compounds which include π bonding. A full set of examples is gathered in Table 17.12.

M-M links for the heavier elements of the Group are found for all elements in the organic compounds $R_2M\text{-}MR_2$, although the Bi examples are unstable. The methylarsenic, *cacodyl*, $Me_2As\text{-}AsMe_2$, has been known since 1760. The corresponding hydrides are less

CATENATION IN GROUP 15 ELEMENTS; HYDRAZINE, THE AZIDE ION AND OTHER NITROGEN-RICH SPECIES

The Group 15 elements show a significant tendency to catenation, if less markedly than does the carbon Group. Polynuclear compounds are discussed in Sections 18.3 and 4. Simpler chain compounds of nitrogen involve both single bonds, as in hydrazine and its derivatives, $R_2N\text{-}NR_2$, where R = H, F or organic groups, and multiple bonds as in the diazenes RN=NR (R=F or organic groups, R=H is unstable unless coordinated—see Section 16.10). Nitrogen is found in a chain of three atoms in hydrazoic acid, HN_3, and in the azides. These are prepared from amide and nitrous oxide:

$$2NaNH_2 + N_2O \rightarrow NaN_3 + NaOH + NH_3$$

The azide ion is linear, and isoelectronic with CO_2. The acid, called hydrazoic acid, is bent at the nitrogen bonded to the substituent, with the NNH angle equal to 114° and the bond distances indicating the presence of single and triple bonds (Fig. 17.31a). Organic azides, RN_3, have similar structures. In the N_3^- ion the two N-N distances are equal at 116 pm, showing delocalization of the π bonding. The amino-diazonium cation, $H_2N_3^+$ has been recently prepared and can be thought of as a protonated hydrazoic acid with both hydrogens residing on one nitrogen, and single and triple bonds between the nitrogens (Fig. 17.31b).

Bond lengths of 121 pm for the central bond and 143 pm for the outer ones, and angles at the central N of 109°, are found in N_4H_4, which is a sublimable solid containing a chain of 4N atoms, $H_2N\text{-}N=N\text{-}NH_2$. A few organic derivatives R_4N_4 are also reported, together with one or two longer chain organic compounds, but all are relatively unstable.

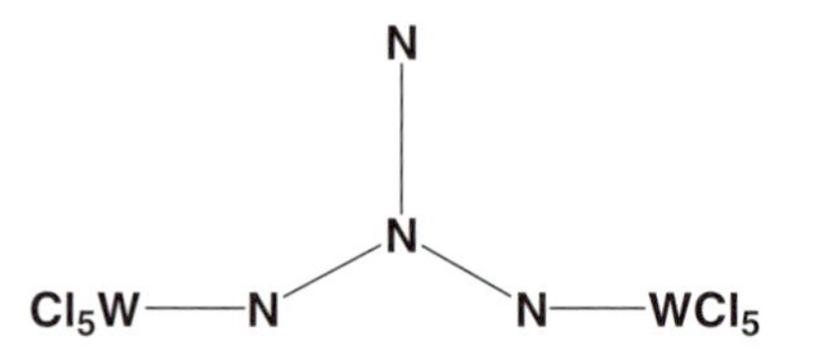

The branched chain isomer of tetrazene is reported, where the planar tetrazene completes an octahedron around each W. Only this coordinated form is known, but this is the first example of a branched nitrogen chain.

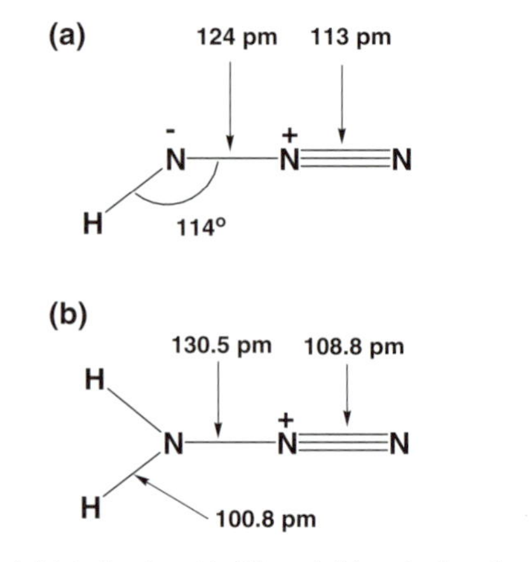

FIG. 17.31 Structures of (a) hydrazoic acid, HN_3 and (b) aminodiazonium cation, $H_2N_3^+$. Only one resonance form is shown for each

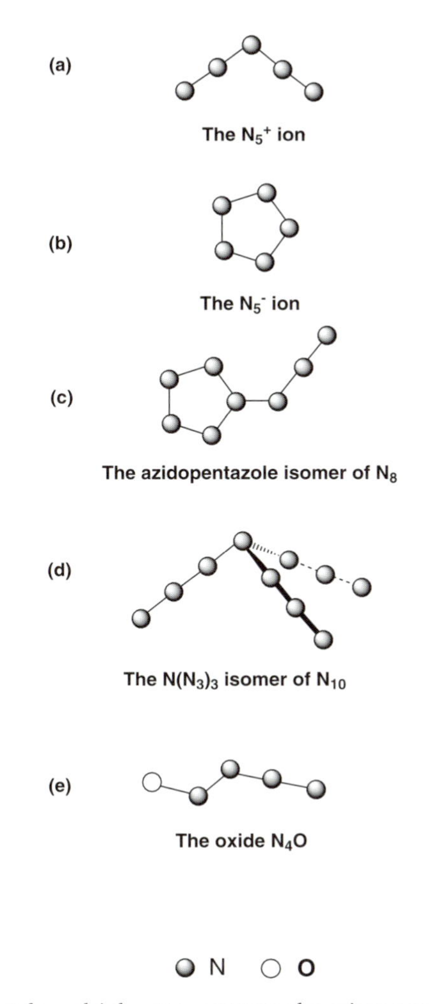

Compounds which are extremely nitrogen-rich are attracting interest as high energy materials, which have potential applications as rocket propellants or explosives. Such materials obtain their energy from the very high, positive heat of formation, and the very high stability of the nitrogen-nitrogen triple bond in N_2, which is invariably one of the major decomposition products. Perhaps the most dramatic example of such a species is the N_5^+ ion (a), which can be prepared as its SbF_6^- salt by reaction of $N_2F^+SbF_6^-$ and HN_3, and has a bent structure. Indeed, one can envisage new allotropes of nitrogen, such as $N_5^+N_3^-$ (which is yet to be prepared). Calculations have suggested that the most stable N_n molecules are those based on the five-membered pentazole ring, and thus N_5^- (b), is thought to be as aromatic as the isoelectronic cyclopentadienide $C_5H_5^-$ ion. For N_8, the most stable isomer is predicted to be azidopentazole (c), whereas for N_{10} pyramidal $N(N_3)_3$ [(d), analogous to NCl_3] might be expected. While such species containing only nitrogen atoms, have not yet been prepared, examples of nitrogen-rich species from other Group 15 elements have been synthesised by substitution of halide ions by azide, and thus species such as $As(N_3)_3$, $As(N_3)_6^-$ and $As(N_3)_4^+$, are analogous to $AsCl_3$, $AsCl_6^-$ and $AsCl_4^+$, respectively. Azides of heavy metal ions, such as $Pd(N_3)_2$ are also explosive. Another compound containing a nitrogen chain is the unusual oxide N_4O (e), prepared by reaction of sodium azide with NOCl, and is formulated as nitrosyl azide.

References

N_5^+: *J. Amer. Chem. Soc.* **123**, 2001, 6308–6313.
Covalent azides: *Chemische Berichte* **130**, 1997, 443–451.
$As(N_3)_6^-$: *Angewandte Chemie*, **39**, 2000, 2108–2109.
Polynitrogen compounds: *Angewandte Chemie* **38**, 1999, 2536–2538.

stable and limited to P_2H_4 and As_2H_4, and all the halides P_2X_4 are also known. Longer chains, rings, nets and open and closed clusters are found for P, and to some extent for As, and are reviewed in Section 18.3.

Trends within the Group are similar to those observed in the boron and carbon Groups. The acidic character of the oxides, and the stability of the V state, decrease in going from nitrogen to bismuth, so that bismuth has only a handful of compounds in the V state and these are unstable and strongly oxidizing. Antimony(V) and arsenic(V) are moderately oxidizing. Phosphorus(V) is very stable, while nitrogen is again oxidizing in the V state, reflecting the differences in formulae and coordination compared with phosphorus. The III state increases in stability as the V state becomes unstable: P(III) and As(III) are reducing, Sb(III) is mildly reducing and Bi(III) is stable. The III state also becomes increasingly stable in cationic forms for antimony and bismuth. Bi^{3+} exists in the salts of strong acids, such as the fluoride, or triflate ($CF_3SO_3^-$); an X-ray structure determination of the bismuth salt of the latter shows that the $[Bi(H_2O)_9]^{3+}$ cation has a tricapped trigonal prism structure. Both elements exist in solution, and in many salts, as the oxycation, MO^+.

17.6.2 The V state

All the oxides of the V state are known and their structures, where known, are shown in Fig. 17.32. N_2O_5 is usually made by dehydrating nitric acid, but a cleaner synthesis results

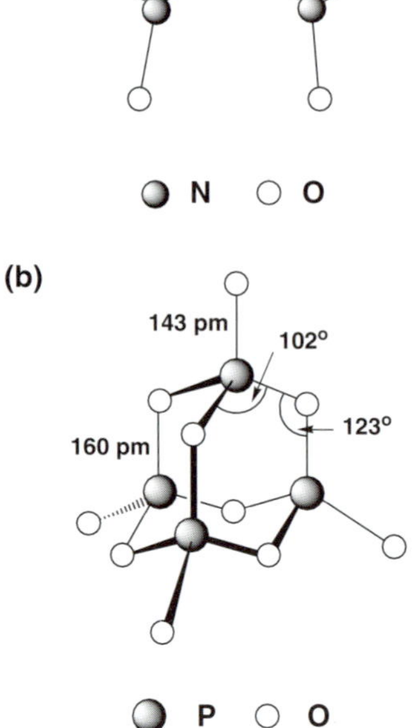

FIG. 17.32 Structures of pentoxides, M_2O_5, of the elements of the nitrogen Group: (a) N_2O_5, (b) P_4O_{10}

from the action of FNO_2 with excess $LiNO_3$. The solid is ionized with a planar nitrate ion and a linear nitronium cation, NO_2^+. Linear nitronium, or nearly so, is found with other stable anions: in $NO_2^+ClO_4^-$, the ONO angle is 175°. In the gas phase, the N_2O_5 molecule exists with the structure of Fig. 17.32. Higher oxides of nitrogen, NO_3 and N_2O_6, have been reported from the reaction of ozone on N_2O_5 but little is known of them.

Phosphorus burns in an excess of air to give the pentoxide which has the molecular formula, P_4O_{10}, and, in the vapour, the tetrahedral structure shown in Fig. 17.32b. This form is also found in the liquid and solid, but prolonged heating of either gives polymeric forms. Both 12- and 20-membered rings of PO_4 tetrahedra linked by corners have been identified. The environments of each phosphorus atom in P_4O_{10} and in the polymeric forms are similar. Each phosphorus atom is linked tetrahedrally to four oxygen atoms, three of which are shared with three other phosphorus atoms, while the fourth link is a P=O. The pentasulfide exists in the same P_4S_{10} form as the pentoxide (Fig. 17.32b) and the mixed oxide-sulfides $P_4O_6S_4$ (with terminal P=S) and $P_4O_4S_6$ (with terminal P=O), completing a satisfying series of symmetric molecules. $P_4S_9N^-$ is also isoelectronic and isostructural, with the N^- in an edge-bridging position.

The pentoxides, M_2O_5, of arsenic, antimony and bismuth are made by oxidizing the element or the trioxide. Increasingly powerful oxidizing agents are needed from arsenic to bismuth, and Bi_2O_5 is not obtained in a pure stoichiometric form. As_2O_5 is a polymeric structure consisting of AsO_4 tetrahedra and AsO_6 octahedra sharing corners. The structures of Sb_2O_5 and Bi_2O_5 are not known but are probably based on MO_6 octahedra like their anions. All three lose oxygen readily on heating to give the trioxides.

Oxyacids and oxyanions of the V state are a very important class of compounds in the chemistry of this Group. The nitrogen, phosphorus and arsenic compounds are included in Table 17.13 p.466. Antimony and bismuth do not form acids in the V state. The oxyanions may be made by reaction of the pentoxides in alkali or by oxidation of the trioxides in an alkaline medium. The bismuthates are strongly oxidizing, the best known being the sodium salt, $NaBiO_3$, which is used to identify manganese in qualitative analysis by oxidizing it to permanganate. The antimonates are oxidizing, but more stable than the bismuthates, and are octahedral ions, $Sb(OH)_6^-$. This is in contrast to the tetrahedral coordination to oxygen shown in the phosphates and arsenates.

Both phosphorus and arsenic form acids in the V state, both the mononuclear acid, H_3MO_4, and polymeric acids. The polyphosphoric acids include chains, rings and more complex structures, all formed from PO_4 tetrahedra sharing oxygen atoms. A wide variety of polyphosphate ions also exists, and structures of the first three members are indicated in Fig. 17.36e, f and g p.467. The detailed structure of the fourth member, $P_4O_{13}^{6-}$, isolated as the $[Co(NH_3)_6]^{3+}$ salt, shows a significant alternation of P-O bond lengths in the chain. Tripolyphosphates are used in detergents as they are excellent sequestering agents, though use has diminished as worries about eutrophication have grown (see Section 20.3.1). These pyro-, meta- and other polyphosphates are stable and hydrolyse only slowly to the ortho-acid, H_3PO_4. Polyarsenates also exist in similar forms, but these are much less stable and readily revert to H_3AsO_4.

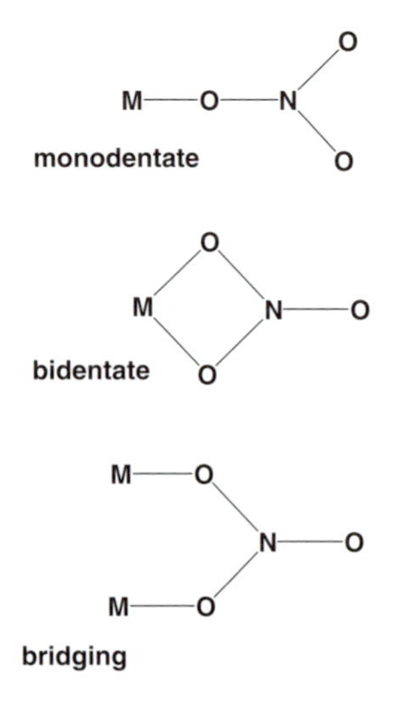

Bonding modes for the nitrate (NO_3^-) ion

Nitrogen in the V state forms nitric acid, HNO_3, and the nitrate ion, NO_3^-. Here, the coordination number of only three to oxygen, and the planar structure, reflect the presence of p_π–p_π bonding between the first row elements, nitrogen and oxygen. Although usually found as the free anion, nitrate does sometimes coordinate to metals. It occurs as a monodentate, bidentate or bridging ligand. A second, unexpected, anion has recently been reported, NO_4^{3-}. A structural study of the sodium salt shows a tetrahedral structure with a NO distance of 139 pm, compared with 122 pm in NO_3^- (Fig. 17.33) and 122 pm for N-O and 141 pm for N-OH in the isolated $HONO_2$ molecule. The N-O bond order in NO_4^{3-} is thus near to one, suggesting a relatively weak interaction, and a description in terms of semi-polar bonds is probably the most appropriate, compare Section 18.9.

The stable V state halides are limited to the pentafluorides, PF_5, AsF_5, SbF_5 and BiF_5, together with PCl_5, $SbCl_5$ and PBr_5. This reflects the decreasing stability of the V state from phosphorus to bismuth. $AsCl_5$ has only recently been identified as a product of the UV photolysis of $AsCl_3$ and Cl_2 at −105°C. It decomposes at −50°C. This behaviour is an

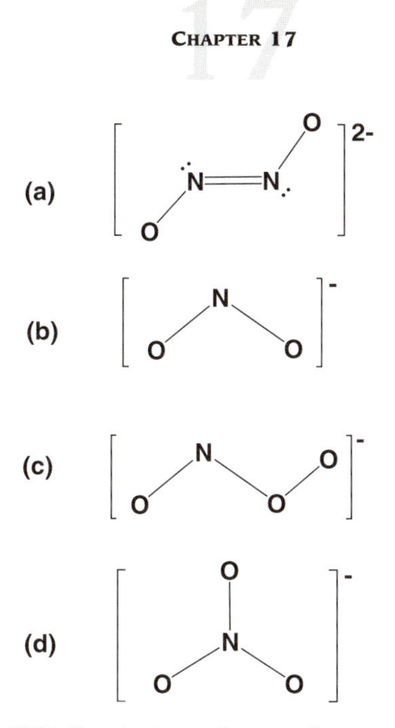

FIG. 17.33 The structures of some nitrogen oxyanions: (a) hyponitrite, (b) nitrite (O-N-O angle 116 to 132°, N-O bond length 113 to 123 pm in different salts), (c) pernitrite, (d) nitrate (N-O bond length 121.8 pm, O-N-O angle 120°)

example of the 'middle element anomaly' already discussed. All the structures so far determined show that the pentahalides are trigonal bipyramidal in the gas phase, but these structures alter in the solid, reflecting the instability of 5-coordination in crystal lattices. PCl_5 ionizes in the solid to $PCl_4^+PCl_6^-$ while PBr_5 ionizes to $PBr_4^+Br^-$. The cations are tetrahedral and PCl_6^- is octahedral. In addition there is another metastable modification of solid PCl_5 which has been found by an X-ray diffraction study to be $(PCl_4^+)_2(PCl_6^-)(Cl^-)$, with a significant interaction between the PCl_4^+ and Cl^- ions. This form slowly reverts to the normal $PCl_4^+PCl_6^-$ form on standing at room temperature. SbF_5 also attains a more stable configuration in the solid, this time by becoming 6-coordinated through sharing fluorine atoms between two antimony atoms in a chain structure. Except for BiF_5, all the pentafluorides readily accept F^- to form the stable octahedral anion, MF_6^-. PF_5, especially, is a strong Lewis acid and forms PF_5.D complexes with a wide variety of nitrogen and oxygen donors (D). Antimony also forms the dimeric ion $Sb_2F_{11}^-$, consisting of two SbF_6 octahedra sharing a corner.

The pentachlorides are similar but weaker acceptors. They do accept a further chloride ion, and $SbCl_6^-$ in particular is well established and stable: PCl_6^- is mentioned above. It has recently been shown that $AsCl_6^-$ may also be prepared if it is stabilized by a large cation, thus $Et_4N^+AsCl_6^-$ has been prepared. All these pentahalides may be made by direct combination or by halogenation of the trihalides, MX_3. The pentafluorides, and PCl_5, are relatively stable, but PBr_5 and $SbCl_5$ readily lose halogen at room temperature (to give the trihalides) and are strong halogenating agents, as is BiF_5 which is by far the most reactive pentafluoride. A number of mixed pentahalides, such as PF_3Cl_2, are also known. They are formed by treating the trihalide with a different halogen:

$$PF_3 + Cl_2 \rightarrow PF_3Cl_2$$

These are similar to the pentahalides in many ways. For example, PF_3Cl_2 is covalent on formation but passes over into the ionic form $PCl_4^+PF_6^-$ on standing. Another example is AsF_3Cl_2 which also appears to be ionized to $AsCl_4^+AsF_6^-$, showing that the second ionic component of the unstable $AsCl_5$ exists. It has also been shown that $AsClF_4$ is quite stable but that, though $AsCl_4F$ exists at low temperatures, it readily loses Cl_2 to give $AsCl_4^+$ AsF_6^- and $AsCl_3$. $AsCl_3F_2$ is not accessible from $AsCl_3$, but results from the treatment of $AsCl_2F_3$ with $CaCl_2$. Like PCl_3F_2, the Cl are equatorial in the trigonal bipyramid. Overall, As(V) with 4 or 5 bonded Cl is unstable with respect to As(III) plus Cl_2, though it can be stabilized by suitable donors or in the presence of suitable counter-ions. The set of AsX_4^+ ions was completed by the discovery of AsI_4^+ as the $AlCl_4^-$ salt. The cation $P_2I_5^+$, with a P-P bond, has been prepared by reaction of PI_3 with AlI_3, or by reaction of P_2I_4 with I_2 and AlI_3. The cation can be viewed as an analogue of the PI_4^+ ion in which an I is replaced by a PI_2 group.

Although nitrogen cannot form a pentahalide, as only four valency orbitals are present in the second level, it is interesting that the cation NF_4^+ exists showing that nitrogen(V) can bond to fluorine, in a species where only σ interactions are possible. The oxychloro cation of nitrogen(V), $(NOCl_2)^+$, has been made as the $[AsF_6]^-$ or $[SbCl_6]^-$ salt. The structure is a flattened pyramid with O-N-Cl = 119°.

In their covalent forms, the mixed pentahalides PX_nY_{5-n}, such as PF_3Cl_2, have structures in which the most electronegative halogens occupy the two axial positions. Although PH_5 is unknown, mixed hydride-fluorides of P(V), are established e.g. PHF_4 and PH_2F_3. These form anions, HPF_5^- and $H_2PF_4^-$, related to PF_6^-.

Nitrogen, in the V state, forms the oxyfluoride NOF_3 from NF_3 and O_2 or from the elements. It readily transfers F^-, e.g. to form $(NOF_2)^+(BF_4)^-$. The structure is pyramidal: see remarks above about the formulation of H_3NO_4. Phosphorus and arsenic form a range of oxyhalides in the V state. Three compounds of phosphorus are known, POX_3 where X = F, Cl or Br, and for arsenic $AsOF_3$ and $AsOCl_3$. $POCl_3$ may be made from PCl_3 or from the pentachloride and pentoxide:

$$PCl_3 + \tfrac{1}{2}O_2 \rightarrow POCl_3$$
$$\text{or } P_4O_{10} + 6PCl_5 \rightarrow 10POCl_3$$

The other phosphorus compounds are made from the oxychloride. $AsOF_3$ is made by the action of fluorine on a mixture of $AsCl_3$ and As_2O_3 while $AsCl_3+O_3$ gives $AsOCl_3$ which decomposes at −30°C. All these are tetrahedral $X_3M{=}O$, with p_π–d_π bonding between the O and M atoms (Fig. 4.23). Phosphorus gives the corresponding sulfur and selenium compounds, PSX_3 and $PSeX_3$, again illustrating the marked stability of 4-coordinated P(V).

Three pentasulfides are found in this Group. P_4S_{10} has the same structure as P_4O_{10}—and there is also a compound $P_4O_6S_4$ which again has the same structure, with the oxygen atoms bridging along the edges of the tetrahedron, and a sulfur atom attached directly to each phosphorus. The structures of As_2S_5 and Sb_2S_5 are unknown. All three sulfides may be formed by direct reaction between the elements, and arsenic and antimony pentasulfides are also formed by the action of H_2S on As(V) or Sb(V) in solution.

PHOSPHONITRILIC HALIDES AND PHOSPHAZENES

One final important class of phosphorus (V) compounds is that of the phosphonitrilic halides. If PCl_5 is heated with ammonium chloride, compounds of the formula $(PNCl_2)_x$ result:

$$PCl_5 + NH_4Cl \rightarrow (PNCl_2)_3 + (PNCl_2)_4 + (PNCl_2)_x + HCl$$

The corresponding bromides may be made in a similar reaction, and the chlorines may also be replaced by groups such as F, NCS, or CH_3 and other alkyl groups, either by substitution reactions, or by using the appropriate starting materials. When $x=3$ or 4, the six- or eight-membered rings shown in Fig. 17.34, are formed. Similar rings have been identified for x values up to 17 in the case of chlorides and fluorides and for $x=6$ for the bromides. In addition, for large values of x, linear polymers are formed, of accurate formula $Cl(PNCl_2)_xPCl_4$. In the ring compounds, the trimer and pentamer are planar, while the tetramer and hexamer are puckered. The nature of the bonding in the rings, and also in the chain compounds, is not yet clearly determined but probably involves π bonding between nitrogen p orbitals and phosphorus d orbitals. In the trimeric chlorides, it has been suggested that this π bonding involves a strong interaction above and below the plane of the ring, as in benzene, and also a weaker interaction in the plane of the ring. In the nonplanar tetramer, this second type of π bonding can make a stronger contribution. In polymers, with OR groups such as OCH_2CF_3 replacing the halides, useful properties are found, especially resistance to oxidation and burning. These polyphosphazenes are finding application in special performance rubbers, gaskets and insulating materials. As these polymers are compatible with tissues, they have potential value in biomedical devices.

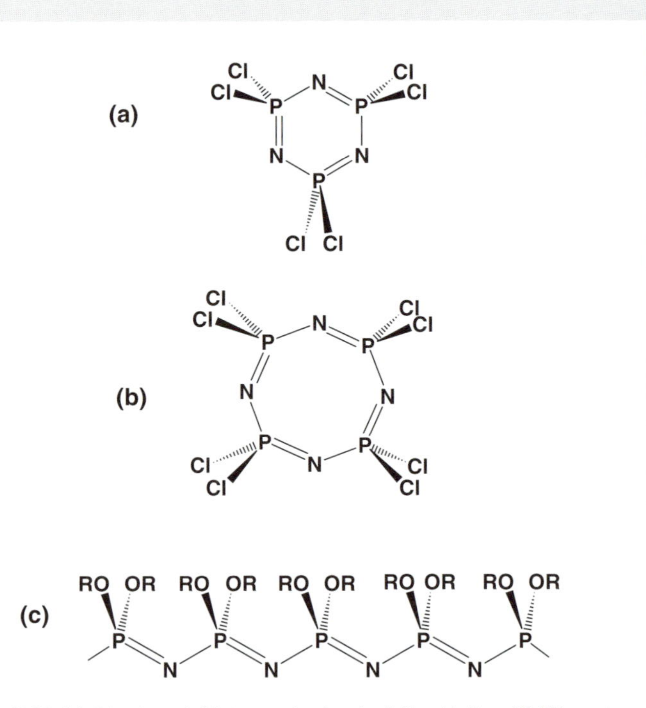

FIG. 17.34 (a) Trimeric and (b) tetrameric phosphonitrilic chlorides $(Cl_2PN)_n$ and (c) a polyphosphazene

Reference

Inorganic polymer, I. MANNERS, Angewandte Chemie **35**, 1996, 1602–1621.

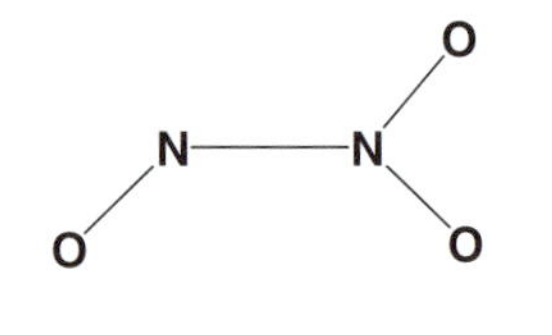

The structure of N_2O_3

17.6.3 The III state

The III state is reducing for nitrogen, phosphorus, and arsenic, and the stable state for antimony and bismuth. Among the oxygen compounds, all the oxides, M_2O_3, and all the oxyanions are known, but the free acids of the III state are found only for nitrogen, phosphorus, and possibly arsenic. This points to an increase in basicity down the Group, as expected.

The oxide of nitrogen, N_2O_3, is found as a deep blue solid or liquid, melting at about −100°C. It is formed by mixing equimolar proportions of NO and NO_2, and reverts to these

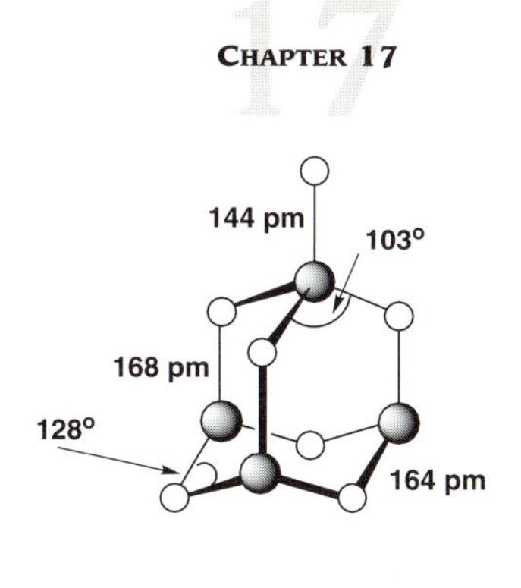

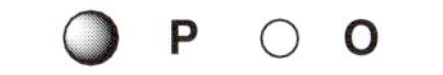

FIG. 17.35 The structure of P_4O_7

two components in the gas phase at room temperature. An X-ray structure determination, carried out at low temperature, showed the compound to be 'nitrosyl nitrite' with an N-N bond. N_2O_3, or an equimolar mixture of NO and NO_2, gives nitrous acid when dissolved in water, and nitrites when dissolved in alkali.

Phosphorus(III) oxide results when phosphorus is burned in a deficiency of air. In the vapour phase, it has the formula P_4O_6, and a structure derived from that of P_4O_{10} by removing the terminal oxygen atoms, see Fig. 17.32. The phosphorus(III) oxide is acidic and reducing, and dissolves in water to give phosphorous acid. We note here that there is now a complete series of mixed P(III)/P(V) oxides, P_4O_x with $x = 7$ (Fig. 17.35, compare Fig. 17.32b), 8 and 9, where terminal oxygens are progressively added to P_4O_6 giving all intermediate structures between that and P_4O_{10}.

When arsenic is burned in air, the only product is the trioxide, which has the formula As_4O_6, and a similar structure to P_4O_6, in both the gas phase and in the solid. A second form also occurs in the solid, but this structure is not known. Arsenic trioxide is acidic.

Antimony and bismuth also burn in air to give only the trioxides. Antimony trioxide has the form, Sb_4O_6, both in the gas and in the solid, and there is a second solid form. This has a structure consisting of long double chains made up of ···-O-Sb-O-Sb-O-··· single chains linked together through an oxygen atom on each antimony. Antimony trioxide is amphoteric. Bi_2O_3 is yellow (all the other compounds are white) and exists in a number of solid forms. These are not known in detail, but some at least contain BiO_6 units in a distorted prism arrangement. Bismuth trioxide is basic only. Antimony and bismuth, in the III state, commonly exist in solution and in their salts as the MO^+ ion, as already noted.

The oxyanions and acids of phosphorus(III) and arsenic(III) are included in Table 17.13 and Fig. 17.36. All the structures known contain 4-coordinated phosphorus or arsenic, and the III oxidation state results from the presence of a direct P-H or As-H bond (which, of course, does not ionize to give a proton). Although the stable form of phosphorous acid is the tetrahedral form shown in Fig. 17.36, there is some evidence from exchange studies (compare Section 2.3) for the transient existence of the pyramidal $P(OH)_3$ form. Organic derivatives of this form, $P(OR)_3$, are well known. Phosphorous acid, and the phosphites, are reducing, and also disproportionate readily as the oxidation state free energy diagram, Fig. 17.27, shows:

$$4H_3PO_3 \rightarrow 3H_3PO_4 + PH_3$$

Arsenites are also mildly reducing with a potential, in acid, of 0.56 V with respect to arsenic(V) acid, so that arsenites are rather weaker reducing agents than iron(II) in acid solution.

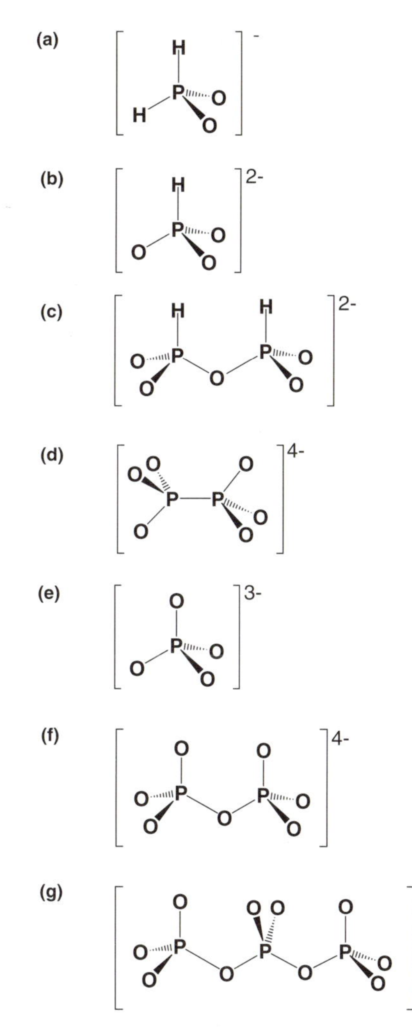

FIG. 17.36 The structures of phosphorus oxyanions: (a) hypophosphite, (b) phosphite, (c) pyrophosphite, (d) hypophosphate, (e) orthophosphate, (f) pyrophosphate, (g) tripolyphosphate

One interesting bismuth(III) compound is the complex oxycation $[Bi_6O_4(OH)_4]^{6+}$ whose structure contains six Bi atoms in an octahedron with 4O and 4OH groups triply bridging the triangular faces in a regular manner. This is formed by Bi_2O_3 in perchloric acid. Analogous Sn, V and Ce cluster ions of this $M_6O_4(OH)_4$ formula are found.

Arsenic, antimony and bismuth all form trisulfides, M_2S_3. The arsenic compound exists as As_4S_6, while the antimony and bismuth sulfides have polymeric chain structures. All are formed by the action of H_2S on solutions of the element in the III state.

The tale of the lower phosphorus sulfides is more complicated, as the main series have P-P bonds and may be seen as derived from the P_4 tetrahedron. The lowest S content is found in P_4S_3 (Fig. 17.37a) where S atoms are inserted in three of the six edges of P_4. The mixed P/As analogue, PAs_3S_3 has the P at the unique apex position and an As_3 triangle in the base. Further edge insertion is found in the two isomers of P_4S_4, where the remaining two P-P bonds are either contiguous or at opposite edges of the tetrahedron. In P_4S_5, S is inserted in an edge (Fig. 17.37b), while P_4S_7 sees edge insertion and terminal addition (Fig. 17.37c). The remaining well-established compound is P_4S_9 where the structure is like the oxide, with one terminal S removed from P_4S_{10} (Fig. 17.32b). We note that, as S is isoelectronic with P^- or PH, the species P_7H_3 and P_7^{3-} (and also As_7^{3-}) have structures analogous to Fig. 17.37a (compare Section 18.2). Arsenic also forms a sulfide, As_4S_3, and a selenide, As_4Se_3, with the P_4S_3 structure. A third arsenic sulfide, As_4S_4 (called *realgar*), is

TABLE 17.13 Oxyacids and oxyanions of the nitrogen Group

Nitrogen

The nitrogen acids and anions all show nitrogen 2- or 3-coordinated to oxygen and all (except hyponitrous acid) have p_π–p_π bonding between N and O.

Acid	Anion	
$H_2N_2O_2$ hyponitrous	$N_2O_2^{2-}$ hyponitrite	Reduction of nitrite by sodium amalgam. Weak acid. Readily decomposes to N_2O.
HNO_2 nitrous	NO_2^- nitrite	Acidify nitrite solution. Free acid known only in gas phase. Weak acid. Aqueous solution decomposes reversibly, $3HNO_2 \rightarrow HNO_3 + 2NO + H_2O$.
HOONO Pernitrous	$(OONO)^-$ Pernitrite	$H_2O_2 + HNO_2$. Free acid postulated as reaction intermediate.
HNO_3 nitric	NO_3^- nitrate	Oxidation of NH_3 from Haber process. Strong acid. Powerful oxidizing agent in concentrated solution.

The structures of the anions are shown in Fig. 17.33

Phosphorus

The phosphorus acids and anions all contain 4-coordinate phosphorus. In the phosphorus(V) acids, all four bonds are to oxygen, while P-H and P-P bonds are present in the acids and ions of the I and III states. Various intermediate oxidation states are found in anions which have P-P bonds, often with P-H ones as well. The most striking case is $[P(O)(OH)]_6$. The crystal structure of the caesium salt shows a six-membered, puckered, P_6 ring.

Acid	Anion	
H_3PO_2 Hypophosphorous	$H_2PO_2^-$ hypophosphite	White P plus hydroxide. Monobasic acid, strongly reducing.
H_3PO_3 Phosphorous	HPO_3^{2-} phosphite	Water plus P_2O_3 or PCl_3. Dibasic acid, reducing.
$H_4P_2O_5$ Pyrophosphorous	$H_2P_2O_5^{2-}$ Pyrophosphite	Heat phosphite: dibasic acid with P-O-P link. Reducing.
$H_4P_2O_6$ hypophosphoric	$P_2O_6^{4-}$ hypophosphate	Oxidation of red P, or of P_2I_4, in alkali, gives sodium salt, which gives the acid on treatment with H^+. Tetrabasic acid with a P-P link. Resistant to oxidation to phosphoric acid.
H_3PO_4 (ortho)phosphoric	PO_4^{3-} phosphate	P_2O_5 or PCl_5 plus water. Stable.
Also pyrophosphate $(O_3POPO_3)^{4-}$ polyphosphates $(O_3P[OPO_2]_nOPO_3)^{(4+n)-}$ metaphosphates $(PO_3)_m^{m-}$		Formed by heating orthophosphate. Linear polyphosphates with n up to 5 and cyclic metaphosphates with $m = 3$ to 10 have been individually identified (cf. Fig. 7.2).

The structures of the phosphorus anions are shown in Fig. 17.36.

Arsenic

Acid	Anion	
H_3AsO_3?, or $As_2O_3.xH_2O$ arsenious acids	$HAsO_3^{2-}$ and more complex forms arsenites	Formed from the trioxide or trihalides. The acid does not contain As-H and is weak, but the arsenites are well established in mononuclear and polynuclear forms. The arsenic(III) species are reducing and thermally unstable.
H_3AsO_4 arsenic acid	AsO_4^{3-} arsenate	$As + HNO_3 \rightarrow H_3AsO_4.\frac{1}{2}H_2O$. Tribasic acid and moderately oxidizing. Arsenates are often isomorphous with the corresponding phosphates.
Condensed arsenates		A number of these exist in the solid state but are less stable than polyphosphates and rapidly hydrolyse to AsO_4^{3-}.

As far as they are known, arsenic anions and acids have the same structures as the corresponding phosphates.

Antimony and bismuth give no free acids, though salts of the III and V states are found, and are discussed in the text. Coordination to oxygen is always six, not four as with phosphorus and arsenic.

also known and its structure is shown in Fig. 17.37d. A phosphorus selenide of composition P_2Se_5 has been found to have the structure shown in Fig. 17.37e. In marked contrast to phosphorus and arsenic, nitrogen sulfides are rare and a good example is given by the contrast in stabilities between the very stable N_2O and the transient sulfur analogue N_2S which has only been detected spectroscopically.

The five elements of the nitrogen Group all give all the trihalides, MX_3. The least stable are the nitrogen compounds of which only NF_3 is stable and occurs as the expected pyramidal molecule formed by the reaction of nitrogen with excess fluorine in the presence of copper. NF_3 has almost no donor power and has only a very low dipole moment as

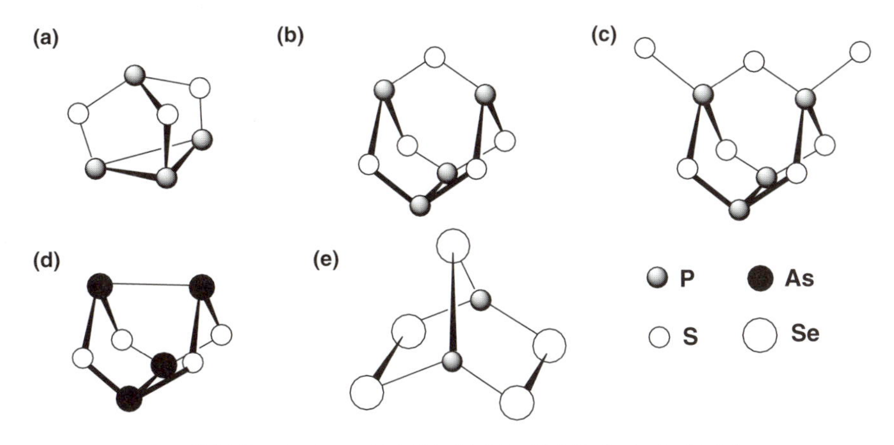

FIG. 17.37 The structures of (a) P_4S_3, (b) P_4S_5, (c) P_4S_7, (d) As_4S_4, (e) P_2Se_5

the strong N-F bond polarizations practically cancel out the effect of the lone pair. Nitrogen trichloride decomposes explosively. NI_3 is typically prepared as an unstable ammoniate such as $NI_3.6NH_3$, which is highly shock sensitive when freed from excess ammonia. The structure of one ammoniate, $NI_3.NH_3$, has been determined crystallographically. It consists of chains of NI_4 tetrahedra formed by sharing corners and the NH_3 molecules are bonded to the nonbridging iodines. NI_3 free from ammonia has been prepared only recently by the reaction of boron nitride with IF:

$$BN + 3IF \longrightarrow NI_3 + BF_3$$

The free NI_3 is rather unstable but can be sublimed as a deep red solid which reacts with ammonia to form the ammoniate $NI_3.3NH_3$.

The other sixteen trihalides of the Group are all relatively stable molecules, with the expected trigonal pyramidal structure in the gas phase. The pyramidal structure is also found in many of the solids, but other solid state structures are also found, particularly among the tri-iodides which adopt layer lattices with the metal atoms octahedrally surrounded by six halogen atoms. PI_3 has three P-I distances of 246 pm (and IPI angles of 102°) and three of 367 pm (angles 60°) showing the P lone pair is still sterically important. In addition to the simple trihalides, MX_3, a wide variety of mixed halides, MX_2Y or MXYZ, are found.

As a ligand in metal complexes, PF_3 (and to a lesser extent PCl_3) resembles CO, with both being moderate σ donors and good π acceptors. Thus, for example, $Ni(PF_3)_4$ is similar to $Ni(CO)_4$, cf. Section 16.2.4.

All the trihalides are readily hydrolysed, giving the oxide, oxyanion, or—in the cases of antimony and bismuth—the oxycation, MO^+. They may act as donor molecules, by virtue of the lone pair, and PF_3 in particular has been widely studied. It is rather less reactive to water and more easily handled than the other trihalides. Protonation of PF_3 (but not the lower basicity AsF_3) can be achieved with the very strong acid HF/SbF_5, giving the PF_3H^+ ion (CO has a similar low proton affinity). Complex ions, such as SbF_5^{2-}, are formed, and SbF_3 also gives the interesting dimeric ion, $Sb_2F_7^-$ (Fig. 17.38). In mixed oxidation state fluorides, such as Sb_8F_{30}, antimony(III) cations are found as well as the $[Sb^VF_6]^-$ anion. This compound contains $[Sb_2F_5]^+$ and $[Sb_3F_7]^{2+}$ together with three anions. The cations are linked through single, nearly linear, F bridges, e.g. $[F_2Sb\text{-}F\text{-}SbF_2]$, and with angles at Sb around 80°. For the heavier halogens, X = Cl, Br or I, three types of complex ion are found: SbX_4^-, SbX_5^{2-} and SbX_6^{3-}. The first two have the AB_4L (Fig. 4.4b) and AB_5L (Fig. 4.5b) structures expected from the presence of the lone pair, but the SbX_6^{3-} species are regular octahedra. In this they parallel the seven-electron-pair MX_6^{2-} species formed by the heavy elements of the oxygen Group (compare Section 17.7.4). In complex ions of formula $M_2X_9^{3-}$, a range of structures is found, from the binuclear (Fig. 17.39a; two octahedra sharing a face) through the tetrameric structure of $Bi_4Cl_{18}^{6-}$ to the polymeric $[Sb_2Cl_9^{2-}]_x$ unit where each antimony carries three terminal Cl atoms, and is linked by three single Cl bridges to three different neighbours in a double-chain. Hydrolysis yields $[Sb_2OCl_6]^{2-}$ with the triply-bridged structure of Fig. 17.39b. A similar structure is found for $[As_2SBr_6]^{2-}$ and for $[Sb_2SCl_6]^{2-}$. The trihalides are common reaction intermediates, and, for example, react with silver salts to give products such as $P(NCO)_3$, and with organometallic reagents to give

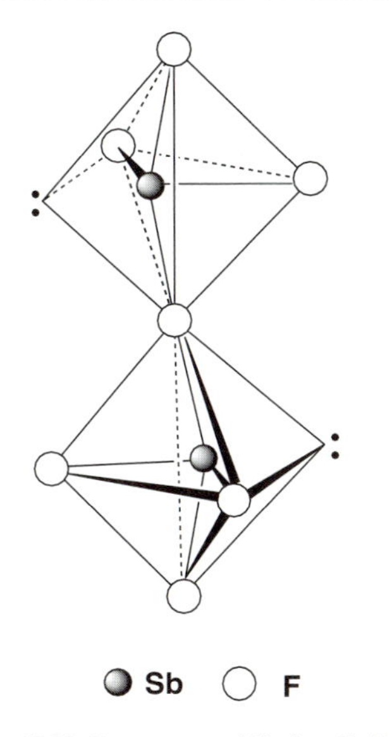

FIG. 17.38 The structure of the ion, $Sb_2F_7^-$

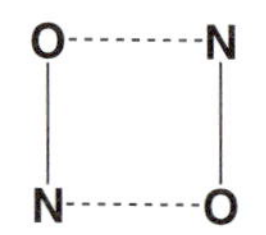

The loosely-bonded NO dimer

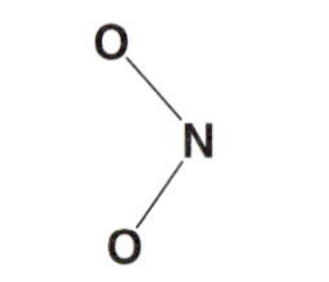

The structure of NO_2

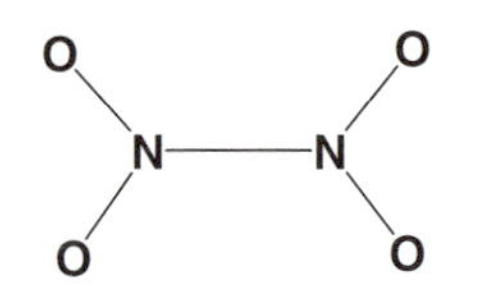

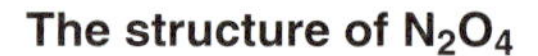

The structure of N_2O_4

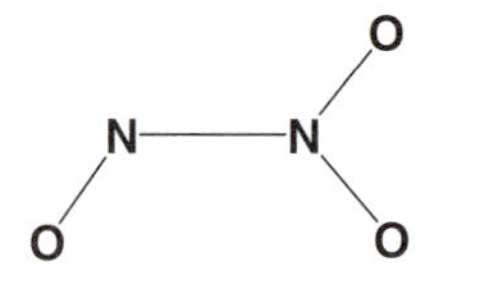

The structure of the $N_2O_3^-$ anion

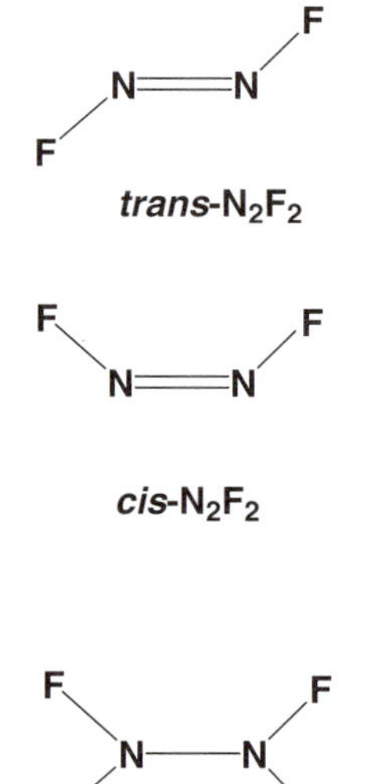

trans-N_2F_2

cis-N_2F_2

The structure of N_2F_4

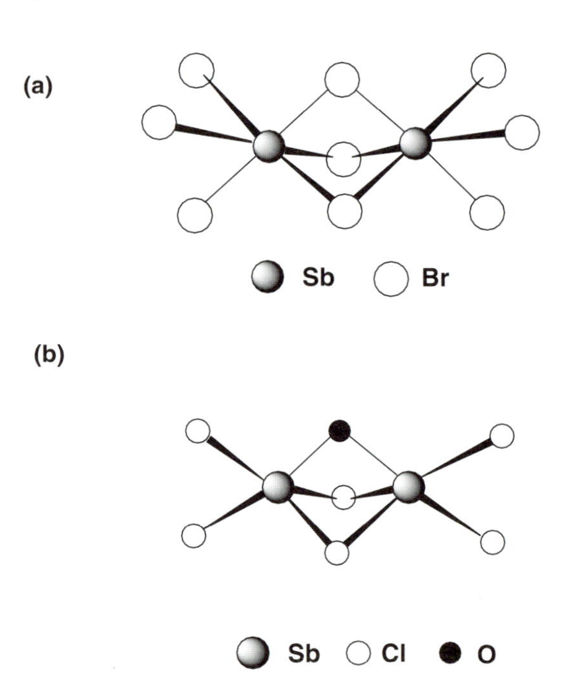

FIG. 17.39 The structures of (a) the $M_2Br_9^{3-}$ ion, (b) the $[Sb_2OCl_6]^{2-}$ ion

a wide variety of organic derivatives, MR_3. Among such analogues we note species like $As(OTeF_5)_3$ and $Sb(CF_3)_3$ where the ligands behave very similarly to F and have a similar extensive chemistry.

17.6.4 Other oxidation states

A number of oxidation states other than V and III are found, especially among the oxides and oxyacids. Nitrogen forms the I, II and IV oxides, N_2O, NO and NO_2 or N_2O_4. Nitrous oxide, N_2O, is formed by heating ammonium nitrate solution or hydroxylamine and is fairly unreactive. It has a π-bonded linear structure, NNO. Nitric oxide has already been discussed from the structural point of view. Although it has an unpaired electron, it shows little tendency to dimerize, though some association to rather loose dimers occurs in the liquid and solid. It rapidly reacts with oxygen to form NO_2. This is also an odd electron molecule but does dimerize readily to dinitrogen tetroxide. The solid is entirely N_2O_4 and this dissociates slightly in the liquid and increasingly in the gas phase until, at 100°C, the vapour contains 90% NO_2. NO_2 is brown and paramagnetic and has an angular structure with ONO = 134°. N_2O_4 is colourless and diamagnetic with a symmetrical structure and a very long N-N bond of 175 pm (compare 145 pm for the N-N single bond length in a molecule like hydrazine).

The compound called *Angeli's salt*, $Na_2N_2O_3.H_2O$, is formally N(II). The anion has the structure shown. Here the unique NO distance is 135 pm and the others are 131 to 132 pm.

When P_2O_3 is heated above 210°C a third oxide, PO_2, is formed, along with red phosphorus. This compound has a vapour density corresponding to P_8O_{18} and it behaves chemically as if it contains both P(V) and P(III). It may also contain P-P bonds as it reacts with iodine to give P_2I_4. Its structure is unknown.

The phosphorus oxide PO, which is isoelectronic to the nitrogen analogue NO, is believed to be the most abundant phosphorus-containing molecule in interstellar clouds. Several metal complexes containing PO as a bridging ligand are known.

Heating either Sb_4O_6 or Sb_2O_5 in air above 900°C gives an oxide of formula SbO_2. This consists of a network of fused SbO_6 octahedra containing both Sb(III) and Sb(V). A corresponding AsO_2 may exist.

There are also two oxyacids of low oxidation states in the Group. These are hyponitrous acid, $H_2N_2O_2$, with nitrogen(I), and hypophosphorous acid, H_3PO_2, with phosphorus(I). These are included in Table 17.13. Nitrogen forms its low oxidation state in hyponitrous acid by p_π-p_π and N-N bonding, while phosphorus in hypophosphorous acid is tetrahedral and the low oxidation state arises from two direct P-H bonds. Intermediate phosphorus oxidation states in oxypolyphosphorus compounds resulting from P-P and P-H bonds are shown in Table 17.13 and Fig. 17.36c, d p.467.

Lower oxidation state nitrogen fluorides are made by direct combination using less fluorine than required for NF_3. N_2F_2 has a planar structure which is most stable in the *trans* form, but which may also occur in the *cis* form. There is a π bond between the two N atoms which have each a lone pair. N_2F_4 is a gas with a skew structure similar to that of hydrazine. In the gas and liquid phases it undergoes reversible dissociation to NF_2:

$$N_2F_4 \rightleftharpoons 2NF_2$$

similar to that of N_2O_4. NF_2 is an angular molecule and contains an unpaired electron. For an odd-electron species, it has fairly high stability resembling NO, NO_2 and ClO_2 in this respect.

Phosphorus forms $X_2P\text{-}PX_2$ lower halides of which P_2I_4 is the most stable. These compounds probably have a skew structure as in N_2F_4. The first phosphorus halide with three linked P atoms, P_3I_5, has been generated in solution by reaction of PI_3, $PSCl_3$ and Zn, and found by nmr spectroscopy to have the structure $I_2P\text{-}P(I)\text{-}PI_2$.

The lower halides of bismuth present a much more complicated picture. When bismuth is dissolved in molten bismuth trichloride, an intensely coloured solution results from which may be isolated a compound with the accurate formula $Bi_{12}Cl_{14}$. This is a complicated structure with 48 Bi atoms and 56 Cl atoms in the unit cell. These are arranged as $4Bi_9^{5+}$, $8BiCl_5^{2-}$ and $2Bi_2Cl_8^{2-}$ units. The $BiCl_5^{2-}$ ion is a square pyramid with Bi(III) and resembles the SbF_5^{2-} ion mentioned earlier. In the structure these units are weakly linked to each other to form a chain. The $Bi_2Cl_8^{2-}$ unit contains Bi(III) and consists of two square pyramids sharing an edge of the base with their apices *trans* to each other. The Bi_9^{5+} unit has six Bi atoms at the corners of a somewhat distorted trigonal prism, and the other three Bi atoms above the rectangular faces. These units are shown in Fig. 17.40. Thus the solid may be written $(Bi_9^{5+})_2(BiCl_5^{2-})_4Bi_2Cl_8^{2-}$. In $Bi_{10}Hf_3Cl_{18}$, the Bi^+ ion has been recognized along with Bi_9^{5+} and $HfCl_6^{2-}$ ions. Further work has produced other cluster compounds of Bi and Sb, and these are reviewed in Section 18.4.

In BiI, an infinite chain structure is found, where Bi atoms are in two environments (Fig. 17.41). In the A chain, Bi is bonded only to three other Bi atoms, and is formally Bi(0). In the B chain, there are four Bi-I bonds, each shared with a second Bi, and thus with the formal oxidation state Bi(II). These Bi atoms form a fifth bond to a bismuth of the A chain.

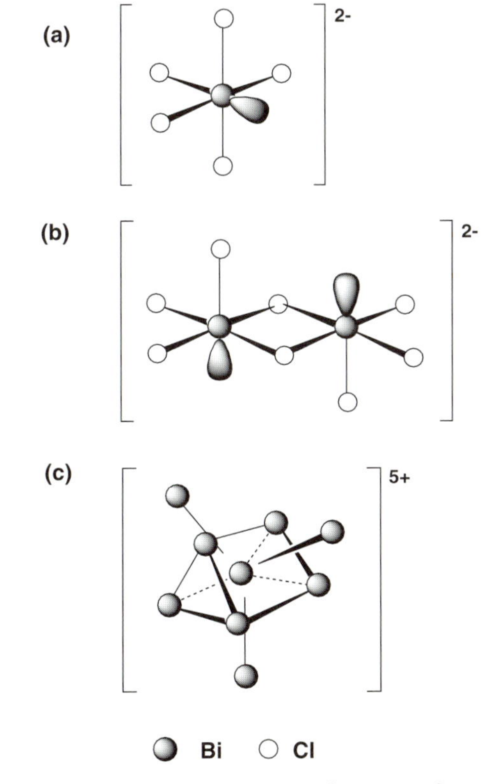

FIG. 17.40 The structures of (a) $BiCl_5^{2-}$, (b) $Bi_2Cl_8^{2-}$ and (c) Bi_9^{5+}

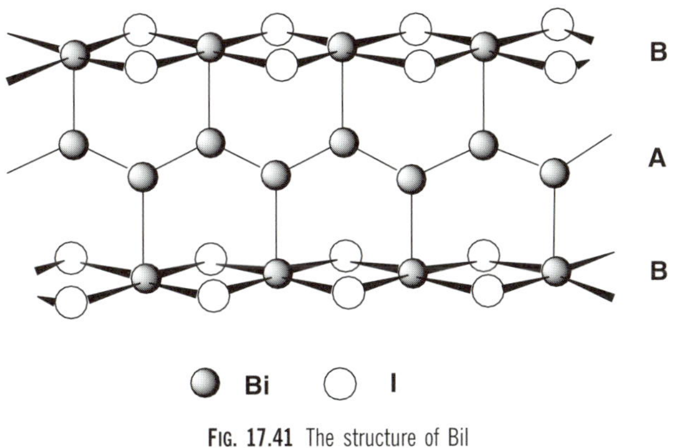

FIG. 17.41 The structure of BiI

In the lower bromides we find BiBr, isostructural with BiI, and also $Bi_{12}Br_{14}$, comparable with $Bi_{12}Cl_{14}$.

Negative oxidation states appear in the hydrides, MH_3, and in the organic compounds, NR_3. The other elements, except possibly arsenic (see Table 2.14), are of lower electronegativity than carbon in alkyl groups, and their organic compounds correspond to positive oxidation states. This distinction is not a useful one and is best regarded—as in the case of carbon chemistry in general—as an accidental result of the definitions. The organic compounds, MR_3, and organohydrides such as R_2PH, behave in a similar way to the hydrides but with the M-C bond stronger than M-H. An extensive organometallic chemistry of this Group exists which requires a textbook of its own for review. Here we note only some direct comparisons with the hydrides. As well as MR_3, analogous to MH_3, ions NR_4^+, PR_4^+, AsR_4^+, SbR_4^+ and BiR_4^+ exist which are tetrahedral and analogous to NH_4^+. The phosphorus analogue of NH_4^+ exists, for example in PH_4I which is prepared from PH_3 and HI. However, the phosphonium halides are relatively unstable and readily decompose to phosphane and hydrogen halide. They are much more covalent than the ammonium salts. The analogous AsH_4^+ and SbH_4^+ ions have also been recently prepared as their SbF_6^- salts by protonation of the parent hydride MH_3 using the superacid HF/SbF_5.

Although no pentavalent hydride, MH_5, exists, the pentaphenyl phosphoranes MPh_5 of P, As, Sb and Bi exist, as does PMe_5.PPh_5 and $AsPh_5$ are trigonal bipyramids in shape but $SbPh_5$ is a square pyramid. $SbPh_5$ reacts with PhLi to give the octahedral $SbPh_6^-$ ion. $BiMe_6^-$ is also known and has the expected octahedral geometry.

A number of mixed chloro-alkyl or chloro-aryl phosphoranes, R_nPX_{5-n} (R = alkyl, aryl; X = halide), are also known, comparable with PH_2F_3 and similar hydride-halides. As a result of recent studies the chemistry of these species has been found to be more complex than first thought. Species of this type can occur in a variety of forms and as a general illustration we take Ph_3PI_2. Reaction of Ph_3P with di-iodine in ether gives the molecular four-coordinate compound Ph_3P-I-I and neither ionic $[Ph_3P\text{-}I^+]I^-$ or 5-coordinate covalent phosphorane $Ph_3P(I)(I)$ which were previously thought to be the stable forms of this compound. Ph_3P-I-I can be thought of as a donor–acceptor compound between Ph_3P and I_2. Analogous compounds Ph_3P-Br-Br and Ph_3As-I-I can also be formed. There appears to be a delicate balance of factors determining whether a covalent or ionic form is produced since $PhPCl_4$ is a molecular compound whilst the corresponding methyl analogue exists as $[MePCl_3]^+Cl^-$. In addition, when R_3P-I-I compounds are dissolved in chloroform they ionize to $[R_3P\text{-}I]^+I^-$.

Nitrogen is found in the –II state in hydrazine and its organic derivatives, R_4N_2, and in the –I state in hydroxylamine, NH_2OH.

17.7 The oxygen Group, ns^2np^4

17.7.1 General properties

Table 17.14 lists some of the properties of the elements and Fig. 17.42 and 17.43 show the variations with Group position of a number of parameters and of the oxidation state free energies:

Oxygen occurs both as the O_2 molecule and as ozone, O_3. O_2 is paramagnetic (Section 3.5) and has a dissociation energy of 489 kJ mol^{-1} It is pale blue in the liquid and solid states. Ozone is usually formed by the action of an electric discharge on O_2, but it can also be formed chemically by the reaction of O_2^+ (dioxygenyl, Section 17.7.2) salts with water in HF at low temperature. Pure O_3 is deep blue as the liquid with m.p. = −250°C and b.p. = −112°C. It is diamagnetic and explodes readily as the decomposition to oxygen is exothermic and easily catalysed:

$$O_3 = \tfrac{3}{2}O_2, \Delta H = -142 \text{ kJ mol}^{-1}$$

OTHER REFERENCES TO THE PROPERTIES OF OXYGEN GROUP ELEMENTS

References to the properties of oxygen Group elements will be found in the following places:

Ionization potentials	Table 2.8
Atomic properties and electron configurations	Table 2.5
Radii	Table 2.10, 2.11, 2.12
Electronegativities	Table 2.14
Redox potentials	Table 6.3

Singlet oxygen is discussed in a Box in Section 3.6. Polysulfides and polyselenides are discussed in fuller detail in Section 18.2 and metal–polysulfur compounds are covered in Section 19.2. Metal dioxygen species are covered in Section 16.9 with structural parameters in Table 16.5.

TABLE 17.14 Properties of the elements of the oxygen Group

Element	Symbol	Oxidation states	Coordination numbers	Availability
Oxygen	O	–II, (–I)	1, 2, (3), (4)	Common
Sulfur	S	–II, (II), IV, VI	2, 4, 6	Common
Selenium	Se	(–II), (II), IV, VI	2, 4, 6	Common
Tellurium	Te	II, IV	6	Common
Polonium	Po	II, IV		Very rare

Ozone has an angular structure with the OOO angle equal to 117°. The bonding in ozone has been described briefly previously in Section 4.10. Of the eighteen valency electrons, four are held in the σ bonds and eight in lone pairs on the two terminal oxygens. Two are present as a lone pair on the centre oxygen, leaving four electrons and the three p orbitals perpendicular to the plane of the molecule to form a π system. The three p orbitals combine to form a bonding, a nonbonding and an antibonding three-centred π orbital and the four electrons occupy the first two. There is thus one bonding π orbital over the three O atoms giving, together with the σ bonds, a bond order of about one and a half. The bond length is 128 pm which agrees with this; O-O for a single bond in H_2O_2 is 149 pm while O=O in O_2 is 121 pm. Adding one more electron to form the O_3^- ion starts to populate the antibonding π^* orbital, and we find the bond length increases to 133 pm with an angle of 114°.

Ozone occurs in the upper atmosphere, where it is formed photochemically from O_2, and has the important property of absorbing harmful middle- and far-UV radiation. The recent 'hole' in the ozone layer, caused by chlorofluorocarbons and related materials, is of significant current concern and this topic is discussed in greater detail in Chapter 20.

Ozone is a strong oxidizing agent, especially in acid solution where the potential is 2.07 V (Table 6.3). It is exceeded in oxidizing power only by fluorine, oxygen difluoride and some radicals.

Since the S-S single bond is strong, sulfur chains form readily, and polysulfur species are one of the major classes of catenated compounds. The S-S bond is also labile, so that a particular chain compound, such as S_6Cl_2, readily redistributes into an equilibrium mixture of different chain lengths, leading to difficulties in characterizing such compounds. The element sulfur itself occurs as chains or rings (which are simply closed chains), and its structural complexities are now understood.

Sulfur shares with phosphorus the ability to form a wide variety of allotropic forms in all three phases. However, many of these varieties of sulfur turn out to be mixtures of long chains and rings. The main interrelations are shown in Fig. 17.44.

Under normal conditions, the thermodynamically stable form of sulfur is the S_8 ring which has the crown structure shown in Fig. 17.45. If we consider a short chain of S atoms

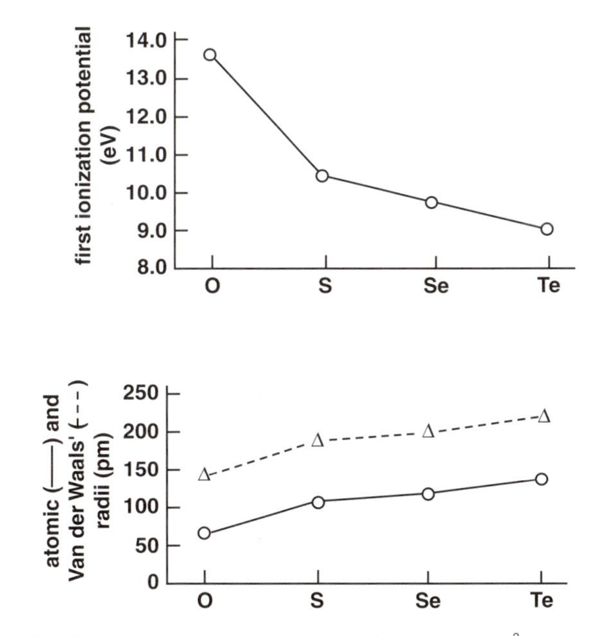

FIG. 17.42 Some properties of the oxygen Group elements. The radii of the anions, X^{2-}, are almost identical with the corresponding Van der Waals' radii

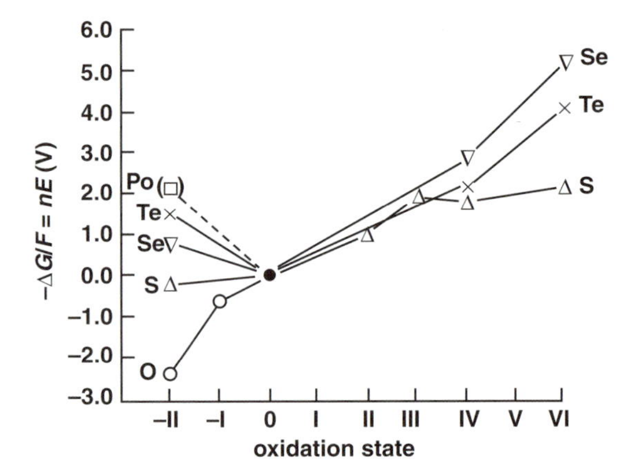

FIG. 17.43 Oxidation state free energies of the oxygen Group elements. Oxygen shows only negative oxidation states. The −II state becomes decreasingly stable from O to Po. The positive oxidation states show the drop in stability of the VI state after S and the tendency for intermediate states to be the more stable for Se and Te. Polonium values are not known

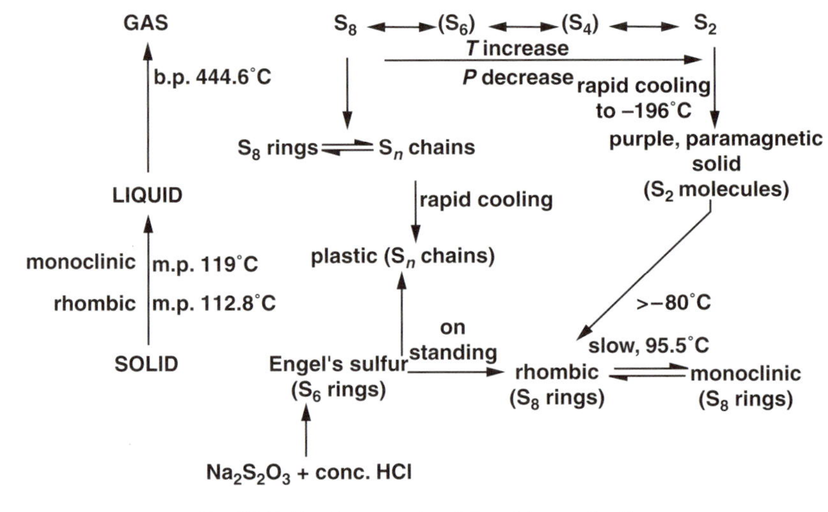

FIG. 17.44 The interconversion of the allotropes of sulfur

(say S-S-S-S-S), then the central S forms two bonds and has two lone pairs. The SSS angle is thus expected to be around 105° and the chain is a zigzag. Further, the arrangement will twist to move lone pairs as far apart as possible, so we find an optimum *dihedral angle* (angle between successive SSS planes), which minimizes lone pair repulsions, of around 85–100°. Putting these preferred angles together, we find that a long chain of S atoms will tend to coil up and 'bite back' on itself. With an arrangement of 8 S atoms, the resulting ring allows an optimum choice of angles.

Several of the allotropes (more specifically polymorphs) of sulfur contain the S_8 ring and differ in the ways in which the rings are packed in the solid. When sulfur is heated to about 160°C, there is a sudden large increase in viscosity and this is ascribed to the $S_8 \rightarrow S_\infty$ change from rings to long chains of S atoms. The S_8 ring is also found in the gas phase, together with smaller units.

A second, long-known, orange-red form of sulfur was first reported by Engel. It has S_6 rings, arranged in the chair form. This, and the many more recently discovered sulfur rings, is treated in Section 18.2.

A further interesting modification of sulfur is the S_2 unit which occurs in the gas at high temperatures. On rapid cooling it condenses to a purple solid, which is paramagnetic like the isoelectronic O_2.

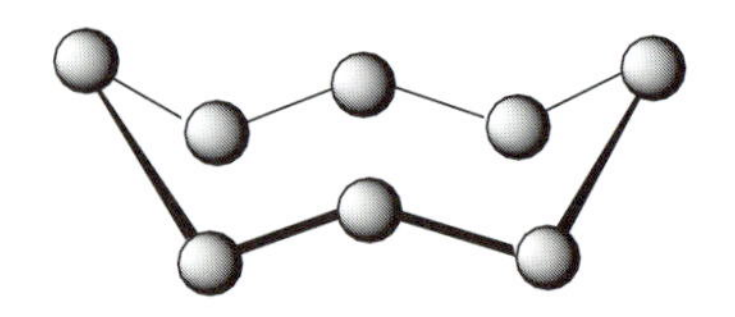

FIG. 17.45 The structure of the S_8 or Se_8 ring

Selenium also forms a number of allotropes. Two different red forms containing Se_8 rings are found but the stable modification is the grey form. This contains infinite spiral chains of selenium atoms with a weak, metallic interaction between atoms in adjoining chains. The chain form of sulfur also contains spiral chains but has no metallic character.

Tellurium has only one form in the solid. This is silvery-white, semi-metallic, and isomorphous with grey selenium, but with rather more metallic interaction as the self-conductivity is higher. In the vapour, selenium and tellurium have a greater tendency than sulfur to exist as diatomic and monatomic species.

All polonium isotopes have relatively short half-lives and this isotope, polonium 210, is the longest-lived with $t_{\frac{1}{2}} = 138$ days. The isotopes all have high activity and the handling problems are severe so that polonium chemistry is not fully explored.

Polonium is a true metal with two allotropic forms, in both of which the coordination number is six. Polonium is found in small amounts in thorium and uranium minerals (where it was discovered by Marie Curie) as one of the decay products of the parent elements, Th or U. It can now be more readily made by bombarding bismuth in a reactor:

$$^{209}_{83}Bi + ^{1}_{0}n \rightarrow ^{210}_{83}Bi \rightarrow ^{210}_{84}Po + ^{0}_{-1}e$$

The trends observed in the nitrogen Group appear in a more marked degree in the chemistry of the oxygen Group. The –II state is well established, not only for oxygen, but also for sulfur and even selenium, and it often occurs as the M^{2-} ion in the compounds of these elements. Apart from the Group state of VI, both the IV and the II states are observed in the Group, the IV state becoming the most stable one for tellurium, and the IV or II state the most stable one for polonium. Apart from the fluorine compounds, oxygen appears only in the –II state, except for the –I state in the peroxides. A variety of shapes is again observed. The VI state is usually octahedral or tetrahedral while the presence of the nonbonding pair in compounds of the IV state gives distorted tetrahedral shapes in the halides, MX_4, and pyramidal or V shapes in oxyhalides or oxides. Coordination numbers higher than six are uncommon but TeF_6 does add F^- to give TeF_7^-.

Oxygen has little in common with the other elements in the Group apart from formal resemblances in the –II states. Polonium shows signs of a distinctive chemistry, similar to that of lead or bismuth, but the difficulties of studying the element mean that there are many gaps in its known chemistry. Tellurium resembles antimony in showing a strong tendency to be 6-coordinate to oxygen, while sulfur and selenium show 4-coordination. Sulfur is stable in the VI state, while selenium(VI) is mildly oxidizing.

The tendency to catenation shown by the elements is continued in the compounds, especially those of sulfur. Oxygen forms O-O links in the peroxides and in O_2F_2, and three- and four-membered chains may be present in the unstable fluorides, O_3F_2 and O_4F_2. Sulfur forms many chain compounds, particularly the dichlorosulfanes, S_xCl_2, where x may be as high as 100, and the polythionates, $(O_3SS_xSO_3)^{2-}$ where the compounds with $x = 1$ to 12 have been isolated. Chain-forming tendencies are slight for selenium and absent in tellurium chemistry.

It is convenient to discuss oxygen chemistry separately from that of the other elements.

17.7.2 Oxygen

Oxygen combines with all elements except the lighter rare gases, and most of the oxides are listed in Tables 13.3 and 17.2. Their properties have been discussed already. The change in acidity from the s element oxides to the oxides of the light p elements will already be familiar to the reader. Oxygen shares with the other first row elements the ability to form p_π–p_π bonds to itself and to the other first row elements, and it is sufficiently reactive and electronegative to form p_π–d_π bonds with the heavier elements.

Dioxygen coordinates to a variety of transition metals in species which have been variously formulated as containing neutral O_2, superoxide O_2^- or peroxide O_2^{2-}. These are discussed in Section 16.9.

As fluorine is more electronegative than oxygen, compounds of the two are oxygen fluorides and are discussed here. The halogen oxides of Cl, Br and I are to be found in Section 17.8. Four oxygen fluorides are well established—OF_2, O_2F_2, O_3F_2 and O_4F_2—and two more have been reported relatively recently—O_5F_2 and O_6F_2. In all of these compounds oxygen is in a formally positive oxidation state and these are the only compounds where this occurs. OF_2 is formed by the action of fluorine on dilute sodium hydroxide solution, while the other five result from the reaction of an electric discharge on an O_2/F_2 mixture. An increasing proportion of O_2 and decreasing temperature are required to make the highest members of the series. Thus O_5F_2 and O_6F_2 were prepared in

OXYGEN AND ITS USES

Oxygen is prepared on a large scale by fractionation of liquid air, and had its first large-scale use in the Bessemer steel-making process where it is used to reduce the carbon content of iron. It has become more widely used in metallurgy to assist the combustion of heavy fuel oils, allowing replacement of coke. It is also used directly in various large-scale organic syntheses such as the formation of ethylene oxide or propylene oxide from the alkenes. Smaller-scale uses, often via peroxides, are in bleaching, biological and medical work, and sewage treatment.

a discharge at –213°C using a 5/1 mixture of O_2 and F_2. The use of higher proportions of oxygen does not give O_7 or higher species at this temperature but it might be possible to make these by working at a still lower temperature. All six compounds are very volatile with boiling points well below room temperature.

OF_2 is the most stable of the six oxygen fluorides. It does not react with H_2, CH_4 or CO on mixing, although such mixtures react explosively on sparking. Mixtures of OF_2 and halogens explode at room temperature. OF_2 reacts slowly with water:

$$OF_2 + H_2O \rightarrow O_2 + 2HF$$

and is readily hydrolysed by base:

$$OF_2 + 2OH^- \rightarrow O_2 + 2F^- + H_2O$$

The structure of OF_2 is V-shaped, like H_2O, with the FOF angle = 103.2° and the bond length, O-F = 141.8 pm.

The other five oxygen fluorides are much less stable. O_2F_2 decomposes at its boiling point of –57°C, O_3F_2 at –157°C, O_4F_2 at –170°C, and O_5F_2 and O_6F_2 are stable only to –200°C. Indeed, O_6F_2 can explode if it is warmed rapidly up to –185°C. All are red or red-brown in colour. Electron spin resonance studies on O_2F_2 and O_3F_2 have shown that these contain the OF radical to the extent of 0.1% for O_2F_2 and 5% for O_3F_2. It is likely that the higher members also contain free radicals, and these may account for the colour. O_2F_2 has a skew structure similar to that of hydrogen peroxide and the others may be chains, FO_nF.

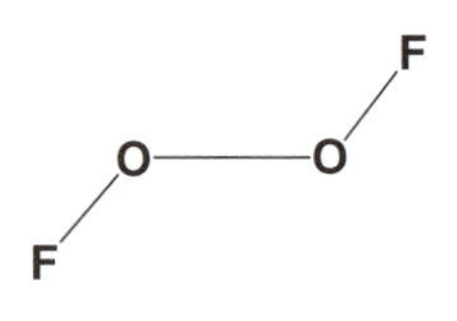

When O_2F_2 is reacted at low temperatures with molecules which will accept F^-, such as BF_3 or PF_5, oxygenyl compounds result:

$$O_2F_2 + BF_3 \rightarrow (O_2)BF_4 + \tfrac{1}{2}F_2(\text{at} - 126°\text{C})$$

Oxygenyl tetrafluoroborate decomposes at room temperature to BF_3, O_2 and fluorine. The oxygenyl group may be replaced by the nitronium ion, NO_2^+:

$$O_2BF_4 + NO_2 \rightarrow (NO_2)BF_4 + O_2$$

The oxygenyl ion may also be formed directly from gaseous oxygen by reaction with the strongly oxidizing platinum hexafluoride molecule:

$$O_2 + PtF_6 \rightarrow (O_2)^+(PtF_6)^-$$

Other $O_2^+(MF_6)^-$ species have M = P, As, Sb, Bi, Nb, Ru, Rh, Pd or Au (note the unusual Au(V) state). $O_2^+(M_2F_{11})^-$ for M = Sb, Bi, Nb and Ta are also known. Recently $O_2^+(GeF_5)^-$ was reported. In all these species the O-O stretching lies in the range 1825–1865 cm^{-1}, reflecting a stronger bond than in O_2.

Oxygen forms two compounds with hydrogen, water and hydrogen peroxide, H_2O_2. Water, and its solvent properties, is discussed in Chapter 6 and hydrogen-bonding in O-H systems is included in Section 9.7 and Figs. 9.1 and 9.2.

Pure H_2O_2 is a pale blue liquid, m.p. –0.89°C, b.p. 150.2°C, with a high dielectric constant and is similar to water in its properties as an ionizing solvent. It is, however, a strong oxidizing agent and readily decomposes in the presence of catalytic amounts of heavy metal ions:

$$H_2O_2 \rightarrow H_2O + \tfrac{1}{2}O_2$$

Accordingly, stabilizers, such as EDTA, which are good complexing agents for these metal ions, need to be added to prevent catalytic decomposition. Hydrogen peroxide has the skew structure shown in Fig. 9.7b.

Finally, it must be noted that the terms, peracid (better, peroxoacid) and peroxide, are properly applied only to compounds which contain the -O-O- group, which may be regarded

THE PREPARATION OF HYDROGEN PEROXIDE

Hydrogen peroxide may be prepared by acidifying an ionic peroxide solution (Section 10.2) or, on a large scale, by electrolytic oxidation of a sulfate system:

$$2HSO_4^- \rightarrow S_2O_8^{2-} + 2H^+ + 2e^-$$
$$S_2O_8^{2-} + 2H_3O^+ \rightarrow 2H_2SO_4 + H_2O_2$$

(The intermediate is called persulfate.) However, most hydrogen peroxide today is manufactured by the anthraquinone autoxidation process. In this a functionalized anthraquinone dissolved in an organic solvent is first reduced (using hydrogen and a catalyst, typically palladium) to give the corresponding hydroquinone which is then oxidized using air, giving hydrogen peroxide and regenerating the anthraquinone, shown in Fig. 17.46. The hydrogen peroxide is extracted with water in a counter-current process and may be concentrated by fractionation.

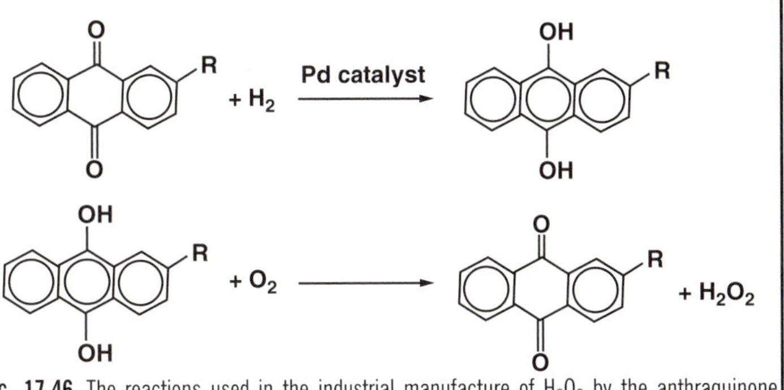

FIG. 17.46 The reactions used in the industrial manufacture of H_2O_2 by the anthraquinone process

PEROXOACIDS AND DERIVATIVES

There are a number of covalent peroxy species, acids or oxyanions, which may be regarded as derived from the normal oxygen compounds by replacing -O- by -O-O-; just as H-O-H is related to H-O-O-H. The best known examples are permonosulfuric acid (Caro's acid) H_2SO_5—which occurs as an intermediate in the persulfate oxidation above—perdisulfuric acid (persulfuric) $H_2S_2O_8$, perphosphoric acid, H_3PO_5, and perdicarbonic acid, $H_2C_2O_6$. The structures of the sulfuric acids have been definitely established and are related to the oxygen compounds as shown in Fig. 17.47. An X-ray structure determination has recently been carried out on explosive crystals of Caro's acid, confirming the expected tetrahedral geometry about sulfur and the presence of an S-O-O-H linkage. The molecules are linked by hydrogen bonding involving both the OH and the OOH groups—S=O···HOS and S=O···HOOS. The O-O bond length is 146.4 pm, compared with 145.8 pm in H_2O_2. The other acids are probably $(HO)_2(HOO)P{=}O$ and HO_2C-O-O-CO_2H, and the latter has been structurally characterized by an X-ray study as its cyclohexyl ester, CyO_2C-O-O-CO_2Cy. The structure of the peroxydiphosphate ion, $[P_2O_8]^{4-}$, is similar to that of peroxydisulfate but with the two PO_3 units lying *trans* across the planar P-O-O-P link. The overall structure is C_{2h}. The O-O length is 149 pm.

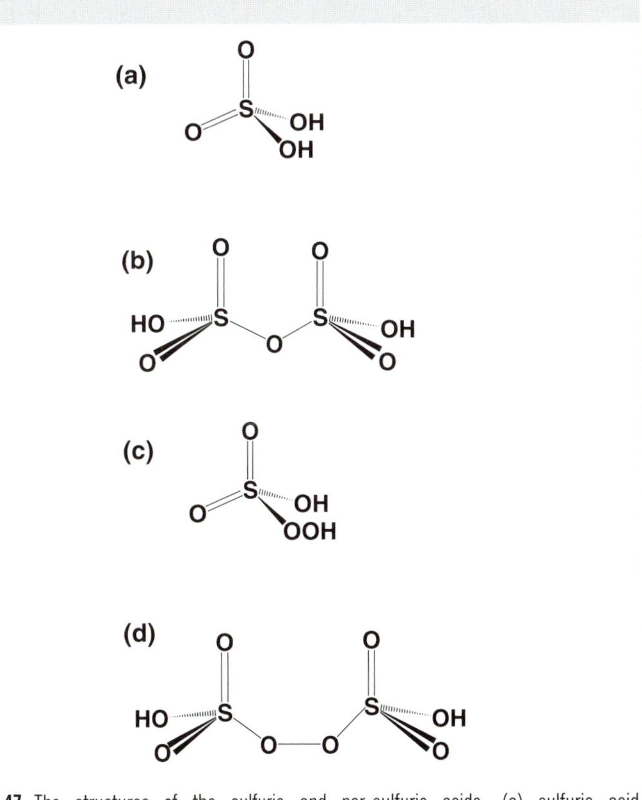

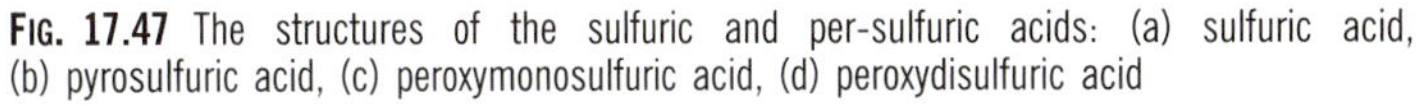
FIG. 17.47 The structures of the sulfuric and per-sulfuric acids: (a) sulfuric acid, (b) pyrosulfuric acid, (c) peroxymonosulfuric acid, (d) peroxydisulfuric acid

A number of other compounds are commonly termed peroxoacids (or salts thereof) but most are only simple acids with hydrogen peroxide of crystallization. For example, the commercial 'percarbonate' is, in fact, $Na_2CO_3.1.5H_2O_2$. While free percarbonic acid, H_2CO_4, is unknown, the monopotassium salt $K[H(O_2)CO_2].H_2O_2$ does contain the peroxocarbonate anion (plus a hydrogen peroxide of crystallization), where the OH group of hydrogen carbonate has been replaced by an OOH group. The peroxocarbonate ion is also known to exist in transition metal complexes such as:

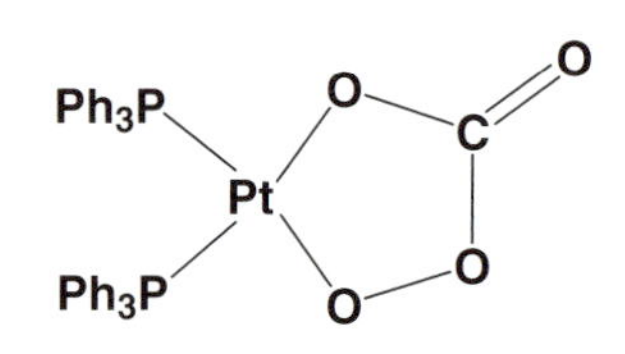

Sodium perborate has also been established as a true peroxy compound with B-O-O linkages, see Section 17.4.1.

Ionic peroxides are discussed in Chapter 10 and transition metal peroxy compounds in Chapter 16.

as derived from hydrogen peroxide. In the older literature, and in much technical literature, some higher oxides such as MnO_2 are termed peroxides. This usage is incorrect on the modern convention of nomenclature.

17.7.3 The other elements: the VI state

The trioxides, MO_3, of the VI state are formed by sulfur, selenium, tellurium and polonium, and all these elements (except possibly polonium) form oxyanions. The acids are included in Table 17.15. Sulfur and selenium are 4-coordinated to oxygen in the acid H_2MO_4 and in the anion $(MO_4)^{2-}$. In the Na_2SeO_4/Na_2O reaction system, new selenium oxyanions SeO_6^{6-} (with SeO_4^{2-} in $Na_6Se_2O_9$) and SeO_5^{4-} (in Na_4SeO_5) have been obtained. Tellurium forms the dibasic, 6-coordinated acid $Te(OH)_6$, and two series of 6-coordinated tellurates, $TeO(OH)_5^-$ and $TeO_2(OH)_4^{2-}$. In addition, the compound $Rb_6(Te_2O_9)$ has been shown to contain equal numbers of tetrahedral $[TeO_4]^{2-}$ and trigonal bipyramidal $[TeO_5]^{4-}$ ions. Sulfur, selenium and tellurium all burn in air to form the dioxide, MO_2, and the trioxides are made by oxidizing these. Sulfuric acid, H_2SO_4, is formed by dissolving SO_3 in water, but selenic acid, H_2SeO_4, is more readily made by oxidizing the selenium(IV) acid, H_2SeO_3, made by dissolving SeO_2. Sulfuric acid and the sulfates are stable, while selenic acid, the selenates, telluric acid and the tellurates are all oxidizing agents, although they are typically slow in reacting.

Only sulfur gives a condensed acid in the VI state, and it forms only the binuclear species, pyrosulfuric acid $H_2S_2O_7$. Condensed anions include $S_3O_{10}^{2-}$ and $S_5O_{16}^{2-}$ as well as $M_2O_7^{2-}$ ($M = S$, Se), but tellurium gives no polyanions at all. As condensation is also restricted in the IV state to binuclear compounds of sulfur and selenium, it will be seen that the tendency to condense is much slighter in this Group than it was in the nitrogen Group.

Oxidizing power of the VI state increases down the Group, and polonium(VI) is strongly oxidizing, so that the existence of the oxide and oxyanions is in some doubt.

OCCURRENCE AND USES OF S, Se AND Te

Sulfur occurs as the free element, especially in volcanic areas around hot springs, and was one of the elements known in ancient times. Its main current source is from the desulfurization of oils and natural gas, mining in volcanic deposits, as a byproduct of extracting sulfide ores, and by the Frasch process where it is recovered from underground deposits by hot water under pressure. Its major use is in sulfuric acid manufacture, with minor consumption in many industries, especially rubber manufacture. Sulfuric acid is made by burning S to SO_2, followed by catalytic oxidation to SO_3 and solution in sulfuric acid and dilution. More than half the sulfuric acid is used in fertilizer manufacture, with other uses spread over more than a hundred industries.

Selenium and tellurium are byproducts of copper extraction. Selenium is used in glass, as a decolorizer and to produce red and yellow colours. It is used in photoconductors and electronic devices, and is the photoreceptor which is basic to xerographic photocopying. Tellurium finds uses in metallurgy and alloys.

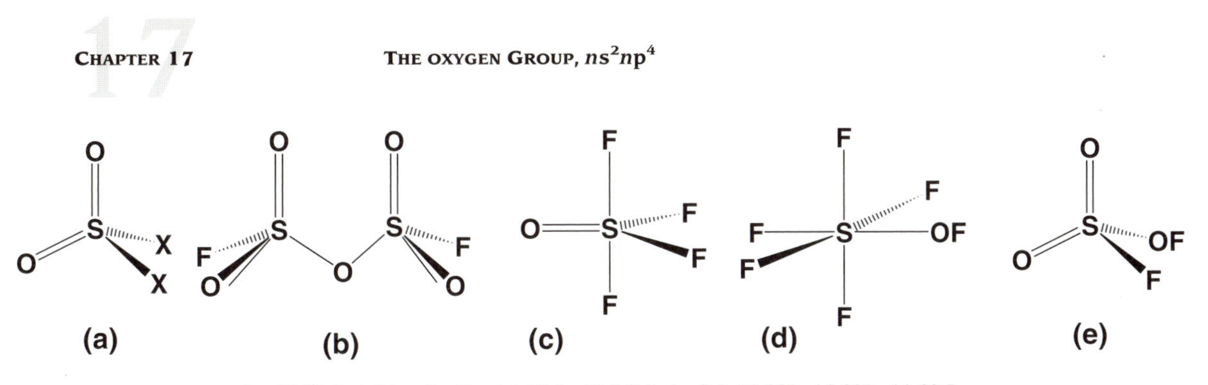

FIG. 17.48 Oxyhalides of sulfur; (a) SO_2X_2, (b) $S_2O_5F_2$ (*or* Cl_2), (c) SOF_4, (d) SOF_6, (e) SO_3F_2

The other main representatives of the VI state are the hexafluorides, MF_6. These compounds are all relatively stable but with reactivity increasing from sulfur to tellurium. SF_6 is extremely stable and inert, both thermally and chemically, and is used as an inert dielectric medium in a number of electrical devices. However, in the presence of oxygen an electrical discharge can generate species such as F_5S-O-SF_5 and the peroxy analogue F_5S-O-O-SF_5. SeF_6 is more reactive and TeF_6 is completely hydrolysed after 24 hours contact with water. The hexafluorides are more stable, and much less reactive as fluorinating agents, than the tetrafluorides of these elements. All the hexafluorides result from the direct reaction of the elements, and dimeric molecules S_2F_{10} or Se_2F_{10} are also found (but not Te_2F_{10}). These compounds are rather more reactive than the hexafluorides, but still reasonably stable. The coordination is octahedral, F_5M-MF_5. They hydrolyse readily, with fission of the M-M bond, and S_2F_{10} reacts with chlorine to give the mixed hexahalide, SF_5Cl. Addition of ClF to SeF_4 gives SeF_5Cl and tellurium forms similar species, TeF_5X for X = Cl, OF or OCl. Pseudohalide compounds such as SF_5CN are also known. Tracer studies reveal the existence of a volatile polonium fluoride, probably PoF_6.

TeF_6 adds F^- to form an 8-coordinated anion and also reacts with other Lewis bases, such as amines, to give 8-coordinated adducts like $(Me_3N)_2TeF_6$.

The VI state is also found in sulfur and selenium oxyhalides, and in halosulfonic acids. Sulfur gives sulfuryl halides, SO_2X_2, with fluorine and chlorine, and also mixed compounds SO_2FCl and SO_2FBr. Only one selenium analogue, SeO_2F_2, is found. If one OH group in sulfuric acid (Fig. 17.47a) is replaced by F, fluorosulfonic acid results. The anion has been characterized and shows the expected tetrahedral configuration around S with the S-F bond distance of 165 pm and the other three S-O distances equal to 143 pm: the SOF bond angle is 108°. Sulfur also gives a number of more complex oxyhalides, some of which are shown in Fig. 17.48. The oxyhalides are formed from halogen and dioxide, or from the halosulfonates. SO_2F_2 is inert chemically but the other compounds are much more reactive and hydrolyse readily.

Tellurium, in particular, forms a number of oxyfluorides of complex formula such as $Te_5O_4F_{22}$ or $Te_7O_6F_{30}$. Structural studies show that many of these are derived from TeF_6 by successive replacement of F by the $OTeF_5$ group. There is evidence for $F_4Te(OTeF_5)_2$, $F_2Te(OTeF_5)_4$, $FTe(OTeF_5)_5$ and $Te(OTeF_5)_6$ (e.g. Fig. 17.49a). The structures are octahedral, and *cis* and *trans* isomers of both the F_4TeO_2 and F_2TeO_4 species are known. Related species are $(TeF_5)_2O$ and $(TeOF_4)_2$ which is a pair of octahedra linked by two oxygen bridges (Fig. 17.49b, c). The selenium analogue is also found. (These structures contrast with the transition metal analogues, such as WOF_4, which are linked by fluorine bridges.)

The $OTeF_5$ and $OSeF_5$ groups are found to bond to other atoms. Thus we find $V(OTeF_5)_6$ with an octahedral VO_6 configuration. Similar species include $I(OSeF_5)_5$, $B(OTeF_5)_3$ and $Xe(OTeF_5)_n$ for $n = 2$, 4 and 6. There is a range of angles at O from XeOTe (125°), TeOTe (139°) to VOTe (170°) which suggests an increasing degree of M-O-Te π bonding. The simple species F_5MOMF_5 is found for M = S, Se and Te, each with similar MOM angles of 142–5°, and in each case the fluorines are eclipsed, again arguing for a p_π–d_π interaction of the oxygen lone pairs.

As this field develops, it is becoming apparent that the $OSeF_5$ and $OTeF_5$ groups are behaving very like F, as ligands of high electronegativity. Many species occur in which

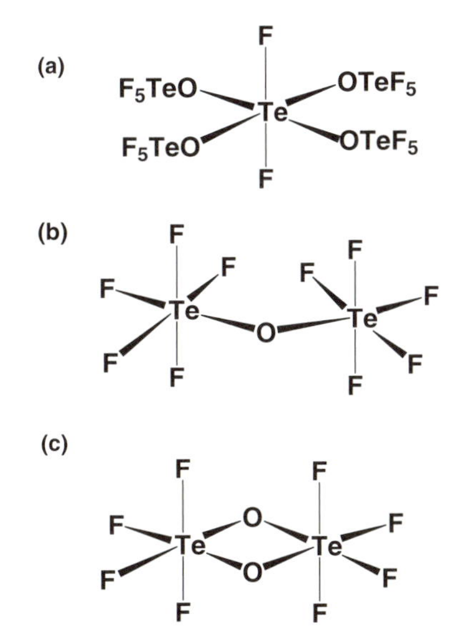

FIG. 17.49 (a) $F_2Te(OTeF_5)_4$, (b) $(F_5Te)_2O$, (c) $(F_4Te)_2(O)_2$

stepwise replacement of F is found, as above and also in oxyfluorides like $OXe(OTeF_5)_4$ (cf. $OXeF_4$), as ionic species, e.g. $Na^+(OSeF_5)^-$, and halo-species as $ClW(OTeF_5)_5$.

Three halosulfonic acids are known, FSO_3H, $ClSO_3H$, and $BrSO_3H$. The structures are tetrahedral and are formed from sulfuric acid by replacing an OH group by a halogen atom. They are strong monobasic acids but only fluorosulfonic acid is stable and forms stable salts, fluorosulfonates FSO_3^-, which are similar in structure and solubilities to the perchlorates.

Hexavalent aryl tellurium compounds are also known, such as Ph_6Te, Ph_5TeCl and Ph_5Te^+. The latter has a square pyramidal geometry, analogous to the structure of the isoelectronic Ph_5Sb.

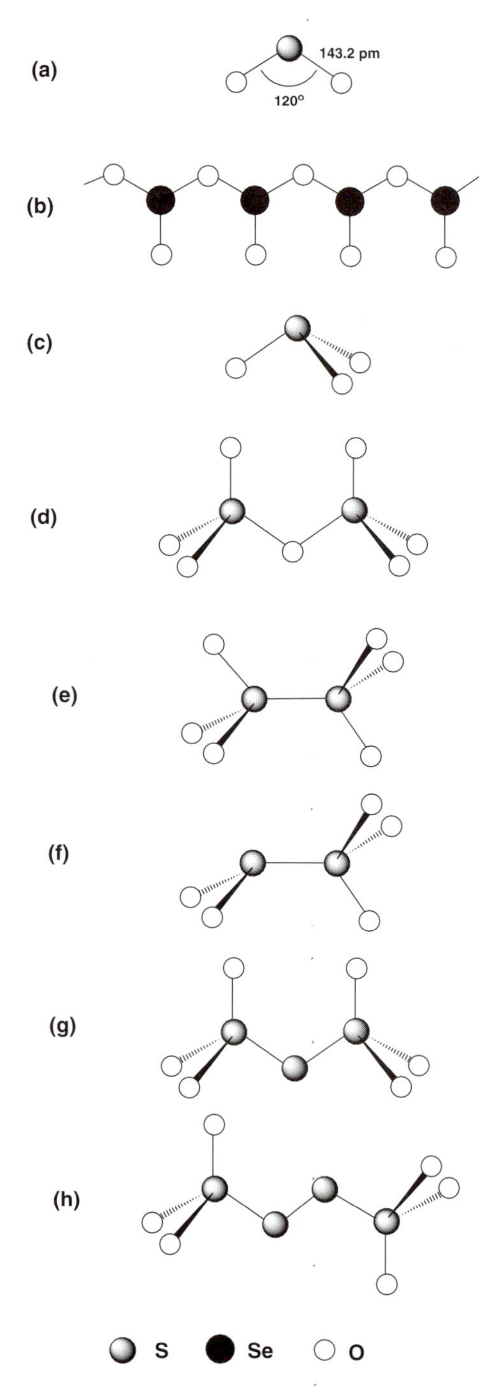

FIG. 17.50 Oxygen compounds of sulfur and selenium: (a) SO_2, (b) SeO_2, (c) S(or Se)O_3^{2-} (sulfite, selenite), (d) $S_2O_7^{2-}$ (pyrosulfate), (e) $S_2O_6^{2-}$ (dithionate), (f) $S_2O_5^{2-}$ (pyrosulfite), (g) $S_3O_6^{2-}$ (trithionate), (h) $S_4O_6^{2-}$ (tetrathionate; related $S_nO_6^{2-}$ have different numbers of sulfur atoms in a chain between the two terminal SO_3^{2-} groups)

17.7.4 The IV state

The IV state is represented by oxides, MO_2, halides, MX_4, oxyhalides, MOX_2, acids and anions, for all four elements. Sulfur(IV) is reducing, selenium(IV) is mildly reducing (going to Se(VI)) and also weakly oxidizing (going to the element), tellurium(IV) is the most stable state of tellurium, while polonium(IV) is weakly oxidizing (going to Po(II)). In addition, a number of complexes are known.

The oxides are given in Table 17.2a and the acids and anions in Table 17.15. The dioxides result from direct reaction of the elements. SO_2 and SeO_2 dissolve in water to give the acids, but TeO_2 and PoO_2 are insoluble and the parent acids do not exist. Tellurium and selenium oxyanions result from the solution of the oxides in bases. Salts of one condensed acid exist: these are the pyrosulfites, $S_2O_5^{2-}$, which have the unsymmetrical structure with a S-S bond, $^-O_3SSO_2^-$.

Figure 17.50 gives the structures of some of the oxides and oxyions. The IV state oxides and oxyanions have an unshared electron pair (in monomeric structures) and thus have unsymmetrical structures. The stable form of the dioxide shows an interesting transition from monomeric covalent molecule, SO_2, through polymeric covalent, SeO_2, to ionic forms for TeO_2 and PoO_2. Sulfur, selenium and tellurium dioxides are acidic and dissolve in bases. PoO_2 appears to be more basic than acidic and dissolves in acids as well as forming polonites with strong bases.

One compound intermediate between the IV and VI oxides is reported. This is Se_2O_5, formed by controlled heating of SeO_3. This compound is conducting in the fused state and it has been suggested that it is $SeO^{2+}SeO_4^{2-}$, a salt of Se(IV) and Se(VI).

The known tetrahalides are listed in Table 17.2b. The missing ones are SBr_4, SI_4 and SeI_4 while SCl_4 and $SeBr_4$ are also unstable, SCl_4 decomposing at −3°C. The tellurium tetrahalides are markedly more stable, even TeI_4 being stable up to 100°C. $TeCl_4$, $TeBr_4$ and $SeCl_4$ are all stable up to 200°C and the tellurium compounds to 400°C. All the tetrafluorides are known and these are rather more stable than the other tetrahalides to thermal decomposition. They are much more reactive than the hexafluorides and act as strong, though selective, fluorinating agents. All four polonium tetrahalides are known and resemble the tellurium analogues, though they seem to be rather less stable. The structures of most tetrahalides are known. SF_4, SeF_4, $TeBr_4$ and $TeCl_4$, are all found as the distorted tetrahedron derived from the trigonal bipyramid with one equatorial position occupied by an unshared pair of electrons (see Section 4.2). Tellurium tetrafluoride forms a polymeric structure in which square pyramids are linked into chains by sharing corners, $TeF_3(F_{2/2})$. TeI_4 is a tetramer consisting of four TeI_6 octahedra sharing edges. Thus the lone pair is still sterically active. SCl_4 and $SeCl_4$ exist only in the solid and vaporize as $MCl_2 + Cl_2$ — the sulfur compound at −30°C and the selenium one at 196°C. Raman spectra suggest the solid is $MCl_3^+\ Cl^-$ and this is supported by crystallographic evidence for EBr_3^+ and EI_3^+ (E = S, Se) in addition to more complex cations like $Se_2I_4^{2+}$.

The selenium and tellurium, and probably the polonium, tetrahalides readily add one or two halide ions to give complex anions such as SeF_5^- and MX_6^{2-} (M = Se, Te, Po; X = any halogen). These hexahalo(IV) complex ions are interesting from the structural point of view as they contain seven electron pairs. Structural determinations on a number of compounds show that the MX_6^{2-} group is octahedral, but the group may be regular or distorted. The electron pair repulsion theory, discussed in Chapter 4, would imply that the nonbonded pair occupied a spatial position and should lead to a distorted octahedron. It is possible,

SO_2 AS A LIGAND

Sulfur dioxide is a very versatile ligand, able to coordinate to metal centres in many different ways, either through oxygen or sulfur atoms, π-bonded, or as a bridging ligand either through sulfur, or sulfur and oxygen. It is also chemically reactive, and can act either as an oxidant, reductant, or small reactive molecule which can undergo insertion reactions e.g. into metal–carbon bonds. As an example of a homoleptic SO_2 complex, containing only SO_2 ligands, oxidation of Ni with AsF_5 in liquid SO_2 gives $[Ni(SO_2)_6]^{2+}$, with 6 O-bonded SO_2 ligands arranged octahedrally around the metal centre.

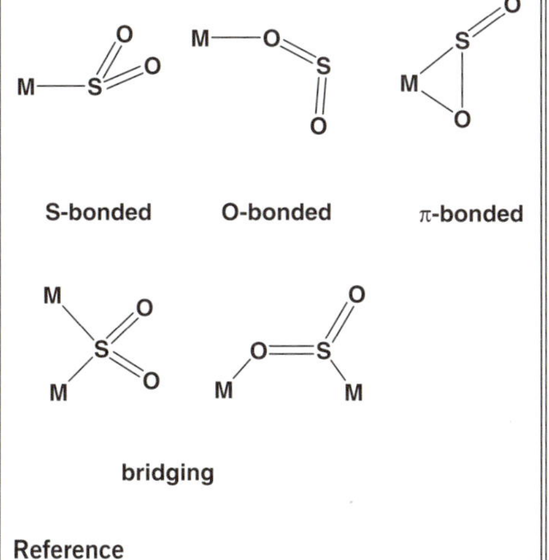

Reference

G.J. KUBAS, *Accounts of Chemical Research* **27**, 1994, 183.

however, that with large, heavy, central atoms, the nonbonded pair might be accommodated in an inner orbital rather than in a particular spatial direction. Structural studies so far reported indicate that two groups of structures are found. In the hexafluorides of the IV state, MF_6^{2-}, as in the isoelectronic IF_6^- and XeF_6, the structure is distorted octahedral, indicating a sterically active lone pair. However, in the heavier halides like MCl_6^{2-}, the structure is regular octahedral. It is proposed here that the dominant factor is repulsion between halogen atoms which is minimized in the regular structure. In this form, the lone pair would have to occupy an s orbital presumably.

The structures of pentafluoro anions, $[EF_5]^-$ for E = S, Se, Te and various cations, are all square pyramids with the lone pair to complete the octahedron. The TeF_5^- units link into chains by weak Te-F···Te bonding while the Se analogue shows weakly bonded tetramers. As in TeF_4, the E atoms lie below the base of the square pyramid, that is, the base F atoms are bent up away from the lone pair position. It will be recalled from Section 13.6 that the opposite distortion occurs when there are five ligands and no lone pair. Selenium forms an interesting pair of dinuclear complexes in the IV state. $Se_2Cl_{10}^{2-}$ has two octahedra sharing an edge, while $Se_2Cl_9^-$ has two octahedra sharing a face (alternatively seen as two Se atoms linked respectively by 2 or 3 bridging Cl).

Oxyhalides of the IV state are formed only by sulfur and selenium and have the formula MOX_2. Thionyl halides, SOF_2, $SOCl_2$, $SOBr_2$ and SOFCl are known while the selenyl fluoride, chloride and bromide exist. Thionyl chloride is made from SO_2 and phosphorus pentachloride:

$$SO_2 + PCl_5 \rightarrow SOCl_2 + POCl_3$$

The other thionyl halides are derived from the chloride. Selenyl chloride is obtained from SeO_2 and $SeCl_4$. These oxyhalides are stable near room temperature but decompose on heating to a mixture of dioxide, halogen and lower halides. SOF_2 is relatively stable to water, but all the other compounds hydrolyse violently. These compounds have an unshared electron pair on the central element and are pyramidal in structure (Fig. 17.51). In $SeOF_2$, FSeF = 92.6° and OSeF = 105.5°. The isoelectronic nitrogen analogue of $SOCl_2$, i.e., $NSCl_2^-$, can be made and has the expected pyramidal shape.

The oxyhalides act as weak donor molecules through the lone pairs on the oxygen atoms, and also as acceptors using d orbitals on the sulfur or selenium. $SeOX_3^-$ complex ions are known for X = F, Cl and Br, and SOF_3^- is also known and has been predicted to have the expected structure based on a pseudo-trigonal bipyramid, with the sulfur lone pair and oxygen atom in the equatorial plane. Ions MO_2F^- are known for S, Se and Te, and there are structural differences between the chalcogens. FSO_2^- is a discrete ionic species, $FSeO_2^-$ shows intermolecular bonding through bridging fluorides, while TeO_2F^- (formed from KF and TeO_2) contains the trimeric anion $Te_3O_6F_3^{3-}$. This contains a puckered six-membered Te_3O_3 ring, and stereochemically active tellurium lone pairs, giving pseudo-trigonal bipyramidal coordination at Te. For selenium and tellurium the anions $MO_2F_2^{2-}$ are known, together with $TeOF_4^{2-}$. In $SeOCl_2$.2py (py = pyridine) the structure is a square pyramid with a sterically active lone pair.

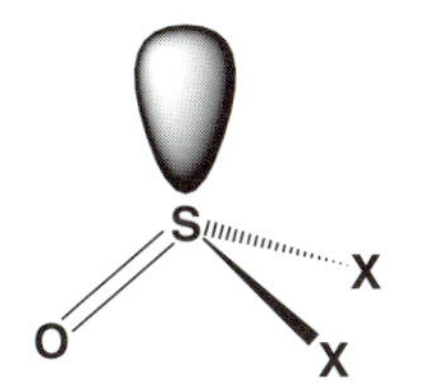

FIG. 17.51 The structure of thionyl halides, SOX_2

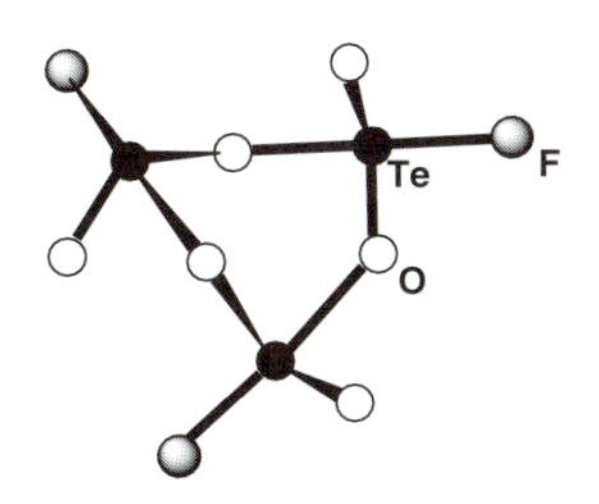

Stucture of the $[Te_3O_6F_3]^{3-}$ ion

17.7.5 The II state: the −II state

If poly-sulfur compounds are excluded, the II state is represented by a more limited range of compounds than the IV or VI states, but it is more fully developed than the I state of the nitrogen Group. Tellurium and polonium form readily oxidized monoxides, MO, and a number of dihalides, MX_2, exist. No difluorides occur (apart from a possible SF_2) but the general pattern of stability of the other dihalides resembles that of the tetrahalides, stability decreasing from tellurium to sulfur and from chloride to iodide. $TeCl_2$, $TeBr_2$, $PoCl_2$ and $PoBr_2$ are stable. $SeCl_2$ has only recently been obtained in a pure form (as a red oil) by the reaction of Se with SO_2Cl_2. $SeCl_2$ is stable for 24 hours in some solutions, but it decomposes readily in the vapour phase, as does $SeBr_2$. SCl_2, the only sulfur dihalide, decomposes at 60°C and is unstable with respect to dissociation to $S_2Cl_2 + Cl_2$. No di-iodide is known although there is a lower iodide of tellurium, $(TeI)_n$. The lower iodide of sulfur has been shown to be S_2I_2: there is no evidence for SI_2.

TABLE 17.15 Oxyacids and oxyanions of sulfur selenium and tellurium

Acid	Anion	Comments
Sulfur		
Sulfur shows coordination numbers up to four, the most common being tetrahedral; S-S bonds are common among the lower acids.		
H_2SO_4 sulfuric acid	HSO_4^- and SO_4^{2-} sulfate	Stable, strong, dibasic acid: formed from SO_3 and water. Structure, Fig. 17.47.
$H_2S_2O_7$ pyrosulfuric acid	$HS_2O_7^-$ and $S_2O_7^{2-}$ pyrosulfate	Strong, dibasic acid: formed from SO_3 and H_2SO_4 (anions by heating HSO_4^-), loses SO_3 on heating. Sulfonating agent. Structure, Fig. 17.47. Also anions $S_3O_{10}^{2-}$ and $S_5O_{16}^{2-}$.
H_2SO_3 sulfurous acid	HSO_3^- and SO_3^{2-} sulfites	Existence of free acid doubtful. SO_2+H_2O gives a solution containing the anions but this loses SO_2 on dehydration. Reducing and weak, dibasic acid. Structure, pyramidal SO_3^{2-} ion; lone pair on S.
	$S_2O_5^{2-}$ pyrosulfite	No free acid. Formed by heating HSO_3^- or by SO_2+SO_3^{2-}.
Polythionic acids		
$H_2S_2O_6$ dithionic acid	$S_2O_6^{2-}$ dithionates	Acid stable only in dilute solution, anions stable. Formed by oxidation of sulfites and stable to further oxidation, or to reduction. Strong, dibasic acid, Structure, Fig. 17.50.
	$S_nO_6^{2-}$ ($n=3$ to 6) polythionates	Free acids cannot be isolated; anions formed by reaction of SO_2 and H_2S or arsenite. Unstable and readily lose sulfur, reducing Structures contain chains of S atoms, Fig. 17.50.
$H_2S_2O_4$ dithionous acid (or hyposulfurous)	$S_2O_4^{2-}$ dithionite	Acid prepared by zinc reduction of sulfurous acid solution, and salts (called also hyposulfites or hydrosulfites) prepared by zinc reduction of sulfites. Unstable in acid solution, but salts are stable in solid or alkaline media, powerful reducing agents. Decompose to sulfite and thiosulfate. Structure, Fig. 17.52 contains S-S link.
	$S_2O_3^{2-}$ thiosulfate	Prepared in alkaline media by action of S with sulfites. Perfectly stable in absence of acid, but gives sulfur in acid media. Mild reducing agent, as in action with I_2 which gives tetrathionate, $S_4O_6^{2-}$. Structure, p.483, derived from sulfate.
	SO_2^{2-} sulfoxylate	Best known as the cobalt salt from $CoS_2O_4+NH_3 \rightarrow CoSO_2+(NH_4)_2SO_3$. The zinc salt may also exist. Unstable and reducing, structure probably V-shaped.
The peroxy acids H_2SO_5 and $H_2S_2O_8$, corresponding to sulfuric acid and pyrosulfuric acid, also exist. Structures are shown in Fig. 17.47.		
Selenium		
Selenium commonly shows a coordination number of four: a smaller number of selenium acids than sulfuric acids is found as the Se-Se bond is weaker.		
H_2SeO_4 selenic acid	$HSeO_4^-$ and SeO_4^{2-} selenates	Formed by oxidation of selenites. Strong dibasic acid, oxidizing. Similar structures to sulfur compounds.
	$Se_2O_7^{2-}$ pyroselenate	No acid, formed by heating selenates.
H_2SeO_3 selenous acid	$HSeO_3^-$ and SeO_3^{2-} selenite	Selenium dioxide solution. Similar to S species, but less reducing and more oxidizing. Structure contains pyramidal SeO_3^{2-} ion.
Chain anions of selenium have not been found, but selenium (and tellurium) may form part of the polythionate chain, as in $SeS_4O_6^{2-}$ and $TeS_4O_6^{2-}$, where the Se or Te atoms occupy the central position in the chain		
Tellurium		
Tellurium, like the preceding elements in this period, is 6-coordinate to oxygen.		
H_6TeO_6 telluric acid	$TeO(OH)_5^-$ and $TeO_2(OH)_4^{2-}$ tellurates	Prepared by strong oxidation of Te or TeO_2. Structure is octahedral $Te(OH)_6$, and only two of the protons are sufficiently acidic to be ionized, and then only weakly. The acid and salts are strong oxidizing agents.
	tellurites	TeO_2 is insoluble and no acid of the IV state is formed. Tellurites, and polytellurites, are formed by fusing TeO_2 with metal oxides.

Different forms of S_2F_2

trans **chain form**

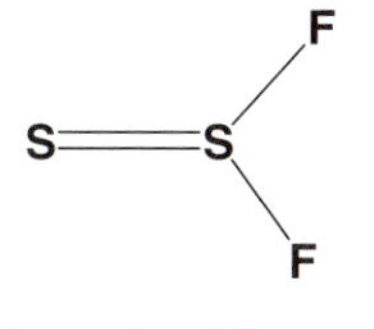

branched form

After considerable confusion, the properties of the fluorides of low oxidation state sulfur are now fairly clear. SF_2, S_2F_4 and two isomers of S_2F_2 exist. SF_2 readily disproportionates into SF_4 and S_8 (probably driven by the high S-S bond energy). It is made from SCl_2 plus KF or HgF_2 under careful conditions: this system also readily yields S_2F_2. SF_2 is a V-shaped molecule with bond angle 98°. It forms the unsymmetrical dimer F_3S-SF.

S_2F_2 exists in the *trans* chain and in the branched form. In the FSSF form, SS is 189 pm, similar to the distances in S_2 and shorter than HSSH or ClSSCl. In the SSF_2 form, SS is reduced further to 186 pm, the FSS angle is 108° and the FSF angle is 93°, arguing some increased electron density in SS.

These species are stable at low temperatures in an inert environment. The FSSF form readily converts to the more stable SSF_2, which can be heated above 200°C before decomposition. However, decomposition is rapid in presence of species like HF.

The complex, $SeBr_2L$, where L is tetramethylthiourea, is an example of the rare T-coordination. The BrSeBr angle is 175° and the S atom of the ligand completes the T. A number of planar halide anions, such as $SeBr_4^{2-}$ and $Se_2Br_6^{2-}$, are known as are mixed oxidation state Se(II)-Se(IV) anions, such as $Se_2Br_8^{2-}$ and $Se_3Br_{10}^{2-}$.

The –II state is found in the hydrides, H_2M, and in the anions, M^{2-}. With the more active metals, sulfur, selenium and tellurium all form compounds which are largely ionic, although they have increasing metallic, and alloy-like, properties as the electronegativity of the metal increases. The tendency to form anions increases from the nitrogen Group (where the evidence for M^{3-} ions, apart from N^{3-}, is limited), through the oxygen Group (where the M^{2-} ion is more widely found), to the halogens where the X^- ion is the most stable form for all the elements. This trend in anionic behaviour reflects the increasing electronegativity of the elements, and the greater ease of forming singly charged species over doubly or triply charged ones.

17.7.6 Compounds with an S-S or Se-Se bond

Many compounds with S-S bonds are known, together with a smaller range of Se-Se species and an even smaller range of Te-Te species. In this section we cover those compounds with one such bond, together with all the oxygen polysulfur compounds. The remaining catenated species with three or more bonded S or Se are discussed in Section 18.2.

Disulfides, S_2^{2-}, are formed by reacting sulfides with S and addition of acids yields H_2S_2. This has a skew structure similar to H_2O_2 but with angles at S of 92°. Reacting Cl_2 or Br_2 with molten sulfur gives S_2Cl_2 or S_2Br_2 which are more stable than the corresponding SX_2 and also have the skew X-S-S-X structure with SSX angles about 103°. Selenium analogues, Se_2X_2 (X = F, Cl, Br), have similar structures. Photolysis of the FE=EF species gives the isomeric form F_2E-E (E = O, S or Se). H_2Te_2 has recently been found to be stable in the gas phase.

Among the oxides, only S_2O contains a S-S bond. This compound was, for a long time, reported as SO but the most recent work has established the existence of S_2O and suggests that SO is a mixture of S_2O and SO_2. S_2O is formed by the action of an electric discharge on sulfur dioxide and it is unstable at room temperature. The structure SSO is proposed.

A variety of oxyacids and oxyanions with S-S bonds exists, among which are the polythionic acids, $H_2(O_3S\text{-}S_n\text{-}SO_3)$ where n varies from 0 to 12, and a miscellaneous group of compounds including thiosulfate, dithionite and pyrosulfite. All these compounds are included in Table 17.15. In sulfur-metabolizing organisms, it would be expected that species representing partial oxidation of sulfur would occur on the path to sulfite and sulfate. It is intriguing to find that chromatographic investigations show the presence of all the polythionates from $n = 1$ to 20 in cultures of *Thiobacillus ferrooxidans*.

In the polythionic acids, there is a marked difference in stability between dithionic acid, $H_2S_2O_6$, and the higher members. As dithionic acid contains no sulfur atom which is bonded only to sulfur, it is much more stable than the other polythionic acids which do contain such sulfurs, see Table 17.15. Dithionates result from the oxidation of sulfites:

$$2SO_3^{2-} + MnO_2 + 4H_3O^+ \rightarrow Mn^{2+} + S_2O_6^{2-} + 6H_2O$$

and the parent acid may be recovered on acidification. Dithionic acid and the dithionates are moderately stable, and the acid is a strong acid. The structure, $O_3S\text{-}SO_3$, has an approximately tetrahedral arrangement at each sulfur, with π bonding between sulfur and the oxygen atoms.

The reaction of H_2S and SO_2 gives a mixture of the polythionates from $S_3O_6^{2-}$ to $S_{14}O_6^{2-}$, while specific preparations for each member also exist, as in the preparation of tetrathionate in volumetric analysis by oxidation of thiosulfate by iodine:

$$2S_2O_3^{2-} + I_2 \rightarrow S_4O_6^{2-} + 2I^-$$

These compounds have very unstable parent acids, which readily decompose to sulfur and sulfur dioxide, but the anions are somewhat more stable. The structures are all established as $O_3S\text{-}S_n\text{-}SO_3$ with sulfur chains which resemble those in sulfur polyanions. As with the other examples of sulfur chain compounds, each polythionic acid readily disproportionates to an equilibrium mixture of all the others.

Thiosulfate, $S_2O_3^{2-}$ is formed by the reaction of sulfite with sulfur. The free acid is unstable but the alkali metal salts are well known in photography, volumetric analysis, in other applications such as paper making and as an antidote for cyanide poisoning, since thiosulfate can convert cyanide to nontoxic products via the reaction:

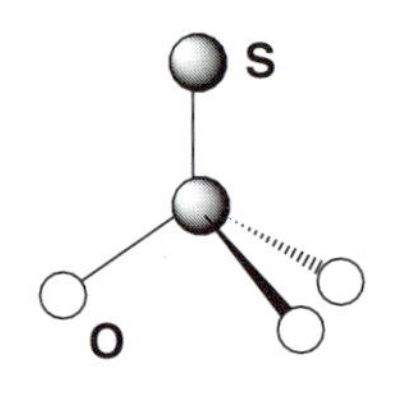

The structure of the thiosulfate anion, $S_2O_3^{2-}$

$$CN^- + S_2O_3^{2-} \rightarrow SCN^- + SO_3^{2-}$$

The thiosulfate ion has been found to have the expected tetrahedral geometry, analogous to sulfate, but with one oxygen replaced by a sulfur atom, and with sulfur–sulfur and sulfur–oxygen bond distances of 201 and 147 pm respectively. Exchange studies with radioactive sulfur have also demonstrated the presence of two types of sulfur atom. The structure of the ion with one added H has been determined by Raman spectroscopy supported by calculations. In $[HS_2O_3]^-$ the H is attached to the S, giving the linkage $[H\text{-}S\text{-}SO_3]^-$ showing that the formal S=S bond in $[S_2O_3]^{2-}$ is easily attacked.

Dithionite ions, $S_2O_4^{2-}$, result from the reduction of sulfite with zinc dust. The free acid is unknown, and the salts are used in alkaline solution as reducing agents. Dithionite (also called hyposulfite or hydrosulfite) decomposes readily:

$$2S_2O_4^{2-} + H_2O \rightarrow S_2O_3^{2-} + 2HSO_3^-$$

and the solutions are also oxidized readily in air. The dithionite ion has the unusual structure shown in Fig. 17.52. The oxygen atoms are in the eclipsed configuration and the S-S bond is very long. The S-O bond lengths show that π bonding exists between the sulfur and oxygen atoms, and there is also a lone pair on each sulfur atom. The S atoms must thus make use of d orbitals and it is proposed that the unusual configuration and the long S-S distance arise from the use of a d orbital.

17.7.7 Sulfur–nitrogen ring compounds

Sulfur and nitrogen form a number of ring compounds, of which the best known is tetrasulfur tetranitride, S_4N_4. This is best formed by reacting a 1:1 mixture of SCl_2 and SCl_4 with ammonia. It is a yellow-orange solid which is not oxidized in air, although it can be detonated by shock. The structure is shown in Fig. 17.53. The four nitrogen atoms lie in a plane, and the four atoms of sulfur form a flattened tetrahedron, interpenetrating the N_4 square. Alternatively, the structure may be regarded as an S_8 ring with every second sulfur replaced by a nitrogen. S-S distances are fairly short, 259 pm, showing that there is some weak bonding across the puckered ring between opposite sulfur atoms. The selenium analogue, Se_4N_4, has a similar structure with Se-N = 178 pm and the long Se···Se distance = 274 pm. In the $S_4N_4^{2+}$ cation, the loss of electrons removes the cross-ring bonds. The cation is found in two forms, planar and boat-shaped, where all the S-N bond lengths are equal and the charge is delocalized. Similar planar $S_4N_3^+$ and $S_5N_5^+$ cations with equal bonds are known.

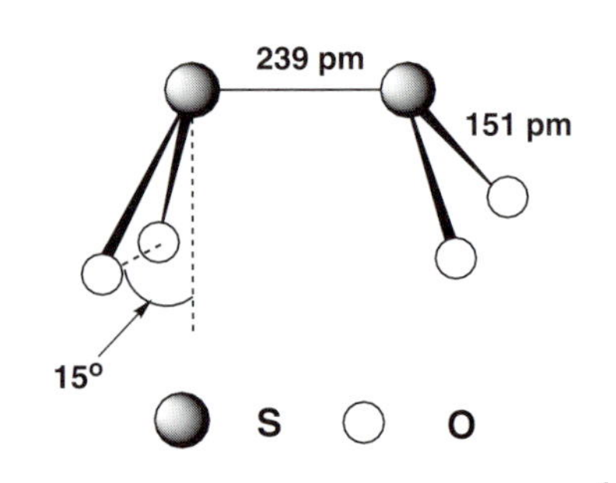

FIG. 17.52 The structure of the dithionite ion, $S_2O_4^{2-}$

A number of other ring compounds exist, including $S_4N_5^-$ where the additional N bridges the two S atoms at the top of the molecule, $F_4S_4N_4$ where the substituent F atoms are on

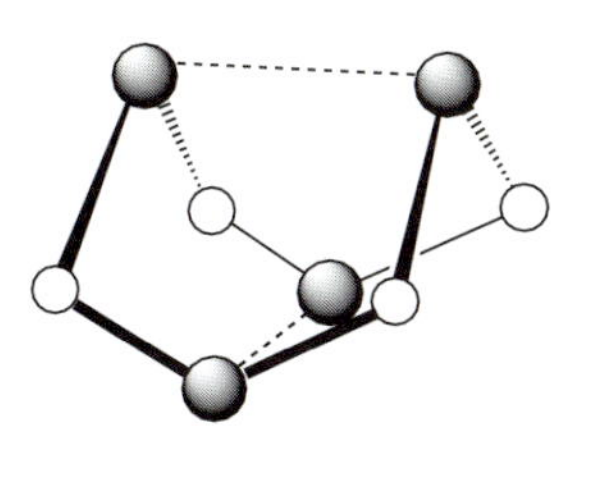

FIG. 17.53 The structure of tetrasulfur tetranitride

the sulfur atoms of the S_4N_4 ring, and $S_4N_4H_4$ with the H atoms on the nitrogens. A different eight-membered ring is found in S_7NH, again with H bonded to the nitrogen. This hydrogen is fairly acidic and can be replaced by a number of metals. Different ring sizes are also found, as in $S_3N_3Cl_3$ and S_4N_3Cl. In S_4N_2 the structure is a 'half-chair' where there is a planar S-N-S-N-S section with the last S linking the two ends and lying above the plane.

Related to these species is polymeric $(SN)_x$ which has metallic conducting properties, and is a semiconductor at 0.26 K. This compound oxidizes too readily for large-scale use, but it has directed interest towards other 'covalent metals', and it is possible that polymers including S-F or S-O units to protect the SN backbone may be more resistant to oxygen.

17.8 The fluorine Group, ns^2np^5 (the halogens)

17.8.1 General properties

Table 17.16 lists some of the properties of the elements, Fig. 17.54 shows the variations with Group position of a number of important parameters, and Fig. 17.55 gives the oxidation state free energies.

Apart from astatine, the elements of the Group are all well known and occur mainly as salts containing the halide anion. The free elements are too reactive to exist in nature. Free fluorine, F_2, is outstandingly reactive and can be handled only in extremely dry glass systems (otherwise it reacts with glass to give SiF_4), in Teflon, or in apparatus made of metals which form a protective layer of fluoride on the surface, such as copper, nickel or steel. Provided these conditions are rigorously adhered to, fluorine can be handled quite readily in the laboratory.

F_2 is made by the electrolysis of KF in anhydrous HF. Because of its extreme reactivity, it is almost always used *in situ*, either to make directly products such as UF_6 used in atomic fuel manufacture or volatile element fluorides such as WF_6 used to deposit metal coatings, or to produce more controllable fluorine compounds such as ClF_3 or BrF_5 which in turn are used to form fluorides. Of fluorine compounds, the widest use is of the very inert

OTHER REFERENCES TO THE PROPERTIES OF THE HALOGENS

Reference to properties of the halogens are included in the following tables and figures:

Ionization potentials	Table 2.8
Electron affinities	Table 2.9
Atomic properties and electron configurations	Table 2.5
Radii	Tables 2.10, 2.11, 2.12
Electronegativities	Table 2.14
Redox potentials	Table 6.3
Uses as strongly acidic media	Sections 6.9 and 18.5
Interhalogens as nonaqueous solvents	Section 6.10

TABLE 17.16 Properties of the elements of the fluorine Group

Element	*Symbol*	*Oxidation states*	*Coordination numbers*	*Availability*
Fluorine	F	−I	1, (2)	Common
Chlorine	Cl	−I, I, III, V, VII	1, 2, 3, 4	Common
Bromine	Br	−I, I, III, V, VII	1, 2, 3, 5	Common
Iodine	I	−I, I, III, V, VII	1, 2, 3, 4, 5, 6, 7	Common
Astatine	At	−I, I, III?, V		Very rare

THE CHEMISTRY OF ASTATINE

Astatine exists only as radioactive isotopes, all of which are very short-lived. Work has been done using either ^{211}At ($t_{\frac{1}{2}} = 7.2$ h) or ^{210}At ($t_{\frac{1}{2}} = 8.3$ h) and the very high activity necessitates working in 10^{-14}M solutions, and following reactions by coprecipitation with iodine compounds. The chemistry of astatine is therefore little known. Preparation is by the action of α-particles on bismuth, for example:

$$^{209}_{83}Bi + ^{4}_{2}He \rightarrow ^{211}_{85}At + 2^{1}_{0}n$$

Coprecipitation with iodine compounds indicates the existence of the oxidation states shown in Table 17.16. The negative state is probably the At^- ion and the positive states, the oxyions, AtO^-, AtO_2^- and AtO_3^-. Astatine appears to differ from iodine in not giving a VII state, in which it parallels the behaviour of the other heavy elements in its period, and in not forming a cation, At^+. Two interhalogens are known for astatine, AtI and AtBr, and there is evidence for polyhalide ions including astatine.

It is possible to synthesize organic astatine compounds, $RAtO_2$, R_2AtCl and $RAtCl_2$, in the III and V states. The compounds in the commonest state, RAt, correspond to the well-known organic halides in the – I state. A very wide range of organic groups R have been explored, including biologically important molecules, such as steroids. These have the potential for specific radiotherapy by taking At to the site of a tumour. A more detailed account of the applications of inorganic compounds in medicine is given in Chapter 20.

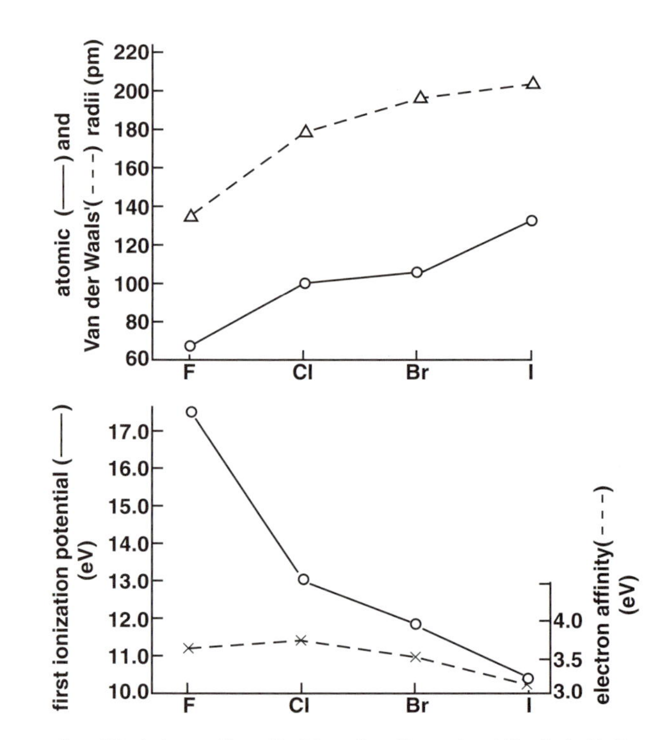

FIG. 17.54 Some properties of the halogens. The radii of the anions, X, are almost identical with the corresponding van der Waals' radii. Notice also, that the electron affinities do not follow a regular trend from fluorine to iodine

fluorocarbons used in all sorts of applications from high-performance lubricants to nonstick frying pans. Simpler members are used as refrigeration liquids and propellants, mostly mixed fluorochloromethanes or ethanes.

Chlorine is made by electrolysis of brines and isolated either as Cl_2, as aqueous base solutions containing OCl^- initially and converted to chlorate or perchlorate, or converted with hydrogen to HCl. Principal uses of the element or the oxychloro compounds are in bleaching (wood pulp for paper, textiles) or sterilization (water supply, swimming pools, some stages of sewage treatment, domestic bleaches). Organochlorine compounds in industry are often produced by the action of Cl_2 over metal chloride catalysts, especially Fe or Cu. These products, such as vinyl chloride, $CH_2{=}CHCl$, are themselves used largely in polymer formation giving the plastics of commerce (e.g. polyvinyl chloride, PVC). Cl_2 is used to make Br_2 from bromides, found in brines or from seawater in general. Bromine is

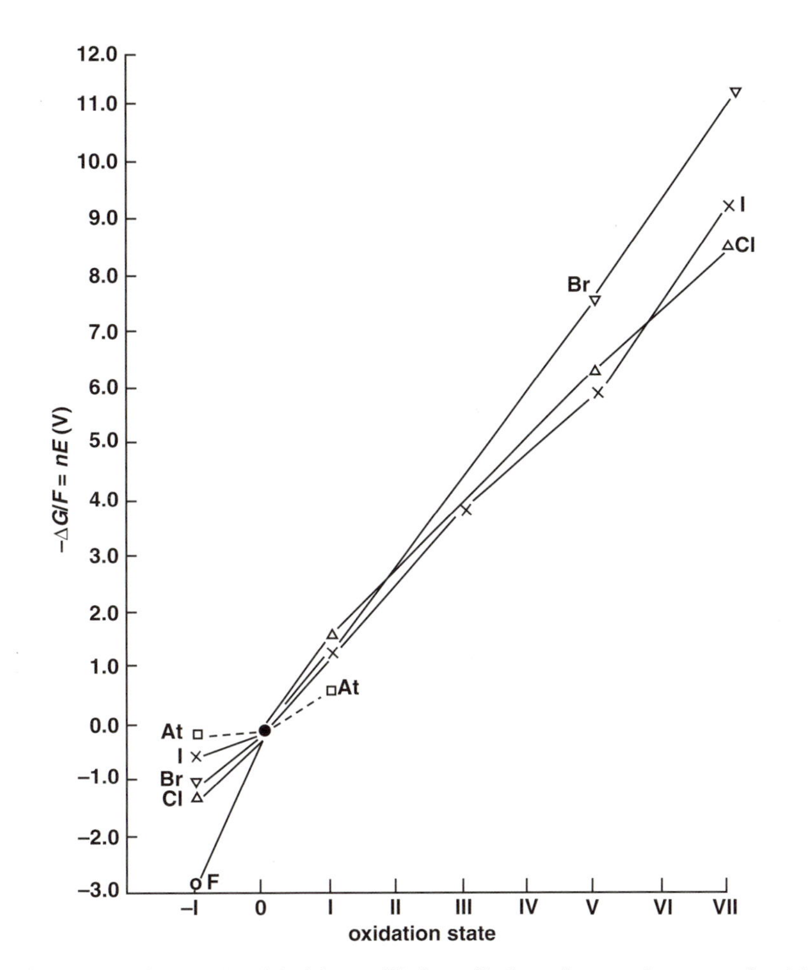

FIG. 17.55 Oxidation state free energies of the halogens. This Group, like the earlier ones, shows a negative state which decreases in stability from the first to the last member of the Group. The positive states are all fairly similar: bromine has the least stable V and VII states while I^{V} is more stable relative to I^{VII} than is Cl^{V} relative to Cl^{VII}

mainly used to prepare organobromine compounds which are used as insecticides, as fire retardants for fibres, as an anti-knock agent in petrol (ethylene dibromide), and in some applications paralleling chlorine in water treatment and sterilization. Iodine is recovered from iodide salts and brines by chlorine oxidation, often with an intermediate stage of iodide concentration. It finds a wide range of uses, mainly as organoiodine compounds, in pharmaceuticals, photography, pigments, sterilization, dyestuffs and rubber manufacture.

Fluorine is the most electronegative element and can therefore exist only in the –I state. The –I state is also the most common and stable state for the other elements, although positive states up to the Group state of VII occur. The positive states are largely found in oxyions and in compounds with other halogens, but iodine does appear to exist in certain systems as the coordinated I^+ cation.

The reactivity of the elements decreases from fluorine to iodine. Fluorine is the most reactive of all the elements and forms compounds with all elements except helium, neon, and argon. In all cases, except with the other rare gases, oxygen and nitrogen, fluorine compounds result from direct, uncatalysed reactions between the elements: the exceptions react in the presence of metal catalysts such as nickel or copper. Chlorine and bromine also combine directly, though less vigorously, with most elements while iodine is less reactive and does not react with some elements such as sulfur.

Oxygen fluorides are discussed in Section 17.7.2.

17.8.2 The positive oxidation states

The positive oxidation states occur in the compounds of chlorine, bromine and iodine with oxygen and fluorine, and of the heavier halogens with the lighter ones. The oxides of the elements are shown in Table 17.17.

TABLE 17.17 The oxides of the halogens

Average oxidation state	*Chlorine*	*Bromine*	*Iodine*
I	Cl_2O (b. 2°)	Br_2O (d. −16°)	
II	$Cl\text{-}ClO_2$		
III		$Br\text{-}O\text{-}BrO_2$ (d. −40)°)	
IV	ClO_2 (b. 11°)	$Br\text{-}O\text{-}BrO_3$	I_2O_4 (d. 130°) I_4O_9 (d. >100°)
V		Br_2O_5 (d. −20°)	I_2O_5 (d. >300°)
VI	Cl_2O_6 (b. 203°)	BrO_3 or Br_3O_8 (d. 20°)	
VII	Cl_2O_7 (b. 80°)	Br_2O_7(?)	

(b = boils d = decomposes, temperatures in °C)

Stability of the oxides is greatest for iodine, then chlorine, with bromine oxides the least stable. The higher oxides are rather more stable than the lower ones. Typical preparations are:

$$2Cl_2 + 2HgO \rightarrow Cl_2O + HgCl_2.HgO \text{ (also for } Br_2O)$$
$$2KClO_3 + 2H_2C_2O_4 \rightarrow 2ClO_2 + 2CO_2 + K_2C_2O_4 + 2H_2O$$

(also with sulfuric acid: the method using oxalic acid gives the very explosive ClO_2 safely diluted with carbon dioxide). ClO_3, BrO_3 and I_4O_9 are prepared by the action of ozone while Cl_2O_7 and I_2O_5 are the result of dehydrating the corresponding acids.

Iodine oxides are typically isolated from complex mixtures, are polymeric, and decompose to iodine and I_2O_5 on heating. Many iodine oxide phases been poorly characterized, and some doubt must be cast upon the authenticity of some materials in the absence of good characterization data. The most stable of all the oxides is iodine pentoxide, I_2O_5, which is obtained as a stable white solid by heating HIO_3 at 200°C. It dissolves in water to re-form iodic acid, and is a strong oxidizing agent which finds one use in the estimation of carbon monoxide:

$$I_2O_5 + 5CO \rightarrow I_2 + 5CO_2$$

The iodine is estimated in the usual way.

Iodine trioxide, with empirical composition IO_3, is one recent example of an iodine oxide material which has been fully characterized by an X-ray structure determination. The material crystallizes as I_4O_{12} units, containing two IO_6 octahedra (with iodine in the VII oxidation state) and two trigonal pyramidal iodines (with iodine in the V oxidation state). I_4O_{12} can thus be viewed as a mixed anhydride of two molecules of H_5IO_6 [iodine(VII)] and two of HIO_3 [iodine(V)] (Fig. 17.56a).

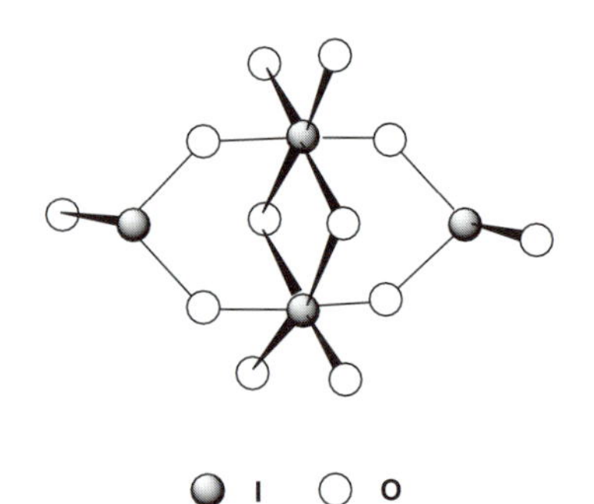

FIG. 17.56a The structure of the I_4O_{12} unit in iodine trioxide, IO_3. In the solid these units are linked together through bridging oxygens from the IO_6 octahedra to the IO_3 trigonal pyramids

I_2O_4 contains chains of IO units which are cross-linked by IO_3 units. The structure of I_4O_9 is less certain but it has been formulated as $I(IO_3)_3$. I_2O_7 has been reported as an orange polymeric solid formed by dehydrating HIO_4 with oleum.

The most stable chlorine oxide is the heptoxide which is the anhydride of perchloric acid, $HClO_4$. It is a strong oxidizing agent and dissolves readily in water to give perchloric acid. It has the structure $O_3Cl\text{-}O\text{-}ClO_3$ with tetrahedral chlorine. Cl_2O_6 is a red oil which melts at 4°C. It is dimeric in carbon tetrachloride solution but the pure liquid may contain some of the monomer, ClO_3. In the solid, it is ionic, $ClO_2^+ClO_4^-$. The ClO_2^+ ion is known in a few other compounds with large anions such as $ClO_2^+GeF_5^-$. It readily decomposes to ClO_2 and oxygen and reacts explosively with organic materials. Like the other chlorine oxides, it can be detonated by shock. In this, it is more sensitive than the heptoxide but more stable than Cl_2O and ClO_2. Chlorine dioxide is a yellow gas with an odd number of electrons. It detonates readily but is gradually being used on a large scale in industrial processes, for example as a more environmentally friendly bleaching agent. In these processes it is made *in situ* and always kept well diluted with an inert gas. Notwithstanding the extremely hazardous nature

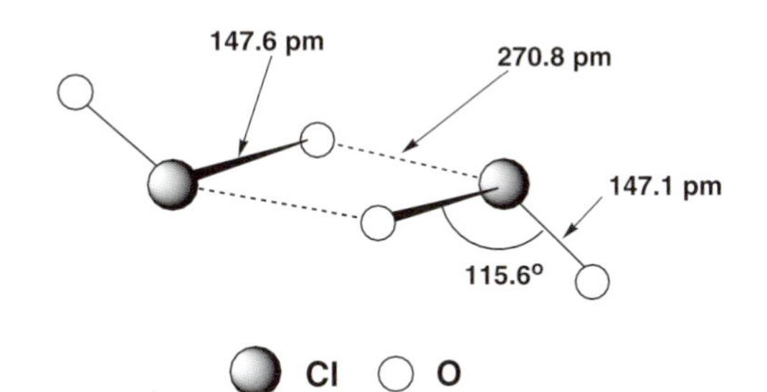

FIG. 17.56b The structure of chlorine dioxide, ClO_2, in the solid state. The ClO_2 molecules dimerize *via* a weak interaction

The structures of some bromine oxides

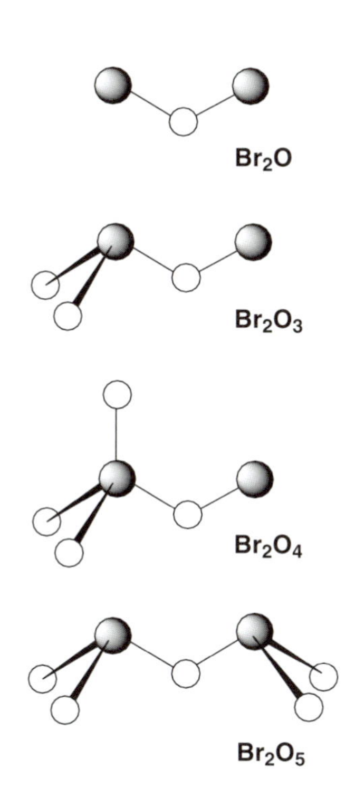

of ClO_2, its molecular structure has been determined in the solid and gaseous states. In the gas phase ClO_2 is an angular molecule with an O-Cl-O angle of 118°. In the solid state, however, ClO_2 is dimerized, Fig. 17.56b, though the interactions are relatively weak. ClO_2 dissolves freely in water to give initially the hydrate, $ClO_2.8H_2O$. On exposure to visible or ultraviolet radiation this gives an HCl and $HClO_4$ solution.

Dichlorine monoxide, Cl_2O is an orange gas which dissolves in water to give a solution containing hypochlorous acid HOCl. It is a symmetrical angular molecule with a Cl-O-Cl angle of 110°. It is powerful oxidizing agent and highly explosive—indeed, most manipulations of Cl_2O are carried out with a substantial wall between the experimenter and the compound.

The mixed oxidation state chlorine oxide, chloryl chloride, Cl-ClO_2 has also been recently described. This compound decomposes to ClO_2 and Cl_2 in the gas phase. In solid matrices, Cl-ClO_2 can be isomerized to Cl-O-Cl-O and Cl-O-O-Cl.

The bromine oxides are the least stable of all the halogen oxides and all decompose below room temperature, through there is some indication that they may be less explosive than the chlorine analogues. Br_2O is dark brown, Br_2O_3 and Br_2O_5 are orange and colourless, respectively. It is not clear whether the next oxide has the formula BrO_3 or Br_3O_8. This is also colourless and it decomposes in vacuum with the evolution of Br_2O, leaving a white solid which could possibly be Br_2O_7. Reaction of ozone with bromine gives Br_2O_3, formulated as bromine bromate, or, under longer reaction times, dibromine pentoxide, Br_2O_5.

One of the possible compounds having the composition BrO_2 has been recently shown to be bromine perbromate, Br-O-BrO_3. The bromine oxides dissolve in water or alkali to give mixtures of the oxyanions.

The oxyacids of the halogens are shown in Table 17.18. Most are obtainable only in solution, although salts of nearly all can be isolated. HOF has also recently been established.

Hypochlorous acid, HOCl, occurs to an appreciable extent, 30%, in solutions of chlorine in water, but only traces of HOBr and no HOI are found in bromine or iodine solutions. All three halogens give hypohalite on solution in alkali:

$$X_2 + 2OH^- \rightleftharpoons XO^- + X^- + H_2O \tag{17.1}$$

$K = [X^-][XO^-]/X_2 = 7 \times 10^{15}$ for Cl, 2×10^8 for Br, 30 for I but the hypohalites readily disproportionate to halide and halate:

$$3XO^- \rightarrow 2X^- + XO_3^- \tag{17.2}$$

with equilibrium constants of 10^{27} for Cl, 10^{15} for Br and 10^{20} for I. Thus the actual products depend on the rates of these two competing reactions. For the case of chlorine, the formation of hypochlorite by reaction (1) is rapid, while the disproportionation by reaction (2) is slow at room temperatures so that the main products of dissolution of chlorine in alkali are chloride and hypochlorite. For bromine, reactions (1) and (2) are both fast at room temperature so that the products are bromide, hypobromite, and bromate, the proportion of bromate being reduced if the reactions occur at 0°C. In the case

TABLE 17.18 Oxyacids of the halogens

Type	*Name*		*Stability*
HOX	Hypohalous	X = F, Cl, Br, I	F > Cl > Br > I. All are unstable and are known only in solution.
HOXO	Halous	X = Cl	$HBrO_2$ possibly exists also.
$HOXO_2$	Halic	X = Cl, Br, I	Cl < Br < I. Chloric and bromic in solution only but iodic acid can be isolated as a solid.
$HOXO_3$	Perhalic	X = Cl, Br, I	Free perchloric, perbromic and periodic acids occur.
Also $(HO)_5IO$ and $H_4I_2O_9$ forms of periodic acid			

of iodine, reaction (2) is very fast and iodine dissolves in alkali at all temperatures to give iodide and iodate quantitatively:

$$3I_2 + 6OH^- \rightarrow IO_3^- + 5I^- + 3H_2O$$

The only halous acid definitely established is chlorous acid, $HClO_2$, though salts containing the bromite ion BrO_2^- are now well defined. An X-ray structure determination on sodium bromite trihydrate confirms the presence of an angular BrO_2^- ion with an O-Br-O angle of 105.3°. $HClO_2$ does not occur in any of the disproportionation reactions above and is formed by acidification of chlorites. The latter are themselves formed by reaction of ClO_2 with bases:

$$2ClO_2 + 2OH^- \rightarrow ClO_2^- + ClO_3^- + H_2O$$

Chlorites are relatively stable in alkaline solution and are used as bleaches. In acid, chlorous acid rapidly disproportionates to chloride, chlorate and chlorine dioxide.

For the halic acids, stability is greatest for iodine. Salts of all three acids are well-known and stable, with a pyramidal structure. The IO_4^{3-} ion, containing iodine(V), has recently been reported in the solid-state compound $Ag_4(UO_2)_4(IO_3)_2(IO_4)_2O_2$. The IO_4^{3-} ions have a sawhorse geometry, derived from a trigonal bipyramid with one equatorial position empty, and occupied by a lone pair of electrons. The halic acids are stronger acids than the lower ones and are weaker oxidizing agents.

All perhalates exist and perchloric acid and perchlorates are well known. $HClO_4$ is the only oxychlorine acid which can be prepared in the free state. It, and perchlorates, although strong oxidizing agents, are the least strongly oxidizing of all the oxychlorine compounds. The perchlorates of many elements exist. The ion is tetrahedral and has the important property of being weakly coordinated by cations. It is thus very useful in the preparation of complexes, as metal ions may be introduced to a reaction as the perchlorates, with the assurance that the perchlorate group will probably remain uncoordinated—contrast this with the behaviour of ions such as halide, carbonate or nitrite which are often found as ligands in the complexes, e.g.:

$$Co^{2+} + NH_3 + Cl^- \xrightarrow{\text{oxidize}} Co(NH_3)_6^{3+} + Co(NH_3)_5Cl^{2+} +, \text{ etc.}$$

$$\text{but } Co^{2+} + NH_3 + ClO_4^- \xrightarrow{\text{oxidize}} Co(NH_3)_6^{3+} \text{ only}$$

After eluding attempts to make it for many years, perbromic acid and its salts were synthesized in 1968. Perbromate, BrO_4^-, was prepared electrolytically, or alternatively by oxidation with XeF_2, but the most convenient synthesis was found to be oxidation of bromate in alkali by molecular fluorine. Acidification yields perbromic acid which is a strong monobasic acid, stable in solutions up to about 6 M. $KBrO_4$ contains tetrahedral BrO_4^- ions, and this species is predominant in solution with no evidence of a second form as found in periodates. The electrode potentials have been assessed for the reaction

$$XO_4^- + 2H^+ + 2e^- \rightarrow XO_3^- + H_2O$$

as 1.23 V for X = Cl, 1.76 V for X = Br and 1.64 V for X = I. Perbromate is a somewhat stronger oxidant than perchlorate or periodate, but its oxidizing reactions are sluggish. Thus, the oxidizing power is not the reason for the difficulties found in the synthesis of perbromate. It seems that the preparation from Br(V) requires the surmounting of an activation barrier, and any process proceeding by one-electron additions might have failed because of the instability of the intermediate species.

Thus, although the long-standing anomaly about perbromate no longer exists, it is clear that bromine(VII) is less stable in compounds with oxygen than either chlorine(VII) or iodine(VII). To this extent, bromine still reflects the middle element anomaly.

Periodic acid and periodates occur, like perrhenates, in 4-, 5- and 6-coordinated forms. Oxidation of iodine in sodium hydroxide solution gives the periodate, $Na_2H_3IO_6$, and the three silver salts, $AgIO_4$, Ag_5IO_6 and Ag_3IO_5, may be precipitated from solutions of this sodium salt under various conditions. Deliquescent white crystals of the acid H_5IO_6 may be obtained from the silver salt and this loses water in two stages:

$$(HO)_5IO \xrightarrow{80^\circ C} H_4I_2O_9 \xrightarrow{100^\circ C} (HO)IO_3$$

Salts, such as $K_4I_2O_9$, of the binuclear acid may be obtained. The $I_2O_9^{2-}$ ion has the $O_3IO_3IO_3$ structure, of two IO_6 octahedra sharing one triangular face. In the periodates, the IO_6 group is octahedral and the IO_4 group is tetrahedral, as expected. In the related ion $[HO_2I_2O_8]^{4-}$, there are two IO_6 octahedra sharing an edge with distance I-OH = 190 pm, I-O (terminal) = 181 pm and I-O (bridge) = 202 pm. In K_3IO_5, the IO_5^{3-} ion has a square pyramidal structure. The iodine oxyanions thus continue the trend, already observed in the earlier members of this period, to become 6-coordinated to oxygen.

It should be noted that halates, and oxyhalides generally, are potentially explosive in contact with oxidizable materials.

Positive oxidation states for halogens are also found in the interhalogen compounds, where the lighter halogen is in the –I state and the heavier one in a positive state. A similar situation pertains for the mixed polyhalide ions. Table 17.19 lists these compounds.

All the interhalogens of type AB are known, although IF and BrF are very unstable. All the AF_3 and AF_5 types also occur, although, again, some are unstable. Only IF_7 is found in the VII oxidation state, and the only other higher interhalogen is iodine trichloride. All are made by direct combination of the elements under suitable conditions, apart from IF_7 which results from the fluorination of IF_5. All the compounds are liquids or volatile solids except ClF, which boils at –100°C. Most boiling points fall between 0°C and 100°C.

The halogen fluorides are all very reactive and act as strong fluorinating agents. Reactivity is highest for chlorine trifluoride, which fluorinates as strongly as elemental fluorine. Reactivity falls from the chlorine to the bromine and iodine fluorides, and also falls off as the number of fluorine atoms in the molecule increases. BrF_3 and IF_5 are particularly useful as fluorinating agents for the production of fluorides of elements in intermediate oxidation states. These two interhalogens, along with the two iodine chlorides ICl and ICl_3, undergo self-ionization and are useful as solvent systems, as discussed in Chapter 6. The anions of these systems, BrF_4^-, IF_6^-, ICl_2^- and ICl_4^-, respectively, are among the polyhalide ions in the table. The cations, BrF_2^+, IF_4^+, I^+ and ICl_2^+, are less familiar but may be isolated by adding halide ion acceptors to the interhalogen, for example:

$$BrF_3 + SbF_5 \longrightarrow (BrF_2)^+(SbF_6)^-$$

TABLE 17.19 Interhalogens and polyhalide ions

AB	AB_3	AB_5	AB_7	A_3^-		AB_4	AB_6^-	AB_8^-	A_n^-	A_n^{2-}	A_2^+	A_3^+	A_5^+	AB_2^+	AB_4^+	AB_6^+
ClF	ClF_3	(ClF_5)	IF_7	Br_3^-	ICl_2^-	ClF_4^-	ClF_6^-	IF_8^-	I_5^-	I_8^{2-}	Br_2^+	Cl_3^+	Br_5^+	ClF_2^+	ClF_4^+	ClF_6^+
BrF	BrF_3	BrF_5		I_3^-	$ClBr_2^-$	BrF_4^-	BrF_6^-		I_7^-		I_2^+	Br_3^+	I_5^+	BrF_2^+	BrF_4^+	BrF_6^+
$(IF)_n$	$(IF_3)_n$	IF_5		ClF_2^-	IBr_2^-	IF_4^-	IF_6^-		I_9^-		also	I_3^+	(also I_7^+)	FCl_2^+	IF_4^+	IF_6^+
BrCl	I_2Cl_6			BrF_2^-	$IBrF^-$	ICl_4^-			$I_2Cl_3^-$		I_4^{2+}			$BrCl_2^+$		
ICl				$BrCl_2^-$	$IBrCl^-$	ICl_3F^-			$I_2Cl_2Br^-$					ICl_2^+		
IBr				BrI_2^-		(IF_6^{3-})								IBr_2^+		

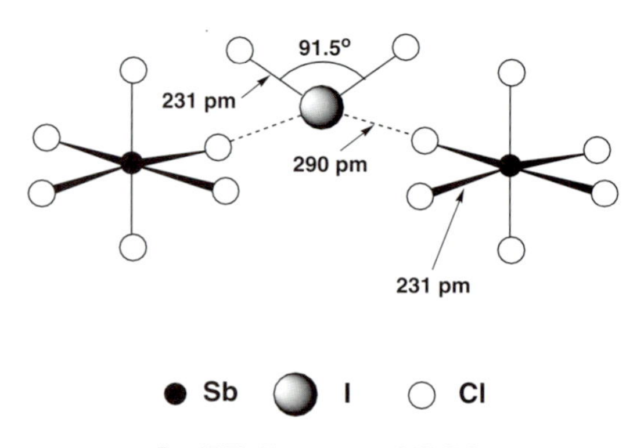

FIG. 17.57 The structure of ICl_2SbCl_6

Although these compounds are formulated as ions, structural studies show that there may be an interaction between the cation and anion. For example, ICl_2SbCl_6, (from ICl_3 and $SbCl_5$) consists of $SbCl_6$ octahedra and angular ICl_2 units, but with a weak coordination of two of the chlorines in an octahedron to two different iodines to give a chain structure, Fig. 17.57. This structure seems to be quite general, $BrF_2^+SbF_6^-$ adopting a very similar form to that shown in Fig. 17.57. In both cases, the AB_2 unit may be described as the cation, AB_2^+, forming two further weak bridges to the anion, SbX_6^-, or as covalent with a very distorted square planar AB_4 unit. The two A-B distances are sufficiently different to make the former description the more acceptable. For FCl_2^+ the vibrational spectrum and calculation points to the asymmetric V-shape structure with the linkage Cl-Cl-F.

The other interhalogens of the AB type have properties and reactivities which are roughly the average of the properties of the constituent halogens. Bonding in AB interhalogens has been discussed in Chapter 3.

The interhalogens AB_3 are predicted by VSEPR to be planar T-shaped, but there are significant variations in the solid state and depending on the central halogen. Thus, in the solid, ClF_3 forms a molecular lattice, while BrF_3 interacts with a bromine from another molecule, such that the Br is planar quadrilateral, and reminiscent of the anion BrF_4^-. However, in the recently determined structure of solid IF_3, the T-shaped IF_3 molecule interacts with iodines of two other IF_3 molecules, giving a pentagonal planar geometry around each iodine atom, reminiscent of the anion IF_5^{2-}. This results in IF_3 having a polymeric structure, which accounts for its low solubility.

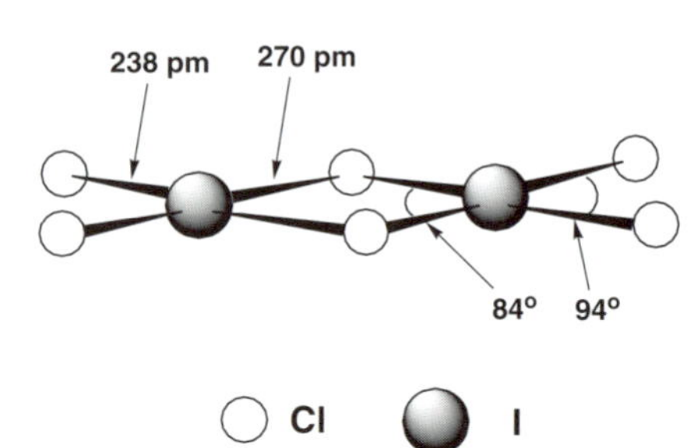

FIG. 17.58 The structure of iodine trichloride, I_2Cl_6

BrF_5 and probably the other pentafluorides, are square pyramids of fluorine atoms with the bromine atom just below the plane of the four fluorines of the base. IF_7 is approximately a pentagonal bipyramid. The remaining compound is iodine trichloride which has the dimeric structure shown in Fig. 17.58.

A variety of polyhalide ions is known. These are all prepared by the general method of adding halogen or interhalogen to a solution of the halide. The examples with three halogen atoms are linear with the heaviest atom in the middle in the mixed types (although it is not known whether $ClBr_2^-$ and BrI_2^- are symmetrical or obey this rule and have structures I-I-Br or Br-Br-Cl). The three-halogen ion, $ClBrI^-$, forms a disordered crystal so its structure is not definitely known. The data are best fitted by the arrangement $Br\text{-}I\text{-}Cl^-$ with I-Cl longer than Br-I (291 pm and 251 pm respectively). These anions are AB_2L_3 structures, while the three-atom cations like ClF_2^+ have an electron pair less and are V-shaped. Established atom sequences include $F\text{-}Cl\text{-}F^+$ and $Cl\text{-}Cl\text{-}F^+$. The AB_4^- compounds are probably all planar ions, with two lone pairs in the remaining octahedral positions. The structures of ICl_4^- and BrF_4^- have been determined and they are planar. The structures of two AB_4^+ cations, IF_4^+ and ClF_4^+, with one pair of electrons fewer, have also been determined. These have the same form as SF_4, which is isoelectronic, and is a trigonal bipyramid with a lone pair in an equatorial position. IF_6^+ and IF_6^-, which again differ by one unshared electron pair, present an interesting structural pair. The cation, which has only the six-bond pairs around the iodine, is a regular octahedron as expected. The anion, with a lone pair in addition, is an example of AB_6L, and resembles some other fluorides with this same number of

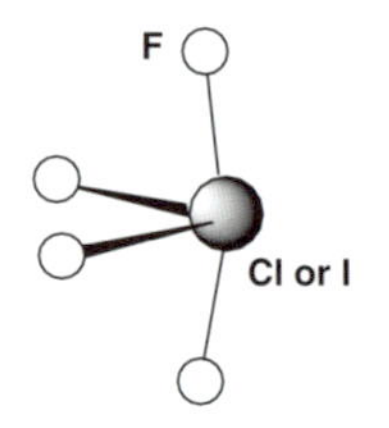

The shape of the ClF_4^+ and IF_4^+ cations

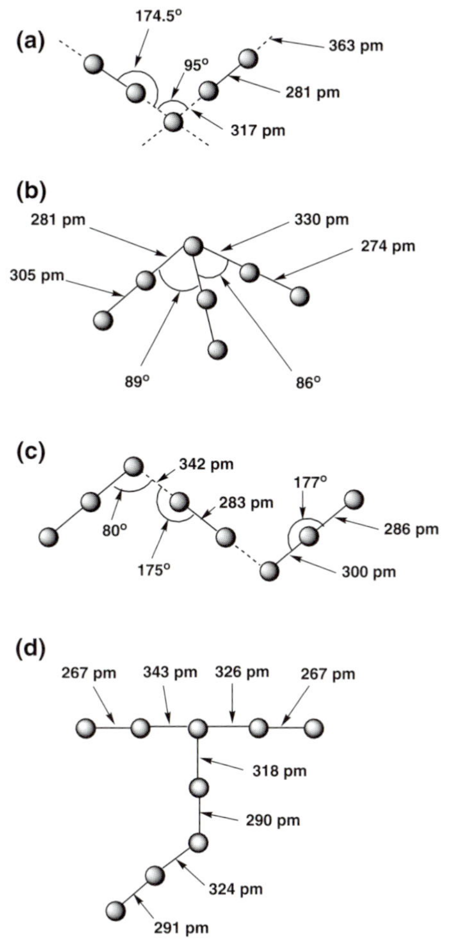

electrons in forming a distorted octahedron. That is, the lone pair is sterically active, as in XeF_6 or SeF_6^{2-}. In contrast, however, a recent X-ray structure determination on $Cs^+BrF_6^-$ shows that the BrF_6^- ion in this species is an almost perfect regular octahedron. In the analogous $[Me_4N]^+[ClF_6]^-$ the vibrational spectrum is very similar to that of BrF_6^- so the Cl species also has a regular octahedral geometry. In these cases the lone pair on the central atom is sterically inactive and presumably lies in an s orbital which is poorly screened from the nucleus.

The stucture of IF_8^- has been determined and found to have a square antiprismatic geometry which is overall very similar to the structure of the isoelectronic species XeF_8^{2-} (see Section 17.9.4).

The other type of polyhalide is limited to iodine and the structures of I_5^-, I_7^-, I_8^{2-} and I_9^- are generally irregular chains. These compounds are only stable in presence of large cations such as caesium or ammonium and alkylammonium ions. The structures are shown in Fig. 17.59. The shorter I-I distances are similar to those in iodine but the longer ones correspond to only weak interactions, and all the polyiodides may be regarded as composed of I^-, I_2 and I_3^- groups weakly bonded together. None of the polyiodides survive in solution and they go to iodide, iodine and tri-iodide. The recently discovered mixed species $I_2Cl_3^-$ and $I_2Cl_2Br^-$ have bent structures like I_5^-.

FIG. 17.59 Structures of polyiodide ions: (a) I_5^-, (b) I_7^-, (c) I_8^{2-}, (d) I_9^-

There are also a number of compounds which contain both halogen–oxygen and halogen–halogen bonds. These are listed in Table 17.20, along with their ions. Syntheses include fluorination of the corresponding anion.

$$\text{e.g.} \quad ClO_4^- + HSO_3F \rightarrow ClO_3F$$

reaction of the interhalogen with OF_2

$$ClF_5 + OF_2 \rightarrow ClOF_3$$

or oxidations as

$$ClO_2F + ClO_2 \rightarrow ClO_3F$$

Cations are formed by reactions with fluoride ion acceptors such as MF_5 or BF_3 while addition of F^- gives anions. Most of the species are strong fluorinating and oxidizing agents though perchloryl fluoride, ClO_3F, is relatively unreactive. The species are relatively stable thermally except for ClOF, the bromine compounds usually being the least stable.

The structures of selected oxyfluoro-halogens are shown in Fig. 17.60 and are those predicted by VSEPR considerations. For IO_2F_3 the structure is the trimer shown in Fig. 17.61. Polymerization also occurs in place of cation formation when MF_5 species are added to IO_2F_3.

TABLE 17.20 Halogen oxyfluorides and their ions

Oxidation state									
VII		*Oxypentafluoride*			*Dioxytrifluoride*			*Perhalylfluoride*	
		$ClOF_5$		$ClO_2F_2^+$	ClO_2F_3	$ClO_2F_4^-$		ClO_3F(a)	
					BrO_2F_3			BrO_3F	
		IOF_5	IOF_6^-	(b)$[IO_2F_3]_3$		$IO_2F_4^-$		$(IO_3F)_n$	$(n = 4?)$
V		*Oxytrifluoride*						*Halylfluoride*	
	$ClOF_2^+$	$ClOF_3$	$ClOF_4^-$				ClO_2^+	ClO_2F	$ClO_2F_2^-$
	$BrOF_2^+$	$BrOF_3$	$BrOF_4^-$				BrO_2^+	BrO_2F	$BrO_2F_2^-$
	IOF_2^+	IOF_3	IOF_4^-				$(IO_2^+)_n$	IO_2F	$IO_2F_2^-$
III					*Oxyfluoride*				
					[ClOF] (unstable)				

The commonly used trivial names are indicated.
(a) The isomer $(FO)ClO_2$, chloryl hypofluorite, exists but is less stable.
(b) $IO_2F_3.MF_5$ etc. are oxygen-bridged polymers.

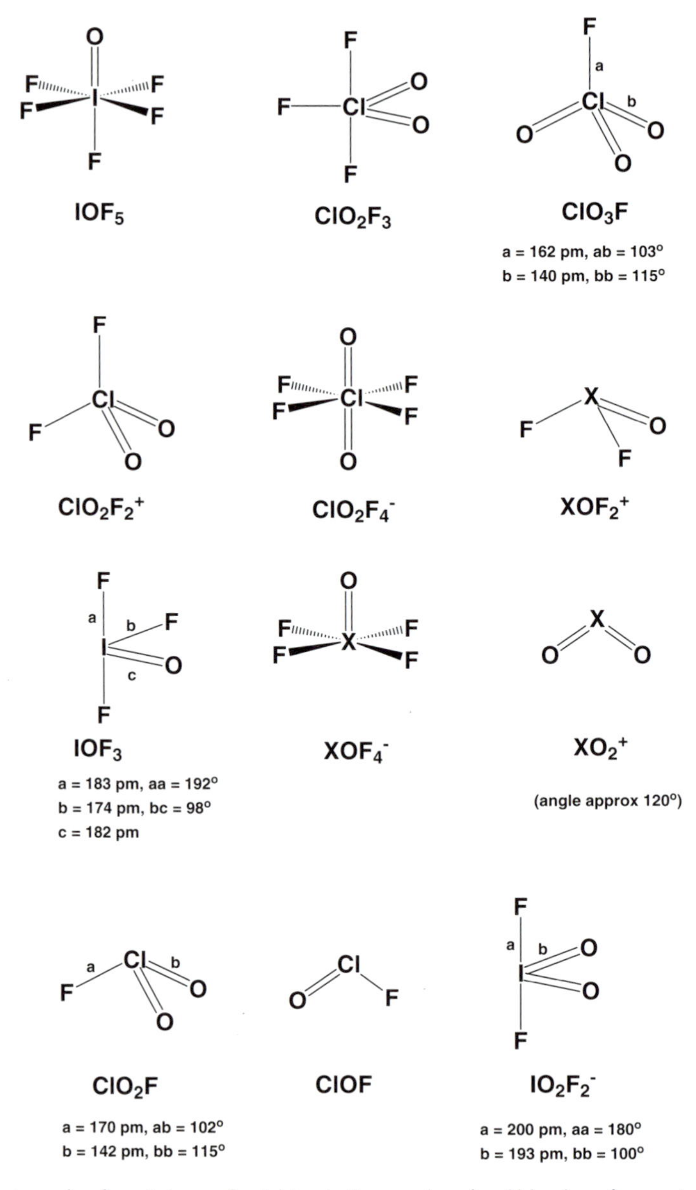

FIG. 17.60 Structures of oxyfluoro-halogens. Exact determinations are those for which values of parameters are given. The remainder are compatible with nmr and vibrational spectra

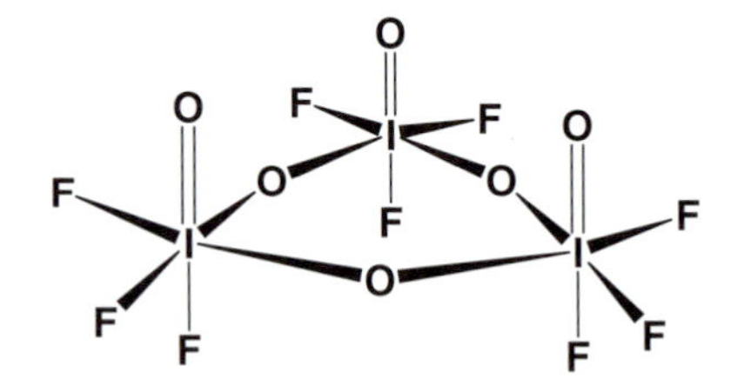

FIG. 17.61 IO_2F_3. The suggested structure is a *cis* bridged trimer

The structure of IOF_6^- is of particular interest since it is a seven-electron pair species and one which therefore could adopt one (or more) of a number of possible structures close in energy—see Section 4.2.7 for a discussion on the VSEPR of such species. An X-ray crystallographic study has shown IOF_6^- to have a pentagonal bipyramidal structure with the oxygen atom in an apical position.

Other electronegative groups may be bonded to halogens in a similar way to F and O. Thus treatment of BrF_5 with sodium nitrate yields $BrOF_4^-$, but if $LiNO_3$ is used, $BrOF_3$ is evolved and bromine(I) nitrate remains, presumably because the small lithium ion is insufficient to stabilize the large $BrOF_4^-$ anion. The structure is covalent, $OBrONO_2$, and the Cl analogue is the same. Action of O_3 yields O_2BrONO_2, a bromine(V) compound. It is probable that other oxyion products, especially with Cl or Br in positive oxidation states, are similar covalent species.

Although most of the halogen compounds in positive oxidation states are best described as covalent, there is limited evidence available to support an ionic formulation, especially for iodine(III) compounds such as the acetate, $I(OCOCH_3)_3$, and phosphate, IPO_4, and the fluorosulfonate, $I(SO_3F)_3$. If a saturated solution of iodine triacetate in acetic anhydride is

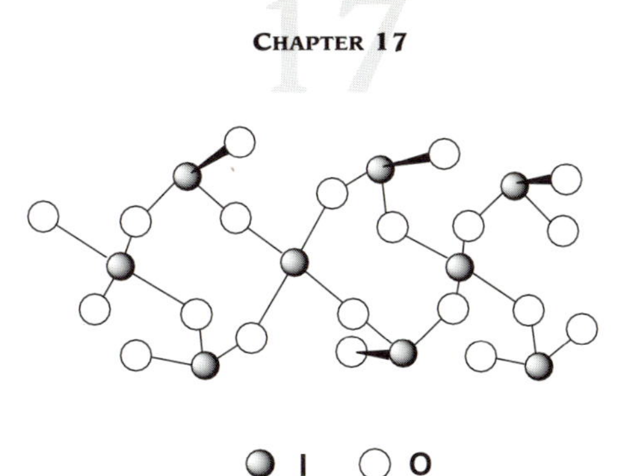

FIG. 17.62 The structure of the polymeric cation $(I_3O_6)_n^{n+}$. Two types of iodines are observed: square-planar iodine(III) and pyramidal iodine(V) bridged by oxygen atoms, forming a chain-like structure

electrolysed, iodine is found at the cathode and, when a silver cathode is used, AgI is formed and current is used as required by the equation

$$I^{3+} + 3e^- + Ag \rightarrow AgI$$

Although no other structural evidence is available, these observations indicate that the compounds could be formulated as containing the I^{3+} cation, though such an ion would be expected to interact strongly with the anion and the solids are not to be regarded as simple salts. It is also possible to formulate the oxides I_4O_9 and I_2O_4 as $I^{3+}(IO_3^-)_3$ and $(IO)^+(IO_3^-)$. A related polymeric iodine-oxygen cation $(I_3O_6)_n^{n+}$ has recently been described which has the structure shown in Fig. 17.62.

A large number of well-characterized compounds are found which contain the I^+ ion stabilized by coordination. For example, the pyridine complexes $(Ipy_2)^+X^-$ are known for a wide variety of anions, and other lone pair donors coordinate similarly. In all these complexes the iodine appears at the cathode on electrolysis. Chlorine and bromine form analogous, coordinated X^+ species. In these compounds, the arrangement of the halogen and ligands is linear, $L\text{-}X\text{-}L^+$.

One system which was thought for several years to contain the I^+ cation has recently been reformulated. This is the blue solution formed by iodine dissolved in oleum and other strong acids (compare Section 6.9). The blue species is also formed by the oxidation of I_2 in IF_5. It was shown, mainly by work in HSO_3F, that the blue species is not I^+ but I_2^+. Magnetic, conductivity and freezing point depression measurements all fit better for the I_2^+ species. Addition of further iodine gives rise to the I_3^+ and I_5^+ ions. Both the I_2^+ and the I_5^+ ions tend to disproportionate to the I_3^+ species. Detailed study of the freezing-point depression shows the equilibrium

$$2I_2^+ \rightleftharpoons I_4^{2+}$$

which lies well to the right at low temperatures. This I_4^{2+} ion can also be synthesized directly in superacid systems.

Work with superacid systems has similarly demonstrated the existence of Br_2^+, Br_3^+ and Br_5^+. The structure of the latter ion has been determined and found to be a zigzag chain of bromines, as shown in Fig. 17.63. In contrast, the most base-sensitive ions, Cl_2^+ and ClF^+, cannot be formed, even in superacid media, contrary to initial reports. Cl_2^+ can, however, be detected spectroscopically in the gas phase. When Cl_2 is oxidized with IrF_6 the Cl_4^+ ion is produced (as the IrF_6^- salt). This chemistry is analogous to the synthesis of $O_2^+PtF_6^-$ (by oxidation of O_2 with PtF_6) and '$XePtF_6$' (of unknown structure, from Xe and PtF_6). The X-ray structure of Cl_4^+ shows a rectangular ion, with two short (194 pm) and two longer (293.6 pm) Cl-Cl distances. The ion can be envisaged as formed from Cl_2 and Cl_2^+ by interaction of their π^* systems (a side-on complex with electron donation from Cl_2 to Cl_2^+, which stabilizes the latter), and there is a strong similarity with the analogous trapezoidal-shaped $Cl_2O_2^+$ ion formed by oxidation of Cl_2 with $O_2^+SbF_6^-$. The Cl_3^+ ion is also known, and has been isolated as salts with various anions.

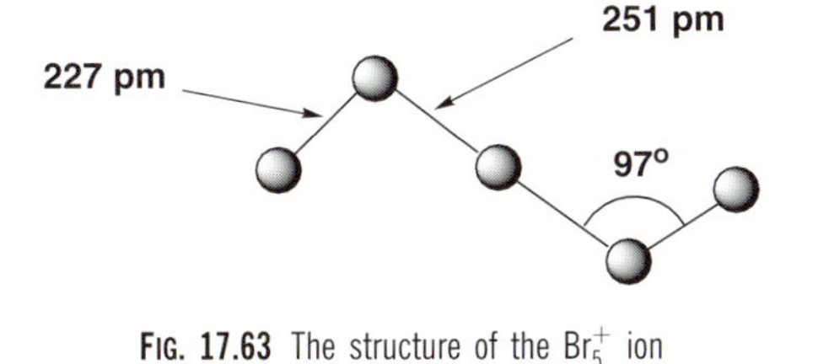

FIG. 17.63 The structure of the Br_5^+ ion

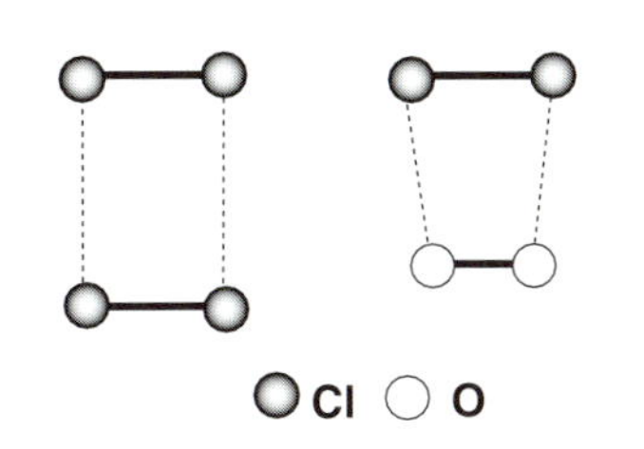

The shapes of the Cl_4^+ and $Cl_2O_2^+$ ions

It is interesting that a structural study of Br_2^+ gives Br-Br = 213 pm, compared with 227 pm in neutral Br_2. This shortening is expected as the electron which is removed in forming the ion comes from an antibonding orbital.

17.8.3 The −I oxidation state

In this state, the chemistry of the halogens is well known and many compounds have already been discussed in the earlier sections. Ionic compounds with X^- are formed with the s elements, except beryllium, and in the II and III oxidation states of the transition elements. The change from ionic to covalent character comes further to the left in the Periodic Table for the heavier halogens than for fluorine, as expected from the electronegativities. The region of change is marked by the occurrence of polymeric structures such as AlF_3, which is a giant molecule with Al-F bonds which can be described as intermediate between ionic and covalent.

In the covalent halides, the main differences between the compounds of the different halogens may be ascribed to the differences in size and reactivity of fluorine compared to

PSEUDOHALOGENS OR HALOGENOIDS

A number of univalent radicals are found which resemble the halogens in many of their properties, and the name pseudohalogen has been given to these. For example, consider the cyanide ion, CN^-. This resembles the halides in the following respects:

(i) it occurs as $(CN)_2$—cyanogen—and forms an HX acid, HCN
(ii) it forms insoluble salts with Ag^+, Hg^+ and Pb^{2+}
(iii) it also gives complex ions of similar formulae to the halogens, e.g. $Co(CN)_6^{3-}$ or $Hg(CN)_4^{2-}$
(iv) it forms covalent compounds and ionic compounds with similar ranges of elements as the halogens
(v) it gives 'interhalogen' compounds such as ClCN or ICN.

The analogy should not be pressed too far. Thus, the CN^- ion has different donor and acceptor properties from the halogens, so its transition metal complexes differ in stability and reactions.

Other radicals with similar properties include cyanate, OCN, thiocyanate SCN, selenocyanate SeCN, and azido-carbondisulfide, $SCSN_3$: all these form R_2 molecules. In addition, the ions azide, N_3^-, and tellurocyanide, $TeCN^-$, act as pseudohalides although no molecule, R_2, is formed. Despite their explosive nature, the halogen azides, XN_3, have found application in the synthesis of organic azides and related nitrogen compounds. Bromine azide has a zigzag molecular structure, Br-N-N-N, with Br-N = 190 pm, the central N-N = 123 pm and the outer N-N = 113 pm. The BrNN angle is 110° and the NNN one = 171°—all parameters very similar to those of BrNCO (Br-N = 186 pm, BrNC = 118°, NCO = 172°). In contrast, the solid-state structure of IN_3 consists of I-N-I-N chains with each azide bridging two iodines through a single terminal nitrogen. Other iodine azide compounds are the $I(N_3)_2^+$ cation (analogous to ICl_2^+ and $I_2N_3^+$, which has been proposed to have an I-I-N-N-N chain structure.

the heavier halides. This is often shown in the formation of hexavalent fluorides and tetravalent halides of the rest of the Group, or, similarly, by the formation of 6-coordinated instead of 4-coordinated complexes. Examples include the formation of SF_6 but only SCl_4, or of CoF_6^{3-} but only $CoCl_4^-$.

In this oxidation state, the halogens show their strongest resemblance and fluorine fits into place as the most reactive of all. This high reactivity derives in part from the relative weakness of the F-F bond in fluorine (similar effects are found for the O-O and N-N single bonds in hydrogen peroxide and hydrazine). The heat of dissociation of F_2 is only 129.3 kJ mol^{-1}, compared with 237.8 kJ mol^{-1} for Cl_2, 188.9 kJ mol^{-1} for Br_2 and 147.9 kJ mol^{-1} for I_2. A value extrapolated from those of the heavier halogens would be about twice the observed F_2 value. The decrease is considerable when it is recalled that most element–hydrogen bond strengths are in the order F > Cl > Br > I.

17.9 The helium Group

The elements of the helium Group are termed the rare gases, or the inert or noble (implying unreactive) gases. None of these terms is now particularly appropriate. The elements are rare only by comparison with the very abundant components of the atmosphere, oxygen and nitrogen. In terms of absolute composition of the crust and atmosphere, the lighter elements of this Group are common. Neither are the gases inert, as has recently been shown. Probably the term 'noble gases' is least unsatisfactory but no general agreement has been reached as yet. The IUPAC recommended name is rare gas and this will be used here.

The rare gases occur as minor components of the atmosphere, ranging in abundance from argon (0.9% by volume) to xenon (9 parts per million). Helium also occurs in natural hydrocarbon gases in some oilfields and is found occluded in some rocks. In both cases, this helium probably arises from α-particles emitted during radioactive decay. The heaviest member, radon, is radioactive and is found in uranium and thorium minerals, where it is produced in the course of the decay of the heavy elements. The other elements are usually produced by fractional distillation of liquid air. The main properties are given in Table 17.21. The low boiling points and heats of vaporization reflect the very low interatomic forces between these monatomic elements: the rise in these values with atomic weight shows the increasing polarizability of the larger electronic clouds.

There is increasing public concern over the harmful effects of radon gas, which is found naturally, accumulating in mines and the basements of houses, etc.

TABLE 17.21 Properties of the rare gases

Element	*Symbol*	*B.p.* (K)	*Heat of vaporization* (kJ mol^{-1})	*Ionization potential* (kJ mol^{-1})	*Uses*
Helium	He	4.18	0.092	2371	Refrigerant at low temperatures: airships
Neon	Ne	27.1	1.84	2080	Lighting
Argon	Ar	87.3	6.27	1520	Inert atmosphere for chemical and technical applications
Krypton	Kr	120.3	9.66	1359	
Xenon	Xe	166.1	13.68	1170	
Radon	Rn	208.2	17.99	1037	Radiotherapy

The main isotope of helium is helium-4, and if this is cooled below 2.178 K surprising properties appear. In this form, called helium-II, the viscosity is too low to be detected, the liquid becomes superfluid, and it appears to flow in thin films without friction and is able to flow uphill from one vessel to another. No full theoretical explanation of these phenomena is yet available.

Until 1962, all attempts to form compounds of the rare gases had failed. Transient species, such as HHe, had been observed in electric discharges but these had very short lifetimes. The rare gases were also found in solids, such as $3C_6H_4(OH)_2.0.74Kr$, but these are not true compounds but *clathrates.*

CLATHRATE COMPOUNDS

A clathrate is formed when a compound crystallizes in a rather open 'cage' lattice which can trap suitably sized atoms or molecules within them. An example is provided by *para*-quinol (p-$C_6H_4(OH)_2$; p-dihydroxybenzene) which, when crystallized under a high pressure of rare gas, forms an open, hydrogen-bonded cage structure which holds the rare gas atoms in compounds like the krypton one above, or like $3C_6H_4(OH)_2.0.88Xe$. When the quinol is dissolved or melted, the rare gas escapes. That the clathrates are not true compounds is shown by the large variety of atoms and molecules which may enter the cages. Not only are quinol clathrates formed by krypton and xenon, but also by O_2, NO, methanol and many others. The only requirement is that the clathrated species should be small enough to fit the cages and not so small that it can diffuse out: thus helium and neon are too small to form clathrates with p-quinol. Other compounds give clathrates with the rare gases. In particular, the reported hydrates of the rare gases are clathrates of these elements in ice, which crystallizes in an unusually open cage form. Although all clathrates do not involve hydrogen-bonded species, for example the benzene clathrate, $Ni(CN)_2.NH_3.C_6H_6$, clathrate formation by hydrogen-bonded molecules is common, as open structures are more readily formed.

17.9.1 Xenon compounds

All other attempts to form rare gas compounds, including many studies of possible donor action to yield compounds such as $Xe \rightarrow BF_3$, failed until 1962. Then Bartlett reported that xenon reacted with PtF_6 to form a compound which he formulated as $Xe^+(PtF_6)^-$. He was led to try this reaction after his discovery of $O_2^+(PtF_6)^-$—see Section 15.8—by the consideration that the ionization potential of xenon was close to that (914 kJ mol^{-1}) of the O_2 molecule, so that if PtF_6 could oxidize O_2 to O_2^+, there was the chance of its oxidizing Xe to Xe^+. Further exploration of this field was extremely rapid. A fuller investigation of the reaction led to the discovery of XeF_4 in the second half of 1962. Interest was then concentrated on simple fluorides, oxides, oxyfluorides and species present in aqueous solution. The compounds of these classes are listed in Table 17.22. The existence of $XeCl_2$ and $XeCl_4$ has been indicated in Mössbauer experiments using ICl_2^- or ICl_4^- as sources.

TABLE 17.22 Simple rare gas compounds

Oxidation state	Fluorides	Oxides	Oxyfluorides	Acids and salts
II	KrF_2			
	XeF_2			
	RnF_2			
IV	XeF_4		($XeOF_2$)	
VI	XeF_6	XeO_3	$XeOF_4$	$HXeO_4^-$
			XeO_2F_2	$Xe(OH)_6$?
				Ba_3XeO_6 (and other salts)
VIII		XeO_4	XeO_3F_2	$HXeO_6^{3-}$
			(XeO_2F_4)	Ba_2XeO_6 (and similar salts)

Compounds in brackets are unstable at room temperature. $XeCl_2$ exists at low temperatures. $XeCl_4$ and $XeBr_2$ exist transiently.

These compounds decompose below room temperature, but $XeCl_2$ was found at 20 K in the products formed by photolysis or by passing Xe and Cl_2 through a microwave discharge. Spectroscopic studies suggest that $XeCl_2$ is linear, though more recent studies have questioned the very nature of this molecule. It has been suggested that $XeCl_2$ is in fact a van der Waals molecule either linear, of the type Xe$\cdots$Cl-Cl (analogous to the known van der Waals molecules $HeCl_2$, $NeCl_2$ and $ArCl_2$), or a T-shaped molecule where the Xe atom bridges the Cl-Cl bond.

Liquid xenon and supercritical xenon (see Section 6.11) are finding increasing use as 'inert' solvents in which oxidative addition reactions (see Sections 13.9 and 15.7) of highly reactive metal complexes with the C-H bonds of alkanes (usually thought to be fairly inert themselves!) can be investigated. As with the earlier discussion on noncoordinating anions (Section 17.5.5), a completely noncoordinating solvent is also somewhat of a 'Holy Grail' in chemistry. This is demonstrated by the detection, at low temperatures, of short-lived donor complexes of the type X-M$(CO)_5$, where X is either Kr or Xe, and M is either Cr or W. The strength of the W-Xe bond in Xe-W$(CO)_5$ has been determined to be around 35 kJ mol^{-1} and the complex has a lifetime of about 1.5 minutes at 170 K in liquid Xe.

While the xenon–metal bonding in the compounds above is rather weak, much stronger bonding is involved in the remarkable species $[AuXe_4]^{2+}$ (formed by reduction of AuF_3 with elemental xenon) which has four xenon atoms acting as ligands to an Au^{2+} centre. The complex is square-planar with a Au-Xe bond length of about 274 pm; given the generally poor donor ligand abilities of the rare gas atoms, it will be interesting to see if new discoveries in the area of transition metal–xenon species arise in the future.

17.9.2 Preparation and properties of simple compounds

Preparations of the fluorides are all by direct reaction under different conditions. Thus a 1:4 mixture of xenon and fluorine passed through a nickel tube at 400°C gives XeF_2, a 1:5 ratio heated for an hour at 13 atmospheres in a nickel can at 400°C gives XeF_4, while heating xenon in excess fluorine at 200 atmospheres pressure gives XeF_6. Since these original routes were discovered, new routes have been added. In fact, exposing a mixture of xenon and fluorine to sunlight provides a route to XeF_2 which crystallizes on the walls of the reaction vessel. The reaction of Xe and F_2 over a hot filament gives good yields and purity of XeF_6, and XeF_4 can be correspondingly prepared from O_2F_2 and Xe. In the presence of a fluoride ion acceptor such as BF_3, AgF_2 is also a strong enough oxidant to convert xenon gas to XeF_2.

Most other compounds result on hydrolysis of the fluorides:

$$XeF_6 + H_2O \rightarrow XeOF_4 + 2HF$$

with an excess of water,

$$XeF_6 + 3H_2O \rightarrow XeO_3 + 6HF$$
$$\text{or } XeF_4 + H_2O \rightarrow Xe + O_2 + XeO_3 + HF$$

Here, about half the Xe(IV) disproportionates to Xe and Xe(VI) while the other half oxidizes water to oxygen, forming xenon. Xenon trioxide is hydrolysed in water, probably according to the equilibrium

$$XeO_3 \rightleftharpoons Xe(OH)_6$$

Interaction of XeO_3 and $XeOF_4$ gives XeO_2.

The xenon(IV) oxyfluoride, $XeOF_2$, is formed in a low temperature matrix by reacting Xe with OF_2, or by the low temperature hydrolysis of XeF_4. At about −20°C, it disproportionates

$$XeOF_2 \rightarrow XeO_2F_2 + XeF_2$$

XeO_2F_2 itself gives XeF_2+O_2 on standing at room temperature.

Solution of xenon trioxide, or hydrolysis of the hexafluoride in acid, gives xenates such as Ba_2XeO_6. In neutral or alkaline solution, xenates rapidly disproportionate to perxenates, xenon(VIII). Thus an overall reaction such as

$$2XeF_6 + 4Na^+ + 16OH^- \rightarrow Na_4XeO_6 + Xe + O_2 + 12F^- + 8H_2O$$

is observed. The addition of acid to a perxenate yields the tetroxide

$$Ba_2XeO_6 + 2H_2SO_4 \rightarrow XeO_4 + 2BaSO_4 + 2H_2O$$

The reaction mixture must be kept well cooled as explosive decomposition of XeO_4 readily occurs. At room temperature, decomposition to $Xe+O_2$ occurs rapidly.

Reaction of XeF_6 with solid Na_4XeO_6 yields the xenon(VIII) oxyfluoride, XeO_3F_2, together with much $XeOF_4$. XeO_3F_2 is volatile and was characterized by its mass spectrum. $XeOF_4$ itself shows some interesting reactions. With CsF, it forms $XeOF_5^-$ which has a distorted octahedral structure (compare XeF_6 below). With an excess of $XeOF_4$ a second product forms, $[(XeOF_4)_3F]^-$ where the central F is bonded to the three Xe atoms in a shallow pyramid (angle 116.5°) which a very weak bond of 262 pm which contrasts with 190 pm for the Xe-F bond within the $XeOF_4$ units. An improved procedure for the synthesis of $XeOF_4$ from XeF_6 employs readily prepared POF_3 as the oxidizing agent. This avoids the generation of explosive XeO_3 and the byproduct PF_5 is readily removed from the $XeOF_4$ product because of its higher volatility.

The fluorides are all strong oxidizing and fluorinating agents, and most reactions give free xenon and oxidized products:

$$XeF_4 + 4KI \rightarrow Xe + 2I_2 + 4KF$$

$$XeF_2 + H_2O \rightarrow Xe + \tfrac{1}{2}O_2 + 2HF$$

$$XeF_4 + Pt \rightarrow Xe + PtF_4$$

$$XeF_4 + 2SF_4 \rightarrow Xe + 2SF_6$$

The three xenon fluorides are all formed in exothermic reactions (heats of formation of the gases are about 85, 230 and 335 kJ mol^{-1} for XeF_2, XeF_4 and XeF_6 respectively). The trioxide is endothermic by 402 kJ mol^{-1}— largely due to the high dissociation energy of O_2. The bond energies of the Xe-F bonds in the fluorides range from 120 to 134 kJ mol^{-1} from the difluoride to the hexafluoride, a difference which is not far outside the experimental errors. The Xe-O energy is about 85 kJ mol^{-1}.

The fluorides are all white volatile solids with the volatility increasing from the difluoride to the hexafluoride. The trioxide is also white but nonvolatile, while the tetroxide, XeO_4, is a yellow, volatile solid which is unstable at room temperature. The oxyfluorides are also white, volatile solids.

While all the early compounds of Xe showed bonds only to F or O, an increasing number of Xe-N, Xe-C and Xe-Cl bonded species are now known, and many are stable (though chemically reactive) compounds. Often, electronegative groups (often fluorines) are required on the ligands to stabilize them, and this is nicely illustrated by the wide range of compounds with C_6F_5-Xe groups discussed later, and shown in Fig. 17.64.

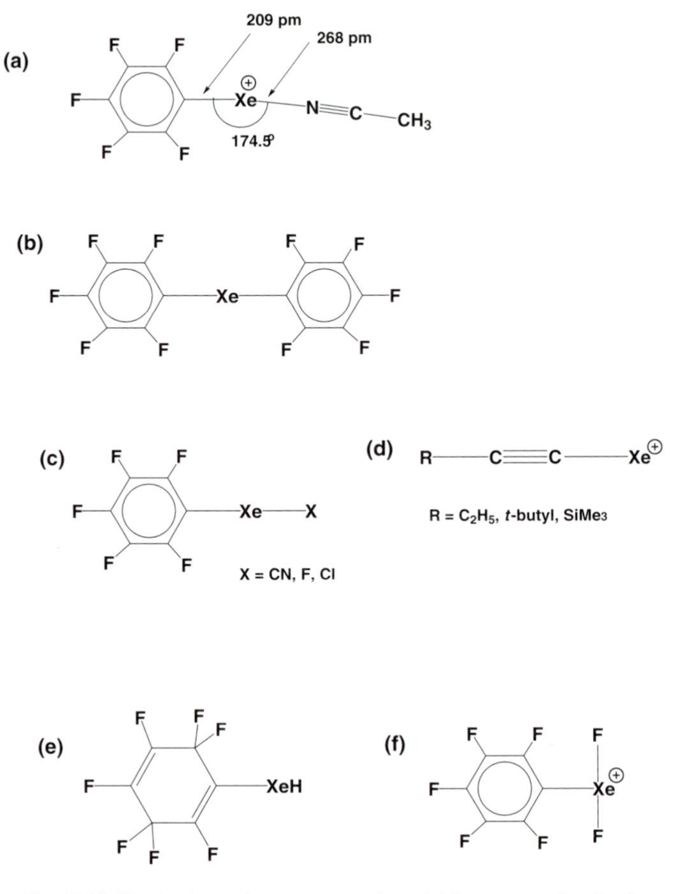

FIG. 17.64 The structures of some compounds containing xenon–carbon bonds

Using matrix isolation techniques at low temperatures, the range of rare gas species which are detectable (though not necessarily stable) can be remarkably extended to include neutral molecules with Xe-H, Xe-I, Xe-Br, Xe-S, Kr-H, Kr-Cl and Kr-C bonds. As an illustration, the species H-Xe-OH has been formed (in a low temperature matrix) from water (dissociated to H and OH radicals using laser radiation) and xenon atoms.

As an illustration of an Xe-N bonded species we note the XeF_2 derivative, $FXeN(SO_2F)_2$, which has a linear F-Xe-N unit and a pyramidal N, bonded to the Xe and to 2 $S(O)_2F$ groups. A number of nitrile–krypton or –xenon compounds of the type $RCN\text{-}NgF^+$ (Ng = Kr, Xe) are also known. In these, the ligand again generally contains an electron-withdrawing fluorinated R group, such as CF_3 or C_2F_5. The unsubstituted krypton ion $HCN\text{-}KrF^+$ has been reported and, in addition, a theoretical study has suggested that the argon analogue, $HCN\text{-}ArF^+$, should be stable and experimentally accessible. These rare gas(II) compounds have the linear structures predicted by VSEPR. Using the fact that xenon–oxygen compounds are less strongly oxidizing than analogous fluorine compounds (in the same xenon oxidation state), compounds containing Xe(VI)-N and Xe(VIII)-N bonds have been prepared in the compounds $O_3Xe\text{-}NCCH_3$ and $O_4Xe\text{-}NCCH_3$ respectively.

In recent years there has been an upsurge of interest in the synthesis of compounds containing xenon–carbon bonds. In these, the organic group generally has a high level of fluorine substitution, and a significant number of the compounds contain the pentafluorophenyl (C_6F_5) group. Some of the xenon compounds of this type which have been prepared are shown in Fig. 17.64. Other than the xenon(IV) compound in Fig. 17.64f, the compounds are of xenon(II) and have a fairly linear geometry at xenon, as illustrated by the acetonitrile complex (Fig. 17.64a), where the C-Xe-N bond angle is 174.5°. Other compounds include ones with two pentafluorophenyl groups (Fig. 17.64b), with cyanide, fluoride or chloride substituents (Fig. 17.64c), or alkynyl or alkenyl groups (Fig. 17.64d and e).

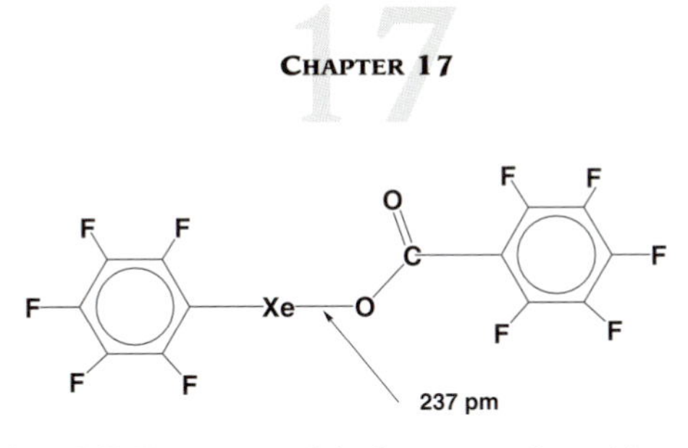

FIG. 17.65 The structure of the first compound containing a xenon–carboxylate group

The first carbon-bonded compounds of xenon in oxidation state IV have recently been prepared. $[C_6F_5XeF_2]^+$, with a T-shaped geometry (Fig. 17.64f) has been prepared by the reaction of XeF_4 with $C_6F_5BF_2$ while reaction of XeF_4 with cyanide gives $[XeF_4CN]^-$. However no compounds of xenon(VI)–carbon bonds have been reported, probably due to the increased oxidizing power of the xenon in combination with oxidizable organic groups. It is also interesting to note that a pentafluorophenyl xenon derivative is the first example to contain a xenon–carboxylate group, in C_6F_5-Xe-OC(O)C_6F_5, Fig. 17.65.

17.9.3 Structures

The structures of most of the xenon compounds are those predicted by the simple electron pair considerations outlined in Chapter 4, and correspond to the structures of isoelectronic iodine compounds. Some structures are shown in Fig 17.66.

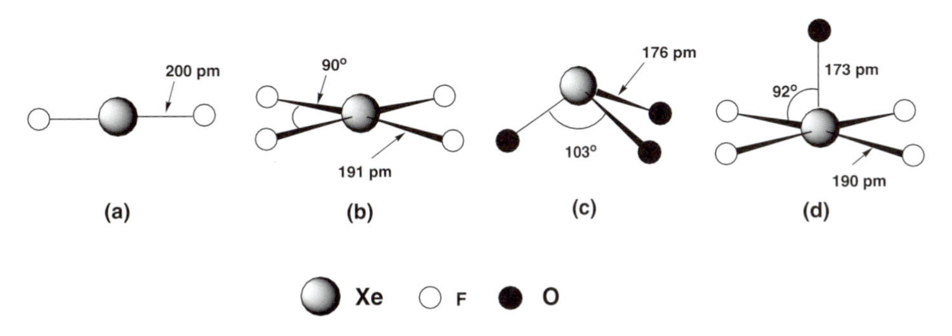

FIG. 17.66 The structures of xenon compounds: (a) XeF_2, (b) XeF_4, (c) XeO_3, (d) $XeOF_4$. These structures are the same as those of the iodine analogues ICl_2^-, IF_4^- and IO_3^-

XeO_2F_2 has a structure like $IO_2F_2^-$ (Fig. 4.6) with a nearly linear F-Xe-F arrangement and an OXeO angle of 106°. XeF is 190 pm and XeO equals 171 pm. XeO_4 is probably a regular tetrahedron, XeO_6^{4-} is octahedral, while the spectrum of $XeOF_2$ shows it is a further example of the T-shape expected for AB_3L_2 species:

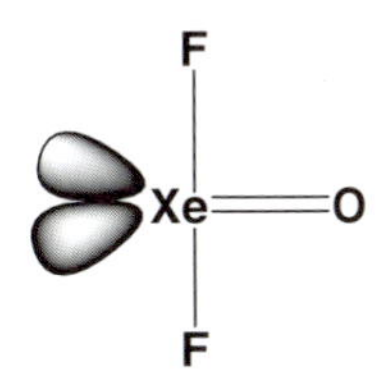

The most interesting structural problem is that presented by XeF_6. A monomer would have the AB_6L structure and would not be expected to be a regular octahedron. In the gas phase, a distorted structure is clearly indicated and three different molecular shapes may be present in equilibrium. However, XeF_6 appears to exist in a polymeric form in the solid and liquid phases. In solution, nmr studies indicate a tetramer, $(XeF_6)_4$, with equal interaction at room temperature between all the F atoms and the four Xe atoms. This can happen only if the fluorines are rapidly exchanging.

The structure of the solid is very complex with the unit cell containing 144 XeF_6 units. These are present as 24 tetramers and 8 hexamers. The configuration of both aggregates is most simply described as XeF_5^+ and F^-. The XeF_5^+ units are square pyramids and are linked together by bridges $\cdots F_5Xe^+ \cdots F^- \cdots XeF_5^+ \cdots$. In the tetramers, the F^- bridges pairs of XeF_5^+ units to form a puckered ring, with unsymmetric bridges of 223 and 260 pm Xe-F^- distances and Xe(F^-)Xe angles of 121°. In the hexamer, each F^- bridges three XeF_5^+ units lying symmetrically with Xe-F^- equal to 256 pm and an angle of 119°. (Compare these distances with the bonded XeF values given in Fig. 17.68d.) Such bridging would, of course, allow the ready fluorine exchange observed in solution. We note that XeF_5^+ is an AB_5L, species and the square pyramid is the shape expected for a sterically active lone pair.

17.9.4 Reactions with fluorides

Xenon fluorides and oxyfluorides react with the fluorides of many elements to give a variety of species. For example, in BrF_5 as solvent

$$XeF_2 + SbF_5 \longrightarrow nXeF_2.mSbF_5$$
$$(n : m = 1:1,\ 1:2,\ 2:1,\ 1:1.5,\ 1:6)$$

By direct reaction, or using SbF_5 as solvent, the species $XeF_4.SbF_5$, $XeF_4.2SbF_5$, $XeOF_4.SbF_5$, $XeOF_4.2SbF_5$ and $XeO_2F_2.2SbF_5$ may be formed. While SbF_5 forms the widest range of species, many other MF_3, MF_4, MF_5 and MOF_4 species react similarly, especially with XeF_2. Under strong fluorinating conditions, $XeF_6.AuF_5$ and $2XeF_6.AuF_5$ are formed (containing the unusual Au(V) oxidation state).

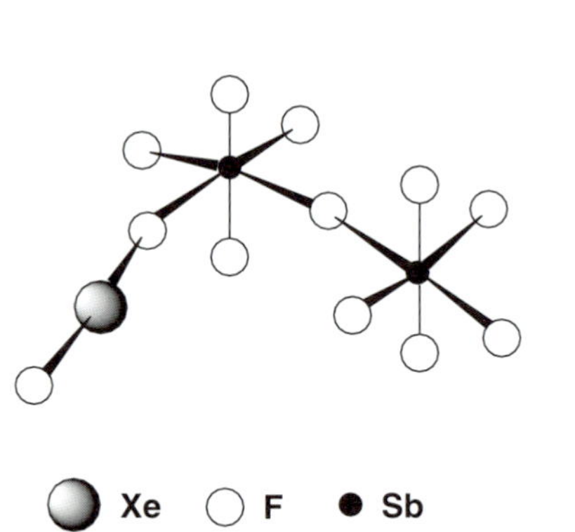

FIG. 17.67 Structure of the adduct $XeF_2.2SbF_5$ The structure is intermediate between that expected for a full covalent Xe-F-Sb bridge and that corresponding to $XeF^+Sb_2F_{11}^-$. Mean non-bridging Sb-F bond length = 183 pm.

The crystal structures of a number of these species have been determined, and they are best described as containing the fluoroxenon cation formed by transferring one F^-. The cation is then weakly bonded to the anion through a fluorine bridge. The 2:1 adducts contain more complex cations with Xe-F-Xe bridges while 1:2 adducts contain dimeric anions such as $Sb_2F_{11}^-$ whose structure is two octahedra linked by a shared F (Fig. 17.67) Further, weaker interactions occur in some crystals, so that the species contain the cations listed in Table 17.23. The structures of some examples are included in Fig. 17.68. The vibrational spectra are compatible with T-shaped XeF_3^+, pyramidal XeO_2F^+, the AB_4L type with equatorial oxygen for $XeOF_3^+$, and square-pyramidal $XeO_2F_3^-$. The crystal structure shows XeO_3F^- is a polymer of XeO_3 units (similar to xenon trioxide) bridged through F^- (see Fig. 17.68e). The more complex xenon(VI) ion, $Xe_2F_{11}^+$ consists of two square pyramids joined through one of the base corners and this is linked to the AuF_6^- ion in $(Xe_2F_{11})^+(AuF_6)^-$ by two more shared fluorines giving an Xe_2Au triangle with an F in each edge. The Au-F···Xe links are very unsymmetrical. A further example of the tendency of Xe(VI) to form complex structures is provided by $XeF_5^+AsF_6^-$. The XeF_5^+ ion is linked to the AsF_6^- one in a very unsymmetric bridge Xe···F-As (Xe···F = 265 pm, F-As = 173 pm) and this $XeF_5^+AsF_6^-$ unit forms a weak dimer by two pairs of $Xe(F)_2As$ bridges with long Xe-F distances of 270 and 281 pm.

All these weak Xe···F interactions should be compared with the van der Waals non-bonding contact distance of 350 pm.

TABLE 17.23 Some rare gas ions

Oxidation state					
II	KrF^+	$Kr_2F_3^+$			
	XeF^+	$Xe_2F_3^+$			
	RnF^+	$Rn_2F_3^+$(?)			
IV	XeF_3^+	$XeOF_3^-$			
	XeF_5^-				
VI	XeF_5^+	$Xe_2F_{11}^+$	$XeOF_3^+$	KrO_2F^+	XeO_3F^-
	XeF_7^-		$XeOF_5^-$	XeO_2F^+	XeO_3Cl^-
	XeF_8^{2-}			$XeO_2F_3^-$	

Notes 1. See text for structures: note especially that there are weak F bonds between cations and fluoro-anions. 2. Countercations are usually large M^+ e.g. Rb^+, Cs^+, NO^+. 3. Counteranions include MF_6^- (M = P?, As, Sb, V?, Nb, Ta, Ru, Os, Ir, Pt and Au);
MF_4^- (M = B, Al); MF_6^{2-} (M = Ge, Sn, Ta, Ir, Pd); $M_2F_{11}^-$ (M = As, Sb, Nb, Ta, Ir, Pt);
MOF_4^- (M = W).

Crystal structures show that the XeF_8^{2-} ion is a regular square antiprism (compare Fig. 15.5b), in which the lone pair of electrons on the xenon has no detectable effect on the shape of the ion. XeF_7^- has a capped octahedral geometry, with a lengthened bond to the capping fluorine. A crystal structure determination has also been carried out on the XeF_5^- ion, and it was found to have a pentagonal planar structure, with the two stereochemically active lone pairs adopting trans-axial positions.

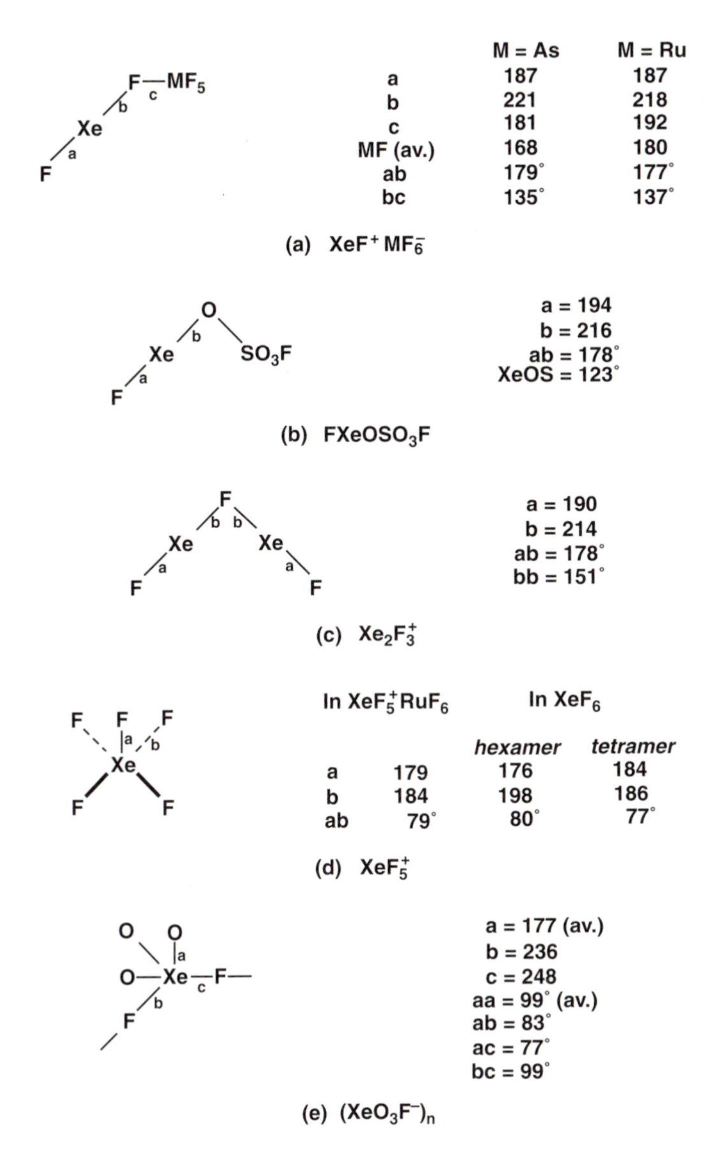

FIG. 17.68 Structures of some xenon–fluorine species. (All bond lengths are in pm)

One of the few compounds to have a bond to Xe from an element other than O or F is formed in the reaction

$$O_2BF_4 + Xe \xrightarrow{-100^\circ C} BXeF_3 \xrightarrow{-30^\circ C} Xe + BF_3$$

A planar structure based on XeF_2 with a BF_2 unit replacing one F is indicated.

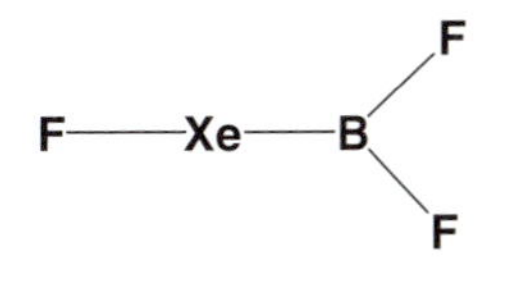

In reactions with strong acids, HOY, stepwise substitution of F in XeF_2 occurs: e.g.

$$XeF_2 + HOY \rightarrow FXeOY \rightarrow Xe(OY)_2$$

This is known for Y = TeF_5, SeF_5, SO_2F, SO_2CF_3, SO_2CH_3, ClO_2 and CF_3CO and possibly for NO_2. Some analogous Xe(IV) and Xe(VI) species also occur. A related synthesis uses the anhydride $(YO)_2O$ as in

$$XeF_2 + P_2O_3F_4 \rightarrow FXeOPOF_2 \rightarrow Xe(OPOF_2)_2$$

The structure of FXeOY is not dissimilar to the $FX_e^+MF_6^-$ species, as shown in Fig. 17.68 a, b. Similarly, $Xe(OY)_2$ has a linear bridge as in F_5SeO-Xe-$OSeF_5$.

In the IV state analogous compounds are also becoming established. $Xe(OTeF_5)_4$ is square planar with Xe-O = 203 pm. The VI state is represented by $O_2Xe(OTeF_5)_2$ and $OXe(OTeF_5)_4$. The first shows the expected five-electron pair structure with equatorial oxygens with distinctly weak double bond repulsion (Xe=O = 173 pm and an O=Xe =O angle of 106°). The axial single bond, Xe-$OTeF_5$, distance is 202 pm.

Reaction of xenon gas with $XeF^+Sb_2F_{11}^-$ in SbF_5 solvent gives rise to a green colouration, and it has been shown that the presence of HF is necessary for the reaction to proceed (no reaction was observed when pure HF-free SbF_5 solvent was used). From the solution, the dark green dixenon cation Xe_2^+ crystallizes (with the $Sb_4F_{21}^-$ ion), with a long Xe-Xe bond length of 308.7 pm. The Xe_2^+ cation is isoelectronic with the ion I_2^-, which is also dark green.

17.9.5 Other rare gas species

The only simple compound of krypton is KrF_2 (earlier reported as KrF_4). This is thermodynamically unstable and the Kr-F bond energy of about 50 kJ mol^{-1} is the lowest known element–fluorine value. KrF_2 should therefore find use as a very reactive fluorinating agent. The compound is linear, with Kr-F 189 pm, and it decomposes at room temperature at the rate of 10% an hour. Its chemistry is little explored but the 1:1, 1:2 and 2:1 adducts with MF_5 (M = Sb, Nb and Ta) are reported. These are formulated as the KrF^+ or $Kr_2F_3^+$ salts of MF_6^- or $M_2F_{11}^-$, as with the xenon analogues. These species are rather more stable than KrF_2. Evidence for compounds containing Kr-O bonds comes from $KrO_2F^+Sb_2F_{11}^-$, made like the Xe analogue, and $Kr(OTeF_5)_2$, prepared by the reaction of KrF_2 with $B(OTeF_5)_3$.

KrF_2, often in conjunction with xenon fluorides, finds application as a powerful fluorinating agent. Thus treatment of Ln(III) compounds yields the Ln(IV) species LnF_4, LnF_7^{3-} and $LnOF_2$, not only for the readily oxidized Ce, but also for Ln = Pr, Nd, Tb and Dy (compare Section 11.4).

Extrapolating from Kr through Xe would indicate that radon should form a range of compounds. However, all of the isotopes are radioactive, and not much work has been reported since the initial studies some years ago. RnF_2 and the RnF^+ cation (in the form of salts with SbF_6^-, BiF_6^- and TaF_6^- anions) are well-established. A number of other compounds which might be expected are higher fluorides, the oxide RnO_3 and ions such as RnO_3F^- but there is some uncertainty as to their authenticity.

It is suggested from the ionization potentials (Table 17.21) that argon, neon and helium are unlikely to form stable compounds corresponding to the xenon and krypton compounds above. However, by using low temperature matrix isolation techniques, (which effectively trap any unstable species formed) the neutral H-Ar-F molecule has recently been identified, and is believed to have predominantly covalent H-Ar and F-Ar bonds, in the ground state. The HArF molecule is stable *only* at low temperatures in a matrix. This work represents an important milestone in the possible future chemistry of argon, since previously the only species known involving argon were van der Waals adducts or species formed in an excited state. The question then arises as to whether other species, for example ArH^+ or ArF^+, might be obtainable; recent calculations have suggested that the latter might be stable enough to be isolated. In such species, the argon atom can be considered to be acting as an electron pair donor to the extremely strong Lewis acid F^+. In this context it is interesting that a number of XHe^+ species (X = a range of common electron deficient radicals) have been identified in interstellar clouds. However, while species such as XeF^+ can be formed by fluoride ion abstraction from XeF_2 (plus a powerful Lewis acid such as SbF_5), the same route is not available for ArF^+ due to the extreme instability of ArF_2. Therefore, routes analogous to the synthesis of NF_4^+ or ClF_6^+, (which have no stable corresponding NF_5 or ClF_7 from which F^- could be removed) must be used.

A calculation reported in 1987, using He as the most difficult case for compound formation, suggested that a number of species might be sufficiently stable to allow isolation, including $(HeCCHe)^+$ and other similar acetylene derivatives. Also quite stable could be the species HeBeO which could be formed by implanting BeO molecules (found in the vapour above solid BeO) into a helium matrix at low temperatures. Argon, krypton and xenon analogues have been recently reported.

17.10 Bonding in Main Group compounds: the use of d orbitals

The bonding in xenon compounds has been the subject of some controversy which raises the wider question of the degree to which the valence shell d orbitals participate in bonding in any compound where more than four electron pairs surround a Main Group atom. The case of XeF_2 is the simplest to discuss. The two ligands occupy axial positions and three lone pairs occupy equatorial positions in a trigonal bipyramid. To accommodate these five electron pairs requires five orbitals which are formed from the s, the three p and a d orbital on the central atom. But it has been objected that, in the case of xenon, the energy gap between the p and d orbitals is very large, equal to about 960 kJ mol^{-1}, and it is unlikely that the bond energies are sufficient to compensate for the energy required to make use of a d orbital in such a scheme. Instead, three-centre σ bonds between xenon and fluorine are proposed. The molecular axis is taken as the z-axis, as in Fig. 17.69, and the relevant orbitals are the xenon p_z orbital, which contains two electrons, and the p_z orbital on each fluorine which hold one electron each (the other six valency electrons on xenon and fluorine fill the s, p_x and p_y orbitals on each atom, all nonbonding). The three p_z orbitals can be combined to give three σ orbitals, centred on the three atoms, one of which is bonding, one nonbonding and one antibonding (Fig. 17.69). The four electrons fill the bonding and nonbonding orbitals to give an overall bonding effect. The position is similar to that in the three-centre B-H-B bond in the boranes, except that there are four electrons instead of two to be accommodated (compare Section 9.5). The cases of XeF_4 and XeF_6 can be explained similarly using xenon p_x (or p_x and p_y for XeF_6) orbitals as well as p_z in the two cases.

Similar descriptions apply to the interhalogen compounds and ions such as ICl_2^-, ICl_4^-, BrF_5, and so on. All these species can be described either in terms of full electron pair bonds, plus nonbonding pairs, by using one or two of the d orbitals, or the use of d orbitals may be avoided by using polycentred molecular orbitals at the price of reducing the bond order. It has been reported that calculations using s and p orbitals only are very successful in reproducing the bond angles and bond lengths found in interhalogen compounds.

The theory using three-centred orbitals formed by the p orbitals would predict that XeF_6, and the isoelectronic IF_6^-, should be regular octahedra while the electron pair repulsion theory suggests that these species with seven pairs around the central atoms should be distorted. It is now established that XeF_6, IF_6^- and TeF_6^{2-} are distorted octahedra, while the isoelectronic MX_6^{2-} species (where M = Se, Te, Po; X = Cl, Br, I but *not* F) are undistorted. This again shows the delicate balance which must exist between the various energies making up these two different structures.

Recent theoretical work on the general problem of d orbital contributions to bonding in Main Group compounds has made considerable progress. This is mainly because use of large computers has allowed work to be carried out with fewer initial assumptions and approximations.

(1) In compounds with second row and heavier elements bonded to highly electronegative elements like oxygen and fluorine, there is evidence of substantial d orbital participation in the bonding orbitals. For example, in PF_3, the population of the phosphorus orbitals is calculated to be:

σ bonding	3s 1.51 electrons	3p 1.05 electrons	3d 0.13 electrons
π bonding		3p 0.84 electrons	3d 0.62 electrons

Thus, the use of the d orbital gives a small stabilization to the σ bonds, but makes a significant contribution to π bonds.

Notice that this contribution is made in a molecule with only four electron pairs on the phosphorus, that is where there are enough s and p orbitals to hold all the valency

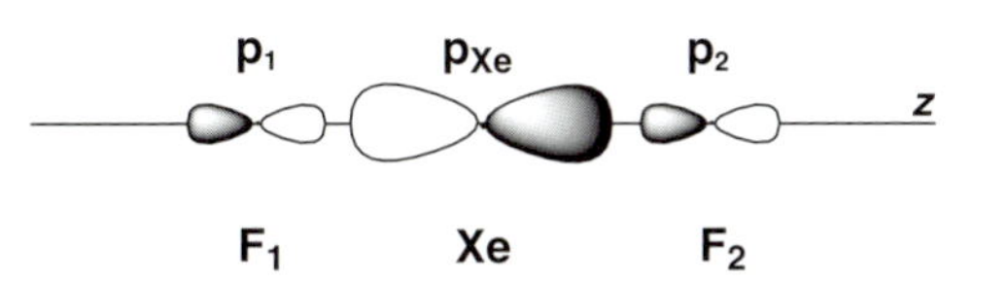

FIG. 17.69 The formation of three-centre bonds in XeF_2: the constituent alomic orbitals. If we define the positive direction of the p orbital as for the Xe one drawn, the three-centre combinations give the following orbitals bonding $\psi = -p_1 + cp_{Xe} - p_2$ nonbonding $\psi = -p_1 + p_2$ antibonding $\psi = -p_1 - c'p_{Xe} + p_2$ where the constants c and c' are of similar size and the expressions are to be regarded as normalized. The bonding orbital and the nonbonding orbital both contain two electrons. As only one pair of bonding electrons exists between three atoms, the Xe-F bond order is only one-half. XeF_4 may be described similarly using p orbitals in both the x and y directions

electrons. In higher oxidation state fluorides and oxides, such as PF_5, OPF_3, SF_6 or XeF_6, there is distinct d orbital involvement in the bonding. This would seem to be supported experimentally by the existence of IF_7 and nonoctahedral MF_6L species.

The test of the calculation is its agreement with experimentally determined ionization or promotion energies (showing the differences between orbital energies) and the agreement between calculated and experimental dipole moments (which reflect the distribution and spatial density of the electron clouds).

(2) In compounds where the bonded atoms are of lower electronegativity, like C or H, only a very low d orbital population, of the order of a few percent, is found for the central atom. This has little effect on the calculated energies, but does greatly improve the agreement between calculated and observed dipole moments. That is, the introduction of d orbitals into the calculations has, as its main benefit, a better description of the dispersion of the electrons in space.

For further discussion of bonding in compounds of the Main Group elements, see Section 18.9.

Problems

The systematic chemistry given in this chapter is best assimilated by working through it in as many ways as possible. Compare behaviour within the same Period, within the same Group, with other species of the same oxidation state or of the same valence electron configuration. See the remarks on the transition element problems and also correlate particularly with Chapters 3 and 4.
When consulting alternative sources on systematic chemistry, you should check out 'Notes to the reader' at the beginning of Appendix A, Further Reading, together with the comments on the different sections of this list.

1 Illustrations of the topics which could be reviewed are:

(a) each oxidation state (properties, stability)
(b) each coordination number (electron pair counting, stability, extent of distribution)
(c) polymeric species
(d) element–element bonding
(e) p_π–p_π bonding
(f) d orbital participation in Main Group chemistry
(g) distinguishing behaviour of first element in a Group.

2 The first dissociation constants K_1 of some oxyacids are listed below. Compare with their structures (see also Table 6.2). Find out the values for the other oxyacids of the main Group elements and discuss any anomalies.

	HNO_2	H_3PO_4	$H_4P_2O_7$	H_3AsO_3	H_3AsO_3
K_1	10^{-3}	7×10^{-3}	10^{-1}	6×10^{-10}	5×10^{-3} (mol l^{-1})

3 Compare the oxidation state free energy diagram for Cl with that of Mn. Discuss the similarities and differences and correlate with the chemistry. Extend this comparison to the other halogens and to rhenium. Carry out a similar analysis for other Groups.

4 Plot the first ionization potential against the atomic radius for the p elements across each of the periods. Comment on trends and anomalies. If the sum of the potentials involving all the valence electrons is plotted similarly, do the same trends emerge?

5 Write an essay on Main Group species with ring structures, including the structures of the elements as well as compounds. Compare ring compounds with the corresponding chain species.

6 Survey the structures found for the chlorides of the p elements. Compare and contrast with those of the same formula found for (a) s elements, (b) d elements, (c) f elements.

7 Recently, $NaPO_3$ has been isolated at low temperatures as a monomeric species. Discuss its likely structure. What compound or compounds would you expect it to form on warming to room temperature?

8 The structural parameters of some sulfuryl halides (compare Fig. 17.48a) and thionyl halides (Fig. 17.51) are tabulated below. Discuss the variations in bond lengths and angles in terms of VSEPR theory and the relative electron-withdrawing effects of the halogens.

	S=O (pm)	OSO (°)	S-F (pm)	XSX (°)	S-Cl(Br) (pm)
SO_2F_2	138.6	125	151.4	99	
SO_2FCl	140.7	123	153.8	99	196.4
$SOCl_2$	141.8	122		77	198.0
$SOCl_2$	143.9			96	206.8
$SOBr_2$	145			99	(225)

9 $Se_2O_2F_8$ is prepared via $SeOF_4$. Predict possible structures for the monomer and dimer. Compare $Se_2O_2F_8$ with other oxyhalides and discuss whether Se-O-Se or Se-F-Se bridging is the more likely. How do your predictions match with the following observed parameters:

For $Se_2F_{10}O$, Se-O = 178 pm, SeOSe = 98°,
Se···Se = 267 pm
For $Se_2F_8O_2$, Se-O = 170 pm, SeOSe = 142°,
Se···Se = 321 pm

10 What structure would you expect for the $Si_4O_{13}^{10-}$ ion, recently isolated as the silver salt? Compare this with the species of overall formula $Si_4O_{12}^{8-}$, $Si_4O_{11}^{6-}$ and $Si_4O_{10}^{4-}$ (compare Section 18.6). What formulae would you find for each of the phosphorus analogues?

11 A bulky cation, I_8^{2-}, has an almost planar structure

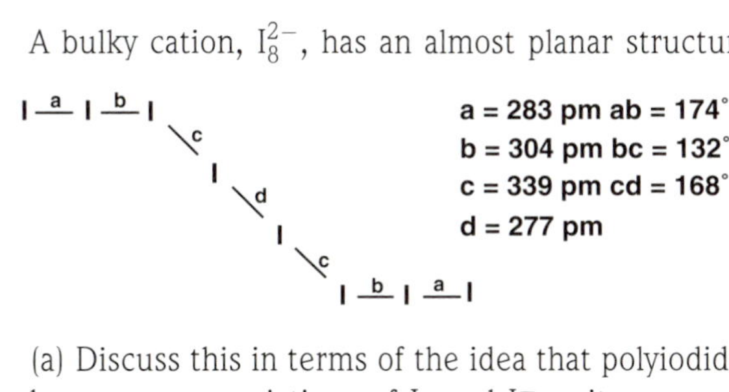

(a) Discuss this in terms of the idea that polyiodides can be seen as associations of I_2 and I_3^- units.

(b) Compare with the structure of Cs_2I_8 shown in Fig. 17.59.

12 In the gas phase, ClF_3O shows the following parameters: lengths (pm) Cl-O = 141, Cl-F(1) = 160, Cl-F(2) = 171 angles (°) OC1F(1) = 109, OClF(2) = 95, F(1)ClF(2) = 88 and F(2)ClF(2) = 171

Discuss these values in terms of the structure predicted by the VSEPR approach. Compare the values with the related compounds shown in Fig. 17.60.

13 The ClO_2^+ ion is bent, with an angle of 119° and Cl-O lengths of 141 pm. In ClO_2 the bond is longer, 148 pm, but the angle is similar, 118°. Discuss these observations in terms of the expected bonding.

14 N_2O_5, PCl_5 and Cl_2O_6 share the property of being covalent in the gas phase, but forming an ionic solid. Find other examples, from both the d and p blocks. Discuss the various reasons accounting for this behaviour.

15 Discuss the molecular parameters listed below:

	M-F$_{eq}$	M-F$_{ax}$	Angle (°)
ClF_3	160 pm	170 pm	87.5
BrF_3	172 pm	181 pm	86.2
XeF_3^+	184 pm	191 pm	81

16 One compound containing the XeF_3^+ ion has the SbF_6^- counterion. Discuss the likely preparation. As the Sb-F distance is 191 pm and there is one of the anion F atoms at 241 pm from Xe, discuss secondary interactions in this compound and compare with related compounds.

17 Discuss the parameters found for SeF_4 and $SeOF_2$:
for SeF_4
Se-F(a) = 177.1 pm, F-Se-F angle(a) = 100.6°
Se-F(b) = 168.2 pm, F-Se-F angle(b) = 169.2°
for $SeOF_2$
Se-O = 157.6 pm, O-Se-F angle = 105°
Se-F = 172.95 pm, F-Se-F angle = 92°

CHAPTER 18

18.1 THE FORMATION OF BONDS BETWEEN LIKE MAIN GROUP ATOMS 506
18.2 POLYSULFUR AND POLYSELENIUM RINGS AND CHAINS 506
18.2.1 Polysulfur rings 507
18.2.2 Polysulfur ions 508
18.2.3 X-S_n-X species 509
18.2.4 Polyselenium and polytellurium rings and chains 510
18.3 NETS AND LINKED RINGS 512
18.3.1 Polyphosphorus compounds and related species 512
18.3.2 Polyarsenic compounds and other analogues 514
18.3.3 Mixed systems 515
18.4 CLUSTER COMPOUNDS OF THE p BLOCK ELEMENTS 516
18.4.1 General 516
18.4.2 Skeletal electron pairs 516
18.4.3 Boron subhalides 518
18.4.4 Naked metal cluster ions 518
18.4.5 Group 14 prismanes and related structures 520
18.5 POLYNUCLEAR IONS AND THE ACID STRENGTH OF PREPARATION MEDIA 521
18.6 SILICATES, ALUMINOSILICATES AND RELATED MATERIALS 523
18.6.1 Simple silicate anions: SiO_4^{4-} and $Si_2O_7^{6-}$ 523
18.6.2 Rings and chains: $[SiO_3]_n^{2-}$ 525
18.6.3 Sheet structures: $Si_4O_{10}^{4-}$ 526
18.6.4 Three-dimensional structures 527
18.6.5 Zeolites 527
18.6.6 Metal phosphates and metal phosphonates 529
18.7 MULTIPLE BONDS INVOLVING HEAVIER MAIN GROUP ELEMENTS 530
18.8 COMMENTARY ON VSEPR 531
18.8.1 Experimental electron densities 531
18.8.2 Limiting cases 532
18.8.3 AB_2 dihalides 532
18.8.4 The problem of s^2 configurations 533
18.9 BONDING IN COMPOUNDS OF THE HEAVIER MAIN GROUP ELEMENTS 534

Selected Topics in Main Group Chemistry and Bonding

In this chapter, some aspects of the chemistry of the elements of the p block raised in earlier chapters (particularly Chapters 4, 8 and 17) are discussed in a little more depth. Some leading references are cited to allow further reading (see Appendix A).

As for Chapter 16, only a limited and arbitrary selection of topics is possible. There are many other areas of current intense interest and development, and these are indicated in the general reading list given in Appendix A.

18.1 The formation of bonds between like Main Group atoms

The chemistry of carbon is so enormously rich, extensive and diverse that it is easy to think that the property of forming bonds between like atoms, called *catenation*, is unique to carbon. Other long-chain species have been known for a long time, for example the plastic form of sulfur is S_∞, but because of the difficulties of handling and identifying such materials, they have often been dismissed as obscure amorphous deposits. More refined, modern experimental approaches have greatly changed this picture. Crystal structures may be determined much more readily, so that solid compounds are more readily identified. For noncrystalline compounds, spectroscopic methods are now much more powerful and give excellent structural information.

In Chapter 17 we have already seen examples of species with groups of like atoms bonded together. The forms of the elements themselves cover the main types. Small finite molecules are represented by the halogens, X_2, while S and Se give *chains* or *rings* (Fig. 17.45) formed by 2-coordinate atoms. Chains are also found in some of the borides and carbides. Graphite (Fig. 5.14c) and bismuth (Fig. 17.28) are examples of layer structures, while full three-dimensional clusters include boron and the borides of Fig. 17.7, white phosphorus, P_4 (Fig. 17.29), the recently discovered form of molecular carbon, C_{60} (Fig. 5.14d), and in infinite extension, diamond (Fig. 5.14a).

Chapter 17 was generally limited to simple cases. Molecules with a single E-E bond are found for the majority of elements while chains E_nR_{2n+2} or rings E_nR_{2n} (R = H, organic group) are represented by the Group 14 compounds E = Si, Ge or Sn.

From Table 17.3, we see that most single-bond energies are quite high, and similar for the majority of Main Group elements within a factor of two. (Single bonds N-N, O-O or F-F are unusually weak.) While there is a general tendency for bond strength to fall with mass the changes are not abrupt. It is no surprise to find that E-E links form for many of the Main Group elements. The resulting compounds range from diatomic molecules for the halogens, to rings, chains, networks and open or closed clusters for the elements of the remaining p Groups. The following sections give illustrations.

18.2 Polysulfur and polyselenium rings and chains

The survey of sulfur allotropes in Section 17.7 included the classical S_6 and S_8 rings and the long chains of plastic sulfur, together with the analogous Se_8 and Se_∞. Discussion also included compounds with single S-S or Se-Se bonds, and some chain oxygen species like

See also Section 9.5.

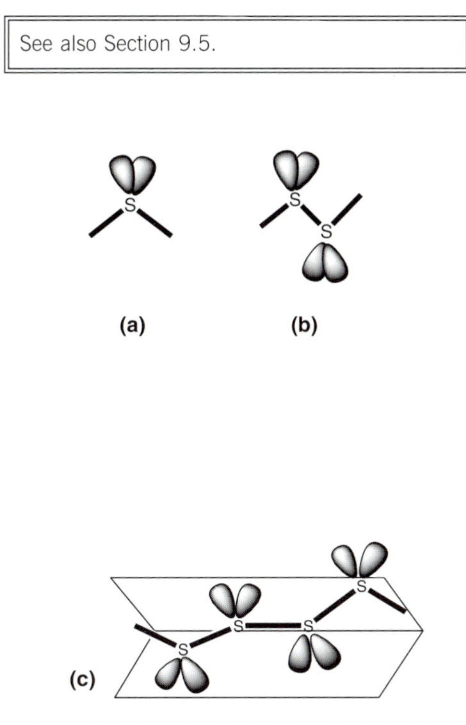

FIG. 18.1 Component units of polysulfides. A divalent S with two lone pairs (a) will tend to form a skew arrangement with its neighbour (b). For a chain, there will be an optimum dihedral angle between successive SSS planes (c)

the polythionates $[(O_3S)\text{-}S_n\text{-}(SO_3)]^{2-}$, for n up to about 20. These species were selected rather arbitrarily to represent the phenomenon of polysulfur and polyselenium chemistry. In this section, we develop this theme further.

If sulfur is compared with its neighbours, oxygen and chlorine, we can see that it has a much greater potential to form extended arrays of -S-S-S- links. Chlorine, with one electron more, has a fully satisfied electron configuration in diatomic Cl-Cl, and the halogens only form longer chains to a limited extent, with weaker interactions, and mainly involving iodine (compare Tables 17.19 and 17.20, and Fig. 17.59). Oxygen differs from S in its ability to form a stable double bond in O=O, and species with O_n chains readily break up with formation of O=O. Thus polyoxygen species are limited to $n=3$ and 4 (Section 17.7.1). In contrast, sulfur has one electron less than Cl, and any S=S is unstable with respect to -S-S-, encouraging chain formation. The S_2 molecule does occur in the gas at high temperatures and gives a purple, paramagnetic solid on rapid cooling, but it reverts to S_8 when allowed to warm to room temperature.

Sulfur chains are built from divalent S atoms with two lone pairs, and will extend in a preferred conformation which minimizes lone-pair/lone-pair interaction as indicated in Fig. 18.1. The long-known S_8 ring attains the optimum array of angles, and the S_6 ring of Engel's sulfur is more strained and readily converts (Section 17.7.1). Clearly, if the ring opens out into a chain, there are no constraints but the chain ends are reactive so that only very long chains survive, as in plastic sulfur where the chains probably coil into a helix.

From this, it is reasonable to find that the further polysulfur species fall into three groups (Table 18.1):

(a) Rings, with those larger than 8-membered more stable
(b) Anions S_n^{2-} and cations S_n^{x+} where $x=1$ or 2
(c) $XS\text{-}S_n\text{-}SX$ where monovalent X groups take up the terminal valencies. An example is the polythionates with $X=SO_3^-$.

18.2.1 Polysulfur rings

Development of a range of new ring compounds depended on the discovery of syntheses giving specific rings, rather than an intractable mixture. One such reaction is between polysulfur hydrides and polysulfur chlorides, e.g.

$$H_2S_8 + S_4Cl_2 \rightarrow S_{12} + 2HCl \tag{1}$$

This procedure gives a convenient route to S_6 and also yields the new rings S_{12}, S_{18} and S_{20}. A modification involves the cleavage of an S_5 unit from the stable, six-membered MS_5 rings formed by a number of transition metals. This route leads to rings with odd numbers of S atoms

$$(C_5H_5)_2TiS_5 + S_2Cl_2 \rightarrow S_7 + (C_5H_5)_2TiCl_2 \tag{2}$$

Use of S_4Cl_2 or S_6Cl_2 in (2) gives S_9 and S_{11} respectively. An attempt to make a ring containing the SO_2 unit by using SO_2Cl_2 in (2) yielded instead S_{10}.

Further, more convenient, routes have been found for some of these rings. Knowing the properties of the isolated species, it has been possible to identify them in classical melts and mixtures. Thus S_{12} is soluble in CS_2, but 150 times less so than S_8, and can thus be separated. All the polysulfur rings are relatively unstable to heat and light, with S_{12}, S_{18} and S_{20} the most stable after S_8. S_6 is very sensitive and a saturated solution of S_6 gives S_{12} after a short exposure to light. The odd-membered rings are also all very reactive, with S_7 polymerizing as low as 45°C.

It is convenient here to mention S_8O, formed by the reaction

$$H_2S_x + SOCl_2 \rightarrow S_8O + 2HCl$$

The species is stable in the dark at −20°C, but soon forms SO_2 and sulfur on warming. The structure is that of the S_8 ring with the SO bond angled towards the crown. The S_nO series has been extended to include $n=6$, 7, 9 and 10—all with similar preparations, stabilities and structures to S_8. There is also one disubstituted species, S_7O_2, formed as dark orange crystals by further oxidation of S_7O with CF_3COOH. It decomposes at 60°C, evolving SO_2, and calculations show the most likely structure to be a chair with the two O atoms 1, 3 or 1, 4.

18.2.2 Polysulfur ions

When sulfur is dissolved in strong acids (see Section 6.9), ring cations are formed. Typical are the long-known coloured solutions of sulfur in oleum, which are red, yellow or blue according to concentration. In such systems, S_4^{2+}, S_8^{2+} and S_{16}^{2+} have been identified. In addition, radical ions S_n^+ give paramagnetic properties. Of these S_4^+ and S_8^+ (formed respectively from S_8^{2+} and S_{16}^{2+}) are reasonably well established.

Table 18.1 summarizes some properties of the polysulfur species, and Fig. 18.2 indicates some of the rings.

The structure of the ion S_8^{2+} has been determined (Fig. 18.3); the bond lengths and angles are very similar to S_8, but the shape changes from the open crown to a half-closed form and the 283 pm S-S distance across the ring corresponds to a weak transannular bond. If we compare S_8 and S_8^{2+} we see that the two missing electrons are balanced by the extra S-S link. This process continues in S_4N_4 with only 44 valency electrons compared with 48 in S_8. As the structure of Fig. 17.53 shows there are now two transannular S-S interactions and the structure is closed up. If two electrons are added to a polysulfur chain, to give an anion S_n^{2-}, the chain ends are no longer radicals, and such anions are expected to be stable

TABLE 18.1 Properties of polysulfur species

	Colour	*M.pt.* (°C)	*S-S* (pm)	*SSS* (°)	*Dihedral* (°)	
S_2	Purple		189.0			Stable in gas at 800°C, paramagnetic. S_3-S_5 also in gas
S_6	Orange-red	50d	205.7	102.2	74.5	Unstable to light
S_7	Yellow	39	Partial structure reported			Unstable to heat and light, polymerizes 45°C
S_8	Yellow	119	206.0	108.0	98.3	Stable, several crystal modifications
S_9	Deep yellow	ca. 50d	205.2			
S_{10}	Yellow		203.3 207.8	103 110	75 79	Two different S environments in crystal
S_{11}						No details available
S_{12}	Yellow	148	205.3	106.6	86.1	Most stable after S_8
S_{14}	Intense yellow	117(d)	205.3	107	93.1	Stable in solid or CS_2 solution at room temperature. Decomposes on heating to S_7 and S_8
S_{18}	Lemon	128	205.9	106.3	84.4	Stable in absence of light
S_{20}	Pale yellow	124	204.7	106.5	83.0	Stable solid, unstable in solution
S chains			206.6	106	85.3	Mixtures of different chain lengths — averaging up to 10^5 S atoms per chain in melt
S_4^{2+}	Pale yellow		198			Square
S_8^{2+}	Deep blue		204.8	108	90	Figure 18.3
S_{19}^{2+}	Red		187 to 221			
S_3^{2-}	Blue			103		Chain
S_5^{2-*}		Terminal	205	109	76	Chain
		Central	207	106	69	

d = decomposes.
*also S_4^{2-} and S_6^{2} (chains). Values for S-S distances and for angles are mean values for the larger molecules.

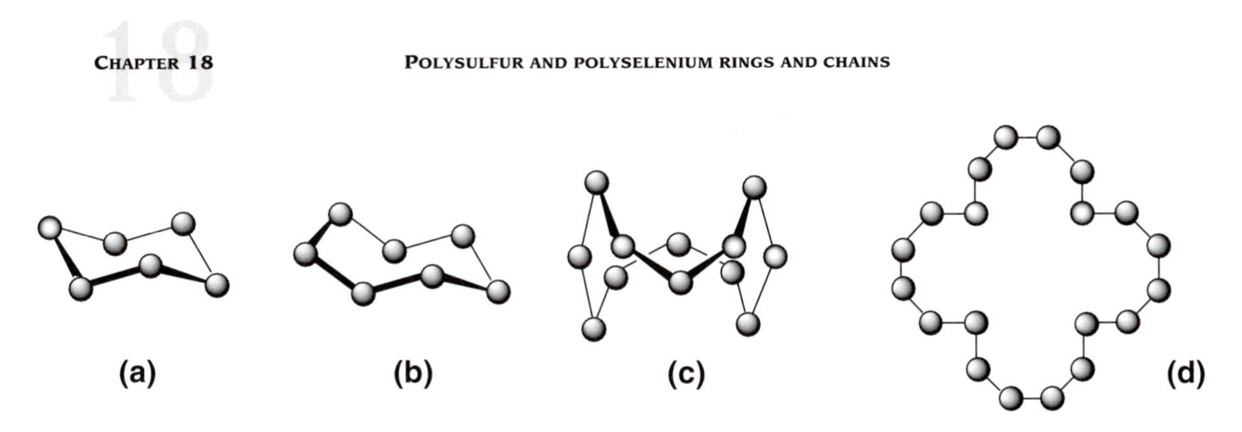

FIG. 18.2 Some sulfur rings: (a) S_6, (b) S_7, (c) S_{12} and (d) S_{20}

ULTRAMARINES

In the precious stone ultramarine, highly coloured radical anions S_2^- and S_3^- substitute for chloride ions in the cages of the host zeolite mineral sodalite (see Section 18.6.5). The ratio of these two sulfur species determines the actual colour, from blue (S_3^-) to green (S_2^-). Such sulfur species have to be generated simultaneously with the formation of the zeolite host, because the free sulfur anions on their own are highly unstable, whereas in the ultramarines the colour centres are remarkably stable.

Other related coloured ions are also known. The S_6^- ion, with a six-membered ring in a cyclohexane-like chair conformation, has been isolated as its Ph_4P^+ salt, and is orange-red in colour. For the heavier chalcogens, Se_2^- has been incorporated into a sodalite material giving a brilliant red solid, which may eventually find application as a pigment as a replacement for the current Cd(S,Se) red-yellow pigments which are often used, but are being phased out. Se_2, Se_2^{2-}, Te_2 and Te_2^-, which are all coloured, can also be incorporated as colour centres in zeolite cages.

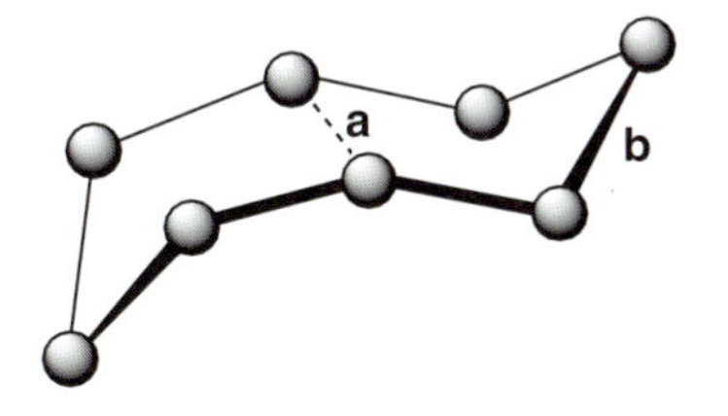

FIG. 18.3 The structure of the S_8^{2+} ion. For S_8, the S-S distance is 204 pm, for S_8^{2+}, a = 283 pm and b = 204 pm. (For the Se analogues, the distances are 234 pm in the molecule and 284 and 232 pm in the cation)

chains. (Alternatively, we can see that adding two electrons opens a ring by breaking one bond.) They are generally made by boiling S^{2-} solutions with sulfur and are stabilized by large cations, as in Cs_2S_6.

18.2.3 X-S_n-X species

Compounds with X = halogen are well known, and the disulfur members ($n = 2$) are included in Section 17.7.6. Longer chains are found in the chloro- and bromo-polysulfanes, S_nCl_2 and S_nBr_2. In the case of the chlorosulfanes, there is evidence that n may be as high as 100. However, individual members of the series have been isolated only up to $n = 8$ for both the chloro- and bromosulfanes. These compounds are all formed by dissolving sulfur in S_2Cl_2. The latter results from the reaction of Cl_2 on molten sulfur. It has a skew Cl-S-S-Cl structure, similar to H_2O_2, with an SSCl angle of 103°. The higher chlorosulfanes probably have chain structures similar to those of the anions. Individual compounds may be isolated by careful distillation or by chromatography. Mild conditions are necessary, as individual members readily disproportionate into mixtures of various chain lengths.

The hydrides, X = H, called *sulfanes*, are formed either from the polysulfides or from the chlorosulfanes:

$$S_n^{2-} + 2HCl \rightarrow H_2S_n + 2Cl^-$$
$$S_nCl_2 + 2H_2S \rightarrow H_2S_{n+2} + 2HCl$$

Compounds up to $n = 8$ have been isolated and the existence of higher members has been demonstrated by chromatographic studies. All are yellow and range from gaseous H_2S to increasingly viscous liquids as the chain length increases. The sulfanes may be interconverted by heating—indeed, any one member readily converts to an equilibrium mixture of the others and all ultimately revert to sulfur and H_2S, although slowly. The structures are not known but are probably chains like the polysulfide ions. The sulfanes

plus chlorosulfanes give syntheses of specific rings as in Equations (1) and (3), which gave the first clean synthesis of Engel's sulfur:

$$H_2S_2 + S_4Cl_2 \rightarrow S_6 + 2HCl \tag{3}$$

Compounds containing mixed sulfur–selenium chains, such as HSSeSeSH and HSSSeSe-SeSSH, have been prepared by similar reactions, starting from H_2S and Se_2Cl_2.

A third important group has X = a transition metal plus ligands. Both chain compounds and rings where the two ends of the S_n chain bond to the same M, are well known and these compounds are reviewed in Section 18.7. Equation (2) illustrates their use to synthesize particular rings.

18.2.4 Polyselenium and polytellurium rings and chains

As in sulfur, recent studies using chromatography for separation, and spectroscopy or crystallography for characterization, have established the existence of a number of selenium rings of different sizes. In the vapour of molten selenium, the main components are Se_5 and Se_6, whereas Se_7 and Se_8 are minor. Se_5, however, cannot be isolated as it converts very rapidly to the larger rings (S_5 behaves similarly). In solution, there is an equilibrium, $2Se_7 \rightleftharpoons Se_6 + Se_8$, which can be demonstrated by high pressure liquid chromatography. Separation of Se_6 and Se_7 is possible but difficult, as they readily convert to stable Se_8. Perhaps the easiest route will be by synthesis, as in

$$(C_5H_5)_2TiSe_5 + Se_2Cl_2 \rightarrow Se_7 + (C_5H_5)_2TiCl_2 \tag{4}$$

which has been established. A crystal structure determination shows Se_6 has a chair form like S_6 (compare Fig. 18.2a), and Se_7 has probably the structure of Fig. 18.2b. Se_8 is stable and known in several crystal modifications which all contain the crown ring (compare Fig. 17.45) packed in different ways. There is evidence for the existence of Se_9 and Se_{10} rings at very low pressures over subliming selenium, and rings such as Se_{16} have been proposed as intermediates in the conversion of Se_8 to the grey form Se_∞. None of these larger ring molecules has been isolated, though they might be more accessible via direct syntheses analogous to Equations (1) or (4). However, selenium differs from sulfur in the higher stability of Se_∞ relative to the rings, and the large-ring chemistry of selenium may remain limited. The established bond lengths and angles are listed in Table 18.2.

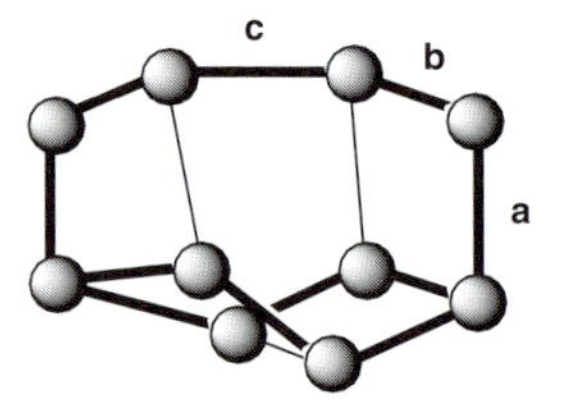

FIG. 18.4 The structure of Se_{10}^{2+}. The bonds from the 3-coordinate Se atoms a are longer (241–246 pm) than their neighbours b (223–227 pm) with an alternation along the chain making c = 236 pm. The lighter links are shorter than nonbonded distance at 330–350 pm

The very low dihedral angle means greatly increased interaction between neighbouring lone pairs in Se_6, and accounts for its ready transformation into Se_8. Note the somewhat smaller Se-Se-Se and larger dihedral angles of the stable Se_∞ and Se_8 forms compared with their sulfur analogues.

The M_4^{2+} ions (M = Se, Te)

Selenium, and also tellurium, forms cations in strong acid media and also in melts (compare Section 6.9). Yellow Se_4^{2+}, green Se_8^{2+} and red Se_{10}^{2+} have been isolated and their structures are known. There is also an indication of Se_n^{2+} ($n = 2$, 12 and 16) in melts. Te_n^{2+} species are known for $n = 2$, 4, 6, 7 and 8. The M_4^{2+} species are planar, with Se-Se 234 pm, and Te-Te 266 pm. The structure of one isomer of Te_7^{2+}, in $Te_7(AsF_6)_2$ (which was formed from $Te_4(AsF_6)_2$ and $Fe(CO)_5$ in liquid SO_2), consists of infinite chains of six-membered Te rings connected through bridging Te atoms. Another compound containing a Te_7^{2+} cation, $Te_7(WOBr_4)_2$, has a substantially different structure, consisting of two Te_4 rings which share a common Te; these units then link together in a complex double chain formed by Te-Te linkages. While Se_8^{2+} has the same structure as S_8^{2+} (see Fig. 18.3), Te_8^{2+} has the structure of two five-membered ring boats sharing an edge, or alternatively described as an

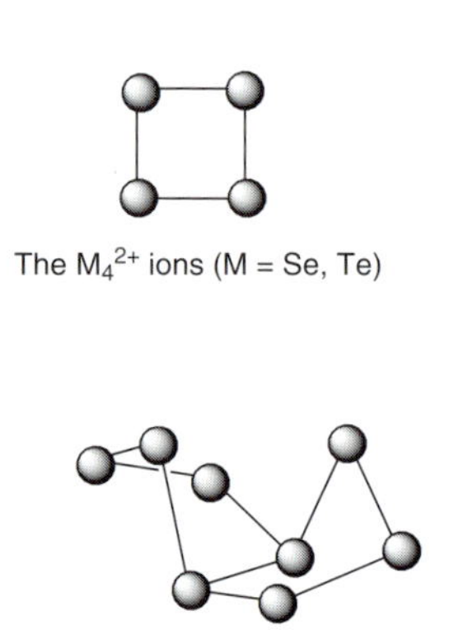

The Te_8^{2+} cation

TABLE 18.2 Some parameters of polyselenium species

	Se-Se (pm)	*SeSeSe* (°)	*Dihedral* (°)
Se_6	236	101	76
Se_8	234	106	101
Se_∞	237	103	101

TABLE 18.3 Some parameters of polyselenium anions in salts with large counterions

	Se-Se (inner) (pm)	*Se-Se* (outer) (pm)	*SeSeSe* (°)	Torsion (°)
Se_5^{2-}	230	235	107–109	98
Se_6^{2-}	230	234–236	109–113	82–98
Se_7^{2-}	230	233–235	109–113	82–98

8-membered ring which is nearer in structure to the crown form of S_8, and with a weak transannular link. The ring Te-Te distances are in the range 270–278 pm and the transannular Te···Te distance is 299 pm. These units are further weakly associated (Te···Te = 342 pm) into chains of rings.

The Se_{10}^{2+} ion is interesting. It forms a structure where a Se_6 ring in a boat-like form is joined 'prow to stern' by a chain of four Se atoms, shown schematically in Fig. 18.4. A range of distances are seen, implying bonding interactions of quite varying strengths. As with Se_∞, we see the evidence for weak additional interactions, (compare the Van der Waals' distance of 380–400 pm) a step along the path to a delocalized cluster.

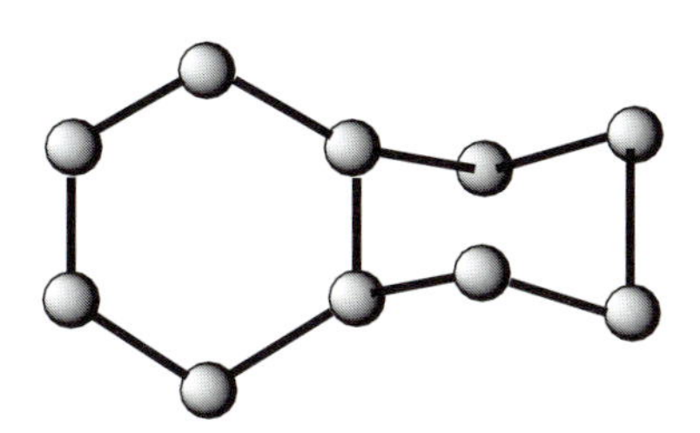

FIG. 18.5 The structure of the Se_{10}^{2-} ion. The structure can be viewed as equivalent to the carbon analogue decalin, with two fused cyclohexane-type chain rings, but with each selenium having two extra valence electrons. The geometry about the two central Se atoms is derived from a trigonal bipyramid, with the two lone pairs on each Se in equatorial positions

In liquid ammonia, selenium forms polyanions Se_n^{2-}. The species with $n = 3,4$ and 6 have been isolated while Se_2^{2-} exists only in equilibrium with Se^{2-} and Se_3^{2-}. The structure of Se_6^{2-}, isolated as the black $(Ph_4P)^+$ salt, is a kinked chain with Se-Se distances of 299 pm (terminal) and 231 pm (inner). Angles at Se are in the range 105–109°. Slightly different values are found when the counter-ion is the larger Rb^+ or Cs^+ stabilized by crown ethers or cryptands (Table 18.3).

As an example of a structurally more complex anion, Se_{10}^{2-} has been isolated in the presence of a bulky cation and has the bicyclic structure shown in Fig. 18.5 (contrast the cation, Fig. 18.4).

$Ph_3Te_3^-$; A TELLURIUM ANALOGUE OF THE WELL-KNOWN TRIIODIDE ION

The $Ph_3Te_3^-$ ion, which has the structure shown, has recently been prepared by several routes, initially by reduction of PhTeTePh with a samarium(II) complex, and subsequently by reduction with $NaBH_4$. The blood-red anion is air-sensitive, and is a direct tellurium analogue of the well-known I_3^- ion. As predicted by VSEPR, the Te_3 chain is slightly bent, with a TeTeTe angle of 172.9°. In contrast, there are a few examples of diorganyltritellurides, RTeTeTeR, which have the expected bent shape. The $Ph_3Te_3^-$ ion is asymmetric in the solid state, suggesting a PhTeTePh unit bonded to $TePh^-$, similar to the I_3^- ion, composed of I_2 plus I^-, which can also crystallize in asymmetric (or symmetric) forms, dependent on the countercation.

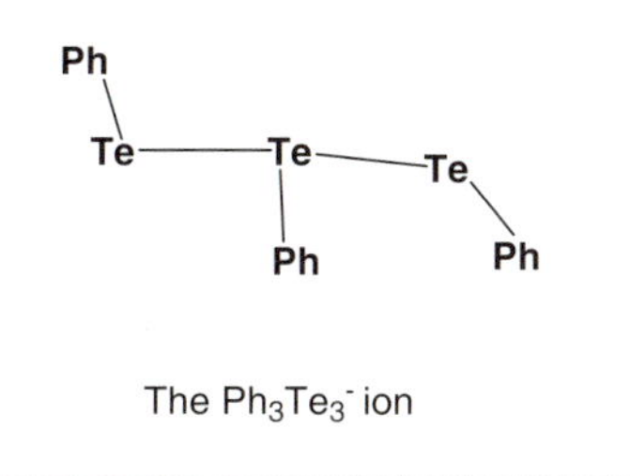

The $Ph_3Te_3^-$ ion

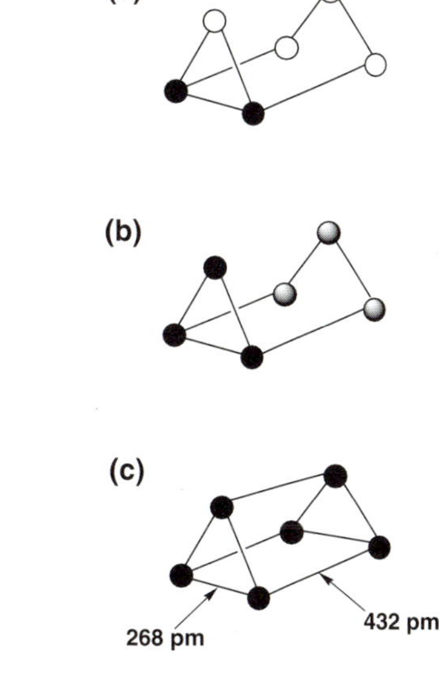

FIG. 18.6 Structures of (a) $Te_2Se_4^{2+}$ (b) $Te_3S_3^{2+}$ (c) Te_6^{4+}. Compared with S_6 (Fig. 18.2a), these show the effect of removing electrons and compensating by added bonds

There also exist mixed-element rings between S, Se and Te such as Se_nS_{8-n} ($n = 1, 2, 3$), Se_2S_{10}, and Se_7TeCl_2. Mixed cations include $Te_2Se_8^{2+}$, which has the same structure as Se_{10}^{2+} (Fig. 18.4) with the two Te atoms in the 3-coordinate positions. $Te_2Se_2^{2+}$ is square with alternate atoms. There are two mixed species which have new structures: $Te_3S_3^{2+}$ (Fig. 18.6b) and $Te_2Se_4^{2+}$ (Fig. 18.6a) have a boat structure with one transannular bond. These can be related to S_6 in the chair form, where removal of the two electrons in the cations requires a further bond. Again, this structure could be regarded in another way as resulting from the P_4S_3 type by removing one S atom (compare Fig. 17.37a).

Tellurium forms the only 4+ ion so far found, Te_6^{4+}, which has the structure of Fig. 18.6c. Here, instead of the removal of two electrons adding one more bond to the Fig. 18.6a structure, a more symmetric structure is adopted, but the bonds linking the two end triangles are distinctly long and weak.

A simple mixed element anion, TeS_3^{2-}, is a pyramid with Te-S = 233 pm and STeS = 106°. The more complex TeS_{10}^{2-} ion is an example of a simple linked-ring species (compare next section). It has two chair-form TeS_5 rings linked by the common Te with Te-S = 270 pm, S-S averaging 205 pm, the STeS angle is 89°, and angles at S in the range 104–109°.

18.3 Nets and linked rings

When more than two valencies are available, structures may grow in two or three dimensions. For cases where both 2- and 3-valent atoms are available, we find linked rings (more 2- than 3-valencies) or open cages or networks (more 3- than 2-valencies). Such behaviour is found in phosphorus and related compounds reviewed here.

18.3.1 Polyphosphorus compounds and related species

In Section 18.2, compounds containing chains of S or Se atoms have been described. If we move one place to the left in the Periodic Table, we see that a suitable structural unit is formed by taking an electron from one of the lone pairs of S, leaving P with one lone pair and three valence electrons (Fig. 18.7a). This unit has greater flexibility, and polyphosphorus compounds show more variety of structure than polysulfur species. The simplest closed unit using the building element of Fig. 18.7a is the P_4 tetrahedron found in white phosphorus (Fig. 17.29a), and other ways of combining trivalent P units are the linked opened tetrahedra of Fig. 17.29c, or the layer structure of black phosphorus thought to be similar to Fig. 17.28. An interesting range of mixed tetrahedra incorporating such P units is found for metal derivatives, as in $PCo_3(CO)_9$.

Further development of polyphosphorus species requires some saturation of the trivalency. A P-X unit (Fig. 18.7b) has two remaining valencies as does a P^- unit (Fig. 18.7c) which is isoelectronic with S (Fig. 18.1a). For X = halogen, examples are largely limited to the P_2X_4 species discussed in Section 17.6.4, where the last two X atoms terminate the chain.

However, when X = H, giving *polyphosphanes*, a large group of compounds has now built up. Phosphanes of formula $H_2P\text{-}(PH)_n\text{-}PH_2$ are hydrocarbon analogues, with the two extra hydrogens allowing chain termination. Straight chains up to $n = 3$ are established while branched chains become preferred for higher n values. Thus a second pentaphosphane, $(H_2P)_2P(PH)PH_2$ has been identified, and branched chain isomers of compounds with up to 9P atoms are known. In a second series, $(PH)_n$ for $n = 3$ to 10, rings are formed. In addition to the arrangement of the phosphorus skeleton, geometrical isomers are possible depending on the arrangement of successive lone pairs. A simple example is P_2H_4 which can give *cis, trans* or *skew* forms.

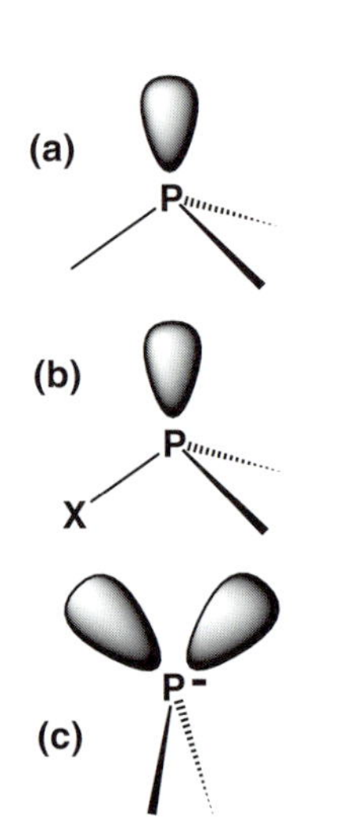

FIG. 18.7 Some building blocks for polyphosphorus compounds: (a) a single P atom which may build in three dimensions, (b) a P-X unit which may form chains, (c) the P^- unit isoelectronic with S

The hydrides are synthesized by the long-known hydrolysis of Ca_3P_2, which gives a mixture of hydrides which can be interconverted by gentle heating (compare Section 9.5), for example

$$2P_2H_4 \rightarrow P_3H_5 + PH_3$$

Reaction of a hydride with $LiPH_2$ or an alkyl-lithium gives the Li salt which is more stable to further reaction or disproportionation. Thus in monoglyme at −20°C

$$9P_2H_4 + 3LiPH_2 \rightarrow Li_3P_7 + 14PH_3$$

The addition of an acid converts the Li salt to the free phosphane P_7H_3. Another example is the ion $[HP\text{-}PH\text{-}PH]^{2-}$, which has been isolated from the reaction of P_4 with sodium in liquid

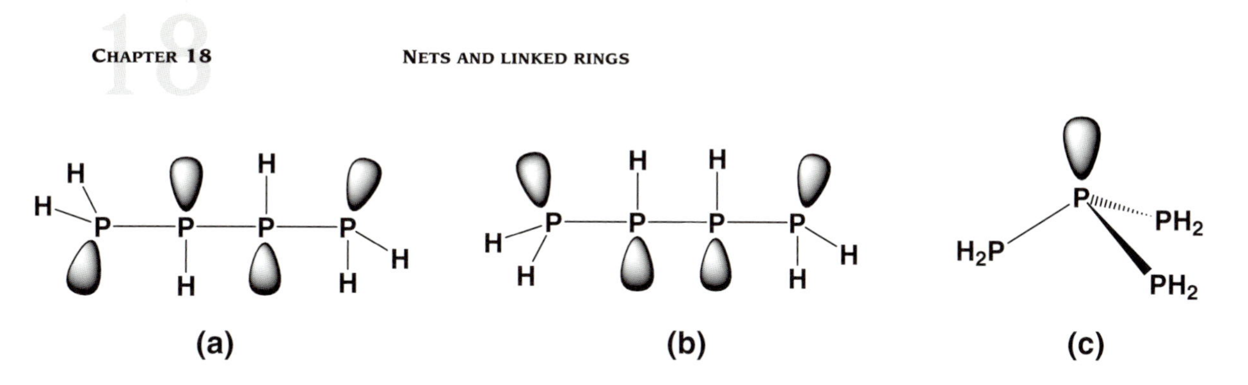

FIG. 18.8 The isomers of P_4H_6. (a) and (b) show the geometrical isomers of n-P_4H_6 and (c) is the branched chain isomer

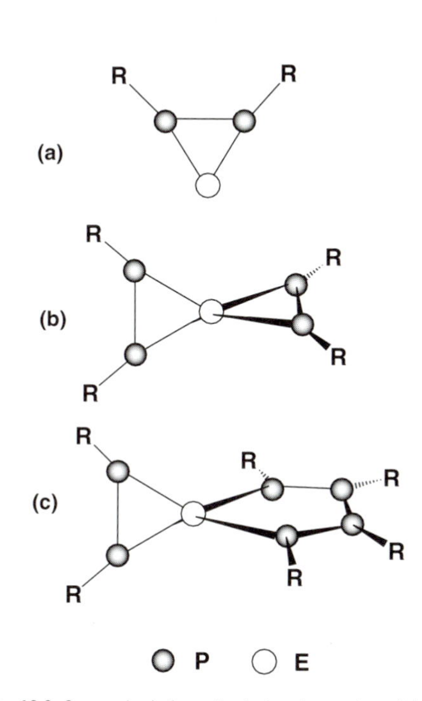

FIG. 18.9 Some mixed-element polyphosphorus rings. (a) The smallest ring, $(RP)_2E$, known for many E groups (see text). (b) A *spiro* structure formed by E = Si or Ge (c) a spiro structure as part of a larger ring in $Ge(PR)_6$. R = *tert*-butyl or similar bulky group

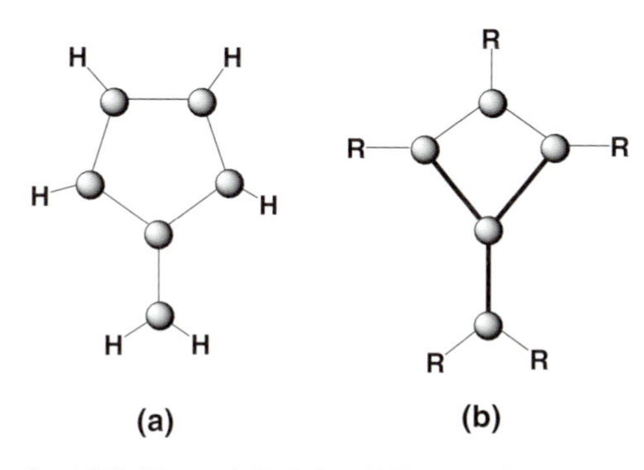

FIG. 18.10 Rings and sidechains. (a) The stable form of P_6H_6, (b) a four-membered ring stabilized by $C(CH_3)_3$ groups in $(RP)_5$

ammonia as $[Na(NH_3)_3(P_3H_3)]^-$, in which the two end PH groups chelate the $Na(NH_3)_3^+$ cation. The $P_3H_3^{2-}$ ion can be viewed as the triphosphane H_2P-PH-PH_2 which has been doubly deprotonated.

These are very striking examples of the development of a new area of chemistry out of a field which was seen to be very difficult. The hydrides are very air-sensitive, readily interconvert, the higher molecular weight species are amorphous, and no solvents were readily available. Because of these difficulties, the initial steps were very slow, with only PH_3, P_2H_4 and P_4 well established. The very richness of the chemistry made interpretation and characterization more difficult.

With much more powerful techniques, including mass spectroscopy and particularly modern ^{31}P nmr carried out at low temperatures to avoid interconversion, the first few additional compounds were characterized. Then it was possible to probe further and identify new species when they occurred in the presence of known ones. Synthetic strategies were further improved and could be monitored, allowing characterization of more compounds, until the whole field began to develop rapidly.

To give a flavour of the chemistry, a few examples are selected. The known hydrides are listed in Table 18.4.

P_4H_6. The two geometrical isomers of the straight chain have been identified, together with the branched chain isomer Fig. 18.8.

In Fig. 18.8a and b, all substituents on the two central P atoms are different, and thus they are chiral centres—(a) has *d* and *l* forms while (b) is *meso*. Different preparation routes give different proportions of (a), (b) and (c), allowing their identification in mixtures. Replacing H by bulky substituents favours the less-hindered isomer. Thus, $P_4(t\text{-Bu})_4H_2$ has alternating *tert*-butyl groups in an (a) structure. Branching is preferred as chain length increases. Thus the P_6H_8 samples characterized so far are H_2P-$P(PH_2)$-PH-PH-PH_2, H_2P-PH-$P(PH_2)$-PH-PH_2, and H_2P-$P(PH_2)$-$P(PH_2)$-PH_2 but not H_2P-PH-PH-PH-PH-PH_2.

Simple rings, P_nH_n, and *planar rings P_n^{x-}.* The smallest ring, P_3H_3, has been identified, but it very easily rearranges, forming P_5H_5. Its organic derivatives, P_3R_3, are stable and readily synthesized. The lone pair on the P atoms may act as donor, as in $(RP)_3Cr(CO)_5$ where the phosphorus ring occupies one position in the octahedron around Cr. In $(RP)_3[Cr(CO)_5]_2$, two different P atoms donate to the two Cr atoms and so the ring links the two octahedra.

A large variety of mixed-element three-membered rings is known of the general type of Fig. 18.9a, where E may be CR_2, C=CR_2 or related species, SiR_2, GeR_2, SnR_2, BNR_2, NR, AsR, SbR, S or Se. For E = Si, Ge or Sn, two rings participate in the *spiro* unit of Fig. 18.9b.

Larger rings are nonplanar, and the five-membered one is relatively stable. Above $n = 5$, there is a strong tendency to form rings with side-chains: thus P_6H_6 is as shown in Fig. 18.10a. The four-membered ring size is not favoured, and is found only when bulky substituents are present, as in Fig. 18.10b. Such stabilized rings are also found in the sequence of molecules $P_nBu^t_{n-2}$ for $n = 8$, 9 and 10 which consist of two P_4 rings slightly bent across a diagonal. For $n = 8$ the link is directly between an apex P atom of each ring and for $n = 9$ and 10 there are, respectively, a PBu^t and a PBu^t-PBu^t bridging unit. Larger rings are also found with hetero-elements. To quote only one example, we find the mixed-ring *spiro* skeleton of Fig. 18.9c in $GeP_6(CMe_3)_6$.

The above rings correspond to the saturated hydrocarbons like cyclopentane. It was surprising, initially, to find a different formula type in P_5^-. This was formed in the reaction of white phosphorus with alkali metals in diglyme, giving for example $[Na(diglyme)_x]^+P_5^-$, and also with $LiPH_2$ in THF giving $[Li(THF)_x]^+P_5^-$. Its properties suggest that P_5^- is a planar ion with delocalized charge, isoelectronic with cyclopentadienide, $C_5H_5^-$ (compare Section 16.4). That is, the P_5^- ion is analogous to aromatic hydrocarbon ions. Strong support for this view came from the isolation of the novel structure

$$(C_5Me_5)Cr(P_5)Cr(C_5Me_5)$$

where the two (C_5Me_5) rings and the P_5 ring are all planar and parallel, making a triple-decker sandwich exactly analogous to that of Fig. 16.7c. The mixed ring, P_4CH^-, has been characterized and is also planar with delocalized charge.

A second member of the aromatic series is found in $(C_5Me_5)Mo(P_6)Mo(C_5Me_5)$, the analogous triple-decker molecule which contains the planar P_6 ring analogous to benzene. This compound was prepared by reacting white phosphorus, P_4, with a molybdenum precursor. Thus the P_6 ring formed spontaneously in the course of the reaction, arguing a significant stability in this environment.

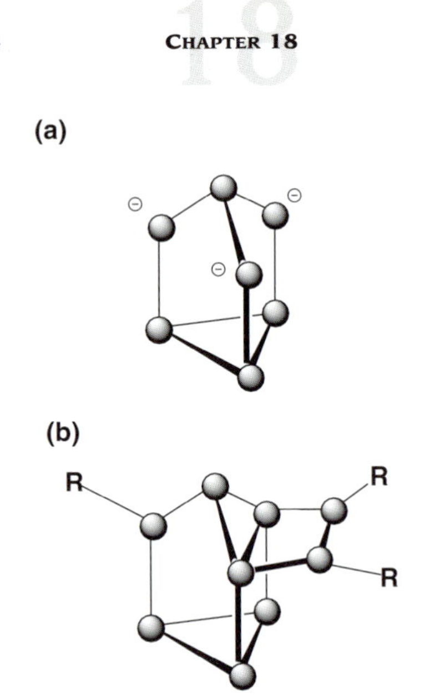

FIG. 18.11 The structure of (a) P_7^{3-} and (b) R_3P_9

Networks or open clusters. Perhaps the most intriguing polyphosphorus compounds are those where the number of H atoms is less than the number of P atoms. These are built of frameworks involving both trivalent and divalent units, Fig. 18.7. Out of the wide range of compounds, we may select one in particular as examplar. The P_7^{3-} cluster is the starting point. This is isoelectronic with P_4S_3 (Fig. 17.37a) and has the same structure with P^- replacing S (Fig. 18.11). Then more complex structures are built up by adding P_x units at the P^- positions, converting some of the divalent bridge P atoms to trivalent ones.

The simplest example is found in the R_3P_9 molecules where a P-P unit bonds to two of the bridging positions (Fig. 18.11b). With a bulky R group, this compound can be crystallized and the structure determined. It is likely that P_9H_3 and P_9^{3-} have the same framework.

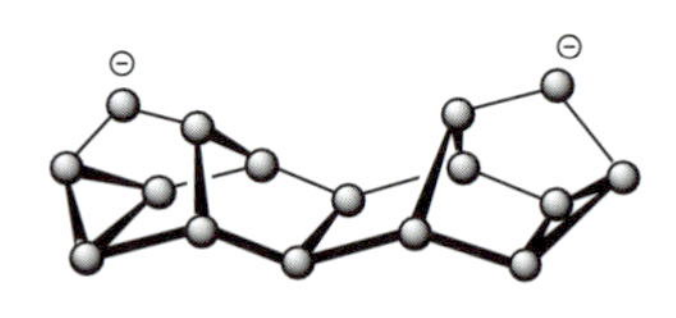

FIG. 18.12 The P_{16}^{2-} ion

Larger molecules may be built up if units are linked together. In the P_{16}^{2-} structure, two P_7 units are linked using a common P-P unit bonded to two of the bridging P^- positions in each P_7 (Fig. 18.12). The third bridge position in each P_7 retains its negative charge. Thus the whole P_7 unit acts as a divalent building block, essentially bonding through a P-P edge involving two of the P^- positions.

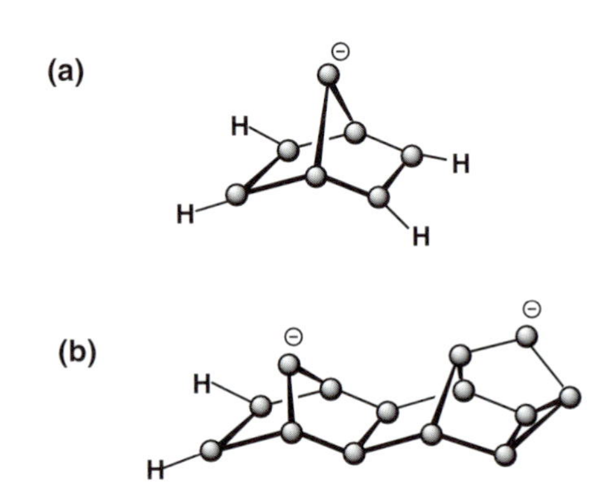

FIG. 18.13 More open polyphosphorus framework ions: (a) $P_7H_4^-$, (b) $P_{14}H_2^{2-}$

By opening up the P_7 skeleton of Fig. 18.11a—by adding two electrons—we find the structural unit of the anion $P_7H_4^-$ (Fig. 18.13a) which results from breaking one of the bonds in the base P_3 triangle. This open P_7 unit then links with a closed one in the ion $P_{14}H_2^{2-}$ (Fig. 18.13b). This is built up by removing two H atoms from one edge of Fig. 18.13a, and using these two positions to join to two bridge positions of Fig. 18.11a.

As final examples, we find the elegant structures of P_{21}^{3-} (Fig. 18.14) and P_{26}^{4-} (Fig. 18.15), which have two end units of the Fig. 18.11a form joined by one and two, respectively, units of the Fig. 18.13a form. As Table 18.4 shows, there is a further wide variety of polyphosphorus compounds. As far as current experimental evidence goes, the main structural theme is one of relatively open clusters, linked together by network units, as in the above examples. Structures will generally become more closed as the H/P ratio decreases.

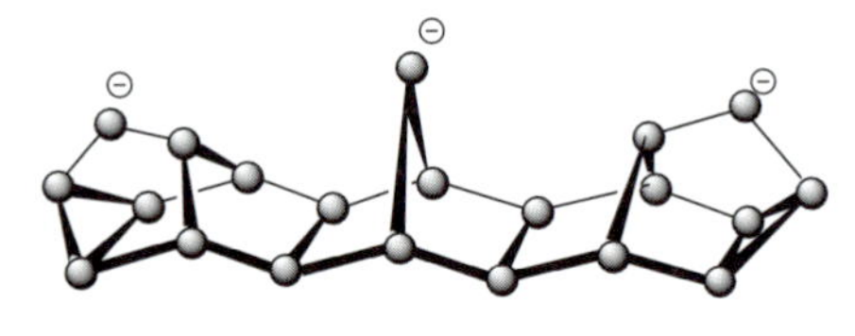

FIG. 18.14 The P_{21}^{3-} ion

18.3.2 Polyarsenic compounds and other analogues

Included in Table 18.4 are the polyarsenic compounds of similar formula to the polyphosphanes. Structural studies are less developed—in particular, there is no useful arsenic nmr nucleus equivalent to ^{31}P, so this powerful technique is not available. While it is likely that some structure types will be common to P and As, it is already clear that there are differences. Thus, both elements form E_8R_6 with $R = CMe_3$, but with different structures, as shown in Fig. 18.16.

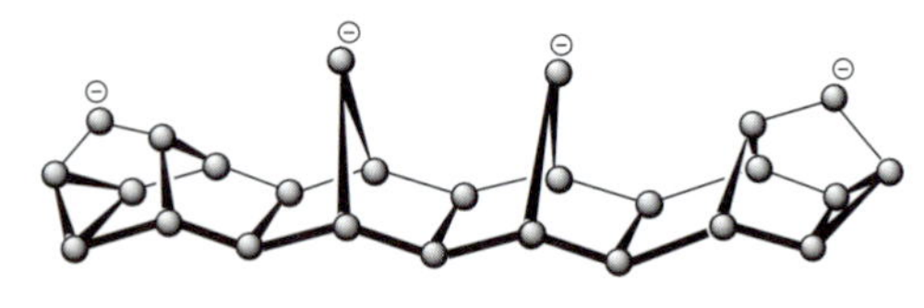

FIG. 18.15 The extended polyphosphorus ion, P_{26}^{4-}

In general, the As-H bond is weaker than the P-H one, so that there are fewer hydrogen-rich arsenic compounds and more use of organic derivatives is to be expected in developing the field. Rings are established for $(AsR)_x$ in the cases $R = Me$, CMe_3 or Ph, $x = 3$, 4, 5 or 6.

TABLE 18.4 Polyphosphorus and polyarsenic compounds

Formula	*E = P; R = H* (*or the corresponding anions of P or As*)	*E = As; R = C(CH$_3$)$_3$*
E_nR_{n+2}	$n = 1$ to 9	
E_nR_n	$n = 3$ to 10	
E_nR_{n-2}	$n = 4$ to 12	$n = 4$ to 9
E_nR_{n-4}	$n = 5$ to 13	$n = 6$ to 13
E_nR_{n-6}	$n = 7$ to 15	$n = 8$ to 16
E_nR_{n-8}	$n = 10$ to 17	$n = 10$ to 14, 20
E_nR_{n-10}	$n = 12$ to 20	$n = 12$, 13
E_nR_{n-12}	$n = 13$ to 20	$n = 14$, 16
E_nR_{n-14}	$n = 15$ to 21	$n = 16$
E_nR_{n-16}	$n = 17$, 19, 20 to 22	
E_nR_{n-18}	$n = 19$ to 22	
The ion P_{26}^{4-} is the first case of the E_nR_{n-22} family		

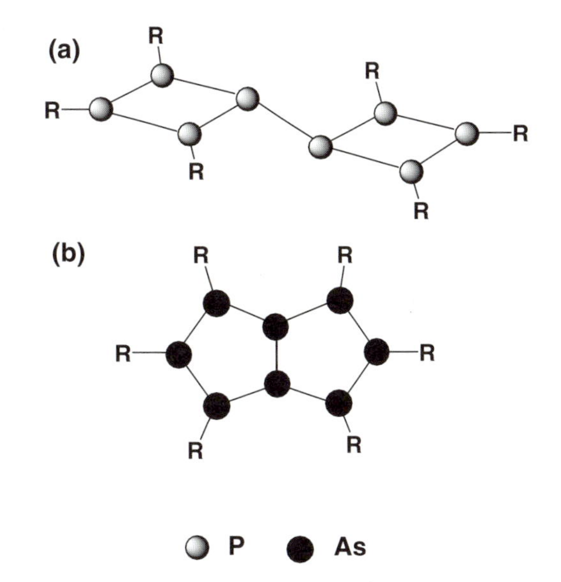

FIG. 18.16 Contrasting structures of polyphosphorus and polyarsenic compounds: (a) P_8R_8 (b) As_8R_8 ($R = CMe_3$)

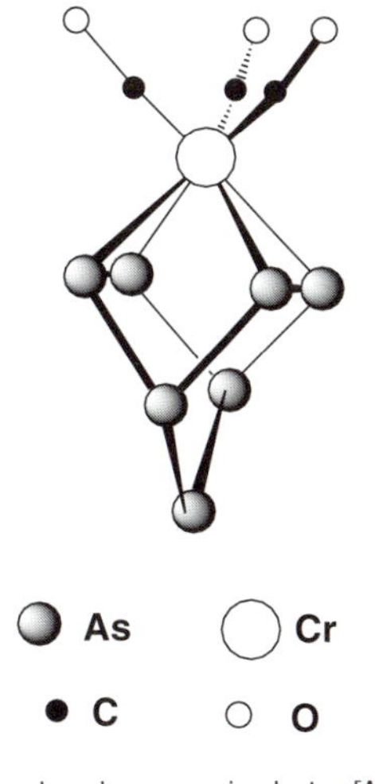

FIG. 18.17 The chromium–arsenic cluster $[As_7Cr(CO)_3]^{3-}$

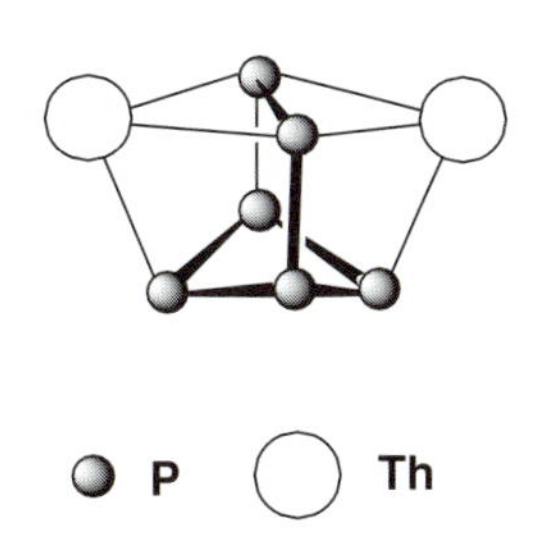

FIG. 18.18 The thorium–phosphorus core of the cluster $Cp''_2Th_2P_6$. Each thorium atom also bears two Cp″ ligands (not shown) where $Cp'' = \eta^5$-1, 3-di-*tert*-butylcyclopentadienyl

The structure of $(PhAs)_6$ shows a chair formation for the As_6 ring. The $(PhSb)_6$ ring is similar, and 4- or 5-membered antimony rings are also found.

18.3.3 Mixed systems

We have already indicated the close relation between the polyphosphorus anions and mixed P-S compounds. There seems to be no reason why an extensive chemistry of polyphosphorus–sulfur should not exist, and this should also extend to As and Se, together with mixed species of three (compare PAs_3S_3) or all four of these elements.

Simple mixed hydrides of phosphorus and Si or Ge, such as GeH_3PH_2 or $(SiH_3)_2PH$, have long been known, and a range of mixed Si-P skeletons is emerging, particularly those where a divalent R_2Si unit replaces P^-. Thus we find $P_4(SiMe_2)_3$ where the Me_2Si units lie in the bridging positions of Fig. 18.11. In $P_4(SiMe_2)_6$ the $SiMe_2$ units bridge all six edges of the phosphorus tetrahedron in a structure analogous to P_4O_6 (Fig. 17.35). In such structures, together with many compounds where the $SiMe_3$ group is terminally bonded to P (acting as a bulky substituent), an extensive chemistry of mixed phosphorus–silicon compounds is growing rapidly. Again, there seems no reason why we should not find similar compounds involving As and Ge.

As an extension of this we note that metals can also be incorporated in such cluster systems, and the polyphosphorus, -arsenic or other element cluster can be viewed, in one sense, as a ligand coordinating to the metal. Examples of recently prepared species are shown in Fig. 18.17 and Fig. 18.18 for an arsenic and a phosphorus cluster containing a chromium and two thorium atoms, respectively. Related polysulfur and -selenium species also act as very good ligands towards metal centres and we develop this theme further in Section 19.2.

18.4 Cluster compounds of the p block elements

When valencies of 3 or more are available, the atoms form three-dimensional units, open or closed clusters. Clusters may be described in terms of localized bonds along the edges, or of delocalized bonds. For closed clusters, as in this section, the latter description is more valid.

18.4.1 General

Closed clusters have been seen for some borides (Fig. 17.7 and Table 17.5), for the E_4 tetrahedral form of P, As and Sb (Fig. 17.29a), and for the Bi_9^{5+} ion which is one component of $Bi_{12}Cl_{14}$ (Fig. 17.40c). In these species, the structure is closed, often a regular solid with equilateral triangular faces. Such regular structures are also found for the boron hydride ions of formula $(BH)_n^{2-}$ (compare Section 9.6), while all the other boron hydrides, and related species like the carboranes, are fragments of such regular structures formed by removing one or more apices. Other electron deficient compounds behave similarly: for example, methyl-lithium crystallizes as $(LiMe)_4$ with a Li_4 tetrahedron bridged on each face by a methyl group, while a chair structure is found in $(EtLi)_6$.

In addition to these classes of compound, there is a further, steadily growing, range of cluster compounds formed by p block elements. The structures are regular solids, or nearly so, and we shall look at two different types; (a) the sub-halides of boron, and (b) the 'naked metal' clusters of the later members of the carbon, nitrogen and oxygen Groups.

It is convenient first to review one general approach to clusters, which involves the rationalization of structures in terms of the electron count. While there is significant theoretical support for the approach, it is best seen as a model and rationalization of observed structures. It involves counting the electrons in a cluster, allocating certain of them to bonds to external substituents, and assigning the rest to hold the skeleton together. The expected skeletal structure is derived from the number of cluster atoms and the number of these electrons. The scheme was first developed for boron hydrides, but has now been extended to all types of transition element and Main Group clusters.

18.4.2 Skeletal electron pairs

The *skeletal electron pair* rationalization postulates that $(N+1)$ such pairs are required to give a stable regular cluster with N vertices. The regular clusters are those with all the faces

TABLE 18.5 Shapes and skeletal electron pairs

No. of skeletal electron pairs $N+1$	*Shape and no. of atoms in the cluster*			
	closo N	*nido* $N-1$	*arachno* $N-2$	*capped* $N+1$
5	4 Tetrahedron	3 Triangle		5 Trigonal bipyramid
6	5 Trigonal bipyramid	4 Tetrahedron	3 Triangle	6 Bicapped tetrahedron
7	6 Octahedron	5 Square pyramid	4 Square	7 Capped octahedron
8	7 Pentagonal bipyramid	6 Pentagonal pyramid	5 Irregular pyramid	

A *butterfly* structure is two triangles sharing an edge. The triangles are usually not coplanar. The *capping* atom is placed regularly over a face. Thus, if the face is a regular triangle, the new faces formed by the capping atom are all regular triangles.

equilateral triangles. The structures expected for some of the simpler cases of $(N+1)$ skeletal electron pairs and N atoms are listed in Table 18.5 which is a fuller version of Table 9.6. The skeletal electron pair (SEP) method is also commonly talked of as applying Wade's rules, after one of the early developers of the approach.

The rules for counting skeletal pairs are simplest for the s and p elements. Let us start with the $(BH)_n^{2-}$ ions mentioned in Section 9.6. For these, we note that each B-H bond points outwards from the cluster, and has all the properties of a normal two-electron bond. Thus we allocate one H per B, together with 2 electrons and an outward-pointing orbital on B, to form these external B-H links. The remaining electrons, and the remaining three orbitals per boron, remain to bond the cluster. For N boron atoms, $N+1$ of these orbitals are stable and bonding, so the cluster forms if there are $N+1$ skeletal electron pairs to fill them.

If there are fewer than N atoms, the structure is formed by removing an appropriate number of vertices from the regular figure. Thus, for 7 electron pairs, $N=6$, and the structures are (i) an octahedron if there are 6 atoms in the cluster (ii) a square pyramid if there are only 5 atoms, and (iii) a square or a butterfly if there are only 4 atoms. Likewise, if there are more than N atoms the extra ones fit over one or more faces of the regular solid in a *capping* position.

To describe these forms, a general terminology is used. The regular figure is called *closo* (simply indicating closed), a structure lacking one apex is called *nido* (from Greek indicating nest-shaped), one lacking two apices is *arachno* (Greek spider, or web-like), one lacking three apices is *hypho* (Greek, net). Extra atoms give *mono-*, *bi-*, *tri-*, etc. *capped* structures. Table 18.5 shows examples.

The skeletal electron pair counting rules for different types of cluster are summarized as follows.

Boron hydrides, carboranes and similar species. Regard the hydride as $(BH)_nH_x$ so that any hydrogens in excess of 1 per B contribute to the cluster. Such extra hydrogens are found bonded to the basic B_n cluster, often face- or edge-bridging, or bonded terminally to B atoms at open faces. Any charges are added or subtracted from the total count, and for mixed-element clusters such as the carboranes, each element contributes its valency electrons, and each is assumed to bear one terminal hydrogen. Thus, $C_2B_3H_7$ has $(2\times4) + (3\times3) + 7 = 24$ electrons, of which 4 are needed for 2CH and 6 for 3BH terminal groups. This leaves 7 skeletal pairs and the structure should be a square pyramid. Similar treatments apply to other additional elements. Note that the electron counting process gives no prediction of where heteroatoms are placed, nor does it define where any additional hydrogens are bonded. However, it should be emphasized that the skeletal electron pair theory is very successful in rationalizing literally hundreds of boron clusters.

Transition metal clusters. An account of the application of SEP or Wade's rules to metal carbonyl clusters is given in Section 16.8 with a number of examples. Some additional considerations are summarized here. A guide to the structures is the realization that a group such as $M(CO)_3$ is similar to BH in the orbitals available for bonding a cluster. BH has a centrally pointing orbital (for example the sp hybrid opposite to that forming the B-H bond) together with two p orbitals at right angles to this, tangential to the cluster. The $M(CO)_3$ unit has a centrally pointing orbital (say d_{z^2}) together with two tangential ones (say d_{xz} and d_{yz}) of similar energies to the HB ones. Such a situation is called *isolobal*, and isolobal groups often behave in similar ways, and substitute one for another. Note that there do not need to be 3 CO groups present on each metal; those of the 12 electrons not needed for M-CO bonding, remain nonbonding: 12 electrons per M are subtracted whatever the structure. Thus, $Rh_6(CO)_{16}$ has $(6\times9)+(16\times2) = 86$ electrons, less 12 per Rh leaves 7 skeletal pairs, and the structure is octahedral. In the actual structure, there are two terminal CO per Rh and the other four CO face bridge every second face.

Other ligands contribute electrons appropriately: H or halogens provide 1, other donors like PF_3 give 2, while NO is usually a three-electron source. Encapsulated atoms, such as C in $Ru_6(CO)_{17}C$, contribute all their valency electrons. Thus $Fe_5(CO)_{15}C$ counts (5×8) + $(15\times2) + 4 = 74$ less (12×5) gives 7 skeletal electron pairs, and the predicted structure is a square pyramid, as found.

The skeletal electron counting approach becomes less useful for transition metal clusters of more than 8 or 9 atoms, and has limitations even with smaller clusters. However, it is generally successful with 3 to 8 atom species, and it is also very valuable for dealing with the metalloboranes which may be regarded as mixed species between the boranes and the metal clusters.

Other cases. The application of electron counting to Main Group clusters is based on the postulate that there is an outward pointing lone pair on each cluster atom (equivalent to the two electrons of the BH unit in the boranes) which does not contribute to the bonding. There are also mixed cases, such as metal clusters incorporating Main Group atoms, for which the approach has moderate success.

Overall, skeletal electron pair counting has had a remarkable success over a very wide range of compounds. It has played an important role, not only in ordering many structures and highlighting anomalies for further study, but also in emphasizing the underlying similarities between classes of compound which seem, at first sight, to be quite different.

18.4.3 Boron subhalides

As well as B_2X_4 and the lower fluorides, discussed in Section 17.4.4, boron forms $(BCl)_n$ for $n = 4$, 8, 9, 10, 11 and 12, $(BBr)_n$ for $n = 7$ to 10, and $(BI)_n$ for $n = 8$ and 9, together with a few related compounds like mixed chloride-bromides.

The structures of three of the compounds have been determined. Each boron bears one terminal chlorine atom: B_4Cl_4 has a regular B_4 tetrahedron; in B_8Cl_8, the boron atoms form a dodecahedron, while in B_9Cl_9 the nine boron atoms lie at the apices of a tricapped trigonal prism. All these are regular arrangements for these numbers of atoms, and it is assumed that the remaining compounds are also regular. We note that the boron subhalides are all short of one electron pair for a closed cluster but the extent of electron donation from the halides is unknown.

The chlorides were prepared from B_2Cl_4 (itself prepared by a radio-frequency discharge on BCl_3) by heating to 1000°C, which gives a mixture of all n species. One specific synthesis is of B_9Cl_9 by treating $B_9H_9^{2-}$ with $SOCl_2$. B_2Br_4 is prepared similarly, and this disproportionates on standing at room temperature to give the tribromide and a mixture of the sub-bromides. Particular members of the group can also be made from the chloride by halogen exchange with Al_2Br_6. B_8I_8 and B_9I_9 were formed when B_2I_4 was melted at 100°C. Characterization depends heavily on mass spectra, helped by the fact that B and all the halogens are polyisotopic, so their combinations give characteristic intensity patterns.

Only a limited study of reactions has so far been made. Treatment of B_9Br_9 with $SnMe_4$ gives a number of methyl derivatives $B_9Br_{9-x}Me_x$ for $x = 1$ to 6, with apparently the same B_9 skeleton.

One link between the halides and the carboranes is the compound $(BCl)_{10}(CH)_2$, which is icosahedral, like the isoelectronic $B_{12}H_{12}^{2-}$. These compounds, taken with the hydrides and binary borides, show boron has a strong tendency to form clusters, whose exploration will undoubtedly continue to attract interest.

18.4.4 Naked metal cluster ions

The first observations of compounds which are now known to be polynuclear anions came a century ago when it was first observed that lead and other heavier p block elements gave a sequence of colours when they were treated with solutions of sodium in liquid ammonia. Later work up to the 1930s showed that such solutions give the polysulfur and polyselenium chain anions, and suggested the other elements gave more condensed species.

For example, germanium, tin and lead give polyanions in liquid ammonia. The tin or lead dihalides react with sodium in ammonia to give, first, the insoluble Na_4M species, and these add extra metal atoms to give intensely coloured solutions of Na_4Sn_9 or Na_4Pb_9. This latter compound was formulated as an ionic species with Pb_9^{4-} ions, on the basis of conductance and electrolytic experiments. For example, electrolysis was found to deposit 2.25 lead equivalents for each faraday passed. The compounds are heavily ammoniated and decompose when the ammonia is removed. GeI_2 behaved similarly. When ethylene-diamine (en) was added to alloys of the p elements with alkali metals, products could be

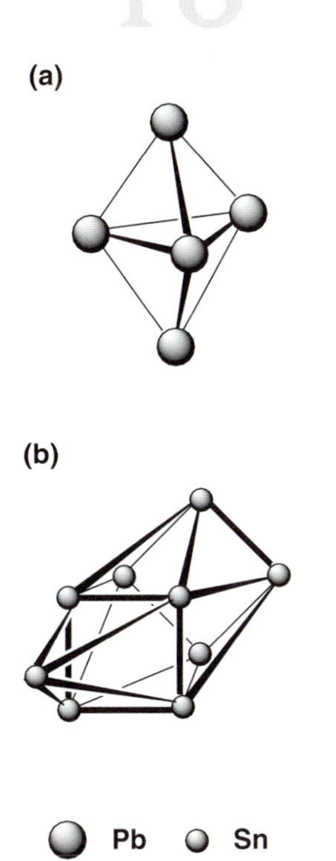

FIG. 18.19 The structure of (Na crypt)$^+$-stabilized (a) Pb_5^{2-} and (b) Sn_9^{4-}. (a) is a trigonal bipyramid with equatorial Pb-Pb = 300 pm and axial-equatorial Pb-Pb = 323 pm. (b) is a square antiprism capped on one square face with Sn-Sn distances of 293 to 302 pm, except that the edges of the capped face are expanded to about 325 pm

isolated, such as $Na_4(en)_5Ge_9$, since the solvating interaction was strong enough to avoid loss of the en. It is probable that, if the cation is in close contact with the large polynuclear anion, reverse electron transfer occurs and a metallic alloy, such as Na_4Ge_9, forms once the solvent NH_3 is removed, and this was avoided in the en complex cation. Crystal structure determinations on such complexes in the 1970s, at last gave some certainty to the field.

By this time, the cryptates and crown ethers (Fig. 10.5) had been evolved specifically to form stable complexes with cations such as the alkali metals. By using cryptate-stabilized alkali metal ions (see Section 10.5), polyanions have been isolated as solids. Crystal structures are known for Na(crypt)$^+$ salts of red Pb_5^{2-} (Fig. 18.19a), orange Sn_5^{2-} and dark red Sn_9^{4-} (Fig. 18.19b). The green Pb_9^{4-} species, strongly indicated by conductivity in solution, has not yet been isolated: perhaps this requires use of an even larger cation.

One particularly striking product resulted from the reaction of an alloy of composition KGe, which gave $K(crypt)_6(Ge_9)^{2-}(Ge_9)^{4-}$.2.5en, where the difference of two electrons between the two germanium clusters gives different structures. A second unexpected species is the odd-electron Sn_9^{3-} cluster. Table 18.6 lists established clusters.

Most of the structures have been determined crystallographically, or by comparison of spectroscopic properties with those of established compounds. The mixed Sn/Pb and Sn/Ge nine-atom clusters were characterized by ^{119}Sn and ^{205}Pb nmr and all ten species occur in the mixture. Interestingly, the Pb_9^{4-} species is seen here. It is noteworthy that some ions with high charge/metal ratio indicated in the solution experiments, such as M_3^{3-} and M_5^{3-} for M = As, Sb or Bi, have also not been isolated as solids. These ions may need an even larger cation than the crypt species for stability, as may the preparation of clusters with more than the present upper limit to cluster size of about 10 atoms.

An alternative synthesis derives from the early observation that Bi dissolved in $BiCl_3$ gives the cluster Bi_9^{5+} (Fig. 17.40c) as part of a mixture of products (see Section 17.6.4). Other fused salts provide similar media, and $AlCl_3$ melts have been generally used. Addition of NaCl increases the basicity of such melts and this type of nonaqueous system has proved to be a source of cluster ions, as well as other unusual species (compare Section 18.9). Using $NaAlCl_4$ melts, two more bismuth clusters, Bi_5^{3+} and Bi_8^{2+} have been prepared. These ions are probably trigonal bipyramidal and square anti-prismatic structures respectively. Reinvestigation of bismuth species in liquid ammonia, stabilized by large solvated alkali metal cations (compare Section 10.5) has led to the isolation of the cluster anion, Bi_4^{2-}, isoelectronic with the known Te_4^{2+} and Se_4^{2+}, with a square of metal atoms.

While the structures of the clusters are clearly related to the number of electrons, there are obviously finely balanced factors as shown by the two different structures of the 40-electron species. If we assume that there is one outward pointing lone pair on each atom, the remaining electrons, listed as *cluster electron pairs* in Table 18.6 may be used in the skeletal electron pair approach. The 18-electron, M_4^{2-} compounds have 5 skeletal pairs and we predict a *closo* figure, a tetrahedron. Similarly, the 22-electron M_5 (Figs 18.19a and 18.20a), the 38-electron M_9 (Fig. 18.20b), and the 42-electron $TlSn_9^{3-}$ clusters are *closo* figures as expected from the 6, 10 or 11 skeletal pairs respectively.

The 20-electron M_4 species have 6 skeletal electron pairs, so a 5-atom cluster would be a trigonal bipyramid, but there are only four M atoms so the structure is a trigonal bipyramid less one vertex—which is a tetrahedron. Likewise, all the 40-electron M_9 species should give the *nido* 10-vertex figure, corresponding to the observed capped square antiprism Figs 18.19b and 18.21. Thus the Bi_9^{5+} structure is the anomalous one, and we note that its trigonal prism is much more elongated than those found for the 38-electron *closo* structures. The odd-electron Sn_9^{3-} adopts the 10-pair rather than the 11-pair configuration.

The square 22-electron M_4 species correspond to an *arachno* octahedron. Alternatively, we can describe this as the result of adding two electron pairs to the 18-electron tetrahedron, breaking two bonds and leaving a square. The square antiprism Bi_8^{2+} is likewise the *arachno* 10-vertex structure.

The closed clusters may be correlated with more open species, as indicated above for the M_4 case. Thus the various E_6 species found for S, Se and Te, may be seen as resulting from

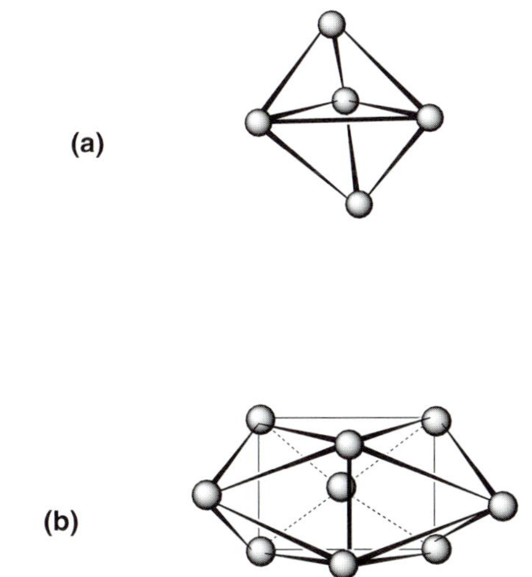

FIG. 18.20 Structures of the tin clusters (a) Sn_5^{2-} and (b) Sn_9^{3-}. The 38-electron Ge_9^{2-} ion is of similar structure to (b)

TABLE 18.6 Some clusters

Examples	*Total electrons*	*Cluster electron pairs*	*Shape*
Ge_4^{2-}, Sn_4^{2-}	18	5	Elongated tetrahedron
P_4, As_4, Sb_4, $Sn_2Bi_2^{2-}$ $Pb_2Sb_2^{2-}$	20	6	Tetrahedron
$Tl_2Te_2^{2-}$	20	6	Butterfly
As_4^{2-}, Sb_4^{2-}, Bi_4^{2-}, Se_4^{2+}, Te_4^{2+}	22	7	Square
Sn_5^{2-}, Pb_5^{2-}, Bi_5^{3+}	22	6	Trigonal bipyramid
Bi_8^{2+}	38	11	Square antiprism
Ge_9^{2-}, $TlSn_8^{3-}$	38	10	Tricapped trigonal prism
Sn_9^{3-}, Ge_9^{3-}, Pb_9^{3-}	39	$10\frac{1}{2}$	Tricapped trigonal prism
Bi_9^{5+}	40	11	Tricapped trigonal prism
Si_9^{4-}, Ge_9^{4-}, Sn_9^{4-}, Pb_9^{4-} $Sn_xGe_{9-x}^{4-}$, $Sn_xPb_{9-x}^{4-}$, Sn_8Sb^{3-}, Sn_8Tl^{5-}	40	11	Capped square antiprism
$TlSn_9^{3-}$	42	11	Bicapped square antiprism

the successive opening of an octahedron, by adding electrons: six pairs removes six of the twelve edges, leaving a chair (Fig. 18.6), for example.

It is also possible to make connections between the clusters and the extended structures of solids. In many of the mixed phases, called *Zintl* phases, formed between electropositive and less electropositive elements, cluster units may be identified as building blocks of the three-dimensional arrays. Tetrahedra are common, but the other isolated clusters are rare. Instead, more extended linkages occur giving rise to puckered layers or extended nets.

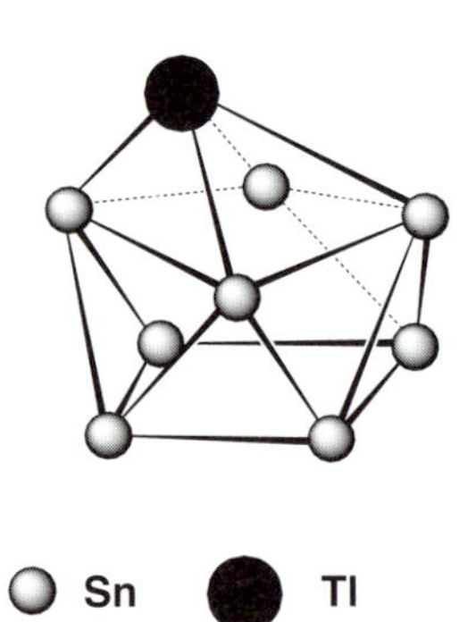

FIG. 18.21 The capped square antiprismatic structure of $TlSn_8^{5-}$, and Sn_9^{4-}. Compare Figure 18.19b

18.4.5 Group 14 prismanes and related structures

An interesting development in organic chemistry was the synthesis of closed carbon frameworks such as tetrahedrane (C_4H_4) or cubane (C_8H_8) with C-C bonds forming each edge of a regular solid. Naturally, making analogous structures has been taken up as a challenge by chemists working with Si, Ge or Sn.

Theoretical studies indicated that a range of closed $(EH)_n$ structures should be stable and that the preferred shape was the *n*-prismane. Prismanes have two regular plane faces in eclipsed positions joined by apex—apex bonds forming a series of square or rectangular faces. We have already met the trigonal (or triangular) prism (3-prismane) and a cube is a special case of a square prism (4-prismane).

The first syntheses of Si, Ge and Sn prismanes were reported almost simultaneously in 1988–9. The key was to use a large bulky substitutent, R, such as *tert*-butyl $(CH_3)_3C$-,

$(Me_3Si)_3C$- or a polysubstituted phenyl. The common syntheses involved removing halogen from REX_3 by using an active metal species like an organolithium. Such reductive dehalogenation of $RGeCl_3$ gave the trigonal prism $[RGe]_6$ (Fig. 18.22a), for R = $CH(SiMe_3)_2$, and the cubane $[RGe]_8$ (Fig. 18.22b) when R was CEt_2Me. The alternative product shown in Fig. 18.22c was also isolated from a similar dehalogenation of $Bu^tGeBr_2GeBr_2Bu^t$. This may be seen as an intermediate *en route* to the cubane, with the two square faces twisted apart. The Si analogues were prepared by similar routes.

In an alternative synthesis, thermolysis of the cyclotristannane $[(2,6\text{-}Et_2C_6H_3)_2Sn]_3$ at 200°C gave not only the cubane (Fig. 18.22b) but also the pentagonal prismane (Fig. 18.22d) and the propellane (Fig. 18.22e). This can be seen as three triangles sharing a common edge. Typical parameters found for representative *n*-prismane clusters are listed in Table 18.7. Note that all the E-E bonds are somewhat longer than indicated by the values of Table 2.10a which are derived from simple E-E species like $H_3Ge\text{-}GeH_3$. They are comparable with the lengthened bonds found in simple R_6E_2 compounds with bulky ligands.

The structures are essentially ideal ones. In the tin propellane (Fig. 18.22e) the outer bonds are normal for Sn-Sn (284–287 pm) while the central bond is very long (Sn-Sn = 336.7 pm). Further structures found for silicon include the striking tetrahedrane Si_4R_4, by making R the 'supersilyl' highly bulky $SiBu^t_3$ ligand, and the mixed-element cubane $(Bu^tSiP)_4$, the analogue of Fig. 18.22b with alternating Si and P in a distorted cage which can be described as a large P_4 tetrahedron interpenetrating a smaller Si_4 one.

18.5 Polynuclear ions and the acid strength of preparation media

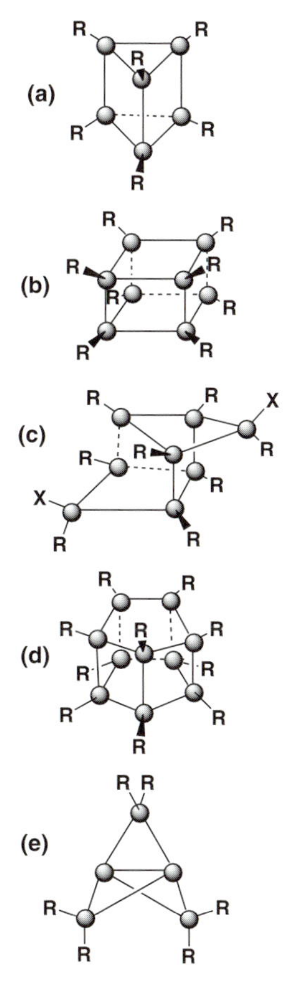

FIG. 18.22 Some cluster structures for germanium and tin

In Chapter 17 and in Sections 18.2, 18.3 and 18.4, we have seen a wide range of examples of polynuclear ions. Almost all the preparations of these ions involve nonaqueous media such as liquid ammonia (Section 6.7), superacids (Section 6.9) or $AlCl_3$ and other melts. Polynuclear cations are very sensitive to base attack and require to be prepared in media of high acidity; conversely, polynuclear anions need strongly basic media. The media also act as vehicles for strong reducing agents (alkali metals in liquid ammonia) or oxidizing agents (F_2 in liquid HF or $S_2O_6F_2$ in HSO_3F), but the acidity appears to be the dominant factor.

Sufficient work has been done to allow these factors to be drawn together in an interesting picture which underlines some of the similarities between various nonaqueous solvents.

As an example, we can consider the iodine cations I_2^+, I_3^+ and I_5^+ (Section 17.8). We use the Hammett acidity function, H_0, as a measure of the relative acidity in the superacid media (the larger the negative value of H_0, the higher the acidity). First, it may be noted that I_x^+ may be prepared in HSO_3F by oxidation of I_2 with $S_2O_6F_2$ in an appropriate stoichiometry: thus 3:1 gives I_3^+ and 5:1 gives I_5^+. However, the species that is recovered depends also on the acidity. The full equation for the formation of I_2^+ is

$$S_2O_6F_2 + 2I_2 \rightarrow 2I_2^+ + 2SO_3F^- \quad (18.1)$$

but the I_2^+ slowly transforms to I_3^+ plus $I(SO_3F)_3$. Now, SO_3F^- is the base in HSO_3F according to the self-dissociation (compare Chapter 6):

$$2HSO_3F \rightleftharpoons H_2SO_3F^+ + SO_3F^- \quad (18.2)$$

Thus the SO_3F^- formed in Equation (18.1) reduces the acidity of the medium. In fact, the H_0 value for HSO_3F is −15.1, while the SO_3F^- will reduce H_0 to about −14.1. If, on the other hand, SbF_5 is added this removes SO_3F^- (see Equation (A2) in Section 6.9), and I_2^+ is very stable (H_0 −16 to −18 depending on SbF_5 concentration). Therefore, the higher the ratio of positive charge to iodine atoms in the ion (the higher the formal oxidation state) the more susceptible the ion to base attack and the higher the required acidity, so I_3^+ is stable in (relatively!) low acidity with charge ratio 1:3, whereas I_2^+ (ratio 1:2) needs higher acidity.

In the early reports of polyiodine cations, the medium was 100% H_2SO_4 (H_0 = −11.9), and I_3^+ and I_5^+ were stable while I_2^+ readily converted to I_3^+. In 60% oleum (i.e. 60% added SO_3 forming $H_2S_2O_7$), with H_0 = −14.8, I_2^+ was stable. Conversely, in H_2SO_4/HSO_4^-

TABLE 18.7 Parameters of representative Group 14 prismanes

Figure 18.22	*ER*	*E-E* (pm) *prism face*	*E-E*(pm) *square face*	*Angles* (°)
(a)	$GeCH(SiMe_3)_2$	258.0	252.1	60° and 90°
(b)	$GeC_6H_4Et_2$	247.8–250.3	Same	88.9–91.1°
(b)	$SnC_6H_4Et_2$	285.4	Same	88–92°
(d)	$SnC_6H_4Et_2$	285.6	285.6	108°, 90°

($H_0 = -11.9$), only I_3^+ was stable. Thus, while oxidant is needed to convert I(0) in I_2 to the fractional positive oxidation state in the I_x^+ ions, the ion recovered depends on the acidity.

Similar results are obtained using HF as a solvent ($H_0 = -15.1$) with acidity varied by adding the base F^- or GeF_4 (which removes F^- by forming GeF_6^{2-}). Since HF has a very small self-dissociation, the acidity is extremely sensitive to $[F^-]$. Thus reaction of F_2 in HF with I_2 gives I_3^+ as the F^- formed reduces H_0 to around –10.5. When GeF_4 was added (H_0 about –5), the product was I_2^+.

The result of increasing basicity is the disproportionation of the cations into a cation of lower charge/atoms ratio plus a covalent IX_3 species:

$$8I_2^+ + 3X^- \rightarrow 5I_3^+ + IX_3$$

$$7I_3^+ + 3X^- \rightarrow 4I_5^+ + IX_3$$

$$3I_5^+ + 3X^- \rightarrow 7I_2 + IX_3 \ (X = F^-, SO_3F^- \text{ or } HSO_4^-).$$

A different form of acidic medium is provided by molten $AlCl_3$, and the acidity may be adjusted by adding varying proportions of KCl (providing the base Cl^- or $AlCl_4^-$). Thus the iodine species formed in the acid melt $2AlCl_3 + 1KCl$ is I_2^+, while I_3^+ is stable in the neutral melt $1AlCl_3 + 1KCl$ (and this melt also gave $I_5^+AlCl_4^-$ at the appropriate stoichiometry).

Finally, a nonbasic medium like SO_2 or AsF_3 allows the formation of the higher polyiodides:

$$I_2 + AsF_5 \rightarrow I_3^+AsF_6^- + AsF_3 \text{ in } SO_2$$

$$I_2 + SbF_5 \rightarrow I_5^+SbF_6^- + SbF_3 \text{ in } AsF_3$$

$$I_2 + 3SbF_5 \rightarrow I_2^+Sb_2F_{11}^- + SbF_3 \text{ in } SO_2$$

because none of the species SO_2, AsF_3 or SbF_3 are sufficiently basic to attack the I_3^+ or I_5^+. The last equation underlines a further contribution—that of specific and usually large counterions in stabilizing a particular cation. This contribution is significant in many cases but is not as dominant as the acidity.

These ideas may be further reinforced if we look at the smaller halogens. Since the charge density per atom is higher in, say, Br_3^+ than in I_3^+, the acidity required to stabilize the bromine species will be higher than for the iodine analogue, while chlorine species will be even more demanding. It is found that Br_3^+ is stable in the highly acidic $HSO_3F/SO_3/SbF_5$ system ($H_0 = -19$), but is only marginally stable at $H_0 = -13.8$ where I_3^+ is stable. Br_2^+ is observed at $H_0 = -19$ but is not stable. No polychlorine cation survives in solution in even the most acidic media, though Cl^{4+} and Cl_3^+ have been isolated as solids. Cl_2^+ is not known.

The similar situation for the chalcogen polycations in sulfuric acid systems is summarized in Table 18.8. The H_0 values range from –10 in 95% H_2SO_4 to –14.1 in an oleum with 40% SO_3.

All these observations on polyatomic cations reinforce the general discussion in Chapter 6 on the similarities between different solvent systems. For example we see the exact parallel

TABLE 18.8 Values of H_0 where X_n^{2+} cations are stable

Formula	X_6^{4+}	X_4^{2+}	X_8^{2+}	X_n^{2+}
Oxidation state	0.67	0.5	–0.25	<0.25
X = Te	–13	–11		
X = Se		–11.9	–11.9	Se_{10}^{2+} at –10
X = S		–14.1	–13.2	Se_{19}^{2+} at –12.5

between the oxidant F_2 in liquid HF, giving the solvent base F^-, and the oxidant $(SO_3F)_2$ in HSO_3F, giving the base SO_3F^-. We see the parallel between these protonic acid solvents and the nonprotonic $AlCl_3$ melt.

Finally we note the difference between such 'participating' solvents and SO_2, which is functioning as a noninteracting medium. It should be noted, of course, that acidity is not the only factor in these reactions. Clearly the intrinsic properties of the reagents are significant —especially where there are small differences in charge density, as in the formation of I_3^+ versus I_5^+, or in the contrasting formation of Se_{10}^{2+} and S_{19}^{2+} for low charge density. For the isolation of solids, the presence of a counterion of appropriate size may be vital.

The considerations above apply also to simpler cations which are unstable in less acidic media, such as water. Thus U^{3+} may be prepared by reacting U with HF, whose acidity is enhanced by adding BF_3, whereas a more basic system containing F^- gives UF_4. U^{3+} was also prepared in an acid 2:1 $AlCl_3$/KCl melt and disproportionated to U plus U(IV) in a 1:1 melt. Similarly, Ti^{2+} (as $Ti[HF]_6^{2+}$) is formed from HF/SbF_5 and $TiCl_2$ in $AlCl_3$. The formation of $Sm^{2+}GeF_6^{2-}$ by reduction of Sm^{3+} in HF with added GeF_4 reflects the effects both of acidity and of a suitable precipitating anion.

All these examples indicate how the control of acidity (supplemented by choice of medium, oxidant and counterions) may lead to the directed synthesis of low-oxidation-state, polynuclear cations. An exactly similar approach in highly basic media would apply to polynuclear anions.

18.6 Silicates, aluminosilicates and related materials

Having discussed a number of ring and chain species formed by the p-block elements, we now turn our attention to another very important class of species formed by these elements—the silicates. The Si-O-Si linkage forms very readily by elimination of water between two Si(OH) groups and a very wide range of silicates is found. In almost all cases the coordination number of silicon with respect to oxygen is four, and thus the vast majority of silicates are built up from SiO_4 tetrahedra. The only exceptions occur in materials formed under conditions of extreme pressure and temperature—such as in meteorite impacts—where the SiO_2 mineral known as *stishovite*, having silicon in 6-coordination, can be found. Our discussion of the silicates will begin with a survey of the various silicate species formed and will then move on to more complex materials having network structures, such as the aluminosilicates, including the very important *zeolite* class of materials, and finally to the related aluminophosphates. One recurring theme in the structural chemistry of silicates is that highly complex structures can be built up from the basic SiO_4 tetrahedron sharing zero, one, two, three or four vertices with other SiO_4 (or MO_4) tetrahedra.

18.6.1 Simple silicate anions: SiO_4^{4-} and $Si_2O_7^{6-}$

The simplest type of silicate contains the isolated SiO_4^{4-} ion shown in Fig. 18.23a. A very common mode of representation of the SiO_4 unit is as a tetrahedral SiO_4 framework unit which has the oxygen atoms at the vertices, as shown in Fig. 18.23a, and this representation will be used extensively to simplify the structures of the more complex silicates discussed later. An example of a material containing the SiO_4^{4-} ion is Mg_2SiO_4 where the magnesium ions are surrounded octahedrally by oxygen atoms. The magnesium ions may be readily replaced by other divalent cations of about 80-pm diameter, such as Fe^{2+} and Mn^{2+}. As a result, natural materials of this type have a wide range of compositions. The mineral olivine is a naturally occurring example of such a magnesium silicate in which about one in ten of the Mg^{2+} ions is replaced by a Fe^{2+} ion. *Garnets* also contain isolated SiO_4^{4-} ions.

A less simple silicate which contains discrete silicate ions is the rather rare scandium silicate $Sc_2Si_2O_7$, known as *thortveitite*, containing the unusual dinuclear ion $Si_2O_7^{6-}$. This ion is formed by two SiO_4 tetrahedra sharing a common oxygen atom, as shown in Fig. 18.23b. The scandium ions in this mineral have octahedral coordination.

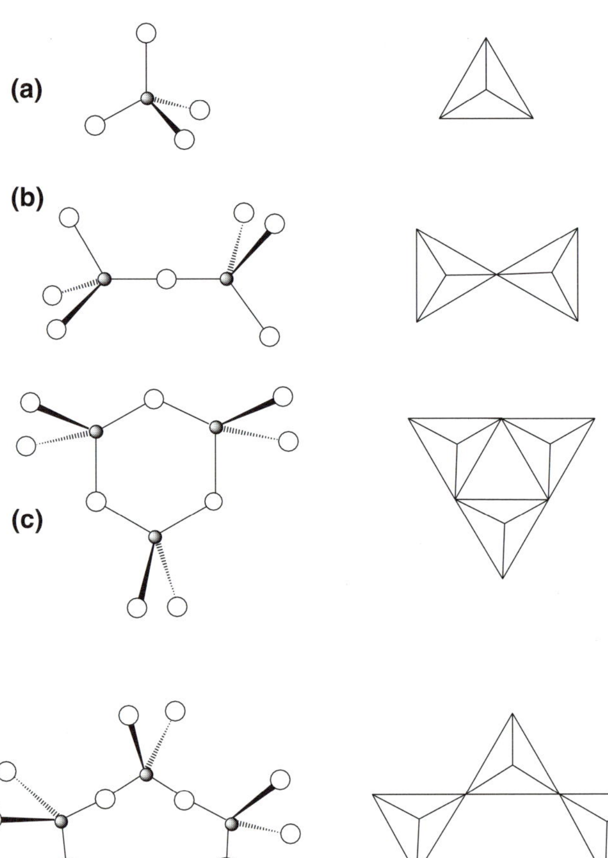

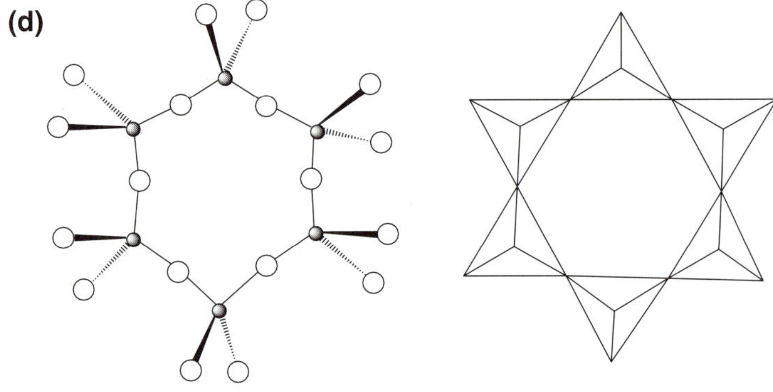

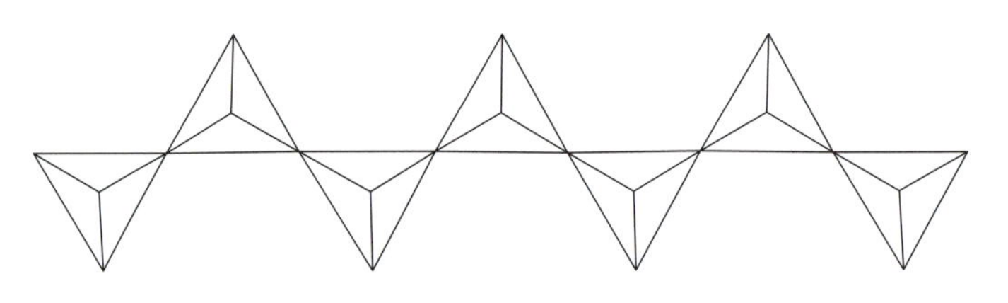

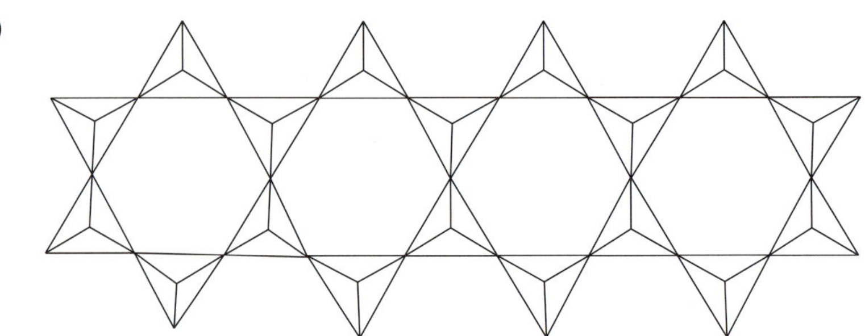

FIG. 18.23 Representations of silicate anions: (a) SiO_4^{4-}, (b) $Si_2O_7^{6-}$, (c) $Si_3O_9^{6-}$, (d) $Si_6O_{18}^{12-}$, (e) an SiO_3^{2-} chain, (f) an $Si_4O_{11}^{6-}$ double chain and (g) an $Si_4O_{10}^{4-}$ sheet

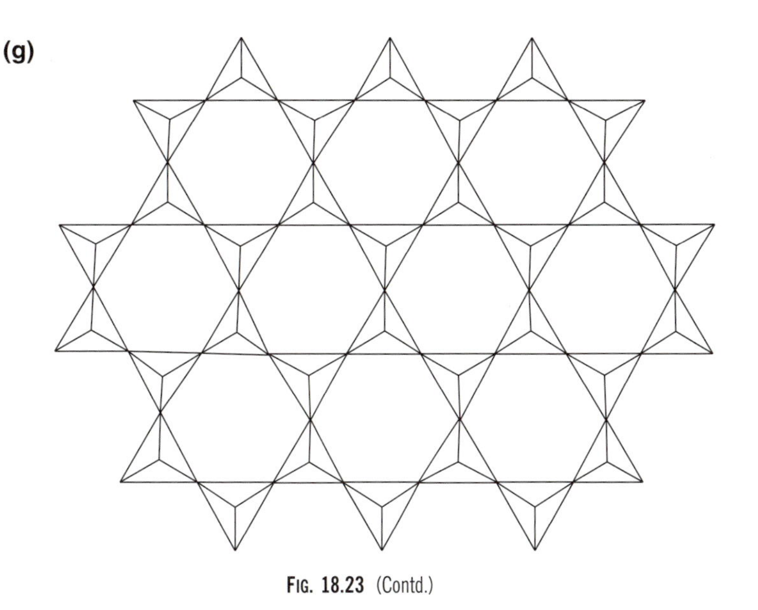

FIG. 18.23 (Contd.)

At this point it is convenient to mention the calculation of charge on a silicate ion—this can be simply done since each terminal oxygen contributes one negative charge while each shared oxygen contributes zero charge. The charge on the Si_2O_7 species is thus confirmed as 6—since the six terminal oxygen each contribute 1—while the single shared oxygen does not contribute to the charge.

18.6.2 Rings and chains: $[SiO_3]_n^{2-}$

If each SiO_4 tetrahedron shares an oxygen with each of two neighbouring tetrahedra, then ring or chain silicate species can result. Two different ring sizes are known. The smaller ring contains three SiO_4 tetrahedra linked together as a six-membered ring $Si_3O_9^{6-}$ (Fig. 18.23c) which occurs in, for example, *benitoite*, $BaTiSi_3O_9$, while the larger ring has six linked tetrahedra in the ion $Si_6O_{18}^{12-}$ (Fig. 18.23d) found in the mineral *beryl*, $Be_3Al_2Si_6O_{18}$. In the structure of these minerals the rings are arranged in sheets with their planes parallel and the metal ions lie between the sheets, binding the parallel rings together by electrostatic forces. The Ba, Ti and Al atoms are coordinated to six O atoms while the smaller Be atom is 4-coordinated to O.

In the green variety of beryl, known as *emerald*, the green Cr^{3+} ions substitute some of the Al^{3+} ions.

The chain structure shown in Fig. 18.23e is found in a very wide range of minerals and these are known collectively as the *pyroxene* group. Examples are $CaMgSiO_3$ and $CaMg(SiO_3)_2$, known as the mineral *diopside*. In these compounds the silicate chains lie parallel to each other and the cations lie between the chains, binding them together by electrostatic forces. The magnesium ion are 6-coordinated while the larger calcium ions are 8-coordinated.

Two $[SiO_3]_n^{2-}$ silicate chains can also be linked together to form *double chains* (Fig. 18.23f) which correspond to the formula $Si_4O_{11}^{6-}$. Materials containing these double-chain silicates are also relatively common and are represented by the class of minerals called *amphiboles* which includes most *asbestos* minerals. One example is $Ca_2Mg_5(Si_4O_{11})_2(OH)_2$ in which the magnesium ions are coordinated to six oxygens, three from the silicate and three from the hydroxyl groups. The presence of hydroxyl groups in the structure is a characteristic feature of the amphibole series of minerals. *Tremolite* is the mineral to which the name 'asbestos' was first given although the term is now applied more widely. As with all minerals, the above composition is an idealized one and other ions of a similar size may be present. In particular, some of the magnesium may be replaced by iron and when the iron content rises to about 2% the mineral is termed *actinolite*.

Minerals having chain silicate ions, particularly the double-chain silicates, typically adopt fibrous forms—this is readily explained since the bonds between the silicate chains are weaker than the very strong covalent Si-O bonds within the chains themselves and so the mineral tends to cleave parallel to the long silicate chains. This relationship between the structure of the silicate and the properties of the minerals is particularly well demonstrated by the next class of silicate materials.

18.6.3 Sheet structures: $Si_4O_{10}^{4-}$

We continue our survey through the structures adopted by silicate materials by considering structures formed when SiO_4 tetrahedra share three vertices with adjacent tetrahedra. When this occurs the sheet silicate structure of Fig. 18.23g results, having the empirical formula $Si_4O_{10}^{4-}$. Again, a wide variety of minerals containing sheet silicates are known and, like the amphiboles, these usually contain hydroxyl groups, as in the well-known mineral *talc*, $Mg_3(OH)_2Si_4O_{10}$ (Fig. 18.24). The weak forces between the silicate sheets result in the material cleaving parallel to the sheets, similar to graphite (Section 5.8) and MoS_2 (Section 15.4.3), resulting in the lubricating properties of these layered materials.

We have already mentioned that it is very common in naturally occurring silicate minerals to find that one element can substitute for another of the same size, resulting in a wide range of mineral compositions. A very important substitution, which occurs extensively in nature and which can be achieved readily in synthetic silicate materials, is the substitution of aluminium for silicon. Aluminium, which is the same size as silicon, readily substitutes but, since aluminium is only a 3+ ion, an additional cation is required for charge balance. Thus an Si^{4+} in the silicate framework can be substituted by the combination of an Al^{3+} with an additional (nonframework) K^+ ion. Substitution of this type happens in the *mica* group of minerals, such as in *phlogopite* $KMg_3(OH)_2Si_3AlO_{10}$ (Fig. 18.25). In this sheet-silicate mineral, one in four silicon atoms of the talc structure is replaced by an aluminium atom plus a potassium atom. In micas the additional cations reside between the silicate sheets and help to bind them together by electrostatic forces. Though these binding forces are still much weaker than the bonds within the sheets themselves, micas are harder than talc but still preferentially cleave parallel to the sheets.

In addition to the micas, the main minerals in the group of sheet-silicate structures are the clay minerals such as the *kaolins*, used in ceramics, *vermiculite*, used as a soil conditioner, and *bentonite* which is used as a binder and adsorbent.

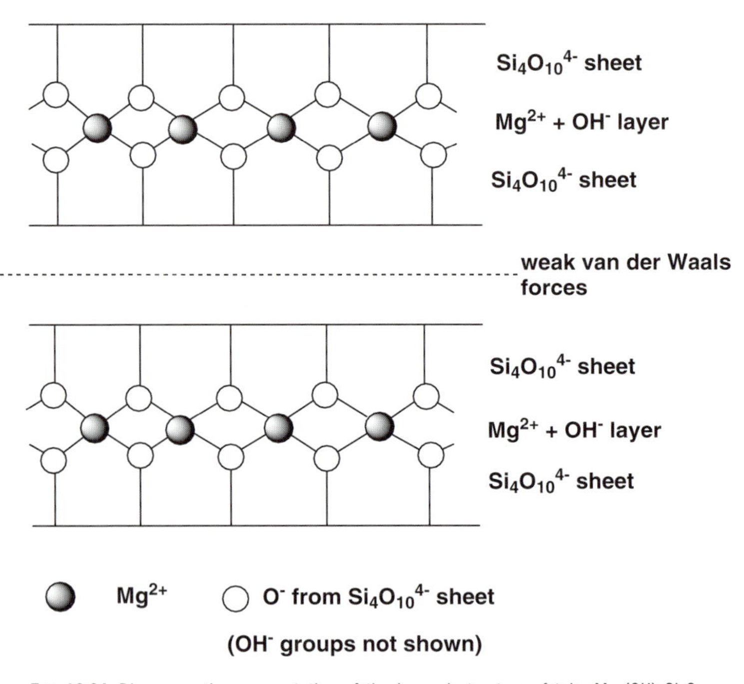

FIG. 18.24 Diagrammatic representation of the layered structure of talc, $Mg_3(OH)_2Si_4O_{10}$

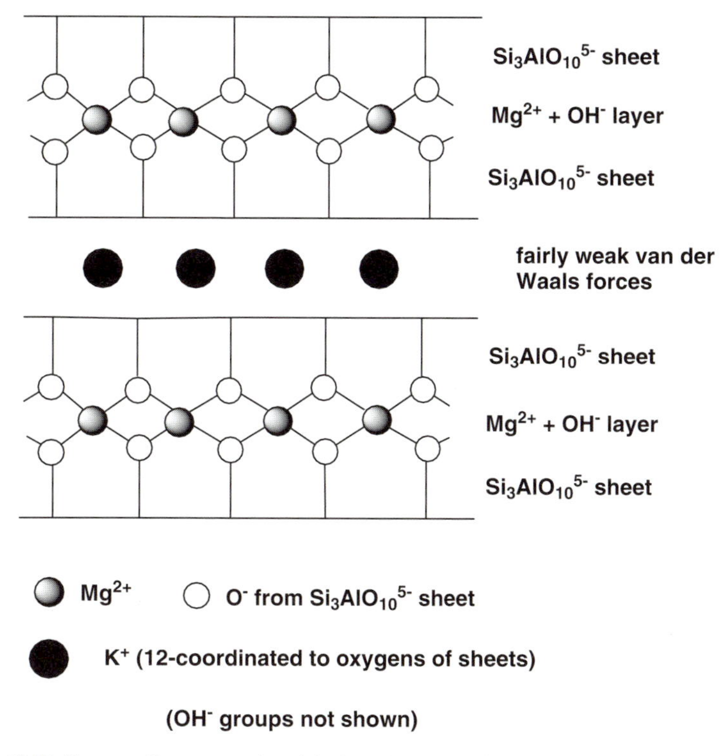

FIG. 18.25 Diagrammatic representation of the layered structure of phlogopite mica, $KMg_3(OH)_2Si_3AlO_{10}$

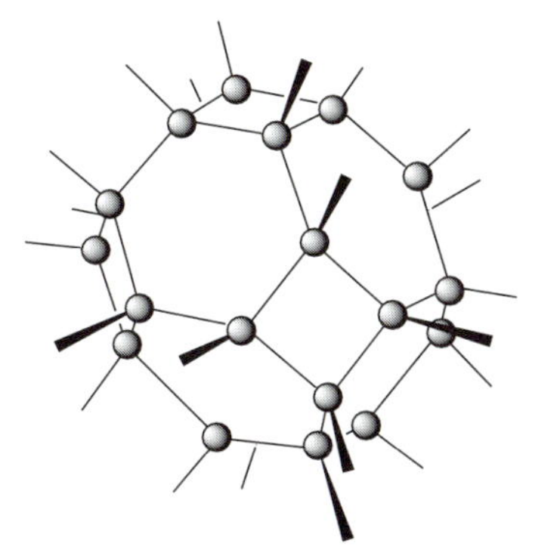

FIG. 18.26 The structure of the sodalite cage, the building block of a number of zeolites. The structure is based on a truncated octahedron. The shaded circles mark the positions of the silicon or aluminium atoms, hence each edge of the polyhedron is bridged by an oxygen atom, forming the usual M-O-M (M = Si or Al) linkages. In addition, each silicon or aluminium atom bears one terminal oxygen

18.6.4 Three-dimensional structures

If SiO_4 tetrahedra share all four oxygen atoms with adjacent tetrahedra, the infinite three-dimensional network structure of silica, SiO_2, results (see Fig. 5.3c). Since all of the oxygen atoms are shared, and there are no terminal Si-O^- groups, there is no net charge and the strongly bonded Si-O covalent network results in the crystalline forms of SiO_2 (e.g. *quartz* or *cristobalite*) which are very hard materials.

As for the sheet silicates, it is again possible to replace some of the silicon atoms by aluminium atoms, with additional cations to balance the charge, and when this occurs three-dimensional framework aluminosilicates result. These structures are represented by the important *feldspar* group of minerals which are the most abundant of all the rock-forming minerals. Examples of feldspars include *orthoclase*, $KAlSi_3O_8$ (often found as large pink crystals in granites), and the continuous solid–solution series formed between the end-members of *albite* ($NaAlSi_3O_8$) and the aluminium-rich *anorthite* ($CaAl_2Si_2O_8$).

18.6.5 Zeolites

A second series of framework aluminosilicates, of significant importance to the chemist, are the *zeolites*. The name zeolite comes from the Greek word for 'boiling stone'. Naturally occurring zeolites are heavily hydrated minerals which, when heated, release their water giving the impression of 'boiling'! While many naturally occurring zeolites are known, these materials can also be synthesized industrially and in the laboratory, and since they have a range of industrial applications, we will discuss them in some detail.

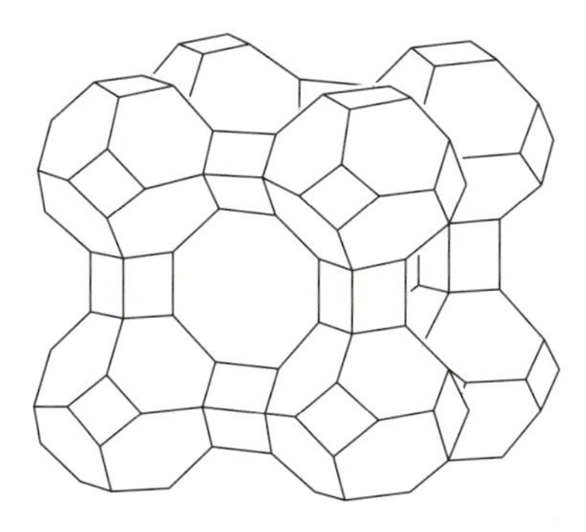

FIG. 18.27 The structure of zeolite A showing how the sodalite units join via linking of their square faces. The vertices mark the position of the Si or Al atoms

Zeolites are typified by having aluminosilicate frameworks with very open structures and large, but highly regular, channels and cavities present in them. A common structural unit in a number of zeolites is the sodalite cage found in the zeolite mineral of the same name. The sodalite cage is depicted as a framework structure in Fig. 18.26 in which the vertices represent the positions of the aluminium or silicon atoms. Conceptually, this building block can be viewed as an octahedron which has had each vertex 'sliced off', yielding a polyhedron with both square and hexagonal faces. A number of different zeolites can then be built up from the sodalite cage building blocks. When sodalite cages are joined *via* their

SYNTHESIS OF ZEOLITES

The synthesis of zeolites in the laboratory can be a relatively straightforward procedure in which an aqueous aluminate/silicate mixture, formed from SiO_2, Al_2O_3 and alkali, (e.g. NaOH), is heated, often under pressure, to form the zeolite. It is believed that the open structure forms around a hydrated cation and one of the common techniques for obtaining zeolites with different cage sizes is to use a large organic cation, such as the *tetra*-butyl ammonium cation, as a 'template' around which the new zeolite structure forms. The material is then heated to a high temperature in air or oxygen to 'burn off' the organic template, leaving behind the desired zeolite. By using different *tetra*-alkyl ammonium cations, different sized cavities can be formed in the zeolite. In a similar fashion it has been possible to prepare zeolites containing no aluminium whatsoever. These materials, consisting entirely of SiO_2 but with very open zeolite structures, are called *silicalite* zeolites.

hexagonal faces, the mineral *faujasite*, $Na_2[Al_2Si_5O_{14}].10H_2O$, results whereas when they join in a cubic pattern *via* the square faces, as shown in Fig. 18.27, the synthetic zeolite A is formed which has the composition $Na_{12}[Al_{12}Si_{12}O_{48}].27H_2O$. This arrangement of sodalite cages results in a very large central cavity, called the α-cage, with an opening diameter of 420 pm. These large cages ordinarily contain the water molecules of hydration, together with the cations (e.g. Na^+). An important characteristic of zeolites is that this water can be readily removed upon heating *without destroying the open framework structure*. The resulting dehydrated zeolites then have a very strong affinity for water and one of the most important applications of these materials is as drying agents.

The cations present inside the cages and channels of a zeolite can also be readily exchanged for other cations. One well-known application of this occurs in commercial water softeners in which the hard-water causing Mg^{2+} and Ca^{2+} ions are exchanged with the Na^+ ions of the zeolite, thereby softening the water. The recent concerns over the use of polyphosphates (another class of inorganic chain species) as water softeners has led to their replacement by zeolites in many countries. However, recent research might appear to suggest that the overall environmental impact of zeolite manufacture and use is similar to that of phosphates: the environmental influences of phosphates and zeolites are discussed in greater detail in Chapter 20.

CATALYTIC APPLICATIONS OF ZEOLITES

The synthetic zeolite ZSM-5 is an excellent catalyst widely used in a number of industrial processes. A good example is the manufacture of synthetic gasoline from methanol. Since methanol is readily formed from any carbon source through the catalysed conversion of synthesis gas (CO plus H_2), this route provides a source of gasoline from nonpetroleum feedstuffs and is therefore an increasingly important chemical technology, given the decline in world oil reserves. This process is carried out on a large scale in New Zealand and produces about one third of the country's gasoline requirements. As a result of the regular structure of the zeolite, the process is highly selective in favour of straight-chain hydrocarbons, such as octane. Branched-chain species, which do not fit into the zeolite channels, are selected against—the net result is a high octane gasoline. The acid form of zeolites is used in this type of reaction (protons are the catalytically active species involved in hydrocarbon synthesis, cracking and rearrangement reactions). Protons are produced by modifying the ammonium salt of the zeolite by ion-exchange, replacing the Na^+ ion with NH_4^+, followed by strong heating to drive off NH_3:

$$Na^+[\text{zeolite}] + NH_4^+ \rightarrow NH_4^+[\text{zeolite}] + Na^+$$

$$NH_4^+[\text{zeolite}] \rightarrow H^+[\text{zeolite}] + NH_3$$

These solid acid catalysts are of great importance as replacements for existing acid catalysts, such as concentrated sulfuric acid, used industrially in a number of processes. The zeolites have advantages in that the acidity is contained within the molecular pores of the solid, resulting in simple handling of the catalyst. In addition, the zeolite can be very easily filtered from the reaction vessel for recycling.

Reference

Syn-fuels from Boiling Stones, G.J. HUTCHINGS, *Chemistry in Britain*, 1987, 762.

The reader is referred to Sections 7.2 and 17.6 for polyphosphates, and to Chapter 20 for a discussion of the environmental chemistry of these species. Similar polymeric species are found for the borates (Section 17.4) and a very extensive series of synthetic polyoxometallate ions are known, almost beginning to rival the silicates in number and complexity. A brief summary is presented in Sections 14.3 and 15.4.

The replacement of the alkali metal ions of a zeolite by appropriate transition metals, such as rhodium, chromium or vanadium, is of interest for the development of new types of catalysts. It is in catalysis that zeolites are perhaps of the greatest interest to the chemist since the presence of large cages and channels within the structure offers the opportunity to use these as a kind of 'molecular-sized reaction vessel'. Since zeolites are crystalline materials, they have very regularly sized cavities and these allow the design of very selective catalysts. The ability of zeolites to incorporate molecules of one type into their internal cavities, while excluding others, has led to the commonly used term 'molecular sieves' being coined for them.

The regular cavities in zeolites lead to their use in gas separation processes. Other novel uses of zeolites are certain to be developed in the not too distant future and the search for new zeolites which have different types of framework structures is an important part of the study of the chemistry of these interesting materials.

This overview of silicate structures only gives an introduction to the rich variety of structures which are possible for these materials. The large number of structures arises from the different ways in which the framework SiO_4 tetrahedra can link up and from the possibility of replacing some of the silicon atoms by others of similar size, especially aluminium. We note, however, that such linked polyoxo species are not restricted to the silicates. Polyphosphates also form important ring and chain species—chain phosphates are used as builders in detergents and are found in ATP, the energy source of life.

18.6.6 Metal phosphates and metal phosphonates

Metal phosphates are a very well-known class of materials which occur naturally. However, there has been a recent resurgence of interest in the structural chemistry of these (and other) materials since novel types of structures have been found. One of the simplest, yet structurally most interesting, materials is aluminium phosphate, $AlPO_4$. Since two silicons are isoelectronic to one aluminium plus one phosphorus, $AlPO_4$ is isoelectronic with 2 SiO_2 and has similar structural properties. As with silicates in the previous discussions, an imbalance between the number of Al and P atoms in the framework will create charged frameworks, in a similar manner to the formation of aluminosilicates. Aluminophosphates having structures similar to those of aluminosilicate zeolites are known and it is anticipated that such phosphate materials will lead to the development of new, improved and more selective catalysts. Alternative framework atoms can be introduced, such as gallium which produces the material $GaPO_4$, known as *cloverite*, named as a result of the very large clover-shaped cavities present in the structure.

The ability to synthesize highly regular solids with interesting structures has led to the investigation of a wide range of metal phosphate materials. A particularly interesting recent example is the vanadium phosphate $[(CH_3)_2NH_2]K_4[V_{10}O_{10}(H_2O)_2(OH)_4(PO_4)_7].4H_2O$, formed in high yield from a 'one-pot' mixture of simple reactants. This compound contains a novel chiral double-helix structure—a kind of inorganic DNA!

When one of the oxygen atoms of the phosphate ion PO_4^{3-} is replaced by an organic group (e.g. a methyl or a phenyl), then a phosphonate anion, RPO_3^{2-}, results. Whereas aluminophosphates have three-dimensional structures, similar to network silicates such as SiO_2 or zeolites, the structures of metal phosphonates resembles those of sheet silicates (clays and micas) (see Fig. 18.28). Such layered structures, like clays, are of interest for their catalytic and ion-exchange properties. Clay minerals have been found to swell by the intercalation between their layers of various small molecules such as alcohols. Metal phosphonates also display the same behaviours but have a very distinct advantage in that the organic substituent on phosphorus may be readily varied, providing the potential for the preparation of a very large range of derivatives with fine-tuning of different chemical properties. Layered materials are not restricted to phosphonates, however, and acid phosphates, in which one P-O group is protonated, such as in the zirconium compound $Zr(HPO_4)_2$, also have layered structures and are of interest as solid acid catalysts.

It can be seen from the above discussion that novel materials, having interesting structures and useful applications, can be formed from very simple inorganic precursors. It is these possibilities, together with the large number of inorganic elements available for study, which has set about a renaissance in the chemistry of materials.

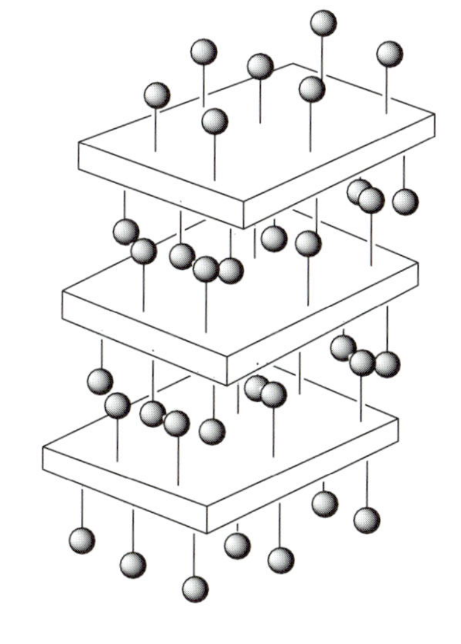

FIG. 18.28 Schematic diagram of a layered metal phosphanate. The 'slabs' represent the metal and oxygen part of the structure with the organic substituents bonded to the phosphorus atoms depicted by shaded circles

18.7 Multiple bonds involving heavier Main Group elements

The discussion in Section 17.2 about the restriction of multiple bonds to first row elements using 2p orbitals is generally valid for species which are stable at room temperature, and with ordinary ligands. There has been evidence both for unsaturated species and low-valent intermediates (such as Me_2Si or $H_2Si{=}SiH_2$) as the reactive species in reactions such as pyrolysis, and such compounds are realistically postulated as reaction intermediates. The isolation of stable compounds with double bonds has also been accomplished, but requires special conditions.

The properties of the $R_2E{=}ER_2$ products are shown in Table 18.9. The structures show some interesting features. The E-E distance is less than the sum of single bond radii, but the shortening is very variable, ranging from about 2% to values similar to that for a C–C double bond. The $R_2E{=}ER_2$ species also differ from C analogues in being increasingly nonplanar as E gets heavier. The angle given in Table 18.9 is that between the R_2E plane and the all-planar position. These structures imply that the R_2E unit retains some lone-pair character in the dimer.

Similar features are found for the heavier elements of the nitrogen Group. The chemistry of RP=C species is particularly well explored and P≡C compounds are reported. Mixed species both within and between Groups are established. Examples include RSb = AsR, $R_2Ge{=}ER$ (E = N, P), 'heavy ketone' analogues $R_2M{=}X$ (M = Ge, Sn; X = S, Se, Te) and a C=Te bond in $F_2C{=}Te$. As described in the box, stable compounds again generally result when R and R′ are large groups which hinder polymerization or other further reaction. As expected, the lack of bulky shielding groups results in very rapid dimerization of $F_2C{=}Te$ to a four-membered Te-C-Te-C ring above −196°C. The B=B system has been established and there is even evidence for a triple bond between N and Si in C_6H_5NSi formed transiently in the gas phase.

Finally, we note that Main Group–transition element multiple bonds are also known. Metal carbenes, $L_xM{=}CR_2$, and carbynes, $L_xM{\equiv}CR$, have been established for a long time, and metal oxo, M=O, and metal nitrido, M≡N, complexes are likewise very well established, for example the nitrido complexes $MNCl_4^-$ (M = Mo, W, Re, Ru and Os). In recent years there has been increased interest in the synthesis and characterization of

STABILIZING HEAVY MAIN GROUP ELEMENT COMPOUNDS

Compounds of the heavy Main Group elements with, for example, E-H or E=E bonds tend to be unstable. The use of bulky organic groups can lead to a dramatic stabilization of the E-H or E=E bond, which is now sterically protected. As an example of the stabilization of a heavy Main Group element hydride, the sterically crowded bismuth halide R_2BiCl can be reduced with $LiAlH_4$ (a hydride source) to give R_2BiH, which contains a pyramidal Bi atom:

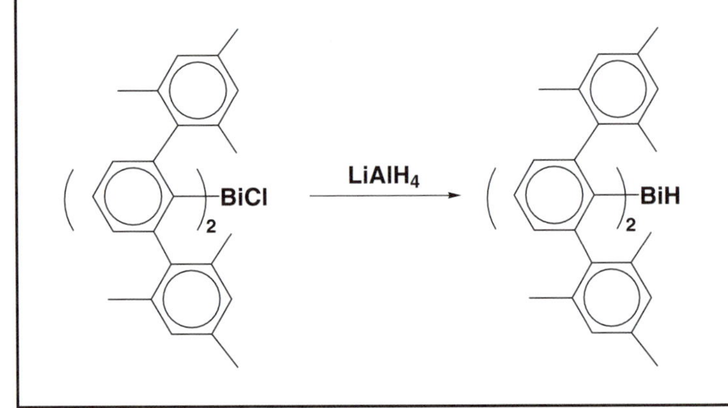

A second example is the reaction sequence:

$$R_2ECl_2 + LiAr \rightarrow (R_2E)_3 \xrightarrow{\text{photolysis}} R_2E{=}ER_2$$

Here, E is Si, Ge or Sn, and R is a sterically hindered ligand like Me_3C-, $(Me_3Si)_{3-n}CH_n$- (n=usually 0 or 1) or substituted aryl rings such as the one shown in the bismuth example, or with methyl groups in *ortho* positions. With smaller R groups, the cyclic trimer or larger rings, $(ER_2)_n$, form the stable products, illustrating the necessity for bulky groups to stabilize the E=E bonds. In such systems it is also possible to produce the divalent monomer, R_2E, which may be in equilibrium with double-bonded $R_2E{=}ER_2$. The product formed depends both on R and E, and also on the reductant and the activating energy.

Particularly noteworthy is the recent synthesis of a compound containing a Pb=Pb double bond, the last remaining double bond between two Group 14 elements to be synthesized, by reaction of $PbCl_2$ with RMgBr, where R is the sterically bulky 2,4,6-tri-iso-propylphenyl group. The initial PbR_2 product dimerizes to form the diplumbene $R_2Pb{=}PbR_2$.

TABLE 18.9 Parameters for multiply bonded species

E	*E-E* (pm)	*Single E-E bond length* (pm)	*Non-planar angle* (°)
Si	213–216	232	0–18
Ge	221–235	244	15–32
Sn	276	281	41
Pb	305	—	44, 51
P	200–203	221	
As	222–225	243	
P=As	212	232	

Note. For E = Si, Ge, Sn, values refer to $R_2E = ER_2$; for E = P, As values are for RE=ER. Range of values are for a variety of hindered ligands R.

heavy element analogues of these compounds. This is illustrated by the recent synthesis of the first examples of complexes containing M≡P triple bonds, Section 16.10.

Calculations suggest that, while rings involving only single bonds, e.g. $(R_2E)_n$ or $(RE)_n$, are the most stable, both the double-bonded species and the monomers, R_2E or RE, are not substantially less stable, so that steric effects may be enough to tip the balance.

These compounds generally show reactions similar to those of classical unsaturated species, particularly addition reactions across the double bond by H-X or R-X.

18.8 Commentary on VSEPR

As Chapter 17 shows, the Valence Shell Electron Pair Repulsion approach is very successful in predicting and rationalizing the shapes of molecules, despite its very simple premises. As with all successful models, it has been subject to further development and to criticism, and its potential and limits have become well defined. To illustrate its successful application, we note that rare gas chemistry, and almost every halogen oxyfluoro compound listed in Table 17.20 and Fig. 17.60, are post-VSEPR and the structures were found to be as predicted. In these, and many other cases, the prediction has been extremely good, *within the limits of the theory*. For example, VSEPR does not predict the preferred linkage isomer or the degree of polymerization (for example the formation of tetrameric XeF_6 in the solid) but it will predict the structure corresponding to each specific formulation.

Many improvements to VSEPR have been suggested, but most would only marginally extend its scope at the price of the essential simplicity of the approach. It is becoming clear that it is probably best to settle for a straightforward version of VSEPR, and leave the finer details of structures to be properly treated at a more sophisticated level.

Indeed, in order to preserve the enormous value of the VSEPR approach as a rule-of-thumb prediction, it might even by appropriate to reduce the weight placed on the rationalizing ideas and reformulate the guideline: *the shape of a polyatomic species is the one which would result if the shape is governed by the repulsion of electron-rich regions of valency electrons: viz., the bond pairs, the lone pairs and multiple bonds, as set out in Tables 4.1, 4.2 and 4.4.*

This is not simply a quibble; it means that VSEPR may be preserved even if quite different explanations appear from deeper treatments of the shapes. For example, it is clear that simple ideas of steric hindrance would also account for the shapes of Table 4.1, as would a model of repulsion (or attraction!) by the electron clouds of the ligands. In the reading list of Appendix A are references to a number of such alternative approaches which the reader should assess.

It is useful to examine three areas in a little more detail.

18.8.1 Experimental electron densities

A major premise of VSEPR is that lone pairs are sterically active, that is that they occupy a definite direction in space and are not spherically symmetric. As valency electron distribution is not measurable by standard structural techniques, the presence of

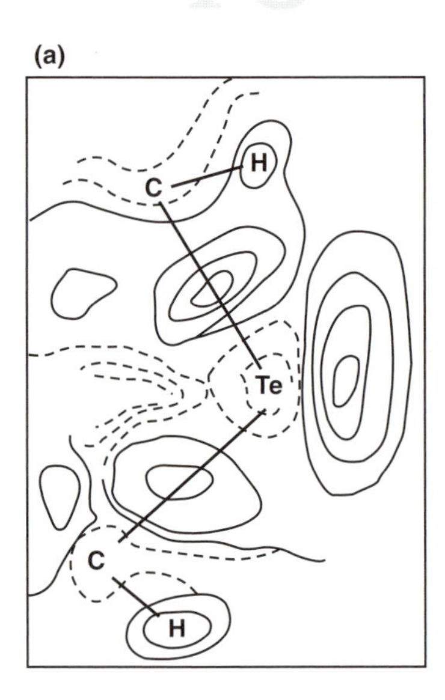

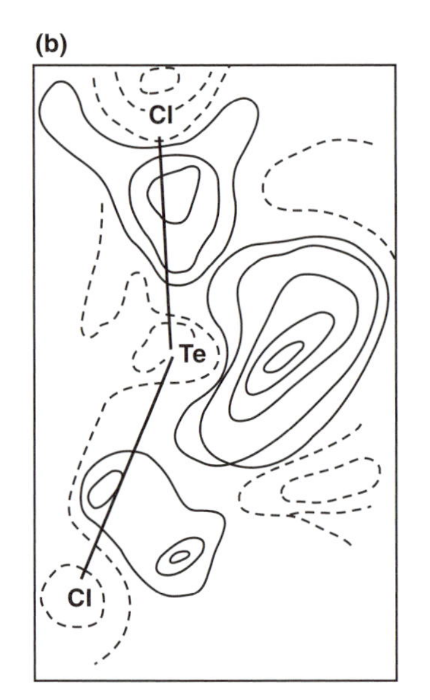

Fig. 18.29 Representation of the electron density difference distribution in Me_2TeCl_2: (a) in the equatorial plane (b) in the axial plane

sterically active lone pairs has had to be deduced less directly. For example, the existence of a dipole rules out a planar structure for NH_3 or a linear structure for GeF_2. The most common evidence is the atom positions from diffraction experiments. Where these match VSEPR predictions, the postulated lone pair position is supported, though not proved.

The recent development of methods of determining the density distribution of valence shell electrons (see Section 19.1) has allowed the location of lone pair electron densities. The experiments are difficult, so the number of examples is small, but the results are basically in accord with VSEPR predictions. One case which is particularly striking, because the difficulties are enhanced with heavy atoms, is a study of Me_2TeCl_2. VSEPR makes this an $AB_2B'_2L$ case where the more electronegative Cl atoms should be in the apex positions and the two Me groups and the lone pair in the equatorial position. Fig. 18.29 shows schematic electron density difference maps (see Section 19.1 for explanation) in the $TeMe_2$ plane and in the plane through ClTeCl bisecting the CTeC angle. The position of the lone pair is exactly as expected.

Such studies thus amount to direct observation of sterically active lone pairs, completing the indirect evidence from atom positions.

18.8.2 Limiting cases

(1) VSEPR is at the limits of its usefulness (i) for d^n configurations which are better treated as in Chapter 13, and (ii) for more than six electron pairs. In such species, the energy difference between alternative possible configurations is small, and the predictive power of the theory fades out. For example, although IF_7 is a pentagonal bipyramid, as expected, the very small angle of 72° in the equatorial plane is relieved by distortion. While this is expected from the basic VSEPR rationalization, the difference between a distorted pentagonal bipyramid and, say, a distorted capped octahedron, is small and almost a matter of semantics only.

In the particular case of AB_6L compounds, where VSEPR predicts that the structure will be distorted from a regular octahedron by the presence of the lone pair, both distorted and regular structures are seen (Section 17.7.4). In structures determined so far, distorted octahedra are usually seen when B is a highly electronegative ligand such as F. However, the balance is clearly a very fine one since $[BrF_6]^-$ is a regular octahedron while $[IF_6]^-$ is distorted. When less electronegative ligands like Cl, Br or I are present, the structures reported so far are regular octahedra. It is likely that, as the use of very large cations is explored, more dichotomies will be found. A number of explanations, including a simple steric one, have been put forward. From a simple approach to VSEPR, it is probably better to note that this case marks one limit of applicability of the system.

(2) Molecules like $(SiH_3)_3N$ (Fig. 17.21) and related compounds with bonds between Si and N or O violate the predictions. The AB_3L nitrogen is planar not pyramidal, and the AB_2L_2 oxygen is linear, or at least has a much wider angle than the VSEPR prediction of 'less than tetrahedral'. These cases were explained by additional π-bonding (Fig. 17.23) and the function of the VSEPR prediction is essentially to highlight a compound class where an extra effect occurs.

(3) Some anomalous structures involve fewer than six electron pairs, as in the symmetrical sandwich structure found for $(Ph_5C_5)_2Sn$, where the lone pair on tin should lead to a bent structure. In such cases the explanation is a simple steric one. This is part of a more general series of observations that very bulky groups, like *tert*-butyl or $(Me_3Si)_3C$-, force unusual stereochemistries. Similarly, rigid ligands with a specific coordination site may force unusual coordinations, such as the square planar Si or Ge in their porphyrin complexes. The overriding effects of such constraints apply, not only to VSEPR, but to other structural approaches.

18.8.3 AB_2 dihalides

Steric effects cannot account for some anomalous AB_2 dihalide structures. Normal VSEPR predictions apply to relatively covalent species: the AB_2, dihalides of Zn, Cd and Hg are linear, and the AB_2L, dihalides of the carbon Group are bent (with angle sequences such as 105°, 101° and 97° for CF_2, SiF_2 and GeF_2 respectively, paralleling Table 4.3 changes).

However, work on the triatomic MX_2 species found in the gas phase above heated alkaline earth element dihalides shows the following pattern of angles (°):

	Be	Mg	Ca	Sr	Ba
MF_2	Linear	Linear	140	108	100
MCl_2	Linear	Linear	Bent	120	ca. 100
MBr_2	Linear	Linear	Linear	Bent	Bent
MI_2	Linear	Linear	Linear	Linear	Bent

These results arise from three different types of experiment: (i) detection of polar molecules by electric field deflection and mass spectrometry in a high-temperature sample; (ii) electron diffraction of gas samples; (iii) infrared study of isolated MX_2 species deposited from a high-temperature beam into a krypton matrix at 20 K. For all these dihalides, a simple VSEPR model predicts a linear structure, as does a simple electrostatic model, $X^-M^{2+}X^-$. It has been suggested that the outer electron shell of the metal core is distorted by the polarizing field of the two halide ligands, resulting in the formation of localized electronic charge in positions spatially opposed to the halides. The bent shapes of many of the AB_2 dihalides can thus be accounted for by modification of the VSEPR treatment. Other explanations have involved the participation of d orbitals, which involve less excitation energy for the larger atoms. It may be noted that the alternative approach to molecular geometry through Valence Bond theory proposed by Smith (see references, Appendix A) does account for these MX_2 cases, in addition to covering the VSEPR-accurate structures.

18.8.4 The problem of s^2 configurations

In Section 4.5, we note that molecules with bond angles close to 90° could be discussed in terms of bonds involving the p orbitals, leaving nonbonding electrons in the s orbital. Further, the photoelectron spectrum of water suggests that the most stable molecular orbital is very close in energy to the unperturbed oxygen 2s orbital (Figs 4.20 and 4.21). These examples represent the general case where VSEPR states that there is at least one sterically active lone pair, and bonding theory places an essentially nonbonding electron pair in an s orbital which is spherically symmetric. Allied to this, we might add the complaint that while there is evidence of one sterically active lone pair, there is no real evidence for two.

Such a problem arises because ideas at different levels of sophistication are being mixed. VSEPR is essentially an empirical model supported by a relatively low-level rationalization. For example, VSEPR arrives at the AB_2L_2 electron count for H_2O and deduces an angle less than tetrahedral. While the relation with AB_4, e.g. for CH_4, is taken to imply some type of sp^3 hybrid, that explanation is not intrinsic to VSEPR. We can take VSEPR at a basic level of sophistication as predicting an angle less than $109\frac{1}{2}^\circ$ for AB_2L_2 without requiring any further interpretation along the lines of 'and the two nonbonding pairs are in sp^3 hybrids'. The shape prediction is at one level while bonding theory discussion is at another, and the difficulty arises because these two are mixed in a way that is not logically required (even if almost irresistible for the chemist!)

A response at the level of bonding theory is to note that the s orbital will only be completely independent if this is imposed by other constraints (for example, by symmetry). Thus, in the water molecule, the (s + s) combination of the H orbitals has the same symmetry, in the $\mathbf{C_{2v}}$ point group of the molecule (see Appendix C), as the oxygen 2s orbital. Thus, some interaction will occur, even if the energy difference means this will be very small.

In general, there will not be an absolutely unperturbed s^2 configuration, and only minor contribution is needed to remove the spherical symmetry. This would allow a reasonable description of PH_3, for example, as containing a lone pair with at least some excess electron density concentrated in the direction away from the three P-H bonds. For two lone pairs, it is not necessary that they be in equivalent orbitals, and it is difficult to see how this could be distinguished by experiment.

It should be emphasized that all the exceptions and anomalies discussed in this section amount to only a tiny fraction of the number of structures where VSEPR predictions are fulfilled.

From a broader viewpoint, it is best to keep the two processes, (i) prediction/rationalization of shape by VSEPR and (ii) description of the bonding by VB or MO or other approach, quite separate, remembering that the strength of VSEPR is in its simple and qualitative approach. Other accounts of molecular shape are available, and it may be that future developments—based perhaps on greatly increased computing power—will supersede VSEPR. At present, it is valuable at its own level, and with its now well-defined limits.

18.9 Bonding in compounds of the heavier Main Group elements

In Chapter 17, and in the earlier discussion in Chapter 4, the general picture was developed that the heavier Main Group elements differed from the second row, B, C, N, O and F by (i) not forming π orbitals from their p orbitals, and (ii) using their valence shell, but higher-energy, d orbitals to allow expansion of the octet. Although double bonds between the heavier elements are found (Section 18.7), the π contribution to the bond strength is only about half the σ contribution (contrast C, N and O where the π contribution is equal to or larger than the σ). While there is significant shortening, compared with the single bond, in molecules like $R_2Si{=}SiR_2$ or $RP{=}PR$ this will reflect not only p π contributions but also contributions from d orbitals and from reductions in repulsion energy. Thus, it is still sound to regard π-bonding by p orbitals as characteristic of the light elements only.

The degree of involvement of the Main Group element d orbitals has been made clearer by substantial theoretical investigation over the last two or three decades. Take a representative example, PF_2H_3, where the F atoms are axial and the H atoms equatorial in a trigonal bipyramid. Regard the planar PH_3 unit as bonded by the s, p_x and p_y orbitals on P, leaving the p_z orbital pointing towards the two F atoms. If three-centred orbitals are formed with the p_z orbitals on each F and on P (compare Fig. 17.69) then PF_2H_3 would be constructed without using d orbitals. However, adding some contribution from the d_{z^2} orbital on P to the nonbonding $(-p_1 + p_2)$ combination of Fig. 17.68 creates a $(-p_1 + d_{z^2} + p_2)$ orbital which is now bonding in character (and an out-of-phase antibonding equivalent). As the originally nonbonded $(-p_1 + p_2)$ orbital was occupied, the effect of adding d contribution is to stabilize these electrons and increase the total bonding. This does not require the d contribution to be as high as the p one. The calculation arrives at a contribution of about 25% by the use of d orbitals to the overall energy, with a d orbital population of around a quarter of an electron.

A further case where d orbitals have commonly been invoked is where it seems that double bonds to O are needed, as in $F_3P{=}O$; compare Figs 4.23 and 4.24. A calculation on H_3PO suggests that including a d orbital contribution shortens the bond length by about 15 pm compared with a single P-O bond, but with a d orbital population of only about 0.5 electron. As there are two equivalent d orbitals available, a more appropriate description is a partial triple bond. There is no suggestion of full d orbital participation, and the description as a semipolar bond, $H_3P^+{-}O^-$ is preferred, with the d orbital involved in partial back-donation to reduce the charge separation. The semipolar description is the only one appropriate for H_3NO.

Thus, in all Main Group oxygen compounds which are formally written with E=O bonds, the better description is probably that of semipolar bonds modified by back-donation. This does involve some use of d orbitals, but does not amount to full d participation. For NO_4^{3-} and NOF_3, the semipolar description is the most useful one, and the bond order for N-O is unity.

In summary, Main Group compounds which cannot be accounted for by up to four two-centre two-electron bonds, are probably best described in terms of multicentre bonds involving p orbitals, and in terms of semipolar bonds where appropriate, rather than by the older descriptions (e.g. sp^3d^2 hybrids) implying d orbital populations of several electrons. Addition of some d character, of up to one electron, substantially improves the energy, bond length and other parameters. None of these approaches seems, at present, to be particularly useful for species with coordination numbers greater than six, or with AB_6L compounds. Overall, the limited contribution of d orbital occupation reflects the penalty in activation energy required to populate them.

In a further reorientation of thinking about Main Group bonding, the emphasis on substantial hybridization between s and p orbitals is seen to be more appropriate for the 2s and 2p orbitals than for those of higher quantum number. As the 2s and 2p orbitals are less extended, and also localized in roughly the same region of space, hybridization is more effective. At the same time, the shorter distances mean that repulsions become more important, and the ability to remove lone pair density from bonding directions by hybridization is significant. For the more diffuse 3p orbitals, hybridization with the relatively compact 3s orbital is less effective, and the larger size means that repulsions are less important: similarly for higher quantum numbers. Thus the low bond angles found for compounds like PH_3 or H_2Se are the 'normal' ones, and the larger angles found for NH_3 or H_2O are the exceptional ones.

Finally, we note that relativistic effects (Section 16.13) increasingly separate the s and p energy levels as the quantum number increases, stabilizing 6s especially. Thus the pattern of the behaviour of the elements of a Main Group (see Section 17.3) re-emerges, although with different emphases in the theoretical explanations:

(i) The first Group member is unique, as 2s–2p hybridization is significant, repulsion effects are important because of the small size, and participation of higher level orbitals in bonding is negligible.

(ii) The remaining elements show the reverse behaviour. Participation of the d orbitals with the same principal quantum number is not as high as implied by earlier theories.

(iii) The heaviest member of the Group has distinct behaviour, in significant part because relativistic stabilization of the 6s orbital reduces its role in bonding.

CHAPTER 19

19.1 ELECTRON DENSITY DETERMINATIONS 536
19.2 METAL–POLYCHALCOGENIDE COMPOUNDS 538
19.2.1 Metal–polysulfur compounds 538
19.2.2 Metal–polyselenide and –telluride complexes 539
19.3 FULLERENES, NANOTUBES AND CARBON 'ONIONS'—NEW FORMS OF ELEMENTAL CARBON 541
19.3.1 Fullerenes and their metal derivatives 541
19.3.2 Carbon nanotubes and giant fullerenes 543
19.3.3 Polyhedral structures formed by other materials: transition metal chalcogenides and metallacarbohedranes 543
19.4 DENDRIMERIC MOLECULES 544
19.4.1 Introduction and dendrimer synthesis 544
19.4.2 Applications 545

General Topics

In this chapter, we review four topics which tie in discussions in different parts of the preceding text. The first is the study of the distribution of the valency electrons which is basic to the whole idea of chemical bonds. In addition, it underlies the experimental ionic radii (Section 2.15.2), and has given evidence for sterically active lone pairs (Section 18.8.1). Current work involves both experimental and theoretical contributions.

The second topic is chosen to reflect the strong current interest in experimental fields which span transition metal and Main Group chemistries. As we have discussed transition metal–sulfur species involving single S atoms (e.g. Sections 14.6.3 and 16.8.3) and also polysulfur rings (Section 18.2), we have chosen to link these by surveying rings containing transition metals and sulfur, together with some discussion on related selenium and tellurium species.

The third topic is designed to illustrate the very recent advances which have been made in 'materials chemistry' and to demonstrate that materials such as carbon, long thought only to exist in the diamond and graphite allotropes (see Section 5.9), can actually form a whole range of polyhedral structural forms. This 'molecular carbon', as will be seen, can form organometallic complexes with transition metals (see Section 16.2 to 16.5). The existence of other nominally layered materials, such as MoS_2 (Section 15.4) and boron nitride which also adopt novel polyhedral structures, suggests that this area of chemistry will become increasingly important.

The fourth topic, concerning dendrimeric molecules, has been chosen because of the recent interest in this topic, and because this represents a new type of molecular material. Examples of dendrimeric molecules also come from many areas of chemistry, including coordination and organometallic chemistry, phosphorus chemistry, as well as the multitude of organic dendrimers which are known.

19.1 Electron density determinations

The improvement in X-ray and neutron diffraction methods since about 1970 has allowed the determination of the distribution of the electron density of bonds, lone pairs and other features. This has been paralleled by improvements in the power and sophistication of molecular orbital calculations. Each approach, through experiment, or through calculation, is hedged by difficulties, but methods have been refined to the point where there is good agreement in key cases and the results may be used with some confidence.

For the experimental approach we note (Section 7.4) that X-rays are scattered by electrons, and the electron density may be calculated from the scattering pattern. The major contribution comes from the inner core electrons, which are highly concentrated, while the valency electrons give only a residual effect which has to be separated from the heavy core scattering and other contributions like those arising from thermal motion. The normal X-ray structure determination refines the atom positions to the centres of electron density, and therefore produces positions which are slightly biased by the distribution of the bond and lone pair electrons. However, accurate atom positions may be measured by neutron scattering, which is a nuclear process, not an electronic one. Alternatively, X-ray scattering at large angles depends only on the central core, and this can be used to determine atom positions.

The majority of X-ray experiments, while fully adequate for structure determination, are not sufficiently refined for electron density determination. Careful work, at low temperatures, and obtaining good high angle data (which is intrinsically less accurate), or carried out in parallel with neutron diffraction, has allowed electron density determination of a good number of molecules.

In calculations, the valency electron density is the small difference between the total electron density in the molecule and that arising from the *promolecule*, which is made up of free atoms with spherically symmetric electron density placed at the positions of the nuclei. Although computing power has increased enormously, an accurate electron density calculation requires a very good degree of approximation in the calculation (indicated by the number and type of *basis functions* used to approximate the true wave function. As the computing time goes up as the fourth power of the number of basis functions, there are clearly limits.) One important criterion for good electron density calculations (which need not handicap total energy calculation, for example) is that the chosen function should be an equally good approximation in all regions of the molecule to avoid introducing artefacts.

However, it is now well understood how to get the best results for both experiment and calculation, and each may be used as a test of the other. The best function to evaluate is the *density difference* which is found by subtracting, either the free atom densities from the observed one to get the *deformation density*, or else the inner shell density to get the *valence density*. A number of ways of creating such density difference maps are used (see the references in Appendix A). Such difference maps accumulate all the errors in both the total density and the free atom or core densities, so only major features are significant.

When an electron density difference map is plotted for an X–Y bond, we find a result which may be idealized as in Fig. 19.1. A bond takes electron density away from the atoms to concentrate it in the bonding region and the plot is the difference in electron density between the free atoms and the molecule. Thus, a zone of negative electron density appears close to the atom positions and a peak of electron density appears in the bond region. Minor density differences are seen on the remote sides of the atoms. Contours are usually plotted at intervals of electron density of 0.05 electrons per cubic ångström—perhaps 0.01 for more refined results—and experimental errors are usually in the range –0.03–0.05 e Å^{-3} in general regions, but are much higher very close to the nucleus.

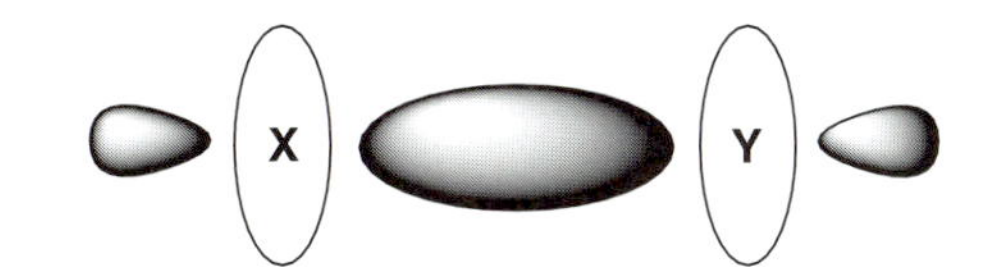

FIG. 19.1 The idealized electron density distribution plot for a single bond. Dark shading, positive density, no shading, negative density

Figure 19.2 shows the electron density difference through a Na^+CN^- pair in a crystal of $NaCN.2H_2O$, measured at 150 K. We see the very high charge density in the CN triple bond, the negative contours around N and C showing where electrons have been removed, and electron density on the remote sides of C and N corresponding to lone pairs. The lone pair on C is less tightly bound, and we know the cyanide ion always donates through the C when it forms complexes. Finally, we note the entirely negative, and almost spherical, distribution of electron density around the sodium ion, corresponding to the loss of the electron in forming the ion. Note that the peak electron density in the triple bond is about 0.5 e Å^{-3}.

As a second example, Fig. 19.3 shows the electron density difference around Co, and along one Co-N bond in the octahedral d^6 complex ion $Co(NO_2)_6^{3-}$. The plane shown passes through two opposite N atoms and bisects the equatorial plane so the remaining four NO_2 groups are above and below the figure. The two O atoms on the N shown are also out of plane. The figure shows the region close to the N where there is extra electron density, corresponding to lone pair donation, with shift of electron density from the remote side of the N. The most interesting feature is that electron density has increased along two axes at 45° to the Co-N bond directions, and has been removed from the Co-N direction in the region near the Co, exactly as expected for electrons moving from the e_g orbitals into the t_{2g} orbitals. When the plot in three dimensions is considered, there are eight zones, one in each quadrant (of which Fig. 19.3 shows a cross section of four) where there is an increase of electron density as expected for the filled t_{2g}^6 configuration of Co(III).

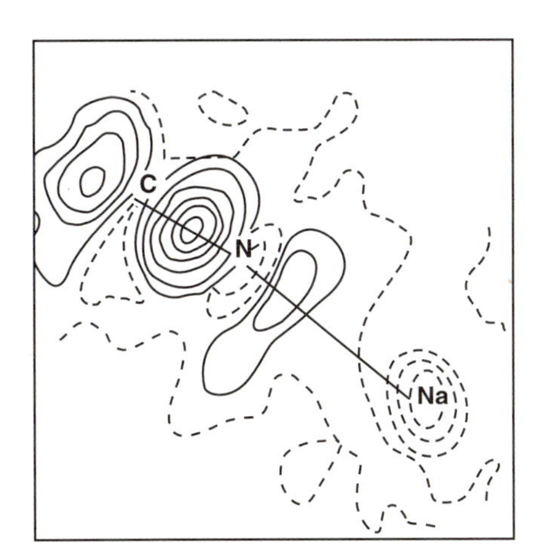

FIG. 19.2 Electron density difference map for a Na-N-C vector in $NaCN.2H_2O$

A very similar situation was seen, though less clearly, in an earlier study on $[Co(NH_3)_6][Co(CN)_6]$ where both the Co(III) complex ions showed a loss of charge density near Co along the bond directions and a gain in regions at 45° to these, though the maxima were only about 0.3 e Å^{-3}. It is interesting to find that the changes in electron density were approximate ellipsoids which were much more elongated at right angles to the bond for Co-CN than for Co-NH_3, matching the π back-donation postulated for such ligands (Fig. 19.4).

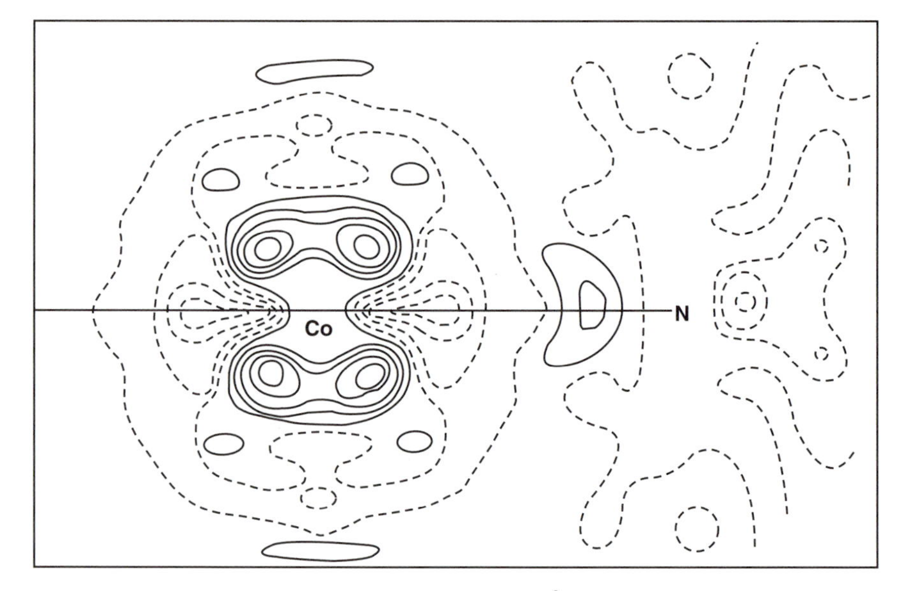

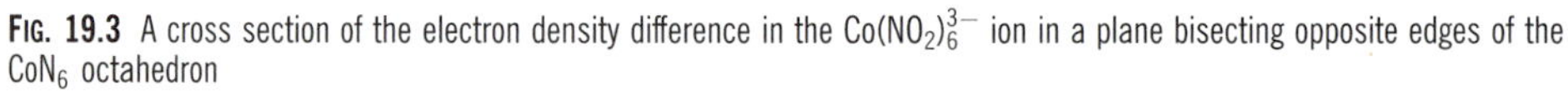
FIG. 19.3 A cross section of the electron density difference in the $Co(NO_2)_6^{3-}$ ion in a plane bisecting opposite edges of the CoN_6 octahedron

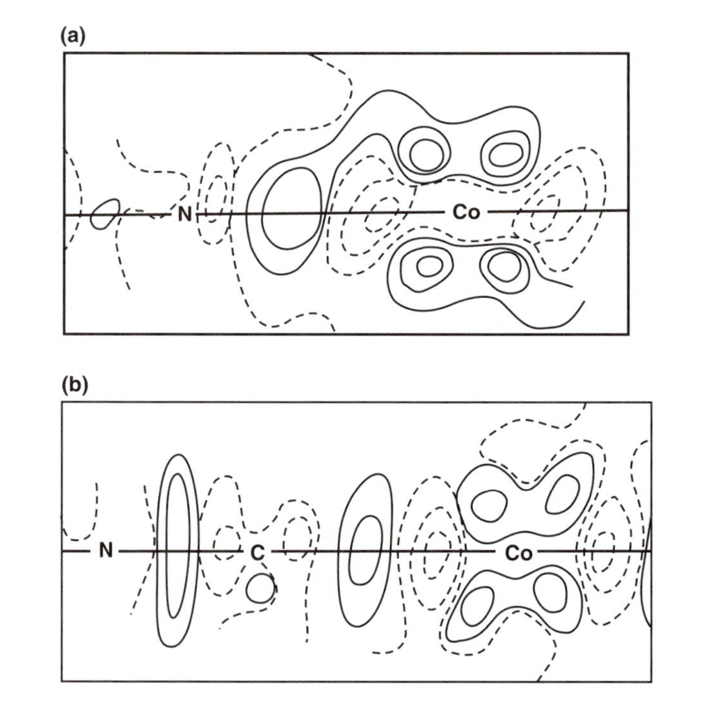

FIG. 19.4 The electron density differences in (a) the cation and (b) the anion of $[Co(NH_3)_6]^{3+}[Co(CN)_6]^{3-}$

19.2 Metal–polychalcogenide compounds

19.2.1 Metal–polysulfur compounds

Just as sulfur itself forms rings reflecting its preferred bond and dihedral angles, so do we find rings containing polysulfur units bonded to metal atoms. As both the metals and the sulfur chain are reasonably flexible in their steric demands, a wide variety of compounds is found. The metal can either be part of the ring system (which is the most common situation), or coordinated to it.

In compounds containing an S-S unit, a range of coordination modes is found. Three-membered rings, describable also as sideways bonded S-S, are common and are analogous

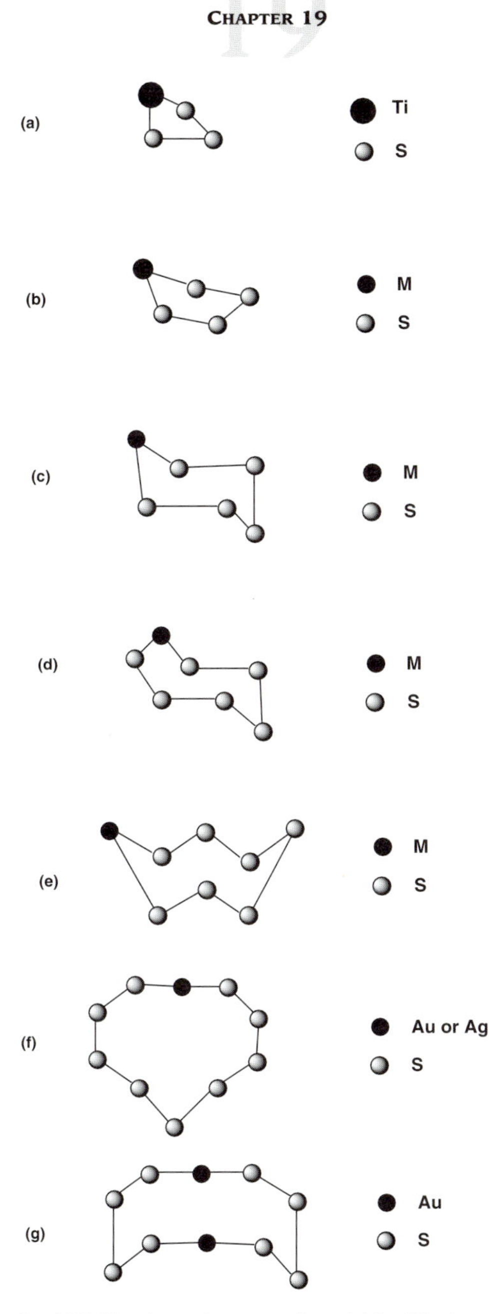

FIG. 19.5 Ring shapes in compounds containing MS_x rings: (a) four-membered ring in $(Me_5C_5)_2TiS_3$, (b) five-membered rings in half-chair form, (c) the chair form of the six-membered rings, (d) seven-membered ring, (e) eight-membered ring, (f) the ten-membered ring of AuS_9^- (g) $Au_2S_8^{2-}$

The very wide range of M-S species includes the binary sulfides, often forming layer lattices, the thioanions which are the sulfur equivalent of oxyanions, the M-S-M species of the types discussed particularly in Sections 14.6, 15.4 and 15.10, and the species containing S-S links discussed above, together with many cases of mixtures of the above, as in thioanions joined by polysulfur bridges.

to the peroxides. Other coordination modes include two metal atoms linked *trans* across an S-S unit or in a double-sideways mode where each M bonds to each S.

Rings of varying sizes are formed. There are only a few examples of a four-membered ring, illustrated by $(Me_5C_5)_2TiS_3$ where the ring is bent with the middle S atom 49° above the STiS plane (Fig. 19.5a). Five-membered rings are more common, and examples include $M(S_4)_2^{2-}$ for M = Ni, Pd, Zn or Hg where the central MS_4 configuration is tetrahedral, and $(C_5H_5)_2MS_4$ for M = Mo or W. Rings are not confined to transition metals, and $Sn(S_4)_3^{2-}$ has three SnS_4 five-membered rings. These MS_4 rings are in the half-chair form of Fig. 19.5b.

The best known of all metal–sulfur ring compounds is $(C_5H_5)_2TiS_5$ which is prepared from $(C_5H_5)_2TiCl_2$ by reaction with Li_2S_2 and sulfur. It is used to form other sulfur species, as in the syntheses of some of the parent polysulfur rings (compare Section 17.7). The Zr, Hf and V analogues of the Ti compounds are known, and other examples of the MS_5 ring include $(NH_3)_2Cr(S_5)_2$ and $Pt(S_5)_3^{2-}$. These rings have the chair structure (Fig. 19.5c), which is also found for the parent S_6.

Larger rings are found including the seven-membered MS_6 in $[M(S_6)_2]^{2-}$ for M = Zn, Cd, Hg, eight-membered in $(R_3P)_3MS_7$ for M = Ru or Os, and also in the Ti_2S_6 ring of $[(C_5H_5)Ti]_2(S_3)_2$. The MS_6 ring has the extended chair form of S_7 and $Ti(S_3)Ti(S_3)$ rings have the crown form of S_8 (Figs 19.5d and e, compare Fig. 17.45).

In the above rings, the metal atoms remain in tetrahedral or octahedral coordination. The preference of Au and Ag for linear 2-coordination is accommodated by the 10-membered rings of MS_9^-, and $Au_2S_8^{2-}$, shown in Figs 19.5f and 19.5g. The large ring size allows approximately linear sections through the M atoms.

Much more complex polysulfides are known. A relatively open structure is found in $M_2S_{20}^{4-}$ (M = Cu or Ag) in which there are two MS_6 rings linked by two S_4 units joined *trans* in an overall linked chain. Similarly, $Bi_2S_{34}^{4-}$ has two Bi atoms linked by a six-membered chain and each bearing two S_7 units in eight-membered BiS_7 rings, all crowns. Most polymetal–polysulfur species tend to assume more condensed forms. A good example is $Re_4S_4(S_4)_6^{4-}$ where the central Re_4S_4 core has an S-S-S chain linking each possible pair of Re atoms. Another interesting rhenium example displaying a diverse range of metal–sulfur bonding types is the complex $[Re_2S_{16}]^{2-}$, shown in Fig. 19.6. The complex contains two Re atoms linked by an Re-Re bond and bridged by two S^{2-} ions as well as by two S_3^{2-} groups. In addition, each Re atom is part of an ReS_4 ring system and overall displays a distorted octahedral geometry. Clearly, the ultimate end of progressive condensation is the formation of the metal sulfide, and it has been suggested that the polysulfur metal compounds are involved in the mobilization of metals in the geochemical formation of metal sulfide minerals.

Compounds are also known where the S_8 molecule is coordinated to a transition metal centre. The S_8 molecule has sulfur atoms each with two lone pairs, and it therefore might be expected to act as a multidentate ligand, forming numerous coordination complexes. Reality is rather different to this, however, with only a very small number of complexes being known. Two of the best characterized complexes are where the S_8 molecule is coordinated to silver(I) in $[Ag(S_8)_2]^+AsF_6^-$, and as axially coordinating ligands in the dinuclear rhodium(II) complexes $[Rh_2(O_2CCF_3)_4]_n(S_8)_m$ (n:m = 1:1 and 3:2). In the 1:1 derivative, S_8 molecules are bonded in the 1,3-positions to dinuclear $Rh_2(O_2CCF_3)_4$ units, such that there is a zigzag chain of alternating S_8 and $Rh_2(O_2CCF_3)_4$ species.

19.2.2 Metal–polyselenide and –telluride complexes

Related polyselenide and polytelluride complexes showing many of the structural types of the sulfur analogues have also been described. This research area continues to be an actively studied one and one which continues to turn up an extensive range of diverse compounds. However, the chemistry of the higher chalcogenides often differs significantly from the sulfur chemistry. Multinuclear nmr spectroscopy is a very useful tool for the study of the selenium and tellurium complexes since both selenium (^{77}Se, nuclear spin $\frac{1}{2}$, 7.6% natural abundance) and tellurium (^{125}Te, nuclear spin $\frac{1}{2}$, 7% natural abundance) are readily accessible nuclei, and this technique can be used to provide information on solution-state structures of these complexes.

GAS-PHASE METAL–POLYSULFUR CHEMISTRY

A final area of metal–polysulfide chemistry occurs in the gas phase. In this area, metal ions are generated by laser ablation techniques (where a short, high-powered laser pulse produces ions from a solid surface, e.g. a metal), and the products of their reactions are investigated by mass spectrometry. Mass spectrometers (see Section 7.9) are now sufficiently well developed that chemistry can be carried out in the gas phase, because individual types of ions can be selected, trapped, reacted, fragmented, neutralized and reionized. Theoretical calculations can then be used to predict the most likely structures for the new species. Gas-phase chemistry is also unaffected by the presence of solvent, and crystal packing forces, which can often influence the types of products formed. However, it is found that coordination numbers in the gas phase are typically much lower than in the condensed (solid or solution) phase.

One type of study carried out concerns reactions between naked M^+ metal ions and S_8: while very few metals form M^+ ions in the solid state, this is the predominant ion in the gas phase. Thus, comparisons between all elements in the same oxidation state are possible in the gas phase, but not in condensed phases. Ag^+ reacts with S_8 to give AgS_4^+, AgS_8^+ and AgS_{16}^+, the last of which has been described earlier in this section. However, it has been found that Ca^+ reacts with S_8 more rapidly than most transition metal ions, giving CaS_3^+.

Alternatively, metal-sulfide ions can be generated in the gas phase by laser ablation of the metal sulfide, and their reactions then investigated. For example, laser ablation of CoS results in detection of over 80 $[Co_xS_y]^-$ anions; the anions are generally reactive towards ligands with soft donor atoms (e.g. H_2S and phosphanes R_3P).

References

Gas-phase Coordination Chemistry Investigated by Laser Ablation Mass Spectrometry, K. J. FISHER, *Progress in Inorganic Chemistry* **50**, 2001, 343–432.

Gas-phase Coordination Chemistry of Transition Metal Ions, K. FISHER, *Chemistry in Australia*, October 1996, 439–441.

Gas-phase Inorganic Chemistry: How Is It Relevant to Condensed-phase Inorganic Chemistry?, K. J. FISHER, I. G. DANCE and G. D. WILLETT, *Rapid Communications in Mass Spectrometry* **10**, 1996, 106–109.

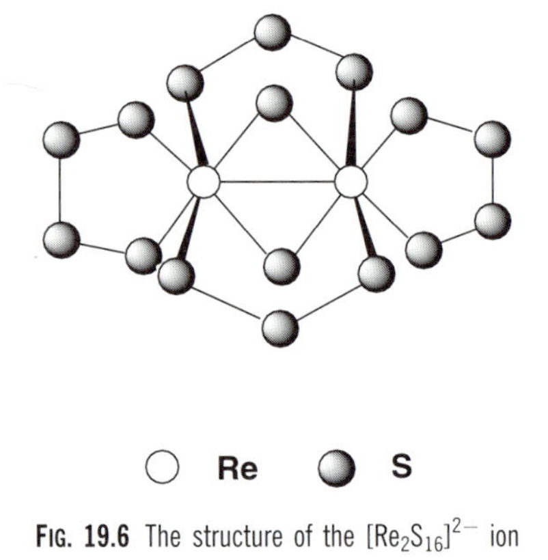

FIG. 19.6 The structure of the $[Re_2S_{16}]^{2-}$ ion

Examples of metal–polyselenide complexes include $[M(Se_4)_2]^{2-}$ (M = e.g. Ni, Pd, Mn, Zn, Cd and Hg), $[Pt(Se_4)_3]^{2-}$ and $[Pt(Se_4)_2]^{2-}$ which all contain five-membered MSe_4 rings analogous to the sulfur complexes (Fig. 19.5b). Metal–polytelluride complexes appear to generate the most 'unconventional' bonding patterns, such as in the complexes $[AgTe_7]^{3-}$ and $[HgTe_7]^{2-}$, shown in Fig. 19.7, where the metal shows a trigonal planar coordination geometry. The polytelluride ligand is an unusual $[\eta^3\text{-}Te_7]^{4-}$ ligand with the metals assigned to their usual Ag(I) and Hg(II) oxidation states in which trigonal planar coordination has a well-defined precedent.

Somewhat similar, in many regards, to the polychalcogenide metal complexes are metal complexes of mixed Se/S/N compounds. The compound S_4N_4 has been known for many years (it was first reported in 1835) and is a well-known explosive (see Section 17.7.7). However, by coordination to metal centres the S_4N_4 (and related Se_4N_4) species are stabilized. The first derivative prepared was the iridium complex shown in Fig. 19.8. Since then quite a diverse range of metal-stabilized unusual chalcogen-nitrides have been synthesized, and the interested reader is referred to the additional reading in Appendix A.

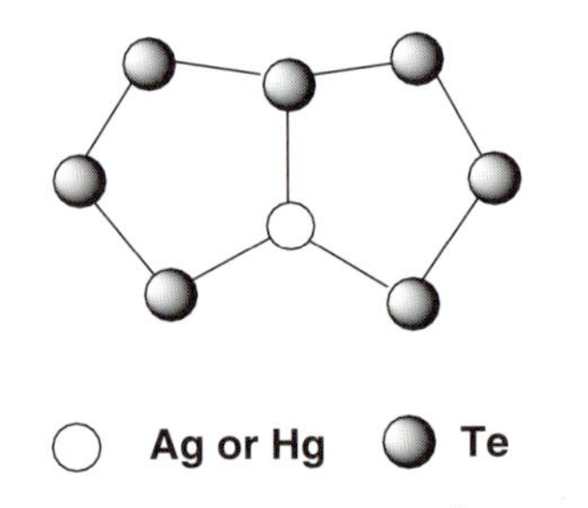

FIG. 19.7 The structure of the ions $[AgTe_7]^{3-}$ and $[HgTe_7]^{2-}$

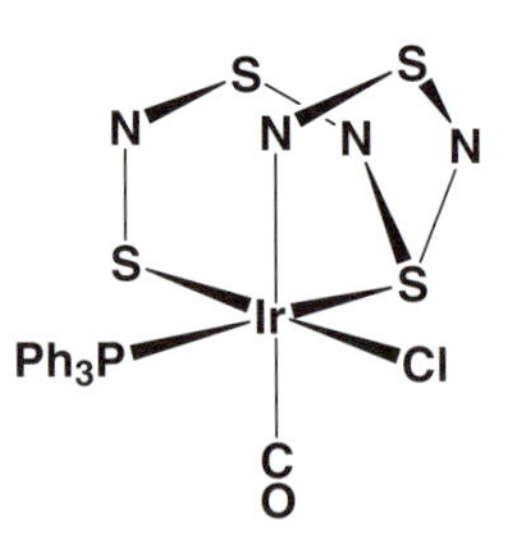

FIG. 19.8 The structure of the sulfur nitride iridium complex $IrCl(CO)(PPh_3)(S_4N_4)$

19.3 Fullerenes, nanotubes and carbon 'onions'—new forms of elemental carbon

19.3.1 Fullerenes and their metal derivatives

It was believed for many years that there were only two well-defined allotropes of the element carbon—the three-dimensional diamond and the planar-sheet structure of graphite. However, the chemistry of carbon has recently taken on a new dimension with the discovery of molecular allotropes such as C_{60} (see Section 17.5 and Fig. 5.14d).

The discovery of C_{60} was a rather serendipitous one. The investigators were looking at carbon-clustering experiments in the gas phase aimed at simulating the chemistry of molecules observed in cold, dark clouds in interstellar space. (The chemistry of outer space is a fascinating subject in itself and a whole range of molecules, such as the polyalkynyl cyanides, HC_xN where x is 5, 7 or 9, have been identified. A very brief overview is given in Section 8.7.) The C_{60}^+ molecular ion was regularly observed as a strong peak in the mass spectra and this led to a closed-shell spheroidal structure being proposed for the C_{60} molecule, as illustrated in Fig. 5.15d. (The name Buckminsterfullerene was coined for this new molecule after R. Buckminster-Fuller, the architect who first designed geodesic domes.) Not until 1990 was it found that carbon soots could contain relatively large amounts of fullerenes, predominantly C_{60} but also a significant amount of C_{70} and larger fullerenes. The fullerenes can be recovered from the soot simply by extracting it with an aromatic hydrocarbon, and chromatography can be used to separate C_{60} from C_{70}.

The availability of macroscopic amounts of C_{60} turned it, almost overnight, from what was essentially a curiosity to a material with an enormous number of potential applications. The synthetic procedure for making fullerenes is deceptively simple and essentially involves striking an arc between two graphite electrodes in a low-pressure helium atmosphere. In this regard it is highly surprising that the fullerenes were not discovered much earlier. The procedure is also highly suitable for use in the undergraduate chemistry laboratory and reference to this is made in Appendix A.

Since their discovery fullerenes have also turned up in some rather extraordinary places! C_{60} and C_{70} have been detected in *fulgurite*, a glassy rock formed where lightning strikes the ground—presumably the intense conditions of the lightning strike provide sufficient energy for the fullerenes to form. Fullerenes have also been detected in an impact crater on a spacecraft which had been in orbit for almost six years. The actual mechanism of formation of these carbon polyhedra is currently the subject of intensive research.

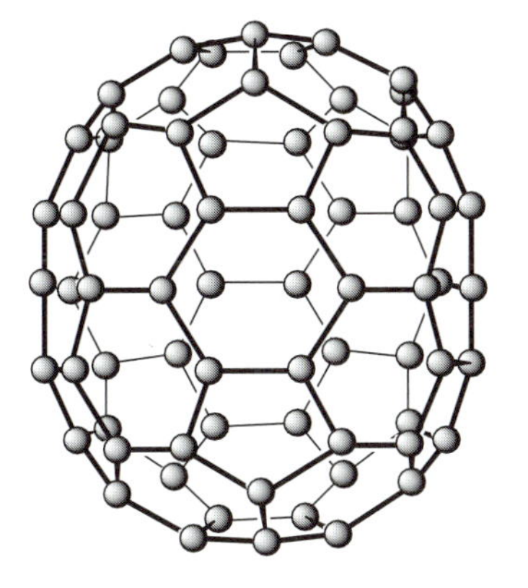

FIG. 19.9 The structure of the fullerene C_{70}

The structures of the fullerenes are now well established as there have been many crystallographic determinations of the structures of both the parent fullerenes and their derivatives. The structure of the C_{70} molecule, given in Fig. 19.9, is similar to that of C_{60} except that it is elongated by insertion of additional hexagons. Again it contains the twelve pentagons necessary to form a closed polyhedron. It was originally envisaged that C_{60} was a kind of 'three-dimensional' benzene, or graphite, with fully aromatic properties. The amount of aromatic character is still under scientific debate, through it appears from ^{3}He nmr data of helium atoms trapped within the fullerene cage that C_{70} has the greater amount of aromatic character, based on the diamagnetic shifts of the ^{3}He resonance. In fact the C_{60} molecule contains 'shorter' and 'longer' C-C bonds rather than having all bonds equal which would be expected if the molecule were fully conjugated. This influences the chemistry of these molecules as described later.

Not surprisingly, chemists, physicists and materials scientists all around the world rapidly began investigating the properties of the fullerenes almost as if a completely new element had been discovered. Given the pace of developments in this field, we can give only a brief mention of this chemistry here. We do not even attempt to apologise for this since it clearly indicates the vigorous activity which is occurring in this area of chemistry today. A selection of review articles from the recent literature is included in Appendix A and almost any journal in the current literature, particularly *Science* and *Nature*, will have the latest findings.

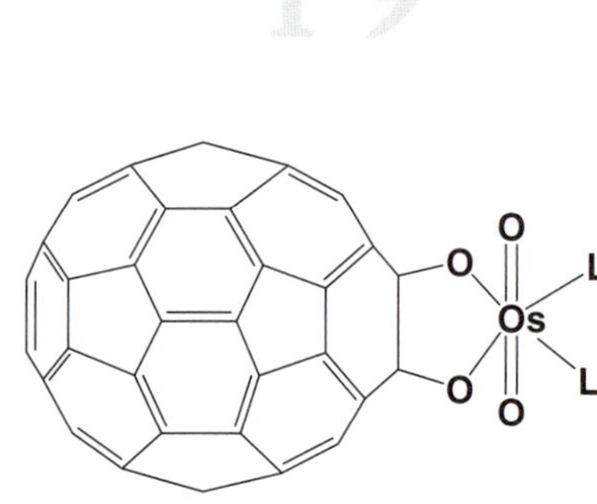

FIG. 19.10 The structure of the osmylated C_{60} derivative formed on reaction of C_{60} with OsO_4 and pyridine *t*-butyl pyridine

C_{60} contains both 6:6 and 6:5 ring junctions but no 5:5 ring junctions—a soccerball provides an extremely useful model!

See Section 16.4 for a discussion on metal-alkene complexes.

Of particular interest are fluorinated derivatives, such as the perfluorinated $C_{60}F_{60}$. Due to the excellent lubricating properties both of graphite and fluorinated polymers, such as Teflon, it was envisaged that $C_{60}F_{60}$ would be a 'molecular lubricant', though whether or not such applications arise remains to be seen.

The coordination chemistry of the fullerenes was one of the earliest areas studied and it also serves a very useful purpose in structurally characterizing these molecules. The problem lies with the high symmetry of the fullerene molecules which in turn produces rotational disorder of the molecules in the crystal lattice. Making a metal derivative disrupts the symmetry, thereby reducing the probability of disorder.

The first metal derivatives prepared were osmylated complexes formed by addition of OsO_4 and pyridine to the C_{60} framework. The complex formed is shown schematically in Fig. 19.10. Very high selectivity is typically observed, with the osmium adding to a 6:6 ring junction where the highest degree of carbon–carbon double bond character exists. This chemistry has also been used to form the first example of an optically active element since the fullerene C_{82} is chiral (it contains either a left-handed or a right-handed helical structure). By using OsO_4 with a chiral pyridine ligand, a kinetic resolution of the different enantiomers is achieved. One enantiomer reacts quicker with one form of the base, giving diasteroisomers which can then be separated by physical methods. The osmium can then be removed by reduction with $SnCl_2$, giving the resolved enantiomers of C_{82}.

A range of organometallic derivatives of fullerenes has also been synthesized and in all of these the general properties of the complexes point towards the fullerenes behaving more like electron deficient alkenes than electron rich benezene-like ligands. For example, the electron rich zero-valent platinum-ethylene complex $Pt(\eta^2\text{-}C_2H_4)(PPh_3)_2$ reacts with C_{60} forming the $\eta^2\text{-}C_{60}$ complex $Pt(\eta^2\text{-}C_{60})(PPh_3)_2$ in which again the platinum bonds to the more reactive 6:6 ring junction. Quite a wide range of complexes with other metals and fullerenes has been reported. Perhaps the most compelling evidence for the electron deficient character of C_{60} comes from ruthenium derivatives. The pentamethylcyclopentadienyl complex $[Cp^*Ru(NCCH_3)_3]^+$ contains labile methyl cyanide ligands, which are readily substituted by benzene and most other arenes to form η^6-arene complexes:

$$[Cp^*Ru(NCCH_3)_3]^+ + \text{arene} \rightarrow [Cp^*Ru(\eta^6\text{-arene})]^+ + 3CH_3CN$$

Compare this hybrid sandwich complex with both ferrocene, Cp_2Fe, and dibenzenechromium, $(\eta^2\text{-}C_6H_6)_2Cr$ (Section 16.4). However, when C_{60} is reacted in the same way only one of the three methyl cyanide ligands is displaced from each ruthenium and three rutheniums add to each C_{60}:

$$\text{excess } [Cp^*Ru(NCCH_3)_3]^+ + C_{60} \rightarrow [Cp^*Ru(NCCH_3)_2]_3C_{60}^{3+} + 3CH_3CN$$

This is more reminiscent of the reaction of $[Cp^*Ru(NCCH_3)_3]^+$ with electron deficient alkenes which also only substitute a single methyl cyanide ligand from the ruthenium.

In addition to these 'traditional' complexes of fullerenes, another completely different class of metal derivative has been found, though these are much less well studied at the present time. It has been found that the vaporization of a composite rod composed of graphite and a metal oxide, for example La_2O_3, under similar conditions as for the preparation of fullerenes, leads to a mixture of fullerenes together with species in which the lanthanum (or other metal atom) is trapped within the fullerene cage. These derivatives have been termed *endohedral* complexes (in order to form a distinction with the organometallic and osmylated derivatives which are *exohedral*). The terminology La@C_{60} has been coined to describe the endohedral lanthanum complex of C_{60}. The properties of these endohedral complexes appear to be somewhat different from those of the parent fullerenes, presumably because the metal causes a significant perturbation in the electronic structure of the fullerene. Larger fullerenes also allow the encapsulation of two or more metal atoms, and in some cases, in combination with other nonmetal atoms. Examples include Y_2@C_{82} and Sc_3N@C_{80} which has a trigonal planar Sc_3N unit within the fullerene cage and Sc_2C_2@C_{84}, which has a C_2 unit bridging 2 Sc atoms within the cage. Even reactive (N) and nonreactive (noble gas) atoms can be encapsulated in fullerenes, as illustrated by the dramatic N@C_{60}, where the inert inner surface of the fullerene cage

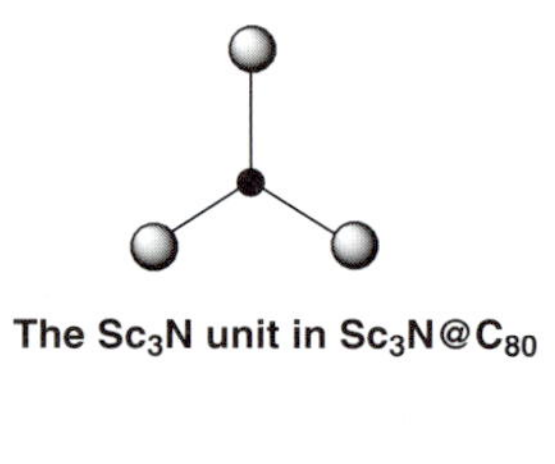

The Sc_3N unit in Sc_3N@C_{80}

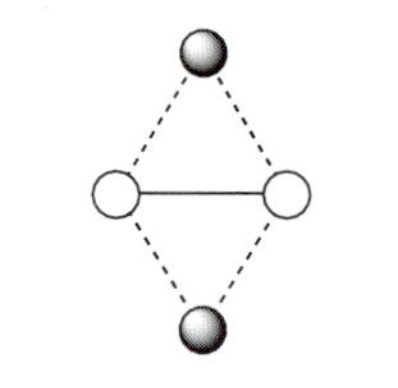

The Sc_2C_2 unit in Sc_2C_2@C_{84}

protects by isolation the highly reactive nitrogen atom. Further surprises are certain to occur in fullerene chemistry.

Another class of metal derivatives of the fullerenes are the fulleride salts formed by reaction of the fullerenes, e.g. C_{60}, with alkali metals. The many vacant molecular orbitals of the C_{60} molecule allow it to accept electrons (from the alkali metals) forming compounds containing the fulleride anions, e.g. M_3C_{60} and M_6C_{60}. The former of these is of current interest since it has been found to superconduct (Section 16.1) at low temperatures. The solid-state structure of M_3C_{60} is also worthy of comment. C_{60} itself, as might be expected of a highly symmetrical, pseudo-spherical molecule, packs together quite efficiently and the compound crystallizes in a face-centered cubic lattice (Section 5.6). It will be recalled that for every atom in a close-packed lattice there are two tetrahedral holes and one octahedral hole per lattice unit, in this case a C_{60} molecule. The structure of M_3C_{60} can therefore be derived based on a close-packed C_{60} array with all of the octahedral and tetrahedral holes filled by metal ions.

19.3.2 Carbon nanotubes and giant fullerenes

In addition to the simple fullerenes C_{60}, C_{70}, and their larger analogues described above, several other completely new forms of elemental carbon have been discovered. Variations of the procedure for forming fullerenes have led to carbon nanotubes, commonly known as buckytubes. These can be thought of as being formed from a number of sheets of graphite folded round on themselves to form 'nested' cylinders, the ends of which are closed with hemispherical fullerene-like caps, as shown in Fig. 19.11. Yet again, 12 pentagons provide the curvature at the fullerene-like ends of the tubes. A number of novel applications can be envisaged—on heating in air in the presence of lead, the caps are oxidized away, opening the tubes which then act as 'nanopipettes' and fill with the molten lead. Such materials chemistry has enormous potential for the fabrication of nanowires which could be used in electronic devices and the like.

Another form of carbon, related to the fullerenes, is the giant nested closed-shell fullerene structures which have been termed 'carbon onions'. It has been found possible to encapsulate moisture-sensitive materials, such as LaC_2, inside these giant structures, thereby protecting them from atmospheric moisture. In this regard these materials resemble the endohedral fullerene complexes described earlier.

19.3.3 Polyhedral structures formed by other materials: transition metal chalcogenides and metallacarbohedranes

The recent discovery that elemental carbon forms molecular species naturally led to the investigation of other materials. The question was asked: 'If graphite, which nominally adopts a layered structure, can be converted into polyhedral forms, such as fullerenes and nanotubes, can the same be done for other layered materials?' Molybdenum and tungsten dichalcogenides MX_2 ($X = S$ or Se) are materials which normally adopt layered structures. These are typified by having strong bonding interactions in two dimensions (in the layer) and weaker interactions in the third direction perpendicular to the layers, as typified by

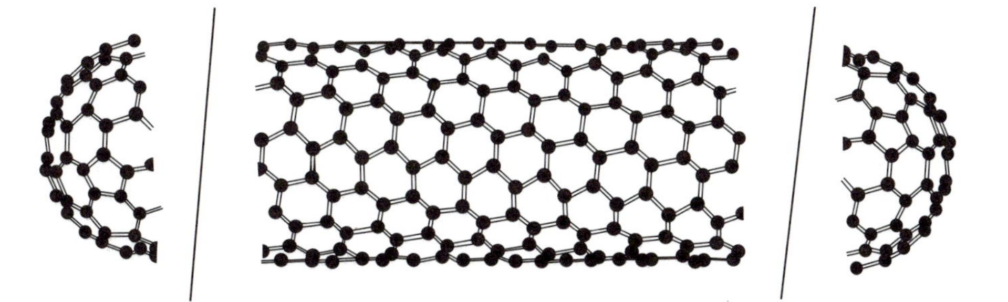

FIG. 19.11 Structure of a single carbon nanotube ('buckytube'). The structure can be considered as a graphite sheet folded to form a cylinder, with fullerene-like hemispherical caps at both ends to eliminate all 'dangling' bonds. Carbon nanotubes prepared in the laboratory typically consist of many concentric, or 'nested', nanotubes. Similar structures are observed for other layered materials, e.g. WS_2

the layered silicates talc and mica (Section 18.6). The structure of MoS_2 consists of close-packed sulfide layers with the Mo atoms lying in trigonal prismatic holes between these layers. These slabs then stack by weak forces, as mentioned in Section 15.4. This results in the material finding application as a lubricant. Recent research has found that the MX_2 compounds also form nanotube and nested polyhedral structures akin to those formed by carbon. A recent report has described the formation of similar structures by boron nitride, another layered material isoelectronic and isostructural with graphite.

Polyhedral cluster species containing a metal and carbon have also recently been prepared by the laser vaporization of a metal in a hydrocarbon atmosphere. This has been accomplished for a range of metals including Ti, Zr, V, Cr, Fe and Cu and the resulting clusters have been termed metallacarbohedranes or 'met-cars'. The most stable of these is the M_8C_{12} cluster. There is general agreement that these M_8C_{12} species form closed-shell polyhedra but the actual structures have been the subject of much recent research and a number of structures have been suggested for them, including a distorted pentagonal dodecahedron containing twelve pentagonal faces, each with two metal and three carbon atoms. Other species which have been detected include the $M_{14}C_{13}$ cluster which has been proposed to have a face-centred cubic structure, essentially a $3 \times 3 \times 3 (= 27)$ atom fragment of the cubic sodium chloride lattice (Fig. 5.la). It is noteworthy that this solid-state structure is adopted by a wide number of binary metal carbides (Sections 5.1 and 5.6). Preliminary studies indicate that the M_8C_{12} met-cars display a coordination chemistry, and species of the type $M_8C_{12}(H_2O)_n$ (n = 1–8) have been detected in which each metal atom can coordinate a donor ligand, in this case water.

Compounds such as these are changing the way that chemists think of materials, and this research is opening up a whole new and exciting area of chemistry. New discoveries and applications of these materials are certain to follow.

19.4 Dendrimeric molecules

19.4.1 Introduction and dendrimer synthesis

Recent years have seen an explosion of interest in a new area of chemistry, that of *dendrimeric* molecules. The word 'dendrimer' comes from the Greek word for branched; the characteristic features of a dendrimer are a globular structure, formed from a high degree of repeated, ordered branching, coming from a central core. Dendrimers can be considered to be a new class of polymeric molecule, with precisely defined size, shape and chemical properties. Dendrimers can also be considered to be a link between small molecular species and linear polymers, which are classified by their entangled polymer chains and relatively wide molecular weight distribution. The first types of dendrimers were constructed from organic building blocks, but recently there have been many examples of dendrimers constructed from phosphorus, organometallic or coordination complex building blocks.

In the *convergent synthesis* method, larger fragments are initially synthesized, and then coupled to the core.

A schematic representation of a dendrimeric molecule is shown in Fig. 19.12. The dendrimer begins with a *core* molecular unit, around which the dendrimer molecule is constructed. To produce a dendrimer, the core must have at least two reactive functional groups. In the *divergent synthesis* method, the dendrimer is then built up from the core outwards, by reaction cycles involving addition of one or more reagents, in repeatable synthesis steps. In order for the synthesis to be successful, each reaction step must occur with very high yields. These reaction cycles thus produce different *generations* of the dendrimer, in the form of concentric layers of repeating structures. The surface of the dendrimer is coated with functional groups (X in Fig. 19.12), which either allow the next generation of the dendrimer to be constructed around it, or alternatively, allow the binding of, e.g., metal complexes, etc.

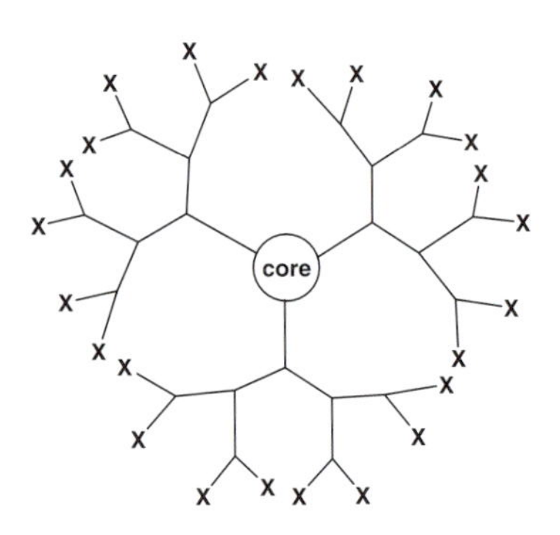

FIG. 19.12 A schematic diagram of a dendrimer molecule. The surface of the molecule is derivatized with functional groups, X

The above points can be illustrated by the synthesis of a small dendrimeric phosphane. Phosphanes are excellent ligands for transition metal centres, and there are potential advantages in using a dendrimeric ligand over smaller analogues, because of the well-defined structure and high molecular weight of a dendrimer. The synthesis, Fig. 19.13, begins with the core species $PhPH_2$. By carrying out two cycles of sequential radical

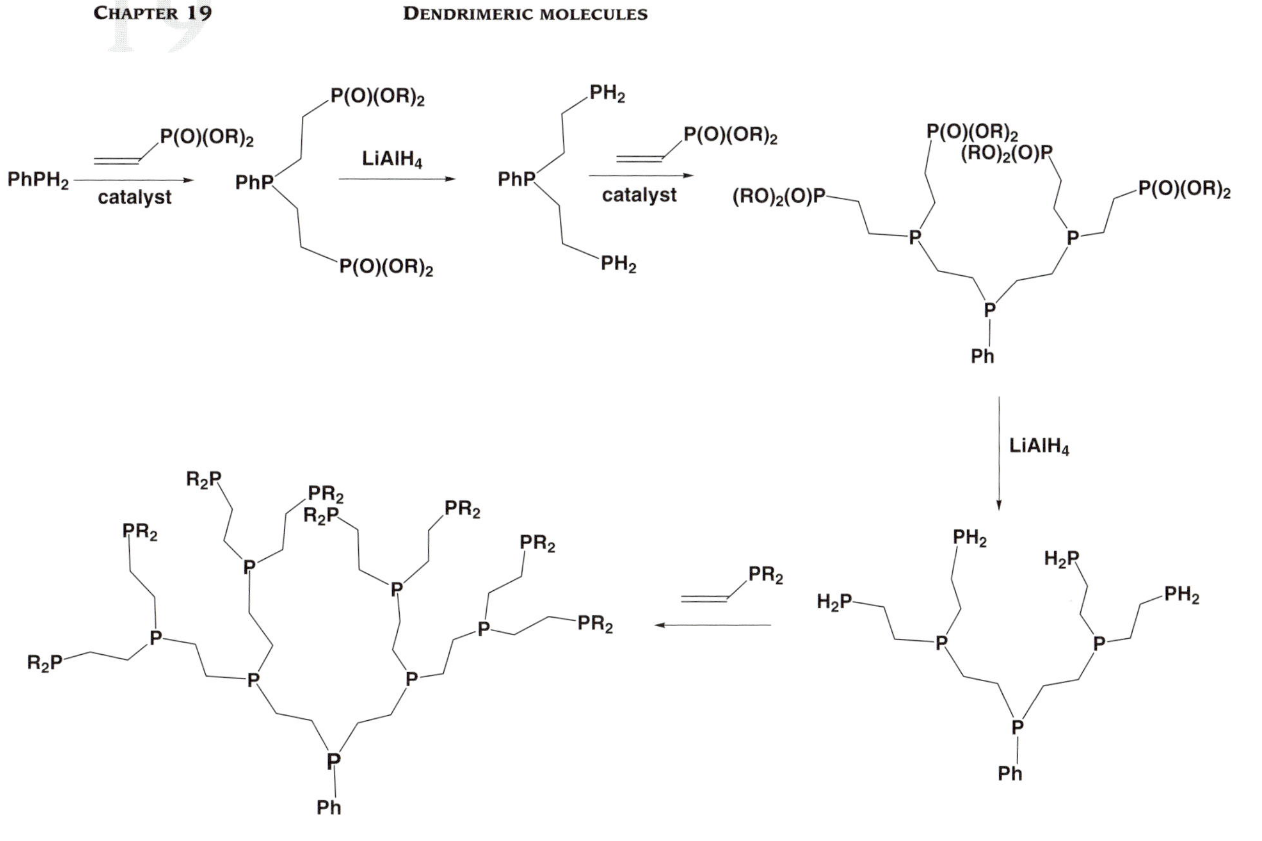

FIG. 19.13 The synthesis of a dendrimeric phosphane

catalyzed addition of the P-H bonds to a vinylphosphonate, $CH_2=CH\text{-}P(O)(OR)_2$, followed by reduction with $LiAlH_4$ (Section 9.5), a small dendrimeric primary phosphane, with terminal PH_2 groups, can be synthesized. To convert this into a dendrimeric tertiary phosphane, the dendrimer PH_2 groups are finally reacted with $CH_2=CH\text{-}PR_2$ (R = Ph or Et) to give the final dendrimeric phosphane with terminal PR_2 groups.

19.4.2 Applications

The high degree of order of a dendrimer, together with its extensive branching, gives a complex, but ordered, three-dimensional structure. Their use as catalyst supports is one area in which this precise molecular definition is potentially of great benefit. Traditional polymers can be used as soluble catalyst supports. However, the random distribution of catalytic sites and difficulty of removing these materials from the reaction medium limits their effectiveness. It is hoped that dendrimer supported catalysts will retain all of the advantages of homogeneous catalysts (high activity, precisely defined active sites) with the principal advantage of heterogeneous catalysts, which is the ease of recovery from the reaction medium. With dendrimers, the large molecular size and generally spherical shape facilitates recovery of the soluble supported catalyst by simple techniques such as ultrafiltration. As an example, the dendrimeric phosphane ligand of Fig. 19.13, containing surface PR_2 groups, has been reacted with $[Pd(MeCN)_4]^{2+}$ cations to give a palladium-functionalized dendrimer which catalyses the electrochemical reduction of CO_2 to CO, with an activity and selectivity which matches that of analogous monomeric catalysts.

CHAPTER 20

20.1 BIOLOGICAL INORGANIC CHEMISTRY 546
20.1.1 Introduction 546
20.1.2 The s elements in biochemistry 546
20.1.3 The p elements in biochemistry 547
20.1.4 The transition metals in biochemistry 549
20.1.4.1 Structural materials and iron storage 549
20.1.4.2 Haemoglobin and related compound 550
20.1.4.3 Ferredoxins: biological electron-transfer agents 551
20.1.4.4 Enzymes nature's catalysts 551
20.1.4.5 Nitogenase enzymes 553
20.1.4.6 Other transition-metal-containing systems 554

20.2 MEDICINAL INORGANIC CHEMISTRY 556
20.2.1 Overview 556
20.2.2 Inorganic diagnostic (imaging) agents in medicine 556
20.2.3 Technetium imaging agents 557
20.2.4 Magnetic resonance imaging 558
20.2.5 Boron neutron capture therapy (BNCT) 559
20.2.6 Platinum-based drugs in cancer chemotherapy 559
20.2.7 Therapeutic gold drugs 560
20.2.8 Chelation therapy for metal poisoning 560

20.3 ENVIRONMENTAL INORGANIC CHEMISTRY 561
20.3.1 Phosphates in detergents: eutrophication 562
20.3.2 Chlorofluorocarbons: the ozone hole 563
20.3.3 The greenhouse effect 564
20.3.4 Acid rain 566
20.3.5 Other inorganic environmental problems 566

Biological, Medicinal and Environmental Inorganic Chemistry

In this chapter we have selected three topics of current interest regarding 'natural' systems where inorganic chemistry plays a significant part. One of these topics, biological inorganic chemistry or bio-inorganic chemistry, is effectively a sub-discipline on its own. The fact that nature utilizes inorganic chemistry to a large extent and with very great effect suggests that we can learn a great deal by studying the way in which nature solves a particular problem. After all, nature has had millions of years to solve problems—our own efforts, by comparison, are often dwarfed into insignificance. Often, nature's solution to a problem involves the use of a metal ion of some sort with various ligands bonded to it. This chapter covers in some detail the inorganic chemistry found and utilized in biological systems.

Having given an overview of inorganic biochemistry, we then move on to cases where nature's solutions break down and drugs are necessary for the maintenance of well-being. Increasingly, drugs made of inorganic compounds are finding application in the treatment of diseases and we have selected several topics, which include elements from various parts of the Periodic Table, to illustrate the general use of inorganic medicinal compounds for both diagnostic and therapeutic purposes.

In the third and final part of this chapter we turn our attention to environmental inorganic chemistry. This is a field which is currently of great importance and concern, and consequently we feel that an inorganic chemistry textbook should at least present some discussion on this area.

20.1 Biological inorganic chemistry

20.1.1 Introduction

In the preceding chapters we have surveyed the chemistry of the elements, starting from relatively simple compounds, such as oxides and halides, and have seen that the Periodic Table can be used to largely predict the properties of the elements and their compounds. Natural systems also use a very diverse range of elements. Table 20.1 shows the elements in the Periodic Table which are utilized by biological systems. Many, such as potassium, phosphorus and sulfur, are ubiquitous whilst elements such as selenium, molybdenum, copper and many others are essential to life in trace amounts. The role of many of these elements is well understood though for others educated speculation is necessary. In the first section we describe several aspects of biological inorganic chemistry, using the Periodic Table as a framework for this discussion. The use of model complexes, to mimic biological systems, is a very important experimental method and we refer to this at several points in the discussion.

20.1.2 The s elements in biochemistry

The ions Na^+, K^+, Mg^{2+} and Ca^{2+} are of great importance in biochemistry. Potassium is an essential plant nutrient and the major production of potassium compounds is for use in fertilizers. In animals Na^+ and K^+ ions are mobile throughout the body and participate in many cell functions, such as nerve impulse transmissions, which depend on the ratio of Na^+ to K^+ and the concentration gradient across the cell membrane. The ions are thought to traverse the cell membrane by means of channels whose surface contains

TABLE 20.1 The Periodic Table showing the naturally occurring elements required by living systems

Group number

1	2	3	4	5	6	7	8	9	10	11	12	13	14	15	16	17	18
H																	*He*
Li	Be											**B**	*C*	*N*	*O*	**F**	Ne
Na	*Mg*											Al	**Si**	*P*	*S*	*Cl*	Ar
K	*Ca*	Sc	Ti	**V**	**Cr**	**Mn**	**Fe**	**Co**	**Ni**	**Cu**	**Zn**	Ga	Ge	<u>As</u>	**Se**	<u>*Br*</u>	Kr
Rb	Sr	Y	Zr	Nb	**Mo**	Tc	Ru	Rh	Pd	Ag	Cd	In	<u>*Sn*</u>	Sb	Te	**I**	Xe
Cs	Ba	Ln	Hf	Ta	W	Re	Os	Ir	Pt	Au	Hg	Tl	Pb	Bi	Po	At	Rn
Fr	Ra	Ac	Th	Pa	U												

XX Bulk biological elements
YY Trace elements believed to be essential for plants or animals
<u>*ZZ*</u> Possible essential trace elements

donor groups in an array similar to the multidentate donors of Fig. 10.5. The 6-oxygen donors (like the crown ether, Fig. 10.5a) have a cavity size which matches the size of K^+ better than Na^+ while a similar 5-oxygen donor favours Na^+ over K^+ so the channels through the cell membrane may select for K^+ or Na^+ under different cell conditions. This allows the build-up of concentration gradients between the inside and outside of the cell. It also results in the creation of charge differences since the total number of cations in the cell changes and anions transfer less readily or not at all. Thus, a nerve cell starts with a higher concentration of K^+ inside the cell than outside and this concentration gradient is maintained by a corresponding negative membrane potential of −70 mV. On stimulation the membrane becomes more permeable to Na^+ ions which flow from the higher external concentration into the lower internal one faster than the K^+ can now move out, causing a brief period when the cell loses its large internal negative charge. Thus charge may be transferred down a line of cells as each stimulates the next. The concentrations of Na^+ and K^+ are restored by chemical action via the hydrolysis of adenosine triphosphate (ATP) by the Na^+/K^+ *ATP-ase* enzyme.

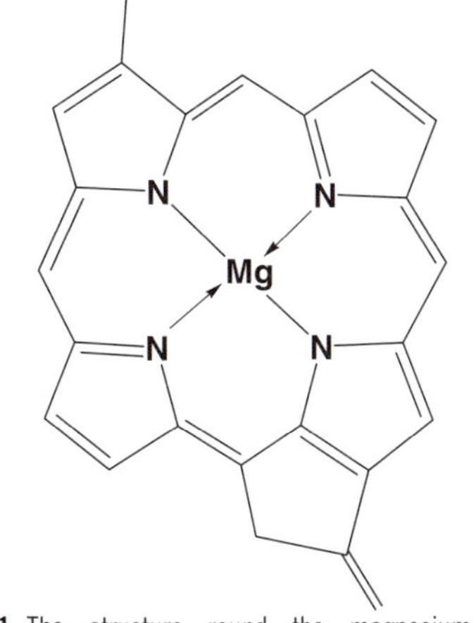

FIG. 20.1 The structure round the magnesium ion in chlorophll-a

Similar processes occur for Ca^{2+} and Mg^{2+} ions. For these doubly charged ions major functions include the triggering of sets of biochemical transformations, probably by coordination to donor atoms and changing the configuration of macromolecules or by bringing the reacting groups closer together. Thus Ca^{2+} is accumulated in exchange for Na^+ in heart mitochondria and contraction of the heart muscle is triggered by the release of this Ca^{2+}. Proposed mechanisms for PO_4^{3-} transfer have involved Mg^{2+} ions, whose coordination changes from one phosphate group to another, in molecules such as ATP and its congeners. This reaction is critical to the pathway by which the energy obtained from the oxidation of food compounds is stored and utilized. In many mechanisms Ca^{2+} and Mg^{2+} have opposing effects just as in the Na^+/K^+ changes in nerves.

Magnesium is found in chlorophylls, the pigments which are responsible for nearly all the conversion of CO_2 and H_2O into organic molecules using the energy of sunlight, i.e. photosynthesis. This is of great importance when the overall carbon balance of the planet is considered. Plants act as sinks for the CO_2 added to the atmosphere as a result of anthropogenic activity (see Section 20.3.3). The central coordination of magnesium in chlorophyll is to four nitrogen atoms in a macrocyclic *porphyrin* ring, as shown in Fig. 20.1. This environment is very similar to that found for iron in haemoglobin (see Section 20.1.4.2) or for cobalt in vitamin B_{12} coenzyme (see Fig. 20.7). Chlorophyll absorbs radiation at both the red and at the blue to near-ultraviolet ends of the spectrum. Radiation between these wavelengths is not absorbed thus giving the green colour which is characteristic of chlorophyll-containing plants.

STRUCTURAL ROLES OF CALCIUM AND MAGNESIUM

Ca^{2+} and Mg^{2+} also have more localized functions. For example, calcium is the main cation accompanying phosphate in bones and teeth. Nature uses inorganic materials as structural components to great effect, often as composite materials with proteins, etc., and we can learn much about the design of new synthetic materials by studying natural materials. Calcium-based biominerals predominate over the other Group 2 metals because of the low solubility of many calcium salts (such as carbonates, sulfates and phosphates) together with the relatively high calcium levels in extracellular fluids. By comparison, magnesium salts are typically more soluble than their calcium analogues and no simple magnesium biominerals have yet been found. Living things also utilize many other solid-state inorganic compounds for structural and other functional means.

20.1.3 The p elements in biochemistry

The p elements have a diverse range of roles in biological systems ranging from carbon, nitrogen, oxygen, sulfur and hydrogen, the building blocks of organic molecules, via phosphates and sulfates as structural components, to trace elements such as selenium and silicon.

Phosphate plays an extremely important role in the energy changes of cell processes and also in buffering acidity changes. The classical interconversion between adenosine triphosphate (ATP) and the diphosphate (ADP) plus phosphate was one of the earliest biochemical processes to be worked out. The role of chloride, sulfate, phosphate and carbonate/bicarbonate in controlling the ion balance and flow of protons is ubiquitous in living systems. These processes are fully described in biochemistry texts. Here we focus on two examples of p element roles of current interest.

The role of silicon is biology is a very interesting one. Amorphous silica is used as a structural component of the exoskeletons of radiolaria and diatoms, and has also been found in various grasses. Silicon may have an important role to play—it has been found that rats and chicks when deprived of silicon in their diet show reduced weight gains and it has been suggested that the major role of silicon could be in the minimization of aluminium toxicity. Aluminium is a toxic element largely due to its high charge/size ratio which results in it forming much more stable complexes than Ca^{2+} or Mg^{2+} with hard oxygen donor groups (such as phosphates). Aluminium has been linked with Alzheimer's disease in

NITRIC OXIDE (NO): A SIMPLE INORGANIC MOLECULE WITH SEVERAL BIOCHEMICAL FUNCTIONS

The small molecule nitric oxide, NO, and its role in biological systems has recently attracted a substantial amount of interest. It may appear as a surprise to the reader that NO is involved in biochemistry at all—it is a highly reactive, toxic gas. Nevertheless, nature uses this molecule in a variety of elegant ways, including neurotransmission, blood clotting and blood pressure control, and it also plays a role in the immune system. Since NO is a small diatomic molecule, which is uncharged, its transport is very different to the channels or other specific transport systems used for other species, such as K^+ or Na^+, and diffusion is its primary mode of movement. During muscle relaxation (which is one of the keys to understanding high blood pressure) NO is produced by endothelial cells on the insides of arteries. An enzyme, called *NO-synthase*, is responsible for the production of NO in the body by the oxidation of the amino acid arginine (Fig. 20.2). The NO then diffuses into the muscle cells where it binds to the iron atom which is at the active site of the enzyme responsible for controlling muscle relaxation, forming an Fe-NO complex (see Section 16.2.5). In fact, now that the role of NO in blood pressure control has come to light, it explains the longstanding success of using glyceryl trinitrate as a drug for the treatment of angina (narrowing of the heart arteries). In the body glyceryl trinitrate is converted into NO which then causes dilation of the arteries. Another drug used for vasodilation (widening of the arteries) in surgery is the NO-coordination complex sodium nitroprusside, $Na_2[Fe(CN)_5NO]$ (see Table 20.3, p.556). What glyceryl trinitrate and sodium nitroprusside have in common is the potential to generate NO in the body.

In a completely different role, NO is involved in the body's immune system, for example to kill any alien bacteria which have invaded the body. Certain cells in the defence system, called *macrophages*, kill bacteria by injecting NO into them. Once the NO has had its toxic effect it is oxidized to NO_2^- and NO_3^- and excreted from the body.

NO also acts as a messenger molecule in the brain—electrical stimulation of certain cells at synapses results in the production of NO and it has been suggested that it may play a role in the memory function of the brain. Research reported in 1993 has also found that NO may have antiviral properties.

It is clear that these recent discoveries indicate that a small reactive molecule has very important biochemical roles to play and is a further example of the elegance of nature in utilizing certain inorganic compounds. It is certainly possible that other new biochemical roles for this simple molecule will be determined in the future. The interested reader is referred to Appendix A for a number of recent articles on this topic which reflect on the current interest.

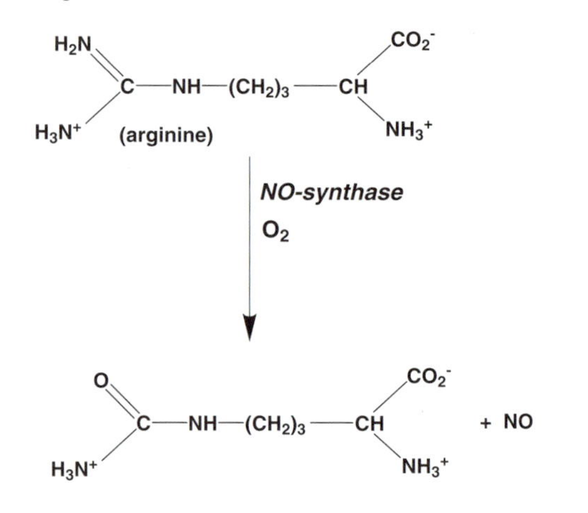

FIG. 20.2 The reaction by which nitric oxide is produced in the body

humans, though the exact causes have not yet been fully determined. Human uptake of aluminium can come from various sources including water treatment (flocculation) chemicals and in particular, aluminium saucepans (especially when used for cooking acidic foodstuffs such as rhubarb—the combination of high acidity and good complexing agents for aluminium, such as citrates, results in relatively large amounts of soluble, and thus available, aluminium). In the presence of silicon, hydroxyaluminosilicate species may form, thereby reducing the bioavailability of the toxic aluminium. It is possible that silicon starvation in laboratory animals increases the availability of aluminium, with its consequent toxic effects.

20.1.4 The transition metals in biochemistry

When the roles of the various transition elements in biology are studied it is clear that the range of functions is even more diverse than with the p or s elements. This can be readily understood in terms of the general chemistry of these elements where typically a range of coordination geometries and oxidation states are accessible for each element allowing many more 'options' in a biological system. A good illustration comes from enzymes—many of the most active of these contain a metal, or metals, at the active site. The presence of the metal allows small molecules (such as CO_2, H_2O, CH_4, etc.) to be selectively coordinated, and subsequently reacted, and good parallels come from a study of coordination and organometallic complexes where many of the fundamental chemical steps have been studied in detail. While we have not attempted to give a comprehensive survey, selected examples of transition elements in biological systems have been chosen to illustrate some of the chemistry. A particular emphasis is placed on iron, since it is the most abundant metallic element in the human body and plays a crucial role in the biochemistry of all living organisms.

20.1.4.1 Structural materials and iron storage

As with the s and p Groups of elements, the transition metals, in particular iron but also manganese, are found in bioinorganic minerals. Several types of *magnetotactic* bacteria contain solid-state, magnetic, mixed-valence oxide or sulfide crystals, which are used as a means of navigation in the ambient magnetic field of the Earth. Most bacteria contain *magnetite*, Fe_3O_4, whereas some, growing in sulfur-rich regions such as geothermal areas, contain the mineral *greigite* Fe_3S_4. The single crystals, as shown in Fig. 20.3, are aligned in chains. Other bacteria are able to mineralize a fairly diverse range of inorganic materials such as cadmium, zinc, copper and lead sulfides.

Related to the deposition of single-phase mineral particles within organisms is the iron storage protein *ferritin* which contains solid iron oxide as nanometre-sized particles. Ferritin has been colourfully called 'biological rust' and the composition of the active site approximates

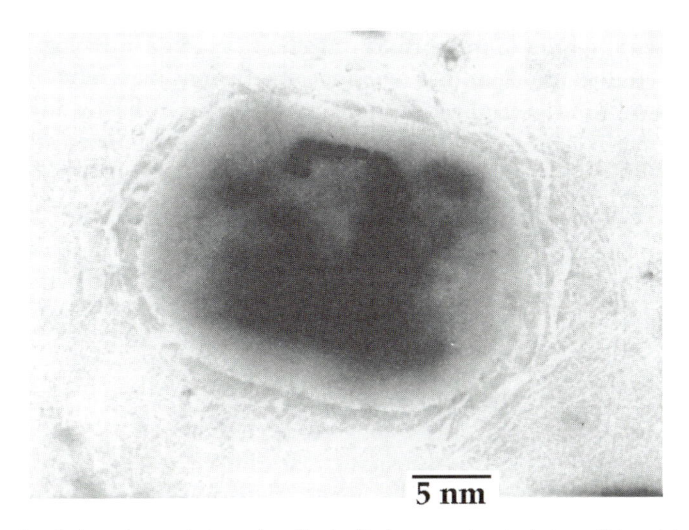

FIG. 20.3 A scanning electron microscopic image (see Chapter 7) of a coccus-type bacterium which contains a chain of nine dark magnetite (Fe_3O_4) single crystals. The bacterium utilizes these microscopic 'bar-magnets' for orientation in the Earth's magnetic field. (Taken with permission from S. MANN, *Journal of the Chemical Society, Dalton Transactions,* 1993, 1.)

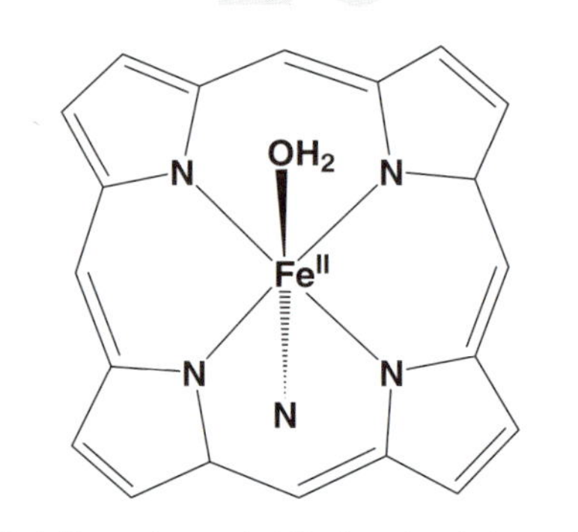

FIG. 20.4 The environment of the iron atom in haemoglobin. The four N atoms of the porphyrin ring are coplanar with the iron. The fifth position on the iron is occupied by a nitrogen atom from a long side-chain on one of the rings, leaving the sixth site in the octahedron around the iron to hold an oxygen molecule, a water molecule or some other group. Side-chains and links to the rest of the protein are attached to the outer carbon atoms

$[FeO(OH)]_8[FeO(OPO_3H_2)]$. The body needs to store and mobilize substantial amounts of iron in a form which prevents the precipitation of iron hydroxide species. Various iron carboxylate complexes have been synthesized, with different degrees of complexity, some of which are models for ferritin. An example is the complex $Fe_{11}O_6(OH)_6(O_2CPh)_{15}$ in which the iron atoms are triply bridged by O or OH and are also linked by the carboxylate groups Fe-O-CPh-O-Fe in a cubane-like structure which lacks some of the edges.

20.1.4.2 Haemoglobin and related compounds

Perhaps the most famous complex of iron in a biological system is the complex *haem* which exists in haemoglobin. The central porphyrin ring system is shown in Fig. 20.4. Side chains are attached to the porphyrin skeleton and an imidazole ring on one of these is coordinated to a fifth position on the iron atom. A water molecule occupies the sixth position, as shown in Fig. 20.5a. This water molecule may be replaced by an O_2 molecule and this process is reversible (Fig. 20.5b), providing the mechanism for the transport of oxygen by red blood cells to various parts of the body (compare Section 16.7). The oxygen site on the iron may also be occupied by CN^-, CO or PF_3 and the coordination in these cases is strong and irreversible. This is one reason for the poisonous nature of these substances (although cyanide, in particular, influences other reactions in the body as well). Iron also occurs in myoglobin which is used to store oxygen in muscles.

It has been disputed for a long time whether the oxygen was carried in these respiratory proteins in an 'end-on' or 'side-on' configuration. Both coordination modes are well known in coordination complexes of dioxygen, as described in Section 16.9. Haemoglobin and myoglobin are large molecules whose structures have not been determined in the detail required to see the O_2 coordination. When simpler molecules (containing the basic porphyrin unit of Fig. 20.4) are used as model complexes, the difficulty has been to obtain a species which reacts *reversibly* with O_2 as do the natural proteins. One approach is to protect the oxygen site on the Fe by using the so-called 'picket-fence' porphyrins. In these, substituents are placed on the C bridges between the C_4N rings of the porphyrin which lie above the FeN_4 ring, as indicated schematically in Fig. 20.6. It was found that such molecules did oxygenate reversibly and that crystalline dioxygen complexes could be isolated. Although the crystals were marginal for X-ray work, an end-on unit with an angle at

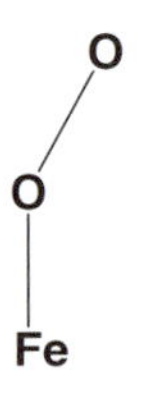

O of about 130° was suggested. Later, and even more demanding, structural work showed an angle at O of 156° in haemoglobin itself. The O-O stretching frequencies of the picket-fence model compounds were in the range 1140–1165cm^{-1}, similar to those for

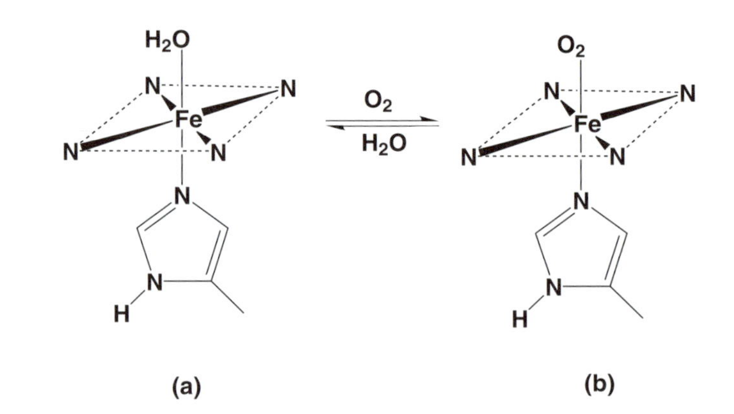

FIG. 20.5 The mode of oxygen carriage by haemoglobin: (a) with a water molecule which is reversibly replaced by an oxygen molecule (b)

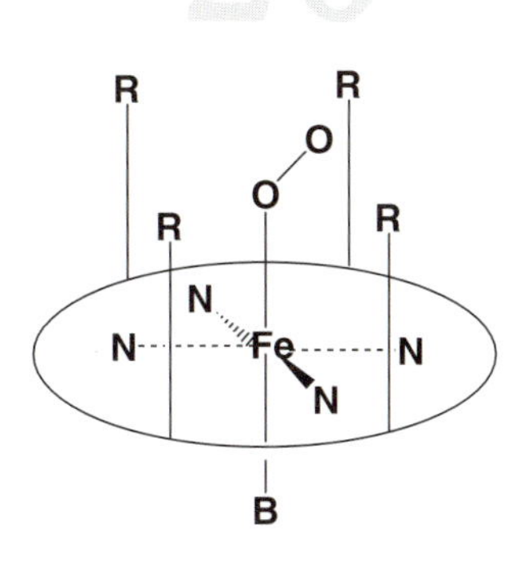

FIG. 20.6 'Picket-fence' porphyrin (diagrammatic). R = NHC(O)CMe$_3$, B = base, e.g. THF. Compare with Fig. 20.4

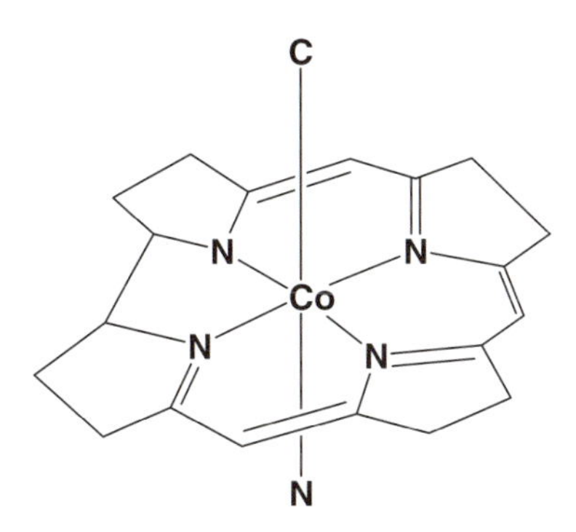

FIG. 20.7 The geometry about the cobalt in vitaminB_{12}. Note that the cobalt atom is bonded to a CH_2 group of an adenosine group and is thus an example of a naturally occurring organo-metallic complex. Compare with the structures of chlorophyll (Fig. 20.1) and haemoglobin (Fig. 20.4)

myoglobin and the cobalt analogue of haemoglobin. This has shown that the end-on, or superoxo (see Section 16.10), description is the best for all the oxygenated respiratory proteins and that model compounds can be put to good use in understanding the structures of the inorganic components of highly complex biological molecules.

Similar, though not identical, to the role of iron in haemoglobin is the natural cobalt(III) complex, vitamin B_{12}. The cobalt is situated in the middle of a porphyrin-type structure (Fig. 20.7) and coordinated by the four ring nitrogens and a fifth nitrogen from a side-chain group. The sixth site, completing the octahedron, is the active site and a number of derivatives are known with different groups occupying this site including CN^-, hydrogen and organo-metallic compounds containing a direct σ cobalt–aliphatic carbon bond.

Iron is also found in various cytochrome pigments. In these a chain of electron-transfer occurs which links dehydrogenation of alcohols or fatty acids with the conversion of O_2 to H_2O via a series of oxidation–reduction steps involving Fe(II)/Fe(III) (or Cu(I)/Cu(II)) conversions in various cytochromes.

20.1.4.3 Ferredoxins: biological electron-transfer agents

A further important group of iron-containing natural products are the *ferredoxins* which contain Fe and S atoms and are used in many organisms as electron-transfer reagents. Two basic structural units (Fig. 20.8) have been observed, (a) Fe_2S_2 and (b) Fe_4S_4. These structures are linked to the protein skeleton by further Fe-S bonds to the sulfur-containing amino acid cysteine, $HSCH_2CH(NH_2)CO_2H$. For example, in one isolated ferredoxin (from *Peptococcus aerogenes*) there are two well-separated Fe_4S_4 units each linked to four positions, again well separated, by the cysteine bonds. It is possible that the electron-transfer functions via the formation of Fe–Fe bonds across the face diagonals of the (b) units. Many iron–sulfur compounds have been synthesized as inorganic models for such systems.

In the preceding section the increasingly important role of NO in biochemical processes was discussed and it has been found that NO can damage cells by attacking and breaking up the Fe-S clusters present in important proteins, as shown schematically in Fig. 20.9.

20.1.4.4 Enzymes—nature's catalysts

A wide range of enzymes are known to occur in biological systems, and these accomplish a large number of chemical transformation steps rapidly and with high reactant and product specificity. Many enzymes find industrial and household applications, such as lipase enzymes in detergents and glucose isomerase for the conversion of glucose to fructose in the food industry. Enzymes which occur in thermophilic, or 'heat-loving', bacteria found in geothermal regions, are also of practical importance since they function most efficiently at high temperatures. In marked contrast most other enzymes are denatured at high temperatures. Because of their specificity for chemical transformations, chemists are

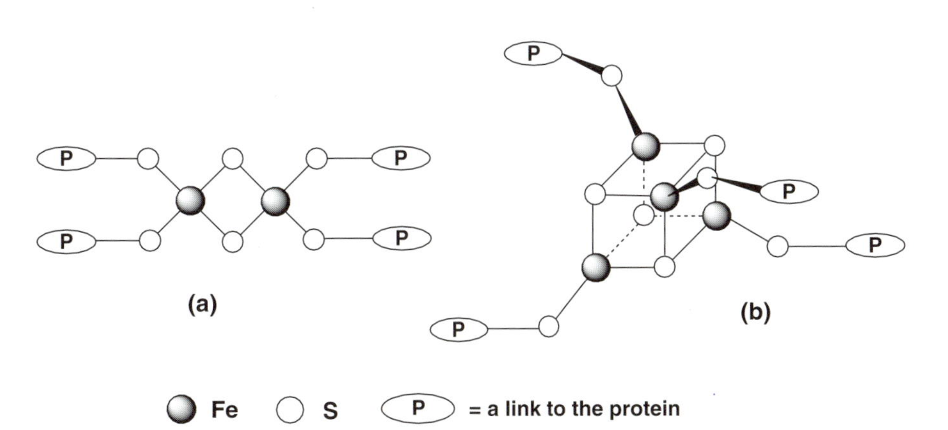

FIG. 20.8 Two iron-sulfur binding units found in ferredoxins, nitrogenase and other iron proteins

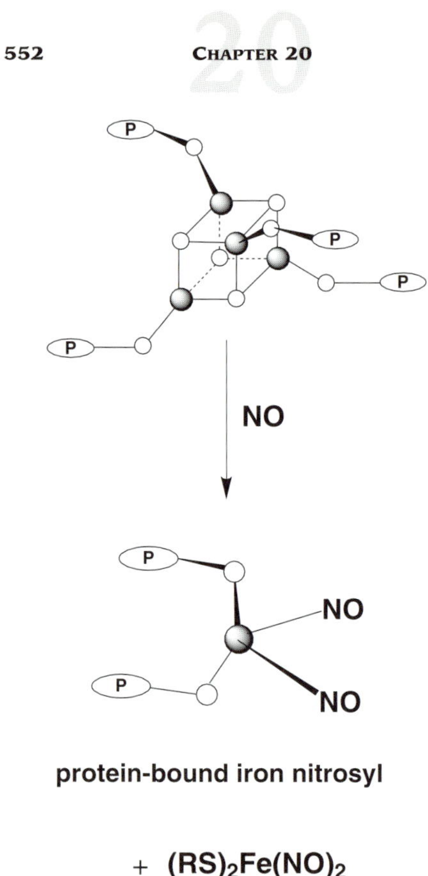

FIG. 20.9 Reaction of NO with iron–sulfur clusters. This is believed to be the mechanism by which NO is utilized in biological systems to destroy invading bacteria

becoming increasingly aware of the utility of thermophilic bacteria in synthetic procedures, such as in making enantiomerically pure materials.

Many enzymes contain a metal, or a number of metals, at the 'active site' where the chemical transformation is accomplished. Table 20.2 shows the metal atoms present in a number of different enzymes, together with their functions. The determination of the structure of the metal-containing active site of an enzyme poses significant challenges since it represents only a small part of the protein structure as a whole. Nevertheless, the active sites of a number of enzymes have been determined. The study of the properties (electrochemical, spectroscopic) of model transition-metal coordination complexes again has a significant role to play in these studies. Learning about the structures of enzyme active sites may lead to the design of better synthetic catalysts and good examples are the ongoing searches to try and find practical catalysts for nitrogen fixation (see Sections 16.9 and 20.1.4.5) and for artificial photosynthesis. If this can be accomplished, we will be readily able to convert the abundant nitrogen in the atmosphere into chemical feedstocks (the Haber process, Section 17.6.1, can already do this but only at high temperatures and pressures using a catalyst) and harness the sun's energy directly into stored chemical energy.

As a recent example of an enzyme where the structure of the active site has been largely determined by a series of elegant experiments, we have chosen to describe a hydroxylase enzyme which catalyses the conversion of methane to methanol in methanotropic bacteria. These bacteria derive their carbon and energy solely from methane via the reaction:

$$CH_4 + O_2 + H^+ + NADH \rightarrow CH_3OH + H_2O + NAD^+$$

The study of the biochemical processes occurring in such bacteria is prompted by several factors, including the implication of methane as one of the greenhouse gases in global warming (sec Section 20.3.3), together with the possibility of using such bacteria to convert readily available natural gas into chemical feedstocks.

While spectroscopic techniques, together with biochemical methods, have given a substantial amount of information on the structure of the active site of the hydroxylase enzymes, X-ray crystallography (Section 7.4) is the most powerful technique, giving direct three-dimensional structural information. However, obtaining crystals of such large biomolecules can be painstaking work—more than 1000 crystallization trials were attempted before a suitable crystal of the hydroxylase enzyme was obtained! Even with this, since the molecule is so large, the resolution of the structure is only around 200 pm.

There are two identical active sites in the enzyme which are dinuclear, hydroxide-bridged units, the structure of which is shown in Fig. 20.10. The two iron(III) atoms are bridged by a hydroxide ligand, by a semibridging carboxylate ligand from an amino acid residue (Glu144) of the protein, and by another ligand, suggested to be an acetate, which has come from the ammonium acetate buffer used in the solution from which the enzyme crystals were grown. It has been proposed that another ligand (other than acetate) bridging the two iron atoms is present in the native enzyme. One possibility is the related formate ion, HCO_2^-, since this is produced by metabolic oxidation of methanol by the bacteria.

TABLE 20.2 The functions of some metal-containing enzymes

Metal ion	*Small molecule reactant*	*Examples*
Co (in B_{12} cofactor)	Glycols, ribose	Rearrangements, reduction
Zn	CO_2, H_2O	Carbonic anhydrase
Zn, Mg	Phosphate esters	Alkaline phosphatase
Fe, Mn	RNA	Acid phosphatase
Mo, Fe	N_2	Nitrogenase
Ni, Fe	CH_4, H_2	Methanogenesis, hydrogenase
Fe	O_2	Cytochrome P-450
Fe, Se	H_2O_2/Cl^-, Br^-, I^-	Catalase, peroxidase
Ni	H_2O/urea	Urease

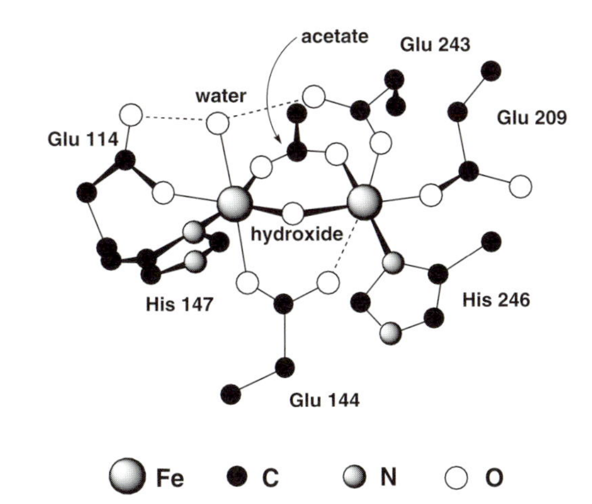

FIG. 20.10 The structure of the dinuclear iron centre in the active site of a methane hydroxylase enzyme. The coordinated amino acid residues are labelled by His = histidine, Glu = glutamine

20.1.4.5 Nitrogenase enzymes

Molybdenum-containing enzymes are responsible for nitrogen fixation in which dinitrogen in the atmosphere is converted to ammonia. Nature thus accomplishes (at about 15°C and 0.8 atmosphere pressure of N_2) what we have been able to accomplish only at high temperatures and pressures using a catalyst. Such nitrogen-assimilating enzymes, the *nitrogenases*, occur in root nodules of clovers and other legumes. Given the increased demand for fertilizers, together with the increased cost of the energy required for high energy industrial syntheses of ammonia, a major research effort is under way to understand, reproduce or model the N_2 fixation process. A discussion on dinitrogen complexes of the transition metals, with emphasis on the chemistry relevant to nitrogen fixation, is given in Section 16.10.

Much lead-up work has established that there are two major proteins in the nitrogenases. One (a) is the nitrogen-binding protein (with a molecular weight around 230 000). This was found by spectroscopic and other methods to contain 2 Mo atoms, some 28 S atoms and 24–32 Fe atoms, and it is probable that the structural units are Fe_4S_4 or Fe_4MoS_4 cubanes. The second (b) is a smaller protein containing Fe and no Mo which carries electrons and contains Fe_4S_4 clusters.

In a major triumph of X-ray crystallography the structure of three FeMo nitrogenases was achieved at 27 pm resolution, the best so far. This, coupled with much other data, establishes that the Fe protein (b) (M_r about 60 000) contains two identical protein subunits bridged by an Fe_4S_4 cubane unit (compare Fig. 20.8). This functions as a one-electron donor for the FeMo protein, which is the major site for the transformation of N_2, by adding overall 8 H^+ and 8e to give NH_3 plus H_2.

The FeMo protein (a) is a tetramer containing two FeMo cofactor units and two units termed P-clusters. The two cofactors are 700 pm apart and so do not interact directly, contrary to earlier ideas. Each P-cluster is separated from the nearest cofactor by only 190 pm and is probably involved in the fixation steps. These metallic centres are surrounded by the protein chain which provides the S, N and other coordinating atoms found in the structure.

The metal centres of the P-clusters consist of two Fe_4S_4 cubanes placed side by side and linked by two S bridges between Fe atoms of the cubane. All these S atoms are part of cysteine components of the proteins which wrap around these centres.

The FeMo cofactor has the structure shown in Fig. 20.11b which compares interestingly with the model compound of Fig. 20.11a which has been prepared earlier (see below). The cofactor consists of two cubane fragments, Fe_4S_3 and Fe_3MoS_3, which are each missing one S corner. The two fragments are bridged by linking the three Fe atoms thus exposed—two by S and the third by a group which may be O or NH. The Mo is 6-coordinate: to the three S of the cubane, to an N from a histidine and by two O from a homocitrate residue.

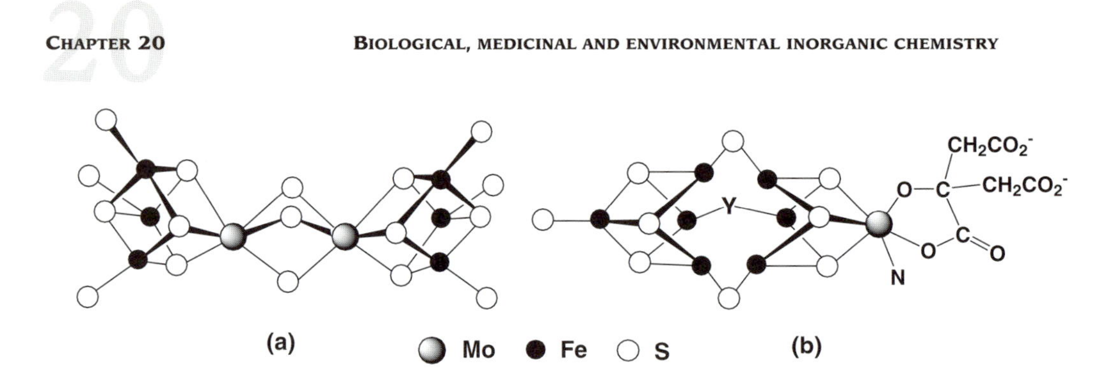

FIG. 20.11 (a) The heavy atom arrangement in the ion $[Fe_6Mo_2S_8(SEt)_9]^{3-}$ (b) Structure of the FeMo cofactor (Y may be O or NH)

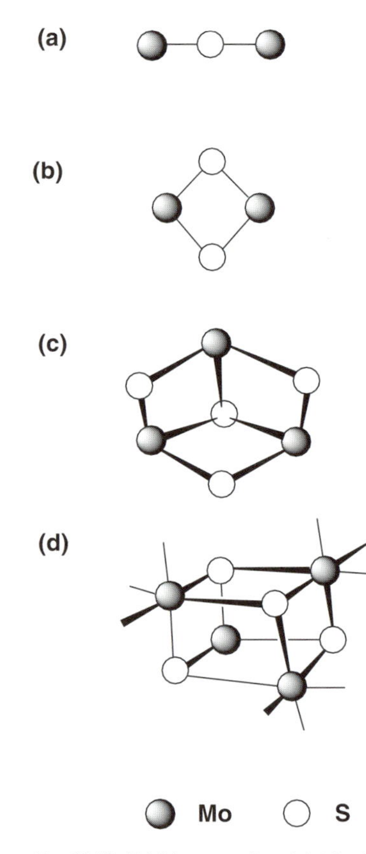

FIG. 20.12 Molybdenum–sulfur skeletal units

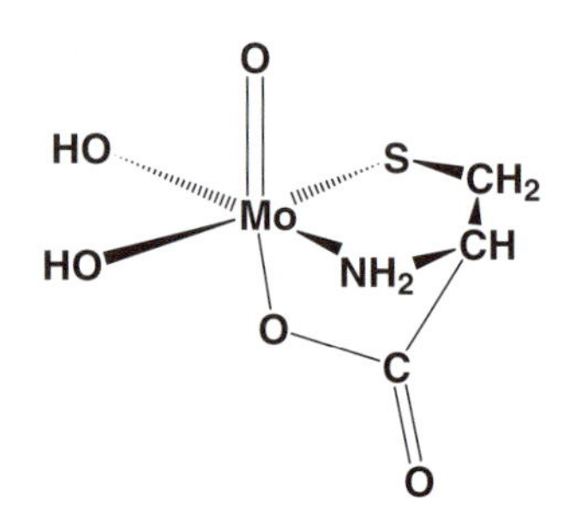

FIG. 20.13 A molybdenum–cysteine model complex for nitrogenase studies

The role of the Mo is thought to be critical because uptake of N_2 seems to depend on its presence in the large majority of systems studied. The determination of the crystal structure is a major advance and will guide further studies. However, discoveries of vanadium-containing nitrogenases, and even of all-Fe species, show there is a long way to go before full understanding is achieved.

The exact mode of action between N_2 and nitrogenase is still undefined. The strategy was developed of synthesizing transition-metal clusters and exploring their chemistry as models for the natural process. This has so far been the main contribution of inorganic chemistry to this side of the nitrogen fixation problem. This chemistry of iron–sulfur clusters is described in Section 20.1.4.3 and for Mo species in Section 15.4. In some ways tungsten is more manageable than molybdenum and its chemistry has been studied as a parallel, creating a major area of the chemistry of these elements. This work has led to many intriguing and interesting compounds but not yet to a detailed scheme for nitrogen fixation.

To illustrate this field we start with an experiment where MoS_2 was treated with CN^- in aqueous solution. One theory of the pre-life evolution of important biomolecules emphasizes the role of cyanide in its polymeric forms, so such species may have been significant in the evolution of Mo-containing enzymes. Four products were isolated where there were increasing numbers of Mo and S atoms bound together and each Mo was made up to 6-coordination by cyanide groups. The skeletons were as shown in Fig. 20.12. These occurred in the ions (a) $[Mo_2S(CN)_{12}]^{6-}$, (b) $[Mo_2S_2(CN)_8]^{n-}$ for $n = 6$ or 8, (c) $[Mo_3S_4(CN)_9]^{5-}$ and (d) $[Mo_4S_4(CN)_{12}]^{-}$.

The skeleton in (d) is a cubane cluster, a common structural unit in transition metal–sulfur cluster chemistry. The other three skeletons are formally fragments of the cubane structure: (b) is a face and (c) is formed by removing one metal corner. Similar structures are widespread both in synthetic compounds and in biological molecules. The units of Fig. 20.12 may be linked together in a wide variety of ways, including double cubanes sharing a face in $[M_6S_6X_{18}]^{6-}$ species and many more open structures. One illustration is given in Fig. 20.11a which parallels the nitrogenases in containing both Fe, Mo and S atoms in the cluster unit. The two cubanes are linked by three bridging S atoms through the Mo corners: note that the Mo atoms are 6-coordinate while the Fe atoms are tetrahedral.

To give an illustration of the experiments directed towards directly mimicking the natural fixation process we note also the following approach. Because of the similarity to ferredoxins (Section 20.1.4.3) and because no N_2 uptake occurs in the absence of molybdenum, a model system was studied consisting of $[(\text{cysteine-S})Fe]_4S_4$ as the electron-transfer agent and the molybdenum–cysteine complex shown in Fig. 20.13 to react with N_2. In this system conversion of N_2 to NH_3 at about 1% of the nitrogenase rate was achieved and the postulated pathway is via diimine and hydrazine.

20.1.4.6 Other transition-metal-containing systems

Over 80 zinc-containing enzymes have been discovered and among the more important ones are *carbonic anhydrase* and *carboxypeptidase A*. In these enzymes the principal role of the zinc is to act as a Lewis acid. For example, in carboxypeptidase A, an enzyme which catalyses the hydrolysis of the terminal peptide bond at the carboxyl end of a protein, the zinc coordinates to the carbonyl oxygen. This polarizes it more strongly, promoting nucleophilic attack at the carbon by the oxygen atom of a Zn-OH group. A similar nucleophilic Zn-OH group undergoes attack at the C=O bond of carbonic anhydrase, thereby greatly speeding up the normally slow hydration of carbon dioxide:

$$CO_{2(aq)} + H_2O \rightarrow H_2CO_3$$

Copper is one of the more important elements in biological systems, being the third most abundant metal ion in the body after iron and zinc. It is particularly concerned in the uptake of inorganic sulfur into organic molecules and copper(I)/copper(II) changes are involved in redox transfer systems in cytochromes in a similar way to iron(II)/(III) ones. Intensely blue copper(II)-containing proteins are typical and are involved in oxidation steps whereas copper(I) is found in haemocyanins.

Vanadium has also been found to be a biologically essential element and perhaps one of the best known occurrences is in the bromo- and iodo-peroxidase enzymes found in seaweeds. Several species of tunicates, better known as 'sea-squirts', contain relatively high concentrations of vanadium in their blood. The vanadium has been shown to exist as V(III) though small amounts of V(IV) as VO^{2+} are also detectable. Current research is directed at establishing the detailed role of vanadium in such organisms.

Nickel occurs in the enzyme urease which catalyses the hydrolysis of urea:

$$H_2NC(O)NH_2 + H_2O \rightarrow CO_2 + 2NH_3$$

While urease was the first enzyme ever to be crystallized, the X-ray structure has only recently been carried out. This showed the active site to contain two Ni centres in accord with previous evidence obtained using other techniques.

Many other elements are found in trace amounts in enzymes and other biological molecules, and while we cannot cover them here the interested reader is referred to the various reviews given in Appendix A.

Manganese is also extensively utilized by biological systems. As an example we quote the Mn-O-OH cluster species implicated in photosynthesis. Discussion of model complexes synthesized in relation to these biological systems is given in Section 14.5 to which the reader is referred.

It is finally worth noting that the biological utilization of transition metals is believed to have varied with time. The very primitive forms of life on Earth lived in a high energy, reducing environment with a planetary atmosphere of N_2, H_2S and CH_4 prevalent. It is believed that the metals which were largely employed by organisms living under these conditions employed sulfide-based metal clusters at the active sites with the metals ranging from iron, cobalt and nickel to molybdenum and vanadium, all in their lower oxidation states. At this stage metals such as copper, which forms highly insoluble sulfides, were generally not available to biological systems in any significant amounts. On the advent of photosynthesis, large-scale introduction of O_2 into the atmosphere occurred, changing it irrevocably (this can be thought of as the biggest single act of 'air pollution' the planet has observed). As a result, H_2S became oxidized to SO_4^{2-} and the sulfide-based chemistry of cobalt and nickel became less important. Oxygen-containing structural units such as the dinuclear Fe-O-Fe (present in the methane hydroxylase enzyme discussed previously in this section) began to take on an increased role in biological systems. It is believed that the inorganic biochemistry of cobalt and nickel is becoming generally of lesser importance. With the exceptions of hydrogenase enzymes and urease, nickel-based biochemistry is largely restricted to the *archaebacteria*, very primitive forms of life occurring in sulfur-rich geothermal regions. These bacteria are believed to be largely unchanged since the earliest forms of life existed. Interestingly, archaebacteria use very little copper in their metabolism. The introduction of an oxidizing planetary atmosphere mobilized copper into the environment and allowed organisms to develop copper biochemistry which is believed to have come relatively recently in evolutionary history.

20.2 Medicinal inorganic chemistry

20.2.1 Overview

The treatment of human (and animal) ailments is one of the major objectives for many researchers and medical practitioners around the world. Although the majority of new drugs result from research on organic compounds, inorganic derivatives are finding an increased role in this field. The large number of inorganic elements available, together with the chemical complexity of inorganic compounds when compared to organic ones, clearly indicates that a great many more combinations of elements are possible with inorganic systems. Of course, only a small number of these are going to be synthesized and of these only a miniscule fraction will have any useful medicinal action. Nevertheless, the search for new drugs among inorganic compounds is a search which is likely to have very fruitful outcomes. Many new drugs are based on metals which are not utilized by natural systems.

While there are a number of inorganic compounds used in medicine (summarized in Table 20.3), some of which will be highlighted in this chapter, the use of inorganic compounds has had a long but somewhat chequered history. In the past many 'treatments' using toxic inorganic compounds, such as mercury salts and even radioactive radium 'tonics', invariably did the patient more harm than good! An organoarsenic compound, arsphenamine, is considered to be the first of the modern chemotherapeutic agents, introduced in 1910 for the treatment of syphilis.

While we do not have space for more than a brief overview, we have chosen a selection of topics to illustrate some of the more important uses of inorganic compounds in medicine. Very broadly speaking these 'drugs' fall into two categories. Firstly, diagnostic agents can be used to investigate ailments without attempting to effect any treatment. As examples of these we discuss the use of radioactive technetium compounds and nonradioactive, but strongly paramagnetic, gadolinium compounds as imaging agents. The second category of drugs are used for therapeutic purposes and the examples chosen include the use of platinum and boron compounds in cancer chemotherapy, and gold drugs in the treatment of rheumatoid arthritis. The use of lithium salts in the treatment of hypertension and chelating agents for heavy metal toxicity represent some other examples of therapeutic inorganic-based drugs.

20.2.2 Inorganic diagnostic (imaging) agents in medicine

Perhaps the most well-known 'imaging' agent is a simple inorganic salt, barium sulfate ($BaSO_4$), which is widely used as a 'barium meal' for imaging stomach ulcers and the like.

TABLE 20.3 Selected example of inorganic medicines and imaging agents in use today

Element	*Trade name*	*Compound*	*Use*
Li	Camcolit	Li_2CO_3	Treatment of manic depression
Mg	Magnesia	MgO	Antacid, laxative
Al	Alludrox	$Al(OH)_3$	Antacid
Ca	Settlers	$CaCO_3$	Antacid
Fe	Nipride	$Na_2[Fe(CN)_5NO]$	Hypertensive, vasodilator
Zn	Calamine	ZnO	Skin ointment
Se	Selsun	SeS_2	Antiseborrhoeic (shampoos)
Sr	Sensodyne	$Sr(acetate)_2$	Toothpastes
Zr	Deodorants	Zr(IV) glycinate	Antiperspirant
Tc	Ceretec	^{99m}Tc propyleneamine oxime	Diagnostic imaging
Ag	Flamazine	Ag sulfadiazine	Antibacterial
Ba	Baridol	$BaSO_4$	X-ray contrast
Gd	Magnevist	$[Gd(DTPA)]\ (meglumine)_2$	MRI contrast
Pt	Cisplatin	*Cis*-$[PtCl_2(NH_3)_2\}$	Anticancer
Au	Myocrisin	$Na_2Au(I)$ thiomalate	Antiarthritic
Bi	De-Nol	Bi(III) citrate	Antacid, antiulcer

Although barium, like other heavy metal ions, is highly toxic the extreme insolubility of $BaSO_4$ gives it a very low toxicity and it is the opaqueness of this material to X-rays which imparts its usefulness in this application.

In general, however, most imaging agents are based on a γ-emitting nuclide which localizes in a specific organ of the body after being injected. The structure of the organ under examination can then be obtained by looking at the radiation emitted by the radioisotope using a scintillation camera.

REQUIREMENTS OF A RADIOACTIVE IMAGING AGENT

(i) Low toxicity.
(ii) A highly specific biodistribution so that the organ in question can be clearly imaged—this is clearly important to minimize the dose of imaging agent required.
(iii) Low radiation dose—this means that a pure γ-emitter is required. At first sight it may appear that the injection of radioactive material into the body is a very dangerous procedure but in fact modern imaging agents expose the patient to little more radiation than a routine X-ray.
(iv) Short isotope half-life so that the patient's long-term exposure to radiation is minimized, though if the half-life is too short this makes transport of the imaging agent rather difficult.

20.2.3 Technetium imaging agents

Technetium provides an excellent, and the most widely used, example of a modern imaging agent. Various technetium complexes, containing different ligands, are commercially available to image different organs of the body. In other words, the coordination sphere of the technetium can be modified to alter its biodistribution. An overview of the chemistry of technetium is given in Section 15.5.

The medically useful isotope of technetium is the metastable ^{99m}Tc. This is manufactured at the point of use by means of a parent–daughter nucleidic pair in which the relatively long-lived parent ^{99}Mo spontaneously decays to the ^{99m}Tc daughter which is then recovered for imaging. The ^{99}Mo is made in a nuclear reactor by thermal neutron fission of ^{235}U:

$$^{235}U + n \rightarrow {}^{99}Zr \xrightarrow[\beta^-]{30s} {}^{99}Nb \xrightarrow[\beta^-]{3\,min} {}^{99}Mo$$

Alternatively, thermal neutron activation of a MoO_3 target also gives ^{99}Mo. The ^{99}Mo is then adsorbed as $^{99}MoO_4^{2-}$ (molybdate) onto an alumina column to form a ^{99m}Tc 'generator' which is transported to the point of use. The ^{99}Mo decays on the column with a 66-hour half-life, generating ^{99m}Tc:

$$^{99}Mo \rightarrow {}^{99m}Tc + \beta^-$$

The technetium is produced in the generator in the form of $^{99m}TcO_4^-$ which, on account of its much lower affinity for the alumina column (compared to MoO_4^{2-}), is readily eluted with sterile saline solution. A schematic diagram of a Mo–Tc generator is given in Fig. 20.14. The ^{99m}Tc decays (half-life of 6 hours) to 'stable' ^{99}Tc (half-life 2×10^5 years) emitting a 140 keV γ-ray which can be easily imaged. The useful lifetime of the technetium generator is therefore limited to about two half-lives of the parent, i.e. about 130 hours.

The $^{99m}TcO_4^-$ solution so produced can either be used directly, to image the thyroid gland, the brain, the kidneys or salivary glands, or it can be converted into a different technetium complex which will image different parts of the body. This is accomplished by adding the $^{99m}TcO_4^-$ solution to a pre-made sterile 'kit' containing the appropriate ligand and any other reagents necessary (such as a reducing agent, to reduce the technetium from oxidation state VII to a lower oxidation state, such as III or IV). It is this versatility, together with the excellent characteristics of the ^{99m}Tc isotope, which has made technetium the most widely used radio-imaging agent available. Some other examples of technetium imaging agents and their applications include the cationic Tc(III) complexes of the type $TcL_2X_2^+$ (where L is a diphosphane or a diarsane—see Appendix B) which image the heart, and technetium phosphonate complexes which are used to image bone tumours.

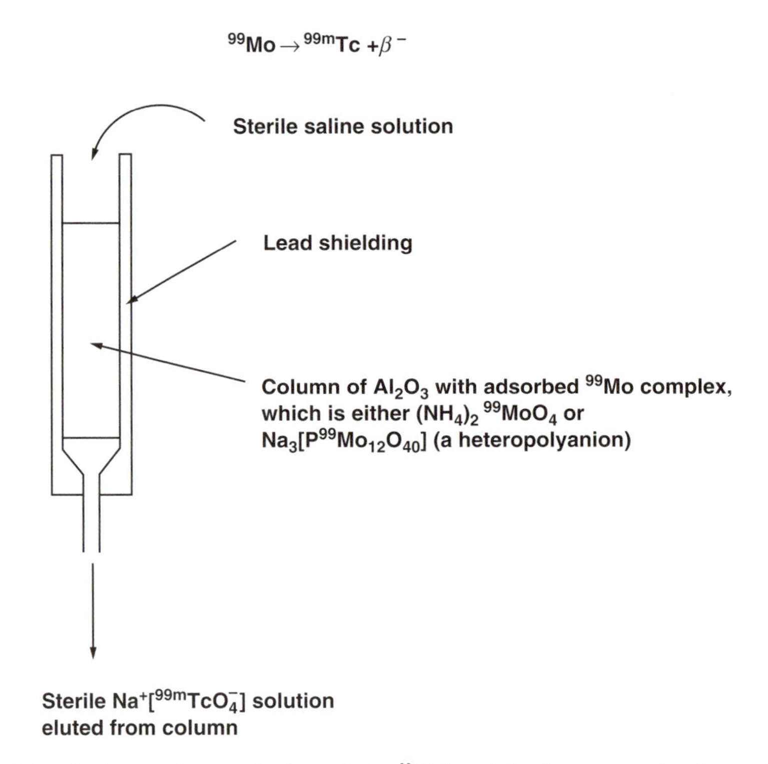

FIG. 20.14 Schematic diagram of a generator for producing $^{99m}TcO_4^-$ solution for use as an imaging agent. The ^{99}Mo complex is strongly adsorbed to the column but the daughter $^{99m}TcO_4^-$ only weakly adsorbs and is readily eluted from the column using saline solution

In addition to technetium imaging agents other radionucleides are also used for specific applications. Examples include thallium-201, gallium-67, iodine-123 and astatine-211.

20.2.4 Magnetic resonance imaging

Another method of imaging bodily organs uses the paramagnetism of certain metal ions, in particular Fe^{3+}, Mn^{2+} and Gd^{3+}, in a technique known as magnetic resonance imaging (MRI). The basic principles of magnetic resonance involving protons (and other nmr active nuclei) have already been introduced in Section 7.8. The imaging technique is based on the fact that the hydrogen atoms of water molecules in different tissues will 'relax' at different rates—the magnetic resonance experiment excites hydrogen atoms into a higher energy state from which they relax back to their original state. Introduction of paramagnetic metal ions into the desired tissues causes the water molecules to relax much more quickly, thereby allowing the desired tissue to be imaged more readily.

Some of the most widely used imaging agents utilize gadolinium complexes (see Chapter 11 for the chemistry of the lanthanide elements). Since simple gadolinium compounds are quite toxic it is necessary to complex the Gd in a stable form which will not break down under physiological conditions but which still has a site at which tissue water molecules can coordinate in order to impart the relaxation effect in magnetic resonance imaging. One example is the diethylenetriamine penta-acetate complex (Fig. 20.15) which is highly stable in water, of low toxicity and approved for use in MRI. An illustration of the type of visual image which can be obtained using the MRI technique enhanced by a gadolinium imaging agent is depicted in Fig. 20.16 which shows an MRI scan of a rat which has had a tumour implanted into its thigh. It is clear that this noninvasive detection of abnormalities such as tumours has become a relatively straightforward process for medical personnel.

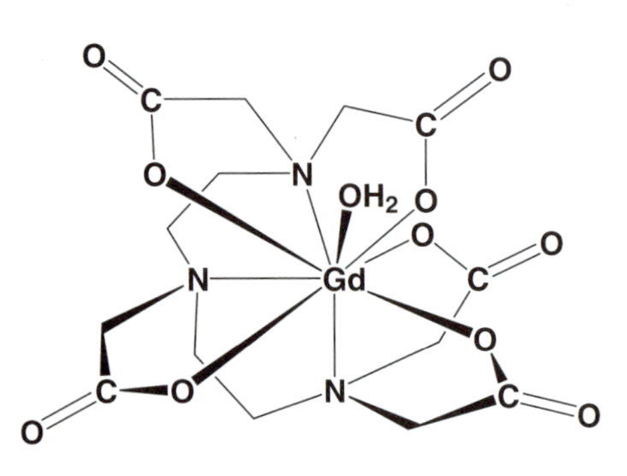

FIG. 20.15 The paramagnetic gadolinium(III) complex $[Gd(H_2O)(DTPA)]^{2-}$ (DTPA = diethylenetriamine penta-acetate) used in magnetic resonance imaging

The discovery that transition metal (and Main Group) compounds have biological activity and can be used as drugs has opened up a new field of research. In the next section we will look at a couple of specific examples—the use of boron and platinum compounds in the treatment of cancer and the use of gold drugs for arthritis.

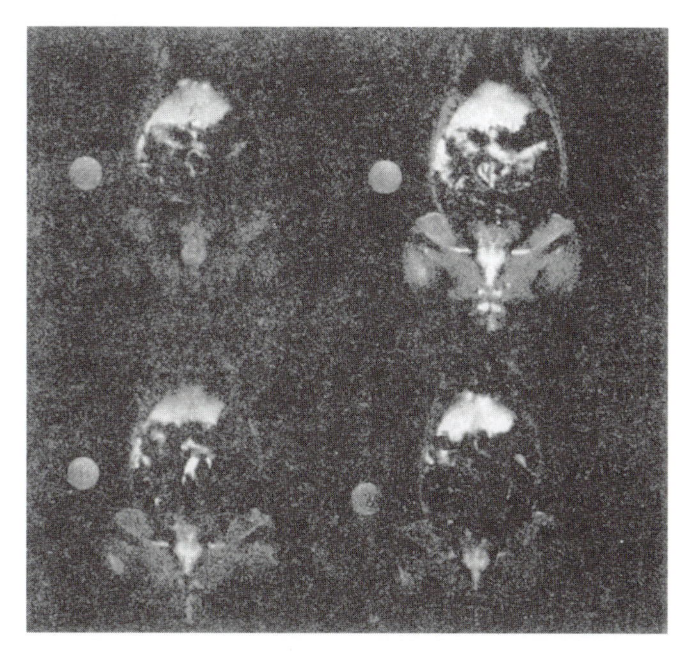

FIG. 20.16 An MRI image of a rat taken at (left to right, top to bottom) 0, 3, 100 and 450 minutes after injecting a gadolinium imaging agent. The presence of an implanted tumor in the rat's thigh can be seen

20.2.5 Boron neutron capture therapy (BNCT)

An example of the use of an inorganic drug in cancer treatment is boron neutron capture therapy (BNCT) in which a boron compound, with a specific biodistribution, becomes localized in a tumour. Irradiation of the tumour with neutrons, which are readily absorbed by the boron atoms, results in the emission of high energy α-particles within the tumour, destroying it. This treatment causes only minimal damage to the surrounding healthy bodily tissues since the penetrative power of α-particles is quite low.

The method relies on the capture of neutrons by boron-10 nuclei which make up about 20% of naturally occurring boron. The role of the chemist in the development of improved versions of this treatment lies in the discovery of new boron compounds which localize specifically in the tumour, again to minimize damage to healthy tissues. Among the early candidates were simple inorganic boron compounds, such as borates (Section 17.4.2), though recent clinical trials have turned to the polyhedral borane clusters, such as those shown in Fig. 20.17. Such closed shell, or *closo*, boranes are relatively stable and have high boron contents. Further developments in the chemistry of these species may lead to new agents for selective BNCT. The reader is referred to Section 9.6 for an introduction to boron-hydride clusters.

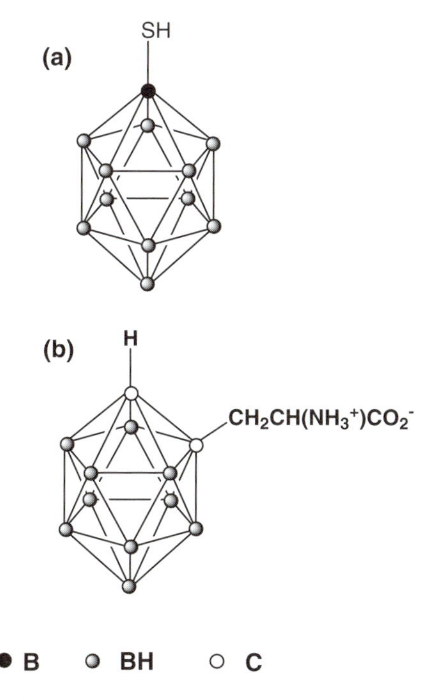

FIG. 20.17 Examples of (a) a polyhedral thiol-containing borane, and (b) a carborane used in boron neutron capture therapy for the treatment of cancer

20.2.6 Platinum-based drugs in cancer chemotherapy

The most widely used drug in the Western world and Japan for the treatment of cancer is a simple inorganic coordination complex, *cis*-$[PtCl_2(NH_3)_2]$, commonly known as *cisplatin.* This drug is particularly effective in the treatment of solid tumours, especially of the genito-urinary tract. Like a number of other important scientific discoveries, the observation that cisplatin kills tumour cells was discovered by accident. In fact the *trans* isomer of $PtCl_2(NH_3)_2$ is inactive against tumour cells and it is thought that the mode of action of cisplatin is the substitution at the N(7) site of guanosine (abbreviated G) in DNA in place of the Cl ligands. The chloride ligands of cisplatin are replaced slowly by water molecules and the resulting aquo complex *cis*-$[Pt(OH_2)_2(NH_3)_2]^{2+}$ is more reactive than cisplatin. It is believed that this aquo complex forms inside the tumour cells. The role of the $(NH_3)_2Pt$ unit is thus to bridge across two neighbouring G residues of DNA, thereby interrupting the replication of the DNA of the tumour cell. It turns out that in cisplatin the Pt-Cl bonds are of just about the right strength since studies on related platinum complexes have shown that, if the platinum–ligand bonds are too strong, then binding of the platinum to the N(7) of guanosine does not occur.

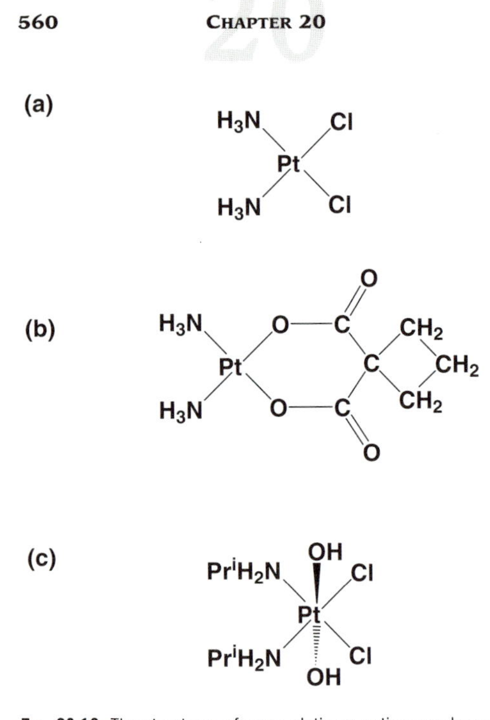

FIG. 20.18 The structures of some platinum anticancer drugs in clinical use: (a) cisplatin (b) carboplatin (c) iproplatin

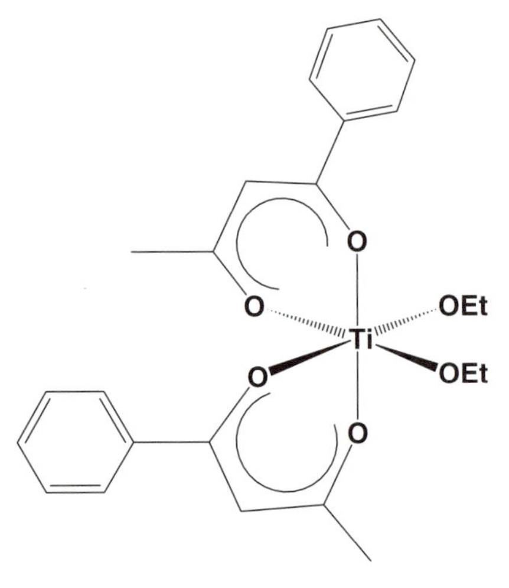

FIG. 20.19 The titanium anticancer drug Budotitane

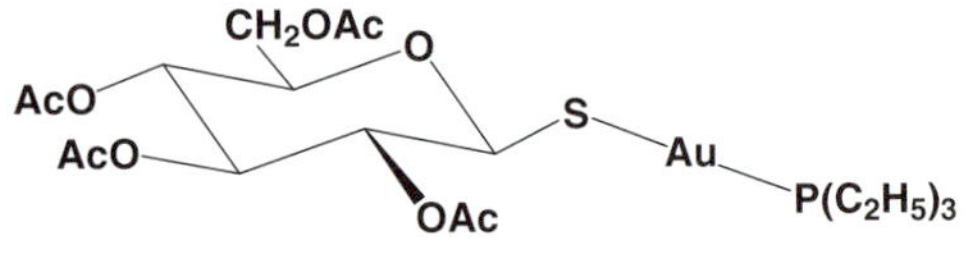

FIG. 20.20 The antiarthritic gold drug Auranofin

Despite the effectiveness of cisplatin in treating certain types of tumours, it has substantial drawbacks in that it is rather toxic towards healthy cells as well as tumour cells and this produces the rather severe side effects often associated with cancer chemotherapy. As a result of this there has been a fairly intensive search for other platinum complexes showing improved anticancer activity but with reduced side effects to the patient. In particular, variations on the amine and the anionic groups (in place of chloride) have been intensively studied, resulting in a number of second-generation anticancer drugs reaching clinical use. Figure 20.18 shows the structure of cisplatin (for comparison) together with some other platinum anticancer drugs in clinical use. In these compounds it is apparent that the amine needs to have at least one hydrogen atom, suggesting that hydrogen bonding plays an important role, and the most promising substitutes for ammonia seem to be diamines, such as cyclohexane-1,2-diamine 1,2-$C_6H_{10}(NH_2)_2$, while the best replacements for Cl may be dicarboxylates, such as cyclobutane-1,1-dicarboxylate in *carboplatin*, Fig. 20.18b.

While we have concentrated here solely on platinum-based anticancer drugs, anticancer activity is not restricted to platinum. A wide range of other metal complexes have been found to exhibit good anticancer activity and a number have reached the clinical trials stage. Examples include the titanium benzoylacetonate complex *Budotitane* (Fig. 20.19) and the cyclopentadienyl titanium complex $(\eta^5\text{-}C_5H_5)_2TiCl_2$.

20.2.7 Therapeutic gold drugs

Gold has been used in medicine for quite some time and the Chinese employed it in medicine as far back as 2500 BC. A number of gold drugs are particularly effective in the treatment of arthritis though their mechanism of action is not well understood. Examples of older drugs of this type include gold(I) thiomalate and gold(I) thioglucose. These materials have been found to contain linearly 2-coordinated gold(I), the most common coordination geometry for gold(I), though they are polymeric in the solid state. However, it is only with the advent of powerful new instrumental techniques (Chapter 7) that the structures of some of these drugs have been established. It is believed that on introduction into the body the polymeric structures react with thiol groups of proteins and the gold is subsequently distributed around the body.

A more recent example of an orally administered antiarthritic gold drug is Auranofin (Fig. 20.20), which contains a triethylphosphane ligand together with an acetylated glucose thiolate ligand. Both the phosphane and the thiolate are 'soft' ligands and show a strong affinity for the soft gold(I) centre, making this drug much more stable than its predecessors. In the body the acetate groups of the thioglucose are hydrolysed off and ultimately the phosphane is displaced from the gold and excreted as triethylphosphane oxide. The gold becomes bound to the thiol groups of blood proteins. There is currently much interest in determining the mechanisms of action of gold in the treatment of arthritis since this will hopefully lead to improved drugs being developed.

20.2.8 Chelation therapy for metal poisoning

For the final part of this section we view the topic of medicinal inorganic chemistry from a slightly different angle in which a drug has to be administered to negate the effects of poisoning by a metal. In this case the aim of the drug is such that the metal and the drug will combine to form a complex which is then much more readily excreted from the body than is the free metal ion. Hence, the principles of coordination chemistry (see Chapter 13) and ligand design (Appendix B) are of great importance in designing drugs to treat poisoning by specific metals. For the 'soft' polarizable metals, such as thallium, mercury and lead, etc., sulfur or other soft donor atom ligands are required whereas for the hard, oxophilic metals, such as the lanthanides and actinides, oxygen (and nitrogen) donors are necessary in order to form stable complexes. By forming chelate complexes (Section 13.7) optimal stability for a given metal can be attained, hence the term *chelation therapy*.

As a specific example, we will look at the developement of drugs for the treatment of plutonium poisoning. Plutonium (Section 12.6) is the most toxic element in the Periodic Table—the lethal dose lies at the *microgram* level. Relatively few people are likely to come

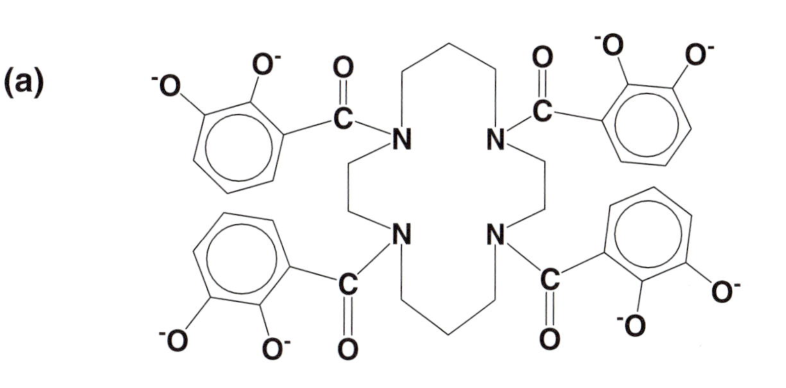

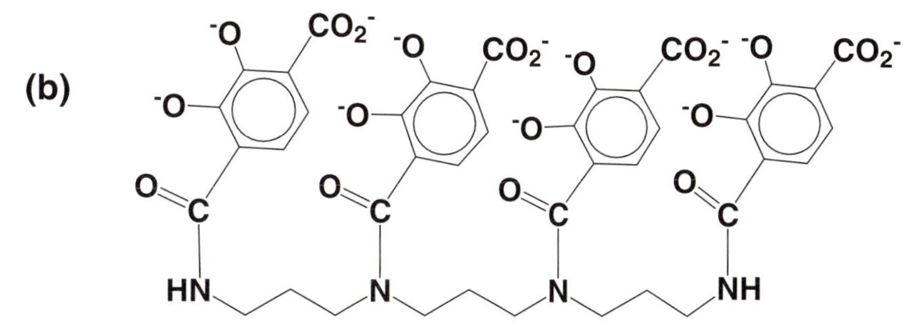

FIG. 20.21 Two highly selective ligands used in plutonium chelation therapy

into contact with plutonium but the extreme toxicity of this element makes it essential that effective treatments are available.

Plutonium readily migrates and concentrates in bone and it is essential that as much of the plutonium as possible is removed before this occurs. A good complexing agent for plutonium is diethylenetriamine penta-acetate (DTPA). This ligand, which is related to EDTA, another excellent complexing agent (see Appendix B), forms stable plutonium complexes analogous to the DTPA-gadolinium complex shown in Fig. 20.15 and used in magnetic resonance imaging. DTPA is potentially an octadentate ligand which can satisfy the coordination number and donor ligand requirements of the oxophilic plutonium ion. The effectiveness of DTPA is shown by the fact that up to 90% of ingested plutonium is excreted within a week when DTPA is administered to the patient within 30 minutes of poisoning occurring. However, given the extreme toxicity of the element, it is naturally desirable to develop much more selective ligands for plutonium chelation therapy.

Two examples of ligands developed for use in chelation therapy in cases of plutonium are shown in Fig. 20.21. These 1,2-dihydroxybenzene, or *catechol*, ligands can bond strongly to a plutonium ion and, like DTPA, are octadentate. The stability constant for the formation of the Pu^{4+} complex of the flexible ligand in Fig. 20.21b has been reported as 10^{52}, making it the most effective, yet nontoxic, ligand for plutonium chelation therapy yet discovered.

20.3 Environmental inorganic chemistry

There is now an increased awareness of the impact of human activities on our environment. Phenomena such as 'the ozone hole', 'the greenhouse effect', 'acid rain', 'eutrophication' (of lakes and rivers) and many others are terms which are in the consciousness of the general public. There is much inorganic chemistry behind these phenomena. Therefore we feel that it is fitting that an inorganic chemistry textbook should reflect on the inorganic chemistry of these current environmental issues. It is our aim in this section to give a general overview of a selected number of relevant environmental topics in which inorganic chemistry is involved to a significant extent. Readers who wish to study the topics in greater depth are encouraged to consult the specialized texts given in Appendix A.

20.3.1 Phosphates in detergents: eutrophication

Phosphates have been utilized as a 'builder' in detergent formulations since around the 1930s. Their role is to complex any hard-water-forming ions present (specifically Ca^{2+} and Mg^{2+}) and hold them in a complexed form in solution. This prevents the formation of solid 'scum' on the water and enables the surfactant to work much more efficiently, preventing soil from being redeposited on the item being washed. Because of their ability to hold Ca^{2+} and Mg^{2+} ions in solution, phosphates have also been used for the prevention of scale formation in industrial cooling water systems. The most widely used phosphate builder is sodium tripolyphosphate, $Na_5P_3O_{10}$, containing a P-O-P-O-P chain (see Section 17.6.2 and Fig. 17.36). The performance/cost ratio of sodium tripolyphosphate is excellent. However, a serious question mark lurks over the use of phosphates in detergents as a result of the phenomenon of *eutrophication*.

Since phosphate is an essential nutrient for plant growth, and is often the limiting reagent, the supply of phosphate to the environment from detergents (and other sources, especially the heavy use of phosphate fertilizers in many areas) may encourage excessive plant growth. This may occur particularly in situations where there is a comparatively low throughput of water, such as lakes and rivers. The large amount of plant biomass produced subsequently decays, recycling carbon dioxide plus the nutrient elements nitrogen, phosphorus and potassium to continue the cycle. Eutrophication has been around for millions of years—it is responsible for the vast peat and coal deposits present on the Earth—yet phosphate has been critiqued for accelerating the problem. As a consequence of this, alternative builders were sought in the 1980s to replace the phosphates used in detergents.

One of the earliest replacements for phosphate was nitrilotris(acetate), $N(CH_2CO_2)_3^{3-}$, known as NTA. However, studies have suggested that this compound is carcinogenic and it has consequently been discontinued as a builder. The discovery that zeolites were efficient ion-exchangers indicated that these materials would be a good replacement for phosphates and they have become the single largest replacement builder in detergents where phosphates are prohibited. A discussion on the chemistry of zeolites is given in Section 18.6. The introduction of zeolite A was initially viewed to be a complete solution to the phosphate eutrophication problem.

More recent studies, however, have suggested that zeolites themselves may have environmental problems associated with them. When the performance/cost ratio of zeolites is compared to that of phosphates, zeolites are not quite as good. In addition, since zeolites are solid materials, they are used as a suspension (unlike tripolyphosphate which is completely soluble in water). Other chemicals need to be added as a 'co-builder' in order to assist in the dispersion of the zeolite particles, to control the pH and to prevent dirt redeposition. Long-chain organic polymers bearing large numbers of carboxylic acid groups are one such class of compound employed for this purpose. Thick brown algal mats have developed off the Adriatic coast and research has suggested that these are caused by the use of zeolites and polycarboxylic acids in detergents. On the other hand, studies have suggested that phosphates are causing destruction of coral reefs off the northeast coast of Australia. The evidence seems to suggest that both types of builder have some significant detrimental effect on the environment.

A recent study has suggested that the *overall* environmental impact of phosphate versus zeolite builders should be considered in order to obtain a fair comparison. The mining, manufacturing, use and disposal operations all clearly have their own environmental impact which must be included in the equation when considering the relative merits of the two detergent builders. When this is done, it has been suggested that zeolites and phosphates are similar in overall environmental impact. This is due, in part, to the nature of bauxite mining, the source of aluminium for the zeolites, which produces large quantities of alkaline wastes, termed 'red muds'. In addition, most bauxite is mined in Australia so significant amounts of energy must be employed to transport it to the point of use. On the other hand, the mining of phosphates produces large quantities of gypsum ($CaSO_4$) which is sometimes discharged into the ocean, smothering the ocean floor and releasing toxic heavy metals originally present in the phosphate rock.

Given that the overall environmental impact of zeolites seems to be comparable to phosphates, it would appear that there is little to choose between them. However, technology is available to recycle phosphates from sewage and convert them to a form which can be re-used in detergents. This is achieved very simply by the addition of calcium ions to the effluent stream and recovery of the insoluble calcium phosphate produced. However, at present, there is no known technology to recover zeolites from sewage.

It is clear that the phosphate debate is still going on and is likely to do so for some time to come. Ideally, one would like to dispense with both zeolites and phosphates as builders. However, it is clear from the above case histories that any replacement must be thoroughly vetted before introduction since detrimental effects may not be discovered until it is too late.

20.3.2 Chlorofluorocarbons: the ozone hole

Examples of CFCs:

CCl_3F	CFC-11
CCl_2F_2	CFC-12
C_2ClF_5F	CFC-115

At this point it is worth noting that ozone in the *lower* atmosphere is highly undesirable—ozone is a toxic, powerful oxidant which causes rubber and other polymers to perish and causes major health problems.

Chlorofluorocarbons, otherwise known as CFCs, are 'inert' compounds with a number of applications: as propellants of aerosol cans, blowing agents for polyurethane foams and fluids in refrigeration units. When they were introduced they were thought to be extremely nontoxic and inert, and as a result found large-scale use. Worldwide production of CFCs reached levels of around half a million metric tons per year. It was first thought that due to the stability of these compounds large quantities could be released into the atmosphere with no effect. However, in 1974 it was suggested that CFCs are responsible for the destruction of stratospheric ozone, O_3 (see Section 17.7.1). Ozone in the upper atmosphere is highly beneficial to life on earth since it filters out hard (i.e. short wavelength) ultraviolet radiation. Such radiation is responsible for causing cancer and other ailments, including eye cataracts.

In the upper atmosphere high energy radiation causes cleavage of the carbon–chlorine bonds of CFCs, generating chlorine radicals:

$$\text{e.g. } CF_2Cl_2 \xrightarrow{h\nu} Cl\cdot + CF_2Cl\cdot$$

These chlorine radicals then react with ozone to produce dioxygen and ClO radicals which then react with a number of species present in the upper atmosphere, such as oxygen atoms (produced by the photolysis of ozone: $O_3 \rightarrow O_2 + O\cdot$) or with nitric oxide, NO.

$$Cl\cdot + O_3 \rightarrow O_2 + ClO\cdot$$

The net effect is to set up a series of chain reactions, the overall result of which is the destruction of the beneficial ozone in the upper atmosphere. Examples of reactions occurring are:

$$ClO\cdot + O\cdot \rightarrow Cl\cdot + O_2$$

then

$$Cl\cdot + O_3 \rightarrow ClO\cdot + O_2$$

the net effect of which is the ozone-destroying reaction

$$O_3 + O\cdot \rightarrow 2O_2$$

and the regeneration of Cl· to repeat the cycle. Similarly, ozone can react with NO, again with the net destruction of ozone:

$$O_3 + NO \rightarrow NO_2 + O_2$$

Perhaps the most well known incidence of ozone depletion is the Antarctic ozone hole which is characterized by up to a 50% depletion in polar stratospheric ozone in the late Antarctic winter and early spring. Under normal conditions the reaction of atmospheric

NO_2 with ClO radicals serves to limit the amount of Cl atoms available for ozone destruction:

$$ClO\cdot + NO_2 \rightarrow ClONO_2 \text{ (chlorine nitrate)}$$

However, with temperatures below −70°C in the cold polar stratospheric clouds, NO_2 is frozen out as compounds such as nitric acid hydrate, $HNO_3.3H_2O$. In addition, $ClONO_2$ can generate, in the ice crystals, various ozone-destroying chlorine species, as shown in the following reactions:

$$ClONO_2 + H_2O \rightarrow HOCl + HNO_3$$
$$ClONO_2 + HCl \rightarrow Cl_2 + HNO_3 \xrightarrow{h\nu} 2Cl\cdot$$
$$HOCl \xrightarrow{h\nu} HO\cdot + Cl\cdot$$

Bromine has also been implicated in the destruction of atmospheric ozone. *Halons* are compounds related to CFCs but which contain bromine atoms. Examples include $CBrF_3$ (Halon-1301) and $C_2Br_2F_4$ (Halon-2402). Such compounds have been used in fire extinguishers since they generate halogen radicals which are very efficient at preventing combustion. However, the major source of bromine in the atmosphere remains uncertain since large quantities of methyl bromide (CH_3Br) are released by natural sources.

The very real potential for the massive destruction of the Earth's protective ozone layer has led to measures to eliminate the production of CFCs. As a result, the inception of the Montreal Protocol (on substances that deplete the ozone layer) in 1987 called for a number of measures including a cutback in CFC production to 50% of 1986 levels by 1998 and a complete ban on the production of CFCs in the USA by the end of 1995.

The Montreal Protocol has led to a massive search for new replacements for CFCs which do not damage the ozone layer. Among the more promising candidates are materials which contain hydrogen atoms in addition to fluorine (and sometimes chlorine). The C-H bonds are much more readily broken than are C-Cl or C-F bonds and so these compounds can degrade much more rapidly in the atmosphere, particularly by reaction with OH radicals, typically causing only 1 to 10% of the ozone destruction of CFCs. A good example is HFC-134a (CH_2FCF_3) where the HFC stands for hydrofluorocarbon. This HFC is now being manufactured for use as a replacement of CFC-12 in refrigeration systems.

One of the major roles of inorganic chemistry in the search for new CFC replacements has been in the development of catalytic processes for their manufacture. The catalytic chemistry used in one of the possible routes for the manufacture of HFC-134a is shown in Fig. 20.22. Other studies are concerned with the gas-phase radical chemistry of CFCs and HCFCs, the range of experiments which can be conducted and the information that can be ascertained from them (see Appendix A). The introduction of the Montreal Protocol is hopefully the start of many similar agreements for the minimization of the causes of other current environmental problems. The ultimate replacement compounds will have no C-Cl bonds (e.g. $C_2F_4H_2$) but these are less accessible and more expensive at present.

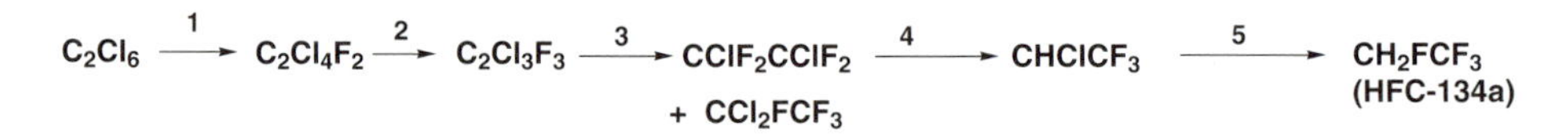

FIG. 20.22 Scheme for the synthesis of the CFC replacement CH_2FCF_3 (HFC-134a). Steps 1 to 3 use a Cr_2O_3 catalyst while steps 4 and 5, which involve hydrogenolysis, use a supported palladium catalyst

20.3.3 The greenhouse effect

The greenhouse effect is an environmental problem of particular concern. Solar radiation (visible light) is absorbed by the Earth which reaches thermal equilibrium by re-radiation in the infrared region. In the greenhouse effect this infrared radiation is absorbed by the so-called 'greenhouse gases' of the atmosphere and lower energy radiation re-emitted towards the Earth. As might be interpreted from the name, it is the same principle which accounts for the warming action of glass greenhouses and the effect on the Earth is similar. The principal greenhouse gases are water and carbon dioxide. The latter is causing the greatest concern due to the increase of CO_2 in the atmosphere from the almost exponential increase in the use of carbon-based fuels.

Measurements over the last 30 years have indicated that the amount of CO_2 in the atmosphere is increasing at the rate of about 1 part per million (ppm) per year. It has been predicted that CO_2 levels will double by the middle of the 21st century and this will produce, as a result of the greenhouse effect, an overall increase in the surface temperature of the Earth predicted to be between 1.5 and 4.5°C. Though this may not sound significant, it will cause melting of substantial portions of the polar ice caps and expansion of the oceans, causing massive destruction of coastal areas close to or below sea level. As a good comparison, a temperature rise of this magnitude is what has occurred since the last ice age. What adds to the problem is that one of the 'sinks' for carbon dioxide are forests which, through the process of photosynthesis, fix vast amounts of CO_2 as organic carbon. The current world policy of deforestation can only add to the problem.

Evidence for the increase in atmospheric CO_2 levels comes from the analysis of air trapped in ice cores from the polar regions (in addition to direct measurements). These experiments suggest that the composition of the air at the time of the last ice age contained CO_2 levels 25% less than pre-industrial levels (it is generally accepted that the first significant human input of CO_2 into the atmosphere started at the time of the industrial revolution).

In addition to water and carbon dioxide, other trace constituents in the atmosphere also contribute to global warming. Among these are nitrous oxide, N_2O (see Section 17.6.4), the chlorofluorocarbons discussed earlier in this chapter, and methane. Along with CO_2, human activities are also increasing the CH_4 content of the atmosphere. This is due to factors such as natural gas spillages, evolution of gas from landfill sites and methane emitted from the digestive tracts of cattle, among others. Methanotropic bacteria, discussed in Section 20.4.4, utilize methane as their source of carbon and energy, and may find use in the bioremediation of land, such as in landfill sites where substantial quantities of methane can be produced.

Various factors also operate to decrease the CO_2 content of the atmosphere. One of these, the photosynthetic conversion of CO_2 to organic carbon, has already been mentioned. Another vast sink of carbon are the oceans—these contain about sixty times the amount of CO_2 found in the atmosphere. However, due to relatively poor mixing of the atmosphere with the oceans, transfer of CO_2 is relatively slow and while the ultimate end for atmospheric CO_2 is likely to be carbonate sediments (such as $CaCO_3$) deposited in the oceans, it is as yet unclear what the effects of increased CO_2 input into the atmosphere are likely to be. Much effort is being spent in computer-modelling of the greenhouse effect. However, the interrelationships between a large number of parameters are difficult to model and predictions are likely to be approximate at the very best.

It is noteworthy that the next, and final, environmental aspect which we will consider, the phenomenon of acid rain, may have a counteractive influence on the greenhouse effect due to the formation of light-reflecting sulfuric acid clouds in the atmosphere. This would clearly be a completely unacceptable solution and emissions of CO_2 must surely be severely curbed in the near future. Perhaps the forthcoming decline in carbon fossil fuel resources may effect a change to a 'hydrogen economy' rather than the current 'carbon economy' with which the world operates. Such an enforced change would have a beneficial effect on CO_2 emissions.

> In the *hydrogen economy* hydrogen is the basic fuel in place of oil or coal. Pollution is greatly reduced if the hydrogen is prepared using hydroelectric, solar, wind or tidal power, and is used in efficient electrochemical devices such as fuel cells.

Finally, it is worth casting our eyes a little further afield for a look at one of our solar system neighbours—Venus. It is generally considered that Venus is a case history of a 'runaway' greenhouse effect. As a result of greenhouse emissions the surface of the planet is believed to have heated up, driving off any liquid volatiles into the atmosphere to act further as greenhouse gases. The surface of Venus is a thoroughly inhospitable place—temperatures of several hundred degrees, the rocks glowing a dull red colour with sulfuric acid clouds lurking overhead. Though Venus is closer to the Sun than Earth, as a result of a runaway greenhouse effect the temperature far exceeds that expected.

20.3.4 Acid rain

One of the consequences of industrial activity is that large quantities of sulfur dioxide and nitrogen oxides are introduced into the atmosphere. The principal source of SO_2 comes from the combustion of fossil fuels which contain various sulfur compounds. Similarly, nitrogen oxides come from emissions from various industrial processes as well as automobile exhaust emissions. In the atmosphere these gases are oxidized and converted to sulfuric and nitric acids:

$$SO_2 + \tfrac{1}{2}O_2 + H_2O \rightarrow H_2SO_4$$
$$2NO_2 + \tfrac{1}{2}O_2 + H_2O \rightarrow 2HNO_3$$

These acids (when combined with other acids such as hydrochloric acid, also added to the atmosphere by emissions from industrial processes) render precipitation acidic—*acid rain*.

Acid rain is an international problem. Emissions resulting from one country's activities may be swept by weather systems to deposit acid rain up to several thousand kilometres away. Perhaps the classical example of this is the occurrence of acid rain in Scandinavia, originating from emissions in the densely industrialized regions of Western Europe.

There are many damaging effects of acid rain including destruction of forests which may occur directly (as a result of the low pH) or indirectly (since decreased pH increases leaching of Al^{3+} from soils and rocks) with concomitant detrimental effects on plant health. Acid rain also produces an increased incidence of human respiratory ailments, acidification of lakes and general corrosion of limestone and steel items.

Various moves are underway to curb acid rain. Naturally, tackling the major source of SO_2 and NO_2 emissions is the logical way to proceed. There is an increased awareness of the need to use relatively sulfur-free coal and oil in power stations and to fit scrubbing systems to minimize SO_2 emissions. For automobile emissions catalytic converters provide significant reductions in emissions of pollutants. Catalytic converters typically use a platinum metal-based catalyst (see Section 15.8) and serve a dual role. First, they ensure complete oxidation of hydrocarbons and carbon monoxide, and second, they aim to reduce (to dinitrogen) any nitrogen oxides which have been produced in the combustion process. The development of new, improved catalytic systems, using cheaper platinum group metals, should decrease the cost of such converters and make them more widely available. Again, movement from a carbon-based economy to cleaner fuel sources (such as those based on elemental hydrogen) should provide future solutions, especially when coupled to low-pollution energy sources (such as solar energy or, ultimately, hydrogen fusion power).

> This will also have benefits on other more localized environmental problems, such as photochemical smog, to which hydrocarbons in the atmosphere are a major contributor.

20.3.5 Other inorganic environmental problems

In concluding this discussion on environmental chemistry we will look, rather superficially, at some of the major environmental problems which concern inorganic chemistry. There are many others. Indeed, the impact of any human activity on the global environment needs to be much more closely regulated. Lead from petrol is one such example. For many years lead has been added to petrols as an antiknocking agent in the form of an organic derivative, tetraethyllead. Naturally, the lead enters the environment and studies have found cases of impaired learning in children living near busy roads.

Heavy metals in the environment must be closely monitored. Biological systems are able to accumulate heavy metals, especially in fatty tissues, and with humans residing at the top of the food chain this can result in significant human uptakes of these toxic metals. Perhaps the classical example comes from mercury poisoning in Japan where mercury-laden fish were consumed, resulting in a large number of deaths from 'Minamata' disease. The increased consumption of fish such as tuna, which accumulate mercury and other heavy metals, is also increasing the human uptake of this element. At the time of writing, there is a debate on the relative contribution to human mercury uptake from the mercury present in amalgam dental fillings. It is worth noting that mercury in inorganic forms has a much lower toxicity when compared with organometallic derivatives, such as CH_3Hg^+, which can be produced by biomethylation of Hg^{2+} by certain microorganisms.

Finally, our planet, as a result of continued population growth and industrialization, has an ever-increasing demand for energy. The continued headlong consumption of declining fossil fuel reserves is not a solution, as we have seen from the earlier discussions on the greenhouse effect and acid rain. Fossil fuels are avoided by using hydroelectric or geothermal generation, but these sources are limited and carry their own problems. A small, but steadily increasing, proportion of electricity is currently being produced by non-polluting methods especially solar cells and wind or tide generators. Such technologies are being steadily developed and improved.

Nuclear power, based on nuclear *fission*, is a relatively clean source of energy when operated carefully and safely. However, accidents such as Three Mile Island and Chernobyl have generated significant distrust of this energy source, and this is reinforced by the problem of long-term disposal of radioactive waste and the ever-growing threat of nuclear terrorism using stolen uranium or plutonium. In the longer term, power from hydrogen nuclear *fusion* may be the best answer. The level of radioactivity produced for a given power output is predicted to be very much lower and more acceptable. However, the advent of readily available fusion power is unlikely to lie in the near future.

It is easy to despair when faced with such problems, or to yearn for a return to a simpler age. However, although earlier ages may have been simpler, they were not golden! Descriptions of pollution in most mediaeval cities make modern problems seem marginal by comparison. The 'idyllic' rural life of early farmers, or even more so of early hunter-gatherers, was characterized by life expectancies of less than 30 years and high mortality of infants and in childbirth.

Problems appear to pile up in part because we currently have the knowledge to see that there is indeed a problem. Conflict and uncertainty arise because we are attempting to understand very complex systems. Information is becoming firmer and the people of Earth are moving slowly towards solutions.

The first problem is to understand and develop the science then the technology and engineering in order to determine and provide adequate and acceptable solutions. The second problem is to develop the social and political understanding, and then the social consensus, to choose good solutions. While everyone can contribute to the second problem, the challenge to scientists, engineers and technologists is to contribute to the solution of both.

Appendix A
Further Reading

(1) Books

TEXTBOOKS AT HONOURS LEVEL

Advanced Inorganic Chemistry, F. A. COTTON, G. WILKINSON, C. A. MURILLO AND M. BOCHMANN, 6th Edition, John Wiley, 1999.

This is the most widely-used Honours textbook in inorganic chemistry. Books with alternative approaches and emphases worth consulting are:

Inorganic Chemistry, D. F. SHRIVER AND P. W. ATKINS, Oxford University Press, 3rd Edition, 1999.

Inorganic Chemistry: Principles of Structure and Reactivity, J. E. HUHEEY, E. A. KEITER AND R. L. KEITER, Benjamin Cummings, 4th Edition, 1997.

Chemistry of the Elements, N. N. GREENWOOD AND A. EARNSHAW, Butterworth-Heinemann, 2nd Edition, 1998.

Holleman-Wiberg Inorganic Chemistry, Ed. N. WIBERG, Harcourt, 2001.

Inorganic Chemistry, A. G. SHARPE, Longman, 3rd Edition, 1992.

Concise Inorganic Chemistry, J. D. LEE, Blackwells, 5th Edition, 1997.

Principles of Inorganic Chemistry, W. L. JOLLY, McGraw-Hill, International (and revised) Edition, 1985.

Inorganic Chemistry, A Unified Approach, W. W. PORTERFIELD, Addison-Wesley, 1984.

The student wishing to undertake a fuller study of inorganic chemistry, or of a particular field, should consult several of the above texts. These will give further references which may be followed up in detail.

A good library will also contain a number of older textbooks which should not be neglected. The experimental facts will be unaltered, and it is often easier to grasp a theory by following its evolution from an earlier form. It is very important to check the date of publication, and to be aware that most of modern chemistry has evolved very rapidly over the past thirty years.

OTHER TEXTS

There are a number of useful general titles which are less wide-ranging than the Honours texts. We also note sources for a number of areas which are not covered in detail in this text.

NOTE TO THE READER

You may wish to consult other sources for a number of reasons, and it helps to be clear about your aims. Nothing is more confusing and frustrating than turning up an explanation in great depth or a review in great detail when only minor clarification is sought.

(1) Explanation of points not understood in text or lectures is best sought, in the first place, from other sources written at a similar level. So often, a different form of words or different approach will clarify a problem. It is surprising how often the process of consulting source A, then B, then C, then A again solves some puzzling point.
If you suspect there is a lack of basic information, then lower-level sources are useful. It often pays to revise a little in school textbooks.
(2) A search for supplementary information, explanations in greater depth or a desire to know where a topic leads are best satisfied by first reading one or more of the advanced texts, Honours or specialist (Section 1). Follow this by consulting the appropriate parts of the multi-volume works listed. These will give the desired material or refer in turn to specialist sources.
(3) To find much fuller information on a limited topic (such as 'chromium chemistry' or 'metal–metal bonds' or 'inorganic biochemistry') the specialist references or reviews should be consulted first. A recent review will normally yield more than enough information for an essay topic, for example, Suitable sources are given in Section 2, while many individual articles are listed in Section 3.
(4) For complete information on a topic (not usually required from an undergraduate), specialized reviews or multi-volume references (Section 1) will provide a terminus. A convenient source for updating these is provided by the appropriate *Specialist Periodical Reports* and *Annual Reports* published by the Royal Society of Chemistry (Section 2). For even fuller or more recent information the next step is a search through *Chemical Abstracts* (typically now done electronically through tools such as *SciFinder Scholar* produced by the American Chemical Society; see Section 4), the *Science Citation Index* or the *Web of Science* to bring the survey up to date. These searches, when done electronically, are now quick and simple, allowing rapid access to large amounts of information. Completely up-to-date information involves scanning the databases mentioned above, plus the latest issues of the major journals.
(5) Note also the various detailed compilations of data, including those listed in Section 3.
(6) For more detailed guides to the chemical literature see *Use of the Chemical Literature*, R. T. BOTTLE, Butterworths, 3rd Edition, 1979, or *Guide to Basic Information Sources in Chemistry*, A. ANTONY, Wiley, 1980; *How to Find Chemical Information: A Guide for Practicing Chemists, Teachers and Students*, R. E. MAIZELL, Wiley, 2nd Edition, 1987.
Electronic access to chemical information via the World Wide Web is gaining in popularity and developments in this area are proceeding with a rapid pace. A good introduction to chemical information on the internet is provided in the September 1995 issue of *Chemistry in Britain* see Section 4.
(7) As an aid to writing, *The Chemist's English*, R. SCHOENFELD, VCH, 3rd Edition, 1989, may be consulted with profit and entertainment.

A series of chemistry primers have been published by Oxford University Press. These cover a range of topics and provide an excellent, concise introduction. The following titles have been published in the year indicated:

The Mechanisms of Reactions at Transition Metal Sites, R. A. HENDERSON (1993).

Energy Levels in Atoms and Molecules, W. G. RICHARDS AND P. R. SCOTT (1994).

Chemical Bonding, M. J. WINTER (1994).

d-Block Chemistry, M. J. WINTER (1994).

Cluster Molecules of the p-Block Elements, C. E. HOUSECROFT (1994).

Organometallics 1: Complexes with Transition Metal Carbon σ Bonds, M. BOCHMANN (1994).

Organometallics 2: Complexes with Transition Metal Carbon π Bonds, M. BOCHMANN (1994).

Essentials of Inorganic Chemistry 1, D. M. P. MINGOS (1995).

Biocoordination Chemistry, D. E. FENTON (1995).

Inorganic Materials Chemistry, M. T. WELLER (1995).

Metal–Metal Bonded Carbonyl Dimers and Clusters, C. E. HOUSECROFT (1996).

Inorganic Chemistry in Biology, P. C. WILKINS AND R. G. WILKINS (1997).

Periodicity and the s- and p-Block Elements, N. C. NORMAN (1997).

Inorganic Spectroscopic Methods, A. K. BRISDON (1998).

Essentials of Inorganic Chemistry 2, D. M. P. MINGOS (1998).

Crystal Structure Determination, W. CLEGG (1998).

The Heavier d-Block Metals – Aspects of Inorganic and Coordination Chemistry, C. E. HOUSECROFT (1999).

Coordination Chemistry of Macrocyclic Compounds, E. C. CONSTABLE (1999).

The f Elements, N. KALTSOYANNIS AND P. SCOTT (1999).

NMR Spectroscopy in Inorganic Chemistry, J. A. IGGO (1999).

Non-Aqueous Solvents, J. R. CHIPPERFIELD (1999).

Supramolecular Chemistry, P. BEER, P. GALE AND D. SMITH (1999).

Foundations of Inorganic Chemistry, M. WINTER AND J. ANDREW (2000).

Introduction to Molecular Symmetry, J. S. OGDEN (2001).

Applied Organometallic Chemistry and Catalysis, R. WHYMAN (2001).

The Royal Society of Chemistry has also produced a series of Tutorial Chemistry Texts, single-topic short books with an emphasis on worked examples and problems. Inorganic chemistry titles so far are:

d- and f-Block Chemistry, C. J. JONES (2000).

Main Group Chemistry, W. HENDERSON (2000).

Reactions and Characterization of Solids, S. E. DANN (2000).

Organotransition Metal Chemistry, A. F. HILL (2001).

Titles scheduled for future publication include *Atomic Structure and Periodicity*, *Bioinorganic Chemistry,* and *Lanthanide and Actinide Elements*.

See also:

Basic Inorganic Chemistry, F. A. COTTON, G. WILKINSON AND P. L. GAUS, John Wiley, 3rd Edition, 1994.

Main Group Chemistry, A. G. MASSEY, Wiley, 2nd Edition, 2000.

Essential Trends in Inorganic Chemistry, D. M. P. MINGOS, Oxford University Press, 1998.

Basic Principles of Inorganic Chemistry: Making the Connections, B. MURPHY, C. MURPHY AND B. J. HATHAWAY, Royal Society of Chemistry Paperback Series, 1998.

Physical Inorganic Chemistry, S. F. A. KETTLE, Oxford University Press, 1997.

Concepts and Models of Inorganic Chemistry, B. E. DOUGLAS, D. H. McDANIEL AND J. J. ALEXANDER, Wiley, 3rd Edition, 1994.

Descriptive Inorganic Chemistry, G. RAYNER-CANHAM, Freeman, 2nd Edition, 2000.

Highlights in Inorganic Chemistry, Eds. G. MEYER, D. NAUMANN AND L. WESEMANN, Wiley-VCH, 2001.

Metals and Ligand Reactivity. An Introduction to the Organic Chemistry of Metal Complexes, E. C. CONSTABLE, VCH, 2nd Edition, 1996.

Transition Metal Chemistry: The Valence Shell in d-Block Chemistry, M. GERLOCH AND E. C. CONSTABLE, VCH, 1994.

Chemical Approaches to the Synthesis of Inorganic Materials, C. N. R. RAO, Wiley, 1995.

Synthesis of Inorganic Materials, U. SCHUBERT AND N. HÜSING, Wiley-VCH, 2000.

Inorganic Substances A Prelude to the Study of Descriptive Inorganic Chemistry, D. W. SMITH, Cambridge University Press, 1990.

The Periodic Table of the Elements, R. J. PUDDEPHATT AND P. K. MONAGHAN, Oxford University Press, 2nd Edition, 1986.

Inorganic Energetics, W. E. DASENT, Cambridge University Press, 2nd Edition, 1982.

Some Thermodynamic Aspects of Inorganic Chemistry, D. A. JOHNSON, Cambridge University Press, 2nd Edition, 1982.

Modern Aspects of Inorganic Chemistry, H. J. EMELEUS AND J. S. ANDERSON, 4th Edition with A. G. SHARPE, Routledge and Kegan Paul, 4th Edition, 1973.

Though now somewhat dated, this classical text which focuses on significant topics is still well worth consulting.

Geochemistry and related topics

The Elements on Earth; Inorganic Chemistry in the Environment, P. A. COX, Oxford University Press, 1995.

Inorganic Geochemistry, P. HENDERSON, Pergamon Press, 1982.

Inorganic Chemistry and the Earth, J. E. FERGUSSON, Pergamon Press, 1982. A short volume which overviews chemical resources, extraction, uses, and environmental effects.

Safety in the chemical laboratory

Hazards in the Chemical Laboratory, Ed. L. BRETHERICK, Royal Society of Chemistry, 4th Edition, 1986.

Handbook of Reactive Chemical Hazards, L. BRETHERICK, Butterworth-Heinemann, 3rd Edition, 1995.

General data

CRC Handbook of Chemistry and Physics, Ed. R. WEAST, CRC Press. Reissued annually with intermittent revision. Check the actual date of the specific data which you use, as this may be much older than the date of the edition consulted.

SI Chemical Data, G. AYLWARD AND T. FINDLAY, Wiley, 5th edition, 2002.

Handbook of Inorganic Compounds, D. L. PERRY, CRC Press, 1995. Has data for more than 3000 selected inorganic compounds.

Dictionary of Chemical Terminology, D. KRYT, Elsevier, 1980.

The Elements, J. EMSLEY, Oxford University Press, 3rd Edition, 1998.

Practical inorganic chemistry

See *Inorganic Experiments*, J. D. WOOLLINS, VCH, 1995.

Synthetic Methods of Organometallic and Inorganic Chemistry, Ed. W. A. HERRMANN, Thieme, 1997. This is an eight-volume series detailing preparations for many common inorganic and organometallic substances.

Advanced Practical Inorganic and Metalorganic Chemistry, R. J. ERRINGTON, Nelson Thornes, 1997.

Synthesis of Organometallic Compounds. A Practical Guide, Ed. S. KOMIYA, Wiley, 1997.

The Manipulation of Air-Sensitive Compounds, D. F. SHRIVER AND M. A. DREZDZON, Wiley-Interscience, 2nd Edition, 1986.

Synthesis and Technique in Inorganic Chemistry, R. J. ANGELICI, W. B. Saunders, 1969.

Inorganic Syntheses. This multivolume series (Wiley, 1939–) provides detailed, checked preparations of selected common inorganic and organometallic compounds.

Technical and industrial

To expand on the details given in the text, a good starting point for any technical query is

Kirk-Othmer Encyclopedia of Chemical Technology, Executive Editor J. I. KROSCHWITZ, Wiley, 4th Edition, 1991–1998.

McGraw Hill Dictionary of Scientific and Technical Terms, 5th Edition, 1994.

Industrial Inorganic Chemicals and Products, Wiley-VCH, 1998.

Inorganic Chemistry: An Industrial and Environmental Perspective, T. W. SWADDLE, Academic Press, 1996.

Industrial Inorganic Chemistry, W. BÜCHNER, R. SCHLIEBS, G. WINTER AND K. H. BÜCHEL, translated by R. TERRELL, VCH, 1989. An excellent, well-written overview of the manufacture and uses of inorganic materials, including ceramics, glasses, etc.

Industrial Chemistry, E. STOCCHI, Ellis Horwood, 1990.

An Introduction to Industrial Chemistry, Ed. C. A. HEATON, Blackie, 2nd Edition, 1991. Of interest to those with a leaning towards chemical engineering, economics, etc.

Catalytic Chemistry, B. C. GATES, Wiley, 1992.

Industrial Applications to Homogeneous Catalysis, Eds. A. MORTREUX AND F. PETIT, Catalysis by Metal Complexes Series, D. Reidel Publishing Company, 1988. See also other volumes in this series.

For those interested in household applications of chemicals. *Chemistry in the Marketplace*, B. SELINGER, Harcourt Brace Publishers, 5th Edition, 1998, provides an excellent overview.

Another, often useful, source is the latest edition of any of the major encyclopedias.

For current information it is worth scanning recent issues of *Chemical* & *Engineering News* or *Chemtech*. A useful guide to current world production of inorganic chemicals is given by the annual 'Facts and Figures for the Chemical Industry' published in *Chemical* & *Engineering News*, usually in June or July.

For more detail, see:

The Modern Inorganic Chemicals Industry, Royal Society of Chemistry Special Publication 31, 1977 (reprinted 1986).

Speciality Inorganic Chemicals, Royal Society of Chemistry Special Publication 40, 1981.

Fine Chemicals for the Electronic Industry, Ed. P. BAMFIELD, Royal Society of Chemistry, Special Publication 60, 1986.

Syn-fuels from Boiling Stones, G. J. HUTCHINGS, *Chemistry in Britain*, 1987, 762 (use of zeolites as catalysts).

Recent Achievements, Trends and Prospects in Homogeneous Catalysis, F. J. WALLER, *J. Molecular Catalysis*, 1986. Review Issue, 43–61: see also the preceding article on Heterogeneous Catalysis by Metals.

Ziegler-Natta Catalysis, H. SINN AND W. KAMISKY, *Advances in Organometallic Chemistry* **18**, 1980, 99–143.

Trends and Opportunities for Organometallic Chemistry in Industry, G. W. PARSHALL, *Organometallics* **6**, 1987, 687–692.

Organometallic chemistry

Several parts of organometallic chemistry are covered briefly in the text, and the survey of organolanthanide chemistry in Section 16.5 gives a more detailed impression of one, small but finite, part of the field. The literature of organometallic chemistry matches that of inorganic chemistry in volume.

Organometallic Compounds, G. E. COATES, Chapman & Hall: all editions are worth consulting. 4th Edition, Volume 1, The Main Group Elements, with B. J. AYLETT AND K. WADE, Part 1, 1980; Part 2, 1979: Volume 2, The Transition Elements, with M. L. H. GREEN AND D. M. P. MINGOS, 1982.

Comprehensive Organometallic Chemistry, Eds. G. WILKINSON, F. G. A. STONE, AND E. W. ABEL, Pergamon Press, 1982. This is a multi-volume survey with many co-authors.

Comprehensive Organometallic Chemistry II (Elsevier, 1995) is a fourteen volume update of the organometallic literature from 1982 to 1994.

The Organometallic Chemistry of the Transition Metals, R. H. CRABTREE, Wiley-Interscience, 2nd Edition, 1993.

Organometallics, C. ELSCHENBROICH AND A. SALZER, VCH, 1989.

Trends and Opportunities for Organometallic Chemistry in Industry, G. W. PARSHALL, *Organometallics* **6**, 1987, 687–692.

The Close Ties Between Organometallic Chemistry, Surface Science, and the Solid State, R. HOFFMANN, S. D. HOFFMANN, S. D. WIJEYESEKEVA AND S.-S. SUNG, *Pure and Applied Chemistry* **58**, 1986, 481–94.

Organometallic Chemistry, Coordination Chemistry, Main Group Chemistry, Where are the Frontiers at Present? J. G. RIESS, *J. Organometallic Chemistry* **281**, 1985, 1–14.

Finally, it is worth looking at the following title, for aesthetic reasons as well as for its wide range of simply presented ideas.

Marvels of the Molecule, L. SALEM, illustrated by C. RATTRAY, VCH Publishers, 1987. A splendidly illustrated sweep from atoms to solids and giant biological molecules.

Comprehensive surveys and multi-volume works

While a full and detailed literature search will not normally be asked of an undergraduate student, the techniques for undertaking such a search are an important part of the chemist's armoury. The multi-volume works listed below present a starting-point which should then be brought up to date using *Chemical Abstracts* and the current literature.

An excellent, and up-to-date, entry point to the inorganic literature is the eight-volume series *Encyclopedia of Inorganic Chemistry*, Editor in Chief R. B. KING, Wiley, 1994. This provides a comprehensive coverage of a wide range of inorganic chemistry topics and provides a substantial number of references to the primary journal literature.

Comprehensive Inorganic Chemistry, Eds. J. C. BAILOR, H. J. EMELEUS, R. S. NYHOLM AND A. F. TROTMAN-DICKENSON, Pergamon Press, 1974, consists of five volumes made up of essay reviews covering all the elements systematically, together with a number of broad topics such as organo-transition metal chemistry. The literature is scanned to the date of the manuscript (note that some terminate as early as 1969) with adequate keys to the older literature.

M.T.P. International Review of Science: Inorganic Chemistry, Series 2, General Ed., H. J. EMELEUS (10 volumes edited by M. F. LAPPERT, D. B. SOWERBY, V. GUTMANN, B. J. AYLETT, D. W. A. SHARP, M. MAYS, K. W. BAGNALL, A. G. MADDOCK, M. L. TOBE and L. E. J. ROBERTS respectively), Butterworths 1975, contains more concentrated reviews.

The above two are to some extent complementary. Unfortunately, both are rather out-of-date and initially promised regular revisions have not yet appeared.

Comprehensive Coordination Chemistry, Eds. G. WILKINSON, R. D. GILLARD AND J. A. MCCLEVERTY, Pergamon Press, 1987. A major multi-volume work. There is also much inorganic chemistry in *Comprehensive Organometallic Chemistry*, cited earlier.

At the time of writing, *Comprehensive Coordination Chemistry II*, providing an update on the recent literature, is under preparation.

The problem of producing a fully comprehensive treatise are immense, due to the very rapid modern expansion of inorganic chemistry. A major effort to produce an up-do-date comprehensive treatise stems from *Handbuch der anorganische Chemie*, L. GMELIN, 1924 onwards.

First published as a multi-volume work, the Gmelin Institute is now pursuing an extensive programme of up-dating by supplementary volumes, for example, *Tellurium Compounds*, 8th Edition, main series, Supplement Volume B, 1977. This handbook now runs to literally hundreds of volumes, is expensive, not held by every library, and is mainly in German. However, for a comprehensive search it is an excellent starting point if the topic is covered by a recent volume, so it is well worth consulting your library catalogue for it.

For information on specific compounds the *Dictionary of Inorganic Compounds* (Executive Editor J. E. MACINTYRE, Chapman & Hall, 1992) and the *Dictionary of Organometallic Compounds* (Chapman & Hall 1984, with four supplements) can be very useful. The corresponding *Dictionary of Organic Compounds* can be useful when searching for information on ligands etc.

(2) Reviews and Journals

More specific information may be obtained from reviews, and ultimately from papers in the journals. There are now many review series published which survey fairly specific areas of chemistry, such as organometallic chemistry, fluorine chemistry, transition metal compounds, spectroscopy, and so on. Series which cover all of inorganic chemistry are

Advances in Inorganic Chemistry and Radiochemistry, Academic Press, approximately annual, 1959 onwards: title changed to *Advances in Inorganic Chemistry* from Volume 31, 1987.

Progress in Inorganic Chemistry, Wiley, 1959 onwards.

Both these series give accumulated contents lists in the latest volume, and every tenth volume in each series has a full accumulated index.

Coordination Chemistry Reviews, Ed. A. B. P. LEVER, Elsevier, 1966 onwards. Covers both Main Group and transition-metal coordination compounds, with some emphasis on methods.

Also important as initial sources of fuller information are *Quarterly Reviews of the Chemical Society*, 1949–1971, titled *Chemical Society Reviews*, 1972 onwards. The American Chemical Society publishes *Accounts of Chemical Research* and *Chemtech*, and general reviews also appear in the German Chemical Society's *Angewandte Chemie* (see the International Edition, which is in English).

These set out to publish articles on any part of chemistry, aimed at the general chemical reader. The student will find a number of useful articles, especially in the earlier volumes.

Chemistry in Britain, Royal Society of Chemistry.

Chemical and Engineering News, American Chemical Society. These are the regular news journals of these societies, which often include survey articles, particularly on topics of current interest, and on industrial activities. Similar informal journals are produced by a number of other Societies, e.g. *Chemistry in Australia, Chemistry in New Zealand, Canadian Chemical News*. Other sources are *Education in Chemistry* (Royal Society of Chemistry) and *Journal of Chemical Education* (American Chemical Society).

JOURNALS

The ultimate source of chemical information is the journal article where working chemists publish the results of their investigations, after their papers have been vetted by referees. A journal reader is assumed to be knowledgeable in the field, and modern publication is under pressure to be concise. Students should consult journal articles only after surveying the information available in textbooks and reviews. Often, the most useful parts of a paper for the student are the introduction, setting the background and outlining earlier work, and the discussion–conclusion section. Many journals publish inorganic chemistry. The two most important English language titles are *Inorganic Chemistry* (American Chemical Society) and *Journal of the Chemical Society, Dalton Transactions* (Royal Society of Chemistry). Other journals covering inorganic chemistry are *Inorganica Chimica Acta, Polyhedron, European Journal of Inorganic Chemistry, Organometallics* and *Journal of Organometallic Chemistry*, to name but a few.

(3) Bibliographies for Particular Sections of the Text

In addition to the appropriate sections of the advanced texts listed above, more information about particular themes may be gained from the following books and articles. This list is not intended to be exhaustive, but is designed to give a lead in to published material in each area. More specialized and advanced references may be traced through the given titles.

SI UNITS AND CHAPTER 1

For a fuller account of units and nomenclature see:

1979 manual of symbols and terminology for physicochemical quantities and units, *Pure and Applied Chemistry* **51**, 1979, 1–41.

A Dictionary of Scientific Units. H. G. JERRARD AND D. R. McNEILL, Chapman & Hall, 1986.

Nomenclature of Inorganic Chemistry II: Recommendations 2000, Eds. J. A. McCLEVERTY AND N. G. CONNELLY, Royal Society of Chemistry, 2001.

Inorganic Chemical Nomenclature – Principles and Practice, B. P. BLOCK, W. H. POWELL AND W. C. FERNELIUS, Oxford University Press, 1998.

For an interesting account see:

Origin of the Names of Chemical Elements, V. RINGNES, *Journal of Chemical Education* **66**, 1989, 731–738. For naming of the superheavy elements see references for Chapter 16.

For examples of the range of inorganic chemistry, we quote:

Silicon and Silicones, About Stone-age Tools, Antique Pottery, Modern Ceramics, Computers, Space Materials, and How They All Got That Way, E. G. ROCHOW, Springer-Verlag, 1987.

Pinpointing the Past, M. COWELL, *Chemistry in Britain* **28**, 1992, 892–896.

For an interesting discussion on clay minerals see *Clay Minerals and the Origin of Life*, A. G. CAIRNS-SMITH AND H. HARTMAN, Cambridge University Press, 1986.

ATOMIC PROPERTIES (Chapter 2)

Table Talk, D. H. ROUVRAY (coordinator), *Chemistry in Britain*, 1994, 371–386. A historical account of the development of the Periodic Table on its 125th anniversary.

Evolution of the Modern Periodic table, G. T. SEABORG, *Journal of the Chemical Society, Dalton Transactions*, 1996, 3899–3907.

Energy Levels in Atoms and Molecules, W. G. RICHARDS AND P. R. SCOTT, Oxford Chemistry Primer, Oxford University Press, 1994. See also: *Atomic Spectra*, T. P. SOFTLEY (1994) in the same series.

Principles of Atomic Orbitals, N. N. GREENWOOD, Royal Society of Chemistry Monographs for Teachers No. 8, 3rd Edition, 1980. See also the references in Chapter 3.

Relative atomic masses (atomic weights): *Pure and Applied Chemistry* Revised values appear every even year in this IUPAC publication.

Ionisation Energies Revisited, N. C. PYPER AND M. BERRY, *Education in Chemistry*, September 1990, 135–137.

The Periodicity of Electron Affinity, R. T. MEYERS, *Journal of Chemical Education* **67**, 1990, 307–308.

Revised Effective Ionic Radii in Halides and Chalcogenides, R. D. SHANNON, *Acta Crystallographica* **A32**, 1976, 751–767: O. JOHNSON, *Inorganic Chemistry* **12**, 1973, 780.

Soft-sphere Ionic Radii for Alkali and Halogenide Ions, L. PAULING, *Journal of the Chemical Society, Dalton Transactions*, 1980, 645 and references therein.

Absolute Electronegativity and Hardness, R. G. PEARSON, *Chemistry in Britain*, May 1991, 444–447.

Principles of Electronegativity, R. T. SANDERSON, *Journal of Chemical Education* **65**, 1988, 112–118.

Electronegativities of Elements in Valence States, Y. ZHANG, *Inorganic Chemistry* **21**, 1982, 3886–9; Applications to Strengths of Lewis Acids, 3889–93.

The 'Inert-Pair' Effect on Electronegativity, R. T. SANDERSON, *Inorganic Chemistry* **25**, 1986, 1856–8; also *Journal of the American Chemical Society*, **105**, 1983, 2259–61 and references therein.

An Electronegativity Scale Based upon Geometry Changes on Ionisation, P. H. BLUSTIN AND W. T. RAYNES, *Journal of the Chemical Society, Dalton Transactions,* 1981, 1237.

The Covalent Potential: A Simple and Useful Measure of the Valence-State Electronegativity for Correlating Molecular Energetics, Y.-R. LUO AND S. W. BENSON, *Accounts of Chemical Research* **25**, 1992, 375–381

Bent's Rule: Energetics, Electronegativity, and the Structures of Nonmetal Fluorides. J. E. Huheey *Inorganic Chemistry* **20**, 1981, 4033–4035.

MOLECULAR SHAPES AND BONDING (Chapters 3 and 4)

Chemical Bonding and Molecular Geometry, R. J. GILLESPIE, Oxford University Press, 2001.

The Shape and Structure of Molecules, C. A. COULSON revised by R. MCWEENY, Clarendon, 2nd Edition, 1982.

Valence, C. A. COULSON, Oxford University Press, 3rd Edition, 1979, by R. MCWEENY. This is a very clear account of atomic and molecular structure, ranging over inorganic and organic molecules. The general student will find it useful if he or she is willing to 'read round' the more detailed sections.

See also the following:

The Chemical Bond in Inorganic Chemistry, I. D. BROWN, Oxford University Press, 2001.

Molecular Geometry, A. RODGER AND P. M. RODGER, Butterworth-Heinemann, 1995.

An Introduction to Molecular Orbitals, Y. JEAN, F. VOLATRON AND J. K. BURDETT, Oxford University Press, 1993.

Bonding and Structure: Structural Principles in Inorganic and Organic Chemistry, N. W. ALCOCK, Ellis Horwood, 1990.

Chemical Bonding Theory, B. WEBSTER, Blackwell, 1990.

The Chemical Bond, J. N. MURRELL, S. F. A. KETTLE, AND J. M. TEDDER, Wiley, 2nd edition, 1985.

Symmetry and Structure, S. F. A. KETTLE, Wiley, 1985.

Valency, M. F. O'DWYER, J. E. KENT AND R. D. BROWN, Springer, 2nd Edition, 1978, reprinted 1986.

The Nature of the Chemical Bond, L. PAULING, Cornell University Press (Oxford University Press in UK), 3rd Edition, 1960. This is Pauling's classical text on chemical bonding, which should be dipped into by every chemist. See also *The Chemical Bond*, Oxford University Press, 1967, in which Pauling gives a shortened and updated survey.

Lewis Structures, Formal Charge, and Oxidation Numbers. A More User-Friendly Approach, J. E. PACKER AND S. D. WOODGATE, *Journal of Chemical Education* **68**, 1991, 456–458. A good summary account of the rules for working out Lewis structures of molecules.

The Simplest Molecule, I. R. MCNAB, *Chemistry in Britain*, 1992, 538–542. A discussion of the bonding in H_2^+.

Describing Electron Distribution in the Hydrogen Molecule. A New Approach. C. J. WILLIS, *Journal of Chemical Education* **68**, 1991 743–747.

The Relative Energies of Molecular Orbitals for Second-Row Homonuclear Diatomic Molecules. The Effect of s–p Mixing, A. HAIM, *Journal of Chemical Education* **68**, 1991, 737–738.

The Three Forms of Molecular Oxygen, M. LAING, *Journal of Chemical Education* **66**, 1989, 453–454.

The Significance of the Bond Angle in Sulfur Dioxide, G. H. PURSER, *Journal of Chemical Education* **66**, 1989, 710–713.

Bonding Considerations of the Nitrate Anion, G. R. WILLEY, *Education in Chemistry*, 1889, 78–82.

The electron pair repulsion theory was introduced by N. V. Sidgwick and H. M. Powell, *Proceedings of the Royal Society* **176A**, 1940, 153, and developed extensively by R. J. Gillespie and R. S. Nyholm, as in *Quarterly Reviews*, **11**, 1957, 261, and more recently in the following articles.

Shaping up with EAN and VSEPR, M. LAING, *Education in Chemistry*, July 1995, 102–105.

The VSEPR Model Revisited, *Chemical Society Reviews*, 1992, 59–69, and Multiple Bonds and the VSEPR Model, *Journal of Chemical Education* **69**, 1992, 116–121.

Electron Domains and the VSEPR Model of Molecular Geometry, R. J. GILLESPIE AND E. A. ROBINSON, *Angewandte Chemie, International Edition* **35**, 1996, 495–514.

Precise data on molecular structure, in the gaseous and condensed states, is to be found in The Chemical Society's Special Publications 11 and 18, *Interatomic Distances and Supplement.*

SOLIDS (Chapter 5)

Inorganic Structural Chemistry, U. MÜLLER, Wiley, 1993.

Structural Inorganic Chemistry, A. F. WELLS, Oxford University Press, 5th Edtion, 1984.

Molecular and Crystal Structure Models, A. WALTON, Ellis Horwood, 1978.

Perovskites—Chemical Chameleons. A. RELLER AND T. WILLIAMS, *Chemistry in Britain*, 1989, 1227–1230.

Lattice energies—a detailed treatment of basic work is given by T. C. WADDINGTON, *Advances in Inorganic Chemistry and Radiochemistry* **1**, 1959, 158–221. For more recent discussions see:

Lattice Enthalpies of Ionic Halides, Hydrides, Oxides, and Sulfides, J. B. HOLBROOK, R. SABRY-GRANT, B. C. SMITH AND T. V. TANDEL, *Journal of Chemical Education* **67**, 1990, 304–306; The Calculation of Lattice Energy: Some Problems and Some Solutions, H. D. B. JENKINS, *Revue de Chimie Minérale* **16**, 1979, 134–150.

A Tetrahedron of Bonding, M. LAING, *Education in Chemistry*, 1993, 160–163.

SOLVENTS (Chapter 6)

Chemical Hardness, R. G. PEARSON, Wiley-VCH, 1997.

Redox Mechanisms in Inorganic Chemistry, A. G. LAPPIN, Ellis Horwood, 1993.

Ions in Solution. Basic Principles of Chemical Interactions, J. BURGESS, Ellis Horwood, 1988.

Recent Advances in the Concept of Hard and Soft Acids and Bases, R. G. PEARSON, *Journal of Chemical Education* **64**, 1987, 561–567. Makes this longstanding theory more quantitative.

An Acidity Scale for Binary Oxides, D.W. SMITH, *Journal of Chemical Education* **64**, 1987, 480–481.

The Usage of the Terms 'Equivalent' and 'Normal', H. M. N. H. IRVING (T. S. WEST), *Pure and Applied Chemistry* **50**, 1978, 325.

See *Modern Aspects of Inorganic Chemistry*, H. J. EMELEUS AND A. G. SHARPE, listed above, for a general survey of nonaqueous solvents.

Nonaqueous Solution Chemistry, O. POPOVYCH AND R. P. T. TOMKINS, Wiley, 1981.

Inorganic Chemistry in Liquid Ammonia, D. NICHOLLS, Elsevier, 1979.

The Chemistry of Aqua Ions: Synthesis, Structure and Reactivity: A Tour Through the Periodic Table of the Elements, D. T. RICHENS, Wiley, 1997.

Coordination Chemistry in and of Sulfur Dioxide, R. MEWS, E. LORK, P. G. WATSON AND B. GÖRTLER, *Coordination Chemistry Reviews* **197**, 2000, 277–320.

Supercritical Chemistry: Synthesis with a Spanner, M. POLIAKOFF AND S. HOWDLE, *Chemistry in Britain*, February 1995, 118–121.

Chemistry Goes Supercritical, T. CLIFFORD AND K. BARTLE, *Chemistry in Britain*, June 1993, 499–502.

EXPERIMENTAL METHODS (Chapter 7)

Spectroscopic Properties of Inorganic and Organometallic Compounds, G. DAVIDSON, Royal Society of Chemistry. This series provides an annual review of literature published in the field.

Inorganic Spectroscopic Methods, A. K. BRISDON, Oxford University Press, 1998.

Structural Methods in Inorganic Chemistry, E. A. V. EBSWORTH, D. W. H. RANKIN, AND S. CRADOCK, Blackwell, 2nd Edition, 1991.

A brief summary of solvent extraction processes can be found in the following:

Liquid-Liquid Extraction: Metals, P. J. BAILES, C. HANSON, AND M. A. HUGHES, *Chemical Engineering*, 1976, 86–94.

For more recent updates see the various sections on solvent extraction and hydrometallurgy in the *Kirk-Othmer Encyclopedia of Chemical Technology.*

The Basics of Crystallography and Diffraction, C. HAMMOND, Oxford University Press, 1997.

Crystal Structure Determination, W. CLEGG, Oxford University Press, 1998.

Molecular Structure: Its Study by Crystal Diffraction, J. C. SPEAKMAN, Royal Society of Chemistry, Monograph for Teachers **30**, 1977.

Handbook of X-ray and Ultraviolet Photoelectron Spectra, Ed., D. BRIGGS, Heyden, 1978.

The Partnership of Gas-Phase and Valence Photoelectron Spectroscopy, W. L. JOLLY, *Accounts of Chemical Research* **16**, 1983, 370–376.

For basic, readable introductions to nmr, see

NMR in Chemistry, W. KEMP, Macmillan, 1986.

Nuclear Magnetic Resonance, P. J. HORE, Oxford Chemistry Primers, Oxford University Press, 1995.

For more advanced students:

Modern NMR Spectroscopy. A Guide for Chemists, J. K. M. SANDERS AND B. K. HUNTER, Oxford University Press, 2nd Edition, 1993. See also the accompanying *Workbook of Chemical Problems*, J. K. M. SANDERS, E. C. CONSTABLE, B. K. HUNTER, AND C. M. PEARCE, Oxford University Press, 2nd Edition, 1993.

NMR and Chemistry. An Introduction to Modern NMR Spectroscopy, J. W. AKITT, Chapman & Hall, 3rd Edition, 1992.

State of the Art for Solids, R. K. HARRIS, *Chemistry in Britain*, 1993, 601–604. A good introduction to the technique and applications of solid-state nmr spectroscopy.

For mass spectrometry, consult first:

Electrospray Mass Spectrometry Applied to Inorganic and Organometallic Chemistry, R. COLTON, A. D'AGOSTINO AND J.C. TRAEGER, *Mass Spectrometry Reviews* **14**, 1995, 79–106.

Scanning Tunneling Microscopy, C. M. LIEBER, *Chemical & Engineering News Special Report*, April 18, 1994, 28–43, See also:

Applications of Scanning Tunneling Microscopy to Inorganic Chemistry, X. L. WU AND C. M. LIEBER, *Progress in Inorganic Chemistry* **39**, 1991, 431–510.

Scanning Tunneling Microscopy and Atomic Force Microscopy in Organic Chemistry, J. FROMMER, *Angewandte Chemie, International Edition* **31**, 1992, 1298–1328.

Probing Surfaces, J. LECKENBY, *Chemistry in Britain*, March 1995, 212.

GENERAL PROPERTIES OF THE ELEMENTS (Chapter 8)

Nucleogenesis

Supernova 1987A—New Evidence for Nucleosynthesis, C. H. ATWOOD, *Chemistry in Britain,* May 1990, 423–426.

Stellar Alchemy—The Origin of the Chemical Elements, E. B. NORMAN, *Journal of Chemical Education* **71**, 1994, 813–820.

See also *Inorganic Geochemistry,* P. HENDERSON, Pergamon Press, 1982, for an overview on nucleosynthesis.

Nobel Lecture, W. A. FOWLER, *Science* **226**, 1984, 922–935.

For interstellar chemistry

Semistable Molecules in the Laboratory and in Space, H. W. KROTO, *Chemical Society Reviews* **11**, 1982, 435–491.

Chemistry of the Solar System, H, E. SUESS, Wiley, 1987.

General

Thermochemistry of Inorganic Fluorine Compounds, A. A. WOOLF, *Advances in Inorganic Chemistry and Radiochemistry* **24**, 1981, 1–56.

Conditions for Stability of Oxidation States Derived from Photo–electron Spectra and Inductive Quantum Chemistry, C. K. JØRGENSEN, *Zeitung für anorganische und allgemeine Chemie* **540/541**, 1986, 91–105. A very broad sweep through the whole of the Periodic Table arranging oxidation states by their Kossel numbers (Kossel number is Z minus the ionic charge) and comparing with the ionization energy from photoelectron spectra. Article is in English.

Oxidation Potentials, W. M. LATIMER, Prentice–Hall, 2nd Edition, 1952. This is a complete survey of oxidation potential data and still the standard source (though more recent values are available for many of the more difficult determinations).

A Graphical Method of Representing the Free Energies of Oxidation–Reduction Systems, E. A. V. EBSWORTH, *Education in Chemistry* **1**, 1964, 123.

For Properties and extraction of elements

See appropriate sections in the *Kirk–Othmer Encyclopedia of Chemical Technology.*

See also:

The Extraction of Metals from Ores Using Bacteria, D. K. EWART AND M. N. HUGHES, *Advances in Inorganic Chemistry* **36**, 1991, 103–135.

Principles of the Extraction of Metals, D. J. G. IVES, Royal Institute of Chemistry Monograph.

HYDROGEN (Chapter 9)

Liquid Water—The Story Unfolds, M. C. R. SYMONS, *Chemistry in Britain*, 1989, 491–494.

Water: From Clusters to the Bulk, R. LUDWIG, *Angewandte Chemie, International Edition* **40**, 2001, 1808–1827.

Active MgH_2–Mg Systems for Reversible Chemical Energy Storage, B. BOGDANOVI'C, A. RITTER, AND B. SPLIETHOFF, *Angewandte Chemie, International Edition* **29**, 1990, 223–234.

Taking Stock: The Astonishing Development of Boron Hydride Cluster Chemistry (Ludwig Mond Lecture), N. N. GREENWOOD, *Chemical Society Reviews* **21**, 1992, 49–57.

The Hydrides of Aluminium, Gallium, Indium, and Thallium, A Re-evaluation, A. J. DOWNS AND C. R. PULHAM, *Chemical Society Reviews* 1994, 175–184.

New Developments in the Chemistry of Organoaluminium and Organogallium Hydrides, A. H. COWLEY, F. P. GABBAÏ, H. S. ISOM AND A. DECKEN, *Journal of Organometallic Chemistry* **500**, 1995, 81–88.

The Hunting of the Gallium Hydrides, A. J. DOWNS AND C. R. PULHAM, *Advances in Inorganic Chemistry* **41**, 1994, 171.

Recent Developments in the Chemistry of Alane (AlH_3) and Gallane (GaH_3), C. L. RASTON, *Journal of Organometallic Chemistry* **475**, 1994, 15–24.

Rings, Clusters and Polymers of Main Group and Transition Elements, H. W. ROESKY, Elsevier, 1989. *See also: Boranes and Metalloboranes,* C. E. HOUSECROFT, Ellis Horwood, 1990.

A New Stage in the Development of Transition Metal Alumohydrides, B. M. BULYCHEV, *Polyhedron* **9**, 1990, 387–408.

Very Strong Hydrogen Bonds, J. EMSLEY, *Chemical Society Reviews* **9**, 1980, 91–124.

's' ELEMENTS (Chapter 10)

Alkali and Alkaline Earth Metal Cryptates, D. PARKER, *Advances in Inorganic Chemistry and Radiochemistry* **27**, 1983, 1–26.

Electrides, Negatively Charged Metal ions and Related Phenomena, J. L. DYE, *Progress in Inorganic Chemistry* **32**, 1984, 327–441.

First Electride Crystal Structure, S. B. DAWES, D. L. WARD, R. H. HUANG, AND J. L. DYE. *Journal of the American Chemical Society* **108**, 1986, 3534–3535 (in the compound $Cs(18\text{-crown-}6)_2^+ \cdot e^-$).

Aqueous Solution Chemistry of Beryllium, L. ALDERIGHI, P. GANS, S. MIDOLLINI AND A. VACCA, *Advances in Inorganic Chemistry* **50**, 2000, 109–172.

To be or not to Be—The Story of Beryllium Toxicity, D. N. SKILLETER, *Chemistry in Britain*, 1990, 26–30, Toxicology, chemistry and applications of Be and its compounds.

Beryllium Coordination Chemistry, C. Y. WONG AND J. D. WOOLLINS, *Coordination Chemistry Reviews* **130**, 1994, 243–273.

The Great Radium Scandal, R. M. MACKLIS, *Scientific American*, August 1993, 78–83. An interesting account of the use of a radioactive radium-laced medicine in the 1920s.

Structures of Organo Alkali Metal Complexes and Related Compounds, E. WEISS, *Angewandte Chemie, International Edition* **32**, 1993, 1501–1523.

Organomagnesium Chemistry: Nearly a Hundred Years but Still Fascinating, F. BICKELHAUPT, *Journal of Organometallic Chemistry* **475**, 1994, 1–14.

Strontium—a neglected element, J. W. NICHOLSON AND L. R. PIERCE, *Education in Chemistry*, May 1995, 74–76.

'f' ELEMENTS (Chapters 11 and 12)

Lanthanides and Actinides, S. COTTON, Macmillan, 1991.

Recent Advances in the Chemistry of Scandium, S. A. COTTON, *Polyhedron* **18**, 1999, 1691–1715.

Transuranium Elements – A Half Century, Eds. L. R. MORSS AND J. FUGER, Oxford University Press, 1998.

The Elements Beyond Uranium, G. T. SEABORG AND W. D. LOVELAND, Wiley, 1990.

Transuranium Elements: Past, Present and Future, G. T. SEABORG, *Accounts of Chemical Research* **28**, 1995, 257–264.

The Discovery of the Rare Earth Element, C. H. EVANS, *Chemistry in Britain*, 1989, 880–882.

Application of Lanthanide Reagents in Organic Synthesis, G. A. MOLANDER, *Chemical Reviews* **92**, 1992, 29–68.

Reduced Halides of the Rare Earth Elements, G. MEYER, *Chemical Reviews* **88**, 1988, 93–107.

Preparation and Purification of Actinide Metals, J. C. SPIRLET, J. R. PETERSON, AND L. B. ASPREY, *Advances in Inorganic Chemistry* **31**, 1987, 1–41.

Plutonium—The Element of Surprise, G. R. CHOPPIN AND B. E. STOUT, *Chemistry in Britain,* 1991, 1126–1129.

The Most Useful Actinide Isotope: Americium-241, J. D. NAVRATIL, W. W. SCHULTZ, AND G. T. SEABORG, *Journal of Chemical Education* **67**, 1990, 15–16.

The Chemistry of Berkelium, J. P. PETERSON AND D. E. HOBART, *Advances in Inorganic Chemistry and Radiochemistry* **28**, 1984, 29–64.

Actinide Alkoxide Chemistry, W. G. VAN DER SLUYS AND A. P. SATTELBERGER, *Chemical Reviews* **90**, 1990, 1027–1040.

TRANSITION ELEMENTS (Chapters 13, 14 and 15)

General references

Reaction Mechanisms of Inorganic and Organometallic Systems, R. B. JORDAN, Oxford University Press, 2nd Edition, 1998.

Inorganic Reaction Mechanisms, M. L. TOBE AND J. BURGESS, Longman, 1999.

Kinetics and Mechanism of Reactions of Transition Metal Complexes, R. G. WILKINS, VCH, 2nd Edition, 1991.

Ligand Field Theory and Its Applications, B. N. FIGGIS AND M. A. HITCHMAN, Wiley, 1999.

Electronic Structure and Properties of Transition Metal Compounds: Introduction to the Theory, I. B. BERSUKER, Wiley, 1996.

Metal-Ligand Bond Distances in First-Row Transition Metal Coordination Compounds: Coordination Number, Oxidation State, and Specific Ligand Effects, R. S. FEE, R. A. KRUSE, AND W. M. STRUB, *Inorganic Chemistry* **37**, 1998, 5369–5375.

Determination and Use of Stability Constants, A. E. MARTELL AND R. J. MOTEKAITIS, Wiley, 2nd Edition, 1992.

A Millennial Overview of Transition Metal Chemistry, F. A. COTTON, *Journal of the Chemical Society, Dalton Transactions*, 2000, 1961–1968.

Chemistry of Transition Metal Cyanide Compounds: Modern Perspectives, K. R. DUNBAR AND R. A. HEINTZ, *Progress in Inorganic Chemistry* **45**, 1997, 283–392.

Alkoxo and Aryloxo Derivatives of Metals, D. C. BRADLEY, R. C. MEHROTRA, I. P. ROTHWELL AND A. SINGH, Academic Press, 2001. See also Recent Trends in Metal Alkoxide Chemistry, R. C. MEHROTRA AND A. SINGH, *Progress in Inorganic Chemistry* **46**, 1997, 239–454.

Early transition metals

Molybdenum. An Outline of its Chemistry and Uses, Eds. E. R. BRAITHWAITE AND J. HABER, Studies in Inorganic Chemistry, Volume 19, Elsevier, 1994.

Early Transition Metal Clusters with pi-Donor Ligands, Ed. M. H. CHISHOLM, Wiley, 1995.

Synthesis of Complex Metal Oxides by Novel Routes, C. N. R. RAO AND J. GOPALAKRISHANAN, *Accounts of Chemical Research* **20**, 1987, 228–235.

Solid State Structures of the Binary Fluorides of the Transition Metals, A. J. EDWARDS, *Advances in Inorganic Chemistry and Radiochemistry* **27**, 1983, 83–112.

Preparation and Reactions of Oxide Fluorides of the Transition Metals, the Lanthanides, and the Actinides, J. H. HOLLOWAY AND D. LAYCOCK, *Advances in Inorganic Chemistry and Radiochemistry* **28**, 1984, 73–100.

The Chemistry and Spectroscopy of Mixed-Valence Compounds, R. J. H. CLARK, *Chemical Society Reviews* **13**, 1984, 219–244.

Vanadium Peroxide Complexes, A. BUTLER, M. J. CLAGUE, AND G. E. MEISTER, *Chemical Reviews* **94**, 1994, 625–638.

Recent Developments in Chromium Chemistry, D. A. HOUSE, *Advances in Inorganic Chemistry* **44**, 1997, 341–374.

Protonation, Oligomerization, and Condensation Reactions of Vanadate(V), Molybdate(VI), and Tungstate(VI), J. J. CRUYWAGEN, *Advances in Inorganic Chemistry* **49**, 2000, 127–182.

Polyoxometalate Chemistry: An Old Field with New Dimensions in Several Disciplines, M. T. POPE AND A. MÜLLER, *Angewandte Chemie, International Edition* **30**, 1991, 34–48.

Peroxo and Superoxo Complexes of Chromium, Molybdenum, and Tungsten, M. H. DICKMAN AND M. T. POPE, *Chemical Reviews* **94**, 1994, 569–584.

The Active Sites in Manganese-Containing Metalloproteins and Inorganic Model Complexes, K. WIEGHARDT, *Angewandte Chemie, International Edition* **28**, 1989, 1153–1172.

The Coordination Chemistry of Technetium, J. BALDAS, *Advances in Inorganic Chemistry* **41**, 1994, 1.

Organorhenium Oxides, W. A. HERRMANN AND F. E. KÜHN, *Accounts of Chemical Research* **30**, 1997, 169–180.

Late transition metals

Chemistry of the Platinum Group Metals: Recent Developments, Ed. F. R. HARTLEY, Studies in Inorganic Chemistry, Volume 11, Elsevier, 1991.

The Chemistry of Ruthenium, E. A. SEDDON AND K. R. SEDDON, Topics in Inorganic and General Chemistry, Volume 19, Elsevier, 1984.

Ferric Iodide as a Nonexistent Compound, K. B. YOON AND J. K. KOCHI, *Inorganic Chemistry* **29**, 1990, 869–874.

Nickel: An Element with Wide Application in Industrial Homogeneous Catalysis, W. KEIM, *Angewandte Chemie, International Edition* **29**, 1990, 235–244.

High Oxidation State Organometallic Chemistry, A Challenge—the Example of Rhenium, W. A. HERRMANN, *Angewandte Chemie, International Edition* **27**, 1988, 1297–1313.

Homogeneous Catalysis with Compounds of Rhodium and Iridium, R. S. DICKSON, D. Reidel, 1985.

One-Dimensional Inorganic Platinum-Chain Electrical Conductors, J. M. WILLIAMS, *Advances in Inorganic Chemistry and Radiochemistry* **26**, 1983, 235–268.

Coordination Chemistry of Halocarbons, R. J. KULAWIEC AND R. H. CRABTREE, *Coordination Chemistry Reviews* **99**, 1990, 89–115.

Gold: Chemistry, Biochemistry and Technology, Ed. H. SCHMIDBAUR, Wiley, 1999.

Homo- and Heteronuclear Cluster Compounds of Gold, K. P. HALL. AND D. M. P. MINGOS, *Progress in Inorganic Chemistry* **32**, 1984, 237–325.

Fluorides of Copper, Silver, Gold, and Palladium, B. G. MÜLLER, *Angewandte Chemie, International Edition* **26**, 1987, 1081–1097.

The Crystal Engineering of Non-Molecular Metal Compounds with Anionic Chalcogenide Ligands E^{2-} and RE^{-}, I. G. DANCE. in *Perspectives in Inorganic Chemistry*, Eds. A. F. WILLIAMS, C. FLORIANI, AND A. E. MERBACH, VCH, 1992, 165–181.

TRANSITION METAL TOPICS (Chapter 16)

Superconductors

Nature **329**, 1987, 763 and *Chemistry in Britain,* 1987, 962–966 give summaries to that time. For review articles see the following:

Room Temperature Superconductors, A. W. SLEIGHT, *Accounts of Chemical Research* **28**, 1995, 103–108.

Chemistry in Britain, September 1994. This issue is devoted to superconductors, with particular emphasis on the cuprate systems.

Structure, Composition and Properties of High-Temperature Cuprate Superconductors, J. K. BURDETT, *Perspectives in Coordination Chemistry* 1992, 293–319.

The New Superconductors, *Chemical Engineering News Special Report,* December 21, 1992, 24–41.

Superconductors Beyond 1–2–3, R. J. CAVA, *Scientific American*, August 1990, 24–31.

Structural Chemistry and the Local Charge Picture of Copper Oxide Superconductors, R. J. CAVA, *Science* **247**, 1990, 656–662.

Structural Aspects of High-Temperature Cuprate Superconductors, C. N. R. RAO AND B. RAVEAU, *Accounts of Chemical Research* **22**, 1989, 106–113.

Chemical Aspects of Solution Routes to Perovskite-Phase Mixed-Metal Oxides from Metal-Organic Precursors, C. D. CHANDLER, C. ROGER, AND M. J. HAMPDEN-SMITH, *Chemical Reviews* **93**, 1993, 1205–1241.

Transition metal carbonyl compounds

100 years of Metal Carbonyls, E. W. ABEL, *Education in Chemistry*, March 1992, 46–49. A very readable historical summary of the developments in this field.

Highly Reduced Metal Carbonyl Anions: Synthesis, Characterisation, and Chemical Properties, J. E. ELLIS, *Advances in Organometallic Chemistry* **31**, 1990, 1–51.

Organometallic compounds

The Organometallic Chemistry of the Transition Metals, R. H. CRABTREE, Wiley, 2nd Edition, 1993.

Organometallics, C. ELSCHENBROICH AND A. SALZER, VCH, 1989.

Organometallic Chemistry. An Overview, J. S. THAYER, VCH, 1987.

An Introduction to Organometallic Chemistry, A. W. PARKINS AND R. C. POLLER, Macmillan, 1986.

Metal–Carbon and Metal–Metal Bonds as Ligands in Transition-Metal Chemistry: the Isolobal Connection, F. G. A. STONE, *Angewandte Chemie, International Edition* **23**, 1984, 89–99.

Organometallic Chemistry of Alkenes and Alkynes, H. WERNER, *Journal of Organometallic Chemistry* **475**, 1994, 45–55.

Cyclopentadienyl compounds

Metallocenes. Synthesis – Reactivity – Applications. Eds. A. TOGNI AND R. L. HALTERMAN, Wiley, 1998. See also *Ferrocenes*, Eds. A. TOGNI AND T. HAYASHI, VCH, 1995.

Metallocenes Come of Age, *Chemistry in Britain,* February 1994, 87–88.

Bulky or Supracyclopentadienyl Derivatives in Organometallic Chemistry, C. JANIAK, AND H. SCHUMANN, *Advances in Organometallic Chemistry* **33**, 1991, 291–393.

Organometallic chemistry of the lanthanides and actinides

Organometallic Chemistry of the Lanthanides, C. J. SCHAVERIEN, *Advances in Oganometallic Chemistry* **36**, 1994, 283–362.

Zero Oxidation State Compounds of Scandium, Yttrium, and the Lanthanides, F. G. A. CLOKE, *Chemical Society Reviews* **22**, 1993, 17–24.

Monocyclopentadienyl Halide Complexes of the d- and f-Block Elements, R. POLI, *Chemical Reviews* **91**, 1991, 509–551.

Multiple metal–metal bonds

Multiple Bonds Between Metal Atoms, F. A. COTTON AND R. A. WALTON, Clarendon Press, 1993.

Highlights from Recent Work on Metal–Metal Bonds, F. A. COTTON, *Inorganic Chemistry* **37**, 1998, 5710.

Metal clusters

Introduction to Cluster Chemistry, D. M. P. MINGOS AND D. J. WALES, Inorganic and Organometallic Chemistry Series, Prentice Hall, 1990.

The Chemistry of Metal Cluster Complexes, D. F. SHRIVER, H. D. KAESZ, AND R. D. ADAMS, VCH, 1990.

Metal Clusters Revisited, J. LEWIS, *Chemistry in Britain,* 1988, 795–800.

Metal Clusters in Catalysis, Eds. B. C. GATES, L. GUCZI AND H. KNÖZINGER, Elsevier, 1986.

Arene-Cluster Compounds, B. F. G. JOHNSON, *Journal of Organometallic Chemistry* **475**, 1994, 31–43.

Clusters and Colloids: From Theory to Applications, Ed. G. SCHMID, VCH, 1994.

The Role of Big Metal Clusters in Nanoscience, G. SCHMID, *Journal of the Chemical Society, Dalton Transactions*, 1998, 1077–1082.

The Application of Au_{55} Clusters as Quantum Dots, U. SIMON, G. SCHÖN AND G. SCHMID, *Angewandte Chemie, International Edition* **32**, 1993, 250–253.

Transition Metal Carbonyl Cluster Chemistry, P. J. DYSON AND J. S. McINDOE, Gordon and Breach, 2000.

High Nuclearity Carbonyl Clusters: Their Synthesis and Reactivity, M. D. VARGAS AND J. N. NICHOLLS, *Advances in Inorganic Chemistry and Radiochemistry* **30**, 1986, 123–222.

Large Clusters and Colloids. Metals in the Embryonic State. G. SCHMID, *Chemical Reviews* **92**, 1992, 1709–1727.

For a good example of a giant cluster see: A New Copper Selenide Cluster with PPh_3 Ligands: $[Cu_{146}Se_{73}(PPh_3)_{30}]$, H. KRAUTSCHEID, D. FENSKE, G. BAUM, AND M. SEMMELMANN, *Angewandte Chemie, International Edition* **32**, 1993, 1303–1305.

Bonding in Molecular Clusters and Their Relationship to Bulk Metals, D. M. P. MINGOS, *Chemical Society Reviews* **15**, 1886, 31–61.

How Chemistry and Physics Meet in the Solid State, R. HOFFMANN, *Angewandte Chemie, International Edition* **26**, 1987, 846–878. Bond theory and band theory—how they are reconciled.

Metal–dioxygen complexes

For comprehensive reviews on metal–dioxygen complexes and closely related topics, see the May 1994 issue of *Chemical Reviews* which is entirely devoted to this topic.

The Structure and Reactivity of Dioxygen Complexes of the Transition Metals, M. H. GUBELMANN AND A. F. WILLIAMS, *Structure and Bonding* **55**, 1983, 1–65.

Mechanistic and Kinetic Aspects of Transition Metal Oxygen Chemistry, A. BAKAC, *Progress in Inorganic Chemistry* **43**, 1995, 267–352.

Metal–nitrogen complexes

The Discovery of $[Ru(NH_3)_5N_2]^{2+}$. A Case of Serendipity and the Scientific Method, C. V. SENOFF, *Journal of Chemical Education* **67**, 1990, 368–370.

The Chemistry of Nitrogen Fixation and Models for the Reactions of Nitrogenase, R. A. HENDERSON, G. J. LEIGH, AND C. J. PICKETT, *Advances in Inorganic Chemistry and Radiochemistry* **27**, 1983, 198–292.

Metal–hydrogen complexes

Recent Advances in Hydride Chemistry, Eds. M. PERUZZINI AND R. POLI, Elsevier, 2001.

Coordination Chemistry of Dihydrogen, D. M. HEINEKEY AND W. J. OLDHAM, JR, *Chemical Reviews* **93**, 1993, 913–926.

Dihydrogen Complexes: Some Structural and Chemical Studies, R. H. CRABTREE, *Accounts of Chemical Research* **23**, 1990, 95–101.

Molecular Hydrogen Complexes: Coordination of a σ Bond to Transition Metals, G. J. KUBAS, *Accounts of Chemical Research* **21**, 1988, 120–128.

Superheavy elements

The Elements Beyond Uranium, G. T. SEABORG AND W. T. LOVELAND, Wiley, 1990.

Voyage to Superheavy Island, Y. T. OGANESSIAN, V. K. UTYONKOV AND K. J. MOODY, *Scientific American*, January 2000, 45–49.

After the Actinides, Then What? S. A. COTTON, *Chemical Society Reviews*, 1996, 219–227.

Electronic Structure and Properties of the Transactinides and Their Compounds, V. G. PERSHINA, *Chemical Reviews* **96**, 1996, 1977–2010.

Searching for the Transactinides, S. A. COTTON, *Education in Chemistry*, May 1995, 67–70.

Transuranium Elements: Past, Present and Future, G. T. SEABORG, *Accounts of Chemical Research* **28**, 1995, 257–264.

The Heaviest Elements, D. C. HOFFMAN, *Chemical & Engineering News*, Special Report, 2 May 1994, 24–34. An excellent account of the methods by which the superheavy elements are synthesized and their chemistry studied.

Ionization Potentials of Seaborgium, E. JOHNSON, V. PERSHINA AND B. FRICKE, *Journal of Physical Chemistry A* **103**, 1999, 8458–8462.

Solution Chemistry of Element 106. Theoretical Predictions of Hydrolysis of Group 6 Cations Mo, W and Sg, V. PERSHINA AND J. V. KRATZ, *Inorganic Chemistry* **40**, 2001, 776–780.

Relativity

Strong Closed-Shell Interactions in Inorganic Chemistry, P. PYYKKÖ, *Chemical Reviews* **97**, 1997, 597–636.

Relativistic Effects in Inorganic and Organometallic Chemistry, N. KALTSOYANNIS, *Journal of the Chemical Society, Dalton Transactions*, 1997, 1–11.

Relativistic Effects in Structural Chemistry, P. PYYKKÖ, *Chemical Reviews* **88**, 1988, 563–594.

Relativistic Effects on Periodic Trends, P. PYYKKÖ, in *The Effects of Relativity in Atoms, Molecules and the Solid-State,* Eds. S. WILSON, I. P. GRANT, AND B. L. GYORFFY, Plenum, 1990.

Why is Mercury Liquid? Or, Why Do Relativistic Effects Not Get Into Chemistry Textbooks? L. J. NORRBY, *Journal of Chemical Education* **68**, 1991, 110 –113.

p ELEMENTS (Chapter 17)

Main Group Chemistry, A. G. MASSEY, Wiley, 2nd Edition, 2000.

Main Group Chemistry, Tutorial Chemistry Text series, W. HENDERSON, Royal Society of Chemistry, 2000.

Main Group Element Chemistry at the Millennium, N. N. GREENWOOD, *Journal of the Chemical Society, Dalton Transactions*, 2001, 2055–2066. For an older review by the same author see: The Resurgence of Main Group Element Chemistry, N. N. GREENWOOD, *Journal of the Chemical Society, Dalton Transactions*, 1991, 565–573.

For inorganic polymers, largely involving main group elements, see:

Main-Group-Based Rings and Polymers, D. P. GATES AND I. MANNERS, *Journal of the Chemical Society, Dalton Transactions,* 1997, 2525–2532.

Polymers and the Periodic Table: Recent Developments in Inorganic Polymer Science, I. MANNERS, *Angewandte Chemie, International Edition* **35**, 1996, 1602–1621.

The Metallic Face of Boron T. P. FEHLNER, *Advances in Inorganic Chemistry* **35**, 1990, 199–233.

Boron Chemistry at the Millenium, Ed. R. B. KING, Elsevier, 1999.

Contemporary Boron Chemistry, Eds. M. G. DAVIDSON, A. K. HUGHES, T.B. MARDER AND K. WADE, Royal Society of Chemistry, 2000.

Coordination Chemistry of Aluminium, Ed. G. H. ROBINSON, VCH, 1993.

Advances in Thallium Aqueous Solution Chemistry, J. GLASER, *Advances in Inorganic Chemistry* **43** 1995, 1–78.

Chemistry of Aluminium, Gallium, Indium, and Thallium, Ed. A. J. DOWNS, Blackie, 1993.

Reactions of Group 13 Alkyls with Dioxygen and Elemental Chalcogens: From Carelessness to Chemistry, A.R. BARRON, *Chemical Society Reviews,* **22**, 1993, 93–99.

The Chemistry of GALLEX—Measurement of Solar Neutrinos with a Radiochemical Gallium Detector, E. HENRICH AND K. H. EBERT, *Angewandte Chemie, International Edition* **31**, 1992, 1283–1297.

The Race for Glittering Prizes, *Chemistry in Britain* **28**, 1992, 686–687. Fabrication of diamond films.

Silicon Nitride—From Powder Synthesis to Ceramic Materials, H. LANGE, G. WÖTTING, AND G. WINTER, *Angewandte Chemie, International Edition* **30**, 1991, 1579–1597.

Where are the Lone-pair Electrons in Subvalent Fourth-Group Compounds? S.-W. NG AND J. J. ZUCKERMAN, *Advances in Inorganic Chemistry and Radiochemistry* **29**, 1985, 297–326.

On the Surprising Kinetic Stability of Carbonic Acid (H_2CO_3), T. LOERTING, C. TAUTERMANN, R. T. KROEMER, I. KOHL, A. HALLBRUCKER, E. MAYER AND K. R. LIEDL, *Angewandte Chemie, International edition* **39**, 2000, 892–894.

Phosgene and Related Carbonyl Halides, T. A. RYAN, C. RYAN, E. A. SEDDON AND K. R. SEDDON, Elsevier, 1996.

Organotin Chemistry, A. G. DAVIES, VCH, 1997.

The Nitrogen Fluorides and Some Related Compounds, H. J. EMELÉUS, J. M. SHREEVE, AND R. D. VERMA, *Advances in Inorganic Chemistry* **33**, 1989, 139–196.

Solid-State Chemistry with Nonmetal Nitrides, W. SCHNICK, *Angewandte Chemie, International Edition*, **32**, 1993, 806–818.

The Shocking History of Phosphorus. A Biography of the Devil's Element, J. EMSLEY, Pan, 2001.

Organobismuth Chemistry, Eds. H. SUZUKI AND Y. MATANO, Elsevier, 2001.

The Chemistry of Arsenic, Antimony and Bismuth, Ed. N. C. NORMAN, Blackie, 1998.

The Stereochemistry of Sb(III) Halides and Some Related Compounds, J. F. SAWYER AND R. J. GILLESPIE, *Progress in Inorganic Chemistry* **34**, 65–113. Notes VSEPR relationships including lone-pair effects.

The Structures of the Group 15 Element(III) Halides and Halogenoanions, G. A. FISHER AND N. C. NORMAN, *Advances in Inorganic Chemistry* **41**, 1994, 223–271.

Recent Aspects of the Structure and Reactivity of Cyclophosphazenes, V. CHANDRASEKHAR AND K. R. J. THOMAS, *Structure and Bonding* **81**, Springer-Verlag, 1993, 41–113.

Applications of Hydrogen Peroxide and Derivatives, C. W. JONES, Royal Society of Chemistry, 1999.

The True Allotropes of Sulphur, G. RAYNER-CANHAM AND J. KETTLE, *Education in Chemistry*, 1991, 49–51.

There is No Such Thing as H_2SO_3, M. LAING, *Education in Chemistry*, 1993, 140.

Thiosulfate. An Interesting Sulfur Oxoanion That is Useful in Both Medicine and Industry—But is Implicated in Corrosion, S. W. DHAWALE, *Journal of Chemical Education* **70**, 1993, 12–14.

Developments in Chalcogen–Halide Chemistry, B. KREBS AND F.-P. AHLERS, *Advances in Inorganic Chemistry* **35**, 1990, 235–317.

Fluorine: The First Hundred Years (1886–1986) Eds. R. E. BANKS, D. W. A. SHARP AND J. C. TATLOW, Elsevier, 1987.

Fluorine Chemistry: *A Comprehensive Treatment*, Ed. M. HOWE-GRANT, Wiley, 1995.

Fluorine Chemistry at the Millenium: *Fascinated by Fluorine*, Ed. R. E. BANKS, Elsevier, 2000.

Astatine: Organonuclear Chemistry and Biomedical Applications, I. BROWN, *Advances in Inorganic Chemistry* **31**, 1987, 43–88.

The Chemistry of Iodine Azide, K. DEHNICKE, *Angewandte Chemie, International Edition* **18**, 1979, 507–514. See also: The Chemistry of the Halogen Azides, *Advances in Inorganic Chemistry and Radiochemistry* **26**, 1983, 201–234 (includes characterization and uses in organic and inorganic synthesis). See also: Covalent Inorganic Azides, I. C. TORNIEPORTH-OETTING AND T. M. KLAPÖTKE, *Angewandte Chemie, International Edition* **34**, 1995, 511–520.

A Renaissance in Noble Gas Chemistry, K. O. CHRISTE, *Angewandte Chemie, International Edition* **40**, 2001, 1419–1421.

Recent Advances in Noble Gas Chemistry, J. H. HOLLOWAY AND E. G. HOPE, *Advances in Inorganic Chemistry* **46**, 1999, 51–100.

A Noble Cause, G. M. R. GRANT, *Chemistry in Britain* 1994, 388–390.

One or Several Pioneers? The Discovery of Noble-Gas Compounds, P. LASZLO AND G. J. SCHROBILGEN, *Angewandte Chemie, International Edition* **27**, 1988, 479–489.

Radon: Not So Noble, J. D. LEE AND T. E. EDMONDS, *Education in Chemistry,* 1991, 152–154. See also: The Menace Under the Floorboards, A. F. GARDNER, R. S. GILLETT AND P. S. PHILLIPS, *Chemistry in Britain*, 1992, 344–348.

MAIN GROUP TOPICS (Chapter 18)

Chains, rings, nets and clusters

Cluster Molecules of the p-Block Elements, C. E. HOUSECROFT, Oxford University Press, 1994.

Cluster Chemistry, G. GONZÁLEZ-MORAGA, Springer-Verlag, 1993.

The Chemistry of Inorganic Homo- and Heterocycles, I. HAIDUC AND D. B. SOWERBY, Academic Press, 1987 (deals with Main Group species).

Main-Group-Based Rings and Polymers, D. P. GATES AND I. MANNERS, *Journal of the Chemical Society, Dalton Transactions*, 1997, 2525–2532.

Recent Advances in the Understanding of the Syntheses, Structures, Bonding and Energetics of the Homopolyatomic Cations of Groups 16 and 17, S. BROWNRIDGE, I. KROSSING, J. PASSMORE, H. D. B. JENKINS AND H. K. ROOBOTTOM, *Coordination Chemistry Reviews* **197**, 2000, 397–481.

Synthesis and Reactions of Phosphorus-rich Silylphosphines, G. FRITZ, *Advances in Inorganic Chemistry* **31**, 1987, 171–243.

Carbosilanes, G. FRITZ, *Angewandte Chemie*, *International Edition* **26**, 1987, 1111–1132.

Chemistry of the Polyhedral Boron Halides and the Diboron Tetrahalides, J. A. MORRISON, *Chemical Reviews* **91**, 1991, 35–48.

Early Carboranes and Their Structural Legacy, R. E. WILLIAMS, *Advances in Organometallic Chemistry* **36**, 1994, 1–55.

Open-Chain Polyphosphorus Hydrides (Phosphanes), M. BAUDLER AND K. GLINKA, *Chemical Reviews* **94**, 1994, 1273–1297.

Monocyclic and Polycyclic Phosphanes, M. BAUDLER AND K. GLINKA, *Chemical Reviews* **93**, 1993, 1623–1667. A comprehensive review on the structural units found in polyphosphorus species.

Clusters of Phosphorus: A Theoretical Investigation, M. HÄSER, U. SCHNEIDER AND R. AHLRICHS, *Journal of the American Chemical Society* **114**, 1992, 9551–9559.

Polyatomic Zintl Anions of the Post-transition Elements, J. D. CORBETT, *Chemical Reviews* **85**, 1985, 383–397.

The Polyborane, Carborane, Carbocation Continuum: Architectural Patterns, R. E. WILLIAMS, *Chemical Reviews* **92**, 1992, 177–207. Other reviews on boranes and carboranes are to be found in the same issue.

For Skeletal Electron Pairs see K. WADE, *Advances in Inorganic Chemistry and Radiochemistry* **18**, 1976, 1–66.

Solvents

Stabilisation of Unusual Cationic Species in Protonic Superacids and Acid Melts, T. A. O'DONNELL, *Chemical Society Reviews* **16**, 1987, 1–43.

Superacids, G. A. OLAH, G. K. S. PRAKASH AND J. SOMMER, Wiley, 1985.

Multiple bonds

Metal Element Triple Bonds of the Heavier Group 15 Elements, M. SCHEER, *Coordination Chemistry Reviews* **163**, 1997, 271–286.

π-Bonding and the Lone Pair Effect in Multiple Bonds Between Heavier Main Group Elements, P. P. POWER, *Chemical Reviews* **99**, 1999, 3463–3503.

Homonuclear Multiple Bonding in Heavier Main Group Elements, P. P. POWER, *Journal of the Chemical Society, Dalton Transactions*, 1998, 2939–2951.

Strained-Ring and Double-Bond Systems Consisting of the Group 14 Elements Si, Ge, and Sn, T. TSUMURAYA, S. A. BATCHELLER AND S. MASAMUNE, *Angewandte Chemie, International Edition* **30**, 1991, 902–939.

Unsaturated Molecules Containing Main Group Metals, M. VIETH, *Angewandte Chemie*, *International Edition* **26**, 1987, 1–14.

Organoelement Compounds with Al-Al, Ga-Ga, and In-In Bonds, W. UHL, *Angewandte Chemie, International Edition* **32**, 1993, 1386–1397.

The Chemistry of the Silicon–Silicon Double Bond, R. WEST, *Angewandte Chemie, International Edition* **26**, 1987, 1201–1211.

Transition Metal Complexes of Silylenes, Silenes, Disilenes, and Related Species, P. D. LICKISS, *Chemical Society Reviews* **21**, 1992, 271–279.

Multiply Bonded Germanium Species. Recent Developments, J. BARRAU, J. ESCUDIÉ AND J. SATGÉ, *Chemical Reviews* **90**, 1990, 283–319.

The Chemistry of Diphosphenes and Their Heavy Congeners: Synthesis, Structure, and Reactivity, L. WEBER, *Chemical Reviews* **92**, 1992, 1839–1906.

The Syntheses, Properties, and Reactivities of Stable Compounds Featuring Double Bonds Between Heavier Group 14 and 15 Elements, A. H. COWLEY AND N. C. NORMAN, *Progress in Inorganic Chemistry* **34**, 1986, 1–63.

Double Bonds Between Phosphorus and Carbon, R. APPEL AND F. KNOLL, *Advances in Inorganic Chemistry* **33**, 1989, 259–361.

Phosphoalkynes: New Building Blocks in Synthetic Chemistry, M. REGITZ, *Chemical Reviews* **90**, 1990, 191–213.

Synthesis of Di-, Tri-, and Polyphosphane and Phosphene Transition-Metal Complexes, A.-M. CAMINADE, J.-P. MAJORAL AND R. MATHIEU, *Chemical Reviews* **91**, 1991, 575–612.

Unconventional Multiple Bonds: Coordination Compounds of Unstable Vth Main Group Ligands, G. HUTTNER, *Pure and Applied Chemistry* **58**, 1986, 585–596.

VSEPR

Experimental observation of the Tellurium(IV) Bonding and Lone-pair Electron Density in Dimethyltellurium Dichloride by X-ray Diffraction Techniques, R. F. ZIOLO AND J. M. TROUP, *Journal of the American Chemical Society* **105**, 1983, 229–234.

A New Electrostatic Model of Molecular Shapes, J. L. BILLS AND S. P. STEED, *Inorganic Chemistry* **22**, 1983, 2401–2405.

The Valence Bond Interpretation of Molecular Geometry, D. W. SMITH, *Journal of Chemical Education* **57**, 1980, 106–109.

Directional Character, Strength, and Nature of the Hydrogen Bond in Gas-phase Dimers, A. C. LEGON AND D. J. MILLER, *Accounts of Chemical Research* **20**, 1987,39–46. 'dimer' = Mol···HX: H-bond is found in the direction expected for the lone pair position. This gives a different type of evidence for sterically active lone pairs.

A Theoretical Study of the Linear Versus Bent Geometry for Several MX_2 Molecules: MgF_2, CaH_2, CaF_2, CeO_2, and $YbCl_2$, R. L. DEKOCK, M. A. PETERSON, L. K. TIMMER, E. J. BAERENDS AND P. VERNOOIJS, *Polyhedron* **9**, 1990, 1919–1934.

Core Distortions and Geometries of the Difluorides and Dihydrides of Ca, Sr and Ba, I. BYTHEWAY, R. J. GILLESPIE, T.-H. TANG AND R. F. W. BADER, *Inorganic Chemistry* **34**, 1995, 2407.

Molecules with Hydride or Alkyl Ligands and Including d^0 Transition Metal Centers: Problem Cases for the Simple VSEPR Model, G. S. MCGRADY AND A. J. DOWNS, *Coordination Chemistry Reviews* **197**, 2000, 95–124. See also other reviews in this issue of the journal for VSEPR.

Bonding

Chemical Bonding in Higher Main Group Elements, W. KUTZELNIGG, *Angewandte Chemie, International Edition* **8**, 1969, 54–68.

The Rise and Decline of d Orbitals. Bonding in Hypervalent Compounds, O. J. CURNOW, *Chemistry in New Zealand,* 1996, September, 10–14; A Simple Qualitative Molecular-Orbital/Valence-Bond Description of the Bonding in Main Group "Hypervalent" Molecules, O. J. CURNOW, *Journal of Chemical Education* **75**, 1998, 910–915. (A hypervalent compound is essentially one where the octet is exceeded.) See also: The Chemistry of Hypervalent Compounds, J. I. MUSHER, *Angewandte Chemie, International Edition* **8**, 1969, 54–68.

Studies of Silicon-Phosphorus Bonding, K. J. DYKEMA, T. N. TRUONG AND M. S. GORDON, *Journal of the American Chemical Society* **107**, 1985, 4535–4541. An example of a characteristic *ab initio* calculation with interesting diagrams of calculated bond densities.

Silicates

For a more detailed account of silicate structures and for a more geochemical perspective see *Introduction to Mineral Sciences*, A. PUTNIS, Cambridge University Press, 1992.

Curiosity, Chance, Paradox and Perspective in the Chemistry of Materials, J. M. THOMAS, *Journal of the Chemical Society, Dalton Transactions,* 1991, 555–563. An account of developments in materials chemistry.

Solid Acid Catalysts, J. M. THOMAS, *Scientific American*, April 1992, 82–88. Use of clays, zeolites and polyoxometallate anions in catalysis.

Zeolite Molecular Sieves, B. M. LOWE, *Education in Chemistry* 1992, 15–18.

Catalysis in Intracrystalline Space, G. HUTCHINGS, *Chemistry in Britain,* 1992, 1006–1009. See also 991–994 of the same issue.

Zeolitic and Layered Materials, S. L. SUIB, *Chemical Reviews* **93**,1993, 803–826.

Metal-Phosphonate Chemistry, A. CLEARFIELD, *Progress in Inorganic Chemistry* **47**, 1998, 371–510.

Layered Metal Phosphates and Phosphonates: From Crystals to Monolayers, G. CAO, H.-G. HONG AND T. E. MALLOUK, *Accounts of Chemical Research* **25**, 1992, 420–427.

An Inorganic Double Helix: Hydrothermal Synthesis, Structure, and Magnetism of Chiral $[(CH_3)_2NH_2]K_4[V_{10}O_{10}(H_2O)_2(OH)_4(PO_4)_7]\cdot4H_2O$, V. SOGHOMONIAN, Q. CHEN, R. C. HAUSHALTER, J. ZUBIETA AND C. J. O'CONNOR, *Science* **259**, 1993, 1596–1599.

GENERAL TOPICS (Chapter 19)

Electron densities

Electron Density Distribution in Inorganic Compounds, K. TORIUMI AND Y. SAITO, *Advances in Inorganic Chemistry and Radiochemistry* **27**, 1983, 28–79.

Experimental Electron Densities and Chemical Bonding, P. COPPENS, *Angewandte Chemie, International Edition* **16**, 1977, 32–40.

Metal–sulfur and –selenium rings

Polysulphide Complexes of Metals, A. MULLER AND E. DIEMANN, *Advances in Inorganic Chemistry* **31**, 1987, 89–122.

Transition Metal Polysulphides: Coordination Compounds with Purely Inorganic Chelate Ligands, M. DRAGANJAC AND T. B. RAUCHFUSS, *Angewandte Chemie*, *International Edition* **24**, 1985, 742–757.

New Developments in the Coordination Chemistry of Inorganic Selenide and Telluride Ligands, L. C. ROOF AND J. W. KOLIS, *Chemical Reviews* **93**, 1993, 1037–1080. A comprehensive review illustrating the diversity in the coordination chemistry of these systems.

Caged Explosives: Metal-Stabilised Chalcogen Nitrides, P. F. KELLY, A. M. Z. SLAWIN, D. J. WILLAMS AND J. D. WOOLLINS, *Chemical Society Reviews* **21**, 1992, 245–252.

Fullerenes and related species

Fullerenes: Chemistry, Physics and Technology, Eds. K. M. KADISH AND R. S. RUOFF, Wiley-Interscience, 2000.

Lecture Notes on Fullerene Chemistry: *A Handbook For Chemists*, R. TAYLOR, Imperial College Press, 1999.

The Fullerenes. New Horizons for the Chemistry, Physics and Astrophysics of Carbon, Eds. H. W. KROTO AND R. M. WALTON, Royal Society, 1993.

Simple Generation of C_{60} (Buckminsterfullerene), D. W. LACOE, W. T. POTTER AND D. TEETERS, *Journal of Chemical Education* **69**, 1992, 663.

C_{60}: Buckminsterfullerene, The Celestial Sphere that Fell to Earth, H. W. KROTO, *Angewandte Chemie*, *International Edition* **31**, 1992, 111–246 A very readable and personal account of the discovery of the fullerenes.

Accounts of Chemical Research **25**, 1992, 97–175 A special issue on buckminsterfullerenes including sections on metal derivatives and fullerides.

The Chemistry of Fullerenes, R. TAYLOR AND R. M. WALTON, *Nature* **363**, 1993, 685–693.

Organometallic Derivatives of Fullerenes, J. R. BOWSER, *Advances in Organometallic Chemistry* **36**, 1994, 57–94.

Inorganic Nanoclusters with Fullerene-Like Structure and Nanotubes, R. TENNE, *Progress in Inorganic Chemistry* **50**, 2001, 269–316.

Organometallic Complexes of Fullerenes, A. H. H. STEPHENS AND M. L. H. GREEN, *Advances in Inorganic Chemistry* **44**, 1997, 1–44.

Atoms in Carbon Cages: The Structure and Properties of Endohedral Fullerenes, D. S. BETHUNE, R. D. JOHNSON, J. R. SALEM, M. S. DE VRIES AND C. S. YANNONI, *Nature* **366**, 1993, 123–128.

BIOLOGICAL, MEDICINAL AND ENVIRONMENTAL INORGANIC CHEMISTRY (Chapter 20)

Inorganic Biochemistry

Bioinorganic Chemistry: Inorganic Elements in the Chemistry of Life, W. KAIM AND B. SCHWEDERSKI, Wiley, 1995.

Principles of Inorganic Biochemistry, S. J. LIPPARD AND J. M. BERG, University Science Books, 1994.

Inorganic Biochemistry, J. A. COWAN, 2nd edition, Wiley-VCH, 1997.

The Biological Chemistry of the Elements. The Inorganic Chemistry of Life, J. J. R. FRAUSTO DA SILVA AND R. J. P. WILLIAMS, Clarendon Press, 1991.

Missing Information in Bio-Inorganic Chemistry, R. J. P. WILLIAMS, *Coordination Chemistry Reviews* **79**, 1987, 175–193.

Bio-inorganic Chemistry: Its Conceptual Evolution, R. J. P. WILLIAMS, *Coordination Chemistry Reviews* **100**, 1990, 573.

The Chemical Elements of Life, R. J. P. WILLIAMS, *Journal of the Chemical Society*, *Dalton Transactions*, 1991, 539–546.

Biomineralization – Principles and Concepts in Bioinorganic Materials Chemistry, S. MANN, Oxford University Press, 2001.

Biomineralization: The Hard Part of Bioinorganic Chemistry! S. MANN, *Journal of the Chemical Society*, *Dalton Transactions*, 1993, 1–9. See also Solid-State Bioinorganic Chemistry: Mechanisms and Models of Biomineralization, S. MANN AND C. C. PERRY, *Advances in Inorganic Chemistry* **36**, 1991, 137–200.

The Role of Silicon in Biology, J. D. BIRCHALL, *Chemistry in Britain*, 1990, 141–144.

Nitric Oxide in Biology: Its Role as a Ligand, R. J. P. WILLIAMS, *Chemical Society Reviews*, 1996, 77–83.

Nitric Oxide—Small but Powerful, A. R. BUTLER, *Education in Chemistry*, 1993, 120.

The Surprising Life of Nitric Oxide, P. L. FELDMAN, O. W. GRIFFITH AND D. J. STUEHR, *Chemical* & *Engineering News*, Special Report, December 1993, 26–38.

Biological Roles of Nitric Oxide, S. H. SNYDER AND D. S. BREDT, *Scientific American*, May 1992, 28–35.

Chemistry Relevant to the Biological Effects of Nitric Oxide and Metallonitrosyls, M. J. CLARKE AND J. B. GAUL, *Structure and Bonding* **81**, Springer-Verlag, 1993, 147–181.

For iron–sulfur proteins, see volume 38 of *Advances in Inorganic Chemistry*, published in 1992. The entire issue is devoted to this topic.

Les Séismes Moléculaires de L'Hémoglobine, J.-L. MARTIN AND J.-C. LAMBY, *La Recherche* **24**, 1993, 572–575.

For a whole issue of *Chemical Reviews* dedicated to Bioinorganic Enzymology, see volume 96, issue 7 (1996).

Iron Carriers and Iron Proteins, Ed. T. M. LEOHR, Wiley, 1989.

Synthetic Models for Hemoglobin and Myoglobin, J. P. COLLMAN AND L. FU, *Accounts of Chemical Research* **32**, 1999, 455–463.

Determining the Structure of a Hydroxylase Enzyme That Catalyses the Conversion of Methane to Methanol in Methanotropic Bacteria, A. C. ROSENZWEIG AND S. J. LIPPARD, *Accounts of Chemical Research* **27**, 1994, 229–236.

For model complexes see also: Oxo- and Hydroxo-Bridged Diiron Complexes: A Chemical Perspective on a Biological unit, D. M. KURTZ, JR., *Chemical Reviews* **90**, 1990, 585–606.

Bioinorganic Chemistry of Vanadium, D. REHDER, *Angewandte Chemie*, *International Edition* **30**, 1991, 148–167. See also: Vanadium: A Biologically Relevant Element, R. WIEVER AND K. KUSTIN, *Advances in Inorganic Chemistry* **35**, 1990, 81–115.

Intrinsic Properties of Zinc(II) Ion Pertinent to Zinc Enzymes, E. KIMURA AND T. KOIKE, *Advances in Inorganic Chemistry* **44**, 1997, 229–263.

The Bioinorganic Chemistry of Nickel, Ed. J. R. LANCASTER, VCH, 1998.

The Bioinorganic Chemistry of Magnesium, Ed. J. A. COWAN, VCH, 1995.

The Chemistry of Synthetic Fe-Mo-S Clusters and Their Relevance to the Structure and Function of the Fe-Mo-S Center in Nitrogenase, S. M. MALINAK AND D. COUCOUVANIS, *Progress in Inorganic Chemistry* **49**, 2001, 599.

Reactions of Small Molecules at Transition Metal Sites: Studies Relevant to Nitrogenase, An Organometallic Enzyme, R. L. RICHARDS, *Coordination Chemistry Reviews* **154**, 1996, 83–98.

Molybdenum Enzymes, Ed. T. SPIRO, Wiley, 1985. See also The Mo-, V-, and Fe-Based Nitrogenase Systems of Azotobacter, R. E. EADY, *Advances in Inorganic Chemistry* **36**, 1991, 77–102.

Metals in the Nitrogenases, R. R. EADY AND G. J. LEIGH, *Journal of the Chemical Society, Dalton Transactions*, 1994, 2739–2747

Medicinal Inorganic Chemistry

Medicinal Inorganic Chemistry, Z. GUO AND P. J. SADLER, *Advances in Inorganic Chemistry* **49**, 2001, 183–306.

Inorganic Chemistry and Drug Design, P.J. SADLER, *Advances in Inorganic Chemistry* **36**, 1991, 1–48. See also 49–75 for a review on the use of lithium compounds in medicine.

Targeting Metal Complexes, D. PARKER, *Chemistry in Britain*, 1994, 818–822.

Medical Diagnostic Imaging with Complexes of ^{99m}Tc, M. J. CLARKE AND L. PODBIELSKI, *Coordination Chemistry Reviews* **78**, 1987, 253–330.

Nuclear Medicine and Positron Emission Tomography: an Overview, T. J. MCCARTHY, S. W. SCHWARZ, AND M. J. WELCH, *Journal of Chemical Education* **71**, 1994, 830–836.

Technetium Chemistry and Technetium Radiopharmaceuticals, E. DEUTSCH, K. LIBSON, S. JURASSIN AND L. F. LINDOY, *Progress in Inorganic Chemistry* **31**, 1984, 75–139. See also Coordination Compounds in Nuclear Medicine, S. JURISSON, D. BERNING, W. JIA AND D. MA, *Chemical Reviews* **93**, 1993, 1137–1156; Technetium Radiopharmaceuticals–Fundamentals, Synthesis, Structure, and Development, K. SCHWOCHAU, *Angewandte Chemie, International Edition* **33**, 1994, 2258–2267.

Inorganic Chemistry and Medicine, P. J. SADLER, *Education in Chemistry*, May 1992, 80–83. An excellent concise summary of this topic.

Metal Compounds in Therapy and Diagnosis, M. J. ABRAMS AND B.A. MURRER, *Science* **261**, 1993, 725–730

Trace Element Medicine and Chelation Therapy, D. M. TAYLOR AND D. R. WILLIAMS, Royal Society of Chemistry, 1995

The Chemistry of Contrast Agents in Medical Magnetic Resonance Imaging, Eds. A. E. MERBACH AND I. TOTH, Wiley, 2001.

Metal Compounds in Cancer Chemotherapy, I. HAIDUC AND C. SILVESTRU, *Coordination Chemistry Reviews* **99**, 1990, 253–296.

Non-Platinum Group Metal Antitumor Agents: History, Current Status, and Perspectives, P. KÖPF-MAIER AND H. KÖPF-MAIER, *Chemical Reviews* **87**, 1987, 1137–1152.

Boron Neutron Capture Therapy for Cancer, R. F. BARTH, A. H. SOLOWAY AND R. G. FAIRCHILD, *Scientific American*, October 1990, 100–107

The Role of Chemistry in the Development of Boron Neutron Capture Therapy of Cancer, M. F. HAWTHORNE, *Angewandte Chemie, International Edition* **32**, 1993, 950–984.

The Chemistry of the Gold Drugs Used in the Treatment of Rheumatoid Arthritis, D. H. BROWN AND W. E. SMITH, *Chemical Society Reviews* **9**, 1980, 217–240.

MRI Primer, W. OLDENDORF AND W. OLDENDORF, JR. Raven Press, 1991. Theory, methods and applications of magnetic resonance imaging.

The Chemistry of Chelating Agents in Medical Sciences, R. A. BULMAN, *Structure and Bonding* **67**, Springer-Verlag, 1987.

Environmental Inorganic Chemistry

Environmental Chemistry, S. E. MANAHAN, Lewis Publishers, 5th Edition, 1991.

Environmental Chemistry, I. WILLIAMS, Wiley, 2001.

Environmental Chemistry. A Global Perspective, G. W. van LOON AND S. J. DUFFY, Oxford University Press, 2000.

Chemistry of Atmospheres, R. P. WAYNE, Oxford University Press, 2000.

For discussion on the relative environmental impact of phosphates versus zeolites, see *The Phosphate Report* published by Landbank Environmental Research and Consulting, 1994. A summary is given in *New Scientist,* 5th February 1994, 10. See also *Science* **263**, 1994, 1086, *New Scientist,* 17 April 1993, 7 and Phosphates go Full Cycle, M. BURKE, *Chemistry in Britain*, March 2002, 460.

Ozone Depletion: 20 Years After The Alarm, *Chemical & Engineering News*, August 15, 1994, 8–13. A good historical account of the discovery of ozone depletion, and subsequent measures taken.

Polar Stratospheric Clouds and Ozone Depletion, O. B. TOON AND R. P. TURCO, *Scientific American*, June 1991, 40–47.

Towards a Laboratory Strategy for the Study of Heterogeneous Catalysis in Stratospheric Ozone Depletion, M. R. S. MECOUSTRA AND A. B. HORN, *Chemistry Society Reviews,* 1994, 195–204.

New Routes to Alternative Halocarbons, G. WEBB AND J. WINFIELD, *Chemistry in Britain,* 1992, 996–997.

Environmental Issues—CFC Alternatives, R. TWEDDLE, *Education in Chemistry*, January 1995, 17–19.

Making Sure That Hydrofluorocarbons are "Ozone Friendly", J. S. FRANCISCO AND M. M. MARICQ, *Accounts of Chemical Research* **29**, 1996, 391–397.

Atmospheric Lifetime, Its Application and Its Determination: CFC-Substitutes as a Case Study, A. R. RAVISHANKARA AND E. R. LOVEJOY, *Journal of the Chemical Society, Faraday Transactions* **90**, 1994, 2159–2169.

Ocean Carbon Cycle, J. I. SARMIENTO, *Chemical & Engineering News,* Special Report, May 1993, 30–43.

Aluminium: A Neurotoxic Product of Acid Rain, R. B. MARTIN, *Accounts of Chemical Research* **27**, 1994, 204–210.

The Sulfur Problem: Cleaning Up Industrial Feedstocks, D. STIRLING, Royal Society of Chemistry, 2000.

GROUP THEORY (Appendix C)

Introductions to Group Theory can be found in the following:

Beginning Group Theory for Chemistry, P. H. WALTON, Oxford University Press, 1998.

Molecular Symmetry and Group Theory, R. L. CARTER, Wiley, 1998.

Molecular Symmetry and Group Theory: A Programmed Introduction to Chemical Applications, A. VINCENT, Wiley, 2nd Edition, 2001.

Group Theory for Chemists, G. DAVIDSON, Macmillan, 1991.

Chemical Applications of Group Theory, F. A. COTTON, Interscience, 2nd Edition, 1971.

(4) Electronic Access to Chemical Information

The World Wide Web can be used to find information on almost any topic, and this task is greatly facilitated by the use of a search engine. In this section, we give details solely on a limited selection of websites which the reader might find useful, including sources of chemical information, chemical societies, and publishers of other chemistry textbooks. To find links to other chemistry websites on the Internet the site http://www.liv.ac.uk/Chemistry/Links/intro.html is useful.

General websites with chemical information

Web Elements (http://www.webelements.com) has an electronic Periodic Table, together with much chemical and physical information on the elements.

Visual Elements Periodic Table (http://www.chemsoc.org/viselements): a visual interpretation of the chemical elements in Periodic Table format, supported by excellent supplementary information and data, from the Royal Society of Chemistry.

Websites of selected chemical societies

Royal Society of Chemistry: http://www.rsc.org

American Chemical Society: http://www.acs.org

See also the website of the International Union of Pure and Applied Chemistry (IUPAC): http://iupac.chemsoc.org. This also has links to other chemistry websites.

Publishers

For information on the latest inorganic chemistry and other textbooks, see:

Nelson Thornes: http://www.nelsonthornes.com

Wiley: http://www.wiley.com

Elsevier: http://www.elsevier.nl

Oxford University Press: http://www.oup.co.uk

Academic Press: http://www.apnet.com

Wiley-VCH: http://wiley-vch.de

CRC Press: http://www.crcpress.com

Addison Wesley & Benjamin Cummings: http://www2.awl.com

Royal Society of Chemistry: http://www.rsc.org

Freeman: http://www.whfreeman.co.uk

Appendix B

Some Common Polydentate Ligands

Ligand	*Formula*	*Mode of Coordination*
BIDENTATE LIGANDS		
Ethylenediamine (en)	$H_2NCH_2CH_2NH_2$	Through both N atoms giving a five-membered ring (Figure 13.13a)
Dicarboxylic acids (and S analogues)	$(CH_2)_n(COOH)_2$ ($n=0, 1, 2$ etc)	A proton is lost from each COOH group and the two O^- atoms co-ordinate giving a $(5+n)$ ring (compare Figure 13.17 for $n=0$)
Acetylacetone (acac)	$CH_3C(OH)=CHCOCH_3$ (in enol form)	Enol proton lost and coordination is through both O atoms (Figures 11.2, 12.5, 14.13)
(and other β-diketones: also as a monodentate ligand in keto form)		
8-quinolinol (oxine or 8-hydroxyquinoline)		Proton lost and coordinates through O and N giving five-membered ring (Figure 13.13d)
Biuret	$H_2NCONHCONH_2$	One NH_2 proton lost and coordinates through two outer N atoms to give a six membered ring
Dimethylglyoxime (DMG)	$CH_3C(=NOH)C(=NOH)CH_3$	One proton lost and coordinates through both N atoms giving five-membered ring. Further six-membered rings formed by hydrogen bonding between NOH and ON (Figure 14.30)
Salicylaldehyde (salic)		Hydroxyl proton lost and bonds through both O atoms giving six-membered ring (Figure 10.4)
(Similarly other disubstituted benzenes. For catechol, 1,2-dihydroxybenzene, $C_6H_4(OH)_2$, bonding is through 2O giving a five-membered ring—compare Figure 20.21 for a polynuclear example. In the salicylaldimines, $C_6H_4(OH)(CH=NR)$ for R = H, alkyl or OH, bonding is through O and N giving a six-membered ring (Figure 13.13e).)		
2:2′-dipyridyl (bipy)		Through both N atoms giving five-membered ring (Figure 14.7)
(Also numerous substituted dipyridyls and analogues)		

Ligand	*Formula*	*Mode of Coordination*
1:10-phenanthroline (phenan)	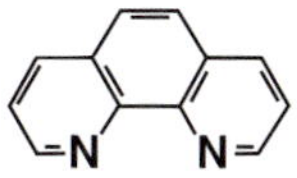	As above
Dithiols	$HS(CH_2)_nSH$; $HSCH_2(SH)CH_2R$; *o*-$(C_6H_4)(SH)_2$ 1,2 unsaturated thiols $RC(SH){=}CR(SH)$ and dithioketones $R{-}C({=}S){-}C({=}S){-}R$	Loss of one proton and coordination through two S atoms generally giving four-, five-, or six-membered rings
Diphosphanes (diphos) and diarsanes (diars)	$R_2MCH_2CH_2MR_2$ or 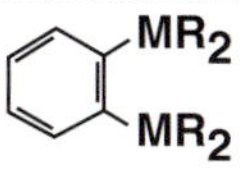 M = P or As	Through two P or two As (or P and As) giving five-membered rings
Dimethoxyethane (DME) (glyme)	$MeOCH_2CH_2OMe$	Through two O atoms giving a five-membered ring
TRIDENTATE LIGANDS		
Diethylenetriamine (dien)	$H_2NCH_2CH_2NHCH_2CH_2NH_2$	Through three N atoms to the same M ion giving two five-membered rings (Figure 13.13c)
2:6-bis (α-pyridyl) pyridine (terpyridine or terpy)	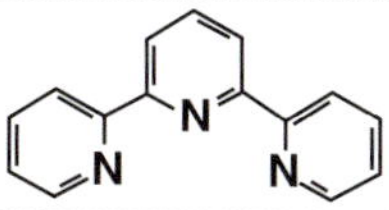	To one metal through three N atoms giving two five-membered rings
Oxydiacetic acid (oda)	$O{=}C(OH){-}CH_2{-}O{-}CH_2{-}C(OH){=}O$	Loss of two protons from OH and coordination through these two O and the central one to give two five-membered rings
Triarsanes (triars)	e.g. $Me_2As(CH_2)_3As(Me)(CH_2)_3AsMe_2$	To one metal atom through three As atoms giving two six-membered rings
Similar molecules with three As, three P or three (As+P) atoms		
QUADRIDENTATE LIGANDS		
Porphyrins and phthalocyanins	Compare Figures 11.5, 20.1 and 20.4	Through four N atoms forming square-planar coordination to M
Bis(acetylacetone) ethylenediamine (acacen) (a Schiff's base)	$MeC(OH){=}CHC(Me){=}NCH_2CH_2N{=}C(Me)CH{=}C(OH)Me$	Two protons lost (from OH) and bonds through two O and two N atoms to an M ion forming two six-membered rings and one five-membered ring (compare Figure 13.14 for a related ligand)
Triethylenetetramine Compare also cyclic analogues, Figure 14.31.	$H_2NCH_2CH_2NHCH_2CH_2NHCH_2CH_2NH_2$	Through four N atoms forming four five-membered rings around a metal atom
Triaminotriethylamine (tren)	$N(CH_2CH_2NR_2)_3$ where R = H or alkyl	Through four N atoms as above
Also similar tetrarsanes and tetraphosphanes where N is replaced by As or P		
Tris (*o*-diphenylarsano-phenyl)arsane (QAS) Also the phosphane where the central As is replaced by P	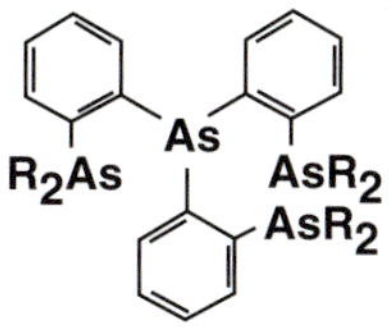	Through four As atoms giving four five-membered rings. The geometric requirements of the ligand give it a pyramidal coordination

Ligand	Formula	Mode of Coordination
QUINQUEDENTATE LIGANDS		
Tetraethylenepentamine (tetren)	$H_2NCH_2CH_2NHCH_2CH_2NHCH_2CH_2NHCH_2CH_2NH_2$	To one metal through all five N atoms
The Schiff's base	$CH{=}NCH_2CH_2NHCH_2CH_2N{=}CH$ (with OH and HO substituents on the terminal CH groups)	Loses two protons and bonds through three N and two O atoms. (Compare Figure 13.14 for the corresponding tetradentate ligand)
SEXADENTATE LIGANDS		
Ethylenediaminetetra acetic acid (EDTA)	$(HO_2CCH_2)_2NCH_2CH_2N(CH_2CO_2H)_2$	Loses four H^+ and bonds through four O and two N atoms to give six five-membered rings (Figure 10.5)
Pentaethylenehexamine	$H_2NCH_2CH_2(NHCH_2CH_2)_4CH_2CH_2NH_2$	To one metal through all six N atoms
Schiff's bases e.g.	$C_6H_4{=}NCH_2CH_2SCH_2CH_2SCH_2CH_2N{=}C_6H_4$ (with OH and HO substituents on the terminal C_6H_4 groups)	Loses two protons and bonds to one metal through two O, two S and two N atoms giving five rings
Also the corresponding compounds with NH in place of S, or with pyridine in place of the $-C_6H_4$-OH ring. (Compare Figure 13.14 for a tetradentate analogue.)		
An octadentate ligand Diethylenetriamine pentaacetic acid (Compare EDTA)	$(HOOCCH_2)_2NCH_2CH_2N(CH_2COOH)CH_2CH_2N(CH_2COOH)_2$	Potentially 8-coordinating through 5O and 3N atoms. For use in plutonium chelation therapy see Section 20.2.8
POLYDENTATE LIGAND FAMILIES		
Several families of ligands with the same unit structure and of increasing size are in use. The size, coordination number, and ring size may be optimized for bonding to a particular species. Commonly used are		
Linear polyethers (polyglymes—compare the bidentate glyme above)	$MeO(CH_2CH_2O)_n$ Me (diglyme, DIME; triglyme, etc)	Bond through $(n+1)$ O atoms giving five-membered rings
Cyclic polyethers Crown ethers	$(CH_2CH_2O)_n$ and related molecules	n-dentate ligands bonding through O (Figure 10.6a)
Cryptates	NZ_3N species where Z is $(CH_2CH_2O)_n$ or S,Se, etc., analogues	Bonding through N and O,S,Se to give multiple rings, usually five membered (Figure 10.6b)

Appendix C Molecular Symmetry and Point Groups

The symmetry of a molecule or ion is a property of fundamental importance in more advanced study of its properties. For example, the choice of atomic orbitals which may be combined into molecular σ or π orbitals is restricted by symmetry considerations. Thus, the basic reason why the combinations indicated in Fig. 3.11 are not allowed is that the two orbitals in each pair belong to different symmetry classes in a diatomic molecule. Similarly, the number and type of transitions expected in the electronic or vibrational spectra of a molecule depend fundamentally on the molecular symmetry. For example, in Section 7.7, the predictions of the number of fundamentals in the infrared and Raman spectra which are indicated there are derived solely from the consideration of the molecular symmetry.

While it is not our purpose, in an introductory text, to discuss these topics it is useful for the student to be able to determine the symmetry of a molecule at an early stage. This may be done quite simply, as explained below, and can conveniently be tackled in conjunction with the determination of molecular shapes discussed in Chapter 4. Familiarity with the nomenclature of symmetry, and the ability to determine the formal symmetry of a molecule, are the necessary first steps to a fuller understanding of bonding and spectroscopy.

There are two steps in the process: first we define symmetry elements and symmetry operations and then the sum of the symmetry elements is used to determine the point group to which the molecule belongs.

Symmetry elements and symmetry operations

A symmetry *operation* is some transformation of the molecule, such as a rotation or a reflection, which leaves the molecule in a configuration in space which is indistinguishable from its initial configuration. A symmetry *element* is that point, line or plane in the molecule about which the symmetry operation takes place. The number of symmetry elements and operations which apply to single, real molecules is quite small and these are detailed in Table C.1.

These operations will be more readily understood from a few examples. It is very much easier to follow the description using a molecular model and it would be well worth the reader's while to make models and carry out the symmetry operations described.

Consider first the ammonia molecule which is pyramidal in shape, Fig. C.1. There is a three-fold axis passing through the nitrogen atom and perpendicular to the plane containing the three hydrogens. Rotation about this axis by $360°/3 = 120°$ leaves the N unchanged and moves each H into the position of the next. The resulting configuration is indistinguishable (though different) from the original one. The molecule may also be rotated twice about this axis by 120° to produce another indistinguishable configuration, but if the molecule is rotated three times successively by 120° it is returned to the original configuration. This symmetry element is labelled C_3, and the operations are distinguished as C_3, C_3^2, and C_3^3. As the last produces the original configuration, we may write

TABLE C.1 Symmetry elements and symmetry operations

Element	*Symbol*	*Operation*
Identity	E	Leaves each particle in its original position
n-fold axis (proper axis)	C_n	Rotation about the axis by 360°/n, or by some multiple of this. Only $n = 1, 2, 3, 4, 5, 6$ and ∞ need be considered for real molecules.
Plane	σ	Reflection in the plane
Centre	i	Inversion through the centre
n-fold alternating axis (improper axis)	S_n	Rotation by 360°/n (or by a multiple) followed by reflection in a plane perpendicular to the axis.

$$C_3^3 = E$$

Similarly, $C_3^4 = C_3$, $C_3^5 = C_3^2$ and so on.

In the NH_3 molecule, there are also three planes of symmetry. Each one contains the N atom and one H, and bisects the angle between the other two H atoms. Reflection in this plane leaves the N and contained H unaltered, and exchanges the other two H atoms. Clearly, if this reflection operation is repeated, the original configuration is restored. That is, $\sigma^2 = E$.

The symmetry elements of the ammonia molecule are thus E, C_3, and the three planes. It is the convention to align the molecule so that the axis of symmetry is vertical. Then a symmetry plane which contains this axis is a *vertical plane*, symbol σ_v, while a plane perpendicular to the axis is a *horizontal plane* with the symbol σ_h. Thus the symmetry planes in ammonia are vertical planes.

If more than one symmetry axis is present, the one of highest order (highest value of n) is termed the *principal axis,* and this is the one which is placed vertically. There are a few cases where a molecule has improper axes but no proper axis, and then the alternating axis of highest order is chosen as the principal axis.

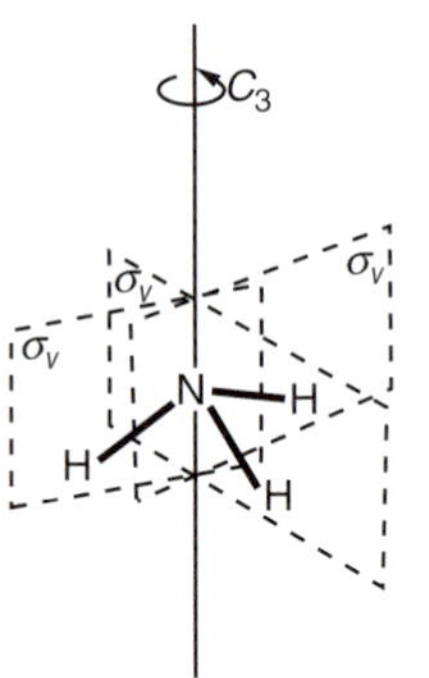

FIG. C.1 Symmetry elements of NH_3. Ammonia belongs to the $\mathbf{C_{3v}}$ point group. The C_3 axis passes through N and the midpoint of the triangle defined by the three H atoms. There are three vertical planes, σ_v, each containing C_3 and one N-H bond and bisecting the opposite HNH angle

As a further example, consider BF_3 which is planar, Fig. C.2. This molecule has the C_3 axis through the B atom, and the three vertical planes, each containing one BF bond, which correspond to the symmetry elements found in NH_3. In addition, there are three C_2 axes, one along each B-F bond. A rotation of 360°/2 about such an axis leaves the contained F and B unchanged, and interchanges the other two F atoms. Furthermore, the plane of the molecule is a symmetry plane as reflection in it leaves all the atoms unchanged. By the definitions above, the principal axis is the C_3 axis, as this is the axis of highest order. Thus the plane of the molecule is a horizontal plane, σ_h, while the planes through the BF bonds are vertical planes. Further, as there is a C_3 axis and a horizontal plane of symmetry, there is necessarily an S_3 axis coincident with the C_3 one. Thus, the symmetry elements of BF_3 are E, C_3, $3C_2$, σ_h, $3\sigma_v$ and S_3.

A centre of symmetry (or an inversion centre) is a point in a molecule such that, if an atom is moved from its position, through the centre of inversion for an equal distance in the same direction, it lands in the position of an identical atom. For example, the centre of an octahedron or of a *trans*-MA_2B_4 species is a centre of symmetry. There is also a centre at the mid-point of the C-C bond in ethane in the staggered configuration. On the other hand, there is no inversion centre in a tetrahedron, or in ethane in the eclipsed configuration.

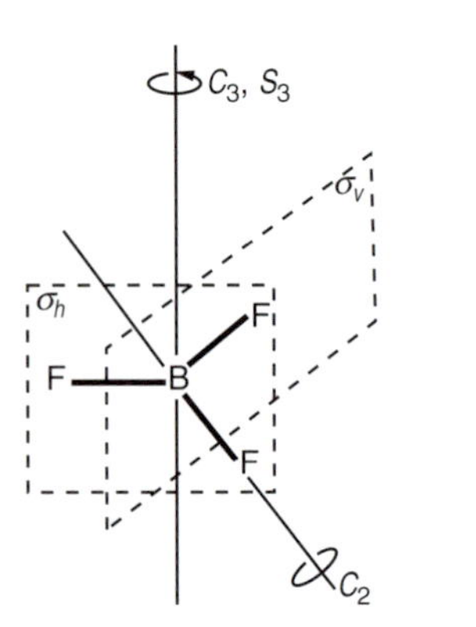

FIG. C.2 Symmetry elements of BF_3. Boron trifluoride belongs to the point group $\mathbf{D_{3h}}$. This has the elements, C_3 (through B perpendicular to the plane of the molecule) and three σ_v (containing one B-F bond and bisecting the opposite angle) similar to those in $\mathbf{C_{3v}}$ (Figure C.1) together with a horizontal plane of symmetry, σ_h (which is the plane of the molecule), an S_3 axis coincident with C_3, and three C_2 axes, one along each B-F bond. Only one of the three vertical planes and one of the two-fold axes is shown

While an alternating axis is necessarily present when both a proper axis and a horizontal plane are present, as in the case of the S_3 axis in BF_3, the alternating axis is an independent symmetry element and it may be present when neither the proper axis nor the plane exist as symmetry elements. An example is provided by ethane in the staggered configuration. If the C-C bond is taken as an axis, rotation by 60° followed by reflection in a plane perpendicular to the C-C bond and containing its mid-point will exchange the two carbons and the hydrogens. Thus there is an S_6 axis along the C-C bond while the highest order proper axis is only C_3 along this bond. There is no horizontal plane of symmetry in staggered ethane. It may be noted that a centre of inversion, i, is equivalent to an S_2 axis.

Point groups

When dealing with a crystal, with extended repeated units, extra symmetry operations become possible which transform one molecule into the neighbouring position which, in an infinitely extended lattice, also gives rise to an indistinguishable configuration. We shall not discuss such elements here.

The complete list of symmetry operations which may be performed on a molecule serves to define the *point group* to which the molecule belongs. Conversely, knowing the point group is equivalent to knowing all the symmetry operations. Thus the fact that ammonia may be subjected to the operations E, C_3, C_3^2 and three different σ_v reflections determines that $\mathbf{NH_3}$ belongs to the $\mathbf{C_{3v}}$ point group.

The point groups are named by symbols, which are related to those for the symmetry elements, and are usually distinguished in print by the use of bold type. The point groups which span molecules are the following:

(a) $\mathbf{C_n}$ to which belong molecules with only a proper axis of symmetry C_n.

Note that $\mathbf{C_1}$ contains only $E = C_1$, and is the point group in which molecules with no symmetry at all are placed.

(b) $\mathbf{C_{nv}}$ to which belong molecules with a C_n axis and n vertical planes only.

Note that, if there is one plane, there must be n as the n-fold rotation about the axis transforms one plane into the next.

(c) $\mathbf{C_{nh}}$ to which belong molecules with only a C_n axis, a horizontal plane, σ_h, and the resulting S_n axis. Note that $\mathbf{C_{1h}} \equiv \mathbf{C_{1v}}$ is more commonly labelled $\mathbf{C_s}$.

(d) $\mathbf{D_n}$ is the point group to which belong molecules containing only one C_n axis together with n C_2 axes.

Note that again, if there is one C_2 axis there must be n of them.

(e) $\mathbf{D_{nd}}$ covers molecules with one C_n and n C_2 axes, together with n vertical planes, and the S_{2n} axis which these imply. Note that the subscript is **d** (for dihedral) and not **v**.

(f) $\mathbf{D_{nh}}$ to which belong molecules with C_n and n C_2 axes, together with a horizontal plane. These elements imply the presense of an S_n axis and n vertical planes.

In chemical structures, axes of higher order than six are extremely rare, thus the point groups specified under (a) to (f) have $\mathbf{n} = 1$, 2, 3, 4, 5 and 6, apart from linear molecules where $\mathbf{n}$ is infinity.

(g) $\mathbf{S_n}$. A few cases exist where molecules have no planes of symmetry and the highest order axis is S_n, together with the $C_{n/2}$ which this implies. Only even values of n are found. $\mathbf{S_2}$ is the group with one S_2 axis, which is equivalent to i, and this group is usually labelled $\mathbf{C_i}$.

(h) $\mathbf{T_d}$, $\mathbf{O_h}$, $\mathbf{I_h}$ are respectively the labels for groups with full tetrahedral, octahedral or icosahedral symmetry. These have large numbers of symmetry elements, listed in Table C.2, and it is impossible not to recognize them. There are a few related groups, such as **T** which has the axes but not the planes of a full tetrahedron, but these are very rarely encountered.

In the discussion which follows, the groups $\mathbf{S_4}$, $\mathbf{S_6}$, $\mathbf{S_8}$ from (g) and **T**, $\mathbf{T_h}$ and **O** which are related to those in (h) are omitted as they are seldom represented by real molecules in their most stable configurations.

Although the more symmetric groups are defined by quite large numbers of symmetry operations, the fact that the presence of the element automatically implies the presence of all the operations of that element (including those giving rise to related elements such as $C_4^2 = C_2$) means that the point group may be decided by considering the symmetry elements. Furthermore, as the presence of certain elements implies the presence of others (as the C_3 and σ_h in BF_3 implied the S_3), the point group to which a molecule belongs may be *diagnosed* by looking for a small number of elements *in a particular order*. If the diagnostic elements are present, then the other symmetry elements, and all the corresponding symmetry operations, are necessarily present.

It then becomes a simple matter to determine the point groups to which a molecule belongs by looking for symmetry elements in a fixed order. This order is:

(1) Find the principal axis.
(2) Look for n C_2 axes perpendicular to the principal axis.
(3) Look for horizontal planes.
(4) Look for vertical planes.
(5) If none of the above are present, check for a centre of inversion.

It is important that the search is always carried out in this order or higher symmetry groups may be missed because too much weight is put on minor elements. Notice, for example, that $\mathbf{D_{nh}}$ groups have vertical as well as horizontal planes of symmetry so that, if vertical planes are looked for before the horizontal plane is excluded, $\mathbf{D_{nh}}$ might be mistakenly assigned as $\mathbf{D_{nd}}$.

TABLE C.2 Some common point groups

Point group	*Diagnostic elements*	*Other elements*	*Examples*
$\mathbf{C_1}$	E only		SiHClBrI
$\mathbf{C_s}$	E, σ only		SiH_2ClBr
$\mathbf{C_i}$	E, i only		*trans*-HClBrSiSiBrClH
$\mathbf{C_2}$	E, C_2 only		H_2O_2 (non-planar)
$\mathbf{C_{2v}}$	E, C_2, $2\sigma_v$		H_2O, SiH_2Cl_2
$\mathbf{C_{3v}}$	E, C_3, $3\sigma_v$		NH_3, $SiHCl_3$
$\mathbf{C_{4v}}$	E, C_4, $4\sigma_v$	$C_2 = C_4^2$	BrF_5, SF_5Cl
$\mathbf{C_{2h}}$	E, C_2, σ_h	i	*trans*-$C_6H_2Cl_2Br_2$
$\mathbf{C_{3h}}$	E, C_3, σ_h		$B(OH)_3$ in form (H–O, O–H, B, O, H)
$\mathbf{D_2}$	E, C_2, $2C_2$		
$\mathbf{D_{2d}}$	E, C_2, $2C_2$, $2\sigma_v$	S_4	$H_2C{=}C{=}CH_2$
$\mathbf{D_{3d}}$	E, C_3, $3C_2$, $3\sigma_v$	i, S_6	C_2H_6, Si_2Cl_6 (staggered)
$\mathbf{D_{4d}}$	E, C_4, $4C_2$, $4\sigma_v$	$C_2 = C_4^2$, S_8	S_8 (puckered ring)
$\mathbf{D_{5d}}$	E, C_5, $5C_2$, $5\sigma_v$	i, S_{10}	$(C_5H_5)_2Fe$ (staggered)
$\mathbf{D_{2h}}$	E, C_2, $2C_2$, σ_h	i, $2\sigma_v$	B_2Cl_4, *trans*-$A_2B_2C_2M$
$\mathbf{D_{3h}}$	E, C_3, $3C_2$, σ_h	S_3, $3\sigma_v$	BF_3, PF_5
$\mathbf{D_{4h}}$	E, C_4, $4C_2$, σ_h	i, S_4, C_2, $4\sigma_v$	$PtCl_4^{2-}$, *trans*-A_2B_4M
$\mathbf{D_{5h}}$	E, C_5, $5C_2$, σ_h	S_5, $5\sigma_v$	C_5H_5, $(C_5H_5)_2Ru$ (eclipsed)
$\mathbf{D_{6h}}$	E, C_6, $6C_2$, σ_h	i, S_6, S_3, C_3, C_2, $6\sigma_v$	C_6H_6
$\mathbf{T_d}$	E, $4C_3$, $3C_2$, $3S_4$, and $6\sigma_v$		SiH_4, $GeCl_4$, $TiCl_4$
$\mathbf{O_h}$	E, $3C_4$, $4C_3$, $6C_2$, i, $3S_4$, $4S_6$, $3\sigma_h$, $6\sigma_v$		SF_6, ML_6, PF_6^-
$\mathbf{I_h}$	E, $6C_5$, $10C_3$, $15C_2$, i, $10S_6$, $6S_{10}$, 15σ		$B_{12}H_{12}^{2-}$, B_{12} (Fig. 17.7d)

The process of diagnosing the point group may then be set out as follows:

(A) A molecule of high symmetry?	Assign as $\mathbf{T_d}$ (tetrahedron), $\mathbf{O_h}$ (octahedron) or $\mathbf{I_h}$ (icosahedron) after checking that all the requisite symmetry elements are present from Table C.2.
(B) Find the axis of highest order, C_n	If present, go to (C). If absent assign as $\mathbf{C_s}$ (plane only), $\mathbf{C_i}$ (centre of inversion, $i \equiv S_2$, only) or $\mathbf{C_1}$ (no element).
(C) Look for n C_2 axes perpendicular to the principal axis	If present, go to (F): if none, go to (D).
(D) Look for horizontal plane	If present, assign as $\mathbf{C_{nh}}$, if absent, go to (E).
(E) If no horizontal plane, look for n vertical planes.	If present, assign as $\mathbf{C_{nv}}$: if no planes then assign as $\mathbf{C_n}$.
(F) If C_n and n C_2 axes are present, look for	
(i) a horizontal plane	Assign as $\mathbf{D_{nh}}$.
(ii) if no horizontal plane, look for n vertical planes	Assign as $\mathbf{D_{nd}}$.
(iii) If no planes at all	Assign as $\mathbf{D_n}$.

A network form of this search is shown in Fig C.3 and Table C.2 lists all the symmetry elements of the chemically important point groups. It will be seen that, while the simpler

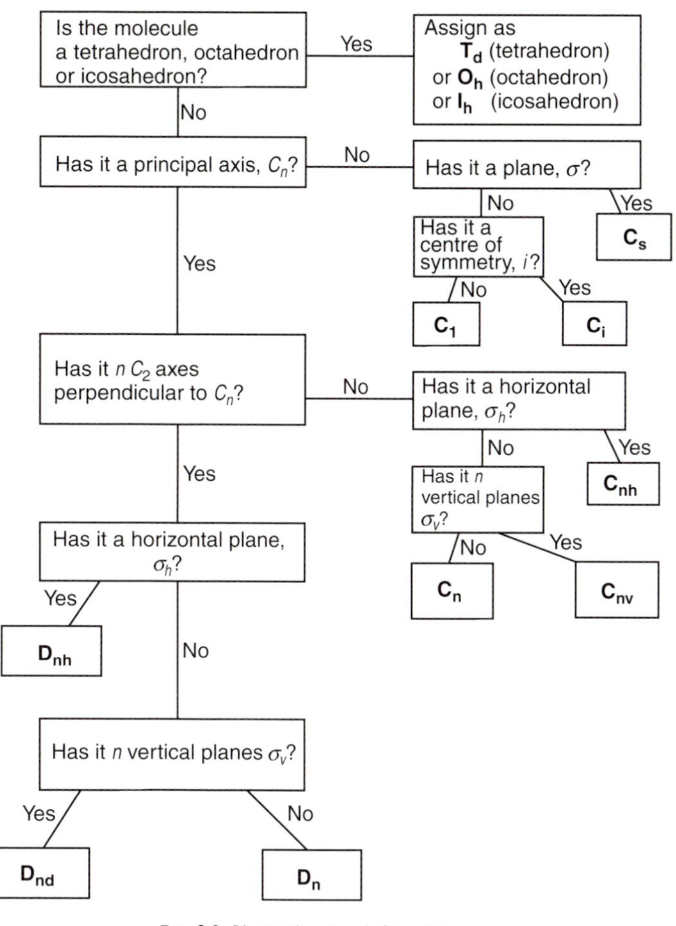

FIG. C.3 Diagnostic network for point groups

groups contain only the diagnostic elements, the more symmetric groups have a considerable number of consequent elements.

A linear molecule fits naturally into the above scheme once it is realized that the molecular axis is an infinity-fold one, because any rotation of the molecule about this axis—however small—produces an indistinguishable configuration. Thus the principal axis is C_∞ and a molecule like N_2 or CO_2 has an infinite number of C_2 axes, perpendicular to the principal axis, passing through the centre of the molecule (mid-point of N-N or through the C in O-C-O). Thus the point group is of the *D* type and, as there is a symmetry plane, σ_h through the midpoint and perpendicular to the molecule axis, these molecules belong to the $\mathbf{D_{\infty h}}$ group. Similar analysis shows that nonsymmetric linear molecules like NO or N_2O (arranged N-N-O) are members of the $\mathbf{C_{\infty v}}$ group.

For further use of the point group, it is necessary to refer to its *character table* which is effectively a summary of all the symmetry properties implied by the point group. For use of the character tables, the reader should refer to one of the more advanced treatments indicated in the references in Appendix A. However, for convenience of reference, we append the characters tables of those point groups which occur most commonly, Table C.3.

From the list, it may be seen that the character table is divided into sections. The symbol for the point group comes first, and then there is listed along the top row all the symmetry operations of the group. The main body of the table shows all the possible ways in which a function might transform under the operations of the group. Here, if the operation transforms the function into itself we find a 1, if the function is reversed we find -1, while if the function is mixed with others we find some other symbol. For example, think of the p orbitals on N in ammonia (that is, the unperturbed atomic orbitals before any combination or hybridization is carried out). Ammonia belongs to the point group $\mathbf{C_{3v}}$ and the p_z orbital will be taken to coincide with the C_3 axis. It can be seen that, if the operation C_3, rotation by 120° about the z axis, is carried out, the p_z orbital is unaltered. Similarly the operations E or reflection in any of the σ_v planes leaves the p_z orbital unaltered. Thus we would write for p_z.

E	C_3	$3\sigma_v$
1	1	1

corresponding to the row labelled A_1. On the other hand, rotation of the p_x orbital about the C_3 axis (i.e. the z axis) by 120° would bring it to a position which must be expressed as a combination of the p_x and p_y orbitals, $-p_x \cos 60 + p_y \sin 60$. Similarly, C_3 operating on p_y mixes it with p_x. Thus the p_x and p_y orbitals are degenerate in the $\mathbf{C_{3v}}$ point group and show the characters indicated in the table opposite E in the left-hand column. In general, the left-hand column in the character table gives symbols which summarize the symmetry properties of the different possible classes of functions in that point group. The exact nomenclature need not concern us, but note that nondegenerate classes are labelled A or B, doubly degenerate ones E and triply degenerate ones T (or F in the older nomenclature).

Beware of confusing the rotational functions R_x, etc., in the third column with Raman activity. When analysing vibrational spectra, Raman activity is shown by a tensor entry in the *last* column.

The final two columns give functions which transform as the various classes. In the third column are vectors in the x, y or z directions and rotations around each axis. The vectors are sometimes represented by translations, T_x, T_y, T_z and sometimes simply as x, y, z which can stand, for ex: nple, for p orbitals. The character tables in Table C.3 give examples of each. In the last column are the second order products, such as the d orbitals.

TABLE C.3 Character tables for common point groups

$\mathbf{C_1}$	E	
A	1	All functions

$\mathbf{C_s}$	E	σ_h		
A'	1	1	T_x, T_y, R_z	x^2, y^2, z^2, xy
A''	1	-1	T_z, R_x, R_y	yz, xz

$\mathbf{C_i}$	E	i		
A_g	1	1	R_x, R_y, R_z	$x^2, y^2, z^2, xy, xz, yz$
A_u	1	-1	T_x, T_y, T_z	

$\mathbf{C_2}$	E	C_2		
A	1	1	T_z, R_z	x^2, y^2, z^2, xy
B	1	-1	T_x, T_y, R_x, R_y	yz, xz

$\mathbf{C_{2v}}$	E	C_2	$\sigma_v(xz)$	$\sigma_v(yz)$		
A_1	1	1	1	1	T_z	x^2, y^2, z^2
A_2	1	1	-1	-1	R_z	xy
B_1	1	-1	1	-1	T_x, R_y	xz
B_2	1	-1	-1	1	T_y, R_x	yz

$\mathbf{C_{3v}}$	E	$2C_3$	$3\sigma_v$		
A_1	1	1	1	T_z	$x^2 + y^2, z^2$
A_2	1	1	-1	R_z	
E	2	-1	0	(T_x, T_y) (R_x, R_y)	$(x^2 - y^2, xy)$ (xz, yz)

$\mathbf{C_{4v}}$	E	$2C_4$	C_2	$2\sigma_v$	$2\sigma_d$		
A_1	1	1	1	1	1	T_z	x^2+y^2, z^2
A_2	1	1	1	−1	−1	R_x	
B_1	1	−1	1	1	−1		x^2-y^2
B_2	1	−1	1	−1	1		xy
E	2	0	−2	0	0	(T_x, T_y) (R_x, R_y)	(xz, yz)

$\mathbf{C_{2h}}$	E	C_2	i	σ_h		
A_g	1	1	1	1	R_z	x^2, y^2, z^2, xy
B_g	1	−1	1	−1	R_x, R_y	xz, yz
A_u	1	1	−1	−1	z	
B_u	1	−1	−1	1	x, y	

$\mathbf{D_{2h}}$	E	$C_2(z)$	$C_2(y)$	$C_2(x)$	i	$\sigma(xy)$	$\sigma(xz)$	$\sigma(yz)$		
A_g	1	1	1	1	1	1	1	1		x^2, y^2, z^2
B_{1g}	1	1	−1	−1	1	1	−1	−1	R_z	xy
B_{2g}	1	−1	1	−1	1	−1	1	−1	R_y	xz
B_{3g}	1	−1	−1	1	1	−1	−1	1	R_x	yz
A_u	1	1	1	1	−1	−1	−1	−1		
B_{1u}	1	1	−1	−1	−1	−1	1	1	T_z	
B_{2u}	1	−1	1	−1	−1	1	−1	1	T_y	
B_{3u}	1	−1	−1	1	−1	1	1	−1	T_x	

$\mathbf{D_{3h}}$	E	$2C_3$	$3C_2$	σ_h	$2S_3$	$3\sigma_v$		
A'_1	1	1	1	1	1	1		x^2+y^2, z^2
A'_2	1	1	−1	1	1	−1	R_z	
E'	2	−1	0	2	−1	0	(T_x, T_y)	(x^2-y^2, xy)
A''_1	1	1	1	−1	−1	−1		
A''_2	1	1	−1	−1	−1	1	T_z	
E''	2	−1	0	−2	1	0	(R_x, R_y)	(xz, yz)

$\mathbf{D_{4h}}$	E	$2C_4$	C_2	$2C'_2$	$2C''_2$	i	$2S_4$	σ_h	$2\sigma_v$	$2\sigma_d$		
A_{1g}	1	1	1	1	1	1	1	1	1	1		x^2+y^2, z^2
A_{2g}	1	1	1	−1	−1	1	1	1	−1	−1	R_z	
B_{1g}	1	−1	1	1	−1	1	−1	1	1	−1		x^2-y^2, z^2
B_{2g}	1	−1	1	−1	1	1	−1	1	−1	1		xy
E_g	2	0	−2	0	0	2	0	−2	0	0	(R_x, R_y)	(xz, yz)
A_{1u}	1	1	1	1	1	−1	−1	−1	−1	−1		
A_{2u}	1	1	1	−1	−1	−1	−1	−1	1	1	T_z	
B_{1u}	1	−1	1	1	−1	−1	1	−1	−1	1		
B_{2u}	1	−1	1	−1	1	−1	1	−1	1	−1		
E_u	2	0	−2	0	0	−2	0	2	0	0	(T_x, T_y)	

$\mathbf{D_{5h}}$	E	$2C_5$	$2C_5^2$	$5C_2$	σ_h	$2S_5$	$2S_5^3$	$5\sigma_v$		
A_1'	1	1	1	1	1	1	1	1		x^2+y^2, z^2
A_2'	1	1	1	−1	1	1	1	−1	R_z	
E_1'	2	2 cos 72°	2 cos 144°	0	2	2 cos 72°	2 cos 144°	0	(T_x, T_y)	
E_2'	2	2 cos 144°	2 cos 72°	0	2	2 cos 144°	2 cos 72°	0		(x^2-y^2, xy)
A_1''	1	1	1	1	−1	−1	−1	−1		
A_2''	1	1	1	−1	−1	−1	−1	1	T_z	
E_1''	2	2 cos 72°	2 cos 144°	0	−2	−2 cos 72°	−2 cos 144°	0	(R_x, R_y)	(xz, yz)
E_2''	2	2 cos 144°	2 cos 72°	0	−2	−2 cos 144°	−2 cos 72°	0		

$\mathbf{D_{6h}}$	E	$2C_6$	$2C_3$	C_2	$3C_2'$	$3C_2''$	i	$2S_3$	$2S_6$	σ_h	$3\sigma_d$	$3\sigma_v$		
A_{1g}	1	1	1	1	1	1	1	1	1	1	1	1		x^2+y^2, z^2
A_{2g}	1	1	1	1	−1	−1	1	1	1	1	−1	−1	R_z	
B_{1g}	1	−1	1	−1	1	−1	1	−1	1	−1	1	−1		
B_{2g}	1	−1	1	−1	−1	1	1	−1	1	−1	−1	1		
E_{1g}	2	1	−1	−2	0	0	2	1	−1	−2	0	0	(R_x, R_y)	(xz, yz)
E_{2g}	2	−1	−1	2	0	0	2	−1	−1	2	0	0		(x^2-y^2, xy)
A_{1u}	1	1	1	1	1	1	−1	−1	−1	−1	−1	−1		
A_{2u}	1	1	1	1	−1	−1	−1	−1	−1	−1	1	1	T_z	
B_{1u}	1	−1	1	−1	1	−1	−1	1	−1	1	−1	1		
B_{2u}	1	−1	1	−1	−1	1	−1	1	−1	1	1	−1		
E_{1u}	2	1	−1	−2	0	0	−2	−1	1	2	0	0	(T_x, T_y)	
E_{2u}	2	−1	−1	2	0	0	−2	1	1	−2	0	0		

$\mathbf{D_{2d}}$	E	$2S_4$	C_2	$2C_2'$	$2\sigma_d$			
A_1	1	1	1	1	1			x^2+y^2, z^2
A_2	1	1	1	−1	−1	R_z		
B_1	1	−1	1	1	−1			x^2-y^2
B_2	1	−1	1	−1	1	z		xy
E	2	0	−2	0	0	(x, y)	(R_x, R_y)	(xz, yz)

$\mathbf{D_{3d}}$	E	$2C_3$	$3C_2$	i	$2S_6$	$3\sigma_d$		
A_{1g}	1	1	1	1	1	1		x^2+y^2, z^2
A_{2g}	1	1	−1	1	1	−1	R_z	
E_g	2	−1	0	2	−1	0	(R_x, R_y)	$(x^2-y^2, xy), (xz, yz)$
A_{1u}	1	1	1	−1	−1	−1		
A_{2u}	1	1	−1	−1	−1	1	z	
E_u	2	−1	0	−2	1	0	(x, y)	

$\mathbf{T_d}$	E	$8C_3$	$3C_2$	$6S_4$	$6\sigma_d$		
A_1	1	1	1	1	1		$x^2+y^2+z^2$
A_2	1	1	1	−1	−1		
E	2	−1	2	0	0		$(2z^2-x^2-y^2, x^2-y^2)$
T_1	3	0	−1	1	−1	(R_x, R_y, R_z)	
T_2	3	0	−1	−1	1	(T_x, T_y, T_z)	(xy, xz, yz)

$\mathbf{O_h}$	E	$8C_3$	$6C_2$	$6C_4$	$3C_2$ $(=C_4^2)$	i	$6S_4$	$8S_6$	$3\sigma_h$	$6\sigma_d$		
A_{1g}	1	1	1	1	1	1	1	1	1	1		$x^2+y^2+z^2$
A_{2g}	1	1	−1	−1	1	1	−1	1	1	−1		
E_g	2	−1	0	0	2	2	0	−1	2	0		$(2z^2-x^2-y^2,\ x^2-y^2)$
T_{1g}	3	0	−1	1	−1	3	1	0	−1	−1	(R_x, R_y, R_z)	
T_{2g}	3	0	1	−1	−1	3	−1	0	−1	1		(xz, yz, xy)
A_{1u}	1	1	1	1	1	−1	−1	−1	−1	−1		
A_{2u}	1	1	−1	−1	1	−1	1	−1	−1	1		
E_u	2	−1	0	0	2	−2	0	1	−2	0		
T_{1u}	3	0	−1	1	−1	−3	−1	0	1	1	(T_x, T_y, T_z)	
T_{2u}	3	0	1	−1	−1	−3	1	0	1	−1		

Index

AB solids, coordination in 111–4
AB_2 solids, coordination in 113–4
AB_n molecules, configurations 81–2
AB_nE_x molecules, configurations 82–3
absorbance 168
absorption spectra
  for H_2 24
  for transition metals 281, 304–7
abundance of elements
  in Earth's crust 202
  in Universe 200
acceptor acids
  definition (Lewis theory) 149–50
  in BrF_3 156
acceptor behaviour of boron Group trihalides 440–1
acetamide, solvent properties 152
acetate
  basic beryllium 244
  chromium 329
  copper 346
  molybdenum 366
  multiple bonding in 406–7
acetic acid
  association 230
  solvent properties 153–5
acetylides of s elements 238–9
acid–base pairs 149–50
acidity, definitions for non-aqueous solvents 148–50
acid rain 566
acids 136–8, 148–50, 152
  hard and soft 292–3
  Lewis 149
  Lowry–Bronsted 136, 148
  in non-aqueous solvents 148
  in water 136
  oxy- 138, 466, 482, 488, 496
  p*K* values 138
  protonic concept 136
  solvent system definition 136, 148
  strengths from catalytic properties 137
  'superacids' 155–6, 522–4
actinide contraction 260
actinide elements 257–72
  free energy changes in redox reactions 261
  halides 266
  oxides 266
  VII state 268
actinides, heavier, redox potentials 272
actinium 246, 259–60
  discovery 9
actinolite 525
active metals, discovery 9
$Al_2(CH_3)_6$ structure 225
Al_2O_3 439–40
  (corundum), structure 115
albite 527
alcohols from alkenes (OXO process) 341
alkali halides, lattice energies 117–8
alkali metals 234–44
  borohydride structures 127, 215
  crown ether and cryptate interactions 240–1
  graphite compounds 448
  heats of hydration of halides 133–4
  hydrides 213
  hydroxides, structures 128
  lattice energies of chalcogenides 118
  oxides and ozonides 238
  solutions in ammonia 151–3
  structures of halides 112, 114
alkaline earth metals 234–40
  acetylides 238, 448
  structures of halides 114
alkene polymerization, (Zeigler–Natta process) 442
alkenes as ligands 402
allotropy
  in arsenic and antimony 458
  in phosphorus 458
  in sulfur Group elements 472
allowed transitions 168
$AIM(SO_4)_2.12H_2O$ (alums) 441
α particle bombardment in actinide element synthesis 257
alumina fibres 436
  alumina in electric cells 436

aluminium 435–45
discovery 9
hydride 226
ionization potentials 45
phosphates in catalysis 436
resemblance to beryllium 244
species in solution 440
aluminohydrides 215, 217, 227
aluminosilicates 526–9
alums 328, 441
amalgams (metal solutions in mercury) 387
americum 197–8, 257–61, 269–70
discovery 9
in smoke detectors 270
amine complexes, formation constants 294
ammonia complexes
with cobalt 339, 538
formation constants 294
with osmium 374
with ruthenium 374
ammonia
shape 84–5, 220
solution of alkali metals 151–3, 237
reactions with Main Group elements 519–20
solution of lanthanides 254
solvent properties 149, 151–3
ammonium salts, structures 128
amphiboles 525
amphoteric behaviour
in non-aqueous solvents 151, 154
in zinc 389
analysis of an unknown compound 61
anion
formation in aqueous solution 143
polarizability 119
radii 51–2, 111–4
sodium 240
anorthite 527
antibonding orbitals
in d element compounds 284
π^* 68
σ^* 64
anti-fluorite structure 126
anti-knock compounds 451
anti-tumour activity 559–60
antimony 457
cluster anions 518–20
discovery 9
pentafluoride 462–3
apatite 459
aqua regia 372
aqueous solutions 133–41
acid–base behaviour 136–8
dielectric constant 135
dipole effect in 134
oxidation potentials 140–2
argon 494–6, 502
aromatic compounds of transition elements 402–3
arsenic 457
cluster anions 518
discovery 9
polyarsenic compounds 514
arsenite oxidation by permanganate 57
asbestos 525
astatine 484
discovery 9
atmospheric composition 202
atom
covalent radii 48
properties 48
atomic fission, separation of products by solvent extraction 163
atomic mass 18
atomic number 18, 185
atomic orbitals 32–40, 62–6, 93–7, 185
and molecular shapes 81–5
energy levels 185–8
hybridization 93
and Periodic position 193
atomic radii
in covalent molecules 49
in ionic compounds 50
in metals 52
atomization, heat of 116
atoms
bonding between 63
electronic structures 29
excited states 24
ground-state 23
radioactive 11
relative mass 18, 33–5
structure 18
aufbau process 29
auride ion 385
azide 460

α-benzoinoxime 294
$B_3N_3H_6$ 227
back-bonding 396
Balmer series 24, 26–7
barium 234–40
barium analogues and correlated study areas 15
discovery 9
base 136–8
hard and soft 292–3
in non-aqueous solvents 148
Lewis 149
basic beryllium acetate 244
bastnasite 246
$BaTiSi_3O_9$, structure 525
bauxite 436
$Be(CH_3)_2$ structure 244
$Be_3Al_2Si_6O_{18}$, beryl 525
$Be_4O(OOCCH_3)$, basic beryllium acetate 244
Beer–Lambert law 168
BeH_2 molecular orbitals 97
benzene, delocalized π orbitals 104
berkelium 258, 270–1
discovery 9
berthollides 131
beryl 525
beryllium 243–4
discovery 9
hydride 244
beryllium group, ionic radii 51
β emission in actinide element synthesis 257–8
bidentate ligands 292–6
biotite (mica) 526–7
bipyridyl *see* dipyridyl
bismuth 456–69
cluster species 518–20
discovery 9
subchloride, $Bi_{12}Cl_{14}$ 469
tri-iodide structure 122
blast furnace 334
body-centred cube 124
boiling points and H-bonding 228
Bohr theory 26
bond energies 430, 433
bond lengths 49
for multiple bonds 49
bond orders 70
bonding
in carbonyls 396
covalent 63
in diatomic molecules 63
electron dot representation 60
and molecular orbital formation 62, 67
octet rule 60–1
in polyatomic molecules 91, 101, 218
ionic 110, 213
metallic 121, 215
bonding orbitals 63, 66, 284
bonding types 123
bonds
hydrogen 228
polarization effects in 119
quadruple 407
three-centre 223
borates, structures 439–40
borax 440
borazine 228
boric acid 229
borides, structures 438
Born–Haber cycle 115
borohydrides 222–6
boron 435
clusters 438, 519
discovery 9
Group, uses 435
halides 441
hydrides 222
nitride, analogy to diamond 130
boron trichloride, bonding 61, 95–6
boron trifluoride
ammonia adduct 61, 87
bonding 61, 441

borosilicate glasses 436
boundary contour representations of 2 p orbitals 37
of 3d orbitals 38
Br_2
bond length 49
electronic structure 73
Brackett series 26–7
Bragg equation 164
brass 345, 446
bridging structures 223, 398
bromides *see* halides
bromine 483
discovery 9
positive oxidation states 486
bromine trifluoride, solvent properties 156
Bronsted theory of acids 136, 148
bronze 345, 446
Buckminsterfullerenes ('bucky balls') 541

cacodyl 460
cadmium 385–9
cadmium iodide, structure 122
caesium 235–7
CaF_2 structure 113
calcium 234–40
calcium carbide structure 127
californium 257, 259
capping 5, 517
carbenes 401
carbides 238, 445, 448
carbon 445
bond length in diamond 49
discovery 9
electronic structure 30
reduction by, in metal extraction 203–7
carbon-14, half-life and dating 21
carbon dioxide
bonding 105
use as solvent 158
carbon Group 445
uses 446
covalent radii 49
halides, bond lengths 49
structures 129
carbon monoxide *see* CO *and* carbonyls
carbonate ion, shape and bonding 102–3
carbonic acid 449
carbonyl compounds of transition elements 395–8, 409
nonclassical 397
carbonyl hydrides, transition-element 398
carboranes 224, 403
carboxylic acids, solvent properties 152
carbynes 401
cassiterite 446
catalysis 315, 317, 379, 413, 442
catalysts
aluminium phosphate 436
Monsanto acetic acid process 378
OXO process 341
platinum 379
Ziegler–Natta 442
catenation 506
in carbon Group 447
in hydrides 219
in nitrogen Group 460–1
in oxygen Group 471–3
$CaTiO_3$, structure 115, 126
cation radii 50–2
cation formation in aqueous solution 143
CdI_2, structure 122
cementite 334
CeO_2, structure 114
cerium 246
discovery 9
IV oxidation state 252
Cesium *see* caesium
CH_2, CH_4, formation energies 96
CH_4
photoelectron spectrum 101
shape 95
chain compounds *see* catenation
chalcogenides *see* sulfides
character tables 594
chelate effect 292–3
chelation therapy 560
chemical behaviour and Periodic position 194
chemical shift 171
chemotherapy 559
chlorides *see* halides
chlorine 484
discovery 9
positive oxidation states 485
chlorine trifluoride, shape 83, 85
chlorofluorocarbons and ozone hole 563
chlorophyll, structure 547
chromatography 161
chromium 324
carbonyl 329, 395
discovery 9
dibenzene 403
hydride 216
isomers of ethylenediamine complexes 297
peroxy compounds 326
cis isomers 174, 179, 297
cisplatin 302, 559
clathrate compounds 495
clay minerals 526
Cl_2, bond length 49
close packing 123–5
and common structures 126
cluster compounds
electron pairs in 516
metal 125
metal carbonyl 409
of Main Group elements 516–20
of transition elements 358, 366, 371, 378, 409–13
transition element/sulfur clusters 336, 363, 389, 538
phosphorus 512
CO_2
π orbitals 105
shape 90
cobalt 338
carbonyl compounds 341, 395, 398
discovery 9
electron density in complexes 537
isomers of ethylenediamine complexes 297
coinage metals *see* Cu, Ag, Au
cold fusion 218
colours of transition element complexes (d-d spectra) 279–80
complex hydride ions 213, 217
complexes
eight-coordinate 291, 365
ligand effect on stability 292
octahedral 279–85
seven-coordinate 291
square planar 285–7
square pyramidal 290–1
structures 128
tetragonal 285, 344
tetrahedral 285–6
trigonal bipyramidal 290–1
cone voltage 177–8
conjugate acid–base pairs 136
continuous extraction 163
coordinate bonds and shapes of compounds 86
coordination in AB and AB_2 solids 113
coordination number 55
and radius ratio in ionic compounds 113
eight (complexes) 249, 253, 262, 268, 269, 291, 353, 356, 365, 369
eight (solids) 112, 121, 123, 363
five 289–90, 322, 341, 344
four 289–90, 342–3, 385
fourteen 268
in Main Group compounds 431
nine 248, 250, 262–3, 269, 291, 369
seven 255, 262, 290, 323, 326, 353, 355, 368
six 112, 279–85, 449
ten 250
three 336, 342
twelve (complexes) 268
two 289, 290
copper 344–8
chemistry and correlated study areas 15
discovery 9
hydride 217
in superconductors 393
correlated study areas 15
corundum structure 115
cosmic abundance of the elements 200
cosmic fusion processes 201
cosmochemistry 201
coupling constant 173
covalent bond
dipole effect 120
energy-level diagrams 65, 69

order 70
summary 78
theory 63
covalent hydrides 218
covalent radii 48
covalent solids, solubility 135, 146
covalent structures 128
β-cristobalite, structure 113, 527
critical temperature 11, 393
crown ethers 240, 547
cryptates 240, 547
crystal field stabilization energy (CFSE) 279
in octahedral complexes 282
reaction mechanisms 297
in square planar complexes 286, 288
in tetrahedral complexes 282, 286–7
crystal structure,
complex ions 127
covalent compounds 128
determination 164
ionic compounds 111–5
metallic compounds 121
summary 126
cubane 452
cubic close packing 124–6
cupron 294
curium 257, 270
discovery 9
cyanide complexes
copper 348
iron 337
molybdenum 365, 366
nickel 343
osmium 376
rhenium 370
ruthenium 375
technetium 371
tungsten 365
cyclic (ring) species 506–12
cyclopentadienyl compounds of transition elements 402
cytochromes 419

d^1 configuration, spectrum 281, 306
d^2, d^3, d^8, spectra 306
d^4, d^6, d^9, spectra 305
d^5, spectrum 305
d, orbitals 38–40, 70, 273
bonding effects 107, 452
participation in Main Group chemistry 434, 534
shapes 38–9
and shielding 186
Dalton 131
defect structures 130–1
degenerate energy levels 25
delocalization
of π bonding 101, 105
of σ bonding in diborane 223
delta bond (quadruple bond) 70
in $Re_2Cl_8^{2-}$ ion 406–7
dendrimers 544
detergents 562
deuterium 208
diamagnetism 178
and Hg(I) state 386
diamond structure 129–30
diatomic molecules 63
p orbital combination 66
s orbital combination 63
dibenzene chromium 403
diborane, electron deficient bonding 222–4
'didymium' 248
dielectric properties of water 135
diethylenetriamine 293
diffraction methods of structure determination 164
dihalides of first transition series 348
dihydrogen complexes in transition elements 419
diimines 416
dimethylglyoxime complex of Ni 342
dinitrogen species 415
dioxygen species 413
dipole effect in covalent bonds 120
dipole moment and structure 119
dipyridyl complexes
chromium 326, 328
titanium 318
vanadium 324
discovery of the elements 9
dissociation energies 430
donor bond 87
double bond
lengths 49
metal–carbon 401
metal–metal 406
radii 49
dysprosium 245

e.m.f., and Hg(I) 386
EDTA, Ca and Mg analysis using 240
effective nuclear charge 185
eight-coordination 291
eighteen electron rule 396
einsteinium 257, 270
electrides 240
electrode potentials 139–44
electrolysis, extraction of elements 203–4
electromagnetic spectrum 166
electron affinity 46
and ionic solids 119
electron counting procedure for π electrons 88–90
electron-deficiency 222
electron densities
difference 536–8
experimental 531, 536–7
electron density
in diatomic molecules 64, 67
in hydrogen d orbitals 38–9
in hydrogen p orbitals 37
in hydrogen s orbitals 36
electron
diffraction 165
dual nature of 22–3
exchange energy 188–9
probability density 22, 36
properties 18
spin resonance 175
transfer and oxidation 138, 297
and wave functions 23
electron pairs and molecular structure 81
electronegativity 53–5
Mulliken definition 55
Pauling values 54
Zhang values 54
electron configurations
of elements 30–2
stable 188
electronic energy
of diatomic molecules 65
of hydrogen-like atoms 24–6
of other atoms 28
electronic spectra 24, 168
electronic structures of diatomic molecules 65, 72–4
electrons, unpaired, determination 175
electrospray ionization 177
electrostatic forces and shapes 81
in ionic crystals 110–1
electrostatic theory of transition element complexes 282
elements
abundance and occurrence 200
cosmic abundances 200
crustal abundances 202
in divisions of the Periodic Table 42, 194
electron affinities 46–8
electronegativities 54
electronic configurations 32–4
extraction 203
ionization potentials 45–6
new ultra-heavy 421
nomenclature 43
occurrence 200
origins and discovery 9
elution 161
emerald 525
empirical formula 60
energies, bond 430
energy changes in forming solutions 133, 146
energy level diagram
diatomic molecules 63–4, 70
$Mo(CN)_8^{4-}$ ion 365
octahedral complex 278, 280
polyatomic molecules 99, 100, 103
square planar complex 384
three-centre bond 223
entropy and hydrogen bonding 230
equilibrium reactions and redox potentials 140–1

erbium 247
ethene, polymerization 442
ethylenediamine 293
europium 247
EXAFS technique 166
exchange energy 188
extraction of the elements 203

f electrons
  associated properties 255
  in actinide elements 260
f element contraction *see* actinide contraction, lanthanide contraction
f orbitals
  shapes 39
  and shielding 186
face–centred cubic close packing 124
fast atom bombardment (FAB) mass spectrometry 177
feldspar 527
fermium 257, 271
ferredoxins 551
ferric compounds *see* iron
ferritin 549
ferrocene 403
ferrous compounds *see* iron
Fischer 399
five coordination 290, 323, 341, 344, 454–5
  in square pyramid 290
  in trigonal bipyramid 82, 290
fluids, supercritical as solvents 158
fluorides
  actinide 266, 267, 270–1
  of p elements 427
  preparation in BrF_3 156
  of transition elements 276
  xenon 500
  zinc Group 385
  (*see also* under individual elements)
fluorinating agents, relative reactivity 489
fluorine
  and correlated study areas 15
  discovery 9
  electronic structure 30, 70
  handling 483
fluorine Group elements 483
  covalent radii 49
  ionic radii 50
  properties 483
  Van der Waals radii 50
  uses 483
fluorite (CaF_2) structure 113
fluoro complexes and high coordination numbers 474, 477
fluoroapatite 459
fluorosulfuric acid as solvent 155
forbidden transitions 164, 308
force constants 169
formamide, solvent properties 152
formation constants, stepwise 295
formula determination 61
Fourier transform methods 177
francium 235
  discovery 9
free energy and oxidation potentials 197–8
Frenkel defect 130
Friedel–Crafts catalysts 440
fullerenes 11, 129, 541
fundamental constants, 4
fundamental particles 18, 19
fused salt batteries 234
fusion processes, cosmic 200

gadolinium 246
  electronic structure 31
galena 446
Gallex experiment 441
gallium 436
  discovery 9
gamma-ray resonance 175–6
garnet 523
GeO_2, structure 114
geometrical isomerism 296
germanes 219
germanium 445
  bond lengths of halides 49
  chain compounds 447
  cluster ions 518
  discovery 9
  hydrides 217
  in II state 479
  in semiconductors 446
gold 384
  clusters 411
  discovery 9
  medical uses of gold compounds 560
Goldschmidt ionic radii 51
graphite, structure 129
graphite fluoride 448
graphite oxide 448
greenhouse effect/global warming 564
Grignard reagents 239
Group oxidation states 195, 429
group theory 169, 589

H_2
  electronic structure 65
  energy of formation 21
  total electronic energy 63
H_2^+, electronic structure 65
H_2^-, total electronic energy 63
H_2O
  bond angle 84
  molecular orbitals 92
  shape 83
$H_5O_2^+$, structure 211
Haber process 459–60
haemoglobin 550
hafnium 352
  discovery of 9
hahnium 423
half-life 20, 258–9, 421
halides (*see also* under individual elements)
  of actinide elements 267
  of p elements 429
  of s elements 235
  of transition elements 277
  of zinc Group 386
halogen cations 489
  oxyanions 492
halogens 483
Hammett acidity function 521
hard/soft acids and bases 292–3
hardness of atoms, definition 55
HCl
  electronic structure 73
  photoelectron spectrum 75
heat of atomization 119
heat of hydration 134
heat of solution 134
heavy metals, discovery 9
Heisenberg's Uncertainty Principle 21–3
helium 495–6
  discovery 9
helium Group 494–502
heteronuclear molecules 73–4
heteropolyacids 361
hexagonal close packing 124–6
HF, H-bonded structure 231
Hg_2^{2+} ion 387
high-spin configurations 282
holmium 247
Hund's Rules 30
hybridization of orbitals 93
  in diborane 223
  equivalent 93
  non equivalent 95
hydrates, first transition series 312
hydration energy
  curve for first transition series 304
  effect on redox potential 144
  effect on solubility 133
hydrazine 220
hydrides 211
  covalent 218
  electron-deficient 222
  interstitial 215
  ionic 213
  metallic 215
  mixed-metal 217
  some bond angles in 85
hydrogen
  covalent radius 49
  discovery 9
  energy levels of molecular 65
  as fuel 210, 565
  isotopes 210
  properties 209, 211
  wave equation for atomic 32
hydrogen bonding 228
  and entropy 230

hydrogen bridging 223–4
hydrogen cyanide, solvent properties 148
hydrogen economy 209, 565
hydrogen electrode 140
hydrogen fluoride, solvent properties 149, 152
(*see also* HF)
hydrogen halides 218–20
photoelectron spectra 77
hydrogen peroxide 220, 474
hydrogen sulfide, shape and bonding 85, 93
hydrogen-like atom, wave equation 24
hydrolysis 136
hydrometallation 341
8-hydroxyquinoline 294, 443

I_2 bond length 49
ice, structure 230
icosahedron 7, 438
indium 438
discovery 9
inert-pair effect 433
relativistic contribution 425
infrared spectrometry 167, 229
inner sphere mechanism 299
interhalogen compounds 490–2
interstitial hydrides 313
iodate–iodine reaction 56
iodides *see* halides
iodine 484
and correlated study areas 15
discovery 9
positive oxidation states 489
ion exchange 160, 262
ionic bonding 110
ionic compounds 110, 133
ionic hydrides 213
ionic radii 51–2, 103, 193
from x-ray diffraction analysis 51
ionization energies, Table 45–6
ionization energy (or potential) 44, 116, 190–2
ions, radial density curves 36
iridium 376
carbonyls 395, 409
discovery 9
iron 333–5
biologically important species 549–51
discovery 9
carbonyls 395–7
electronic structure 31
isomerism 178, 296–7, 338
isopolyacids 360–1
isotopes 19
actinide elements 259
astatine 484
carbon 21
hydrogen 210
polonium 471
superheavy elements 421
technetium 366

Jahn–Teller effect 288
Johnson ionic radii 51

K^+ compounds *see* potassium
kaolins 526
kinetics and mechanism 297, 453
krypton 495–6, 500

LCAO method 63
Ladd ionic radii 52
lanthanide contraction 250, 351
relativistic contribution 424
lanthanide elements 246
discovery 9
magnetic moments 253
occurrence and uses 246–7
lanthanum 246
discovery 9
lattice defects 130–1
lattice energy 116
and solubility 133
ligand field contribution 302
lawrencium 258, 271
layer lattice structure 121–2
lead 447
clusters 519
discovery 9
structure 129
Lewis theory of acids 149–50
Lewis theory and covalent bonding 60–2
ligands
hard and soft 292, 376
polydentate, in isomerism studies 296
polydentate, stepwide formation constants 295
ligand field effect and
exchange energies 282
high and low-spin configurations 282
lattice energy 303–4
octahedral complexes 278
spectra 305–7
spectrochemical series 280
square planar complexes 286, 288
stabilization energy 278
structural aspects 278
tetrahedral complexes 286–7
theory 277
light elements, general characteristics
s elements 234
p elements 427
litharge 452
lithium 234, 241
batteries 234
discovery 9
lithium Group ionic radii 51
localized π bonds, steric effects 89, 101
lone pair electrons and shapes 82
lone pairs, effect on shapes of sigma bonded species 83
low-spin configurations 282
Lowry–Bronsted theory of acids 136–8

lutetium 247
Lyman series 26–7

Madelung constant 117–8
magnesium 234, 241–3
discovery 9
^{23}Mg, half-life 20
magnetic measurements and structures 178
magnetic moments 283
magnetic resonance imaging (MRI) 558
magnetite 334, 549
Magnus's green salt 381
Main Group elements
bond energies 430
halides 429
oxides 428
stability of oxidation states 428
sulfides 428–9
use of d orbitals in compounds 502, 534
MALDI (in mass spectrometry) 177
manganese 329
discovery 9
oxygen compounds 122, 329–2
manganese carbonyl 395, 398–9
many-electron atoms, wave function 28
martensite 334
mass spectrometry 176
Meissner effect 391
mendelevium 257, 271–2
mercury 386
discovery 9
metal carbonyls 395, 409
metal clusters 409, 517
metal dihydrogen complexes 419
metal dinitrogen species 415
metal dioxygen species 413
metal–carbon bonds 399
metal–organic species 399, 403, 405
metal solutions in liquid ammonia 151–3
metal–sulfur rings 538
metallic hydrides 215
metallic radii 53
metals, bonding 121
micas, structures 526–7
mischmetal 246
mixed-metal hydrides 217
Mn_2O_3, structure 122
Molar extinction coefficient 168
molecular formulae from analysis 61
molecular orbital bonding theories 63
in d element compounds 279
molecular orbital formation
diatomic molecule 63
polyatomic molecule 92
molecular orbitals, nomenclature 67
molecular sieves *see* zeolites
molecules
π bonds in 101
shapes of AB_n species 81
species with lone pairs 82

molybdenum 359
  discovery 9
molybdenum blue 363
molybdenum carbonyl 366, 395
monazite 246
Mond process 342, 395
Monsanto acetic acid process 378
Mossbauer effect 175
MRI (magnetic resonance imaging) 558
Mulliken definition of electronegativity 53
multiple metal bonds 406

NH_3 460, 470
  bond angles 84–5
  liquid, as solvent 151
  photoelectron spectrum 101
N_2H_4, structure 220
N_2-metal species 341, 375, 415–9
Na compounds *see* sodium
Na^- 240
$NaBH_4$ 215, 227
Natta catalysis 442
neodymium 246
neon 495–6
  electronic structure 30
neptunium 257, 268
  discovery 9
neutron diffraction 165, 229
NiAs, structure 122, 126
nickel 342
  discovery 9
nickel carbonyl 344, 395
nickel dimethylglyoxime complex 342
niobium 354
  discovery 9
  hydride 216
nitrate ion, bonding 107
nitric acid 466, *see also* aqua regia
nitric oxide *see* NO
nitro–nitrito isomerism 298
nitrogen 457
  complexes 341, 375, 415
  discovery 9
  oxidation numbers 55
nitrogen fixation 317, 375, 415
nitrogen Group 457
  organic compounds 469–70
  uses 459
  Van der Waals' radii 50
nitrogenases 553
nitrous oxide, N_2O 468
nmr *see* nuclear magnetic resonance
NO
  electronic structure 74
  ozone hole chemistry 563
  physiological effects 548
NO_2^-
  bonding 104
  shape 60
nobelium 258, 270
noble gases *see* rare gases
noble metals *see* platinum metals, gold
nodal plane 36
nomenclature 13, 42
non-aqueous solvents 145–58
  acidity 148
  redox properties 150
  solvation and solvolysis 146–8
non-equivalent hybrid orbitals 95
nonstoichiometric compounds 130, 319, 334
nuclear magnetic resonance (nmr) 171
    see also MRI
nuclear power 265, 567
nucleophilic reactions 301, 455

O_2
  electronic structure and bonding 72–3
  photoelectron spectrum 76
O_2–metal compounds 413
octahedral complexes 279–85
octahedral sites in close-packed lattices 125–6
octahedron, distortion by unshared pairs 86
olefins
  oxidation to alcohols 341
  polymerization 442
olivine, structure 523
optical isomers 296–7
orbitals
  2 p, boundary contour representation 37
  3 d, boundary contour representation 38–40
  combination of d 69
  combination of p 66–8
  combination of s 63
  hybridization 912
  nomenclature 67
  participation of d 107, 273, 503
  π (pi) 86, 101–5
  shapes 37–40
  σ (sigma) 64, 93, 104
organometallic compounds
    207, 238, 389, 399, 442, 451, 470
origins of the elements 8
ortho-hydrogen 212
osmium 372
  carbonyls 395, 411
  discovery 9
oxidation, definitions 56, 138
oxidation reduction free energies
  actinides 261
  americium 198
  boron Group 439
  carbon Group 447
  chromium 324
  cobalt 338
  first transition series 312
  fluorine Group 485
  free energy diagrams definition 197–9
  iron 333
  manganese 329
  molybdenum 359
  nitrogen Group 457
  oxygen Group 471
  rhenium 366
  technetium 366
  titanium 314
  tungsten 359
  uranium 198
  vanadium 318
oxidation number *see* oxidation states
oxidation potential *see* redox potential
oxidation state stabilities
  of d elements 273
  of p elements 428
oxidation state, definition 56–7
oxidative addition 377
oxides (*see also* under individual elements)
  actinides 266
  lanthanides(IV) 252, 253
  p element 428
  s element 236
  transition element 275
  xenon 497
OXO process 341
oxyacids 138
  of halogens 488
  of nitrogen Group elements 466
  of oxygen Group elements 475, 480
  of xenon 497
oxyanions *see* oxyacids
oxygen 474 *see also* dioxygen, ozone
oxygen Group
  covalent radii 49
  discovery 9
  ionic radii 51
  uses 375
  Van der Waals' radii 50
oxygen species, bond orders 74
oxyhaemoglobin 550
oxyhalides, solvent properties 150
ozone
  bonding 107
  effects in upper atmosphere 202
  electron counting and shape 86
ozone layer 471
ozone hole 563
ozonides 238

p elements
  cluster compounds 516
  general properties 427
  oxidation states 428
p orbitals 23, 37
  boundary contour representations 37
  combination 66–8
  multicentred bonds 535
  π bonds 431–2
  shielding effects 185
palladium 379
  discovery 9
  hydride 216
para-hydrogen 212
paramagnetic compounds 178, 470
Paschen series 26–7

Pauli principle 29
Pauling electronegativity values 54
Pauling *S* parameter 94
Pb compounds *see* lead
pearlite 334
pentaborane 223
perbromates 488
Periodic Table 41, 185, 610
permanganates 330
perovskite
  and correlated study areas 15
  structure 115
  in superconductors 391
peroxo (peroxy) compounds
    238, 315, 320, 326, 415
perrhenic acid 368
pewter 446
Pfund series 27
phosphates 462, 466, 529
phosphane oxides, π-bonding 99
phosphazenes 464
phosphoric acid 466
phosphorus 457
  discovery 9
  hydride 85, 92
phosphorus pentachloride 87
photoelectron spectra
  CH_4 101
  CO 76, 77
  H_2 180
  HX 77
  H_2O 100
  NH_3 101
  O_2 76
photoelectron spectrometry 182–3
photosynthesis 547
physical constants 4
π electrons, electron counting procedure
    88
π bonding 88, 101
  in benzene 103
  in carbon dioxide 105
  in Main Group compounds 431–3
  in nitrite 104
  in nitrate 107
  in octahedral complexes 309
    in ozone 107
  in polyatomic molecules 101–4
  shapes of species containing 86
  in sulfur trioxide 107
  in Transition Element compounds 307
  involving d orbitals in Main Group
    compounds 431–2, 451, 504
PI_3 solid, structure 467
picket-fence porphyrin 550
p*K* values 137
Planck's constant 23, 164
platinum 379
  carbonyl clusters 382
  discovery 9
  isomerism in complexes 296
platinum-based drugs 559
platinum metals as catalysts 377
platinum metals, separation 372
plumbous, plumbic compounds *see* lead
plutonium 257, 268
  discovery 9
  separation 163
point groups 609
polarization 55, 119, 133
pollution 562
polonium 473
  discovery 9
polyatomic molecules
  delocalized bonding in 97
  localized bonding in 93
  π bonding in 101
  shapes 81, 86
polycentred orbitals 97, 105
polydentate ligands 292–6, 586
polyhalides 489
polyhydride Mg/Fe complex 336
polyiodine cations in superacid media 522
polymerization 442, 451
polyphosphanes 512–4
polyphosphazenes 464
polyphosphorus rings, mixed-element 512–4
polysulfides 507
polysulfur species 509
polysulfur–metal compounds 538
polyvanadates 320
porphin (porphyrin) ring 550–1
post-uranium elements, discovery 9
potassium 234
  discovery 9
potassium halides in non-aqueous solvents 147
potentials, oxidation 139
potentials, standard 142
praseodymium 247, 253
probability density, electron 23, 32, 35–6
promethium 247
  discovery 9
1, 3-propanediamine as ligand 294
protactinium 257, 263
  discovery 9
proton 18
  and acids 137
  solvated 211, 231
Prussian blue 337
pseudohalogens 495
pyrolusite 329
pyroxenes 525

quadruple metal bonds 407
quantum numbers 24–5, 29
quantum dots 413
quantum shells 29
quartz 113, 450, 528

radial density function 36
radioactive material, disposal 265
radioactive species 20, 256
radium 235, 236
  discovery 9
radius
  atomic 48
  covalent 48
  hydride ion 209
  ionic 51
  metallic 53
  Periodic position, variation with
    193–4
  Shannon 51, 52
  transition element ions 303
  Van der Waals' 50
radius ratio 111
radon 494–5, 502
  discovery 9
Raman spectra 166
rare earths *see* lanthanides
rare gas Group (helium Group) 494
rare gases, uses 495
reaction mechanisms
  for silicon 454
  for transition metals 297
realgar 465
redox in non-aqueous solvents 150
redox potential 139, 142
  for heavier actinides 271
reduction 56, 137
Reduction elimination 377
Reinecke's salt 327
relativistic effects 424
rhenium 366
  discovery 9
rhenium carbonyls 371, 395
rhenium trioxide, structure 113
rhodium 376
  discovery 9
rhodium carbonyls 377, 395
  complex as catalyst in Monsanto process
    378
ring compounds
  boron–nitrogen 228
  phosphorus–nitrogen 464
  selenium and tellurium 473, 510
  sulfur 478–83, 507
  sulfur–nitrogen 482
rotational spectra 167
rubidium 234
  discovery 9
  suboxides 238
ruthenium 372
  N_2 complexes 375
  discovery 9
ruthenium carbonyls 377, 395, 409
rutile structure 113
Rydberg constants 27

s elements
  in biochemistry 546
  origin and discovery 9
  sources and uses 234

s orbitals 24–5, 36, 62
shielding effects 185
salicylaldoxime 294
samarium 246
'sandwich' molecules *see* cyclopentadiene compounds
Sb species *see* antimony
SbF_5-HSO_3F as a superacid solvent 156
'scandide contraction' 435
scandium 246
scanning electron microscopy SEM 180
scanning tunneling microscopy STM 180–1
Schomaker–Stevenson formula 49
Schottky defect 130
Schrodinger equation 21, 23, 24
selection rules 168–9, 307
selenium 472
discovery 9
oxychloride, solvent properties 152
uses 476
self-consistent field method 28
semiconductors 129, 131
silicon, germanium 446
$YBa_2Cu_3O_6$ structure 392, 393
seven coordination 290
Shannon crystal radii 51, 52
for Zn, Cd, Hg 386
shapes
and skeletal electron pairs (Wade's rules) 224, 517
molecular 79–86
orbital 32, 35–9
shielding effects 185
sigma bond 63, 67, 70, 92, 98
effect of lone pairs on 83
in polyatomic molecules 91–101
sigma metal–carbon bonds 399
silanes 219
silica 113, 446, 450, 528
silicate structures 524–8
silicides 445
silicon 445
bond lengths of halides 49
chains 219–20, 447, 506
discovery 9
hydrides 219
reaction mechanisms 451
in semiconductors 446
silicones 451
silver 382
discovery 9
silver halides
in photography 383
solubilities in nonaqueous solvents 147
SiO_2, structure 113
skeletal electron pairs and shapes 516
Slater's rules 185
S_N1 reactions 299
S_N2 reactions 300, 454
SO_3, bonding 107
sodalite cage structure 428
sodium 234
discovery 9
sodium chloride
structure 112
Born–Haber cycle and 115
soft/hard acids and bases 292–3
solder 446
solid-state electric cells 436
solubility
in nonaqueous solvents 146
in water 133
solutions, metal in ammonia 151
solvated proton 150
solvation and solvolysis in non-aqueous solvents 148
solvent extraction 162, 251–2
solvent system theory of acids and bases 137, 148
solvents, nonaqueous 145–8
solvolysis 148
sources of s elements 234
spectra 24, 26, 255, 280, 304
stellar 29
spectrochemical series 280
spectroscopic methods 166
spectrum
electromagnetic 166
electronic, of transition elements 279, 304
'spin only' magnetic moments 282–3
spin–spin coupling 169
spinel structure 115
spodumene 234
square planar complexes 285–7
square pyramidal complexes 290–1
stabilities of oxidation states 196
stability
in aqueous or nonaqueous environments 146
of transition metal compounds 278
stability constants 293–4
stable electronic configuration 190
standard potentials 138
stannic, stannous compounds *see* tin
steel 313
stepwise formation constants 294
stereochemistry *see* structures, shapes, and individual molecules
stereoisomerism with polydentate ligands 296
steric effects 433
Stock 222
strength of oxyacids 138
strong field configuration 282
strontium 235
structure determination 164
structures
borides 438
carbonyl 398
close packing 124–5, 126
covalent molecules *see* shapes
diamond 129–30
ionic solid 111, 127
layer lattice 121–2, 388
less regular 119, 127
Main Group cluster ions 397
poly-P, -S, -Se species 506
silicates 523
subatomic particles 18
suboxides 238
sulfanes 509
sulfides 465
sulfite ion structure 87
sulfoxides π bonding in 101
sulfur 476
chains and rings 482–3
discovery 9
dioxide as solvent 155
and acid rain 566
trioxide, bonding 107
uses 476
sulfur-metal rings 538
sulfuric acid, solvent properties 149
superacid media 155–6, 521–3
superconductors
copper chemistry and 392
discovery 11
perovskite structures and 392
warm 393
$YBa_2Cu_3O_{7-x}$ structure and 392
supercritical fluids 158
superfluidity in helium 495
superheavy elements 421
isotopes 421–3
nomenclature 423
superoxides 238
superphosphate 459
symmetry 81, 110, 167–8, 589
synergic contributions to bonding 396

talc 526
tantalum 356
discovery 9
technetium 366
discovery 9
medical applications 557
tellurium 472
clusters and polynuclear anions 510–1
discovery 9
uses 476
terbium 247
tetrahydroborates 215, 227
tetragonal distortion in complexes 285, 345
tetraethyl lead 451
tetrahedral complexes 285–7
shapes for covalent species 82
sites in close packing 125–6
tetrahydroaluminates 215, 217, 227
thallium 435
discovery 9
in cluster anions 521
thiocyanate 296, 340, 375, 494
thionyl halides 479

thorium 261
  discovery 9
three-centre bonds 223, 224
thulium 247
tin 445
  bond lengths of halides 49
  clusters 518–9
  discovery 9
  structure 129
titanium 313
  discovery 9
  hydride 131, 216
  oxides, structural relations 315–7
  spectrum of (III) state 281
titanium chloride, structure 87
titanium dioxide, structure and uses 113
trans-directing ligands 301
trans effect 301–2
trans influence 174
trans isomers 174, 177, 297
transition elements
  aromatic compounds 402
  carbonyl hydrides 399
  carbonyls 396, 409
  clusters 409, 517
  complexes 279–96
  dihydrogen complexes 419
  dinitrogen complexes 415
  dioxygen complexes 413
  first series, hydrates 312
  general properties 273, 312, 351
  halides 276, 277, 348
  hydration energy 303
  hydrides 213
  ligand field theory 279
  oxidation free energy diagrams 312
  oxidation states 274
  oxides 275
  production 313
  radii of ions 303
transuranium elements 268
trigonal bipyramidal complexes 84, 290–1
triple bond lengths 49
triple metal bonds 406–7
trisilylamine 451
tritium 210
  half-life 19
tungsten 359
  bronze 363
  carbonyl 366
  discovery 9
  polynuclear anions (polytungstates) 360–1
tunneling microscopy 180–2
Turnbull's blue 337
two-centre bonding
  hybridization 93
  in polyatomic molecules 92
type metal 446

uncertainty principle 22
unpaired electrons, determination 175
uranium 197–8, 257, 264–8
  discovery 9
  separation by solvent extraction 163
uranium-238, half-life 20
uranocene 405

valency 55
valency electrons, definition 31
Van der Waals radii 50
vanadium 318
  carbonyl 324, 395
  discovery 9
vapour phase chromatography 161
vermiculite 526
vibrational spectra 167
vibrational structure in PE spectra 183
vitamin B_{12} 551
VSEPR theory 81, 532

W compounds *see* tungsten
Wacker process 381
Wade's rules 224, 517
warm superconductors 393
water
  solvent properties 133
  treatment of by ion exchange 160
wave equation 23–5, 62
wave function 23
wave mechanics 23
wave number 166
weak field configuration 282
Werner complexes 339
Wilkinson 399
wolfram *see* tungsten
wurtzite structure 112, 389

X-ray absorption edge 166
X-ray diffraction 164–6, 229
X-ray photoelectron spectroscopy 183
X-rays in electron density determination 536
Xe species and correlated study areas 15
xenon 495–502
xenon fluorides, structures 87

ytterbium 247
yttrium 246
yttrium analogues and correlated study areas 15

Z and Z^* 185
Zeise's salt 381, 402
zeolites 527–9
Zhang electronegativity values 54
Ziegler–Natta catalysts 442
  Ti compounds in 314
zinc 385
  discovery 9
  in biochemistry and medicine 552, 554–5
zinc blende structure 112, 389
Zintl phases 520
zircon 352
zirconium 352
  discovery 9
ZnS (wurtzite) structure 112
zone refining 446
ZSM-5 (zeolite) 528

Relative Atomic Masses (based on $C^{12} = 12.000$)

Element	Mass	Element	Mass	Element	Mass	Element	Mass
Actinium*	227.03	Erbium	167.26	Mercury	200.59	Samarium	150.36
Aluminium	26.982	Europium	151.97	Molybdenum	95.94	Scandium	44.956
Americium*	241.06	Fermium*	257	Neodymium	144.24	Selenium	78.96
Antimony	121.76	Fluorine	18.998	Neon	20.180	Silicon	28.086
Argon	39.948	Francium*	223.02	Neptunium*	237.05	Silver	107.87
Arsenic	74.922	Gadolinium	157.25	Nickel	58.69	Sodium	22.990
Astatine*	209.99	Gallium	69.723	Niobium	92.906	Strontium	87.62
Barium	137.33	Germanium	72.64	Nitrogen	14.007	Sulfur	32.065
Berkelium*	247.07	Gold	196.97	Nobelium*	259	Tantalum	180.95
Beryllium	9.0122	Hafnium	178.49	Osmium	190.2	Technetium*	98.906
Bismuth	208.98	Helium	4.0026	Oxygen	15.999	Tellurium	127.60
Boron	10.811	Holmium	164.93	Palladium	106.42	Terbium	158.93
Bromine	79.904	Hydrogen	1.0079	Phosphorus	30.974	Thallium	204.38
Cadmium	112.41	Indium	114.82	Platinum	195.08	Thorium*	232.04
Calcium	40.078	Iodine	126.90	Plutonium*	244.06	Thulium	168.93
Californium*	251.08	Iridium	192.22	Polonium*	209.98	Tin	118.71
Carbon	12.011	Iron	55.845	Potassium	39.098	Titanium	47.867
Cerium	140.12	Krypton	83.80	Praseodymium	140.91	Tungsten	183.85
Cesium	132.91	Lanthanum	138.91	Promethium*	146.92	Uranium*	238.03
Chlorine	35.453	Lawrencium*	262	Protactinium*	231.04	Vanadium	50.942
Chromium	51.996	Lead	207.2	Radium*	226.03	Xenon	131.29
Cobalt	58.933	Lithium	6.941	Radon*	222.02	Ytterbium	173.04
Copper	63.546	Lutetium	174.97	Rhenium	186.21	Yttrium	88.906
Curium*	247.07	Magnesium	24.305	Rhodium	102.91	Zinc	65.39
Dysprosium	162.50	Manganese	54.938	Rubidium	85.468	Zirconium	91.224
Einsteinium*	252	Mendelevium*	258.10	Ruthenium	101.07		

Note: The above values are rounded off to values sufficiently accurate for normal calculations: for accurate values based on the 1999 revision, see Table 2.5. Transfermium elements, see Table 16.8.
*Element with no stable isotope.

Periodic Table of the Elements

	s^1	s^2	$d^n s^x (n = 1 \text{ to } 10; x = 0, 1 \text{ or } 2)$										s^2p^1	s^2p^2	s^2p^3	s^2p^4	s^2p^5	s^2p^6
1*s*						1 **H**		2 **He**										
2*s* 2*p*	3 **Li**	4 **Be**											5 **B**	6 **C**	7 **N**	8 **O**	9 **F**	10 **Ne**
3*s* 3*p*	11 **Na**	12 **Mg**											13 **Al**	14 **Si**	15 **P**	16 **S**	17 **Cl**	18 **Ar**
4*s* 3*d* 4*p*	19 **K**	20 **Ca**	21 **Sc**	22 **Ti**	23 **V**	24 **Cr**	25 **Mn**	26 **Fe**	27 **Co**	28 **Ni**	29 **Cu**	30 **Zn**	31 **Ga**	32 **Ge**	33 **As**	34 **Se**	35 **Br**	36 **Kr**
5*s* 4*d* 5*p*	37 **Rb**	38 **Sr**	39 **Y**	40 **Zr**	41 **Nb**	42 **Mo**	43 **Tc**	44 **Ru**	45 **Rh**	46 **Pd**	47 **Ag**	48 **Cd**	49 **In**	50 **Sn**	51 **Sb**	52 **Te**	53 **I**	54 **Xe**
6*s* (4*f*) 5*d* 6*p*	55 **Cs**	56 **Ba**	57* **La**	72 **Hf**	73 **Ta**	74 **W**	75 **Re**	76 **Os**	77 **Ir**	78 **Pt**	79 **Au**	80 **Hg**	81 **Tl**	82 **Pb**	83 **Bi**	84 **Po**	85 **At**	86 **Rn**
7*s* (5*f*) 6*d*	87 **Fr**	88 **Ra**	89** **Ac**	104 **Rf**	105 **Db**	106 **Sg**	107 **Bh**	108 **Hs**	109 **Mt**	110 **Uun**	111 **Uuu**	112 **Uub**		114 **Uuq**				

	$f^p d^n s^2$ ($p = 1$ to 14, $n = 0$ or 1 (2 for Th))													
*Lanthanide series 4*f*	58 **Ce**	59 **Pr**	60 **Nd**	61 **Pm**	62 **Sm**	63 **Eu**	64 **Gd**	65 **Tb**	66 **Dy**	67 **Ho**	68 **Er**	69 **Tm**	70 **Yb**	71 **Lu**
**Actinide series 5*f*	90 **Th**	91 **Pa**	92 **U**	93 **Np**	94 **Pu**	95 **Am**	96 **Cm**	97 **Bk**	98 **Cf**	99 **Es**	100 **Fm**	101 **Md**	102 **No**	103 **Lr**